Rational (Reciprocal) Function

$$f(x) = \frac{1}{x}$$

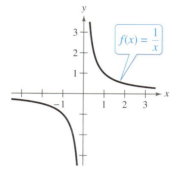

Domain: $(-\infty, 0) \cup (0, \infty)$
Range: $(-\infty, 0) \cup (0, \infty)$
No intercepts
Decreasing on $(-\infty, 0)$ and $(0, \infty)$
Odd function
Origin symmetry
Vertical asymptote: y-axis
Horizontal asymptote: x-axis

Exponential Function

$$f(x) = a^x, \; a > 1$$

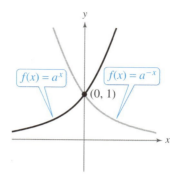

Domain: $(-\infty, \infty)$
Range: $(0, \infty)$
Intercept: $(0, 1)$
Increasing on $(-\infty, \infty)$
   for $f(x) = a^x$
Decreasing on $(-\infty, \infty)$
   for $f(x) = a^{-x}$
Horizontal asymptote: x-axis
Continuous

Logarithmic Function

$$f(x) = \log_a x, \; a > 1$$

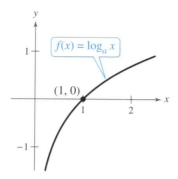

Domain: $(0, \infty)$
Range: $(-\infty, \infty)$
Intercept: $(1, 0)$
Increasing on $(0, \infty)$
Vertical asymptote: y-axis
Continuous
Reflection of graph of $f(x) = a^x$
   in the line $y = x$

SYMMETRY

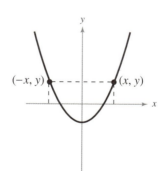

y-Axis Symmetry

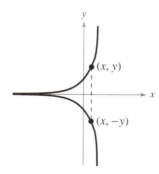

x-Axis Symmetry

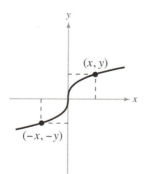

Origin Symmetry

COLLEGE ALGEBRA

10E

with

CalcChat® *and* **CalcView®**

Ron Larson
The Pennsylvania State University
The Behrend College

With the assistance of David C. Falvo
The Pennsylvania State University
The Behrend College

CENGAGE
Learning·

Australia • Brazil • Mexico • Singapore • United Kingdom • United States

College Algebra
with CalcChat and CalcView
Tenth Edition

Ron Larson

Product Director: Terry Boyle

Product Manager: Gary Whalen

Senior Content Developer: Stacy Green

Associate Content Developer: Samantha Lugtu

Product Assistant: Katharine Werring

Media Developer: Lynh Pham

Marketing Manager: Ryan Ahern

Content Project Manager: Jennifer Risden

Manufacturing Planner: Doug Bertke

Production Service: Larson Texts, Inc.

Photo Researcher: Lumina Datamatics

Text Researcher: Lumina Datamatics

Illustrator: Larson Texts, Inc.

Text Designer: Larson Texts, Inc.

Cover Designer: Larson Texts, Inc.

Front Cover Image: betibup33/Shutterstock.com

Back Cover Image: SidorArt/Shutterstock.com

Compositor: Larson Texts, Inc.

For product information and technology assistance, contact us at **Cengage Learning Customer & Sales Support, 1-800-354-9706.**

For permission to use material from this text or product, submit all requests online at **www.cengage.com/permissions.** Further permissions questions can be emailed to **permissionrequest@cengage.com.**

Library of Congress Control Number: 2016944979

Student Edition:
ISBN: 978-1-337-28229-1

Loose-leaf Edition:
ISBN: 978-1-337-29152-1

Cengage Learning
20 Channel Center Street
Boston, MA 02210
USA

Cengage Learning is a leading provider of customized learning solutions with employees residing in nearly 40 different countries and sales in more than 125 countries around the world. Find your local representative at **www.cengage.com.**

Cengage Learning products are represented in Canada by Nelson Education, Ltd.

To learn more about Cengage Learning Solutions, visit **www.cengage.com.**

Purchase any of our products at your local college store or at our preferred online store **www.cengagebrain.com.**

QR Code is a registered trademark of Denso Wave Incorporated

Printed in the United States of America
Print Number: 03   Print Year: 2017

Contents

*Available at the text-specific website *www.cengagebrain.com*

Preface

Welcome to *College Algebra,* Tenth Edition. We are excited to offer you a new edition with even more resources that will help you understand and master algebra. This textbook includes features and resources that continue to make *College Algebra* a valuable learning tool for students and a trustworthy teaching tool for instructors.

College Algebra provides the clear instruction, precise mathematics, and thorough coverage that you expect for your course. Additionally, this new edition provides you with **free** access to three companion websites:

- **CalcView.com**—video solutions to selected exercises
- **CalcChat.com**—worked-out solutions to odd-numbered exercises and access to online tutors
- **LarsonPrecalculus.com**—companion website with resources to supplement your learning

These websites will help enhance and reinforce your understanding of the material presented in this text and prepare you for future mathematics courses. CalcView® and CalcChat® are also available as free mobile apps.

Features

NEW CalcView®

The website *CalcView.com* contains video solutions of selected exercises. Watch instructors progress step-by-step through solutions, providing guidance to help you solve the exercises. The CalcView mobile app is available for free at the Apple® App Store® or Google Play™ store. The app features an embedded QR Code® reader that can be used to scan the on-page codes and go directly to the videos. You can also access the videos at *CalcView.com.*

UPDATED CalcChat®

In each exercise set, be sure to notice the reference to *CalcChat.com.* This website provides free step-by-step solutions to all odd-numbered exercises in many of our textbooks. Additionally, you can chat with a tutor, at no charge, during the hours posted at the site. For over 14 years, hundreds of thousands of students have visited this site for help. The CalcChat mobile app is also available as a free download at the Apple® App Store® or Google Play™ store and features an embedded QR Code® reader.

App Store is a service mark of Apple Inc. Google Play is a trademark of Google Inc.
QR Code is a registered trademark of Denso Wave Incorporated.

REVISED LarsonPrecalculus.com

All companion website features have been updated based on this revision, plus we have added a new Collaborative Project feature. Access to these features is free. You can view and listen to worked-out solutions of Checkpoint problems in English or Spanish, explore examples, download data sets, watch lesson videos, and much more.

NEW Collaborative Project

You can find these extended group projects at *LarsonPrecalculus.com.* Check your understanding of the chapter concepts by solving in-depth, real-life problems. These collaborative projects provide an interesting and engaging way for you and other students to work together and investigate ideas.

REVISED Exercise Sets

The exercise sets have been carefully and extensively examined to ensure they are rigorous and relevant, and include topics our users have suggested. The exercises have been reorganized and titled so you can better see the connections between examples and exercises. Multi-step, real-life exercises reinforce problem-solving skills and mastery of concepts by giving you the opportunity to apply the concepts in real-life situations. Error Analysis exercises have been added throughout the text to help you identify common mistakes.

Chapter Opener

Each Chapter Opener highlights real-life applications used in the examples and exercises.

Section Objectives

A bulleted list of learning objectives provides you the opportunity to preview what will be presented in the upcoming section.

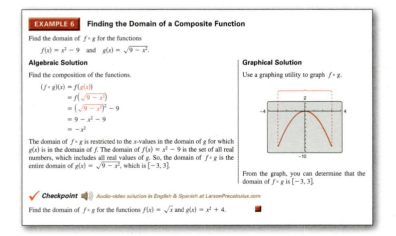

Side-By-Side Examples

Throughout the text, we present solutions to many examples from multiple perspectives—algebraically, graphically, and numerically. The side-by-side format of this pedagogical feature helps you to see that a problem can be solved in more than one way and to see that different methods yield the same result. The side-by-side format also addresses many different learning styles.

Remarks

These hints and tips reinforce or expand upon concepts, help you learn how to study mathematics, caution you about common errors, address special cases, or show alternative or additional steps to a solution of an example.

Checkpoints

Accompanying every example, the Checkpoint problems encourage immediate practice and check your understanding of the concepts presented in the example. View and listen to worked-out solutions of the Checkpoint problems in English or Spanish at *LarsonPrecalculus.com*.

Technology

The technology feature gives suggestions for effectively using tools such as calculators, graphing utilities, and spreadsheet programs to help deepen your understanding of concepts, ease lengthy calculations, and provide alternate solution methods for verifying answers obtained by hand.

Historical Notes

These notes provide helpful information regarding famous mathematicians and their work.

Algebra of Calculus

Throughout the text, special emphasis is given to the algebraic techniques used in calculus. Algebra of Calculus examples and exercises are integrated throughout the text and are identified by the symbol $\int$.

Summarize

The Summarize feature at the end of each section helps you organize the lesson's key concepts into a concise summary, providing you with a valuable study tool.

Vocabulary Exercises

The vocabulary exercises appear at the beginning of the exercise set for each section. These problems help you review previously learned vocabulary terms that you will use in solving the section exercises.

▷ **TECHNOLOGY** You can use a graphing utility to check that a solution is reasonable. One way is to graph the left side of the equation, then graph the right side of the equation, and determine the point of intersection. For instance, in Example 2, if you graph the equations

$$y_1 = 6(x - 1) + 4 \quad \text{The left side}$$

and

$$y_2 = 3(7x + 1) \quad \text{The right side}$$

in the same viewing window, they intersect at $x = -\frac{1}{3}$, as shown in the graph below.

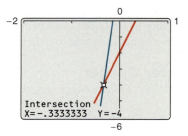

92. **HOW DO YOU SEE IT?** The graph represents the height h of a projectile after t seconds.

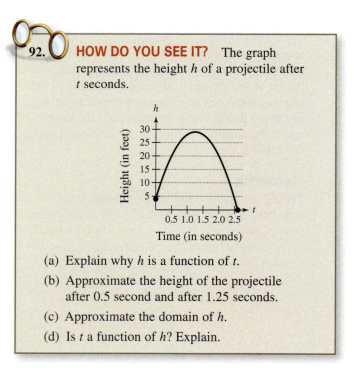

Time (in seconds)

(a) Explain why h is a function of t.

(b) Approximate the height of the projectile after 0.5 second and after 1.25 seconds.

(c) Approximate the domain of h.

(d) Is t a function of h? Explain.

How Do You See It?

The How Do You See It? feature in each section presents a real-life exercise that you will solve by visual inspection using the concepts learned in the lesson. This exercise is excellent for classroom discussion or test preparation.

Project

The projects at the end of selected sections involve in-depth applied exercises in which you will work with large, real-life data sets, often creating or analyzing models. These projects are offered online at *LarsonPrecalculus.com*.

Chapter Summary

The Chapter Summary includes explanations and examples of the objectives taught in each chapter.

Instructor Resources

Annotated Instructor's Edition / ISBN-13: 978-1-337-28230-7
This is the complete student text plus point-of-use annotations for the instructor, including extra projects, classroom activities, teaching strategies, and additional examples. Answers to even-numbered text exercises, Vocabulary Checks, and Explorations are also provided.

Complete Solutions Manual (on instructor companion site)
This manual contains solutions to all exercises from the text, including Chapter Review Exercises and Chapter Tests, and Practice Tests with solutions.

Cengage Learning Testing Powered by Cognero (login.cengage.com)
CLT is a flexible online system that allows you to author, edit, and manage test bank content; create multiple test versions in an instant; and deliver tests from your LMS, your classroom, or wherever you want. This is available online via *www.cengage.com/login.*

Instructor Companion Site
Everything you need for your course in one place! This collection of book-specific lecture and class tools is available online via *www.cengage.com/login.* Access and download PowerPoint® presentations, images, the instructor's manual, and more.

The Test Bank (on instructor companion site)
This contains text-specific multiple-choice and free response test forms.

Lesson Plans (on instructor companion site)
This manual provides suggestions for activities and lessons with notes on time allotment in order to ensure timeliness and efficiency during class.

MindTap for Mathematics
MindTap® is the digital learning solution that helps instructors engage and transform today's students into critical thinkers. Through paths of dynamic assignments and applications that you can personalize, real-time course analytics and an accessible reader, MindTap helps you turn cookie cutter into cutting edge, apathy into engagement, and memorizers into higher-level thinkers.

Enhanced WebAssign®   **ENHANCED WebAssign**
Exclusively from Cengage Learning, Enhanced WebAssign combines the exceptional mathematics content that you know and love with the most powerful online homework solution, WebAssign. Enhanced WebAssign engages students with immediate feedback, rich tutorial content, and interactive, fully customizable e-books (YouBook), helping students to develop a deeper conceptual understanding of their subject matter. Quick Prep and Just In Time exercises provide opportunities for students to review prerequisite skills and content, both at the start of the course and at the beginning of each section. Flexible assignment options give instructors the ability to release assignments conditionally on the basis of students' prerequisite assignment scores. Visit us at **www.cengage.com/ewa** to learn more.

Student Resources

Student Study and Solutions Manual / ISBN-13: 978-1-337-29150-7
This guide offers step-by-step solutions for all odd-numbered text exercises, Chapter Tests, and Cumulative Tests. It also contains Practice Tests.

Note-Taking Guide / ISBN-13: 978-1-337-29151-4
This is an innovative study aid, in the form of a notebook organizer, that helps students develop a section-by-section summary of key concepts.

CengageBrain.com
To access additional course materials, please visit *www.cengagebrain.com.* At the *CengageBrain.com* home page, search for the ISBN of your title (from the back cover of your book) using the search box at the top of the page. This will take you to the product page where these resources can be found.

MindTap for Mathematics
MindTap® provides you with the tools you need to better manage your limited time—you can complete assignments whenever and wherever you are ready to learn with course material specially customized for you by your instructor and streamlined in one proven, easy-to-use interface. With an array of tools and apps—from note taking to flashcards—you'll get a true understanding of course concepts, helping you to achieve better grades and setting the groundwork for your future courses. This access code entitles you to one term of usage.

Enhanced WebAssign®
Enhanced WebAssign (assigned by the instructor) provides you with instant feedback on homework assignments. This online homework system is easy to use and includes helpful links to textbook sections, video examples, and problem-specific tutorials.

Acknowledgments

I would like to thank the many people who have helped me prepare the text and the supplements package. Their encouragement, criticisms, and suggestions have been invaluable.

Thank you to all of the instructors who took the time to review the changes in this edition and to provide suggestions for improving it. Without your help, this book would not be possible.

Reviewers of the Tenth Edition

Gurdial Arora, *Xavier University of Louisiana*
Russell C. Chappell, *Twinsburg High School, Ohio*
Darlene Martin, *Lawson State Community College*
John Fellers, *North Allegheny School District*
Professor Steven Sikes, *Collin College*
Ann Slate, *Surry Community College*
John Elias, *Glenda Dawson High School*
Kathy Wood, *Lansing Catholic High School*
Darin Bauguess, *Surry Community College*
Brianna Kurtz, *Daytona State College*

Reviewers of the Previous Editions

Timothy Andrew Brown, *South Georgia College;* Blair E. Caboot, *Keystone College;* Shannon Cornell, *Amarillo College;* Gayla Dance, *Millsaps College;* Paul Finster, *El Paso Community College;* Paul A. Flasch, *Pima Community College West Campus;* Vadas Gintautas, *Chatham University;* Lorraine A. Hughes, *Mississippi State University;* Shu-Jen Huang, *University of Florida;* Renyetta Johnson, *East Mississippi Community College;* George Keihany, *Fort Valley State University;* Mulatu Lemma, *Savannah State University;* William Mays Jr., *Salem Community College;* Marcella Melby, *University of Minnesota;* Jonathan Prewett, *University of Wyoming;* Denise Reid, *Valdosta State University;* David L. Sonnier, *Lyon College;* David H. Tseng, *Miami Dade College—Kendall Campus;* Kimberly Walters, *Mississippi State University;* Richard Weil, *Brown College;* Solomon Willis, *Cleveland Community College;* Bradley R. Young, *Darton College*

My thanks to Robert Hostetler, The Behrend College, The Pennsylvania State University, and David Heyd, The Behrend College, The Pennsylvania State University, for their significant contributions to previous editions of this text.

I would also like to thank the staff at Larson Texts, Inc. who assisted with proofreading the manuscript, preparing and proofreading the art package, and checking and typesetting the supplements.

On a personal level, I am grateful to my spouse, Deanna Gilbert Larson, for her love, patience, and support. Also, a special thanks goes to R. Scott O'Neil. If you have suggestions for improving this text, please feel free to write to me. Over the past two decades, I have received many useful comments from both instructors and students, and I value these comments very highly.

Ron Larson, Ph.D.
Professor of Mathematics
Penn State University
www.RonLarson.com

P Prerequisites

Autocatalytic Chemical Reaction *(Exercise 84, page 40)*

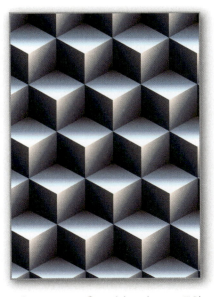

Computer Graphics *(page 56)*

Steel Beam Loading *(Exercise 81, page 33)*

Gallons of Water on Earth *(page 17)*

Change in Temperature *(page 7)*

P.1 Review of Real Numbers and Their Properties

Real numbers can represent many real-life quantities. For example, in Exercises 49–52 on page 13, you will use real numbers to represent the federal surplus or deficit.

- ■ Represent and classify real numbers.
- ■ Order real numbers and use inequalities.
- ■ Find the absolute values of real numbers and find the distance between two real numbers.
- ■ Evaluate algebraic expressions.
- ■ Use the basic rules and properties of algebra.

Real Numbers

Real numbers can describe quantities in everyday life such as age, miles per gallon, and population. Symbols such as

$$-5, 9, 0, \tfrac{4}{3}, 0.666\ldots, 28.21, \sqrt{2}, \pi, \text{ and } \sqrt[3]{-32}$$

represent real numbers. Here are some important **subsets** (each member of a subset B is also a member of a set A) of the real numbers. The three dots, or *ellipsis points*, tell you that the pattern continues indefinitely.

$$\{1, 2, 3, 4, \ldots\} \qquad \text{Set of natural numbers}$$
$$\{0, 1, 2, 3, 4, \ldots\} \qquad \text{Set of whole numbers}$$
$$\{\ldots, -3, -2, -1, 0, 1, 2, 3, \ldots\} \qquad \text{Set of integers}$$

A real number is **rational** when it can be written as the ratio p/q of two integers, where $q \neq 0$. For example, the numbers

$$\tfrac{1}{3} = 0.3333\ldots = 0.\overline{3}, \tfrac{1}{8} = 0.125, \text{ and } \tfrac{125}{111} = 1.126126\ldots = 1.\overline{126}$$

are rational. The decimal representation of a rational number either repeats $\left(\text{as in } \tfrac{173}{55} = 3.1\overline{45}\right)$ or terminates $\left(\text{as in } \tfrac{1}{2} = 0.5\right)$. A real number that cannot be written as the ratio of two integers is **irrational.** The decimal representation of an irrational number neither terminates nor repeats. For example, the numbers

$$\sqrt{2} = 1.4142135\ldots \approx 1.41 \quad \text{and} \quad \pi = 3.1415926\ldots \approx 3.14$$

are irrational. (The symbol $\approx$ means "is approximately equal to.") Figure P.1 shows subsets of the real numbers and their relationships to each other.

```
                  Real
                numbers
                   |
        +----------+----------+
        |                     |
   Irrational            Rational
    numbers               numbers
                             |
                   +---------+---------+
                   |                   |
                Integers          Noninteger
                   |               fractions
                   |             (positive and
                   |               negative)
           +-------+-------+
           |               |
       Negative          Whole
       integers         numbers
                           |
                   +-------+-------+
                   |               |
                Natural          Zero
                numbers
```

Subsets of the real numbers
Figure P.1

EXAMPLE 1   Classifying Real Numbers

Determine which numbers in the set $\left\{-13, -\sqrt{5}, -1, -\tfrac{1}{3}, 0, \tfrac{5}{8}, \sqrt{2}, \pi, 7\right\}$ are (a) natural numbers, (b) whole numbers, (c) integers, (d) rational numbers, and (e) irrational numbers.

Solution

a. Natural numbers: $\{7\}$

b. Whole numbers: $\{0, 7\}$

c. Integers: $\{-13, -1, 0, 7\}$

d. Rational numbers: $\left\{-13, -1, -\tfrac{1}{3}, 0, \tfrac{5}{8}, 7\right\}$

e. Irrational numbers: $\left\{-\sqrt{5}, \sqrt{2}, \pi\right\}$

✓ *Checkpoint* ◄))) *Audio-video solution in English & Spanish at LarsonPrecalculus.com*

Repeat Example 1 for the set $\left\{-\pi, -\tfrac{1}{4}, \tfrac{6}{3}, \tfrac{1}{2}\sqrt{2}, -7.5, -1, 8, -22\right\}$.

Real numbers are represented graphically on the **real number line.** When you draw a point on the real number line that corresponds to a real number, you are **plotting** the real number. The point representing 0 on the real number line is the **origin.** Numbers to the right of 0 are positive, and numbers to the left of 0 are negative, as shown in the figure below. The term **nonnegative** describes a number that is either positive or zero.

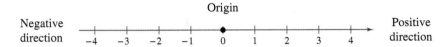

As the next two number lines illustrate, there is a *one-to-one correspondence* between real numbers and points on the real number line.

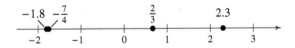

Every real number corresponds to exactly one point on the real number line.

Every point on the real number line corresponds to exactly one real number.

EXAMPLE 2   Plotting Points on the Real Number Line

Plot the real numbers on the real number line.

a. $-\dfrac{7}{4}$

b. 2.3

c. $\dfrac{2}{3}$

d. -1.8

Solution   The figure below shows all four points.

a. The point representing the real number $-\frac{7}{4} = -1.75$ lies between -2 and -1, but closer to -2, on the real number line.

b. The point representing the real number 2.3 lies between 2 and 3, but closer to 2, on the real number line.

c. The point representing the real number $\frac{2}{3} = 0.666 \ldots$ lies between 0 and 1, but closer to 1, on the real number line.

d. The point representing the real number -1.8 lies between -2 and -1, but closer to -2, on the real number line. Note that the point representing -1.8 lies slightly to the left of the point representing $-\frac{7}{4}$.

✓ *Checkpoint* 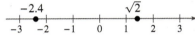 *Audio-video solution in English & Spanish at LarsonPrecalculus.com*

Plot the real numbers on the real number line.

a. $\dfrac{5}{2}$ b. -1.6

c. $-\dfrac{3}{4}$ d. 0.7

Ordering Real Numbers

One important property of real numbers is that they are *ordered*.

> ### Definition of Order on the Real Number Line
>
> If *a* and *b* are real numbers, then *a* is *less than* *b* when $b - a$ is positive. The **inequality** $a < b$ denotes the **order** of *a* and *b*. This relationship can also be described by saying that *b* is *greater than* *a* and writing $b > a$. The inequality $a \le b$ means that *a* is *less than or equal to b*, and the inequality $b \ge a$ means that *b* is *greater than or equal to a*. The symbols $<$, $>$, $\le$, and $\ge$ are *inequality symbols*.

$a < b$ if and only if *a* lies to the left of *b*.

Figure P.2

Geometrically, this definition implies that $a < b$ if and only if *a* lies to the *left* of *b* on the real number line, as shown in Figure P.2.

EXAMPLE 3   **Ordering Real Numbers**

Place the appropriate inequality symbol ($<$ or $>$) between the pair of real numbers.

a. $-3, 0$ **b.** $-2, -4$ **c.** $\frac{1}{4}, \frac{1}{3}$

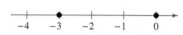

Figure P.3

Figure P.4

Solution

a. On the real number line, -3 lies to the left of 0, as shown in Figure P.3. So, you can say that -3 is *less than* 0, and write $-3 < 0$.

b. On the real number line, -2 lies to the right of -4, as shown in Figure P.4. So, you can say that -2 is *greater than* -4, and write $-2 > -4$.

c. On the real number line, $\frac{1}{4}$ lies to the left of $\frac{1}{3}$, as shown in Figure P.5. So, you can say that $\frac{1}{4}$ is *less than* $\frac{1}{3}$, and write $\frac{1}{4} < \frac{1}{3}$.

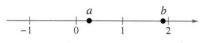

Figure P.5

✓ *Checkpoint* 🔊))) *Audio-video solution in English & Spanish at LarsonPrecalculus.com*

Place the appropriate inequality symbol ($<$ or $>$) between the pair of real numbers.

a. $1, -5$ **b.** $\frac{3}{2}, 7$ **c.** $-\frac{2}{3}, -\frac{3}{4}$

EXAMPLE 4   **Interpreting Inequalities**

See LarsonPrecalculus.com for an interactive version of this type of example.

Describe the subset of real numbers that the inequality represents.

a. $x \le 2$ **b.** $-2 \le x < 3$

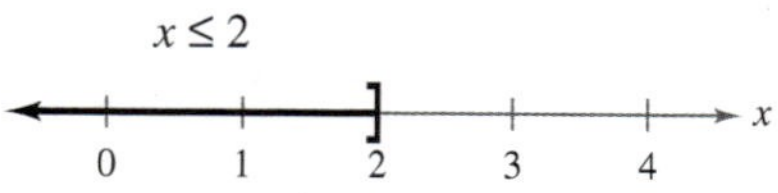

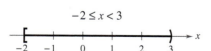

Figure P.6

Solution

a. The inequality $x \le 2$ denotes all real numbers less than or equal to 2, as shown in Figure P.6.

b. The inequality $-2 \le x < 3$ means that $x \ge -2$ *and* $x < 3$. This "double inequality" denotes all real numbers between -2 and 3, including -2 but not including 3, as shown in Figure P.7.

Figure P.7

✓ *Checkpoint* 🔊))) *Audio-video solution in English & Spanish at LarsonPrecalculus.com*

Describe the subset of real numbers that the inequality represents.

a. $x > -3$ **b.** $0 < x \le 4$

Inequalities can describe subsets of real numbers called **intervals.** In the bounded intervals below, the real numbers a and b are the **endpoints** of each interval. The endpoints of a closed interval are included in the interval, whereas the endpoints of an open interval are not included in the interval.

REMARK The reason that the four types of intervals at the right are called *bounded* is that each has a finite length. An interval that does not have a finite length is *unbounded* (see below).

Bounded Intervals on the Real Number Line

Notation	Interval Type	Inequality	Graph
$[a, b]$	Closed	$a \leq x \leq b$	
(a, b)	Open	$a < x < b$	
$[a, b)$		$a \leq x < b$	
$(a, b]$		$a < x \leq b$	

The symbols ∞, **positive infinity,** and $-\infty$, **negative infinity,** do not represent real numbers. They are convenient symbols used to describe the unboundedness of an interval such as $(1, \infty)$ or $(-\infty, 3]$.

REMARK Whenever you write an interval containing ∞ or $-\infty$, always use a parenthesis and never a bracket next to these symbols. This is because ∞ and $-\infty$ are never included in the interval.

Unbounded Intervals on the Real Number Line

Notation	Interval Type	Inequality	Graph
$[a, \infty)$		$x \geq a$	
(a, ∞)	Open	$x > a$	
$(-\infty, b]$		$x \leq b$	
$(-\infty, b)$	Open	$x < b$	
$(-\infty, \infty)$	Entire real line	$-\infty < x < \infty$	

EXAMPLE 5   **Interpreting Intervals**

a. The interval $(-1, 0)$ consists of all real numbers greater than -1 and less than 0.

b. The interval $[2, \infty)$ consists of all real numbers greater than or equal to 2.

✓ *Checkpoint*))) *Audio-video solution in English & Spanish at LarsonPrecalculus.com*

Give a verbal description of the interval $[-2, 5)$.

EXAMPLE 6   **Using Inequalities to Represent Intervals**

a. The inequality $c \leq 2$ can represent the statement "c is at most 2."

b. The inequality $-3 < x \leq 5$ can represent "all x in the interval $(-3, 5]$."

✓ *Checkpoint*))) *Audio-video solution in English & Spanish at LarsonPrecalculus.com*

Use inequality notation to represent the statement "x is less than 4 and at least -2." ■

Absolute Value and Distance

The **absolute value** of a real number is its *magnitude,* or the distance between the origin and the point representing the real number on the real number line.

Definition of Absolute Value

If a is a real number, then the **absolute value** of a is

$$|a| = \begin{cases} a, & a \geq 0 \\ -a, & a < 0 \end{cases}.$$

Notice in this definition that the absolute value of a real number is never negative. For example, if $a = -5$, then $|-5| = -(-5) = 5$. The absolute value of a real number is either positive or zero. Moreover, 0 is the only real number whose absolute value is 0. So, $|0| = 0$.

Properties of Absolute Values

1. $|a| \geq 0$ **2.** $|-a| = |a|$

3. $|ab| = |a||b|$ **4.** $\left|\dfrac{a}{b}\right| = \dfrac{|a|}{|b|}, \quad b \neq 0$

EXAMPLE 7 **Finding Absolute Values**

a. $|-15| = 15$ **b.** $\left|\dfrac{2}{3}\right| = \dfrac{2}{3}$

c. $|-4.3| = 4.3$ **d.** $-|-6| = -(6) = -6$

✓ *Checkpoint* 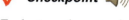 *Audio-video solution in English & Spanish at LarsonPrecalculus.com*

Evaluate each expression.

a. $|1|$ **b.** $-\left|\dfrac{3}{4}\right|$ **c.** $\dfrac{2}{|-3|}$ **d.** $-|0.7|$

EXAMPLE 8 **Evaluating an Absolute Value Expression**

Evaluate $\dfrac{|x|}{x}$ for (a) $x > 0$ and (b) $x < 0$.

Solution

a. If $x > 0$, then x is positive and $|x| = x$. So, $\dfrac{|x|}{x} = \dfrac{x}{x} = 1$.

b. If $x < 0$, then x is negative and $|x| = -x$. So, $\dfrac{|x|}{x} = \dfrac{-x}{x} = -1$.

✓ *Checkpoint* *Audio-video solution in English & Spanish at LarsonPrecalculus.com*

Evaluate $\dfrac{|x + 3|}{x + 3}$ for (a) $x > -3$ and (b) $x < -3$. ■

The **Law of Trichotomy** states that for any two real numbers a and b, *precisely* one of three relationships is possible:

$$a = b, \quad a < b, \quad \text{or} \quad a > b. \qquad \text{Law of Trichotomy}$$

EXAMPLE 9   **Comparing Real Numbers**

Place the appropriate symbol ($<$, $>$, or $=$) between the pair of real numbers.

a. $|-4| \quad \blacksquare \quad |3|$ **b.** $|-10| \quad \blacksquare \quad |10|$ **c.** $-|-7| \quad \blacksquare \quad |-7|$

Solution

a. $|-4| > |3|$ because $|-4| = 4$ and $|3| = 3$, and 4 is greater than 3.
b. $|-10| = |10|$ because $|-10| = 10$ and $|10| = 10$.
c. $-|-7| < |-7|$ because $-|-7| = -7$ and $|-7| = 7$, and -7 is less than 7.

✓ *Checkpoint*  *Audio-video solution in English & Spanish at LarsonPrecalculus.com*

Place the appropriate symbol ($<$, $>$, or $=$) between the pair of real numbers.

a. $|-3| \quad \blacksquare \quad |4|$
b. $-|-4| \quad \blacksquare \quad -|4|$
c. $|-3| \quad \blacksquare \quad -|-3|$

Absolute value can be used to find the distance between two points on the real number line. For example, the distance between -3 and 4 is

$$|-3 - 4| = |-7|$$
$$= 7$$

The distance between -3 and 4 is 7.
Figure P.8

as shown in Figure P.8.

Distance Between Two Points on the Real Number Line

Let a and b be real numbers. The **distance between a and b** is

$$d(a, b) = |b - a| = |a - b|.$$

EXAMPLE 10   **Finding a Distance**

Find the distance between -25 and 13.

Solution

The distance between -25 and 13 is

$$|-25 - 13| = |-38| = 38. \qquad \text{Distance between } -25 \text{ and } 13$$

The distance can also be found as follows.

$$|13 - (-25)| = |38| = 38 \qquad \text{Distance between } -25 \text{ and } 13$$

✓ *Checkpoint* 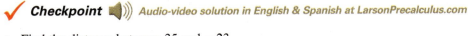 *Audio-video solution in English & Spanish at LarsonPrecalculus.com*

a. Find the distance between 35 and -23.
b. Find the distance between -35 and -23.
c. Find the distance between 35 and 23.

One application of finding the distance between two points on the real number line is finding a change in temperature.

Algebraic Expressions

One characteristic of algebra is the use of letters to represent numbers. The letters are **variables,** and combinations of letters and numbers are **algebraic expressions.** Here are a few examples of algebraic expressions.

$$5x, \qquad 2x - 3, \qquad \frac{4}{x^2 + 2}, \qquad 7x + y$$

> ### Definition of an Algebraic Expression
> An **algebraic expression** is a collection of letters (**variables**) and real numbers (**constants**) combined using the operations of addition, subtraction, multiplication, division, and exponentiation.

The **terms** of an algebraic expression are those parts that are separated by *addition*. For example, $x^2 - 5x + 8 = x^2 + (-5x) + 8$ has three terms: x^2 and $-5x$ are the **variable terms** and 8 is the **constant term.** For terms such as x^2, $-5x$, and 8, the numerical factor is the **coefficient.** Here, the coefficients are 1, -5, and 8.

EXAMPLE 11 Identifying Terms and Coefficients

Algebraic Expression	Terms	Coefficients
a. $5x - \dfrac{1}{7}$	$5x, -\dfrac{1}{7}$	$5, -\dfrac{1}{7}$
b. $2x^2 - 6x + 9$	$2x^2, -6x, 9$	$2, -6, 9$
c. $\dfrac{3}{x} + \dfrac{1}{2}x^4 - y$	$\dfrac{3}{x}, \dfrac{1}{2}x^4, -y$	$3, \dfrac{1}{2}, -1$

✓ *Checkpoint*  *Audio-video solution in English & Spanish at LarsonPrecalculus.com*

Identify the terms and coefficients of $-2x + 4$. ■

The **Substitution Principle** states, "If $a = b$, then b can replace a in any expression involving a." Use the Substitution Principle to **evaluate** an algebraic expression by substituting numerical values for each of the variables in the expression. The next example illustrates this.

EXAMPLE 12 Evaluating Algebraic Expressions

Expression	Value of Variable	Substitute.	Value of Expression
a. $-3x + 5$	$x = 3$	$-3(3) + 5$	$-9 + 5 = -4$
b. $3x^2 + 2x - 1$	$x = -1$	$3(-1)^2 + 2(-1) - 1$	$3 - 2 - 1 = 0$
c. $\dfrac{2x}{x + 1}$	$x = -3$	$\dfrac{2(-3)}{-3 + 1}$	$\dfrac{-6}{-2} = 3$

Note that you must substitute the value for *each* occurrence of the variable.

✓ *Checkpoint*  *Audio-video solution in English & Spanish at LarsonPrecalculus.com*

Evaluate $4x - 5$ when $x = 0$. ■

Basic Rules of Algebra

There are four arithmetic operations with real numbers: *addition, multiplication, subtraction,* and *division,* denoted by the symbols $+$, $\times$ or $\cdot$, $-$, and $\div$ or $/$, respectively. Of these, addition and multiplication are the two primary operations. Subtraction and division are the inverse operations of addition and multiplication, respectively.

Definitions of Subtraction and Division

Subtraction: Add the opposite.   **Division:** Multiply by the reciprocal.

$$a - b = a + (-b) \qquad\qquad \text{If } b \neq 0, \text{ then } a/b = a\left(\frac{1}{b}\right) = \frac{a}{b}.$$

In these definitions, $-b$ is the **additive inverse** (or opposite) of b, and $1/b$ is the **multiplicative inverse** (or reciprocal) of b. In the fractional form a/b, a is the **numerator** of the fraction and b is the **denominator.**

The properties of real numbers below are true for variables and algebraic expressions as well as for real numbers, so they are often called the **Basic Rules of Algebra.** Formulate a verbal description of each of these properties. For example, the first property states that *the order in which two real numbers are added does not affect their sum.*

Basic Rules of Algebra

Let a, b, and c be real numbers, variables, or algebraic expressions.

Property		**Example**
Commutative Property of Addition:	$a + b = b + a$	$4x + x^2 = x^2 + 4x$
Commutative Property of Multiplication:	$ab = ba$	$(4 - x)x^2 = x^2(4 - x)$
Associative Property of Addition:	$(a + b) + c = a + (b + c)$	$(x + 5) + x^2 = x + (5 + x^2)$
Associative Property of Multiplication:	$(ab)c = a(bc)$	$(2x \cdot 3y)(8) = (2x)(3y \cdot 8)$
Distributive Properties:	$a(b + c) = ab + ac$	$3x(5 + 2x) = 3x \cdot 5 + 3x \cdot 2x$
	$(a + b)c = ac + bc$	$(y + 8)y = y \cdot y + 8 \cdot y$
Additive Identity Property:	$a + 0 = a$	$5y^2 + 0 = 5y^2$
Multiplicative Identity Property:	$a \cdot 1 = a$	$(4x^2)(1) = 4x^2$
Additive Inverse Property:	$a + (-a) = 0$	$5x^3 + (-5x^3) = 0$
Multiplicative Inverse Property:	$a \cdot \dfrac{1}{a} = 1, \quad a \neq 0$	$(x^2 + 4)\left(\dfrac{1}{x^2 + 4}\right) = 1$

Subtraction is defined as "adding the opposite," so the Distributive Properties are also true for subtraction. For example, the "subtraction form" of $a(b + c) = ab + ac$ is $a(b - c) = ab - ac$. Note that the operations of subtraction and division are neither commutative nor associative. The examples

$$7 - 3 \neq 3 - 7 \quad \text{and} \quad 20 \div 4 \neq 4 \div 20$$

show that subtraction and division are not commutative. Similarly

$$5 - (3 - 2) \neq (5 - 3) - 2 \quad \text{and} \quad 16 \div (4 \div 2) \neq (16 \div 4) \div 2$$

demonstrate that subtraction and division are not associative.

EXAMPLE 13   **Identifying Rules of Algebra**

Identify the rule of algebra illustrated by the statement.

a. $(5x^3)2 = 2(5x^3)$ **b.** $(4x + 3) - (4x + 3) = 0$

c. $7x \cdot \dfrac{1}{7x} = 1, \quad x \neq 0$ **d.** $(2 + 5x^2) + x^2 = 2 + (5x^2 + x^2)$

Solution

a. This statement illustrates the Commutative Property of Multiplication. In other words, you obtain the same result whether you multiply $5x^3$ by 2, or 2 by $5x^3$.

b. This statement illustrates the Additive Inverse Property. In terms of subtraction, this property states that when any expression is subtracted from itself, the result is 0.

c. This statement illustrates the Multiplicative Inverse Property. Note that x must be a nonzero number. The reciprocal of x is undefined when x is 0.

d. This statement illustrates the Associative Property of Addition. In other words, to form the sum $2 + 5x^2 + x^2$, it does not matter whether 2 and $5x^2$, or $5x^2$ and x^2 are added first.

✓ **Checkpoint** 🔊))) *Audio-video solution in English & Spanish at LarsonPrecalculus.com*

Identify the rule of algebra illustrated by the statement.

a. $x + 9 = 9 + x$ **b.** $5(x^3 \cdot 2) = (5x^3)2$ **c.** $(2 + 5x^2)y^2 = 2 \cdot y^2 + 5x^2 \cdot y^2$

• • **REMARK** Notice the difference between the *opposite of a number* and a *negative number*. If a is already negative, then its opposite, $-a$, is positive. For example, if $a = -5$, then

$$-a = -(-5) = 5.$$

▷

Properties of Negation and Equality

Let a, b, and c be real numbers, variables, or algebraic expressions.

Property	Example
1. $(-1)a = -a$	$(-1)7 = -7$
2. $-(-a) = a$	$-(-6) = 6$
3. $(-a)b = -(ab) = a(-b)$	$(-5)3 = -(5 \cdot 3) = 5(-3)$
4. $(-a)(-b) = ab$	$(-2)(-x) = 2x$
5. $-(a + b) = (-a) + (-b)$	$-(x + 8) = (-x) + (-8)$
	$\quad\quad\quad\quad = -x - 8$
6. If $a = b$, then $a \pm c = b \pm c$.	$\frac{1}{2} + 3 = 0.5 + 3$
7. If $a = b$, then $ac = bc$.	$4^2 \cdot 2 = 16 \cdot 2$
8. If $a \pm c = b \pm c$, then $a = b$.	$1.4 - 1 = \frac{7}{5} - 1 \implies 1.4 = \frac{7}{5}$
9. If $ac = bc$ and $c \neq 0$, then $a = b$.	$3x = 3 \cdot 4 \implies x = 4$

• • **REMARK** The "or" in the Zero-Factor Property includes the possibility that either or both factors may be zero. This is an *inclusive or,* and it is generally the way the word "or" is used in mathematics.

▷

Properties of Zero

Let a and b be real numbers, variables, or algebraic expressions.

1. $a + 0 = a$ and $a - 0 = a$ **2.** $a \cdot 0 = 0$

3. $\dfrac{0}{a} = 0, \quad a \neq 0$ **4.** $\dfrac{a}{0}$ is undefined.

5. Zero-Factor Property: If $ab = 0$, then $a = 0$ or $b = 0$.

Properties and Operations of Fractions

Let a, b, c, and d be real numbers, variables, or algebraic expressions such that $b \neq 0$ and $d \neq 0$.

1. **Equivalent Fractions:** $\dfrac{a}{b} = \dfrac{c}{d}$ if and only if $ad = bc$.

2. **Rules of Signs:** $-\dfrac{a}{b} = \dfrac{-a}{b} = \dfrac{a}{-b}$ and $\dfrac{-a}{-b} = \dfrac{a}{b}$

3. **Generate Equivalent Fractions:** $\dfrac{a}{b} = \dfrac{ac}{bc}$, $\quad c \neq 0$

4. **Add or Subtract with Like Denominators:** $\dfrac{a}{b} \pm \dfrac{c}{b} = \dfrac{a \pm c}{b}$

5. **Add or Subtract with Unlike Denominators:** $\dfrac{a}{b} \pm \dfrac{c}{d} = \dfrac{ad \pm bc}{bd}$

6. **Multiply Fractions:** $\dfrac{a}{b} \cdot \dfrac{c}{d} = \dfrac{ac}{bd}$

7. **Divide Fractions:** $\dfrac{a}{b} \div \dfrac{c}{d} = \dfrac{a}{b} \cdot \dfrac{d}{c} = \dfrac{ad}{bc}$, $\quad c \neq 0$

• • • • • • • • • • • • • • • • ▷

• •**REMARK** In Property 1, the phrase "if and only if" implies two statements. One statement is: If $a/b = c/d$, then $ad = bc$. The other statement is: If $ad = bc$, where $b \neq 0$ and $d \neq 0$, then $a/b = c/d$.

EXAMPLE 14 **Properties and Operations of Fractions**

a. $\dfrac{x}{5} = \dfrac{3 \cdot x}{3 \cdot 5} = \dfrac{3x}{15}$
   b. $\dfrac{7}{x} \div \dfrac{3}{2} = \dfrac{7}{x} \cdot \dfrac{2}{3} = \dfrac{14}{3x}$

✓ **Checkpoint** *Audio-video solution in English & Spanish at LarsonPrecalculus.com*

a. Multiply fractions: $\dfrac{3}{5} \cdot \dfrac{x}{6}$ b. Add fractions: $\dfrac{x}{10} + \dfrac{2x}{5}$ ■

If a, b, and c are integers such that $ab = c$, then a and b are **factors** or **divisors** of c. A **prime number** is an integer that has exactly two positive factors—itself and 1—such as 2, 3, 5, 7, and 11. The numbers 4, 6, 8, 9, and 10 are **composite** because each can be written as the product of two or more prime numbers. The **Fundamental Theorem of Arithmetic** states that every positive integer greater than 1 is a prime number or can be written as the product of prime numbers in precisely one way (disregarding order). For example, the *prime factorization* of 24 is $24 = 2 \cdot 2 \cdot 2 \cdot 3$.

• •**REMARK** The number 1 is neither prime nor composite.
• • • • • • • • • • • • • • • • ▷

Summarize (Section P.1)

1. Explain how to represent and classify real numbers *(pages 2 and 3)*. For examples of representing and classifying real numbers, see Examples 1 and 2.

2. Explain how to order real numbers and use inequalities *(pages 4 and 5)*. For examples of ordering real numbers and using inequalities, see Examples 3–6.

3. State the definition of the absolute value of a real number *(page 6)*. For examples of using absolute value, see Examples 7–10.

4. Explain how to evaluate an algebraic expression *(page 8)*. For examples involving algebraic expressions, see Examples 11 and 12.

5. State the basic rules and properties of algebra *(pages 9–11)*. For examples involving the basic rules and properties of algebra, see Examples 13 and 14.

P.1 Exercises

See CalcChat.com for tutorial help and worked-out solutions to odd-numbered exercises.

Vocabulary: Fill in the blanks.

1. The decimal representation of an _____ number neither terminates nor repeats.
2. The point representing 0 on the real number line is the _____.
3. The distance between the origin and a point representing a real number on the real number line is the _____ _____ of the real number.
4. A number that can be written as the product of two or more prime numbers is a _____ number.
5. The _____ of an algebraic expression are those parts that are separated by addition.
6. The _____ _____ states that if $ab = 0$, then $a = 0$ or $b = 0$.

Skills and Applications

 Classifying Real Numbers In Exercises 7–10, determine which numbers in the set are (a) natural numbers, (b) whole numbers, (c) integers, (d) rational numbers, and (e) irrational numbers.

7. $\left\{-9, -\frac{7}{2}, 5, \frac{2}{3}, \sqrt{2}, 0, 1, -4, 2, -11\right\}$
8. $\left\{\sqrt{5}, -7, -\frac{7}{3}, 0, 3.14, \frac{5}{4}, -3, 12, 5\right\}$
9. $\{2.01, 0.\overline{6}, -13, 0.010110111\ldots, 1, -6\}$
10. $\left\{25, -17, -\frac{12}{5}, \sqrt{9}, 3.12, \frac{1}{2}\pi, 7, -11.1, 13\right\}$

Plotting Points on the Real Number Line In Exercises 11 and 12, plot the real numbers on the real number line.

11. (a) 3   (b) $\frac{7}{2}$ (c) $-\frac{5}{2}$ (d) -5.2
12. (a) 8.5   (b) $\frac{4}{3}$ (c) -4.75 (d) $-\frac{8}{3}$

 Plotting and Ordering Real Numbers In Exercises 13–16, plot the two real numbers on the real number line. Then place the appropriate inequality symbol (< or >) between them.

13. $-4, -8$
14. $1, \frac{16}{3}$
15. $\frac{5}{6}, \frac{2}{3}$
16. $-\frac{8}{7}, -\frac{3}{7}$

Interpreting an Inequality or an Interval In Exercises 17–24, (a) give a verbal description of the subset of real numbers represented by the inequality or the interval, (b) sketch the subset on the real number line, and (c) state whether the subset is bounded or unbounded.

17. $x \le 5$
18. $x < 0$
19. $-2 < x < 2$
20. $0 < x \le 6$
21. $[4, \infty)$
22. $(-\infty, 2)$
23. $[-5, 2)$
24. $(-1, 2]$

Using Inequality and Interval Notation In Exercises 25–28, use inequality notation and interval notation to describe the set.

25. y is nonnegative.
26. y is no more than 25.
27. t is at least 10 and at most 22.
28. k is less than 5 but no less than -3.

 Evaluating an Absolute Value Expression In Exercises 29–38, evaluate the expression.

29. $|-10|$
30. $|0|$
31. $|3 - 8|$
32. $|6 - 2|$
33. $|-1| - |-2|$
34. $-3 - |-3|$
35. $5|-5|$
36. $-4|-4|$
37. $\dfrac{|x + 2|}{x + 2}, \quad x < -2$
38. $\dfrac{|x - 1|}{x - 1}, \quad x > 1$

 Comparing Real Numbers In Exercises 39–42, place the appropriate symbol (<, >, or =) between the pair of real numbers.

39. $|-4|$ ▧ $|4|$
40. -5 ▧ $-|5|$
41. $-|-6|$ ▧ $|-6|$
42. $-|-2|$ ▧ $-|2|$

 Finding a Distance In Exercises 43–46, find the distance between a and b.

43. $a = 126, b = 75$
44. $a = -20, b = 30$
45. $a = -\frac{5}{2}, b = 0$
46. $a = -\frac{1}{4}, b = -\frac{11}{4}$

Using Absolute Value Notation In Exercises 47 and 48, use absolute value notation to represent the situation.

47. The distance between x and 5 is no more than 3.
48. The distance between x and -10 is at least 6.

The symbol ▦ and a red exercise number indicates that a video solution can be seen at *CalcView.com*.

• • Federal Deficit • • • • • • • • • • • • • • • •

In Exercises 49–52, use the bar graph, which shows the receipts of the federal government (in billions of dollars) for selected years from 2008 through 2014. In each exercise, you are given the expenditures of the federal government. Find the magnitude of the surplus or deficit for the year. *(Source: U.S. Office of Management and Budget)*

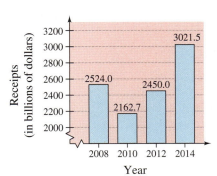

Year	Receipts, R	Expenditures, E	$\lvert R - E \rvert$
49. 2008		$2982.5 billion	
50. 2010		$3457.1 billion	
51. 2012		$3537.0 billion	
52. 2014		$3506.1 billion	

 Identifying Terms and Coefficients In Exercises 53–58, identify the terms. Then identify the coefficients of the variable terms of the expression.

53. $7x + 4$

54. $2x - 3$

55. $6x^3 - 5x$

56. $4x^3 + 0.5x - 5$

57. $3\sqrt{3}x^2 + 1$

58. $2\sqrt{2}x^2 - 3$

 Evaluating an Algebraic Expression In Exercises 59–64, evaluate the expression for each value of x. (If not possible, state the reason.)

59. $4x - 6$ (a) $x = -1$ (b) $x = 0$

60. $9 - 7x$ (a) $x = -3$ (b) $x = 3$

61. $x^2 - 3x + 2$ (a) $x = 0$ (b) $x = -1$

62. $-x^2 + 5x - 4$ (a) $x = -1$ (b) $x = 1$

63. $\dfrac{x + 1}{x - 1}$ (a) $x = 1$ (b) $x = -1$

64. $\dfrac{x - 2}{x + 2}$ (a) $x = 2$ (b) $x = -2$

Identifying Rules of Algebra In Exercises 65–68, identify the rule(s) of algebra illustrated by the statement.

65. $\dfrac{1}{h + 6}(h + 6) = 1, \quad h \neq -6$

66. $(x + 3) - (x + 3) = 0$

67. $x(3y) = (x \cdot 3)y = (3x)y$

68. $\frac{1}{7}(7 \cdot 12) = \left(\frac{1}{7} \cdot 7\right)12 = 1 \cdot 12 = 12$

Operations with Fractions In Exercises 69–72, perform the operation. (Write fractional answers in simplest form.)

69. $\dfrac{2x}{3} - \dfrac{x}{4}$

70. $\dfrac{3x}{4} + \dfrac{x}{5}$

71. $\dfrac{3x}{10} \cdot \dfrac{5}{6}$

72. $\dfrac{2x}{3} \div \dfrac{6}{7}$

Exploration

True or False? In Exercises 73–75, determine whether the statement is true or false. Justify your answer.

73. Every nonnegative number is positive.

74. If $a > 0$ and $b < 0$, then $ab > 0$.

75. If $a < 0$ and $b < 0$, then $ab > 0$.

76. HOW DO YOU SEE IT? Match each description with its graph. Which types of real numbers shown in Figure P.1 on page 2 may be included in a range of prices? a range of lengths? Explain.

(i)

1.87 1.88 1.89 1.90 1.91 1.92 1.93

(ii)

1.87 1.88 1.89 1.90 1.91 1.92 1.93

(a) The price of an item is within $0.03 of $1.90.

(b) The distance between the prongs of an electric plug may not differ from 1.9 centimeters by more than 0.03 centimeter.

77. Conjecture

(a) Use a calculator to complete the table.

n	0.0001	0.01	1	100	10,000
$\dfrac{5}{n}$					

(b) Use the result from part (a) to make a conjecture about the value of $5/n$ as n (i) approaches 0, and (ii) increases without bound.

P.2 Exponents and Radicals

■ Use properties of exponents.
■ Use scientific notation to represent real numbers.
■ Use properties of radicals.
■ Simplify and combine radical expressions.
■ Rationalize denominators and numerators.
■ Use properties of rational exponents.

Integer Exponents and Their Properties

Repeated *multiplication* can be written in **exponential form.**

Repeated Multiplication	Exponential Form
$a \cdot a \cdot a \cdot a \cdot a$	a^5
$(-4)(-4)(-4)$	$(-4)^3$
$(2x)(2x)(2x)(2x)$	$(2x)^4$

Real numbers and algebraic expressions are often written with exponents and radicals. For example, in Exercise 69 on page 25, you will use an expression involving rational exponents to find the times required for a funnel to empty for different water heights.

Exponential Notation

If a is a real number and n is a positive integer, then

$$a^n = \underbrace{a \cdot a \cdot a \cdot \cdots \cdot a}_{n \text{ factors}}$$

where n is the **exponent** and a is the **base.** You read a^n as "a to the nth **power.**"

An exponent can also be negative or zero. Properties 3 and 4 below show how to use negative and zero exponents.

Properties of Exponents

Let a and b be real numbers, variables, or algebraic expressions, and let m and n be integers. (All denominators and bases are nonzero.)

Property	Example
1. $a^m a^n = a^{m+n}$	$3^2 \cdot 3^4 = 3^{2+4} = 3^6 = 729$
2. $\dfrac{a^m}{a^n} = a^{m-n}$	$\dfrac{x^7}{x^4} = x^{7-4} = x^3$
3. $a^{-n} = \dfrac{1}{a^n} = \left(\dfrac{1}{a}\right)^n$	$y^{-4} = \dfrac{1}{y^4} = \left(\dfrac{1}{y}\right)^4$
4. $a^0 = 1$	$(x^2 + 1)^0 = 1$
5. $(ab)^m = a^m b^m$	$(5x)^3 = 5^3 x^3 = 125x^3$
6. $(a^m)^n = a^{mn}$	$(y^3)^{-4} = y^{3(-4)} = y^{-12} = \dfrac{1}{y^{12}}$
7. $\left(\dfrac{a}{b}\right)^m = \dfrac{a^m}{b^m}$	$\left(\dfrac{2}{x}\right)^3 = \dfrac{2^3}{x^3} = \dfrac{8}{x^3}$
8. $\lvert a^2 \rvert = \lvert a \rvert^2 = a^2$	$\lvert (-2)^2 \rvert = \lvert -2 \rvert^2 = 2^2 = 4 = (-2)^2$

The properties of exponents listed on the preceding page apply to *all* integers m and n, not just to positive integers, as shown in Examples 1–4.

It is important to recognize the difference between expressions such as $(-2)^4$ and -2^4. In $(-2)^4$, the parentheses tell you that the exponent applies to the negative sign as well as to the 2, but in $-2^4 = -(2^4)$, the exponent applies only to the 2. So, $(-2)^4 = 16$ and $-2^4 = -16$.

EXAMPLE 1   Evaluating Exponential Expressions

a. $(-5)^2 = (-5)(-5) = 25$ Negative sign is part of the base.

b. $-5^2 = -(5)(5) = -25$ Negative sign is *not* part of the base.

c. $2 \cdot 2^4 = 2^{1+4} = 2^5 = 32$ Property 1

d. $\dfrac{4^4}{4^6} = 4^{4-6} = 4^{-2} = \dfrac{1}{4^2} = \dfrac{1}{16}$ Properties 2 and 3

✓ *Checkpoint*  *Audio-video solution in English & Spanish at LarsonPrecalculus.com*

Evaluate each expression.

a. -3^4 **b.** $(-3)^4$

c. $3^2 \cdot 3$ **d.** $\dfrac{3^5}{3^8}$

▷ **TECHNOLOGY** When using a calculator to evaluate exponential expressions, it is important to know when to use parentheses because the calculator follows the order of operations. For example, here is how you would evaluate $(-2)^4$ on a graphing utility.

⟮ ⟮ (−) ⟯ 2 ⟯ ⟮ ^ ⟯ 4 ⟮ENTER⟯

The display will be 16. If you omit the parentheses, the display will be -16.

EXAMPLE 2   Evaluating Algebraic Expressions

Evaluate each algebraic expression when $x = 3$.

a. $5x^{-2}$ **b.** $\dfrac{1}{3}(-x)^3$

Solution

a. When $x = 3$, the expression $5x^{-2}$ has a value of

$$5x^{-2} = 5(3)^{-2} = \frac{5}{3^2} = \frac{5}{9}.$$

b. When $x = 3$, the expression $\dfrac{1}{3}(-x)^3$ has a value of

$$\frac{1}{3}(-x)^3 = \frac{1}{3}(-3)^3 = \frac{1}{3}(-27) = -9.$$

✓ *Checkpoint*  *Audio-video solution in English & Spanish at LarsonPrecalculus.com*

Evaluate each algebraic expression when $x = 4$.

a. $-x^{-2}$ **b.** $\dfrac{1}{4}(-x)^4$

EXAMPLE 3   **Using Properties of Exponents**

Use the properties of exponents to simplify each expression.

a. $(-3ab^4)(4ab^{-3})$ **b.** $(2xy^2)^3$ **c.** $3a(-4a^2)^0$ **d.** $\left(\dfrac{5x^3}{y}\right)^2$

Solution

a. $(-3ab^4)(4ab^{-3}) = (-3)(4)(a)(a)(b^4)(b^{-3}) = -12a^2b$

b. $(2xy^2)^3 = 2^3(x)^3(y^2)^3 = 8x^3y^6$

c. $3a(-4a^2)^0 = 3a(1) = 3a$

d. $\left(\dfrac{5x^3}{y}\right)^2 = \dfrac{5^2(x^3)^2}{y^2} = \dfrac{25x^6}{y^2}$

✓ **Checkpoint** ◀))) *Audio-video solution in English & Spanish at LarsonPrecalculus.com*

Use the properties of exponents to simplify each expression.

a. $(2x^{-2}y^3)(-x^4y)$ **b.** $(4a^2b^3)^0$ **c.** $(-5z)^3(z^2)$ **d.** $\left(\dfrac{3x^4}{x^2y^2}\right)^2$

EXAMPLE 4   **Rewriting with Positive Exponents**

a. $x^{-1} = \dfrac{1}{x}$ Property 3

b. $\dfrac{1}{3x^{-2}} = \dfrac{1(x^2)}{3}$ Property 3 (The exponent -2 does not apply to 3.)

   $= \dfrac{x^2}{3}$ Simplify.

c. $\dfrac{12a^3b^{-4}}{4a^{-2}b} = \dfrac{12a^3 \cdot a^2}{4b \cdot b^4}$ Property 3

   $= \dfrac{3a^5}{b^5}$ Property 1

d. $\left(\dfrac{3x^2}{y}\right)^{-2} = \dfrac{3^{-2}(x^2)^{-2}}{y^{-2}}$ Properties 5 and 7

   $= \dfrac{3^{-2}x^{-4}}{y^{-2}}$ Property 6

   $= \dfrac{y^2}{3^2x^4}$ Property 3

   $= \dfrac{y^2}{9x^4}$ Simplify.

✓ **Checkpoint** ◀))) *Audio-video solution in English & Spanish at LarsonPrecalculus.com*

Rewrite each expression with positive exponents. Simplify, if possible.

a. $2a^{-2}$ **b.** $\dfrac{3a^{-3}b^4}{15ab^{-1}}$

c. $\left(\dfrac{x}{10}\right)^{-1}$ **d.** $(-2x^2)^3(4x^3)^{-1}$

• • **REMARK**   Rarely in algebra
is there only one way to solve a
problem. Do not be concerned
when the steps you use to solve
a problem are not exactly the
same as the steps presented in
this text. It is important to use
steps that you understand and,
of course, steps that are justified
by the rules of algebra. For
example, the fractional form of
Property 3 is

$$\left(\dfrac{a}{b}\right)^{-m} = \left(\dfrac{b}{a}\right)^m.$$

So, you might prefer the steps
below for Example 4(d).

$$\left(\dfrac{3x^2}{y}\right)^{-2} = \left(\dfrac{y}{3x^2}\right)^2 = \dfrac{y^2}{9x^4}$$

There are about 366 billion billion gallons of water on Earth. It is convenient to write such a number in scientific notation.

Scientific Notation

Exponents provide an efficient way of writing and computing with very large (or very small) numbers. For example, there are about 366 billion billion gallons of water on Earth—that is, 366 followed by 18 zeros.

   366,000,000,000,000,000,000

It is convenient to write such numbers in **scientific notation.** This notation has the form $\pm c \times 10^n$, where $1 \le c < 10$ and n is an integer. So, the number of gallons of water on Earth, written in scientific notation, is

   $3.66 \times 100{,}000{,}000{,}000{,}000{,}000{,}000 = 3.66 \times 10^{20}.$

The *positive* exponent 20 tells you that the number is *large* (10 or more) and that the decimal point has been moved 20 places. A *negative* exponent tells you that the number is *small* (less than 1). For example, the mass (in grams) of one electron is approximately

   $9.1 \times 10^{-28} = 0.00000000000000000000000000091.$

   28 decimal places

EXAMPLE 5   Scientific Notation

a. $0.0000782 = 7.82 \times 10^{-5}$

b. $836{,}100{,}000 = 8.361 \times 10^{8}$

✓ *Checkpoint* *Audio-video solution in English & Spanish at LarsonPrecalculus.com*

Write 45,850 in scientific notation.

EXAMPLE 6   Decimal Notation

a. $-9.36 \times 10^{-6} = -0.00000936$

b. $1.345 \times 10^{2} = 134.5$

✓ *Checkpoint* *Audio-video solution in English & Spanish at LarsonPrecalculus.com*

Write -2.718×10^{-3} in decimal notation.

EXAMPLE 7   Using Scientific Notation

Evaluate $\dfrac{(2{,}400{,}000{,}000)(0.0000045)}{(0.00003)(1500)}$.

Solution   Begin by rewriting each number in scientific notation. Then simplify.

$$\frac{(2{,}400{,}000{,}000)(0.0000045)}{(0.00003)(1500)} = \frac{(2.4 \times 10^{9})(4.5 \times 10^{-6})}{(3.0 \times 10^{-5})(1.5 \times 10^{3})}$$

$$= \frac{(2.4)(4.5)(10^{3})}{(4.5)(10^{-2})}$$

$$= (2.4)(10^{5})$$

$$= 240{,}000$$

✓ *Checkpoint* *Audio-video solution in English & Spanish at LarsonPrecalculus.com*

Evaluate $(24{,}000{,}000{,}000)(0.00000012)(300{,}000)$.

▷ **TECHNOLOGY** Most calculators automatically switch to scientific notation when showing large (or small) numbers that exceed the display range.

   To enter numbers in scientific notation, your calculator should have an exponential entry key labeled

   $\boxed{\text{EE}}$ or $\boxed{\text{EXP}}$.

Consult the user's guide for instructions on keystrokes and how your calculator displays numbers in scientific notation.

Radicals and Their Properties

A **square root** of a number is one of its two equal factors. For example, 5 is a square root of 25 because 5 is one of the two equal factors of 25. In a similar way, a **cube root** of a number is one of its three equal factors, as in $125 = 5^3$.

Definition of *n*th Root of a Number

Let a and b be real numbers and let $n \geq 2$ be a positive integer. If

$$a = b^n$$

then b is an ***n*th root of *a*.** If $n = 2$, then the root is a **square root.** If $n = 3$, then the root is a **cube root.**

Some numbers have more than one *n*th root. For example, both 5 and -5 are square roots of 25. The *principal square* root of 25, written as $\sqrt{25}$, is the positive root, 5.

Principal *n*th Root of a Number

Let a be a real number that has at least one *n*th root. The **principal *n*th root of *a*** is the *n*th root that has the same sign as a. It is denoted by a **radical symbol**

$$\sqrt[n]{a}. \qquad \text{Principal } n\text{th root}$$

The positive integer $n \geq 2$ is the **index** of the radical, and the number a is the **radicand.** When $n = 2$, omit the index and write $\sqrt{a}$ rather than $\sqrt[2]{a}$. (The plural of index is *indices*.)

A common misunderstanding is that the square root sign implies both negative and positive roots. This is not correct. The square root sign implies only a positive root. When a negative root is needed, you must use the negative sign with the square root sign.

$$\text{Incorrect: } \sqrt{4} = \pm 2 \quad \text{✗} \qquad \text{Correct: } -\sqrt{4} = -2 \quad \text{and} \quad \sqrt{4} = 2$$

EXAMPLE 8   **Evaluating Radical Expressions**

a. $\sqrt{36} = 6$ because $6^2 = 36$.

b. $-\sqrt{36} = -6$ because $-\left(\sqrt{36}\right) = -\left(\sqrt{6^2}\right) = -(6) = -6$.

c. $\sqrt[3]{\dfrac{125}{64}} = \dfrac{5}{4}$ because $\left(\dfrac{5}{4}\right)^3 = \dfrac{5^3}{4^3} = \dfrac{125}{64}$.

d. $\sqrt[5]{-32} = -2$ because $(-2)^5 = -32$.

e. $\sqrt[4]{-81}$ is not a real number because no real number raised to the fourth power produces -81.

✓ **Checkpoint**  *Audio-video solution in English & Spanish at LarsonPrecalculus.com*

Evaluate each expression, if possible.

a. $-\sqrt{144}$ b. $\sqrt{-144}$

c. $\sqrt{\dfrac{25}{64}}$ d. $-\sqrt[3]{\dfrac{8}{27}}$

Here are some generalizations about the nth roots of real numbers.

Generalizations About nth Roots of Real Numbers			
Real Number a	Integer $n > 0$	Root(s) of a	Example
$a > 0$	n is even.	$\sqrt[n]{a},\ -\sqrt[n]{a}$	$\sqrt[4]{81} = 3,\ -\sqrt[4]{81} = -3$
$a > 0$ or $a < 0$	n is odd.	$\sqrt[n]{a}$	$\sqrt[3]{-8} = -2$
$a < 0$	n is even.	No real roots	$\sqrt{-4}$ is not a real number.
$a = 0$	n is even or odd.	$\sqrt[n]{0} = 0$	$\sqrt[5]{0} = 0$

Integers such as 1, 4, 9, 16, 25, and 36 are **perfect squares** because they have integer square roots. Similarly, integers such as 1, 8, 27, 64, and 125 are **perfect cubes** because they have integer cube roots.

Properties of Radicals

Let a and b be real numbers, variables, or algebraic expressions such that the roots below are real numbers, and let m and n be positive integers.

Property	Example
1. $\sqrt[n]{a^m} = \left(\sqrt[n]{a}\right)^m$	$\sqrt[3]{8^2} = \left(\sqrt[3]{8}\right)^2 = (2)^2 = 4$
2. $\sqrt[n]{a} \cdot \sqrt[n]{b} = \sqrt[n]{ab}$	$\sqrt{5} \cdot \sqrt{7} = \sqrt{5 \cdot 7} = \sqrt{35}$
3. $\dfrac{\sqrt[n]{a}}{\sqrt[n]{b}} = \sqrt[n]{\dfrac{a}{b}},\quad b \neq 0$	$\dfrac{\sqrt[4]{27}}{\sqrt[4]{9}} = \sqrt[4]{\dfrac{27}{9}} = \sqrt[4]{3}$
4. $\sqrt[m]{\sqrt[n]{a}} = \sqrt[mn]{a}$	$\sqrt[3]{\sqrt{10}} = \sqrt[6]{10}$
5. $\left(\sqrt[n]{a}\right)^n = a$	$\left(\sqrt{3}\right)^2 = 3$
6. For n even, $\sqrt[n]{a^n} = \lvert a \rvert$.	$\sqrt{(-12)^2} = \lvert -12 \rvert = 12$
For n odd, $\sqrt[n]{a^n} = a$.	$\sqrt[3]{(-12)^3} = -12$

A common use of Property 6 is $\sqrt{a^2} = \lvert a \rvert$.

EXAMPLE 9   **Using Properties of Radicals**

Use the properties of radicals to simplify each expression.

a. $\sqrt{8} \cdot \sqrt{2}$ **b.** $\left(\sqrt[3]{5}\right)^3$

c. $\sqrt[3]{x^3}$ **d.** $\sqrt[6]{y^6}$

Solution

a. $\sqrt{8} \cdot \sqrt{2} = \sqrt{8 \cdot 2} = \sqrt{16} = 4$ **b.** $\left(\sqrt[3]{5}\right)^3 = 5$

c. $\sqrt[3]{x^3} = x$ **d.** $\sqrt[6]{y^6} = \lvert y \rvert$

✓ *Checkpoint* *Audio-video solution in English & Spanish at LarsonPrecalculus.com*

Use the properties of radicals to simplify each expression.

a. $\dfrac{\sqrt{125}}{\sqrt{5}}$ **b.** $\sqrt[3]{125^2}$ **c.** $\sqrt[3]{x^2} \cdot \sqrt[3]{x}$ **d.** $\sqrt{\sqrt{x}}$

Simplifying Radical Expressions

An expression involving radicals is in **simplest form** when the three conditions below are satisfied.

1. All possible factors are removed from the radical.

2. All fractions have radical-free denominators (a process called *rationalizing the denominator* accomplishes this).

3. The index of the radical is reduced.

To simplify a radical, factor the radicand into factors whose exponents are multiples of the index. Write the roots of these factors outside the radical. The "leftover" factors make up the new radicand.

> **REMARK** When you simplify a radical, it is important that both the original and the simplified expressions are defined for the same values of the variable. For instance, in Example 10(c), $\sqrt{75x^3}$ and $5x\sqrt{3x}$ are both defined only for nonnegative values of x. Similarly, in Example 10(e), $\sqrt[4]{(5x)^4}$ and $5|x|$ are both defined for all real values of x.

EXAMPLE 10   **Simplifying Radical Expressions**

Perfect cube   Leftover factor

a. $\sqrt[3]{24} = \sqrt[3]{8 \cdot 3} = \sqrt[3]{2^3 \cdot 3} = 2\sqrt[3]{3}$

Perfect 4th power   Leftover factor

b. $\sqrt[4]{48} = \sqrt[4]{16 \cdot 3} = \sqrt[4]{2^4 \cdot 3} = 2\sqrt[4]{3}$

c. $\sqrt{75x^3} = \sqrt{25x^2 \cdot 3x} = \sqrt{(5x)^2 \cdot 3x} = 5x\sqrt{3x}$

d. $\sqrt[3]{24a^4} = \sqrt[3]{8a^3 \cdot 3a} = \sqrt[3]{(2a)^3 \cdot 3a} = 2a\sqrt[3]{3a}$

e. $\sqrt[4]{(5x)^4} = |5x| = 5|x|$

✓ **Checkpoint** 🔊)) *Audio-video solution in English & Spanish at LarsonPrecalculus.com*

Simplify each radical expression.

a. $\sqrt{32}$ **b.** $\sqrt[3]{250}$ **c.** $\sqrt{24a^5}$ **d.** $\sqrt[3]{-135x^3}$

Radical expressions can be combined (added or subtracted) when they are **like radicals**—that is, when they have the same index and radicand. For example, $\sqrt{2}$, $3\sqrt{2}$, and $\frac{1}{2}\sqrt{2}$ are like radicals, but $\sqrt{3}$ and $\sqrt{2}$ are unlike radicals. To determine whether two radicals can be combined, first simplify each radical.

EXAMPLE 11   **Combining Radical Expressions**

a. $2\sqrt{48} - 3\sqrt{27} = 2\sqrt{16 \cdot 3} - 3\sqrt{9 \cdot 3}$ Find square factors.

$= 8\sqrt{3} - 9\sqrt{3}$ Find square roots and multiply by coefficients.

$= (8 - 9)\sqrt{3}$ Combine like radicals.

$= -\sqrt{3}$ Simplify.

b. $\sqrt[3]{16x} - \sqrt[3]{54x^4} = \sqrt[3]{8 \cdot 2x} - \sqrt[3]{27x^3 \cdot 2x}$ Find cube factors.

$= 2\sqrt[3]{2x} - 3x\sqrt[3]{2x}$ Find cube roots.

$= (2 - 3x)\sqrt[3]{2x}$ Combine like radicals.

✓ **Checkpoint** 🔊)) *Audio-video solution in English & Spanish at LarsonPrecalculus.com*

Simplify each radical expression.

a. $3\sqrt{8} + \sqrt{18}$ **b.** $\sqrt[3]{81x^5} - \sqrt[3]{24x^2}$

Rationalizing Denominators and Numerators

To rationalize a denominator or numerator of the form $a - b\sqrt{m}$ or $a + b\sqrt{m}$, multiply both numerator and denominator by a **conjugate:** $a + b\sqrt{m}$ and $a - b\sqrt{m}$ are conjugates of each other. If $a = 0$, then the rationalizing factor for $\sqrt{m}$ is itself, $\sqrt{m}$. For cube roots, choose a rationalizing factor that produces a perfect cube radicand.

EXAMPLE 12   Rationalizing Single-Term Denominators

a. $\dfrac{5}{2\sqrt{3}} = \dfrac{5}{2\sqrt{3}} \cdot \dfrac{\sqrt{3}}{\sqrt{3}}$ $\sqrt{3}$ is rationalizing factor.

$\qquad\quad = \dfrac{5\sqrt{3}}{2(3)}$ Multiply.

$\qquad\quad = \dfrac{5\sqrt{3}}{6}$ Simplify.

b. $\dfrac{2}{\sqrt[3]{5}} = \dfrac{2}{\sqrt[3]{5}} \cdot \dfrac{\sqrt[3]{5^2}}{\sqrt[3]{5^2}}$ $\sqrt[3]{5^2}$ is rationalizing factor.

$\qquad\quad = \dfrac{2\sqrt[3]{5^2}}{\sqrt[3]{5^3}}$ Multiply.

$\qquad\quad = \dfrac{2\sqrt[3]{25}}{5}$ Simplify.

✓ *Checkpoint* ◀))) *Audio-video solution in English & Spanish at LarsonPrecalculus.com*

Rationalize the denominator of each expression.

a. $\dfrac{5}{3\sqrt{2}}$ **b.** $\dfrac{1}{\sqrt[3]{25}}$

EXAMPLE 13   Rationalizing a Denominator with Two Terms

$\dfrac{2}{3 + \sqrt{7}} = \dfrac{2}{3 + \sqrt{7}} \cdot \dfrac{3 - \sqrt{7}}{3 - \sqrt{7}}$ Multiply numerator and denominator by conjugate of denominator.

$\qquad\quad = \dfrac{2(3 - \sqrt{7})}{3(3 - \sqrt{7}) + \sqrt{7}(3 - \sqrt{7})}$ Distributive Property

$\qquad\quad = \dfrac{2(3 - \sqrt{7})}{3(3) - 3(\sqrt{7}) + \sqrt{7}(3) - \sqrt{7}(\sqrt{7})}$ Distributive Property

$\qquad\quad = \dfrac{2(3 - \sqrt{7})}{(3)^2 - (\sqrt{7})^2}$ Simplify.

$\qquad\quad = \dfrac{2(3 - \sqrt{7})}{2}$ Simplify.

$\qquad\quad = 3 - \sqrt{7}$ Divide out common factor.

✓ *Checkpoint* ◀))) *Audio-video solution in English & Spanish at LarsonPrecalculus.com*

Rationalize the denominator: $\dfrac{8}{\sqrt{6} - \sqrt{2}}$.   ◼

Sometimes it is necessary to rationalize the numerator of an expression. For instance, in Section P.5 you will use the technique shown in Example 14 on the next page to rationalize the numerator of an expression from calculus.

EXAMPLE 14 **Rationalizing a Numerator**

REMARK Do not confuse the expression $\sqrt{5} + \sqrt{7}$ with the expression $\sqrt{5 + 7}$. In general, $\sqrt{x + y}$ does not equal $\sqrt{x} + \sqrt{y}$. Similarly, $\sqrt{x^2 + y^2}$ does not equal $x + y$.

$$\frac{\sqrt{5} - \sqrt{7}}{2} = \frac{\sqrt{5} - \sqrt{7}}{2} \cdot \frac{\sqrt{5} + \sqrt{7}}{\sqrt{5} + \sqrt{7}}$$ Multiply numerator and denominator by conjugate of numerator.

$$= \frac{(\sqrt{5})^2 - (\sqrt{7})^2}{2(\sqrt{5} + \sqrt{7})}$$ Simplify.

$$= \frac{5 - 7}{2(\sqrt{5} + \sqrt{7})}$$ Property 5 of radicals

$$= \frac{-2}{2(\sqrt{5} + \sqrt{7})}$$ Simplify.

$$= \frac{-1}{\sqrt{5} + \sqrt{7}}$$ Divide out common factor.

✓ *Checkpoint* *Audio-video solution in English & Spanish at LarsonPrecalculus.com*

Rationalize the numerator: $\dfrac{2 - \sqrt{2}}{3}$.

Rational Exponents and Their Properties

Definition of Rational Exponents

If a is a real number and n is a positive integer such that the principal nth root of a exists, then $a^{1/n}$ is defined as

$$a^{1/n} = \sqrt[n]{a}.$$

Moreover, if m is a positive integer, then

$$a^{m/n} = (a^{1/n})^m.$$

$1/n$ and m/n are called **rational exponents** of a.

REMARK If m and n have no common factors, then it is also true that $a^{m/n} = (a^m)^{1/n}$.

The numerator of a rational exponent denotes the *power* to which the base is raised, and the denominator denotes the *index* or the *root* to be taken.

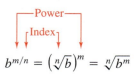

$$b^{m/n} = \left(\sqrt[n]{b}\right)^m = \sqrt[n]{b^m}$$

When you are working with rational exponents, the properties of integer exponents still apply. For example, $2^{1/2}2^{1/3} = 2^{(1/2)+(1/3)} = 2^{5/6}$.

EXAMPLE 15 **Changing From Radical to Exponential Form**

a. $\sqrt{3} = 3^{1/2}$

b. $\sqrt{(3xy)^5} = \sqrt[2]{(3xy)^5} = (3xy)^{5/2}$

c. $2x\sqrt[4]{x^3} = (2x)(x^{3/4}) = 2x^{1+(3/4)} = 2x^{7/4}$

✓ *Checkpoint* *Audio-video solution in English & Spanish at LarsonPrecalculus.com*

Write (a) $\sqrt[3]{27}$, (b) $\sqrt{x^3y^5z}$, and (c) $3x\sqrt[3]{x^2}$ in exponential form.

▷ **TECHNOLOGY** There are four
methods of evaluating radicals on
most graphing utilities. For square
roots, you can use the *square root
key* $\boxed{\sqrt{}}$. For cube roots, you can
use the *cube root key* $\boxed{\sqrt[3]{}}$. For
other roots, first convert the radical
to exponential form and then use
the *exponential key* $\boxed{\wedge}$, or use the
*x*th root key $\boxed{\sqrt[x]{}}$ (or menu choice).
Consult the user's guide for
your graphing utility for specific
keystrokes.

EXAMPLE 16   **Changing From Exponential to Radical Form**

See LarsonPrecalculus.com for an interactive version of this type of example.

a. $(x^2 + y^2)^{3/2} = \left(\sqrt{x^2 + y^2}\right)^3 = \sqrt{(x^2 + y^2)^3}$

b. $2y^{3/4}z^{1/4} = 2(y^3z)^{1/4} = 2\sqrt[4]{y^3z}$

c. $a^{-3/2} = \dfrac{1}{a^{3/2}} = \dfrac{1}{\sqrt{a^3}}$

d. $x^{0.2} = x^{1/5} = \sqrt[5]{x}$

✓ *Checkpoint* ◀))) *Audio-video solution in English & Spanish at LarsonPrecalculus.com*

Write each expression in radical form.

a. $(x^2 - 7)^{-1/2}$ **b.** $-3b^{1/3}c^{2/3}$

c. $a^{0.75}$ **d.** $(x^2)^{2/5}$ ◼

Rational exponents are useful for evaluating roots of numbers on a calculator, for
reducing the index of a radical, and for simplifying expressions in calculus.

EXAMPLE 17   **Simplifying with Rational Exponents**

a. $(-32)^{-4/5} = \left(\sqrt[5]{-32}\right)^{-4} = (-2)^{-4} = \dfrac{1}{(-2)^4} = \dfrac{1}{16}$

b. $(-5x^{5/3})(3x^{-3/4}) = -15x^{(5/3)-(3/4)} = -15x^{11/12}, \quad x \neq 0$

c. $\sqrt[9]{a^3} = a^{3/9} = a^{1/3} = \sqrt[3]{a}$ Reduce index.

d. $\sqrt[3]{\sqrt{125}} = \sqrt[6]{125} = \sqrt[6]{(5)^3} = 5^{3/6} = 5^{1/2} = \sqrt{5}$

e. $(2x - 1)^{4/3}(2x - 1)^{-1/3} = (2x - 1)^{(4/3)-(1/3)} = 2x - 1, \quad x \neq \dfrac{1}{2}$

✓ *Checkpoint* ◀))) *Audio-video solution in English & Spanish at LarsonPrecalculus.com*

▷ **REMARK** The expression in
Example 17(b) is not defined
when $x = 0$ because $0^{-3/4}$ is
not a real number. Similarly, the
expression in Example 17(e) is
not defined when $x = \frac{1}{2}$ because

$$\left(2 \cdot \tfrac{1}{2} - 1\right)^{-1/3} = (0)^{-1/3}$$

is not a real number.

Simplify each expression.

a. $(-125)^{-2/3}$ **b.** $(4x^2y^{3/2})(-3x^{-1/3}y^{-3/5})$

c. $\sqrt[3]{\sqrt[4]{27}}$ **d.** $(3x + 2)^{5/2}(3x + 2)^{-1/2}$ ◼

Summarize   **(Section P.2)**

1. Make a list of the properties of exponents *(page 14)*. For examples that use
   these properties, see Examples 1–4.

2. State the definition of scientific notation *(page 17)*. For examples involving
   scientific notation, see Examples 5–7.

3. Make a list of the properties of radicals *(page 19)*. For examples involving
   radicals, see Examples 8 and 9.

4. Explain how to simplify a radical expression *(page 20)*. For examples of
   simplifying radical expressions, see Examples 10 and 11.

5. Explain how to rationalize a denominator or a numerator *(page 21)*. For
   examples of rationalizing denominators and numerators, see Examples 12–14.

6. State the definition of a rational exponent *(page 22)*. For examples involving
   rational exponents, see Examples 15–17.

P.2 Exercises

See CalcChat.com for tutorial help and worked-out solutions to odd-numbered exercises.

Vocabulary: Fill in the blanks.

1. In the exponential form a^n, n is the _____ and a is the _____.
2. A convenient way of writing very large or very small numbers is _____ _____.
3. One of the two equal factors of a number is a _____ _____ of the number.
4. In the radical form $\sqrt[n]{a}$, the positive integer n is the _____ of the radical and the number a is the _____.
5. Radical expressions can be combined (added or subtracted) when they are _____ _____.
6. The expressions $a + b\sqrt{m}$ and $a - b\sqrt{m}$ are _____ of each other.
7. The process used to create a radical-free denominator is known as _____ the denominator.
8. In the expression $b^{m/n}$, m denotes the _____ to which the base is raised and n denotes the _____ or root to be taken.

Skills and Applications

 Evaluating Exponential Expressions In Exercises 9–14, evaluate each expression.

9. (a) $5 \cdot 5^3$ (b) $\dfrac{5^2}{5^4}$

10. (a) $(3^3)^0$ (b) -3^2

11. (a) $(2^3 \cdot 3^2)^2$ (b) $\left(-\dfrac{3}{5}\right)^3 \left(\dfrac{5}{3}\right)^2$

12. (a) $\dfrac{3}{3^{-4}}$ (b) $48(-4)^{-3}$

13. (a) $\dfrac{4 \cdot 3^{-2}}{2^{-2} \cdot 3^{-1}}$ (b) $(-2)^0$

14. (a) $3^{-1} + 2^{-2}$ (b) $(3^{-2})^2$

Evaluating an Algebraic Expression In Exercises 15–20, evaluate the expression for the given value of x.

15. $-3x^3$, $x = 2$
16. $7x^{-2}$, $x = 4$
17. $6x^0$, $x = 10$
18. $2x^3$, $x = -3$
19. $-3x^4$, $x = -2$
20. $12(-x)^3$, $x = -\frac{1}{3}$

 Using Properties of Exponents In Exercises 21–26, simplify each expression.

21. (a) $(5z)^3$ (b) $5x^4(x^2)$
22. (a) $(-2x)^2$ (b) $(4x^3)^0$
23. (a) $6y^2(2y^0)^2$ (b) $(-z)^3(3z^4)$
24. (a) $\dfrac{7x^2}{x^3}$ (b) $\dfrac{12(x+y)^3}{9(x+y)}$
25. (a) $\left(\dfrac{4}{y}\right)^3 \left(\dfrac{3}{y}\right)^4$ (b) $\left(\dfrac{b^{-2}}{a^{-2}}\right)\left(\dfrac{b}{a}\right)^2$
26. (a) $[(x^2y^{-2})^{-1}]^{-1}$ (b) $(5x^2z^6)^3(5x^2z^6)^{-3}$

Rewriting with Positive Exponents In Exercises 27–30, rewrite each expression with positive exponents. Simplify, if possible.

27. (a) $(x + 5)^0$ (b) $(2x^2)^{-2}$
28. (a) $(4y^{-2})(8y^4)$ (b) $(z + 2)^{-3}(z + 2)^{-1}$
29. (a) $\left(\dfrac{x^{-3}y^4}{5}\right)^{-3}$ (b) $\left(\dfrac{a^{-2}}{b^{-2}}\right)\left(\dfrac{b}{a}\right)^3$
30. (a) $\dfrac{3^n \cdot 3^{2n}}{3^{3n} \cdot 3^2}$ (b) $\dfrac{x^2 \cdot x^n}{x^3 \cdot x^n}$

 Scientific Notation In Exercises 31 and 32, write the number in scientific notation.

31. 10,250.4 32. -0.000125

Decimal Notation In Exercises 33–36, write the number in decimal notation.

33. 3.14×10^{-4} 34. -2.058×10^6
35. Light year: 9.46×10^{12} kilometers
36. Diameter of a human hair: 9.0×10^{-6} meter

Using Scientific Notation In Exercises 37 and 38, evaluate each expression without using a calculator.

37. (a) $(2.0 \times 10^9)(3.4 \times 10^{-4})$
    (b) $(1.2 \times 10^7)(5.0 \times 10^{-3})$
38. (a) $\dfrac{6.0 \times 10^8}{3.0 \times 10^{-3}}$ (b) $\dfrac{2.5 \times 10^{-3}}{5.0 \times 10^2}$

Evaluating Radical Expressions In Exercises 39 and 40, evaluate each expression without using a calculator.

39. (a) $\sqrt{9}$ (b) $\sqrt[3]{\frac{27}{8}}$ 40. (a) $\sqrt[3]{27}$ (b) $\left(\sqrt{36}\right)^3$

Using Properties of Radicals In Exercises 41 and 42, use the properties of radicals to simplify each expression.

41. (a) $\left(\sqrt[5]{2}\right)^5$ (b) $\sqrt[5]{32x^5}$
42. (a) $\sqrt{12} \cdot \sqrt{3}$ (b) $\sqrt[4]{(3x^2)^4}$

 Simplifying Radical Expressions In Exercises 43–50, simplify each radical expression.

43. (a) $\sqrt{20}$ (b) $\sqrt[3]{128}$
44. (a) $\sqrt[3]{\frac{16}{27}}$ (b) $\sqrt{\frac{75}{4}}$
45. (a) $\sqrt{72x^3}$ (b) $\sqrt{54xy^4}$
46. (a) $\sqrt{\frac{18^2}{z^3}}$ (b) $\sqrt{\frac{32a^4}{b^2}}$
47. (a) $\sqrt[3]{16x^5}$ (b) $\sqrt{75x^2y^{-4}}$
48. (a) $\sqrt[4]{3x^4y^2}$ (b) $\sqrt[5]{160x^8z^4}$
49. (a) $2\sqrt{20x^2} + 5\sqrt{125x^2}$
    (b) $8\sqrt{147x} - 3\sqrt{48x}$
50. (a) $3\sqrt[3]{54x^3} + \sqrt[3]{16x^3}$
    (b) $\sqrt[3]{64x} - \sqrt[3]{27x^4}$

 Rationalizing a Denominator In Exercises 51–54, rationalize the denominator of the expression. Then simplify your answer.

51. $\dfrac{1}{\sqrt{3}}$ 52. $\dfrac{8}{\sqrt[3]{2}}$

53. $\dfrac{5}{\sqrt{14} - 2}$ 54. $\dfrac{3}{\sqrt{5} + \sqrt{6}}$

ƒ Rationalizing a Numerator In Exercises 55 and 56, rationalize the numerator of the expression. Then simplify your answer.

55. $\dfrac{\sqrt{5} + \sqrt{3}}{3}$ 56. $\dfrac{\sqrt{7} - 3}{4}$

 Writing Exponential and Radical Forms In Exercises 57–60, fill in the missing form of the expression.

Radical Form	Rational Exponent Form
57. $\sqrt[3]{64}$	
58. $x^2\sqrt{x}$	
59.	$3x^{-2/3}$
60.	$a^{0.4}$

 Simplifying Expressions In Exercises 61–68, simplify each expression.

61. (a) $32^{-3/5}$ (b) $\left(\frac{16}{81}\right)^{-3/4}$
62. (a) $100^{-3/2}$ (b) $\left(\frac{9}{4}\right)^{-1/2}$

63. (a) $\sqrt[4]{3^2}$ (b) $\sqrt[6]{(x+1)^4}$
64. (a) $\sqrt[6]{x^3}$ (b) $\sqrt[4]{(3x^2)^4}$
65. (a) $\sqrt{\sqrt{32}}$ (b) $\sqrt{\sqrt[4]{2x}}$
66. (a) $\sqrt{\sqrt{243(x+1)}}$
    (b) $\sqrt{\sqrt[3]{10a^7b}}$
67. (a) $(x-1)^{1/3}(x-1)^{2/3}$
    (b) $(x-1)^{1/3}(x-1)^{-4/3}$
68. (a) $(4x+3)^{5/2}(4x+3)^{-5/3}$
    (b) $(4x+3)^{-5/2}(4x+3)^{2/3}$

• • 69. **Mathematical Modeling** • • • • • • •

A funnel is filled with water to a height of h centimeters. The formula

$t = 0.03[12^{5/2} - (12 - h)^{5/2}], \quad 0 \le h \le 12$

represents the amount of time t (in seconds) that it will take for the funnel to empty. Use the *table* feature of a graphing utility to find the times required for the funnel to empty for water heights of $h = 0, h = 1, h = 2, \ldots, h = 12$ centimeters.

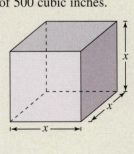

 70. **HOW DO YOU SEE IT?** Package A is a cube with a volume of 500 cubic inches. Package B is a cube with a volume of 250 cubic inches. Is the length x of a side of package A greater than, less than, or equal to twice the length of a side of package B? Explain.

Exploration

True or False? In Exercises 71–74, determine whether the statement is true or false. Justify your answer.

71. $\dfrac{x^{k+1}}{x} = x^k$ 72. $(a^n)^k = a^{n^k}$

73. $(a+b)^2 = a^2 + b^2$

74. $\dfrac{a}{\sqrt{b}} = \dfrac{a^2}{\left(\sqrt{b}\right)^2} = \dfrac{a^2}{b}$

P.3 Polynomials and Special Products

Polynomials have many real-life applications. For example, in Exercise 81 on page 33, you will work with polynomials that model uniformly distributed safe loads for steel beams.

■ Write polynomials in standard form.
■ Add, subtract, and multiply polynomials.
■ Use special products to multiply polynomials.
■ Use polynomials to solve real-life problems.

Polynomials

One of the most common types of algebraic expressions is the **polynomial.** Some examples are $2x + 5$, $3x^4 - 7x^2 + 2x + 4$, and $5x^2y^2 - xy + 3$. The first two are *polynomials in x* and the third is a *polynomial in x and y.* The terms of a polynomial in x have the form ax^k, where a is the **coefficient** and k is the **degree** of the term. For example, the polynomial $2x^3 - 5x^2 + 1 = 2x^3 + (-5)x^2 + (0)x + 1$ has coefficients 2, -5, 0, and 1.

Definition of a Polynomial in x

Let $a_0, a_1, a_2, \ldots, a_n$ be real numbers and let n be a nonnegative integer. A polynomial in x is an expression of the form

$$a_n x^n + a_{n-1}x^{n-1} + \cdots + a_1 x + a_0$$

where $a_n \neq 0$. The polynomial is of **degree** n, a_n is the **leading coefficient,** and a_0 is the **constant term.**

Polynomials with one, two, and three terms are **monomials, binomials,** and **trinomials,** respectively. A polynomial written with descending powers of x is in **standard form.**

EXAMPLE 1   **Writing Polynomials in Standard Form**

Polynomial	Standard Form	Degree	Leading Coefficient
a. $4x^2 - 5x^7 - 2 + 3x$	$-5x^7 + 4x^2 + 3x - 2$	7	-5
b. $4 - 9x^2$	$-9x^2 + 4$	2	-9
c. 8	8 or $8x^0$	0	8

✓ **Checkpoint** 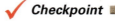 *Audio-video solution in English & Spanish at LarsonPrecalculus.com*

Write the polynomial $6 - 7x^3 + 2x$ in standard form. Then identify the degree and leading coefficient of the polynomial. ■

A polynomial that has all zero coefficients is called the **zero polynomial,** denoted by 0. No degree is assigned to the zero polynomial. For polynomials in more than one variable, the degree of a *term* is the sum of the exponents of the variables in the term. The degree of the *polynomial* is the highest degree of its terms. For example, the degree of the polynomial $-2x^3y^6 + 4xy - x^7y^4$ is 11 because the sum of the exponents in the last term is the greatest. The leading coefficient of the polynomial is the coefficient of the highest-degree term. Expressions are not polynomials when a variable is underneath a radical or when a polynomial expression (with degree greater than 0) is in the denominator of a term. For example, the expressions $x^3 - \sqrt{3x} = x^3 - (3x)^{1/2}$ and $x^2 + (5/x) = x^2 + 5x^{-1}$ are not polynomials.

Operations with Polynomials

You can add and subtract polynomials in much the same way you add and subtract real numbers. Add or subtract the *like terms* (terms having the same variables to the same powers) by adding or subtracting their coefficients. For example, $-3xy^2$ and $5xy^2$ are like terms and their sum is

$$-3xy^2 + 5xy^2 = (-3 + 5)xy^2 = 2xy^2.$$

EXAMPLE 2 **Adding or Subtracting Polynomials**

a. $(5x^3 - 7x^2 - 3) + (x^3 + 2x^2 - x + 8)$

$= (5x^3 + x^3) + (-7x^2 + 2x^2) + (-x) + (-3 + 8)$ Group like terms.

$= 6x^3 - 5x^2 - x + 5$ Combine like terms.

b. $(7x^4 - x^2 - 4x + 2) - (3x^4 - 4x^2 + 3x)$

$= 7x^4 - x^2 - 4x + 2 - 3x^4 + 4x^2 - 3x$ Distributive Property

$= (7x^4 - 3x^4) + (-x^2 + 4x^2) + (-4x - 3x) + 2$ Group like terms.

$= 4x^4 + 3x^2 - 7x + 2$ Combine like terms.

✓ **Checkpoint** 🔊))) *Audio-video solution in English & Spanish at LarsonPrecalculus.com*

Find the difference $(2x^3 - x + 3) - (x^2 - 2x - 3)$ and write the resulting polynomial in standard form.

> **REMARK** When a negative sign precedes an expression inside parentheses, remember to distribute the negative sign to each term inside the parentheses. In other words, multiply each term by -1.
>
> $-(3x^4 - 4x^2 + 3x)$
> $= -3x^4 + 4x^2 - 3x$

To find the *product* of two polynomials, use the right and left Distributive Properties. For example, you can find the product of $3x - 2$ and $5x + 7$ by first treating $5x + 7$ as a single quantity.

$(3x - 2)(5x + 7) = 3x(5x + 7) - 2(5x + 7)$

$= (3x)(5x) + (3x)(7) - (2)(5x) - (2)(7)$

$= 15x^2 + 21x - 10x - 14$

| Product of First terms | Product of Outer terms | Product of Inner terms | Product of Last terms |

$= 15x^2 + 11x - 14$

Note that when using the **FOIL Method** above (which can be used only to multiply two binomials), some of the terms in the product may be like terms that can be combined into one term.

EXAMPLE 3 **Finding a Product by the FOIL Method**

Use the FOIL Method to find the product of $2x - 4$ and $x + 5$.

Solution

$$(2x - 4)(x + 5) = 2x^2 + 10x - 4x - 20 = 2x^2 + 6x - 20$$

✓ **Checkpoint** 🔊))) *Audio-video solution in English & Spanish at LarsonPrecalculus.com*

Use the FOIL Method to find the product of $3x - 1$ and $x - 5$.

When multiplying two polynomials, be sure to multiply *each* term of one polynomial by *each* term of the other. A vertical arrangement can be helpful.

EXAMPLE 4   **A Vertical Arrangement for Multiplication**

Multiply $x^2 - 2x + 2$ by $x^2 + 2x + 2$ using a vertical arrangement.

Solution

$$
\begin{array}{ll}
x^2 - 2x + 2 & \text{Write in standard form.} \\
\underline{\times\ x^2 + 2x + 2} & \text{Write in standard form.} \\
2x^2 - 4x + 4 \qquad \Longleftarrow & 2(x^2 - 2x + 2) \\
2x^3 - 4x^2 + 4x \qquad \Longleftarrow & 2x(x^2 - 2x + 2) \\
\underline{x^4 - 2x^3 + 2x^2 } \qquad \Longleftarrow & x^2(x^2 - 2x + 2) \\
x^4 + 0x^3 + 0x^2 + 0x + 4 = x^4 + 4 & \text{Combine like terms.}
\end{array}
$$

So, $(x^2 - 2x + 2)(x^2 + 2x + 2) = x^4 + 4$.

✔ *Checkpoint* 🔊))) *Audio-video solution in English & Spanish at LarsonPrecalculus.com*

Multiply $x^2 + 2x + 3$ by $x^2 - 2x + 3$ using a vertical arrangement. ■

Special Products

Some binomial products have special forms that occur frequently in algebra. You do not need to memorize these formulas because you can use the Distributive Property to multiply. However, becoming familiar with these formulas will enable you to manipulate the algebra more quickly.

Special Products

Let u and v be real numbers, variables, or algebraic expressions.

Special Product	**Example**
Sum and Difference of Same Terms	
$(u + v)(u - v) = u^2 - v^2$	$(x + 4)(x - 4) = x^2 - 4^2$
	$\qquad\qquad\qquad = x^2 - 16$
Square of a Binomial	
$(u + v)^2 = u^2 + 2uv + v^2$	$(x + 3)^2 = x^2 + 2(x)(3) + 3^2$
	$\qquad\qquad = x^2 + 6x + 9$
$(u - v)^2 = u^2 - 2uv + v^2$	$(3x - 2)^2 = (3x)^2 - 2(3x)(2) + 2^2$
	$\qquad\qquad = 9x^2 - 12x + 4$
Cube of a Binomial	
$(u + v)^3 = u^3 + 3u^2v + 3uv^2 + v^3$	$(x + 2)^3 = x^3 + 3x^2(2) + 3x(2^2) + 2^3$
	$\qquad\qquad = x^3 + 6x^2 + 12x + 8$
$(u - v)^3 = u^3 - 3u^2v + 3uv^2 - v^3$	$(x - 1)^3 = x^3 - 3x^2(1) + 3x(1^2) - 1^3$
	$\qquad\qquad = x^3 - 3x^2 + 3x - 1$

EXAMPLE 5 **Sum and Difference of Same Terms**

Find the product of $5x + 9$ and $5x - 9$.

Solution

The product of a sum and a difference of the *same* two terms has no middle term and takes the form $(u + v)(u - v) = u^2 - v^2$.

$$(5x + 9)(5x - 9) = (5x)^2 - 9^2 = 25x^2 - 81$$

✓ *Checkpoint* ◀))) *Audio-video solution in English & Spanish at LarsonPrecalculus.com*

Find the product of $3x - 2$ and $3x + 2$.

· · REMARK When squaring a binomial, note that the resulting middle term is always *twice* the product of the two terms of the binomial. ▷

EXAMPLE 6 **Square of a Binomial**

Find $(6x - 5)^2$.

Solution

The square of the binomial $u - v$ is $(u - v)^2 = u^2 - 2uv + v^2$.

$$(6x - 5)^2 = (6x)^2 - 2(6x)(5) + 5^2 = 36x^2 - 60x + 25$$

✓ *Checkpoint* ◀))) *Audio-video solution in English & Spanish at LarsonPrecalculus.com*

Find $(x + 10)^2$.

EXAMPLE 7 **Cube of a Binomial**

Find $(3x + 2)^3$.

Solution

The cube of the binomial $u + v$ is $(u + v)^3 = u^3 + 3u^2v + 3uv^2 + v^3$. Note the *decreasing* powers of u and the *increasing* powers of v. Letting $u = 3x$ and $v = 2$,

$$(3x + 2)^3 = (3x)^3 + 3(3x)^2(2) + 3(3x)(2^2) + 2^3 = 27x^3 + 54x^2 + 36x + 8.$$

✓ *Checkpoint* ◀))) *Audio-video solution in English & Spanish at LarsonPrecalculus.com*

Find $(4x - 1)^3$.

EXAMPLE 8 **Multiplying Two Trinomials**

See LarsonPrecalculus.com for an interactive version of this type of example.

Find the product of $x + y - 2$ and $x + y + 2$.

Solution

One way to find this product is to group $x + y$ and form a special product.

Difference   Sum

$$(x + y - 2)(x + y + 2) = [(x + y) - 2][(x + y) + 2]$$

$$= (x + y)^2 - 2^2 \qquad \text{Sum and difference of same terms}$$

$$= x^2 + 2xy + y^2 - 4 \qquad \text{Square of a binomial}$$

✓ *Checkpoint* ◀))) *Audio-video solution in English & Spanish at LarsonPrecalculus.com*

Find the product of $x - 2 + 3y$ and $x - 2 - 3y$.

Application

EXAMPLE 9   **Finding the Volume of a Box**

An open box is made by cutting squares from the corners of a piece of metal that is 20 inches by 16 inches, as shown in the figure. The edge of each cut-out square is x inches. Find the volume of the box in terms of x. Then find the volume of the box when $x = 1$, $x = 2$, and $x = 3$.

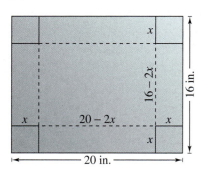

Solution

The volume of a rectangular box is equal to the product of its length, width, and height. From the figure, the length is $20 - 2x$, the width is $16 - 2x$, and the height is x. So, the volume of the box is

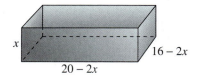

$$\text{Volume} = (20 - 2x)(16 - 2x)(x)$$

$$= (320 - 72x + 4x^2)(x)$$

$$= 320x - 72x^2 + 4x^3.$$

When $x = 1$ inch, the volume of the box is

$$\text{Volume} = 320(1) - 72(1)^2 + 4(1)^3$$

$$= 252 \text{ cubic inches.}$$

When $x = 2$ inches, the volume of the box is

$$\text{Volume} = 320(2) - 72(2)^2 + 4(2)^3$$

$$= 384 \text{ cubic inches.}$$

When $x = 3$ inches, the volume of the box is

$$\text{Volume} = 320(3) - 72(3)^2 + 4(3)^3$$

$$= 420 \text{ cubic inches.}$$

 ✓ *Checkpoint* ))) *Audio-video solution in English & Spanish at LarsonPrecalculus.com*

In Example 9, find the volume of the box in terms of x when the piece of metal is 12 inches by 10 inches. Then find the volume when $x = 2$ and $x = 3$.

Summarize   (Section P.3)

1. State the definition of a polynomial in x and explain what is meant by the standard form of a polynomial *(page 26)*. For an example of writing polynomials in standard form, see Example 1.

2. Explain how to add and subtract polynomials *(page 27)*. For an example of adding and subtracting polynomials, see Example 2.

3. Explain the FOIL Method *(page 27)*. For an example of finding a product using the FOIL Method, see Example 3.

4. Explain how to find binomial products that have special forms *(page 28)*. For examples of binomial products that have special forms, see Examples 5–8.

P.3 Exercises

See **CalcChat.com** for tutorial help and worked-out solutions to odd-numbered exercises.

Vocabulary: Fill in the blanks.

1. For the polynomial $a_n x^n + a_{n-1} x^{n-1} + \cdots + a_1 x + a_0$, $a_n \neq 0$, the degree is _____, the leading coefficient is _____, and the constant term is _____.

2. A polynomial with one term is a _____, while a polynomial with two terms is a _____ and a polynomial with three terms is a _____.

3. To add or subtract polynomials, add or subtract the _____ _____ by adding or subtracting their coefficients.

4. The letters in "FOIL" stand for F _____, O _____, I _____, and L _____.

Skills and Applications

Writing Polynomials in Standard Form In Exercises 5–10, (a) write the polynomial in standard form, (b) identify the degree and leading coefficient of the polynomial, and (c) state whether the polynomial is a monomial, a binomial, or a trinomial.

5. $7x$

6. 3

7. $14x - \frac{1}{2}x^5$

8. $3 + 2x$

9. $1 + 6x^4 - 4x^5$

10. $-y + 25y^2 + 1$

Identifying Polynomials In Exercises 11–16, determine whether the expression is a polynomial. If so, write the polynomial in standard form.

11. $2x - 3x^3 + 8$

12. $5x^4 - 2x^2 + x^{-2}$

13. $\dfrac{3x + 4}{x}$

14. $\dfrac{x^2 + 2x - 3}{2}$

15. $y^2 - y^4 + y^3$

16. $y^4 - \sqrt{y}$

Adding or Subtracting Polynomials In Exercises 17–24, add or subtract and write the result in standard form.

17. $(6x + 5) - (8x + 15)$

18. $(2x^2 + 1) - (x^2 - 2x + 1)$

19. $(t^3 - 1) + (6t^3 - 5t)$

20. $(4y^2 - 3) + (-7y^2 + 9)$

21. $(15x^2 - 6) + (-8.3x^3 - 14.7x^2 - 17)$

22. $(15.6w^4 - 14w - 17.4) + (16.9w^4 - 9.2w + 13)$

23. $5z - [3z - (10z + 8)]$

24. $(y^3 + 1) - [(y^2 + 1) + (3y - 7)]$

Multiplying Polynomials In Exercises 25–38, multiply the polynomials.

25. $3x(x^2 - 2x + 1)$

26. $y^2(4y^2 + 2y - 3)$

27. $-5z(3z - 1)$

28. $-3x(5x + 2)$

29. $(1.5t^2 + 5)(-3t)$

30. $(2 - 3.5y)(2y^3)$

31. $-2x(0.1x + 17)$

32. $6y(5 - \frac{3}{8}y)$

33. $(x + 7)(x + 5)$

34. $(x - 8)(x + 4)$

35. $(3x - 5)(2x + 1)$

36. $(7x - 2)(4x - 3)$

37. $(x^2 - x + 2)(x^2 + x + 1)$

38. $(2x^2 - x + 4)(x^2 + 3x + 2)$

Finding Special Products In Exercises 39–62, find the special product.

39. $(x + 10)(x - 10)$

40. $(2x + 3)(2x - 3)$

41. $(x + 2y)(x - 2y)$

42. $(4a + 5b)(4a - 5b)$

43. $(2x + 3)^2$

44. $(5 - 8x)^2$

45. $(4x^3 - 3)^2$

46. $(8x + 3)^2$

47. $(x + 3)^3$

48. $(x - 2)^3$

49. $(2x - y)^3$

50. $(3x + 2y)^3$

51. $\left(\frac{1}{5}x - 3\right)\left(\frac{1}{5}x + 3\right)$

52. $(1.5x - 4)(1.5x + 4)$

53. $(-6x + 3y)(-6x - 3y)$

54. $(3a^3 - 4b^2)(3a^3 + 4b^2)$

55. $\left(\frac{1}{4}x - 5\right)^2$

56. $(2.4x + 3)^2$

57. $[(x - 3) + y]^2$

58. $[(x + 1) - y]^2$

59. $[(m - 3) + n][(m - 3) - n]$

60. $[(x - 3y) + z][(x - 3y) - z]$

61. $(u + 2)(u - 2)(u^2 + 4)$

62. $(x + y)(x - y)(x^2 + y^2)$

Operations with Polynomials In Exercises 63–66, perform the operation.

63. Subtract $4x^2 - 5$ from $-3x^3 + x^2 + 9$.

64. Subtract $-7t^4 + 5t^2 - 1$ from $2t^4 - 10t^3 - 4t$.

65. Multiply $y^2 + 3y - 5$ by $y^2 - 6y + 4$.

66. Multiply $x^2 + 4x - 1$ by $x^2 - x + 3$.

Finding a Product In Exercises 67–70, find the product. (The expressions are not polynomials, but the formulas can still be used.)

67. $\left(\sqrt{x} + \sqrt{y}\right)\left(\sqrt{x} - \sqrt{y}\right)$ 68. $\left(5 + \sqrt{x}\right)\left(5 - \sqrt{x}\right)$

69. $\left(x - \sqrt{5}\right)^2$ 70. $\left(x + \sqrt{3}\right)^2$

71. **Cost, Revenue, and Profit** An electronics manufacturer can produce and sell x MP3 players per week. The total cost C (in dollars) of producing x MP3 players is $C = 93x + 35{,}000$, and the total revenue R (in dollars) is $R = 135x$.

(a) Find the profit P in terms of x.

(b) Find the profit obtained by selling 5000 MP3 players per week.

72. **Compound Interest** An investment of $500 compounded annually for 2 years at an interest rate r (in decimal form) yields an amount of $500(1 + r)^2$.

(a) Write this polynomial in standard form.

(b) Use a calculator to evaluate the polynomial for the values of r given in the table.

r	$2\frac{1}{2}\%$	3%	4%	$4\frac{1}{2}\%$	5%
$500(1 + r)^2$					

(c) What conclusion can you make from the table?

73. **Genetics** In deer, the gene N is for normal coloring and the gene a is for albino. Any gene combination with an N results in normal coloring. The Punnett square shows the possible gene combinations of an offspring and the resulting colors when both parents have the gene combination Na.

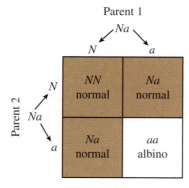

Parent 1

(a) What percent of the possible gene combinations result in albino coloring?

(b) Each parent's gene combination is represented by the polynomial $0.5N + 0.5a$. The product $(0.5N + 0.5a)^2$ represents the possible gene combinations of an offspring. Find this product.

(c) The coefficient of each term of the polynomial you wrote in part (b) is the probability (in decimal form) of the offspring having that gene combination. Use this polynomial to confirm your answer in part (a). Explain.

74. **Construction Management** A square-shaped foundation for a building with 100-foot sides is reduced by x feet on one side and extended by x feet on an adjacent side.

(a) The area of the new foundation is represented by $(100 - x)(100 + x)$. Find this product.

(b) Does the area of the foundation increase, decrease, or stay the same? Explain.

(c) Use the polynomial in part (a) to find the area of the new foundation when $x = 21$.

Geometry In Exercises 75–78, find the area of the shaded region in terms of x. Write your result as a polynomial in standard form.

75.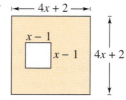

76.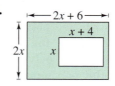

77.

78.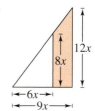

79. **Volume of a Box** A take-out fast-food restaurant is constructing an open box by cutting squares from the corners of the piece of cardboard shown in the figure. The edge of each cut-out square is x centimeters.

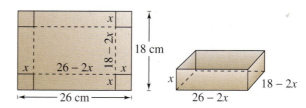

(a) Find the volume of the box in terms of x.

(b) Find the volume when $x = 1$, $x = 2$, and $x = 3$.

80. **Volume of a Box** An overnight shipping company designs a closed box by cutting along the solid lines and folding along the broken lines on the rectangular piece of corrugated cardboard shown in the figure.

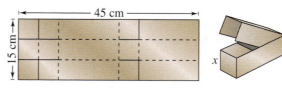

(a) Find the volume of the shipping box in terms of x.

(b) Find the volume when $x = 3$, $x = 5$, and $x = 7$.

81. Engineering

A one-inch-wide steel beam has a uniformly distributed load. When the span of the beam is x feet and its depth is 6 inches, the safe load S (in pounds) is approximately $S_6 = (0.06x^2 - 2.42x + 38.71)^2$. When the depth is 8 inches, the safe load is approximately $S_8 = (0.08x^2 - 3.30x + 51.93)^2$.

(a) Approximate the difference of the safe loads for these two beams when the span is 12 feet.

(b) How does the difference of the safe loads change as the span increases?

82. Stopping Distance The stopping distance of an automobile is the distance traveled during the driver's reaction time plus the distance traveled after the driver applies the brakes. In an experiment, researchers measured these distances (in feet) when the automobile was traveling at a speed of x miles per hour on dry, level pavement, as shown in the bar graph. The distance traveled during the reaction time R was

$$R = 1.1x$$

and the braking distance B was

$$B = 0.0475x^2 - 0.001x + 0.23.$$

(a) Determine the polynomial that represents the total stopping distance T.

(b) Use the result of part (a) to estimate the total stopping distance when $x = 30$, $x = 40$, and $x = 55$ miles per hour.

(c) Use the bar graph to make a statement about the total stopping distance required for increasing speeds.

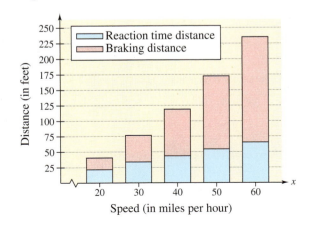

Exploration

True or False? In Exercises 83–86, determine whether the statement is true or false. Justify your answer.

83. The product of two binomials is always a second-degree polynomial.

84. The sum of two second-degree polynomials is always a second-degree polynomial.

85. The sum of two binomials is always a binomial.

86. The leading coefficient of the product of two polynomials is always the product of the leading coefficients of the two polynomials.

87. Degree of a Product Find the degree of the product of two polynomials of degrees m and n.

88. Degree of a Sum Find the degree of the sum of two polynomials of degrees m and n, where $m < n$.

89. Error Analysis Describe the error.

$$(x - 3)^2 = x^2 + 9 \quad ✗$$

90. HOW DO YOU SEE IT? An open box has a length of $(52 - 2x)$ inches, a width of $(42 - 2x)$ inches, and a height of x inches, as shown.

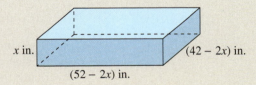

(a) Describe a way that you could make the box from a rectangular piece of cardboard. Give the original dimensions of the cardboard.

(b) What degree is the polynomial that represents the volume of the box? Explain your reasoning.

(c) Describe a procedure for finding the value of x (to the nearest tenth of an inch) that yields the maximum possible volume of the box.

91. Think About It When the polynomial

$$-x^3 + 3x^2 + 2x - 1$$

is subtracted from an unknown polynomial, the difference is $5x^2 + 8$. Find the unknown polynomial.

92. Logical Reasoning Verify that $(x + y)^2$ is not equal to $x^2 + y^2$ by letting $x = 3$ and $y = 4$ and evaluating both expressions. Are there any values of x and y for which $(x + y)^2$ and $x^2 + y^2$ are equal? Explain.

P.4 Factoring Polynomials

- ◼ Factor out common factors from polynomials.
- ◼ Factor special polynomial forms.
- ◼ Factor trinomials as the product of two binomials.
- ◼ Factor polynomials by grouping.

Polynomial factoring has many real-life applications. For example, in Exercise 84 on page 40, you will use polynomial factoring to write an alternative form of an expression that models the rate of change of an autocatalytic chemical reaction.

Polynomials with Common Factors

The process of writing a polynomial as a product is called **factoring.** It is an important tool for solving equations and for simplifying rational expressions.

Unless noted otherwise, when you are asked to factor a polynomial, assume that you are looking for factors that have integer coefficients. If a polynomial does not factor using integer coefficients, then it is **prime** or **irreducible over the integers.** For example, the polynomial $x^2 - 3$ is irreducible over the integers. Over the *real numbers*, this polynomial factors as

$$x^2 - 3 = (x + \sqrt{3})(x - \sqrt{3}).$$

A polynomial is **completely factored** when each of its factors is prime. For example,

$$x^3 - x^2 + 4x - 4 = (x - 1)(x^2 + 4) \qquad \text{Completely factored}$$

is completely factored, but

$$x^3 - x^2 - 4x + 4 = (x - 1)(x^2 - 4) \qquad \text{Not completely factored}$$

is not completely factored. Its complete factorization is

$$x^3 - x^2 - 4x + 4 = (x - 1)(x + 2)(x - 2).$$

The simplest type of factoring involves a polynomial that can be written as the product of a monomial and another polynomial. The technique used here is the Distributive Property, $a(b + c) = ab + ac$, in the *reverse* direction.

$$ab + ac = a(b + c) \qquad \text{\textit{a} is a common factor.}$$

Factoring out any common factors is the first step in completely factoring a polynomial.

EXAMPLE 1 **Factoring Out Common Factors**

Factor each expression.

a. $6x^3 - 4x$ **b.** $-4x^2 + 12x - 16$ **c.** $(x - 2)(2x) + (x - 2)(3)$

Solution

a. $6x^3 - 4x = 2x(3x^2) - 2x(2) \qquad \text{2\textit{x} is a common factor.}$

$\qquad\qquad\;\; = 2x(3x^2 - 2)$

b. $-4x^2 + 12x - 16 = -4(x^2) + (-4)(-3x) + (-4)4 \qquad \text{−4 is a common factor.}$

$\qquad\qquad\qquad\qquad\;\; = -4(x^2 - 3x + 4)$

c. $(x - 2)(2x) + (x - 2)(3) = (x - 2)(2x + 3) \qquad \text{(\textit{x} − 2) is a common factor.}$

✓ *Checkpoint* ◀))) *Audio-video solution in English & Spanish at LarsonPrecalculus.com*

Factor each expression.

a. $5x^3 - 15x^2$ **b.** $-3 + 6x - 12x^3$ **c.** $(x + 1)(x^2) - (x + 1)(2)$ ◼

Factoring Special Polynomial Forms

Some polynomials have special forms that arise from the special product forms on page 28. You should learn to recognize these forms.

Factoring Special Polynomial Forms

Factored Form	Example

Difference of Two Squares

$u^2 - v^2 = (u + v)(u - v)$ $\qquad\qquad$ $9x^2 - 4 = (3x)^2 - 2^2 = (3x + 2)(3x - 2)$

Perfect Square Trinomial

$u^2 + 2uv + v^2 = (u + v)^2$ $\qquad\qquad$ $x^2 + 6x + 9 = x^2 + 2(x)(3) + 3^2 = (x + 3)^2$

$u^2 - 2uv + v^2 = (u - v)^2$ $\qquad\qquad$ $x^2 - 6x + 9 = x^2 - 2(x)(3) + 3^2 = (x - 3)^2$

Sum or Difference of Two Cubes

$u^3 + v^3 = (u + v)(u^2 - uv + v^2)$ $\qquad\qquad$ $x^3 + 8 = x^3 + 2^3 = (x + 2)(x^2 - 2x + 4)$

$u^3 - v^3 = (u - v)(u^2 + uv + v^2)$ $\qquad\qquad$ $27x^3 - 1 = (3x)^3 - 1^3 = (3x - 1)(9x^2 + 3x + 1)$

The factored form of the difference of two squares is always a set of **conjugate pairs.**

$$u^2 - v^2 = (u + v)(u - v) \qquad\qquad \text{Conjugate pairs}$$

Difference   Opposite signs

To recognize perfect square terms, look for coefficients that are squares of integers and variables raised to *even powers.*

EXAMPLE 2 **Factoring Out a Common Factor First**

$3 - 12x^2 = 3(1 - 4x^2)$ $\qquad\qquad$ 3 is a common factor.

$\qquad\qquad = 3[1^2 - (2x)^2]$ $\qquad\qquad$ Rewrite $1 - 4x^2$ as the difference of two squares.

$\qquad\qquad = 3(1 + 2x)(1 - 2x)$ $\qquad\qquad$ Factor.

REMARK In Example 2, note that the first step in factoring a polynomial is to check for any common factors. Once you have removed any common factors, it is often possible to recognize patterns that were not immediately obvious.

 Checkpoint *Audio-video solution in English & Spanish at LarsonPrecalculus.com*

Factor $100 - 4y^2$.

EXAMPLE 3 **Factoring the Difference of Two Squares**

a. $(x + 2)^2 - y^2 = [(x + 2) + y][(x + 2) - y]$

$\qquad\qquad = (x + 2 + y)(x + 2 - y)$

b. $16x^4 - 81 = (4x^2)^2 - 9^2$ $\qquad\qquad$ Rewrite as the difference of two squares.

$\qquad\qquad = (4x^2 + 9)(4x^2 - 9)$ $\qquad\qquad$ Factor.

$\qquad\qquad = (4x^2 + 9)[(2x)^2 - 3^2]$ $\qquad\qquad$ Rewrite $4x^2 - 9$ as the difference of two squares.

$\qquad\qquad = (4x^2 + 9)(2x + 3)(2x - 3)$ $\qquad\qquad$ Factor.

 Checkpoint *Audio-video solution in English & Spanish at LarsonPrecalculus.com*

Factor $(x - 1)^2 - 9y^4$.

A **perfect square trinomial** is the square of a binomial, and it has the form

$$u^2 + 2uv + v^2 = (u + v)^2 \quad \text{or} \quad u^2 - 2uv + v^2 = (u - v)^2.$$

Like signs   Like signs

Note that the first and last terms are squares and the middle term is twice the product of u and v.

EXAMPLE 4   **Factoring Perfect Square Trinomials**

Factor each trinomial.

a. $x^2 - 10x + 25$ **b.** $16x^2 + 24x + 9$

Solution

a. $x^2 - 10x + 25 = x^2 - 2(x)(5) + 5^2 = (x - 5)^2$

b. $16x^2 + 24x + 9 = (4x)^2 + 2(4x)(3) + 3^2 = (4x + 3)^2$

✓ **Checkpoint** ◀)) *Audio-video solution in English & Spanish at LarsonPrecalculus.com*

Factor $9x^2 - 30x + 25$.

The next two formulas show the sum and difference of two cubes. Pay special attention to the signs of the terms.

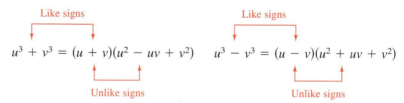

Like signs   Like signs

$$u^3 + v^3 = (u + v)(u^2 - uv + v^2) \quad u^3 - v^3 = (u - v)(u^2 + uv + v^2)$$

Unlike signs   Unlike signs

EXAMPLE 5   **Factoring the Difference of Two Cubes**

$$
\begin{aligned}
x^3 - 27 &= x^3 - 3^3 && \text{Rewrite 27 as } 3^3. \\
&= (x - 3)(x^2 + 3x + 9) && \text{Factor.}
\end{aligned}
$$

✓ **Checkpoint** ◀)) *Audio-video solution in English & Spanish at LarsonPrecalculus.com*

Factor $64x^3 - 1$.

EXAMPLE 6   **Factoring the Sum of Two Cubes**

$$
\begin{aligned}
\textbf{a.} \quad y^3 + 8 &= y^3 + 2^3 && \text{Rewrite 8 as } 2^3. \\
&= (y + 2)(y^2 - 2y + 4) && \text{Factor.} \\
\textbf{b.} \quad 3x^3 + 192 &= 3(x^3 + 64) && \text{3 is a common factor.} \\
&= 3(x^3 + 4^3) && \text{Rewrite 64 as } 4^3. \\
&= 3(x + 4)(x^2 - 4x + 16) && \text{Factor.}
\end{aligned}
$$

✓ **Checkpoint** ◀)) *Audio-video solution in English & Spanish at LarsonPrecalculus.com*

Factor each expression.

a. $x^3 + 216$ **b.** $5y^3 + 135$

Trinomials with Binomial Factors

To factor a trinomial of the form $ax^2 + bx + c$, use the pattern below.

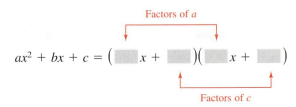

$$ax^2 + bx + c = \left(\boxed{}\, x + \boxed{}\right)\left(\boxed{}\, x + \boxed{}\right)$$

The goal is to find a combination of factors of a and c such that the sum of the outer and inner products is the middle term bx. For example, for the trinomial $6x^2 + 17x + 5$, you can write all possible factorizations and determine which one has outer and inner products whose sum is $17x$.

$$(6x + 5)(x + 1),\ (6x + 1)(x + 5),\ (2x + 1)(3x + 5),\ (2x + 5)(3x + 1)$$

The correct factorization is $(2x + 5)(3x + 1)$ because the sum of the outer (O) and inner (I) products is $17x$.

$$(2x + 5)(3x + 1) = 6x^2 + 2x + 15x + 5 = 6x^2 + 17x + 5$$

REMARK Factoring a trinomial can involve trial and error. However, it is relatively easy to check your answer by multiplying the factors. The product should be the original trinomial. For instance, in Example 7, verify that $(x - 3)(x - 4) = x^2 - 7x + 12$.

EXAMPLE 7   Factoring a Trinomial: Leading Coefficient Is 1

Factor $x^2 - 7x + 12$.

Solution   For this trinomial, $a = 1$, $b = -7$, and $c = 12$. Because b is negative and c is positive, both factors of 12 must be negative. So, the possible factorizations of $x^2 - 7x + 12$ are

$$(x - 1)(x - 12),\quad (x - 2)(x - 6),\quad \text{and}\quad (x - 3)(x - 4).$$

Testing the middle term, you will find the correct factorization to be

$$x^2 - 7x + 12 = (x - 3)(x - 4). \qquad \text{O + I} = -4x - 3x = -7x$$

✓ **Checkpoint** ◀))) *Audio-video solution in English & Spanish at LarsonPrecalculus.com*

Factor $x^2 + x - 6$.

EXAMPLE 8   Factoring a Trinomial: Leading Coefficient Is Not 1

See LarsonPrecalculus.com for an interactive version of this type of example.

Factor $2x^2 + x - 15$.

Solution   For this trinomial, $a = 2$, $b = 1$, and $c = -15$. Because c is negative, its factors must have unlike signs. The eight possible factorizations are below.

$$(2x - 1)(x + 15)\quad (2x + 1)(x - 15)\quad (2x - 3)(x + 5)\quad (2x + 3)(x - 5)$$
$$(2x - 5)(x + 3)\quad (2x + 5)(x - 3)\quad (2x - 15)(x + 1)\quad (2x + 15)(x - 1)$$

Testing the middle term, you will find the correct factorization to be

$$2x^2 + x - 15 = (2x - 5)(x + 3). \qquad \text{O + I} = 6x - 5x = x$$

✓ **Checkpoint** ◀))) *Audio-video solution in English & Spanish at LarsonPrecalculus.com*

Factor $2x^2 - 5x + 3$.

Factoring by Grouping

Sometimes, polynomials with more than three terms can be **factored by grouping.**

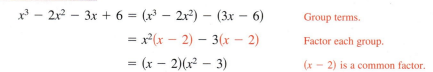

$$x^3 - 2x^2 - 3x + 6 = (x^3 - 2x^2) - (3x - 6) \qquad \text{Group terms.}$$
$$= x^2(x - 2) - 3(x - 2) \qquad \text{Factor each group.}$$
$$= (x - 2)(x^2 - 3) \qquad (x - 2) \text{ is a common factor.}$$

✓ *Checkpoint* *Audio-video solution in English & Spanish at LarsonPrecalculus.com*

Factor $x^3 + x^2 - 5x - 5$.

> **REMARK** Sometimes, more than one grouping will work. For instance, another way to factor the polynomial in Example 9 is
>
> $$x^3 - 2x^2 - 3x + 6$$
> $$= (x^3 - 3x) - (2x^2 - 6)$$
> $$= x(x^2 - 3) - 2(x^2 - 3)$$
> $$= (x^2 - 3)(x - 2).$$
>
> Notice that this is the same result as in Example 9.

Factoring by grouping can eliminate some of the trial and error involved in factoring a trinomial. To factor a trinomial of the form $ax^2 + bx + c$ by grouping, choose factors of the product ac that sum to b and use these factors to rewrite the middle term. Example 10 illustrates this technique.

EXAMPLE 10 **Factoring a Trinomial by Grouping**

In the trinomial $2x^2 + 5x - 3$, $a = 2$ and $c = -3$, so the product ac is -6. Now, -6 factors as $(6)(-1)$ and $6 + (-1) = 5 = b$. So, rewrite the middle term as $5x = 6x - x$ and factor by grouping.

$$2x^2 + 5x - 3 = 2x^2 + 6x - x - 3 \qquad \text{Rewrite middle term.}$$
$$= (2x^2 + 6x) - (x + 3) \qquad \text{Group terms.}$$
$$= 2x(x + 3) - (x + 3) \qquad \text{Factor } 2x^2 + 6x.$$
$$= (x + 3)(2x - 1) \qquad (x + 3) \text{ is a common factor.}$$

✓ *Checkpoint* *Audio-video solution in English & Spanish at LarsonPrecalculus.com*

Use factoring by grouping to factor $2x^2 + 5x - 12$.

Guidelines for Factoring Polynomials

1. Factor out any common factors using the Distributive Property.
2. Factor according to one of the special polynomial forms.
3. Factor as $ax^2 + bx + c = (mx + r)(nx + s)$.
4. Factor by grouping.

Summarize (Section P.4)

1. Explain what it means to completely factor a polynomial *(page 34)*. For an example of factoring out common factors, see Example 1.
2. Make a list of the special polynomial forms of factoring *(page 35)*. For examples of factoring these special forms, see Examples 2–6.
3. Explain how to factor a trinomial of the form $ax^2 + bx + c$ *(page 37)*. For examples of factoring trinomials of this form, see Examples 7 and 8.
4. Explain how to factor a polynomial by grouping *(page 38)*. For examples of factoring by grouping, see Examples 9 and 10.

P.4 Exercises

See CalcChat.com for tutorial help and worked-out solutions to odd-numbered exercises.

Vocabulary: Fill in the blanks.

1. The process of writing a polynomial as a product is called _____.

2. A polynomial is _____ _____ when each of its factors is prime.

3. A _____ _____ _____ is the square of a binomial, and it has the form $u^2 + 2uv + v^2$ or $u^2 - 2uv + v^2$.

4. Sometimes, polynomials with more than three terms can be factored by _____.

Skills and Applications

 Factoring Out a Common Factor In Exercises 5–8, factor out the common factor.

5. $2x^3 - 6x$

6. $3z^3 - 6z^2 + 9z$

7. $3x(x - 5) + 8(x - 5)$

8. $(x + 3)^2 - 4(x + 3)$

 Factoring the Difference of Two Squares In Exercises 9–18, completely factor the difference of two squares.

9. $x^2 - 81$

10. $x^2 - 64$

11. $25y^2 - 4$

12. $4y^2 - 49$

13. $64 - 9z^2$

14. $81 - 36z^2$

15. $(x - 1)^2 - 4$

16. $25 - (z + 5)^2$

17. $81u^4 - 1$

18. $x^4 - 16y^4$

 Factoring a Perfect Square Trinomial In Exercises 19–24, factor the perfect square trinomial.

19. $x^2 - 4x + 4$

20. $4t^2 + 4t + 1$

21. $25z^2 - 30z + 9$

22. $36y^2 + 84y + 49$

23. $4y^2 - 12y + 9$

24. $9u^2 + 24uv + 16v^2$

 Factoring the Sum or Difference of Two Cubes In Exercises 25–32, factor the sum or difference of two cubes.

25. $x^3 - 8$

26. $x^3 + 125$

27. $8t^3 - 1$

28. $27z^3 + 1$

29. $27x^3 + 8$

30. $64y^3 - 125$

31. $u^3 + 27v^3$

32. $(x + 2)^3 - y^3$

 Factoring a Trinomial In Exercises 33–42, factor the trinomial.

33. $x^2 + x - 2$

34. $x^2 + 5x + 6$

35. $s^2 - 5s + 6$

36. $t^2 - t - 6$

37. $3x^2 + 10x - 8$

38. $2x^2 - 3x - 27$

39. $5x^2 + 31x + 6$

40. $8x^2 + 51x + 18$

41. $-5y^2 - 8y + 4$

42. $-6z^2 + 17z + 3$

 Factoring by Grouping In Exercises 43–48, factor by grouping.

43. $x^3 - x^2 + 2x - 2$

44. $x^3 + 5x^2 - 5x - 25$

45. $2x^3 - x^2 - 6x + 3$

46. $3x^3 + x^2 - 15x - 5$

47. $3x^5 + 6x^3 - 2x^2 - 4$

48. $8x^5 - 6x^2 + 12x^3 - 9$

 Factoring a Trinomial by Grouping In Exercises 49–52, factor the trinomial by grouping.

49. $2x^2 + 9x + 9$

50. $6x^2 + x - 2$

51. $6x^2 - x - 15$

52. $12x^2 - 13x + 1$

Factoring Completely In Exercises 53–70, completely factor the expression.

53. $6x^2 - 54$

54. $12x^2 - 48$

55. $x^3 - x^2$

56. $x^3 - 16x$

57. $1 - 4x + 4x^2$

58. $-9x^2 + 6x - 1$

59. $2x^2 + 4x - 2x^3$

60. $9x^2 + 12x - 3x^3$

61. $(x^2 + 3)^2 - 16x^2$

62. $(x^2 + 8)^2 - 36x^2$

63. $2x^3 + x^2 - 8x - 4$

64. $3x^3 + x^2 - 27x - 9$

65. $2x(3x + 1) + (3x + 1)^2$

66. $4x(2x - 1) + (2x - 1)^2$

67. $2(x - 2)(x + 1)^2 - 3(x - 2)^2(x + 1)$

68. $2(x + 1)(x - 3)^2 - 3(x + 1)^2(x - 3)$

69. $5(2x + 1)^2(x + 1)^2 + (2x + 1)(x + 1)^3$

70. $7(3x + 2)^2(1 - x)^2 + (3x + 2)(1 - x^3)$

Fractional Coefficients In Exercises 71–76, completely factor the expression. (*Hint:* The factors will contain fractional coefficients.)

71. $16x^2 - \frac{1}{9}$

72. $\frac{4}{25}y^2 - 64$

73. $z^2 + z + \frac{1}{4}$

74. $9y^2 - \frac{3}{2}y + \frac{1}{16}$

75. $y^3 + \frac{8}{27}$

76. $x^3 - \frac{27}{64}$

Geometric Modeling In Exercises 77–80, draw a "geometric factoring model" to represent the factorization. For example, a factoring model for $2x^2 + 3x + 1 = (2x + 1)(x + 1)$ is shown below.

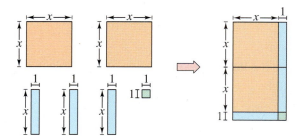

77. $x^2 + 3x + 2 = (x + 2)(x + 1)$
78. $x^2 + 4x + 3 = (x + 3)(x + 1)$
79. $2x^2 + 7x + 3 = (2x + 1)(x + 3)$
80. $3x^2 + 7x + 2 = (3x + 1)(x + 2)$

Geometry In Exercises 81 and 82, write an expression in factored form for the area of the shaded portion of the figure.

81.

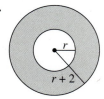

82.

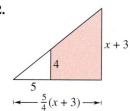

83. Geometry The cylindrical shell shown in the figure has a volume of

$$V = \pi R^2 h - \pi r^2 h.$$

(a) Factor the expression for the volume.

(b) From the result of part (a), show that the volume is

$$2\pi(\text{average radius})(\text{thickness of the shell})h.$$

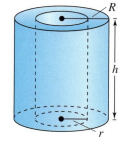

84. Chemistry

The rate of change of an autocatalytic chemical reaction is

$kQx - kx^2$

where Q is the amount of the original substance, x is the amount of substance formed, and k is a constant of proportionality. Factor the expression.

Factoring a Trinomial In Exercises 85 and 86, find all values of b for which the trinomial is factorable.

85. $x^2 + bx - 15$ **86.** $x^2 + bx + 24$

Factoring a Trinomial In Exercises 87 and 88, find two integer values of c such that the trinomial is factorable. (There are many correct answers.)

87. $2x^2 + 5x + c$ **88.** $3x^2 - x + c$

Exploration

True or False? In Exercises 89 and 90, determine whether the statement is true or false. Justify your answer.

89. The difference of two perfect squares can be factored as the product of conjugate pairs.

90. A perfect square trinomial can always be factored as the square of a binomial.

91. Error Analysis Describe the error.

$$9x^2 - 9x - 54 = (3x + 6)(3x - 9)$$
$$= 3(x + 2)(x - 3)$$

92. Think About It Is $(3x - 6)(x + 1)$ completely factored? Explain.

Factoring with Variables in the Exponents In Exercises 93 and 94, factor the expression as completely as possible.

93. $x^{2n} - y^{2n}$ **94.** $x^{3n} + y^{3n}$

95. Think About It Give an example of a polynomial that is prime.

96. HOW DO YOU SEE IT? The figure shows a large square with an area of a^2 that contains a smaller square with an area of b^2.

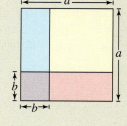

(a) Describe the regions that represent $a^2 - b^2$. How can you rearrange these regions to show that $a^2 - b^2 = (a - b)(a + b)$?

(b) How can you use the figure to show that $(a - b)^2 = a^2 - 2ab + b^2$?

(c) Draw another figure to show that $(a + b)^2 = a^2 + 2ab + b^2$. Explain how the figure shows this.

97. Difference of Two Sixth Powers Rewrite $u^6 - v^6$ as the difference of two squares. Then find a formula for completely factoring $u^6 - v^6$. Use your formula to completely factor $x^6 - 1$ and $x^6 - 64$.

P.5 Rational Expressions

Rational expressions have many real-life applications. For example, in Exercise 71 on page 49, you will work with a rational expression that models the temperature of food in a refrigerator.

- ■ Find domains of algebraic expressions.
- ■ Simplify rational expressions.
- ■ Add, subtract, multiply, and divide rational expressions.
- ■ Simplify complex fractions and rewrite difference quotients.

Domain of an Algebraic Expression

The set of real numbers for which an algebraic expression is defined is the **domain** of the expression. Two algebraic expressions are **equivalent** when they have the same domain and yield the same values for all numbers in their domain. For example,

$$(x + 1) + (x + 2) \quad \text{and} \quad 2x + 3$$

are equivalent because

$$
\begin{aligned}
(x + 1) + (x + 2) &= x + 1 + x + 2 \\
&= x + x + 1 + 2 \\
&= 2x + 3.
\end{aligned}
$$

EXAMPLE 1   Finding Domains of Algebraic Expressions

a. The domain of the polynomial

$$2x^3 + 3x + 4$$

is the set of all real numbers. In fact, the domain of any polynomial is the set of all real numbers, unless the domain is specifically restricted.

b. The domain of the radical expression

$$\sqrt{x - 2}$$

is the set of real numbers greater than or equal to 2, because the square root of a negative number is not a real number.

c. The domain of the expression

$$\frac{x + 2}{x - 3}$$

is the set of all real numbers except $x = 3$, which would result in division by zero, which is undefined.

✓ *Checkpoint*  *Audio-video solution in English & Spanish at LarsonPrecalculus.com*

Find the domain of each expression.

a. $4x^3 + 3, \quad x \geq 0$ **b.** $\sqrt{x + 7}$ **c.** $\dfrac{1 - x}{x}$ ■

The quotient of two algebraic expressions is a *fractional expression*. Moreover, the quotient of two *polynomials* such as

$$\frac{1}{x}, \quad \frac{2x - 1}{x + 1}, \quad \text{or} \quad \frac{x^2 - 1}{x^2 + 1}$$

is a **rational expression.**

Simplifying Rational Expressions

Recall that a fraction is in simplest form when its numerator and denominator have no factors in common other than ± 1. To write a fraction in simplest form, divide out common factors.

$$\frac{a \cdot \cancel{c}}{b \cdot \cancel{c}} = \frac{a}{b}, \quad c \neq 0$$

The key to success in simplifying rational expressions lies in your ability to *factor* polynomials. When simplifying rational expressions, factor each polynomial completely to determine whether the numerator and denominator have factors in common.

EXAMPLE 2 **Simplifying a Rational Expression**

$$\frac{x^2 + 4x - 12}{3x - 6} = \frac{(x + 6)\cancel{(x - 2)}}{3\cancel{(x - 2)}} \qquad \text{Factor completely.}$$

$$= \frac{x + 6}{3}, \quad x \neq 2 \qquad \text{Divide out common factor.}$$

Note that the original expression is undefined when $x = 2$ (because division by zero is undefined). To make the simplified expression *equivalent* to the original expression, you must restrict the domain of the simplified expression by excluding the value of $x = 2$.

✓ **Checkpoint** 🔊))) *Audio-video solution in English & Spanish at LarsonPrecalculus.com*

Write $\dfrac{4x + 12}{x^2 - 3x - 18}$ in simplest form. ◼

Sometimes it may be necessary to change the sign of a factor by factoring out (-1) to simplify a rational expression, as shown in Example 3.

EXAMPLE 3 **Simplifying a Rational Expression**

$$\frac{12 + x - x^2}{2x^2 - 9x + 4} = \frac{(4 - x)(3 + x)}{(2x - 1)(x - 4)} \qquad \text{Factor completely.}$$

$$= \frac{-\cancel{(x - 4)}(3 + x)}{(2x - 1)\cancel{(x - 4)}} \qquad (4 - x) = -(x - 4)$$

$$= -\frac{3 + x}{2x - 1}, \quad x \neq 4 \qquad \text{Divide out common factor.}$$

✓ **Checkpoint** 🔊))) *Audio-video solution in English & Spanish at LarsonPrecalculus.com*

Write $\dfrac{3x^2 - x - 2}{5 - 4x - x^2}$ in simplest form. ◼

In this text, the domain is usually not listed with a rational expression. It is *implied* that the real numbers that make the denominator zero are excluded from the domain. Also, when performing operations with rational expressions, this text follows the convention of listing *by the simplified expression* all values of x that must be specifically excluded from the domain to make the domains of the simplified and original expressions agree. Example 3, for instance, lists the restriction $x \neq 4$ with the simplified expression to make the two domains agree. Note that the value $x = \frac{1}{2}$ is excluded from *both* domains, so it is not necessary to list this value.

REMARK In Example 2, do not make the mistake of trying to simplify further by dividing out terms.

$$\frac{x + 6}{3} = \frac{x + \cancel{6}}{\cancel{3}}$$

$$= x + 2$$

To simplify fractions, divide out common *factors,* not terms. To learn about other common errors, see Appendix A.

Operations with Rational Expressions

To multiply or divide rational expressions, use the properties of fractions discussed in Section P.1. Recall that to divide fractions, you invert the divisor and multiply.

EXAMPLE 4 **Multiplying Rational Expressions**

$$\frac{2x^2 + x - 6}{x^2 + 4x - 5} \cdot \frac{x^3 - 3x^2 + 2x}{4x^2 - 6x} = \frac{(2x - 3)(x + 2)}{(x + 5)(x - 1)} \cdot \frac{x(x - 2)(x - 1)}{2x(2x - 3)}$$

$$= \frac{(x + 2)(x - 2)}{2(x + 5)}, \quad x \neq 0, \ x \neq 1, \ x \neq \frac{3}{2}$$

✓ *Checkpoint*))) *Audio-video solution in English & Spanish at LarsonPrecalculus.com*

Multiply and simplify: $\dfrac{15x^2 + 5x}{x^3 - 3x^2 - 18x} \cdot \dfrac{x^2 - 2x - 15}{3x^2 - 8x - 3}$.

> **• • REMARK** Note that Example 4 lists the restrictions $x \neq 0$, $x \neq 1$, and $x \neq \frac{3}{2}$ with the simplified expression to make the two domains agree. Also note that the value $x = -5$ is excluded from both domains, so it is not necessary to list this value.

EXAMPLE 5 **Dividing Rational Expressions**

$$\frac{x^3 - 8}{x^2 - 4} \div \frac{x^2 + 2x + 4}{x^3 + 8} = \frac{x^3 - 8}{x^2 - 4} \cdot \frac{x^3 + 8}{x^2 + 2x + 4} \qquad \textcolor{red}{\text{Invert and multiply.}}$$

$$= \frac{(x - 2)(x^2 + 2x + 4)}{(x + 2)(x - 2)} \cdot \frac{(x + 2)(x^2 - 2x + 4)}{(x^2 + 2x + 4)}$$

$$= x^2 - 2x + 4, \quad x \neq \pm 2 \qquad \textcolor{red}{\text{Divide out common factors.}}$$

✓ *Checkpoint*))) *Audio-video solution in English & Spanish at LarsonPrecalculus.com*

Divide and simplify: $\dfrac{x^3 - 1}{x^2 - 1} \div \dfrac{x^2 + x + 1}{x^2 + 2x + 1}$.

To add or subtract rational expressions, use the LCD (least common denominator) method or the *basic definition*

$$\frac{a}{b} \pm \frac{c}{d} = \frac{ad \pm bc}{bd}, \quad b \neq 0, d \neq 0. \qquad \textcolor{red}{\text{Basic definition}}$$

This definition provides an efficient way of adding or subtracting two fractions that have no common factors in their denominators.

EXAMPLE 6 **Subtracting Rational Expressions**

$$\frac{x}{x - 3} - \frac{2}{3x + 4} = \frac{x(3x + 4) - 2(x - 3)}{(x - 3)(3x + 4)} \qquad \textcolor{red}{\text{Basic definition}}$$

$$= \frac{3x^2 + 4x - 2x + 6}{(x - 3)(3x + 4)} \qquad \textcolor{red}{\text{Distributive Property}}$$

$$= \frac{3x^2 + 2x + 6}{(x - 3)(3x + 4)} \qquad \textcolor{red}{\text{Combine like terms.}}$$

> **• • REMARK** When subtracting rational expressions, remember to distribute the negative sign to *all* the terms in the quantity that is being subtracted.

✓ *Checkpoint*))) *Audio-video solution in English & Spanish at LarsonPrecalculus.com*

Subtract and simplify: $\dfrac{x}{2x - 1} - \dfrac{1}{x + 2}$.

For three or more fractions, or for fractions with a repeated factor in the denominators, the LCD method works well. Recall that the least common denominator of several fractions consists of the product of all prime factors in the denominators, with each factor given the highest power of its occurrence in any denominator. Here is a numerical example.

$$\frac{1}{6} + \frac{3}{4} - \frac{2}{3} = \frac{1 \cdot 2}{6 \cdot 2} + \frac{3 \cdot 3}{4 \cdot 3} - \frac{2 \cdot 4}{3 \cdot 4} \qquad \text{The LCD is 12.}$$

$$= \frac{2}{12} + \frac{9}{12} - \frac{8}{12}$$

$$= \frac{3}{12}$$

$$= \frac{1}{4}$$

Sometimes, the numerator of the answer has a factor in common with the denominator. In such cases, simplify the answer, as shown in the example above.

EXAMPLE 7   Combining Rational Expressions: The LCD Method

See LarsonPrecalculus.com for an interactive version of this type of example.

Perform the operations and simplify.

$$\frac{3}{x - 1} - \frac{2}{x} + \frac{x + 3}{x^2 - 1}$$

Solution   Use the factored denominators

$$(x - 1), \quad x, \quad \text{and} \quad (x + 1)(x - 1)$$

to determine that the LCD is $x(x + 1)(x - 1)$.

$$\frac{3}{x - 1} - \frac{2}{x} + \frac{x + 3}{(x + 1)(x - 1)}$$

$$= \frac{3(x)(x + 1)}{x(x + 1)(x - 1)} - \frac{2(x + 1)(x - 1)}{x(x + 1)(x - 1)} + \frac{(x + 3)(x)}{x(x + 1)(x - 1)}$$

$$= \frac{3(x)(x + 1) - 2(x + 1)(x - 1) + (x + 3)(x)}{x(x + 1)(x - 1)}$$

$$= \frac{3x^2 + 3x - 2x^2 + 2 + x^2 + 3x}{x(x + 1)(x - 1)} \qquad \text{Multiply.}$$

$$= \frac{(3x^2 - 2x^2 + x^2) + (3x + 3x) + 2}{x(x + 1)(x - 1)} \qquad \text{Group like terms.}$$

$$= \frac{2x^2 + 6x + 2}{x(x + 1)(x - 1)} \qquad \text{Combine like terms.}$$

$$= \frac{2(x^2 + 3x + 1)}{x(x + 1)(x - 1)} \qquad \text{Factor.}$$

✓ **Checkpoint** ◀)) *Audio-video solution in English & Spanish at LarsonPrecalculus.com*

Perform the operations and simplify.

$$\frac{4}{x} - \frac{x + 5}{x^2 - 4} + \frac{4}{x + 2}$$

Complex Fractions and the Difference Quotient

Complex fractions are fractional expressions with separate fractions in the numerator, denominator, or both. Here are two examples.

$$\frac{\left(\dfrac{1}{x}\right)}{x^2 + 1} \quad \text{and} \quad \frac{\left(\dfrac{1}{x}\right)}{\left(\dfrac{1}{x^2 + 1}\right)}$$

One way to simplify a complex fraction is to combine the fractions in the numerator into a single fraction and then combine the fractions in the denominator into a single fraction. Then invert the denominator and multiply. Example 8 shows this method.

EXAMPLE 8 **Simplifying a Complex Fraction**

$$\frac{\left(\dfrac{2}{x} - 3\right)}{\left(1 - \dfrac{1}{x - 1}\right)} = \frac{\left[\dfrac{2 - 3(x)}{x}\right]}{\left[\dfrac{1(x - 1) - 1}{x - 1}\right]} \qquad \text{Combine fractions.}$$

$$= \frac{\left(\dfrac{2 - 3x}{x}\right)}{\left(\dfrac{x - 2}{x - 1}\right)} \qquad \text{Simplify.}$$

$$= \frac{2 - 3x}{x} \cdot \frac{x - 1}{x - 2} \qquad \text{Invert and multiply.}$$

$$= \frac{(2 - 3x)(x - 1)}{x(x - 2)}, \quad x \neq 1$$

✓ **Checkpoint** ◁))) *Audio-video solution in English & Spanish at LarsonPrecalculus.com*

Simplify the complex fraction $\dfrac{\left(\dfrac{1}{x + 2} + 1\right)}{\left(\dfrac{x}{3} - 1\right)}$.

Another way to simplify a complex fraction is to multiply its numerator and denominator by the LCD of all fractions in its numerator and denominator. This method, applied to the fraction in Example 8, is shown below. Notice that both methods yield the same result.

$$\frac{\left(\dfrac{2}{x} - 3\right)}{\left(1 - \dfrac{1}{x - 1}\right)} \cdot \frac{x(x - 1)}{x(x - 1)} = \frac{\left(\dfrac{2}{x}\right)(x)(x - 1) - (3)(x)(x - 1)}{(1)(x)(x - 1) - \left(\dfrac{1}{x - 1}\right)(x)(x - 1)} \qquad \begin{array}{l}\text{LCD is}\\ x(x - 1).\end{array}$$

$$= \frac{2(x - 1) - 3x(x - 1)}{x(x - 1) - x} \qquad \text{Simplify.}$$

$$= \frac{(2 - 3x)(x - 1)}{x(x - 2)}, \quad x \neq 1 \qquad \text{Factor.}$$

The next three examples illustrate some methods for simplifying rational expressions involving negative exponents and radicals. These types of expressions occur frequently in calculus.

To simplify an expression with negative exponents, one method is to begin by factoring out the common factor with the *lesser* exponent. Remember that when factoring, you *subtract* exponents. For example, in $3x^{-5/2} + 2x^{-3/2}$, the lesser exponent is $-\frac{5}{2}$ and the common factor is $x^{-5/2}$.

$$3x^{-5/2} + 2x^{-3/2} = x^{-5/2}[3 + 2x^{-3/2-(-5/2)}]$$
$$= x^{-5/2}(3 + 2x^1)$$
$$= \frac{3 + 2x}{x^{5/2}}$$

EXAMPLE 9 **Simplifying an Expression**

Simplify $x(1 - 2x)^{-3/2} + (1 - 2x)^{-1/2}$.

Solution Begin by factoring out the common factor with the lesser exponent.

$$x(1 - 2x)^{-3/2} + (1 - 2x)^{-1/2} = (1 - 2x)^{-3/2}[x + (1 - 2x)^{(-1/2)-(-3/2)}]$$
$$= (1 - 2x)^{-3/2}[x + (1 - 2x)^1]$$
$$= \frac{1 - x}{(1 - 2x)^{3/2}}$$

✓ **Checkpoint**)))) *Audio-video solution in English & Spanish at LarsonPrecalculus.com*

Simplify $(x - 1)^{-1/3} - x(x - 1)^{-4/3}$.

The next example shows a complex fraction with a negative exponent and a second method for simplifying an expression with negative exponents.

EXAMPLE 10 **Simplifying an Expression**

$$\frac{\left[\dfrac{1}{(4 - x^2)^{-1/2}} + \dfrac{x^2}{(4 - x^2)^{1/2}}\right]}{4 - x^2} = \frac{(4 - x^2)^{1/2} + x^2(4 - x^2)^{-1/2}}{4 - x^2}$$

$$= \frac{(4 - x^2)^{1/2} + x^2(4 - x^2)^{-1/2}}{4 - x^2} \cdot \frac{(4 - x^2)^{1/2}}{(4 - x^2)^{1/2}}$$

$$= \frac{(4 - x^2)^1 + x^2(4 - x^2)^0}{(4 - x^2)^{3/2}}$$

$$= \frac{4 - x^2 + x^2}{(4 - x^2)^{3/2}}$$

$$= \frac{4}{(4 - x^2)^{3/2}}$$

✓ **Checkpoint**)))) *Audio-video solution in English & Spanish at LarsonPrecalculus.com*

Simplify

$$\frac{\left[\dfrac{x^2}{(x^2 - 2)^{1/2}} + \dfrac{1}{(x^2 - 2)^{-1/2}}\right]}{x^2 - 2}.$$

Difference quotients, such as

$$\frac{\sqrt{x + h} - \sqrt{x}}{h}$$

occur frequently in calculus. Often, they need to be rewritten in an equivalent form, as shown in Example 11.

EXAMPLE 11 **Rewriting a Difference Quotient**

Rewrite the difference quotient

$$\frac{\sqrt{x + h} - \sqrt{x}}{h}$$

by rationalizing its numerator.

Solution

$$\frac{\sqrt{x + h} - \sqrt{x}}{h} = \frac{\sqrt{x + h} - \sqrt{x}}{h} \cdot \frac{\sqrt{x + h} + \sqrt{x}}{\sqrt{x + h} + \sqrt{x}}$$

$$= \frac{\left(\sqrt{x + h}\right)^2 - \left(\sqrt{x}\right)^2}{h\left(\sqrt{x + h} + \sqrt{x}\right)}$$

$$= \frac{x + h - x}{h\left(\sqrt{x + h} + \sqrt{x}\right)}$$

$$= \frac{h}{h\left(\sqrt{x + h} + \sqrt{x}\right)}$$

$$= \frac{1}{\sqrt{x + h} + \sqrt{x}}, \quad h \neq 0$$

✓ *Checkpoint* *Audio-video solution in English & Spanish at LarsonPrecalculus.com*

Rewrite the difference quotient

$$\frac{\sqrt{9 + h} - 3}{h}$$

by rationalizing its numerator.

Summarize (Section P.5)

1. State the definition of the domain of an algebraic expression *(page 41)*. For an example of finding domains of algebraic expressions, see Example 1.

2. State the definition of a rational expression and explain how to simplify a rational expression *(pages 41 and 42)*. For examples of simplifying rational expressions, see Examples 2 and 3.

3. Explain how to multiply, divide, add, and subtract rational expressions *(page 43)*. For examples of operations with rational expressions, see Examples 4–7.

4. State the definition of a complex fraction *(page 45)*. For an example of simplifying a complex fraction, see Example 8.

5. Explain how to rewrite a difference quotient *(page 47)*. For an example of rewriting a difference quotient, see Example 11.

P.5 Exercises

See **CalcChat.com** for tutorial help and worked-out solutions to odd-numbered exercises.

Vocabulary: Fill in the blanks.

1. The set of real numbers for which an algebraic expression is defined is the _____ of the expression.
2. The quotient of two algebraic expressions is a fractional expression, and the quotient of two polynomials is a _____ _____.
3. Fractional expressions with separate fractions in the numerator, denominator, or both are _____ fractions.
4. Two algebraic expressions that have the same domain and yield the same values for all numbers in their domains are _____.

Skills and Applications

 Finding the Domain of an Algebraic Expression In Exercises 5–16, find the domain of the expression.

5. $3x^2 - 4x + 7$

6. $6x^2 - 9, \quad x > 0$

7. $\dfrac{1}{3 - x}$

8. $\dfrac{1}{x + 5}$

9. $\dfrac{x + 6}{3x + 2}$

10. $\dfrac{x - 4}{1 - 2x}$

11. $\dfrac{x^2 - 5x + 6}{x^2 + 6x + 8}$

12. $\dfrac{x^2 - 1}{x^2 + 3x - 10}$

13. $\sqrt{x - 7}$

14. $\sqrt{2x - 5}$

15. $\dfrac{1}{\sqrt{x - 3}}$

16. $\dfrac{1}{\sqrt{x + 2}}$

 Simplifying a Rational Expression In Exercises 17–30, write the rational expression in simplest form.

17. $\dfrac{15x^2}{10x}$

18. $\dfrac{18y^2}{60y^5}$

19. $\dfrac{x - 5}{10 - 2x}$

20. $\dfrac{12 - 4x}{x - 3}$

21. $\dfrac{y^2 - 16}{y + 4}$

22. $\dfrac{x^2 - 25}{5 - x}$

23. $\dfrac{6y + 9y^2}{12y + 8}$

24. $\dfrac{4y - 8y^2}{10y - 5}$

25. $\dfrac{x^2 + 4x - 5}{x^2 + 8x + 15}$

26. $\dfrac{x^2 + 8x - 20}{x^2 + 11x + 10}$

27. $\dfrac{x^2 - x - 2}{10 - 3x - x^2}$

28. $\dfrac{4 + 3x - x^2}{2x^2 - 7x - 4}$

29. $\dfrac{x^2 - 16}{x^3 + x^2 - 16x - 16}$

30. $\dfrac{x^2 - 1}{x^3 + x^2 + 9x + 9}$

31. **Error Analysis** Describe the error.

$$\dfrac{5x^3}{2x^3 + 4} = \dfrac{5\cancel{x^3}}{2\cancel{x^3} + 4} = \dfrac{5}{2 + 4} = \dfrac{5}{6} \quad \textcolor{red}{\bigtimes}$$

32. **Evaluating a Rational Expression** Complete the table. What can you conclude?

x	0	1	2	3	4	5	6
$\dfrac{x - 3}{x^2 - x - 6}$							
$\dfrac{1}{x + 2}$							

 Multiplying or Dividing Rational Expressions In Exercises 33–38, perform the multiplication or division and simplify.

33. $\dfrac{5}{x - 1} \cdot \dfrac{x - 1}{25(x - 2)}$

34. $\dfrac{r}{r - 1} \div \dfrac{r^2}{r^2 - 1}$

35. $\dfrac{x^2 - 4}{12} \div \dfrac{2 - x}{2x + 4}$

36. $\dfrac{t^2 - t - 6}{t^2 + 6t + 9} \cdot \dfrac{t + 3}{t^2 - 4}$

37. $\dfrac{x^2 + xy - 2y^2}{x^3 + x^2y} \cdot \dfrac{x}{x^2 + 3xy + 2y^2}$

38. $\dfrac{x^2 - 14x + 49}{x^2 - 49} \div \dfrac{3x - 21}{x + 7}$

 Adding or Subtracting Rational Expressions In Exercises 39–46, perform the addition or subtraction and simplify.

39. $\dfrac{x - 1}{x + 2} - \dfrac{x - 4}{x + 2}$

40. $\dfrac{2x - 1}{x + 3} + \dfrac{1 - x}{x + 3}$

41. $\dfrac{1}{3x + 2} + \dfrac{x}{x + 1}$

42. $\dfrac{x}{x + 4} - \dfrac{6}{x - 1}$

43. $\dfrac{3}{2x + 4} - \dfrac{x}{x + 2}$

44. $\dfrac{2}{x^2 - 9} + \dfrac{4}{x + 3}$

45. $-\dfrac{1}{x} + \dfrac{2}{x^2 + 1} + \dfrac{1}{x^3 + x}$

46. $\dfrac{2}{x + 1} + \dfrac{2}{x - 1} + \dfrac{1}{x^2 - 1}$

Error Analysis In Exercises 47 and 48, describe the error.

47. $\dfrac{x+4}{x+2} - \dfrac{3x-8}{x+2} = \dfrac{x+4-3x-8}{x+2}$

$= \dfrac{-2x-4}{x+2}$

$= \dfrac{-2(x+2)}{x+2}$

$= -2, \quad x \neq -2$

48. $\dfrac{6-x}{x(x+2)} + \dfrac{x+2}{x^2} + \dfrac{8}{x^2(x+2)}$

$= \dfrac{6-x+(x+2)^2+8}{x^2(x+2)}$

$= \dfrac{6-x+x^2+4x+4+8}{x^2(x+2)}$

$= \dfrac{x^2+3x+18}{x^2(x+2)}$

 Simplifying a Complex Fraction In Exercises 49–54, simplify the complex fraction.

49. $\dfrac{\left(\dfrac{x}{2}-1\right)}{x-2}$

50. $\dfrac{x+5}{\left(\dfrac{x}{5}-5\right)}$

51. $\dfrac{\left[\dfrac{x^2}{(x+1)^2}\right]}{\left[\dfrac{x}{(x+1)^3}\right]}$

52. $\dfrac{\left(\dfrac{x^2-1}{x}\right)}{\left[\dfrac{(x-1)^2}{x}\right]}$

53. $\dfrac{\left(\sqrt{x}-\dfrac{1}{2\sqrt{x}}\right)}{\sqrt{x}}$

54. $\dfrac{\left(\dfrac{t^2}{\sqrt{t^2+1}}-\sqrt{t^2+1}\right)}{t^2}$

Factoring an Expression In Exercises 55–58, factor the expression by factoring out the common factor with the lesser exponent.

55. $x^2(x^2+3)^{-4} + (x^2+3)^3$
56. $2x(x-5)^{-3} - 4x^2(x-5)^{-4}$
57. $2x^2(x-1)^{1/2} - 5(x-1)^{-1/2}$
58. $4x^3(x+1)^{-3/2} - x(x+1)^{-1/2}$

 Simplifying an Expression In Exercises 59 and 60, simplify the expression.

59. $\dfrac{3x^{1/3}-x^{-2/3}}{3x^{-2/3}}$

60. $\dfrac{-x^3(1-x^2)^{-1/2} - 2x(1-x^2)^{1/2}}{x^4}$

 Simplifying a Difference Quotient In Exercises 61–64, simplify the difference quotient.

61. $\dfrac{\left(\dfrac{1}{x+h}-\dfrac{1}{x}\right)}{h}$

62. $\dfrac{\left[\dfrac{1}{(x+h)^2}-\dfrac{1}{x^2}\right]}{h}$

63. $\dfrac{\left(\dfrac{1}{x+h-4}-\dfrac{1}{x-4}\right)}{h}$

64. $\dfrac{\left(\dfrac{x+h}{x+h+1}-\dfrac{x}{x+1}\right)}{h}$

 **Rewriting a Difference Quotient** In Exercises 65–70, rewrite the difference quotient by rationalizing the numerator.

65. $\dfrac{\sqrt{x+2}-\sqrt{x}}{2}$

66. $\dfrac{\sqrt{z-3}-\sqrt{z}}{-3}$

67. $\dfrac{\sqrt{t+3}-\sqrt{3}}{t}$

68. $\dfrac{\sqrt{x+5}-\sqrt{5}}{x}$

69. $\dfrac{\sqrt{x+h+1}-\sqrt{x+1}}{h}$

70. $\dfrac{\sqrt{x+h-2}-\sqrt{x-2}}{h}$

71. Refrigeration

After placing food (at room temperature) in a refrigerator, the time required for the food to cool depends on the amount of food, the air circulation in the refrigerator, the original temperature of the food, and the temperature of the refrigerator. The model that gives the temperature of food that has an original temperature of 75°F and is placed in a 40°F refrigerator is

$$T = 10\left(\dfrac{4t^2+16t+75}{t^2+4t+10}\right)$$

where T is the temperature (in degrees Fahrenheit) and t is the time (in hours).

(a) Complete the table.

t	0	2	4	6	8	10	12
T							

t	14	16	18	20	22
T					

(b) What value of T does the mathematical model appear to be approaching?

72. Rate A copier copies at a rate of 50 pages per minute.

(a) Find the time required to copy one page.

(b) Find the time required to copy x pages.

(c) Find the time required to copy 120 pages.

Probability **In Exercises 73 and 74, consider an experiment in which a marble is tossed into a box whose base is shown in the figure. The probability that the marble will come to rest in the shaded portion of the base is equal to the ratio of the shaded area to the total area of the figure. Find the probability.**

73.

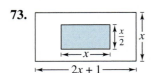

74.

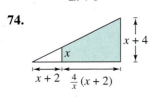

75. Interactive Money Management The table shows the numbers of U.S. households (in millions) using online banking and mobile banking from 2011 through 2014. (*Source: Fiserv, Inc.*)

Year	Online Banking	Mobile Banking
2011	79	18
2012	81	24
2013	83	30
2014	86	35

Mathematical models for the data are

Number using online banking $= \dfrac{-2.9709t + 70.517}{-0.0474t + 1}$

Number using mobile banking $= \dfrac{0.661t^2 - 47}{0.007t^2 + 1}$

where t represents the year, with $t = 11$ corresponding to 2011.

(a) Using the models, create a table showing the numbers of households using online banking and the numbers of households using mobile banking for the given years.

(b) Compare the values from the models with the actual data.

(c) Determine a model for the ratio of the number of households using mobile banking to the number of households using online banking.

(d) Use the model from part (c) to find the ratios for the given years. Interpret your results.

76. Finance The formula that approximates the annual interest rate r of a monthly installment loan is

$$r = \dfrac{24(NM - P)}{N} \div \left(P + \dfrac{NM}{12}\right)$$

where N is the total number of payments, M is the monthly payment, and P is the amount financed.

(a) Approximate the annual interest rate for a five-year car loan of \$28,000 that has monthly payments of \$525.

(b) Simplify the expression for the annual interest rate r, and then rework part (a).

77. Electrical Engineering The formula for the total resistance R_T (in ohms) of two resistors connected in parallel is

$$R_T = \dfrac{1}{\left(\dfrac{1}{R_1} + \dfrac{1}{R_2}\right)}$$

where R_1 and R_2 are the resistance values of the first and second resistors, respectively. Simplify the expression for the total resistance R_T.

78. HOW DO YOU SEE IT? The mathematical model

$$P = 100\left(\dfrac{t^2 - t + 1}{t^2 + 1}\right), \quad t \geq 0$$

gives the percent P of the normal level of oxygen in a pond, where t is the time (in weeks) after organic waste is dumped into the pond. The bar graph shows the situation. What conclusions can you draw from the bar graph?

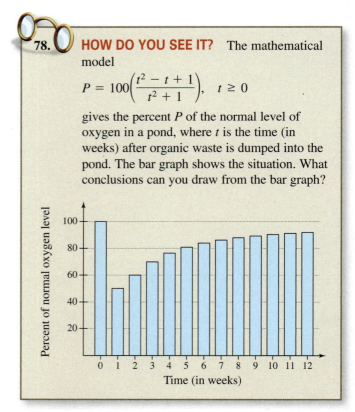

Exploration

True or False? **In Exercises 79 and 80, determine whether the statement is true or false. Justify your answer.**

79. $\dfrac{x^{2n} - 1^{2n}}{x^n - 1^n} = x^n + 1^n$

80. $\dfrac{x^2 - 3x + 2}{x - 1} = x - 2$, for all values of x

P.6 The Rectangular Coordinate System and Graphs

The Cartesian plane can help you visualize relationships between two variables. For example, in Exercise 37 on page 58, given how far north and west one city is from another, plotting points to represent the cities can help you visualize these distances and determine the flying distance between the cities.

■ Plot points in the Cartesian plane.
■ Use the Distance Formula to find the distance between two points.
■ Use the Midpoint Formula to find the midpoint of a line segment.
■ Use a coordinate plane to model and solve real-life problems.

The Cartesian Plane

Just as you can represent real numbers by points on a real number line, you can represent ordered pairs of real numbers by points in a plane called the **rectangular coordinate system,** or the **Cartesian plane,** named after the French mathematician René Descartes (1596–1650).

Two real number lines intersecting at right angles form the Cartesian plane, as shown in Figure P.9. The horizontal real number line is usually called the **x-axis,** and the vertical real number line is usually called the **y-axis.** The point of intersection of these two axes is the **origin,** and the two axes divide the plane into four **quadrants.**

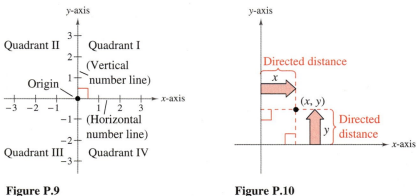

Figure P.9

Figure P.10

Each point in the plane corresponds to an **ordered pair** (x, y) of real numbers x and y, called **coordinates** of the point. The **x-coordinate** represents the directed distance from the y-axis to the point, and the **y-coordinate** represents the directed distance from the x-axis to the point, as shown in Figure P.10.

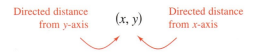

Directed distance from y-axis $\quad (x, y) \quad$ Directed distance from x-axis

The notation (x, y) denotes both a point in the plane and an open interval on the real number line. The context will tell you which meaning is intended.

Figure P.11

EXAMPLE 1 Plotting Points in the Cartesian Plane

Plot the points $(-1, 2)$, $(3, 4)$, $(0, 0)$, $(3, 0)$, and $(-2, -3)$.

Solution To plot the point $(-1, 2)$, imagine a vertical line through -1 on the x-axis and a horizontal line through 2 on the y-axis. The intersection of these two lines is the point $(-1, 2)$. Plot the other four points in a similar way, as shown in Figure P.11.

✓ **Checkpoint**))) Audio-video solution in English & Spanish at LarsonPrecalculus.com

Plot the points $(-3, 2)$, $(4, -2)$, $(3, 1)$, $(0, -2)$, and $(-1, -2)$.

The beauty of a rectangular coordinate system is that it allows you to *see* relationships between two variables. It would be difficult to overestimate the importance of Descartes's introduction of coordinates in the plane. Today, his ideas are in common use in virtually every scientific and business-related field.

EXAMPLE 2 Sketching a Scatter Plot

The table shows the numbers N (in millions) of subscribers to a cellular telecommunication service in the United States from 2005 through 2014, where t represents the year. Sketch a scatter plot of the data. *(Source: CTIA-The Wireless Association)*

Solution   To sketch a *scatter plot* of the data shown in the table, represent each pair of values by an ordered pair (t, N) and plot the resulting points. For example, let $(2005, 207.9)$ represent the first pair of values. Note that in the scatter plot below, the break in the t-axis indicates omission of the years before 2005, and the break in the N-axis indicates omission of the numbers less than 150 million.

Year, t	Subscribers, N
2005	207.9
2006	233.0
2007	255.4
2008	270.3
2009	285.6
2010	296.3
2011	316.0
2012	326.5
2013	335.7
2014	355.4

Spreadsheet at LarsonPrecalculus.com

Subscribers to a Cellular Telecommunication Service

✓ *Checkpoint* ◀))) *Audio-video solution in English & Spanish at LarsonPrecalculus.com*

The table shows the numbers N (in thousands) of cellular telecommunication service employees in the United States from 2005 through 2014, where t represents the year. Sketch a scatter plot of the data. *(Source: CTIA-The Wireless Association)*

t	N
2005	233.1
2006	253.8
2007	266.8
2008	268.5
2009	249.2
2010	250.4
2011	238.1
2012	230.1
2013	230.4
2014	232.2

Spreadsheet at LarsonPrecalculus.com

▷ **TECHNOLOGY** The scatter plot in Example 2 is only one way to represent the data graphically. You could also represent the data using a bar graph or a line graph. Use a graphing utility to represent the data given in Example 2 graphically.

In Example 2, you could let $t = 1$ represent the year 2005. In that case, there would not be a break in the horizontal axis, and the labels 1 through 10 (instead of 2005 through 2014) would be on the tick marks.

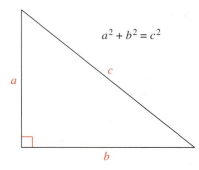

$$a^2 + b^2 = c^2$$

Figure P.12

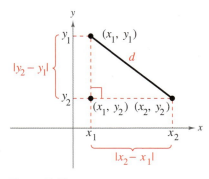

Figure P.13

The Distance Formula

Recall from the Pythagorean Theorem that, for a right triangle with hypotenuse of length c and sides of lengths a and b, you have

$$a^2 + b^2 = c^2 \qquad \text{Pythagorean Theorem}$$

as shown in Figure P.12. (The converse is also true. That is, if $a^2 + b^2 = c^2$, then the triangle is a right triangle.)

Using the points (x_1, y_1) and (x_2, y_2), you can form a right triangle, as shown in Figure P.13. The length of the hypotenuse of the right triangle is the distance d between the two points. The length of the vertical side of the triangle is $|y_2 - y_1|$ and the length of the horizontal side is $|x_2 - x_1|$. By the Pythagorean Theorem,

$$d^2 = |x_2 - x_1|^2 + |y_2 - y_1|^2$$
$$d = \sqrt{|x_2 - x_1|^2 + |y_2 - y_1|^2}$$
$$= \sqrt{(x_2 - x_1)^2 + (y_2 - y_1)^2}.$$

This result is the **Distance Formula.**

The Distance Formula

The distance d between the points (x_1, y_1) and (x_2, y_2) in the plane is

$$d = \sqrt{(x_2 - x_1)^2 + (y_2 - y_1)^2}.$$

EXAMPLE 3   **Finding a Distance**

Find the distance between the points

$$(-2, 1) \quad \text{and} \quad (3, 4).$$

Algebraic Solution

Let $(x_1, y_1) = (-2, 1)$ and $(x_2, y_2) = (3, 4)$. Then apply the Distance Formula.

$$d = \sqrt{(x_2 - x_1)^2 + (y_2 - y_1)^2} \qquad \text{Distance Formula}$$
$$= \sqrt{[3 - (-2)]^2 + (4 - 1)^2} \qquad \text{Substitute for } x_1, y_1, x_2, \text{ and } y_2.$$
$$= \sqrt{(5)^2 + (3)^2} \qquad \text{Simplify.}$$
$$= \sqrt{34} \qquad \text{Simplify.}$$
$$\approx 5.83 \qquad \text{Use a calculator.}$$

So, the distance between the points is about 5.83 units.

Check

$$d^2 \overset{?}{=} 5^2 + 3^2 \qquad \text{Pythagorean Theorem}$$
$$\left(\sqrt{34}\right)^2 \overset{?}{=} 5^2 + 3^2 \qquad \text{Substitute for } d.$$
$$34 = 34 \qquad \text{Distance checks. } ✓$$

Graphical Solution

Use centimeter graph paper to plot the points $A(-2, 1)$ and $B(3, 4)$. Carefully sketch the line segment from A to B. Then use a centimeter ruler to measure the length of the segment.

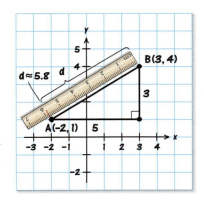

The line segment measures about 5.8 centimeters. So, the distance between the points is about 5.8 units.

 Checkpoint ◀))) *Audio-video solution in English & Spanish at LarsonPrecalculus.com*

Find the distance between the points $(3, 1)$ and $(-3, 0)$.

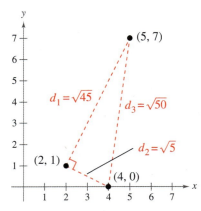

Figure P.14

EXAMPLE 4 Verifying a Right Triangle

Show that the points

$(2, 1)$, $(4, 0)$, and $(5, 7)$

are vertices of a right triangle.

Solution The three points are plotted in Figure P.14. Using the Distance Formula, the lengths of the three sides are

$$d_1 = \sqrt{(5-2)^2 + (7-1)^2} = \sqrt{9+36} = \sqrt{45},$$

$$d_2 = \sqrt{(4-2)^2 + (0-1)^2} = \sqrt{4+1} = \sqrt{5}, \text{ and}$$

$$d_3 = \sqrt{(5-4)^2 + (7-0)^2} = \sqrt{1+49} = \sqrt{50}.$$

Because $(d_1)^2 + (d_2)^2 = 45 + 5 = 50 = (d_3)^2$, you can conclude by the converse of the Pythagorean Theorem that the triangle is a right triangle.

 ✓ **Checkpoint** ◀))) *Audio-video solution in English & Spanish at LarsonPrecalculus.com*

Show that the points $(2, -1)$, $(5, 5)$, and $(6, -3)$ are vertices of a right triangle. ■

The Midpoint Formula

To find the **midpoint** of the line segment that joins two points in a coordinate plane, find the average values of the respective coordinates of the two endpoints using the **Midpoint Formula.**

The Midpoint Formula

The midpoint of the line segment joining the points (x_1, y_1) and (x_2, y_2) is

$$\text{Midpoint} = \left(\frac{x_1 + x_2}{2}, \frac{y_1 + y_2}{2} \right).$$

For a proof of the Midpoint Formula, see Proofs in Mathematics on page 66.

EXAMPLE 5 Finding the Midpoint of a Line Segment

Find the midpoint of the line segment joining the points

$(-5, -3)$ and $(9, 3)$.

Solution Let $(x_1, y_1) = (-5, -3)$ and $(x_2, y_2) = (9, 3)$.

$$\text{Midpoint} = \left(\frac{x_1 + x_2}{2}, \frac{y_1 + y_2}{2} \right) \qquad \textcolor{red}{\text{Midpoint Formula}}$$

$$= \left(\frac{-5 + 9}{2}, \frac{-3 + 3}{2} \right) \qquad \textcolor{red}{\text{Substitute for } x_1, y_1, x_2, \text{ and } y_2.}$$

$$= (2, 0) \qquad \textcolor{red}{\text{Simplify.}}$$

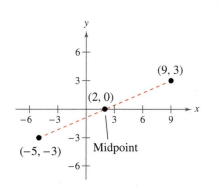

Figure P.15

The midpoint of the line segment is $(2, 0)$, as shown in Figure P.15.

✓ **Checkpoint** ◀))) *Audio-video solution in English & Spanish at LarsonPrecalculus.com*

Find the midpoint of the line segment joining the points $(-2, 8)$ and $(4, -10)$. ■

Applications

EXAMPLE 6 **Finding the Length of a Pass**

A football quarterback throws a pass from the 28-yard line, 40 yards from the sideline. A wide receiver catches the pass on the 5-yard line, 20 yards from the same sideline, as shown in Figure P.16. How long is the pass?

Solution The length of the pass is the distance between the points $(40, 28)$ and $(20, 5)$.

$$d = \sqrt{(x_2 - x_1)^2 + (y_2 - y_1)^2} \qquad \text{Distance Formula}$$

$$= \sqrt{(40 - 20)^2 + (28 - 5)^2} \qquad \text{Substitute for } x_1, y_1, x_2, \text{ and } y_2.$$

$$= \sqrt{20^2 + 23^2} \qquad \text{Simplify.}$$

$$= \sqrt{400 + 529} \qquad \text{Simplify.}$$

$$= \sqrt{929} \qquad \text{Simplify.}$$

$$\approx 30 \qquad \text{Use a calculator.}$$

So, the pass is about 30 yards long.

✓ **Checkpoint** Audio-video solution in English & Spanish at LarsonPrecalculus.com

A football quarterback throws a pass from the 10-yard line, 10 yards from the sideline. A wide receiver catches the pass on the 32-yard line, 25 yards from the same sideline. How long is the pass?

In Example 6, the scale along the goal line does not normally appear on a football field. However, when you use coordinate geometry to solve real-life problems, you are free to place the coordinate system in any way that helps you solve the problem.

EXAMPLE 7 **Estimating Annual Sales**

Starbucks Corporation had annual sales of approximately $13.3 billion in 2012 and $16.4 billion in 2014. Without knowing any additional information, what would you estimate the 2013 sales to have been? *(Source: Starbucks Corporation)*

Solution Assuming that sales followed a linear pattern, you can estimate the 2013 sales by finding the midpoint of the line segment connecting the points $(2012, 13.3)$ and $(2014, 16.4)$.

$$\text{Midpoint} = \left(\frac{x_1 + x_2}{2}, \frac{y_1 + y_2}{2} \right) \qquad \text{Midpoint Formula}$$

$$= \left(\frac{2012 + 2014}{2}, \frac{13.3 + 16.4}{2} \right) \qquad \text{Substitute for } x_1, x_2, y_1, \text{ and } y_2.$$

$$= (2013, 14.85) \qquad \text{Simplify.}$$

So, you would estimate the 2013 sales to have been about $14.85 billion, as shown in Figure P.17. (The actual 2013 sales were about $14.89 billion.)

✓ **Checkpoint** Audio-video solution in English & Spanish at LarsonPrecalculus.com

Yahoo! Inc. had annual revenues of approximately $5.0 billon in 2012 and $4.6 billion in 2014. Without knowing any additional information, what would you estimate the 2013 revenue to have been? *(Source: Yahoo! Inc.)*

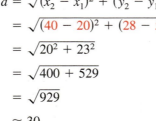

Football Pass

Distance (in yards)

Distance (in yards)

Figure P.16

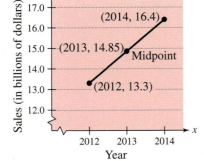

Starbucks Corporation Sales

Sales (in billions of dollars)

Year

Figure P.17

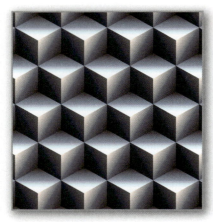

Much of computer graphics, including this computer-generated tessellation, consists of transformations of points in a coordinate plane. Example 8 illustrates one type of transformation called a translation. Other types include reflections, rotations, and stretches.

Translating Points in the Plane

See LarsonPrecalculus.com for an interactive version of this type of example.

The triangle in Figure P.18 has vertices at the points $(-1, 2)$, $(1, -2)$, and $(2, 3)$. Shift the triangle three units to the right and two units up and find the coordinates of the vertices of the shifted triangle shown in Figure P.19.

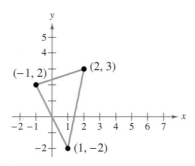

Figure P.18

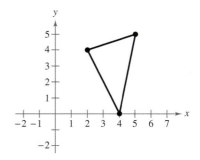

Figure P.19

Solution To shift the vertices three units to the right, add 3 to each of the x-coordinates. To shift the vertices two units up, add 2 to each of the y-coordinates.

Original Point	Translated Point
$(-1, 2)$	$(-1 + 3, 2 + 2) = (2, 4)$
$(1, -2)$	$(1 + 3, -2 + 2) = (4, 0)$
$(2, 3)$	$(2 + 3, 3 + 2) = (5, 5)$

 Checkpoint *Audio-video solution in English & Spanish at LarsonPrecalculus.com*

Find the coordinates of the vertices of the parallelogram shown after translating it two units to the left and four units down.

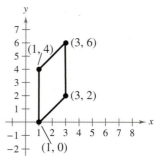

The figures in Example 8 were not really essential to the solution. Nevertheless, you should develop the habit of including sketches with your solutions because they serve as useful problem-solving tools.

Summarize (Section P.6)

1. Describe the Cartesian plane *(page 51)*. For examples of plotting points in the Cartesian plane, see Examples 1 and 2.

2. State the Distance Formula *(page 53)*. For examples of using the Distance Formula to find the distance between two points, see Examples 3 and 4.

3. State the Midpoint Formula *(page 54)*. For an example of using the Midpoint Formula to find the midpoint of a line segment, see Example 5.

4. Describe examples of how to use a coordinate plane to model and solve real-life problems *(pages 55 and 56, Examples 6–8)*.

P.6 Exercises

See **CalcChat.com** for tutorial help and worked-out solutions to odd-numbered exercises.

Vocabulary: Fill in the blanks.

1. An ordered pair of real numbers can be represented in a plane called the rectangular coordinate system or the _____ plane.
2. The point of intersection of the x- and y-axes is the _____, and the two axes divide the coordinate plane into four _____.
3. The _____ _____ is derived from the Pythagorean Theorem.
4. Finding the average values of the respective coordinates of the two endpoints of a line segment in a coordinate plane is also known as using the _____ _____.

Skills and Applications

 Plotting Points in the Cartesian Plane
In Exercises 5 and 6, plot the points.

5. $(2, 4), (3, -1), (-6, 2), (-4, 0), (-1, -8), (1.5, -3.5)$
6. $(1, -5), (-2, -7), (3, 3), (-2, 4), (0, 5), \left(\frac{2}{3}, \frac{5}{2}\right)$

Finding the Coordinates of a Point   **In Exercises 7 and 8, find the coordinates of the point.**

7. The point is three units to the left of the y-axis and four units above the x-axis.
8. The point is on the x-axis and 12 units to the left of the y-axis.

 Determining Quadrant(s) for a Point
In Exercises 9–14, determine the quadrant(s) in which (x, y) could be located.

9. $x > 0$ and $y < 0$
10. $x < 0$ and $y < 0$
11. $x = -4$ and $y > 0$
12. $x < 0$ and $y = 7$
13. $x + y = 0, x \neq 0, y \neq 0$
14. $xy > 0$

 Sketching a Scatter Plot   **In Exercises 15 and 16, sketch a scatter plot of the data shown in the table.**

15. The table shows the number y of Wal-Mart stores for each year x from 2007 through 2014. *(Source: Wal-Mart Stores, Inc.)*

Year, x	Number of Stores, y
2007	7262
2008	7720
2009	8416
2010	8970
2011	10,130
2012	10,773
2013	10,942
2014	11,453

Spreadsheet at LarsonPrecalculus.com

16. The table shows the lowest temperature on record y (in degrees Fahrenheit) in Duluth, Minnesota, for each month x, where $x = 1$ represents January. *(Source: NOAA)*

Month, x	Temperature, y
1	-39
2	-39
3	-29
4	-5
5	17
6	27
7	35
8	32
9	22
10	8
11	-23
12	-34

Spreadsheet at LarsonPrecalculus.com

 Finding a Distance   **In Exercises 17–22, find the distance between the points.**

17. $(-2, 6), (3, -6)$
18. $(8, 5), (0, 20)$
19. $(1, 4), (-5, -1)$
20. $(1, 3), (3, -2)$
21. $\left(\frac{1}{2}, \frac{4}{3}\right), (2, -1)$
22. $(9.5, -2.6), (-3.9, 8.2)$

 Verifying a Right Triangle   **In Exercises 23 and 24, (a) find the length of each side of the right triangle, and (b) show that these lengths satisfy the Pythagorean Theorem.**

23.

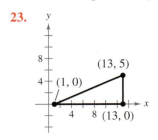

24.

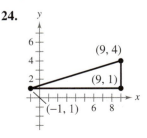

Verifying a Polygon   In Exercises 25–28, show that the points form the vertices of the polygon.

25. Right triangle: $(4, 0)$, $(2, 1)$, $(-1, -5)$

26. Right triangle: $(-1, 3)$, $(3, 5)$, $(5, 1)$

27. Isosceles triangle: $(1, -3)$, $(3, 2)$, $(-2, 4)$

28. Isosceles triangle: $(2, 3)$, $(4, 9)$, $(-2, 7)$

Plotting, Distance, and Midpoint   In Exercises 29–36, (a) plot the points, (b) find the distance between the points, and (c) find the midpoint of the line segment joining the points.

29. $(6, -3)$, $(6, 5)$

30. $(1, 4)$, $(8, 4)$

31. $(1, 1)$, $(9, 7)$

32. $(1, 12)$, $(6, 0)$

33. $(-1, 2)$, $(5, 4)$

34. $(2, 10)$, $(10, 2)$

35. $(-16.8, 12.3)$, $(5.6, 4.9)$

36. $\left(\frac{1}{2}, 1\right)$, $\left(-\frac{5}{2}, \frac{4}{3}\right)$

• • **37. Flying Distance** • • • • • • • • • • • •

An airplane flies from Naples, Italy, in a straight line to Rome, Italy, which is 120 kilometers north and 150 kilometers west of Naples. How far does the plane fly?

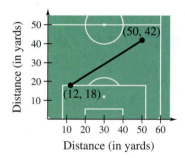

38. Sports   A soccer player passes the ball from a point that is 18 yards from the endline and 12 yards from the sideline. A teammate who is 42 yards from the same endline and 50 yards from the same sideline receives the pass. (See figure.) How long is the pass?

Distance (in yards)

39. Sales   The Coca-Cola Company had sales of \$35,123 million in 2010 and \$45,998 million in 2014. Use the Midpoint Formula to estimate the sales in 2012. Assume that the sales followed a linear pattern. *(Source: The Coca-Cola Company)*

40. Revenue per Share   The revenue per share for Twitter, Inc. was \$1.17 in 2013 and \$3.25 in 2015. Use the Midpoint Formula to estimate the revenue per share in 2014. Assume that the revenue per share followed a linear pattern. *(Source: Twitter, Inc.)*

Translating Points in the Plane   In Exercises 41–44, find the coordinates of the vertices of the polygon after the given translation to a new position in the plane.

41.

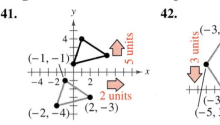

42.

43. Original coordinates of vertices: $(-7, -2)$, $(-2, 2)$, $(-2, -4)$, $(-7, -4)$

Shift: eight units up, four units to the right

44. Original coordinates of vertices: $(5, 8)$, $(3, 6)$, $(7, 6)$

Shift: 6 units down, 10 units to the left

45. Minimum Wage   Use the graph below, which shows the minimum wages in the United States (in dollars) from 1950 through 2015. *(Source: U.S. Department of Labor)*

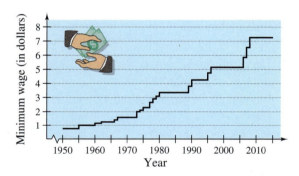

(a) Which decade shows the greatest increase in the minimum wage?

(b) Approximate the percent increases in the minimum wage from 1985 to 2000 and from 2000 to 2015.

(c) Use the percent increase from 2000 to 2015 to predict the minimum wage in 2030.

(d) Do you believe that your prediction in part (c) is reasonable? Explain.

46. Exam Scores   The table shows the mathematics entrance test scores x and the final examination scores y in an algebra course for a sample of 10 students.

x	22	29	35	40	44	48	53	58	65	76
y	53	74	57	66	79	90	76	93	83	99

(a) Sketch a scatter plot of the data.

(b) Find the entrance test score of any student with a final exam score in the 80s.

(c) Does a higher entrance test score imply a higher final exam score? Explain.

Exploration

True or False?   In Exercises 47–50, determine whether the statement is true or false. Justify your answer.

47. If the point (x, y) is in Quadrant II, then the point $(2x, -3y)$ is in Quadrant III.

48. To divide a line segment into 16 equal parts, you have to use the Midpoint Formula 16 times.

49. The points $(-8, 4)$, $(2, 11)$, and $(-5, 1)$ represent the vertices of an isosceles triangle.

50. If four points represent the vertices of a polygon, and the four side lengths are equal, then the polygon must be a square.

51. Think About It When plotting points on the rectangular coordinate system, when should you use different scales for the x- and y-axes? Explain.

52. Think About It What is the y-coordinate of any point on the x-axis? What is the x-coordinate of any point on the y-axis?

53. Using the Midpoint Formula A line segment has (x_1, y_1) as one endpoint and (x_m, y_m) as its midpoint. Find the other endpoint (x_2, y_2) of the line segment in terms of x_1, y_1, x_m, and y_m.

54. Using the Midpoint Formula Use the result of Exercise 53 to find the endpoint (x_2, y_2) of each line segment with the given endpoint (x_1, y_1) and midpoint (x_m, y_m).

(a) $(x_1, y_1) = (1, -2)$
$(x_m, y_m) = (4, -1)$
(b) $(x_1, y_1) = (-5, 11)$
$(x_m, y_m) = (2, 4)$

55. Using the Midpoint Formula Use the Midpoint Formula three times to find the three points that divide the line segment joining (x_1, y_1) and (x_2, y_2) into four equal parts.

56. Using the Midpoint Formula Use the result of Exercise 55 to find the points that divide each line segment joining the given points into four equal parts.

(a) $(x_1, y_1) = (1, -2)$
$(x_2, y_2) = (4, -1)$
(b) $(x_1, y_1) = (-2, -3)$
$(x_2, y_2) = (0, 0)$

57. Proof Prove that the diagonals of the parallelogram in the figure intersect at their midpoints.

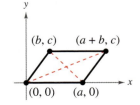

58. HOW DO YOU SEE IT? Use the plot of the point (x_0, y_0) in the figure. Match the transformation of the point with the correct plot. Explain. [The plots are labeled (i), (ii), (iii), and (iv).]

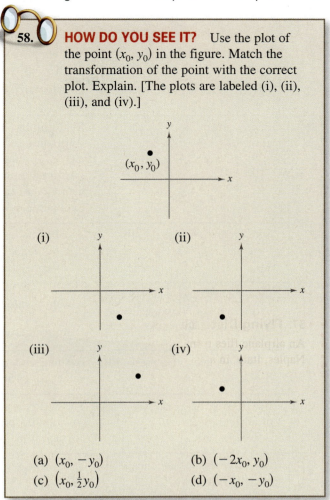

(a) $(x_0, -y_0)$ (b) $(-2x_0, y_0)$
(c) $\left(x_0, \frac{1}{2}y_0\right)$ (d) $(-x_0, -y_0)$

59. Collinear Points Three or more points are collinear when they all lie on the same line. Use the steps below to determine whether the set of points $\{A(2, 3), B(2, 6), C(6, 3)\}$ and the set of points $\{A(8, 3), B(5, 2), C(2, 1)\}$ are collinear.

(a) For each set of points, use the Distance Formula to find the distances from A to B, from B to C, and from A to C. What relationship exists among these distances for each set of points?

(b) Plot each set of points in the Cartesian plane. Do all the points of either set appear to lie on the same line?

(c) Compare your conclusions from part (a) with the conclusions you made from the graphs in part (b). Make a general statement about how to use the Distance Formula to determine collinearity.

60. Make a Conjecture

(a) Use the result of Exercise 58(a) to make a conjecture about the new location of a point when the sign of the y-coordinate is changed.

(b) Use the result of Exercise 58(d) to make a conjecture about the new location of a point when the signs of both x- and y-coordinates are changed.

Chapter Summary

	What Did You Learn?	Explanation/Examples	Review Exercises
Section P.1	Represent and classify real numbers (*p. 2*).	**Real numbers:** set of all rational and irrational numbers **Rational numbers:** real numbers that can be written as the ratio of two integers **Irrational numbers:** real numbers that cannot be written as the ratio of two integers Real numbers can be represented on the real number line.	1, 2
	Order real numbers and use inequalities (*p. 4*).	$a < b$: a is less than b. $a > b$: a is greater than b. $a \le b$: a is less than or equal to b. $a \ge b$: a is greater than or equal to b.	3–8
	Find the absolute values of real numbers and find the distance between two real numbers (*p. 6*).	**Absolute value of a:** $\|a\| = \begin{cases} a, & a \ge 0 \\ -a, & a < 0 \end{cases}$ **Distance between a and b:** $d(a, b) = \|b - a\| = \|a - b\|$	9–12
	Evaluate algebraic expressions (*p. 8*).	Evaluate an algebraic expression by substituting numerical values for each of the variables in the expression.	13–16
	Use the basic rules and properties of algebra (*p. 9*).	The basic rules of algebra, the properties of negation and equality, the properties of zero, and the properties and operations of fractions can be used to perform operations.	17–30
Section P.2	Use properties of exponents (*p. 14*).	**1.** $a^m a^n = a^{m+n}$ **2.** $a^m/a^n = a^{m-n}$ **3.** $a^{-n} = (1/a)^n$ **4.** $a^0 = 1$ **5.** $(ab)^m = a^m b^m$ **6.** $(a^m)^n = a^{mn}$ **7.** $(a/b)^m = a^m/b^m$ **8.** $\|a^2\| = a^2$	31–38
	Use scientific notation to represent real numbers (*p. 17*).	A number written in scientific notation has the form $\pm c \times 10^n$, where $1 \le c < 10$ and n is an integer.	39–42
	Use properties of radicals (*p. 18*) to simplify and combine radical expressions (*p. 20*).	**1.** $\sqrt[n]{a^m} = (\sqrt[n]{a})^m$ **2.** $\sqrt[n]{a} \cdot \sqrt[n]{b} = \sqrt[n]{ab}$ **3.** $\sqrt[n]{a}/\sqrt[n]{b} = \sqrt[n]{a/b},\ b \ne 0$ **4.** $\sqrt[m]{\sqrt[n]{a}} = \sqrt[mn]{a}$ **5.** $(\sqrt[n]{a})^n = a$ **6.** *n* even: $\sqrt[n]{a^n} = \|a\|$, *n* odd: $\sqrt[n]{a^n} = a$ A radical expression is in simplest form when all possible factors are removed from the radical, all fractions have radical-free denominators, and the index of the radical is reduced. Radical expressions can be combined when they are like radicals.	43–50
	Rationalize denominators and numerators (*p. 21*).	To rationalize a denominator or numerator of the form $a - b\sqrt{m}$ or $a + b\sqrt{m}$, multiply both numerator and denominator by a conjugate.	51–56
	Use properties of rational exponents (*p. 22*).	If a is a real number and n is a positive integer such that the principal nth root of a exists, then $a^{1/n}$ is defined as $a^{1/n} = \sqrt[n]{a}$, where $1/n$ is the rational exponent of a.	57–60
Section P.3	Write polynomials in standard form (*p. 26*), and add, subtract, and multiply polynomials (*p. 27*).	A polynomial written with descending powers of x is in standard form. To add and subtract polynomials, add or subtract the like terms. To find the product of two binomials, use the FOIL Method.	61–70, 75, 76

	What Did You Learn?	**Explanation/Examples**	**Review Exercises**
Section P.3	Use special products to multiply polynomials *(p. 28)*.	**Sum and difference of same terms:** $(u + v)(u - v) = u^2 - v^2$ **Square of a binomial:** $(u + v)^2 = u^2 + 2uv + v^2$ $\qquad\qquad (u - v)^2 = u^2 - 2uv + v^2$ **Cube of a binomial:** $(u + v)^3 = u^3 + 3u^2v + 3uv^2 + v^3$ $\qquad\qquad (u - v)^3 = u^3 - 3u^2v + 3uv^2 - v^3$	71–74
	Use polynomials to solve real-life problems *(p. 30)*.	Polynomials can be used to find the volume of a box. (See Example 9.)	77–80
Section P.4	Factor out common factors from polynomials *(p. 34)*.	The process of writing a polynomial as a product is called factoring. Factoring out any common factors is the first step in completely factoring a polynomial.	81, 82
	Factor special polynomial forms *(p. 35)*.	**Difference of two squares:** $u^2 - v^2 = (u + v)(u - v)$ **Perfect square trinomial:** $u^2 + 2uv + v^2 = (u + v)^2$ $\qquad\qquad u^2 - 2uv + v^2 = (u - v)^2$ **Sum or difference** $u^3 + v^3 = (u + v)(u^2 - uv + v^2)$ **of two cubes:** $\quad u^3 - v^3 = (u - v)(u^2 + uv + v^2)$	83–86
	Factor trinomials as the product of two binomials *(p. 37)*.	$ax^2 + bx + c = (\;\blacksquare x + \blacksquare\;)(\;\blacksquare x + \blacksquare\;)$ Factors of a — Factors of c	87, 88
	Factor polynomials by grouping *(p. 38)*.	Polynomials with more than three terms can sometimes be factored by grouping. (See Examples 9 and 10.)	89, 90
Section P.5	Find domains of algebraic expressions *(p. 41)*.	The set of real numbers for which an algebraic expression is defined is the domain of the expression.	91–94
	Simplify rational expressions *(p. 42)*.	When simplifying rational expressions, factor each polynomial completely to determine whether the numerator and denominator have factors in common.	95, 96
	Add, subtract, multiply, and divide rational expressions *(p. 43)*.	To add or subtract, use the LCD method or the basic definition $\dfrac{a}{b} \pm \dfrac{c}{d} = \dfrac{ad \pm bc}{bd}$, $b \neq 0$, $d \neq 0$. To multiply or divide, use the properties of fractions.	97–100
	Simplify complex fractions and rewrite difference quotients *(p. 45)*.	One way to simplify a complex fraction is to combine the fractions in the numerator into a single fraction and then combine the fractions in the denominator into a single fraction. Then invert the denominator and multiply.	101–104
Section P.6	Plot points in the Cartesian plane *(p. 51)*.	For an ordered pair (x, y), the x-coordinate is the directed distance from the y-axis to the point, and the y-coordinate is the directed distance from the x-axis to the point.	105–108
	Use the Distance Formula *(p. 53)* and the Midpoint Formula *(p. 54)*.	**Distance Formula:** $d = \sqrt{(x_2 - x_1)^2 + (y_2 - y_1)^2}$ **Midpoint Formula:** Midpoint $= \left(\dfrac{x_1 + x_2}{2}, \dfrac{y_1 + y_2}{2} \right)$	109–112
	Use a coordinate plane to model and solve real-life problems *(p. 55)*.	The coordinate plane can be used to estimate the annual sales of a company. (See Example 7.)	113–116

Review Exercises
See CalcChat.com for tutorial help and worked-out solutions to odd-numbered exercises.

P.1 **Classifying Real Numbers** In Exercises 1 and 2, determine which numbers in the set are (a) natural numbers, (b) whole numbers, (c) integers, (d) rational numbers, and (e) irrational numbers.

1. $\left\{ 11, -14, -\frac{8}{9}, \frac{5}{2}, \sqrt{6}, 0.4 \right\}$
2. $\left\{ \sqrt{15}, -22, -\frac{10}{3}, 0, 5.2, \frac{3}{7} \right\}$

Plotting and Ordering Real Numbers In Exercises 3 and 4, plot the two real numbers on the real number line. Then place the appropriate inequality symbol (< or >) between them.

3. (a) $\frac{5}{4}$ (b) $\frac{7}{8}$ 4. (a) $-\frac{9}{25}$ (b) $-\frac{5}{7}$

Interpreting an Inequality or an Interval In Exercises 5–8, (a) give a verbal description of the subset of real numbers represented by the inequality or the interval, (b) sketch the subset on the real number line, and (c) state whether the subset is bounded or unbounded.

5. $x \geq 6$ 6. $-4 < x < 4$
7. $[-3, 4)$ 8. $[2, \infty)$

Finding a Distance In Exercises 9 and 10, find the distance between a and b.

9. $a = -74, b = 48$
10. $a = -112, b = -6$

Using Absolute Value Notation In Exercises 11 and 12, use absolute value notation to describe the situation.

11. The distance between x and 7 is at least 4.
12. The distance between x and 25 is no more than 10.

Evaluating an Algebraic Expression In Exercises 13–16, evaluate the expression for each value of x. (If not possible, state the reason.)

13. $12x - 7$ (a) $x = 0$ (b) $x = -1$
14. $x^2 - 6x + 5$ (a) $x = -2$ (b) $x = 2$
15. $-x^2 + x - 1$ (a) $x = 1$ (b) $x = -1$
16. $\dfrac{x}{x - 3}$ (a) $x = -3$ (b) $x = 3$

Identifying Rules of Algebra In Exercises 17–22, identify the rule(s) of algebra illustrated by the statement.

17. $0 + (a - 5) = a - 5$
18. $1 \cdot (3x + 4) = 3x + 4$
19. $2x + (3x - 10) = (2x + 3x) - 10$
20. $4(t + 2) = 4 \cdot t + 4 \cdot 2$

21. $(t^2 + 1) + 3 = 3 + (t^2 + 1)$
22. $\dfrac{2}{y + 4} \cdot \dfrac{y + 4}{2} = 1, \quad y \neq -4$

Performing Operations In Exercises 23–30, perform the operation(s). (Write fractional answers in simplest form.)

23. $-6 + 6$ 24. $2 - (-3)$
25. $(-8)(-4)$ 26. $5(20 + 7)$
27. $\dfrac{x}{5} + \dfrac{7x}{12}$ 28. $\dfrac{x}{2} - \dfrac{2x}{5}$
29. $\dfrac{3x}{10} \cdot \dfrac{5}{3}$ 30. $\dfrac{9}{x} \div \dfrac{1}{6}$

P.2 **Using Properties of Exponents** In Exercises 31–34, simplify each expression.

31. (a) $3x^2(4x^3)^3$ (b) $\dfrac{5y^6}{10y}$
32. (a) $(3a)^2(6a^3)$ (b) $\dfrac{36x^5}{9x^{10}}$
33. (a) $(-2z)^3$ (b) $\dfrac{(8y)^0}{y^2}$
34. (a) $[(x + 2)^2]^3$ (b) $\dfrac{40(b - 3)^5}{75(b - 3)^2}$

Rewriting with Positive Exponents In Exercises 35–38, rewrite each expression with positive exponents. Simplify, if possible.

35. (a) $\dfrac{a^2}{b^{-2}}$ (b) $(a^2 b^4)(3ab^{-2})$
36. (a) $\dfrac{6^2 u^3 v^{-3}}{12u^{-2}v}$ (b) $\dfrac{3^{-4}m^{-1}n^{-3}}{9^{-2}mn^{-3}}$
37. (a) $\dfrac{(5a)^{-2}}{(5a)^2}$ (b) $\dfrac{4(x^{-1})^{-3}}{4^{-2}(x^{-1})^{-1}}$
38. (a) $(x + y^{-1})^{-1}$ (b) $\left(\dfrac{x^{-3}}{y} \right)\left(\dfrac{x}{y} \right)^{-1}$

Scientific Notation In Exercises 39 and 40, write the number in scientific notation.

39. Sales for Nautilus, Inc. in 2014: \$274,400,000 (*Source: Nautilus, Inc.*)
40. Number of meters in 1 foot: 0.3048

Decimal Notation In Exercises 41 and 42, write the number in decimal notation.

41. Distance between the sun and Jupiter: 4.84×10^8 miles
42. Ratio of day to year: 2.74×10^{-3}

Simplifying Radical Expressions   In Exercises 43–48, simplify each radical expression.

43. (a) $\sqrt[3]{27^2}$ (b) $\sqrt{49^3}$

44. (a) $\sqrt[3]{\frac{64}{125}}$ (b) $\sqrt{\frac{81}{100}}$

45. (a) $\left(\sqrt[3]{216}\right)^3$ (b) $\sqrt[4]{32^4}$

46. (a) $\sqrt[3]{\frac{2x^3}{27}}$ (b) $\sqrt[5]{64x^6}$

47. (a) $\sqrt{12x^3} + \sqrt{3x}$ (b) $\sqrt{27x^3} - \sqrt{3x^3}$

48. (a) $\sqrt{8x^3} + \sqrt{2x}$ (b) $\sqrt{18x^5} - \sqrt{8x^3}$

49. Writing   Explain why $\sqrt{5u} + \sqrt{3u} \neq 2\sqrt{2u}$.

50. Engineering   The rectangular cross section of a wooden beam cut from a log of diameter 24 inches (see figure) will have a maximum strength when its width w and height h are

$$w = 8\sqrt{3} \quad \text{and} \quad h = \sqrt{24^2 - \left(8\sqrt{3}\right)^2}.$$

Find the area of the rectangular cross section and write the answer in simplest form.

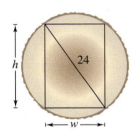

Rationalizing a Denominator   In Exercises 51–54, rationalize the denominator of the expression. Then simplify your answer.

51. $\dfrac{3}{4\sqrt{3}}$ **52.** $\dfrac{12}{\sqrt[3]{4}}$

53. $\dfrac{1}{2 - \sqrt{3}}$ **54.** $\dfrac{1}{\sqrt{5} + 1}$

∫ Rationalizing a Numerator   In Exercises 55 and 56, rationalize the numerator of the expression. Then simplify your answer.

55. $\dfrac{\sqrt{7} + 1}{2}$ **56.** $\dfrac{\sqrt{2} - \sqrt{11}}{3}$

Simplifying an Expression   In Exercises 57–60, simplify the expression.

57. $16^{3/2}$ **58.** $64^{-2/3}$

59. $(3x^{2/5})(2x^{1/2})$ **60.** $(x - 1)^{1/3}(x - 1)^{-1/4}$

P.3 **Writing Polynomials in Standard Form**   In Exercises 61–64, write the polynomial in standard form. Identify the degree and leading coefficient.

61. $3 - 11x^2$ **62.** $3x^3 - 5x^5 + x - 4$

63. $-4 - 12x^2$ **64.** $12x - 7x^2 + 6$

Adding or Subtracting Polynomials   In Exercises 65 and 66, perform the operation and write the result in standard form.

65. $-(3x^2 + 2x) + (1 - 5x)$

66. $8y - [2y^2 - (3y - 8)]$

Multiplying Polynomials   In Exercises 67–74, find the product.

67. $2x(x^2 - 5x + 6)$

68. $(3x^3 - 1.5x^2 + 4)(-3x)$

69. $(3x - 6)(5x + 1)$ **70.** $(x + 2)(x^2 - 2)$

71. $(6x + 5)(6x - 5)$

72. $(3\sqrt{5} + 2x)(3\sqrt{5} - 2x)$

73. $(2x - 3)^2$ **74.** $(x - 4)^3$

Operations with Polynomials   In Exercises 75 and 76, perform the operation.

75. Multiply $x^2 + x + 5$ and $x^2 - 7x - 2$.

76. Subtract $9x^4 - 11x^2 + 16$ from $6x^4 - 20x^2 - x + 3$.

77. Compound Interest   An investment of $2500 compounded annually for 2 years at an interest rate r (in decimal form) yields an amount of $2500(1 + r)^2$. Write this polynomial in standard form.

78. Surface Area   The surface area S of a right circular cylinder is $S = 2\pi r^2 + 2\pi rh$.

(a) Draw a right circular cylinder of radius r and height h. Use your drawing to explain how to obtain the surface area formula.

(b) Find the surface area when the radius is 6 inches and the height is 8 inches.

79. Geometry   Find a polynomial that represents the total number of square feet for the floor plan shown in the figure.

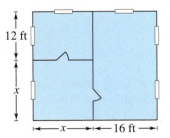

80. Geometry   Use the area model to write two different expressions for the area. Then equate the two expressions and name the algebraic property illustrated.

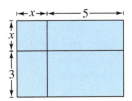

P.4 Factoring Completely In Exercises 81–90, completely factor the expression.

81. $x^3 - x$

82. $x(x - 3) + 4(x - 3)$

83. $25x^2 - 49$

84. $36x^2 - 81$

85. $x^3 - 64$

86. $8x^3 + 27$

87. $2x^2 + 21x + 10$

88. $3x^2 + 14x + 8$

89. $x^3 - x^2 + 2x - 2$

90. $x^3 - 4x^2 + 2x - 8$

P.5 Finding the Domain of an Algebraic Expression In Exercises 91–94, find the domain of the expression.

91. $\dfrac{1}{x + 1}$

92. $\dfrac{1}{x + 6}$

93. $\sqrt{x + 2}$

94. $\sqrt{x + 4}$

Simplifying a Rational Expression In Exercises 95 and 96, write the rational expression in simplest form.

95. $\dfrac{x^2 - 64}{5(3x + 24)}$

96. $\dfrac{x^3 + 27}{x^2 + x - 6}$

Operations with Rational Expressions In Exercises 97–100, perform the operation and simplify.

97. $\dfrac{x^2 - 7x + 12}{x^2 + 8x + 16} \cdot \dfrac{x + 4}{x^2 - 9}$

98. $\dfrac{2x - 6}{3x + 6} \div \dfrac{3 - x}{2x + 4}$

99. $\dfrac{2}{x - 3} + \dfrac{6}{3 - x}$

100. $\dfrac{3x}{x - 4} - \dfrac{2}{4 - x}$

Simplifying a Complex Fraction In Exercises 101 and 102, simplify the complex fraction.

101. $\dfrac{\left[\dfrac{3a}{(a^2/x) - 1}\right]}{\left(\dfrac{a}{x} - 1\right)}$

102. $\dfrac{\left(\dfrac{1}{2x - 3} - \dfrac{1}{2x + 3}\right)}{\left(\dfrac{1}{2x} - \dfrac{1}{2x + 3}\right)}$

∫ Simplifying a Difference Quotient In Exercises 103 and 104, simplify the difference quotient.

103. $\dfrac{\left[\dfrac{1}{2(x + h)} - \dfrac{1}{2x}\right]}{h}$

104. $\dfrac{\left(\dfrac{1}{x + h - 3} - \dfrac{1}{x - 3}\right)}{h}$

P.6 Plotting Points in the Cartesian Plane In Exercises 105 and 106, plot the points.

105. $(5, 5), (-2, 0), (-3, 6), (-1, -7)$

106. $(0, 6), (8, 1), (5, -4), (-3, -3)$

Determining Quadrant(s) for a Point In Exercises 107 and 108, determine the quadrant(s) in which (x, y) could be located.

107. $x > 0$ and $y = -2$

108. $xy = 4$

Plotting, Distance, and Midpoint In Exercises 109–112, (a) plot the points, (b) find the distance between the points, and (c) find the midpoint of the line segment joining the points.

109. $(-3, 8), (1, 5)$

110. $(-2, 6), (4, -3)$

111. $(5.6, 0), (0, 8.2)$

112. $(1.8, 7.4), (-0.6, -14.5)$

Translating Points in the Plane In Exercises 113 and 114, find the coordinates of the vertices of the polygon after the given translation to a new position in the plane.

113. Original coordinates of vertices:

$(4, 8), (6, 8), (4, 3), (6, 3)$

Shift: eight units down, four units to the left

114. Original coordinates of vertices:

$(0, 1), (3, 3), (0, 5), (-3, 3)$

Shift: three units up, two units to the left

115. **Sales** Barnes & Noble had annual sales of $6.8 billion in 2013 and $6.1 billion in 2015. Use the Midpoint Formula to estimate the sales in 2014. Assume that the annual sales followed a linear pattern. *(Source: Barnes & Noble, Inc.)*

116. **Meteorology** The apparent temperature is a measure of relative discomfort to a person from heat and high humidity. The table shows the actual temperatures x (in degrees Fahrenheit) versus the apparent temperatures y (in degrees Fahrenheit) for a relative humidity of 75%.

x	70	75	80	85	90	95	100
y	70	77	85	95	109	130	150

(a) Sketch a scatter plot of the data shown in the table.

(b) Find the change in the apparent temperature when the actual temperature changes from 70°F to 100°F.

Exploration

True or False? In Exercises 117 and 118, determine whether the statement is true or false. Justify your answer.

117. A binomial sum squared is equal to the sum of the terms squared.

118. $x^n - y^n$ factors as conjugates for all values of n.

Chapter Test

See **CalcChat.com** for tutorial help and worked-out solutions to odd-numbered exercises.

Take this test as you would take a test in class. When you are finished, check your work against the answers given in the back of the book.

1. Place the appropriate inequality symbol ($<$ or $>$) between the real numbers $-\frac{10}{3}$ and $-\frac{5}{3}$.

2. Find the distance between the real numbers $-\frac{7}{4}$ and $\frac{5}{4}$.

3. Identify the rule of algebra illustrated by $(5 - x) + 0 = 5 - x$.

In Exercises 4 and 5, evaluate each expression without using a calculator.

4. (a) $\left(-\dfrac{3}{5}\right)^3$ (b) $\left(\dfrac{3^2}{2}\right)^{-3}$

   (c) $\dfrac{5^3 \cdot 7^{-1}}{5^2 \cdot 7}$ (d) $(2^3)^{-2}$

5. (a) $\sqrt{5} \cdot \sqrt{125}$ (b) $\dfrac{\sqrt{27}}{\sqrt{2}}$

   (c) $\dfrac{5.4 \times 10^8}{3 \times 10^3}$ (d) $(4.0 \times 10^8)(2.4 \times 10^{-3})$

In Exercises 6 and 7, simplify each expression.

6. (a) $3z^2(2z^3)^2$ (b) $(u - 2)^{-4}(u - 2)^{-3}$ (c) $\left(\dfrac{x^{-2}y^2}{3}\right)^{-1}$

7. (a) $9z\sqrt{8z} - 3\sqrt{2z^3}$ (b) $(4x^{3/5})(x^{1/3})$ (c) $\sqrt[3]{\dfrac{16}{v^5}}$

8. Write the polynomial $3 - 2x^5 + 3x^3 - x^4$ in standard form. Identify the degree and leading coefficient.

In Exercises 9–12, perform the operation and simplify.

9. $(x^2 + 3) - [3x + (8 - x^2)]$ 10. $\left(x + \sqrt{5}\right)\left(x - \sqrt{5}\right)$

11. $\dfrac{5x}{x - 4} + \dfrac{20}{4 - x}$ 12. $\dfrac{\left[\dfrac{x^3}{(x - 1)^3}\right]}{\left[\dfrac{x}{(x - 1)^4}\right]}$

13. Completely factor (a) $2x^4 - 3x^3 - 2x^2$ and (b) $x^3 + 2x^2 - 4x - 8$.

14. Rationalize each denominator and simplify. (a) $\dfrac{16}{\sqrt[3]{16}}$ (b) $\dfrac{4}{1 - \sqrt{2}}$

15. Find the domain of $\dfrac{6 - x}{1 - x}$.

16. Multiply and simplify: $\dfrac{y^2 + 8y + 16}{2y - 4} \cdot \dfrac{8y - 16}{(y + 4)^3}$.

17. A T-shirt company can produce and sell x T-shirts per day. The total cost C (in dollars) for producing x T-shirts is $C = 1480 + 6x$, and the total revenue R (in dollars) is $R = 15x$. Find the profit obtained by selling 225 T-shirts per day.

18. Plot the points $(-2, 5)$ and $(6, 0)$. Then find the distance between the points and the midpoint of the line segment joining the points.

19. Write an expression for the area of the shaded region in the figure at the left, and simplify the result.

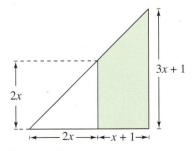

Figure for 19

Proofs in Mathematics ▪ ▪ ▪ ▪ ▪ ▪ ▪ ▪ ▪ ▪ ▪ ▪ ▪ ▪ ▪ ▪

What does the word *proof* mean to you? In mathematics, the word *proof* means a valid argument. When you prove a statement or theorem, you must use facts, definitions, and accepted properties in a logical order. You can also use previously proved theorems in your proof. For example, the proof of the Midpoint Formula below uses the Distance Formula. There are several different proof methods, which you will see in later chapters.

> ### The Midpoint Formula *(p.54)*
> The midpoint of the line segment joining the points (x_1, y_1) and (x_2, y_2) is
> $$\text{Midpoint} = \left(\frac{x_1 + x_2}{2}, \frac{y_1 + y_2}{2}\right).$$

THE CARTESIAN PLANE

The Cartesian plane is named after French mathematician René Descartes (1596–1650). According to some accounts, while Descartes was lying in bed, he noticed a fly buzzing around on the ceiling. He realized that he could describe the fly's position by its distance from the bedroon walls. This led to the development of the Cartesian plane. Descartes felt that using a coordinate plane could facilitate descriptions of the positions of objects.

Proof

Using the figure, you must show that $d_1 = d_2$ and $d_1 + d_2 = d_3$.

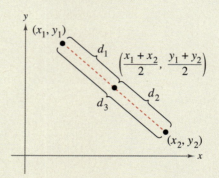

By the Distance Formula, you obtain

$$d_1 = \sqrt{\left(\frac{x_1 + x_2}{2} - x_1\right)^2 + \left(\frac{y_1 + y_2}{2} - y_1\right)^2}$$

$$= \sqrt{\left(\frac{x_2 - x_1}{2}\right)^2 + \left(\frac{y_2 - y_1}{2}\right)^2}$$

$$= \frac{1}{2}\sqrt{(x_2 - x_1)^2 + (y_2 - y_1)^2},$$

$$d_2 = \sqrt{\left(x_2 - \frac{x_1 + x_2}{2}\right)^2 + \left(y_2 - \frac{y_1 + y_2}{2}\right)^2}$$

$$= \sqrt{\left(\frac{x_2 - x_1}{2}\right)^2 + \left(\frac{y_2 - y_1}{2}\right)^2}$$

$$= \frac{1}{2}\sqrt{(x_2 - x_1)^2 + (y_2 - y_1)^2},$$

and

$$d_3 = \sqrt{(x_2 - x_1)^2 + (y_2 - y_1)^2}.$$

So, it follows that $d_1 = d_2$ and $d_1 + d_2 = d_3$. ◾

P.S. Problem Solving ■ ■ ■ ■ ■ ■ ■ ■ ■ ■ ■ ■ ■ ■ ■

1. Volume and Density The USATF states that the men's and women's shots for track and field competition must comply with the specifications below. *(Source: USATF)*

	Men's	Women's
Weight	7.26 kg	4.0 kg
Diameter (minimum)	110 mm	95 mm
Diameter (maximum)	130 mm	110 mm

(a) Find the maximum and minimum volumes of both the men's and women's shots.

(b) To find the *density* of an object, divide its mass (weight) by its volume. Find the maximum and minimum densities of both the men's and women's shots.

(c) A shot is sometimes made out of iron. When a ball of cork has the same volume as an iron shot, do you think they would have the same density? Explain.

2. Proof Find an example for which

$$|a - b| > |a| - |b|$$

and an example for which

$$|a - b| = |a| - |b|.$$

Then prove that

$$|a - b| \ge |a| - |b|$$

for all a, b.

3. Significant Digits The accuracy of an approximation to a number is related to how many significant digits there are in the approximation. Write a definition of significant digits and illustrate the concept with examples.

4. Stained Glass Window A stained glass window is in the shape of a rectangle with a semicircular arch (see figure). The width of the window is 2 feet and the perimeter is approximately 13.14 feet. Find the least amount of glass required to construct the window.

←— 2 ft —→

5. Heartbeats The life expectancies at birth in 2013 for men and women were 76.4 years and 81.2 years, respectively. Assuming an average healthy heart rate of 70 beats per minute, find the numbers of beats in a lifetime for a man and for a woman. *(Source: National Center for Health Statistics)*

6. Population The table shows the census population y (in millions) of the United States for each census year x from 1960 through 2010. *(Source: U.S. Census Bureau)*

DATA Year, x	Population, y
1960	179.32
1970	203.30
1980	226.54
1990	248.72
2000	281.42
2010	308.75

Spreadsheet at LarsonPrecalculus.com

(a) Sketch a scatter plot of the data. Describe any trends in the data.

(b) Find the increase in population from each census year to the next.

(c) Over which decade did the population increase the most? the least?

(d) Find the percent increase in population from each census year to the next.

(e) Over which decade was the percent increase the greatest? the least?

7. Declining Balances Method Find the annual depreciation rate r (in decimal form) from the bar graph below. To find r by the declining balances method, use the formula

$$r = 1 - \left(\frac{S}{C}\right)^{1/n}$$

where n is the useful life of the item (in years), S is the salvage value (in dollars), and C is the original cost (in dollars).

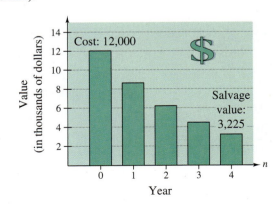

8. Planetary Distance and Period Johannes Kepler (1571–1630), a well-known German astronomer, discovered a relationship between the average distance of a planet from the sun and the time (or period) it takes the planet to orbit the sun. People then knew that planets that are closer to the sun take less time to complete an orbit than planets that are farther from the sun. Kepler discovered that the distance and period are related by an exact mathematical formula.

The table shows the average distances x (in astronomical units) and periods y (in years) for the five planets that are closest to the sun. By completing the table, can you rediscover Kepler's relationship? Write a paragraph that summarizes your conclusions.

Planet	Mercury	Venus	Earth	Mars	Jupiter
x	0.387	0.723	1.000	1.524	5.203
$\sqrt{x}$					
y	0.241	0.615	1.000	1.881	11.862
$\sqrt[3]{y}$					

9. Surface Area of a Box A mathematical model for the volume V (in cubic inches) of the box shown below is

$$V = 2x^3 + x^2 - 8x - 4$$

where x is in inches. Find an expression for the surface area of the box. Then find the surface area when $x = 6$ inches.

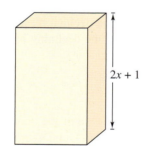

$2x + 1$

10. Proof Prove that

$$\left(\frac{2x_1 + x_2}{3}, \frac{2y_1 + y_2}{3} \right)$$

is one of the points of trisection of the line segment joining (x_1, y_1) and (x_2, y_2). Find the midpoint of the line segment joining

$$\left(\frac{2x_1 + x_2}{3}, \frac{2y_1 + y_2}{3} \right)$$

and (x_2, y_2) to find the second point of trisection.

11. Points of Trisection Use the results of Exercise 10 to find the points of trisection of the line segment joining each pair of points.

(a) $(1, -2)$, $(4, 1)$

(b) $(-2, -3)$, $(0, 0)$

12. Nonequivalent and Equivalent Equations Verify that $y_1 \neq y_2$ by letting $x = 0$ and evaluating y_1 and y_2.

$$y_1 = 2x\sqrt{1 - x^2} - \frac{x^3}{\sqrt{1 - x^2}}$$

$$y_2 = \frac{2 - 3x^2}{\sqrt{1 - x^2}}$$

Change y_2 so that $y_1 = y_2$.

13. Weight of a Golf Ball A major feature of Epcot Center at Disney World is Spaceship Earth. The building is spherically shaped and weighs 1.6×10^7 pounds, which is equal in weight to 1.58×10^8 golf balls. Use these values to find the approximate weight (in pounds) of one golf ball. Then convert the weight to ounces. *(Source: Disney.com)*

14. Misleading Graphs Although graphs can help visualize relationships between two variables, they can also be misleading. The graphs shown below represent the same data points.

(a) Which of the two graphs is misleading, and why? Discuss other ways in which graphs can be misleading.

(b) Why would it be beneficial for someone to use a misleading graph?

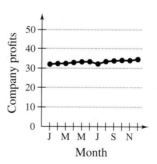

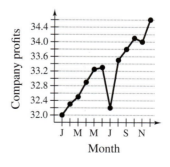

1 Equations, Inequalities, and Mathematical Modeling

Electrical Circuit *(Exercise 87, page 120)*

Compound Interest
(Example 9, page 127)

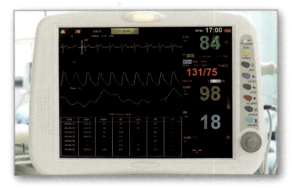

Blood Oxygen Level *(Exercise 117, page 112)*

Cross-Country Running
(Example 6, page 86)

Population Statistics *(Exercise 77, page 80)*

1.1 Graphs of Equations

The graph of an equation can help you visualize relationships between real-life quantities. For example, in Exercise 77 on page 80, you will use a graph to analyze life expectancy.

- ■ Sketch graphs of equations.
- ■ Identify *x*- and *y*-intercepts of graphs of equations.
- ■ Use symmetry to sketch graphs of equations.
- ■ Write equations of circles.
- ■ Use graphs of equations to solve real-life problems.

The Graph of an Equation

In Section P.6, you used a coordinate system to graphically represent the relationship between two quantities as points in a coordinate plane.

Frequently, a relationship between two quantities is expressed as an **equation in two variables.** For example, $y = 7 - 3x$ is an equation in x and y. An ordered pair (a, b) is a **solution** or **solution point** of an equation in x and y when the substitutions $x = a$ and $y = b$ result in a true statement. For example, $(1, 4)$ is a solution of $y = 7 - 3x$ because $4 = 7 - 3(1)$ is a true statement.

In this section, you will review some basic procedures for sketching the graph of an equation in two variables. The **graph of an equation** is the set of all points that are solutions of the equation.

EXAMPLE 1 Determining Solution Points

Determine whether (a) $(2, 13)$ and (b) $(-1, -3)$ lie on the graph of $y = 10x - 7$.

Solution

a. $y = 10x - 7$ Write original equation.

$13 \overset{?}{=} 10(2) - 7$ Substitute 2 for x and 13 for y.

$13 = 13$ $(2, 13)$ is a solution. ✔

The point $(2, 13)$ *does* lie on the graph of $y = 10x - 7$ because it is a solution point of the equation.

b. $y = 10x - 7$ Write original equation.

$-3 \overset{?}{=} 10(-1) - 7$ Substitute -1 for x and -3 for y.

$-3 \neq -17$ $(-1, -3)$ is not a solution.

The point $(-1, -3)$ *does not* lie on the graph of $y = 10x - 7$ because it is *not* a solution point of the equation.

✔ *Checkpoint* 🔊))) *Audio-video solution in English & Spanish at LarsonPrecalculus.com*

Determine whether (a) $(3, -5)$ and (b) $(-2, 26)$ lie on the graph of $y = 14 - 6x$. ■

The basic technique used for sketching the graph of an equation is the **point-plotting method.**

▷ **ALGEBRA HELP** When evaluating an expression or an equation, remember to follow the Basic Rules of Algebra. To review these rules, see Section P.1.

The Point-Plotting Method of Graphing

1. When possible, isolate one of the variables.
2. Construct a table of values showing several solution points.
3. Plot these points in a rectangular coordinate system.
4. Connect the points with a smooth curve or line.

It is important to use negative values, zero, and positive values for x (if possible) when constructing a table.

EXAMPLE 2   **Sketching the Graph of an Equation**

Sketch the graph of

$3x + y = 7.$

Solution

First, isolate the variable y.

$y = -3x + 7$ Solve equation for y.

Next, construct a table of values that consists of several solution points of the equation. For example, when $x = -3$,

$y = -3(-3) + 7 = 16$

which implies that $(-3, 16)$ is a solution point of the equation.

x	$y = -3x + 7$	(x, y)
-3	16	$(-3, 16)$
-2	13	$(-2, 13)$
-1	10	$(-1, 10)$
0	7	$(0, 7)$
1	4	$(1, 4)$
2	1	$(2, 1)$
3	-2	$(3, -2)$

From the table, it follows that

$(-3, 16), \ (-2, 13), \ (-1, 10), \ (0, 7), \ (1, 4), \ (2, 1), \ \text{and} \ (3, -2)$

are solution points of the equation. Plot these points and connect them with a line, as shown below.

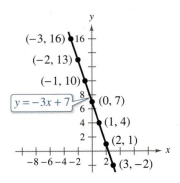

✓ *Checkpoint* 🔊))) *Audio-video solution in English & Spanish at LarsonPrecalculus.com*

Sketch the graph of each equation.

a. $3x + y = 2$

b. $-2x + y = 1$

EXAMPLE 3   **Sketching the Graph of an Equation**

See LarsonPrecalculus.com for an interactive version of this type of example.

Sketch the graph of

$$y = x^2 - 2.$$

Solution

The equation is already solved for y, so begin by constructing a table of values.

x	-2	-1	0	1	2	3
$y = x^2 - 2$	2	-1	-2	-1	2	7
(x, y)	$(-2, 2)$	$(-1, -1)$	$(0, -2)$	$(1, -1)$	$(2, 2)$	$(3, 7)$

Next, plot the points given in the table, as shown in Figure 1.1. Finally, connect the points with a smooth curve, as shown in Figure 1.2.

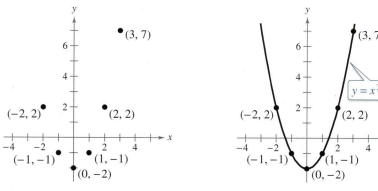

Figure 1.1   Figure 1.2

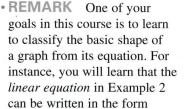

REMARK   One of your goals in this course is to learn to classify the basic shape of a graph from its equation. For instance, you will learn that the *linear equation* in Example 2 can be written in the form

$$y = mx + b$$

and its graph is a line. Similarly, the *quadratic equation* in Example 3 has the form

$$y = ax^2 + bx + c$$

and its graph is a parabola.

✓ *Checkpoint* *Audio-video solution in English & Spanish at LarsonPrecalculus.com*

Sketch the graph of each equation.

a. $y = x^2 + 3$ **b.** $y = 1 - x^2$

The point-plotting method demonstrated in Examples 2 and 3 is straightforward, but it has shortcomings. For instance, with too few solution points, it is possible to misrepresent the graph of an equation. To illustrate, when you only plot the four points

$$(-2, 2), (-1, -1), (1, -1), \quad \text{and} \quad (2, 2)$$

in Example 3, any one of the three graphs below is reasonable.

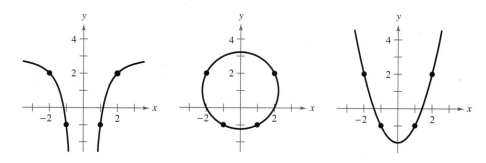

To graph an equation involving x and y on a graphing utility, use the procedure below.

1. If necessary, rewrite the equation so that y is isolated on the left side.

2. Enter the equation in the graphing utility.

3. Determine a *viewing window* that shows all important features of the graph.

4. Graph the equation.

Intercepts of a Graph

Solution points of an equation that have zero as either the x-coordinate or the y-coordinate are called **intercepts.** They are the points at which the graph intersects or touches the x- or y-axis. It is possible for a graph to have no intercepts, one intercept, or several intercepts, as shown in the graphs below.

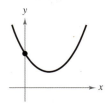

No x-intercepts
One y-intercept

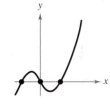

Three x-intercepts
One y-intercept

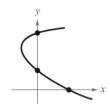

One x-intercept
Two y-intercepts

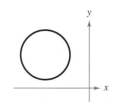

No intercepts

Note that an x-intercept can be written as the ordered pair $(a, 0)$ and a y-intercept can be written as the ordered pair $(0, b)$. Sometimes it is convenient to denote the x-intercept as the x-coordinate a of the point $(a, 0)$, or the y-intercept as the y-coordinate b of the point $(0, b)$. Unless it is necessary to make a distinction, the term *intercept* will refer to either the point or the coordinate.

EXAMPLE 4   **Identifying *x*- and *y*-Intercepts**

Identify the x- and y-intercepts of the graph of

$$y = x^3 + 1$$

shown at the right.

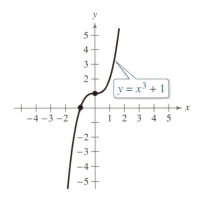

$y = x^3 + 1$

Solution

From the figure, the graph of the equation

$$y = x^3 + 1$$

has an x-intercept (where y is zero) at $(-1, 0)$ and a y-intercept (where x is zero) at $(0, 1)$.

✓ *Checkpoint* *Audio-video solution in English & Spanish at LarsonPrecalculus.com*

Identify the x- and y-intercepts of the graph of

$$y = -x^2 - 5x$$

shown at the right.

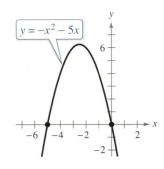

$y = -x^2 - 5x$

Symmetry

Graphs of equations can have **symmetry** with respect to one of the coordinate axes or with respect to the origin. Symmetry with respect to the x-axis means that when you fold the Cartesian plane along the x-axis, the portion of the graph above the x-axis coincides with the portion below the x-axis. Symmetry with respect to the y-axis or the origin can be described in a similar manner. The graphs below show these three types of symmetry.

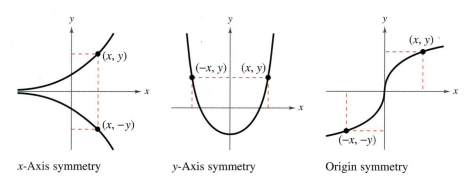

| x-Axis symmetry | y-Axis symmetry | Origin symmetry |

Knowing the symmetry of a graph *before* attempting to sketch it is helpful, because then you need only half as many solution points to sketch the graph. Graphical and algebraic tests for these three basic types of symmetry are described below.

Graphical Tests for Symmetry

1. A graph is **symmetric with respect to the x-axis** if, whenever (x, y) is on the graph, $(x, -y)$ is also on the graph.

2. A graph is **symmetric with respect to the y-axis** if, whenever (x, y) is on the graph, $(-x, y)$ is also on the graph.

3. A graph is **symmetric with respect to the origin** if, whenever (x, y) is on the graph, $(-x, -y)$ is also on the graph.

For example, the graph of $y = x^2 - 2$ is symmetric with respect to the y-axis because (x, y) and $(-x, y)$ are on the graph of $y = x^2 - 2$. (See the table below and Figure 1.3.)

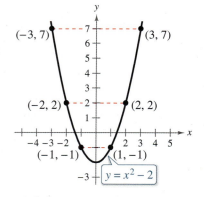

y-Axis symmetry
Figure 1.3

x	-3	-2	-1	1	2	3
y	7	2	-1	-1	2	7
(x, y)	$(-3, 7)$	$(-2, 2)$	$(-1, -1)$	$(1, -1)$	$(2, 2)$	$(3, 7)$

Algebraic Tests for Symmetry

1. The graph of an equation is symmetric with respect to the x-axis when replacing y with $-y$ yields an equivalent equation.

2. The graph of an equation is symmetric with respect to the y-axis when replacing x with $-x$ yields an equivalent equation.

3. The graph of an equation is symmetric with respect to the origin when replacing x with $-x$ and y with $-y$ yields an equivalent equation.

EXAMPLE 5 **Testing for Symmetry**

Test $y = 2x^3$ for symmetry with respect to both axes and the origin.

Solution

x-Axis: $y = 2x^3$ Write original equation.

$-y = 2x^3$ Replace y with $-y$. Result is *not* an equivalent equation.

y-Axis: $y = 2x^3$ Write original equation.

$y = 2(-x)^3$ Replace x with $-x$.

$y = -2x^3$ Simplify. Result is *not* an equivalent equation.

Origin: $y = 2x^3$ Write original equation.

$-y = 2(-x)^3$ Replace y with $-y$ and x with $-x$.

$-y = -2x^3$ Simplify.

$y = 2x^3$ Simplify. Result is an equivalent equation.

Of the three tests for symmetry, the test for origin symmetry is the only one satisfied. So, the graph of $y = 2x^3$ is symmetric with respect to the origin (see Figure 1.4).

✓ *Checkpoint* 🔊))) *Audio-video solution in English & Spanish at LarsonPrecalculus.com*

Test $y^2 = 6 - x$ for symmetry with respect to both axes and the origin.

EXAMPLE 6 **Using Symmetry as a Sketching Aid**

Use symmetry to sketch the graph of $x - y^2 = 1$.

Solution   Of the three tests for symmetry, the test for x-axis symmetry is the only one satisfied, because $x - (-y)^2 = 1$ is equivalent to $x - y^2 = 1$. So, the graph is symmetric with respect to the x-axis. Find solution points above (or below) the x-axis and then use symmetry to obtain the graph, as shown in Figure 1.5.

✓ *Checkpoint* 🔊))) *Audio-video solution in English & Spanish at LarsonPrecalculus.com*

Use symmetry to sketch the graph of $y = x^2 - 4$.

EXAMPLE 7 **Sketching the Graph of an Equation**

Sketch the graph of $y = |x - 1|$.

Solution   This equation fails all three tests for symmetry, so its graph is not symmetric with respect to either axis or to the origin. The absolute value bars tell you that y is always nonnegative. Construct a table of values. Then plot and connect the points, as shown in Figure 1.6. Notice from the table that $x = 0$ when $y = 1$. So, the y-intercept is $(0, 1)$. Similarly, $y = 0$ when $x = 1$. So, the x-intercept is $(1, 0)$.

x	-2	-1	0	1	2	3	4
$y = |x - 1|$	3	2	1	0	1	2	3
(x, y)	$(-2, 3)$	$(-1, 2)$	$(0, 1)$	$(1, 0)$	$(2, 1)$	$(3, 2)$	$(4, 3)$

✓ *Checkpoint* 🔊))) *Audio-video solution in English & Spanish at LarsonPrecalculus.com*

Sketch the graph of $y = |x - 2|$.

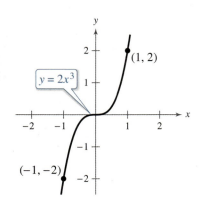

Figure 1.4

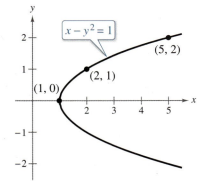

Figure 1.5

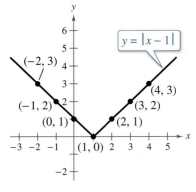

Figure 1.6

Circles

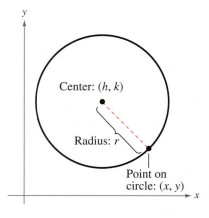

A **circle** is a set of points (x, y) in a plane that are the same distance r from a point called the center, (h, k), as shown at the right. By the Distance Formula,

$$\sqrt{(x - h)^2 + (y - k)^2} = r.$$

By squaring each side of this equation, you obtain the **standard form of the equation of a circle.** For example, for a circle with its center at $(h, k) = (1, 3)$ and radius $r = 4$,

$$\sqrt{(x - 1)^2 + (y - 3)^2} = 4 \qquad \text{Substitute for } h, k, \text{ and } r.$$
$$(x - 1)^2 + (y - 3)^2 = 16. \qquad \text{Square each side.}$$

Standard Form of the Equation of a Circle

A point (x, y) lies on the circle of **radius** r and **center** (h, k) if and only if

$$(x - h)^2 + (y - k)^2 = r^2.$$

From this result, the standard form of the equation of a circle with radius r *and center at the origin,* $(h, k) = (0, 0)$, is

$$x^2 + y^2 = r^2. \qquad \text{Circle with radius } r \text{ and center at origin}$$

EXAMPLE 8 **Writing the Equation of a Circle**

The point $(3, 4)$ lies on a circle whose center is at $(-1, 2)$, as shown in Figure 1.7. Write the standard form of the equation of this circle.

Solution

The radius of the circle is the distance between $(-1, 2)$ and $(3, 4)$.

$$r = \sqrt{(x - h)^2 + (y - k)^2} \qquad \text{Distance Formula}$$
$$= \sqrt{[3 - (-1)]^2 + (4 - 2)^2} \qquad \text{Substitute for } x, y, h, \text{ and } k.$$
$$= \sqrt{4^2 + 2^2} \qquad \text{Simplify.}$$
$$= \sqrt{16 + 4} \qquad \text{Simplify.}$$
$$= \sqrt{20} \qquad \text{Radius}$$

Using $(h, k) = (-1, 2)$ and $r = \sqrt{20}$, the equation of the circle is

$$(x - h)^2 + (y - k)^2 = r^2 \qquad \text{Equation of circle}$$
$$[x - (-1)]^2 + (y - 2)^2 = \left(\sqrt{20}\right)^2 \qquad \text{Substitute for } h, k, \text{ and } r.$$
$$(x + 1)^2 + (y - 2)^2 = 20. \qquad \text{Standard form}$$

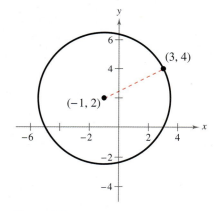

Figure 1.7

✓ *Checkpoint*))) *Audio-video solution in English & Spanish at LarsonPrecalculus.com*

The point $(1, -2)$ lies on a circle whose center is at $(-3, -5)$. Write the standard form of the equation of this circle. ◼

To find h and k from the standard form of the equation of a circle, you may want to rewrite one or both of the quantities in parentheses. For example, $x + 1 = x - (-1)$.

Application

In this course, you will learn that there are many ways to approach a problem. Example 9 illustrates three common approaches.

A numerical approach: Construct and use a table.

A graphical approach: Draw and use a graph.

An algebraic approach: Use the rules of algebra.

EXAMPLE 9   Maximum Weight

The maximum weight y (in pounds) for a man in the United States Marine Corps can be approximated by the mathematical model

$$y = 0.040x^2 - 0.11x + 3.9, \quad 58 \le x \le 80$$

where x is the man's height (in inches).   *(Source: U.S. Department of Defense)*

a. Construct a table of values that shows the maximum weights for men with heights of 62, 64, 66, 68, 70, 72, 74, and 76 inches.

b. Use the table of values to sketch a graph of the model. Then use the graph to estimate *graphically* the maximum weight for a man whose height is 71 inches.

c. Use the model to confirm *algebraically* the estimate you found in part (b).

Solution

a. Use a calculator to construct a table, as shown at the left.

b. Use the table of values to sketch the graph of the equation, as shown in Figure 1.8. From the graph, you can estimate that a height of 71 inches corresponds to a weight of about 198 pounds.

c. To confirm algebraically the estimate you found in part (b), substitute 71 for x in the model.

$$y = 0.040(71)^2 - 0.11(71) + 3.9$$
$$\approx 197.7$$

So, the graphical estimate of 198 pounds is fairly good.

✓ **Checkpoint** 🔊)) *Audio-video solution in English & Spanish at LarsonPrecalculus.com*

Use Figure 1.8 to estimate *graphically* the maximum weight for a man whose height is 75 inches. Then confirm the estimate *algebraically.*

REMARK You should develop the habit of using at least two approaches to solve every problem. This helps build your intuition and helps you check that your answers are reasonable.

DATA	Height, x	Weight, y
	62	150.8
	64	160.7
	66	170.9
	68	181.4
	70	192.2
	72	203.3
	74	214.8
	76	226.6

Spreadsheet at LarsonPrecalculus.com

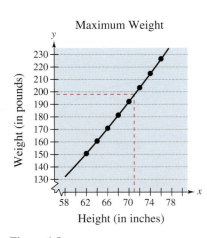

Figure 1.8

Summarize   (Section 1.1)

1. Explain how to sketch the graph of an equation *(page 70)*. For examples of sketching graphs of equations, see Examples 2 and 3.

2. Explain how to identify the x- and y-intercepts of a graph *(page 73)*. For an example of identifying x- and y-intercepts, see Example 4.

3. Explain how to use symmetry to graph an equation *(page 74)*. For an example of using symmetry to graph an equation, see Example 6.

4. State the standard form of the equation of a circle *(page 76)*. For an example of writing the standard form of the equation of a circle, see Example 8.

5. Describe an example of how to use the graph of an equation to solve a real-life problem *(page 77, Example 9)*.

1.1 Exercises

See **CalcChat.com** for tutorial help and worked-out solutions to odd-numbered exercises.

Vocabulary: Fill in the blanks.

1. An ordered pair (a, b) is a _____ of an equation in x and y when the substitutions $x = a$ and $y = b$ result in a true statement.

2. The set of all solution points of an equation is the _____ of the equation.

3. The points at which a graph intersects or touches an axis are the _____ of the graph.

4. A graph is symmetric with respect to the _____ if, whenever (x, y) is on the graph, $(-x, y)$ is also on the graph.

5. The equation $(x - h)^2 + (y - k)^2 = r^2$ is the standard form of the equation of a _____ with center _____ and radius _____.

6. When you construct and use a table to solve a problem, you are using a _____ approach.

Skills and Applications

 Determining Solution Points In Exercises 7–14, determine whether each point lies on the graph of the equation.

Equation	Points
7. $y = \sqrt{x + 4}$	(a) $(0, 2)$ (b) $(5, 3)$
8. $y = \sqrt{5 - x}$	(a) $(1, 2)$ (b) $(5, 0)$
9. $y = x^2 - 3x + 2$	(a) $(2, 0)$ (b) $(-2, 8)$
10. $y = 3 - 2x^2$	(a) $(-1, 1)$ (b) $(-2, 11)$
11. $y = 4 - |x - 2|$	(a) $(1, 5)$ (b) $(6, 0)$
12. $y = |x - 1| + 2$	(a) $(2, 3)$ (b) $(-1, 0)$
13. $x^2 + y^2 = 20$	(a) $(3, -2)$ (b) $(-4, 2)$
14. $2x^2 + 5y^2 = 8$	(a) $(6, 0)$ (b) $(0, 4)$

 Sketching the Graph of an Equation In Exercises 15–18, complete the table. Use the resulting solution points to sketch the graph of the equation.

15. $y = -2x + 5$

x	-1	0	1	2	$\frac{5}{2}$
y					
(x, y)					

16. $y + 1 = \frac{3}{4}x$

x	-2	0	1	$\frac{4}{3}$	2
y					
(x, y)					

17. $y + 3x = x^2$

x	-1	0	1	2	3
y					
(x, y)					

18. $y = 5 - x^2$

x	-2	-1	0	1	2
y					
(x, y)					

 Identifying x- and y-Intercepts In Exercises 19–24, identify the x- and y-intercepts of the graph.

19. $y = (x - 3)^2$

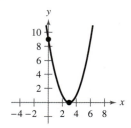

20. $y = 16 - 4x^2$

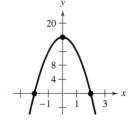

21. $y = |x + 2|$

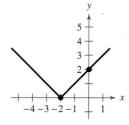

22. $y^2 = 4 - x$

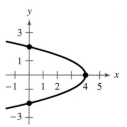

23. $y = 2 - 2x^3$

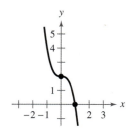

24. $y = x^3 - 4x$

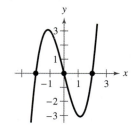

The symbol ▦ and a red exercise number indicates that a video solution can be seen at *CalcView.com*.

Testing for Symmetry In Exercises 25–32, use the algebraic tests to check for symmetry with respect to both axes and the origin.

25. $x^2 - y = 0$ **26.** $x - y^2 = 0$

27. $y = x^3$ **28.** $y = x^4 - x^2 + 3$

29. $y = \dfrac{x}{x^2 + 1}$ **30.** $y = \dfrac{1}{x^2 + 1}$

31. $xy^2 + 10 = 0$ **32.** $xy = 4$

 Using Symmetry as a Sketching Aid In Exercises 33–36, assume that the graph has the given type of symmetry. Complete the graph of the equation. To print an enlarged copy of the graph, go to *MathGraphs.com*.

33.
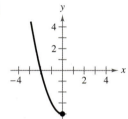
y-Axis symmetry

34.
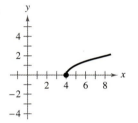
x-Axis symmetry

35.
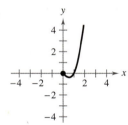
Origin symmetry

36.
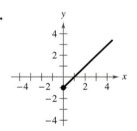
y-Axis symmetry

 Sketching the Graph of an Equation In Exercises 37–48, test for symmetry and graph the equation. Then identify any intercepts.

37. $y = -3x + 1$ **38.** $y = 2x - 3$

39. $y = x^2 - 2x$ **40.** $y = -x^2 - 2x$

41. $y = x^3 + 3$ **42.** $y = x^3 - 1$

43. $y = \sqrt{x - 3}$ **44.** $y = \sqrt{1 - x}$

45. $y = |x - 6|$ **46.** $y = 1 - |x|$

47. $x = y^2 - 1$ **48.** $x = y^2 - 5$

 Using Technology In Exercises 49–58, use a graphing utility to graph the equation. Use a standard setting. Approximate any intercepts.

49. $y = 5 - \frac{1}{2}x$ **50.** $y = \frac{2}{3}x - 1$

51. $y = x^2 - 4x + 3$

52. $y = x^2 + x - 2$

The symbol indicates an exercise or a part of an exercise in which you are instructed to use a graphing utility.

53. $y = \dfrac{2x}{x - 1}$ **54.** $y = \dfrac{4}{x^2 + 1}$

55. $y = \sqrt[3]{x + 1}$ **56.** $y = x\sqrt{x + 6}$

57. $y = |x + 3|$ **58.** $y = 2 - |x|$

 Writing the Equation of a Circle In Exercises 59–66, write the standard form of the equation of the circle with the given characteristics.

59. Center: $(0, 0)$; Radius: 3

60. Center: $(0, 0)$; Radius: 7

61. Center: $(-4, 5)$; Radius: 2

62. Center: $(1, -3)$; Radius: $\sqrt{11}$

63. Center: $(3, 8)$; Solution point: $(-9, 13)$

64. Center: $(-2, -6)$; Solution point: $(1, -10)$

65. Endpoints of a diameter: $(3, 2), (-9, -8)$

66. Endpoints of a diameter: $(11, -5), (3, 15)$

Sketching a Circle In Exercises 67–72, find the center and radius of the circle with the given equation. Then sketch the circle.

67. $x^2 + y^2 = 25$ **68.** $x^2 + y^2 = 36$

69. $(x - 1)^2 + (y + 3)^2 = 9$

70. $x^2 + (y - 1)^2 = 1$

71. $\left(x - \frac{1}{2}\right)^2 + \left(y - \frac{1}{2}\right)^2 = \frac{9}{4}$

72. $(x - 2)^2 + (y + 3)^2 = \frac{16}{9}$

73. Depreciation A hospital purchases a new magnetic resonance imaging (MRI) machine for $1.2 million. The depreciated value y (reduced value) after t years is given by $y = 1,200,000 - 80,000t$, $0 \le t \le 10$. Sketch the graph of the equation.

74. Depreciation You purchase an all-terrain vehicle (ATV) for $9500. The depreciated value y (reduced value) after t years is given by $y = 9500 - 1000t$, $0 \le t \le 6$. Sketch the graph of the equation.

75. Geometry A regulation NFL playing field of length x and width y has a perimeter of $346\frac{2}{3}$ or $\frac{1040}{3}$ yards.

(a) Draw a rectangle that gives a visual representation of the problem. Use the specified variables to label the sides of the rectangle.

(b) Show that the width of the rectangle is $y = \frac{520}{3} - x$ and its area is $A = x\left(\frac{520}{3} - x\right)$.

(c) Use a graphing utility to graph the area equation. Be sure to adjust your window settings.

(d) From the graph in part (c), estimate the dimensions of the rectangle that yield a maximum area.

(e) Use your school's library, the Internet, or some other reference source to find the actual dimensions and area of a regulation NFL playing field and compare your findings with the results of part (d).

76. Architecture   The arch support of a bridge is modeled by $y = -0.0012x^2 + 300$, where x and y are measured in feet and the x-axis represents the ground.

(a) Use a graphing utility to graph the equation.

(b) Identify one x-intercept of the graph. Explain how to use the intercept and the symmetry of the graph to find the width of the arch support.

77. Population Statistics

The table shows the life expectancies of a child (at birth) in the United States for selected years from 1940 through 2010.   *(Source: U.S. National Center for Health Statistics)*

DATA	Year	Life Expectancy, y
	1940	62.9
	1950	68.2
	1960	69.7
	1970	70.8
	1980	73.7
	1990	75.4
	2000	76.8
	2010	78.7

A model for the life expectancy during this period is

$$y = \frac{63.6 + 0.97t}{1 + 0.01t}, \quad 0 \le t \le 70$$

where y represents the life expectancy and t is the time in years, with $t = 0$ corresponding to 1940.

(a) Use a graphing utility to graph the data from the table and the model in the same viewing window. How well does the model fit the data? Explain.

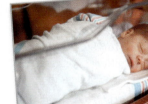

(b) Determine the life expectancy in 1990 both graphically and algebraically.

(c) Use the graph to determine the year when life expectancy was approximately 70.1. Verify your answer algebraically.

(d) Identify the y-intercept of the graph of the model. What does it represent in the context of the problem?

(e) Do you think this model can be used to predict the life expectancy of a child 50 years from now? Explain.

78. Electronics   The resistance y (in ohms) of 1000 feet of solid copper wire at 68 degrees Fahrenheit is

$$y = \frac{10,370}{x^2}$$

where x is the diameter of the wire in mils (0.001 inch).

(a) Complete the table.

x	5	10	20	30	40	50
y						

x	60	70	80	90	100
y					

(b) Use the table of values in part (a) to sketch a graph of the model. Then use your graph to estimate the resistance when $x = 85.5$.

(c) Use the model to confirm algebraically the estimate you found in part (b).

(d) What can you conclude about the relationship between the diameter of the copper wire and the resistance?

Exploration

True or False?   In Exercises 79–81, determine whether the statement is true or false. Justify your answer.

79. The graph of a linear equation cannot be symmetric with respect to the origin.

80. The graph of a linear equation can have either no x-intercepts or only one x-intercept.

81. A circle can have a total of zero, one, two, three, or four x- and y-intercepts.

82.   HOW DO YOU SEE IT?   The graph shows the circle with the equation $x^2 + y^2 = 1$. Describe the types of symmetry that you observe.

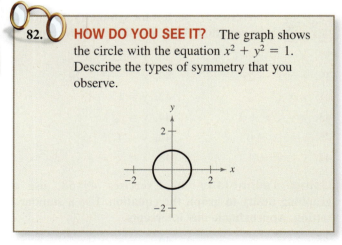

83. Think About It   Find a and b when the graph of $y = ax^2 + bx^3$ is symmetric with respect to (a) the y-axis and (b) the origin. (There are many correct answers.)

1.2 Linear Equations in One Variable

- Identify different types of equations.
- Solve linear equations in one variable.
- Solve rational equations that lead to linear equations.
- Find x- and y-intercepts of graphs of equations algebraically.
- Use linear equations to model and solve real-life problems.

Linear equations have many real-life applications, such as in forensics. For example, in Exercises 67 and 68 on page 88, you will use linear equations to determine height from femur length.

Equations and Solutions of Equations

An **equation** in x is a statement that two algebraic expressions are equal. For example,

$$3x - 5 = 7, \quad x^2 - x - 6 = 0, \quad \text{and} \quad \sqrt{2x} = 4$$

are equations. To **solve** an equation in x means to find all values of x for which the equation is true. Such values are **solutions.** For example, $x = 4$ is a solution of the equation

$$3x - 5 = 7$$

because

$$3(4) - 5 = 7$$

is a true statement.

The solutions of an equation depend on the kinds of numbers being considered. For example, in the set of rational numbers, $x^2 = 10$ has no solution because there is no rational number whose square is 10. However, in the set of real numbers, the equation has the two solutions

$$x = \sqrt{10} \quad \text{and} \quad x = -\sqrt{10}.$$

An equation that is true for *every* real number in the domain of the variable is an **identity.** For example,

$$x^2 - 9 = (x + 3)(x - 3) \qquad \text{Identity}$$

is an identity because it is a true statement for any real value of x. The equation

$$\frac{x}{3x^2} = \frac{1}{3x} \qquad \text{Identity}$$

is an identity because it is true for any nonzero real value of x.

An equation that is true for just *some* (but not all) of the real numbers in the domain of the variable is a **conditional equation.** For example, the equation

$$x^2 - 9 = 0 \qquad \text{Conditional equation}$$

is conditional because $x = 3$ and $x = -3$ are the only values in the domain that satisfy the equation. The equation

$$2x + 4 = 6$$

is conditional because $x = 1$ is the only value in the domain that satisfies the equation.

A **contradiction** is an equation that is false for *every* real number in the domain of the variable. For example, the equation

$$2x - 4 = 2x + 1 \qquad \text{Contradiction}$$

is a contradiction because there are no real values of x for which the equation is true.

Linear Equations in One Variable

> ### Definition of a Linear Equation in One Variable
> A **linear equation in one variable** x is an equation that can be written in the standard form
>
> $$ax + b = 0$$
>
> where a and b are real numbers with $a \neq 0$.

A linear equation in one variable has exactly one solution. To see this, consider the steps below. (Remember that $a \neq 0$.)

$ax + b = 0$	Write original equation.
$ax = -b$	Subtract b from each side.
$x = -\dfrac{b}{a}$	Divide each side by a.

So, the equation

$$ax + b = 0$$

has exactly one solution,

$$x = -\frac{b}{a}.$$

The above suggests that to solve a conditional equation in x, you isolate x on one side of the equation using a sequence of **equivalent equations,** each having the same solution as the original equation. The operations that yield equivalent equations come from the properties of equality reviewed in Section P.1.

HISTORICAL NOTE

This ancient Egyptian papyrus, discovered in 1858, contains one of the earliest examples of mathematical writing in existence. The papyrus itself dates back to around 1650 B.C., but it is actually a copy of writings from two centuries earlier. The algebraic equations on the papyrus were written in words. Diophantus, a Greek who lived around A.D. 250, is often called the Father of Algebra. He was the first to use abbreviated word forms in equations.

> ### Generating Equivalent Equations
> An equation can be transformed into an *equivalent equation* by one or more of the steps listed below.
>
> | | Given Equation | Equivalent Equation |
> |---|---|---|
> | **1.** Remove symbols of grouping, combine like terms, or simplify fractions on one or both sides of the equation. | $2x - x = 4$ | $x = 4$ |
> | **2.** Add (or subtract) the same quantity to (from) *each* side of the equation. | $x + 1 = 6$ | $x = 5$ |
> | **3.** Multiply (or divide) *each* side of the equation by the same *nonzero* quantity. | $2x = 6$ | $x = 3$ |
> | **4.** Interchange the two sides of the equation. | $2 = x$ | $x = 2$ |

In Examples 1 and 2, you will use these steps to solve linear equations in one variable x.

EXAMPLE 1 **Solving Linear Equations**

a. $3x - 6 = 0$ Original equation

      $3x = 6$ Add 6 to each side.

      $x = 2$ Divide each side by 3.

b. $5x + 4 = 3x - 8$ Original equation

    $2x + 4 = -8$ Subtract $3x$ from each side.

      $2x = -12$ Subtract 4 from each side.

       $x = -6$ Divide each side by 2.

✓ *Checkpoint* *Audio-video solution in English & Spanish at LarsonPrecalculus.com*

Solve each equation.

a. $7 - 2x = 15$

b. $7x - 9 = 5x + 7$

After solving an equation, you should check each solution in the original equation. For instance, here is a check of the solution in Example 1(a).

    $3x - 6 = 0$ Write original equation.

  $3(2) - 6 \stackrel{?}{=} 0$ Substitute 2 for x.

        $0 = 0$ Solution checks. ✓

Check the solution in Example 1(b) on your own.

▷ **TECHNOLOGY** You can use a graphing utility to check that a solution is reasonable. One way is to graph the left side of the equation, then graph the right side of the equation, and determine the point of intersection. For instance, in Example 2, if you graph the equations

$y_1 = 6(x - 1) + 4$ The left side

and

$y_2 = 3(7x + 1)$ The right side

in the same viewing window, they intersect at $x = -\frac{1}{3}$, as shown in the graph below.

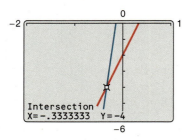

EXAMPLE 2 **Solving a Linear Equation**

Solve $6(x - 1) + 4 = 3(7x + 1)$ and check your solution.

Solution

   $6(x - 1) + 4 = 3(7x + 1)$ Write original equation.

    $6x - 6 + 4 = 21x + 3$ Distributive Property

       $6x - 2 = 21x + 3$ Simplify.

            $-5 = 15x$ Simplify.

              $x = -\frac{1}{3}$ Interchange sides and divide each side by 15.

Check

     $6(x - 1) + 4 = 3(7x + 1)$ Write original equation.

  $6\left(-\frac{1}{3} - 1\right) + 4 \stackrel{?}{=} 3\left[7\left(-\frac{1}{3}\right) + 1\right]$ Substitute $-\frac{1}{3}$ for x.

     $6\left(-\frac{4}{3}\right) + 4 \stackrel{?}{=} 3\left[-\frac{7}{3} + 1\right]$ Simplify.

        $-8 + 4 \stackrel{?}{=} 3\left(-\frac{4}{3}\right)$ Simplify.

            $-4 = -4$ Solution checks. ✓

✓ *Checkpoint* *Audio-video solution in English & Spanish at LarsonPrecalculus.com*

Solve $4(x + 2) - 12 = 5(x - 6)$ and check your solution.

Rational Equations That Lead to Linear Equations

A **rational equation** involves one or more rational expressions. To solve a rational equation, multiply every term by the least common denominator (LCD) of all the terms. This clears the original equation of fractions and produces a simpler equation.

> **REMARK** An equation with a single fraction on each side can be cleared of denominators by *cross multiplying*. To do this, multiply the left numerator by the right denominator and the right numerator by the left denominator.
>
> $$\frac{a}{b} = \frac{c}{d} \qquad \text{Original equation}$$
>
> $$ad = cb \qquad \text{Cross multiply.}$$

EXAMPLE 3   Solving a Rational Equation

Solve $\dfrac{x}{3} + \dfrac{3x}{4} = 2$.

Solution

$$\frac{x}{3} + \frac{3x}{4} = 2 \qquad \text{Write original equation.}$$

$$(12)\frac{x}{3} + (12)\frac{3x}{4} = (12)2 \qquad \text{Multiply each term by the LCD.}$$

$$4x + 9x = 24 \qquad \text{Simplify.}$$

$$13x = 24 \qquad \text{Combine like terms.}$$

$$x = \frac{24}{13} \qquad \text{Divide each side by 13.}$$

The solution is $x = \frac{24}{13}$. Check this in the original equation.

✓ **Checkpoint**)))   *Audio-video solution in English & Spanish at LarsonPrecalculus.com*

Solve $\dfrac{4x}{9} - \dfrac{1}{3} = x + \dfrac{5}{3}$.   ■

When multiplying or dividing an equation by a *variable expression,* it is possible to introduce an **extraneous solution,** which is a solution that does not satisfy the original equation.

EXAMPLE 4   An Equation with an Extraneous Solution

See LarsonPrecalculus.com for an interactive version of this type of example.

Solve $\dfrac{1}{x-2} = \dfrac{3}{x+2} - \dfrac{6x}{x^2-4}$.

> ▷ **ALGEBRA HELP** To review how to find the LCD (least common denominator) of several rational expressions, see Section P.5.

Solution   The LCD is $x^2 - 4 = (x+2)(x-2)$. Multiply each term by the LCD.

$$\frac{1}{x-2}(x+2)(x-2) = \frac{3}{x+2}(x+2)(x-2) - \frac{6x}{x^2-4}(x+2)(x-2)$$

$$x + 2 = 3(x-2) - 6x, \quad x \neq \pm 2$$

$$x + 2 = 3x - 6 - 6x$$

$$x + 2 = -3x - 6$$

$$4x = -8 \implies x = -2 \qquad \text{Extraneous solution}$$

In the original equation, $x = -2$ yields a denominator of zero. So, $x = -2$ is an extraneous solution, and the original equation has *no solution.*

✓ **Checkpoint** )))   *Audio-video solution in English & Spanish at LarsonPrecalculus.com*

Solve $\dfrac{3x}{x-4} = 5 + \dfrac{12}{x-4}$.   ■

Finding Intercepts Algebraically

In Section 1.1, you learned to find x- and y-intercepts using a graphical approach. All points on the x-axis have a y-coordinate equal to zero, and all points on the y-axis have an x-coordinate equal to zero. This suggests an algebraic approach to finding x- and y-intercepts.

Finding Intercepts Algebraically

1. To find x-intercepts, set y equal to zero and solve the equation for x.

2. To find y-intercepts, set x equal to zero and solve the equation for y.

EXAMPLE 5   **Finding Intercepts Algebraically**

Find the x- and y-intercepts of the graph of each equation algebraically.

a. $y = 4x + 1$ **b.** $3x + 2y = 6$

Solution

a. To find the x-intercept, set y equal to zero and solve for x.

$y = 4x + 1$	Write original equation.
$0 = 4x + 1$	Substitute 0 for y.
$-1 = 4x$	Subtract 1 from each side.
$-\frac{1}{4} = x$	Divide each side by 4.

So, the x-intercept is $\left(-\frac{1}{4}, 0\right)$.

To find the y-intercept, set x equal to zero and solve for y.

$y = 4x + 1$	Write original equation.
$y = 4(0) + 1$	Substitute 0 for x.
$y = 1$	Simplify.

So, the y-intercept is $(0, 1)$.

b. To find the x-intercept, set y equal to zero and solve for x.

$3x + 2y = 6$	Write original equation.
$3x + 2(0) = 6$	Substitute 0 for y.
$3x = 6$	Simplify.
$x = 2$	Divide each side by 3.

So, the x-intercept is $(2, 0)$.

To find the y-intercept, set x equal to zero and solve for y.

$3x + 2y = 6$	Write original equation.
$3(0) + 2y = 6$	Substitute 0 for x.
$2y = 6$	Simplify.
$y = 3$	Divide each side by 2.

So, the y-intercept is $(0, 3)$.

✓ *Checkpoint*  *Audio-video solution in English & Spanish at LarsonPrecalculus.com*

Find the x- and y-intercepts of the graph of each equation algebraically.

a. $y = -3x - 2$ **b.** $5x + 3y = 15$

Application

EXAMPLE 6 **Female Participants in Cross-Country Running**

The number y (in thousands) of female participants in high school cross-country running in the United States from 2007 through 2014 can be approximated by the linear model

$$y = 4.33t + 205.5, \quad -3 \le t \le 4$$

where t represents the year, with $t = 0$ corresponding to 2010. (a) Find algebraically and interpret the y-intercept of the graph of the linear model shown in Figure 1.9. (b) Use the linear model to predict the year in which there will be 250,000 female participants. *(Source: National Federation of State High School Associations)*

Solution

a. To find the y-intercept, let $t = 0$ and solve for y.

$y = 4.33t + 205.5$	Write original equation.
$= 4.33(0) + 205.5$	Substitute 0 for t.
$= 205.5$	Simplify.

So, the y-intercept is $(0, 205.5)$. This means that, according to the model, there were about 205,500 female participants in 2010.

b. Let $y = 250$ and solve for t.

$y = 4.33t + 205.5$	Write original equation.
$250 = 4.33t + 205.5$	Substitute 250 for y.
$44.5 = 4.33t$	Subtract 205.5 from each side.
$10 \approx t$	Divide each side by 4.33.

Because $t = 0$ represents 2010, $t = 10$ must represent 2020. This means that, according to the model, there will be 250,000 female participants in 2020.

Female Participants in
Cross-Country Running

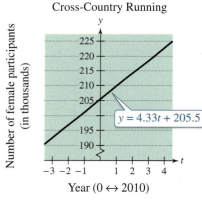

Year (0 ↔ 2010)

Figure 1.9

✓ **Checkpoint** 🔊))) *Audio-video solution in English & Spanish at LarsonPrecalculus.com*

The number y (in thousands) of male participants in high school lacrosse in the United States from 2008 through 2015 can be approximated by the linear model

$$y = 3.66t + 91.4, \quad -2 \le t \le 5$$

where t represents the year, with $t = 0$ corresponding to 2010. (a) Find algebraically and interpret the y-intercept of the graph of the linear model. (b) Use the linear model to predict the year in which there will be 128,000 male participants. *(Source: National Federation of State High School Associations)*

Summarize   (Section 1.2)

1. State the definitions of an identity, a conditional equation, and a contradiction *(page 81)*.

2. State the definition of a linear equation in one variable and list the four steps that can be used to form an equivalent equation *(page 82)*. For examples of solving linear equations, see Examples 1 and 2.

3. Explain how to solve a rational equation *(page 84)*. For examples of solving rational equations, see Examples 3 and 4.

4. Explain how to find intercepts algebraically *(page 85)*. For an example of finding intercepts algebraically, see Example 5.

5. Describe a real-life application involving a linear equation *(page 86, Example 6)*.

1.2 Exercises

See **CalcChat.com** for tutorial help and worked-out solutions to odd-numbered exercises.

Vocabulary: Fill in the blanks.

1. An _____ is a statement that equates two algebraic expressions.
2. There are three types of equations: _____, _____ equations, and _____.
3. A linear equation in one variable x is an equation that can be written in the standard form _____.
4. An _____ equation has the same solution(s) as the original equation.
5. A _____ equation is an equation that involves one or more rational expressions.
6. An _____ solution is a solution that does not satisfy the original equation.

Skills and Applications

Classifying an Equation In Exercises 7–14, determine whether the equation is an identity, a conditional equation, or a contradiction.

7. $3(x - 1) = 3x - 3$ 8. $2(x + 1) = 2x - 1$
9. $2(x - 1) = 3x + 1$ 10. $4(x + 2) = 2x + 2$
11. $3(x + 2) = 3x + 2$ 12. $5(x + 2) = 5x + 10$
13. $2(x + 3) - 5 = 2x + 1$
14. $3(x - 1) + 2 = 4x - 2$

 Solving a Linear Equation In Exercises 15–26, solve the equation and check your solution. (If not possible, explain why.)

15. $x + 11 = 15$ 16. $7 - x = 19$
17. $7 - 2x = 25$ 18. $7x + 2 = 23$
19. $3x - 5 = 2x + 7$ 20. $5x + 3 = 6 - 2x$
21. $4y + 2 - 5y = 7 - 6y$ 22. $5y + 1 = 8y - 5 + 6y$
23. $x - 3(2x + 3) = 8 - 5x$
24. $9x - 10 = 5x + 2(2x - 5)$
25. $0.25x + 0.75(10 - x) = 3$
26. $0.60x + 0.40(100 - x) = 50$

 Solving a Rational Equation In Exercises 27–44, solve the equation and check your solution. (If not possible, explain why.)

27. $\dfrac{3x}{8} - \dfrac{4x}{3} = 4$ 28. $\dfrac{2x}{5} + 5x = \dfrac{4}{3}$

29. $\dfrac{5x}{4} + \dfrac{1}{2} = x - \dfrac{1}{2}$ 30. $\dfrac{x}{5} - \dfrac{x}{2} = 3 + \dfrac{3x}{10}$

31. $\dfrac{5x - 4}{5x + 4} = \dfrac{2}{3}$ 32. $\dfrac{10x + 3}{5x + 6} = \dfrac{1}{2}$

33. $10 - \dfrac{13}{x} = 4 + \dfrac{5}{x}$ 34. $\dfrac{15}{x} - 4 = \dfrac{6}{x} + 3$

35. $3 = 2 + \dfrac{2}{z + 2}$ 36. $\dfrac{1}{x} + \dfrac{2}{x - 5} = 0$

37. $\dfrac{x}{x + 4} + \dfrac{4}{x + 4} + 2 = 0$

38. $\dfrac{7}{2x + 1} - \dfrac{8x}{2x - 1} = -4$

39. $\dfrac{2}{(x - 4)(x - 2)} = \dfrac{1}{x - 4} + \dfrac{2}{x - 2}$

40. $\dfrac{12}{(x - 1)(x + 3)} = \dfrac{3}{x - 1} + \dfrac{2}{x + 3}$

41. $\dfrac{1}{x - 3} + \dfrac{1}{x + 3} = \dfrac{10}{x^2 - 9}$

42. $\dfrac{1}{x - 2} + \dfrac{3}{x + 3} = \dfrac{4}{x^2 + x - 6}$

43. $\dfrac{3}{x^2 - 3x} + \dfrac{4}{x} = \dfrac{1}{x - 3}$

44. $\dfrac{6}{x} - \dfrac{2}{x + 3} = \dfrac{3(x + 5)}{x^2 + 3x}$

 Finding Intercepts Algebraically In Exercises 45–54, find the x- and y-intercepts of the graph of the equation algebraically.

45. $y = 12 - 5x$ 46. $y = 16 - 3x$
47. $y = -3(2x + 1)$ 48. $y = 5 - (6 - x)$
49. $2x + 3y = 10$ 50. $4x - 5y = 12$
51. $4y - 0.75x + 1.2 = 0$ 52. $3y + 2.5x - 3.4 = 0$
53. $\dfrac{2x}{5} + 8 - 3y = 0$ 54. $\dfrac{8x}{3} + 50 - 2y = 0$

Using Technology In Exercises 55–60, use a graphing utility to graph the equation and approximate any x-intercepts. Then set $y = 0$ and solve the resulting equation. Compare the result with the graph's x-intercept.

55. $y = 2(x - 1) - 4$ 56. $y = \frac{4}{3}x + 2$
57. $y = 20 - (3x - 10)$ 58. $y = 10 + 2(x - 2)$
59. $y = -38 + 5(9 - x)$ 60. $y = 6x - 6\left(\frac{16}{11} + x\right)$

Solving an Equation   **In Exercises 61–64, solve the equation. (Round your solution to three decimal places.)**

61. $0.275x + 0.725(500 - x) = 300$

62. $2.763 - 4.5(2.1x - 5.1432) = 6.32x + 5$

63. $\dfrac{2}{7.398} - \dfrac{4.405}{x} = \dfrac{1}{x}$ **64.** $\dfrac{3}{6.350} - \dfrac{6}{x} = 18$

65. Geometry The surface area S of the circular cylinder shown in the figure is

$$S = 2\pi(25) + 2\pi(5h).$$

Find the height h of the cylinder when the surface area is 471 square feet. Use 3.14 for π.

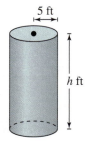
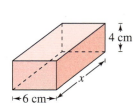

Figure for 65   Figure for 66

66. Geometry The surface area S of the rectangular solid shown in the figure is $S = 2(24) + 2(4x) + 2(6x)$. Find the length x of the solid when the surface area is 248 square centimeters.

• • **Forensics** • • • • • • • • • • • • • • • • • •

In Exercises 67 and 68, use the following information. The relationship between the length of an adult's femur (thigh bone) and the height of the adult can be approximated by the linear equations

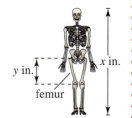

$y = 0.514x - 14.75$ Female

$y = 0.532x - 17.03$ Male

where y is the length of the femur in inches and x is the height of the adult in inches (see figure).

67. A crime scene investigator discovers a femur belonging to an adult human female. The bone is 18 inches long. Estimate the height of the female.

68. Officials search a forest for a missing man who is 6 feet 3 inches tall. They find an adult male femur that is 23 inches long. Is it possible that the femur belongs to the missing man?

69. Population The population y (in thousands) of Raleigh, North Carolina, from 2000 to 2014 can be approximated by the model $y = 11.09t + 293.4$, $0 \le t \le 14$, where t represents the year, with $t = 0$ corresponding to 2000 (see figure). *(Source: U.S. Census Bureau)*

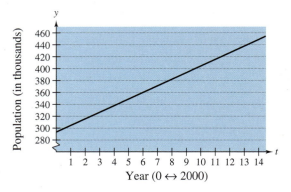

(a) Graphically estimate the y-intercept of the graph.

(b) Find algebraically and interpret the y-intercept of the graph.

(c) Use the model to predict the year in which the population will be 538,000. Does your answer seem reasonable? Explain.

70. Population The population y (in thousands) of Buffalo, New York, from 2000 to 2014 can be approximated by the model $y = -2.60t + 291.7$, $0 \le t \le 14$, where t represents the year, with $t = 0$ corresponding to 2000 (see figure). *(Source: U.S. Census Bureau)*

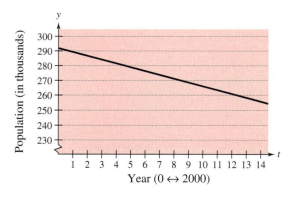

(a) Graphically estimate the y-intercept of the graph.

(b) Find algebraically and interpret the y-intercept of the graph.

(c) Use the model to predict the year in which the population will be 239,000. Does your answer seem reasonable? Explain.

71. Operating Cost A delivery company has a fleet of vans. The annual operating cost C (in dollars) per van is $C = 0.37m + 2600$, where m is the number of miles traveled by a van in a year. What number of miles yields an annual operating cost of $10,000?

72. Flood Control   A river is 8 feet above its flood stage. The water is receding at a rate of 3 inches per hour. Write a mathematical model that shows the number of feet above flood stage after t hours. Assuming the water continually recedes at this rate, when will the river be 1 foot above its flood stage?

Exploration

True or False?   In Exercises 73–78, determine whether the statement is true or false. Justify your answer.

73. The equation $x(3 - x) = 10$ is a linear equation.

74. The equation $2x + 3 = x$ is a linear equation.

75. The equation $2(x - 3) + 1 = 2x - 5$ has no solution.

76. The equation $2(x + 3) = 3x + 3$ has no solution.

77. The equation $3(x - 1) - 2 = 3x - 6$ is an identity.

78. The equation $2 - \dfrac{1}{x - 2} = \dfrac{3}{x - 2}$ has no solution because $x = 2$ is an extraneous solution.

79. Think About It

(a) Complete the table.

x	-1	0	1	2	3	4
$3.2x - 5.8$						

(b) Use the table in part (a) to determine the interval in which the solution of the equation $3.2x - 5.8 = 0$ is located. Explain.

(c) Complete the table.

x	1.5	1.6	1.7	1.8	1.9	2.0
$3.2x - 5.8$						

(d) Use the table in part (c) to determine the interval in which the solution of the equation $3.2x - 5.8 = 0$ is located. Explain how to use this procedure to approximate the solution to any desired degree of accuracy.

(e) Use the procedure in parts (a)–(d) to approximate the solution of the equation $0.3(x - 1.5) - 2 = 0$, accurate to two decimal places.

80. Think About It   Are $\dfrac{3x + 2}{5} = 7$ and $x + 9 = 20$ equivalent equations? Explain.

81. Graphical Reasoning

(a) Use a graphing utility to graph the equation $y = 3x - 6$.

(b) Use the result of part (a) to estimate the x-intercept.

(c) Explain how the x-intercept is related to the solution of $3x - 6 = 0$, as shown in Example 1(a).

82.   HOW DO YOU SEE IT?   Use the information below about a possible tax credit for a family consisting of two adults and two children (see figure).

Earned income:

E

Subsidy (a grant of money):

$S = 10,000 - \frac{1}{2}E, \quad 0 \le E \le 20,000$

Total income:

$T = E + S$

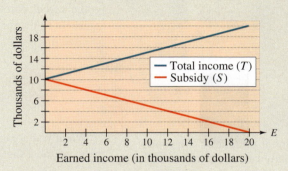

(a) Graphically estimate the intercepts of the red line. Then interpret the meaning of each intercept.

(b) Explain how to solve part (a) algebraically.

(c) Graphically estimate the earned income for which the total income is $14,000.

(d) Explain how to solve part (c) algebraically.

83. Finding Intercepts   Consider the linear equation

$$y = ax + b$$

where a and b are real numbers.

(a) What is the x-intercept of the graph of the equation when $a \ne 0$?

(b) What is the y-intercept of the graph of the equation?

(c) Use your results from parts (a) and (b) to find the x- and y-intercepts of the graph of $y = 5x + 10$.

84. Finding Intercepts   Consider the linear equation

$$ax + by = c$$

where a, b, and c are real numbers.

(a) What is the x-intercept of the graph of the equation when $a \ne 0$?

(b) What is the y-intercept of the graph of the equation when $b \ne 0$?

(c) Use your results from parts (a) and (b) to find the x- and y-intercepts of the graph of $2x + 7y = 11$.

1.3 Modeling with Linear Equations

Linear equations can model many real-life situations. For example, in Exercise 57 on page 99, you will use a linear model to determine how many gallons of gasoline to add to a gasoline-oil mixture to bring the mixture to the desired concentration for a chainsaw engine.

■ **Write and use mathematical models to solve real-life problems.**
■ **Solve mixture problems.**
■ **Use common formulas to solve real-life problems.**

Using Mathematical Models

In this section, you will use algebra to solve problems that occur in real-life situations. The process of translating phrases or sentences into algebraic expressions or equations is called **mathematical modeling.**

A good approach to mathematical modeling is to use two stages. Begin by using the verbal description of the problem to form a *verbal model*. Then, after assigning labels to the quantities in the verbal model, form a *mathematical model* or *algebraic equation*.

$$\boxed{\text{Verbal description}} \Rightarrow \boxed{\text{Verbal model}} \Rightarrow \boxed{\text{Algebraic equation}}$$

When you are constructing a verbal model, it is helpful to look for a *hidden equality*—a statement that two algebraic expressions are equal.

EXAMPLE 1   Using a Verbal Model

You accept a job with an annual income of $32,300. This includes your salary and a year-end bonus of $500. You are paid twice a month. What is your gross pay (pay before taxes) for each paycheck?

Solution   There are 12 months in a year and you are paid twice a month, so you receive 24 paychecks during the year.

Verbal model:	$\boxed{\text{Income for year}}$ = $\boxed{\text{24 paychecks}}$ · $\boxed{\text{Amount of each paycheck}}$ + $\boxed{\text{Bonus}}$

Labels:	Income for year = 32,300	(dollars)
	Amount of each paycheck = x	(dollars)
	Bonus = 500	(dollars)

Equation: $32{,}300 = 24x + 500$

The algebraic equation for this problem is a linear equation in one variable x, whose solution is shown below.

$32{,}300 = 24x + 500$	Write original equation.
$31{,}800 = 24x$	Subtract 500 from each side.
$1325 = x$	Divide each side by 24.

So, your gross pay for each paycheck is $1325.

✓ *Checkpoint*  *Audio-video solution in English & Spanish at LarsonPrecalculus.com*

You accept a job with an annual income of $58,400. This includes your salary and a $1200 year-end bonus. You are paid weekly. What is your salary per pay period? ■

A fundamental step in writing a mathematical model to represent a real-life problem is translating key words and phrases into algebraic expressions and equations. The list below gives several examples.

Translating Key Words and Phrases

Key Words and Phrases	Verbal Description	Algebraic Expression or Equation
Addition:		
Sum, plus, increased by, more than, total of	• The sum of 5 and x • Seven more than y	$5 + x$ or $x + 5$ $7 + y$ or $y + 7$
Subtraction:		
Difference, minus, less than, decreased by, subtracted from, reduced by	• The difference of 4 and b • Three less than z	$4 - b$ $z - 3$
Multiplication:		
Product, multiplied by, twice, times, percent of	• Two times x • Three percent of t	$2x$ $0.03t$
Division:		
Quotient, divided by, ratio, per	• The ratio of x to 8	$\dfrac{x}{8}$
Equality:		
Equals, equal to, is, are, was, will be, represents	• The sale price S is \$10 less than the list price L.	$S = L - 10$

EXAMPLE 2   **Finding a Percent Raise**

You accept a job that pays \$10 per hour. After a two-month probationary period, your hourly wage will increase to \$11 per hour. What percent raise will you receive after the two-month period?

Solution

Verbal model: Raise $=$ Percent $\cdot$ Old wage

Labels: Old wage $= 10$ (dollars per hour)
    Raise $= 11 - 10 = 1$ (dollars per hour)
    Percent $= r$ (in decimal form)

Equation: $1 = r \cdot 10$

  $\dfrac{1}{10} = r$ Divide each side by 10.

  $0.1 = r$ Rewrite the fraction as a decimal.

You will receive a raise of 0.1 or 10%.

✓ *Checkpoint*  *Audio-video solution in English & Spanish at LarsonPrecalculus.com*

You buy stock at \$15 per share. You sell the stock at \$18 per share. What is the percent increase in the stock's value?

EXAMPLE 3 **Finding a Percent of Annual Income**

Your family has an annual income of $57,000 and these monthly expenses: mortgage ($1100), car payment ($375), food ($900), utilities ($240), and credit cards ($220). What percent of your family's annual income does the total amount of the monthly expenses represent?

Solution The total amount of your family's monthly expenses is $2835. The total monthly expenses for 1 year are $34,020.

Verbal model:

| Monthly expenses | = | Percent | · | Income |

Labels: Income = 57,000 (dollars)
Monthly expenses = 34,020 (dollars)
Percent = r (in decimal form)

Equation: $34,020 = r \cdot 57,000$

$\dfrac{34,020}{57,000} = r$ Divide each side by 57,000.

$0.597 \approx r$ Use a calculator.

Your family's monthly expenses are approximately 0.597 or 59.7% of your family's annual income.

✓ **Checkpoint** ◄))) *Audio-video solution in English & Spanish at LarsonPrecalculus.com*

Your family has annual loan payments equal to 28% of its annual income. During the year, the loan payments total $17,920. What is your family's annual income?

EXAMPLE 4 **Finding the Dimensions of a Room**

A rectangular kitchen is twice as long as it is wide, and its perimeter is 84 feet. Find the dimensions of the kitchen.

Solution For this problem, it helps to draw a diagram, as shown in Figure 1.10.

Verbal model: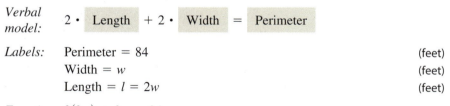

| 2 · | Length | + 2 · | Width | = | Perimeter |

Labels: Perimeter = 84 (feet)
Width = w (feet)
Length = $l = 2w$ (feet)

Equation: $2(2w) + 2w = 84$

$6w = 84$ Combine like terms.

$w = 14$ Divide each side by 6.

The length is twice the width, so

$l = 2w$
$= 2(14)$
$= 28.$

The dimensions of the kitchen are 14 feet by 28 feet.

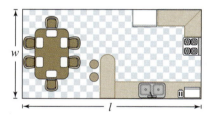

Figure 1.10

✓ **Checkpoint** ◄))) *Audio-video solution in English & Spanish at LarsonPrecalculus.com*

A rectangular family room is 3 times as long as it is wide, and its perimeter is 112 feet. Find the dimensions of the family room. ■

•• REMARK Writing units for each of the labels in a real-life problem helps you determine the unit(s) for the answer. This is called *unit analysis.* When the same unit of measure occurs in the numerator and denominator of an expression, divide out the unit. For example, unit analysis verifies that time in the formula below is in hours.

$$\text{Time} = \frac{\text{distance}}{\text{rate}}$$

$$= \frac{\text{miles}}{\dfrac{\text{miles}}{\text{hour}}}$$

$$= \text{miles} \cdot \frac{\text{hours}}{\text{miles}}$$

$$= \text{hours}$$

EXAMPLE 5   Estimating Travel Time

A plane flies nonstop from Portland, Oregon, to Atlanta, Georgia, a distance of about 2170 miles. After 3 hours in the air, the plane flies over Topeka, Kansas (a distance of about 1440 miles from Portland). Assuming the plane flies at a constant speed, how long does the entire trip take?

Solution

Verbal model:  | Distance | = | Rate | · | Time |

Labels:

Distance = 2170   (miles)

Time = t (hours)

Rate $= \dfrac{\text{distance to Topeka}}{\text{time to Topeka}} = \dfrac{1440}{3}$ (miles per hour)

Equation: $\quad 2170 = \dfrac{1440}{3}t$

$\quad\quad\quad\quad 2170 = 480t$ Simplify.

$\quad\quad\quad\quad \dfrac{2170}{480} = t$ Divide each side by 480.

$\quad\quad\quad\quad 4.52 \approx t$ Use a calculator.

The entire trip takes about 4.52 hours, or about 4 hours and 31 minutes.

✓ **Checkpoint** 🔊))) *Audio-video solution in English & Spanish at LarsonPrecalculus.com*

A boat travels at a constant speed to an island 14 miles away. It takes 0.5 hour to travel the first 5 miles. How long does the entire trip take?

EXAMPLE 6   Determining the Height of a Building

You measure the shadow cast by the Aon Center in Chicago, Illinois, and find that it is 142 feet long, as shown in Figure 1.11. Then you measure the shadow cast by a nearby four-foot post and find that it is 6 inches long. Determine the building's height.

Solution   To solve this problem, use the result from geometry that the ratios of corresponding sides of similar triangles are equal.

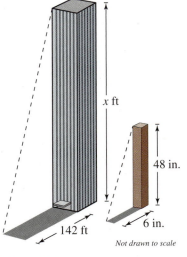

x ft

48 in.

142 ft   6 in.

Not drawn to scale

Figure 1.11

Verbal model: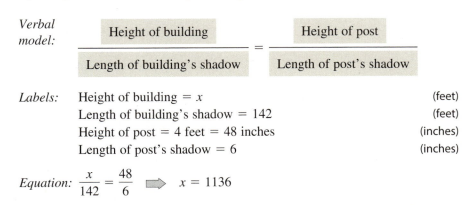

$$\frac{\text{Height of building}}{\text{Length of building's shadow}} = \frac{\text{Height of post}}{\text{Length of post's shadow}}$$

Labels:

Height of building = x (feet)

Length of building's shadow = 142   (feet)

Height of post = 4 feet = 48 inches   (inches)

Length of post's shadow = 6   (inches)

Equation: $\dfrac{x}{142} = \dfrac{48}{6} \implies x = 1136$

So, the height of the Aon Center is 1136 feet.

✓ **Checkpoint** 🔊))) *Audio-video solution in English & Spanish at LarsonPrecalculus.com*

You measure the shadow cast by a building and find that it is 55 feet long. Then you measure the shadow cast by a nearby four-foot post and find that it is 1.8 feet long. Determine the building's height. ◼

Mixture Problems

Problems that involve two or more rates are called **mixture problems.**

EXAMPLE 7 A Simple Interest Problem

You invested a total of \$10,000 at $4\frac{1}{2}\%$ and $5\frac{1}{2}\%$ simple interest. During one year, the two accounts earned \$508.75. How much did you invest in each account?

Solution Let x represent the amount invested at $4\frac{1}{2}\%$. Then the amount invested at $5\frac{1}{2}\%$ is $10,000 - x$.

Verbal model:	Interest from $4\frac{1}{2}\%$	+	Interest from $5\frac{1}{2}\%$	=	Total interest

Labels: Interest from $4\frac{1}{2}\%$ = Prt = $(x)(0.045)(1)$ (dollars)
   Interest from $5\frac{1}{2}\%$ = Prt = $(10,000 - x)(0.055)(1)$ (dollars)
   Total interest = 508.75   (dollars)

Equation: $0.045x + 0.055(10,000 - x) = 508.76$

$$-0.01x = -41.25 \implies x = 4125$$

So, you invested \$4125 at $4\frac{1}{2}\%$ and $10,000 - x = \$5875$ at $5\frac{1}{2}\%$.

 Checkpoint *Audio-video solution in English & Spanish at LarsonPrecalculus.com*

You invested a total of \$5000 at $2\frac{1}{2}\%$ and $3\frac{1}{2}\%$ simple interest. During one year, the two accounts earned \$151.25. How much did you invest in each account?

> **REMARK** Example 7 uses the simple interest formula $I = Prt$, where I is the interest, P is the principal (original deposit), r is the annual interest rate (in decimal form), and t is the time in years. Notice that in this example the amount invested, \$10,000, is separated into two parts, x and $10,000 - x$.

EXAMPLE 8 An Inventory Problem

A store has \$30,000 of inventory in 24-inch and 50-inch televisions. The profit on a 24-inch television is 22% and the profit on a 50-inch television is 40%. The profit on the entire stock is 35%. How much was invested in each type of television?

Solution Let x represent the amount invested in 24-inch televisions. Then the amount invested in 50-inch televisions is $30,000 - x$.

Verbal model:	Profit from 24-inch televisions	+	Profit from 50-inch televisions	=	Total profit

Labels: Inventory of 24-inch televisions = x (dollars)
   Inventory of 50-inch televisions = $30,000 - x$ (dollars)
   Profit from 24-inch televisions = $0.22x$ (dollars)
   Profit from 50-inch televisions = $0.40(30,000 - x)$ (dollars)
   Total profit = $0.35(30,000) = 10,500$ (dollars)

Equation: $0.22x + 0.40(30,000 - x) = 10,500$

$$-0.18x = -1500 \implies x \approx 8333.33$$

So, \$8333.33 was invested in 24-inch televisions and $30,000 - x = \$21,666.67$ was invested in 50-inch televisions.

 Checkpoint *Audio-video solution in English & Spanish at LarsonPrecalculus.com*

In Example 8, the profit on a 24-inch television is 24% and the profit on a 50-inch television is 42%. The profit on the entire stock is 36%. How much was invested in each type of television?

Common Formulas

A **literal equation** is an equation that contains more than one variable. Many common types of geometric, scientific, and investment problems use ready-made literal equations, or **formulas.** Knowing these formulas will help you translate and solve a wide variety of real-life applications.

Common Formulas for Area *A*, Perimeter *P*, Circumference *C*, and Volume *V*

Square	*Rectangle*	*Circle*	*Triangle*
$A = s^2$	$A = lw$	$A = \pi r^2$	$A = \dfrac{1}{2}bh$
$P = 4s$	$P = 2l + 2w$	$C = 2\pi r$	$P = a + b + c$

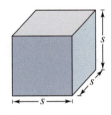 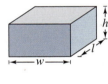 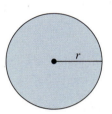 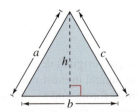

Cube	*Rectangular Solid*	*Circular Cylinder*	*Sphere*
$V = s^3$	$V = lwh$	$V = \pi r^2 h$	$V = \dfrac{4}{3}\pi r^3$

 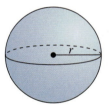

Miscellaneous Common Formulas

Temperature:

$$F = \frac{9}{5}C + 32 \qquad F = \text{degrees Fahrenheit, } C = \text{degrees Celsius}$$

$$C = \frac{5}{9}(F - 32)$$

Simple Interest:

$$I = Prt \qquad I = \text{interest, } P = \text{principal (original deposit),}$$
$$r = \text{annual interest rate (in decimal form), } t = \text{time in years}$$

Compound Interest:

$$A = P\left(1 + \frac{r}{n}\right)^{nt} \qquad A = \text{balance, } P = \text{principal (original deposit), } r = \text{annual interest rate (in decimal form),}$$
$$n = \text{compoundings (number of times interest is calculated) per year, } t = \text{time in years}$$

Distance:

$$d = rt \qquad d = \text{distance traveled, } r = \text{rate, } t = \text{time}$$

When solving an applied problem, you may find it helpful to rewrite a common formula. For example, the formula for the perimeter of a rectangle, $P = 2l + 2w$, can be solved for w as $w = \frac{1}{2}(P - 2l)$.

EXAMPLE 9 **Using a Formula**

See LarsonPrecalculus.com for an interactive version of this type of example.

The cylindrical can shown below has a volume of 200 cubic centimeters (cm³). Find the height of the can.

Solution

The formula for the *volume of a cylinder* is $V = \pi r^2 h$. To find the height of the can, solve for h.

$$h = \frac{V}{\pi r^2}$$

Then, using $V = 200$ and $r = 4$, find the height.

$$h = \frac{200}{\pi(4)^2} \qquad \text{Substitute 200 for } V \text{ and 4 for } r.$$

$$= \frac{200}{16\pi} \qquad \text{Simplify denominator.}$$

$$\approx 3.98 \qquad \text{Use a calculator.}$$

So, the height of the can is about 3.98 centimeters.

Check

$$V = \pi r^2 h$$
$$\approx \pi(4)^2(3.98)$$
$$\approx 200$$

 Checkpoint  *Audio-video solution in English & Spanish at LarsonPrecalculus.com*

A cylindrical container has a volume of 84 cubic inches and a radius of 3 inches. Find the height of the container.

Summarize (Section 1.3)

1. Describe the process of mathematical modeling *(page 90)*. For examples of writing and using mathematical models, see Examples 1–6.

2. Explain how to solve a mixture problem *(page 94)*. For examples of solving mixture problems, see Examples 7 and 8.

3. State some common formulas used to solve real-life problems *(page 95)*. For an example that uses a volume formula, see Example 9.

1.3 Exercises

See **CalcChat.com** for tutorial help and worked-out solutions to odd-numbered exercises.

Vocabulary: Fill in the blanks.

1. The process of translating phrases or sentences into algebraic expressions or equations is called _____ _____.

2. A good approach to mathematical modeling is a two-stage approach, using a verbal description to form a _____ _____, and then, after assigning labels to the quantities, forming an _____ _____.

Skills and Applications

 Writing a Verbal Description In Exercises 3–12, write a verbal description of the algebraic expression without using the variable.

3. $y + 2$

4. $x - 8$

5. $\dfrac{t}{6}$

6. $\dfrac{1}{3}u$

7. $\dfrac{z - 2}{3}$

8. $\dfrac{x + 9}{5}$

9. $-2(d + 5)$

10. $10y(y - 3)$

11. $\dfrac{3(x - 2)}{x}$

12. $\dfrac{(r + 3)(2 - r)}{3r}$

 Writing an Algebraic Expression In Exercises 13–24, write an algebraic expression for the verbal description.

13. The sum of two consecutive natural numbers

14. The product of two consecutive natural numbers

15. The product of two consecutive odd integers, the first of which is $2n - 1$

16. The sum of the squares of two consecutive even integers, the first of which is $2n$

17. The distance a car travels in t hours at a rate of 55 miles per hour

18. The travel time for a plane traveling at a rate of r kilometers per hour for 900 kilometers

19. The amount of acid in x liters of a 20% acid solution

20. The sale price of an item with a 33% discount on its list price L

21. The perimeter of a rectangle with a width x and a length that is twice the width

22. The area of a triangle with a base that is 16 inches and a height that is h inches

23. The total cost of producing x units for which the fixed costs are $2500 and the cost per unit is $40

24. The total revenue obtained by selling x units at $12.99 per unit

Translating a Statement In Exercises 25–28, translate the statement into an algebraic expression or equation.

25. The discount d is 30% of the list price L.

26. The amount A of water in q quarts of a liquid is 72% of the liquid.

27. The number N represents p percent of 672.

28. The sales for this month S_2 are 20% greater than the sales from last month S_1.

Writing an Expression In Exercises 29 and 30, write an expression for the area of the figure.

29.

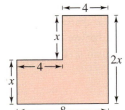

30.

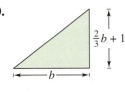

 Number Problems In Exercises 31–36, write a mathematical model for the problem and solve.

31. The sum of two consecutive natural numbers is 525. Find the numbers.

32. The sum of three consecutive natural numbers is 804. Find the numbers.

33. One positive number is 5 times another number. The difference between the two numbers is 148. Find the numbers.

34. One positive number is $\frac{1}{5}$ of another number. The difference between the two numbers is 76. Find the numbers.

35. Find two consecutive integers whose product is 5 less than the square of the smaller number.

36. Find two consecutive natural numbers such that the difference of their reciprocals is $\frac{1}{4}$ the reciprocal of the smaller number.

37. Finance A salesperson's weekly paycheck is 15% less than a second salesperson's paycheck. The two paychecks total $1125. Find the amount of each paycheck.

38. Discount The sale price of a swimming pool after a 16.5% discount is $1210.75. Find the original list price of the pool.

39. Finance A family has annual loan payments equal to 32% of their annual income. During the year, the loan payments total $15,680. What is the family's annual income?

40. Finance A family has a monthly mortgage payment of $760, which is 16% of their monthly income. What is the family's monthly income?

41. Dimensions A rectangular room is 1.5 times as long as it is wide, and its perimeter is 25 meters.

(a) Draw a diagram that gives a visual representation of the problem. Let l represent the length and let w represent the width.

(b) Write l in terms of w and write an equation for the perimeter in terms of w.

(c) Find the dimensions of the room.

42. Dimensions A rectangular picture frame has a perimeter of 3 meters. The height of the frame is $\frac{2}{3}$ times its width. Find the dimensions of the picture frame.

43. Course Grade To get an A in a course, you must have an average of at least 90% on four tests worth 100 points each. Your scores so far are 87, 92, and 84. What must you score on the fourth test to get an A in the course?

44. Course Grade You are taking a course that has four tests. The first three tests are worth 100 points each and the fourth test is worth 200 points. To get an A in the course, you must have an average of at least 90% on the four tests. Your scores so far are 87, 92, and 84. What must you score on the fourth test to get an A in the course?

45. Travel Time You are driving on a Canadian freeway to a town that is 500 kilometers from your home. After 30 minutes, you pass a freeway exit that you know is 50 kilometers from your home. Assuming that you continue at the same constant speed, how long does the entire trip take?

46. Average Speed A truck driver travels at an average speed of 55 miles per hour on a 200-mile trip to pick up a load of freight. On the return trip (with the truck fully loaded), the average speed is 40 miles per hour. What is the average speed for the round trip?

47. Physics Light travels at the speed of approximately 3.0×10^8 meters per second. Find the time in minutes required for light to travel from the sun to Earth (an approximate distance of 1.5×10^{11} meters).

48. Radio Waves Radio waves travel at the same speed as light, approximately 3.0×10^8 meters per second. Find the time required for a radio wave to travel from Mission Control in Houston to NASA astronauts on the surface of the moon 3.84×10^8 meters away.

49. Height of a Building You measure the shadow cast by One Liberty Place in Philadelphia, Pennsylvania, and find that it is 105 feet long. Then you measure the shadow cast by a nearby three-foot post and find that it is 4 inches long. Determine the building's height.

50. Height of a Tree You measure a tree's shadow and find that it is 8 meters long. Then you measure the shadow of a nearby two-meter lamppost and find that it is 75 centimeters long. (See figure.) How tall is the tree?

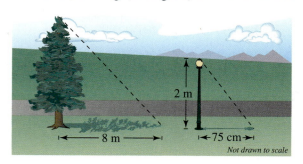

Not drawn to scale

51. Flagpole Height A person who is 6 feet tall walks away from a flagpole toward the tip of the shadow of the flagpole. When the person is 30 feet from the flagpole, the tips of the person's shadow and the shadow cast by the flagpole coincide at a point 5 feet in front of the person.

(a) Draw a diagram that gives a visual representation of the problem. Let h represent the height of the flagpole.

(b) Find the height of the flagpole.

52. Shadow Length A person who is 6 feet tall walks away from a 50-foot tower toward the tip of the tower's shadow. At a distance of 32 feet from the tower, the person's shadow begins to emerge beyond the tower's shadow. How much farther must the person walk to be completely out of the tower's shadow?

53. Simple Interest You invested a total of $12,000 at $4\frac{1}{2}$% and 5% simple interest. During one year, the two accounts earned $580. How much did you invest in each account?

54. Simple Interest You invested a total of $25,000 at 3% and $4\frac{1}{2}$% simple interest. During one year, the two accounts earned $900. How much did you invest in each account?

55. Inventory A nursery has $40,000 of inventory in dogwood trees and red maple trees. The profit on a dogwood tree is 25% and the profit on a red maple tree is 17%. The profit for the entire stock is 20%. How much was invested in each type of tree?

56. Inventory An automobile dealer has $600,000 of inventory in minivans and alternative-fueled vehicles. The profit on a minivan is 24% and the profit on an alternative-fueled vehicle is 28%. The profit for the entire stock is 25%. How much was invested in each type of vehicle?

· · 57. Mixture Problem · · · · · · · · · · · · ·

A forester is making a gasoline-oil mixture for a chainsaw engine. The forester has 2 gallons of a mixture that is 32 parts gasoline and 1 part oil. How many gallons of gasoline should the forester add to bring the mixture to 50 parts gasoline and 1 part oil?

58. Mixture Problem A grocer mixes peanuts that cost $1.49 per pound and walnuts that cost $2.69 per pound to make 100 pounds of a mixture that costs $2.21 per pound. How much of each kind of nut is in the mixture?

59. Area of a Triangle

Solve for h: $A = \frac{1}{2}bh$.

60. Volume of a Rectangular Prism

Solve for l: $V = lwh$.

61. Markup

Solve for C: $S = C + RC$.

62. Discount

Solve for L: $S = L - RL$.

63. Investment at Simple Interest

Solve for r: $A = P + Prt$.

64. Area of a Trapezoid

Solve for b: $A = \frac{1}{2}(a + b)h$.

65. Physiology The average body temperature of a person is 98.6°F. What is this temperature in degrees Celsius?

66. Physiology The average body temperature of a dog is 101.3°F. What is this temperature in degrees Celsius?

67. Chemistry The melting point of francium is 27°C. What is this temperature in degrees Fahrenheit?

68. Chemistry The boiling point of bromine is 58.8°C. What is this temperature in degrees Fahrenheit?

69. Volume of a Billiard Ball A billiard ball has a volume of 5.96 cubic inches. Find the radius of the billiard ball.

70. Length of a Tank The diameter of a cylindrical propane gas tank is 4 feet. The total volume of the tank is 603.2 cubic feet. Find the length of the tank.

71. Physics You have a uniform beam of length L with a fulcrum x feet from one end (see figure). Objects with weights W_1 and W_2 are placed at opposite ends of the beam. The beam balances when $W_1x = W_2(L - x)$.

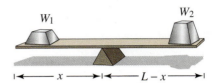

Find x such that each beam will balance.

(a) Two children weighing 50 pounds and 75 pounds are playing on a seesaw that is 10 feet long.

(b) A person weighing 200 pounds is attempting to move a 550-pound rock with a bar that is 5 feet long.

72. HOW DO YOU SEE IT? To determine a building's height, you measure the shadows cast by the building and a nearby four-foot post (see figure).

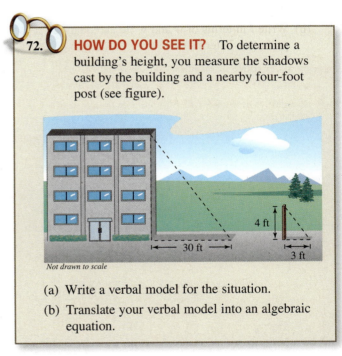

Not drawn to scale

(a) Write a verbal model for the situation.

(b) Translate your verbal model into an algebraic equation.

Exploration

True or False? In Exercises 73–76, determine whether the statement is true or false. Justify your answer.

73. "8 less than z cubed divided by the difference of z squared and 9" can be written as $(z^3 - 8)/(z - 9)^2$.

74. The expression $x^3/(x - 4)^2$ can be described as "x cubed divided by the square of the difference of x and 4".

75. The area of a circle with a radius of 2 inches is less than the area of a square with a side length of 4 inches.

76. The volume of a cube with a side length of 9.5 inches is greater than the volume of a sphere with a radius of 5.9 inches.

1.4 Quadratic Equations and Applications

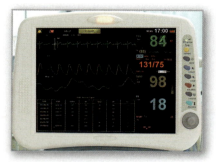

Quadratic equations have many real-life applications. For example, in Exercise 117 on page 112, you will use a quadratic equation to analyze a patient's blood oxygen level.

- Solve quadratic equations by factoring.
- Solve quadratic equations by extracting square roots.
- Solve quadratic equations by completing the square.
- Use the Quadratic Formula to solve quadratic equations.
- Use quadratic equations to model and solve real-life problems.

Solving Quadratic Equations by Factoring

A **quadratic equation** in x is an equation that can be written in the general form

$$ax^2 + bx + c = 0$$

where a, b, and c are real numbers with $a \neq 0$. A quadratic equation in x is also called a **second-degree polynomial equation** in x.

In this section, you will study four methods for solving quadratic equations: *factoring, extracting square roots, completing the square,* and the *Quadratic Formula.* The first method is based on the Zero-Factor Property from Section P.1.

If $ab = 0$, then $a = 0$ or $b = 0$.   Zero-Factor Property

To use this property, rewrite the left side of the general form of a quadratic equation as the product of two linear factors. Then find the solutions of the quadratic equation by setting each linear factor equal to zero.

EXAMPLE 1   Solving Quadratic Equations by Factoring

a.
$$2x^2 + 9x + 7 = 3$$ Original equation
$$2x^2 + 9x + 4 = 0$$ Write in general form.
$$(2x + 1)(x + 4) = 0$$ Factor.
$$2x + 1 = 0 \implies x = -\tfrac{1}{2}$$ Set 1st factor equal to 0 and solve.
$$x + 4 = 0 \implies x = -4$$ Set 2nd factor equal to 0 and solve.

The solutions are $x = -\tfrac{1}{2}$ and $x = -4$. Check these in the original equation.

b.
$$6x^2 - 3x = 0$$ Original equation
$$3x(2x - 1) = 0$$ Factor.
$$3x = 0 \implies x = 0$$ Set 1st factor equal to 0 and solve.
$$2x - 1 = 0 \implies x = \tfrac{1}{2}$$ Set 2nd factor equal to 0 and solve.

The solutions are $x = 0$ and $x = \tfrac{1}{2}$. Check these in the original equation.

✓ *Checkpoint* 🔊)) *Audio-video solution in English & Spanish at LarsonPrecalculus.com*

Solve $2x^2 - 3x + 1 = 6$ by factoring.   ■

The Zero-Factor Property applies *only* to equations written in general form (in which the right side of the equation is zero). So, collect all terms on one side *before* factoring. For example, in the equation $(x - 5)(x + 2) = 8$, it is *incorrect* to set each factor equal to 8. To solve this equation, first multiply the binomials on the left side of the equation. Then subtract 8 from each side. After simplifying the equation, factor the left side and use the Zero-Factor Property to find the solutions. Solve this equation correctly on your own. Then check the solutions in the original equation.

▷ **TECHNOLOGY** You can use a graphing utility to check graphically the real solutions of a quadratic equation. Begin by writing the equation in general form. Then set y equal to the left side and graph the resulting equation. The x-intercepts of the graph represent the real solutions of the original equation. Use the *zero* or *root* feature of the graphing utility to approximate the x-intercepts of the graph. For example, to check the solutions of $6x^2 - 3x = 0$, graph $y = 6x^2 - 3x$, and find that the x-intercepts are $(0, 0)$ and $\left(\frac{1}{2}, 0\right)$, as shown in the graph below. These x-intercepts represent the solutions $x = 0$ and $x = \frac{1}{2}$, as found in Example 1(b).

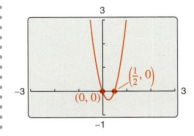

Extracting Square Roots

Consider a quadratic equation of the form $u^2 = d$, where $d > 0$ and u is an algebraic expression. Factoring shows that this equation has two solutions.

$$u^2 = d \qquad \text{Write original equation.}$$
$$u^2 - d = 0 \qquad \text{Write in general form.}$$
$$\left(u + \sqrt{d}\right)\left(u - \sqrt{d}\right) = 0 \qquad \text{Factor.}$$
$$u + \sqrt{d} = 0 \implies u = -\sqrt{d} \qquad \text{Set 1st factor equal to 0 and solve.}$$
$$u - \sqrt{d} = 0 \implies u = \sqrt{d} \qquad \text{Set 2nd factor equal to 0 and solve.}$$

The two solutions differ only in sign, so you can write the solutions together, using a "plus or minus sign," as

$$u = \pm\sqrt{d}.$$

This form of the solution is read as "u is equal to plus or minus the square root of d." Solving an equation of the form $u^2 = d$ without going through the steps of factoring is called **extracting square roots.**

Extracting Square Roots

The equation $u^2 = d$, where $d > 0$, has exactly two solutions:

$$u = \sqrt{d} \quad \text{and} \quad u = -\sqrt{d}.$$

These solutions can also be written as

$$u = \pm\sqrt{d}.$$

EXAMPLE 2   **Extracting Square Roots**

Solve each equation by extracting square roots.

a. $4x^2 = 12$

b. $(x - 3)^2 = 7$

Solution

a. $4x^2 = 12$ Write original equation.

$\qquad x^2 = 3$ Divide each side by 4.

$\qquad x = \pm\sqrt{3}$ Extract square roots.

The solutions are $x = \sqrt{3}$ and $x = -\sqrt{3}$. Check these in the original equation.

b. $(x - 3)^2 = 7$ Write original equation.

$\qquad x - 3 = \pm\sqrt{7}$ Extract square roots.

$\qquad x = 3 \pm \sqrt{7}$ Add 3 to each side.

The solutions are $x = 3 \pm \sqrt{7}$. Check these in the original equation.

✓ **Checkpoint** 🔊))) Audio-video solution in English & Spanish at LarsonPrecalculus.com

Solve each equation by extracting square roots.

a. $3x^2 = 36$

b. $(x - 1)^2 = 10$

Completing the Square

The general form of the equation $(x - 3)^2 = 7$ from Example 2(b) is $x^2 - 6x + 2 = 0$. This equation is equivalent to the original, so it has the same two solutions

$$x = 3 \pm \sqrt{7}.$$

However, the left side of the equation is not factorable, and you cannot find its solutions unless you rewrite the equation so it can be solved by extracting square roots. You can do this using a method called **completing the square.**

Completing the Square

To **complete the square** for the expression $x^2 + bx$, add $(b/2)^2$, which is the square of half the coefficient of x. Consequently,

$$x^2 + bx + \left(\frac{b}{2}\right)^2 = \left(x + \frac{b}{2}\right)^2.$$

When solving quadratic equations by completing the square, you must add $(b/2)^2$ to *each side* in order to maintain equality.

EXAMPLE 3 **Completing the Square: Leading Coefficient Is 1**

Solve

$$x^2 + 2x - 6 = 0$$

by completing the square.

Solution

$x^2 + 2x - 6 = 0$	Write original equation.
$x^2 + 2x = 6$	Add 6 to each side.
$x^2 + 2x + 1^2 = 6 + 1^2$	Add 1^2 to each side.

$$(\text{Half of 2})^2$$

$(x + 1)^2 = 7$	Simplify.
$x + 1 = \pm\sqrt{7}$	Extract square roots.
$x = -1 \pm \sqrt{7}$	Subtract 1 from each side.

The solutions are $x = -1 \pm \sqrt{7}$.

Check $x = -1 + \sqrt{7}$

$x^2 + 2x - 6 = 0$	Write original equation.
$\left(-1 + \sqrt{7}\right)^2 + 2\left(-1 + \sqrt{7}\right) - 6 \stackrel{?}{=} 0$	Substitute $-1 + \sqrt{7}$ for x.
$8 - 2\sqrt{7} - 2 + 2\sqrt{7} - 6 \stackrel{?}{=} 0$	Multiply.
$8 - 2 - 6 = 0$	Solution checks. ✓

Check $x = -1 - \sqrt{7}$ on your own.

✓ *Checkpoint* 🔊))) *Audio-video solution in English & Spanish at LarsonPrecalculus.com*

Solve

$$x^2 - 4x - 1 = 0$$

by completing the square.

When the leading coefficient is *not* 1, divide each side of the equation by the leading coefficient *before* completing the square.

EXAMPLE 4 **Completing the Square: Leading Coefficient Is Not 1**

Solve $2x^2 + 8x + 3 = 0$ by completing the square.

Solution

$2x^2 + 8x + 3 = 0$	Write original equation.
$2x^2 + 8x = -3$	Subtract 3 from each side.
$x^2 + 4x = -\dfrac{3}{2}$	Divide each side by 2.
$x^2 + 4x + 2^2 = -\dfrac{3}{2} + 2^2$	Add 2^2 to each side.

(Half of 4)^2

$(x + 2)^2 = \dfrac{5}{2}$	Simplify.
$x + 2 = \pm\sqrt{\dfrac{5}{2}}$	Extract square roots.
$x + 2 = \pm\dfrac{\sqrt{10}}{2}$	Rationalize denominator.
$x = -2 \pm \dfrac{\sqrt{10}}{2}$	Subtract 2 from each side.

▷ **ALGEBRA HELP** To review rationalizing denominators, see Section P.2.

✓ **Checkpoint** ◆))) *Audio-video solution in English & Spanish at LarsonPrecalculus.com*

Solve $2x^2 - 4x + 1 = 0$ by completing the square.

EXAMPLE 5 **Completing the Square: Leading Coefficient Is Not 1**

Solve $3x^2 - 4x - 5 = 0$ by completing the square.

Solution

$3x^2 - 4x - 5 = 0$	Write original equation.
$3x^2 - 4x = 5$	Add 5 to each side.
$x^2 - \dfrac{4}{3}x = \dfrac{5}{3}$	Divide each side by 3.
$x^2 - \dfrac{4}{3}x + \left(-\dfrac{2}{3}\right)^2 = \dfrac{5}{3} + \left(-\dfrac{2}{3}\right)^2$	Add $\left(-\dfrac{2}{3}\right)^2$ to each side.
$\left(x - \dfrac{2}{3}\right)^2 = \dfrac{19}{9}$	Simplify.
$x - \dfrac{2}{3} = \pm\dfrac{\sqrt{19}}{3}$	Extract square roots.
$x = \dfrac{2}{3} \pm \dfrac{\sqrt{19}}{3}$	Add $\dfrac{2}{3}$ to each side.

✓ **Checkpoint** ◆))) *Audio-video solution in English & Spanish at LarsonPrecalculus.com*

Solve $3x^2 - 10x - 2 = 0$ by completing the square.

The Quadratic Formula

Often in mathematics you learn the long way of solving a problem first. Then, you use the longer method to develop shorter techniques. The long way stresses understanding and the short way stresses efficiency.

For example, completing the square is a "long way" of solving a quadratic equation. When you use completing the square to solve quadratic equations, you must complete the square for *each* equation separately. In the derivation below, you complete the square *once* for the general form of a quadratic equation

$$ax^2 + bx + c = 0$$

to obtain the **Quadratic Formula**—a shortcut for solving quadratic equations.

$ax^2 + bx + c = 0$	General form, $a \neq 0$
$ax^2 + bx = -c$	Subtract c from each side.
$x^2 + \dfrac{b}{a}x = -\dfrac{c}{a}$	Divide each side by a.
$x^2 + \dfrac{b}{a}x + \left(\dfrac{b}{2a}\right)^2 = -\dfrac{c}{a} + \left(\dfrac{b}{2a}\right)^2$	Complete the square.

$$\left(\text{Half of } \frac{b}{a}\right)^2$$

$\left(x + \dfrac{b}{2a}\right)^2 = \dfrac{b^2 - 4ac}{4a^2}$	Simplify.
$x + \dfrac{b}{2a} = \pm\sqrt{\dfrac{b^2 - 4ac}{4a^2}}$	Extract square roots.
$x = -\dfrac{b}{2a} \pm \dfrac{\sqrt{b^2 - 4ac}}{2|a|}$	Solutions

Note that $\pm 2|a|$ represents the same numbers as $\pm 2a$, so the formula simplifies to

$$x = \frac{-b \pm \sqrt{b^2 - 4ac}}{2a}.$$

The Quadratic Formula

The solutions of a quadratic equation in the general form

$$ax^2 + bx + c = 0, \quad a \neq 0$$

are given by the **Quadratic Formula**

$$x = \frac{-b \pm \sqrt{b^2 - 4ac}}{2a}.$$

The Quadratic Formula is one of the most important formulas in algebra. It is possible to solve every quadratic equation by completing the square or using the Quadratic Formula. You should learn the verbal statement of the Quadratic Formula:

"Negative b, plus or minus the square root of b squared minus $4ac$, all divided by $2a$."

In the Quadratic Formula, the quantity under the radical sign, $b^2 - 4ac$, is the **discriminant** of the quadratic equation $ax^2 + bx + c = 0$. It can be used to determine the number of real solutions of a quadratic equation.

> **Solutions of a Quadratic Equation**
>
> The solutions of a quadratic equation
>
> $$ax^2 + bx + c = 0, a \neq 0$$
>
> can be classified in three ways. If the discriminant $b^2 - 4ac$ is
>
> **1.** *positive*, then the quadratic equation has *two* distinct real solutions and its graph has *two* x-intercepts.
>
> **2.** *zero*, then the quadratic equation has *one* repeated real solution and its graph has *one* x-intercept.
>
> **3.** *negative*, then the quadratic equation has *no* real solutions and its graph has *no* x-intercepts.

If the discriminant of a quadratic equation is negative, then its square root is not a real number and the Quadratic Formula yields two complex solutions. You will study complex solutions in Section 1.5.

When you use the Quadratic Formula, remember that *before* applying the formula, you must first write the quadratic equation in general form.

EXAMPLE 6 **Using the Quadratic Formula**

See LarsonPrecalculus.com for an interactive version of this type of example.

Use the Quadratic Formula to solve

$$x^2 + 3x = 9.$$

Solution The general form of the equation is $x^2 + 3x - 9 = 0$. The discriminant is $b^2 - 4ac = 9 + 36 = 45$, which is positive. So, the equation has two real solutions.

$$x^2 + 3x - 9 = 0 \qquad \text{Write in general form.}$$

$$x = \frac{-b \pm \sqrt{b^2 - 4ac}}{2a} \qquad \text{Quadratic Formula}$$

$$x = \frac{-3 \pm \sqrt{(3)^2 - 4(1)(-9)}}{2(1)} \qquad \text{Substitute } a = 1, b = 3, \text{ and } c = -9.$$

$$x = \frac{-3 \pm \sqrt{45}}{2} \qquad \text{Simplify.}$$

$$x = \frac{-3 \pm 3\sqrt{5}}{2} \qquad \text{Simplify.}$$

The two solutions are

$$x = \frac{-3 + 3\sqrt{5}}{2}$$

and

$$x = \frac{-3 - 3\sqrt{5}}{2}.$$

Check these in the original equation.

✓ *Checkpoint* ◄))) Audio-video solution in English & Spanish at LarsonPrecalculus.com

Use the Quadratic Formula to solve

$$3x^2 + 2x = 10.$$

Applications

Quadratic equations are often used in problems dealing with area. Here is a relatively simple example.

A square room has an area of 144 square feet. Find the dimensions of the room.

To solve this problem, let x represent the length of each side of the room. Then use the formula for the area of a square to write and solve the equation

$$x^2 = 144$$

and conclude that each side of the room is 12 feet long. Note that although the equation $x^2 = 144$ has two solutions, $x = -12$ and $x = 12$, the negative solution does not make sense in the context of the problem, so choose the positive solution.

EXAMPLE 7   **Finding the Dimensions of a Room**

A rectangular sunroom is 3 feet longer than it is wide (see figure) and has an area of 154 square feet. Find the dimensions of the room.

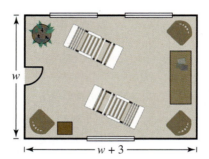

Solution

Verbal model: | Width of room | $\cdot$ | Length of room | $=$ | Area of room |

Labels:	Width of room $= w$	(feet)
	Length of room $= w + 3$	(feet)
	Area of room $= 154$	(square feet)

Equation:

$$w(w + 3) = 154$$
$$w^2 + 3w - 154 = 0 \qquad \text{Write in general form.}$$
$$(w - 11)(w + 14) = 0 \qquad \text{Factor.}$$
$$w - 11 = 0 \implies w = 11 \qquad \text{Set 1st factor equal to 0 and solve.}$$
$$w + 14 = 0 \implies w = -14 \qquad \text{Set 2nd factor equal to 0 and solve.}$$

Choosing the positive value, the width is 11 feet and the length is $11 + 3 = 14$ feet.

Check

The length of 14 feet is 3 more than the width of 11 feet. ✓
The area of the sunroom is $11(14) = 154$ square feet. ✓

✓ **Checkpoint** 🔊))) *Audio-video solution in English & Spanish at LarsonPrecalculus.com*

A rectangular kitchen is 6 feet longer than it is wide and has an area of 112 square feet. Find the dimensions of the kitchen.

1.5 Exercises   See **CalcChat.com** for tutorial help and worked-out solutions to odd-numbered exercises.

Vocabulary: Fill in the blanks.

1. A _____ number has the form $a + bi$, where $a \neq 0$, $b = 0$.
2. An _____ number has the form $a + bi$, where $a \neq 0$, $b \neq 0$.
3. A _____ _____ number has the form $a + bi$, where $a = 0$, $b \neq 0$.
4. The imaginary unit i is defined as $i =$ _____, where $i^2 =$ _____.
5. When a is a positive real number, the _____ _____ root of $-a$ is defined as $\sqrt{-a} = \sqrt{a}i$.
6. The numbers $a + bi$ and $a - bi$ are called _____ _____, and their product is a real number $a^2 + b^2$.

Skills and Applications

Equality of Complex Numbers In Exercises 7–10, find real numbers a and b such that the equation is true.

7. $a + bi = 9 + 8i$
8. $a + bi = 10 - 5i$
9. $(a - 2) + (b + 1)i = 6 + 5i$
10. $(a + 2) + (b - 3)i = 4 + 7i$

 Writing a Complex Number in Standard Form In Exercises 11–22, write the complex number in standard form.

11. $2 + \sqrt{-25}$
12. $4 + \sqrt{-49}$
13. $1 - \sqrt{-12}$
14. $2 - \sqrt{-18}$
15. $\sqrt{-40}$
16. $\sqrt{-27}$
17. 23
18. 50
19. $-6i + i^2$
20. $-2i^2 + 4i$
21. $\sqrt{-0.04}$
22. $\sqrt{-0.0025}$

 Adding or Subtracting Complex Numbers In Exercises 23–30, perform the operation and write the result in standard form.

23. $(5 + i) + (2 + 3i)$
24. $(13 - 2i) + (-5 + 6i)$
25. $(9 - i) - (8 - i)$
26. $(3 + 2i) - (6 + 13i)$
27. $\left(-2 + \sqrt{-8}\right) + \left(5 - \sqrt{-50}\right)$
28. $\left(8 + \sqrt{-18}\right) - \left(4 + 3\sqrt{2}i\right)$
29. $13i - (14 - 7i)$
30. $25 + (-10 + 11i) + 15i$

 Multiplying Complex Numbers In Exercises 31–38, perform the operation and write the result in standard form.

31. $(1 + i)(3 - 2i)$
32. $(7 - 2i)(3 - 5i)$
33. $12i(1 - 9i)$
34. $-8i(9 + 4i)$
35. $\left(\sqrt{2} + 3i\right)\left(\sqrt{2} - 3i\right)$
36. $\left(4 + \sqrt{7}i\right)\left(4 - \sqrt{7}i\right)$
37. $(6 + 7i)^2$
38. $(5 - 4i)^2$

Multiplying Conjugates In Exercises 39–46, write the complex conjugate of the complex number. Then multiply the number by its complex conjugate.

39. $9 + 2i$
40. $8 - 10i$
41. $-1 - \sqrt{5}i$
42. $-3 + \sqrt{2}i$
43. $\sqrt{-20}$
44. $\sqrt{-15}$
45. $\sqrt{6}$
46. $1 + \sqrt{8}$

 A Quotient of Complex Numbers in Standard Form In Exercises 47–54, write the quotient in standard form.

47. $\dfrac{2}{4 - 5i}$
48. $\dfrac{13}{1 - i}$
49. $\dfrac{5 + i}{5 - i}$
50. $\dfrac{6 - 7i}{1 - 2i}$
51. $\dfrac{9 - 4i}{i}$
52. $\dfrac{8 + 16i}{2i}$
53. $\dfrac{3i}{(4 - 5i)^2}$
54. $\dfrac{5i}{(2 + 3i)^2}$

 Performing Operations with Complex Numbers In Exercises 55–58, perform the operation and write the result in standard form.

55. $\dfrac{2}{1 + i} - \dfrac{3}{1 - i}$
56. $\dfrac{2i}{2 + i} + \dfrac{5}{2 - i}$
57. $\dfrac{i}{3 - 2i} + \dfrac{2i}{3 + 8i}$
58. $\dfrac{1 + i}{i} - \dfrac{3}{4 - i}$

 Writing a Complex Number in Standard Form In Exercises 59–66, write the complex number in standard form.

59. $\sqrt{-6}\sqrt{-2}$
60. $\sqrt{-5}\sqrt{-10}$
61. $\left(\sqrt{-15}\right)^2$
62. $\left(\sqrt{-75}\right)^2$
63. $\sqrt{-8} + \sqrt{-50}$
64. $\sqrt{-45} - \sqrt{-5}$
65. $\left(3 + \sqrt{-5}\right)\left(7 - \sqrt{-10}\right)$
66. $\left(2 - \sqrt{-6}\right)^2$

Complex Solutions of a Quadratic Equation In Exercises 67–76, use the Quadratic Formula to solve the quadratic equation.

67. $x^2 - 2x + 2 = 0$ **68.** $x^2 + 6x + 10 = 0$

69. $4x^2 + 16x + 17 = 0$ **70.** $9x^2 - 6x + 37 = 0$

71. $4x^2 + 16x + 21 = 0$ **72.** $16t^2 - 4t + 3 = 0$

73. $\frac{3}{2}x^2 - 6x + 9 = 0$ **74.** $\frac{7}{8}x^2 - \frac{3}{4}x + \frac{5}{16} = 0$

75. $1.4x^2 - 2x + 10 = 0$ **76.** $4.5x^2 - 3x + 12 = 0$

Simplifying a Complex Number In Exercises 77–86, simplify the complex number and write it in standard form.

77. $-6i^3 + i^2$ **78.** $4i^2 - 2i^3$

79. $-14i^5$ **80.** $(-i)^3$

81. $\left(\sqrt{-72}\right)^3$ **82.** $\left(\sqrt{-2}\right)^6$

83. $\dfrac{1}{i^3}$ **84.** $\dfrac{1}{(2i)^3}$

85. $(3i)^4$ **86.** $(-i)^6$

87. Impedance of a Circuit

The opposition to current in an electrical circuit is called its impedance. The impedance z in a parallel circuit with two pathways satisfies the equation

$$\frac{1}{z} = \frac{1}{z_1} + \frac{1}{z_2}$$

where z_1 is the impedance (in ohms) of pathway 1 and z_2 is the impedance (in ohms) of pathway 2.

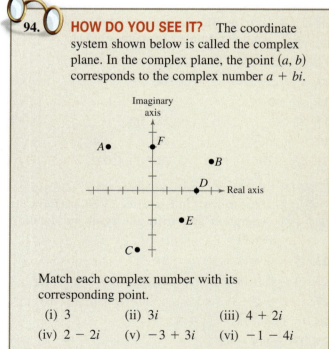

(a) The impedance of each pathway in a parallel circuit is found by adding the impedances of all components in the pathway. Use the table to find z_1 and z_2.

	Resistor	Inductor	Capacitor
Symbol	$a\ \Omega$	$b\ \Omega$	$c\ \Omega$
Impedance	a	bi	$-ci$

(b) Find the impedance z.

88. Cube of a Complex Number Cube each complex number.

(a) $-1 + \sqrt{3}i$ (b) $-1 - \sqrt{3}i$

Exploration

True or False? In Exercises 89–92, determine whether the statement is true or false. Justify your answer.

89. The sum of two complex numbers is always a real number.

90. There is no complex number that is equal to its complex conjugate.

91. $-i\sqrt{6}$ is a solution of $x^4 - x^2 + 14 = 56$.

92. $i^{44} + i^{150} - i^{74} - i^{109} + i^{61} = -1$

93. Pattern Recognition Find the missing values.

$i^1 = i$ $i^2 = -1$ $i^3 = -i$ $i^4 = 1$

$i^5 = \rule{1cm}{0.4pt}$ $i^6 = \rule{1cm}{0.4pt}$ $i^7 = \rule{1cm}{0.4pt}$ $i^8 = \rule{1cm}{0.4pt}$

$i^9 = \rule{1cm}{0.4pt}$ $i^{10} = \rule{1cm}{0.4pt}$ $i^{11} = \rule{1cm}{0.4pt}$ $i^{12} = \rule{1cm}{0.4pt}$

What pattern do you see? Write a brief description of how you would find i raised to any positive integer power.

94. HOW DO YOU SEE IT? The coordinate system shown below is called the complex plane. In the complex plane, the point (a, b) corresponds to the complex number $a + bi$.

Imaginary axis

$A\bullet$ $\bullet F$

$\bullet B$

$\bullet D$ → Real axis

$\bullet E$

$C\bullet$

Match each complex number with its corresponding point.

(i) 3 (ii) $3i$ (iii) $4 + 2i$

(iv) $2 - 2i$ (v) $-3 + 3i$ (vi) $-1 - 4i$

95. Error Analysis Describe the error.

$$\sqrt{-6}\sqrt{-6} = \sqrt{(-6)(-6)} = \sqrt{36} = 6 \quad \text{✗}$$

96. Proof Prove that the complex conjugate of the product of two complex numbers $a_1 + b_1 i$ and $a_2 + b_2 i$ is the product of their complex conjugates.

97. Proof Prove that the complex conjugate of the sum of two complex numbers $a_1 + b_1 i$ and $a_2 + b_2 i$ is the sum of their complex conjugates.

1.6   Other Types of Equations

- Solve polynomial equations of degree three or greater.
- Solve radical equations.
- Solve rational equations and absolute value equations.
- Use nonlinear and nonquadratic models to solve real-life problems.

Polynomial Equations

In this section, you will extend the techniques for solving equations to nonlinear and nonquadratic equations. At this point in the text, you have only four basic methods for solving nonlinear equations—*factoring, extracting square roots, completing the square,* and the *Quadratic Formula.* So, the main goal of this section is to learn to *rewrite* nonlinear equations in a form that enables you to apply one of these methods.

Example 1 shows how to use factoring to solve a **polynomial equation,** which is an equation that can be written in the general form

$$a_n x^n + a_{n-1} x^{n-1} + \cdots + a_2 x^2 + a_1 x + a_0 = 0.$$

Polynomial equations, radical equations, rational equations, and absolute value equations have many real-life applications. For example, in Exercise 101 on page 129, you will use a radical equation to analyze the relationship between the pressure and temperature of saturated steam.

EXAMPLE 1   **Solving a Polynomial Equation by Factoring**

Solve $3x^4 = 48x^2$ and check your solution(s).

Solution   First write the polynomial equation in general form. Then factor the polynomial, set each factor equal to zero, and solve.

$3x^4 = 48x^2$	Write original equation.
$3x^4 - 48x^2 = 0$	Write in general form.
$3x^2(x^2 - 16) = 0$	Factor out common factor.
$3x^2(x + 4)(x - 4) = 0$	Factor completely.
$3x^2 = 0 \implies x = 0$	Set 1st factor equal to 0 and solve.
$x + 4 = 0 \implies x = -4$	Set 2nd factor equal to 0 and solve.
$x - 4 = 0 \implies x = 4$	Set 3rd factor equal to 0 and solve.

Check these solutions by substituting in the original equation.

Check

$$3(0)^4 \overset{?}{=} 48(0)^2 \implies 0 = 0 \qquad \text{0 checks. } ✔$$
$$3(-4)^4 \overset{?}{=} 48(-4)^2 \implies 768 = 768 \qquad -4 \text{ checks. } ✔$$
$$3(4)^4 \overset{?}{=} 48(4)^2 \implies 768 = 768 \qquad 4 \text{ checks. } ✔$$

So, the solutions are $x = 0$, $x = -4$, and $x = 4$.

✔ **Checkpoint** ◀))) *Audio-video solution in English & Spanish at LarsonPrecalculus.com*

Solve $9x^4 - 12x^2 = 0$ and check your solution(s). ◼

A common mistake when solving an equation such as that in Example 1 is to divide each side of the equation by the variable factor x^2. This loses the solution $x = 0$. When solving a polynomial equation, always write the equation in general form, then factor the polynomial and set each factor equal to zero. Do not divide each side of an equation by a variable factor in an attempt to simplify the equation.

▷ **ALGEBRA HELP** To review factoring polynomials, see Section P.4.

EXAMPLE 2 **Solving a Polynomial Equation by Factoring**

Solve $x^3 - 3x^2 + 3x - 9 = 0$.

Solution

$x^3 - 3x^2 + 3x - 9 = 0$	Write original equation.
$x^2(x - 3) + 3(x - 3) = 0$	Group terms and factor.
$(x - 3)(x^2 + 3) = 0$	$(x - 3)$ is a common factor.
$x - 3 = 0 \implies x = 3$	Set 1st factor equal to 0 and solve.
$x^2 + 3 = 0 \implies x = \pm\sqrt{3}i$	Set 2nd factor equal to 0 and solve.

The solutions are $x = 3$, $x = \sqrt{3}i$, and $x = -\sqrt{3}i$. Check these in the original equation.

✓ **Checkpoint** ◀))) *Audio-video solution in English & Spanish at LarsonPrecalculus.com*

Solve each equation.

a. $x^3 - 5x^2 - 2x + 10 = 0$

b. $6x^3 - 27x^2 - 54x = 0$

Occasionally, mathematical models involve equations that are of **quadratic type.** In general, an equation is of quadratic type when it can be written in the form

$$au^2 + bu + c = 0$$

where $a \neq 0$ and u is an algebraic expression.

▷ **TECHNOLOGY** You can use a graphing utility to check graphically the solutions of the equation in Example 2. Graph the equation

$$y = x^3 - 3x^2 + 3x - 9.$$

Then use the *zero* or *root* feature to approximate any x-intercepts. As shown in the graph below, the x-intercept of the graph occurs at $(3, 0)$, confirming the *real* solution $x = 3$ found in Example 2.

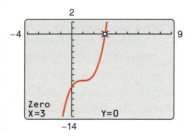

Use a graphing utility to check the solutions found in Example 3.

EXAMPLE 3 **Solving an Equation of Quadratic Type**

Solve $x^4 - 3x^2 + 2 = 0$.

Solution

This equation is of quadratic type with $u = x^2$.

$x^4 - 3x^2 + 2 = 0$	Write original equation.
$(x^2)^2 - 3(x^2) + 2 = 0$	Quadratic form
$u^2 - 3u + 2 = 0$	$u = x^2$
$(u - 1)(u - 2) = 0$	Factor.
$u - 1 = 0 \implies u = 1$	Set 1st factor equal to 0 and solve.
$u - 2 = 0 \implies u = 2$	Set 2nd factor equal to 0 and solve.

Next, replace u with x^2 and solve for x in each equation.

$u = 1$	$u = 2$
$x^2 = 1$	$x^2 = 2$
$x = \pm 1$	$x = \pm\sqrt{2}$

The solutions are $x = 1$, $x = -1$, $x = \sqrt{2}$, and $x = -\sqrt{2}$. Check these in the original equation.

✓ **Checkpoint** ◀))) *Audio-video solution in English & Spanish at LarsonPrecalculus.com*

Solve each equation.

a. $x^4 - 7x^2 + 12 = 0$

b. $9x^4 - 37x^2 + 4 = 0$

Radical Equations

A **radical equation** is an equation that involves one or more radical expressions. Examples 4 and 5 demonstrate how to solve radical equations.

EXAMPLE 4   **Solving Radical Equations**

a.

$\sqrt{2x + 7} - x = 2$	Original equation
$\sqrt{2x + 7} = x + 2$	Isolate radical.
$2x + 7 = x^2 + 4x + 4$	Square each side.
$0 = x^2 + 2x - 3$	Write in general form.
$0 = (x + 3)(x - 1)$	Factor.
$x + 3 = 0 \implies x = -3$	Set 1st factor equal to 0 and solve.
$x - 1 = 0 \implies x = 1$	Set 2nd factor equal to 0 and solve.

Checking these values shows that the only solution is $x = 1$.

b.

$\sqrt{2x - 5} - \sqrt{x - 3} = 1$	Original equation
$\sqrt{2x - 5} = \sqrt{x - 3} + 1$	Isolate $\sqrt{2x - 5}$.
$2x - 5 = x - 3 + 2\sqrt{x - 3} + 1$	Square each side.
$x - 3 = 2\sqrt{x - 3}$	Isolate $2\sqrt{x - 3}$.
$x^2 - 6x + 9 = 4(x - 3)$	Square each side.
$x^2 - 10x + 21 = 0$	Write in general form.
$(x - 3)(x - 7) = 0$	Factor.
$x - 3 = 0 \implies x = 3$	Set 1st factor equal to 0 and solve.
$x - 7 = 0 \implies x = 7$	Set 2nd factor equal to 0 and solve.

The solutions are $x = 3$ and $x = 7$. Check these in the original equation.

✓ *Checkpoint* *Audio-video solution in English & Spanish at LarsonPrecalculus.com*

Solve $-\sqrt{40 - 9x} + 2 = x$.

EXAMPLE 5   **Solving an Equation Involving a Rational Exponent**

Solve $(x - 4)^{2/3} = 25$.

Solution

$(x - 4)^{2/3} = 25$	Write original equation.
$\sqrt[3]{(x - 4)^2} = 25$	Rewrite in radical form.
$(x - 4)^2 = 15{,}625$	Cube each side.
$x - 4 = \pm 125$	Extract square roots.
$x = 129, \ x = -121$	Add 4 to each side.

The solutions are $x = 129$ and $x = -121$. Check these in the original equation.

✓ *Checkpoint* *Audio-video solution in English & Spanish at LarsonPrecalculus.com*

Solve $(x - 5)^{2/3} = 16$.

Rational Equations and Absolute Value Equations

In Section 1.2, you learned how to solve rational equations. Recall that the first step is to multiply each term of the equation by the least common denominator (LCD).

EXAMPLE 6 **Solving a Rational Equation**

Solve $\dfrac{2}{x} = \dfrac{3}{x-2} - 1$ and check your solution(s).

Solution For this equation, the LCD of the three terms is $x(x-2)$, so begin by multiplying each term of the equation by this expression.

$$\frac{2}{x} = \frac{3}{x-2} - 1 \qquad \text{Write original equation.}$$

$$x(x-2)\frac{2}{x} = x(x-2)\frac{3}{x-2} - x(x-2)(1) \qquad \text{Multiply each term by the LCD.}$$

$$2(x-2) = 3x - x(x-2), \quad x \neq 0, 2 \qquad \text{Simplify.}$$

$$2x - 4 = -x^2 + 5x \qquad \text{Simplify.}$$

$$x^2 - 3x - 4 = 0 \qquad \text{Write in general form.}$$

$$(x-4)(x+1) = 0 \qquad \text{Factor.}$$

$$x - 4 = 0 \implies x = 4 \qquad \text{Set 1st factor equal to 0 and solve.}$$

$$x + 1 = 0 \implies x = -1 \qquad \text{Set 2nd factor equal to 0 and solve.}$$

Both $x = 4$ and $x = -1$ are possible solutions. Multiplying each side of an equation by a variable expression can introduce extraneous solutions, so it is important to check your solutions.

Check $x = 4$

$$\frac{2}{x} = \frac{3}{x-2} - 1 \qquad \text{Write original equation.}$$

$$\frac{2}{4} \stackrel{?}{=} \frac{3}{4-2} - 1 \qquad \text{Substitute 4 for } x.$$

$$\frac{1}{2} \stackrel{?}{=} \frac{3}{2} - 1 \qquad \text{Simplify.}$$

$$\frac{1}{2} = \frac{1}{2} \qquad \text{4 checks. } \checkmark$$

Check $x = -1$

$$\frac{2}{x} = \frac{3}{x-2} - 1 \qquad \text{Write original equation.}$$

$$\frac{2}{-1} \stackrel{?}{=} \frac{3}{-1-2} - 1 \qquad \text{Substitute } -1 \text{ for } x.$$

$$-2 \stackrel{?}{=} -1 - 1 \qquad \text{Simplify.}$$

$$-2 = -2 \qquad -1 \text{ checks. } \checkmark$$

So, the solutions are $x = 4$ and $x = -1$.

REMARK Notice that the values $x = 0$ and $x = 2$ are excluded because they result in division by zero in the original equation.

✓ *Checkpoint* ◀))) 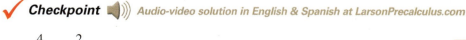 *Audio-video solution in English & Spanish at LarsonPrecalculus.com*

Solve $\dfrac{4}{x} + \dfrac{2}{x+3} = -3$ and check your solution(s). ■

An **absolute value equation** is an equation that involves one or more absolute value expressions. To solve an absolute value equation, remember that the expression inside the absolute value bars can be positive or negative. This results in *two* separate equations, each of which must be solved. For example, the equation

$$|x - 2| = 3$$

results in the two equations

$$x - 2 = 3$$

and

$$-(x - 2) = 3$$

which implies that the original equation has two solutions: $x = 5$ and $x = -1$.

▷ **ALGEBRA HELP**   To review the definition of absolute value, see Section P.1.

EXAMPLE 7   **Solving an Absolute Value Equation**

See LarsonPrecalculus.com for an interactive version of this type of example.

Solve $|x^2 - 3x| = -4x + 6$ and check your solution(s).

Solution   Solve the two equations below.

First Equation

$$x^2 - 3x = -4x + 6 \qquad \text{Use positive expression.}$$
$$x^2 + x - 6 = 0 \qquad \text{Write in general form.}$$
$$(x + 3)(x - 2) = 0 \qquad \text{Factor.}$$
$$x + 3 = 0 \implies x = -3 \qquad \text{Set 1st factor equal to 0 and solve.}$$
$$x - 2 = 0 \implies x = 2 \qquad \text{Set 2nd factor equal to 0 and solve.}$$

Second Equation

$$-(x^2 - 3x) = -4x + 6 \qquad \text{Use negative expression.}$$
$$x^2 - 7x + 6 = 0 \qquad \text{Write in general form.}$$
$$(x - 1)(x - 6) = 0 \qquad \text{Factor.}$$
$$x - 1 = 0 \implies x = 1 \qquad \text{Set 1st factor equal to 0 and solve.}$$
$$x - 6 = 0 \implies x = 6 \qquad \text{Set 2nd factor equal to 0 and solve.}$$

Check

$$|(-3)^2 - 3(-3)| \stackrel{?}{=} -4(-3) + 6 \qquad \text{Substitute } -3 \text{ for } x.$$
$$18 = 18 \qquad -3 \text{ checks. } ✔$$
$$|(2)^2 - 3(2)| \stackrel{?}{=} -4(2) + 6 \qquad \text{Substitute 2 for } x.$$
$$2 \neq -2 \qquad 2 \text{ does not check.}$$
$$|(1)^2 - 3(1)| \stackrel{?}{=} -4(1) + 6 \qquad \text{Substitute 1 for } x.$$
$$2 = 2 \qquad 1 \text{ checks. } ✔$$
$$|(6)^2 - 3(6)| \stackrel{?}{=} -4(6) + 6 \qquad \text{Substitute 6 for } x.$$
$$18 \neq -18 \qquad 6 \text{ does not check.}$$

The solutions are $x = -3$ and $x = 1$.

✓ *Checkpoint*  Audio-video solution in English & Spanish at LarsonPrecalculus.com

Solve $|x^2 + 4x| = 7x + 18$ and check your solution(s).

Applications

It would be virtually impossible to categorize all of the different types of applications that involve nonlinear and nonquadratic models. However, you will see a variety of applications in the next two examples and in the exercises.

EXAMPLE 8 **Reduced Rates**

A ski club charters a bus for a ski trip at a cost of $480. To lower the bus fare per skier, the club invites nonmembers to go on the trip. After 5 nonmembers join the trip, the fare per skier decreases by $4.80. How many club members are going on the trip?

Solution

Verbal model:

| Cost per skier | $\cdot$ | Number of skiers | $=$ | Cost of trip |

Labels:

Cost of trip $= 480$ (dollars)

Number of ski club members $= x$ (people)

Number of skiers $= x + 5$ (people)

Original cost per member $= \dfrac{480}{x}$ (dollars per person)

Cost per skier $= \dfrac{480}{x} - 4.80$ (dollars per person)

Equation:

$$\left(\frac{480}{x} - 4.80\right)(x + 5) = 480$$

$$\left(\frac{480 - 4.8x}{x}\right)(x + 5) = 480 \qquad \text{Rewrite first factor.}$$

$$(480 - 4.8x)(x + 5) = 480x, \ x \neq 0 \qquad \text{Multiply each side by } x.$$

$$480x + 2400 - 4.8x^2 - 24x = 480x \qquad \text{Multiply.}$$

$$-4.8x^2 - 24x + 2400 = 0 \qquad \text{Write in general form.}$$

$$x^2 + 5x - 500 = 0 \qquad \text{Divide each side by } -4.8.$$

$$(x + 25)(x - 20) = 0 \qquad \text{Factor.}$$

$$x + 25 = 0 \implies x = -25 \qquad \text{Set 1st factor equal to 0 and solve.}$$

$$x - 20 = 0 \implies x = 20 \qquad \text{Set 2nd factor equal to 0 and solve.}$$

Only the positive value of x makes sense in the context of the problem, so 20 ski club members are going on the trip. Check this in the original statement of the problem.

Check

$$\left(\frac{480}{20} - 4.80\right)(20 + 5) \overset{?}{=} 480 \qquad \text{Substitute 20 for } x.$$

$$(24 - 4.80)25 \overset{?}{=} 480 \qquad \text{Simplify.}$$

$$480 = 480 \qquad \text{20 checks. } \checkmark$$

✓ *Checkpoint*  *Audio-video solution in English & Spanish at LarsonPrecalculus.com*

A high school charters a bus for $560 to take a group of students to an observatory. When 8 more students join the trip, the cost per student decreases by $3.50. How many students were in the original group?

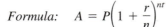

Compound Interest

When you were born, your grandparents deposited $5000 in a long-term investment in which the interest was compounded quarterly. On your 25th birthday, the value of the investment is $25,062.59. What is the annual interest rate for this investment?

Solution

Formula: $\quad A = P\left(1 + \dfrac{r}{n}\right)^{nt}$

Labels:

Balance $= A = 25{,}062.59$		(dollars)
Principal $= P = 5000$		(dollars)
Time $= t = 25$		(years)
Compoundings per year $= n = 4$		(compoundings per year)
Annual interest rate $= r$		(percent in decimal form)

Equation:

$25{,}062.59 = 5000\left(1 + \dfrac{r}{4}\right)^{4(25)}$ 　　Substitute.

$\dfrac{25{,}062.59}{5000} = \left(1 + \dfrac{r}{4}\right)^{100}$ 　　Divide each side by 5000.

$5.0125 \approx \left(1 + \dfrac{r}{4}\right)^{100}$ 　　Use a calculator.

$(5.0125)^{1/100} \approx 1 + \dfrac{r}{4}$ 　　Raise each side to reciprocal power.

$1.01625 \approx 1 + \dfrac{r}{4}$ 　　Use a calculator.

$0.01625 \approx \dfrac{r}{4}$ 　　Subtract 1 from each side.

$0.065 \approx r$ 　　Multiply each side by 4.

The annual interest rate is about 0.065, or 6.5%. Check this in the formula using $r = 0.065$ and the values of A, P, t, and n listed above.

 Checkpoint  Audio-video solution in English & Spanish at LarsonPrecalculus.com

You deposit $2500 in a long-term investment in which the interest is compounded monthly. After 5 years, the balance is $3544.06. What is the annual interest rate?

> **·REMARK** Recall that the formula for interest that is compounded *n* times per year is
>
> $$A = P\left(1 + \dfrac{r}{n}\right)^{nt}.$$
>
> In this formula, A is the balance in the account, P is the principal (or original deposit), r is the annual interest rate (in decimal form), n is the number of compoundings per year, and t is the time in years.

With compound interest, it is beneficial to begin saving for retirement as soon as possible. Use the formula above to verify that investing $10,000 at 5.5% compounded quarterly yields $123,389.89 after 46 years, but only $60,656.10 after 33 years.

Summarize　(Section 1.6)

1. Explain how to solve a polynomial equation of degree three or greater by factoring *(page 121)*. For examples of solving polynomial equations by factoring, see Examples 1–3.

2. Explain how to solve a radical equation *(page 123)*. For an example of solving radical equations, see Example 4.

3. Explain how to solve rational equations and absolute value equations *(pages 124 and 125)*. For examples of solving these types of equations, see Examples 6 and 7.

4. Describe real-life applications that use nonlinear and nonquadratic models *(pages 126 and 127, Examples 8 and 9)*.

1.6 Exercises

See CalcChat.com for tutorial help and worked-out solutions to odd-numbered exercises.

Vocabulary: Fill in the blanks.

1. The general form of a _____ equation in x is $a_n x^n + a_{n-1} x^{n-1} + \cdots + a_2 x^2 + a_1 x + a_0 = 0$.

2. An equation is of _____ _____ when it can be written in the form $au^2 + bu + c = 0$.

3. A _____ equation is an equation that involves one or more radical expressions.

4. An _____ _____ equation is an equation that involves one or more absolute value expressions.

Skills and Applications

 Solving a Polynomial Equation In Exercises 5–16, solve the equation. Check your solutions.

5. $6x^4 - 54x^2 = 0$ 6. $36x^3 - 100x = 0$

7. $5x^3 + 30x^2 + 45x = 0$ 8. $9x^4 - 24x^3 + 16x^2 = 0$

9. $x^4 - 81 = 0$ 10. $x^6 - 64 = 0$

11. $x^3 + 512 = 0$ 12. $27x^3 - 343 = 0$

13. $x^3 - 3x^2 - x = -3$ 14. $x^3 + 2x^2 + 3x = -6$

15. $x^4 - x^3 + x = 1$ 16. $x^4 + 2x^3 - 8x = 16$

 Solving an Equation of Quadratic Type In Exercises 17–30, solve the equation. Check your solutions.

17. $x^4 - 4x^2 + 3 = 0$ 18. $x^4 - 13x^2 + 36 = 0$

19. $4x^4 - 65x^2 + 16 = 0$ 20. $36t^4 + 29t^2 - 7 = 0$

21. $x^6 + 7x^3 - 8 = 0$ 22. $x^6 + 3x^3 + 2 = 0$

23. $\dfrac{1}{x^2} + \dfrac{8}{x} + 15 = 0$ 24. $1 + \dfrac{3}{x} = -\dfrac{2}{x^2}$

25. $2\left(\dfrac{x}{x+2}\right)^2 - 3\left(\dfrac{x}{x+2}\right) - 2 = 0$

26. $6\left(\dfrac{x}{x+1}\right)^2 + 5\left(\dfrac{x}{x+1}\right) - 6 = 0$

27. $2x + 9\sqrt{x} = 5$ 28. $6x - 7\sqrt{x} - 3 = 0$

29. $9t^{2/3} + 24t^{1/3} = -16$ 30. $3x^{1/3} + 2x^{2/3} = 5$

 Solving a Radical Equation In Exercises 31–44, solve the equation, if possible. Check your solutions.

31. $\sqrt{5x} - 10 = 0$ 32. $6 - 2\sqrt{x} = 0$

33. $\sqrt{x+8} - 5 = 0$ 34. $\sqrt{3x+1} = 7$

35. $4 + \sqrt[3]{2x-9} = 0$ 36. $\sqrt[3]{12-x} - 3 = 0$

37. $\sqrt{x+8} = 2 + x$ 38. $2x = \sqrt{-5x+24} - 3$

39. $\sqrt{x-3} + 1 = \sqrt{x}$ 40. $\sqrt{x} + \sqrt{x-24} = 2$

41. $2\sqrt{x+1} - \sqrt{2x+3} = 1$

42. $4\sqrt{x-3} - \sqrt{6x-17} = 3$

43. $\sqrt{4\sqrt{4x+9}} = \sqrt{8x+2}$

44. $\sqrt{16 + 9\sqrt{x}} = 4 + \sqrt{x}$

 Solving an Equation Involving a Rational Exponent In Exercises 45–50, solve the equation. Check your solutions.

45. $(x-5)^{3/2} = 8$ 46. $(x+2)^{2/3} = 9$

47. $(x^2-5)^{3/2} = 27$ 48. $(x^2 - x - 22)^{3/2} = 27$

49. $3x(x-1)^{1/2} + 2(x-1)^{3/2} = 0$

50. $4x^2(x-1)^{1/3} + 6x(x-1)^{4/3} = 0$

 Solving a Rational Equation In Exercises 51–58, solve the equation. Check your solutions.

51. $x = \dfrac{3}{x} + \dfrac{1}{2}$ 52. $\dfrac{4}{x} - \dfrac{5}{3} = \dfrac{x}{6}$

53. $\dfrac{1}{x} - \dfrac{1}{x+1} = 3$ 54. $\dfrac{4}{x+1} - \dfrac{3}{x+2} = 1$

55. $3 - \dfrac{14}{x} - \dfrac{5}{x^2} = 0$ 56. $5 = \dfrac{18}{x} + \dfrac{8}{x^2}$

57. $\dfrac{x}{x^2-4} + \dfrac{1}{x+2} = 3$ 58. $\dfrac{x+1}{3} - \dfrac{x+1}{x+2} = 0$

 Solving an Absolute Value Equation In Exercises 59–64, solve the equation. Check your solutions.

59. $|2x - 5| = 11$ 60. $|3x + 2| = 7$

61. $|x| = x^2 + x - 24$ 62. $|x^2 + 6x| = 3x + 18$

63. $|x + 1| = x^2 - 5$ 64. $|x - 15| = x^2 - 15x$

Using Technology In Exercises 65–74, (a) use a graphing utility to graph the equation, (b) use the graph to approximate any x-intercepts, (c) set $y = 0$ and solve the resulting equation, and (d) compare the result of part (c) with the x-intercept(s) of the graph.

65. $y = x^3 - 2x^2 - 3x$ 66. $y = 2x^4 - 15x^3 + 18x^2$

67. $y = x^4 - 10x^2 + 9$ 68. $y = x^4 - 29x^2 + 100$

69. $y = \sqrt{11x - 30} - x$ 70. $y = 2x - \sqrt{15 - 4x}$

71. $y = \dfrac{1}{x} - \dfrac{4}{x-1} - 1$ 72. $y = x + \dfrac{9}{x+1} - 5$

73. $y = |x + 1| - 2$ 74. $y = |x - 2| - 3$

Solving an Equation   In Exercises 75–82, find the real solution(s) of the equation. (Round your answer(s) to three decimal places.)

75. $3.2x^4 - 1.5x^2 = 2.1$

76. $0.1x^4 - 2.4x^2 = 3.6$

77. $7.08x^6 + 4.15x^3 = 9.6$

78. $5.25x^6 - 0.2x^3 = 1.55$

79. $1.8x - 6\sqrt{x} = 5.6$

80. $5.3x + 3.1 = 9.8\sqrt{x}$

81. $4x^{2/3} + 8x^{1/3} = -3.6$

82. $8.4x^{2/3} - 1.2x^{1/3} = 24$

 Think About It   In Exercises 83–92, write an equation that has the given solutions. (There are many correct answers.)

83. $-4, 7$

84. $0, 2, 9$

85. $-\frac{7}{3}, \frac{6}{7}$

86. $-\frac{1}{8}, -\frac{4}{5}$

87. $\sqrt{3}, -\sqrt{3}, 4$

88. $2\sqrt{7}, -\sqrt{7}$

89. $i, -i$

90. $2i, -2i$

91. $-1, 1, i, -i$

92. $4i, -4i, 6, -6$

93. **Reduced Rates**   A college charters a bus for $1700 to take a group of students to see a Broadway production. When 6 more students join the trip, the cost per student decreases by $7.50. How many students were in the original group?

94. **Reduced Rates**   Three students plan to divide the rent for an apartment equally. When they add a fourth student, the cost per student decreases by $150 per month. What is the total monthly rent for the apartment?

95. **Average Speed**   An airline runs a commuter flight between Portland, Oregon, and Seattle, Washington, which are 145 miles apart. An increase of 40 miles per hour in the average speed of the plane decreases the travel time by 12 minutes. What initial average speed results in this decrease in travel time?

96. **Average Speed**   A family drives 1080 miles to a vacation lodge. On the return trip, it takes the family $2\frac{1}{2}$ hours longer, traveling at an average speed that is 6 miles per hour slower, to drive the same distance. Determine the average speed on the way to the lodge.

97. **Compound Interest**   You deposit $2500 in a long-term investment fund in which the interest is compounded monthly. After 5 years, the balance is $2694.58. What is the annual interest rate?

98. **Compound Interest**   You deposit $6000 in a long-term investment fund in which the interest is compounded quarterly. After 5 years, the balance is $7734.27. What is the annual interest rate?

99. **Airline Passengers**   An airline offers daily flights between Chicago and Denver. The total monthly cost C (in millions of dollars) of these flights can be modeled by

$$C = \sqrt{0.2x + 1}$$

where x is the number of passengers (in thousands). The total cost of the flights for June is 2.5 million dollars. How many passengers flew in June?

100. **Nursing**   The number of registered nursing graduates N (in thousands) in the United States from 2003 to 2014 can be modeled by

$$N = \sqrt{734.024 + 1839.666t}, \quad 3 \le t \le 14$$

where t represents the year, with $t = 3$ corresponding to 2003.   *(Source: National Council of State Boards of Nursing)*

(a) In which year did the number of registered nursing graduates reach 135,000?

(b) Use the model to predict when the number of registered nursing graduates will reach 200,000. Is this prediction reasonable? Explain.

101. **Saturated Steam**

The temperature T (in degrees Fahrenheit) of saturated steam increases as pressure increases. This relationship can be approximated by the model

$$T = 75.82 - 2.11x + 43.51\sqrt{x}$$

$$5 \le x \le 40$$

where x is the absolute pressure (in pounds per square inch).

(a) The temperature of saturated steam at sea level is 212°F. Find the absolute pressure at this temperature.

(b) Use a graphing utility to verify your solution for part (a).

102. **Voting-Age Population**   The total voting-age population P (in millions) in the United States from 1990 through 2014 can be modeled by

$$P = \frac{184.64 + 0.7524t^2}{1 + 0.0028t^2}, \quad 0 \le t \le 24$$

where t represents the year, with $t = 0$ corresponding to 1990.   *(Source: U.S. Census Bureau)*

(a) In which year did the total voting-age population reach 210 million?

(b) Use the model to predict when the total voting-age population will reach 260 million. Is this prediction reasonable? Explain.

103. Demand The demand for a video game can be modeled by

$$p = 40 - \sqrt{0.01x + 1}$$

where x is the number of units demanded per day and p is the price per unit. Find the demand when the price is \$37.55.

104. Power Line A power station is on one side of a river that is $\frac{3}{4}$ mile wide, and a factory is 8 miles downstream on the other side of the river. It costs \$24 per foot to run power lines over land and \$30 per foot to run them under water. The project's cost is \$1,098,662.40. Find the length x labeled in the figure.

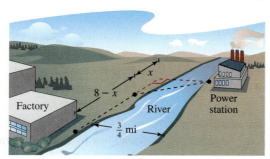

105. Tiling a Floor Working alone, you can tile a floor in t hours. When you work with a friend, the time y (in hours) it takes to tile the floor satisfies the equation

$$\frac{1}{t} + \frac{1}{t + 3} = \frac{1}{y}.$$

Find the time it takes you to tile the floor working alone when you and your friend can tile the floor in 2 hours working together.

106. Painting a Fence Working alone, you can paint a fence in t hours. Working with a friend, the time y (in hours) it takes to paint the fence satisfies the equation

$$\frac{1}{t} + \frac{1}{t + 2} = \frac{1}{y}.$$

Find the time it takes you to paint the fence working alone when you and your friend can paint the fence in 3 hours working together.

107. Potential Energy of a Spring The distance d a spring is stretched can be calculated using the formula

$$d = \sqrt{\frac{2U}{k}}$$

where U is the potential energy and k is the spring constant. Solve the formula for U.

108. Circumference of an Ellipse The circumference C of an ellipse with major axis length $2a$ and minor axis length $2b$ can be approximated using the formula below. Solve the formula for a.

$$C \approx 2\pi \sqrt{\frac{a^2 + b^2}{2}}$$

Exploration

True or False? **In Exercises 109 and 110, determine whether the statement is true or false. Justify your answer.**

109. An equation can never have more than one extraneous solution.

110. The equation $\sqrt{x + 10} - \sqrt{x - 10} = 0$ has no solution.

111. Distance Find x such that the distance between $(3, -5)$ and $(x, 7)$ is 13.

112. Distance Find y such that the distance between $(10, y)$ and $(4, -3)$ is 10.

113. Error Analysis Describe the error(s).

$$\sqrt{3}x = \sqrt{7x + 4}$$
$$3x^2 = 7x + 4$$
$$x = \frac{-7 \pm \sqrt{7^2 - 4(3)(4)}}{2(3)}$$
$$x = -1 \text{ and } x = -\frac{4}{3}$$

114. **HOW DO YOU SEE IT?** The figure shows a glass cube partially filled with water.

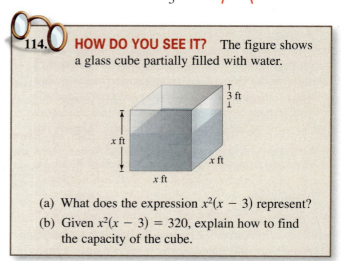

(a) What does the expression $x^2(x - 3)$ represent?

(b) Given $x^2(x - 3) = 320$, explain how to find the capacity of the cube.

Solving an Absolute Value Equation **In Exercises 115 and 116, consider an equation of the form $x + |x - a| = b$, where a and b are constants.**

115. Find a and b when the solution of the equation is $x = 9$. (There are many correct answers.)

116. Write a short paragraph listing the steps required to solve this absolute value equation and explain why it is important to check your solutions.

Solving a Radical Equation **In Exercises 117 and 118, consider an equation of the form $x + \sqrt{x - a} = b$, where a and b are constants.**

117. Find a and b when the solution of the equation is $x = 20$. (There are many correct answers.)

118. Write a short paragraph listing the steps required to solve this radical equation and explain why it is important to check your solutions.

1.7 Linear Inequalities in One Variable

Linear inequalities have many real-life applications. For example, in Exercise 104 on page 139, you will use an absolute value inequality to describe the distance between two locations.

■ **Represent solutions of linear inequalities in one variable.**
■ **Use properties of inequalities to write equivalent inequalities.**
■ **Solve linear inequalities in one variable.**
■ **Solve absolute value inequalities.**
■ **Use linear inequalities to model and solve real-life problems.**

Introduction

Simple inequalities were discussed in Section P.1. There, the inequality symbols $<$, $\le$, $>$, and $\ge$ were used to compare two numbers and to denote subsets of real numbers. For example, the simple inequality

$$x \ge 3$$

denotes all real numbers x that are greater than or equal to 3.

Now, you will expand your work with inequalities to include more involved statements such as

$$5x - 7 < 3x + 9 \quad \text{and} \quad -3 \le 6x - 1 < 3.$$

As with an equation, you **solve an inequality** in the variable x by finding all values of x for which the inequality is true. Such values are **solutions** that **satisfy** the inequality. The set of all real numbers that are solutions of an inequality is the **solution set** of the inequality. For example, the solution set of

$$x + 1 < 4$$

is all real numbers that are less than 3.

The set of all points on the real number line that represents the solution set is the **graph of the inequality.** Graphs of many types of inequalities consist of intervals on the real number line. See Section P.1 to review the nine basic types of intervals on the real number line. Note that each type of interval can be classified as *bounded* or *unbounded*.

EXAMPLE 1   Intervals and Inequalities

Write an inequality that represents each interval. Then state whether the interval is bounded or unbounded.

a. $(-3, 5]$ **b.** $(-3, \infty)$

c. $[0, 2]$ **d.** $(-\infty, \infty)$

Solution

a. $(-3, 5]$ corresponds to $-3 < x \le 5$.   Bounded

b. $(-3, \infty)$ corresponds to $x > -3$.   Unbounded

c. $[0, 2]$ corresponds to $0 \le x \le 2$.   Bounded

d. $(-\infty, \infty)$ corresponds to $-\infty < x < \infty$.   Unbounded

✓ *Checkpoint* 🔊))) *Audio-video solution in English & Spanish at LarsonPrecalculus.com*

Write an inequality that represents each interval. Then state whether the interval is bounded or unbounded.

a. $[-1, 3]$ **b.** $(-1, 6)$

c. $(-\infty, 4)$ **d.** $[0, \infty)$

Properties of Inequalities

The procedures for solving linear inequalities in one variable are similar to those for solving linear equations. To isolate the variable, use the **properties of inequalities.** These properties are similar to the properties of equality, but there are two important exceptions. When you multiply or divide each side of an inequality by a negative number, you must reverse the direction of the inequality symbol. Here is an example.

$$-2 < 5 \qquad \text{Original inequality}$$
$$(-3)(-2) > (-3)(5) \qquad \text{Multiply each side by } -3 \text{ and reverse inequality symbol.}$$
$$6 > -15 \qquad \text{Simplify.}$$

Notice that when you do not reverse the inequality symbol in the example above, you obtain the false statement

$$6 < -15. \qquad \text{False statement}$$

Two inequalities that have the same solution set are **equivalent.** For example, the inequalities

$$x + 2 < 5$$

and

$$x < 3$$

are equivalent. To obtain the second inequality from the first, subtract 2 from each side of the inequality. The list below describes operations used to write equivalent inequalities.

Properties of Inequalities

Let a, b, c, and d be real numbers.

1. Transitive Property

$$a < b \text{ and } b < c \implies a < c$$

2. Addition of Inequalities

$$a < b \text{ and } c < d \implies a + c < b + d$$

3. Addition of a Constant

$$a < b \implies a + c < b + c$$

4. Multiplication by a Constant

$$\text{For } c > 0, a < b \implies ac < bc$$
$$\text{For } c < 0, a < b \implies ac > bc \qquad \text{Reverse the inequality symbol.}$$

Each of the properties above is true when you replace the symbol $<$ with $\leq$ and you replace the symbol $>$ with $\geq$. For example, another form of the multiplication property is shown below.

$$\text{For } c > 0, a \leq b \implies ac \leq bc$$
$$\text{For } c < 0, a \leq b \implies ac \geq bc$$

On your own, verify each of the properties of inequalities by using several examples with real numbers.

Solving Linear Inequalities in One Variable

The simplest type of inequality to solve is a **linear inequality** in one variable. For example, $2x + 3 > 4$ is a linear inequality in x.

> **EXAMPLE 2** **Solving a Linear Inequality**

Solve $5x - 7 > 3x + 9$. Then graph the solution set.

Solution

$$5x - 7 > 3x + 9 \qquad \text{Write original inequality.}$$
$$2x - 7 > 9 \qquad \text{Subtract } 3x \text{ from each side.}$$
$$2x > 16 \qquad \text{Add 7 to each side.}$$
$$x > 8 \qquad \text{Divide each side by 2.}$$

The solution set is all real numbers that are greater than 8, denoted by $(8, \infty)$. The graph of this solution set is shown below. Note that a parenthesis at 8 on the real number line indicates that 8 *is not* part of the solution set.

Solution interval: $(8, \infty)$

✓ **Checkpoint** 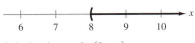 *Audio-video solution in English & Spanish at LarsonPrecalculus.com*

Solve $7x - 3 \le 2x + 7$. Then graph the solution set.

> **EXAMPLE 3** **Solving a Linear Inequality**

See LarsonPrecalculus.com for an interactive version of this type of example.

Solve $1 - \frac{3}{2}x \ge x - 4$.

Algebraic Solution

$$1 - \frac{3}{2}x \ge x - 4 \qquad \text{Write original inequality.}$$
$$2 - 3x \ge 2x - 8 \qquad \text{Multiply each side by 2.}$$
$$2 - 5x \ge -8 \qquad \text{Subtract } 2x \text{ from each side.}$$
$$-5x \ge -10 \qquad \text{Subtract 2 from each side.}$$
$$x \le 2 \qquad \text{Divide each side by } -5 \text{ and reverse the inequality symbol.}$$

The solution set is all real numbers that are less than or equal to 2, denoted by $(-\infty, 2]$. The graph of this solution set is shown below. Note that a bracket at 2 on the real number line indicates that 2 *is* part of the solution set.

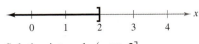

Solution interval: $(-\infty, 2]$

Graphical Solution

Use a graphing utility to graph $y_1 = 1 - \frac{3}{2}x$ and $y_2 = x - 4$ in the same viewing window.

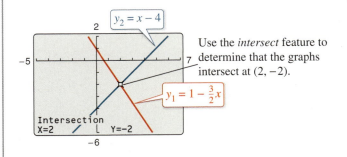

Use the *intersect* feature to determine that the graphs intersect at $(2, -2)$.

The graph of y_1 lies above the graph of y_2 to the left of their point of intersection, which implies that $y_1 \ge y_2$ for all $x \le 2$.

✓ **Checkpoint**   *Audio-video solution in English & Spanish at LarsonPrecalculus.com*

Solve $2 - \frac{5}{3}x > x - 6$ (a) algebraically and (b) graphically.

Sometimes it is possible to write two inequalities as a **double inequality.** For example, you can write the two inequalities

$$-4 \le 5x - 2$$

and

$$5x - 2 < 7$$

as

$$-4 \le 5x - 2 < 7. \qquad \text{Double inequality}$$

This form allows you to solve the two inequalities together, as demonstrated in Example 4.

EXAMPLE 4 **Solving a Double Inequality**

Solve $-3 \le 6x - 1 < 3$. Then graph the solution set.

Solution One way to solve this double inequality is to isolate x as the middle term.

$-3 \le 6x - 1 < 3$	Write original inequality.
$-3 + 1 \le 6x - 1 + 1 < 3 + 1$	Add 1 to each part.
$-2 \le 6x < 4$	Simplify.
$\dfrac{-2}{6} \le \dfrac{6x}{6} < \dfrac{4}{6}$	Divide each part by 6.
$-\dfrac{1}{3} \le x < \dfrac{2}{3}$	Simplify.

The solution set is all real numbers that are greater than or equal to $-\frac{1}{3}$ and less than $\frac{2}{3}$, denoted by $\left[-\frac{1}{3}, \frac{2}{3}\right)$. The graph of this solution set is shown below.

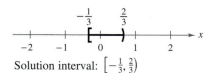

Solution interval: $\left[-\frac{1}{3}, \frac{2}{3}\right)$

 ✔ *Checkpoint* ◀))) *Audio-video solution in English & Spanish at LarsonPrecalculus.com*

Solve $1 < 2x + 7 < 11$. Then graph the solution set.

Another way to solve the double inequality in Example 4 is to solve it in two parts.

$-3 \le 6x - 1$	and	$6x - 1 < 3$
$-2 \le 6x$		$6x < 4$
$-\dfrac{1}{3} \le x$		$x < \dfrac{2}{3}$

The solution set consists of all real numbers that satisfy *both* inequalities. In other words, the solution set is the set of all values of x for which

$$-\frac{1}{3} \le x < \frac{2}{3}.$$

When combining two inequalities to form a double inequality, be sure that the inequalities satisfy the Transitive Property. For example, it is *incorrect* to combine the inequalities $3 < x$ and $x \le -1$ as $3 < x \le -1$. This "inequality" is wrong because 3 is not less than -1.

Absolute Value Inequalities

▷ **TECHNOLOGY** A graphing
utility can help you identify the
solution set of an inequality.
For instance, to find the
solution set of $|x - 5| < 2$
(see Example 5a), rewrite the
inequality as $|x - 5| - 2 < 0$,
enter

$$Y1 = \text{abs}(X - 5) - 2$$

and press the *graph* key. The
graph should resemble the one
shown below.

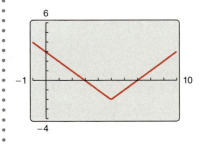

Notice that the graph is below
the x-axis on the interval $(3, 7)$.

> **Solving an Absolute Value Inequality**
>
> Let u be an algebraic expression and let a be a real number such that $a > 0$.
>
> **1.** $|u| < a$ if and only if $-a < u < a$.
>
> **2.** $|u| \le a$ if and only if $-a \le u \le a$.
>
> **3.** $|u| > a$ if and only if $u < -a$ or $u > a$.
>
> **4.** $|u| \ge a$ if and only if $u \le -a$ or $u \ge a$.

EXAMPLE 5   **Solving Absolute Value Inequalities**

Solve each inequality. Then graph the solution set.

a. $|x - 5| < 2$ **b.** $|x + 3| \ge 7$

Solution

a. $|x - 5| < 2$ Write original inequality.

    $-2 < x - 5 < 2$ Write related double inequality.

   $-2 + 5 < x - 5 + 5 < 2 + 5$ Add 5 to each part.

       $3 < x < 7$ Simplify.

The solution set is all real numbers that are greater than 3 and less than 7, denoted
by $(3, 7)$. The graph of this solution set is shown below. Note that the graph of the
inequality can be described as all real numbers less than two units from 5.

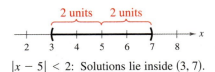

$|x - 5| < 2$: Solutions lie inside $(3, 7)$.

b. $|x + 3| \ge 7$ Write original inequality.

    $x + 3 \le -7$ or $x + 3 \ge 7$ Write related inequalities.

$x + 3 - 3 \le -7 - 3$ $x + 3 - 3 \ge 7 - 3$ Subtract 3 from each side.

      $x \le -10$ $x \ge 4$ Simplify.

The solution set is all real numbers that are less than or equal to -10 *or* greater than
or equal to 4, denoted by $(-\infty, -10] \cup [4, \infty)$. The symbol $\cup$ is the *union* symbol,
which denotes the combining of two sets. The graph of this solution set is shown
below. Note that the graph of the inequality can be described as all real numbers at
least seven units from -3.

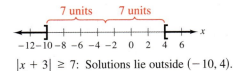

$|x + 3| \ge 7$: Solutions lie outside $(-10, 4)$.

✓ *Checkpoint* 🔊))) *Audio-video solution in English & Spanish at LarsonPrecalculus.com*

Solve $|x - 20| \le 4$. Then graph the solution set.

Applications

EXAMPLE 6   **Comparative Shopping**

A car sharing company offers two plans, as shown in Figure 1.13. How many hours must you use a car in one month for plan B to cost more than plan A?

Solution   Let h represent the number of hours you use the car. Write and solve an inequality.

$$10.25h + 8 > 8.75h + 50$$
$$1.5h > 42$$
$$h > 28$$

Plan B costs more when you use the car for more than 28 hours in one month.

 ✓ **Checkpoint** *Audio-video solution in English & Spanish at LarsonPrecalculus.com*

Rework Example 6 when plan A costs $25 per month plus $10 per hour.

Car Sharing Company

Plan A:
$50.00 per month
plus $8.75 per hour

Plan B:
$8.00 per month
plus $10.25 per hour

Figure 1.13

EXAMPLE 7   **Accuracy of a Measurement**

You buy a bag of chocolates that cost $9.89 per pound. The scale used to weigh the bag is accurate to within $\frac{1}{32}$ pound. According to the scale, the bag weighs $\frac{1}{2}$ pound and costs $4.95. How much might you have been undercharged or overcharged?

Solution   Let x represent the actual weight of the bag. The difference of the actual weight and the weight shown on the scale is at most $\frac{1}{32}$ pound. That is, $\left| x - \frac{1}{2} \right| \le \frac{1}{32}$. Solve this inequality.

$$-\frac{1}{32} \le x - \frac{1}{2} \le \frac{1}{32}$$
$$\frac{15}{32} \le x \le \frac{17}{32}$$

The least the bag can weigh is $\frac{15}{32}$ pound, which would have cost $4.64. The most the bag can weigh is $\frac{17}{32}$ pound, which would have cost $5.25. So, you might have been overcharged by as much as $0.31 or undercharged by as much as $0.30.

✓ **Checkpoint** *Audio-video solution in English & Spanish at LarsonPrecalculus.com*

Rework Example 7 when the scale is accurate to within $\frac{1}{64}$ pound.

Summarize   (Section 1.7)

1. Explain how to use inequalities to represent intervals *(page 131)*. For an example of writing inequalities that represent intervals, see Example 1.

2. State the properties of inequalities *(page 132)*.

3. Explain how to solve a linear inequality in one variable *(page 133)*. For examples of solving linear inequalities in one variable, see Examples 2–4.

4. Explain how to solve an absolute value inequality *(page 135)*. For an example of solving absolute value inequalities, see Example 5.

5. Describe real-life applications of linear inequalities in one variable *(page 136, Examples 6 and 7)*.

1.7 Exercises

See CalcChat.com for tutorial help and worked-out solutions to odd-numbered exercises.

Vocabulary: Fill in the blanks.

1. The set of all real numbers that are solutions of an inequality is the _____ _____ of the inequality.

2. The set of all points on the real number line that represents the solution set of an inequality is the _____ of the inequality.

3. It is sometimes possible to write two inequalities as a _____ inequality.

4. The symbol $\cup$ is the _____ symbol, which denotes the combining of two sets.

Skills and Applications

 Intervals and Inequalities In Exercises 5–12, write an inequality that represents the interval. Then state whether the interval is bounded or unbounded.

5. $[-2, 6)$ 6. $(-7, 4)$

7. $[-1, 5]$ 8. $(2, 10]$

9. $(11, \infty)$ 10. $[-5, \infty)$

11. $(-\infty, -2)$ 12. $(-\infty, 7]$

 Solving a Linear Inequality In Exercises 13–30, solve the inequality. Then graph the solution set.

13. $4x < 12$ 14. $10x < -40$

15. $-2x > -3$ 16. $-6x > 15$

17. $x - 5 \geq 7$ 18. $x + 7 \leq 12$

19. $2x + 7 < 3 + 4x$ 20. $3x + 1 \geq 2 + x$

21. $3x - 4 \geq 4 - 5x$ 22. $6x - 4 \leq 2 + 8x$

23. $4 - 2x < 3(3 - x)$ 24. $4(x + 1) < 2x + 3$

25. $\frac{3}{4}x - 6 \leq x - 7$ 26. $3 + \frac{2}{7}x > x - 2$

27. $\frac{1}{2}(8x + 1) \geq 3x + \frac{5}{2}$ 28. $9x - 1 < \frac{3}{4}(16x - 2)$

29. $3.6x + 11 \geq -3.4$ 30. $15.6 - 1.3x < -5.2$

 Solving a Double Inequality In Exercises 31–42, solve the inequality. Then graph the solution set.

31. $1 < 2x + 3 < 9$ 32. $-9 \leq -2x - 7 < 5$

33. $0 < 3(x + 7) \leq 20$ 34. $-1 \leq -(x - 4) < 7$

35. $-4 < \frac{2x - 3}{3} < 4$ 36. $0 \leq \frac{x + 3}{2} < 5$

37. $-1 < \frac{-x - 2}{3} \leq 1$ 38. $-1 \leq \frac{-3x + 5}{7} \leq 2$

39. $\frac{3}{4} > x + 1 > \frac{1}{4}$

40. $-1 < 2 - \frac{x}{3} < 1$

41. $3.2 \leq 0.4x - 1 \leq 4.4$

42. $1.6 < 0.3x + 1 < 2.8$

 Solving an Absolute Value Inequality In Exercises 43–58, solve the inequality. Then graph the solution set. (Some inequalities have no solution.)

43. $|x| < 5$ 44. $|x| \geq 8$

45. $\left|\frac{x}{2}\right| > 1$ 46. $\left|\frac{x}{3}\right| < 2$

47. $|x - 5| < -1$ 48. $|x - 7| < -5$

49. $|x - 20| \leq 6$ 50. $|x - 8| \geq 0$

51. $|7 - 2x| \geq 9$ 52. $|1 - 2x| < 5$

53. $\left|\frac{x - 3}{2}\right| \geq 4$ 54. $\left|1 - \frac{2x}{3}\right| < 1$

55. $|9 - 2x| - 2 < -1$ 56. $|x + 14| + 3 > 17$

57. $2|x + 10| \geq 9$ 58. $3|4 - 5x| \leq 9$

Using Technology In Exercises 59–68, use a graphing utility to graph the inequality and identify the solution set.

59. $7x > 21$ 60. $-4x \leq 9$

61. $8 - 3x \geq 2$ 62. $20 < 6x - 1$

63. $4(x - 3) \leq 8 - x$ 64. $3(x + 1) < x + 7$

65. $|x - 8| \leq 14$ 66. $|2x + 9| > 13$

67. $2|x + 7| \geq 13$ 68. $\frac{1}{2}|x + 1| \leq 3$

Using Technology In Exercises 69–74, use a graphing utility to graph the equation. Use the graph to approximate the values of x that satisfy each inequality.

Equation	Inequalities	
69. $y = 3x - 1$	(a) $y \geq 2$	(b) $y \leq 0$
70. $y = \frac{2}{3}x + 1$	(a) $y \leq 5$	(b) $y \geq 0$
71. $y = -\frac{1}{2}x + 2$	(a) $0 \leq y \leq 3$	(b) $y \geq 0$
72. $y = -3x + 8$	(a) $-1 \leq y \leq 3$	(b) $y \leq 0$
73. $y = |x - 3|$	(a) $y \leq 2$	(b) $y \geq 4$
74. $y = \left|\frac{1}{2}x + 1\right|$	(a) $y \leq 4$	(b) $y \geq 1$

75. **Think About It** The graph of $|x - 5| < 3$ can be described as all real numbers less than three units from 5. Give a similar description of $|x - 10| < 8$.

76. Think About It The graph of $|x - 2| > 5$ can be described as all real numbers more than five units from 2. Give a similar description of $|x - 8| > 4$.

Using Absolute Value In Exercises 77–84, use absolute value notation to define the interval (or pair of intervals) on the real number line.

77.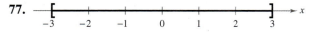

78. (number line with open parenthesis at -2 going left, open parenthesis at 2 going right)

79. (number line from 4 to 14, arrow left from 5, bracket at 10)

80. (number line from -7 to 3, bracket from -5 to 3)

81. All real numbers at least three units from 7

82. All real numbers more than five units from 8

83. All real numbers less than four units from -3

84. All real numbers no more than seven units from -6

Writing an Inequality In Exercises 85–88, write an inequality to describe the situation.

85. During a trading day, the price P of a stock is no less than $7.25 and no more than $7.75.

86. During a month, a person's weight w is greater than 180 pounds but less than 185.5 pounds.

87. The expected return r on an investment is no more than 8%.

88. The expected net income I of a company is no less than $239 million.

Physiology One formula that relates a person's maximum heart rate r (in beats per minute) to the person's age A (in years) is

$$r = 220 - A.$$

In Exercises 89 and 90, determine the interval in which the person's heart rate is from 50% to 85% of the maximum heart rate. *(Source: American Heart Association)*

89. a 20-year-old

90. a 40-year-old

91. Job Offers You are considering two job offers. The first job pays $13.50 per hour. The second job pays $9.00 per hour plus $0.75 per unit produced per hour. How many units must you produce per hour for the second job to pay more per hour than the first job?

92. Job Offers You are considering two job offers. The first job pays $13.75 per hour. The second job pays $10.00 per hour plus $1.25 per unit produced per hour. How many units must you produce per hour for the second job to pay more than the first job?

93. Investment What annual interest rates yield a balance of more than $2000 on a 10-year investment of $1000? $[A = P(1 + rt)]$

94. Investment What annual interest rates yield a balance of more than $750 on a 5-year investment of $500? $[A = P(1 + rt)]$

95. Cost, Revenue, and Profit The revenue from selling x units of a product is $R = 115.95x$. The cost of producing x units is $C = 95x + 750$. To obtain a profit, the revenue must be greater than the cost. For what values of x does this product return a profit?

96. Cost, Revenue, and Profit The revenue from selling x units of a product is $R = 24.55x$. The cost of producing x units is $C = 15.4x + 150,000$. To obtain a profit, the revenue must be greater than the cost. For what values of x does this product return a profit?

97. Daily Sales A doughnut shop sells a dozen doughnuts for $7.95. Beyond the fixed costs (rent, utilities, and insurance) of $165 per day, it costs $1.45 for enough materials and labor to produce a dozen doughnuts. The daily profit from doughnut sales varies between $400 and $1200. Between what levels (in dozens of doughnuts) do the daily sales vary?

98. Weight Loss Program A person enrolls in a diet and exercise program that guarantees a loss of at least $1\frac{1}{2}$ pounds per week. The person's weight at the beginning of the program is 164 pounds. Find the maximum number of weeks before the person attains a goal weight of 128 pounds.

99. GPA An equation that relates the college grade-point averages y and high school grade-point averages x of the students at a college is $y = 0.692x + 0.988$.

(a) Use a graphing utility to graph the model.

(b) Use the graph to estimate the values of x that predict a college grade-point average of at least 3.0.

(c) Verify your estimate from part (b) algebraically.

(d) List other factors that may influence college GPA.

100. Weightlifting The 6RM load for a weightlifting exercise is the maximum weight at which a person can perform six repetitions. An equation that relates an athlete's 6RM bench press load x (in kilograms) and the athlete's 6RM barbell curl load y (in kilograms) is $y = 0.33x + 6.20$. *(Source: Journal of Sports Science & Medicine)*

(a) Use a graphing utility to graph the model.

(b) Use the graph to estimate the values of x that predict a 6RM barbell curl load of no more than 80 kilograms.

(c) Verify your estimate from part (b) algebraically.

(d) List other factors that may influence an athlete's 6RM barbell curl load.

101. Chemists' Wages   The mean hourly wage W (in dollars) of chemists in the United States from 2000 through 2014 can be modeled by

$$W = 0.903t + 26.08, \quad 0 \le t \le 14$$

where t represents the year, with $t = 0$ corresponding to 2000.   *(Source: U.S. Bureau of Labor Statistics)*

(a) According to the model, when was the mean hourly wage at least $30, but no more than $32?

(b) Use the model to predict when the mean hourly wage will exceed $45.

102. Milk Production   Milk production M (in billions of pounds) in the United States from 2000 through 2014 can be modeled by

$$M = 3.00t + 163.3, \quad 0 \le t \le 14$$

where t represents the year, with $t = 0$ corresponding to 2000.   *(Source: U.S. Department of Agriculture)*

(a) According to the model, when was the annual milk production greater than 180 billion pounds, but no more than 190 billion pounds?

(b) Use the model to predict when milk production will exceed 230 billion pounds.

103. Time Study   The times required to perform a task in a manufacturing process by approximately two-thirds of the workers in a study satisfy the inequality

$$|t - 15.6| \le 1.9$$

where t is time in minutes. Determine the interval in which these times lie.

104. Geography

A geographic information system reports that the distance between two locations is 206 meters. The system is accurate to within 3 meters.

(a) Write an absolute value inequality for the possible distances between the locations.

(b) Graph the solution set.

105. Accuracy of Measurement   You buy 6 T-bone steaks that cost $8.99 per pound. The weight that is listed on the package is 5.72 pounds. The scale that weighed the package is accurate to within $\frac{1}{32}$ pound. How much might you be undercharged or overcharged?

106. Accuracy of Measurement   You stop at a self-service gas station to buy 15 gallons of 87-octane gasoline at $2.22 per gallon. The gas pump is accurate to within $\frac{1}{10}$ gallon. How much might you be undercharged or overcharged?

107. Geometry   The side length of a square is 10.4 inches with a possible error of $\frac{1}{16}$ inch. Determine the interval containing the possible areas of the square.

108. Geometry   The side length of a square is 24.2 centimeters with a possible error of 0.25 centimeter. Determine the interval containing the possible areas of the square.

Exploration

True or False?   In Exercises 109–112, determine whether the statement is true or false. Justify your answer.

109. If a, b, and c are real numbers, and $a < b$, then $a + c < b + c$.

110. If $a, b,$ and c are real numbers, and $a \le b$, then $ac \le bc$.

111. If $-10 \le x \le 8$, then $-10 \ge -x$ and $-x \ge -8$.

112. If $-2 < x < -1$, then $1 < -x < 2$.

113. Think About It   Give an example of an inequality whose solution set is $(-\infty, \infty)$.

114.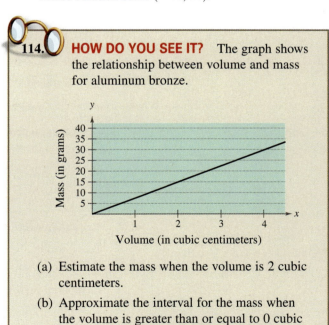

HOW DO YOU SEE IT?   The graph shows the relationship between volume and mass for aluminum bronze.

(a) Estimate the mass when the volume is 2 cubic centimeters.

(b) Approximate the interval for the mass when the volume is greater than or equal to 0 cubic centimeters and less than 4 cubic centimeters.

115. Think About It   Find sets of values of a, b, and c such that $0 \le x \le 10$ is a solution of the inequality $|ax - b| \le c$.

1.8 Other Types of Inequalities

Nonlinear inequalities have many real-life applications. For example, in Exercises 67 and 68 on page 148, you will use a polynomial inequality to model the height of a projectile.

■ Solve polynomial inequalities.
■ Solve rational inequalities.
■ Use nonlinear inequalities to model and solve real-life problems.

Polynomial Inequalities

To solve a polynomial inequality such as $x^2 - 2x - 3 < 0$, use the fact that a polynomial can change signs only at its *zeros* (the x-values that make the polynomial equal to zero). Between two consecutive zeros, a polynomial must be entirely positive or entirely negative. This means that when the real zeros of a polynomial are put in order, they divide the real number line into intervals in which the polynomial has no sign changes. These zeros are the **key numbers** of the inequality, and the resulting open intervals are the **test intervals** for the inequality. For example, the polynomial $x^2 - 2x - 3$ factors as

$$x^2 - 2x - 3 = (x + 1)(x - 3)$$

so it has two zeros,

$$x = -1 \quad \text{and} \quad x = 3.$$

These zeros divide the real number line into three test intervals:

$$(-\infty, -1), \quad (-1, 3), \quad \text{and} \quad (3, \infty). \qquad \text{(See figure below.)}$$

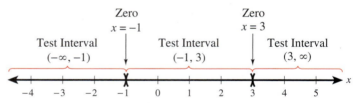

Three test intervals for $x^2 - 2x - 3$

To solve the inequality $x^2 - 2x - 3 < 0$, you need to test only one value from each of these test intervals. When a value from a test interval satisfies the original inequality, you can conclude that the interval is a solution of the inequality.

Use the same basic approach, generalized below, to find the solution set of any polynomial inequality.

REMARK The solution set of

$$x^2 - 2x - 3 < 0$$

discussed above, is the open interval $(-1, 3)$. Use Step 3 to verify this. By choosing the representative x-values $x = -2$, $x = 0$, and $x = 4$, you will find that the value of the polynomial is negative only in $(-1, 3)$.

Test Intervals for a Polynomial Inequality

To determine the intervals on which the values of a polynomial are entirely negative or entirely positive, use the steps below.

1. Find all real zeros of the polynomial, and arrange the zeros in increasing order. These zeros are the key numbers of the inequality.

2. Use the key numbers of the inequality to determine the test intervals.

3. Choose one representative x-value in each test interval and evaluate the polynomial at that value. When the value of the polynomial is negative, the polynomial has negative values for every x-value in the interval. When the value of the polynomial is positive, the polynomial has positive values for every x-value in the interval.

EXAMPLE 1 **Solving a Polynomial Inequality**

Solve $x^2 - x - 6 < 0$. Then graph the solution set.

▷ **ALGEBRA HELP** To review
the techniques for factoring
polynomials, see Section P.4.

Solution Factoring the polynomial

$$x^2 - x - 6 = (x + 2)(x - 3)$$

shows that the key numbers are $x = -2$ and $x = 3$. So, the inequality's test intervals are

$$(-\infty, -2), \quad (-2, 3), \quad \text{and} \quad (3, \infty) \qquad \text{Test intervals.}$$

In each test interval, choose a representative x-value and evaluate the polynomial.

Test Interval	x-Value	Polynomial Value	Conclusion
$(-\infty, -2)$	$x = -3$	$(-3)^2 - (-3) - 6 = 6$	Positive
$(-2, 3)$	$x = 0$	$(0)^2 - (0) - 6 = -6$	Negative
$(3, \infty)$	$x = 4$	$(4)^2 - (4) - 6 = 6$	Positive

The inequality is satisfied for all x-values in $(-2, 3)$. This implies that the solution set of the inequality

$$x^2 - x - 6 < 0$$

is the interval $(-2, 3)$, as shown on the number line below. Note that the original inequality contains a "less than" symbol. This means that the solution set does not contain the endpoints of the test interval $(-2, 3)$.

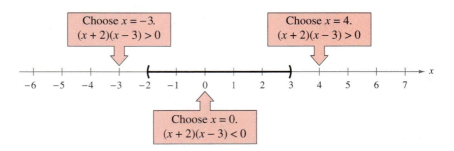

Choose $x = -3$.
$(x + 2)(x - 3) > 0$

Choose $x = 4$.
$(x + 2)(x - 3) > 0$

Choose $x = 0$.
$(x + 2)(x - 3) < 0$

✓ *Checkpoint* 🔊)) *Audio-video solution in English & Spanish at LarsonPrecalculus.com*

Solve $x^2 - x - 20 < 0$. Then graph the solution set. ■

As with linear inequalities, you can check the reasonableness of a solution by substituting x-values into the original inequality. For instance, to check the solution found in Example 1, substitute several x-values from the interval $(-2, 3)$ into the inequality

$$x^2 - x - 6 < 0.$$

Regardless of which x-values you choose, the inequality should be satisfied.
    You can also use a graph to check the result of Example 1. Sketch the graph of

$$y = x^2 - x - 6$$

as shown in Figure 1.14. Notice that the graph is below the x-axis on the interval $(-2, 3)$.
    In Example 1, the polynomial inequality is in general form (with the polynomial on one side and zero on the other). Whenever this is not the case, you should begin by writing the inequality in general form.

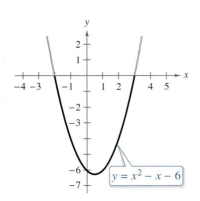

$y = x^2 - x - 6$

Figure 1.14

EXAMPLE 2   **Solving a Polynomial Inequality**

See LarsonPrecalculus.com for an interactive version of this type of example.

Solve $4x^2 - 5x > 6$.

Algebraic Solution

$4x^2 - 5x - 6 > 0$ Write in general form.

$(x - 2)(4x + 3) > 0$ Factor.

Key numbers: $x = -\frac{3}{4}$, $x = 2$

Test intervals: $\left(-\infty, -\frac{3}{4}\right)$, $\left(-\frac{3}{4}, 2\right)$, $(2, \infty)$

Test: Is $(x - 2)(4x + 3) > 0$?

Testing these intervals shows that the polynomial $4x^2 - 5x - 6$ is positive on the open intervals $\left(-\infty, -\frac{3}{4}\right)$ and $(2, \infty)$. So, the solution set of the inequality is $\left(-\infty, -\frac{3}{4}\right) \cup (2, \infty)$.

Graphical Solution

First write the polynomial inequality $4x^2 - 5x > 6$ as $4x^2 - 5x - 6 > 0$. Then use a graphing utility to graph $y = 4x^2 - 5x - 6$.

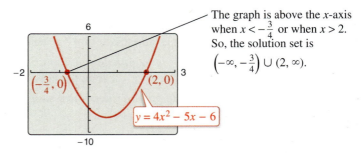

The graph is above the x-axis when $x < -\frac{3}{4}$ or when $x > 2$. So, the solution set is $\left(-\infty, -\frac{3}{4}\right) \cup (2, \infty)$.

$\left(-\frac{3}{4}, 0\right)$ $(2, 0)$

$y = 4x^2 - 5x - 6$

✓ *Checkpoint* 🔊 *Audio-video solution in English & Spanish at LarsonPrecalculus.com*

Solve $2x^2 + 3x < 5$ (a) algebraically and (b) graphically.

EXAMPLE 3   **Solving a Polynomial Inequality**

Solve $2x^3 - 3x^2 - 32x > -48$. Then graph the solution set.

Solution

$2x^3 - 3x^2 - 32x + 48 > 0$ Write in general form.

$(x - 4)(x + 4)(2x - 3) > 0$ Factor by grouping.

The key numbers are $x = -4$, $x = \frac{3}{2}$, and $x = 4$, and the test intervals are $(-\infty, -4)$, $\left(-4, \frac{3}{2}\right)$, $\left(\frac{3}{2}, 4\right)$, and $(4, \infty)$.

Test Interval	x-Value	Polynomial Value	Conclusion
$(-\infty, -4)$	$x = -5$	$2(-5)^3 - 3(-5)^2 - 32(-5) + 48 = -117$	Negative
$\left(-4, \frac{3}{2}\right)$	$x = 0$	$2(0)^3 - 3(0)^2 - 32(0) + 48 = 48$	Positive
$\left(\frac{3}{2}, 4\right)$	$x = 2$	$2(2)^3 - 3(2)^2 - 32(2) + 48 = -12$	Negative
$(4, \infty)$	$x = 5$	$2(5)^3 - 3(5)^2 - 32(5) + 48 = 63$	Positive

The inequality is satisfied on the open intervals $\left(-4, \frac{3}{2}\right)$ and $(4, \infty)$. So, the solution set is $\left(-4, \frac{3}{2}\right) \cup (4, \infty)$, as shown on the number line below.

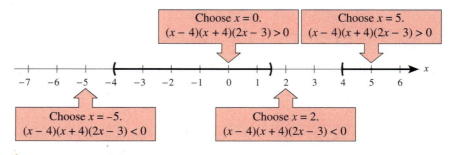

Choose $x = 0$.
$(x - 4)(x + 4)(2x - 3) > 0$

Choose $x = 5$.
$(x - 4)(x + 4)(2x - 3) > 0$

Choose $x = -5$.
$(x - 4)(x + 4)(2x - 3) < 0$

Choose $x = 2$.
$(x - 4)(x + 4)(2x - 3) < 0$

✓ *Checkpoint* 🔊 *Audio-video solution in English & Spanish at LarsonPrecalculus.com*

Solve $3x^3 - x^2 - 12x > -4$. Then graph the solution set.

You may find it easier to determine the sign of a polynomial from its *factored* form. For instance, in Example 2, when you substitute the test value $x = 1$ into the factored form

$$(x - 2)(4x + 3)$$

the sign pattern of the factors is

$$(-)(+)$$

which yields a negative result. Use factored forms to determine the signs of the polynomials in other examples in this section.

When solving a polynomial inequality, be sure to account for the inequality symbol. For instance, in Example 2, note that the original inequality symbol is "greater than" and the solution consists of two open intervals. If the original inequality had been

$$4x^2 - 5x \geq 6$$

then the solution set would have been

$$\left(-\infty, -\tfrac{3}{4}\right] \cup [2, \infty).$$

Each of the polynomial inequalities in Examples 1, 2, and 3 has a solution set that consists of a single interval or the union of two intervals. When solving the exercises for this section, watch for unusual solution sets, as illustrated in Example 4.

EXAMPLE 4   Unusual Solution Sets

a. The solution set of

$$x^2 + 2x + 4 > 0$$

consists of the entire set of real numbers, $(-\infty, \infty)$. In other words, the value of the quadratic polynomial $x^2 + 2x + 4$ is positive for every real value of x.

b. The solution set of

$$x^2 + 2x + 1 \leq 0$$

consists of the single real number $\{-1\}$, because the inequality has only one key number, $x = -1$, and it is the only value that satisfies the inequality.

c. The solution set of

$$x^2 + 3x + 5 < 0$$

is empty. In other words, $x^2 + 3x + 5$ is not less than zero for any value of x.

d. The solution set of

$$x^2 - 4x + 4 > 0$$

consists of all real numbers except $x = 2$. This solution set can be written in interval notation as

$$(-\infty, 2) \cup (2, \infty).$$

✓ *Checkpoint* ◀))) *Audio-video solution in English & Spanish at LarsonPrecalculus.com*

What is unusual about the solution set of each inequality?

a. $x^2 + 6x + 9 < 0$

b. $x^2 + 4x + 4 \leq 0$

c. $x^2 - 6x + 9 > 0$

d. $x^2 - 2x + 1 \geq 0$

Rational Inequalities

The concepts of key numbers and test intervals can be extended to rational inequalities. Use the fact that the value of a rational expression can change sign at its *zeros* (the x-values for which its numerator is zero) and at its *undefined values* (the x-values for which its denominator is zero). These two types of numbers make up the *key numbers* of a rational inequality. When solving a rational inequality, begin by writing the inequality in general form, that is, with zero on the right side of the inequality.

EXAMPLE 5 **Solving a Rational Inequality**

Solve $\dfrac{2x - 7}{x - 5} \le 3$. Then graph the solution set.

Solution

$$\frac{2x - 7}{x - 5} \le 3 \qquad \text{Write original inequality.}$$

$$\frac{2x - 7}{x - 5} - 3 \le 0 \qquad \text{Write in general form.}$$

$$\frac{2x - 7 - 3x + 15}{x - 5} \le 0 \qquad \text{Find the LCD and subtract fractions.}$$

$$\frac{-x + 8}{x - 5} \le 0 \qquad \text{Simplify.}$$

Key numbers: $x = 5$, $x = 8$ Zeros and undefined values of rational expression

Test intervals: $(-\infty, 5), (5, 8), (8, \infty)$

Test: Is $\dfrac{-x + 8}{x - 5} \le 0$?

Testing these intervals, as shown in the figure below, the inequality is satisfied on the open intervals $(-\infty, 5)$ and $(8, \infty)$. Moreover,

$$\frac{-x + 8}{x - 5} = 0$$

when $x = 8$, so the solution set is $(-\infty, 5) \cup [8, \infty)$. (Be sure to use a bracket to signify that $x = 8$ is included in the solution set.)

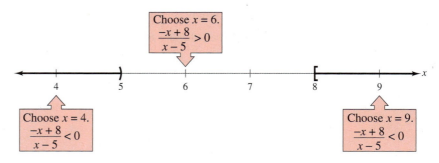

✓ Checkpoint 🔊))) *Audio-video solution in English & Spanish at LarsonPrecalculus.com*

Solve each inequality. Then graph the solution set.

a. $\dfrac{x - 2}{x - 3} \ge -3$

b. $\dfrac{4x - 1}{x - 6} > 3$

•• **REMARK** By writing 3 as $\frac{3}{1}$, you should be able to see that the least common denominator is $(x - 5)(1) = x - 5$. So, rewriting the general form as

$$\frac{2x - 7}{x - 5} - \frac{3(x - 5)}{x - 5} \le 0$$

and subtracting gives the result shown.

Applications

One common application of inequalities comes from business and involves profit, revenue, and cost. The formula that relates these three quantities is

$$\boxed{\text{Profit}} \;=\; \boxed{\text{Revenue}} \;-\; \boxed{\text{Cost}}$$

$$P = R - C.$$

EXAMPLE 6   Profit from a Product

The marketing department of a calculator manufacturer determines that the demand for a new model of calculator is

$$p = 100 - 0.00001x, \quad 0 \le x \le 10{,}000{,}000 \qquad \text{Demand equation}$$

where p is the price per calculator (in dollars) and x represents the number of calculators sold. (According to this model, no one would be willing to pay \$100 for the calculator. At the other extreme, the company could not *give* away more than 10 million calculators.) The revenue for selling x calculators is

$$R = xp = x(100 - 0.00001x). \qquad \text{Revenue equation}$$

The total cost of producing x calculators is \$10 per calculator plus a one-time development cost of \$2,500,000. So, the total cost is

$$C = 10x + 2{,}500{,}000. \qquad \text{Cost equation}$$

What prices can the company charge per calculator to obtain a profit of at least \$190,000,000?

Solution

Verbal model: $\boxed{\text{Profit}} \;=\; \boxed{\text{Revenue}} \;-\; \boxed{\text{Cost}}$

Equation: $P = R - C$

$$P = 100x - 0.00001x^2 - (10x + 2{,}500{,}000)$$

$$P = -0.00001x^2 + 90x - 2{,}500{,}000$$

To answer the question, solve the inequality

$$P \ge 190{,}000{,}000$$

$$-0.00001x^2 + 90x - 2{,}500{,}000 \ge 190{,}000{,}000.$$

Write the inequality in general form, find the key numbers and the test intervals, and then test a value in each test interval to find that the solution is

$$3{,}500{,}000 \le x \le 5{,}500{,}000$$

as shown in Figure 1.15. Substituting the x-values in the original demand equation shows that prices of

$$\$45.00 \le p \le \$65.00$$

yield a profit of at least \$190,000,000.

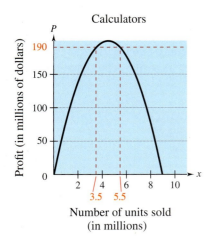

Figure 1.15

Calculators

Number of units sold (in millions)

Profit (in millions of dollars)

✓ *Checkpoint* *Audio-video solution in English & Spanish at LarsonPrecalculus.com*

The revenue and cost equations for a product are

$$R = x(60 - 0.0001x) \quad \text{and} \quad C = 12x + 1{,}800{,}000$$

where R and C are measured in dollars and x represents the number of units sold. How many units must be sold to obtain a profit of at least \$3,600,000?

Another common application of inequalities is finding the domain of an expression that involves a square root, as shown in Example 7.

EXAMPLE 7   **Finding the Domain of an Expression**

Find the domain of $\sqrt{64 - 4x^2}$.

Algebraic Solution

Recall that the domain of an expression is the set of all x-values for which the expression is defined. The expression $\sqrt{64 - 4x^2}$ is defined only when $64 - 4x^2$ is nonnegative, so the inequality $64 - 4x^2 \geq 0$ gives the domain.

$$64 - 4x^2 \geq 0 \qquad \text{Write in general form.}$$
$$16 - x^2 \geq 0 \qquad \text{Divide each side by 4.}$$
$$(4 - x)(4 + x) \geq 0 \qquad \text{Write in factored form.}$$

The inequality has two key numbers: $x = -4$ and $x = 4$. Use these two numbers to test the inequality.

Key numbers: $x = -4, x = 4$

Test intervals: $(-\infty, -4), (-4, 4), (4, \infty)$

Test: Is $(4 - x)(4 + x) \geq 0$?

A test shows that the inequality is satisfied in the *closed interval* $[-4, 4]$. So, the domain of the expression $\sqrt{64 - 4x^2}$ is the closed interval $[-4, 4]$.

Graphical Solution

Begin by sketching the graph of the equation $y = \sqrt{64 - 4x^2}$, as shown below. The graph shows that the x-values extend from -4 to 4 (including -4 and 4). So, the domain of the expression $\sqrt{64 - 4x^2}$ is the closed interval $[-4, 4]$.

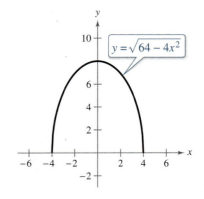

$y = \sqrt{64 - 4x^2}$

✓ **Checkpoint** ◀))) *Audio-video solution in English & Spanish at LarsonPrecalculus.com*

Find the domain of $\sqrt{x^2 - 7x + 10}$.

You can check the reasonableness of the solution to Example 7 by choosing a representative x-value in the interval and evaluating the radical expression at that value. When you substitute any number from the closed interval $[-4, 4]$ into the expression $\sqrt{64 - 4x^2}$, you obtain a nonnegative number under the radical symbol that simplifies to a real number. When you substitute any number from the intervals $(-\infty, -4)$ and $(4, \infty)$, you obtain a complex number. A visual representation of the intervals is shown below.

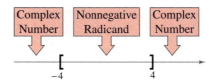

Summarize   (Section 1.8)

1. Explain how to solve a polynomial inequality *(page 140)*. For examples of solving polynomial inequalities, see Examples 1–4.

2. Explain how to solve a rational inequality *(page 144)*. For an example of solving a rational inequality, see Example 5.

3. Describe applications of polynomial inequalities *(pages 145 and 146, Examples 6 and 7)*.

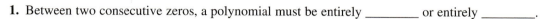

1.8 Exercises

See **CalcChat.com** for tutorial help and worked-out solutions to odd-numbered exercises.

Vocabulary: Fill in the blanks.

1. Between two consecutive zeros, a polynomial must be entirely _____ or entirely _____.
2. To solve a polynomial inequality, find the _____ numbers of the inequality, and use these numbers to create _____ _____ for the inequality.
3. A rational expression can change sign at its _____ and its _____ _____.
4. The formula that relates cost, revenue, and profit is _____.

Skills and Applications

 Checking Solutions In Exercises 5–8, determine whether each value of x is a solution of the inequality.

Inequality	Values
5. $x^2 - 3 < 0$	(a) $x = 3$ (b) $x = 0$ (c) $x = \frac{3}{2}$ (d) $x = -5$
6. $x^2 - 2x - 8 \geq 0$	(a) $x = -2$ (b) $x = 0$ (c) $x = -4$ (d) $x = 1$
7. $\dfrac{x + 2}{x - 4} \geq 3$	(a) $x = 5$ (b) $x = 4$ (c) $x = -\frac{9}{2}$ (d) $x = \frac{9}{2}$
8. $\dfrac{3x^2}{x^2 + 4} < 1$	(a) $x = -2$ (b) $x = -1$ (c) $x = 0$ (d) $x = 3$

Finding Key Numbers In Exercises 9–12, find the key numbers of the inequality.

9. $x^2 - 3x - 18 > 0$ 10. $9x^3 - 25x^2 \leq 0$

11. $\dfrac{1}{x - 5} + 1 \geq 0$ 12. $\dfrac{x}{x + 2} - \dfrac{2}{x - 1} < 0$

 Solving a Polynomial Inequality In Exercises 13–36, solve the inequality. Then graph the solution set.

13. $2x^2 + 4x < 0$ 14. $3x^2 - 9x \geq 0$

15. $x^2 < 9$ 16. $x^2 \leq 25$

17. $(x + 2)^2 \leq 25$ 18. $(x - 3)^2 \geq 1$

19. $x^2 + 6x + 1 \geq -7$ 20. $x^2 - 8x + 2 < 11$

21. $x^2 + x < 6$ 22. $x^2 + 2x > 3$

23. $x^2 < 3 - 2x$ 24. $x^2 > 2x + 8$

25. $3x^2 - 11x > 20$ 26. $-2x^2 + 6x \leq -15$

27. $x^3 - 3x^2 - x + 3 > 0$ 28. $x^3 + 2x^2 - 4x \leq 8$

29. $-x^3 + 7x^2 + 9x > 63$

30. $2x^3 + 13x^2 - 8x \geq 52$

31. $4x^3 - 6x^2 < 0$ 32. $4x^3 - 12x^2 > 0$

33. $x^3 - 4x \geq 0$ 34. $2x^3 - x^4 \leq 0$

35. $(x - 1)^2(x + 2)^3 \geq 0$ 36. $x^4(x - 3) \leq 0$

Unusual Solution Sets In Exercises 37–40, explain what is unusual about the solution set of the inequality.

37. $4x^2 - 4x + 1 \leq 0$ 38. $x^2 + 3x + 8 > 0$

39. $x^2 - 6x + 12 \leq 0$ 40. $x^2 - 8x + 16 > 0$

 Solving a Rational Inequality In Exercises 41–52, solve the inequality. Then graph the solution set.

41. $\dfrac{4x - 1}{x} > 0$ 42. $\dfrac{x^2 - 1}{x} < 0$

43. $\dfrac{3x + 5}{x - 1} < 2$ 44. $\dfrac{x + 12}{x + 2} \geq 3$

45. $\dfrac{2}{x + 5} > \dfrac{1}{x - 3}$ 46. $\dfrac{5}{x - 6} > \dfrac{3}{x + 2}$

47. $\dfrac{1}{x - 3} \leq \dfrac{9}{4x + 3}$ 48. $\dfrac{1}{x} \geq \dfrac{1}{x + 3}$

49. $\dfrac{x^2 + 2x}{x^2 - 9} \leq 0$ 50. $\dfrac{x^2 + x - 6}{x} \geq 0$

51. $\dfrac{3}{x - 1} + \dfrac{2x}{x + 1} > -1$ 52. $\dfrac{3x}{x - 1} \leq \dfrac{x}{x + 4} + 3$

Using Technology In Exercises 53–60, use a graphing utility to graph the equation. Use the graph to approximate the values of x that satisfy each inequality.

Equation	Inequalities
53. $y = -x^2 + 2x + 3$	(a) $y \leq 0$ (b) $y \geq 3$
54. $y = \frac{1}{2}x^2 - 2x + 1$	(a) $y \leq 0$ (b) $y \geq 7$
55. $y = \frac{1}{8}x^3 - \frac{1}{2}x$	(a) $y \geq 0$ (b) $y \leq 6$
56. $y = x^3 - x^2 - 16x + 16$	(a) $y \leq 0$ (b) $y \geq 36$
57. $y = \dfrac{3x}{x - 2}$	(a) $y \leq 0$ (b) $y \geq 6$
58. $y = \dfrac{2(x - 2)}{x + 1}$	(a) $y \leq 0$ (b) $y \geq 8$
59. $y = \dfrac{2x^2}{x^2 + 4}$	(a) $y \geq 1$ (b) $y \leq 2$
60. $y = \dfrac{5x}{x^2 + 4}$	(a) $y \geq 1$ (b) $y \leq 0$

Solving an Inequality In Exercises 61–66, solve the inequality. (Round your answers to two decimal places.)

61. $0.3x^2 + 6.26 < 10.8$ **62.** $-1.3x^2 + 3.78 > 2.12$

63. $-0.5x^2 + 12.5x + 1.6 > 0$

64. $1.2x^2 + 4.8x + 3.1 < 5.3$

65. $\dfrac{1}{2.3x - 5.2} > 3.4$ **66.** $\dfrac{2}{3.1x - 3.7} > 5.8$

Height of a Projectile

In Exercises 67 and 68, use the position equation

$$s = -16t^2 + v_0 t + s_0$$

where s represents the height of an object (in feet), v_0 represents the initial velocity of the object (in feet per second), s_0 represents the initial height of the object (in feet), and t represents the time (in seconds).

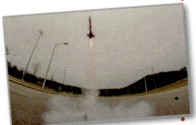

67. A projectile is fired straight upward from ground level ($s_0 = 0$) with an initial velocity of 160 feet per second.

(a) At what instant will it be back at ground level?

(b) When will the height exceed 384 feet?

68. A projectile is fired straight upward from ground level ($s_0 = 0$) with an initial velocity of 128 feet per second.

(a) At what instant will it be back at ground level?

(b) When will the height be less than 128 feet?

69. Cost, Revenue, and Profit The revenue and cost equations for a product are $R = x(75 - 0.0005x)$ and $C = 30x + 250,000$, where R and C are measured in dollars and x represents the number of units sold. How many units must be sold to obtain a profit of at least $750,000? What is the price per unit?

70. Cost, Revenue, and Profit The revenue and cost equations for a product are $R = x(50 - 0.0002x)$ and $C = 12x + 150,000$, where R and C are measured in dollars and x represents the number of units sold. How many units must be sold to obtain a profit of at least $1,650,000? What is the price per unit?

Finding the Domain of an Expression In Exercises 71–76, find the domain of the expression. Use a graphing utility to verify your result.

71. $\sqrt{4 - x^2}$ **72.** $\sqrt{x^2 - 9}$

73. $\sqrt{x^2 - 9x + 20}$ **74.** $\sqrt{49 - x^2}$

75. $\sqrt{\dfrac{x}{x^2 - 2x - 35}}$ **76.** $\sqrt{\dfrac{x}{x^2 - 9}}$

77. School Enrollment The table shows the numbers N (in millions) of students enrolled in elementary and secondary schools in the United States from 2005 through 2014. (*Source: National Center for Education Statistics*)

DATA	Year	Number, N
	2005	49.11
	2006	49.32
	2007	49.29
	2008	49.27
	2009	49.36
	2010	49.48
	2011	49.52
	2012	49.77
	2013	49.94
	2014	49.99

Spreadsheet at LarsonPrecalculus.com

(a) Use a graphing utility to create a scatter plot of the data. Let t represent the year, with $t = 5$ corresponding to 2005.

(b) Use the *regression* feature of the graphing utility to find a *quartic* model for the data. (A quartic model has the form $at^4 + bt^3 + ct^2 + dt + e$, where a, b, c, d, and e are constant and t is variable.)

(c) Graph the model and the scatter plot in the same viewing window. How well does the model fit the data?

(d) According to the model, after 2014, when did the number of students enrolled in elementary and secondary schools fall below 48 million?

(e) Is the model valid for long-term predictions of student enrollment? Explain.

78. Safe Load The maximum safe load uniformly distributed over a one-foot section of a two-inch-wide wooden beam can be approximated by the model

$$\text{Load} = 168.5d^2 - 472.1$$

where d is the depth of the beam.

(a) Evaluate the model for $d = 4$, $d = 6$, $d = 8$, $d = 10$, and $d = 12$. Use the results to create a bar graph.

(b) Determine the minimum depth of the beam that will safely support a load of 2000 pounds.

79. Geometry A rectangular playing field with a perimeter of 100 meters is to have an area of at least 500 square meters. Within what bounds must the length of the rectangle lie?

80. Geometry   A rectangular parking lot with a perimeter of 440 feet is to have an area of at least 8000 square feet. Within what bounds must the length of the rectangle lie?

81. Resistors   When two resistors of resistances R_1 and R_2 are connected in parallel (see figure), the total resistance R satisfies the equation

$$\frac{1}{R} = \frac{1}{R_1} + \frac{1}{R_2}.$$

Find R_1 for a parallel circuit in which $R_2 = 2$ ohms and R must be at least 1 ohm.

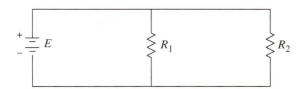

82. Teachers' Salaries   The table shows the mean salaries S (in thousands of dollars) of public school classroom teachers in the United States from 2002 through 2013.

DATA	Year	Salary, S
	2002	44.7
	2003	45.7
	2004	46.5
	2005	47.5
	2006	49.1
	2007	51.1
	2008	52.8
	2009	54.3
	2010	55.2
	2011	56.1
	2012	55.4
	2013	56.4

Spreadsheet at LarsonPrecalculus.com

A model that approximates these data is

$$S = \frac{40.32 + 3.53t}{1 + 0.039t}, \quad 2 \le t \le 13$$

where t represents the year, with $t = 2$ corresponding to 2002. (*Source: National Center for Education Statistics*)

(a) Use a graphing utility to create a scatter plot of the data. Then graph the model in the same viewing window.

(b) How well does the model fit the data? Explain.

(c) Use the model to predict when the salary for classroom teachers will exceed \$65,000.

(d) Is the model valid for long-term predictions of classroom teacher salaries? Explain.

Exploration

True or False?   In Exercises 83 and 84, determine whether the statement is true or false. Justify your answer.

83. The zeros of the polynomial

$$x^3 - 2x^2 - 11x + 12 = (x + 3)(x - 1)(x - 4)$$

divide the real number line into three test intervals.

84. The solution set of the inequality $\frac{3}{2}x^2 + 3x + 6 \ge 0$ is the entire set of real numbers.

85. Graphical Reasoning   Use a graphing utility to verify the results in Example 4. For instance, the graph of $y = x^2 + 2x + 4$ is shown below. Notice that the y-values are greater than 0 for all values of x, as stated in Example 4(a). Use the graphing utility to graph $y = x^2 + 2x + 1$, $y = x^2 + 3x + 5$, and $y = x^2 - 4x + 4$. Explain how you can use the graphs to verify the results of parts (b), (c), and (d) of Example 4.

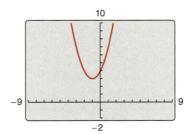

86. HOW DO YOU SEE IT?   Consider the polynomial

$$(x - a)(x - b)$$

and the real number line shown below.

(a) Identify the points on the line at which the polynomial is zero.

(b) For each of the three subintervals of the real number line, write the sign of each factor and the sign of the product.

(c) At what x-values does the polynomial change signs?

Conjecture   In Exercises 87–90, (a) find the interval(s) for b such that the equation has at least one real solution and (b) write a conjecture about the interval(s) based on the values of the coefficients.

87. $x^2 + bx + 9 = 0$

88. $x^2 + bx - 9 = 0$

89. $3x^2 + bx + 10 = 0$

90. $2x^2 + bx + 5 = 0$

Chapter Summary

	What Did You Learn?	Explanation/Examples	Review Exercises
Section 1.1	Sketch graphs of equations (p. 70), and identify x- and y-intercepts of graphs of equations (p. 73).	To graph an equation, construct a table of values, plot the points, and connect the points with a smooth curve or line. Intercepts are points where a graph intersects or touches the x- or y-axis.	1–4
	Use symmetry to sketch graphs of equations (p. 74).	Graphs can have symmetry with respect to one of the coordinate axes or with respect to the origin.	5–12
	Write equations of circles (p. 76).	A point (x, y) lies on the circle of radius r and center (h, k) if and only if $(x - h)^2 + (y - k)^2 = r^2$.	13–18
	Use graphs of equations to solve real-life problems (p. 77).	The graph of an equation can be used to estimate the maximum weight for a man in the U.S. Marine Corps. (See Example 9.)	19, 20
Section 1.2	Identify different types of equations (p. 81).	**Identity:** true for *every* real number in the domain **Conditional equation:** true for just *some* (but not all) of the real numbers in the domain **Contradiction:** false for *every* real number in the domain	21–24
	Solve linear equations in one variable (p. 82), and solve rational equations that lead to linear equations (p. 84).	**Linear equation in one variable:** an equation that can be written in the standard form $ax + b = 0$, where a and b are real numbers with $a \neq 0$ To solve a rational equation, multiply every term by the LCD.	25–32
	Find x- and y-intercepts algebraically (p. 85).	To find x-intercepts, set y equal to zero and solve for x. To find y-intercepts, set x equal to zero and solve for y.	33–38
	Use linear equations to model and solve real-life problems (p. 86).	A linear equation can be used to model the number of female participants in cross-country running. (See Example 6.)	39, 40
Section 1.3	Use mathematical models to solve real-life problems (p. 90).	Mathematical models can be used to find a percent raise and a building's height. (See Examples 2 and 6.)	41–44
	Solve mixture problems (p. 94).	Mixture problems include simple interest problems and inventory problems. (See Examples 7 and 8.)	45, 46
	Use common formulas to solve real-life problems (p. 95).	A literal equation contains more than one variable. A formula is an example of a literal equation. (See Example 9.)	47, 48
Section 1.4	Solve quadratic equations by factoring (p. 100).	The method of factoring is based on the Zero-Factor Property, which states if $ab = 0$, then $a = 0$ or $b = 0$.	49, 50
	Solve quadratic equations by extracting square roots (p. 101).	The equation $u^2 = d$, where $d > 0$, has exactly two solutions: $u = \sqrt{d}$ and $u = -\sqrt{d}$.	51–54
	Solve quadratic equations by completing the square (p. 102) and using the Quadratic Formula (p. 104).	To complete the square for $x^2 + bx$, add $(b/2)^2$. **Quadratic Formula:** $x = \dfrac{-b \pm \sqrt{b^2 - 4ac}}{2a}$	55–58
	Use quadratic equations to model and solve real-life problems (p. 106).	A quadratic equation can be used to model the numbers of monthly active Facebook users worldwide from 2009 through 2015. (See Example 9.)	59, 60

	What Did You Learn?	**Explanation/Examples**	**Review Exercises**
Section 1.5	Use the imaginary unit i to write complex numbers (p. 114), and add, subtract, and multiply complex numbers (p. 115).	When a and b are real numbers, $a + bi$ is a complex number. **Sum:** $(a + bi) + (c + di) = (a + c) + (b + d)i$ **Difference:** $(a + bi) - (c + di) = (a - c) + (b - d)i$ Use the Distributive Property or the FOIL method to multiply.	61–70
	Use complex conjugates to write the quotient of two complex numbers in standard form (p. 117).	To write $(a + bi)/(c + di)$ in standard form, where c and d are not both zero, multiply the numerator and denominator by the complex conjugate of the denominator, $c - di$.	71–76
	Find complex solutions of quadratic equations (p. 118).	When a is a positive real number, the principal square root of $-a$ is defined as $\sqrt{-a} = \sqrt{a}i$.	77–80
Section 1.6	Solve polynomial equations of degree three or greater (p. 121).	Some polynomial equations of degree three or greater can be solved by factoring.	81–86
	Solve radical equations (p. 123).	Solving radical equations usually involves squaring or cubing each side of the equation.	87–90
	Solve rational equations and absolute value equations (p. 124).	To solve a rational equation, first multiply each side of the equation by the LCD of all terms in the equation. To solve an absolute value equation, remember that the expression inside the absolute value bars can be positive or negative.	91–96
	Use nonlinear and nonquadratic models to solve real-life problems (p. 126).	Nonlinear and nonquadratic models can be used to find the number of ski club members going on a ski trip and the annual interest rate for an investment. (See Examples 8 and 9.)	97, 98
Section 1.7	Represent solutions of linear inequalities in one variable (p. 131).	**Bounded** **Unbounded** $[-1, 2) \rightarrow -1 \le x < 2$ $(3, \infty) \rightarrow x > 3$ $[-4, 5] \rightarrow -4 \le x \le 5$ $(-\infty, \infty) \rightarrow -\infty < x < \infty$	99–102
	Use properties of inequalities to write equivalent inequalities (p. 132) and solve linear inequalities in one variable (p. 133).	Solving linear inequalities is similar to solving linear equations. Use the properties of inequalities to isolate the variable. Remember to reverse the inequality symbol when you multiply or divide by a negative number.	103–106
	Solve absolute value inequalities (p. 135).	Let u be an algebraic expression and let a be a real number such that $a > 0$. **1.** $|u| < a$ if and only if $-a < u < a$. **2.** $|u| \le a$ if and only if $-a \le u \le a$. **3.** $|u| > a$ if and only if $u < -a$ or $u > a$. **4.** $|u| \ge a$ if and only if $u \le -a$ or $u \ge a$.	107, 108
	Use linear inequalities to model and solve real-life problems (p. 136).	A linear inequality can be used to determine the accuracy of a measurement. (See Example 7.)	109, 110
Section 1.8	Solve polynomial (p. 140) and rational (p. 144) inequalities.	Use the concepts of key numbers and test intervals to solve both polynomial and rational inequalities.	111–116
	Use nonlinear inequalities to model and solve real-life problems (p. 145).	A common application of nonlinear inequalities involves profit P, revenue R, and cost C. (See Example 6.)	117, 118

Review Exercises See CalcChat.com for tutorial help and worked-out solutions to odd-numbered exercises.

1.1 **Sketching the Graph of an Equation** In Exercises 1 and 2, construct a table of values that consists of several solution points of the equation. Use the resulting solution points to sketch the graph of the equation.

1. $y = -4x + 1$ **2.** $y = x^2 + 2x$

Identifying x- and y-Intercepts In Exercises 3 and 4, identify the x- and y-intercepts of the graph.

3. $y = (x - 3)^2 - 4$ **4.** $y = |x + 1| - 3$

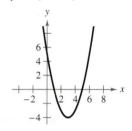
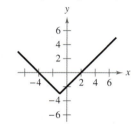

Testing for Symmetry In Exercises 5–12, use the algebraic tests to check for symmetry with respect to both axes and the origin. Then sketch the graph of the equation.

5. $y = -3x + 7$ **6.** $x = -8$

7. $x = y^2 - 5$ **8.** $y = 3x^3$

9. $y = -x^4 + 6x^2$ **10.** $y = x^4 + x^3 - 1$

11. $y = \dfrac{3}{x}$ **12.** $y = |x| - 4$

Sketching the Graph of a Circle In Exercises 13–16, find the center and radius of the circle with the given equation. Then sketch the circle.

13. $x^2 + y^2 = 9$ **14.** $x^2 + y^2 = 4$

15. $(x + 2)^2 + y^2 = 16$ **16.** $x^2 + (y - 8)^2 = 81$

17. Writing the Equation of a Circle Write the standard form of the equation of the circle for which the endpoints of a diameter are $(0, 0)$ and $(4, -6)$.

18. Writing the Equation of a Circle Write the standard form of the equation of the circle for which the endpoints of a diameter are $(-2, -3)$ and $(4, -10)$.

19. Sales The sales S (in millions of dollars) for Jazz Pharmaceuticals for the years 2006 through 2014 can be approximated by the model

$$S = 27.215t^2 - 409.06t + 1563.6, \quad 6 \le t \le 14$$

where t represents the year, with $t = 6$ corresponding to 2006. (*Source: Jazz Pharmaceuticals*)

(a) Sketch a graph of the model.

(b) Use the graph to estimate the year in which the sales were $200 million.

20. Physics The force F (in pounds) required to stretch a spring x inches from its natural length (see figure) is

$$F = \frac{5}{4}x, \quad 0 \le x \le 20.$$

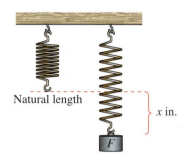

(a) Use the model to complete the table.

x	0	4	8	12	16	20
Force, F						

(b) Sketch a graph of the model.

(c) Use the graph to estimate the force necessary to stretch the spring 10 inches.

1.2 **Classifying an Equation** In Exercises 21–24, determine whether the equation is an identity, a conditional equation, or a contradiction.

21. $2(x - 2) = 2x - 4$ **22.** $2(x + 3) = 2x - 2$

23. $3(x - 2) + 2x = 2(x + 3)$

24. $5(x - 1) - 2x = 3x - 5$

Solving an Equation In Exercises 25–32, solve the equation and check your solution. (If not possible, explain why.)

25. $8x - 5 = 3x + 20$ **26.** $7x + 3 = 3x - 17$

27. $2(x + 5) - 7 = x + 9$ **28.** $7(x - 4) = 1 - (x + 9)$

29. $\dfrac{x}{5} - 3 = \dfrac{x}{3} + 1$ **30.** $\dfrac{4x - 3}{6} + \dfrac{x}{4} = x - 2$

31. $3 + \dfrac{2}{x - 5} = \dfrac{2x}{x - 5}$

32. $\dfrac{1}{x^2 + 3x - 18} - \dfrac{3}{x + 6} = \dfrac{4}{x - 3}$

Finding Intercepts Algebraically In Exercises 33–38, find the x- and y-intercepts of the graph of the equation algebraically.

33. $y = 3x - 1$ **34.** $y = -5x + 6$

35. $y = 2(x - 4)$ **36.** $y = 4(7x + 1)$

37. $y = -\frac{1}{2}x + \frac{2}{3}$ **38.** $y = \frac{3}{4}x - \frac{1}{4}$

39. Geometry The surface area S of the cylinder shown in the figure is approximated by the equation $S = 2(3.14)(3)^2 + 2(3.14)(3)h$. The surface area is 244.92 square inches. Find the height h of the cylinder.

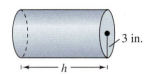

40. Temperature The Fahrenheit and Celsius temperature scales are related by the equation $C = \frac{5}{9}F - \frac{160}{9}$. Find the Fahrenheit temperature that corresponds to 100° Celsius.

1.3

41. Profit In 2014, LinkedIn's total revenue was 45.2% more than it was in 2013. The total revenue for the two years was $3.75 billion. Find the revenue for each year. *(Source: LinkedIn Corp.)*

42. Discount The sale price of a digital camera after a 20% discount is $340. Find the original price.

43. Business Venture You are starting a small business. You have 9 investors who are willing to share equally in the venture. When you add 3 more investors, each person's share decreases by $2500. What is the total investment required to start the business?

44. Average Speed You commute 56 miles one way to work. The trip to work takes 10 minutes longer than the trip home. Your average speed on the trip home is 8 miles per hour faster. What is your average speed on the trip home?

45. Mixture Problem A car radiator contains 10 liters of a 30% antifreeze solution. How many liters should you replace with pure antifreeze to get a 50% antifreeze solution?

46. Simple Interest You invest $6000 at $4\frac{1}{2}\%$ and $5\frac{1}{2}\%$ simple interest. During the first year, the two accounts earn $305. Find the initial investment for each account.

47. Volume of a Cone The volume V of a cone can be calculated by the formula $V = \frac{1}{3}\pi r^2 h$, where r is the radius and h is the height. Solve for h.

48. Kinetic Energy The kinetic energy E of an object can be calculated by the formula $E = \frac{1}{2}mv^2$, where m is the mass and v is the velocity. Solve for m.

1.4 Choosing a Method In Exercises 49–58, solve the equation using any convenient method.

49. $15 + x - 2x^2 = 0$ **50.** $2x^2 - x - 28 = 0$

51. $6 = 3x^2$ **52.** $16x^2 = 25$

53. $(x + 13)^2 = 25$ **54.** $(x - 5)^2 = 30$

55. $x^2 + 12x = -25$ **56.** $9x^2 - 12x = 14$

57. $-2x^2 - 5x + 27 = 0$

58. $-20 - 3x + 3x^2 = 0$

59. Simply Supported Beam A simply supported 20-foot beam supports a uniformly distributed load of 1000 pounds per foot. The bending moment M (in foot-pounds) x feet from one end of the beam is given by $M = 500x(20 - x)$.

(a) Where is the bending moment zero?

(b) Use a graphing utility to graph the equation.

(c) Use the graph to determine the point on the beam where the bending moment is the greatest.

60. Sports You throw a softball straight up into the air at a velocity of 30 feet per second. You release the softball at a height of 5.8 feet and catch it when it falls back to a height of 6.2 feet.

(a) Use the position equation to write a mathematical model for the height of the softball.

(b) What is the height of the softball after 1 second?

(c) How many seconds is the softball in the air?

1.5 Writing a Complex Number in Standard Form In Exercises 61–64, write the complex number in standard form.

61. $4 + \sqrt{-9}$ **62.** $3 + \sqrt{-16}$

63. $i^2 + 3i$ **64.** $-5i + i^2$

Performing Operations with Complex Numbers In Exercises 65–70, perform the operation and write the result in standard form.

65. $(6 - 4i) + (-9 + i)$ **66.** $(7 - 2i) - (3 - 8i)$

67. $-3i(-2 + 5i)$ **68.** $(4 + i)(3 - 10i)$

69. $(1 + 7i)(1 - 7i)$ **70.** $(5 - 9i)^2$

Quotient of Complex Numbers in Standard Form In Exercises 71–74, write the quotient in standard form.

71. $\dfrac{4}{1 - 2i}$ **72.** $\dfrac{6 - 5i}{i}$

73. $\dfrac{3 + 2i}{5 + i}$ **74.** $\dfrac{7i}{(3 + 2i)^2}$

Performing Operations with Complex Numbers In Exercises 75 and 76, perform the operation and write the result in standard form.

75. $\dfrac{4}{2 - 3i} + \dfrac{2}{1 + i}$ **76.** $\dfrac{1}{2 + i} - \dfrac{5}{1 + 4i}$

Complex Solutions of a Quadratic Equation In Exercises 77–80, use the Quadratic Formula to solve the quadratic equation.

77. $x^2 - 2x + 10 = 0$

78. $x^2 + 6x + 34 = 0$

79. $4x^2 + 4x + 7 = 0$

80. $6x^2 + 3x + 27 = 0$

1.6 Solving an Equation **In Exercises 81–96, solve the equation. Check your solutions.**

81. $5x^4 - 12x^3 = 0$ 82. $4x^3 - 6x^2 = 0$

83. $x^3 - 7x^2 + 4x - 28 = 0$

84. $9x^4 + 27x^3 - 4x^2 - 12x = 0$

85. $x^6 - 7x^3 - 8 = 0$ 86. $x^4 - 13x^2 - 48 = 0$

87. $\sqrt{2x + 3} + \sqrt{x - 2} = 2$

88. $5\sqrt{x} - \sqrt{x - 1} = 6$

89. $(x - 1)^{2/3} - 25 = 0$ 90. $(x + 2)^{3/4} = 27$

91. $\dfrac{5}{x} = 1 + \dfrac{3}{x + 2}$ 92. $\dfrac{6}{x} + \dfrac{8}{x + 5} = 3$

93. $|x - 5| = 10$ 94. $|2x + 3| = 7$

95. $|x^2 - 3| = 2x$ 96. $|x^2 - 6| = x$

97. **Demand** The demand equation for a hair dryer is $p = 42 - \sqrt{0.001x + 2}$, where x is the number of units demanded per day and p is the price per unit. Find the demand when the price is set at $29.95.

98. **Newspapers** The average number N (in thousands) of daily newspapers in circulation in the United States from 2000 through 2014 can be approximated by the model $N = 56{,}146 - 314.980t^{3/2}$, $0 \le t \le 14$, where t represents the year, with $t = 0$ corresponding to 2000. The table shows the average daily number of newspapers for selected years. *(Source: Editor and Publisher International Yearbook)*

Year	Newspapers, N
2000	55,773
2002	55,186
2004	54,626
2006	52,329
2008	48,597
2010	44,100
2012	43,433
2014	40,420

Spreadsheet at LarsonPrecalculus.com

(a) Use a graphing utility to plot the data and graph the model in the same viewing window. How well does the model fit the data?

(b) Use the graph in part (a) to estimate the year in which there was an average of 50,000,000 daily newspapers in circulation.

(c) Use the model to verify algebraically the estimate from part (b).

1.7 Intervals and Inequalities **In Exercises 99–102, write an inequality that represents the interval. Then state whether the interval is bounded or unbounded.**

99. $(-7, 2]$ 100. $(4, \infty)$

101. $(-\infty, -10]$ 102. $[-2, 2]$

Solving an Inequality **In Exercises 103–108, solve the inequality. Then graph the solution set.**

103. $3(x + 2) + 7 < 2x - 5$

104. $2(x + 7) - 4 \ge 5(x - 3)$

105. $4(5 - 2x) \le \frac{1}{2}(8 - x)$ 106. $\frac{1}{2}(3 - x) > \frac{1}{3}(2 - 3x)$

107. $|x + 6| < 5$ 108. $\frac{2}{3}|3 - x| \ge 4$

109. **Cost, Revenue, and Profit** The revenue for selling x units of a product is $R = 125.33x$. The cost of producing x units is $C = 92x + 1200$. To obtain a profit, the revenue must be greater than the cost. For what values of x does this product return a profit?

110. **Geometry** The side length of a square is 19.3 centimeters with a possible error of 0.5 centimeter. Determine the interval containing the possible areas of the square.

1.8 Solving an Inequality **In Exercises 111–116, solve the inequality. Then graph the solution set.**

111. $x^2 - 6x - 27 < 0$ 112. $x^2 - 2x \ge 3$

113. $5x^3 - 45x < 0$ 114. $2x^3 - 5x^2 - 3x \ge 0$

115. $\dfrac{2}{x + 1} \le \dfrac{3}{x - 1}$ 116. $\dfrac{x - 5}{3 - x} < 0$

117. **Investment** An investment of P dollars at interest rate r (in decimal form) compounded annually increases to an amount $A = P(1 + r)^2$ in 2 years. An investment of $5000 increases to an amount greater than $5500 in 2 years. The interest rate must be greater than what percent?

118. **Biology** A biologist introduces 200 ladybugs into a crop field. The population P of the ladybugs can be approximated by the model $P = [1000(1 + 3t)]/(5 + t)$, where t is the time in days. Find the time required for the population to increase to at least 2000 ladybugs.

Exploration

True or False? **In Exercises 119 and 120, determine whether the statement is true or false. Justify your answer.**

119. $\sqrt{-18}\sqrt{-2} = \sqrt{(-18)(-2)}$

120. The equation $325x^2 - 717x + 398 = 0$ has no solution.

121. **Writing** Explain why it is essential to check your solutions to radical, absolute value, and rational equations.

122. **Error Analysis** Describe the error below.

$$|11x + 4| \ge 26$$
$$11x + 4 \le 26 \quad \text{or} \quad 11x + 4 \ge 26$$
$$11x \le 22 \qquad\qquad 11x \ge 22$$
$$x \le 2 \qquad\qquad x \ge 2$$

Take this test as you would take a test in class. When you are finished, check your work against the answers given in the back of the book.

In Exercises 1–6, use the algebraic tests to check for symmetry with respect to both axes and the origin. Then sketch the graph of the equation. Identify any x- and y-intercepts.

1. $y = 4 - \frac{3}{4}x$

2. $y = 4 - \frac{3}{4}|x|$

3. $y = 4 - (x - 2)^2$

4. $y = x - x^3$

5. $y = \sqrt{5 - x}$

6. $(x - 3)^2 + y^2 = 9$

In Exercises 7–12, solve the equation and check your solution. (If not possible, explain why.)

7. $\frac{2}{3}(x - 1) + \frac{1}{4}x = 10$

8. $(x - 4)(x + 2) = 7$

9. $\frac{x - 2}{x + 2} + \frac{4}{x + 2} + 4 = 0$

10. $x^4 + x^2 - 6 = 0$

11. $2\sqrt{x} - \sqrt{2x + 1} = 1$

12. $|3x - 1| = 7$

In Exercises 13–16, solve the inequality. Then graph the solution set.

13. $-3 \le 2(x + 4) < 14$

14. $\frac{2}{x} > \frac{5}{x + 6}$

15. $2x^2 + 5x > 12$

16. $|3x + 5| \ge 10$

17. Perform each operation and write the result in standard form.

   (a) $\sqrt{-16} - 2(7 + 2i)$

   (b) $(5 - i)(3 + 4i)$

18. Write the quotient in standard form: $\frac{8}{1 + 2i}$.

19. Solve $2x^2 - 6x + 5 = 0$.

20. The sales S (in billions of dollars) for Oracle from 2005 through 2014 can be approximated by the model

   $S = 3.2205t - 3.908, \quad 5 \le t \le 14$

   where t represents the year, with $t = 5$ corresponding to 2005. *(Source: Oracle Corp.)*

   (a) Sketch a graph of the model.

   (b) Use the graph in part (a) to predict the sales in 2017.

   (c) Use the model to verify algebraically your prediction from part (b).

21. A basketball has a volume of about 455.9 cubic inches. Find the radius of the basketball (accurate to three decimal places).

22. On the first part of a 350-kilometer trip, a salesperson travels 2 hours and 15 minutes at an average speed of 100 kilometers per hour. The salesperson needs to arrive at the destination in another hour and 20 minutes. Find the average speed required for the remainder of the trip.

23. The area of the ellipse in the figure is $A = \pi ab$. Find a and b such that the area of the ellipse equals the area of the circle when a and b satisfy the constraint $a + b = 100$.

Figure for 23

Proofs in Mathematics ■ ■ ■ ■ ■ ■ ■ ■ ■ ■ ■ ■ ■ ■

Conditional Statements

Many theorems are written in the **if-then** form "if p, then q," which is denoted by

$p \rightarrow q$ Conditional statement

where p is the **hypothesis** and q is the **conclusion.** Here are some other ways to express the conditional statement $p \rightarrow q$.

p implies q. p only if q. p is sufficient for q.

Conditional statements can be either true or false. The conditional statement $p \rightarrow q$ is false only when p is true and q is false. To show that a conditional statement is true, you must prove that the conclusion follows for all cases that fulfill the hypothesis. To show that a conditional statement is false, you need to describe only a single **counterexample** that shows that the statement is not always true.

For instance, $x = -4$ is a counterexample that shows that the statement below is false.

If $x^2 = 16$, then $x = 4$.

The hypothesis "$x^2 = 16$" is true because $(-4)^2 = 16$. However, the conclusion "$x = 4$" is false. This implies that the given conditional statement is false.

For the conditional statement $p \rightarrow q$, there are three important associated conditional statements.

1. The **converse** of $p \rightarrow q$: $q \rightarrow p$

2. The **inverse** of $p \rightarrow q$: $\sim p \rightarrow \sim q$

3. The **contrapositive** of $p \rightarrow q$: $\sim q \rightarrow \sim p$

The symbol $\sim$ means the **negation** of a statement. For example, the negation of "The engine is running" is "The engine is not running."

> **EXAMPLE**   **Writing the Converse, Inverse, and Contrapositive**

Write the converse, inverse, and contrapositive of the conditional statement "If I get a B on my test, then I will pass the course."

Solution

Converse: If I pass the course, then I got a B on my test.

Inverse: If I do not get a B on my test, then I will not pass the course.

Contrapositive: If I do not pass the course, then I did not get a B on my test. ■

In the example above, notice that neither the converse nor the inverse is logically equivalent to the original conditional statement. On the other hand, the contrapositive *is* logically equivalent to the original conditional statement.

P.S. Problem Solving

1. **Time and Distance** Let x represent the time (in seconds), and let y represent the distance (in feet) between you and a tree. Sketch a possible graph that shows how x and y are related when you are walking toward the tree.

2. **Sum of the First n Natural Numbers**

   (a) Find each sum.

   $$1 + 2 + 3 + 4 + 5 = $$

   $$1 + 2 + 3 + 4 + 5 + 6 + 7 + 8 = $$

   $$1 + 2 + 3 + 4 + 5 + 6$$
   $$+ 7 + 8 + 9 + 10 = $$

   (b) Use the formula below for the sum of the first n natural numbers to verify your answers to part (a).

   $$1 + 2 + 3 + \cdots + n = \frac{1}{2}n(n + 1)$$

   (c) Use the formula in part (b) to find n such that the sum of the first n natural numbers is 210.

3. **Area of an Ellipse** The area of an ellipse is equal to $A = \pi ab$ (see figure). For the ellipse below, $a + b = 20$.

   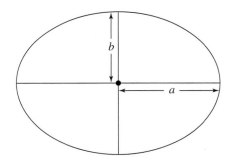

   (a) Show that

   $$A = \pi a(20 - a).$$

   (b) Complete the table.

a	4	7	10	13	16
A					

   (c) Find two values of a such that $A = 300$.

   (d) Use a graphing utility to graph the area equation.

   (e) Find the a-intercepts of the graph of the area equation. What do these values represent?

   (f) What is the maximum area? What values of a and b yield the maximum area?

4. **Using a Graph to Solve an Inequality** Use the graph of

   $$y = x^4 - x^3 - 6x^2 + 4x + 8$$

   to solve the inequality

   $$x^4 - x^3 - 6x^2 + 4x + 8 > 0.$$

5. **Wind Pressure** A building code requires that a building be able to withstand a specific amount of wind pressure. The pressure P (in pounds per square foot) from wind blowing at s miles per hour is given by

   $$P = 0.00256s^2.$$

   (a) A structural engineer is designing a library. The building is required to withstand wind pressure of 20 pounds per square foot. Under this requirement, how fast must the wind blow to produce excessive stress on the building?

   (b) To be safe, the engineer designs the library so that it can withstand wind pressure of 40 pounds per square foot. Does this mean that the library can survive wind blowing at twice the speed you found in part (a)? Justify your answer.

   (c) Use the pressure formula to explain why even a relatively small increase in the wind speed could have potentially serious effects on a building.

6. **Water Height** For a bathtub with a rectangular base, Toricelli's Law implies that the height h of water in the tub t seconds after it begins draining is given by

   $$h = \left(\sqrt{h_0} - \frac{2\pi d^2 \sqrt{3}}{lw} t \right)^2$$

   where l and w are the tub's length and width, d is the diameter of the drain, and h_0 is the water's initial height. (All measurements are in inches.) You completely fill a tub with water. The tub is 60 inches long by 30 inches wide by 25 inches high and has a drain with a two-inch diameter.

   (a) Find the time it takes for the tub to go from being full to half-full.

   (b) Find the time it takes for the tub to go from being half-full to empty.

   (c) Based on your results in parts (a) and (b), what general statement can you make about the speed at which the water drains?

7. **Sum of Squares**

   (a) Consider the sum of squares $x^2 + 9$. If the sum can be factored, then there are integers m and n such that $x^2 + 9 = (x + m)(x + n)$. Write two equations relating the sum and the product of m and n to the coefficients in $x^2 + 9$.

   (b) Show that there are no integers m and n that satisfy both equations you wrote in part (a). What can you conclude?

8. **Finding Values** Use the equation $4\sqrt{x} = 2x + k$ to find three different values of k such that the equation has two solutions, one solution, and no solution. Describe the process you used to find the values.

9. Pythagorean Triples A Pythagorean Triple is a group of three integers, such as 3, 4, and 5, that could be the lengths of the sides of a right triangle.

(a) Find two other Pythagorean Triples.

(b) Notice that $3 \cdot 4 \cdot 5 = 60$. Is the product of the three numbers in each Pythagorean Triple evenly divisible by 3? by 4? by 5?

(c) Write a conjecture involving Pythagorean Triples and divisibility by 60.

10. Sums and Products of Solutions Determine the solutions x_1 and x_2 of each quadratic equation. Use the values of x_1 and x_2 to fill in the boxes.

Equation	x_1, x_2	$x_1 + x_2$	$x_1 \cdot x_2$
(a) $x^2 - x - 6 = 0$			
(b) $2x^2 + 5x - 3 = 0$			
(c) $4x^2 - 9 = 0$			
(d) $x^2 - 10x + 34 = 0$			

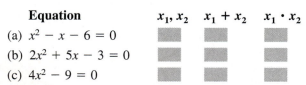

11. Proof The solutions of a quadratic equation are

$$x = \frac{-b \pm \sqrt{b^2 - 4ac}}{2a}.$$

(a) Prove that the sum of the solutions is

$$S = -\frac{b}{a}.$$

(b) Prove that the product of the solutions is

$$P = \frac{c}{a}.$$

12. Principal Cube Root

(a) The principal cube root of 125, $\sqrt[3]{125}$, is 5. Evaluate the expression x^3 for each value of x.

$\quad$ (i) $x = \dfrac{-5 + 5\sqrt{3}i}{2}$

$\quad$ (ii) $x = \dfrac{-5 - 5\sqrt{3}i}{2}$

(b) The principal cube root of 27, $\sqrt[3]{27}$, is 3. Evaluate the expression x^3 for each value of x.

$\quad$ (i) $x = \dfrac{-3 + 3\sqrt{3}i}{2}$

$\quad$ (ii) $x = \dfrac{-3 - 3\sqrt{3}i}{2}$

(c) Use the results of parts (a) and (b) to list possible cube roots of (i) 1, (ii) 8, and (iii) 64. Verify your results algebraically.

13. Multiplicative Inverse of a Complex Number The multiplicative inverse of a complex number z is a complex number z_m such that $z \cdot z_m = 1$. Find the multiplicative inverse of each complex number.

(a) $z = 1 + i$

(b) $z = 3 - i$

(c) $z = -2 + 8i$

14. Proof Prove that the product of a complex number $a + bi$ and its complex conjugate is a real number.

15. The Mandelbrot Set A **fractal** is a geometric figure that consists of a pattern that is repeated infinitely on a smaller and smaller scale. The most famous fractal is the **Mandelbrot Set,** named after the Polish-born mathematician Benoit Mandelbrot (1924–2010). To draw the Mandelbrot Set, consider the sequence of numbers below.

$$c, \ c^2 + c, \ (c^2 + c)^2 + c, \ [(c^2 + c)^2 + c]^2 + c, \ldots$$

The behavior of this sequence depends on the value of the complex number c. If the sequence is bounded (the absolute value of each number in the sequence,

$$|a + bi| = \sqrt{a^2 + b^2}$$

is less than some fixed number N), then the complex number c is in the Mandelbrot Set, and if the sequence is unbounded (the absolute value of the terms of the sequence become infinitely large), then the complex number c is not in the Mandelbrot Set. Determine whether the complex number c is in the Mandelbrot Set.

(a) $c = i$

(b) $c = 1 + i$

(c) $c = -2$

The figure below shows a graph of the Mandelbrot Set, where the horizontal and vertical axes represent the real and imaginary parts of c, respectively.

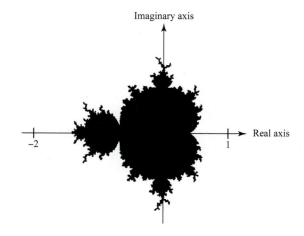

2 Functions and Their Graphs

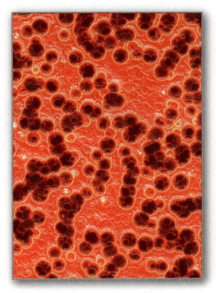

Bacteria *(Example 8, page 218)*

Snowstorm *(Exercise 47, page 204)*

Average Speed *(Example 7, page 192)*

Alternative-Fuel Stations
(Example 10, page 180)

Americans with Disabilities Act *(page 166)*

2.1   Linear Equations in Two Variables

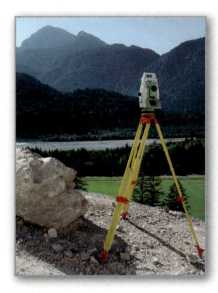

Linear equations in two variables can help you model and solve real-life problems. For example, in Exercise 90 on page 171, you will use a surveyor's measurements to find a linear equation that models a mountain road.

■ Use slope to graph linear equations in two variables.
■ Find the slope of a line given two points on the line.
■ Write linear equations in two variables.
■ Use slope to identify parallel and perpendicular lines.
■ Use slope and linear equations in two variables to model and solve real-life problems.

Using Slope

The simplest mathematical model for relating two variables is the **linear equation in two variables** $y = mx + b$. The equation is called *linear* because its graph is a line. (In mathematics, the term *line* means *straight line*.) By letting $x = 0$, you obtain

$$y = m(0) + b = b.$$

So, the line crosses the y-axis at $y = b$, as shown in the figures below. In other words, the y-intercept is $(0, b)$. The steepness, or *slope,* of the line is m.

$$y = mx + b$$

Slope ———┘   └——— y-Intercept

The **slope** of a nonvertical line is the number of units the line rises (or falls) vertically for each unit of horizontal change from left to right, as shown below.

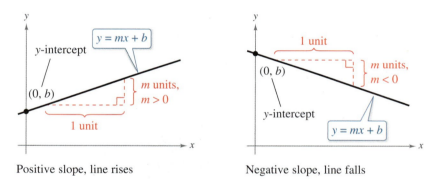

Positive slope, line rises   Negative slope, line falls

A linear equation written in **slope-intercept form** has the form $y = mx + b$.

The Slope-Intercept Form of the Equation of a Line

The graph of the equation

$$y = mx + b$$

is a line whose slope is m and whose y-intercept is $(0, b)$.

Once you determine the slope and the y-intercept of a line, it is relatively simple to sketch its graph. In the next example, note that none of the lines is vertical. A vertical line has an equation of the form

$$x = a. \qquad \text{Vertical line}$$

The equation of a vertical line cannot be written in the form $y = mx + b$ because the slope of a vertical line is undefined (see Figure 2.1).

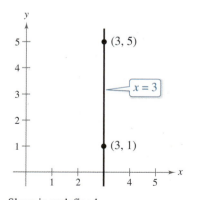

Slope is undefined.
Figure 2.1

> **EXAMPLE 1**   **Graphing Linear Equations**

See LarsonPrecalculus.com for an interactive version of this type of example.

Sketch the graph of each linear equation.

a. $y = 2x + 1$

b. $y = 2$

c. $x + y = 2$

Solution

a. Because $b = 1$, the y-intercept is $(0, 1)$. Moreover, the slope is $m = 2$, so the line *rises* two units for each unit the line moves to the right (see figure).

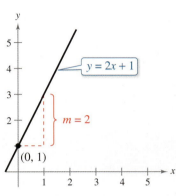

When m is positive, the line rises.

b. By writing this equation in the form $y = (0)x + 2$, you find that the y-intercept is $(0, 2)$ and the slope is $m = 0$. A slope of 0 implies that the line is horizontal—that is, it does not rise *or* fall (see figure).

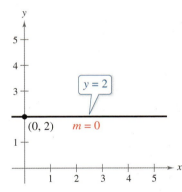

When m is 0, the line is horizontal.

c. By writing this equation in slope-intercept form

$$x + y = 2 \qquad \text{Write original equation.}$$

$$y = -x + 2 \qquad \text{Subtract } x \text{ from each side.}$$

$$y = (-1)x + 2 \qquad \text{Write in slope-intercept form.}$$

you find that the y-intercept is $(0, 2)$. Moreover, the slope is $m = -1$, so the line *falls* one unit for each unit the line moves to the right (see figure).

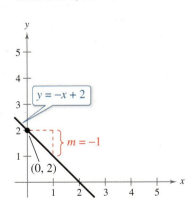

When m is negative, the line falls.

✓ *Checkpoint* 🔊))) *Audio-video solution in English & Spanish at LarsonPrecalculus.com*

Sketch the graph of each linear equation.

a. $y = 3x + 2$ **b.** $y = -3$ **c.** $4x + y = 5$

Finding the Slope of a Line

Given an equation of a line, you can find its slope by writing the equation in slope-intercept form. When you are not given an equation, you can still find the slope by using two points on the line. For example, consider the line passing through the points (x_1, y_1) and (x_2, y_2) in the figure below.

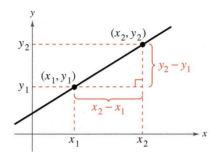

As you move from left to right along this line, a change of $(y_2 - y_1)$ units in the vertical direction corresponds to a change of $(x_2 - x_1)$ units in the horizontal direction.

$$y_2 - y_1 = \text{change in } y = \text{rise}$$

and

$$x_2 - x_1 = \text{change in } x = \text{run}$$

The ratio of $(y_2 - y_1)$ to $(x_2 - x_1)$ represents the slope of the line that passes through the points (x_1, y_1) and (x_2, y_2).

$$\text{Slope} = \frac{\text{change in } y}{\text{change in } x} = \frac{\text{rise}}{\text{run}} = \frac{y_2 - y_1}{x_2 - x_1}$$

The Slope of a Line Passing Through Two Points

The **slope** m of the nonvertical line through (x_1, y_1) and (x_2, y_2) is

$$m = \frac{y_2 - y_1}{x_2 - x_1}$$

where $x_1 \neq x_2$.

When using the formula for slope, the *order of subtraction* is important. Given two points on a line, you are free to label either one of them as (x_1, y_1) and the other as (x_2, y_2). However, once you do this, you must form the numerator and denominator using the same order of subtraction.

$$m = \frac{y_2 - y_1}{x_2 - x_1} \qquad m = \frac{y_1 - y_2}{x_1 - x_2} \qquad m = \frac{y_2 - y_1}{x_1 - x_2} \quad \times$$

Correct   Correct   Incorrect

For example, the slope of the line passing through the points $(3, 4)$ and $(5, 7)$ can be calculated as

$$m = \frac{7 - 4}{5 - 3} = \frac{3}{2}$$

or as

$$m = \frac{4 - 7}{3 - 5} = \frac{-3}{-2} = \frac{3}{2}.$$

EXAMPLE 2 **Finding the Slope of a Line Through Two Points**

Find the slope of the line passing through each pair of points.

a. $(-2, 0)$ and $(3, 1)$ **b.** $(-1, 2)$ and $(2, 2)$

c. $(0, 4)$ and $(1, -1)$ **d.** $(3, 4)$ and $(3, 1)$

Solution

a. Letting $(x_1, y_1) = (-2, 0)$ and $(x_2, y_2) = (3, 1)$, you find that the slope is

$$m = \frac{y_2 - y_1}{x_2 - x_1} = \frac{1 - 0}{3 - (-2)} = \frac{1}{5}.$$ See Figure 2.2.

b. The slope of the line passing through $(-1, 2)$ and $(2, 2)$ is

$$m = \frac{2 - 2}{2 - (-1)} = \frac{0}{3} = 0.$$ See Figure 2.3.

c. The slope of the line passing through $(0, 4)$ and $(1, -1)$ is

$$m = \frac{-1 - 4}{1 - 0} = \frac{-5}{1} = -5.$$ See Figure 2.4.

d. The slope of the line passing through $(3, 4)$ and $(3, 1)$ is

$$m = \frac{1 - 4}{3 - 3} = \frac{-3}{0}.$$ 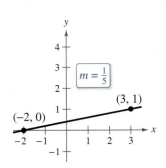 See Figure 2.5.

Division by 0 is undefined, so the slope is undefined and the line is vertical.

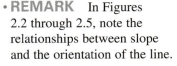

•• **REMARK** In Figures
2.2 through 2.5, note the
relationships between slope
and the orientation of the line.

a. Positive slope: line rises
from left to right

b. Zero slope: line is horizontal

c. Negative slope: line falls
from left to right

d. Undefined slope: line is
vertical

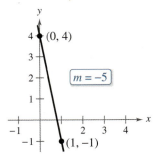

Figure 2.2

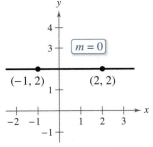

Figure 2.3

Figure 2.4

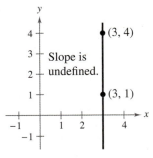

Figure 2.5

✓ *Checkpoint* ◄))) *Audio-video solution in English & Spanish at LarsonPrecalculus.com*

Find the slope of the line passing through each pair of points.

a. $(-5, -6)$ and $(2, 8)$ **b.** $(4, 2)$ and $(2, 5)$

c. $(0, 0)$ and $(0, -6)$ **d.** $(0, -1)$ and $(3, -1)$

Writing Linear Equations in Two Variables

If (x_1, y_1) is a point on a line of slope m and (x, y) is *any other* point on the line, then

$$\frac{y - y_1}{x - x_1} = m.$$

This equation in the variables x and y can be rewritten in the **point-slope form** of the equation of a line

$$y - y_1 = m(x - x_1).$$

> ### Point-Slope Form of the Equation of a Line
>
> The equation of the line with slope m passing through the point (x_1, y_1) is
>
> $$y - y_1 = m(x - x_1).$$

The point-slope form is useful for *finding* the equation of a line. You should remember this form.

EXAMPLE 3   **Using the Point-Slope Form**

Find the slope-intercept form of the equation of the line that has a slope of 3 and passes through the point $(1, -2)$.

Solution   Use the point-slope form with $m = 3$ and $(x_1, y_1) = (1, -2)$.

$y - y_1 = m(x - x_1)$	Point-slope form
$y - (-2) = 3(x - 1)$	Substitute for m, x_1, and y_1.
$y + 2 = 3x - 3$	Simplify.
$y = 3x - 5$	Write in slope-intercept form.

The slope-intercept form of the equation of the line is $y = 3x - 5$. Figure 2.6 shows the graph of this equation.

✓ *Checkpoint* 🔊))) *Audio-video solution in English & Spanish at LarsonPrecalculus.com*

Find the slope-intercept form of the equation of the line that has the given slope and passes through the given point.

a. $m = 2$, $(3, -7)$

b. $m = -\frac{2}{3}$, $(1, 1)$

c. $m = 0$, $(1, 1)$

The point-slope form can be used to find an equation of the line passing through two points (x_1, y_1) and (x_2, y_2). To do this, first find the slope of the line.

$$m = \frac{y_2 - y_1}{x_2 - x_1}, \quad x_1 \neq x_2$$

Then use the point-slope form to obtain the equation.

$$y - y_1 = \frac{y_2 - y_1}{x_2 - x_1}(x - x_1) \qquad \text{Two-point form}$$

This is sometimes called the **two-point form** of the equation of a line.

Figure 2.6

• • REMARK When you find an equation of the line that passes through two given points, you only need to substitute the coordinates of one of the points in the point-slope form. It does not matter which point you choose because both points will yield the same result.

Parallel and Perpendicular Lines

Slope can tell you whether two nonvertical lines in a plane are parallel, perpendicular, or neither.

Parallel and Perpendicular Lines

1. Two distinct nonvertical lines are **parallel** if and only if their slopes are equal. That is,

$$m_1 = m_2.$$

2. Two nonvertical lines are **perpendicular** if and only if their slopes are negative reciprocals of each other. That is,

$$m_1 = \frac{-1}{m_2}.$$

EXAMPLE 4   **Finding Parallel and Perpendicular Lines**

Find the slope-intercept form of the equations of the lines that pass through the point $(2, -1)$ and are (a) parallel to and (b) perpendicular to the line $2x - 3y = 5$.

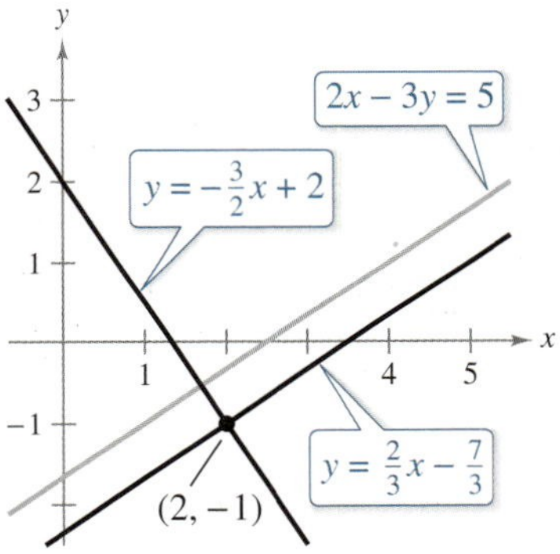

Solution   Write the equation of the given line in slope-intercept form.

$2x - 3y = 5$	Write original equation.
$-3y = -2x + 5$	Subtract $2x$ from each side.
$y = \frac{2}{3}x - \frac{5}{3}$	Write in slope-intercept form.

Notice that the line has a slope of $m = \frac{2}{3}$.

a. Any line parallel to the given line must also have a slope of $\frac{2}{3}$. Use the point-slope form with $m = \frac{2}{3}$ and $(x_1, y_1) = (2, -1)$.

$y - (-1) = \frac{2}{3}(x - 2)$	Write in point-slope form.
$3(y + 1) = 2(x - 2)$	Multiply each side by 3.
$3y + 3 = 2x - 4$	Distributive Property
$y = \frac{2}{3}x - \frac{7}{3}$	Write in slope-intercept form.

Notice the similarity between the slope-intercept form of this equation and the slope-intercept form of the given equation.

b. Any line perpendicular to the given line must have a slope of $-\frac{3}{2}$ (because $-\frac{3}{2}$ is the negative reciprocal of $\frac{2}{3}$). Use the point-slope form with $m = -\frac{3}{2}$ and $(x_1, y_1) = (2, -1)$.

$y - (-1) = -\frac{3}{2}(x - 2)$	Write in point-slope form.
$2(y + 1) = -3(x - 2)$	Multiply each side by 2.
$2y + 2 = -3x + 6$	Distributive Property
$y = -\frac{3}{2}x + 2$	Write in slope-intercept form.

The graphs of all three equations are shown in Figure 2.7.

✓ **Checkpoint** *Audio-video solution in English & Spanish at LarsonPrecalculus.com*

Find the slope-intercept form of the equations of the lines that pass through the point $(-4, 1)$ and are (a) parallel to and (b) perpendicular to the line $5x - 3y = 8$. ∎

Figure 2.7

$2x - 3y = 5$

$y = -\frac{3}{2}x + 2$

$y = \frac{2}{3}x - \frac{7}{3}$

$(2, -1)$

▷ **TECHNOLOGY** On a graphing utility, lines will not appear to have the correct slope unless you use a viewing window that has a square setting. For instance, graph the lines in Example 4 using the standard setting $-10 \le x \le 10$ and $-10 \le y \le 10$. Then reset the viewing window with the square setting $-9 \le x \le 9$ and $-6 \le y \le 6$. On which setting do the lines $y = \frac{2}{3}x - \frac{5}{3}$ and $y = -\frac{3}{2}x + 2$ appear to be perpendicular?

Applications

In real-life problems, the slope of a line can be interpreted as either a *ratio* or a *rate*. When the *x*-axis and *y*-axis have the same unit of measure, the slope has no units and is a **ratio.** When the *x*-axis and *y*-axis have different units of measure, the slope is a **rate** or **rate of change.**

EXAMPLE 5   Using Slope as a Ratio

The maximum recommended slope of a wheelchair ramp is $\frac{1}{12}$. A business installs a wheelchair ramp that rises 22 inches over a horizontal length of 24 feet. Is the ramp steeper than recommended?   *(Source: ADA Standards for Accessible Design)*

Solution   The horizontal length of the ramp is 24 feet or 12(24) = 288 inches (see figure). So, the slope of the ramp is

$$\text{Slope} = \frac{\text{vertical change}}{\text{horizontal change}} = \frac{22 \text{ in.}}{288 \text{ in.}} \approx 0.076.$$

Because $\frac{1}{12} \approx 0.083$, the slope of the ramp is not steeper than recommended.

The Americans with Disabilities Act (ADA) became law on July 26, 1990. It is the most comprehensive formulation of rights for persons with disabilities in U.S. (and world) history.

22 in.

24 ft

 Checkpoint)) *Audio-video solution in English & Spanish at LarsonPrecalculus.com*

The business in Example 5 installs a second ramp that rises 36 inches over a horizontal length of 32 feet. Is the ramp steeper than recommended?

EXAMPLE 6   Using Slope as a Rate of Change

A kitchen appliance manufacturing company determines that the total cost C (in dollars) of producing x units of a blender is given by

$$C = 25x + 3500.\qquad \text{Cost equation}$$

Interpret the *y*-intercept and slope of this line.

Solution   The *y*-intercept (0, 3500) tells you that the cost of producing 0 units is $3500. This is the *fixed cost* of production—it includes costs that must be paid regardless of the number of units produced. The slope of $m = 25$ tells you that the cost of producing each unit is $25, as shown in Figure 2.8. Economists call the cost per unit the *marginal cost.* When the production increases by one unit, the "margin," or extra amount of cost, is $25. So, the cost increases at a rate of $25 per unit.

 Checkpoint)) *Audio-video solution in English & Spanish at LarsonPrecalculus.com*

An accounting firm determines that the value V (in dollars) of a copier t years after its purchase is given by

$$V = -300t + 1500.$$

Interpret the *y*-intercept and slope of this line.

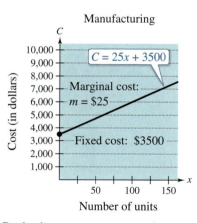

Manufacturing

$C = 25x + 3500$

Marginal cost: $m = \$25$

Fixed cost: $3500

Number of units

Production cost

Figure 2.8

Businesses can deduct most of their expenses in the same year they occur. One exception is the cost of property that has a useful life of more than 1 year. Such costs must be *depreciated* (decreased in value) over the useful life of the property. Depreciating the *same amount* each year is called *linear* or *straight-line depreciation*. The *book value* is the difference between the original value and the total amount of depreciation accumulated to date.

EXAMPLE 7   **Straight-Line Depreciation**

A college purchased exercise equipment worth $12,000 for the new campus fitness center. The equipment has a useful life of 8 years. The salvage value at the end of 8 years is $2000. Write a linear equation that describes the book value of the equipment each year.

Solution   Let V represent the value of the equipment at the end of year t. Represent the initial value of the equipment by the data point $(0, 12,000)$ and the salvage value of the equipment by the data point $(8, 2000)$. The slope of the line is

$$m = \frac{2000 - 12{,}000}{8 - 0} = -\$1250$$

which represents the annual depreciation in *dollars per year*. Using the point-slope form, write an equation of the line.

$$V - 12{,}000 = -1250(t - 0) \qquad \text{Write in point-slope form.}$$
$$V = -1250t + 12{,}000 \qquad \text{Write in slope-intercept form.}$$

The table shows the book value at the end of each year, and Figure 2.9 shows the graph of the equation.

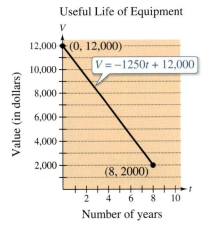

Useful Life of Equipment

$V = -1250t + 12{,}000$

Straight-line depreciation
Figure 2.9

Year, t	Value, V
0	12,000
1	10,750
2	9500
3	8250
4	7000
5	5750
6	4500
7	3250
8	2000

✓ *Checkpoint* 🔊)) *Audio-video solution in English & Spanish at LarsonPrecalculus.com*

A manufacturing firm purchases a machine worth $24,750. The machine has a useful life of 6 years. After 6 years, the machine will have to be discarded and replaced, because it will have no salvage value. Write a linear equation that describes the book value of the machine each year.

In many real-life applications, the two data points that determine the line are often given in a disguised form. Note how the data points are described in Example 7.

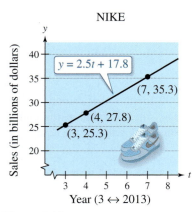

Figure 2.10

EXAMPLE 8 **Predicting Sales**

The sales for NIKE were approximately $25.3 billion in 2013 and $27.8 billion in 2014. Using only this information, write a linear equation that gives the sales in terms of the year. Then predict the sales in 2017.   *(Source: NIKE Inc.)*

Solution   Let $t = 3$ represent 2013. Then the two given values are represented by the data points $(3, 25.3)$ and $(4, 27.8)$ The slope of the line through these points is

$$m = \frac{27.8 - 25.3}{4 - 3} = 2.5.$$

Use the point-slope form to write an equation that relates the sales y and the year t.

$$y - 25.3 = 2.5(t - 3) \qquad \text{Write in point-slope form.}$$
$$y = 2.5t + 17.8 \qquad \text{Write in slope-intercept form.}$$

According to this equation, the sales in 2017 will be

$$y = 2.5(7) + 17.8 = 17.5 + 17.8 = \$35.3 \text{ billion. (See Figure 2.10.)}$$

✓ *Checkpoint* ◀))) *Audio-video solution in English & Spanish at LarsonPrecalculus.com*

The sales for Foot Locker were approximately $6.5 billion in 2013 and $7.2 billion in 2014. Repeat Example 8 using this information.   *(Source: Foot Locker)* ◼

The prediction method illustrated in Example 8 is called **linear extrapolation.** Note in Figure 2.11 that an extrapolated point does not lie between the given points. When the estimated point lies between two given points, as shown in Figure 2.12, the procedure is called **linear interpolation.**

The slope of a vertical line is undefined, so its equation cannot be written in slope-intercept form. However, every line has an equation that can be written in the **general form** $Ax + By + C = 0$, where A and B are not both zero.

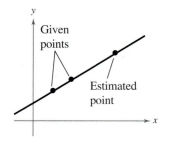

Linear extrapolation
Figure 2.11

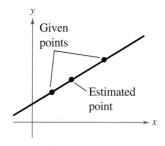

Linear interpolation
Figure 2.12

Summary of Equations of Lines

1. General form: $Ax + By + C = 0$
2. Vertical line: $x = a$
3. Horizontal line: $y = b$
4. Slope-intercept form: $y = mx + b$
5. Point-slope form: $y - y_1 = m(x - x_1)$
6. Two-point form: $y - y_1 = \dfrac{y_2 - y_1}{x_2 - x_1}(x - x_1)$

Summarize   (Section 2.1)

1. Explain how to use slope to graph a linear equation in two variables *(page 160)* and how to find the slope of a line passing through two points *(page 162)*. For examples of using and finding slopes, see Examples 1 and 2.

2. State the point-slope form of the equation of a line *(page 164)*. For an example of using point-slope form, see Example 3.

3. Explain how to use slope to identify parallel and perpendicular lines *(page 165)*. For an example of finding parallel and perpendicular lines, see Example 4.

4. Describe examples of how to use slope and linear equations in two variables to model and solve real-life problems *(pages 166–168, Examples 5–8)*.

2.1 Exercises

See CalcChat.com for tutorial help and worked-out solutions to odd-numbered exercises.

Vocabulary: Fill in the blanks.

1. The simplest mathematical model for relating two variables is the _____ equation in two variables $y = mx + b$.

2. For a line, the ratio of the change in y to the change in x is the _____ of the line.

3. The _____-_____ form of the equation of a line with slope m passing through the point (x_1, y_1) is $y - y_1 = m(x - x_1)$.

4. Two distinct nonvertical lines are _____ if and only if their slopes are equal.

5. Two nonvertical lines are _____ if and only if their slopes are negative reciprocals of each other.

6. When the x-axis and y-axis have different units of measure, the slope can be interpreted as a _____.

7. _____ _____ is the prediction method used to estimate a point on a line when the point does not lie between the given points.

8. Every line has an equation that can be written in _____ form.

Skills and Applications

Identifying Lines In Exercises 9 and 10, identify the line that has each slope.

9. (a) $m = \frac{2}{3}$
   (b) m is undefined.
   (c) $m = -2$

10. (a) $m = 0$
    (b) $m = -\frac{3}{4}$
    (c) $m = 1$

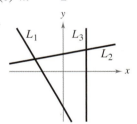

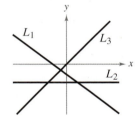

Sketching Lines In Exercises 11 and 12, sketch the lines through the point with the given slopes on the same set of coordinate axes.

Point	Slopes
11. $(2, 3)$	(a) 0 (b) 1
	(c) 2 (d) -3
12. $(-4, 1)$	(a) 3 (b) -3
	(c) $\frac{1}{2}$ (d) Undefined

Estimating the Slope of a Line In Exercises 13 and 14, estimate the slope of the line.

13.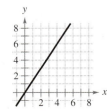

14.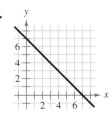

Graphing a Linear Equation In Exercises 15–24, find the slope and y-intercept (if possible) of the line. Sketch the line.

15. $y = 5x + 3$
16. $y = -x - 10$
17. $y = -\frac{3}{4}x - 1$
18. $y = \frac{2}{3}x + 2$
19. $y - 5 = 0$
20. $x + 4 = 0$
21. $5x - 2 = 0$
22. $3y + 5 = 0$
23. $7x - 6y = 30$
24. $2x + 3y = 9$

Finding the Slope of a Line Through Two Points In Exercises 25–34, find the slope of the line passing through the pair of points.

25. $(0, 9), (6, 0)$
26. $(10, 0), (0, -5)$
27. $(-3, -2), (1, 6)$
28. $(2, -1), (-2, 1)$
29. $(5, -7), (8, -7)$
30. $(-2, 1), (-4, -5)$
31. $(-6, -1), (-6, 4)$
32. $(0, -10), (-4, 0)$
33. $(4.8, 3.1), (-5.2, 1.6)$
34. $\left(\frac{11}{2}, -\frac{4}{3}\right), \left(-\frac{3}{2}, -\frac{1}{3}\right)$

Using the Slope and a Point In Exercises 35–42, use the slope of the line and the point on the line to find three additional points through which the line passes. (There are many correct answers.)

35. $m = 0, \quad (5, 7)$
36. $m = 0, \quad (3, -2)$
37. $m = 2, \quad (-5, 4)$
38. $m = -2, \quad (0, -9)$
39. $m = -\frac{1}{3}, \quad (4, 5)$
40. $m = \frac{1}{4}, \quad (3, -4)$
41. m is undefined, $\quad (-4, 3)$
42. m is undefined, $\quad (2, 14)$

Using the Point-Slope Form In Exercises 43–54, find the slope-intercept form of the equation of the line that has the given slope and passes through the given point. Sketch the line.

43. $m = 3$, $(0, -2)$
44. $m = -1$, $(0, 10)$
45. $m = -2$, $(-3, 6)$
46. $m = 4$, $(0, 0)$
47. $m = -\frac{1}{3}$, $(4, 0)$
48. $m = \frac{1}{4}$, $(8, 2)$
49. $m = -\frac{1}{2}$, $(2, -3)$
50. $m = \frac{3}{4}$, $(-2, -5)$
51. $m = 0$, $\left(4, \frac{5}{2}\right)$
52. $m = 6$, $\left(2, \frac{3}{2}\right)$
53. $m = 5$, $(-5.1, 1.8)$
54. $m = 0$, $(-2.5, 3.25)$

Finding an Equation of a Line In Exercises 55–64, find an equation of the line passing through the pair of points. Sketch the line.

55. $(5, -1), (-5, 5)$
56. $(4, 3), (-4, -4)$
57. $(-7, 2), (-7, 5)$
58. $(-6, -3), (2, -3)$
59. $\left(2, \frac{1}{2}\right), \left(\frac{1}{2}, \frac{5}{4}\right)$
60. $(1, 1), \left(6, -\frac{2}{3}\right)$
61. $(1, 0.6), (-2, -0.6)$
62. $(-8, 0.6), (2, -2.4)$
63. $(2, -1), \left(\frac{1}{3}, -1\right)$
64. $\left(\frac{7}{3}, -8\right), \left(\frac{7}{3}, 1\right)$

Parallel and Perpendicular Lines In Exercises 65–68, determine whether the lines are parallel, perpendicular, or neither.

65. L_1: $y = -\frac{2}{3}x - 3$
    L_2: $y = -\frac{2}{3}x + 4$
66. L_1: $y = \frac{1}{4}x - 1$
    L_2: $y = 4x + 7$
67. L_1: $y = \frac{1}{2}x - 3$
    L_2: $y = -\frac{1}{2}x + 1$
68. L_1: $y = -\frac{4}{5}x - 5$
    L_2: $y = \frac{5}{4}x + 1$

Parallel and Perpendicular Lines In Exercises 69–72, determine whether the lines L_1 and L_2 passing through the pairs of points are parallel, perpendicular, or neither.

69. L_1: $(0, -1), (5, 9)$
    L_2: $(0, 3), (4, 1)$
70. L_1: $(-2, -1), (1, 5)$
    L_2: $(1, 3), (5, -5)$
71. L_1: $(-6, -3), (2, -3)$
    L_2: $\left(3, -\frac{1}{2}\right), \left(6, -\frac{1}{2}\right)$
72. L_1: $(4, 8), (-4, 2)$
    L_2: $(3, -5), \left(-1, \frac{1}{3}\right)$

Finding Parallel and Perpendicular Lines In Exercises 73–80, find equations of the lines that pass through the given point and are (a) parallel to and (b) perpendicular to the given line.

73. $4x - 2y = 3$, $(2, 1)$
74. $x + y = 7$, $(-3, 2)$
75. $3x + 4y = 7$, $\left(-\frac{2}{3}, \frac{7}{8}\right)$
76. $5x + 3y = 0$, $\left(\frac{7}{8}, \frac{3}{4}\right)$
77. $y + 5 = 0$, $(-2, 4)$
78. $x - 4 = 0$, $(3, -2)$
79. $x - y = 4$, $(2.5, 6.8)$
80. $6x + 2y = 9$, $(-3.9, -1.4)$

Using Intercept Form In Exercises 81–86, use the *intercept form* to find the general form of the equation of the line with the given intercepts. The intercept form of the equation of a line with intercepts $(a, 0)$ and $(0, b)$ is

$$\frac{x}{a} + \frac{y}{b} = 1, \quad a \neq 0, \quad b \neq 0.$$

81. x-intercept: $(3, 0)$
    y-intercept: $(0, 5)$
82. x-intercept: $(-3, 0)$
    y-intercept: $(0, 4)$
83. x-intercept: $\left(-\frac{1}{6}, 0\right)$
    y-intercept: $\left(0, -\frac{2}{3}\right)$
84. x-intercept: $\left(\frac{2}{3}, 0\right)$
    y-intercept: $(0, -2)$
85. Point on line: $(1, 2)$
    x-intercept: $(c, 0)$, $c \neq 0$
    y-intercept: $(0, c)$, $c \neq 0$
86. Point on line: $(-3, 4)$
    x-intercept: $(d, 0)$, $d \neq 0$
    y-intercept: $(0, d)$, $d \neq 0$

87. **Sales** The slopes of lines representing annual sales y in terms of time x in years are given below. Use the slopes to interpret any change in annual sales for a one-year increase in time.
    (a) The line has a slope of $m = 135$.
    (b) The line has a slope of $m = 0$.
    (c) The line has a slope of $m = -40$.

88. **Sales** The graph shows the sales (in billions of dollars) for Apple Inc. in the years 2009 through 2015. (*Source: Apple Inc.*)

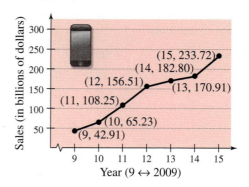

(a) Use the slopes of the line segments to determine the years in which the sales showed the greatest increase and the least increase.

(b) Find the slope of the line segment connecting the points for the years 2009 and 2015.

(c) Interpret the meaning of the slope in part (b) in the context of the problem.

89. Road Grade   You are driving on a road that has a 6% uphill grade. This means that the slope of the road is $\frac{6}{100}$. Approximate the amount of vertical change in your position when you drive 200 feet.

90. Road Grade

From the top of a mountain road, a surveyor takes several horizontal measurements x and several vertical measurements y, as shown in the table (x and y are measured in feet).

x	300	600	900	1200
y	-25	-50	-75	-100

x	1500	1800	2100
y	-125	-150	-175

(a) Sketch a scatter plot of the data.

(b) Use a straightedge to sketch the line that you think best fits the data.

(c) Find an equation for the line you sketched in part (b).

(d) Interpret the meaning of the slope of the line in part (c) in the context of the problem.

(e) The surveyor needs to put up a road sign that indicates the steepness of the road. For example, a surveyor would put up a sign that states "8% grade" on a road with a downhill grade that has a slope of $-\frac{8}{100}$. What should the sign state for the road in this problem?

Rate of Change   **In Exercises 91 and 92, you are given the dollar value of a product in 2016 and the rate at which the value of the product is expected to change during the next 5 years. Use this information to write a linear equation that gives the dollar value V of the product in terms of the year t. (Let $t = 16$ represent 2016.)**

	2016 Value	Rate
91.	$3000	$150 decrease per year
92.	$200	$6.50 increase per year

93. Cost   The cost C of producing n computer laptop bags is given by

$$C = 1.25n + 15{,}750, \quad n > 0.$$

Explain what the C-intercept and the slope represent.

94. Monthly Salary   A pharmaceutical salesperson receives a monthly salary of $5000 plus a commission of 7% of sales. Write a linear equation for the salesperson's monthly wage W in terms of monthly sales S.

95. Depreciation   A sandwich shop purchases a used pizza oven for $875. After 5 years, the oven will have to be discarded and replaced. Write a linear equation giving the value V of the equipment during the 5 years it will be in use.

96. Depreciation   A school district purchases a high-volume printer, copier, and scanner for $24,000. After 10 years, the equipment will have to be replaced. Its value at that time is expected to be $2000. Write a linear equation giving the value V of the equipment during the 10 years it will be in use.

97. Temperature Conversion   Write a linear equation that expresses the relationship between the temperature in degrees Celsius C and degrees Fahrenheit F. Use the fact that water freezes at 0°C (32°F) and boils at 100°C (212°F).

98. Neurology   The average weight of a male child's brain is 970 grams at age 1 and 1270 grams at age 3. *(Source: American Neurological Association)*

(a) Assuming that the relationship between brain weight y and age t is linear, write a linear model for the data.

(b) What is the slope and what does it tell you about brain weight?

(c) Use your model to estimate the average brain weight at age 2.

(d) Use your school's library, the Internet, or some other reference source to find the actual average brain weight at age 2. How close was your estimate?

(e) Do you think your model could be used to determine the average brain weight of an adult? Explain.

99. Cost, Revenue, and Profit   A roofing contractor purchases a shingle delivery truck with a shingle elevator for $42,000. The vehicle requires an average expenditure of $9.50 per hour for fuel and maintenance, and the operator is paid $11.50 per hour.

(a) Write a linear equation giving the total cost C of operating this equipment for t hours. (Include the purchase cost of the equipment.)

(b) Assuming that customers are charged $45 per hour of machine use, write an equation for the revenue R obtained from t hours of use.

(c) Use the formula for profit $P = R - C$ to write an equation for the profit obtained from t hours of use.

(d) Use the result of part (c) to find the break-even point—that is, the number of hours this equipment must be used to yield a profit of 0 dollars.

100. Geometry The length and width of a rectangular garden are 15 meters and 10 meters, respectively. A walkway of width x surrounds the garden.

    (a) Draw a diagram that gives a visual representation of the problem.

    (b) Write the equation for the perimeter y of the walkway in terms of x.

    (c) Use a graphing utility to graph the equation for the perimeter.

    (d) Determine the slope of the graph in part (c). For each additional one-meter increase in the width of the walkway, determine the increase in its perimeter.

Exploration

True or False? **In Exercises 101 and 102, determine whether the statement is true or false. Justify your answer.**

101. A line with a slope of $-\frac{5}{7}$ is steeper than a line with a slope of $-\frac{6}{7}$.

102. The line through $(-8, 2)$ and $(-1, 4)$ and the line through $(0, -4)$ and $(-7, 7)$ are parallel.

103. Right Triangle Explain how you can use slope to show that the points $A(-1, 5)$, $B(3, 7)$, and $C(5, 3)$ are the vertices of a right triangle.

104. Vertical Line Explain why the slope of a vertical line is undefined.

105. Error Analysis Describe the error.

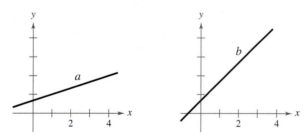

Line b has a greater slope than line a. ✗

106. Perpendicular Segments Find d_1 and d_2 in terms of m_1 and m_2, respectively (see figure). Then use the Pythagorean Theorem to find a relationship between m_1 and m_2.

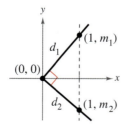

107. Think About It Is it possible for two lines with positive slopes to be perpendicular? Explain.

108. Slope and Steepness The slopes of two lines are -4 and $\frac{5}{2}$. Which is steeper? Explain.

109. Comparing Slopes Use a graphing utility to compare the slopes of the lines $y = mx$, where $m = 0.5$, 1, 2, and 4. Which line rises most quickly? Now, let $m = -0.5$, -1, -2, and -4. Which line falls most quickly? Use a square setting to obtain a true geometric perspective. What can you conclude about the slope and the "rate" at which the line rises or falls?

110. HOW DO YOU SEE IT? Match the description of the situation with its graph. Also determine the slope and y-intercept of each graph and interpret the slope and y-intercept in the context of the situation. [The graphs are labeled (i), (ii), (iii), and (iv).]

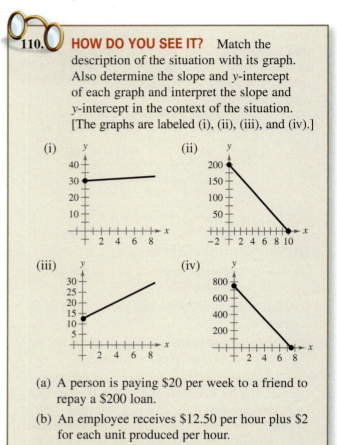

    (a) A person is paying \$20 per week to a friend to repay a \$200 loan.

    (b) An employee receives \$12.50 per hour plus \$2 for each unit produced per hour.

    (c) A sales representative receives \$30 per day for food plus \$0.32 for each mile traveled.

    (d) A computer that was purchased for \$750 depreciates \$100 per year.

Finding a Relationship for Equidistance **In Exercises 111–114, find a relationship between x and y such that (x, y) is equidistant (the same distance) from the two points.**

111. $(4, -1)$, $(-2, 3)$ **112.** $(6, 5)$, $(1, -8)$

113. $\left(3, \frac{5}{2}\right)$, $(-7, 1)$ **114.** $\left(-\frac{1}{2}, -4\right)$, $\left(\frac{7}{2}, \frac{5}{4}\right)$

Project: Bachelor's Degrees To work an extended application analyzing the numbers of bachelor's degrees earned by women in the United States from 2002 through 2013, visit this text's website at *LarsonPrecalculus.com.* (*Source: National Center for Education Statistics*)

2.2 Functions

Functions are used to model and solve real-life problems. For example, in Exercise 70 on page 185, you will use a function that models the force of water against the face of a dam.

■ **Determine whether relations between two variables are functions, and use function notation.**
■ **Find the domains of functions.**
■ **Use functions to model and solve real-life problems.**
■ **Evaluate difference quotients.**

Introduction to Functions and Function Notation

Many everyday phenomena involve two quantities that are related to each other by some rule of correspondence. The mathematical term for such a rule of correspondence is a **relation.** In mathematics, equations and formulas often represent relations. For example, the simple interest I earned on \$1000 for 1 year is related to the annual interest rate r by the formula $I = 1000r$.

The formula $I = 1000r$ represents a special kind of relation that matches each item from one set with *exactly one* item from a different set. Such a relation is a **function.**

Definition of Function

A **function** f from a set A to a set B is a relation that assigns to each element x in the set A exactly one element y in the set B. The set A is the **domain** (or set of inputs) of the function f, and the set B contains the **range** (or set of outputs).

To help understand this definition, look at the function below, which relates the time of day to the temperature.

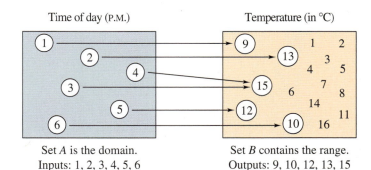

Set A is the domain.   Set B contains the range.
Inputs: 1, 2, 3, 4, 5, 6   Outputs: 9, 10, 12, 13, 15

The ordered pairs below can represent this function. The first coordinate (x-value) is the input and the second coordinate (y-value) is the output.

$$\{(1, 9), (2, 13), (3, 15), (4, 15), (5, 12), (6, 10)\}$$

Characteristics of a Function from Set A to Set B

1. Each element in A must be matched with an element in B.

2. Some elements in B may not be matched with any element in A.

3. Two or more elements in A may be matched with the same element in B.

4. An element in A (the domain) cannot be matched with two different elements in B.

Here are four common ways to represent functions.

> **Four Ways to Represent a Function**
>
> 1. *Verbally* by a sentence that describes how the input variable is related to the output variable
>
> 2. *Numerically* by a table or a list of ordered pairs that matches input values with output values
>
> 3. *Graphically* by points in a coordinate plane in which the horizontal positions represent the input values and the vertical positions represent the output values
>
> 4. *Algebraically* by an equation in two variables

To determine whether a relation is a function, you must decide whether each input value is matched with exactly one output value. When any input value is matched with two or more output values, the relation is not a function.

EXAMPLE 1   **Testing for Functions**

Determine whether the relation represents y as a function of x.

a. The input value x is the number of representatives from a state, and the output value y is the number of senators.

b.

Input, x	Output, y
2	11
2	10
3	8
4	5
5	1

c.

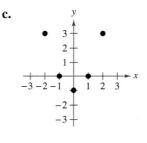

Solution

a. This verbal description *does* describe y as a function of x. Regardless of the value of x, the value of y is always 2. This is an example of a *constant function*.

b. This table *does not* describe y as a function of x. The input value 2 is matched with two different y-values.

c. The graph *does* describe y as a function of x. Each input value is matched with exactly one output value.

✓ **Checkpoint**))) *Audio-video solution in English & Spanish at LarsonPrecalculus.com*

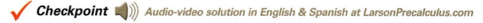

Determine whether the relation represents y as a function of x.

a. *Domain, x*   *Range, y*

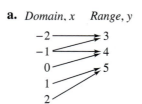

b.

Input, x	0	1	2	3	4
Output, y	-4	-2	0	2	4

Representing functions by sets of ordered pairs is common in *discrete mathematics*. In algebra, however, it is more common to represent functions by equations or formulas involving two variables. For example, the equation

$$y = x^2 \qquad \text{\textcolor{red}{y is a function of x.}}$$

represents the variable y as a function of the variable x. In this equation, x is the **independent variable** and y is the **dependent variable.** The domain of the function is the set of all values taken on by the independent variable x, and the range of the function is the set of all values taken on by the dependent variable y.

EXAMPLE 2   **Testing for Functions Represented Algebraically**

See LarsonPrecalculus.com for an interactive version of this type of example.

Determine whether each equation represents y as a function of x.

a. $x^2 + y = 1$
b. $-x + y^2 = 1$

Solution   To determine whether y is a function of x, solve for y in terms of x.

a. Solving for y yields

$$x^2 + y = 1 \qquad \text{\textcolor{red}{Write original equation.}}$$
$$y = 1 - x^2. \qquad \text{\textcolor{red}{Solve for y.}}$$

To each value of x there corresponds exactly one value of y. So, y is a function of x.

b. Solving for y yields

$$-x + y^2 = 1 \qquad \text{\textcolor{red}{Write original equation.}}$$
$$y^2 = 1 + x \qquad \text{\textcolor{red}{Add x to each side.}}$$
$$y = \pm\sqrt{1 + x}. \qquad \text{\textcolor{red}{Solve for y.}}$$

The $\pm$ indicates that to a given value of x there correspond two values of y. So, y is not a function of x.

✓ **Checkpoint**  *Audio-video solution in English & Spanish at LarsonPrecalculus.com*

Determine whether each equation represents y as a function of x.

a. $x^2 + y^2 = 8$ **b.** $y - 4x^2 = 36$ ■

When using an equation to represent a function, it is convenient to name the function for easy reference. For example, the equation $y = 1 - x^2$ describes y as a function of x. By renaming this function "f," you can write the input, output, and equation using **function notation.**

Input	Output	Equation
x	$f(x)$	$f(x) = 1 - x^2$

The symbol $f(x)$ is read as *the value of f at x* or simply *f of x.* The symbol $f(x)$ corresponds to the y-value for a given x. So, $y = f(x)$. Keep in mind that f is the *name* of the function, whereas $f(x)$ is the *value* of the function at x. For example, the function $f(x) = 3 - 2x$ has *function values* denoted by $f(-1)$, $f(0)$, $f(2)$, and so on. To find these values, substitute the specified input values into the given equation.

For $x = -1$, $f(-1) = 3 - 2(-1) = 3 + 2 = 5.$
For $x = 0$, $f(0) = 3 - 2(0) = 3 - 0 = 3.$
For $x = 2$, $f(2) = 3 - 2(2) = 3 - 4 = -1.$

HISTORICAL NOTE

Many consider Leonhard Euler (1707–1783), a Swiss mathematician, to be the most prolific and productive mathematician in history. One of his greatest influences on mathematics was his use of symbols, or notation. Euler introduced the function notation $y = f(x)$.

Although it is often convenient to use f as a function name and x as the independent variable, other letters may be used as well. For example,

$$f(x) = x^2 - 4x + 7, \quad f(t) = t^2 - 4t + 7, \quad \text{and} \quad g(s) = s^2 - 4s + 7$$

all define the same function. In fact, the role of the independent variable is that of a "placeholder." Consequently, the function can be described by

$$f(\quad) = (\quad)^2 - 4(\quad) + 7.$$

EXAMPLE 3   Evaluating a Function

Let $g(x) = -x^2 + 4x + 1$. Find each function value.

a. $g(2)$ **b.** $g(t)$ **c.** $g(x + 2)$

Solution

a. Replace x with 2 in $g(x) = -x^2 + 4x + 1$.

$$g(2) = -(2)^2 + 4(2) + 1$$
$$= -4 + 8 + 1$$
$$= 5$$

b. Replace x with t.

$$g(t) = -(t)^2 + 4(t) + 1$$
$$= -t^2 + 4t + 1$$

c. Replace x with $x + 2$.

$$g(x + 2) = -(x + 2)^2 + 4(x + 2) + 1$$
$$= -(x^2 + 4x + 4) + 4x + 8 + 1$$
$$= -x^2 - 4x - 4 + 4x + 8 + 1$$
$$= -x^2 + 5$$

$\cdots$ **REMARK**   In Example 3(c), note that $g(x + 2)$ is not equal to $g(x) + g(2)$. In general, $g(u + v) \neq g(u) + g(v)$. $\triangleright$

✓ *Checkpoint* 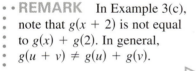 *Audio-video solution in English & Spanish at LarsonPrecalculus.com*

Let $f(x) = 10 - 3x^2$. Find each function value.

a. $f(2)$ **b.** $f(-4)$ **c.** $f(x - 1)$ ◼

A function defined by two or more equations over a specified domain is called a **piecewise-defined function.**

EXAMPLE 4   A Piecewise-Defined Function

Evaluate the function when $x = -1, 0,$ and 1.

$$f(x) = \begin{cases} x^2 + 1, & x < 0 \\ x - 1, & x \geq 0 \end{cases}$$

Solution   Because $x = -1$ is less than 0, use $f(x) = x^2 + 1$ to obtain $f(-1) = (-1)^2 + 1 = 2$. For $x = 0$, use $f(x) = x - 1$ to obtain $f(0) = (0) - 1 = -1$. For $x = 1$, use $f(x) = x - 1$ to obtain $f(1) = (1) - 1 = 0$.

✓ *Checkpoint* 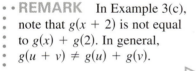 *Audio-video solution in English & Spanish at LarsonPrecalculus.com*

Evaluate the function given in Example 4 when $x = -2, 2,$ and 3.   ◼

EXAMPLE 5   Finding Values for Which $f(x) = 0$

Find all real values of x for which $f(x) = 0$.

a. $f(x) = -2x + 10$ **b.** $f(x) = x^2 - 5x + 6$

Solution   For each function, set $f(x) = 0$ and solve for x.

a. $-2x + 10 = 0$ Set $f(x)$ equal to 0.

$\qquad -2x = -10$ Subtract 10 from each side.

$\qquad\qquad x = 5$ Divide each side by -2.

So, $f(x) = 0$ when $x = 5$.

b. $x^2 - 5x + 6 = 0$ Set $f(x)$ equal to 0.

$\quad (x - 2)(x - 3) = 0$ Factor.

$\qquad x - 2 = 0 \implies x = 2$ Set 1st factor equal to 0 and solve.

$\qquad x - 3 = 0 \implies x = 3$ Set 2nd factor equal to 0 and solve.

So, $f(x) = 0$ when $x = 2$ or $x = 3$.

✓ *Checkpoint* *Audio-video solution in English & Spanish at LarsonPrecalculus.com*

Find all real values of x for which $f(x) = 0$, where $f(x) = x^2 - 16$.

EXAMPLE 6   Finding Values for Which $f(x) = g(x)$

Find the values of x for which $f(x) = g(x)$.

a. $f(x) = x^2 + 1$ and $g(x) = 3x - x^2$

b. $f(x) = x^2 - 1$ and $g(x) = -x^2 + x + 2$

Solution

a. $x^2 + 1 = 3x - x^2$ Set $f(x)$ equal to $g(x)$.

$\quad 2x^2 - 3x + 1 = 0$ Write in general form.

$\ (2x - 1)(x - 1) = 0$ Factor.

$\qquad 2x - 1 = 0 \implies x = \frac{1}{2}$ Set 1st factor equal to 0 and solve.

$\qquad x - 1 = 0 \implies x = 1$ Set 2nd factor equal to 0 and solve.

So, $f(x) = g(x)$ when $x = \dfrac{1}{2}$ or $x = 1$.

b. $x^2 - 1 = -x^2 + x + 2$ Set $f(x)$ equal to $g(x)$.

$\quad 2x^2 - x - 3 = 0$ Write in general form.

$\ (2x - 3)(x + 1) = 0$ Factor.

$\qquad 2x - 3 = 0 \implies x = \frac{3}{2}$ Set 1st factor equal to 0 and solve.

$\qquad x + 1 = 0 \implies x = -1$ Set 2nd factor equal to 0 and solve.

So, $f(x) = g(x)$ when $x = \dfrac{3}{2}$ or $x = -1$.

✓ *Checkpoint* *Audio-video solution in English & Spanish at LarsonPrecalculus.com*

Find the values of x for which $f(x) = g(x)$, where $f(x) = x^2 + 6x - 24$ and $g(x) = 4x - x^2$.

The Domain of a Function

The domain of a function can be described explicitly or it can be *implied* by the expression used to define the function. The **implied domain** is the set of all real numbers for which the expression is defined. For example, the function

$$f(x) = \frac{1}{x^2 - 4} \qquad \text{Domain excludes } x\text{-values that result in division by zero.}$$

has an implied domain consisting of all real x other than $x = \pm 2$. These two values are excluded from the domain because division by zero is undefined. Another common type of implied domain is that used to avoid even roots of negative numbers. For example, the function

$$f(x) = \sqrt{x} \qquad \text{Domain excludes } x\text{-values that result in even roots of negative numbers.}$$

is defined only for $x \geq 0$. So, its implied domain is the interval $[0, \infty)$. In general, the domain of a function *excludes* values that cause division by zero *or* that result in the even root of a negative number.

EXAMPLE 7 **Finding the Domains of Functions**

Find the domain of each function.

a. f: $\{(-3, 0), (-1, 4), (0, 2), (2, 2), (4, -1)\}$ **b.** $g(x) = \dfrac{1}{x + 5}$

c. Volume of a sphere: $V = \frac{4}{3}\pi r^3$ **d.** $h(x) = \sqrt{4 - 3x}$

Solution

a. The domain of f consists of all first coordinates in the set of ordered pairs.

$$\text{Domain} = \{-3, -1, 0, 2, 4\}$$

b. Excluding x-values that yield zero in the denominator, the domain of g is the set of all real numbers x except $x = -5$.

c. This function represents the volume of a sphere, so the values of the radius r must be positive. The domain is the set of all real numbers r such that $r > 0$.

d. This function is defined only for x-values for which

$$4 - 3x \geq 0.$$

Using the methods described in Section 1.7, you can conclude that $x \leq \frac{4}{3}$. So, the domain is the interval $\left(-\infty, \frac{4}{3}\right]$.

✓ *Checkpoint*  *Audio-video solution in English & Spanish at LarsonPrecalculus.com*

Find the domain of each function.

a. f: $\{(-2, 2), (-1, 1), (0, 3), (1, 1), (2, 2)\}$ **b.** $g(x) = \dfrac{1}{3 - x}$

c. Circumference of a circle: $C = 2\pi r$ **d.** $h(x) = \sqrt{x - 16}$

In Example 7(c), note that the domain of a function may be implied by the physical context. For example, from the equation

$$V = \frac{4}{3}\pi r^3$$

you have no reason to restrict r to positive values, but the physical context implies that a sphere cannot have a negative or zero radius.

Applications

EXAMPLE 8   **The Dimensions of a Container**

You work in the marketing department of a soft-drink company and are experimenting with a new can for iced tea that is slightly narrower and taller than a standard can. For your experimental can, the ratio of the height to the radius is 4.

a. Write the volume of the can as a function of the radius r.

b. Write the volume of the can as a function of the height h.

$\frac{h}{r} = 4$

Solution

a. $V(r) = \pi r^2 h = \pi r^2(4r) = 4\pi r^3$ Write V as a function of r.

b. $V(h) = \pi r^2 h = \pi\left(\dfrac{h}{4}\right)^2 h = \dfrac{\pi h^3}{16}$ Write V as a function of h.

✓ *Checkpoint* ◀))) *Audio-video solution in English & Spanish at LarsonPrecalculus.com*

For the experimental can described in Example 8, write the *surface area* as a function of (a) the radius r and (b) the height h.

EXAMPLE 9   **The Path of a Baseball**

A batter hits a baseball at a point 3 feet above ground at a velocity of 100 feet per second and an angle of 45°. The path of the baseball is given by the function

$$f(x) = -0.0032x^2 + x + 3$$

where $f(x)$ is the height of the baseball (in feet) and x is the horizontal distance from home plate (in feet). Will the baseball clear a 10-foot fence located 300 feet from home plate?

Algebraic Solution

Find the height of the baseball when $x = 300$.

$f(x) = -0.0032x^2 + x + 3$ Write original function.

$f(300) = -0.0032(300)^2 + 300 + 3$ Substitute 300 for x.

$\quad\quad = 15$ Simplify.

When $x = 300$, the height of the baseball is 15 feet. So, the baseball will clear a 10-foot fence.

Graphical Solution

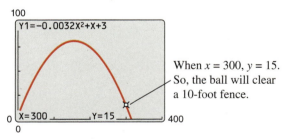

When $x = 300$, $y = 15$. So, the ball will clear a 10-foot fence.

✓ *Checkpoint* ◀))) *Audio-video solution in English & Spanish at LarsonPrecalculus.com*

A second baseman throws a baseball toward the first baseman 60 feet away. The path of the baseball is given by the function

$$f(x) = -0.004x^2 + 0.3x + 6$$

where $f(x)$ is the height of the baseball (in feet) and x is the horizontal distance from the second baseman (in feet). The first baseman can reach 8 feet high. Can the first baseman catch the baseball without jumping?

Flexible-fuel vehicles are designed to operate on gasoline, E85, or a mixture of the two fuels. The concentration of ethanol in E85 fuel ranges from 51% to 83%, depending on where and when the E85 is produced.

EXAMPLE 10 Alternative-Fuel Stations

The number S of fuel stations that sold E85 (a gasoline-ethanol blend) in the United States increased in a linear pattern from 2008 through 2011, and then increased in a different linear pattern from 2012 through 2015, as shown in the bar graph. These two patterns can be approximated by the function

$$S(t) = \begin{cases} 260.8t - 439, & 8 \leq t \leq 11 \\ 151.2t + 714, & 12 \leq t \leq 15 \end{cases}$$

where t represents the year, with $t = 8$ corresponding to 2008. Use this function to approximate the number of stations that sold E85 each year from 2008 to 2015. *(Source: Alternative Fuels Data Center)*

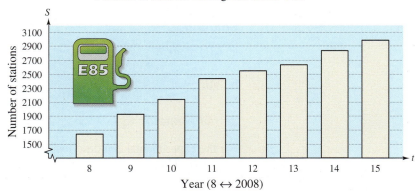

Number of Stations Selling E85 in the U.S.

Solution From 2008 through 2011, use $S(t) = 260.8t - 439$.

1647	1908	2169	2430
2008	2009	2010	2011

From 2012 to 2015, use $S(t) = 151.2t + 714$.

2528	2680	2831	2982
2012	2013	2014	2015

✓ **Checkpoint** ◀))) *Audio-video solution in English & Spanish at LarsonPrecalculus.com*

The number S of fuel stations that sold compressed natural gas in the United States from 2009 to 2015 can be approximated by the function

$$S(t) = \begin{cases} 69t + 151, & 9 \leq t \leq 11 \\ 160t - 803, & 12 \leq t \leq 15 \end{cases}$$

where t represents the year, with $t = 9$ corresponding to 2009. Use this function to approximate the number of stations that sold compressed natural gas each year from 2009 through 2015. *(Source: Alternative Fuels Data Center)*

Difference Quotients

One of the basic definitions in calculus uses the ratio

$$\frac{f(x + h) - f(x)}{h}, \quad h \neq 0.$$

This ratio is a **difference quotient,** as illustrated in Example 11.

EXAMPLE 11   **Evaluating a Difference Quotient**

For $f(x) = x^2 - 4x + 7$, find $\dfrac{f(x + h) - f(x)}{h}$.

Solution

$$\frac{f(x + h) - f(x)}{h} = \frac{[(x + h)^2 - 4(x + h) + 7] - (x^2 - 4x + 7)}{h}$$

$$= \frac{x^2 + 2xh + h^2 - 4x - 4h + 7 - x^2 + 4x - 7}{h}$$

$$= \frac{2xh + h^2 - 4h}{h} = \frac{h(2x + h - 4)}{h} = 2x + h - 4, \quad h \neq 0$$

✓ **Checkpoint** *Audio-video solution in English & Spanish at LarsonPrecalculus.com*

For $f(x) = x^2 + 2x - 3$, find $\dfrac{f(x + h) - f(x)}{h}$.   ■

• • REMARK You may find it easier to calculate the difference quotient in Example 11 by first finding $f(x + h)$, and then substituting the resulting expression into the difference quotient

$$\frac{f(x + h) - f(x)}{h}.$$

Summary of Function Terminology

Function: A **function** is a relationship between two variables such that to each value of the independent variable there corresponds exactly one value of the dependent variable.

Function notation: $y = f(x)$

    f is the *name* of the function.

    y is the **dependent variable.**

    x is the **independent variable.**

    $f(x)$ is the *value of the function at x.*

Domain: The **domain** of a function is the set of all values (inputs) of the independent variable for which the function is defined. If x is in the domain of f, then f is *defined* at x. If x is not in the domain of f, then f is *undefined* at x.

Range: The **range** of a function is the set of all values (outputs) taken on by the dependent variable (that is, the set of all function values).

Implied domain: If f is defined by an algebraic expression and the domain is not specified, then the **implied domain** consists of all real numbers for which the expression is defined.

Summarize   (Section 2.2)

1. State the definition of a function and describe function notation *(pages 173–177)*. For examples of determining functions and using function notation, see Examples 1–6.

2. State the definition of the implied domain of a function *(page 178)*. For an example of finding the domains of functions, see Example 7.

3. Describe examples of how functions can model real-life problems *(pages 179 and 180, Examples 8–10)*.

4. State the definition of a difference quotient *(page 180)*. For an example of evaluating a difference quotient, see Example 11.

2.2 Exercises

See **CalcChat.com** for tutorial help and worked-out solutions to odd-numbered exercises.

Vocabulary: Fill in the blanks.

1. A relation that assigns to each element x from a set of inputs, or _____, exactly one element y in a set of outputs, or _____, is a _____.

2. For an equation that represents y as a function of x, the set of all values taken on by the _____ variable x is the domain, and the set of all values taken on by the _____ variable y is the range.

3. If the domain of the function f is not given, then the set of values of the independent variable for which the expression is defined is the _____ _____.

4. One of the basic definitions in calculus uses the ratio $\dfrac{f(x + h) - f(x)}{h}$, $h \neq 0$. This ratio is a _____ _____.

Skills and Applications

 Testing for Functions In Exercises 5–8, determine whether the relation represents y as a function of x.

5. Domain, x Range, y

$$
\begin{array}{ll}
-2 & 5 \\
-1 & 6 \\
0 & 7 \\
1 & 8 \\
2 &
\end{array}
$$

6. Domain, x Range, y

$$
\begin{array}{ll}
-2 & 0 \\
-1 & 1 \\
0 & 2 \\
1 & \\
2 &
\end{array}
$$

7.

Input, x	10	7	4	7	10
Output, y	3	6	9	12	15

8.

Input, x	-2	0	2	4	6
Output, y	1	1	1	1	1

Testing for Functions In Exercises 9 and 10, which sets of ordered pairs represent functions from A to B? Explain.

9. $A = \{0, 1, 2, 3\}$ and $B = \{-2, -1, 0, 1, 2\}$
   (a) $\{(0, 1), (1, -2), (2, 0), (3, 2)\}$
   (b) $\{(0, -1), (2, 2), (1, -2), (3, 0), (1, 1)\}$
   (c) $\{(0, 0), (1, 0), (2, 0), (3, 0)\}$
   (d) $\{(0, 2), (3, 0), (1, 1)\}$

10. $A = \{a, b, c\}$ and $B = \{0, 1, 2, 3\}$
   (a) $\{(a, 1), (c, 2), (c, 3), (b, 3)\}$
   (b) $\{(a, 1), (b, 2), (c, 3)\}$
   (c) $\{(1, a), (0, a), (2, c), (3, b)\}$
   (d) $\{(c, 0), (b, 0), (a, 3)\}$

 Testing for Functions Represented Algebraically In Exercises 11–18, determine whether the equation represents y as a function of x.

11. $x^2 + y^2 = 4$

12. $x^2 - y = 9$

13. $y = \sqrt{16 - x^2}$

14. $y = \sqrt{x + 5}$

15. $y = 4 - |x|$

16. $|y| = 4 - x$

17. $y = -75$

18. $x - 1 = 0$

 Evaluating a Function In Exercises 19–30, find each function value, if possible.

19. $f(x) = 3x - 5$
   (a) $f(1)$ (b) $f(-3)$ (c) $f(x + 2)$

20. $V(r) = \frac{4}{3}\pi r^3$
   (a) $V(3)$ (b) $V\!\left(\frac{3}{2}\right)$ (c) $V(2r)$

21. $g(t) = 4t^2 - 3t + 5$
   (a) $g(2)$ (b) $g(t - 2)$ (c) $g(t) - g(2)$

22. $h(t) = -t^2 + t + 1$
   (a) $h(2)$ (b) $h(-1)$ (c) $h(x + 1)$

23. $f(y) = 3 - \sqrt{y}$
   (a) $f(4)$ (b) $f(0.25)$ (c) $f(4x^2)$

24. $f(x) = \sqrt{x + 8} + 2$
   (a) $f(-8)$ (b) $f(1)$ (c) $f(x - 8)$

25. $q(x) = 1/(x^2 - 9)$
   (a) $q(0)$ (b) $q(3)$ (c) $q(y + 3)$

26. $q(t) = (2t^2 + 3)/t^2$
   (a) $q(2)$ (b) $q(0)$ (c) $q(-x)$

27. $f(x) = |x|/x$
   (a) $f(2)$ (b) $f(-2)$ (c) $f(x - 1)$

28. $f(x) = |x| + 4$
   (a) $f(2)$ (b) $f(-2)$ (c) $f(x^2)$

29. $f(x) = \begin{cases} 2x + 1, & x < 0 \\ 2x + 2, & x \geq 0 \end{cases}$
   (a) $f(-1)$ (b) $f(0)$ (c) $f(2)$

30. $f(x) = \begin{cases} -3x - 3, & x < -1 \\ x^2 + 2x - 1, & x \geq -1 \end{cases}$
   (a) $f(-2)$ (b) $f(-1)$ (c) $f(1)$

Evaluating a Function In Exercises 31–34, complete the table.

31. $f(x) = -x^2 + 5$

x	-2	-1	0	1	2
$f(x)$					

32. $h(t) = \frac{1}{2}|t + 3|$

t	-5	-4	-3	-2	-1
$h(t)$					

33. $f(x) = \begin{cases} -\frac{1}{2}x + 4, & x \le 0 \\ (x - 2)^2, & x > 0 \end{cases}$

x	-2	-1	0	1	2
$f(x)$					

34. $f(x) = \begin{cases} 9 - x^2, & x < 3 \\ x - 3, & x \ge 3 \end{cases}$

x	1	2	3	4	5
$f(x)$					

 Finding Values for Which $f(x) = 0$ In Exercises 35–42, find all real values of x for which $f(x) = 0$.

35. $f(x) = 15 - 3x$

36. $f(x) = 4x + 6$

37. $f(x) = \dfrac{3x - 4}{5}$

38. $f(x) = \dfrac{12 - x^2}{8}$

39. $f(x) = x^2 - 81$

40. $f(x) = x^2 - 6x - 16$

41. $f(x) = x^3 - x$

42. $f(x) = x^3 - x^2 - 3x + 3$

 Finding Values for Which $f(x) = g(x)$ In Exercises 43–46, find the value(s) of x for which $f(x) = g(x)$.

43. $f(x) = x^2$, $g(x) = x + 2$

44. $f(x) = x^2 + 2x + 1$, $g(x) = 5x + 19$

45. $f(x) = x^4 - 2x^2$, $g(x) = 2x^2$

46. $f(x) = \sqrt{x} - 4$, $g(x) = 2 - x$

 Finding the Domain of a Function In Exercises 47–56, find the domain of the function.

47. $f(x) = 5x^2 + 2x - 1$

48. $g(x) = 1 - 2x^2$

49. $g(y) = \sqrt{y + 6}$

50. $f(t) = \sqrt[3]{t + 4}$

51. $g(x) = \dfrac{1}{x} - \dfrac{3}{x + 2}$

52. $h(x) = \dfrac{6}{x^2 - 4x}$

53. $f(s) = \dfrac{\sqrt{s - 1}}{s - 4}$

54. $f(x) = \dfrac{\sqrt{x + 6}}{6 + x}$

55. $f(x) = \dfrac{x - 4}{\sqrt{x}}$

56. $f(x) = \dfrac{x + 2}{\sqrt{x - 10}}$

57. Maximum Volume An open box of maximum volume is made from a square piece of material 24 centimeters on a side by cutting equal squares from the corners and turning up the sides (see figure).

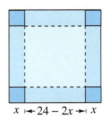 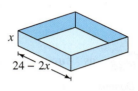

(a) The table shows the volumes V (in cubic centimeters) of the box for various heights x (in centimeters). Use the table to estimate the maximum volume.

Height, x	1	2	3	4	5	6
Volume, V	484	800	972	1024	980	864

(b) Plot the points (x, V) from the table in part (a). Does the relation defined by the ordered pairs represent V as a function of x?

(c) Given that V is a function of x, write the function and determine its domain.

58. Maximum Profit The cost per unit in the production of an MP3 player is \$60. The manufacturer charges \$90 per unit for orders of 100 or less. To encourage large orders, the manufacturer reduces the charge by \$0.15 per MP3 player for each unit ordered in excess of 100 (for example, the charge is reduced to \$87 per MP3 player for an order size of 120).

(a) The table shows the profits P (in dollars) for various numbers of units ordered, x. Use the table to estimate the maximum profit.

Units, x	130	140	150	160	170
Profit, P	3315	3360	3375	3360	3315

(b) Plot the points (x, P) from the table in part (a). Does the relation defined by the ordered pairs represent P as a function of x?

(c) Given that P is a function of x, write the function and determine its domain. (*Note:* $P = R - C$, where R is revenue and C is cost.)

59. Geometry Write the area A of a square as a function of its perimeter P.

60. Geometry Write the area A of a circle as a function of its circumference C.

61. Path of a Ball You throw a baseball to a child 25 feet away. The height y (in feet) of the baseball is given by

$$y = -\frac{1}{10}x^2 + 3x + 6$$

where x is the horizontal distance (in feet) from where you threw the ball. Can the child catch the baseball while holding a baseball glove at a height of 5 feet?

62. Postal Regulations A rectangular package has a combined length and girth (perimeter of a cross section) of 108 inches (see figure).

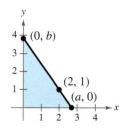

(a) Write the volume V of the package as a function of x. What is the domain of the function?

(b) Use a graphing utility to graph the function. Be sure to use an appropriate window setting.

(c) What dimensions will maximize the volume of the package? Explain.

63. Geometry A right triangle is formed in the first quadrant by the x- and y-axes and a line through the point $(2, 1)$ (see figure). Write the area A of the triangle as a function of x, and determine the domain of the function.

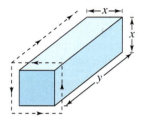

64. Geometry A rectangle is bounded by the x-axis and the semicircle $y = \sqrt{36 - x^2}$ (see figure). Write the area A of the rectangle as a function of x, and graphically determine the domain of the function.

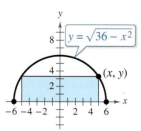

65. Pharmacology The percent p of prescriptions filled with generic drugs at CVS Pharmacies from 2008 through 2014 (see figure) can be approximated by the model

$$p(t) = \begin{cases} 2.77t + 45.2, & 8 \le t \le 11 \\ 1.95t + 55.9, & 12 \le t \le 14 \end{cases}$$

where t represents the year, with $t = 8$ corresponding to 2008. Use this model to find the percent of prescriptions filled with generic drugs in each year from 2008 through 2014. (*Source: CVS Health*)

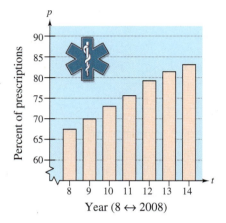

Year (8 ↔ 2008)

66. Median Sale Price The median sale price p (in thousands of dollars) of an existing one-family home in the United States from 2002 through 2014 (see figure) can be approximated by the model

$$p(t) = \begin{cases} -0.757t^2 + 20.80t + 127.2, & 2 \le t \le 6 \\ 3.879t^2 - 82.50t + 605.8, & 7 \le t \le 11 \\ -4.171t^2 + 124.34t - 714.2, & 12 \le t \le 14 \end{cases}$$

where t represents the year, with $t = 2$ corresponding to 2002. Use this model to find the median sale price of an existing one-family home in each year from 2002 through 2014. (*Source: National Association of Realtors*)

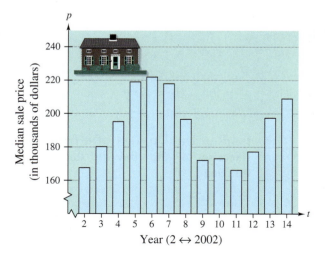

Year (2 ↔ 2002)

67. Cost, Revenue, and Profit   A company produces a product for which the variable cost is $12.30 per unit and the fixed costs are $98,000. The product sells for $17.98. Let x be the number of units produced and sold.

(a) The total cost for a business is the sum of the variable cost and the fixed costs. Write the total cost C as a function of the number of units produced.

(b) Write the revenue R as a function of the number of units sold.

(c) Write the profit P as a function of the number of units sold. (*Note:* $P = R - C$)

68. Average Cost   The inventor of a new game believes that the variable cost for producing the game is $0.95 per unit and the fixed costs are $6000. The inventor sells each game for $1.69. Let x be the number of games produced.

(a) The total cost for a business is the sum of the variable cost and the fixed costs. Write the total cost C as a function of the number of games produced.

(b) Write the average cost per unit $\overline{C} = \dfrac{C}{x}$ as a function of x.

69. Height of a Balloon   A balloon carrying a transmitter ascends vertically from a point 3000 feet from the receiving station.

(a) Draw a diagram that gives a visual representation of the problem. Let h represent the height of the balloon and let d represent the distance between the balloon and the receiving station.

(b) Write the height of the balloon as a function of d. What is the domain of the function?

70. Physics

The function $F(y) = 149.76\sqrt{10}\,y^{5/2}$ estimates the force F (in tons) of water against the face of a dam, where y is the depth of the water (in feet).

(a) Complete the table. What can you conclude from the table?

y	5	10	20	30	40
$F(y)$					

(b) Use the table to approximate the depth at which the force against the dam is 1,000,000 tons.

(c) Find the depth at which the force against the dam is 1,000,000 tons algebraically.

71. Transportation   For groups of 80 or more people, a charter bus company determines the rate per person according to the formula

$$\text{Rate} = 8 - 0.05(n - 80), \quad n \geq 80$$

where the rate is given in dollars and n is the number of people.

(a) Write the revenue R for the bus company as a function of n.

(b) Use the function in part (a) to complete the table. What can you conclude?

n	90	100	110	120	130	140	150
$R(n)$							

72. E-Filing   The table shows the numbers of tax returns (in millions) made through e-file from 2007 through 2014. Let $f(t)$ represent the number of tax returns made through e-file in the year t.   (*Source: eFile*)

DATA	Year	Number of Tax Returns Made Through E-File
	2007	80.0
	2008	89.9
	2009	95.0
	2010	98.7
	2011	112.2
	2012	112.1
	2013	114.4
	2014	125.8

Spreadsheet at LarsonPrecalculus.com

(a) Find $\dfrac{f(2014) - f(2007)}{2014 - 2007}$ and interpret the result in the context of the problem.

(b) Make a scatter plot of the data.

(c) Find a linear model for the data algebraically. Let N represent the number of tax returns made through e-file and let $t = 7$ correspond to 2007.

(d) Use the model found in part (c) to complete the table.

t	7	8	9	10	11	12	13	14
N								

(e) Compare your results from part (d) with the actual data.

(f) Use a graphing utility to find a linear model for the data. Let $x = 7$ correspond to 2007. How does the model you found in part (c) compare with the model given by the graphing utility?

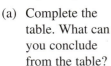

Evaluating a Difference Quotient In Exercises 73–80, find the difference quotient and simplify your answer.

73. $f(x) = x^2 - 2x + 4$, $\dfrac{f(2 + h) - f(2)}{h}$, $h \neq 0$

74. $f(x) = 5x - x^2$, $\dfrac{f(5 + h) - f(5)}{h}$, $h \neq 0$

75. $f(x) = x^3 + 3x$, $\dfrac{f(x + h) - f(x)}{h}$, $h \neq 0$

76. $f(x) = 4x^3 - 2x$, $\dfrac{f(x + h) - f(x)}{h}$, $h \neq 0$

77. $g(x) = \dfrac{1}{x^2}$, $\dfrac{g(x) - g(3)}{x - 3}$, $x \neq 3$

78. $f(t) = \dfrac{1}{t - 2}$, $\dfrac{f(t) - f(1)}{t - 1}$, $t \neq 1$

79. $f(x) = \sqrt{5x}$, $\dfrac{f(x) - f(5)}{x - 5}$, $x \neq 5$

80. $f(x) = x^{2/3} + 1$, $\dfrac{f(x) - f(8)}{x - 8}$, $x \neq 8$

Modeling Data In Exercises 81–84, determine which of the following functions

$$f(x) = cx, \quad g(x) = cx^2, \quad h(x) = c\sqrt{|x|}, \quad \text{and} \quad r(x) = \frac{c}{x}$$

can be used to model the data and determine the value of the constant c that will make the function fit the data in the table.

81.

x	-4	-1	0	1	4
y	-32	-2	0	-2	-32

82.

x	-4	-1	0	1	4
y	-1	$-\frac{1}{4}$	0	$\frac{1}{4}$	1

83.

x	-4	-1	0	1	4
y	-8	-32	Undefined	32	8

84.

x	-4	-1	0	1	4
y	6	3	0	3	6

Exploration

True or False? In Exercises 85–88, determine whether the statement is true or false. Justify your answer.

85. Every relation is a function.

86. Every function is a relation.

87. For the function
$$f(x) = x^4 - 1$$
the domain is $(-\infty, \infty)$ and the range is $(0, \infty)$.

88. The set of ordered pairs $\{(-8, -2), (-6, 0), (-4, 0), (-2, 2), (0, 4), (2, -2)\}$ represents a function.

89. Error Analysis Describe the error.

The functions
$$f(x) = \sqrt{x - 1} \quad \text{and} \quad g(x) = \frac{1}{\sqrt{x - 1}}$$
have the same domain, which is the set of all real numbers x such that $x \geq 1$.

90. Think About It Consider
$$f(x) = \sqrt{x - 2} \quad \text{and} \quad g(x) = \sqrt[3]{x - 2}.$$
Why are the domains of f and g different?

91. Think About It Given $f(x) = x^2$, is f the independent variable? Why or why not?

92. HOW DO YOU SEE IT? The graph represents the height h of a projectile after t seconds.

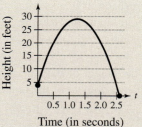

(a) Explain why h is a function of t.

(b) Approximate the height of the projectile after 0.5 second and after 1.25 seconds.

(c) Approximate the domain of h.

(d) Is t a function of h? Explain.

Think About It In Exercises 93 and 94, determine whether the statements use the word *function* in ways that are mathematically correct. Explain.

93. (a) The sales tax on a purchased item is a function of the selling price.

(b) Your score on the next algebra exam is a function of the number of hours you study the night before the exam.

94. (a) The amount in your savings account is a function of your salary.

(b) The speed at which a free-falling baseball strikes the ground is a function of the height from which it was dropped.

2.3 Analyzing Graphs of Functions

Graphs of functions can help you visualize relationships between variables in real life. For example, in Exercise 90 on page 197, you will use the graph of a function to visually represent the temperature in a city over a 24-hour period.

- ■ Use the Vertical Line Test for functions.
- ■ Find the zeros of functions.
- ■ Determine intervals on which functions are increasing or decreasing.
- ■ Determine relative minimum and relative maximum values of functions.
- ■ Determine the average rate of change of a function.
- ■ Identify even and odd functions.

The Graph of a Function

In Section 2.2, you studied functions from an algebraic point of view. In this section, you will study functions from a graphical perspective.

The **graph of a function** f is the collection of ordered pairs $(x, f(x))$ such that x is in the domain of f. As you study this section, remember that

x = the directed distance from the y-axis

$y = f(x)$ = the directed distance from the x-axis

as shown in the figure at the right.

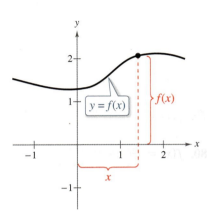

EXAMPLE 1   Finding the Domain and Range of a Function

Use the graph of the function f, shown in Figure 2.13, to find (a) the domain of f, (b) the function values $f(-1)$ and $f(2)$, and (c) the range of f.

Solution

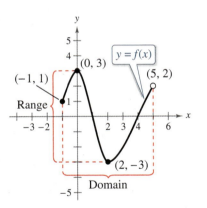

Figure 2.13

a. The closed dot at $(-1, 1)$ indicates that $x = -1$ is in the domain of f, whereas the open dot at $(5, 2)$ indicates that $x = 5$ is not in the domain. So, the domain of f is all x in the interval $[-1, 5)$.

b. One point on the graph of f is $(-1, 1)$, so $f(-1) = 1$. Another point on the graph of f is $(2, -3)$, so $f(2) = -3$.

c. The graph does not extend below $f(2) = -3$ or above $f(0) = 3$, so the range of f is the interval $[-3, 3]$.

✓ **Checkpoint** 🔊)) *Audio-video solution in English & Spanish at LarsonPrecalculus.com*

Use the graph of the function f to find (a) the domain of f, (b) the function values $f(0)$ and $f(3)$, and (c) the range of f.

- • **REMARK** The use of dots (open or closed) at the extreme left and right points of a graph indicates that the graph does not extend beyond these points. If such dots are not on the graph, then assume that the graph extends beyond these points.

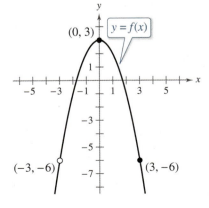

By the definition of a function, at most one y-value corresponds to a given x-value. So, no two points on the graph of a function have the same x-coordinate, or lie on the same vertical line. It follows, then, that a vertical line can intersect the graph of a function at most once. This observation provides a convenient visual test called the **Vertical Line Test** for functions.

Vertical Line Test for Functions

A set of points in a coordinate plane is the graph of y as a function of x if and only if no *vertical* line intersects the graph at more than one point.

EXAMPLE 2   **Vertical Line Test for Functions**

Use the Vertical Line Test to determine whether each graph represents y as a function of x.

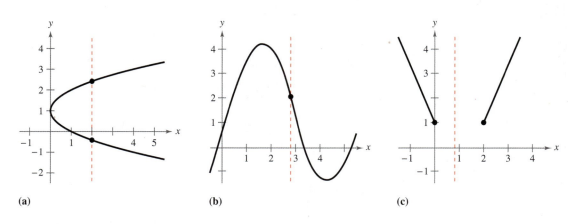

(a)   (b)   (c)

Solution

a. This *is not* a graph of y as a function of x, because there are vertical lines that intersect the graph twice. That is, for a particular input x, there is more than one output y.

b. This *is* a graph of y as a function of x, because every vertical line intersects the graph at most once. That is, for a particular input x, there is at most one output y.

c. This *is* a graph of y as a function of x, because every vertical line intersects the graph at most once. That is, for a particular input x, there is at most one output y. (Note that when a vertical line does not intersect the graph, it simply means that the function is undefined for that particular value of x.)

✓ **Checkpoint** ◀))) *Audio-video solution in English & Spanish at LarsonPrecalculus.com*

Use the Vertical Line Test to determine whether the graph represents y as a function of x.

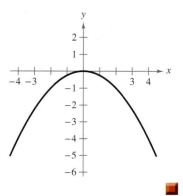

▷ **TECHNOLOGY** Most graphing utilities graph functions of x more easily than other types of equations. For example, the graph shown in (a) above represents the equation $x - (y - 1)^2 = 0$. To duplicate this graph using a graphing utility, you must first solve the equation for y to obtain $y = 1 \pm \sqrt{x}$, and then graph the two equations $y_1 = 1 + \sqrt{x}$ and $y_2 = 1 - \sqrt{x}$ in the same viewing window.

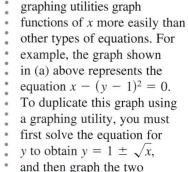

Zeros of a Function

If the graph of a function of x has an x-intercept at $(a, 0)$, then a is a **zero** of the function.

> ### Zeros of a Function
> The **zeros of a function** $y = f(x)$ are the x-values for which $f(x) = 0$.

EXAMPLE 3   **Finding the Zeros of Functions**

Find the zeros of each function algebraically.

a. $f(x) = 3x^2 + x - 10$

b. $g(x) = \sqrt{10 - x^2}$

c. $h(t) = \dfrac{2t - 3}{t + 5}$

Solution   To find the zeros of a function, set the function equal to zero and solve for the independent variable.

a.
$$3x^2 + x - 10 = 0 \qquad \text{Set } f(x) \text{ equal to 0.}$$
$$(3x - 5)(x + 2) = 0 \qquad \text{Factor.}$$
$$3x - 5 = 0 \;\Longrightarrow\; x = \tfrac{5}{3} \qquad \text{Set 1st factor equal to 0 and solve.}$$
$$x + 2 = 0 \;\Longrightarrow\; x = -2 \qquad \text{Set 2nd factor equal to 0 and solve.}$$

The zeros of f are $x = \frac{5}{3}$ and $x = -2$. In Figure 2.14, note that the graph of f has $\left(\frac{5}{3}, 0\right)$ and $(-2, 0)$ as its x-intercepts.

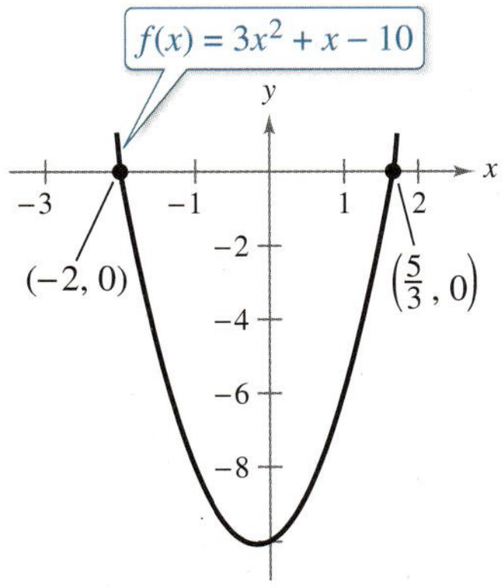

b.
$$\sqrt{10 - x^2} = 0 \qquad \text{Set } g(x) \text{ equal to 0.}$$
$$10 - x^2 = 0 \qquad \text{Square each side.}$$
$$10 = x^2 \qquad \text{Add } x^2 \text{ to each side.}$$
$$\pm\sqrt{10} = x \qquad \text{Extract square roots.}$$

The zeros of g are $x = -\sqrt{10}$ and $x = \sqrt{10}$. In Figure 2.15, note that the graph of g has $\left(-\sqrt{10}, 0\right)$ and $\left(\sqrt{10}, 0\right)$ as its x-intercepts.

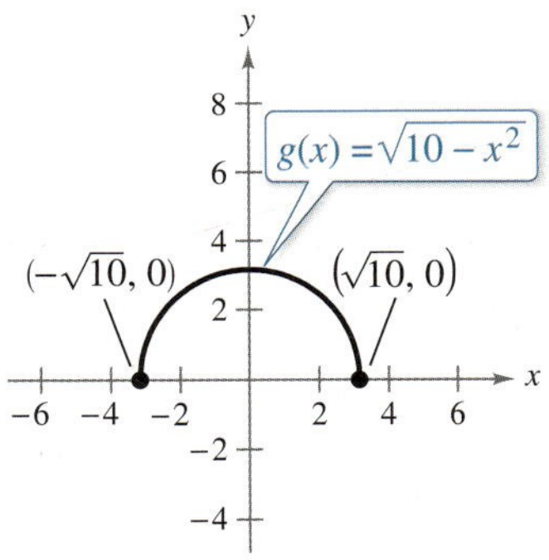

c.
$$\frac{2t - 3}{t + 5} = 0 \qquad \text{Set } h(t) \text{ equal to 0.}$$
$$2t - 3 = 0 \qquad \text{Multiply each side by } t + 5.$$
$$2t = 3 \qquad \text{Add 3 to each side.}$$
$$t = \frac{3}{2} \qquad \text{Divide each side by 2.}$$

The zero of h is $t = \frac{3}{2}$. In Figure 2.16, note that the graph of h has $\left(\frac{3}{2}, 0\right)$ as its t-intercept.

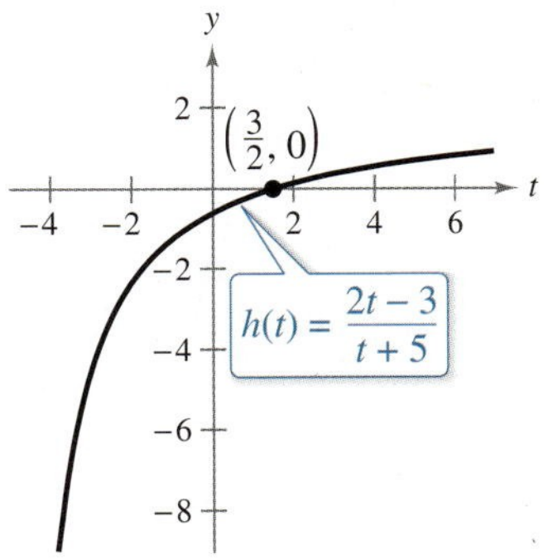

✓ *Checkpoint* 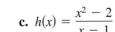 Audio-video solution in English & Spanish at LarsonPrecalculus.com

Find the zeros of each function.

a. $f(x) = 2x^2 + 13x - 24$ **b.** $g(t) = \sqrt{t - 25}$ **c.** $h(x) = \dfrac{x^2 - 2}{x - 1}$ ▪

$f(x) = 3x^2 + x - 10$

Zeros of f: $x = -2$, $x = \frac{5}{3}$
Figure 2.14

$g(x) = \sqrt{10 - x^2}$

Zeros of g: $x = \pm\sqrt{10}$
Figure 2.15

$h(t) = \dfrac{2t - 3}{t + 5}$

Zero of h: $t = \frac{3}{2}$
Figure 2.16

Increasing and Decreasing Functions

The more you know about the graph of a function, the more you know about the function itself. Consider the graph shown in Figure 2.17. As you move from *left to right*, this graph falls from $x = -2$ to $x = 0$, is constant from $x = 0$ to $x = 2$, and rises from $x = 2$ to $x = 4$.

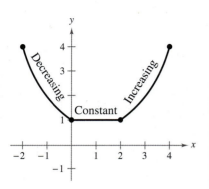

Figure 2.17

Increasing, Decreasing, and Constant Functions

A function f is **increasing** on an interval when, for any x_1 and x_2 in the interval,

$$x_1 < x_2 \quad \text{implies} \quad f(x_1) < f(x_2).$$

A function f is **decreasing** on an interval when, for any x_1 and x_2 in the interval,

$$x_1 < x_2 \quad \text{implies} \quad f(x_1) > f(x_2).$$

A function f is **constant** on an interval when, for any x_1 and x_2 in the interval,

$$f(x_1) = f(x_2).$$

EXAMPLE 4 **Describing Function Behavior**

Determine the open intervals on which each function is increasing, decreasing, or constant.

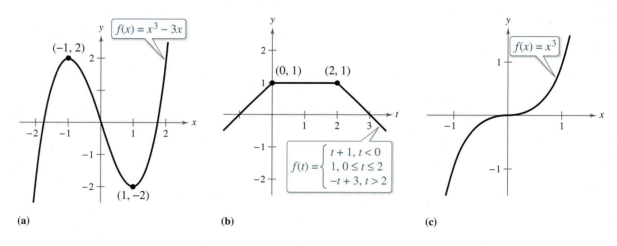

(a)   (b)   (c)

Solution

a. This function is increasing on the interval $(-\infty, -1)$, decreasing on the interval $(-1, 1)$, and increasing on the interval $(1, \infty)$.

b. This function is increasing on the interval $(-\infty, 0)$, constant on the interval $(0, 2)$, and decreasing on the interval $(2, \infty)$.

c. This function may appear to be constant on an interval near $x = 0$, but for all real values of x_1 and x_2, if $x_1 < x_2$, then $(x_1)^3 < (x_2)^3$. So, the function is increasing on the interval $(-\infty, \infty)$.

✓ *Checkpoint*  *Audio-video solution in English & Spanish at LarsonPrecalculus.com*

Graph the function

$$f(x) = x^3 + 3x^2 - 1.$$

Then determine the open intervals on which the function is increasing, decreasing, or constant.

Relative Minimum and Relative Maximum Values

The points at which a function changes its increasing, decreasing, or constant behavior are helpful in determining the **relative minimum** or **relative maximum** values of the function.

• • • • • • • • • • • • • • • • ▷

•• REMARK A relative minimum or relative maximum is also referred to as a local minimum or local maximum.

Definitions of Relative Minimum and Relative Maximum

A function value $f(a)$ is a **relative minimum** of f when there exists an interval (x_1, x_2) that contains a such that

$$x_1 < x < x_2 \quad \text{implies} \quad f(a) \le f(x).$$

A function value $f(a)$ is a **relative maximum** of f when there exists an interval (x_1, x_2) that contains a such that

$$x_1 < x < x_2 \quad \text{implies} \quad f(a) \ge f(x).$$

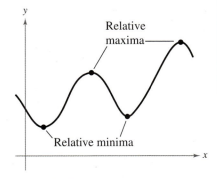

Figure 2.18 shows several different examples of relative minima and relative maxima. In Section 3.1, you will study a technique for finding the *exact point* at which a second-degree polynomial function has a relative minimum or relative maximum. For the time being, however, you can use a graphing utility to find reasonable approximations of these points.

Figure 2.18

EXAMPLE 5   **Approximating a Relative Minimum**

Use a graphing utility to approximate the relative minimum of the function

$$f(x) = 3x^2 - 4x - 2.$$

Solution   The graph of f is shown in Figure 2.19. By using the *zoom* and *trace* features or the *minimum* feature of a graphing utility, you can approximate that the relative minimum of the function occurs at the point

$$(0.67, -3.33).$$

So, the relative minimum is approximately -3.33. Later, in Section 3.1, you will learn how to determine that the exact point at which the relative minimum occurs is $\left(\frac{2}{3}, -\frac{10}{3}\right)$ and the exact relative minimum is $-\frac{10}{3}$.

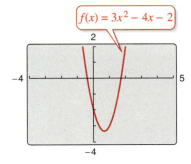

Figure 2.19

✓ *Checkpoint* 🔊))) *Audio-video solution in English & Spanish at LarsonPrecalculus.com*

Use a graphing utility to approximate the relative maximum of the function

$$f(x) = -4x^2 - 7x + 3.$$

You can also use the *table* feature of a graphing utility to numerically approximate the relative minimum of the function in Example 5. Using a table that begins at 0.6 and increments the value of x by 0.01, you can approximate that the minimum of

$$f(x) = 3x^2 - 4x - 2$$

occurs at the point $(0.67, -3.33)$.

▷ **TECHNOLOGY** When you use a graphing utility to approximate the x- and y-values of the point where a relative minimum or relative maximum occurs, the *zoom* feature will often produce graphs that are nearly flat. To overcome this problem, manually change the vertical setting of the viewing window. The graph will stretch vertically when the values of Ymin and Ymax are closer together.

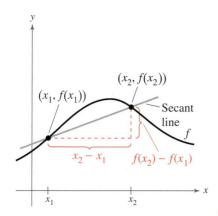

Figure 2.20

Average Rate of Change

In Section 2.1, you learned that the slope of a line can be interpreted as a *rate of change*. For a nonlinear graph, the **average rate of change** between any two points $(x_1, f(x_1))$ and $(x_2, f(x_2))$ is the slope of the line through the two points (see Figure 2.20). The line through the two points is called a **secant line**, and the slope of this line is denoted as m_{sec}.

$$
\text{Average rate of change of } f \text{ from } x_1 \text{ to } x_2 = \frac{f(x_2) - f(x_1)}{x_2 - x_1}
$$

$$
= \frac{\text{change in } y}{\text{change in } x}
$$

$$
= m_{\text{sec}}
$$

EXAMPLE 6 **Average Rate of Change of a Function**

Find the average rates of change of $f(x) = x^3 - 3x$ (a) from $x_1 = -2$ to $x_2 = -1$ and (b) from $x_1 = 0$ to $x_2 = 1$ (see Figure 2.21).

Solution

a. The average rate of change of f from $x_1 = -2$ to $x_2 = -1$ is

$$
\frac{f(x_2) - f(x_1)}{x_2 - x_1} = \frac{f(-1) - f(-2)}{-1 - (-2)} = \frac{2 - (-2)}{1} = 4. \qquad \text{Secant line has positive slope.}
$$

b. The average rate of change of f from $x_1 = 0$ to $x_2 = 1$ is

$$
\frac{f(x_2) - f(x_1)}{x_2 - x_1} = \frac{f(1) - f(0)}{1 - 0} = \frac{-2 - 0}{1} = -2. \qquad \text{Secant line has negative slope.}
$$

 Checkpoint  Audio-video solution in English & Spanish at LarsonPrecalculus.com

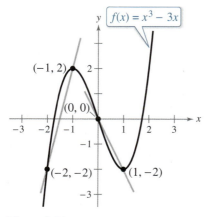

Figure 2.21

Find the average rates of change of $f(x) = x^2 + 2x$ (a) from $x_1 = -3$ to $x_2 = -2$ and (b) from $x_1 = -2$ to $x_2 = 0$.

EXAMPLE 7 **Finding Average Speed**

The distance s (in feet) a moving car is from a stoplight is given by the function

$$
s(t) = 20t^{3/2}
$$

where t is the time (in seconds). Find the average speed of the car (a) from $t_1 = 0$ to $t_2 = 4$ seconds and (b) from $t_1 = 4$ to $t_2 = 9$ seconds.

Solution

a. The average speed of the car from $t_1 = 0$ to $t_2 = 4$ seconds is

$$
\frac{s(t_2) - s(t_1)}{t_2 - t_1} = \frac{s(4) - s(0)}{4 - 0} = \frac{160 - 0}{4} = 40 \text{ feet per second.}
$$

b. The average speed of the car from $t_1 = 4$ to $t_2 = 9$ seconds is

$$
\frac{s(t_2) - s(t_1)}{t_2 - t_1} = \frac{s(9) - s(4)}{9 - 4} = \frac{540 - 160}{5} = 76 \text{ feet per second.}
$$

Average speed is an average rate of change.

Checkpoint Audio-video solution in English & Spanish at LarsonPrecalculus.com

In Example 7, find the average speed of the car (a) from $t_1 = 0$ to $t_2 = 1$ second and (b) from $t_1 = 1$ second to $t_2 = 4$ seconds.

Even and Odd Functions

In Section 1.1, you studied different types of symmetry of a graph. In the terminology of functions, a function is said to be **even** when its graph is symmetric with respect to the y-axis and **odd** when its graph is symmetric with respect to the origin. The symmetry tests in Section 1.1 yield the tests for even and odd functions below.

Tests for Even and Odd Functions

A function $y = f(x)$ is **even** when, for each x in the domain of f, $f(-x) = f(x)$.

A function $y = f(x)$ is **odd** when, for each x in the domain of f, $f(-x) = -f(x)$.

EXAMPLE 8   **Even and Odd Functions**

See LarsonPrecalculus.com for an interactive version of this type of example.

a. The function $g(x) = x^3 - x$ is odd because $g(-x) = -g(x)$, as follows.

$$g(-x) = (-x)^3 - (-x) \qquad \text{Substitute } -x \text{ for } x.$$
$$= -x^3 + x \qquad \text{Simplify.}$$
$$= -(x^3 - x) \qquad \text{Distributive Property}$$
$$= -g(x) \qquad \text{Test for odd function}$$

b. The function $h(x) = x^2 + 1$ is even because $h(-x) = h(x)$, as follows.

$$h(-x) = (-x)^2 + 1 = x^2 + 1 = h(x) \qquad \text{Test for even function}$$

Figure 2.22 shows the graphs and symmetry of these two functions.

✓ *Checkpoint* 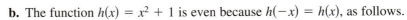 *Audio-video solution in English & Spanish at LarsonPrecalculus.com*

Determine whether each function is even, odd, or neither. Then describe the symmetry.

a. $f(x) = 5 - 3x$ **b.** $g(x) = x^4 - x^2 - 1$ **c.** $h(x) = 2x^3 + 3x$ ▪

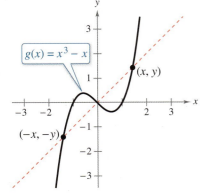

(a) Symmetric to origin: Odd Function

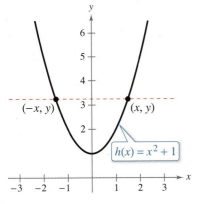

(b) Symmetric to y-axis: Even Function

Figure 2.22

Summarize   (Section 2.3)

1. State the Vertical Line Test for functions *(page 188)*. For an example of using the Vertical Line Test, see Example 2.

2. Explain how to find the zeros of a function *(page 189)*. For an example of finding the zeros of functions, see Example 3.

3. Explain how to determine intervals on which functions are increasing or decreasing *(page 190)*. For an example of describing function behavior, see Example 4.

4. Explain how to determine relative minimum and relative maximum values of functions *(page 191)*. For an example of approximating a relative minimum, see Example 5.

5. Explain how to determine the average rate of change of a function *(page 192)*. For examples of determining average rates of change, see Examples 6 and 7.

6. State the definitions of an even function and an odd function *(page 193)*. For an example of identifying even and odd functions, see Example 8.

2.3 Exercises

See **CalcChat.com** for tutorial help and worked-out solutions to odd-numbered exercises.

Vocabulary: Fill in the blanks.

1. The _____ _____ _____ is used to determine whether a graph represents y as a function of x.

2. The _____ of a function $y = f(x)$ are the values of x for which $f(x) = 0$.

3. A function f is _____ on an interval when, for any x_1 and x_2 in the interval, $x_1 < x_2$ implies $f(x_1) > f(x_2)$.

4. A function value $f(a)$ is a relative _____ of f when there exists an interval (x_1, x_2) containing a such that $x_1 < x < x_2$ implies $f(a) \geq f(x)$.

5. The _____ _____ _____ _____ between any two points $(x_1, f(x_1))$ and $(x_2, f(x_2))$ is the slope of the line through the two points, and this line is called the _____ line.

6. A function f is _____ when, for each x in the domain of f, $f(-x) = -f(x)$.

Skills and Applications

 Domain, Range, and Values of a Function In Exercises 7–10, use the graph of the function to find the domain and range of f and each function value.

7. (a) $f(-1)$ (b) $f(0)$
   (c) $f(1)$ (d) $f(2)$

8. (a) $f(-1)$ (b) $f(0)$
   (c) $f(1)$ (d) $f(3)$

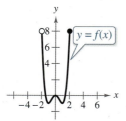

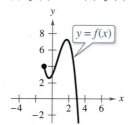

9. (a) $f(2)$ (b) $f(1)$
   (c) $f(3)$ (d) $f(-1)$

10. (a) $f(-2)$ (b) $f(1)$
    (c) $f(0)$ (d) $f(2)$

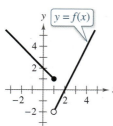

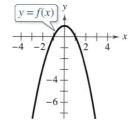

 Vertical Line Test for Functions In Exercises 11–14, use the Vertical Line Test to determine whether the graph represents y as a function of x. To print an enlarged copy of the graph, go to *MathGraphs.com*.

11.

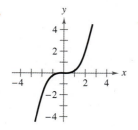

12.

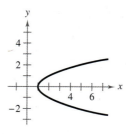

13.

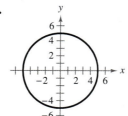

14.

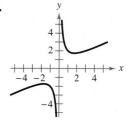

 Finding the Zeros of a Function In Exercises 15–26, find the zeros of the function algebraically.

15. $f(x) = 3x + 18$

16. $f(x) = 15 - 2x$

17. $f(x) = 2x^2 - 7x - 30$

18. $f(x) = 3x^2 + 22x - 16$

19. $f(x) = \dfrac{x + 3}{2x^2 - 6}$

20. $f(x) = \dfrac{x^2 - 9x + 14}{4x}$

21. $f(x) = \frac{1}{3}x^3 - 2x$

22. $f(x) = -25x^4 + 9x^2$

23. $f(x) = x^3 - 4x^2 - 9x + 36$

24. $f(x) = 4x^3 - 24x^2 - x + 6$

25. $f(x) = \sqrt{2x - 1}$

26. $f(x) = \sqrt{3x + 2}$

Graphing and Finding Zeros In Exercises 27–32, (a) use a graphing utility to graph the function and find the zeros of the function and (b) verify your results from part (a) algebraically.

27. $f(x) = x^2 - 6x$

28. $f(x) = 2x^2 - 13x - 7$

29. $f(x) = \sqrt{2x + 11}$

30. $f(x) = \sqrt{3x - 14} - 8$

31. $f(x) = \dfrac{3x - 1}{x - 6}$

32. $f(x) = \dfrac{2x^2 - 9}{3 - x}$

Describing Function Behavior In Exercises 33–40, determine the open intervals on which the function is increasing, decreasing, or constant.

33. $f(x) = -\frac{1}{2}x^3$

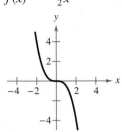

34. $f(x) = x^2 - 4x$

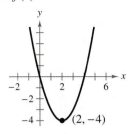

35. $f(x) = \sqrt{x^2 - 1}$

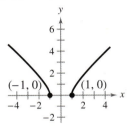

36. $f(x) = x^3 - 3x^2 + 2$

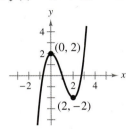

37. $f(x) = |x + 1| + |x - 1|$ **38.** $f(x) = \dfrac{x^2 + x + 1}{x + 1}$

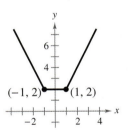

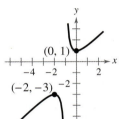

39. $f(x) = \begin{cases} 2x + 1, & x \le -1 \\ x^2 - 2, & x > -1 \end{cases}$

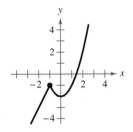

40. $f(x) = \begin{cases} x + 3, & x \le 0 \\ 3, & 0 < x \le 2 \\ 2x + 1, & x > 2 \end{cases}$

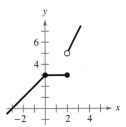

 Describing Function Behavior In Exercises 41–48, use a graphing utility to graph the function and visually determine the open intervals on which the function is increasing, decreasing, or constant. Use a table of values to verify your results.

41. $f(x) = 3$ **42.** $g(x) = x$
43. $g(x) = \frac{1}{2}x^2 - 3$ **44.** $f(x) = 3x^4 - 6x^2$
45. $f(x) = \sqrt{1 - x}$ **46.** $f(x) = x\sqrt{x + 3}$
47. $f(x) = x^{3/2}$ **48.** $f(x) = x^{2/3}$

 Approximating Relative Minima or Maxima In Exercises 49–54, use a graphing utility to approximate (to two decimal places) any relative minima or maxima of the function.

49. $f(x) = x(x + 3)$
50. $f(x) = -x^2 + 3x - 2$
51. $h(x) = x^3 - 6x^2 + 15$
52. $f(x) = x^3 - 3x^2 - x + 1$
53. $h(x) = (x - 1)\sqrt{x}$
54. $g(x) = x\sqrt{4 - x}$

 Graphical Reasoning In Exercises 55–60, graph the function and determine the interval(s) for which $f(x) \ge 0$.

55. $f(x) = 4 - x$ **56.** $f(x) = 4x + 2$
57. $f(x) = 9 - x^2$ **58.** $f(x) = x^2 - 4x$
59. $f(x) = \sqrt{x - 1}$ **60.** $f(x) = |x + 5|$

∫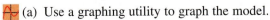 **Average Rate of Change of a Function In Exercises 61–64, find the average rate of change of the function from x_1 to x_2.**

Function	x-Values
61. $f(x) = -2x + 15$	$x_1 = 0, x_2 = 3$
62. $f(x) = x^2 - 2x + 8$	$x_1 = 1, x_2 = 5$
63. $f(x) = x^3 - 3x^2 - x$	$x_1 = -1, x_2 = 2$
64. $f(x) = -x^3 + 6x^2 + x$	$x_1 = 1, x_2 = 6$

65. Research and Development The amounts (in billions of dollars) the U.S. federal government spent on research and development for defense from 2010 through 2014 can be approximated by the model

$$y = 0.5079t^2 - 8.168t + 95.08$$

where t represents the year, with $t = 0$ corresponding to 2010. *(Source: American Association for the Advancement of Science)*

(a) Use a graphing utility to graph the model.

(b) Find the average rate of change of the model from 2010 to 2014. Interpret your answer in the context of the problem.

66. Finding Average Speed Use the information in Example 7 to find the average speed of the car from $t_1 = 0$ to $t_2 = 9$ seconds. Explain why the result is less than the value obtained in part (b) of Example 7.

Physics In Exercises 67–70, (a) use the position equation $s = -16t^2 + v_0 t + s_0$ to write a function that represents the situation, (b) use a graphing utility to graph the function, (c) find the average rate of change of the function from t_1 to t_2, (d) describe the slope of the secant line through t_1 and t_2, (e) find the equation of the secant line through t_1 and t_2, and (f) graph the secant line in the same viewing window as your position function.

67. An object is thrown upward from a height of 6 feet at a velocity of 64 feet per second.

$t_1 = 0, t_2 = 3$

68. An object is thrown upward from a height of 6.5 feet at a velocity of 72 feet per second.

$t_1 = 0, t_2 = 4$

69. An object is thrown upward from ground level at a velocity of 120 feet per second.

$t_1 = 3, t_2 = 5$

70. An object is dropped from a height of 80 feet.

$t_1 = 1, t_2 = 2$

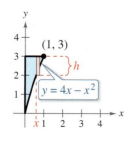

Even, Odd, or Neither? In Exercises 71–76, determine whether the function is even, odd, or neither. Then describe the symmetry.

71. $f(x) = x^6 - 2x^2 + 3$ **72.** $g(x) = x^3 - 5x$

73. $h(x) = x\sqrt{x + 5}$ **74.** $f(x) = x\sqrt{1 - x^2}$

75. $f(s) = 4s^{3/2}$ **76.** $g(s) = 4s^{2/3}$

Even, Odd, or Neither? In Exercises 77–82, sketch a graph of the function and determine whether it is even, odd, or neither. Verify your answer algebraically.

77. $f(x) = -9$ **78.** $f(x) = 5 - 3x$

79. $f(x) = -|x - 5|$ **80.** $h(x) = x^2 - 4$

81. $f(x) = \sqrt[3]{4x}$ **82.** $f(x) = \sqrt[3]{x - 4}$

Height of a Rectangle In Exercises 83 and 84, write the height h of the rectangle as a function of x.

83.
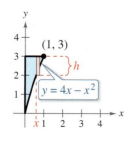

84.
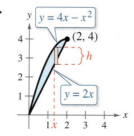

Length of a Rectangle In Exercises 85 and 86, write the length L of the rectangle as a function of y.

85.
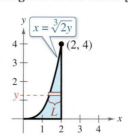

86.
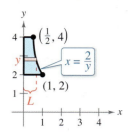

87. Error Analysis Describe the error.

The function $f(x) = 2x^3 - 5$ is odd because $f(-x) = -f(x)$, as follows.

$$f(-x) = 2(-x)^3 - 5$$
$$= -2x^3 - 5$$
$$= -(2x^3 - 5)$$
$$= -f(x)$$

88. Geometry Corners of equal size are cut from a square with sides of length 8 meters (see figure).

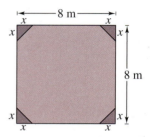

(a) Write the area A of the resulting figure as a function of x. Determine the domain of the function.

(b) Use a graphing utility to graph the area function over its domain. Use the graph to find the range of the function.

(c) Identify the figure that results when x is the maximum value in the domain of the function. What would be the length of each side of the figure?

89. Coordinate Axis Scale Each function described below models the specified data for the years 2006 through 2016, with $t = 6$ corresponding to 2006. Estimate a reasonable scale for the vertical axis (e.g., hundreds, thousands, millions, etc.) of the graph and justify your answer. (There are many correct answers.)

(a) $f(t)$ represents the average salary of college professors.

(b) $f(t)$ represents the U.S. population.

(c) $f(t)$ represents the percent of the civilian workforce that is unemployed.

(d) $f(t)$ represents the number of games a college football team wins.

90. Temperature

The table shows the temperatures y (in degrees Fahrenheit) in a city over a 24-hour period. Let x represent the time of day, where $x = 0$ corresponds to 6 A.M.

Time, x	Temperature, y
0	34
2	50
4	60
6	64
8	63
10	59
12	53
14	46
16	40
18	36
20	34
22	37
24	45

Spreadsheet at LarsonPrecalculus.com

These data can be approximated by the model

$y = 0.026x^3 - 1.03x^2 + 10.2x + 34, \quad 0 \le x \le 24.$

(a) Use a graphing utility to create a scatter plot of the data. Then graph the model in the same viewing window.

(b) How well does the model fit the data?

(c) Use the graph to approximate the times when the temperature was increasing and decreasing.

(d) Use the graph to approximate the maximum and minimum temperatures during this 24-hour period.

(e) Could this model predict the temperatures in the city during the next 24-hour period? Why or why not?

Exploration

True or False? In Exercises 91–93, determine whether the statement is true or false. Justify your answer.

91. A function with a square root cannot have a domain that is the set of real numbers.

92. It is possible for an odd function to have the interval $[0, \infty)$ as its domain.

93. It is impossible for an even function to be increasing on its entire domain.

94. HOW DO YOU SEE IT? Use the graph of the function to answer parts (a)–(e).

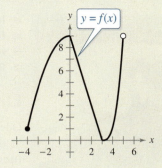

(a) Find the domain and range of f.

(b) Find the zero(s) of f.

(c) Determine the open intervals on which f is increasing, decreasing, or constant.

(d) Approximate any relative minimum or relative maximum values of f.

(e) Is f even, odd, or neither?

Think About It In Exercises 95 and 96, find the coordinates of a second point on the graph of a function f when the given point is on the graph and the function is (a) even and (b) odd.

95. $\left(-\frac{5}{3}, -7\right)$ **96.** $(2a, 2c)$

97. Writing Use a graphing utility to graph each function. Write a paragraph describing any similarities and differences you observe among the graphs.

(a) $y = x$ (b) $y = x^2$ (c) $y = x^3$
(d) $y = x^4$ (e) $y = x^5$ (f) $y = x^6$

98. Graphical Reasoning Graph each of the functions with a graphing utility. Determine whether each function is even, odd, or neither.

$f(x) = x^2 - x^4$ $g(x) = 2x^3 + 1$
$h(x) = x^5 - 2x^3 + x$ $j(x) = 2 - x^6 - x^8$
$k(x) = x^5 - 2x^4 + x - 2$ $p(x) = x^9 + 3x^5 - x^3 + x$

What do you notice about the equations of functions that are odd? What do you notice about the equations of functions that are even? Can you describe a way to identify a function as odd or even by inspecting the equation? Can you describe a way to identify a function as neither odd nor even by inspecting the equation?

99. Even, Odd, or Neither? Determine whether g is even, odd, or neither when f is an even function. Explain.

(a) $g(x) = -f(x)$ (b) $g(x) = f(-x)$
(c) $g(x) = f(x) - 2$ (d) $g(x) = f(x - 2)$

2.4 A Library of Parent Functions

Piecewise-defined functions model many real-life situations. For example, in Exercise 47 on page 204, you will write a piecewise-defined function to model the depth of snow during a snowstorm.

■ Identify and graph linear and squaring functions.
■ Identify and graph cubic, square root, and reciprocal functions.
■ Identify and graph step and other piecewise-defined functions.
■ Recognize graphs of parent functions.

Linear and Squaring Functions

One of the goals of this text is to enable you to recognize the basic shapes of the graphs of different types of functions. For example, you know that the graph of the **linear function** $f(x) = ax + b$ is a line with slope $m = a$ and y-intercept at $(0, b)$. The graph of a linear function has the characteristics below.

- The domain of the function is the set of all real numbers.
- When $m \neq 0$, the range of the function is the set of all real numbers.
- The graph has an x-intercept at $(-b/m, 0)$ and a y-intercept at $(0, b)$.
- The graph is increasing when $m > 0$, decreasing when $m < 0$, and constant when $m = 0$.

EXAMPLE 1   Writing a Linear Function

Write the linear function f for which $f(1) = 3$ and $f(4) = 0$.

Solution   To find the equation of the line that passes through $(x_1, y_1) = (1, 3)$ and $(x_2, y_2) = (4, 0)$, first find the slope of the line.

$$m = \frac{y_2 - y_1}{x_2 - x_1} = \frac{0 - 3}{4 - 1} = \frac{-3}{3} = -1$$

Next, use the point-slope form of the equation of a line.

$y - y_1 = m(x - x_1)$	Point-slope form
$y - 3 = -1(x - 1)$	Substitute for x_1, y_1, and m.
$y = -x + 4$	Simplify.
$f(x) = -x + 4$	Function notation

The figure below shows the graph of this function.

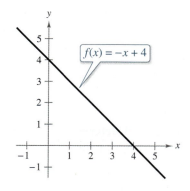

$f(x) = -x + 4$

✓ **Checkpoint** ◀))) *Audio-video solution in English & Spanish at LarsonPrecalculus.com*

Write the linear function f for which $f(-2) = 6$ and $f(4) = -9$.

There are two special types of linear functions, the **constant function** and the **identity function.** A constant function has the form

$$f(x) = c$$

and has a domain of all real numbers with a range consisting of a single real number c. The graph of a constant function is a horizontal line, as shown in Figure 2.23. The identity function has the form

$$f(x) = x.$$

Its domain and range are the set of all real numbers. The identity function has a slope of $m = 1$ and a y-intercept at $(0, 0)$. The graph of the identity function is a line for which each x-coordinate equals the corresponding y-coordinate. The graph is always increasing, as shown in Figure 2.24.

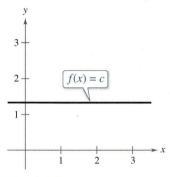

Figure 2.23

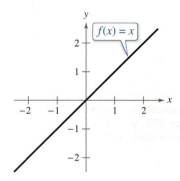

Figure 2.24

The graph of the **squaring function**

$$f(x) = x^2$$

is a U-shaped curve with the characteristics below.

- The domain of the function is the set of all real numbers.
- The range of the function is the set of all nonnegative real numbers.
- The function is even.
- The graph has an intercept at $(0, 0)$.
- The graph is decreasing on the interval $(-\infty, 0)$ and increasing on the interval $(0, \infty)$.
- The graph is symmetric with respect to the y-axis.
- The graph has a relative minimum at $(0, 0)$.

The figure below shows the graph of the squaring function.

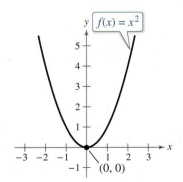

Cubic, Square Root, and Reciprocal Functions

Here are the basic characteristics of the graphs of the **cubic, square root,** and **reciprocal functions.**

1. The graph of the *cubic* function

   $$f(x) = x^3$$

   has the characteristics below.

   • The domain of the function is the set of all real numbers.

   • The range of the function is the set of all real numbers.

   • The function is odd.

   • The graph has an intercept at $(0, 0)$.

   • The graph is increasing on the interval $(-\infty, \infty)$.

   • The graph is symmetric with respect to the origin.

   The figure shows the graph of the cubic function.

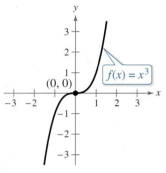

Cubic function

2. The graph of the *square root* function

   $$f(x) = \sqrt{x}$$

   has the characteristics below.

   • The domain of the function is the set of all nonnegative real numbers.

   • The range of the function is the set of all nonnegative real numbers.

   • The graph has an intercept at $(0, 0)$.

   • The graph is increasing on the interval $(0, \infty)$.

   The figure shows the graph of the square root function.

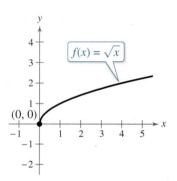

Square root function

3. The graph of the *reciprocal* function

   $$f(x) = \frac{1}{x}$$

   has the characteristics below.

   • The domain of the function is $(-\infty, 0) \cup (0, \infty)$.

   • The range of the function is $(-\infty, 0) \cup (0, \infty)$.

   • The function is odd.

   • The graph does not have any intercepts.

   • The graph is decreasing on the intervals $(-\infty, 0)$ and $(0, \infty)$.

   • The graph is symmetric with respect to the origin.

   The figure shows the graph of the reciprocal function.

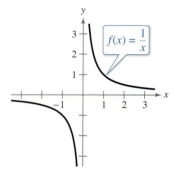

Reciprocal function

Step and Piecewise-Defined Functions

Functions whose graphs resemble sets of stairsteps are known as **step functions.** One common type of step function is the **greatest integer function,** denoted by $[\![x]\!]$ and defined as

$f(x) = [\![x]\!] = $ *the greatest integer less than or equal to x.*

Here are several examples of evaluating the greatest integer function.

$[\![-1]\!] = (\text{greatest integer} \le -1) = -1$

$\left[\!\left[-\tfrac{1}{2}\right]\!\right] = \left(\text{greatest integer} \le -\tfrac{1}{2}\right) = -1$

$\left[\!\left[\tfrac{1}{10}\right]\!\right] = \left(\text{greatest integer} \le \tfrac{1}{10}\right) = 0$

$[\![1.5]\!] = (\text{greatest integer} \le 1.5) = 1$

$[\![1.9]\!] = (\text{greatest integer} \le 1.9) = 1$

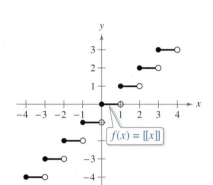

The graph of the greatest integer function

$f(x) = [\![x]\!]$

has the characteristics below, as shown in Figure 2.25.

- The domain of the function is the set of all real numbers.
- The range of the function is the set of all integers.
- The graph has a *y*-intercept at $(0, 0)$ and *x*-intercepts in the interval $[0, 1)$.
- The graph is constant between each pair of consecutive integer values of *x*.
- The graph jumps vertically one unit at each integer value of *x*.

Figure 2.25

▷ TECHNOLOGY Most graphing utilities display graphs in *connected* mode, which works well for graphs that do not have breaks. For graphs that do have breaks, such as the graph of the greatest integer function, it may be better to use *dot* mode. Graph the greatest integer function [often called Int(*x*)] in *connected* and *dot* modes, and compare the two results.

EXAMPLE 2   **Evaluating a Step Function**

Evaluate the function $f(x) = [\![x]\!] + 1$ when $x = -1, 2,$ and $\tfrac{3}{2}$.

Solution   For $x = -1$, the greatest integer ≤ -1 is -1, so

$f(-1) = [\![-1]\!] + 1 = -1 + 1 = 0.$

For $x = 2$, the greatest integer ≤ 2 is 2, so

$f(2) = [\![2]\!] + 1 = 2 + 1 = 3.$

For $x = \tfrac{3}{2}$, the greatest integer $\le \tfrac{3}{2}$ is 1, so

$f\left(\tfrac{3}{2}\right) = \left[\!\left[\tfrac{3}{2}\right]\!\right] + 1 = 1 + 1 = 2.$

Verify your answers by examining the graph of $f(x) = [\![x]\!] + 1$ shown in Figure 2.26.

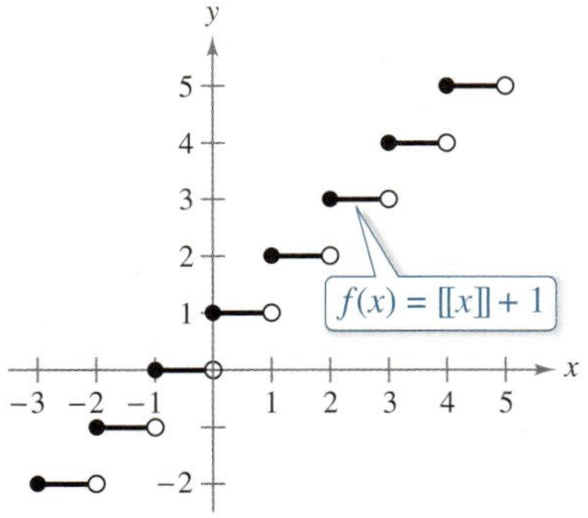

Figure 2.26

✓ **Checkpoint** ◀))) *Audio-video solution in English & Spanish at LarsonPrecalculus.com*

Evaluate the function $f(x) = [\![x + 2]\!]$ when $x = -\tfrac{3}{2}, 1,$ and $-\tfrac{5}{2}$. ■

Recall from Section 2.2 that a piecewise-defined function is defined by two or more equations over a specified domain. To graph a piecewise-defined function, graph each equation separately over the specified domain, as shown in Example 3.

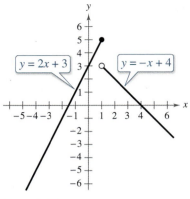

Figure 2.27

EXAMPLE 3 **Graphing a Piecewise-Defined Function**

See LarsonPrecalculus.com for an interactive version of this type of example.

Sketch the graph of $f(x) = \begin{cases} 2x + 3, & x \le 1 \\ -x + 4, & x > 1 \end{cases}$.

Solution This piecewise-defined function consists of two linear functions. At $x = 1$ and to the left of $x = 1$, the graph is the line $y = 2x + 3$, and to the right of $x = 1$, the graph is the line $y = -x + 4$, as shown in Figure 2.27. Notice that the point $(1, 5)$ is a solid dot and the point $(1, 3)$ is an open dot. This is because $f(1) = 2(1) + 3 = 5$.

✓ *Checkpoint* 🔊))) *Audio-video solution in English & Spanish at LarsonPrecalculus.com*

Sketch the graph of $f(x) = \begin{cases} -\frac{1}{2}x - 6, & x \le -4 \\ x + 5, & x > -4 \end{cases}$.

Parent Functions

The graphs below represent the most commonly used functions in algebra. Familiarity with the characteristics of these graphs will help you analyze more complicated graphs obtained from these graphs by the transformations studied in the next section.

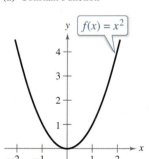

(a) Constant Function

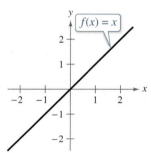

(b) Identity Function

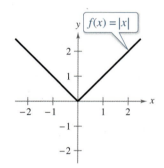

(c) Absolute Value Function

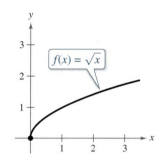

(d) Square Root Function

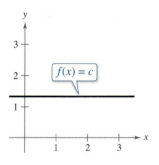

(e) Squaring Function

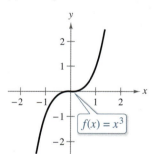

(f) Cubic Function

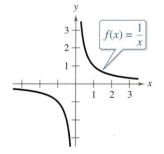

(g) Reciprocal Function

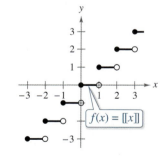

(h) Greatest Integer Function

Summarize (Section 2.4)

1. Explain how to identify and graph linear and squaring functions (*pages 198 and 199*). For an example involving a linear function, see Example 1.

2. Explain how to identify and graph cubic, square root, and reciprocal functions (*page 200*).

3. Explain how to identify and graph step and other piecewise-defined functions (*page 201*). For examples involving these functions, see Examples 2 and 3.

4. Identify and sketch the graphs of parent functions (*page 202*).

2.4 Exercises

See CalcChat.com for tutorial help and worked-out solutions to odd-numbered exercises.

Vocabulary

In Exercises 1–9, write the most specific name of the function.

1. $f(x) = [\![x]\!]$

2. $f(x) = x$

3. $f(x) = 1/x$

4. $f(x) = x^2$

5. $f(x) = \sqrt{x}$

6. $f(x) = c$

7. $f(x) = |x|$

8. $f(x) = x^3$

9. $f(x) = ax + b$

10. Fill in the blank: The constant function and the identity function are two special types of _____ functions.

Skills and Applications

 Writing a Linear Function In Exercises 11–14, (a) write the linear function f that has the given function values and (b) sketch the graph of the function.

11. $f(1) = 4, \quad f(0) = 6$

12. $f(-3) = -8, \quad f(1) = 2$

13. $f\left(\frac{1}{2}\right) = -\frac{5}{3}, \quad f(6) = 2$

14. $f\left(\frac{3}{5}\right) = \frac{1}{2}, \quad f(4) = 9$

 Graphing a Function In Exercises 15–26, use a graphing utility to graph the function. Be sure to choose an appropriate viewing window.

15. $f(x) = 2.5x - 4.25$

16. $f(x) = \frac{5}{6} - \frac{2}{3}x$

17. $g(x) = x^2 + 3$

18. $f(x) = -2x^2 - 1$

19. $f(x) = x^3 - 1$

20. $f(x) = (x - 1)^3 + 2$

21. $f(x) = \sqrt{x} + 4$

22. $h(x) = \sqrt{x + 2} + 3$

23. $f(x) = \dfrac{1}{x - 2}$

24. $k(x) = 3 + \dfrac{1}{x + 3}$

25. $g(x) = |x| - 5$

26. $f(x) = |x - 1|$

Evaluating a Step Function In Exercises 27–30, evaluate the function for the given values.

27. $f(x) = [\![x]\!]$

 (a) $f(2.1)$ (b) $f(2.9)$ (c) $f(-3.1)$ (d) $f\left(\frac{7}{2}\right)$

28. $h(x) = [\![x + 3]\!]$

 (a) $h(-2)$ (b) $h\left(\frac{1}{2}\right)$ (c) $h(4.2)$ (d) $h(-21.6)$

29. $k(x) = [\![2x + 1]\!]$

 (a) $k\left(\frac{1}{3}\right)$ (b) $k(-2.1)$ (c) $k(1.1)$ (d) $k\left(\frac{2}{3}\right)$

30. $g(x) = -7[\![x + 4]\!] + 6$

 (a) $g\left(\frac{1}{8}\right)$ (b) $g(9)$ (c) $g(-4)$ (d) $g\left(\frac{3}{2}\right)$

Graphing a Step Function In Exercises 31–34, sketch the graph of the function.

31. $g(x) = -[\![x]\!]$

32. $g(x) = 4[\![x]\!]$

33. $g(x) = [\![x]\!] - 1$

34. $g(x) = [\![x - 3]\!]$

 Graphing a Piecewise-Defined Function In Exercises 35–40, sketch the graph of the function.

35. $g(x) = \begin{cases} x + 6, & x \le -4 \\ \frac{1}{2}x - 4, & x > -4 \end{cases}$

36. $f(x) = \begin{cases} 4 + x, & x \le 2 \\ x^2 + 2, & x > 2 \end{cases}$

37. $f(x) = \begin{cases} 1 - (x - 1)^2, & x \le 2 \\ \sqrt{x - 2}, & x > 2 \end{cases}$

38. $f(x) = \begin{cases} \sqrt{4 + x}, & x < 0 \\ \sqrt{4 - x}, & x \ge 0 \end{cases}$

39. $h(x) = \begin{cases} 4 - x^2, & x < -2 \\ 3 + x, & -2 \le x < 0 \\ x^2 + 1, & x \ge 0 \end{cases}$

40. $k(x) = \begin{cases} 2x + 1, & x \le -1 \\ 2x^2 - 1, & -1 < x \le 1 \\ 1 - x^2, & x > 1 \end{cases}$

Graphing a Function In Exercises 41 and 42, (a) use a graphing utility to graph the function and (b) state the domain and range of the function.

41. $s(x) = 2\left(\frac{1}{4}x - \left[\!\left[\frac{1}{4}x\right]\!\right]\right)$

42. $k(x) = 4\left(\frac{1}{2}x - \left[\!\left[\frac{1}{2}x\right]\!\right]\right)^2$

43. Wages A mechanic's pay is $14 per hour for regular time and time-and-a-half for overtime. The weekly wage function is

$$W(h) = \begin{cases} 14h, & 0 < h \le 40 \\ 21(h - 40) + 560, & h > 40 \end{cases}$$

where h is the number of hours worked in a week.

(a) Evaluate $W(30)$, $W(40)$, $W(45)$, and $W(50)$.

(b) The company decreases the regular work week to 36 hours. What is the new weekly wage function?

(c) The company increases the mechanic's pay to $16 per hour. What is the new weekly wage function? Use a regular work week of 40 hours.

44. Revenue   The table shows the monthly revenue y (in thousands of dollars) of a landscaping business for each month of the year 2016, with $x = 1$ representing January.

Month, x	Revenue, y
1	5.2
2	5.6
3	6.6
4	8.3
5	11.5
6	15.8
7	12.8
8	10.1
9	8.6
10	6.9
11	4.5
12	2.7

Spreadsheet at LarsonPrecalculus.com

A mathematical model that represents these data is

$$f(x) = \begin{cases} -1.97x + 26.3 \\ 0.505x^2 - 1.47x + 6.3 \end{cases}.$$

(a) Use a graphing utility to graph the model. What is the domain of each part of the piecewise-defined function? How can you tell?

(b) Find $f(5)$ and $f(11)$ and interpret your results in the context of the problem.

(c) How do the values obtained from the model in part (b) compare with the actual data values?

45. Fluid Flow   The intake pipe of a 100-gallon tank has a flow rate of 10 gallons per minute, and two drainpipes have flow rates of 5 gallons per minute each. The figure shows the volume V of fluid in the tank as a function of time t. Determine whether the input pipe and each drainpipe are open or closed in specific subintervals of the 1 hour of time shown in the graph. (There are many correct answers.)

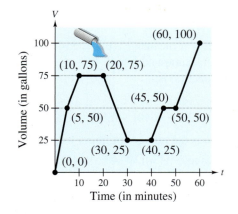

46. Delivery Charges   The cost of mailing a package weighing up to, but not including, 1 pound is \$2.72. Each additional pound or portion of a pound costs \$0.50.

(a) Use the greatest integer function to create a model for the cost C of mailing a package weighing x pounds, where $x > 0$.

(b) Sketch the graph of the function.

47. Snowstorm

During a nine-hour snowstorm, it snows at a rate of 1 inch per hour for the first 2 hours, at a rate of 2 inches per hour for the next 6 hours, and at a rate of 0.5 inch per hour for the final hour. Write and graph a piecewise-defined function that gives the depth of the snow during the snowstorm. How many inches of snow accumulated from the storm?

48. HOW DO YOU SEE IT?   For each graph of f shown below, answer parts (a)–(d).

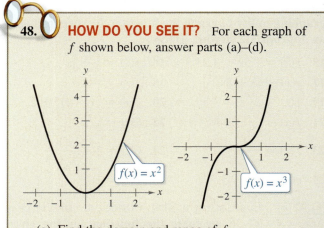

(a) Find the domain and range of f.

(b) Find the x- and y-intercepts of the graph of f.

(c) Determine the open intervals on which f is increasing, decreasing, or constant.

(d) Determine whether f is even, odd, or neither. Then describe the symmetry.

Exploration

True or False?   **In Exercises 49 and 50, determine whether the statement is true or false. Justify your answer.**

49. A piecewise-defined function will always have at least one x-intercept or at least one y-intercept.

50. A linear equation will always have an x-intercept and a y-intercept.

2.5 Transformations of Functions

Transformations of functions model many real-life applications. For example, in Exercise 61 on page 212, you will use a transformation of a function to model the number of horsepower required to overcome wind drag on an automobile.

- Use vertical and horizontal shifts to sketch graphs of functions.
- Use reflections to sketch graphs of functions.
- Use nonrigid transformations to sketch graphs of functions.

Shifting Graphs

Many functions have graphs that are transformations of the parent graphs summarized in Section 2.4. For example, you obtain the graph of

$$h(x) = x^2 + 2$$

by shifting the graph of $f(x) = x^2$ *up* two units, as shown in Figure 2.28. In function notation, h and f are related as follows.

$$h(x) = x^2 + 2 = f(x) + 2 \qquad \text{Upward shift of two units}$$

Similarly, you obtain the graph of

$$g(x) = (x - 2)^2$$

by shifting the graph of $f(x) = x^2$ to the *right* two units, as shown in Figure 2.29. In this case, the functions g and f have the following relationship.

$$g(x) = (x - 2)^2 = f(x - 2) \qquad \text{Right shift of two units}$$

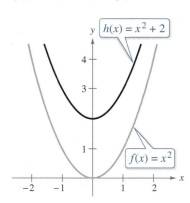

Figure 2.28

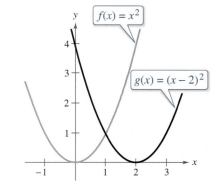

Figure 2.29

The list below summarizes this discussion about horizontal and vertical shifts.

•• REMARK In items 3 and 4, be sure you see that $h(x) = f(x - c)$ corresponds to a *right* shift and $h(x) = f(x + c)$ corresponds to a *left* shift for $c > 0$.

Vertical and Horizontal Shifts

Let c be a positive real number. **Vertical and horizontal shifts** in the graph of $y = f(x)$ are represented as follows.

1. Vertical shift c units *up:* $h(x) = f(x) + c$

2. Vertical shift c units *down:* $h(x) = f(x) - c$

3. Horizontal shift c units to the *right:* $h(x) = f(x - c)$

4. Horizontal shift c units to the *left:* $h(x) = f(x + c)$

Some graphs are obtained from combinations of vertical and horizontal shifts, as demonstrated in Example 1(b). Vertical and horizontal shifts generate a *family of functions,* each with the same shape but at a different location in the plane.

EXAMPLE 1 **Shifting the Graph of a Function**

Use the graph of $f(x) = x^3$ to sketch the graph of each function.

a. $g(x) = x^3 - 1$

b. $h(x) = (x + 2)^3 + 1$

Solution

a. Relative to the graph of $f(x) = x^3$, the graph of

$$g(x) = x^3 - 1$$

is a downward shift of one unit, as shown below.

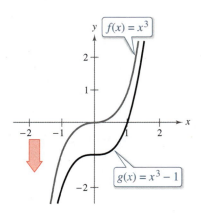

b. Relative to the graph of $f(x) = x^3$, the graph of

$$h(x) = (x + 2)^3 + 1$$

is a left shift of two units and an upward shift of one unit, as shown below.

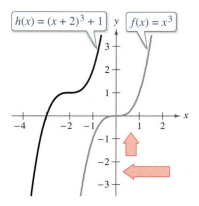

✓ *Checkpoint* ◉))) *Audio-video solution in English & Spanish at LarsonPrecalculus.com*

Use the graph of $f(x) = x^3$ to sketch the graph of each function.

a. $h(x) = x^3 + 5$

b. $g(x) = (x - 3)^3 + 2$

In Example 1(a), note that $g(x) = f(x) - 1$ and in Example 1(b), $h(x) = f(x + 2) + 1$. In Example 1(b), you obtain the same result whether the vertical shift precedes the horizontal shift or the horizontal shift precedes the vertical shift.

Reflecting Graphs

Another common type of transformation is a **reflection.** For example, if you consider the x-axis to be a mirror, then the graph of $h(x) = -x^2$ is the mirror image (or reflection) of the graph of $f(x) = x^2$, as shown in Figure 2.30.

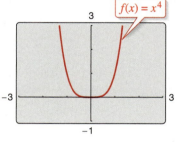

Figure 2.30

> ### Reflections in the Coordinate Axes
>
> **Reflections** in the coordinate axes of the graph of $y = f(x)$ are represented as follows.
>
> **1.** Reflection in the x-axis: $h(x) = -f(x)$
>
> **2.** Reflection in the y-axis: $h(x) = f(-x)$

EXAMPLE 2   **Writing Equations from Graphs**

The graph of the function

$$f(x) = x^4$$

is shown in Figure 2.31. Each graph below is a transformation of the graph of f. Write an equation for the function represented by each graph.

Figure 2.31

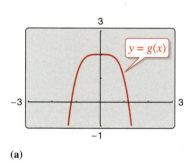
(a)

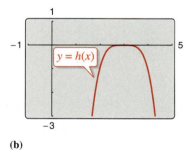
(b)

Solution

a. The graph of g is a reflection in the x-axis *followed by* an upward shift of two units of the graph of $f(x) = x^4$. So, an equation for g is

$$g(x) = -x^4 + 2.$$

b. The graph of h is a right shift of three units *followed by* a reflection in the x-axis of the graph of $f(x) = x^4$. So, an equation for h is

$$h(x) = -(x-3)^4.$$

✓ **Checkpoint** 🔊)) *Audio-video solution in English & Spanish at LarsonPrecalculus.com*

The graph is a transformation of the graph of $f(x) = x^4$. Write an equation for the function represented by the graph.

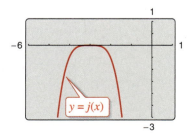

Reflections and Shifts

Compare the graph of each function with the graph of $f(x) = \sqrt{x}$.

a. $g(x) = -\sqrt{x}$ **b.** $h(x) = \sqrt{-x}$ **c.** $k(x) = -\sqrt{x+2}$

Algebraic Solution

a. The graph of g is a reflection of the graph of f in the x-axis because

$$g(x) = -\sqrt{x}$$
$$= -f(x).$$

b. The graph of h is a reflection of the graph of f in the y-axis because

$$h(x) = \sqrt{-x}$$
$$= f(-x).$$

c. The graph of k is a left shift of two units followed by a reflection in the x-axis because

$$k(x) = -\sqrt{x+2}$$
$$= -f(x+2).$$

Graphical Solution

a. Graph f and g on the same set of coordinate axes. The graph of g is a reflection of the graph of f in the x-axis.

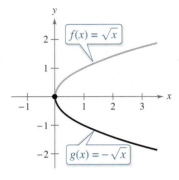

b. Graph f and h on the same set of coordinate axes. The graph of h is a reflection of the graph of f in the y-axis.

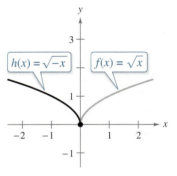

c. Graph f and k on the same set of coordinate axes. The graph of k is a left shift of two units followed by a reflection in the x-axis of the graph of f.

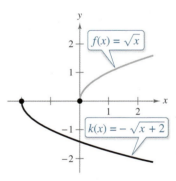

✓ *Checkpoint* ◀))) *Audio-video solution in English & Spanish at LarsonPrecalculus.com*

Compare the graph of each function with the graph of

$$f(x) = \sqrt{x-1}.$$

a. $g(x) = -\sqrt{x-1}$ **b.** $h(x) = \sqrt{-x-1}$

When sketching the graphs of functions involving square roots, remember that you must restrict the domain to exclude negative numbers inside the radical. For instance, here are the domains of the functions in Example 3.

Domain of $g(x) = -\sqrt{x}$: $x \geq 0$

Domain of $h(x) = \sqrt{-x}$: $x \leq 0$

Domain of $k(x) = -\sqrt{x+2}$: $x \geq -2$

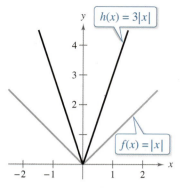

Figure 2.32

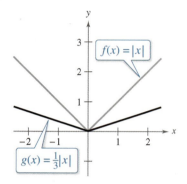

Figure 2.33

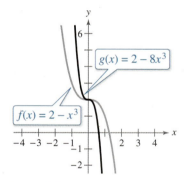

Figure 2.34

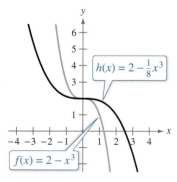

Figure 2.35

Nonrigid Transformations

Horizontal shifts, vertical shifts, and reflections are **rigid transformations** because the basic shape of the graph is unchanged. These transformations change only the *position* of the graph in the coordinate plane. **Nonrigid transformations** are those that cause a *distortion*—a change in the shape of the original graph. For example, a nonrigid transformation of the graph of $y = f(x)$ is represented by $g(x) = cf(x)$, where the transformation is a **vertical stretch** when $c > 1$ and a **vertical shrink** when $0 < c < 1$. Another nonrigid transformation of the graph of $y = f(x)$ is represented by $h(x) = f(cx)$, where the transformation is a **horizontal shrink** when $c > 1$ and a **horizontal stretch** when $0 < c < 1$.

EXAMPLE 4 **Nonrigid Transformations**

Compare the graph of each function with the graph of $f(x) = |x|$.

a. $h(x) = 3|x|$ **b.** $g(x) = \frac{1}{3}|x|$

Solution

a. Relative to the graph of $f(x) = |x|$, the graph of $h(x) = 3|x| = 3f(x)$ is a vertical stretch (each y-value is multiplied by 3). (See Figure 2.32.)

b. Similarly, the graph of $g(x) = \frac{1}{3}|x| = \frac{1}{3}f(x)$ is a vertical shrink $\left(\text{each } y\text{-value is multiplied by } \frac{1}{3}\right)$ of the graph of f. (See Figure 2.33.)

✓ *Checkpoint* *Audio-video solution in English & Spanish at LarsonPrecalculus.com*

Compare the graph of each function with the graph of $f(x) = x^2$.

a. $g(x) = 4x^2$ **b.** $h(x) = \frac{1}{4}x^2$

EXAMPLE 5 **Nonrigid Transformations**

See LarsonPrecalculus.com for an interactive version of this type of example.

Compare the graph of each function with the graph of $f(x) = 2 - x^3$.

a. $g(x) = f(2x)$ **b.** $h(x) = f\left(\frac{1}{2}x\right)$

Solution

a. Relative to the graph of $f(x) = 2 - x^3$, the graph of $g(x) = f(2x) = 2 - (2x)^3 = 2 - 8x^3$ is a horizontal shrink $(c > 1)$. (See Figure 2.34.)

b. Similarly, the graph of $h(x) = f\left(\frac{1}{2}x\right) = 2 - \left(\frac{1}{2}x\right)^3 = 2 - \frac{1}{8}x^3$ is a horizontal stretch $(0 < c < 1)$ of the graph of f. (See Figure 2.35.)

✓ *Checkpoint*  *Audio-video solution in English & Spanish at LarsonPrecalculus.com*

Compare the graph of each function with the graph of $f(x) = x^2 + 3$.

a. $g(x) = f(2x)$ **b.** $h(x) = f\left(\frac{1}{2}x\right)$

Summarize (Section 2.5)

1. Explain how to shift the graph of a function vertically and horizontally (*page 205*). For an example of shifting the graph of a function, see Example 1.

2. Explain how to reflect the graph of a function in the x-axis and in the y-axis (*page 207*). For examples of reflecting graphs of functions, see Examples 2 and 3.

3. Describe nonrigid transformations of the graph of a function (*page 209*). For examples of nonrigid transformations, see Examples 4 and 5.

2.5 Exercises

See CalcChat.com for tutorial help and worked-out solutions to odd-numbered exercises.

Vocabulary

In Exercises 1–3, fill in the blanks.

1. Horizontal shifts, vertical shifts, and reflections are _____ transformations.
2. A reflection in the x-axis of the graph of $y = f(x)$ is represented by $h(x) = $ _____, while a reflection in the y-axis of the graph of $y = f(x)$ is represented by $h(x) = $ _____.
3. A nonrigid transformation of the graph of $y = f(x)$ represented by $g(x) = cf(x)$ is a _____ _____ when $c > 1$ and a _____ _____ when $0 < c < 1$.

4. Match each function h with the transformation it represents, where $c > 0$.
   (a) $h(x) = f(x) + c$ (i) A horizontal shift of f, c units to the right
   (b) $h(x) = f(x) - c$ (ii) A vertical shift of f, c units down
   (c) $h(x) = f(x + c)$ (iii) A horizontal shift of f, c units to the left
   (d) $h(x) = f(x - c)$ (iv) A vertical shift of f, c units up

Skills and Applications

5. **Shifting the Graph of a Function** For each function, sketch the graphs of the function when $c = -2$, -1, 1, and 2 on the same set of coordinate axes.
   (a) $f(x) = |x| + c$ (b) $f(x) = |x - c|$

6. **Shifting the Graph of a Function** For each function, sketch the graphs of the function when $c = -3$, -2, 2, and 3 on the same set of coordinate axes.
   (a) $f(x) = \sqrt{x} + c$ (b) $f(x) = \sqrt{x - c}$

7. **Shifting the Graph of a Function** For each function, sketch the graphs of the function when $c = -4$, -1, 2, and 5 on the same set of coordinate axes.
   (a) $f(x) = [\![x]\!] + c$ (b) $f(x) = [\![x + c]\!]$

8. **Shifting the Graph of a Function** For each function, sketch the graphs of the function when $c = -3$, -2, 1, and 2 on the same set of coordinate axes.
   (a) $f(x) = \begin{cases} x^2 + c, & x < 0 \\ -x^2 + c, & x \ge 0 \end{cases}$

   (b) $f(x) = \begin{cases} (x + c)^2, & x < 0 \\ -(x + c)^2, & x \ge 0 \end{cases}$

Sketching Transformations In Exercises 9 and 10, use the graph of f to sketch each graph. To print an enlarged copy of the graph, go to *MathGraphs.com*.

9. (a) $y = f(-x)$
   (b) $y = f(x) + 4$
   (c) $y = 2f(x)$
   (d) $y = -f(x - 4)$
   (e) $y = f(x) - 3$
   (f) $y = -f(x) - 1$
   (g) $y = f(2x)$

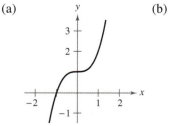

10. (a) $y = f(x - 5)$
    (b) $y = -f(x) + 3$
    (c) $y = \frac{1}{3}f(x)$
    (d) $y = -f(x + 1)$
    (e) $y = f(-x)$
    (f) $y = f(x) - 10$
    (g) $y = f(\frac{1}{3}x)$

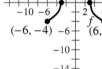

11. **Writing Equations from Graphs** Use the graph of $f(x) = x^2$ to write an equation for the function represented by each graph.
    (a)   (b)

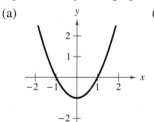   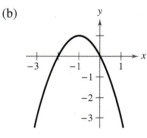

12. **Writing Equations from Graphs** Use the graph of $f(x) = x^3$ to write an equation for the function represented by each graph.
    (a)   (b)

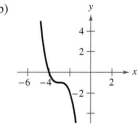

13. Writing Equations from Graphs   Use the graph of $f(x) = |x|$ to write an equation for the function represented by each graph.

(a)

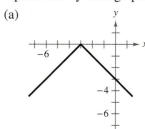

(b)

14. Writing Equations from Graphs   Use the graph of $f(x) = \sqrt{x}$ to write an equation for the function represented by each graph.

(a)

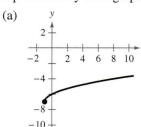

(b)

 Writing Equations from Graphs   In Exercises 15–20, identify the parent function and the transformation represented by the graph. Write an equation for the function represented by the graph.

15.

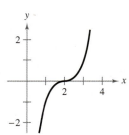

16.

17.

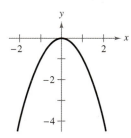

18.

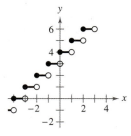

19.

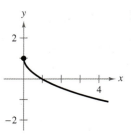

20.

 Describing   Transformations   In Exercises 21–38, g is related to one of the parent functions described in Section 2.4. (a) Identify the parent function f. (b) Describe the sequence of transformations from f to g. (c) Sketch the graph of g. (d) Use function notation to write g in terms of f.

21. $g(x) = x^2 + 6$ **22.** $g(x) = x^2 - 2$

23. $g(x) = -(x - 2)^3$ **24.** $g(x) = -(x + 1)^3$

25. $g(x) = -3 - (x + 1)^2$

26. $g(x) = 4 - (x - 2)^2$

27. $g(x) = |x - 1| + 2$ **28.** $g(x) = |x + 3| - 2$

29. $g(x) = 2\sqrt{x}$ **30.** $g(x) = \frac{1}{2}\sqrt{x}$

31. $g(x) = 2[\![x]\!] - 1$ **32.** $g(x) = -[\![x]\!] + 1$

33. $g(x) = |2x|$ **34.** $g(x) = \left|\frac{1}{2}x\right|$

35. $g(x) = -2x^2 + 1$ **36.** $g(x) = \frac{1}{2}x^2 - 2$

37. $g(x) = 3|x - 1| + 2$

38. $g(x) = -2|x + 1| - 3$

 Writing an Equation from a Description In Exercises 39–46, write an equation for the function whose graph is described.

39. The shape of $f(x) = x^2$, but shifted three units to the right and seven units down

40. The shape of $f(x) = x^2$, but shifted two units to the left, nine units up, and then reflected in the x-axis

41. The shape of $f(x) = x^3$, but shifted 13 units to the right

42. The shape of $f(x) = x^3$, but shifted six units to the left, six units down, and then reflected in the y-axis

43. The shape of $f(x) = |x|$, but shifted 12 units up and then reflected in the x-axis

44. The shape of $f(x) = |x|$, but shifted four units to the left and eight units down

45. The shape of $f(x) = \sqrt{x}$, but shifted six units to the left and then reflected in both the x-axis and the y-axis

46. The shape of $f(x) = \sqrt{x}$, but shifted nine units down and then reflected in both the x-axis and the y-axis

47. Writing Equations from Graphs   Use the graph of $f(x) = x^2$ to write an equation for the function represented by each graph.

(a)

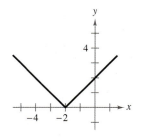

(b)

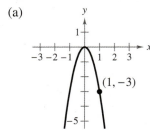

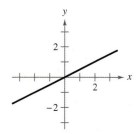

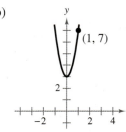

48. Writing Equations from Graphs   Use the graph of

$$f(x) = x^3$$

to write an equation for the function represented by each graph.

(a)
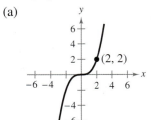

(b)
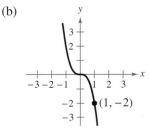

49. Writing Equations from Graphs   Use the graph of

$$f(x) = |x|$$

to write an equation for the function represented by each graph.

(a)

(b)
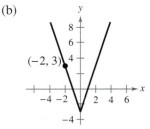

50. Writing Equations from Graphs   Use the graph of

$$f(x) = \sqrt{x}$$

to write an equation for the function represented by each graph.

(a)
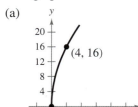

(b)
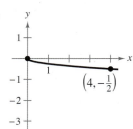

Writing Equations from Graphs   In Exercises 51–56, identify the parent function and the transformation represented by the graph. Write an equation for the function represented by the graph. Then use a graphing utility to verify your answer.

51.
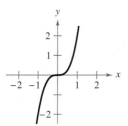

52.
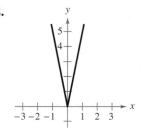

53.
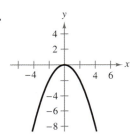

54.
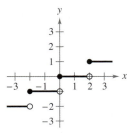

55.
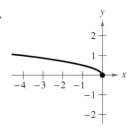

56.
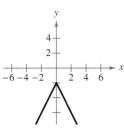

Writing Equations from Graphs   In Exercises 57–60, write an equation for the transformation of the parent function.

57.
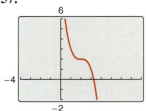

58.
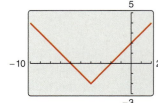

59.
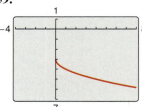

60.
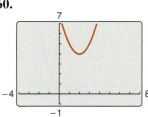

61. Automobile Aerodynamics

The horsepower H required to overcome wind drag on a particular automobile is given by

$$H(x) = 0.00004636x^3$$

where x is the speed of the car (in miles per hour).

(a) Use a graphing utility to graph the function.

(b) Rewrite the horsepower function so that x represents the speed in kilometers per hour. [Find $H(x/1.6)$.] Identify the type of transformation applied to the graph of the horsepower function.

62. Households The number N (in millions) of households in the United States from 2000 through 2014 can be approximated by

$$N(x) = -0.023(x - 33.12)^2 + 131, \quad 0 \le t \le 14$$

where t represents the year, with $t = 0$ corresponding to 2000. (*Source: U.S. Census Bureau*)

(a) Describe the transformation of the parent function $f(x) = x^2$. Then use a graphing utility to graph the function over the specified domain.

(b) Find the average rate of change of the function from 2000 to 2014. Interpret your answer in the context of the problem.

(c) Use the model to predict the number of households in the United States in 2022. Does your answer seem reasonable? Explain.

Exploration

True or False?   **In Exercises 63–66, determine whether the statement is true or false. Justify your answer.**

63. The graph of $y = f(-x)$ is a reflection of the graph of $y = f(x)$ in the x-axis.

64. The graph of $y = -f(x)$ is a reflection of the graph of $y = f(x)$ in the y-axis.

65. The graphs of $f(x) = |x| + 6$ and $f(x) = |-x| + 6$ are identical.

66. If the graph of the parent function $f(x) = x^2$ is shifted six units to the right, three units up, and reflected in the x-axis, then the point $(-2, 19)$ will lie on the graph of the transformation.

67. Finding Points on a Graph   The graph of $y = f(x)$ passes through the points $(0, 1)$, $(1, 2)$, and $(2, 3)$. Find the corresponding points on the graph of $y = f(x + 2) - 1$.

68. Think About It   Two methods of graphing a function are plotting points and translating a parent function as shown in this section. Which method of graphing do you prefer to use for each function? Explain.

(a) $f(x) = 3x^2 - 4x + 1$ (b) $f(x) = 2(x - 1)^2 - 6$

69. Error Analysis   Describe the error.

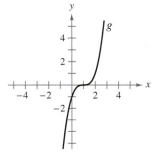

The graph of g is a right shift of one unit of the graph of $f(x) = x^3$. So, an equation for g is $g(x) = (x + 1)^3$.

70.   **HOW DO YOU SEE IT?**   Use the graph of $y = f(x)$ to find the open intervals on which the graph of each transformation is increasing and decreasing. If not possible, state the reason.

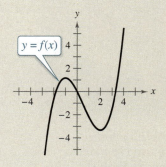

(a) $y = f(-x)$ (b) $y = -f(x)$ (c) $y = \frac{1}{2}f(x)$
(d) $y = -f(x - 1)$ (e) $y = f(x - 2) + 1$

71. Describing Profits   Management originally predicted that the profits from the sales of a new product could be approximated by the graph of the function f shown. The actual profits are represented by the graph of the function g along with a verbal description. Use the concepts of transformations of graphs to write g in terms of f.

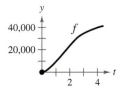

(a) The profits were only three-fourths as large as expected.

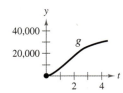

(b) The profits were consistently $10,000 greater than predicted.

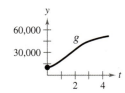

(c) There was a two-year delay in the introduction of the product. After sales began, profits grew as expected.

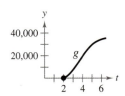

72. Reversing the Order of Transformations   Reverse the order of transformations in Example 2(a). Do you obtain the same graph? Do the same for Example 2(b). Do you obtain the same graph? Explain.

2.6 Combinations of Functions: Composite Functions

■ Add, subtract, multiply, and divide functions.
■ Find the composition of one function with another function.
■ Use combinations and compositions of functions to model and solve real-life problems.

Arithmetic combinations of functions are used to model and solve real-life problems. For example, in Exercise 60 on page 220, you will use arithmetic combinations of functions to analyze numbers of pets in the United States.

Arithmetic Combinations of Functions

Just as two real numbers can be combined by the operations of addition, subtraction, multiplication, and division to form other real numbers, two *functions* can be combined to create new functions. For example, the functions $f(x) = 2x - 3$ and $g(x) = x^2 - 1$ can be combined to form the sum, difference, product, and quotient of f and g.

$$f(x) + g(x) = (2x - 3) + (x^2 - 1) = x^2 + 2x - 4 \qquad \text{Sum}$$

$$f(x) - g(x) = (2x - 3) - (x^2 - 1) = -x^2 + 2x - 2 \qquad \text{Difference}$$

$$f(x)g(x) = (2x - 3)(x^2 - 1) = 2x^3 - 3x^2 - 2x + 3 \qquad \text{Product}$$

$$\frac{f(x)}{g(x)} = \frac{2x - 3}{x^2 - 1}, \quad x \neq \pm 1 \qquad \text{Quotient}$$

The domain of an **arithmetic combination** of functions f and g consists of all real numbers that are common to the domains of f and g. In the case of the quotient $f(x)/g(x)$, there is the further restriction that $g(x) \neq 0$.

Sum, Difference, Product, and Quotient of Functions

Let f and g be two functions with overlapping domains. Then, for all x common to both domains, the *sum, difference, product,* and *quotient* of f and g are defined as follows.

1. Sum: $\qquad (f + g)(x) = f(x) + g(x)$

2. Difference: $\quad (f - g)(x) = f(x) - g(x)$

3. Product: $\qquad (fg)(x) = f(x) \cdot g(x)$

4. Quotient: $\qquad \left(\dfrac{f}{g}\right)(x) = \dfrac{f(x)}{g(x)}, \quad g(x) \neq 0$

EXAMPLE 1 **Finding the Sum of Two Functions**

Given $f(x) = 2x + 1$ and $g(x) = x^2 + 2x - 1$, find $(f + g)(x)$. Then evaluate the sum when $x = 3$.

Solution The sum of f and g is

$$(f + g)(x) = f(x) + g(x) = (2x + 1) + (x^2 + 2x - 1) = x^2 + 4x.$$

When $x = 3$, the value of this sum is

$$(f + g)(3) = 3^2 + 4(3) = 21.$$

✓ *Checkpoint* ◗)) *Audio-video solution in English & Spanish at LarsonPrecalculus.com*

Given $f(x) = x^2$ and $g(x) = 1 - x$, find $(f + g)(x)$. Then evaluate the sum when $x = 2$. ■

> **EXAMPLE 2**   **Finding the Difference of Two Functions**

Given $f(x) = 2x + 1$ and $g(x) = x^2 + 2x - 1$, find $(f - g)(x)$. Then evaluate the difference when $x = 2$.

Solution   The difference of f and g is

$$(f - g)(x) = f(x) - g(x) = (2x + 1) - (x^2 + 2x - 1) = -x^2 + 2.$$

When $x = 2$, the value of this difference is

$$(f - g)(2) = -(2)^2 + 2 = -2.$$

✓ **Checkpoint** 🔊))) *Audio-video solution in English & Spanish at LarsonPrecalculus.com*

Given $f(x) = x^2$ and $g(x) = 1 - x$, find $(f - g)(x)$. Then evaluate the difference when $x = 3$.

> **EXAMPLE 3**   **Finding the Product of Two Functions**

Given $f(x) = x^2$ and $g(x) = x - 3$, find $(fg)(x)$. Then evaluate the product when $x = 4$.

Solution   The product of f and g is

$$(fg)(x) = f(x)g(x) = (x^2)(x - 3) = x^3 - 3x^2.$$

When $x = 4$, the value of this product is

$$(fg)(4) = 4^3 - 3(4)^2 = 16.$$

✓ **Checkpoint** 🔊))) *Audio-video solution in English & Spanish at LarsonPrecalculus.com*

Given $f(x) = x^2$ and $g(x) = 1 - x$, find $(fg)(x)$. Then evaluate the product when $x = 3$.

In Examples 1–3, both f and g have domains that consist of all real numbers. So, the domains of $f + g$, $f - g$, and fg are also the set of all real numbers. Remember to consider any restrictions on the domains of f and g when forming the sum, difference, product, or quotient of f and g.

> **EXAMPLE 4**   **Finding the Quotients of Two Functions**

Find $(f/g)(x)$ and $(g/f)(x)$ for the functions $f(x) = \sqrt{x}$ and $g(x) = \sqrt{4 - x^2}$. Then find the domains of f/g and g/f.

Solution   The quotient of f and g is

$$\left(\frac{f}{g}\right)(x) = \frac{f(x)}{g(x)} = \frac{\sqrt{x}}{\sqrt{4 - x^2}}$$

and the quotient of g and f is

$$\left(\frac{g}{f}\right)(x) = \frac{g(x)}{f(x)} = \frac{\sqrt{4 - x^2}}{\sqrt{x}}.$$

- - **REMARK**   Note that the domain of f/g includes $x = 0$, but not $x = 2$, because $x = 2$ yields a zero in the denominator, whereas the domain of g/f includes $x = 2$, but not $x = 0$, because $x = 0$ yields a zero in the denominator.

The domain of f is $[0, \infty)$ and the domain of g is $[-2, 2]$. The intersection of these domains is $[0, 2]$. So, the domains of f/g and g/f are as follows.

Domain of f/g: $[0, 2)$ Domain of g/f: $(0, 2]$

✓ **Checkpoint** 🔊))) *Audio-video solution in English & Spanish at LarsonPrecalculus.com*

Find $(f/g)(x)$ and $(g/f)(x)$ for the functions $f(x) = \sqrt{x - 3}$ and $g(x) = \sqrt{16 - x^2}$. Then find the domains of f/g and g/f.

Composition of Functions

Another way of combining two functions is to form the **composition** of one with the other. For example, if $f(x) = x^2$ and $g(x) = x + 1$, then the composition of f with g is

$$f(g(x)) = f(x + 1)$$
$$= (x + 1)^2.$$

This composition is denoted as $f \circ g$ and reads as "f composed with g."

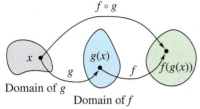

$f \circ g$

Domain of g

Domain of f

Figure 2.36

Definition of Composition of Two Functions

The **composition** of the function f with the function g is

$$(f \circ g)(x) = f(g(x)).$$

The domain of $f \circ g$ is the set of all x in the domain of g such that $g(x)$ is in the domain of f. (See Figure 2.36.)

EXAMPLE 5   **Compositions of Functions**

See LarsonPrecalculus.com for an interactive version of this type of example.

Given $f(x) = x + 2$ and $g(x) = 4 - x^2$, find the following.

a. $(f \circ g)(x)$ **b.** $(g \circ f)(x)$ **c.** $(g \circ f)(-2)$

Solution

a. The composition of f with g is as shown.

$$(f \circ g)(x) = f(g(x)) \qquad \text{Definition of } f \circ g$$
$$= f(4 - x^2) \qquad \text{Definition of } g(x)$$
$$= (4 - x^2) + 2 \qquad \text{Definition of } f(x)$$
$$= -x^2 + 6 \qquad \text{Simplify.}$$

b. The composition of g with f is as shown.

$$(g \circ f)(x) = g(f(x)) \qquad \text{Definition of } g \circ f$$
$$= g(x + 2) \qquad \text{Definition of } f(x)$$
$$= 4 - (x + 2)^2 \qquad \text{Definition of } g(x)$$
$$= 4 - (x^2 + 4x + 4) \qquad \text{Expand.}$$
$$= -x^2 - 4x \qquad \text{Simplify.}$$

Note that, in this case, $(f \circ g)(x) \neq (g \circ f)(x)$.

c. Evaluate the result of part (b) when $x = -2$.

$$(g \circ f)(-2) = -(-2)^2 - 4(-2) \qquad \text{Substitute.}$$
$$= -4 + 8 \qquad \text{Simplify.}$$
$$= 4 \qquad \text{Simplify.}$$

• • **REMARK** The tables of values below help illustrate the composition $(f \circ g)(x)$ in Example 5(a).

x	0	1	2	3
$g(x)$	4	3	0	-5

$g(x)$	4	3	0	-5
$f(g(x))$	6	5	2	-3

x	0	1	2	3
$f(g(x))$	6	5	2	-3

Note that the first two tables are combined (or "composed") to produce the values in the third table.

✓ **Checkpoint** ◄))) *Audio-video solution in English & Spanish at LarsonPrecalculus.com*

Given $f(x) = 2x + 5$ and $g(x) = 4x^2 + 1$, find the following.

a. $(f \circ g)(x)$ **b.** $(g \circ f)(x)$ **c.** $(f \circ g)\left(-\frac{1}{2}\right)$

EXAMPLE 6 **Finding the Domain of a Composite Function**

Find the domain of $f \circ g$ for the functions

$$f(x) = x^2 - 9 \quad \text{and} \quad g(x) = \sqrt{9 - x^2}.$$

Algebraic Solution

Find the composition of the functions.

$$
\begin{aligned}
(f \circ g)(x) &= f(g(x)) \\
&= f\left(\sqrt{9 - x^2}\right) \\
&= \left(\sqrt{9 - x^2}\right)^2 - 9 \\
&= 9 - x^2 - 9 \\
&= -x^2
\end{aligned}
$$

The domain of $f \circ g$ is restricted to the x-values in the domain of g for which $g(x)$ is in the domain of f. The domain of $f(x) = x^2 - 9$ is the set of all real numbers, which includes all real values of g. So, the domain of $f \circ g$ is the entire domain of $g(x) = \sqrt{9 - x^2}$, which is $[-3, 3]$.

Graphical Solution

Use a graphing utility to graph $f \circ g$.

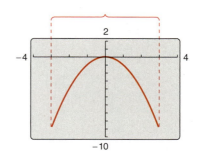

From the graph, you can determine that the domain of $f \circ g$ is $[-3, 3]$.

✓ *Checkpoint*  Audio-video solution in English & Spanish at LarsonPrecalculus.com

Find the domain of $f \circ g$ for the functions $f(x) = \sqrt{x}$ and $g(x) = x^2 + 4$.

In Examples 5 and 6, you formed the composition of two given functions. In calculus, it is also important to be able to identify two functions that make up a given composite function. For example, the function $h(x) = (3x - 5)^3$ is the composition of $f(x) = x^3$ and $g(x) = 3x - 5$. That is,

$$h(x) = (3x - 5)^3 = [g(x)]^3 = f(g(x)).$$

Basically, to "decompose" a composite function, look for an "inner" function and an "outer" function. In the function h above, $g(x) = 3x - 5$ is the inner function and $f(x) = x^3$ is the outer function.

EXAMPLE 7 **Decomposing a Composite Function**

Write the function $h(x) = \dfrac{1}{(x - 2)^2}$ as a composition of two functions.

Solution Consider $g(x) = x - 2$ as the inner function and $f(x) = \dfrac{1}{x^2} = x^{-2}$ as the outer function. Then write

$$
\begin{aligned}
h(x) &= \frac{1}{(x - 2)^2} \\
&= (x - 2)^{-2} \\
&= f(x - 2) \\
&= f(g(x)).
\end{aligned}
$$

✓ *Checkpoint*  Audio-video solution in English & Spanish at LarsonPrecalculus.com

Write the function $h(x) = \dfrac{\sqrt[3]{8 - x}}{5}$ as a composition of two functions.

Application

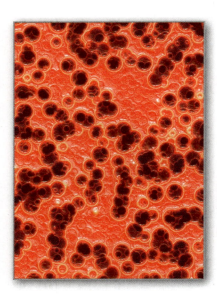

Refrigerated foods can have two types of bacteria: pathogenic bacteria, which can cause foodborne illness, and spoilage bacteria, which give foods an unpleasant look, smell, taste, or texture.

EXAMPLE 8 **Bacteria Count**

The number N of bacteria in a refrigerated food is given by

$$N(T) = 20T^2 - 80T + 500, \quad 2 \le T \le 14$$

where T is the temperature of the food in degrees Celsius. When the food is removed from refrigeration, the temperature of the food is given by

$$T(t) = 4t + 2, \quad 0 \le t \le 3$$

where t is the time in hours.

a. Find and interpret $(N \circ T)(t)$.

b. Find the time when the bacteria count reaches 2000.

Solution

a. $(N \circ T)(t) = N(T(t))$

$$= 20(4t + 2)^2 - 80(4t + 2) + 500$$

$$= 20(16t^2 + 16t + 4) - 320t - 160 + 500$$

$$= 320t^2 + 320t + 80 - 320t - 160 + 500$$

$$= 320t^2 + 420$$

The composite function $N \circ T$ represents the number of bacteria in the food as a function of the amount of time the food has been out of refrigeration.

b. The bacteria count reaches 2000 when $320t^2 + 420 = 2000$. By solving this equation algebraically, you find that the count reaches 2000 when $t \approx 2.2$ hours. Note that the negative solution $t \approx -2.2$ hours is rejected because it is not in the domain of the composite function.

✓ *Checkpoint* ◄))) *Audio-video solution in English & Spanish at LarsonPrecalculus.com*

The number N of bacteria in a refrigerated food is given by

$$N(T) = 8T^2 - 14T + 200, \quad 2 \le T \le 12$$

where T is the temperature of the food in degrees Celsius. When the food is removed from refrigeration, the temperature of the food is given by

$$T(t) = 2t + 2, \quad 0 \le t \le 5$$

where t is the time in hours.

a. Find $(N \circ T)(t)$.

b. Find the time when the bacteria count reaches 1000.

Summarize (Section 2.6)

1. Explain how to add, subtract, multiply, and divide functions *(page 214)*. For examples of finding arithmetic combinations of functions, see Examples 1–4.

2. Explain how to find the composition of one function with another function *(page 216)*. For examples that use compositions of functions, see Examples 5–7.

3. Describe a real-life example that uses a composition of functions *(page 218, Example 8)*.

2.6 Exercises

See **CalcChat.com** for tutorial help and worked-out solutions to odd-numbered exercises.

Vocabulary: Fill in the blanks.

1. Two functions f and g can be combined by the arithmetic operations of _____, _____, _____, and _____ to create new functions.

2. The _____ of the function f with the function g is $(f \circ g)(x) = f(g(x))$.

Skills and Applications

Graphing the Sum of Two Functions In Exercises 3 and 4, use the graphs of f and g to graph $h(x) = (f + g)(x)$. To print an enlarged copy of the graph, go to *MathGraphs.com*.

3.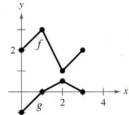

4.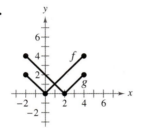

 Finding Arithmetic Combinations of Functions In Exercises 5–12, find (a) $(f + g)(x)$, (b) $(f - g)(x)$, (c) $(fg)(x)$, and (d) $(f/g)(x)$. What is the domain of f/g?

5. $f(x) = x + 2, \quad g(x) = x - 2$

6. $f(x) = 2x - 5, \quad g(x) = 2 - x$

7. $f(x) = x^2, \quad g(x) = 4x - 5$

8. $f(x) = 3x + 1, \quad g(x) = x^2 - 16$

9. $f(x) = x^2 + 6, \quad g(x) = \sqrt{1 - x}$

10. $f(x) = \sqrt{x^2 - 4}, \quad g(x) = \dfrac{x^2}{x^2 + 1}$

11. $f(x) = \dfrac{x}{x + 1}, \quad g(x) = x^3$

12. $f(x) = \dfrac{2}{x}, \quad g(x) = \dfrac{1}{x^2 - 1}$

Evaluating an Arithmetic Combination of Functions In Exercises 13–24, evaluate the function for $f(x) = x + 3$ and $g(x) = x^2 - 2$.

13. $(f + g)(2)$

14. $(f + g)(-1)$

15. $(f - g)(0)$

16. $(f - g)(1)$

17. $(f - g)(3t)$

18. $(f + g)(t - 2)$

19. $(fg)(6)$

20. $(fg)(-6)$

21. $(f/g)(5)$

22. $(f/g)(0)$

23. $(f/g)(-1) - g(3)$

24. $(fg)(5) + f(4)$

 Graphical Reasoning In Exercises 25–28, use a graphing utility to graph f, g, and $f + g$ in the same viewing window. Which function contributes most to the magnitude of the sum when $0 \le x \le 2$? Which function contributes most to the magnitude of the sum when $x > 6$?

25. $f(x) = 3x, \quad g(x) = -\dfrac{x^3}{10}$

26. $f(x) = \dfrac{x}{2}, \quad g(x) = \sqrt{x}$

27. $f(x) = 3x + 2, \quad g(x) = -\sqrt{x + 5}$

28. $f(x) = x^2 - \frac{1}{2}, \quad g(x) = -3x^2 - 1$

 Finding Compositions of Functions In Exercises 29–34, find (a) $f \circ g$, (b) $g \circ f$, and (c) $g \circ g$.

29. $f(x) = x + 8, \quad g(x) = x - 3$

30. $f(x) = -4x, \quad g(x) = x + 7$

31. $f(x) = x^2, \quad g(x) = x - 1$

32. $f(x) = 3x, \quad g(x) = x^4$

33. $f(x) = \sqrt[3]{x - 1}, \quad g(x) = x^3 + 1$

34. $f(x) = x^3, \quad g(x) = \dfrac{1}{x}$

 Finding Domains of Functions and Composite Functions In Exercises 35–42, find (a) $f \circ g$ and (b) $g \circ f$. Find the domain of each function and of each composite function.

35. $f(x) = \sqrt{x + 4}, \quad g(x) = x^2$

36. $f(x) = \sqrt[3]{x - 5}, \quad g(x) = x^3 + 1$

37. $f(x) = x^3, \quad g(x) = x^{2/3}$

38. $f(x) = x^5, \quad g(x) = \sqrt[4]{x}$

39. $f(x) = |x|, \quad g(x) = x + 6$

40. $f(x) = |x - 4|, \quad g(x) = 3 - x$

41. $f(x) = \dfrac{1}{x}, \quad g(x) = x + 3$

42. $f(x) = \dfrac{3}{x^2 - 1}, \quad g(x) = x + 1$

Graphing Combinations of Functions In Exercises 43 and 44, on the same set of coordinate axes, **(a) graph the functions** f, g, **and** $f + g$ **and (b) graph the functions** f, g, **and** $f \circ g$.

43. $f(x) = \frac{1}{2}x$, $g(x) = x - 4$

44. $f(x) = x + 3$, $g(x) = x^2$

 Evaluating Combinations of Functions In Exercises 45–48, use the graphs of f and g to evaluate the functions.

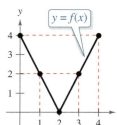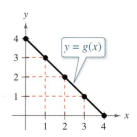

45. (a) $(f + g)(3)$ (b) $(f/g)(2)$
46. (a) $(f - g)(1)$ (b) $(fg)(4)$
47. (a) $(f \circ g)(2)$ (b) $(g \circ f)(2)$
48. (a) $(f \circ g)(1)$ (b) $(g \circ f)(3)$

 Decomposing a Composite Function In Exercises 49–56, find two functions f and g such that $(f \circ g)(x) = h(x)$. (There are many correct answers.)

49. $h(x) = (2x + 1)^2$ 50. $h(x) = (1 - x)^3$

51. $h(x) = \sqrt[3]{x^2 - 4}$ 52. $h(x) = \sqrt{9 - x}$

53. $h(x) = \dfrac{1}{x + 2}$ 54. $h(x) = \dfrac{4}{(5x + 2)^2}$

55. $h(x) = \dfrac{-x^2 + 3}{4 - x^2}$

56. $h(x) = \dfrac{27x^3 + 6x}{10 - 27x^3}$

57. **Stopping Distance** The research and development department of an automobile manufacturer determines that when a driver is required to stop quickly to avoid an accident, the distance (in feet) the car travels during the driver's reaction time is given by $R(x) = \frac{3}{4}x$, where x is the speed of the car in miles per hour. The distance (in feet) the car travels while the driver is braking is given by $B(x) = \frac{1}{15}x^2$.

(a) Find the function that represents the total stopping distance T.

(b) Graph the functions R, B, and T on the same set of coordinate axes for $0 \le x \le 60$.

(c) Which function contributes most to the magnitude of the sum at higher speeds? Explain.

58. **Business** The annual cost C (in thousands of dollars) and revenue R (in thousands of dollars) for a company each year from 2010 through 2016 can be approximated by the models

$$C = 254 - 9t + 1.1t^2 \quad \text{and} \quad R = 341 + 3.2t$$

where t is the year, with $t = 10$ corresponding to 2010.

(a) Write a function P that represents the annual profit of the company.

(b) Use a graphing utility to graph C, R, and P in the same viewing window.

59. **Vital Statistics** Let $b(t)$ be the number of births in the United States in year t, and let $d(t)$ represent the number of deaths in the United States in year t, where $t = 10$ corresponds to 2010.

(a) If $p(t)$ is the population of the United States in year t, find the function $c(t)$ that represents the percent change in the population of the United States.

(b) Interpret $c(16)$.

60. **Pets**

Let $d(t)$ be the number of dogs in the United States in year t, and let $c(t)$ be the number of cats in the United States in year t, where $t = 10$ corresponds to 2010.

(a) Find the function $p(t)$ that represents the total number of dogs and cats in the United States.

(b) Interpret $p(16)$.

(c) Let $n(t)$ represent the population of the United States in year t, where $t = 10$ corresponds to 2010. Find and interpret

$$h(t) = p(t)/n(t).$$

61. **Geometry** A square concrete foundation is a base for a cylindrical tank (see figure).

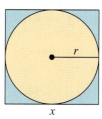

(a) Write the radius r of the tank as a function of the length x of the sides of the square.

(b) Write the area A of the circular base of the tank as a function of the radius r.

(c) Find and interpret $(A \circ r)(x)$.

62. Biology   The number N of bacteria in a refrigerated food is given by

$$N(T) = 10T^2 - 20T + 600, \quad 2 \le T \le 20$$

where T is the temperature of the food in degrees Celsius. When the food is removed from refrigeration, the temperature of the food is given by

$$T(t) = 3t + 2, \quad 0 \le t \le 6$$

where t is the time in hours.

(a) Find and interpret $(N \circ T)(t)$.

(b) Find the bacteria count after 0.5 hour.

(c) Find the time when the bacteria count reaches 1500.

63. Salary   You are a sales representative for a clothing manufacturer. You are paid an annual salary, plus a bonus of 3% of your sales over $500,000. Consider the two functions $f(x) = x - 500,000$ and $g(x) = 0.03x$. When x is greater than $500,000, which of the following represents your bonus? Explain.

(a) $f(g(x))$

(b) $g(f(x))$

64. Consumer Awareness   The suggested retail price of a new hybrid car is p dollars. The dealership advertises a factory rebate of $2000 and a 10% discount.

(a) Write a function R in terms of p giving the cost of the hybrid car after receiving the rebate from the factory.

(b) Write a function S in terms of p giving the cost of the hybrid car after receiving the dealership discount.

(c) Find and interpret $(R \circ S)(p)$ and $(S \circ R)(p)$.

(d) Find $(R \circ S)(25,795)$ and $(S \circ R)(25,795)$. Which yields the lower cost for the hybrid car? Explain.

Exploration

True or False?   In Exercises 65 and 66, determine whether the statement is true or false. Justify your answer.

65. If $f(x) = x + 1$ and $g(x) = 6x$, then

$$(f \circ g)(x) = (g \circ f)(x).$$

66. When you are given two functions f and g and a constant c, you can find $(f \circ g)(c)$ if and only if $g(c)$ is in the domain of f.

Siblings   In Exercises 67 and 68, three siblings are three different ages. The oldest is twice the age of the middle sibling, and the middle sibling is six years older than one-half the age of the youngest.

67. (a) Write a composite function that gives the oldest sibling's age in terms of the youngest. Explain how you arrived at your answer.

(b) If the oldest sibling is 16 years old, find the ages of the other two siblings.

68. (a) Write a composite function that gives the youngest sibling's age in terms of the oldest. Explain how you arrived at your answer.

(b) If the youngest sibling is 2 years old, find the ages of the other two siblings.

69. Proof   Prove that the product of two odd functions is an even function, and that the product of two even functions is an even function.

70. Conjecture   Use examples to hypothesize whether the product of an odd function and an even function is even or odd. Then prove your hypothesis.

71. Writing Functions   Write two unique functions f and g such that $(f \circ g)(x) = (g \circ f)(x)$ and f and g are (a) linear functions and (b) polynomial functions with degrees greater than one.

72. **HOW DO YOU SEE IT?**   The graphs labeled $L_1, L_2, L_3,$ and L_4 represent four different pricing discounts, where p is the original price (in dollars) and S is the sale price (in dollars). Match each function with its graph. Describe the situations in parts (c) and (d).

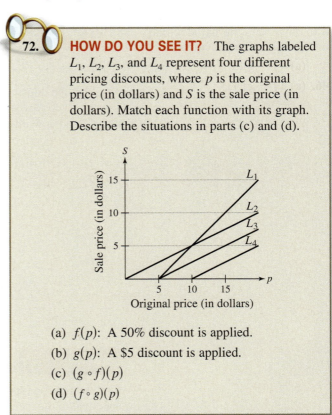

(a) $f(p)$: A 50% discount is applied.

(b) $g(p)$: A $5 discount is applied.

(c) $(g \circ f)(p)$

(d) $(f \circ g)(p)$

73. Proof

(a) Given a function f, prove that g is even and h is odd, where $g(x) = \frac{1}{2}[f(x) + f(-x)]$ and

$$h(x) = \frac{1}{2}[f(x) - f(-x)].$$

(b) Use the result of part (a) to prove that any function can be written as a sum of even and odd functions. [*Hint*: Add the two equations in part (a).]

(c) Use the result of part (b) to write each function as a sum of even and odd functions.

$$f(x) = x^2 - 2x + 1, \quad k(x) = \frac{1}{x + 1}$$

2.7 Inverse Functions

Inverse functions can help you model and solve real-life problems. For example, in Exercise 90 on page 230, you will write an inverse function and use it to determine the percent load interval for a diesel engine.

- Find inverse functions informally and verify that two functions are inverse functions of each other.
- Use graphs to verify that two functions are inverse functions of each other.
- Use the Horizontal Line Test to determine whether functions are one-to-one.
- Find inverse functions algebraically.

Inverse Functions

Recall from Section 2.2 that a set of ordered pairs can represent a function. For example, the function $f(x) = x + 4$ from the set $A = \{1, 2, 3, 4\}$ to the set $B = \{5, 6, 7, 8\}$ can be written as

$$f(x) = x + 4: \{(1, 5), (2, 6), (3, 7), (4, 8)\}.$$

In this case, by interchanging the first and second coordinates of each of the ordered pairs, you form the **inverse function** of f, which is denoted by f^{-1}. It is a function from the set B to the set A, and can be written as

$$f^{-1}(x) = x - 4: \{(5, 1), (6, 2), (7, 3), (8, 4)\}.$$

Note that the domain of f is equal to the range of f^{-1}, and vice versa, as shown in the figure below. Also note that the functions f and f^{-1} have the effect of "undoing" each other. In other words, when you form the composition of f with f^{-1} or the composition of f^{-1} with f, you obtain the identity function.

$$f(f^{-1}(x)) = f(x - 4) = (x - 4) + 4 = x$$
$$f^{-1}(f(x)) = f^{-1}(x + 4) = (x + 4) - 4 = x$$

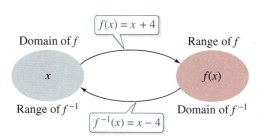

Domain of f $f(x) = x + 4$ Range of f

x $f(x)$

Range of f^{-1} $f^{-1}(x) = x - 4$ Domain of f^{-1}

EXAMPLE 1   **Finding an Inverse Function Informally**

Find the inverse function of $f(x) = 4x$. Then verify that both $f(f^{-1}(x))$ and $f^{-1}(f(x))$ are equal to the identity function.

Solution The function f *multiplies* each input by 4. To "undo" this function, you need to *divide* each input by 4. So, the inverse function of $f(x) = 4x$ is

$$f^{-1}(x) = \frac{x}{4}.$$

Verify that $f(f^{-1}(x)) = x$ and $f^{-1}(f(x)) = x$.

$$f(f^{-1}(x)) = f\left(\frac{x}{4}\right) = 4\left(\frac{x}{4}\right) = x \qquad f^{-1}(f(x)) = f^{-1}(4x) = \frac{4x}{4} = x$$

✓ **Checkpoint** Audio-video solution in English & Spanish at LarsonPrecalculus.com

Find the inverse function of $f(x) = \frac{1}{5}x$. Then verify that both $f(f^{-1}(x))$ and $f^{-1}(f(x))$ are equal to the identity function. ∎

Definition of Inverse Function

Let f and g be two functions such that

$$f(g(x)) = x \qquad \text{for every } x \text{ in the domain of } g$$

and

$$g(f(x)) = x \qquad \text{for every } x \text{ in the domain of } f.$$

Under these conditions, the function g is the **inverse function** of the function f. The function g is denoted by f^{-1} (read "f-inverse"). So,

$$f(f^{-1}(x)) = x \quad \text{and} \quad f^{-1}(f(x)) = x.$$

The domain of f must be equal to the range of f^{-1}, and the range of f must be equal to the domain of f^{-1}.

Do not be confused by the use of -1 to denote the inverse function f^{-1}. In this text, whenever f^{-1} is written, it *always* refers to the inverse function of the function f and *not* to the reciprocal of $f(x)$.

If the function g is the inverse function of the function f, then it must also be true that the function f is the inverse function of the function g. So, it is correct to say that the functions f and g are *inverse functions of each other*.

EXAMPLE 2 Verifying Inverse Functions

Which of the functions is the inverse function of $f(x) = \dfrac{5}{x - 2}$?

$$g(x) = \frac{x - 2}{5} \qquad h(x) = \frac{5}{x} + 2$$

Solution By forming the composition of f with g, you have

$$f(g(x)) = f\left(\frac{x - 2}{5}\right) = \frac{5}{\left(\dfrac{x - 2}{5}\right) - 2} = \frac{25}{x - 12} \neq x.$$

This composition is not equal to the identity function x, so g *is not* the inverse function of f. By forming the composition of f with h, you have

$$f(h(x)) = f\left(\frac{5}{x} + 2\right) = \frac{5}{\left(\dfrac{5}{x} + 2\right) - 2} = \frac{5}{\left(\dfrac{5}{x}\right)} = x.$$

So, it appears that h *is* the inverse function of f. Confirm this by showing that the composition of h with f is also equal to the identity function.

$$h(f(x)) = h\left(\frac{5}{x - 2}\right) = \frac{5}{\left(\dfrac{5}{x - 2}\right)} + 2 = x - 2 + 2 = x$$

Check to see that the domain of f is the same as the range of h and vice versa.

✓ **Checkpoint**))) *Audio-video solution in English & Spanish at LarsonPrecalculus.com*

Which of the functions is the inverse function of $f(x) = \dfrac{x - 4}{7}$?

$$g(x) = 7x + 4 \qquad h(x) = \frac{7}{x - 4}$$

The Graph of an Inverse Function

The graphs of a function f and its inverse function f^{-1} are related to each other in this way: If the point (a, b) lies on the graph of f, then the point (b, a) must lie on the graph of f^{-1}, and vice versa. This means that the graph of f^{-1} is a *reflection* of the graph of f in the line $y = x$, as shown in Figure 2.37.

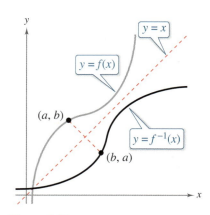

Figure 2.37

EXAMPLE 3 **Verifying Inverse Functions Graphically**

Verify graphically that the functions $f(x) = 2x - 3$ and $g(x) = \frac{1}{2}(x + 3)$ are inverse functions of each other.

Solution Sketch the graphs of f and g on the same rectangular coordinate system, as shown in Figure 2.38. It appears that the graphs are reflections of each other in the line $y = x$. Further verify this reflective property by testing a few points on each graph. Note that for each point (a, b) on the graph of f, the point (b, a) is on the graph of g.

Graph of $f(x) = 2x - 3$	Graph of $g(x) = \frac{1}{2}(x + 3)$
$(-1, -5)$	$(-5, -1)$
$(0, -3)$	$(-3, 0)$
$(1, -1)$	$(-1, 1)$
$(2, 1)$	$(1, 2)$
$(3, 3)$	$(3, 3)$

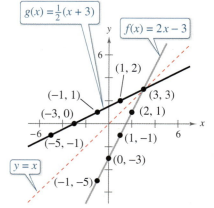

Figure 2.38

The graphs of f and g are reflections of each other in the line $y = x$. So, f and g are inverse functions of each other.

✓ *Checkpoint* 🔊)) *Audio-video solution in English & Spanish at LarsonPrecalculus.com*

Verify graphically that the functions $f(x) = 4x - 1$ and $g(x) = \frac{1}{4}(x + 1)$ are inverse functions of each other.

EXAMPLE 4 **Verifying Inverse Functions Graphically**

Verify graphically that the functions $f(x) = x^2$ ($x \geq 0$) and $g(x) = \sqrt{x}$ are inverse functions of each other.

Solution Sketch the graphs of f and g on the same rectangular coordinate system, as shown in Figure 2.39. It appears that the graphs are reflections of each other in the line $y = x$. Test a few points on each graph.

Graph of $f(x) = x^2$, $x \geq 0$	Graph of $g(x) = \sqrt{x}$
$(0, 0)$	$(0, 0)$
$(1, 1)$	$(1, 1)$
$(2, 4)$	$(4, 2)$
$(3, 9)$	$(9, 3)$

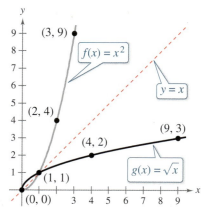

Figure 2.39

The graphs of f and g are reflections of each other in the line $y = x$. So, f and g are inverse functions of each other.

✓ *Checkpoint* 🔊)) *Audio-video solution in English & Spanish at LarsonPrecalculus.com*

Verify graphically that the functions $f(x) = x^2 + 1$ ($x \geq 0$) and $g(x) = \sqrt{x - 1}$ are inverse functions of each other.

One-to-One Functions

The reflective property of the graphs of inverse functions gives you a graphical test for determining whether a function has an inverse function. This test is the **Horizontal Line Test** for inverse functions.

> ### Horizontal Line Test for Inverse Functions
>
> A function f has an inverse function if and only if no *horizontal* line intersects the graph of f at more than one point.

If no horizontal line intersects the graph of f at more than one point, then no y-value corresponds to more than one x-value. This is the essential characteristic of **one-to-one functions.**

> ### One-to-One Functions
>
> A function f is **one-to-one** when each value of the dependent variable corresponds to exactly one value of the independent variable. A function f has an inverse function if and only if f is one-to-one.

Consider the table of values for the function $f(x) = x^2$ on the left. The output $f(x) = 4$ corresponds to two inputs, $x = -2$ and $x = 2$, so f is not one-to-one. In the table on the right, x and y are interchanged. Here $x = 4$ corresponds to both $y = -2$ and $y = 2$, so this table does not represent a function. So, $f(x) = x^2$ is not one-to-one and does not have an inverse function.

x	$f(x) = x^2$
-2	4
-1	1
0	0
1	1
2	4
3	9

x	y
4	-2
1	-1
0	0
1	1
4	2
9	3

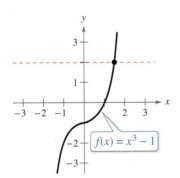

Figure 2.40

EXAMPLE 5　Applying the Horizontal Line Test

See LarsonPrecalculus.com for an interactive version of this type of example.

a. The graph of the function $f(x) = x^3 - 1$ is shown in Figure 2.40. No horizontal line intersects the graph of f at more than one point, so f *is* a one-to-one function and *does* have an inverse function.

b. The graph of the function $f(x) = x^2 - 1$ is shown in Figure 2.41. It is possible to find a horizontal line that intersects the graph of f at more than one point, so f *is not* a one-to-one function and *does not* have an inverse function.

✓ **Checkpoint** 🔊))) *Audio-video solution in English & Spanish at LarsonPrecalculus.com*

Use the graph of f to determine whether the function has an inverse function.

a. $f(x) = \frac{1}{2}(3 - x)$　　**b.** $f(x) = |x|$

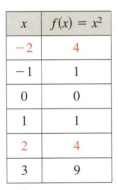

Figure 2.41

Finding Inverse Functions Algebraically

For relatively simple functions (such as the one in Example 1), you can find inverse functions by inspection. For more complicated functions, however, it is best to use the guidelines below. The key step in these guidelines is Step 3—interchanging the roles of x and y. This step corresponds to the fact that inverse functions have ordered pairs with the coordinates reversed.

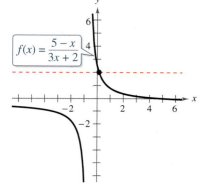

• • **REMARK** Note what happens when you try to find the inverse function of a function that is not one-to-one.

$f(x) = x^2 + 1$ Original function

$y = x^2 + 1$ Replace $f(x)$ with y.

$x = y^2 + 1$ Interchange x and y.

$x - 1 = y^2$ Isolate y-term.

$y = \pm\sqrt{x - 1}$ Solve for y.

You obtain two y-values for each x.

Finding an Inverse Function

1. Use the Horizontal Line Test to decide whether f has an inverse function.

2. In the equation for $f(x)$, replace $f(x)$ with y.

3. Interchange the roles of x and y, and solve for y.

4. Replace y with $f^{-1}(x)$ in the new equation.

5. Verify that f and f^{-1} are inverse functions of each other by showing that the domain of f is equal to the range of f^{-1}, the range of f is equal to the domain of f^{-1}, and $f(f^{-1}(x)) = x$ and $f^{-1}(f(x)) = x$.

EXAMPLE 6   **Finding an Inverse Function Algebraically**

Find the inverse function of

$$f(x) = \frac{5 - x}{3x + 2}.$$

Solution   The graph of f is shown in Figure 2.42. This graph passes the Horizontal Line Test. So, you know that f is one-to-one and has an inverse function.

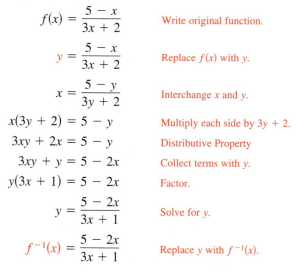

$$f(x) = \frac{5 - x}{3x + 2} \qquad \text{Write original function.}$$

$$y = \frac{5 - x}{3x + 2} \qquad \text{Replace } f(x) \text{ with } y.$$

$$x = \frac{5 - y}{3y + 2} \qquad \text{Interchange } x \text{ and } y.$$

$$x(3y + 2) = 5 - y \qquad \text{Multiply each side by } 3y + 2.$$

$$3xy + 2x = 5 - y \qquad \text{Distributive Property}$$

$$3xy + y = 5 - 2x \qquad \text{Collect terms with } y.$$

$$y(3x + 1) = 5 - 2x \qquad \text{Factor.}$$

$$y = \frac{5 - 2x}{3x + 1} \qquad \text{Solve for } y.$$

$$f^{-1}(x) = \frac{5 - 2x}{3x + 1} \qquad \text{Replace } y \text{ with } f^{-1}(x).$$

Check that $f(f^{-1}(x)) = x$ and $f^{-1}(f(x)) = x$.

✓ **Checkpoint** 🔊)) *Audio-video solution in English & Spanish at LarsonPrecalculus.com*

Find the inverse function of

$$f(x) = \frac{5 - 3x}{x + 2}.$$

Figure 2.42

$f(x) = \dfrac{5 - x}{3x + 2}$

EXAMPLE 7 **Finding an Inverse Function Algebraically**

Find the inverse function of

$$f(x) = \sqrt{2x - 3}.$$

Solution The graph of f is shown in the figure below. This graph passes the Horizontal Line Test. So, you know that f is one-to-one and has an inverse function.

$f(x) = \sqrt{2x - 3}$	Write original function.
$y = \sqrt{2x - 3}$	Replace $f(x)$ with y.
$x = \sqrt{2y - 3}$	Interchange x and y.
$x^2 = 2y - 3$	Square each side.
$2y = x^2 + 3$	Isolate y-term.
$y = \dfrac{x^2 + 3}{2}$	Solve for y.
$f^{-1}(x) = \dfrac{x^2 + 3}{2}, \ x \geq 0$	Replace y with $f^{-1}(x)$.

The graph of f^{-1} in the figure is the reflection of the graph of f in the line $y = x$. Note that the range of f is the interval $[0, \infty)$, which implies that the domain of f^{-1} is the interval $[0, \infty)$. Moreover, the domain of f is the interval $\left[\frac{3}{2}, \infty\right)$, which implies that the range of f^{-1} is the interval $\left[\frac{3}{2}, \infty\right)$. Verify that $f(f^{-1}(x)) = x$ and $f^{-1}(f(x)) = x$.

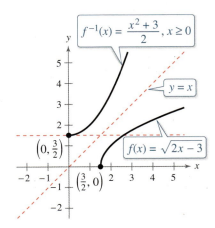

✓ *Checkpoint* Audio-video solution in English & Spanish at LarsonPrecalculus.com

Find the inverse function of

$$f(x) = \sqrt[3]{10 + x}.$$

Summarize (Section 2.7)

1. State the definition of an inverse function *(page 223)*. For examples of finding inverse functions informally and verifying inverse functions, see Examples 1 and 2.

2. Explain how to use graphs to verify that two functions are inverse functions of each other *(page 224)*. For examples of verifying inverse functions graphically, see Examples 3 and 4.

3. Explain how to use the Horizontal Line Test to determine whether a function is one-to-one *(page 225)*. For an example of applying the Horizontal Line Test, see Example 5.

4. Explain how to find an inverse function algebraically *(page 226)*. For examples of finding inverse functions algebraically, see Examples 6 and 7.

2.7 Exercises

See **CalcChat.com** for tutorial help and worked-out solutions to odd-numbered exercises.

Vocabulary: Fill in the blanks.

1. If $f(g(x))$ and $g(f(x))$ both equal x, then the function g is the _____ function of the function f.
2. The inverse function of f is denoted by _____.
3. The domain of f is the _____ of f^{-1}, and the _____ of f^{-1} is the range of f.
4. The graphs of f and f^{-1} are reflections of each other in the line _____.
5. A function f is _____ when each value of the dependent variable corresponds to exactly one value of the independent variable.
6. A graphical test for the existence of an inverse function of f is the _____ Line Test.

Skills and Applications

Finding an Inverse Function Informally
In Exercises 7–14, find the inverse function of f informally. Verify that $f(f^{-1}(x)) = x$ and $f^{-1}(f(x)) = x$.

7. $f(x) = 6x$

8. $f(x) = \dfrac{1}{3}x$

9. $f(x) = 3x + 1$

10. $f(x) = \dfrac{x - 3}{2}$

11. $f(x) = x^2 - 4, \ x \geq 0$
12. $f(x) = x^2 + 2, \ x \geq 0$
13. $f(x) = x^3 + 1$
14. $f(x) = \dfrac{x^5}{4}$

Verifying Inverse Functions In Exercises 15–18, verify that f and g are inverse functions algebraically.

15. $f(x) = \dfrac{x - 9}{4}, \quad g(x) = 4x + 9$

16. $f(x) = -\dfrac{3}{2}x - 4, \quad g(x) = -\dfrac{2x + 8}{3}$

17. $f(x) = \dfrac{x^3}{4}, \quad g(x) = \sqrt[3]{4x}$

18. $f(x) = x^3 + 5, \quad g(x) = \sqrt[3]{x - 5}$

Sketching the Graph of an Inverse Function In Exercises 19 and 20, use the graph of the function to sketch the graph of its inverse function $y = f^{-1}(x)$.

19.

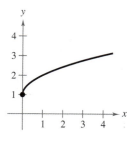

20.

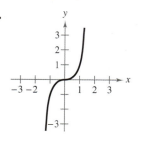

Verifying Inverse Functions In Exercises 21–32, verify that f and g are inverse functions (a) algebraically and (b) graphically.

21. $f(x) = x - 5, \quad g(x) = x + 5$

22. $f(x) = 2x, \quad g(x) = \dfrac{x}{2}$

23. $f(x) = 7x + 1, \quad g(x) = \dfrac{x - 1}{7}$

24. $f(x) = 3 - 4x, \quad g(x) = \dfrac{3 - x}{4}$

25. $f(x) = x^3, \quad g(x) = \sqrt[3]{x}$

26. $f(x) = \dfrac{x^3}{3}, \quad g(x) = \sqrt[3]{3x}$

27. $f(x) = \sqrt{x + 5}, \quad g(x) = x^2 - 5, \quad x \geq 0$
28. $f(x) = 1 - x^3, \quad g(x) = \sqrt[3]{1 - x}$

29. $f(x) = \dfrac{1}{x}, \quad g(x) = \dfrac{1}{x}$

30. $f(x) = \dfrac{1}{1 + x}, \quad x \geq 0, \quad g(x) = \dfrac{1 - x}{x}, \quad 0 < x \leq 1$

31. $f(x) = \dfrac{x - 1}{x + 5}, \quad g(x) = -\dfrac{5x + 1}{x - 1}$

32. $f(x) = \dfrac{x + 3}{x - 2}, \quad g(x) = \dfrac{2x + 3}{x - 1}$

Using a Table to Determine an Inverse Function In Exercises 33 and 34, does the function have an inverse function?

33.

x	-1	0	1	2	3	4
$f(x)$	-2	1	2	1	-2	-6

34.

x	-3	-2	-1	0	2	3
$f(x)$	10	6	4	1	-3	-10

Using a Table to Find an Inverse Function In Exercises 35 and 36, use the table of values for $y = f(x)$ to complete a table for $y = f^{-1}(x)$.

35.

x	-1	0	1	2	3	4
$f(x)$	3	5	7	9	11	13

36.

x	-3	-2	-1	0	1	2
$f(x)$	10	5	0	-5	-10	-15

Applying the Horizontal Line Test In Exercises 37–40, does the function have an inverse function?

37.

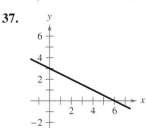

38.

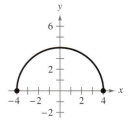

39.

40.

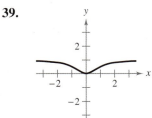

 Applying the Horizontal Line Test In Exercises 41–44, use a graphing utility to graph the function, and use the Horizontal Line Test to determine whether the function has an inverse function.

41. $g(x) = (x + 3)^2 + 2$ **42.** $f(x) = \frac{1}{5}(x + 2)^3$

43. $f(x) = x\sqrt{9 - x^2}$ **44.** $h(x) = |x| - |x - 4|$

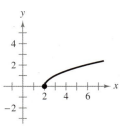 **Finding and Analyzing Inverse Functions** In Exercises 45–54, (a) find the inverse function of f, (b) graph both f and f^{-1} on the same set of coordinate axes, (c) describe the relationship between the graphs of f and f^{-1}, and (d) state the domains and ranges of f and f^{-1}.

45. $f(x) = x^5 - 2$ **46.** $f(x) = x^3 + 8$

47. $f(x) = \sqrt{4 - x^2}$, $0 \le x \le 2$

48. $f(x) = x^2 - 2$, $x \le 0$

49. $f(x) = \dfrac{4}{x}$ **50.** $f(x) = -\dfrac{2}{x}$

51. $f(x) = \dfrac{x + 1}{x - 2}$ **52.** $f(x) = \dfrac{x - 2}{3x + 5}$

53. $f(x) = \sqrt[3]{x - 1}$ **54.** $f(x) = x^{3/5}$

 Finding an Inverse Function In Exercises 55–70, determine whether the function has an inverse function. If it does, find the inverse function.

55. $f(x) = x^4$ **56.** $f(x) = \dfrac{1}{x^2}$

57. $g(x) = \dfrac{x + 1}{6}$ **58.** $f(x) = 3x + 5$

59. $p(x) = -4$ **60.** $f(x) = 0$

61. $f(x) = (x + 3)^2$, $x \ge -3$

62. $q(x) = (x - 5)^2$

63. $f(x) = \begin{cases} x + 3, & x < 0 \\ 6 - x, & x \ge 0 \end{cases}$

64. $f(x) = \begin{cases} -x, & x \le 0 \\ x^2 - 3x, & x > 0 \end{cases}$

65. $h(x) = |x + 1| - 1$

66. $f(x) = |x - 2|$, $x \le 2$

67. $f(x) = \sqrt{2x + 3}$

68. $f(x) = \sqrt{x - 2}$

69. $f(x) = \dfrac{6x + 4}{4x + 5}$

70. $f(x) = \dfrac{5x - 3}{2x + 5}$

Restricting the Domain In Exercises 71–78, restrict the domain of the function f so that the function is one-to-one and has an inverse function. Then find the inverse function f^{-1}. State the domains and ranges of f and f^{-1}. Explain your results. (There are many correct answers.)

71. $f(x) = |x + 2|$ **72.** $f(x) = |x - 5|$

73. $f(x) = (x + 6)^2$ **74.** $f(x) = (x - 4)^2$

75. $f(x) = -2x^2 + 5$

76. $f(x) = \frac{1}{2}x^2 - 1$

77. $f(x) = |x - 4| + 1$

78. $f(x) = -|x - 1| - 2$

Composition with Inverses In Exercises 79–84, use the functions $f(x) = \frac{1}{8}x - 3$ and $g(x) = x^3$ to find the value or function.

79. $(f^{-1} \circ g^{-1})(1)$ **80.** $(g^{-1} \circ f^{-1})(-3)$

81. $(f^{-1} \circ f^{-1})(4)$ **82.** $(g^{-1} \circ g^{-1})(-1)$

83. $(f \circ g)^{-1}$ **84.** $g^{-1} \circ f^{-1}$

Composition with Inverses In Exercises 85–88, use the functions $f(x) = x + 4$ and $g(x) = 2x - 5$ to find the function.

85. $g^{-1} \circ f^{-1}$ **86.** $f^{-1} \circ g^{-1}$

87. $(f \circ g)^{-1}$ **88.** $(g \circ f)^{-1}$

89. Hourly Wage   Your wage is $10.00 per hour plus $0.75 for each unit produced per hour. So, your hourly wage y in terms of the number of units produced x is $y = 10 + 0.75x$.

(a) Find the inverse function. What does each variable represent in the inverse function?

(b) Determine the number of units produced when your hourly wage is $24.25.

90. Diesel Mechanics

The function

$$y = 0.03x^2 + 245.50, \quad 0 < x < 100$$

approximates the exhaust temperature y in degrees Fahrenheit, where x is the percent load for a diesel engine.

(a) Find the inverse function. What does each variable represent in the inverse function?

(b) Use a graphing utility to graph the inverse function.

(c) The exhaust temperature of the engine must not exceed 500 degrees Fahrenheit. What is the percent load interval?

Exploration

True or False?   **In Exercises 91 and 92, determine whether the statement is true or false. Justify your answer.**

91. If f is an even function, then f^{-1} exists.

92. If the inverse function of f exists and the graph of f has a y-intercept, then the y-intercept of f is an x-intercept of f^{-1}.

Creating a Table   **In Exercises 93 and 94, use the graph of the function f to create a table of values for the given points. Then create a second table that can be used to find f^{-1}, and sketch the graph of f^{-1}, if possible.**

93.

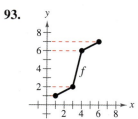

94.

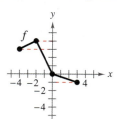

95. Proof   Prove that if f and g are one-to-one functions, then $(f \circ g)^{-1}(x) = (g^{-1} \circ f^{-1})(x)$.

96. Proof   Prove that if f is a one-to-one odd function, then f^{-1} is an odd function.

97. Think About It   The function $f(x) = k(2 - x - x^3)$ has an inverse function, and $f^{-1}(3) = -2$. Find k.

98. Think About It   Consider the functions $f(x) = x + 2$ and $f^{-1}(x) = x - 2$. Evaluate $f(f^{-1}(x))$ and $f^{-1}(f(x))$ for the given values of x. What can you conclude about the functions?

x	-10	0	7	45
$f(f^{-1}(x))$				
$f^{-1}(f(x))$				

99. Think About It   Restrict the domain of

$$f(x) = x^2 + 1$$

to $x \geq 0$. Use a graphing utility to graph the function. Does the restricted function have an inverse function? Explain.

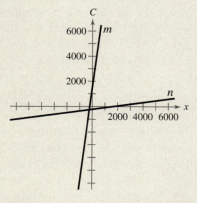

100.   HOW DO YOU SEE IT?   The cost C for a business to make personalized T-shirts is given by

$$C(x) = 7.50x + 1500$$

where x represents the number of T-shirts.

(a) The graphs of C and C^{-1} are shown below. Match each function with its graph.

(b) Explain what $C(x)$ and $C^{-1}(x)$ represent in the context of the problem.

One-to-One Function Representation   **In Exercises 101 and 102, determine whether the situation can be represented by a one-to-one function. If so, write a statement that best describes the inverse function.**

101. The number of miles n a marathon runner has completed in terms of the time t in hours

102. The depth of the tide d at a beach in terms of the time t over a 24-hour period

Chapter Summary

	What Did You Learn?	Explanation/Examples	Review Exercises
Section 2.1	Use slope to graph linear equations in two variables *(p. 160)*.	**The Slope-Intercept Form of the Equation of a Line** The graph of the equation $y = mx + b$ is a line whose slope is m and whose y-intercept is $(0, b)$.	1–4
	Find the slope of a line given two points on the line *(p. 162)*.	The slope m of the nonvertical line through (x_1, y_1) and (x_2, y_2) is $m = (y_2 - y_1)/(x_2 - x_1)$, where $x_1 \neq x_2$.	5, 6
	Write linear equations in two variables *(p. 164)*.	**Point-Slope Form of the Equation of a Line** The equation of the line with slope m passing through the point (x_1, y_1) is $y - y_1 = m(x - x_1)$.	7–10
	Use slope to identify parallel and perpendicular lines *(p. 165)*.	**Parallel lines:** Slopes are equal. **Perpendicular lines:** Slopes are negative reciprocals of each other.	11, 12
	Use slope and linear equations in two variables to model and solve real-life problems *(p. 166)*.	A linear equation in two variables can be used to describe the book value of exercise equipment each year. (See Example 7.)	13, 14
Section 2.2	Determine whether relations between two variables are functions, and use function notation *(p. 173)*.	A function f from a set A (domain) to a set B (range) is a relation that assigns to each element x in the set A exactly one element y in the set B. **Equation:** $f(x) = 5 - x^2$ $f(2)$: $f(2) = 5 - 2^2 = 1$	15–20
	Find the domains of functions *(p. 178)*.	**Domain of $f(x) = 5 - x^2$:** All real numbers	21, 22
	Use functions to model and solve real-life problems *(p. 179)*.	A function can be used to model the path of a baseball. (See Example 9.)	23, 24
	Evaluate difference quotients *(p. 180)*.	**Difference quotient:** $\dfrac{f(x + h) - f(x)}{h}, \quad h \neq 0$	25, 26
Section 2.3	Use the Vertical Line Test for functions *(p. 188)*.	A set of points in a coordinate plane is the graph of y as a function of x if and only if no *vertical* line intersects the graph at more than one point.	27, 28
	Find the zeros of functions *(p. 189)*.	**Zeros of $y = f(x)$:** x-values for which $f(x) = 0$	29–32
	Determine intervals on which functions are increasing or decreasing *(p. 190)*, determine relative minimum and relative maximum values of functions *(p. 191)*, and determine the average rate of change of a function *(p. 192)*.	To determine whether a function is increasing, decreasing, or constant on an interval, determine whether the graph of the function rises, falls, or is constant from left to right. The points at which the behavior of a function changes can help determine relative minimum or relative maximum values. The average rate of change between any two points is the slope of the line (secant line) through the two points.	33–38
	Identify even and odd functions *(p. 193)*.	**Even:** For each x in the domain of f, $f(-x) = f(x)$. **Odd:** For each x in the domain of f, $f(-x) = -f(x)$.	39–42

What Did You Learn?	**Explanation/Examples**	**Review Exercises**
Section 2.4 Identify and graph linear functions *(p. 198)*, squaring functions *(p. 199)*, cubic, square root, and reciprocal functions *(p. 200)*, and step and other piecewise-defined functions *(p. 201)*. Recognize graphs of parent functions *(p. 202)*.	**Linear:** $f(x) = ax + b$ **Squaring:** $f(x) = x^2$ **Square Root:** $f(x) = \sqrt{x}$ **Step:** $f(x) = [\![x]\!]$ The graphs of eight of the most commonly used functions in algebra are shown on page 202.	43–48
Section 2.5 Use vertical and horizontal shifts *(p. 205)*, reflections *(p. 207)*, and nonrigid transformations *(p. 209)* to sketch graphs of functions.	**Vertical shifts:** $h(x) = f(x) + c$ or $h(x) = f(x) - c$ **Horizontal shifts:** $h(x) = f(x - c)$ or $h(x) = f(x + c)$ **Reflection in x-axis:** $h(x) = -f(x)$ **Reflection in y-axis:** $h(x) = f(-x)$ **Nonrigid transformations:** $h(x) = cf(x)$ or $h(x) = f(cx)$	49–58
Section 2.6 Add, subtract, multiply, and divide functions *(p. 214)*.	$(f + g)(x) = f(x) + g(x)$ $(f - g)(x) = f(x) - g(x)$ $(fg)(x) = f(x) \cdot g(x)$ $(f/g)(x) = f(x)/g(x), g(x) \neq 0$	59, 60
Find the composition of one function with another function *(p. 216)*.	The composition of the function f with the function g is $(f \circ g)(x) = f(g(x))$.	61, 62
Use combinations and compositions of functions to model and solve real-life problems *(p. 218)*.	A composite function can be used to represent the number of bacteria in food as a function of the amount of time the food has been out of refrigeration. (See Example 8.)	63, 64
Section 2.7 Find inverse functions informally and verify that two functions are inverse functions of each other *(p. 222)*.	Let f and g be two functions such that $f(g(x)) = x$ for every x in the domain of g and $g(f(x)) = x$ for every x in the domain of f. Under these conditions, the function g is the inverse function of the function f.	65, 66
Use graphs to verify that two functions are inverse functions of each other *(p. 224)*, use the Horizontal Line Test to determine whether functions are one-to-one *(p. 225)*, and find inverse functions algebraically *(p. 226)*.	If the point (a, b) lies on the graph of f, then the point (b, a) must lie on the graph of f^{-1}, and vice versa. In short, the graph of f^{-1} is a reflection of the graph of f in the line $y = x$. **Horizontal Line Test for Inverse Functions** A function f has an inverse function if and only if no *horizontal* line intersects the graph of f at more than one point. To find an inverse function, replace $f(x)$ with y, interchange the roles of x and y, solve for y, and then replace y with $f^{-1}(x)$.	67–72

Review Exercises   <small>See **CalcChat.com** for tutorial help and worked-out solutions to odd-numbered exercises.</small>

2.1 **Graphing a Linear Equation**   **In Exercises 1–4,** find the slope and y-intercept (if possible) of the line. Sketch the line.

1. $y = -\frac{1}{2}x + 1$ **2.** $2x - 3y = 6$

3. $y = 1$ **4.** $x = -6$

Finding the Slope of a Line Through Two Points **In Exercises 5 and 6, find the slope of the line passing through the pair of points.**

5. $(5, -2), (-1, 4)$ **6.** $(-1, 6), (3, -2)$

Using the Point-Slope Form   **In Exercises 7 and 8,** find the slope-intercept form of the equation of the line that has the given slope and passes through the given point. Sketch the line.

7. $m = \frac{1}{3},\quad (6, -5)$ **8.** $m = -\frac{3}{4},\quad (-4, -2)$

Finding an Equation of a Line   **In Exercises 9 and 10,** find an equation of the line passing through the pair of points. Sketch the line.

9. $(-6, 4), (4, 9)$ **10.** $(-9, -3), (-3, -5)$

Finding Parallel and Perpendicular Lines **In Exercises 11 and 12, find equations of the lines that pass through the given point and are (a) parallel to and (b) perpendicular to the given line.**

11. $5x - 4y = 8, \quad (3, -2)$ **12.** $2x + 3y = 5, \quad (-8, 3)$

13. Sales   A discount outlet offers a 20% discount on all items. Write a linear equation giving the sale price S for an item with a list price L.

14. Hourly Wage   A manuscript translator charges a starting fee of \$50 plus \$2.50 per page translated. Write a linear equation for the amount A earned for translating p pages.

2.2 **Testing for Functions Represented Algebraically**   **In Exercises 15–18, determine whether the equation represents y as a function of x.**

15. $16x - y^4 = 0$ **16.** $2x - y - 3 = 0$

17. $y = \sqrt{1 - x}$ **18.** $|y| = x + 2$

Evaluating a Function   **In Exercises 19 and 20, find each function value.**

19. $g(x) = x^{4/3}$

  (a) $g(8)$ (b) $g(t + 1)$ (c) $g(-27)$ (d) $g(-x)$

20. $h(x) = |x - 2|$

  (a) $h(-4)$ (b) $h(-2)$ (c) $h(0)$ (d) $h(-x + 2)$

Finding the Domain of a Function   **In Exercises 21 and 22, find the domain of the function.**

21. $f(x) = \sqrt{25 - x^2}$ **22.** $h(x) = \dfrac{x}{x^2 - x - 6}$

Physics   **In Exercises 23 and 24, the velocity of a ball projected upward from ground level is given by** $v(t) = -32t + 48$, **where t is the time in seconds and v is the velocity in feet per second.**

23. Find the velocity when $t = 1$.

24. Find the time when the ball reaches its maximum height. [*Hint:* Find the time when $v(t) = 0$.]

$\displaystyle\smallint$ **Evaluating a Difference Quotient**   **In Exercises 25 and 26, find the difference quotient and simplify your answer.**

25. $f(x) = 2x^2 + 3x - 1,\quad \dfrac{f(x + h) - f(x)}{h},\quad h \neq 0$

26. $f(x) = x^3 - 5x^2 + x,\quad \dfrac{f(x + h) - f(x)}{h},\quad h \neq 0$

2.3 **Vertical Line Test for Functions**   **In Exercises 27 and 28, use the Vertical Line Test to determine whether the graph represents y as a function of x. To print an enlarged copy of the graph, go to *MathGraphs.com*.**

27.

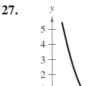

28.

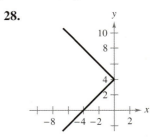

Finding the Zeros of a Function   **In Exercises 29–32, find the zeros of the function algebraically.**

29. $f(x) = 5x^2 + 4x - 1$

30. $f(x) = \dfrac{8x + 3}{11 - x}$

31. $f(x) = \sqrt{2x + 1}$

32. $f(x) = x^3 - x^2$

Describing Function Behavior   **In Exercises 33 and 34, use a graphing utility to graph the function and visually determine the open intervals on which the function is increasing, decreasing, or constant.**

33. $f(x) = |x| + |x + 1|$

34. $f(x) = (x^2 - 4)^2$

Approximating Relative Minima or Maxima In Exercises 35 and 36, use a graphing utility to approximate (to two decimal places) any relative minima or maxima of the function.

35. $f(x) = -x^2 + 2x + 1$ **36.** $f(x) = x^3 - 4x^2 - 1$

Average Rate of Change of a Function In Exercises 37 and 38, find the average rate of change of the function from x_1 to x_2.

37. $f(x) = -x^2 + 8x - 4$, $x_1 = 0, x_2 = 4$
38. $f(x) = x^3 + 2x + 1$, $x_1 = 1, x_2 = 3$

Even, Odd, or Neither? In Exercises 39–42, determine whether the function is even, odd, or neither. Then describe the symmetry.

39. $f(x) = x^5 + 4x - 7$ **40.** $f(x) = x^4 - 20x^2$
41. $f(x) = 2x\sqrt{x^2 + 3}$ **42.** $f(x) = \sqrt[5]{6x^2}$

2.4 **Writing a Linear Function** In Exercises 43 and 44, (a) write the linear function f that has the given function values and (b) sketch the graph of the function.

43. $f(2) = -6$, $f(-1) = 3$
44. $f(0) = -5$, $f(4) = -8$

Graphing a Function In Exercises 45–48, sketch the graph of the function.

45. $g(x) = [\![x]\!] - 2$ **46.** $g(x) = [\![x + 4]\!]$
47. $f(x) = \begin{cases} 5x - 3, & x \geq -1 \\ -4x + 5, & x < -1 \end{cases}$
48. $f(x) = \begin{cases} 2x + 1, & x \leq 2 \\ x^2 + 1, & x > 2 \end{cases}$

2.5 **Describing Transformations** In Exercises 49–58, h is related to one of the parent functions described in this chapter. (a) Identify the parent function f. (b) Describe the sequence of transformations from f to h. (c) Sketch the graph of h. (d) Use function notation to write h in terms of f.

49. $h(x) = x^2 - 9$ **50.** $h(x) = (x - 2)^3 + 2$
51. $h(x) = -\sqrt{x} + 4$ **52.** $h(x) = |x + 3| - 5$
53. $h(x) = -(x + 2)^2 + 3$ **54.** $h(x) = \frac{1}{2}(x - 1)^2 - 2$
55. $h(x) = -[\![x]\!] + 6$ **56.** $h(x) = -\sqrt{x + 1} + 9$
57. $h(x) = 5[\![x - 9]\!]$ **58.** $h(x) = -\frac{1}{3}x^3$

2.6 **Finding Arithmetic Combinations of Functions** In Exercises 59 and 60, find (a) $(f + g)(x)$, (b) $(f - g)(x)$, (c) $(fg)(x)$, and (d) $(f/g)(x)$. What is the domain of f/g?

59. $f(x) = x^2 + 3$, $g(x) = 2x - 1$
60. $f(x) = x^2 - 4$, $g(x) = \sqrt{3 - x}$

Finding Domains of Functions and Composite Functions In Exercises 61 and 62, find (a) $f \circ g$ and (b) $g \circ f$. Find the domain of each function and of each composite function.

61. $f(x) = \frac{1}{3}x - 3$, $g(x) = 3x + 1$
62. $f(x) = \sqrt{x + 1}$, $g(x) = x^2$

Retail In Exercises 63 and 64, the price of a washing machine is x dollars. The function $f(x) = x - 100$ gives the price of the washing machine after a \$100 rebate. The function $g(x) = 0.95x$ gives the price of the washing machine after a 5% discount.

63. Find and interpret $(f \circ g)(x)$.
64. Find and interpret $(g \circ f)(x)$.

2.7 **Finding an Inverse Function Informally** In Exercises 65 and 66, find the inverse function of f informally. Verify that $f(f^{-1}(x)) = x$ and $f^{-1}(f(x)) = x$.

65. $f(x) = \dfrac{x - 4}{5}$ **66.** $f(x) = x^3 - 1$

Applying the Horizontal Line Test In Exercises 67 and 68, use a graphing utility to graph the function, and use the Horizontal Line Test to determine whether the function has an inverse function.

67. $f(x) = (x - 1)^2$ **68.** $h(t) = \dfrac{2}{t - 3}$

Finding and Analyzing Inverse Functions In Exercises 69 and 70, (a) find the inverse function of f, (b) graph both f and f^{-1} on the same set of coordinate axes, (c) describe the relationship between the graphs of f and f^{-1}, and (d) state the domains and ranges of f and f^{-1}.

69. $f(x) = \frac{1}{2}x - 3$ **70.** $f(x) = \sqrt{x + 1}$

Restricting the Domain In Exercises 71 and 72, restrict the domain of the function f to an interval on which the function is increasing, and find f^{-1} on that interval.

71. $f(x) = 2(x - 4)^2$ **72.** $f(x) = |x - 2|$

Exploration

True or False? In Exercises 73 and 74, determine whether the statement is true or false. Justify your answer.

73. Relative to the graph of $f(x) = \sqrt{x}$, the graph of the function $h(x) = -\sqrt{x + 9} - 13$ is shifted 9 units to the left and 13 units down, then reflected in the x-axis.

74. If f and g are two inverse functions, then the domain of g is equal to the range of f.

Chapter Test

See CalcChat.com for tutorial help and worked-out solutions to odd-numbered exercises.

Take this test as you would take a test in class. When you are finished, check your work against the answers given in the back of the book.

In Exercises 1 and 2, find an equation of the line passing through the pair of points. Sketch the line.

1. $(-2, 5), (1, -7)$ **2.** $(-4, -7), \left(1, \frac{4}{3}\right)$

3. Find equations of the lines that pass through the point $(0, 4)$ and are (a) parallel to and (b) perpendicular to the line $5x + 2y = 3$.

In Exercises 4 and 5, find each function value.

4. $f(x) = |x + 2| - 15$

   (a) $f(-8)$ (b) $f(14)$ (c) $f(x - 6)$

5. $f(x) = \dfrac{\sqrt{x + 9}}{x^2 - 81}$

   (a) $f(7)$ (b) $f(-5)$ (c) $f(x - 9)$

In Exercises 6 and 7, find (a) the domain and (b) the zeros of the function.

6. $f(x) = \dfrac{x - 5}{2x^2 - x}$ **7.** $f(x) = 10 - \sqrt{3 - x}$

In Exercises 8–10, (a) use a graphing utility to graph the function, (b) approximate the open intervals on which the function is increasing, decreasing, or constant, and (c) determine whether the function is even, odd, or neither.

8. $f(x) = 2x^6 + 5x^4 - x^2$ **9.** $f(x) = 4x\sqrt{3 - x}$ **10.** $f(x) = |x + 5|$

11. Use a graphing utility to approximate (to two decimal places) any relative minima or maxima of $f(x) = -x^3 + 2x - 1$.

12. Find the average rate of change of $f(x) = -2x^2 + 5x - 3$ from $x_1 = 1$ to $x_2 = 3$.

13. Sketch the graph of $f(x) = \begin{cases} 3x + 7, & x \le -3 \\ 4x^2 - 1, & x > -3 \end{cases}$.

In Exercises 14–16, (a) identify the parent function f in the transformation, (b) describe the sequence of transformations from f to h, and (c) sketch the graph of h.

14. $h(x) = 4[\![x]\!]$

15. $h(x) = -\sqrt{x + 5} + 8$

16. $h(x) = -2(x - 5)^3 + 3$

In Exercises 17 and 18, find (a) $(f + g)(x)$, (b) $(f - g)(x)$, (c) $(fg)(x)$, (d) $(f/g)(x)$, (e) $(f \circ g)(x)$, and (f) $(g \circ f)(x)$.

17. $f(x) = 3x^2 - 7, \quad g(x) = -x^2 - 4x + 5$ **18.** $f(x) = \dfrac{1}{x}, \quad g(x) = 2\sqrt{x}$

In Exercises 19–21, determine whether the function has an inverse function. If it does, find the inverse function.

19. $f(x) = x^3 + 8$ **20.** $f(x) = |x^2 - 3| + 6$ **21.** $f(x) = 3x\sqrt{x}$

22. It costs a company \$58 to produce 6 units of a product and \$78 to produce 10 units. Assuming that the cost function is linear, how much does it cost to produce 25 units?

Cumulative Test for Chapters P–2

See **CalcChat.com** for tutorial help and worked-out solutions to odd-numbered exercises.

Take this test as you would take a test in class. When you are finished, check your work against the answers given in the back of the book.

In Exercises 1 and 2, simplify the expression.

1. $\dfrac{8x^2y^{-3}}{30x^{-1}y^2}$

2. $\sqrt{20x^2y^3}$

In Exercises 3–5, perform the operation(s) and simplify the result.

3. $4x - [2x + 3(2 - x)]$

4. $(x - 2)(x^2 + x - 3)$

5. $\dfrac{3}{s + 5} - \dfrac{2}{s - 3}$

In Exercises 6–8, completely factor the expression.

6. $36 - (x + 1)^2$

7. $x - 5x^2 - 6x^3$

8. $54x^3 + 16$

In Exercises 9 and 10, write an expression for the area of the figure as a polynomial in standard form.

9.

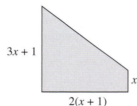

10.

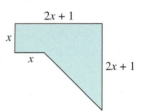

In Exercises 11–13, sketch the graph of the equation.

11. $x - 3y + 12 = 0$

12. $y = x^2 - 9$

13. $y = \sqrt{4 - x}$

In Exercises 14–16, solve the equation and check your solution.

14. $3x - 5 = 6x + 8$

15. $-(x + 3) = 14(x - 6)$

16. $\dfrac{1}{x - 2} = \dfrac{10}{4x + 3}$

In Exercises 17–22, solve the equation using any convenient method.

17. $x^2 - 4x + 3 = 0$

18. $-2x^2 + 4x + 6 = 0$

19. $3x^2 + 9x + 1 = 0$

20. $3x^2 + 5x - 6 = 0$

21. $\frac{2}{3}x^2 = 24$

22. $\frac{1}{2}x^2 - 7 = 25$

In Exercises 23–28, solve the equation, if possible. Check your solutions.

23. $x^4 + 12x^3 + 4x^2 + 48x = 0$

24. $8x^3 - 48x^2 + 72x = 0$

25. $x^{3/2} + 21 = 13$

26. $\sqrt{x + 10} = x - 2$

27. $|2(x - 1)| = 8$

28. $|x - 12| = -2$

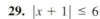

Figure for 34

In Exercises 29–32, solve the inequality. Then graph the solution set.

29. $|x + 1| \leq 6$

30. $|5 + 6x| > 3$

31. $5x^2 + 12x + 7 \geq 0$

32. $-8x^2 + 10x + 3 > 0$

33. Find the slope-intercept form of the equation of the line passing through $\left(-\frac{1}{2}, 1\right)$ and $(3, 8)$.

34. Explain why the graph at the left does not represent y as a function of x.

35. Let $f(x) = \dfrac{x}{x - 2}$. Find each function value, if possible.

(a) $f(6)$ (b) $f(2)$ (c) $f(s + 2)$

In Exercises 36–38, determine whether the function is even, odd, or neither.

36. $f(x) = 5 + \sqrt{4 - x}$

37. $f(x) = 2x^3 - 4x$

38. $f(x) = x^4 + 1$

39. Compare the graph of each function with the graph of $y = \sqrt[3]{x}$. (*Note:* It is not necessary to sketch the graphs.)

(a) $r(x) = \frac{1}{2}\sqrt[3]{x}$ (b) $h(x) = \sqrt[3]{x} + 2$ (c) $g(x) = \sqrt[3]{x + 2}$

In Exercises 40 and 41, find (a) $(f + g)(x)$, (b) $(f - g)(x)$, (c) $(fg)(x)$, and (d) $(f/g)(x)$. What is the domain of f/g?

40. $f(x) = x - 4$, $g(x) = 3x + 1$

41. $f(x) = \sqrt{x - 1}$, $g(x) = x^2 + 1$

In Exercises 42 and 43, find (a) $f \circ g$ and (b) $g \circ f$. Find the domain of each composite function.

42. $f(x) = 2x^2$, $g(x) = \sqrt{x + 6}$

43. $f(x) = x - 2$, $g(x) = |x|$

44. Determine whether $h(x) = 3x - 4$ has an inverse function. If it does, find the inverse function.

45. A group of n people decide to buy a \$36,000 minibus. Each person will pay an equal share of the cost. When three additional people join the group, the cost per person will decrease by \$1000. Find n.

46. For groups of 60 or more people, a charter bus company determines the rate per person according to the formula

$$\text{Rate} = 10 - 0.05(n - 60), \quad n \geq 60$$

where the rate is given in dollars and n is the number of people.

(a) Write the revenue R as a function of n.

(b) Use a graphing utility to graph the revenue function. Use the graph to approximate the number of people that will maximize the revenue.

47. The height of an object thrown upward from a height of 8 feet at a velocity of 36 feet per second can be modeled by $s(t) = -16t^2 + 36t + 8$, where s is the height (in feet) and t is the time (in seconds). Find the average rate of change of the function from $t_1 = 0$ to $t_2 = 2$. Interpret your answer in the context of the problem.

Proofs in Mathematics ■ ■ ■ ■ ■ ■ ■ ■ ■ ■ ■ ■ ■ ■ ■ ■

Biconditional Statements

Recall from the Proofs in Mathematics in Chapter 1 that a conditional statement is a statement of the form "if p, then q." A statement of the form "p if and only if q" is a **biconditional statement.** A biconditional statement, denoted by

$p \leftrightarrow q$ Biconditional statement

is the conjunction of the conditional statement $p \rightarrow q$ and its converse $q \rightarrow p$.

A biconditional statement can be either true or false. To be true, *both* the conditional statement and its converse must be true.

EXAMPLE 1   **Analyzing a Biconditional Statement**

Consider the statement "$x = 3$ if and only if $x^2 = 9$."

a. Is the statement a biconditional statement?

b. Is the statement true?

Solution

a. The statement is a biconditional statement, because it is of the form "p if and only if q."

b. Rewrite the statement as a conditional statement and its converse.

> *Conditional statement:* If $x = 3$, then $x^2 = 9$.

> *Converse:* If $x^2 = 9$, then $x = 3$.

The conditional statement is true, but the converse is false, because x can also equal -3. So, the biconditional statement is false. ■

Knowing how to use biconditional statements is an important tool for reasoning in mathematics.

EXAMPLE 2   **Analyzing a Biconditional Statement**

Determine whether the biconditional statement is true or false. If it is false, provide a counterexample.

> A number is divisible by 5 if and only if it ends in 0.

Solution   Rewrite the biconditional statement as a conditional statement and its converse.

> *Conditional statement:* If a number is divisible by 5, then it ends in 0.
> *Converse:* If a number ends in 0, then it is divisible by 5.

The biconditional statement is false, because the conditional statement is false. A counterexample is the number 15, which is divisible by 5 but does not end in 0. ■

P.S. Problem Solving

1. **Monthly Wages** As a salesperson, you receive a monthly salary of $2000, plus a commission of 7% of sales. You receive an offer for a new job at $2300 per month, plus a commission of 5% of sales.

   (a) Write a linear equation for your current monthly wage W_1 in terms of your monthly sales S.

   (b) Write a linear equation for the monthly wage W_2 of your new job offer in terms of the monthly sales S.

   (c) Use a graphing utility to graph both equations in the same viewing window. Find the point of intersection. What does the point of intersection represent?

   (d) You expect sales of $20,000 per month. Should you change jobs? Explain.

2. **Cellphone Keypad** For the numbers 2 through 9 on a cellphone keypad (see figure), consider two relations: one mapping numbers onto letters, and the other mapping letters onto numbers. Are both relations functions? Explain.

3. **Sums and Differences of Functions** What can be said about the sum and difference of each pair of functions?

   (a) Two even functions

   (b) Two odd functions

   (c) An odd function and an even function

4. **Inverse Functions** The functions

   $$f(x) = x \quad \text{and} \quad g(x) = -x$$

   are their own inverse functions. Graph each function and explain why this is true. Graph other linear functions that are their own inverse functions. Find a formula for a family of linear functions that are their own inverse functions.

5. **Proof** Prove that a function of the form

   $$y = a_{2n}x^{2n} + a_{2n-2}x^{2n-2} + \cdots + a_2x^2 + a_0$$

   is an even function.

6. **Miniature Golf** A golfer is trying to make a hole-in-one on the miniature golf green shown. The golf ball is at the point $(2.5, 2)$ and the hole is at the point $(9.5, 2)$. The golfer wants to bank the ball off the side wall of the green at the point (x, y). Find the coordinates of the point (x, y). Then write an equation for the path of the ball.

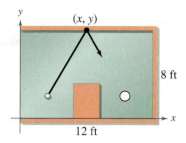

Figure for 6

7. **Titanic** At 2:00 P.M. on April 11, 1912, the *Titanic* left Cobh, Ireland, on her voyage to New York City. At 11:40 P.M. on April 14, the *Titanic* struck an iceberg and sank, having covered only about 2100 miles of the approximately 3400-mile trip.

   (a) What was the total duration of the voyage in hours?

   (b) What was the average speed in miles per hour?

   (c) Write a function relating the distance of the *Titanic* from New York City and the number of hours traveled. Find the domain and range of the function.

   (d) Graph the function in part (c).

8. **Average Rate of Change** Consider the function $f(x) = -x^2 + 4x - 3$. Find the average rate of change of the function from x_1 to x_2.

   (a) $x_1 = 1, x_2 = 2$

   (b) $x_1 = 1, x_2 = 1.5$

   (c) $x_1 = 1, x_2 = 1.25$

   (d) $x_1 = 1, x_2 = 1.125$

   (e) $x_1 = 1, x_2 = 1.0625$

   (f) Does the average rate of change seem to be approaching one value? If so, state the value.

   (g) Find the equations of the secant lines through the points $(x_1, f(x_1))$ and $(x_2, f(x_2))$ for parts (a)–(e).

   (h) Find the equation of the line through the point $(1, f(1))$ using your answer from part (f) as the slope of the line.

9. **Inverse of a Composition** Consider the functions $f(x) = 4x$ and $g(x) = x + 6$.

   (a) Find $(f \circ g)(x)$.

   (b) Find $(f \circ g)^{-1}(x)$.

   (c) Find $f^{-1}(x)$ and $g^{-1}(x)$.

   (d) Find $(g^{-1} \circ f^{-1})(x)$ and compare the result with that of part (b).

   (e) Repeat parts (a) through (d) for $f(x) = x^3 + 1$ and $g(x) = 2x$.

   (f) Write two one-to-one functions f and g, and repeat parts (a) through (d) for these functions.

   (g) Make a conjecture about $(f \circ g)^{-1}(x)$ and $(g^{-1} \circ f^{-1})(x)$.

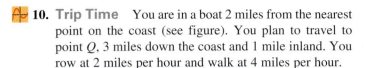

10. Trip Time You are in a boat 2 miles from the nearest point on the coast (see figure). You plan to travel to point Q, 3 miles down the coast and 1 mile inland. You row at 2 miles per hour and walk at 4 miles per hour.

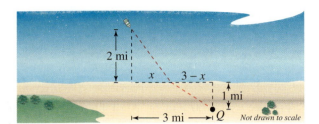

Not drawn to scale

(a) Write the total time T (in hours) of the trip as a function of the distance x (in miles).

(b) Determine the domain of the function.

(c) Use a graphing utility to graph the function. Be sure to choose an appropriate viewing window.

(d) Find the value of x that minimizes T.

(e) Write a brief paragraph interpreting these values.

11. Heaviside Function The **Heaviside function**

$$H(x) = \begin{cases} 1, & x \ge 0 \\ 0, & x < 0 \end{cases}$$

is widely used in engineering applications. (See figure.) To print an enlarged copy of the graph, go to *MathGraphs.com.*

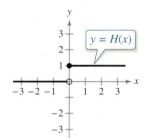

Sketch the graph of each function by hand.

(a) $H(x) - 2$

(b) $H(x - 2)$

(c) $-H(x)$

(d) $H(-x)$

(e) $\frac{1}{2}H(x)$

(f) $-H(x - 2) + 2$

12. Repeated Composition Let $f(x) = \dfrac{1}{1 - x}$.

(a) Find the domain and range of f.

(b) Find $f(f(x))$. What is the domain of this function?

(c) Find $f(f(f(x)))$. Is the graph a line? Why or why not?

13. Associative Property with Compositions Show that the Associative Property holds for compositions of functions—that is,

$$(f \circ (g \circ h))(x) = ((f \circ g) \circ h)(x).$$

14. Graphical Reasoning Use the graph of the function f to sketch the graph of each function. To print an enlarged copy of the graph, go to *MathGraphs.com.*

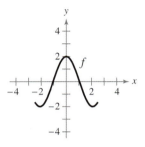

(a) $f(x + 1)$

(b) $f(x) + 1$

(c) $2f(x)$

(d) $f(-x)$

(e) $-f(x)$

(f) $|f(x)|$

(g) $f(|x|)$

15. Graphical Reasoning Use the graphs of f and f^{-1} to complete each table of function values.

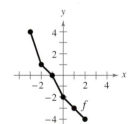 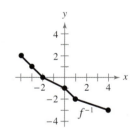

(a)

x	-4	-2	0	4
$(f(f^{-1}(x)))$				

(b)

x	-3	-2	0	1
$(f + f^{-1})(x)$				

(c)

x	-3	-2	0	1
$(f \cdot f^{-1})(x)$				

(d)

x	-4	-3	0	4
$|f^{-1}(x)|$				

3 Polynomial Functions

Candle Making Kits *(Example 12, page 282)*

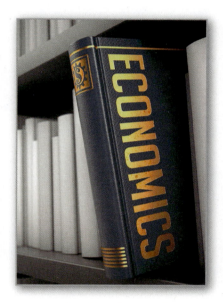

Product Demand
(Example 5, page 291)

Lyme Disease *(Exercise 82, page 272)*

Tree Growth
(Exercise 98, page 263)

Path of a Diver *(Exercise 67, page 249)*

3.1 Quadratic Functions and Models

Quadratic functions have many real-life applications. For example, in Exercise 67 on page 249, you will use a quadratic function that models the path of a diver.

- ■ Analyze graphs of quadratic functions.
- ■ Write quadratic functions in standard form and use the results to sketch their graphs.
- ■ Find minimum and maximum values of quadratic functions in real-life applications.

The Graph of a Quadratic Function

In this and the next section, you will study graphs of polynomial functions. Section 2.4 introduced basic functions such as linear, constant, and squaring functions.

$$f(x) = ax + b \qquad \text{Linear function}$$
$$f(x) = c \qquad \text{Constant function}$$
$$f(x) = x^2 \qquad \text{Squaring function}$$

These are examples of **polynomial functions.**

Definition of a Polynomial Function

Let n be a nonnegative integer and let $a_n, a_{n-1}, \ldots, a_2, a_1, a_0$ be real numbers with $a_n \neq 0$. The function

$$f(x) = a_n x^n + a_{n-1} x^{n-1} + \cdots + a_2 x^2 + a_1 x + a_0$$

is a **polynomial function of x with degree n.**

Polynomial functions are classified by degree. For example, a constant function $f(x) = c$ with $c \neq 0$ has degree 0, and a linear function $f(x) = ax + b$ with $a \neq 0$ has degree 1. In this section, you will study **quadratic functions,** which are second-degree polynomial functions.

For example, each function listed below is a quadratic function.

$$f(x) = x^2 + 6x + 2$$
$$g(x) = 2(x + 1)^2 - 3$$
$$h(x) = 9 + \tfrac{1}{4}x^2$$
$$k(x) = (x - 2)(x + 1)$$

Note that the squaring function is a simple quadratic function.

Definition of a Quadratic Function

Let a, b, and c be real numbers with $a \neq 0$. The function

$$f(x) = ax^2 + bx + c \qquad \text{Quadratic function}$$

is a **quadratic function.**

Time, t	Height, h
0	6
4	774
8	1030
12	774
16	6

Often, quadratic functions can model real-life data. For example, the table at the left shows the heights h (in feet) of a projectile fired from an initial height of 6 feet with an initial velocity of 256 feet per second at selected values of time t (in seconds). A quadratic model for the data in the table is

$$h(t) = -16t^2 + 256t + 6, \quad 0 \le t \le 16.$$

The graph of a quadratic function is a "U"-shaped curve called a **parabola.** Parabolas occur in many real-life applications—including those that involve reflective properties of satellite dishes and flashlight reflectors. You will study these properties in Section 4.3.

All parabolas are symmetric with respect to a line called the **axis of symmetry,** or simply the **axis** of the parabola. The point where the axis intersects the parabola is the **vertex** of the parabola. When the leading coefficient is positive, the graph of

$$f(x) = ax^2 + bx + c$$

is a parabola that opens upward. When the leading coefficient is negative, the graph is a parabola that opens downward. The next two figures show the axes and vertices of parabolas for cases where $a > 0$ and $a < 0$.

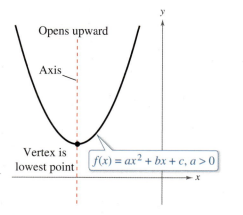

Leading coefficient is positive.

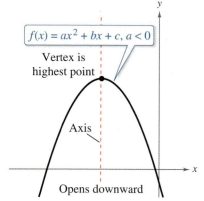

Leading coefficient is negative.

The simplest type of quadratic function is one in which $b = c = 0$. In this case, the function has the form $f(x) = ax^2$. Its graph is a parabola whose vertex is $(0, 0)$. When $a > 0$, the vertex is the point with the *minimum* y-value on the graph, and when $a < 0$, the vertex is the point with the *maximum* y-value on the graph, as shown in the figures below.

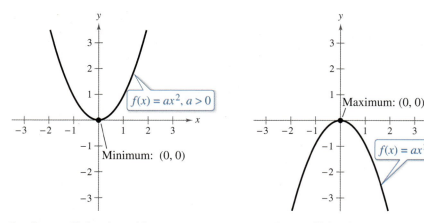

Leading coefficient is positive.

Leading coefficient is negative.

When sketching the graph of $f(x) = ax^2$, it is helpful to use the graph of $y = x^2$ as a reference, as suggested in Section 2.5. There you learned that when $a > 1$, the graph of $y = af(x)$ is a vertical stretch of the graph of $y = f(x)$. When $0 < a < 1$, the graph of $y = af(x)$ is a vertical shrink of the graph of $y = f(x)$. Example 1 demonstrates this again.

| **EXAMPLE 1** | **Sketching Graphs of Quadratic Functions** |

See LarsonPrecalculus.com for an interactive version of this type of example.

Sketch the graph of each quadratic function and compare it with the graph of $y = x^2$.

a. $f(x) = \frac{1}{3}x^2$ **b.** $g(x) = 2x^2$

Solution

a. Compared with $y = x^2$, each output of $f(x) = \frac{1}{3}x^2$ "shrinks" by a factor of $\frac{1}{3}$, producing the broader parabola shown in Figure 3.1.

b. Compared with $y = x^2$, each output of $g(x) = 2x^2$ "stretches" by a factor of 2, producing the narrower parabola shown in Figure 3.2.

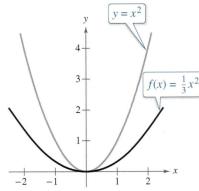

Figure 3.1   **Figure 3.2**

✓ **Checkpoint** 🔊)) *Audio-video solution in English & Spanish at LarsonPrecalculus.com*

Sketch the graph of each quadratic function and compare it with the graph of $y = x^2$.

a. $f(x) = \frac{1}{4}x^2$ **b.** $g(x) = -\frac{1}{6}x^2$ **c.** $h(x) = \frac{5}{2}x^2$ **d.** $k(x) = -4x^2$ ■

In Example 1, note that the coefficient a determines how wide the parabola $f(x) = ax^2$ opens. The smaller the value of $|a|$, the wider the parabola opens.

Recall from Section 2.5 that the graphs of

$$y = f(x \pm c), \quad y = f(x) \pm c, \quad y = f(-x), \quad \text{and} \quad y = -f(x)$$

are rigid transformations of the graph of $y = f(x)$. For example, in the figures below, notice how transformations of the graph of $y = x^2$ can produce the graphs of

$$f(x) = -x^2 + 1 \quad \text{and} \quad g(x) = (x + 2)^2 - 3.$$

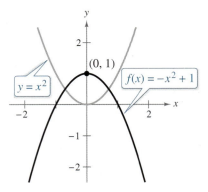

Reflection in x-axis followed by
an upward shift of one unit

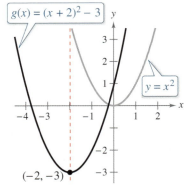

Left shift of two units followed by
a downward shift of three units

The Standard Form of a Quadratic Function

The **standard form** of a quadratic function is $f(x) = a(x - h)^2 + k$. This form is especially convenient for sketching a parabola because it identifies the vertex of the parabola as (h, k).

• • REMARK The standard form of a quadratic function identifies four basic transformations of the graph of $y = x^2$.

a. The factor a produces a vertical stretch or shrink.

b. When $a < 0$, the factor a also produces a reflection in the x-axis.

c. The factor $(x - h)^2$ represents a horizontal shift of h units.

d. The term k represents a vertical shift of k units.

Standard Form of a Quadratic Function

The quadratic function

$$f(x) = a(x - h)^2 + k, \quad a \neq 0$$

is in **standard form**. The graph of f is a parabola whose axis is the vertical line $x = h$ and whose vertex is the point (h, k). When $a > 0$, the parabola opens upward, and when $a < 0$, the parabola opens downward.

To graph a parabola, it is helpful to begin by writing the quadratic function in standard form using the process of completing the square, as illustrated in Example 2. In this example, notice that when completing the square, you *add and subtract* the square of half the coefficient of x within the parentheses instead of adding the value to each side of the equation as is done in Section 1.4.

EXAMPLE 2   Using Standard Form to Graph a Parabola

Sketch the graph of $f(x) = 2x^2 + 8x + 7$. Identify the vertex and the axis of the parabola.

▷ **ALGEBRA HELP**   To review techniques for completing the square, see Section 1.4.

Solution   Begin by writing the quadratic function in standard form. Notice that the first step in completing the square is to factor out any coefficient of x^2 that is not 1.

$$
\begin{aligned}
f(x) &= 2x^2 + 8x + 7 && \text{Write original function.} \\
&= 2(x^2 + 4x) + 7 && \text{Factor 2 out of } x\text{-terms.} \\
&= 2(x^2 + 4x + 4 - 4) + 7 && \text{Add and subtract 4 within parentheses.}
\end{aligned}
$$

$$\underbrace{\qquad}_{(4/2)^2}$$

$$
\begin{aligned}
&= 2(x^2 + 4x + 4) - 2(4) + 7 && \text{Distributive Property} \\
&= 2(x^2 + 4x + 4) - 8 + 7 && \text{Simplify.} \\
&= 2(x + 2)^2 - 1 && \text{Write in standard form.}
\end{aligned}
$$

The graph of f is a parabola that opens upward and has its vertex at $(-2, -1)$. This corresponds to a left shift of two units and a downward shift of one unit relative to the graph of $y = 2x^2$, as shown in the figure. The axis of the parabola is the vertical line through the vertex, $x = -2$, also shown in the figure.

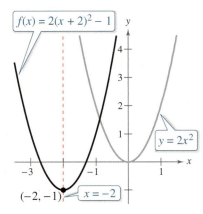

$f(x) = 2(x + 2)^2 - 1$

$y = 2x^2$

$(-2, -1)$ $x = -2$

✓ *Checkpoint* ◀))) *Audio-video solution in English & Spanish at LarsonPrecalculus.com*

Sketch the graph of $f(x) = 3x^2 - 6x + 4$. Identify the vertex and the axis of the parabola.

▷ **ALGEBRA HELP** To review techniques for solving quadratic equations, see Section 1.4.

To find the x-intercepts of the graph of $f(x) = ax^2 + bx + c$, you must solve the equation $ax^2 + bx + c = 0$. When $ax^2 + bx + c$ does not factor, use completing the square or the Quadratic Formula to find the x-intercepts. Remember, however, that a parabola may not have x-intercepts.

EXAMPLE 3 **Finding the Vertex and x-Intercepts of a Parabola**

Sketch the graph of $f(x) = -x^2 + 6x - 8$. Identify the vertex and x-intercepts.

Solution

$$f(x) = -x^2 + 6x - 8 \qquad \text{Write original function.}$$
$$= -(x^2 - 6x) - 8 \qquad \text{Factor } -1 \text{ out of } x\text{-terms.}$$
$$= -(x^2 - 6x + 9 - 9) - 8 \qquad \text{Add and subtract 9 within parentheses.}$$
$$(-6/2)^2$$
$$= -(x^2 - 6x + 9) - (-9) - 8 \qquad \text{Distributive Property}$$
$$= -(x - 3)^2 + 1 \qquad \text{Write in standard form.}$$

The graph of f is a parabola that opens downward with vertex $(3, 1)$. Next, find the x-intercepts of the graph.

$$-(x^2 - 6x + 8) = 0 \qquad \text{Factor out } -1.$$
$$-(x - 2)(x - 4) = 0 \qquad \text{Factor.}$$
$$x - 2 = 0 \implies x = 2 \qquad \text{Set 1st factor equal to 0 and solve.}$$
$$x - 4 = 0 \implies x = 4 \qquad \text{Set 2nd factor equal to 0 and solve.}$$

So, the x-intercepts are $(2, 0)$ and $(4, 0)$, as shown in Figure 3.3.

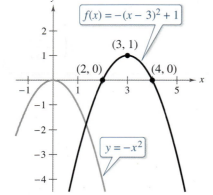
Figure 3.3

✓ **Checkpoint** ◀))) Audio-video solution in English & Spanish at LarsonPrecalculus.com

Sketch the graph of $f(x) = x^2 - 4x + 3$. Identify the vertex and x-intercepts.

EXAMPLE 4 **Writing a Quadratic Function**

Write the standard form of the quadratic function whose graph is a parabola with vertex $(1, 2)$ and that passes through the point $(3, -6)$.

Solution The vertex is $(h, k) = (1, 2)$, so the equation has the form

$$f(x) = a(x - 1)^2 + 2. \qquad \text{Substitute for } h \text{ and } k \text{ in standard form.}$$

The parabola passes through the point $(3, -6)$, so it follows that $f(3) = -6$. So,

$$f(x) = a(x - 1)^2 + 2 \qquad \text{Write in standard form.}$$
$$-6 = a(3 - 1)^2 + 2 \qquad \text{Substitute 3 for } x \text{ and } -6 \text{ for } f(x).$$
$$-6 = 4a + 2 \qquad \text{Simplify.}$$
$$-8 = 4a \qquad \text{Subtract 2 from each side.}$$
$$-2 = a. \qquad \text{Divide each side by 4.}$$

The function in standard form is $f(x) = -2(x - 2)^2 + 2$. Figure 3.4 shows the graph of f.

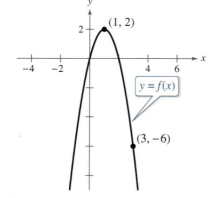
Figure 3.4

✓ **Checkpoint** ◀))) Audio-video solution in English & Spanish at LarsonPrecalculus.com

Write the standard form of the quadratic function whose graph is a parabola with vertex $(-4, 11)$ and that passes through the point $(-6, 15)$.

Finding Minimum and Maximum Values

Many applications involve finding the maximum or minimum value of a quadratic function. By completing the square within the quadratic function $f(x) = ax^2 + bx + c$, you can rewrite the function in standard form (see Exercise 79).

$$f(x) = a\left(x + \frac{b}{2a}\right)^2 + \left(c - \frac{b^2}{4a}\right) \qquad \text{Standard form}$$

So, the vertex of the graph of f is $\left(-\dfrac{b}{2a}, f\left(-\dfrac{b}{2a}\right)\right)$.

Minimum and Maximum Values of Quadratic Functions

Consider the function $f(x) = ax^2 + bx + c$ with vertex $\left(-\dfrac{b}{2a}, f\left(-\dfrac{b}{2a}\right)\right)$.

1. When $a > 0$, f has a *minimum* at $x = -\dfrac{b}{2a}$. The minimum value is $f\left(-\dfrac{b}{2a}\right)$.

2. When $a < 0$, f has a *maximum* at $x = -\dfrac{b}{2a}$. The maximum value is $f\left(-\dfrac{b}{2a}\right)$.

EXAMPLE 5 Maximum Height of a Baseball

The path of a baseball after being hit is modeled by $f(x) = -0.0032x^2 + x + 3$, where $f(x)$ is the height of the baseball (in feet) and x is the horizontal distance from home plate (in feet). What is the maximum height of the baseball?

Algebraic Solution

For this quadratic function, you have

$$f(x) = ax^2 + bx + c = -0.0032x^2 + x + 3$$

which shows that $a = -0.0032$ and $b = 1$. Because $a < 0$, the function has a maximum at $x = -b/(2a)$. So, the baseball reaches its maximum height when it is

$$x = -\frac{b}{2a} = -\frac{1}{2(-0.0032)} = 156.25 \text{ feet}$$

from home plate. At this distance, the maximum height is

$$f(156.25) = -0.0032(156.25)^2 + 156.25 + 3 = 81.125 \text{ feet.}$$

Graphical Solution

The maximum height is $y = 81.125$ feet at $x = 156.25$ feet.

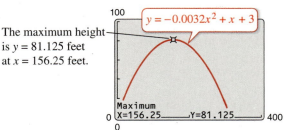

✓ **Checkpoint** ◀))) *Audio-video solution in English & Spanish at LarsonPrecalculus.com*

Rework Example 5 when the path of the baseball is modeled by

$$f(x) = -0.007x^2 + x + 4.$$

Summarize   (Section 3.1)

1. State the definition of a quadratic function and describe its graph *(pages 242–244)*. For an example of sketching graphs of quadratic functions, see Example 1.

2. State the standard form of a quadratic function *(page 245)*. For examples that use the standard form of a quadratic function, see Examples 2–4.

3. Explain how to find the minimum or maximum value of a quadratic function *(page 247)*. For a real-life application, see Example 5.

3.1 Exercises

See **CalcChat.com** for tutorial help and worked-out solutions to odd-numbered exercises.

Vocabulary: Fill in the blanks.

1. Linear, constant, and squaring functions are examples of _____ functions.
2. A polynomial function of x with degree n has the form $f(x) = a_n x^n + a_{n-1}x^{n-1} + \cdots + a_1 x + a_0$ ($a_n \neq 0$), where n is a _____ _____ and $a_n, a_{n-1}, \ldots, a_1, a_0$ are _____ numbers.
3. A _____ function is a second-degree polynomial function, and its graph is called a _____.
4. When the graph of a quadratic function opens downward, its leading coefficient is _____ and the vertex of the graph is a _____.

Skills and Applications

Matching In Exercises 5–8, match the quadratic function with its graph. [The graphs are labeled (a), (b), (c), and (d).]

(a)

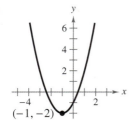

(b)

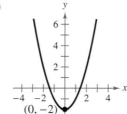

(c)

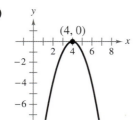

(d)

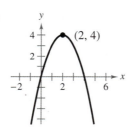

5. $f(x) = x^2 - 2$
6. $f(x) = (x + 1)^2 - 2$
7. $f(x) = -(x - 4)^2$
8. $f(x) = 4 - (x - 2)^2$

 Sketching Graphs of Quadratic Functions In Exercises 9–12, sketch the graph of each quadratic function and compare it with the graph of $y = x^2$.

9. (a) $f(x) = \frac{1}{2}x^2$ (b) $g(x) = -\frac{1}{8}x^2$
   (c) $h(x) = \frac{3}{2}x^2$ (d) $k(x) = -3x^2$
10. (a) $f(x) = x^2 + 1$ (b) $g(x) = x^2 - 1$
   (c) $h(x) = x^2 + 3$ (d) $k(x) = x^2 - 3$
11. (a) $f(x) = (x - 1)^2$ (b) $g(x) = (3x)^2 + 1$
   (c) $h(x) = \left(\frac{1}{3}x\right)^2 - 3$ (d) $k(x) = (x + 3)^2$
12. (a) $f(x) = -\frac{1}{2}(x - 2)^2 + 1$
   (b) $g(x) = \left[\frac{1}{2}(x - 1)\right]^2 - 3$
   (c) $h(x) = -\frac{1}{2}(x + 2)^2 - 1$
   (d) $k(x) = [2(x + 1)]^2 + 4$

 Using Standard Form to Graph a Parabola In Exercises 13–26, write the quadratic function in standard form and sketch its graph. Identify the vertex, axis of symmetry, and x-intercept(s).

13. $f(x) = x^2 - 6x$
14. $g(x) = x^2 - 8x$
15. $h(x) = x^2 - 8x + 16$
16. $g(x) = x^2 + 2x + 1$
17. $f(x) = x^2 - 6x + 2$
18. $f(x) = x^2 + 16x + 61$
19. $f(x) = x^2 - 8x + 21$
20. $f(x) = x^2 + 12x + 40$
21. $f(x) = x^2 - x + \frac{5}{4}$
22. $f(x) = x^2 + 3x + \frac{1}{4}$
23. $f(x) = -x^2 + 2x + 5$
24. $f(x) = -x^2 - 4x + 1$
25. $h(x) = 4x^2 - 4x + 21$
26. $f(x) = 2x^2 - x + 1$

Using Technology In Exercises 27–34, use a graphing utility to graph the quadratic function. Identify the vertex, axis of symmetry, and x-intercept(s). Then check your results algebraically by writing the quadratic function in standard form.

27. $f(x) = -(x^2 + 2x - 3)$
28. $f(x) = -(x^2 + x - 30)$
29. $g(x) = x^2 + 8x + 11$
30. $f(x) = x^2 + 10x + 14$
31. $f(x) = -2x^2 + 12x - 18$
32. $f(x) = -4x^2 + 24x - 41$
33. $g(x) = \frac{1}{2}(x^2 + 4x - 2)$
34. $f(x) = \frac{3}{5}(x^2 + 6x - 5)$

Writing a Quadratic Function In Exercises 35 and 36, write the standard form of the quadratic function whose graph is the parabola shown.

35.

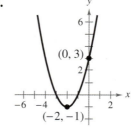

36.

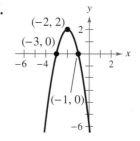

Writing a Quadratic Function In Exercises 37–46, write the standard form of the quadratic function whose graph is a parabola with the given vertex and that passes through the given point.

37. Vertex: $(-2, 5)$; point: $(0, 9)$

38. Vertex: $(-3, -10)$; point: $(0, 8)$

39. Vertex: $(1, -2)$; point: $(-1, 14)$

40. Vertex: $(2, 3)$; point: $(0, 2)$

41. Vertex: $(5, 12)$; point: $(7, 15)$

42. Vertex: $(-2, -2)$; point: $(-1, 0)$

43. Vertex: $\left(-\frac{1}{4}, \frac{3}{2}\right)$; point: $(-2, 0)$

44. Vertex: $\left(\frac{5}{2}, -\frac{3}{4}\right)$; point: $(-2, 4)$

45. Vertex: $\left(-\frac{5}{2}, 0\right)$; point: $\left(-\frac{7}{2}, -\frac{16}{3}\right)$

46. Vertex: $(6, 6)$; point: $\left(\frac{61}{10}, \frac{3}{2}\right)$

Graphical Reasoning In Exercises 47–50, determine the x-intercept(s) of the graph visually. Then find the x-intercept(s) algebraically to confirm your results.

47. $y = x^2 - 2x - 3$

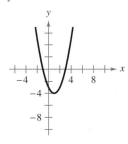

48. $y = x^2 - 4x - 5$

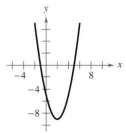

49. $y = 2x^2 + 5x - 3$

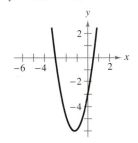

50. $y = -2x^2 + 5x + 3$

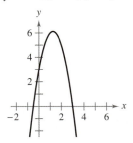

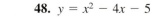

Using Technology In Exercises 51–56, use a graphing utility to graph the quadratic function. Find the x-intercept(s) of the graph and compare them with the solutions of the corresponding quadratic equation when $f(x) = 0$.

51. $f(x) = x^2 - 4x$

52. $f(x) = -2x^2 + 10x$

53. $f(x) = x^2 - 9x + 18$

54. $f(x) = x^2 - 8x - 20$

55. $f(x) = 2x^2 - 7x - 30$

56. $f(x) = \frac{7}{10}(x^2 + 12x - 45)$

Finding Quadratic Functions In Exercises 57–62, find two quadratic functions, one that opens upward and one that opens downward, whose graphs have the given x-intercepts. (There are many correct answers.)

57. $(-3, 0)$, $(3, 0)$

58. $(-5, 0)$, $(5, 0)$

59. $(-1, 0)$, $(4, 0)$

60. $(-2, 0)$, $(3, 0)$

61. $(-3, 0)$, $\left(-\frac{1}{2}, 0\right)$

62. $\left(-\frac{3}{2}, 0\right)$, $(-5, 0)$

Number Problems In Exercises 63–66, find two positive real numbers whose product is a maximum.

63. The sum is 110.

64. The sum is S.

65. The sum of the first and twice the second is 24.

66. The sum of the first and three times the second is 42.

67. Path of a Diver

The path of a diver is modeled by

$$f(x) = -\frac{4}{9}x^2 + \frac{24}{9}x + 12$$

where $f(x)$ is the height (in feet) and x is the horizontal distance (in feet) from the end of the diving board. What is the maximum height of the diver?

68. Height of a Ball The path of a punted football is modeled by

$$f(x) = -\frac{16}{2025}x^2 + \frac{9}{5}x + 1.5$$

where $f(x)$ is the height (in feet) and x is the horizontal distance (in feet) from the point at which the ball is punted.

(a) How high is the ball when it is punted?

(b) What is the maximum height of the punt?

(c) How long is the punt?

69. Minimum Cost A manufacturer of lighting fixtures has daily production costs of $C = 800 - 10x + 0.25x^2$, where C is the total cost (in dollars) and x is the number of units produced. What daily production number yields a minimum cost?

70. Maximum Profit The profit P (in hundreds of dollars) that a company makes depends on the amount x (in hundreds of dollars) the company spends on advertising according to the model $P = 230 + 20x - 0.5x^2$. What expenditure for advertising yields a maximum profit?

71. Maximum Revenue The total revenue R earned (in thousands of dollars) from manufacturing handheld video games is given by $R(p) = -25p^2 + 1200p$, where p is the price per unit (in dollars).

(a) Find the revenues when the prices per unit are $20, $25, and $30.

(b) Find the unit price that yields a maximum revenue. What is the maximum revenue? Explain.

72. Maximum Revenue The total revenue R earned per day (in dollars) from a pet-sitting service is given by $R(p) = -12p^2 + 150p$, where p is the price charged per pet (in dollars).

(a) Find the revenues when the prices per pet are $4, $6, and $8.

(b) Find the unit price that yields a maximum revenue. What is the maximum revenue? Explain.

73. Maximum Area A rancher has 200 feet of fencing to enclose two adjacent rectangular corrals (see figure).

(a) Write the area A of the corrals as a function of x.

(b) What dimensions produce a maximum enclosed area?

74. Maximum Area A Norman window is constructed by adjoining a semicircle to the top of an ordinary rectangular window (see figure). The perimeter of the window is 16 feet.

(a) Write the area A of the window as a function of x.

(b) What dimensions produce a window of maximum area?

Exploration

True or False? **In Exercises 75 and 76, determine whether the statement is true or false. Justify your answer.**

75. The graph of $f(x) = -12x^2 - 1$ has no x-intercepts.

76. The graphs of $f(x) = -4x^2 - 10x + 7$ and $g(x) = 12x^2 + 30x + 1$ have the same axis of symmetry.

Think About It **In Exercises 77 and 78, find the values of b such that the function has the given maximum or minimum value.**

77. $f(x) = -x^2 + bx - 75$; Maximum value: 25

78. $f(x) = x^2 + bx - 25$; Minimum value: -50

79. Verifying the Vertex Write the quadratic function

$$f(x) = ax^2 + bx + c$$

in standard form to verify that the vertex occurs at

$$\left(-\frac{b}{2a}, f\left(-\frac{b}{2a}\right)\right).$$

80. **HOW DO YOU SEE IT?** The graph shows a quadratic function of the form

$$P(t) = at^2 + bt + c$$

which represents the yearly profit for a company, where $P(t)$ is the profit in year t.

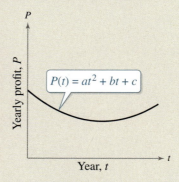

(a) Is the value of a positive, negative, or zero? Explain.

(b) Write an expression in terms of a and b that represents the year t when the company made the least profit.

(c) The company made the same yearly profits in 2008 and 2016. Estimate the year in which the company made the least profit.

81. Proof Assume that the function

$$f(x) = ax^2 + bx + c, \quad a \neq 0$$

has two real zeros. Prove that the x-coordinate of the vertex of the graph is the average of the zeros of f. (*Hint:* Use the Quadratic Formula.)

Project: Height of a Basketball To work an extended application analyzing the height of a dropped basketball, visit this text's website at *LarsonPrecalculus.com*.

3.2 Polynomial Functions of Higher Degree

Polynomial functions have many real-life applications. For example, in Exercise 98 on page 263, you will use a polynomial function to analyze the growth of a red oak tree.

- ■ Use transformations to sketch graphs of polynomial functions.
- ■ Use the Leading Coefficient Test to determine the end behaviors of graphs of polynomial functions.
- ■ Find real zeros of polynomial functions and use them as sketching aids.
- ■ Use the Intermediate Value Theorem to help locate real zeros of polynomial functions.

Graphs of Polynomial Functions

In this section, you will study basic features of the graphs of polynomial functions. One feature is that the graph of a polynomial function is **continuous.** Essentially, this means that the graph of a polynomial function has no breaks, holes, or gaps, as shown in Figure 3.5(a). The graph shown in Figure 3.5(b) is an example of a piecewise-defined function that is not continuous.

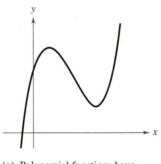

(a) Polynomial functions have continuous graphs.

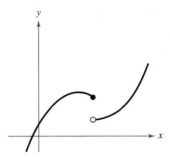

(b) Functions with graphs that are not continuous are not polynomial functions.

Figure 3.5

Another feature of the graph of a polynomial function is that it has only smooth, rounded turns, as shown in Figure 3.6(a). The graph of a polynomial function cannot have a sharp turn, such as the one shown in Figure 3.6(b).

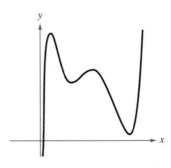

(a) Polynomial functions have graphs with smooth, rounded turns.

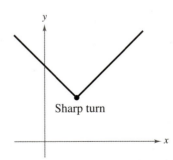

(b) Functions with graphs that have sharp turns are not polynomial functions.

Figure 3.6

Sketching graphs of polynomial functions of degree greater than 2 is often more involved than sketching graphs of polynomial functions of degree 0, 1, or 2. However, using the features presented in this section, along with your knowledge of point plotting, intercepts, and symmetry, you should be able to make reasonably accurate sketches by hand.

REMARK For functions of the form $f(x) = x^n$, if n is even, then the graph of the function is symmetric with respect to the y-axis, and if n is odd, then the graph of the function is symmetric with respect to the origin.

The polynomial functions that have the simplest graphs are monomial functions of the form $f(x) = x^n$, where n is an integer greater than zero. When n is *even*, the graph is similar to the graph of $f(x) = x^2$, and when n is *odd*, the graph is similar to the graph of $f(x) = x^3$, as shown in Figure 3.7. Moreover, the greater the value of n, the flatter the graph near the origin. Polynomial functions of the form $f(x) = x^n$ are often referred to as **power functions.**

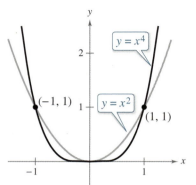

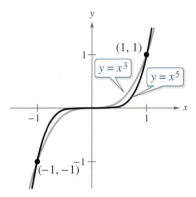

(**a**) When n is even, the graph of $y = x^n$ touches the x-axis at the x-intercept.

(**b**) When n is odd, the graph of $y = x^n$ crosses the x-axis at the x-intercept.

Figure 3.7

EXAMPLE 1 **Sketching Transformations of Monomial Functions**

See LarsonPrecalculus.com for an interactive version of this type of example.

Sketch the graph of each function.

a. $f(x) = -x^5$ **b.** $h(x) = (x + 1)^4$

Solution

a. The degree of $f(x) = -x^5$ is odd, so its graph is similar to the graph of $y = x^3$. In Figure 3.8, note that the negative coefficient has the effect of reflecting the graph in the x-axis.

b. The degree of $h(x) = (x + 1)^4$ is even, so its graph is similar to the graph of $y = x^2$. In Figure 3.9, note that the graph of h is a left shift by one unit of the graph of $y = x^4$.

 **ALGEBRA HELP** To review techniques for shifting, reflecting, stretching, and shrinking graphs, see Section 2.5.

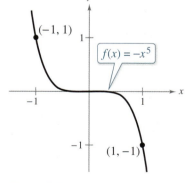

Figure 3.8

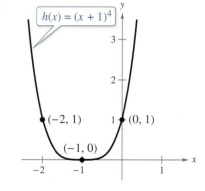

Figure 3.9

✓ **Checkpoint** 🔊 *Audio-video solution in English & Spanish at LarsonPrecalculus.com*

Sketch the graph of each function.

a. $f(x) = (x + 5)^4$ **b.** $g(x) = x^4 - 7$

c. $h(x) = 7 - x^4$ **d.** $k(x) = \frac{1}{4}(x - 3)^4$

The Leading Coefficient Test

In Example 1, note that both graphs eventually rise or fall without bound as x moves to the left or to the right. A polynomial function's degree (even or odd) and its leading coefficient (positive or negative) determine whether the graph of the function eventually rises or falls, as described in the **Leading Coefficient Test.**

Leading Coefficient Test

As x moves without bound to the left or to the right, the graph of the polynomial function

$$f(x) = a_n x^n + \cdots + a_1 x + a_0, \quad a_n \neq 0$$

eventually rises or falls in the manner described below.

1. When n is *odd:*

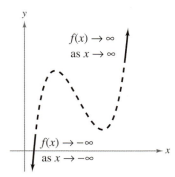

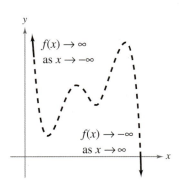

If the leading coefficient is positive $(a_n > 0)$, then the graph falls to the left and rises to the right.

If the leading coefficient is negative $(a_n < 0)$, then the graph rises to the left and falls to the right.

2. When n is *even:*

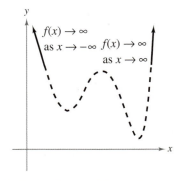

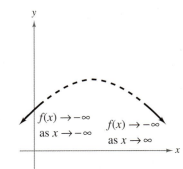

If the leading coefficient is positive $(a_n > 0)$, then the graph rises to the left and to the right.

If the leading coefficient is negative $(a_n < 0)$, then the graph falls to the left and to the right.

The dashed portions of the graphs indicate that the test determines *only* the right-hand and left-hand behavior of the graph.

> **REMARK** The notation "$f(x) \to -\infty$ as $x \to -\infty$" means that the graph falls to the left. The notation "$f(x) \to \infty$ as $x \to \infty$" means that the graph rises to the right. Identify and interpret similar notation for the other two possible types of end behavior given in the Leading Coefficient Test.

As you continue to study polynomial functions and their graphs, you will notice that the degree of a polynomial plays an important role in determining other characteristics of the polynomial function and its graph.

EXAMPLE 2 **Applying the Leading Coefficient Test**

Describe the left-hand and right-hand behavior of the graph of each function.

a. $f(x) = -x^3 + 4x$ **b.** $f(x) = x^4 - 5x^2 + 4$ **c.** $f(x) = x^5 - x$

Solution

a. The degree is odd and the leading coefficient is negative, so the graph rises to the left and falls to the right, as shown in the figure below.

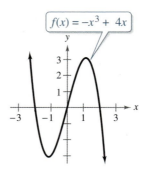

$f(x) = -x^3 + 4x$

b. The degree is even and the leading coefficient is positive, so the graph rises to the left and to the right, as shown in the figure below.

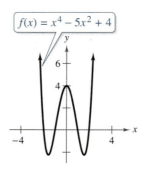

$f(x) = x^4 - 5x^2 + 4$

c. The degree is odd and the leading coefficient is positive, so the graph falls to the left and rises to the right, as shown in the figure below.

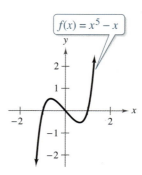

$f(x) = x^5 - x$

✓ *Checkpoint*   *Audio-video solution in English & Spanish at LarsonPrecalculus.com*

Describe the left-hand and right-hand behavior of the graph of each function.

a. $f(x) = \frac{1}{4}x^3 - 2x$ **b.** $f(x) = -3.6x^5 + 5x^3 - 1$

In Example 2, note that the Leading Coefficient Test tells you only whether the graph *eventually* rises or falls to the left or to the right. You must use other tests to determine other characteristics of the graph, such as intercepts and minimum and maximum points.

Real Zeros of Polynomial Functions

It is possible to show that for a polynomial function f of degree n, the two statements below are true.

• • • • • • • • • • • • • • • • ▷

REMARK Remember that the *zeros* of a function of x are the x-values for which the function is zero.

1. The function f has, at most, n real zeros. (You will study this result in detail in the discussion of the Fundamental Theorem of Algebra in Section 3.4.)
2. The graph of f has, at most, $n-1$ turning points. (Turning points, also called relative minima or relative maxima, are points at which the graph changes from increasing to decreasing or vice versa.)

Finding the zeros of a polynomial function is an important problem in algebra. There is a strong interplay between graphical and algebraic approaches to this problem.

Real Zeros of Polynomial Functions

When f is a polynomial function and a is a real number, the statements listed below are equivalent.

1. $x = a$ is a *zero* of the function f.
2. $x = a$ is a *solution* of the polynomial equation $f(x) = 0$.
3. $(x - a)$ is a *factor* of the polynomial $f(x)$.
4. $(a, 0)$ is an *x-intercept* of the graph of f.

EXAMPLE 3 **Finding Real Zeros of a Polynomial Function**

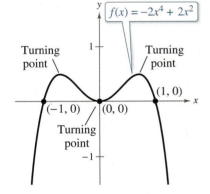

Turning point

Turning point

$f(x) = -2x^4 + 2x^2$

$(1, 0)$

$(-1, 0)$ $(0, 0)$

Turning point

Figure 3.10

Find all real zeros of $f(x) = -2x^4 + 2x^2$. Then determine the maximum possible number of turning points of the graph of the function.

Solution To find the real zeros of the function, set $f(x)$ equal to zero and then solve for x.

$$-2x^4 + 2x^2 = 0 \qquad \text{Set } f(x) \text{ equal to 0.}$$
$$-2x^2(x^2 - 1) = 0 \qquad \text{Remove common monomial factor.}$$
$$-2x^2(x - 1)(x + 1) = 0 \qquad \text{Factor completely.}$$

So, the real zeros are $x = 0$, $x = 1$, and $x = -1$, and the corresponding x-intercepts occur at $(0, 0)$, $(1, 0)$, and $(-1, 0)$. The function is a fourth-degree polynomial, so the graph of f can have at most $4 - 1 = 3$ turning points. In this case, the graph of f has three turning points. Figure 3.10 shows the graph of f.

✓ **Checkpoint** ◀))) *Audio-video solution in English & Spanish at LarsonPrecalculus.com*

Find all real zeros of $f(x) = x^3 - 12x^2 + 36x$. Then determine the maximum possible number of turning points of the graph of the function. ∎

In Example 3, note that the factor $-2x^2$ yields the *repeated* zero $x = 0$. The exponent is even, so the graph touches the x-axis at $x = 0$.

▷ **ALGEBRA HELP** The solution to Example 3 uses polynomial factoring. To review the techniques for factoring polynomials, see Section P.4.

Repeated Zeros

A factor $(x - a)^k$, $k > 1$, yields a **repeated zero** $x = a$ of **multiplicity** k.

1. When k is odd, the graph *crosses* the x-axis at $x = a$.
2. When k is even, the graph *touches* the x-axis (but does not cross the x-axis) at $x = a$.

A polynomial function is in **standard form** when its terms are in descending order of exponents from left to right. To avoid making a mistake when applying the Leading Coefficient Test, rewrite the polynomial function in standard form first, if necessary.

EXAMPLE 4   Sketching the Graph of a Polynomial Function

Sketch the graph of

$$f(x) = -4x^3 + 3x^4.$$

Solution

1. *Rewrite in Standard Form and Apply the Leading Coefficient Test.*   In standard form, the polynomial function is $f(x) = 3x^4 - 4x^3$. The leading coefficient is positive and the degree is even, so you know that the graph eventually rises to the left and to the right (see Figure 3.11).

2. *Find the Real Zeros of the Function.*   Factoring

$$f(x) = 3x^4 - 4x^3 = x^3(3x - 4) \qquad \text{Remove common monomial factor.}$$

shows that the real zeros of f are $x = 0$ and $x = \frac{4}{3}$ (both of odd multiplicity). So, the x-intercepts occur at $(0, 0)$ and $\left(\frac{4}{3}, 0\right)$. Add these points to your graph, as shown in Figure 3.11.

3. *Plot a Few Additional Points.*   To sketch the graph, find a few additional points, as shown in the table.

x	-1	$\frac{1}{2}$	1	$\frac{3}{2}$
$f(x)$	7	$-\frac{5}{16}$	-1	$\frac{27}{16}$

Then plot the points (see Figure 3.12).

4. *Draw the Graph.*   Draw a continuous curve through the points, as shown in Figure 3.12. Both zeros are of odd multiplicity, so you know that the graph should cross the x-axis at $x = 0$ and $x = \frac{4}{3}$. If you are unsure of the shape of a portion of a graph, plot additional points.

▷ **TECHNOLOGY** Example 4 uses an *algebraic approach* to describe the graph of the function. A graphing utility can complement this approach. Remember to find a viewing window that shows all significant features of the graph. For instance, viewing window (a) illustrates all of the significant features of the function in Example 4, but viewing window (b) does not.

(a)

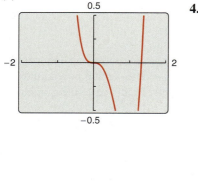

(b)

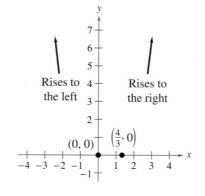

Figure 3.11

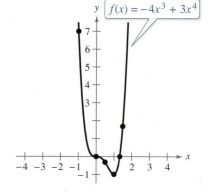

Figure 3.12

✓ *Checkpoint*))) *Audio-video solution in English & Spanish at LarsonPrecalculus.com*

Sketch the graph of

$$f(x) = 2x^3 - 6x^2.$$

EXAMPLE 5 **Sketching the Graph of a Polynomial Function**

Sketch the graph of

$$f(x) = -2x^3 + 6x^2 - \tfrac{9}{2}x.$$

Solution

1. *Apply the Leading Coefficient Test.* The leading coefficient is negative and the degree is odd, so you know that the graph eventually rises to the left and falls to the right (see Figure 3.13).

2. *Find the Real Zeros of the Function.* Factoring

$$\begin{aligned} f(x) &= -2x^3 + 6x^2 - \tfrac{9}{2}x \\ &= -\tfrac{4}{2}x^3 + \tfrac{12}{2}x^2 - \tfrac{9}{2}x \\ &= -\tfrac{1}{2}x(4x^2 - 12x + 9) \\ &= -\tfrac{1}{2}x(2x - 3)^2 \end{aligned}$$

shows that the real zeros of f are

$$x = 0 \text{ (odd multiplicity)} \quad \text{and} \quad x = \tfrac{3}{2} \text{ (even multiplicity)}.$$

So, the x-intercepts occur at $(0, 0)$ and $\left(\tfrac{3}{2}, 0\right)$. Add these points to your graph, as shown in Figure 3.13.

3. *Plot a Few Additional Points.* To sketch the graph, find a few additional points, as shown in the table.

x	$-\tfrac{1}{2}$	$\tfrac{1}{2}$	1	2
$f(x)$	4	-1	$-\tfrac{1}{2}$	-1

Then plot the points (see Figure 3.14).

4. *Draw the Graph.* Draw a continuous curve through the points, as shown in Figure 3.14. From the multiplicities of the zeros, you know that the graph crosses the x-axis at $(0, 0)$ but does not cross the x-axis at $\left(\tfrac{3}{2}, 0\right)$.

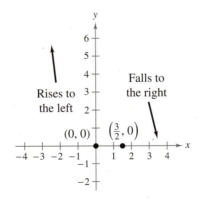

Figure 3.13

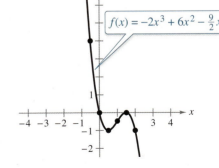

Figure 3.14

· · · · · · · · · · · · · · · · · · ·▷

REMARK Observe in Example 5 that the sign of $f(x)$ is positive to the left of and negative to the right of the zero $x = 0$. Similarly, the sign of $f(x)$ is negative to the left and to the right of the zero $x = \tfrac{3}{2}$. This illustrates that (1) if the zero of a polynomial function is of *odd* multiplicity, then the graph crosses the x-axis at that zero, and (2) if the zero is of *even* multiplicity, then the graph touches the x-axis at that zero. Constructing a table such as the one below may be helpful in graphing polynomial functions.

x	$-\tfrac{1}{2}$	0	$\tfrac{1}{2}$
$f(x)$	4	0	-1
Sign	$+$		$-$

x	1	$\tfrac{3}{2}$	2
$f(x)$	$-\tfrac{1}{2}$	0	-1
Sign	$-$		$-$

✓ ***Checkpoint*** 🔊)) *Audio-video solution in English & Spanish at LarsonPrecalculus.com*

Sketch the graph of

$$f(x) = -\tfrac{1}{4}x^4 + \tfrac{3}{2}x^3 - \tfrac{9}{4}x^2.$$

The Intermediate Value Theorem

The **Intermediate Value Theorem** implies that if

$$(a, f(a)) \quad \text{and} \quad (b, f(b))$$

are two points on the graph of a polynomial function such that $f(a) \neq f(b)$, then for any number d between $f(a)$ and $f(b)$ there must be a number c between a and b such that $f(c) = d$. (See figure below.)

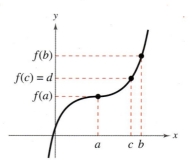

Intermediate Value Theorem

Let a and b be real numbers such that $a < b$. If f is a polynomial function such that $f(a) \neq f(b)$, then, in the interval $[a, b]$, f takes on every value between $f(a)$ and $f(b)$.

REMARK Note that $f(a)$ and $f(b)$ must be of opposite signs in order to guarantee that a zero exists between them. If $f(a)$ and $f(b)$ are of the same sign, then it is inconclusive whether a zero exists between them.

One application of the Intermediate Value Theorem is in helping you locate real zeros of a polynomial function. If there exists a value $x = a$ at which a polynomial function is negative, and another value $x = b$ at which it is positive (or if it is positive when $x = a$ and negative when $x = b$), then the function has at least one real zero between these two values. For example, the function

$$f(x) = x^3 + x^2 + 1$$

is negative when $x = -2$ and positive when $x = -1$. So, it follows from the Intermediate Value Theorem that f must have a real zero somewhere between -2 and -1, as shown in the figure below.

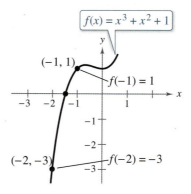

The function f must have a real zero somewhere between -2 and -1.

By continuing this line of reasoning, it is possible to approximate real zeros of a polynomial function to any desired accuracy. Example 6 further demonstrates this concept.

X	Y1
-2	-11
-1	-1
0	**1**
1	1
2	5
3	19
4	49
X=0	

X	Y1
-1	-1
-.9	-.539
-.8	-.152
-.7	**.167**
-.6	.424
-.5	.625
-.4	.776
X=-.7	

EXAMPLE 6 **Using the Intermediate Value Theorem**

Use the Intermediate Value Theorem to approximate the real zero of

$$f(x) = x^3 - x^2 + 1.$$

Solution Begin by computing a few function values.

x	-2	-1	0	1
$f(x)$	-11	-1	1	1

The value $f(-1)$ is negative and $f(0)$ is positive, so by the Intermediate Value Theorem, the function has a real zero between -1 and 0. To pinpoint this zero more closely, divide the interval $[-1, 0]$ into tenths and evaluate the function at each point. When you do this, you will find that

$$f(-0.8) = -0.152$$

and

$$f(-0.7) = 0.167.$$

So, f must have a real zero between -0.8 and -0.7, as shown in the figure. For a more accurate approximation, compute function values between $f(-0.8)$ and $f(-0.7)$ and apply the Intermediate Value Theorem again. Continue this process to verify that

$$x \approx -0.755$$

is an approximation (to the nearest thousandth) of the real zero of f.

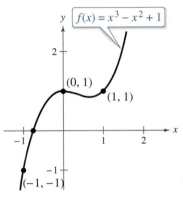

The function f has a real zero between -0.8 and -0.7.

✓ *Checkpoint* *Audio-video solution in English & Spanish at LarsonPrecalculus.com*

Use the Intermediate Value Theorem to approximate the real zero of

$$f(x) = x^3 - 3x^2 - 2.$$

Summarize (Section 3.2)

1. Explain how to use transformations to sketch graphs of polynomial functions *(page 252)*. For an example of sketching transformations of monomial functions, see Example 1.

2. Explain how to apply the Leading Coefficient Test *(page 253)*. For an example of applying the Leading Coefficient Test, see Example 2.

3. Explain how to find real zeros of polynomial functions and use them as sketching aids *(page 255)*. For examples involving finding real zeros of polynomial functions, see Examples 3–5.

4. Explain how to use the Intermediate Value Theorem to help locate real zeros of polynomial functions *(page 258)*. For an example of using the Intermediate Value Theorem, see Example 6.

3.2 Exercises

See **CalcChat.com** for tutorial help and worked-out solutions to odd-numbered exercises.

Vocabulary: Fill in the blanks.

1. The graph of a polynomial function is _____, which means that the graph has no breaks, holes, or gaps.

2. The _____ _____ _____ is used to determine the left-hand and right-hand behavior of the graph of a polynomial function.

3. A polynomial function of degree n has at most _____ real zeros and at most _____ turning points.

4. When $x = a$ is a zero of a polynomial function f, the three statements below are true.
   (a) $x = a$ is a _____ of the polynomial equation $f(x) = 0$.
   (b) _____ is a factor of the polynomial $f(x)$.
   (c) $(a, 0)$ is an _____ of the graph of f.

5. When a real zero $x = a$ of a polynomial function f is of even multiplicity, the graph of f _____ the x-axis at $x = a$, and when it is of odd multiplicity, the graph of f _____ the x-axis at $x = a$.

6. A factor $(x - a)^k$, $k > 1$, yields a _____ _____ $x = a$ of _____ k.

7. A polynomial function is written in _____ form when its terms are written in descending order of exponents from left to right.

8. The _____ _____ Theorem states that if f is a polynomial function such that $f(a) \neq f(b)$, then, in the interval $[a, b]$, f takes on every value between $f(a)$ and $f(b)$.

Skills and Applications

Matching In Exercises 9–14, match the polynomial function with its graph. [The graphs are labeled (a), (b), (c), (d), (e), and (f).]

9. $f(x) = -2x^2 - 5x$

10. $f(x) = 2x^3 - 3x + 1$

11. $f(x) = -\frac{1}{4}x^4 + 3x^2$

12. $f(x) = -\frac{1}{3}x^3 + x^2 - \frac{4}{3}$

13. $f(x) = x^4 + 2x^3$

14. $f(x) = \frac{1}{5}x^5 - 2x^3 + \frac{9}{5}x$

(a)

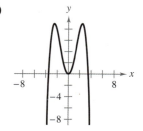

(b)

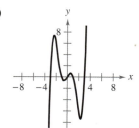

Sketching Transformations of Monomial Functions In Exercises 15–18, sketch the graph of $y = x^n$ and each transformation.

15. $y = x^3$
   (a) $f(x) = (x - 4)^3$
   (b) $f(x) = x^3 - 4$
   (c) $f(x) = -\frac{1}{4}x^3$
   (d) $f(x) = (x - 4)^3 - 4$

16. $y = x^5$
   (a) $f(x) = (x + 1)^5$
   (b) $f(x) = x^5 + 1$
   (c) $f(x) = 1 - \frac{1}{2}x^5$
   (d) $f(x) = -\frac{1}{2}(x + 1)^5$

(c)

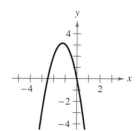

(d)

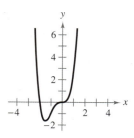

17. $y = x^4$
   (a) $f(x) = (x + 3)^4$
   (b) $f(x) = x^4 - 3$
   (c) $f(x) = 4 - x^4$
   (d) $f(x) = \frac{1}{2}(x - 1)^4$
   (e) $f(x) = (2x)^4 + 1$
   (f) $f(x) = \left(\frac{1}{2}x\right)^4 - 2$

18. $y = x^6$
   (a) $f(x) = (x - 5)^6$
   (b) $f(x) = \frac{1}{8}x^6$
   (c) $f(x) = (x + 3)^6 - 4$
   (d) $f(x) = -\frac{1}{4}x^6 + 1$
   (e) $f(x) = \left(\frac{1}{4}x\right)^6 - 2$
   (f) $f(x) = (2x)^6 - 1$

(e)

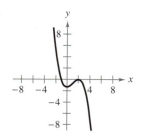

(f)

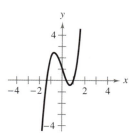

Applying the Leading Coefficient Test In Exercises 19–28, describe the left-hand and right-hand behavior of the graph of the polynomial function.

19. $f(x) = 12x^3 + 4x$ **20.** $f(x) = 2x^2 - 3x + 1$

21. $g(x) = 5 - \frac{7}{2}x - 3x^2$ **22.** $h(x) = 1 - x^6$

23. $h(x) = 6x - 9x^3 + x^2$ **24.** $g(x) = 8 + \frac{1}{4}x^5 - x^4$

25. $f(x) = 9.8x^6 - 1.2x^3$

26. $h(x) = 1 - 0.5x^5 - 2.7x^3$

27. $f(s) = -\frac{7}{8}(s^3 + 5s^2 - 7s + 1)$

28. $h(t) = -\frac{4}{3}(t - 6t^3 + 2t^4 + 9)$

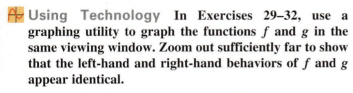 **Using Technology** In Exercises 29–32, use a graphing utility to graph the functions f and g in the same viewing window. Zoom out sufficiently far to show that the left-hand and right-hand behaviors of f and g appear identical.

29. $f(x) = 3x^3 - 9x + 1$, $g(x) = 3x^3$

30. $f(x) = -\frac{1}{3}(x^3 - 3x + 2)$, $g(x) = -\frac{1}{3}x^3$

31. $f(x) = -(x^4 - 4x^3 + 16x)$, $g(x) = -x^4$

32. $f(x) = 3x^4 - 6x^2$, $g(x) = 3x^4$

 Finding Real Zeros of a Polynomial Function In Exercises 33–48, (a) find all real zeros of the polynomial function, (b) determine whether the multiplicity of each zero is even or odd, (c) determine the maximum possible number of turning points of the graph of the function, and (d) use a graphing utility to graph the function and verify your answers.

33. $f(x) = x^2 - 36$ **34.** $f(x) = 81 - x^2$

35. $h(t) = t^2 - 6t + 9$ **36.** $f(x) = x^2 + 10x + 25$

37. $f(x) = \frac{1}{3}x^2 + \frac{1}{3}x - \frac{2}{3}$ **38.** $f(x) = \frac{1}{2}x^2 + \frac{5}{2}x - \frac{3}{2}$

39. $g(x) = 5x(x^2 - 2x - 1)$ **40.** $f(t) = t^2(3t^2 - 10t + 7)$

41. $f(x) = 3x^3 - 12x^2 + 3x$

42. $f(x) = x^4 - x^3 - 30x^2$

43. $g(t) = t^5 - 6t^3 + 9t$ **44.** $f(x) = x^5 + x^3 - 6x$

45. $f(x) = 3x^4 + 9x^2 + 6$ **46.** $f(t) = 2t^4 - 2t^2 - 40$

47. $g(x) = x^3 + 3x^2 - 4x - 12$

48. $f(x) = x^3 - 4x^2 - 25x + 100$

Using Technology In Exercises 49–52, (a) use a graphing utility to graph the function, (b) use the graph to approximate any x-intercepts of the graph, (c) find any real zeros of the function algebraically, and (d) compare the results of part (c) with those of part (b).

49. $y = 4x^3 - 20x^2 + 25x$

50. $y = 4x^3 + 4x^2 - 8x - 8$

51. $y = x^5 - 5x^3 + 4x$ **52.** $y = \frac{1}{5}x^5 - \frac{9}{5}x^3$

 Finding a Polynomial Function In Exercises 53–62, find a polynomial function that has the given zeros. (There are many correct answers.)

53. $0, 7$ **54.** $-2, 5$

55. $0, -2, -4$ **56.** $0, 1, 6$

57. $4, -3, 3, 0$ **58.** $-2, -1, 0, 1, 2$

59. $1 + \sqrt{2}, 1 - \sqrt{2}$ **60.** $4 + \sqrt{3}, 4 - \sqrt{3}$

61. $2, 2 + \sqrt{5}, 2 - \sqrt{5}$ **62.** $3, 2 + \sqrt{7}, 2 - \sqrt{7}$

 Finding a Polynomial Function In Exercises 63–70, find a polynomial of degree n that has the given zero(s). (There are many correct answers.)

Zero(s)	Degree
63. $x = -3$	$n = 2$
64. $x = -\sqrt{2}, \sqrt{2}$	$n = 2$
65. $x = -5, 0, 1$	$n = 3$
66. $x = -2, 6$	$n = 3$
67. $x = -5, 1, 2$	$n = 4$
68. $x = -4, -1$	$n = 4$
69. $x = 0, -\sqrt{3}, \sqrt{3}$	$n = 5$
70. $x = -1, 4, 7, 8$	$n = 5$

 Sketching the Graph of a Polynomial Function In Exercises 71–84, sketch the graph of the function by (a) applying the Leading Coefficient Test, (b) finding the real zeros of the polynomial, (c) plotting sufficient solution points, and (d) drawing a continuous curve through the points.

71. $f(t) = \frac{1}{4}(t^2 - 2t + 15)$ **72.** $g(x) = -x^2 + 10x - 16$

73. $f(x) = x^3 - 25x$ **74.** $g(x) = -9x^2 + x^4$

75. $f(x) = -8 + \frac{1}{2}x^4$ **76.** $f(x) = 8 - x^3$

77. $f(x) = 3x^3 - 15x^2 + 18x$

78. $f(x) = -4x^3 + 4x^2 + 15x$

79. $f(x) = -5x^2 - x^3$ **80.** $f(x) = -48x^2 + 3x^4$

81. $f(x) = 9x^2(x + 2)^3$ **82.** $h(x) = \frac{1}{3}x^3(x - 4)^2$

83. $g(t) = -\frac{1}{4}(t - 2)^2(t + 2)^2$

84. $g(x) = \frac{1}{10}(x + 1)^2(x - 3)^3$

Using Technology In Exercises 85–88, use a graphing utility to graph the function. Use the *zero* or *root* feature to approximate the real zeros of the function. Then determine whether the multiplicity of each zero is even or odd.

85. $f(x) = x^3 - 16x$ **86.** $f(x) = \frac{1}{4}x^4 - 2x^2$

87. $g(x) = \frac{1}{5}(x + 1)^2(x - 3)(2x - 9)$

88. $h(x) = \frac{1}{5}(x + 2)^2(3x - 5)^2$

Using the Intermediate Value Theorem
In Exercises 89–92, (a) use the Intermediate Value Theorem and the *table* feature of a graphing utility to find intervals one unit in length in which the polynomial function is guaranteed to have a zero. (b) Adjust the table to approximate the zeros of the function to the nearest thousandth.

89. $f(x) = x^3 - 3x^2 + 3$

90. $f(x) = 0.11x^3 - 2.07x^2 + 9.81x - 6.88$

91. $g(x) = 3x^4 + 4x^3 - 3$ **92.** $h(x) = x^4 - 10x^2 + 3$

93. Maximum Volume You construct an open box from a square piece of material, 36 inches on a side, by cutting equal squares with sides of length x from the corners and turning up the sides (see figure).

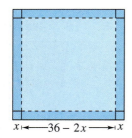
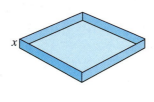

(a) Write a function V that represents the volume of the box.

(b) Determine the domain of the function V.

(c) Use a graphing utility to construct a table that shows the box heights x and the corresponding volumes $V(x)$. Use the table to estimate the dimensions that produce a maximum volume.

(d) Use the graphing utility to graph V and use the graph to estimate the value of x for which $V(x)$ is a maximum. Compare your result with that of part (c).

94. Maximum Volume You construct an open box with locking tabs from a square piece of material, 24 inches on a side, by cutting equal sections from the corners and folding along the dashed lines (see figure).

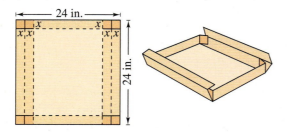

(a) Write a function V that represents the volume of the box.

(b) Determine the domain of the function V.

(c) Sketch a graph of the function and estimate the value of x for which $V(x)$ is a maximum.

95. Revenue The revenue R (in millions of dollars) for a software company from 2003 through 2016 can be modeled by

$$R = 6.212t^3 - 152.87t^2 + 990.2t - 414, \quad 3 \le t \le 16$$

where t represents the year, with $t = 3$ corresponding to 2003.

(a) Use a graphing utility to approximate any relative minima or maxima of the model over its domain.

(b) Use the graphing utility to approximate the intervals on which the revenue for the company is increasing and decreasing over its domain.

(c) Use the results of parts (a) and (b) to describe the company's revenue during this time period.

96. Revenue The revenue R (in millions of dollars) for a construction company from 2003 through 2010 can be modeled by

$$R = 0.1104t^4 - 5.152t^3 + 88.20t^2 - 654.8t + 1907, \quad 7 \le t \le 16$$

where t represents the year, with $t = 7$ corresponding to 2007.

(a) Use a graphing utility to approximate any relative minima or maxima of the model over its domain.

(b) Use the graphing utility to approximate the intervals on which the revenue for the company is increasing and decreasing over its domain.

(c) Use the results of parts (a) and (b) to describe the company's revenue during this time period.

97. Revenue The revenue R (in millions of dollars) for a beverage company is related to its advertising expense by the function

$$R = \frac{1}{100{,}000}(-x^3 + 600x^2), \quad 0 \le x \le 400$$

where x is the amount spent on advertising (in tens of thousands of dollars). Use the graph of this function to estimate the point on the graph at which the function is increasing most rapidly. This point is called the *point of diminishing returns* because any expense above this amount will yield less return per dollar invested in advertising.

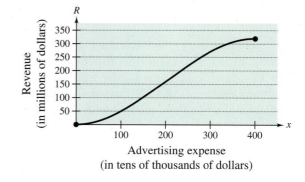

Advertising expense
(in tens of thousands of dollars)

98. Arboriculture

The growth of a red oak tree is approximated by the function

$$G = -0.003t^3 + 0.137t^2 + 0.458t - 0.839,$$
$$2 \le t \le 34$$

where G is the height of the tree (in feet) and t is its age (in years).

(a) Use a graphing utility to graph the function.

(b) Estimate the age of the tree when it is growing most rapidly. This point is called the *point of diminishing returns* because the increase in size will be less with each additional year.

(c) Using calculus, the point of diminishing returns can be found by finding the vertex of the parabola

$$y = -0.009t^2 + 0.274t + 0.458.$$

Find the vertex of this parabola.

(d) Compare your results from parts (b) and (c).

Exploration

True or False? In Exercises 99–102, determine whether the statement is true or false. Justify your answer.

99. If the graph of a polynomial function falls to the right, then its leading coefficient is negative.

100. A fifth-degree polynomial function can have five turning points in its graph.

101. It is possible for a polynomial with an even degree to have a range of $(-\infty, \infty)$.

102. If f is a polynomial function of x such that $f(2) = -6$ and $f(6) = 6$, then f has at most one real zero between $x = 2$ and $x = 6$.

103. Modeling Polynomials Sketch the graph of a fourth-degree polynomial function that has a zero of multiplicity 2 and a negative leading coefficient. Sketch the graph of another polynomial function with the same characteristics except that the leading coefficient is positive.

104. Modeling Polynomials Sketch the graph of a fifth-degree polynomial function that has a zero of multiplicity 2 and a negative leading coefficient. Sketch the graph of another polynomial function with the same characteristics except that the leading coefficient is positive.

105. Graphical Reasoning Sketch the graph of the function $f(x) = x^4$. Explain how the graph of each function g differs (if it does) from the graph of f. Determine whether g is even, odd, or neither.

(a) $g(x) = f(x) + 2$
(b) $g(x) = f(x + 2)$
(c) $g(x) = f(-x)$
(d) $g(x) = -f(x)$
(e) $g(x) = f\left(\frac{1}{2}x\right)$
(f) $g(x) = \frac{1}{2}f(x)$
(g) $g(x) = f(x^{3/4})$
(h) $g(x) = (f \circ f)(x)$

106. HOW DO YOU SEE IT? For each graph, describe a polynomial function that could represent the graph. (Indicate the degree of the function and the sign of its leading coefficient.)

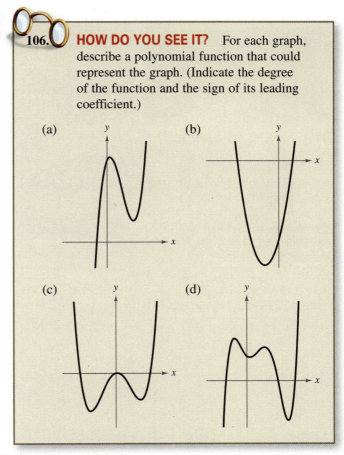

107. Think About It Use a graphing utility to graph the functions

$$y_1 = -\frac{1}{3}(x - 2)^5 + 1 \quad \text{and} \quad y_2 = \frac{3}{5}(x + 2)^5 - 3.$$

(a) Determine whether the graphs of y_1 and y_2 are increasing or decreasing. Explain.

(b) Will the graph of

$$g(x) = a(x - h)^5 + k$$

always be strictly increasing or strictly decreasing? If so, is this behavior determined by a, h, or k? Explain.

(c) Use a graphing utility to graph

$$f(x) = x^5 - 3x^2 + 2x + 1.$$

Use a graph and the result of part (b) to determine whether f can be written in the form $f(x) = a(x - h)^5 + k$. Explain.

3.3 Polynomial and Synthetic Division

■ Use long division to divide polynomials by other polynomials.
■ Use synthetic division to divide polynomials by binomials of the form $(x - k)$.
■ Use the Remainder Theorem and the Factor Theorem.

Long Division of Polynomials

Consider the graph of

$$f(x) = 6x^3 - 19x^2 + 16x - 4$$

shown at the right. Notice that one of the zeros of f is $x = 2$. This means that $(x - 2)$ is a factor of $f(x)$, and there exists a second-degree polynomial $q(x)$ such that

$$f(x) = (x - 2) \cdot q(x).$$

One way to find $q(x)$ is to use **long division,** as illustrated in Example 1.

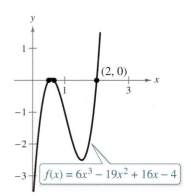

EXAMPLE 1   Long Division of Polynomials

Divide the polynomial $6x^3 - 19x^2 + 16x - 4$ by $x - 2$, and use the result to factor the polynomial completely.

Solution

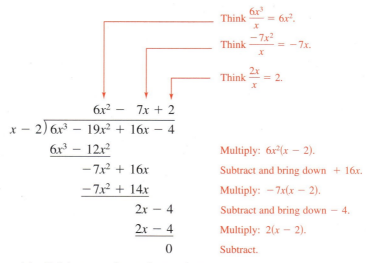

Think $\dfrac{6x^3}{x} = 6x^2$.

Think $\dfrac{-7x^2}{x} = -7x$.

Think $\dfrac{2x}{x} = 2$.

$$
\require{enclose}
\begin{array}{r}
6x^2 - 7x + 2 \\[-3pt]
x - 2 \enclose{longdiv}{6x^3 - 19x^2 + 16x - 4} \\
\end{array}
$$

$6x^3 - 12x^2$	Multiply: $6x^2(x - 2)$.
$-7x^2 + 16x$	Subtract and bring down $+ 16x$.
$-7x^2 + 14x$	Multiply: $-7x(x - 2)$.
$2x - 4$	Subtract and bring down $- 4$.
$2x - 4$	Multiply: $2(x - 2)$.
0	Subtract.

From this division, you have shown that

$$6x^3 - 19x^2 + 16x - 4 = (x - 2)(6x^2 - 7x + 2)$$

and by factoring the quadratic $6x^2 - 7x + 2$, you have

$$6x^3 - 19x^2 + 16x - 4 = (x - 2)(2x - 1)(3x - 2).$$

✓ **Checkpoint** 🔊))) *Audio-video solution in English & Spanish at LarsonPrecalculus.com*

Divide the polynomial $9x^3 + 36x^2 - 49x - 196$ by $x + 4$, and use the result to factor the polynomial completely. ■

One application of synthetic division is in evaluating polynomial functions. For example, in Exercise 82 on page 272, you will use synthetic division to evaluate a polynomial function that models the number of confirmed cases of Lyme disease in Maryland.

•• **REMARK**   Note that in Example 1, the division process requires $-7x^2 + 14x$ to be subtracted from $-7x^2 + 16x$. So, it is implied that

$$\frac{-7x^2 + 16x}{-(-7x^2 + 14x)} = \frac{-7x^2 + 16x}{7x^2 - 14x}$$

and is written as

$$
\begin{array}{r}
-7x^2 + 16x \\
-7x^2 + 14x \\
\hline
2x.
\end{array}
$$

•• **REMARK**   Note that the factorization found in Example 1 agrees with the graph of f above. The three x-intercepts occur at $(2, 0)$, $\left(\frac{1}{2}, 0\right)$, and $\left(\frac{2}{3}, 0\right)$.

In Example 1, $x - 2$ is a factor of the polynomial

$$6x^3 - 19x^2 + 16x - 4$$

and the long division process produces a remainder of zero. Often, long division will produce a nonzero remainder. For example, when you divide $x^2 + 3x + 5$ by $x + 1$, you obtain a remainder of 3.

$$
\begin{array}{r}
x + 2 \quad\longleftarrow \text{ Quotient} \\
\text{Divisor} \longrightarrow \; x + 1 \overline{)\,x^2 + 3x + 5\,} \quad\longleftarrow \text{ Dividend} \\
\underline{x^2 + \; x} \\
2x + 5 \\
\underline{2x + 2} \\
3 \quad\longleftarrow \text{ Remainder}
\end{array}
$$

In fractional form, you can write this result as

$$
\underbrace{\frac{\overbrace{x^3 + 3x + 5}^{\text{Dividend}}}{\underbrace{x + 1}_{\text{Divisor}}}} = \overbrace{x + 2}^{\text{Quotient}} + \frac{\overset{\text{Remainder}}{\downarrow}{3}}{\underbrace{x + 1}_{\text{Divisor}}}.
$$

This implies that

$$x^2 + 3x + 5 = (x + 1)(x + 2) + 3 \qquad \text{Multiply each side by } (x + 1).$$

which illustrates a theorem called the **Division Algorithm.**

The Division Algorithm

If $f(x)$ and $d(x)$ are polynomials such that $d(x) \neq 0$, and the degree of $d(x)$ is less than or equal to the degree of $f(x)$, then there exist unique polynomials $q(x)$ and $r(x)$ such that

$$f(x) = d(x)q(x) + r(x)$$

$$
\begin{array}{cccc}
\uparrow & \uparrow & & \uparrow \\
\text{Dividend} & \text{Quotient} & & \text{Remainder} \\
& \text{Divisor} & &
\end{array}
$$

where $r(x) = 0$ or the degree of $r(x)$ is less than the degree of $d(x)$. If the remainder $r(x)$ is zero, then $d(x)$ *divides evenly* into $f(x)$.

Another way to write the Division Algorithm is

$$\frac{f(x)}{d(x)} = q(x) + \frac{r(x)}{d(x)}.$$

In the Division Algorithm, the rational expression $f(x)/d(x)$ is **improper** because the degree of $f(x)$ is greater than or equal to the degree of $d(x)$. On the other hand, the rational expression $r(x)/d(x)$ is **proper** because the degree of $r(x)$ is less than the degree of $d(x)$.

If necessary, follow these steps before you apply the Division Algorithm.

1. Write the terms of the dividend and divisor in descending powers of the variable.

2. Insert placeholders with zero coefficients for missing powers of the variable.

Note how Examples 2 and 3 apply these steps.

<table>
<tr><td style="background-color:#b5321e;color:white">**EXAMPLE 2**</td><td>**Long Division of Polynomials**</td></tr>
</table>

Divide $x^3 - 1$ by $x - 1$. Check the result.

Solution   There is no x^2-term or x-term in the dividend $x^3 - 1$, so you need to rewrite the dividend as $x^3 + 0x^2 + 0x - 1$ before you apply the Division Algorithm.

$$
\begin{array}{r}
x^2 + x + 1 \\
x - 1 \overline{\smash{)}\, x^3 + 0x^2 + 0x - 1} \\
\underline{x^3 - x^2} \\
x^2 + 0x \\
\underline{x^2 - x} \\
x - 1 \\
\underline{x - 1} \\
0
\end{array}
$$

Multiply: $x^2(x - 1)$.
Subtract and bring down $0x$.
Multiply: $x(x - 1)$.
Subtract and bring down -1.
Multiply: $1(x - 1)$.
Subtract.

So, $x - 1$ divides evenly into $x^3 - 1$, and you can write

$$\frac{x^3 - 1}{x - 1} = x^2 + x + 1, \quad x \neq 1.$$

Check the result by multiplying.

$$
\begin{aligned}
(x - 1)(x^2 + x + 1) &= x^3 + x^2 + x - x^2 - x - 1 \\
&= x^3 - 1
\end{aligned}
$$

✓ *Checkpoint* Audio-video solution in English & Spanish at LarsonPrecalculus.com

Divide $x^3 - 2x^2 - 9$ by $x - 3$. Check the result.

<table>
<tr><td style="background-color:#b5321e;color:white">**EXAMPLE 3**</td><td>**Long Division of Polynomials**</td></tr>
</table>

See LarsonPrecalculus.com for an interactive version of this type of example.

Divide $-5x^2 - 2 + 3x + 2x^4 + 4x^3$ by $2x - 3 + x^2$. Check the result.

Solution   Write the terms of the dividend and divisor in descending powers of x.

$$
\begin{array}{r}
2x^2 + 1 \\
x^2 + 2x - 3 \overline{\smash{)}\, 2x^4 + 4x^3 - 5x^2 + 3x - 2} \\
\underline{2x^4 + 4x^3 - 6x^2} \\
x^2 + 3x - 2 \\
\underline{x^2 + 2x - 3} \\
x + 1
\end{array}
$$

Multiply: $2x^2(x^2 + 2x - 3)$.
Subtract and bring down $3x - 2$.
Multiply: $1(x^2 + 2x - 3)$.
Subtract.

Note that the first subtraction eliminated two terms from the dividend. When this happens, the quotient skips a term. You can write the result as

$$\frac{2x^4 + 4x^3 - 5x^2 + 3x - 2}{x^2 + 2x - 3} = 2x^2 + 1 + \frac{x + 1}{x^2 + 2x - 3}.$$

Check the result by multiplying.

$$
\begin{aligned}
(x^2 + 2x - 3)(2x^2 + 1) + x + 1 &= 2x^4 + x^2 + 4x^3 + 2x - 6x^2 - 3 + x + 1 \\
&= 2x^4 + 4x^3 - 5x^2 + 3x - 2
\end{aligned}
$$

✓ *Checkpoint* Audio-video solution in English & Spanish at LarsonPrecalculus.com

Divide $-x^3 + 9x + 6x^4 - x^2 - 3$ by $1 + 3x$. Check the result.

Synthetic Division

For long division of polynomials by divisors of the form $x - k$, there is a shortcut called **synthetic division.** The pattern for synthetic division of a cubic polynomial is summarized below. (The pattern for higher-degree polynomials is similar.)

Synthetic Division (for a Cubic Polynomial)

To divide $ax^3 + bx^2 + cx + d$ by $x - k$, use this pattern.

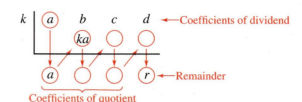

Vertical pattern: Add terms in columns.
Diagonal pattern: Multiply results by k.

This algorithm for synthetic division works only for divisors of the form $x - k$. Remember that $x + k = x - (-k)$.

EXAMPLE 4 **Using Synthetic Division**

Use synthetic division to divide

$$x^4 - 10x^2 - 2x + 4 \quad \text{by} \quad x + 3.$$

Solution Begin by setting up an array. Include a zero for the missing x^3-term in the dividend.

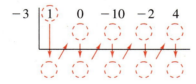

Then, use the synthetic division pattern by adding terms in columns and multiplying the results by -3.

Divisor: $x + 3$ Dividend: $x^4 - 10x^2 - 2x + 4$

$$
\begin{array}{r|rrrrr}
-3 & 1 & 0 & -10 & -2 & 4 \\
   &   & -3 & 9 & 3 & -3 \\
\hline
   & 1 & -3 & -1 & 1 & \boxed{1} \leftarrow \text{Remainder: 1}
\end{array}
$$

Quotient: $x^3 - 3x^2 - x + 1$

So, you have

$$\frac{x^4 - 10x^2 - 2x + 4}{x + 3} = x^3 - 3x^2 - x + 1 + \frac{1}{x + 3}.$$

✓ *Checkpoint* ◀))) Audio-video solution in English & Spanish at LarsonPrecalculus.com

Use synthetic division to divide $5x^3 + 8x^2 - x + 6$ by $x + 2$.

The Remainder and Factor Theorems

The remainder obtained in the synthetic division process has an important interpretation, as described in the **Remainder Theorem.**

The Remainder Theorem

If a polynomial $f(x)$ is divided by $x - k$, then the remainder is

$$r = f(k).$$

For a proof of the Remainder Theorem, see Proofs in Mathematics on page 305.

The Remainder Theorem tells you that synthetic division can be used to evaluate a polynomial function. That is, to evaluate a polynomial $f(x)$ when $x = k$, divide $f(x)$ by $x - k$. The remainder will be $f(k)$, as illustrated in Example 5.

EXAMPLE 5 **Using the Remainder Theorem**

Use the Remainder Theorem to evaluate

$$f(x) = 3x^3 + 8x^2 + 5x - 7$$

when $x = -2$. Check your answer.

Solution Using synthetic division gives the result below.

$$
\begin{array}{r|rrrr}
-2 & 3 & 8 & 5 & -7 \\
   &   & -6 & -4 & -2 \\
\hline
   & 3 & 2 & 1 & -9
\end{array}
$$

The remainder is $r = -9$, so

$$f(-2) = -9. \qquad r = f(k)$$

This means that $(-2, -9)$ is a point on the graph of f. Check this by substituting $x = -2$ in the original function.

Check

$$
\begin{aligned}
f(-2) &= 3(-2)^3 + 8(-2)^2 + 5(-2) - 7 \\
&= 3(-8) + 8(4) - 10 - 7 \\
&= -24 + 32 - 10 - 7 \\
&= -9
\end{aligned}
$$

✓ **Checkpoint** 🔊))) *Audio-video solution in English & Spanish at LarsonPrecalculus.com*

Use the Remainder Theorem to find each function value given

$$f(x) = 4x^3 + 10x^2 - 3x - 8.$$

Check your answer.

a. $f(-1)$ **b.** $f(4)$

c. $f\left(\frac{1}{2}\right)$ **d.** $f(-3)$

▷ **TECHNOLOGY** One way to evaluate a function with your graphing utility is to enter the function in the equation editor and use the *table* feature in *ask* mode. When you enter values in the X column of a table in *ask* mode, the corresponding function values are displayed in the function column.

Another important theorem is the **Factor Theorem,** stated below.

> ### The Factor Theorem
>
> A polynomial $f(x)$ has a factor $(x - k)$ if and only if $f(k) = 0$.

For a proof of the Factor Theorem, see Proofs in Mathematics on page 305.

Using the Factor Theorem, you can test whether a polynomial has $(x - k)$ as a factor by evaluating the polynomial at $x = k$. If the result is 0, then $(x - k)$ is a factor.

EXAMPLE 6   **Factoring a Polynomial: Repeated Division**

Show that $(x - 2)$ and $(x + 3)$ are factors of

$$f(x) = 2x^4 + 7x^3 - 4x^2 - 27x - 18.$$

Then find the remaining factors of $f(x)$.

Algebraic Solution

Using synthetic division with the factor $(x - 2)$ gives the result below.

$$
\begin{array}{r|rrrrr}
2 & 2 & 7 & -4 & -27 & -18 \\
  &   & 4 & 22 & 36 & 18 \\
\hline
  & 2 & 11 & 18 & 9 & 0
\end{array}
$$

⟶ 0 remainder, so $f(2) = 0$ and $(x - 2)$ is a factor.

Take the result of this division and perform synthetic division again using the factor $(x + 3)$.

$$
\begin{array}{r|rrrr}
-3 & 2 & 11 & 18 & 9 \\
   &   & -6 & -15 & -9 \\
\hline
   & 2 & 5 & 3 & 0
\end{array}
$$

⟶ 0 remainder, so $f(-3) = 0$ and $(x + 3)$ is a factor.

$\underbrace{}_{2x^2 + 5x + 3}$

The resulting quadratic expression factors as

$$2x^2 + 5x + 3 = (2x + 3)(x + 1)$$

so the complete factorization of $f(x)$ is

$$f(x) = (x - 2)(x + 3)(2x + 3)(x + 1).$$

Graphical Solution

The graph of $f(x) = 2x^4 + 7x^3 - 4x^2 - 27x - 18$ has four x-intercepts (see figure). These occur at $x = -3$, $x = -\frac{3}{2}$, $x = -1$, and $x = 2$. (Check this algebraically.) This implies that $(x + 3)$, $\left(x + \frac{3}{2}\right)$, $(x + 1)$, and $(x - 2)$ are factors of $f(x)$. [Note that $\left(x + \frac{3}{2}\right)$ and $(2x + 3)$ are equivalent factors because they both yield the same zero, $x = -\frac{3}{2}$.]

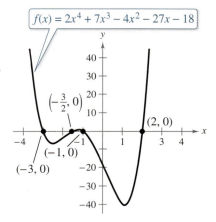

$f(x) = 2x^4 + 7x^3 - 4x^2 - 27x - 18$

✓ *Checkpoint* �)) *Audio-video solution in English & Spanish at LarsonPrecalculus.com*

Show that $(x + 3)$ is a factor of $f(x) = x^3 - 19x - 30$. Then find the remaining factors of $f(x)$.

> ### Summarize   (Section 3.3)
>
> 1. Explain how to use long division to divide two polynomials *(pages 264 and 265)*. For examples of long division of polynomials, see Examples 1–3.
>
> 2. Describe the algorithm for synthetic division *(page 267)*. For an example of synthetic division, see Example 4.
>
> 3. State the Remainder Theorem and the Factor Theorem *(pages 268 and 269)*. For an example of using the Remainder Theorem, see Example 5. For an example of using the Factor Theorem, see Example 6.

3.3 Exercises

See **CalcChat.com** for tutorial help and worked-out solutions to odd-numbered exercises.

Vocabulary

1. Two forms of the Division Algorithm are shown below. Identify and label each term or function.

$$f(x) = d(x)q(x) + r(x) \qquad \frac{f(x)}{d(x)} = q(x) + \frac{r(x)}{d(x)}$$

In Exercises 2–6, fill in the blanks.

2. In the Division Algorithm, the rational expression $r(x)/d(x)$ is _____ because the degree of $r(x)$ is less than the degree of $d(x)$.

3. In the Division Algorithm, the rational expression $f(x)/d(x)$ is _____ because the degree of $f(x)$ is greater than or equal to the degree of $d(x)$.

4. A shortcut for long division of polynomials is _____ _____, in which the divisor must be of the form $x - k$.

5. The _____ Theorem states that a polynomial $f(x)$ has a factor $(x - k)$ if and only if $f(k) = 0$.

6. The _____ Theorem states that if a polynomial $f(x)$ is divided by $x - k$, then the remainder is $r = f(k)$.

Skills and Applications

Using the Division Algorithm In Exercises 7 and 8, use long division to verify that $y_1 = y_2$.

7. $y_1 = \dfrac{x^2}{x + 2}, \quad y_2 = x - 2 + \dfrac{4}{x + 2}$

8. $y_1 = \dfrac{x^3 - 3x^2 + 4x - 1}{x + 3}, \quad y_2 = x^2 - 6x + 22 - \dfrac{67}{x + 3}$

Using Technology In Exercises 9 and 10, (a) use a graphing utility to graph the two equations in the same viewing window, (b) use the graphs to verify that the expressions are equivalent, and (c) use long division to verify the results algebraically.

9. $y_1 = \dfrac{x^2 + 2x - 1}{x + 3}, \quad y_2 = x - 1 + \dfrac{2}{x + 3}$

10. $y_1 = \dfrac{x^4 + x^2 - 1}{x^2 + 1}, \quad y_2 = x^2 - \dfrac{1}{x^2 + 1}$

Long Division of Polynomials In Exercises 11–24, use long division to divide.

11. $(2x^2 + 10x + 12) \div (x + 3)$

12. $(5x^2 - 17x - 12) \div (x - 4)$

13. $(4x^3 - 7x^2 - 11x + 5) \div (4x + 5)$

14. $(6x^3 - 16x^2 + 17x - 6) \div (3x - 2)$

15. $(x^4 + 5x^3 + 6x^2 - x - 2) \div (x + 2)$

16. $(x^3 + 4x^2 - 3x - 12) \div (x - 3)$

17. $(6x + 5) \div (x + 1)$ **18.** $(9x - 4) \div (3x + 2)$

19. $(x^3 - 9) \div (x^2 + 1)$ **20.** $(x^5 + 7) \div (x^4 - 1)$

21. $(3x + 2x^3 - 9 - 8x^2) \div (x^2 + 1)$

22. $(5x^3 - 16 - 20x + x^4) \div (x^2 - x - 3)$

23. $\dfrac{x^4}{(x - 1)^3}$ **24.** $\dfrac{2x^3 - 4x^2 - 15x + 5}{(x - 1)^2}$

 Using Synthetic Division In Exercises 25–44, use synthetic division to divide.

25. $(2x^3 - 10x^2 + 14x - 24) \div (x - 4)$

26. $(5x^3 + 18x^2 + 7x - 6) \div (x + 3)$

27. $(6x^3 + 7x^2 - x + 26) \div (x - 3)$

28. $(2x^3 + 12x^2 + 14x - 3) \div (x + 4)$

29. $(4x^3 - 9x + 8x^2 - 18) \div (x + 2)$

30. $(9x^3 - 16x - 18x^2 + 32) \div (x - 2)$

31. $(-x^3 + 75x - 250) \div (x + 10)$

32. $(3x^3 - 16x^2 - 72) \div (x - 6)$

33. $(x^3 - 3x^2 + 5) \div (x - 4)$

34. $(5x^3 + 6x + 8) \div (x + 2)$

35. $\dfrac{10x^4 - 50x^3 - 800}{x - 6}$ **36.** $\dfrac{x^5 - 13x^4 - 120x + 80}{x + 3}$

37. $\dfrac{x^3 + 512}{x + 8}$ **38.** $\dfrac{x^3 - 729}{x - 9}$

39. $\dfrac{-3x^4}{x - 2}$ **40.** $\dfrac{-2x^5}{x + 2}$

41. $\dfrac{180x - x^4}{x - 6}$ **42.** $\dfrac{5 - 3x + 2x^2 - x^3}{x + 1}$

43. $\dfrac{4x^3 + 16x^2 - 23x - 15}{x + \frac{1}{2}}$ **44.** $\dfrac{3x^3 - 4x^2 + 5}{x - \frac{3}{2}}$

Using the Remainder Theorem In Exercises 45–50, write the function in the form $f(x) = (x - k)q(x) + r$ for the given value of k, and demonstrate that $f(k) = r$.

45. $f(x) = x^3 - x^2 - 10x + 7, \quad k = 3$

46. $f(x) = x^3 - 4x^2 - 10x + 8, \quad k = -2$

47. $f(x) = 15x^4 + 10x^3 - 6x^2 + 14, \quad k = -\frac{2}{3}$

48. $f(x) = 10x^3 - 22x^2 - 3x + 4, \quad k = \frac{1}{5}$

49. $f(x) = -4x^3 + 6x^2 + 12x + 4, \quad k = 1 - \sqrt{3}$

50. $f(x) = -3x^3 + 8x^2 + 10x - 8, \quad k = 2 + \sqrt{2}$

 Using the Remainder Theorem In Exercises 51–54, use the Remainder Theorem and synthetic division to find each function value. Verify your answers using another method.

51. $f(x) = 2x^3 - 7x + 3$

   (a) $f(1)$ (b) $f(-2)$ (c) $f(3)$ (d) $f(2)$

52. $g(x) = 2x^6 + 3x^4 - x^2 + 3$

   (a) $g(2)$ (b) $g(1)$ (c) $g(3)$ (d) $g(-1)$

53. $h(x) = x^3 - 5x^2 - 7x + 4$

   (a) $h(3)$ (b) $h(\frac{1}{2})$ (c) $h(-2)$ (d) $h(-5)$

54. $f(x) = 4x^4 - 16x^3 + 7x^2 + 20$

   (a) $f(1)$ (b) $f(-2)$ (c) $f(5)$ (d) $f(-10)$

Using the Factor Theorem In Exercises 55–62, use synthetic division to show that x is a solution of the third-degree polynomial equation, and use the result to factor the polynomial completely. List all real solutions of the equation.

55. $x^3 + 6x^2 + 11x + 6 = 0, \quad x = -3$

56. $x^3 - 52x - 96 = 0, \quad x = -6$

57. $2x^3 - 15x^2 + 27x - 10 = 0, \quad x = \frac{1}{2}$

58. $48x^3 - 80x^2 + 41x - 6 = 0, \quad x = \frac{2}{3}$

59. $x^3 + 2x^2 - 3x - 6 = 0, \quad x = \sqrt{3}$

60. $x^3 + 2x^2 - 2x - 4 = 0, \quad x = \sqrt{2}$

61. $x^3 - 3x^2 + 2 = 0, \quad x = 1 + \sqrt{3}$

62. $x^3 - x^2 - 13x - 3 = 0, \quad x = 2 - \sqrt{5}$

 Factoring a Polynomial In Exercises 63–70, (a) verify the given factors of $f(x)$, (b) find the remaining factor(s) of $f(x)$, (c) use your results to write the complete factorization of $f(x)$, (d) list all real zeros of f, and (e) confirm your results by using a graphing utility to graph the function.

Function	Factors
63. $f(x) = 2x^3 + x^2 - 5x + 2$	$(x + 2), (x - 1)$
64. $f(x) = 3x^3 - x^2 - 8x - 4$	$(x + 1), (x - 2)$

Function	Factors
65. $f(x) = x^4 - 8x^3 + 9x^2$ $+ 38x - 40$	$(x - 5), (x + 2)$
66. $f(x) = 8x^4 - 14x^3 - 71x^2$ $- 10x + 24$	$(x + 2), (x - 4)$
67. $f(x) = 6x^3 + 41x^2 - 9x - 14$	$(2x + 1), (3x - 2)$
68. $f(x) = 10x^3 - 11x^2 - 72x + 45$	$(2x + 5), (5x - 3)$
69. $f(x) = 2x^3 - x^2 - 10x + 5$	$(2x - 1), (x + \sqrt{5})$
70. $f(x) = x^3 + 3x^2 - 48x - 144$	$(x + 4\sqrt{3}), (x + 3)$

Approximating Zeros In Exercises 71–76, (a) use the *zero* or *root* feature of a graphing utility to approximate the zeros of the function accurate to three decimal places, (b) determine the exact value of one of the zeros, and (c) use synthetic division to verify your result from part (b), and then factor the polynomial completely.

71. $f(x) = x^3 - 2x^2 - 5x + 10$

72. $g(x) = x^3 + 3x^2 - 2x - 6$

73. $h(t) = t^3 - 2t^2 - 7t + 2$

74. $f(s) = s^3 - 12s^2 + 40s - 24$

75. $h(x) = x^5 - 7x^4 + 10x^3 + 14x^2 - 24x$

76. $g(x) = 6x^4 - 11x^3 - 51x^2 + 99x - 27$

Simplifying Rational Expressions In Exercises 77–80, simplify the rational expression by using long division or synthetic division.

77. $\dfrac{x^3 + x^2 - 64x - 64}{x + 8}$

78. $\dfrac{4x^3 - 8x^2 + x + 3}{2x - 3}$

79. $\dfrac{x^4 + 6x^3 + 11x^2 + 6x}{x^2 + 3x + 2}$

80. $\dfrac{x^4 + 9x^3 - 5x^2 - 36x + 4}{x^2 - 4}$

81. Profit A company that produces calculators estimates that the profit P (in dollars) from selling a specific model of calculator is given by

$$P = -152x^3 + 7545x^2 - 169{,}625, \quad 0 \le x \le 45$$

where x is the advertising expense (in tens of thousands of dollars). For this model of calculator, an advertising expense of \$400,000 ($x = 40$) results in a profit of \$2,174,375.

(a) Use a graphing utility to graph the profit function.

(b) Use the graph from part (a) to estimate another amount the company can spend on advertising that results in the same profit.

(c) Use synthetic division to confirm the result of part (b) algebraically.

• • 82. Lyme Disease • • • • • • • • • • • • •

The numbers N of confirmed cases
of Lyme disease in
Maryland from 2007
through 2014 are
shown in the table,
where t represents
the year, with $t = 7$
corresponding to
2007. *(Source:
Centers for Disease Control and Prevention)*

Year, t	Number, N
7	2576
8	1746
9	1466
10	1163
11	938
12	1113
13	801
14	957

Spreadsheet at
LarsonPrecalculus.com

(a) Use a graphing utility to create a scatter plot of
the data.

(b) Use the *regression* feature of the graphing utility
to find a *quartic* model for the data. (A quartic
model has the form $at^4 + bt^3 + ct^2 + dt + e$,
where a, b, c, d, and e are constant and t is
variable.) Graph the model in the same viewing
window as the scatter plot.

(c) Use the model to create a table of estimated values
of N. Compare the model with the original data.

(d) Use synthetic division to confirm algebraically
your estimated value for the year 2014.

Exploration

True or False?   In Exercises 83–86, determine whether
the statement is true or false. Justify your answer.

83. If $(7x + 4)$ is a factor of some polynomial function
$f(x)$, then $\frac{4}{7}$ is a zero of f.

84. $(2x - 1)$ is a factor of the polynomial

$6x^6 + x^5 - 92x^4 + 45x^3 + 184x^2 + 4x - 48.$

85. The rational expression $\dfrac{x^3 + 2x^2 - 7x + 4}{x^2 - 4x - 12}$ is improper.

86. The equation

$$\frac{x^3 - 3x^2 + 4}{x + 1} = x^2 - 4x + 4$$

is true for all values of x.

Think About It   In Exercises 87 and 88, perform the
division. Assume that n is a positive integer.

87. $\dfrac{x^{3n} + 9x^{2n} + 27x^n + 27}{x^n + 3}$

88. $\dfrac{x^{3n} - 3x^{2n} + 5x^n - 6}{x^n - 2}$

89. Error Analysis   Describe the error.

Use synthetic division to find the remainder when
$x^2 + 3x - 5$ is divided by $x + 1$.

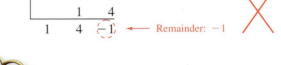

$$1 \ \begin{array}{|rrr} 1 & 3 & -5 \\ & 1 & 4 \\ \hline 1 & 4 & -1 \end{array}$$

← Remainder: -1

90. HOW DO YOU SEE IT?   The graph
below shows a company's estimated profits
for different advertising expenses. The
company's actual profit was \$936,660 for
an advertising expense of \$300,000.

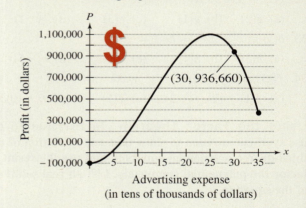

P

(30, 936,660)

Profit (in dollars)

Advertising expense
(in tens of thousands of dollars)

(a) From the graph, it appears that the company
could have obtained the same profit for a lesser
advertising expense. Use the graph to estimate
this expense.

(b) The company's model is

$P = -140.75x^3 + 5348.3x^2 - 76{,}560,$

$0 \le x \le 35$

where P is the profit (in dollars) and x is the
advertising expense (in tens of thousands of
dollars). Explain how you could verify the
lesser expense from part (a) algebraically.

Exploration   In Exercises 91 and 92, find the constant
c such that the denominator will divide evenly into the
numerator.

91. $\dfrac{x^3 + 4x^2 - 3x + c}{x - 5}$

92. $\dfrac{x^5 - 2x^2 + x + c}{x + 2}$

93. Think About It   Find the value of k such that $x - 4$
is a factor of $x^3 - kx^2 + 2kx - 8.$

3.4 Zeros of Polynomial Functions

Finding zeros of polynomial functions is an important part of solving many real-life problems. For example, in Exercise 105 on page 285, you will use the zeros of a polynomial function to redesign a storage bin so that it can hold five times as much food.

- Use the Fundamental Theorem of Algebra to determine numbers of zeros of polynomial functions.
- Find rational zeros of polynomial functions.
- Find complex zeros using conjugate pairs.
- Find zeros of polynomials by factoring.
- Use Descartes's Rule of Signs and the Upper and Lower Bound Rules to find zeros of polynomials.
- Find zeros of polynomials in real-life applications.

The Fundamental Theorem of Algebra

In the complex number system, every nth-degree polynomial function has *precisely n* zeros. This important result is derived from the **Fundamental Theorem of Algebra,** first proved by German mathematician Carl Friedrich Gauss (1777–1855).

The Fundamental Theorem of Algebra

If $f(x)$ is a polynomial of degree n, where $n > 0$, then f has at least one zero in the complex number system.

Using the Fundamental Theorem of Algebra and the equivalence of zeros and factors, you obtain the **Linear Factorization Theorem.**

Linear Factorization Theorem

If $f(x)$ is a polynomial of degree n, where $n > 0$, then $f(x)$ has precisely n linear factors

$$f(x) = a_n(x - c_1)(x - c_2) \cdots (x - c_n)$$

where $c_1, c_2, \ldots, c_n$ are complex numbers.

REMARK Recall that in order to find the zeros of a function f, set $f(x)$ equal to 0 and solve the resulting equation for x. For instance, the function in Example 1(a) has a zero at $x = 2$ because

$$x - 2 = 0$$
$$x = 2.$$

For a proof of the Linear Factorization Theorem, see Proofs in Mathematics on page 306.

Note that the Fundamental Theorem of Algebra and the Linear Factorization Theorem tell you only that the zeros or factors of a polynomial exist, not how to find them. Such theorems are called **existence theorems.**

EXAMPLE 1 Zeros of Polynomial Functions

See LarsonPrecalculus.com for an interactive version of this type of example.

a. The first-degree polynomial function $f(x) = x - 2$ has exactly *one* zero: $x = 2$.

b. The second-degree polynomial function $f(x) = x^2 - 6x + 9 = (x - 3)(x - 3)$ has exactly *two* zeros: $x = 3$ and $x = 3$ (a *repeated zero*).

c. The third-degree polynomial function $f(x) = x^3 + 4x = x(x - 2i)(x + 2i)$ has exactly *three* zeros: $x = 0$, $x = 2i$, and $x = -2i$.

✓ **Checkpoint** 🔊))) *Audio-video solution in English & Spanish at LarsonPrecalculus.com*

Determine the number of zeros of the polynomial function $f(x) = x^4 - 1$. ∎

The Rational Zero Test

The **Rational Zero Test** relates the possible rational zeros of a polynomial (having integer coefficients) to the leading coefficient and to the constant term of the polynomial.

The Rational Zero Test

If the polynomial

$$f(x) = a_n x^n + a_{n-1} x^{n-1} + \cdots + a_2 x^2 + a_1 x + a_0$$

has *integer* coefficients, then every rational zero of f has the form

$$\text{Rational zero} = \frac{p}{q}$$

where p and q have no common factors other than 1, and

$p =$ a factor of the constant term a_0

$q =$ a factor of the leading coefficient a_n.

Although they were not contemporaries, French mathematician Jean Le Rond d'Alembert (1717–1783) worked independently of Carl Friedrich Gauss in trying to prove the Fundamental Theorem of Algebra. His efforts were such that, in France, the Fundamental Theorem of Algebra is frequently known as d'Alembert's Theorem.

To use the Rational Zero Test, you should first list all rational numbers whose numerators are factors of the constant term and whose denominators are factors of the leading coefficient.

$$\text{Possible rational zeros: } \frac{\text{Factors of constant term}}{\text{Factors of leading coefficient}}$$

Having formed this list of *possible rational zeros,* use a trial-and-error method to determine which, if any, are actual zeros of the polynomial. Note that when the leading coefficient is 1, the possible rational zeros are simply the factors of the constant term.

EXAMPLE 2   **Rational Zero Test with Leading Coefficient of 1**

Find (if possible) the rational zeros of

$$f(x) = x^3 + x + 1.$$

Solution   The leading coefficient is 1, so the possible rational zeros are the factors of the constant term.

Possible rational zeros: 1 and -1

Testing these possible zeros shows that neither works.

$$f(1) = (1)^3 + 1 + 1$$
$$= 3$$
$$f(-1) = (-1)^3 + (-1) + 1$$
$$= -1$$

So, the given polynomial has *no* rational zeros. Note from the graph of f in Figure 3.15 that f does have one real zero between -1 and 0. However, by the Rational Zero Test, you know that this real zero is *not* a rational number.

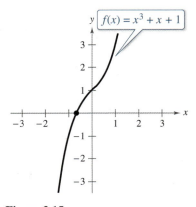

Figure 3.15

✔ *Checkpoint* 🔊))) *Audio-video solution in English & Spanish at LarsonPrecalculus.com*

Find (if possible) the rational zeros of

$$f(x) = x^3 + 2x^2 + 6x - 4.$$

EXAMPLE 3   **Rational Zero Test with Leading Coefficient of 1**

Find the rational zeros of

$$f(x) = x^4 - x^3 + x^2 - 3x - 6.$$

Solution   The leading coefficient is 1, so the possible rational zeros are the factors of the constant term.

Possible rational zeros: $\pm 1, \pm 2, \pm 3, \pm 6$

By applying synthetic division successively, you find that $x = -1$ and $x = 2$ are the only two rational zeros.

$$
\begin{array}{r|rrrrr}
-1 & 1 & -1 & 1 & -3 & -6 \\
   &   & -1 & 2 & -3 & 6 \\
\hline
   & 1 & -2 & 3 & -6 & 0
\end{array}
\longrightarrow \quad \text{0 remainder, so } x = -1 \text{ is a zero.}
$$

$$
\begin{array}{r|rrrr}
2 & 1 & -2 & 3 & -6 \\
  &   & 2 & 0 & 6 \\
\hline
  & 1 & 0 & 3 & 0
\end{array}
\longrightarrow \quad \text{0 remainder, so } x = 2 \text{ is a zero.}
$$

So, $f(x)$ factors as

$$f(x) = (x + 1)(x - 2)(x^2 + 3).$$

The factor $(x^2 + 3)$ produces no real zeros, so $x = -1$ and $x = 2$ are the only *real* zeros of f. The figure below verifies this.

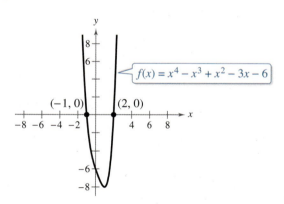

$$f(x) = x^4 - x^3 + x^2 - 3x - 6$$

  Checkpoint Audio-video solution in English & Spanish at LarsonPrecalculus.com

Find the rational zeros of

$$f(x) = x^3 - 15x^2 + 75x - 125.$$

When the leading coefficient of a polynomial is not 1, the number of possible rational zeros can increase dramatically. In such cases, the search can be shortened in several ways.

1. A graphing utility can help to speed up the calculations.

2. A graph can give good estimates of the locations of the zeros.

3. The Intermediate Value Theorem, along with a table of values, can give approximations of the zeros.

4. Synthetic division can be used to test the possible rational zeros.

After finding the first zero, the search becomes simpler by working with the lower-degree polynomial obtained in synthetic division, as shown in Example 3.

REMARK   When there are few possible rational zeros, as in Example 2, it may be quicker to test the zeros by evaluating the function. When there are more possible rational zeros, as in Example 3, it may be quicker to use a different approach to test the zeros, such as using synthetic division or sketching a graph.

Find the rational zeros of $f(x) = 2x^3 + 3x^2 - 8x + 3$.

Solution The leading coefficient is 2 and the constant term is 3.

$$\textit{Possible rational zeros:} \quad \frac{\text{Factors of } 3}{\text{Factors of } 2} = \frac{\pm 1, \pm 3}{\pm 1, \pm 2} = \pm 1, \pm 3, \pm \frac{1}{2}, \pm \frac{3}{2}$$

By synthetic division, $x = 1$ is a rational zero.

$$
\begin{array}{r|rrrr}
1 & 2 & 3 & -8 & 3 \\
  &   & 2 & 5 & -3 \\
\hline
  & 2 & 5 & -3 & 0 \\
\end{array}
$$

So, $f(x)$ factors as

$$
\begin{aligned}
f(x) &= (x - 1)(2x^2 + 5x - 3) \\
     &= (x - 1)(2x - 1)(x + 3)
\end{aligned}
$$

which shows that the rational zeros of f are $x = 1$, $x = \frac{1}{2}$, and $x = -3$.

✓ **Checkpoint** ◀))) Audio-video solution in English & Spanish at LarsonPrecalculus.com

Find the rational zeros of

$$f(x) = 2x^3 + x^2 - 13x + 6.$$

Recall from Section 3.2 that if $x = a$ is a zero of the polynomial function f, then $x = a$ is a solution of the polynomial equation $f(x) = 0$.

Find all real solutions of $-10x^3 + 15x^2 + 16x - 12 = 0$.

Solution The leading coefficient is -10 and the constant term is -12.

$$\textit{Possible rational solutions:} \quad \frac{\text{Factors of } -12}{\text{Factors of } -10} = \frac{\pm 1, \pm 2, \pm 3, \pm 4, \pm 6, \pm 12}{\pm 1, \pm 2, \pm 5, \pm 10}$$

With so many possibilities (32, in fact), it is worth your time to sketch a graph. In Figure 3.16, three reasonable solutions appear to be $x = -\frac{6}{5}$, $x = \frac{1}{2}$, and $x = 2$. Testing these by synthetic division shows that $x = 2$ is the only rational solution. So, you have

$$(x - 2)(-10x^2 - 5x + 6) = 0.$$

Using the Quadratic Formula to solve $-10x^2 - 5x + 6 = 0$, you find that the two additional solutions are irrational numbers.

$$x = \frac{5 + \sqrt{265}}{-20} \approx -1.0639$$

and

$$x = \frac{5 - \sqrt{265}}{-20} \approx 0.5639$$

✓ **Checkpoint** ◀))) Audio-video solution in English & Spanish at LarsonPrecalculus.com

Find all real solutions of

$$-2x^3 - 5x^2 + 15x + 18 = 0.$$

• • • • • • • • • • • • • • • ▷

•• **REMARK** Remember that when you find the rational zeros of a polynomial function with many possible rational zeros, as in Example 4, you must use trial and error. There is no quick algebraic method to determine which of the possibilities is an actual zero; however, sketching a graph may be helpful.

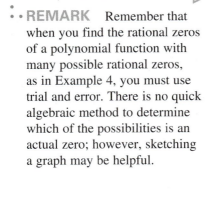

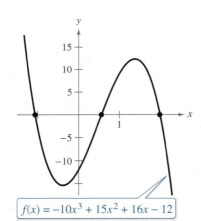

$f(x) = -10x^3 + 15x^2 + 16x - 12$

Figure 3.16

▷ **ALGEBRA HELP** To review the Quadratic Formula, see Section 1.4.

Conjugate Pairs

In Example 1(c), note that the two complex zeros $2i$ and $-2i$ are complex conjugates. That is, they are of the forms $a + bi$ and $a - bi$.

> **Complex Zeros Occur in Conjugate Pairs**
>
> Let f be a polynomial function that has *real coefficients*. If $a + bi$, where $b \neq 0$, is a zero of the function, then the complex conjugate $a - bi$ is also a zero of the function.

Be sure you see that this result is true only when the polynomial function has *real coefficients*. For example, the result applies to the function $f(x) = x^2 + 1$, but not to the function $g(x) = x - i$.

EXAMPLE 6 **Finding a Polynomial Function with Given Zeros**

Find a fourth-degree polynomial function f with real coefficients that has -1, -1, and $3i$ as zeros.

Solution   You are given that $3i$ is a zero of f *and* the polynomial has real coefficients, so you know that the complex conjugate $-3i$ must also be a zero. Using the Linear Factorization Theorem, write $f(x)$ as

$$f(x) = a(x + 1)(x + 1)(x - 3i)(x + 3i).$$

For simplicity, let $a = 1$ to obtain

$$f(x) = (x^2 + 2x + 1)(x^2 + 9) = x^4 + 2x^3 + 10x^2 + 18x + 9.$$

✓ *Checkpoint*  *Audio-video solution in English & Spanish at LarsonPrecalculus.com*

Find a fourth-degree polynomial function f with real coefficients that has 2, -2, and $-7i$ as zeros.

EXAMPLE 7 **Finding a Polynomial Function with Given Zeros**

Find the cubic polynomial function f with real coefficients that has 2 and $1 - i$ as zeros, and $f(1) = 3$.

Solution   You are given that $1 - i$ is a zero of f, so the complex conjugate $1 + i$ is also a zero.

$$\begin{aligned}
f(x) &= a(x - 2)[x - (1 - i)][x - (1 + i)] \\
&= a(x - 2)[(x - 1) + i][(x - 1) - i] \\
&= a(x - 2)[(x - 1)^2 + 1] \\
&= a(x - 2)(x^2 - 2x + 2) \\
&= a(x^3 - 4x^2 + 6x - 4)
\end{aligned}$$

To find the value of a, use the fact that $f(1) = 3$ to obtain

$$a[(1)^3 - 4(1)^2 + 6(1) - 4] = 3.$$

So, $a = -3$ and

$$f(x) = -3(x^3 - 4x^2 + 6x - 4) = -3x^3 + 12x^2 - 18x + 12.$$

✓ *Checkpoint*  *Audio-video solution in English & Spanish at LarsonPrecalculus.com*

Find the *quartic* (fourth-degree) polynomial function f with real coefficients that has 1, -2, and $2i$ as zeros, and $f(-1) = 10$.

Factoring a Polynomial

The Linear Factorization Theorem states that you can write any nth-degree polynomial as the product of n linear factors.

$$f(x) = a_n(x - c_1)(x - c_2)(x - c_3) \cdots (x - c_n)$$

This result includes the possibility that some of the values of c_i are imaginary. The theorem below states that you can write $f(x)$ as the product of linear and quadratic factors with real coefficients. For a proof of this theorem, see Proofs in Mathematics on page 306.

> ### Factors of a Polynomial
>
> Every polynomial of degree $n > 0$ with real coefficients can be written as the product of linear and quadratic factors with real coefficients, where the quadratic factors have no real zeros.

A quadratic factor with no real zeros is *prime* or **irreducible over the reals.** Note that this is not the same as being *irreducible over the rationals*. For example, the quadratic $x^2 + 1 = (x - i)(x + i)$ is irreducible over the reals (and therefore over the rationals). On the other hand, the quadratic $x^2 - 2 = \left(x - \sqrt{2}\right)\left(x + \sqrt{2}\right)$ is irreducible over the rationals but *reducible* over the reals.

▷ **TECHNOLOGY** Another way to find the real zeros of the function in Example 8 is to use a graphing utility to graph the function (see figure).

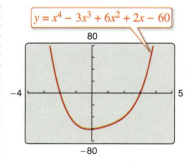

$y = x^4 - 3x^3 + 6x^2 + 2x - 60$

Then use the *zero* or *root* feature of the graphing utility to determine that $x = -2$ and $x = 3$ are the real zeros.

EXAMPLE 8 **Finding the Zeros of a Polynomial Function**

Find all the zeros of $f(x) = x^4 - 3x^3 + 6x^2 + 2x - 60$ given that $1 + 3i$ is a zero of f.

Solution Complex zeros occur in conjugate pairs, so you know that $1 - 3i$ is also a zero of f. This means that both $[x - (1 + 3i)]$ and $[x - (1 - 3i)]$ are factors of $f(x)$. Multiplying these two factors produces

$$[x - (1 + 3i)][x - (1 - 3i)] = [(x - 1) - 3i][(x - 1) + 3i]$$
$$= (x - 1)^2 - 9i^2$$
$$= x^2 - 2x + 10.$$

Using long division, divide $x^2 - 2x + 10$ into $f(x)$.

$$
\begin{array}{r}
x^2 - x - 6 \\
x^2 - 2x + 10 \overline{)\,x^4 - 3x^3 + 6x^2 + 2x - 60} \\
\underline{x^4 - 2x^3 + 10x^2} \\
-x^3 - 4x^2 + 2x \\
\underline{-x^3 + 2x^2 - 10x} \\
-6x^2 + 12x - 60 \\
\underline{-6x^2 + 12x - 60} \\
0
\end{array}
$$

So, you have

$$f(x) = (x^2 - 2x + 10)(x^2 - x - 6) = (x^2 - 2x + 10)(x - 3)(x + 2)$$

and can conclude that the zeros of f are $x = 1 + 3i$, $x = 1 - 3i$, $x = 3$, and $x = -2$.

▷ **ALGEBRA HELP** To review the techniques for polynomial long division, see Section 3.3.

✓ *Checkpoint* ◀))) *Audio-video solution in English & Spanish at LarsonPrecalculus.com*

Find all the zeros of $f(x) = 3x^3 - 2x^2 + 48x - 32$ given that $4i$ is a zero of f. ■

In Example 8, without knowing that $1 + 3i$ is a zero of f, it is still possible to find all the zeros of the function. You can first use synthetic division to find the real zeros -2 and 3. Then, factor the polynomial as

$$(x + 2)(x - 3)(x^2 - 2x + 10).$$

Finally, use the Quadratic Formula to solve $x^2 - 2x + 10 = 0$ to obtain the zeros $1 + 3i$ and $1 - 3i$.

In Example 9, you will find all the zeros, including the imaginary zeros, of a fifth-degree polynomial function.

EXAMPLE 9   Finding the Zeros of a Polynomial Function

Write

$$f(x) = x^5 + x^3 + 2x^2 - 12x + 8$$

as the product of linear factors and list all the zeros of the function.

Solution   The leading coefficient is 1, so the possible rational zeros are the factors of the constant term.

   Possible rational zeros: $\pm 1, \pm 2, \pm 4,$ and ± 8

By synthetic division, $x = 1$ and $x = -2$ are zeros.

$$
\begin{array}{r|rrrrrr}
1 & 1 & 0 & 1 & 2 & -12 & 8 \\
  &   & 1 & 1 & 2 & 4 & -8 \\
\hline
  & 1 & 1 & 2 & 4 & -8 & 0
\end{array}
\longrightarrow \quad \text{1 is a zero.}
$$

$$
\begin{array}{r|rrrrr}
-2 & 1 & 1 & 2 & 4 & -8 \\
   &   & -2 & 2 & -8 & 8 \\
\hline
   & 1 & -1 & 4 & -4 & 0
\end{array}
\longrightarrow \quad \text{-2 is a zero.}
$$

So, you have

$$
\begin{aligned}
f(x) &= x^5 + x^3 + 2x^2 - 12x + 8 \\
&= (x - 1)(x + 2)(x^3 - x^2 + 4x - 4).
\end{aligned}
$$

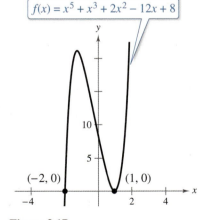

$f(x) = x^5 + x^3 + 2x^2 - 12x + 8$

Factoring by grouping,

$$x^3 - x^2 + 4x - 4 = (x - 1)(x^2 + 4)$$

and by factoring $x^2 + 4$ as

$$x^2 + 4 = (x - 2i)(x + 2i)$$

you obtain

$$f(x) = (x - 1)(x - 1)(x + 2)(x - 2i)(x + 2i)$$

which gives all five zeros of f.

$$x = 1, \quad x = 1, \quad x = -2, \quad x = 2i, \quad \text{and} \quad x = -2i$$

Figure 3.17 shows the graph of f. Notice that the *real* zeros are the only ones that appear as x-intercepts and that the real zero $x = 1$ is repeated.

Figure 3.17

✔ *Checkpoint*   *Audio-video solution in English & Spanish at LarsonPrecalculus.com*

Write

$$f(x) = x^4 + 8x^2 - 9$$

as the product of linear factors and list all the zeros of the function.

Other Tests for Zeros of Polynomials

You know that an nth-degree polynomial function can have *at most n real zeros*. Of course, many nth-degree polynomial functions do not have that many real zeros. For example, $f(x) = x^2 + 1$ has no real zeros, and $f(x) = x^3 + 1$ has only one real zero. The theorem below, called **Descartes's Rule of Signs,** uses variations in sign to analyze the number of real zeros of a polynomial. A **variation in sign** means that two consecutive nonzero coefficients have opposite signs.

Descartes's Rule of Signs

Let $f(x) = a_n x^n + a_{n-1} x^{n-1} + \cdots + a_2 x^2 + a_1 x + a_0$ be a polynomial with real coefficients and $a_0 \neq 0$.

1. The number of *positive real zeros* of f is either equal to the number of variations in sign of $f(x)$ or less than that number by an even integer.

2. The number of *negative real zeros* of f is either equal to the number of variations in sign of $f(-x)$ or less than that number by an even integer.

When using Descartes's Rule of Signs, count a zero of multiplicity k as k zeros. For example, the polynomial $x^3 - 3x + 2$ has two variations in sign, and so it has either two positive or no positive real zeros. This polynomial factors as

$$x^3 - 3x + 2 = (x - 1)(x - 1)(x + 2)$$

so the two positive real zeros are $x = 1$ of multiplicity 2.

EXAMPLE 10   **Using Descartes's Rule of Signs**

Determine the possible numbers of positive and negative real zeros of

$$f(x) = 3x^3 - 5x^2 + 6x - 4.$$

Solution   The original polynomial has *three* variations in sign.

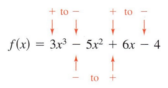

The polynomial

$$f(-x) = 3(-x)^3 - 5(-x)^2 + 6(-x) - 4$$
$$= -3x^3 - 5x^2 - 6x - 4$$

has no variations in sign. So, from Descartes's Rule of Signs, the polynomial

$$f(x) = 3x^3 - 5x^2 + 6x - 4$$

has either three positive real zeros or one positive real zero, and has no negative real zeros. Figure 3.18 shows that the function has only one real zero, $x = 1$.

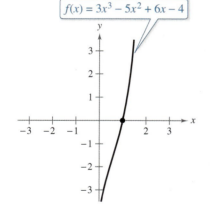

Figure 3.18

✓ *Checkpoint* 🔊))) *Audio-video solution in English & Spanish at LarsonPrecalculus.com*

Determine the possible numbers of positive and negative real zeros of

$$f(x) = 2x^3 + 5x^2 + x + 8.$$

Another test for zeros of a polynomial function is related to the sign pattern in the last row of the synthetic division array. This test can give you an upper or lower bound for the real zeros of f. A real number c is an **upper bound** for the real zeros of f when no zeros are greater than c. Similarly, c is a **lower bound** when no real zeros of f are less than c.

Upper and Lower Bound Rules

Let $f(x)$ be a polynomial with real coefficients and a positive leading coefficient. Divide $f(x)$ by $x - c$ using synthetic division.

1. If $c > 0$ and each number in the last row is either positive or zero, then c is an **upper bound** for the real zeros of f.

2. If $c < 0$ and the numbers in the last row are alternately positive and negative (zero entries count as positive or negative), then c is a **lower bound** for the real zeros of f.

EXAMPLE 11   **Finding Real Zeros of a Polynomial Function**

Find all real zeros of

$$f(x) = 6x^3 - 4x^2 + 3x - 2.$$

Solution   List the possible rational zeros of f.

$$\frac{\text{Factors of } -2}{\text{Factors of } 6} = \frac{\pm 1, \pm 2}{\pm 1, \pm 2, \pm 3, \pm 6} = \pm 1, \pm\frac{1}{2}, \pm\frac{1}{3}, \pm\frac{1}{6}, \pm\frac{2}{3}, \pm 2$$

The original polynomial $f(x)$ has three variations in sign. The polynomial

$$f(-x) = 6(-x)^3 - 4(-x)^2 + 3(-x) - 2$$
$$= -6x^3 - 4x^2 - 3x - 2$$

has no variations in sign. So, by Descartes's Rule of Signs, there are three positive real zeros or one positive real zero, and no negative real zeros. Test $x = 1$.

$$
\begin{array}{r|rrrr}
1 & 6 & -4 & 3 & -2 \\
  &   & 6 & 2 & 5 \\
\hline
  & 6 & 2 & 5 & 3
\end{array}
$$

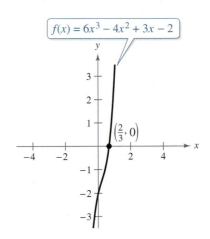

$f(x) = 6x^3 - 4x^2 + 3x - 2$

This shows that $x = 1$ is not a zero. However, the last row has all positive entries, telling you that $x = 1$ is an upper bound for the real zeros. So, restrict the search to zeros between 0 and 1. By trial and error, $x = \frac{2}{3}$ is a zero, and factoring,

$$f(x) = \left(x - \frac{2}{3}\right)(6x^2 + 3).$$

The factor $6x^2 + 3$ has no real zeros, so it follows that $x = \frac{2}{3}$ is the only real zero, as verified in the graph of f at the right.

✓ **Checkpoint** ◀))) *Audio-video solution in English & Spanish at LarsonPrecalculus.com*

Find all real zeros of $f(x) = 8x^3 - 4x^2 + 6x - 3$.

Application

EXAMPLE 12 **Using a Polynomial Model**

You design candle making kits. Each kit contains 25 cubic inches of candle wax and a mold for making a pyramid-shaped candle. You want the height of the candle to be 2 inches less than the length of each side of the candle's square base. What should the dimensions of your candle mold be?

Solution The volume of a pyramid is $V = \frac{1}{3}Bh$, where B is the area of the base and h is the height. The area of the base is x^2 and the height is $(x - 2)$. So, the volume of the pyramid is $V = \frac{1}{3}x^2(x - 2)$. Substitute 25 for the volume and solve for x.

$$25 = \tfrac{1}{3}x^2(x - 2) \qquad \text{\color{red}Substitute 25 for } V.$$

$$75 = x^3 - 2x^2 \qquad \text{\color{red}Multiply each side by 3, and distribute } x^2.$$

$$0 = x^3 - 2x^2 - 75 \qquad \text{\color{red}Write in general form.}$$

The possible rational solutions are $x = \pm 1, \pm 3, \pm 5, \pm 15, \pm 25, \pm 75$. Note that in this case it makes sense to consider only positive x-values. Use synthetic division to test some of the possible solutions and determine that $x = 5$ is a solution.

$$
\begin{array}{r|rrrr}
5 & 1 & -2 & 0 & -75 \\
  &   & 5 & 15 & 75 \\
\hline
  & 1 & 3 & 15 & 0
\end{array}
$$

The other two solutions, which satisfy $x^2 + 3x + 15 = 0$, are imaginary, so discard them and conclude that the base of the candle mold should be 5 inches by 5 inches and the height should be $5 - 2 = 3$ inches.

✓ *Checkpoint* *Audio-video solution in English & Spanish at LarsonPrecalculus.com*

Rework Example 12 when each kit contains 147 cubic inches of candle wax and you want the height of the pyramid-shaped candle to be 2 inches more than the length of each side of the candle's square base.

Before concluding this section, here is an additional hint that can help you find the zeros of a polynomial function. When the terms of $f(x)$ have a common monomial factor, you should factor it out before applying the tests in this section. For example, writing $f(x) = x^4 - 5x^3 + 3x^2 + x = x(x^3 - 5x^2 + 3x + 1)$ shows that $x = 0$ is a zero of f. Obtain the remaining zeros by analyzing the cubic factor.

Summarize   (Section 3.4)

1. State the Fundamental Theorem of Algebra and the Linear Factorization Theorem *(page 273, Example 1)*.

2. Explain how to use the Rational Zero Test *(page 274, Examples 2–5)*.

3. Explain how to use complex conjugates when analyzing a polynomial function *(page 277, Examples 6 and 7)*.

4. Explain how to find the zeros of a polynomial function *(page 278, Examples 8 and 9)*.

5. State Descartes's Rule of Signs and the Upper and Lower Bound Rules *(pages 280 and 281, Examples 10 and 11)*.

6. Describe a real-life application of finding the zeros of a polynomial function *(page 282, Example 12)*.

3.4 Exercises

See CalcChat.com for tutorial help and worked-out solutions to odd-numbered exercises.

Vocabulary: Fill in the blanks.

1. The _____ _____ of _____ states that if $f(x)$ is a polynomial of degree n ($n > 0$), then f has at least one zero in the complex number system.

2. The _____ _____ _____ states that if $f(x)$ is a polynomial of degree n ($n > 0$), then $f(x)$ has precisely n linear factors, $f(x) = a_n(x - c_1)(x - c_2) \cdots (x - c_n)$, where $c_1, c_2, \ldots, c_n$ are complex numbers.

3. The test that gives a list of the possible rational zeros of a polynomial function is the _____ _____ Test.

4. If $a + bi$, where $b \neq 0$, is a complex zero of a polynomial with real coefficients, then so is its _____ _____, $a - bi$.

5. Every polynomial of degree $n > 0$ with real coefficients can be written as the product of _____ and _____ factors with real coefficients, where the _____ factors have no real zeros.

6. A quadratic factor that cannot be factored further as a product of linear factors containing real numbers is _____ over the _____.

7. The theorem that can be used to determine the possible numbers of positive and negative real zeros of a function is called _____ _____ of _____.

8. A real number c is a _____ bound for the real zeros of f when no real zeros are less than c, and is a _____ bound when no real zeros are greater than c.

Skills and Applications

 Zeros of Polynomial Functions In Exercises 9–14, determine the number of zeros of the polynomial function.

9. $f(x) = x^3 + 2x^2 + 1$ 10. $f(x) = x^4 - 3x$

11. $g(x) = x^4 - x^5$ 12. $f(x) = x^3 - x^6$

13. $f(x) = (x + 5)^2$

14. $h(t) = (t - 1)^2 - (t + 1)^2$

Using the Rational Zero Test In Exercises 15–18, use the Rational Zero Test to list the possible rational zeros of f. Verify that the zeros of f shown in the graph are contained in the list.

15. $f(x) = x^3 + 2x^2 - x - 2$

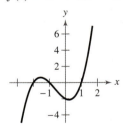

16. $f(x) = x^3 - 4x^2 - 4x + 16$

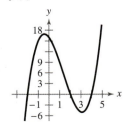

17. $f(x) = 2x^4 - 17x^3 + 35x^2 + 9x - 45$

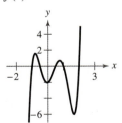

18. $f(x) = 4x^5 - 8x^4 - 5x^3 + 10x^2 + x - 2$

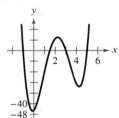

 Using the Rational Zero Test In Exercises 19–28, find (if possible) the rational zeros of the function.

19. $f(x) = x^3 - 7x - 6$ 20. $f(x) = x^3 - 13x + 12$

21. $g(t) = t^3 - 4t^2 + 4$ 22. $h(x) = x^3 - 19x + 30$

23. $h(t) = t^3 + 8t^2 + 13t + 6$

24. $g(x) = x^3 + 8x^2 + 12x + 18$

25. $C(x) = 2x^3 + 3x^2 - 1$

26. $f(x) = 3x^3 - 19x^2 + 33x - 9$

27. $g(x) = 9x^4 - 9x^3 - 58x^2 + 4x + 24$

28. $f(x) = 2x^4 - 15x^3 + 23x^2 + 15x - 25$

Solving a Polynomial Equation In Exercises 29–32, find all real solutions of the polynomial equation.

29. $-5x^3 + 11x^2 - 4x - 2 = 0$

30. $8x^3 + 10x^2 - 15x - 6 = 0$

31. $x^4 + 6x^3 + 3x^2 - 16x + 6 = 0$

32. $x^4 + 8x^3 + 14x^2 - 17x - 42 = 0$

Using the Rational Zero Test In Exercises 33–36, (a) list the possible rational zeros of f, (b) sketch the graph of f so that some of the possible zeros in part (a) can be disregarded, and then (c) determine all real zeros of f.

33. $f(x) = x^3 + x^2 - 4x - 4$

34. $f(x) = -3x^3 + 20x^2 - 36x + 16$

35. $f(x) = -4x^3 + 15x^2 - 8x - 3$

36. $f(x) = 4x^3 - 12x^2 - x + 15$

 Using the Rational Zero Test In Exercises 37–40, (a) list the possible rational zeros of f, (b) use a graphing utility to graph f so that some of the possible zeros in part (a) can be disregarded, and then (c) determine all real zeros of f.

37. $f(x) = -2x^4 + 13x^3 - 21x^2 + 2x + 8$

38. $f(x) = 4x^4 - 17x^2 + 4$

39. $f(x) = 32x^3 - 52x^2 + 17x + 3$

40. $f(x) = 4x^3 + 7x^2 - 11x - 18$

Finding a Polynomial Function with Given Zeros In Exercises 41–46, find a polynomial function with real coefficients that has the given zeros. (There are many correct answers.)

41. $1, 5i$

42. $4, -3i$

43. $2, 2, 1 + i$

44. $-1, 5, 3 - 2i$

45. $\frac{2}{3}, -1, 3 + \sqrt{2}i$

46. $-\frac{5}{2}, -5, 1 + \sqrt{3}i$

Finding a Polynomial Function with Given Zeros In Exercises 47–50, find the polynomial function f with real coefficients that has the given degree, zeros, and solution point.

	Degree	Zeros	Solution Point
47.	4	$-2, 1, i$	$f(0) = -4$
48.	4	$-1, 2, \sqrt{2}i$	$f(1) = 12$
49.	3	$-3, 1 + \sqrt{3}i$	$f(-2) = 12$
50.	3	$-2, 1 - \sqrt{2}i$	$f(-1) = -12$

Factoring a Polynomial In Exercises 51–54, write the polynomial (a) as the product of factors that are irreducible over the *rationals*, (b) as the product of linear and quadratic factors that are irreducible over the *reals*, and (c) in completely factored form.

51. $f(x) = x^4 + 2x^2 - 8$

52. $f(x) = x^4 + 6x^2 - 27$

53. $f(x) = x^4 - 2x^3 - 3x^2 + 12x - 18$
  (*Hint:* One factor is $x^2 - 6$.)

54. $f(x) = x^4 - 3x^3 - x^2 - 12x - 20$
  (*Hint:* One factor is $x^2 + 4$.)

Finding the Zeros of a Polynomial Function In Exercises 55–60, use the given zero to find all the zeros of the function.

Function	Zero
55. $f(x) = x^3 - x^2 + 4x - 4$	$2i$
56. $f(x) = 2x^3 + 3x^2 + 18x + 27$	$3i$
57. $g(x) = x^3 - 8x^2 + 25x - 26$	$3 + 2i$
58. $g(x) = x^3 + 9x^2 + 25x + 17$	$-4 + i$
59. $h(x) = x^4 - 6x^3 + 14x^2 - 18x + 9$	$1 - \sqrt{2}i$
60. $h(x) = x^4 + x^3 - 3x^2 - 13x + 14$	$-2 + \sqrt{3}i$

Finding the Zeros of a Polynomial Function In Exercises 61–72, write the polynomial as the product of linear factors and list all the zeros of the function.

61. $f(x) = x^2 + 36$ **62.** $f(x) = x^2 + 49$

63. $h(x) = x^2 - 2x + 17$ **64.** $g(x) = x^2 + 10x + 17$

65. $f(x) = x^4 - 16$ **66.** $f(y) = y^4 - 256$

67. $f(z) = z^2 - 2z + 2$

68. $h(x) = x^3 - 3x^2 + 4x - 2$

69. $g(x) = x^3 - 3x^2 + x + 5$

70. $f(x) = x^3 - x^2 + x + 39$

71. $g(x) = x^4 - 4x^3 + 8x^2 - 16x + 16$

72. $h(x) = x^4 + 6x^3 + 10x^2 + 6x + 9$

Finding the Zeros of a Polynomial Function In Exercises 73–78, find all the zeros of the function. When there is an extended list of possible rational zeros, use a graphing utility to graph the function in order to disregard any of the possible rational zeros that are obviously not zeros of the function.

73. $f(x) = x^3 + 24x^2 + 214x + 740$

74. $f(s) = 2s^3 - 5s^2 + 12s - 5$

75. $f(x) = 16x^3 - 20x^2 - 4x + 15$

76. $f(x) = 9x^3 - 15x^2 + 11x - 5$

77. $f(x) = 2x^4 + 5x^3 + 4x^2 + 5x + 2$

78. $g(x) = x^5 - 8x^4 + 28x^3 - 56x^2 + 64x - 32$

Using Descartes's Rule of Signs In Exercises 79–86, use Descartes's Rule of Signs to determine the possible numbers of positive and negative real zeros of the function.

79. $g(x) = 2x^3 - 3x^2 - 3$ **80.** $h(x) = 4x^2 - 8x + 3$

81. $h(x) = 2x^3 + 3x^2 + 1$ **82.** $h(x) = 2x^4 - 3x - 2$

83. $g(x) = 6x^4 + 2x^3 - 3x^2 + 2$

84. $f(x) = 4x^3 - 3x^2 - 2x - 1$

85. $f(x) = 5x^3 + x^2 - x + 5$

86. $f(x) = 3x^3 - 2x^2 - x + 3$

Verifying Upper and Lower Bounds In Exercises 87–90, use synthetic division to verify the upper and lower bounds of the real zeros of f.

87. $f(x) = x^3 + 3x^2 - 2x + 1$

    (a) Upper: $x = 1$ (b) Lower: $x = -4$

88. $f(x) = x^3 - 4x^2 + 1$

    (a) Upper: $x = 4$ (b) Lower: $x = -1$

89. $f(x) = x^4 - 4x^3 + 16x - 16$

    (a) Upper: $x = 5$ (b) Lower: $x = -3$

90. $f(x) = 2x^4 - 8x + 3$

    (a) Upper: $x = 3$ (b) Lower: $x = -4$

Finding Real Zeros of a Polynomial Function In Exercises 91–94, find all real zeros of the function.

91. $f(x) = 16x^3 - 12x^2 - 4x + 3$

92. $f(z) = 12z^3 - 4z^2 - 27z + 9$

93. $f(y) = 4y^3 + 3y^2 + 8y + 6$

94. $g(x) = 3x^3 - 2x^2 + 15x - 10$

Finding the Rational Zeros of a Polynomial In Exercises 95–98, find the rational zeros of the polynomial function.

95. $P(x) = x^4 - \frac{25}{4}x^2 + 9 = \frac{1}{4}(4x^4 - 25x^2 + 36)$

96. $f(x) = x^3 - \frac{3}{2}x^2 - \frac{23}{2}x + 6$

    $= \frac{1}{2}(2x^3 - 3x^2 - 23x + 12)$

97. $f(x) = x^3 - \frac{1}{4}x^2 - x + \frac{1}{4} = \frac{1}{4}(4x^3 - x^2 - 4x + 1)$

98. $f(z) = z^3 + \frac{11}{6}z^2 - \frac{1}{2}z - \frac{1}{3} = \frac{1}{6}(6z^3 + 11z^2 - 3z - 2)$

Rational and Irrational Zeros In Exercises 99–102, match the cubic function with the numbers of rational and irrational zeros.

(a) Rational zeros: 0; irrational zeros: 1

(b) Rational zeros: 3; irrational zeros: 0

(c) Rational zeros: 1; irrational zeros: 2

(d) Rational zeros: 1; irrational zeros: 0

99. $f(x) = x^3 - 1$ **100.** $f(x) = x^3 - 2$

101. $f(x) = x^3 - x$ **102.** $f(x) = x^3 - 2x$

103. Geometry You want to make an open box from a rectangular piece of material, 15 centimeters by 9 centimeters, by cutting equal squares from the corners and turning up the sides.

    (a) Let x represent the side length of each of the squares removed. Draw a diagram showing the squares removed from the original piece of material and the resulting dimensions of the open box.

    (b) Use the diagram to write the volume V of the box as a function of x. Determine the domain of the function.

    (c) Sketch the graph of the function and approximate the dimensions of the box that yield a maximum volume.

    (d) Find values of x such that $V = 56$. Which of these values is a physical impossibility in the construction of the box? Explain.

104. Geometry A rectangular package to be sent by a delivery service (see figure) has a combined length and girth (perimeter of a cross section) of 120 inches.

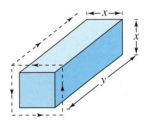

    (a) Use the diagram to write the volume V of the package as a function of x.

    (b) Use a graphing utility to graph the function and approximate the dimensions of the package that yield a maximum volume.

    (c) Find values of x such that $V = 13{,}500$. Which of these values is a physical impossibility in the construction of the package? Explain.

105. Geometry

A bulk food storage bin with dimensions 2 feet by 3 feet by 4 feet needs to be increased in size to hold five times as much food as the current bin.

    (a) Assume each dimension is increased by the same amount. Write a function that represents the volume V of the new bin.

    (b) Find the dimensions of the new bin.

106. Cost The ordering and transportation cost C (in thousands of dollars) for machine parts is given by

$$C(x) = 100\left(\frac{200}{x^2} + \frac{x}{x + 30}\right), \quad x \geq 1$$

where x is the order size (in hundreds). In calculus, it can be shown that the cost is a minimum when

$$3x^3 - 40x^2 - 2400x - 36,000 = 0.$$

Use a graphing utility to approximate the optimal order size to the nearest hundred units.

Exploration

True or False? **In Exercises 107 and 108, decide whether the statement is true or false. Justify your answer.**

107. It is possible for a third-degree polynomial function with integer coefficients to have no real zeros.

108. If $x = -i$ is a zero of the function

$$f(x) = x^3 + ix^2 + ix - 1$$

then $x = i$ must also be a zero of f.

Think About It **In Exercises 109–114, determine (if possible) the zeros of the function g when the function f has zeros at $x = r_1$, $x = r_2$, and $x = r_3$.**

109. $g(x) = -f(x)$

110. $g(x) = 3f(x)$

111. $g(x) = f(x - 5)$

112. $g(x) = f(2x)$

113. $g(x) = 3 + f(x)$

114. $g(x) = f(-x)$

115. Think About It A cubic polynomial function f has real zeros -2, $\frac{1}{2}$, and 3, and its leading coefficient is negative. Write an equation for f and sketch its graph. How many different polynomial functions are possible for f?

116. Think About It Sketch the graph of a fifth-degree polynomial function whose leading coefficient is positive and that has a zero at $x = 3$ of multiplicity 2.

Writing an Equation **In Exercises 117 and 118, the graph of a cubic polynomial function $y = f(x)$ is shown. One of the zeros is $1 + i$. Write an equation for f.**

117.

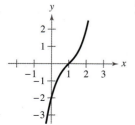

118.

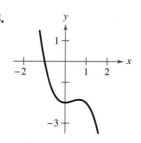

119. Error Analysis Describe the error.

The graph of a quartic (fourth-degree) polynomial $y = f(x)$ is shown. One of the zeros is i.

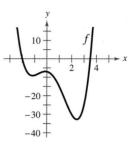

The function is $f(x) = (x + 2)(x - 3.5)(x - i)$. ✗

120. **HOW DO YOU SEE IT?** Use the information in the table to answer each question.

Interval	Value of $f(x)$
$(-\infty, -2)$	Positive
$(-2, 1)$	Negative
$(1, 4)$	Negative
$(4, \infty)$	Positive

(a) What are the three real zeros of the polynomial function f?

(b) What can be said about the behavior of the graph of f at $x = 1$?

(c) What is the least possible degree of f? Explain. Can the degree of f ever be odd? Explain.

(d) Is the leading coefficient of f positive or negative? Explain.

(e) Sketch a graph of a function that exhibits the behavior described in the table.

121. Think About It Let $y = f(x)$ be a quartic (fourth-degree) polynomial with leading coefficient $a = 1$ and

$$f(i) = f(2i) = 0.$$

Write an equation for f.

122. Think About It Let $y = f(x)$ be a cubic polynomial with leading coefficient $a = -1$ and

$$f(2) = f(i) = 0.$$

Write an equation for f.

123. Writing an Equation Write the equation for a quadratic function f (with integer coefficients) that has the given zeros. Assume that b is a positive integer.

(a) $\pm\sqrt{b}i$ (b) $a \pm bi$

3.5   Mathematical Modeling and Variation

Mathematical models have a wide variety of real-life applications. For example, in Exercise 71 on page 297, you will use variation to model ocean temperatures at various depths.

■ Use mathematical models to approximate sets of data points.
■ Use the *regression* feature of a graphing utility to find equations of least squares regression lines.
■ Write mathematical models for direct variation.
■ Write mathematical models for direct variation as an *n*th power.
■ Write mathematical models for inverse variation.
■ Write mathematical models for combined variation.
■ Write mathematical models for joint variation.

Introduction

In this section, you will study two techniques for fitting models to data: *least squares regression* and *direct and inverse variation*.

EXAMPLE 1   Using a Mathematical Model

The table shows the populations *y* (in millions) of the United States from 2008 through 2015.   *(Source: U.S. Census Bureau)*

DATA Year	2008	2009	2010	2011	2012	2013	2014	2015
Population, *y*	304.1	306.8	309.3	311.7	314.1	316.5	318.9	321.2

Spreadsheet at LarsonPrecalculus.com

A linear model that approximates the data is

$$y = 2.43t + 284.9, \quad 8 \leq t \leq 15$$

where *t* represents the year, with *t* = 8 corresponding to 2008. Plot the actual data *and* the model on the same graph. How closely does the model represent the data?

Solution   Figure 3.19 shows the actual data and the model plotted on the same graph. From the graph, it appears that the model is a "good fit" for the actual data. To see how well the model fits, compare the actual values of *y* with the values of *y* found using the model. The values found using the model are labeled *y** in the table below.

t	8	9	10	11	12	13	14	15
y	304.1	306.8	309.3	311.7	314.1	316.5	318.9	321.2
y*	304.3	306.8	309.2	311.6	314.1	316.5	318.9	321.4

✓ *Checkpoint* ◀))) *Audio-video solution in English & Spanish at LarsonPrecalculus.com*

The ordered pairs below give the median sales prices *y* (in thousands of dollars) of new homes sold in a neighborhood from 2009 through 2016.   *(Spreadsheet at LarsonPrecalculus.com)*

DATA
(2009, 179.4)   (2011, 191.0)   (2013, 202.6)   (2015, 214.9)
(2010, 185.4)   (2012, 196.7)   (2014, 208.7)   (2016, 221.4)

A linear model that approximates the data is *y* = 5.96*t* + 125.5, 9 ≤ *t* ≤ 16, where *t* represents the year, with *t* = 9 corresponding to 2009. Plot the actual data *and* the model on the same graph. How closely does the model represent the data?   ■

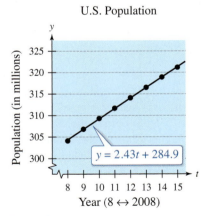

U.S. Population

Population (in millions)

Year (8 ↔ 2008)

Figure 3.19

Least Squares Regression and Graphing Utilities

So far in this text, you have worked with many different types of mathematical models that approximate real-life data. In some instances the model was given (as in Example 1), whereas in other instances you found the model using algebraic techniques or a graphing utility.

To find a model that approximates a set of data most accurately, statisticians use a measure called the **sum of the squared differences,** which is the sum of the squares of the differences between actual data values and model values. The "best-fitting" linear model, called the **least squares regression line,** is the one with the least sum of the squared differences.

Recall that you can approximate this line visually by plotting the data points and drawing the line that appears to best fit the data—or you can enter the data points into a graphing utility or software program and use the *linear regression* feature.

When you use the *regression* feature of a graphing utility or software program, an "*r*-value" may be output. This is the **correlation coefficient** of the data and gives a measure of how well the model fits the data. The closer the value of $|r|$ is to 1, the better the fit.

EXAMPLE 2   **Finding a Least Squares Regression Line**

See LarsonPrecalculus.com for an interactive version of this type of example.

The table shows the numbers E (in millions) of Medicare private health plan enrollees from 2008 through 2015. Construct a scatter plot that represents the data and find the equation of the least squares regression line for the data.   *(Source: U.S. Centers for Medicare and Medicaid Services)*

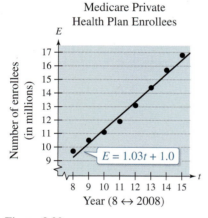

Medicare Private
Health Plan Enrollees

$E = 1.03t + 1.0$

Year (8 ↔ 2008)

Figure 3.20

	Year	Enrollees, *E*
	2008	9.7
	2009	10.5
	2010	11.1
	2011	11.9
	2012	13.1
	2013	14.4
	2014	15.7
	2015	16.8

Spreadsheet at LarsonPrecalculus.com

t	E	E^*
8	9.7	9.2
9	10.5	10.3
10	11.1	11.3
11	11.9	12.3
12	13.1	13.4
13	14.4	14.4
14	15.7	15.4
15	16.8	16.5

Solution   Let $t = 8$ represent 2008. Figure 3.20 shows a scatter plot of the data. Using the *regression* feature of a graphing utility or software program, the equation of the least squares regression line is $E = 1.03t + 1.0$. To check this model, compare the actual E-values with the E-values found using the model, which are labeled E^* in the table at the left. The correlation coefficient for this model is $r \approx 0.992$, so the model is a good fit.

✓ **Checkpoint** 🔊)) *Audio-video solution in English & Spanish at LarsonPrecalculus.com*

The ordered pairs below give the numbers E (in millions) of Medicare Advantage enrollees in health maintenance organization plans from 2008 through 2015. *(Spreadsheet at LarsonPrecalculus.com)* Construct a scatter plot that represents the data and find the equation of the least squares regression line for the data.   *(Source: U.S. Centers for Medicare and Medicaid Services)*

 DATA

(2008, 6.3)	(2010, 7.2)	(2012, 8.5)	(2014, 10.1)
(2009, 6.7)	(2011, 7.7)	(2013, 9.3)	(2015, 10.7)

Direct Variation

There are two basic types of linear models. The more general model has a nonzero y-intercept.

$$y = mx + b, \quad b \neq 0$$

The simpler model

$$y = kx$$

has a y-intercept of zero. In the simpler model, y **varies directly** as x, or is **directly proportional** to x.

Direct Variation

The statements below are equivalent.

1. y **varies directly** as x.
2. y is **directly proportional** to x.
3. $y = kx$ for some nonzero constant k.

k is the **constant of variation** or the **constant of proportionality.**

EXAMPLE 3   Direct Variation

In Pennsylvania, the state income tax is directly proportional to *gross income*. You work in Pennsylvania and your state income tax deduction is $46.05 for a gross monthly income of $1500. Find a mathematical model that gives the Pennsylvania state income tax in terms of gross income.

Solution

Verbal model:

$$\boxed{\text{State income tax}} = \boxed{k} \cdot \boxed{\text{Gross income}}$$

Labels: State income tax $= y$ (dollars)
Gross income $= x$ (dollars)
Income tax rate $= k$ (percent in decimal form)

Equation: $y = kx$

To find the state income tax rate k, substitute the given information into the equation $y = kx$ and solve.

$$y = kx \qquad \text{Write direct variation model.}$$
$$46.05 = k(1500) \qquad \text{Substitute 46.05 for } y \text{ and 1500 for } x.$$
$$0.0307 = k \qquad \text{Divide each side by 1500.}$$

So, the equation (or model) for state income tax in Pennsylvania is

$$y = 0.0307x.$$

In other words, Pennsylvania has a state income tax rate of 3.07% of gross income. Figure 3.21 shows the graph of this equation.

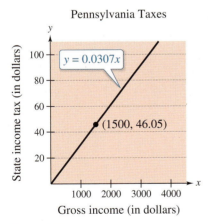

Pennsylvania Taxes

$y = 0.0307x$

(1500, 46.05)

State income tax (in dollars)

Gross income (in dollars)

Figure 3.21

✓ *Checkpoint*  *Audio-video solution in English & Spanish at LarsonPrecalculus.com*

The simple interest on an investment is directly proportional to the amount of the investment. For example, an investment of $2500 earns $187.50 after 1 year. Find a mathematical model that gives the interest I after 1 year in terms of the amount invested P.

Direct Variation as an *n*th Power

Another type of direct variation relates one variable to a *power* of another variable. For example, in the formula for the area of a circle

$$A = \pi r^2$$

the area A is directly proportional to the square of the radius r. Note that for this formula, π is the constant of proportionality.

•• REMARK Note that the direct variation model $y = kx$ is a special case of $y = kx^n$ with $n = 1$.

> ### Direct Variation as an *n*th Power
>
> The statements below are equivalent.
> 1. y **varies directly as the *n*th power** of x.
> 2. y is **directly proportional to the *n*th power** of x.
> 3. $y = kx^n$ for some nonzero constant k.

EXAMPLE 4 **Direct Variation as an *n*th Power**

The distance a ball rolls down an inclined plane is directly proportional to the square of the time it rolls. During the first second, the ball rolls 8 feet. (See Figure 3.22.)

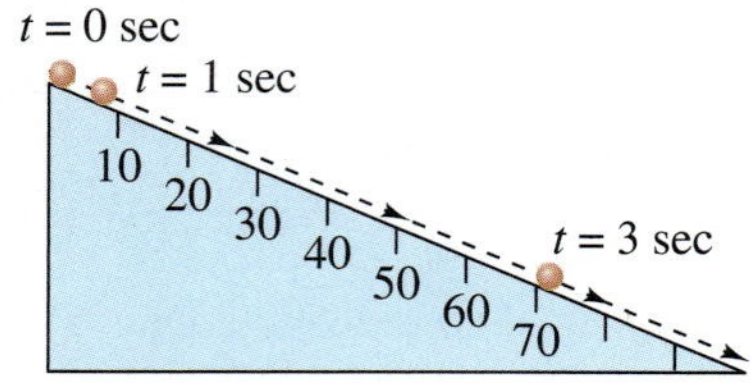

a. Write an equation relating the distance traveled to the time.

b. How far does the ball roll during the first 3 seconds?

$t = 0$ sec
$t = 1$ sec
$t = 3$ sec

Figure 3.22

Solution

a. Letting d be the distance (in feet) the ball rolls and letting t be the time (in seconds), you have

$$d = kt^2.$$

Now, $d = 8$ when $t = 1$, so you have

$$d = kt^2 \qquad \text{Write direct variation model.}$$
$$8 = k(\ )^2 \qquad \text{Substitute 8 for } d \text{ and 1 for } t.$$
$$8 = k \qquad \text{Simplify.}$$

and, the equation relating distance to time is

$$d = 8t^2.$$

b. When $t = 3$, the distance traveled is

$$d = 8(3)^2 \qquad \text{Substitute 3 for } t.$$
$$= 8(9) \qquad \text{Simplify.}$$
$$= 72 \text{ feet.} \qquad \text{Simplify.}$$

So, the ball rolls 72 feet during the first 3 seconds.

✓ *Checkpoint* *Audio-video solution in English & Spanish at LarsonPrecalculus.com*

Neglecting air resistance, the distance s an object falls varies directly as the square of the duration t of the fall. An object falls a distance of 144 feet in 3 seconds. How far does it fall in 6 seconds?

In Examples 3 and 4, the direct variations are such that an *increase* in one variable corresponds to an *increase* in the other variable. You should not, however, assume that this always occurs with direct variation. For example, for the model $y = -3x$, an increase in x results in a *decrease* in y, and yet y is said to vary directly as x.

Inverse Variation

> ### Inverse Variation
>
> The statements below are equivalent.
>
> **1.** y **varies inversely** as x.
>
> **2.** y is **inversely proportional** to x.
>
> **3.** $y = \dfrac{k}{x}$ for some nonzero constant k.

If x and y are related by an equation of the form $y = k/x^n$, then y varies inversely as the nth power of x (or y is inversely proportional to the nth power of x).

EXAMPLE 5 **Inverse Variation**

A company has found that the demand for one of its products varies inversely as the price of the product. When the price is $6.25, the demand is 400 units. Approximate the demand when the price is $5.75.

Solution

Let p be the price and let x be the demand. The demand varies inversely as the price, so you have

$$x = \frac{k}{p}.$$

Now, $x = 400$ when $p = 6.25$, so you have

$x = \dfrac{k}{p}$	Write inverse variation model.
$400 = \dfrac{k}{6.25}$	Substitute 400 for x and 6.25 for p.
$(400)(6.25) = k$	Multiply each side by 6.25.
$2500 = k$	Simplify.

and the equation relating price and demand is

$$x = \frac{2500}{p}.$$

When $p = 5.75$, the demand is

$x = \dfrac{2500}{p}$	Write inverse variation model.
$= \dfrac{2500}{5.75}$	Substitute 5.75 for p.
≈ 435 units.	Simplify.

So, the demand for the product is about 435 units when the price is $5.75.

 Checkpoint *Audio-video solution in English & Spanish at LarsonPrecalculus.com*

The company in Example 5 has found that the demand for another of its products also varies inversely as the price of the product. When the price is $2.75, the demand is 600 units. Approximate the demand when the price is $3.25.

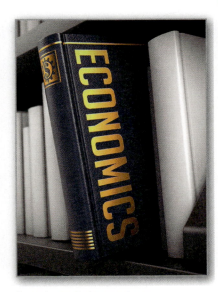

Supply and demand are fundamental concepts in economics. The law of demand states that, all other factors remaining equal, the lower the price of the product, the higher the quantity demanded. The law of supply states that the higher the price of the product, the higher the quantity supplied. *Equilibrium* occurs when the demand and the supply are the same.

Combined Variation

Some applications of variation involve problems with *both* direct and inverse variations in the same model. These types of models have **combined variation.**

EXAMPLE 6 **Combined Variation**

A gas law states that the volume of an enclosed gas varies inversely as the pressure (Figure 3.23) *and* directly as the temperature. The pressure of a gas is 0.75 kilogram per square centimeter when the temperature is 294 K and the volume is 8000 cubic centimeters.

a. Write an equation relating pressure, temperature, and volume.

b. Find the pressure when the temperature is 300 K and the volume is 7000 cubic centimeters.

Solution

a. Volume V varies directly as temperature T and inversely as pressure P, so you have

$$V = \frac{kT}{P}.$$

Now, $P = 0.75$ when $T = 294$ and $V = 8000$, so you have

$$V = \frac{kT}{P} \qquad \text{Write combined variation model.}$$

$$8000 = \frac{k(294)}{0.75} \qquad \text{Substitute 8000 for } V, \text{ 294 for } T, \text{ and 0.75 for } P.$$

$$\frac{6000}{294} = k \qquad \text{Simplify.}$$

$$\frac{1000}{49} = k \qquad \text{Simplify.}$$

and the equation relating pressure, temperature, and volume is

$$V = \frac{1000}{49}\left(\frac{T}{P}\right).$$

b. Isolate P on one side of the equation by multiplying each side by P and dividing each side by V to obtain $P = \dfrac{1000}{49}\left(\dfrac{T}{V}\right)$. When $T = 300$ and $V = 7000$, the pressure is

$$P = \frac{1000}{49}\left(\frac{T}{V}\right) \qquad \text{Combined variation model solved for } P.$$

$$= \frac{1000}{49}\left(\frac{300}{7000}\right) \qquad \text{Substitute 300 for } T \text{ and 7000 for } V.$$

$$= \frac{300}{343} \qquad \text{Simplify.}$$

$$\approx 0.87 \text{ kilogram per square centimeter.} \qquad \text{Simplify.}$$

So, the pressure is about 0.87 kilogram per square centimeter when the temperature is 300 K and the volume is 7000 cubic centimeters.

✓ *Checkpoint*  *Audio-video solution in English & Spanish at LarsonPrecalculus.com*

The resistance of a copper wire carrying an electrical current is directly proportional to its length and inversely proportional to its cross-sectional area. A copper wire with a diameter of 0.0126 inch has a resistance of 64.9 ohms per thousand feet. What length of 0.0201-inch-diameter copper wire will produce a resistance of 33.5 ohms? ■

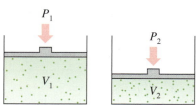

If $P_2 > P_1$, then $V_2 < V_1$.

If the temperature is held constant and pressure increases, then the volume *decreases.*
Figure 3.23

Joint Variation

> **Joint Variation**
>
> The statements below are equivalent.
>
> 1. z **varies jointly** as x and y.
> 2. z is **jointly proportional** to x and y.
> 3. $z = kxy$ for some nonzero constant k.

If x, y, and z are related by an equation of the form $z = kx^n y^m$, then z varies jointly as the nth power of x and the mth power of y.

EXAMPLE 7 **Joint Variation**

The *simple* interest for an investment is jointly proportional to the time and the principal. After one quarter (3 months), the interest on a principal of $5000 is $43.75. (a) Write an equation relating the interest, principal, and time. (b) Find the interest after three quarters.

Solution

a. Interest I (in dollars) is jointly proportional to principal P (in dollars) and time t (in years), so you have

$$I = kPt.$$

For $I = 43.75$, $P = 5000$, and $t = \frac{3}{12} = \frac{1}{4}$, you have $43.75 = k(5000)\left(\frac{1}{4}\right)$, which implies that $k = 4(43.75)/5000 = 0.035$. So, the equation relating interest, principal, and time is

$$I = 0.035Pt$$

which is the familiar equation for simple interest where the constant of proportionality, 0.035, represents an annual interest rate of 3.5%.

b. When $P = \$5000$ and $t = \frac{3}{4}$, the interest is $I = (0.035)(5000)\left(\frac{3}{4}\right) = \$131.25$.

✓ *Checkpoint*   *Audio-video solution in English & Spanish at LarsonPrecalculus.com*

The kinetic energy E of an object varies jointly with the object's mass m and the square of the object's velocity v. An object with a mass of 50 kilograms traveling at 16 meters per second has a kinetic energy of 6400 joules. What is the kinetic energy of an object with a mass of 70 kilograms traveling at 20 meters per second?

> **Summarize** (Section 3.5)
>
> 1. Explain how to use a mathematical model to approximate a set of data points *(page 287)*. For an example of using a mathematical model to approximate a set of data points, see Example 1.
> 2. Explain how to use the *regression* feature of a graphing utility to find the equation of a least squares regression line *(page 288)*. For an example of finding the equation of a least squares regression line, see Example 2.
> 3. Explain how to write mathematical models for direct variation, direct variation as an nth power, inverse variation, combined variation, and joint variation *(pages 289–293)*. For examples of these types of variation, see Examples 3–7.

3.5 Exercises

See **CalcChat.com** for tutorial help and worked-out solutions to odd-numbered exercises.

Vocabulary: **Fill in the blanks.**

1. Two techniques for fitting models to data are direct and inverse _____ and least squares _____.

2. Statisticians use a measure called the _____ of the _____ _____ to find a model that approximates a set of data most accurately.

3. The linear model with the least sum of the squared differences is called the _____ _____ _____ line.

4. An r-value, or _____ _____, of a set of data gives a measure of how well a model fits the data.

5. The direct variation model $y = kx^n$ can be described as "y varies directly as the nth power of x," or "y is _____ _____ to the nth power of x."

6. The mathematical model $y = \dfrac{2}{x}$ is an example of _____ variation.

7. Mathematical models that involve both direct and inverse variation have _____ variation.

8. The joint variation model $z = kxy$ can be described as "z varies jointly as x and y," or "z is _____ _____ to x and y."

Skills and Applications

Mathematical Models **In Exercises 9 and 10,** (a) plot the actual data and the model of the same graph and (b) describe how closely the model represents the data. If the model does not closely represent the data, suggest another type of model that may be a better fit.

9. The ordered pairs below give the civilian noninstitutional U.S. populations y (in millions of people) 16 years of age and over not in the civilian labor force from 2006 through 2014. (*Spreadsheet at LarsonPrecalculus.com*)

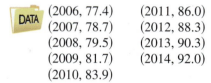

(2006, 77.4)	(2011, 86.0)
(2007, 78.7)	(2012, 88.3)
(2008, 79.5)	(2013, 90.3)
(2009, 81.7)	(2014, 92.0)
(2010, 83.9)	

A model for the data is $y = 1.92t + 65.0$, $6 \leq t \leq 14$, where t represents the years, with $t = 6$ corresponding to 2006. (*Source: U.S. Bureau of Labor Statistics*)

10. The ordered pairs below give the revenues y (in billions of dollars) for Activision Blizzard, Inc., from 2008 through 2014. (*Spreadsheet at LarsonPrecalculus.com*)

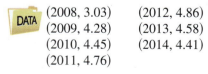

(2008, 3.03)	(2012, 4.86)
(2009, 4.28)	(2013, 4.58)
(2010, 4.45)	(2014, 4.41)
(2011, 4.76)	

A model for the data is $y = 0.184t + 2.32$, $8 \leq t \leq 14$, where t represents the year, with $t = 8$ corresponding to 2008. (*Source: Activision Blizzard, Inc.*)

 Sketching a Line **In Exercises 11–16,** sketch the line that you think best approximates the data in the scatter plot. Then find an equation of the line. To print an enlarged copy of the graph, go to *MathGraphs.com*.

11.
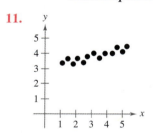

12.
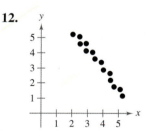

13.
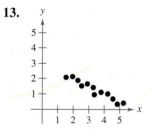

14.
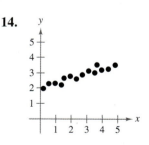

15.
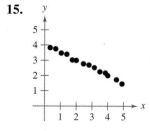

16.
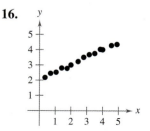

17. Sports The ordered pairs below give the winning times (in seconds) of the women's 100-meter freestyle in the Olympics from 1984 through 2012. *(Spreadsheet at LarsonPrecalculus.com)* *(Source: International Olympic Committee)*

DATA
(1984, 55.92)   (2000, 53.83)
(1988, 54.93)   (2004, 53.84)
(1992, 54.64)   (2008, 53.12)
(1996, 54.50)   (2012, 53.00)

(a) Sketch a scatter plot of the data. Let y represent the winning time (in seconds) and let $t = 84$ represent 1984.

(b) Sketch the line that you think best approximates the data and find an equation of the line.

(c) Use the *regression* feature of a graphing utility to find the equation of the least squares regression line that fits the data.

(d) Compare the linear model you found in part (b) with the linear model you found in part (c).

18. Broadway The ordered pairs below give the starting year and gross ticket sales S (in millions of dollars) for each Broadway season in New York City from 1997 through 2014. *(Spreadsheet at LarsonPrecalculus.com)* *(Source: The Broadway League)*

DATA
(1997, 558)   (2003, 771)   (2009, 1020)
(1998, 588)   (2004, 769)   (2010, 1081)
(1999, 603)   (2005, 862)   (2011, 1139)
(2000, 666)   (2006, 939)   (2012, 1139)
(2001, 643)   (2007, 938)   (2013, 1269)
(2002, 721)   (2008, 943)   (2014, 1365)

(a) Use a graphing utility to create a scatter plot of the data. Let $t = 7$ represent 1997.

(b) Use the *regression* feature of the graphing utility to find the equation of the least squares regression line that fits the data.

(c) Use the graphing utility to graph the scatter plot you created in part (a) and the model you found in part (b) in the same viewing window. How closely does the model represent the data?

(d) Use the model to predict the gross ticket sales during the season starting in 2021.

(e) Interpret the meaning of the slope of the linear model in the context of the problem.

Direct Variation In Exercises 19–24, find a direct variation model that relates y and x.

19. $x = 2$, $y = 14$ 20. $x = 5$, $y = 12$
21. $x = 5$, $y = 1$ 22. $x = -24$, $y = 3$
23. $x = 4$, $y = 8\pi$ 24. $x = \pi$, $y = -1$

Direct Variation as an nth Power In Exercises 25–28, use the given values of k and n to complete the table for the direct variation model $y = kx^n$. Plot the points in a rectangular coordinate system.

x	2	4	6	8	10
$y = kx^n$					

25. $k = 1$, $n = 2$ 26. $k = 2$, $n = 2$
27. $k = \frac{1}{2}$, $n = 3$ 28. $k = \frac{1}{4}$, $n = 3$

Inverse Variation as an nth Power In Exercises 29–32, use the given values of k and n to complete the table for the inverse variation model $y = k/x^n$. Plot the points in a rectangular coordinate system.

x	2	4	6	8	10
$y = k/x^n$					

29. $k = 2$, $n = 1$ 30. $k = 5$, $n = 1$
31. $k = 10$, $n = 2$ 32. $k = 20$, $n = 2$

Think About It In Exercises 33 and 34, use the graph to determine whether y varies directly as some power of x or inversely as some power of x. Explain.

33.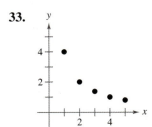

34.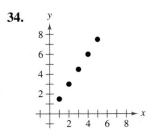

 Determining Variation In Exercises 35–38, determine whether the variation model represented by the ordered pairs (x, y) is of the form $y = kx$ or $y = k/x$, and find k. Then write a model that relates y and x.

35. $(5, 1)$, $\left(10, \frac{1}{2}\right)$, $\left(15, \frac{1}{3}\right)$, $\left(20, \frac{1}{4}\right)$, $\left(25, \frac{1}{5}\right)$
36. $(5, 2)$, $(10, 4)$, $(15, 6)$, $(20, 8)$, $(25, 10)$
37. $(5, -3.5)$, $(10, -7)$, $(15, -10.5)$, $(20, -14)$, $(25, -17.5)$
38. $(5, 24)$, $(10, 12)$, $(15, 8)$, $(20, 6)$, $\left(25, \frac{24}{5}\right)$

Finding a Mathematical Model In Exercises 39–48, find a mathematical model for the verbal statement.

39. A varies directly as the square of r.

40. V varies directly as the cube of l.

41. y varies inversely as the square of x.

42. h varies inversely as the square root of s.

43. F varies directly as g and inversely as r^2.

44. z varies jointly as the square of x and the cube of y.

45. *Newton's Law of Cooling:* The rate of change R of the temperature of an object is directly proportional to the difference between the temperature T of the object and the temperature T_e of the environment.

46. *Boyle's Law:* For a constant temperature, the pressure P of a gas is inversely proportional to the volume V of the gas.

47. *Direct Current:* The electric power P of a direct current circuit is jointly proportional to the voltage V and the electric current I.

48. *Newton's Law of Universal Gravitation:* The gravitational attraction F between two objects of masses m_1 and m_2 is jointly proportional to the masses and inversely proportional to the square of the distance r between the objects.

Describing a Formula In Exercises 49–52, use variation terminology to describe the formula.

49. $y = 2x^2$

50. $t = \dfrac{72}{r}$

51. $A = \frac{1}{2}bh$

52. $K = \frac{1}{2}mv^2$

Finding a Mathematical Model In Exercises 53–60, find a mathematical model that represents the statement. (Determine the constant of proportionality.)

53. y is directly proportional to x. ($y = 54$ when $x = 3$.)

54. A varies directly as r^2. ($A = 9\pi$ when $r = 3$.)

55. y varies inversely as x. ($y = 3$ when $x = 25$.)

56. y is inversely proportional to x^3. ($y = 7$ when $x = 2$.)

57. z varies jointly as x and y. ($z = 64$ when $x = 4$ and $y = 8$.)

58. F is jointly proportional to r and the third power of s. ($F = 4158$ when $r = 11$ and $s = 3$.)

59. P varies directly as x and inversely as the square of y. $\left(P = \frac{28}{3}\text{ when } x = 42 \text{ and } y = 9.\right)$

60. z varies directly as the square of x and inversely as y. ($z = 6$ when $x = 6$ and $y = 4$.)

61. **Simple Interest** The simple interest on an investment is directly proportional to the amount of the investment. An investment of \$3250 earns \$113.75 after 1 year. Find a mathematical model that gives the interest I after 1 year in terms of the amount invested P.

62. **Simple Interest** The simple interest on an investment is directly proportional to the amount of the investment. An investment of \$6500 earns \$211.25 after 1 year. Find a mathematical model that gives the interest I after 1 year in terms of the amount invested P.

63. **Measurement** Use the fact that 13 inches is approximately the same length as 33 centimeters to find a mathematical model that relates centimeters y to inches x. Then use the model to find the numbers of centimeters in 10 inches and 20 inches.

64. **Measurement** Use the fact that 14 gallons is approximately the same amount as 53 liters to find a mathematical model that relates liters y to gallons x. Then use the model to find the numbers of liters in 5 gallons and 25 gallons.

Hooke's Law In Exercises 65–68, use Hooke's Law, which states that the distance a spring stretches (or compresses) from its natural, or equilibrium, length varies directly as the applied force on the spring.

65. A force of 220 newtons stretches a spring 0.12 meter. What force stretches the spring 0.16 meter?

66. A force of 265 newtons stretches a spring 0.15 meter.

(a) What force stretches the spring 0.1 meter?

(b) How far does a force of 90 newtons stretch the spring?

67. The coiled spring of a toy supports the weight of a child. The weight of a 25-pound child compresses the spring a distance of 1.9 inches. The toy does not work properly when a weight compresses the spring more than 3 inches. What is the maximum weight for which the toy works properly?

68. An overhead garage door has two springs, one on each side of the door. A force of 15 pounds is required to stretch each spring 1 foot. Because of a pulley system, the springs stretch only one-half the distance the door travels. The door moves a total of 8 feet, and the springs are at their natural lengths when the door is open. Find the combined lifting force applied to the door by the springs when the door is closed.

69. **Ecology** The diameter of the largest particle that a stream can move is approximately directly proportional to the square of the velocity of the stream. When the velocity is $\frac{1}{4}$ mile per hour, the stream can move coarse sand particles about 0.02 inch in diameter. Approximate the velocity required to carry particles 0.12 inch in diameter.

70. Work The work W required to lift an object varies jointly with the object's mass m and the height h that the object is lifted. The work required to lift a 120-kilogram object 1.8 meters is 2116.8 joules. Find the amount of work required to lift a 100-kilogram object 1.5 meters.

• • 71. Ocean Temperatures • • • • • • • •

The ordered pairs below give the average water temperatures C (in degrees Celsius) at several depths d (in meters) in the Indian Ocean. (*Spreadsheet at LarsonPrecalculus.com*) (*Source: NOAA*)

 (1000, 4.85)   (2500, 1.888)
(1500, 3.525)   (3000, 1.583)
(2000, 2.468)   (3500, 1.422)

(a) Sketch a scatter plot of the data.

(b) Determine whether a direct variation model or an inverse variation model better fits the data.

(c) Find k for each pair of coordinates. Then find the mean value of k to find the constant of proportionality for the model you chose in part (b).

(d) Use your model to approximate the depth at which the water temperature is 3°C.

72. Light Intensity The ordered pairs below give the intensities y (in microwatts per square centimeter) of the light measured by a light probe located x centimeters from a light source. (*Spreadsheet at LarsonPrecalculus.com*)

DATA (30, 0.1881)   (38, 0.1172)   (46, 0.0775)
(34, 0.1543)   (42, 0.0998)   (50, 0.0645)

A model that approximates the data is $y = 171.33/x^2$.

(a) Use a graphing utility to plot the data points and the model in the same viewing window.

(b) Use the model to approximate the light intensity 25 centimeters from the light source.

73. Music The fundamental frequency (in hertz) of a piano string is directly proportional to the square root of its tension and inversely proportional to its length and the square root of its mass density. A string has a frequency of 100 hertz. Find the frequency of a string with each property.

(a) Four times the tension

(b) Twice the length

(c) Four times the tension and twice the length

74. Beam Load The maximum load that a horizontal beam can safely support varies jointly as the width of the beam and the square of its depth and inversely as the length of the beam. Determine how each change affects the beam's maximum load.

(a) Doubling the width

(b) Doubling the depth

(c) Halving the length

(d) Halving the width and doubling the length

Exploration

True or False? **In Exercises 75 and 76, decide whether the statement is true or false. Justify your answer.**

75. If y is directly proportional to x and x is directly proportional to z, then y is directly proportional to z.

76. If y is inversely proportional to x and x is inversely proportional to z, then y is inversely proportional to z.

77. Error Analysis Describe the error.

In the equation for the surface area of a sphere, $S = 4\pi r^2$, the surface area S varies jointly with π and the square of the radius r.

78.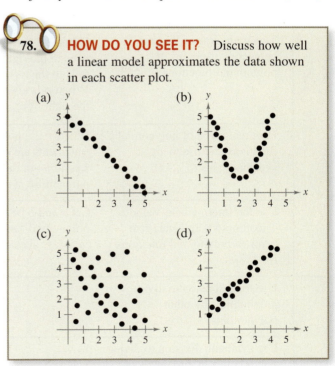

HOW DO YOU SEE IT? Discuss how well a linear model approximates the data shown in each scatter plot.

79. Think About It Let $y = 2x + 2$ and $t = x + 1$. What kind of variation do y and t have? Explain.

Project: Fraud and Identity Theft To work an extended application analyzing the numbers of fraud complaints and identity theft victims in the United States in 2014, visit this text's website at *LarsonPrecalculus.com*. (*Source: U.S. Federal Trade Commission*)

Chapter Summary

	What Did You Learn?	**Explanation/Examples**	**Review Exercises**
Section 3.1	Analyze graphs of quadratic functions *(p. 242)*.	Let a, b, and c be real numbers with $a \neq 0$. The function $f(x) = ax^2 + bx + c$ is a quadratic function. Its graph is a "U"-shaped curve called a parabola.	1, 2
	Write quadratic functions in standard form and use the results to sketch their graphs *(p. 245)*.	The quadratic function $f(x) = a(x - h)^2 + k$, $a \neq 0$, is in standard form. The graph of f is a parabola whose axis is the vertical line $x = h$ and whose vertex is (h, k). When $a > 0$, the parabola opens upward, and when $a < 0$, the parabola opens downward.	3–20
	Find minimum and maximum values of quadratic functions in real-life applications *(p. 247)*.	Consider $f(x) = ax^2 + bx + c$ with vertex $\left(-\dfrac{b}{2a}, f\left(-\dfrac{b}{2a}\right)\right)$. When $a > 0$, f has a *minimum* at $x = -b/(2a)$. When $a < 0$, f has a *maximum* at $x = -b/(2a)$.	21–26
Section 3.2	Use transformations to sketch graphs of polynomial functions *(p. 251)*.	The graph of a polynomial function is continuous (no breaks, holes, or gaps) and has only smooth, rounded turns.	27–32
	Use the Leading Coefficient Test to determine the end behaviors of graphs of polynomial functions *(p. 253)*.	Consider the graph of $f(x) = a_n x^n + \cdots + a_1 x + a_0$, $a_n \neq 0$. **When n is odd:** If $a_n > 0$, then the graph falls to the left and rises to the right. If $a_n < 0$, then the graph rises to the left and falls to the right. **When n is even:** If $a_n > 0$, then the graph rises to the left and to the right. If $a_n < 0$, then the graph falls to the left and to the right.	33–36
	Find real zeros of polynomial functions and use them as sketching aids *(p. 255)*.	When f is a polynomial function and a is a real number, these statements are equivalent: (1) $x = a$ is a *zero* of f, (2) $x = a$ is a *solution* of the equation $f(x) = 0$, (3) $(x - a)$ is a *factor* of the polynomial $f(x)$, and (4) $(a, 0)$ is an *x-intercept* of the graph of f.	37–46
	Use the Intermediate Value Theorem to help locate real zeros of polynomial functions *(p. 258)*.	Let a and b be real numbers such that $a < b$. If f is a polynomial function such that $f(a) \neq f(b)$, then, in the interval $[a, b]$, f takes on every value between $f(a)$ and $f(b)$.	47–50
Section 3.3	Use long division to divide polynomials by other polynomials *(p. 264)*.	Dividend, Quotient, Remainder $$\frac{x^2 + 3x + 5}{x + 1} = x + 2 + \frac{3}{x + 1}$$ Divisor → $x + 1$, $x + 1$ ← Divisor	51–56
	Use synthetic division to divide polynomials by binomials of the form $(x - k)$ *(p. 267)*.	Divisor: $x + 3$ Dividend: $x^4 - 10x^2 - 2x + 4$ $$-3 \begin{array}{|rrrrr} 1 & 0 & -10 & -2 & 4 \\ & -3 & 9 & 3 & -3 \\ \hline 1 & -3 & -1 & 1 & (1) \end{array}$$ ← Remainder: 1 Quotient: $x^3 - 3x^2 - x + 1$	57–60
	Use the Remainder Theorem and the Factor Theorem *(p. 268)*.	**The Remainder Theorem:** If a polynomial $f(x)$ is divided by $x - k$, then the remainder is $r = f(k)$. **The Factor Theorem:** A polynomial $f(x)$ has a factor $(x - k)$ if and only if $f(k) = 0$.	61–68

	What Did You Learn?	Explanation/Examples	Review Exercises
Section 3.4	Use the Fundamental Theorem of Algebra to determine numbers of zeros of polynomial functions *(p. 273)*.	**The Fundamental Theorem of Algebra** If $f(x)$ is a polynomial of degree n, where $n > 0$, then f has at least one zero in the complex number system. **Linear Factorization Theorem** If $f(x)$ is a polynomial of degree n, where $n > 0$, then $f(x)$ has precisely n linear factors $f(x) = a_n(x - c_1)(x - c_2) \cdots (x - c_n)$, where $c_1, c_2, \ldots, c_n$ are complex numbers.	69–74
	Find rational zeros of polynomial functions *(p. 274)*.	The Rational Zero Test relates the possible rational zeros of a polynomial to the leading coefficient and constant term.	75–80
	Find complex zeros using conjugate pairs *(p. 277)*.	**Complex Zeros Occur in Conjugate Pairs** Let f be a polynomial function that has real coefficients. If $a + bi$, where $b \neq 0$, is a zero of the function, then the complex conjugate $a - bi$ is also a zero of the function.	81, 82
	Find zeros of polynomials by factoring *(p. 278)*.	Every polynomial of degree $n > 0$ with real coefficients can be written as the product of linear and quadratic factors with real coefficients, where the quadratic factors have no real zeros.	83–92
	Use Descartes's Rule of Signs and the Upper and Lower Bound Rules to find zeros of polynomials *(p. 280)*.	**Descartes's Rule of Signs** Let $f(x) = a_n x^n + a_{n-1} x^{n-1} + \cdots + a_2 x^2 + a_1 x + a_0$ be a polynomial with real coefficients and $a_0 \neq 0$. **1.** The number of *positive real zeros* of f is either equal to the number of variations in sign of $f(x)$ or less than that number by an even integer. **2.** The number of *negative real zeros* of f is either equal to the number of variations in sign of $f(-x)$ or less than that number by an even integer.	93–96
	Find zeros of polynomials in real-life applications *(p. 282)*.	Zeros of polynomials can be used to find the dimensions of a pyramid-shaped candle mold. (See Example 12.)	97, 98
Section 3.5	Use mathematical models to approximate sets of data points *(p. 287)*.	To see how well a model fits a set of data, compare the actual values of y with the model values. (See Example 1.)	99
	Use the *regression* feature of a graphing utility to find equations of least squares regression lines *(p. 288)*.	The sum of the squared differences is the sum of the squares of the differences between actual data values and model values. The least squares regression line is the linear model with the least sum of the squared differences. The *regression* feature of a graphing utility can be used to find the equation of the least squares regression line. The correlation coefficient (r-value) of the data gives a measure of how well the model fits the data. The closer the value of $|r|$ is to 1, the better the fit.	100
	Write mathematical models for direct variation *(p. 289)*, direct variation as an nth power *(p. 290)*, inverse variation *(p. 291)*, combined variation *(p. 292)*, and joint variation *(p. 293)*.	**Direct variation:** $y = kx$ for some nonzero constant k. **Direct variation as an nth power:** $y = kx^n$ for some nonzero constant k. **Inverse variation:** $y = k/x$ for some nonzero constant k. **Combined variation:** Model has *both* direct and inverse variations. **Joint variation:** $z = kxy$ for some nonzero constant k.	101–106

3.1 Sketching Graphs of Quadratic Functions
In Exercises 1 and 2, sketch the graph of each quadratic function and compare it with the graph of $y = x^2$.

1. (a) $f(x) = 2x^2$
    (b) $g(x) = -2x^2$
    (c) $h(x) = x^2 + 2$
    (d) $k(x) = (x + 2)^2$

2. (a) $f(x) = x^2 - 4$
    (b) $g(x) = 4 - x^2$
    (c) $h(x) = (x - 5)^2 + 3$
    (d) $k(x) = \frac{1}{2}x^2 - 1$

Using Standard Form to Graph a Parabola   In Exercises 3–14, write the quadratic function in standard form and sketch its graph. Identify the vertex, axis of symmetry, and x-intercept(s).

3. $g(x) = x^2 - 2x$ 4. $f(x) = 6x - x^2$

5. $f(x) = x^2 - 6x + 1$ 6. $h(x) = x^2 + 5x - 4$

7. $f(x) = x^2 + 8x + 10$

8. $f(x) = x^2 - 8x + 12$

9. $h(x) = 3 + 4x - x^2$

10. $f(t) = -2t^2 + 4t + 1$

11. $h(x) = 4x^2 + 4x + 13$

12. $f(x) = 4x^2 + 4x + 5$

13. $f(x) = \frac{1}{3}(x^2 + 5x - 4)$

14. $f(x) = \frac{1}{2}(6x^2 - 24x + 22)$

Writing a Quadratic Function   In Exercises 15–20, write the standard form of the quadratic function whose graph is a parabola with the given vertex and that passes through the given point.

15.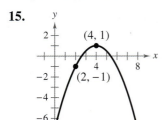

16.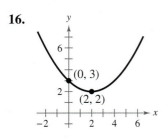

17. Vertex: $(6, 0)$; point: $(3, -9)$

18. Vertex: $(-3, -8)$; point: $(-6, 10)$

19. Vertex: $\left(2, -\frac{5}{2}\right)$; point: $\left(4, \frac{1}{2}\right)$

20. Vertex: $\left(-\frac{1}{4}, 7\right)$; point: $\left(\frac{3}{4}, \frac{49}{8}\right)$

21. **Geometry**   A rectangle is inscribed in the region bounded by the x-axis, the y-axis, and the graph of $x + 2y - 8 = 0$, as shown in the figure.

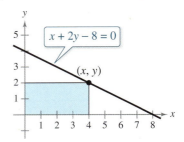

 (a) Write the area A of the rectangle as a function of x.

 (b) Determine the domain of the function in the context of the problem.

 (c) Construct a table showing possible values of x and the corresponding areas of the rectangle. Use the table to estimate the dimensions that produce the maximum area.

 (d) Use a graphing utility to graph the area function. Use the graph to approximate the dimensions that produce the maximum area.

 (e) Write the area function in standard form to find analytically the dimensions that produce the maximum area.

22. **Geometry**   The perimeter of a rectangle is 200 meters.

 (a) Draw a diagram that gives a visual representation of the problem. Let x and y represent the length and width of the rectangle, respectively.

 (b) Write y as a function of x. Use the result to write the area A as a function of x.

 (c) Of all possible rectangles with perimeters of 200 meters, find the dimensions of the one with the maximum area.

23. **Maximum Revenue**   The total revenue R earned (in dollars) from producing a gift box of tea is given by

$$R(p) = -10p^2 + 800p$$

where p is the price per box (in dollars).

 (a) Find the revenues when the prices per box are $20, $25, and $30.

 (b) Find the unit price that yields a maximum revenue. What is the maximum revenue? Explain your results.

24. Maximum Profit A real estate office handles an apartment building that has 50 units. When the rent is $540 per month, all units are occupied. However, for each $30 increase in rent, one unit becomes vacant. Each occupied unit requires an average of $18 per month for service and repairs. What rent should they charge to obtain the maximum profit?

25. Minimum Cost A soft-drink manufacturer has a daily production cost of

$$C = 70{,}000 - 120x + 0.055x^2$$

where C is the total cost (in dollars) and x is the number of units produced. How many units should they produce each day to yield a minimum cost?

26. Maximum Revenue A small theater has a seating capacity of 2000. When the ticket price is $20, attendance is 1500. For each $1 decrease in price, attendance increases by 100.

(a) Write the revenue R of the theater as a function of ticket price x.

(b) What ticket price will yield a maximum revenue? What is the maximum revenue?

3.2 **Sketching a Transformation of a Monomial Function** In Exercises 27–32, sketch the graphs of $y = x^n$ and the transformation.

27. $y = x^3,$ $f(x) = (x - 2)^3$

28. $y = x^3,$ $f(x) = 4x^3$

29. $y = x^4,$ $f(x) = 6 - x^4$

30. $y = x^4,$ $f(x) = 2(x - 8)^4$

31. $y = x^5,$ $f(x) = (x - 5)^5 + 1$

32. $y = x^5,$ $f(x) = \frac{1}{2}x^5 + 3$

Applying the Leading Coefficient Test In Exercises 33–36, describe the left-hand and right-hand behavior of the graph of the polynomial function.

33. $f(x) = -2x^2 - 5x + 12$

34. $f(x) = 4x - \frac{1}{2}x^3$

35. $g(x) = -3x^3 - 8x^4 + x^5$

36. $h(x) = 5 + 9x^6 - 6x^5$

Finding Real Zeros of a Polynomial Function In Exercises 37–42, (a) find all real zeros of the polynomial function, (b) determine whether the multiplicity of each zero is even or odd, (c) determine the maximum possible number of turning points of the graph of the function, and (d) use a graphing utility to graph the function and verify your answers.

37. $f(x) = 3x^2 + 20x - 32$ **38.** $f(x) = x^2 + 12x + 36$

39. $f(t) = t^3 - 3t$ **40.** $f(x) = x^3 - 8x^2$

41. $f(x) = x^4 - 8x^2 - 9$ **42.** $g(x) = x^4 + x^3 - 12x^2$

Sketching the Graph of a Polynomial Function In Exercises 43–46, sketch the graph of the function by (a) applying the Leading Coefficient Test, (b) finding the real zeros of the polynomial, (c) plotting sufficient solution points, and (d) drawing a continuous curve through the points.

43. $f(x) = -x^3 + x^2 - 2$

44. $g(x) = 2x^3 + 4x^2$

45. $f(x) = x(x^3 + x^2 - 5x + 3)$

46. $h(x) = 3x^2 - x^4$

Using the Intermediate Value Theorem In Exercises 47–50, (a) use the Intermediate Value Theorem and the *table* feature of a graphing utility to find intervals one unit in length in which the polynomial function is guaranteed to have a zero. (b) Adjust the table to approximate the zeros of the function to the nearest thousandth. Use the *zero* or *root* feature of the graphing utility to verify your results.

47. $f(x) = 3x^3 - x^2 + 3$

48. $f(x) = 0.25x^3 - 3.65x + 6.12$

49. $f(x) = x^4 - 5x - 1$

50. $f(x) = 7x^4 + 3x^3 - 8x^2 + 2$

3.3 **Long Division of Polynomials** In Exercises 51–56, use long division to divide.

51. $\dfrac{30x^2 - 3x + 8}{5x - 3}$

52. $\dfrac{4x + 7}{3x - 2}$

53. $\dfrac{5x^3 - 21x^2 - 25x - 4}{x^2 - 5x - 1}$

54. $\dfrac{3x^4}{x^2 - 1}$

55. $\dfrac{x^4 - 3x^3 + 4x^2 - 6x + 3}{x^2 + 2}$

56. $\dfrac{6x^4 + 10x^3 + 13x^2 - 5x + 2}{2x^2 - 1}$

Using Synthetic Division In Exercises 57–60, use synthetic division to divide.

57. $\dfrac{2x^3 - 25x^2 + 66x + 48}{x - 8}$

58. $\dfrac{5x^3 + 33x^2 + 50x - 8}{x + 4}$

59. $\dfrac{x^4 - 2x^2 + 9x}{x + 3}$

60. $\dfrac{6x^4 - 4x^3 - 27x^2 + 18x}{x - 2}$

Using the Remainder Theorem In Exercises 61 and 62, use the Remainder Theorem and synthetic division to find each function value.

61. $f(x) = x^4 + 10x^3 - 24x^2 + 20x + 44$

    (a) $f(-3)$ (b) $f(-1)$

62. $g(t) = 2t^5 - 5t^4 - 8t + 20$

    (a) $g(-4)$ (b) $g(\sqrt{2})$

Using the Factor Theorem In Exercises 63 and 64, use synthetic division to determine whether the given values of x are zeros of the function.

63. $f(x) = 20x^4 + 9x^3 - 14x^2 - 3x$

    (a) $x = -1$ (b) $x = \frac{3}{4}$ (c) $x = 0$ (d) $x = 1$

64. $f(x) = 3x^3 - 8x^2 - 20x + 16$

    (a) $x = 4$ (b) $x = -4$ (c) $x = \frac{2}{3}$ (d) $x = -1$

Factoring a Polynomial In Exercises 65–68, (a) verify the given factor(s) of $f(x)$, (b) find the remaining factors of $f(x)$, (c) use your results to write the complete factorization of $f(x)$, (d) list all real zeros of f, and (e) confirm your results by using a graphing utility to graph the function.

Function	Factor(s)
65. $f(x) = x^3 + 4x^2 - 25x - 28$	$(x - 4)$
66. $f(x) = 2x^3 + 11x^2 - 21x - 90$	$(x + 6)$
67. $f(x) = x^4 - 4x^3 - 7x^2 + 22x + 24$	$(x + 2), (x - 3)$
68. $f(x) = x^4 - 11x^3 + 41x^2 - 61x + 30$	$(x - 2), (x - 5)$

3.4 Zeros of Polynomial Functions In Exercises 69–74, determine the number of zeros of the polynomial function.

69. $f(x) = x - 6$ **70.** $g(x) = x^2 - 2x - 8$

71. $h(t) = t^2 - t^5$ **72.** $f(x) = x^8 + x^9$

73. $f(x) = (x - 8)^3$ **74.** $g(t) = (2t - 1)^2 - t^4$

Using the Rational Zero Test In Exercises 75–80, find the rational zeros of the function.

75. $f(x) = x^3 + 3x^2 - 28x - 60$

76. $f(x) = x^3 - 10x^2 + 17x - 8$

77. $f(x) = 3x^3 + 8x^2 - 4x - 16$

78. $f(x) = 4x^3 - 27x^2 + 11x + 42$

79. $f(x) = x^4 + x^3 - 11x^2 + x - 12$

80. $f(x) = 25x^4 + 25x^3 - 154x^2 - 4x + 24$

Finding a Polynomial Function with Given Zeros In Exercises 81 and 82, find a polynomial function with real coefficients that has the given zeros. (There are many correct answers.)

81. $\frac{2}{3}, 4, \sqrt{3}i$

82. $2, -3, 1 - 2i$

Finding the Zeros of a Polynomial Function In Exercises 83 and 84, use the given zero to find all the zeros of the function.

Function	Zero
83. $h(x) = -x^3 + 2x^2 - 16x + 32$	$-4i$
84. $g(x) = 2x^4 - 3x^3 - 13x^2 + 37x - 15$	$2 + i$

Finding the Zeros of a Polynomial Function In Exercises 85–88, write the polynomial as the product of linear factors and list all the zeros of the function.

85. $f(x) = x^3 + 4x^2 - 5x$

86. $g(x) = x^3 - 7x^2 + 36$

87. $g(x) = x^4 + 4x^3 - 3x^2 + 40x + 208$

88. $f(x) = x^4 + 8x^3 + 8x^2 - 72x - 153$

Finding the Zeros of a Polynomial Function In Exercises 89–92, find all the zeros of the function. When there is an extended list of possible rational zeros, use a graphing utility to graph the function in order to disregard any of the possible rational zeros that are obviously not zeros of the function.

89. $f(x) = x^3 - 16x^2 + x - 16$

90. $f(x) = 4x^4 - 12x^3 - 71x^2 - 3x - 18$

91. $g(x) = x^4 - 3x^3 - 14x^2 - 12x - 72$

92. $g(x) = 9x^5 - 27x^4 - 86x^3 + 204x^2 - 40x + 96$

Using Descartes's Rule of Signs In Exercises 93 and 94, use Descartes's Rule of Signs to determine the possible numbers of positive and negative real zeros of the function.

93. $g(x) = 5x^3 + 3x^2 - 6x + 9$

94. $h(x) = -2x^5 + 4x^3 - 2x^2 + 5$

Verifying Upper and Lower Bounds In Exercises 95 and 96, use synthetic division to verify the upper and lower bounds of the real zeros of f.

95. $f(x) = 4x^3 - 3x^2 + 4x - 3$

    (a) Upper: $x = 1$

    (b) Lower: $x = -\frac{1}{4}$

96. $f(x) = 2x^3 - 5x^2 - 14x + 8$

    (a) Upper: $x = 8$

    (b) Lower: $x = -4$

97. Geometry A right cylindrical water bottle has a volume of 36π cubic inches and a height nine inches greater than its radius. Find the dimensions of the water bottle.

98. Geometry A kitchen has a volume of 60 cubic meters. The width of the room is one meter greater than the length and the height is one meter less than the length. Find the dimensions of the room.

3.5

99. Business   The table shows the number of restaurants R operated by Chipotle Mexican Grill, Inc. at the end of each year from 2007 through 2014. (*Source: Chipotle Mexican Grill, Inc.*)

Year	Restaurants, R
2007	704
2008	837
2009	956
2010	1084
2011	1230
2012	1410
2013	1595
2014	1783

DATA Spreadsheet at LarsonPrecalculus.com

A linear model that approximates the data is

$$R = 153.0t - 407, \quad 7 \le t \le 14$$

where t represents the year, with $t = 7$ corresponding to 2007. Plot the actual data and the model on the same graph. How closely does the model represent the data?

100. Agriculture   The table shows the amount B (in millions of pounds) of beef produced on private farms each year from 2007 through 2014.   (*Source: United States Department of Agriculture*)

Year	Pounds, B
2007	102.7
2008	95.9
2009	90.2
2010	84.2
2011	75.0
2012	76.3
2013	70.4
2014	67.9

DATA Spreadsheet at LarsonPrecalculus.com

(a) Use a graphing utility to create a scatter plot of the data. Let t represent the year, with $t = 7$ corresponding to 2007.

(b) Use the *regression* feature of the graphing utility to find the equation of the least squares regression line that fits the data. Then graph the model and the scatter plot you found in part (a) in the same viewing window. How closely does the model represent the data?

(c) Use the model to predict the amount of beef produced on private farms in 2020.

(d) Interpret the meaning of the slope of the linear model in the context of the problem.

101. Measurement   A billboard says that it is 12.5 miles or 20 kilometers to the next gas station. Use this information to find a mathematical model that relates miles x to kilometers y. Then use the model to find the numbers of kilometers in 5 miles and 25 miles.

102. Energy   The power P produced by a wind turbine is directly proportional to the cube of the wind speed S. A wind speed of 27 miles per hour produces a power output of 750 kilowatts. Find the output for a wind speed of 40 miles per hour.

103. Frictional Force   The frictional force F between the tires and the road required to keep a car on a curved section of a highway is directly proportional to the square of the speed s of the car. If the speed of the car is doubled, the force will change by what factor?

104. Demand   A company has found that the daily demand x for its boxes of chocolates is inversely proportional to the price p. When the price is $5, the demand is 800 boxes. Approximate the demand when the price is increased to $6.

105. Travel Time   The travel time between two cities is inversely proportional to the average speed. A train travels between the cities in 3 hours at an average speed of 65 miles per hour. How long does it take to travel between the cities at an average speed of 80 miles per hour?

106. Cost   The cost of constructing a wooden box with a square base varies jointly as the height of the box and the square of the width of the box. Constructing a box of height 16 inches and of width 6 inches costs $28.80. How much does it cost to construct a box of height 14 inches and of width 8 inches?

Exploration

True or False?   **In Exercises 107–109, determine whether the statement is true or false. Justify your answer.**

107. The graph of the function

$$f(x) = 2 + x - x^2 + x^3 - x^4 + x^5 + x^6 - x^7$$

rises to the left and falls to the right.

108. A fourth-degree polynomial with real coefficients can have

$$-5, \quad -8i, \quad 4i, \quad \text{and} \quad 5$$

as its zeros.

109. If y is directly proportional to x, then x is directly proportional to y.

110. Writing   Explain how to determine the maximum or minimum value of a quadratic function.

111. Writing   Explain the connections between factors of a polynomial, zeros of a polynomial function, and solutions of a polynomial equation.

Chapter Test

See **CalcChat.com** for tutorial help and worked-out solutions to odd-numbered exercises.

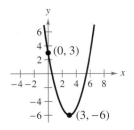

Figure for 3

Take this test as you would take a test in class. When you are finished, check your work against the answers given in the back of the book.

1. Sketch the graph of each quadratic function and compare it with the graph of $y = x^2$.

   (a) $g(x) = -x^2 + 4$ (b) $g(x) = \left(x - \frac{3}{2}\right)^2$

2. Identify the vertex and intercepts of the graph of $f(x) = x^2 - 2x - 3$.

3. Write the standard form of the equation of the parabola shown at the left.

4. The path of a ball is modeled by the function $f(x) = -\frac{1}{20}x^2 + 3x + 5$, where $f(x)$ is the height (in feet) of the ball and x is the horizontal distance (in feet) from where the ball was thrown.

   (a) What is the maximum height of the ball?

   (b) Which number determines the height at which the ball was thrown? Does changing this value change the coordinates of the maximum height of the ball? Explain.

5. Describe the left-hand and right-hand behavior of the graph of the function $h(t) = -\frac{3}{4}t^5 + 2t^2$. Then sketch its graph.

6. Divide using long division.   7. Divide using synthetic division.

   $$\frac{3x^3 + 4x - 1}{x^2 + 1}$$ $$\frac{2x^4 - 3x^2 + 4x - 1}{x + 2}$$

8. Use synthetic division to show that $x = \sqrt{3}$ is a zero of the function

   $$f(x) = 2x^3 - 5x^2 - 6x + 15.$$

   Use the result to factor the polynomial function completely and list all the zeros of the function.

In Exercises 9 and 10, find the rational zeros of the function.

9. $g(t) = 2t^4 - 3t^3 + 16t - 24$ 10. $h(x) = 3x^5 + 2x^4 - 3x - 2$

In Exercises 11 and 12, find a polynomial function with real coefficients that has the given zeros. (There are many correct answers.)

11. $0, 2, 3i$ 12. $1, 1, 2 + \sqrt{3}i$

In Exercises 13 and 14, find all the zeros of the function.

13. $f(x) = 3x^3 + 14x^2 - 7x - 10$ 14. $f(x) = x^4 - 9x^2 - 22x - 24$

In Exercises 15–17, find a mathematical model that represents the statement. (Determine the constant of proportionality.)

15. v varies directly as the square root of s. ($v = 24$ when $s = 16$.)

16. A varies jointly as x and y. ($A = 500$ when $x = 15$ and $y = 8$.)

17. b varies inversely as a. ($b = 32$ when $a = 1.5$.)

18. The table at the left shows the numbers y of insured commercial banks in the United States for the years 2009 through 2014, where t represents the year, with $t = 9$ corresponding to 2009. Use the *regression* feature of a graphing utility to find the equation of the least squares regression line that fits the data. How well does the model represent the data? (*Source: Federal Deposit Insurance Corporation*)

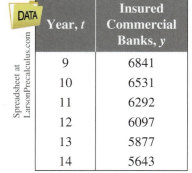

Year, t	Insured Commercial Banks, y
9	6841
10	6531
11	6292
12	6097
13	5877
14	5643

Proofs in Mathematics ■ ■ ■ ■ ■ ■ ■ ■ ■ ■ ■ ■ ■ ■

These two pages contain proofs of four important theorems about polynomial functions. The first two theorems are from Section 3.3, and the second two theorems are from Section 3.4.

The Remainder Theorem *(p. 268)*

If a polynomial $f(x)$ is divided by $x - k$, then the remainder is

$$r = f(k).$$

Proof

Using the Division Algorithm with the divisor $(x - k)$, you have

$$f(x) = (x - k)q(x) + r(x).$$

Either $r(x) = 0$ or the degree of $r(x)$ is less than the degree of $x - k$, so you know that $r(x)$ must be a constant. That is, $r(x) = r$. Now, by evaluating $f(x)$ at $x = k$, you have

$$f(k) = (k - k)q(k) + r$$
$$= (0)q(k) + r$$
$$= r. \qquad ■$$

To be successful in algebra, it is important that you understand the connection among *factors* of a polynomial, *zeros* of a polynomial function, and *solutions* or *roots* of a polynomial equation. The Factor Theorem is the basis for this connection.

The Factor Theorem *(p. 269)*

A polynomial $f(x)$ has a factor $(x - k)$ if and only if $f(k) = 0$.

Proof

Using the Division Algorithm with the factor $(x - k)$, you have

$$f(x) = (x - k)q(x) + r(x).$$

By the Remainder Theorem, $r(x) = r = f(k)$, and you have

$$f(x) = (x - k)q(x) + f(k)$$

where $q(x)$ is a polynomial of lesser degree than $f(x)$. If $f(k) = 0$, then

$$f(x) = (x - k)q(x)$$

and you see that $(x - k)$ is a factor of $f(x)$. Conversely, if $(x - k)$ is a factor of $f(x)$, then division of $f(x)$ by $(x - k)$ yields a remainder of 0. So, by the Remainder Theorem, you have $f(k) = 0$. ■

THE FUNDAMENTAL THEOREM OF ALGEBRA

The Fundamental Theorem of Algebra, which is closely related to the Linear Factorization Theorem, has a long and interesting history. In the early work with polynomial equations, the Fundamental Theorem of Algebra was thought to have been false, because imaginary solutions were not considered. In fact, in the very early work by mathematicians such as Abu al-Khwarizmi (c. 800 A.D.), negative solutions were also not considered.

Once imaginary numbers were considered, several mathematicians attempted to give a general proof of the Fundamental Theorem of Algebra. These included Jean Le Rond d'Alembert (1746), Leonhard Euler (1749), Joseph-Louis Lagrange (1772), and Pierre Simon Laplace (1795). The mathematician usually credited with the first complete and correct proof of the Fundamental Theorem of Algebra is Carl Friedrich Gauss, who published the proof in 1816.

Linear Factorization Theorem   *(p. 273)*

If $f(x)$ is a polynomial of degree n, where $n > 0$, then $f(x)$ has precisely n linear factors

$$f(x) = a_n(x - c_1)(x - c_2) \cdots (x - c_n)$$

where $c_1, c_2, \ldots, c_n$ are complex numbers.

Proof

Using the Fundamental Theorem of Algebra, you know that f must have at least one zero, c_1. Consequently, $(x - c_1)$ is a factor of $f(x)$, and you have

$$f(x) = (x - c_1)f_1(x).$$

If the degree of $f_1(x)$ is greater than zero, then you again apply the Fundamental Theorem of Algebra to conclude that f_1 must have a zero c_2, which implies that

$$f(x) = (x - c_1)(x - c_2)f_2(x).$$

It is clear that the degree of $f_1(x)$ is $n - 1$, that the degree of $f_2(x)$ is $n - 2$, and that you can repeatedly apply the Fundamental Theorem of Algebra n times until you obtain

$$f(x) = a_n(x - c_1)(x - c_2) \cdots (x - c_n)$$

where a_n is the leading coefficient of the polynomial $f(x)$. ∎

Factors of a Polynomial   *(p. 278)*

Every polynomial of degree $n > 0$ with real coefficients can be written as the product of linear and quadratic factors with real coefficients, where the quadratic factors have no real zeros.

Proof

To begin, use the Linear Factorization Theorem to conclude that $f(x)$ can be *completely* factored in the form

$$f(x) = d(x - c_1)(x - c_2)(x - c_3) \cdots (x - c_n).$$

If each c_i is real, then there is nothing more to prove. If any c_i is imaginary ($c_i = a + bi, b \neq 0$), then you know that the conjugate $c_j = a - bi$ is also a zero, because the coefficients of $f(x)$ are real. By multiplying the corresponding factors, you obtain

$$
\begin{aligned}
(x - c_i)(x - c_j) &= [x - (a + bi)][x - (a - bi)] \\
&= [(x - a) - bi][(x - a) + bi] \\
&= (x - a)^2 + b^2 \\
&= x^2 - 2ax + (a^2 + b^2)
\end{aligned}
$$

where each coefficient is real. ∎

P.S. Problem Solving ▪ ▪ ▪ ▪ ▪ ▪ ▪ ▪ ▪ ▪ ▪ ▪ ▪

1. Exploring Zeros of Functions

(a) Find the zeros of each quadratic function $g(x)$.

   (i) $g(x) = x^2 - 4x - 12$

   (ii) $g(x) = x^2 + 5x$

   (iii) $g(x) = x^2 + 3x - 10$

   (iv) $g(x) = x^2 - 4x + 4$

   (v) $g(x) = x^2 - 2x - 6$

   (vi) $g(x) = x^2 + 3x + 4$

(b) For each function in part (a), use a graphing utility to graph $f(x) = (x - 2) \cdot g(x)$. Verify that $(2, 0)$ is an x-intercept of the graph of $f(x)$. Describe any similarities or differences in the behaviors of the six functions at this x-intercept.

(c) For each function in part (b), use the graph of $f(x)$ to approximate the other x-intercepts of the graph.

(d) Describe the connections that you find among the results of parts (a), (b), and (c).

2. Exploring Zeros of Functions

(a) Find the zeros of each quadratic function $g(x)$.

   (i) $g(x) = 2x^2 + 5x - 3$

   (ii) $g(x) = -x^2 - 3x - 2$

(b) For each function in part (a), find the zeros of $f(x) = g\left(\frac{1}{2}x\right)$.

(c) Describe the connection between the results in parts (a) and (b).

3. Building a Quonset Hut Quonset huts were developed during World War II. They were temporary housing structures that could be assembled quickly and easily. A Quonset hut is shaped like a half cylinder. A manufacturer has 600 square feet of material with which to build a Quonset hut.

(a) The formula for the surface area of half a cylinder is $S = \pi r^2 + \pi rl$, where r is the radius and l is the length of the hut. Solve this equation for l when $S = 600$.

(b) The formula for the volume of the hut is $V = \frac{1}{2}\pi r^2 l$. Write the volume V of the Quonset hut as a polynomial function of r.

(c) Use the function you wrote in part (b) to find the maximum volume of a Quonset hut with a surface area of 600 square feet. What are the dimensions of the hut?

4. Verifying the Remainder Theorem Show that if $f(x) = ax^3 + bx^2 + cx + d$, then $f(k) = r$, where $r = ak^3 + bk^2 + ck + d$, using long division. In other words, verify the Remainder Theorem for a third-degree polynomial function.

5. Babylonian Mathematics In 2000 B.C., the Babylonians solved polynomial equations by referring to tables of values. One such table gave the values of $y^3 + y^2$. To be able to use this table, the Babylonians sometimes used the method below to manipulate the equation.

$$ax^3 + bx^2 = c \qquad \text{Original equation}$$

$$\frac{a^3x^3}{b^3} + \frac{a^2x^2}{b^2} = \frac{a^2c}{b^3} \qquad \text{Multiply each side by } \frac{a^2}{b^3}.$$

$$\left(\frac{ax}{b}\right)^3 + \left(\frac{ax}{b}\right)^2 = \frac{a^2c}{b^3} \qquad \text{Rewrite.}$$

Then they would find $(a^2c)/b^3$ in the $y^3 + y^2$ column of the table. They knew that the corresponding y-value was equal to $(ax)/b$, so they could conclude that $x = (by)/a$.

(a) Calculate $y^3 + y^2$ for $y = 1, 2, 3, \ldots, 10$. Record the values in a table.

(b) Use the table from part (a) and the method above to solve each equation.

   (i) $x^3 + x^2 = 252$ (ii) $x^3 + 2x^2 = 288$

   (iii) $3x^3 + x^2 = 90$ (iv) $2x^3 + 5x^2 = 2500$

   (v) $7x^3 + 6x^2 = 1728$ (vi) $10x^3 + 3x^2 = 297$

(c) Using the methods from this chapter, verify your solution of each equation.

6. Zeros of a Cubic Function Can a cubic function with real coefficients have two real zeros and one complex zero? Explain.

7. Sums and Products of Zeros

(a) Complete the table.

Function	Zeros	Sum of Zeros	Product of Zeros
$f_1(x) = x^2 - 5x + 6$			
$f_2(x) = x^3 - 7x + 6$			
$f_3(x) = x^4 + 2x^3 + x^2 + 8x - 12$			
$f_4(x) = x^5 - 3x^4 - 9x^3 + 25x^2 - 6x$			

(b) Use the table to make a conjecture relating the sum of the zeros of a polynomial function to the coefficients of the polynomial function.

(c) Use the table to make a conjecture relating the product of the zeros of a polynomial function to the coefficients of the polynomial function.

8. True or False? Determine whether the statement is true or false. If false, provide one or more reasons why the statement is false and correct the statement. Let

$$f(x) = ax^3 + bx^2 + cx + d, \quad a \neq 0$$

and let $f(2) = -1$. Then

$$\frac{f(x)}{x + 1} = q(x) + \frac{2}{x + 1}$$

where $q(x)$ is a second-degree polynomial.

9. Finding the Equation of a Parabola The parabola shown in the figure has an equation of the form $y = ax^2 + bx + c$. Find the equation of this parabola using each method. (a) Find the equation analytically. (b) Use the *regression* feature of a graphing utility to find the equation.

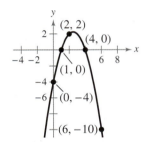

10. Finding the Slope of a Tangent Line One of the fundamental themes of calculus is to find the slope of the tangent line to a curve at a point. To see how this can be done, consider the point $(2, 4)$ on the graph of the quadratic function $f(x) = x^2$, as shown in the figure.

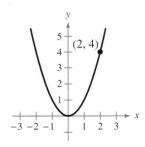

(a) Find the slope m_1 of the line joining $(2, 4)$ and $(3, 9)$.

(b) Find the slope m_2 of the line joining $(2, 4)$ and $(1, 1)$.

(c) Find the slope m_3 of the line joining $(2, 4)$ and $(2.1, 4.41)$.

(d) Find the slope m_h of the line joining $(2, 4)$ and $(2 + h, f(2 + h))$ in terms of the nonzero number h.

(e) Evaluate the slope formula from part (d) for $h = -1$, 1, and 0.1. Compare these values with those in parts (a)–(c).

(f) What can you conclude the slope $m_{\tan}$ of the tangent line at $(2, 4)$ to be? Explain.

11. Maximum Area A rancher plans to fence a rectangular pasture adjacent to a river (see figure). The rancher has 100 meters of fencing, and no fencing is needed along the river.

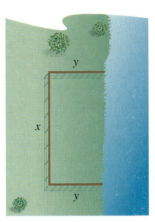

(a) Write the area A of the pasture as a function of x, the length of the side parallel to the river. What is the domain of A?

(b) Graph the function A and estimate the dimensions that yield the maximum area of the pasture.

(c) Find the exact dimensions that yield the maximum area of the pasture by writing the quadratic function in standard form.

12. Maximum and Minimum Area A wire 100 centimeters in length is cut into two pieces. One piece is bent to form a square and the other to form a circle. Let x equal the length of the wire used to form the square.

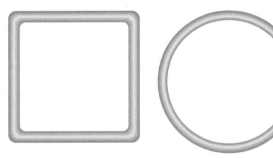

(a) Write the function that represents the combined area of the two figures.

(b) Determine the domain of the function.

(c) Find the value(s) of x that yield a maximum area and a minimum area. Explain.

13. Finding Dimensions At a glassware factory, molten cobalt glass is poured into molds to make paperweights. Each mold is a rectangular prism whose height is 3 inches greater than the length of each side of the square base. A machine pours 20 cubic inches of liquid glass into each mold. What are the dimensions of the mold?

4 Rational Functions and Conics

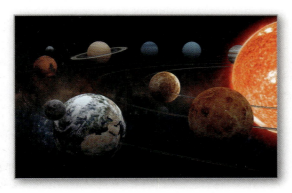

Solar System *(Page 330)*

Satellite Orbit
(Exercise 41, page 348)

Suspension Bridge *(Exercise 31, page 338)*

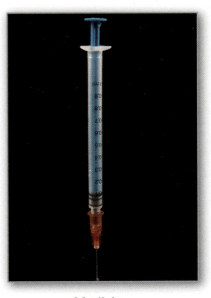

Medicine
(Exercise 90, page 326)

Coal *(Example 4, page 313)*

4.1 Rational Functions and Asymptotes

Rational functions have many real-life applications. For example, in Exercise 41 on page 316, you will use a rational function to determine the cost of supplying recycling bins to the population of a rural township.

■ Find domains of rational functions.
■ Find vertical and horizontal asymptotes of graphs of rational functions.
■ Use rational functions to model and solve real-life problems.

Introduction

A **rational function** is a quotient of polynomial functions. It can be written in the form

$$f(x) = \frac{N(x)}{D(x)}$$

where $N(x)$ and $D(x)$ are polynomials and $D(x)$ is not the zero polynomial.

The *domain* of a rational function of x includes all real numbers except x-values that make the denominator zero. Much of the discussion of rational functions will focus on the behavior of their graphs near x-values excluded from the domain.

EXAMPLE 1 **Finding the Domain of a Rational Function**

See LarsonPrecalculus.com for an interactive version of this type of example.

Find the domain of $f(x) = \dfrac{1}{x}$ and discuss the behavior of f near any excluded x-values.

Solution The denominator is zero when $x = 0$, so the domain of f is all real numbers except $x = 0$. To determine the behavior of f near this excluded value, evaluate $f(x)$ to the left and right of $x = 0$, as shown in the tables below.

x	-1	-0.5	-0.1	-0.01	-0.001	$\to 0$
$f(x)$	-1	-2	-10	-100	-1000	$\to -\infty$

x	$0 \leftarrow$	0.001	0.01	0.1	0.5	1
$f(x)$	$\infty \leftarrow$	1000	100	10	2	1

Note that as x approaches 0 *from the left*, $f(x)$ decreases without bound. In contrast, as x approaches 0 *from the right*, $f(x)$ increases without bound. The graph of f is shown below.

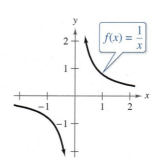

··▷ **REMARK** Recall from Section 2.4 that the rational function

$$f(x) = \frac{1}{x}$$

is also referred to as the *reciprocal function.*

✓ *Checkpoint*  *Audio-video solution in English & Spanish at LarsonPrecalculus.com*

Find the domain of $f(x) = \dfrac{3x}{x - 1}$ and discuss the behavior of f near any excluded x-values.

Vertical and Horizontal Asymptotes

In Example 1, the behavior of f near $x = 0$ is as denoted below.

$$f(x) \to -\infty \text{ as } x \to 0^- \qquad f(x) \to \infty \text{ as } x \to 0^+$$

$f(x)$ decreases without bound as x approaches 0 from the left. $f(x)$ increases without bound as x approaches 0 from the right.

The line $x = 0$ is a **vertical asymptote** of the graph of f, as shown in Figure 4.1. Notice that the graph of f also has a **horizontal asymptote**—the line $y = 0$. The behavior of f near $y = 0$ is as denoted below.

$$f(x) \to 0 \text{ as } x \to -\infty \qquad f(x) \to 0 \text{ as } x \to \infty$$

$f(x)$ approaches 0 as x decreases without bound. $f(x)$ approaches 0 as x increases without bound.

Vertical asymptote: $x = 0$

$f(x) = \dfrac{1}{x}$

Horizontal asymptote: $y = 0$

Figure 4.1

Definitions of Vertical and Horizontal Asymptotes

1. The line $x = a$ is a **vertical asymptote** of the graph of f when

$$f(x) \to \infty \quad \text{or} \quad f(x) \to -\infty$$

as $x \to a$, either from the right or from the left.

2. The line $y = b$ is a **horizontal asymptote** of the graph of f when

$$f(x) \to b$$

as $x \to \infty$ or $x \to -\infty$.

Eventually (as $x \to \infty$ or $x \to -\infty$), the distance between the horizontal asymptote and the points on the graph must approach zero. Figure 4.2 shows the vertical and horizontal asymptotes of the graphs of three rational functions.

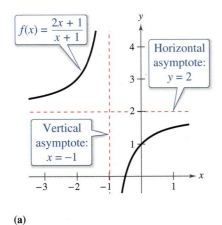

$f(x) = \dfrac{2x + 1}{x + 1}$

Horizontal asymptote: $y = 2$

Vertical asymptote: $x = -1$

(a)

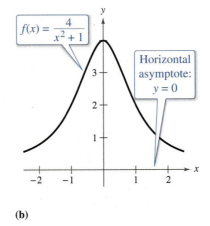

$f(x) = \dfrac{4}{x^2 + 1}$

Horizontal asymptote: $y = 0$

(b)

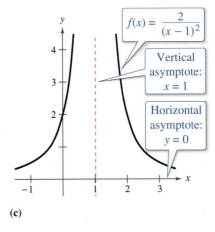

$f(x) = \dfrac{2}{(x - 1)^2}$

Vertical asymptote: $x = 1$

Horizontal asymptote: $y = 0$

(c)

Figure 4.2

Verify numerically the horizontal asymptotes shown in Figure 4.2. For example, to show that the line $y = 2$ is the horizontal asymptote of the graph of

$$f(x) = \frac{2x + 1}{x + 1}$$

create a table that shows the value of $f(x)$ as x increases and decreases without bound.

The graphs of $f(x) = \dfrac{1}{x}$ in Figure 4.1 and $f(x) = \dfrac{2x + 1}{x + 1}$ in Figure 4.2(a) are **hyperbolas.** You will study hyperbolas in Sections 4.3 and 4.4.

> ### Vertical and Horizontal Asymptotes
>
> Let f be the rational function
>
> $$f(x) = \frac{N(x)}{D(x)} = \frac{a_n x^n + a_{n-1}x^{n-1} + \cdots + a_1 x + a_0}{b_m x^m + b_{m-1}x^{m-1} + \cdots + b_1 x + b_0}$$
>
> where $N(x)$ and $D(x)$ have no common factors.
>
> **1.** The graph of f has *vertical* asymptotes at the zeros of $D(x)$.
>
> **2.** The graph of f has at most one *horizontal* asymptote determined by comparing the degrees of $N(x)$ and $D(x)$.
>
>   **a.** When $n < m$, the graph of f has the line $y = 0$ (the x-axis) as a horizontal asymptote.
>
>   **b.** When $n = m$, the graph of f has the line $y = \dfrac{a_n}{b_m}$ (ratio of the leading coefficients) as a horizontal asymptote.
>
>   **c.** When $n > m$, the graph of f has no horizontal asymptote.

EXAMPLE 2   Finding Vertical and Horizontal Asymptotes

Find all vertical and horizontal asymptotes of the graph of each rational function.

a. $f(x) = \dfrac{2x}{3x^2 + 1}$ **b.** $f(x) = \dfrac{2x^2}{x^2 - 1}$

Solution

a. For this rational function, the degree of the numerator is *less than* the degree of the denominator, so the graph has the line $y = 0$ as a horizontal asymptote. To find any vertical asymptotes, set the denominator equal to zero and solve the resulting equation for x. The equation

$$3x^2 + 1 = 0$$

has no real solutions, so the graph has no vertical asymptote. Figure 4.3 shows the graph of the function.

b. For this rational function, the degree of the numerator is *equal* to the degree of the denominator. The leading coefficient of the numerator is 2 and the leading coefficient of the denominator is 1, so the graph has the line $y = 2/1 = 2$ as a horizontal asymptote. To find any vertical asymptotes, set the denominator equal to zero and solve the resulting equation for x.

$$x^2 - 1 = 0 \qquad \text{Set denominator equal to zero.}$$
$$(x + 1)(x - 1) = 0 \qquad \text{Factor.}$$
$$x + 1 = 0 \implies x = -1 \qquad \text{Set 1st factor equal to 0 and solve.}$$
$$x - 1 = 0 \implies x = 1 \qquad \text{Set 2nd factor equal to 0 and solve.}$$

This equation has two real solutions, $x = -1$ and $x = 1$, so the graph has the lines $x = -1$ and $x = 1$ as vertical asymptotes. Figure 4.4 shows the graph of the function.

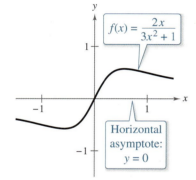

Figure 4.3

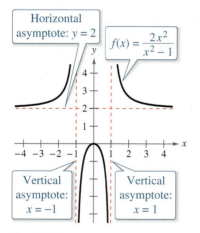

Figure 4.4

✓ *Checkpoint* 🔊))) Audio-video solution in English & Spanish at LarsonPrecalculus.com

Find all vertical and horizontal asymptotes of the graph of $f(x) = \dfrac{5x^2}{x^2 - 1}$. ∎

EXAMPLE 3 **Finding Vertical and Horizontal Asymptotes**

Find all vertical and horizontal asymptotes of the graph of

$$f(x) = \frac{x^2 + x - 2}{x^2 - x - 6}.$$

Solution For this rational function, the degree of the numerator is *equal to* the degree of the denominator. The leading coefficients of the numerator and denominator are both 1, so the graph has the line $y = 1/1 = 1$ as a horizontal asymptote. To find any vertical asymptotes, first factor the numerator and denominator.

$$f(x) = \frac{x^2 + x - 2}{x^2 - x - 6} = \frac{(x - 1)(x + 2)}{(x + 2)(x - 3)} = \frac{x - 1}{x - 3}, \quad x \neq -2$$

Setting the denominator $x - 3$ (of the simplified function) equal to zero shows that the graph has the line $x = 3$ as a vertical asymptote.

 Checkpoint)) *Audio-video solution in English & Spanish at LarsonPrecalculus.com*

Find all vertical and horizontal asymptotes of the graph of

$$f(x) = \frac{3x^2 + 7x - 6}{x^2 + 4x + 3}.$$

· · · · · · · · · · · · · · · · · ▷
· · **REMARK** There is a *hole* in the graph of f at $x = -2$. In Section 4.2, you will sketch the graph of a rational function that has a hole.

Applications

EXAMPLE 4 **Cost-Benefit Model**

A utility company burns coal to generate electricity. The cost C (in dollars) of removing $p\%$ of the smokestack pollutants is given by $C = 80{,}000p/(100 - p)$, $0 \le p < 100$. Sketch the graph of this function. You are a member of a state legislature considering a law that would require utility companies to remove 90% of the pollutants from their smokestack emissions. The current law requires 85% removal. How much additional cost would the utility company incur as a result of the new law?

Solution Figure 4.5 shows the graph of this function. Note that the graph has a vertical asymptote at $p = 100$. The current law requires 85% removal, so the current cost to the utility company is

$$C = \frac{80{,}000(85)}{100 - 85} \approx \$453{,}333. \qquad \text{Evaluate } C \text{ when } p = 85.$$

The cost to remove 90% of the pollutants would be

$$C = \frac{80{,}000(90)}{100 - 90} = \$720{,}000. \qquad \text{Evaluate } C \text{ when } p = 90.$$

So, the new law would require the utility company to spend an additional

$$720{,}000 - 453{,}333 = \$266{,}667. \qquad \text{Subtract 85\% removal cost from 90\% removal cost.}$$

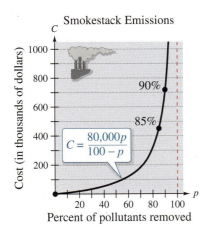
Worldwide, coal is the largest source of energy used to generate electricity.

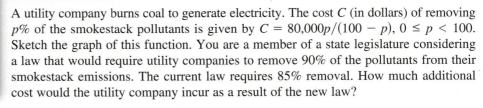

Smokestack Emissions

Cost (in thousands of dollars)

Percent of pollutants removed

$$C = \frac{80{,}000p}{100 - p}$$

Figure 4.5

 Checkpoint)) *Audio-video solution in English & Spanish at LarsonPrecalculus.com*

The cost C (in millions of dollars) of removing $p\%$ of the industrial and municipal pollutants discharged into a river is given by $C = 255p/(100 - p)$, $0 \le p < 100$.

a. Find the costs of removing 20%, 45%, and 80% of the pollutants.

b. According to the model, is it possible to remove 100% of the pollutants? Explain.

Ultraviolet Radiation
Protection

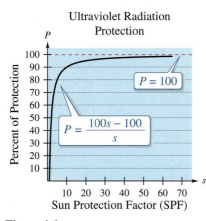

Figure 4.6

EXAMPLE 5 **Ultraviolet Radiation Protection**

The percent P of ultraviolet radiation protection provided by a sunscreen can be modeled by

$$P = \frac{100s - 100}{s}, \quad 1 \le s \le 65$$

where s is the Sun Protection Factor, or SPF. *(Source: U.S. Environmental Protection Agency)*

a. Find the percents of ultraviolet radiation protection provided by sunscreens with SPF values of 15, 30, and 50.

b. If the model were valid for all $s \ge 1$, what would be the horizontal asymptote of the graph of this function, and what would it represent?

Solution

a. When $s = 15$, $P = \dfrac{100(15) - 100}{15}$

$$\approx 93.3\%.$$

When $s = 30$, $P = \dfrac{100(30) - 100}{30}$

$$\approx 96.7\%.$$

When $s = 50$, $P = \dfrac{100(50) - 100}{50}$

$$= 98\%.$$

b. As shown in Figure 4.6, the horizontal asymptote is the line $P = 100$. This line represents 100% protection from ultraviolet radiation.

✓ **Checkpoint**  *Audio-video solution in English & Spanish at LarsonPrecalculus.com*

A business has a cost function of $C = 0.4x + 8000$, where C is measured in dollars and x is the number of units produced. The average cost per unit is given by

$$\overline{C} = \frac{C}{x} = \frac{0.4x + 8000}{x}, \quad x > 0.$$

a. Find the average costs per unit when $x = 1000$, $x = 8000$, $x = 20{,}000$, and $x = 100{,}000$.

b. What is the horizontal asymptote of the graph of this function, and what does it represent?

Summarize (Section 4.1)

1. State the definition of a rational function and explain how to find the domain of a rational function *(page 310)*. For an example of finding the domain of a rational function, see Example 1.

2. Explain how to find the vertical and horizontal asymptotes of the graph of a rational function *(page 311)*. For examples of finding vertical and horizontal asymptotes of graphs of rational functions, see Examples 2 and 3.

3. Describe real-life applications of rational functions *(pages 313 and 314, Examples 4 and 5)*.

4.1 Exercises

See CalcChat.com for tutorial help and worked-out solutions to odd-numbered exercises.

Vocabulary: Fill in the blanks.

1. A _____ _____ is a quotient of polynomial functions.
2. When $f(x) \to \pm\infty$ as $x \to a$ from the left or the right, $x = a$ is a _____ _____ of the graph of f.
3. When $f(x) \to b$ as $x \to \pm\infty$, $y = b$ is a _____ _____ of the graph of f.
4. The graph of $f(x) = 1/x$ is a _____.

Skills and Applications

Finding the Domain of a Rational Function In Exercises 5–12, find the domain of the function and discuss the behavior of f near any excluded x-values.

5. $f(x) = \dfrac{1}{x-1}$

6. $f(x) = \dfrac{4}{x+3}$

7. $f(x) = \dfrac{5x}{x+2}$

8. $f(x) = \dfrac{6x}{x-3}$

9. $f(x) = \dfrac{3x^2}{x^2-1}$

10. $f(x) = \dfrac{2x}{x^2-4}$

11. $f(x) = \dfrac{x^2+3x+2}{x^2-2x+1}$

12. $f(x) = \dfrac{x^2+x-20}{x^2+8x+16}$

Finding Vertical and Horizontal Asymptotes In Exercises 13–28, find all vertical and horizontal asymptotes of the graph of the function.

13. $f(x) = \dfrac{4}{x^2}$

14. $f(x) = \dfrac{1}{(x-2)^3}$

15. $f(x) = \dfrac{5+x}{5-x}$

16. $f(x) = \dfrac{3-7x}{3+2x}$

17. $f(x) = \dfrac{x^3}{x^2-1}$

18. $f(x) = \dfrac{2x^2}{x+1}$

19. $f(x) = \dfrac{-4x^2+1}{x^2+x+3}$

20. $f(x) = \dfrac{3x^2+x-5}{x^2+1}$

21. $f(x) = \dfrac{x-4}{x^2-16}$

22. $f(x) = \dfrac{x+3}{x^2-9}$

23. $f(x) = \dfrac{x^2-1}{x^2-x-6}$

24. $f(x) = \dfrac{x^2-4}{x^2-3x+2}$

25. $f(x) = \dfrac{x^2-3x-4}{2x^2+x-1}$

26. $f(x) = \dfrac{x^2+x-2}{2x^2+5x+2}$

27. $f(x) = \dfrac{6x^2+5x-6}{3x^2-8x+4}$

28. $f(x) = \dfrac{-8x^2+10x+3}{3x^2+10x-8}$

Matching In Exercises 29–36, match the rational function with its graph. [The graphs are labeled (a)–(h).]

(a)
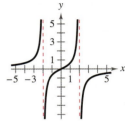

(b)
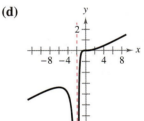

(c)
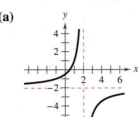

(d)
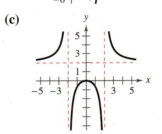

(e)
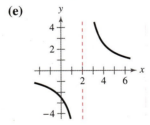

(f)
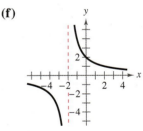

(g)
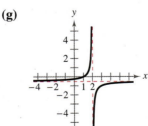

(h)
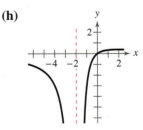

29. $f(x) = \dfrac{4}{x+2}$

30. $f(x) = \dfrac{5}{x-2}$

31. $f(x) = -\dfrac{2x-1}{x-2}$

32. $f(x) = -\dfrac{x-1}{2x-4}$

33. $f(x) = \dfrac{2x^2}{x^2-4}$

34. $f(x) = \dfrac{-2x}{x^2-4}$

35. $f(x) = \dfrac{x^3}{4(x+2)^2}$

36. $f(x) = \dfrac{3x}{(x+2)^2}$

Numerical Analysis In Exercises 37–40, (a) determine the domains of f and g, (b) simplify f and find any vertical asymptotes of the graph of f, (c) complete the table, and (d) explain how the two functions differ.

37. $f(x) = \dfrac{x^2 - 4}{x + 2}$, $g(x) = x - 2$

x	-4	-3	-2.5	-2	-1.5	-1	0
$f(x)$							
$g(x)$							

38. $f(x) = \dfrac{x^2(x + 3)}{x^2 + 3x}$, $g(x) = x$

x	-3	-2	-1	0	1	2	3
$f(x)$							
$g(x)$							

39. $f(x) = \dfrac{2x - 1}{2x^2 - x}$, $g(x) = \dfrac{1}{x}$

x	-1	-0.5	0	0.5	2	3	4
$f(x)$							
$g(x)$							

40. $f(x) = \dfrac{2x - 8}{x^2 - 9x + 20}$, $g(x) = \dfrac{2}{x - 5}$

x	0	1	2	3	4	5	6
$f(x)$							
$g(x)$							

41. Recycling

The cost C (in dollars) of supplying recycling bins to $p\%$ of the population of a rural township is given by

$$C = \frac{25{,}000p}{100 - p}, \quad 0 \le p < 100.$$

(a) Use a graphing utility to graph the cost function.

(b) Find the costs of supplying bins to 15%, 50%, and 90% of the population.

(c) According to the model, is it possible to supply bins to 100% of the population? Explain.

42. Ecology The cost C (in millions of dollars) of removing $p\%$ of the industrial and municipal pollutants discharged into a river is given by

$$C = \frac{225p}{100 - p}, \quad 0 \le p < 100.$$

(a) Use a graphing utility to graph the cost function.

(b) Find the costs of removing 10%, 40%, and 75% of the pollutants.

(c) According to the model, is it possible to remove 100% of the pollutants? Explain.

43. Population Growth The game commission introduces 100 deer into newly acquired state game lands. The population N of the herd is modeled by

$$N = \frac{20(5 + 3t)}{1 + 0.04t}, \quad t \ge 0$$

where t is the time in years.

(a) Use a graphing utility to graph this model.

(b) Find the populations when $t = 5$, $t = 10$, and $t = 25$.

(c) What is the limiting size of the herd as time increases?

44. Biology A science class performs an experiment comparing the quantity of food consumed by a species of moth with the quantity of food supplied. The model for the experimental data is

$$y = \frac{1.568x - 0.001}{6.360x + 1}, \quad x > 0$$

where x is the quantity (in milligrams) of food supplied and y is the quantity (in milligrams) of food consumed.

(a) Use a graphing utility to graph this model.

(b) At what level of consumption will the moth become satiated?

45. Psychology Psychologists have developed mathematical models to predict memory performance as a function of the number of trials n of a certain task. Consider the learning curve

$$P = \frac{0.5 + 0.9(n - 1)}{1 + 0.9(n - 1)}, \quad n > 0$$

where P is the fraction of correct responses after n trials.

(a) Complete the table for this model. What does it suggest?

n	1	2	3	4	5	6	7	8	9	10
P										

(b) According to the model, what is the limiting percent of correct responses as n increases?

46. Physics Experiment Consider a physics laboratory experiment designed to determine an unknown mass. A flexible metal meter stick is clamped to a table with 50 centimeters overhanging the edge (see figure). Known masses M ranging from 200 grams to 2000 grams are attached to the end of the meter stick. For each mass, the meter stick is displaced vertically and then allowed to oscillate. The average time t (in seconds) of one oscillation for each mass is recorded in the table.

DATA	Mass, M	Time, t
	200	0.450
	400	0.597
	600	0.712
	800	0.831
	1000	0.906
	1200	1.003
	1400	1.088
	1600	1.168
	1800	1.218
	2000	1.338

Spreadsheet at LarsonPrecalculus.com

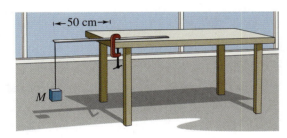

A model for the data that can be used to predict the time of one oscillation is

$$t = \frac{38M + 16{,}965}{10(M + 5000)}.$$

(a) Use this model to create a table showing the predicted time for each of the masses shown in the table above.

(b) Compare the predicted times with the experimental times. What can you conclude?

(c) Use the model to approximate the mass of an object for which $t = 1.056$ seconds.

Exploration

True or False? In Exercises 47–49, determine whether the statement is true or false. Justify your answer.

47. The graph of a polynomial function can have infinitely many vertical asymptotes.

48. $f(x) = x^3 - 2x^2 - 5x + 6$ is a rational function.

49. The graph of a rational function never has two horizontal asymptotes.

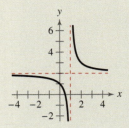

50. **HOW DO YOU SEE IT?** The graph of a rational function

$$f(x) = \frac{N(x)}{D(x)}$$

is shown below. Determine which of the statements about the function is false. Justify your answer.

(a) When $x = 1$, $D(x) = 0$.

(b) The degree of $N(x)$ is equal to the degree of $D(x)$.

(c) The ratio of the leading coefficients of $N(x)$ and $D(x)$ is 1.

Determining Function Behavior In Exercises 51–54, (a) determine the value that the function f approaches as the magnitude of x increases. Is $f(x)$ greater than or less than this value when (b) x is positive and large in magnitude and (c) x is negative and large in magnitude?

51. $f(x) = \dfrac{4x - 1}{x}$

52. $f(x) = \dfrac{2x - 5}{x - 3}$

53. $f(x) = \dfrac{2x - 1}{x^2 + 1}$

54. $f(x) = \dfrac{-3x^2}{x^2 + 6}$

Writing a Rational Function In Exercises 55 and 56, write a rational function f whose graph has the specified characteristics. (There are many correct answers.)

55. Vertical asymptote: None
Horizontal asymptote: $y = 2$

56. Vertical asymptotes: $x = -2$, $x = 1$
Horizontal asymptote: None

57. Error Analysis Describe the error.

A real zero of the numerator of a rational function f is $x = c$. So, $x = c$ must also be a zero of f.

58. Think About It Give an example of a rational function whose domain is the set of all real numbers. Give an example of a rational function whose domain is the set of all real numbers except $x = 15$.

4.2 Graphs of Rational Functions

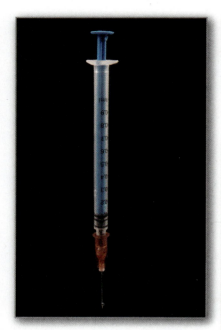

Graphs of rational functions have many real-life applications. For example, in Exercise 90 on page 326, you will use the graph of a rational function to analyze the concentration of a chemical in the bloodstream after injection.

- Sketch graphs of rational functions.
- Sketch graphs of rational functions that have slant asymptotes.
- Use graphs of rational functions to model and solve real-life problems.

Sketching the Graph of a Rational Function

To sketch the graph of a rational function, use the guidelines listed below.

Guidelines for Graphing Rational Functions

Let $f(x) = \dfrac{N(x)}{D(x)}$, where $N(x)$ and $D(x)$ are polynomials and $D(x)$ is not the zero polynomial.

1. Simplify f, if possible. List any restrictions on the domain of f that are not implied by the simplified function.

2. Find and plot the y-intercept (if any) by evaluating $f(0)$.

3. Find the zeros of the numerator (if any). Then plot the corresponding x-intercepts.

4. Find the zeros of the denominator (if any). Then sketch the corresponding vertical asymptotes.

5. Find and sketch the horizontal asymptote (if any) by using the rule for finding the horizontal asymptote of a rational function on page 312.

6. Plot at least one point *between* and one point *beyond* each x-intercept and vertical asymptote.

7. Use smooth curves to complete the graph between and beyond the vertical asymptotes.

You may also want to test for symmetry when graphing rational functions, especially for simple rational functions. Recall from Section 2.4 that the graph of the reciprocal function $f(x) = \dfrac{1}{x}$ is symmetric with respect to the origin.

▷ **TECHNOLOGY** Some graphing utilities have difficulty graphing rational functions with vertical asymptotes. In *connected* mode, the utility may connect parts of the graph that are not supposed to be connected. For example, the graph on the left should consist of two unconnected portions—one to the left of $x = 2$ and the other to the right of $x = 2$. Changing the mode of the graphing utility to *dot* mode eliminates this problem. However, in *dot* mode, the graph is represented as a collection of dots (as shown in the graph on the right) rather than as a smooth curve.

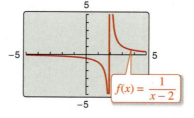

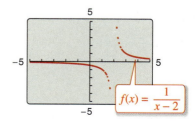

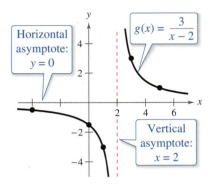

Horizontal asymptote: $y = 0$

$g(x) = \dfrac{3}{x - 2}$

Vertical asymptote: $x = 2$

Figure 4.7

REMARK In Examples 1–5, note that the vertical asymptotes are included in the tables of additional points to emphasize numerically the behavior of the graph of the function.

EXAMPLE 1 **Sketching the Graph of a Rational Function**

See LarsonPrecalculus.com for an interactive version of this type of example.

Sketch the graph of $g(x) = \dfrac{3}{x - 2}$ and state its domain.

Solution

y-intercept:	$\left(0, -\frac{3}{2}\right)$, because $g(0) = -\frac{3}{2}$
x-intercept:	none, because there are no zeros of the numerator
Vertical asymptote:	$x = 2$, zero of denominator
Horizontal asymptote:	$y = 0$, because degree of $N(x) <$ degree of $D(x)$

Additional points:

x	-4	1	2	3	5
$g(x)$	-0.5	-3	Undefined	3	1

By plotting the intercept, asymptotes, and a few additional points, you obtain the graph shown in Figure 4.7. The domain of g is all real numbers except $x = 2$.

✓ *Checkpoint* *Audio-video solution in English & Spanish at LarsonPrecalculus.com*

Sketch the graph of $f(x) = \dfrac{1}{x + 3}$ and state its domain.

Note that the graph of g in Example 1 is a vertical stretch and a right shift of the graph of $f(x) = 1/x$, because

$$g(x) = \frac{3}{x - 2} = 3\left(\frac{1}{x - 2}\right) = 3f(x - 2).$$

EXAMPLE 2 **Sketching the Graph of a Rational Function**

Sketch the graph of $f(x) = \dfrac{2x - 1}{x}$ and state its domain.

Solution

y-intercept:	none, because $x = 0$ is not in the domain
x-intercept:	$\left(\frac{1}{2}, 0\right)$, because $2x - 1 = 0$ when $x = \frac{1}{2}$
Vertical asymptote:	$x = 0$, zero of denominator
Horizontal asymptote:	$y = 2$, because degree of $N(x) =$ degree of $D(x)$

Additional points:

x	-4	-1	0	$\frac{1}{4}$	4
$f(x)$	2.25	3	Undefined	-2	1.75

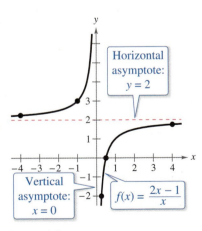

Horizontal asymptote: $y = 2$

Vertical asymptote: $x = 0$

$f(x) = \dfrac{2x - 1}{x}$

Figure 4.8

By plotting the intercept, asymptotes, and a few additional points, you obtain the graph shown in Figure 4.8. The domain of f is all real numbers except $x = 0$.

✓ *Checkpoint* *Audio-video solution in English & Spanish at LarsonPrecalculus.com*

Sketch the graph of $g(x) = \dfrac{3 + 2x}{1 + x}$ and state its domain.

Figure 4.9

Hole at $x = 3$

Figure 4.10

EXAMPLE 3   **Sketching the Graph of a Rational Function**

Sketch the graph of $f(x) = \dfrac{x}{x^2 - x - 2}$ and state its domain.

Solution   Factor the denominator to more easily determine the zeros of the denominator.

$$f(x) = \frac{x}{x^2 - x - 2} = \frac{x}{(x + 1)(x - 2)}$$

y-intercept: $(0, 0)$, because $f(0) = 0$

x-intercept: $(0, 0)$

Vertical asymptotes: $x = -1$, $x = 2$, zeros of denominator

Horizontal asymptote: $y = 0$, because degree of $N(x) <$ degree of $D(x)$

Additional points:

x	-3	-1	-0.5	1	2	3
$f(x)$	-0.3	Undefined	0.4	-0.5	Undefined	0.75

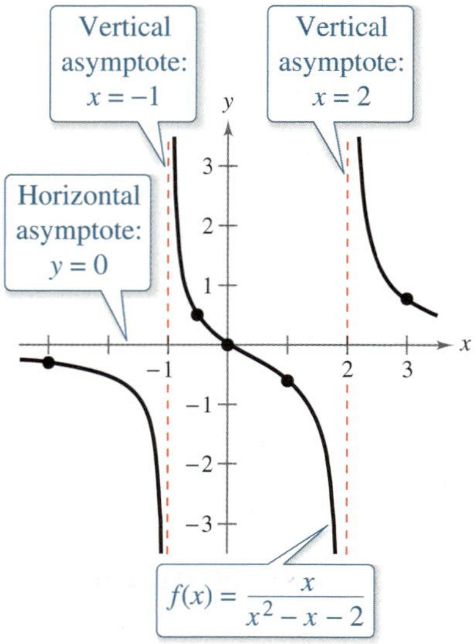

Figure 4.9 shows the graph. The domain of f is all real numbers except $x = -1$ and $x = 2$.

✓ **Checkpoint** 🔊)) *Audio-video solution in English & Spanish at LarsonPrecalculus.com*

Sketch the graph of $f(x) = \dfrac{3x}{x^2 + x - 2}$ and state its domain.

EXAMPLE 4   **Sketching the Graph of a Rational Function**

Sketch the graph of $f(x) = \dfrac{x^2 - 9}{x^2 - 2x - 3}$ and state its domain.

Solution   By factoring the numerator and denominator, you have

$$f(x) = \frac{x^2 - 9}{x^2 - 2x - 3} = \frac{(x - 3)(x + 3)}{(x - 3)(x + 1)} = \frac{x + 3}{x + 1}, \quad x \neq 3.$$

y-intercept: $(0, 3)$, because $f(0) = 3$

x-intercept: $(-3, 0)$, because $x + 3 = 0$ when $x = -3$

Vertical asymptote: $x = -1$, zero of (simplified) denominator

Horizontal asymptote: $y = 1$, because degree of $N(x) =$ degree of $D(x)$

Additional points:

x	-5	-2	-1	-0.5	1	3	4
$f(x)$	0.5	-1	Undefined	5	2	Undefined	1.4

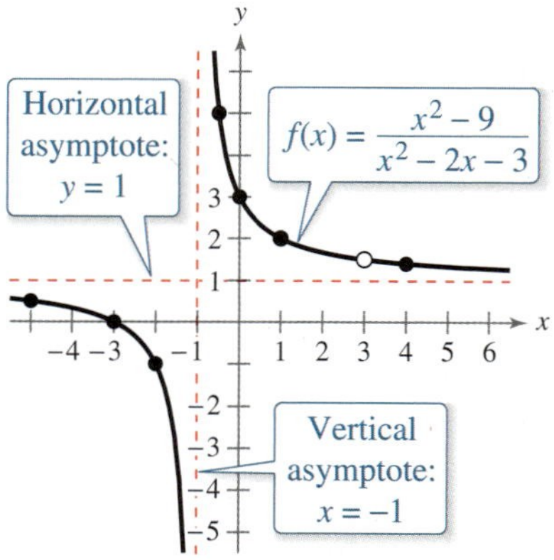

Figure 4.10 shows the graph. Notice that there is a *hole* in the graph at $x = 3$ because the numerator and denominator have a common factor of $x - 3$. The domain of f is all real numbers except $x = -1$ and $x = 3$.

✓ **Checkpoint** 🔊)) *Audio-video solution in English & Spanish at LarsonPrecalculus.com*

Sketch the graph of $f(x) = \dfrac{x^2 - 4}{x^2 - x - 6}$ and state its domain. ■

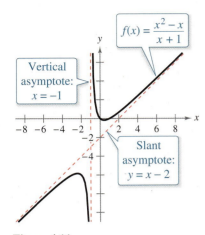

Figure 4.11

Slant Asymptotes

Consider a rational function whose denominator is of degree 1 or greater. If the degree of the numerator is exactly *one more* than the degree of the denominator, then the graph of the function has a **slant** (or **oblique**) **asymptote.** For example, the graph of

$$f(x) = \frac{x^2 - x}{x + 1}$$

has a slant asymptote, as shown in Figure 4.11. To find the equation of a slant asymptote, use long division. For example, by dividing $x + 1$ into $x^2 - x$, you obtain

$$f(x) = \frac{x^2 - x}{x + 1} = x - 2 + \frac{2}{x + 1}.$$

$$\underbrace{}_{\substack{\text{Slant asymptote} \\ (y = x - 2)}}$$

As x increases or decreases without bound, the remainder term $2/(x + 1)$ approaches 0, so the graph of f approaches the line $y = x - 2$, as shown in Figure 4.11.

EXAMPLE 5   **A Rational Function with a Slant Asymptote**

Sketch the graph of $f(x) = \dfrac{x^2 - x - 2}{x - 1}$ and state its domain.

Solution   First write $f(x)$ in two different ways. Factoring the numerator

$$f(x) = \frac{x^2 - x - 2}{x - 1}$$

$$= \frac{(x - 2)(x + 1)}{x - 1}$$

enables you to recognize the x-intercepts. Long division

$$f(x) = \frac{x^2 - x - 2}{x - 1}$$

$$= x - \frac{2}{x - 1}$$

enables you to recognize that the line $y = x$ is a slant asymptote of the graph.

y-intercept: $(0, 2)$, because $f(0) = 2$

x-intercepts: $(2, 0)$ and $(-1, 0)$, because $x - 2 = 0$ when $x = 2$ and $x + 1 = 0$ when $x = -1$

Vertical asymptote: $x = 1$, zero of denominator

Slant asymptote: $y = x$

Additional points:

x	-2	0.5	1	1.5	3
$f(x)$	-1.33	4.5	Undefined	-2.5	2

Figure 4.12 shows the graph. The domain of f is all real numbers except $x = 1$.

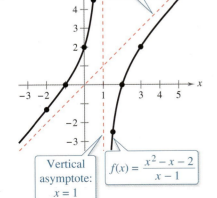

Figure 4.12

✓ *Checkpoint* 🔊)) *Audio-video solution in English & Spanish at LarsonPrecalculus.com*

Sketch the graph of $f(x) = \dfrac{3x^2 + 1}{x}$ and state its domain.

Application

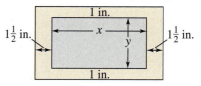

Figure 4.13

EXAMPLE 6   **Finding a Minimum Area**

A rectangular page contains 48 square inches of print. The margins at the top and bottom of the page are each 1 inch deep. The margins on each side are $1\frac{1}{2}$ inches wide. What should the dimensions of the page be to use the least amount of paper?

Graphical Solution

Let A be the area to be minimized. From Figure 4.13, you can write $A = (x + 3)(y + 2)$. The printed area inside the margins is given by $xy = 48$ or $y = 48/x$. To find the minimum area, rewrite the equation for A in terms of just one variable by substituting $48/x$ for y.

$$A = (x + 3)\left(\frac{48}{x} + 2\right)$$

$$= \frac{(x + 3)(48 + 2x)}{x}, \quad x > 0$$

The graph of this rational function is shown below. Because x represents the width of the printed area, you need to consider only the portion of the graph for which x is positive. Use the *minimum* feature of a graphing utility to estimate that the minimum value of A occurs when $x \approx 8.5$ inches. The corresponding value of y is $48/8.5 \approx 5.6$ inches. So, the dimensions should be

$$x + 3 \approx 11.5 \text{ inches} \quad \text{by} \quad y + 2 \approx 7.6 \text{ inches.}$$

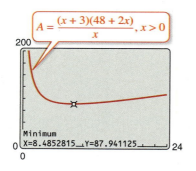

Numerical Solution

Let A be the area to be minimized. From Figure 4.13, you can write $A = (x + 3)(y + 2)$. The printed area inside the margins is given by $xy = 48$ or $y = 48/x$. To find the minimum area, rewrite the equation for A in terms of just one variable by substituting $48/x$ for y.

$$A = (x + 3)\left(\frac{48}{x} + 2\right) = \frac{(x + 3)(48 + 2x)}{x}, \quad x > 0$$

Use the *table* feature of a graphing utility to create a table of values for the function

$$y_1 = \frac{(x + 3)(48 + 2x)}{x}$$

beginning at $x = 1$ and increasing by 1. The minimum value of y_1 occurs when x is somewhere between 8 and 9, as shown in Figure 4.14. To approximate the minimum value of y_1 to one decimal place, change the table to begin at $x = 8$ and increase by 0.1. The minimum value of y_1 occurs when $x \approx 8.5$, as shown in Figure 4.15. The corresponding value of y is $48/8.5 \approx 5.6$ inches. So, the dimensions should be

$$x + 3 \approx 11.5 \text{ inches} \quad \text{by} \quad y + 2 \approx 7.6 \text{ inches.}$$

X	Y₁	
6	90	
7	88.571	
8	**88**	
9	88	
10	88.4	
11	89.091	
12	90	

X=8

Figure 4.14

X	Y₁	
8.2	87.961	
8.3	87.949	
8.4	87.943	
8.5	**87.941**	
8.6	87.944	
8.7	87.952	
8.8	87.964	

X=8.5

Figure 4.15

 **Checkpoint**   *Audio-video solution in English & Spanish at LarsonPrecalculus.com*

Rework Example 6 when the margins on each side are 2 inches wide and the page contains 40 square inches of print.

Summarize   (Section 4.2)

1. Explain how to sketch the graph of a rational function *(page 318)*. For examples of sketching graphs of rational functions, see Examples 1–4.

2. Explain how to determine whether the graph of a rational function has a slant asymptote *(page 321)*. For an example of sketching the graph of a rational function that has a slant asymptote, see Example 5.

3. Describe an example of how to use the graph of a rational function to model and solve a real-life problem *(page 322, Example 6)*.

4.2 Exercises

See CalcChat.com for tutorial help and worked-out solutions to odd-numbered exercises.

Vocabulary: Fill in the blanks.

1. For the rational function $f(x) = N(x)/D(x)$, if the degree of $N(x)$ is exactly one more than the degree of $D(x)$, then the graph of f has a _____ (or oblique) _____.

2. The graph of $g(x) = 3/(x - 2)$ has a _____ asymptote at $x = 2$.

Skills and Applications

 Sketching a Transformation of a Rational Function In Exercises 3–6, use the graph of $f(x) = \dfrac{2}{x}$ to sketch the graph of g.

3. $g(x) = \dfrac{2}{x} + 4$

4. $g(x) = \dfrac{2}{x - 4}$

5. $g(x) = -\dfrac{2}{x}$

6. $g(x) = \dfrac{1}{x + 2}$

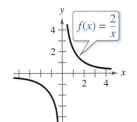

Sketching a Transformation of a Rational Function In Exercises 7–10, use the graph of $f(x) = \dfrac{3}{x^2}$ to sketch the graph of g.

7. $g(x) = \dfrac{3}{x^2} - 1$

8. $g(x) = -\dfrac{3}{x^2}$

9. $g(x) = \dfrac{3}{(x - 1)^2}$

10. $g(x) = \dfrac{1}{x^2}$

Sketching a Transformation of a Rational Function In Exercises 11–14, use the graph of $f(x) = \dfrac{4}{x^3}$ to sketch the graph of g.

11. $g(x) = \dfrac{4}{(x + 2)^3}$

12. $g(x) = \dfrac{4}{x^3} - 2$

13. $g(x) = -\dfrac{4}{x^3}$

14. $g(x) = \dfrac{2}{x^3}$

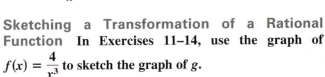

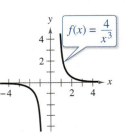

 Sketching the Graph of a Rational Function In Exercises 15–44, (a) state the domain of the function, (b) identify all intercepts, (c) find any vertical or horizontal asymptotes, and (d) plot additional solution points as needed to sketch the graph of the rational function.

15. $f(x) = \dfrac{1}{x + 1}$

16. $f(x) = \dfrac{1}{x - 3}$

17. $h(x) = \dfrac{-1}{x + 4}$

18. $g(x) = \dfrac{1}{6 - x}$

19. $C(x) = \dfrac{2x + 3}{x + 2}$

20. $P(x) = \dfrac{1 - 3x}{1 - x}$

21. $g(x) = \dfrac{1}{x + 2} + 2$

22. $f(x) = \dfrac{3}{x - 1} - 5$

23. $f(x) = \dfrac{x^2}{x^2 + 9}$

24. $f(t) = \dfrac{1 - 2t}{t}$

25. $h(x) = \dfrac{x^2}{x^2 - 9}$

26. $g(x) = \dfrac{x}{x^2 - 9}$

27. $g(s) = \dfrac{4s}{s^2 + 4}$

28. $f(x) = -\dfrac{x}{(x - 2)^2}$

29. $g(x) = \dfrac{4(x + 1)}{x(x - 4)}$

30. $h(x) = \dfrac{2}{x^2(x - 2)}$

31. $f(x) = \dfrac{2x}{x^2 - 3x - 4}$

32. $f(x) = \dfrac{3x}{x^2 + 2x - 3}$

33. $f(x) = \dfrac{5(x + 4)}{x^2 + x - 12}$

34. $f(x) = \dfrac{x^2 + 3x}{x^2 + x - 6}$

35. $f(t) = \dfrac{t^2 - 1}{t - 1}$

36. $f(x) = \dfrac{x^2 - 36}{x + 6}$

37. $h(x) = \dfrac{x^2 - 5x + 4}{x^2 - 4}$

38. $g(x) = \dfrac{x^2 - 2x - 8}{x^2 - 9}$

39. $f(x) = \dfrac{x^2 + 4}{x^2 + 3x - 4}$

40. $f(x) = \dfrac{3(x^2 + 1)}{x^2 + 2x - 15}$

41. $f(x) = \dfrac{2x^2 - 5x + 2}{2x^2 - x - 6}$

42. $f(x) = \dfrac{3x^2 - 8x + 4}{2x^2 - 3x - 2}$

43. $f(x) = \dfrac{2x^2 - 5x - 3}{x^3 - 2x^2 - x + 2}$

44. $f(x) = \dfrac{x^2 - x - 2}{x^3 - 2x^2 - 5x + 6}$

Comparing Graphs of Functions In Exercises 45 and 46, (a) state the domains of f and g, (b) use a graphing utility to graph f and g in the same viewing window, and (c) explain why the graphing utility may not show the difference in the domains of f and g.

45. $f(x) = \dfrac{x^2 - 1}{x + 1}$, $g(x) = x - 1$

46. $f(x) = \dfrac{x - 2}{x^2 - 2x}$, $g(x) = \dfrac{1}{x}$

A Rational Function with a Slant Asymptote In Exercises 47–62, (a) state the domain of the function, (b) identify all intercepts, (c) find any vertical or slant asymptotes, and (d) plot additional solution points as needed to sketch the graph of the rational function.

47. $h(x) = \dfrac{x^2 - 4}{x}$

48. $g(x) = \dfrac{x^2 + 5}{x}$

49. $f(x) = \dfrac{2x^2 + 1}{x}$

50. $f(x) = \dfrac{-x^2 - 2}{x}$

51. $g(x) = \dfrac{x^2 + 1}{x}$

52. $h(x) = \dfrac{x^2}{x - 1}$

53. $f(t) = -\dfrac{t^2 + 1}{t + 5}$

54. $f(x) = \dfrac{x^2 + 1}{x + 1}$

55. $f(x) = \dfrac{x^3}{x^2 - 4}$

56. $g(x) = \dfrac{x^3}{2x^2 - 8}$

57. $f(x) = \dfrac{x^3 - 1}{x^2 - x}$

58. $f(x) = \dfrac{x^4 + x}{x^3}$

59. $f(x) = \dfrac{x^2 - x + 1}{x - 1}$

60. $f(x) = \dfrac{2x^2 - 5x + 5}{x - 2}$

61. $f(x) = \dfrac{2x^3 - x^2 - 2x + 1}{x^2 + 3x + 2}$

62. $f(x) = \dfrac{2x^3 + x^2 - 8x - 4}{x^2 - 3x + 2}$

Using Technology In Exercises 63–66, use a graphing utility to graph the rational function. Give the domain of the function and find any asymptotes. Then zoom out sufficiently far so that the graph appears as a line. Identify the line.

63. $f(x) = \dfrac{x^2 + 2x - 8}{x + 2}$

64. $f(x) = \dfrac{2x^2 + x}{x + 1}$

65. $g(x) = \dfrac{1 + 3x^2 - x^3}{x^2}$

66. $h(x) = \dfrac{12 - 2x - x^2}{2(4 + x)}$

Graphical Reasoning In Exercises 67–74, (a) use the graph to determine any x-intercepts of the graph of the rational function and (b) set $y = 0$ and solve the resulting equation to confirm your result in part (a).

67. $y = \dfrac{x + 1}{x - 3}$

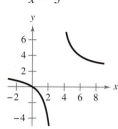

68. $y = \dfrac{2x}{x - 3}$

69. $y = \dfrac{3x}{x + 1}$

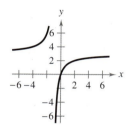

70. $y = \dfrac{x}{x^2 - 4}$

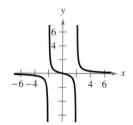

71. $y = \dfrac{1}{x} - x$

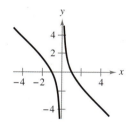

72. $y = x - 3 + \dfrac{2}{x}$

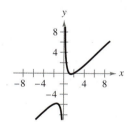

73. $y = x + 2 + \dfrac{1}{x}$

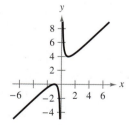

74. $y = x - \dfrac{2}{x + 1}$

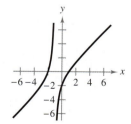

Graphical Reasoning In Exercises 75–78, (a) use a graphing utility to graph the function and determine any x-intercepts of the graph and (b) set $y = 0$ and solve the resulting equation to confirm your result in part (a).

75. $y = \dfrac{1}{x + 5} + \dfrac{4}{x}$

76. $y = 20\left(\dfrac{2}{x + 1} - \dfrac{3}{x}\right)$

77. $y = x - \dfrac{6}{x - 1}$

78. $y = x - \dfrac{16}{x}$

79. **Geometry**   A rectangular region of length x and width y has an area of 600 square meters.

    (a) Write the width y as a function of x.

    (b) Determine the domain of the function based on the physical constraints of the problem.

    (c) Sketch the graph of the function and determine the width of the rectangle when $x = 35$ meters.

80. **Page Design**   A rectangular page contains 64 square inches of print. The margins at the top and bottom of the page are each 1 inch deep. The margins on each side are $1\frac{1}{2}$ inches wide. What should the dimensions of the page be to use the least amount of paper?

81. **Page Design**   A page that is x inches wide and y inches high contains 30 square inches of print. The top and bottom margins are each 1 inch deep and the margins on each side are 2 inches wide (see figure).

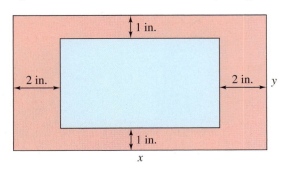

    (a) Show that the total area A of the page is

    $$A = \frac{2x(x + 11)}{x - 4}.$$

    (b) Determine the domain of the function based on the physical constraints of the problem.

    (c) Use a graphing utility to graph the area function and approximate the dimensions of the page that use the least amount of paper. Verify your answer numerically using the *table* feature of the graphing utility.

82. **Concentration of a Mixture**   A 1000-liter tank contains 50 liters of a 25% brine solution. You add x liters of a 75% brine solution to the tank.

    (a) Show that the concentration C, the proportion of brine to total solution, in the final mixture is

    $$C = \frac{3x + 50}{4(x + 50)}.$$

    (b) Determine the domain of the function based on the physical constraints of the problem.

    (c) Sketch the graph of the concentration function.

    (d) As the tank is filled, what happens to the rate at which the concentration of brine is increasing? What percent does the concentration of brine appear to approach?

Using Technology   In Exercises 83–86, use a graphing utility to graph the function and locate any relative maximum or minimum points on the graph.

83. $f(x) = \dfrac{3(x + 1)}{x^2 + x + 1}$

84. $g(x) = \dfrac{6x}{x^2 + x + 1}$

85. $C(x) = x - 2 + \dfrac{32}{x}$

86. $f(x) = x - 4 + \dfrac{16}{x}$

87. **Minimum Cost**   The ordering and transportation cost C (in thousands of dollars) for the components used in manufacturing a product is given by

    $$C = 100\left(\frac{200}{x^2} + \frac{x}{x + 30}\right), \quad x \geq 1$$

    where x is the order size (in hundreds). Use a graphing utility to graph the cost function. From the graph, estimate the order size that minimizes cost.

88. **Minimum Cost**   The cost C (in dollars) of producing x units of a product is given by

    $$C = 0.2x^2 + 10x + 5$$

    and the average cost per unit $\overline{C}$ is given by

    $$\overline{C} = \frac{C}{x} = \frac{0.2x^2 + 10x + 5}{x}, \quad x > 0.$$

    Sketch the graph of the average cost function and estimate the number of units that should be produced to minimize the average cost per unit.

89. **Average Speed**   A driver's average speed is 50 miles per hour on a round trip between two cities 100 miles apart. The average speeds for going and returning were x and y miles per hour, respectively.

    (a) Show that $y = \dfrac{25x}{x - 25}.$

    (b) Determine the vertical and horizontal asymptotes of the graph of the function.

    (c) Use a graphing utility to graph the function.

    (d) Complete the table.

x	30	35	40	45	50	55	60
y							

    (e) Are the results in the table what you expected? Explain.

    (f) Is it possible to average 20 miles per hour in one direction and still average 50 miles per hour on the round trip? Explain.

• • 90. Medicine • • • • • • • • • • • •

The concentration C of a chemical in the bloodstream t hours after injection into muscle tissue is given by

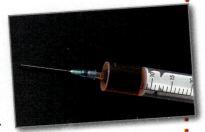

$$C = \frac{3t^2 + t}{t^3 + 50}, \quad t > 0.$$

(a) Determine the horizontal asymptote of the graph of the function and interpret its meaning in the context of the problem.

(b) Use a graphing utility to graph the function and approximate the time when the bloodstream concentration is greatest.

(c) Use the graphing utility to determine when the concentration is less than 0.345.

Exploration

True or False? In Exercises 91–94, determine whether the statement is true or false. Justify your answer.

91. When the graph of a rational function f has a vertical asymptote at $x = 5$, it is possible to sketch the graph without lifting your pencil from the paper.

92. The graph of a rational function can never cross one of its asymptotes.

93. The graph of $f(x) = \dfrac{x^2}{x + 1}$ has a slant asymptote.

94. The graph of every rational function has a horizontal asymptote.

95. **Error Analysis** Describe the error.

The graph of

$$h(x) = \frac{6 - 2x}{3 - x}$$

has the line $x = 3$ as a vertical asymptote because $3 - x = 0$ when $x = 3$.

96. **Writing**

(a) Given a rational function f, how can you determine whether the graph of f has a slant asymptote?

(b) You determine that the graph of a rational function f has a slant asymptote. How do you find it?

97. **Writing and Graphing a Rational Function** Write a rational function satisfying the criteria below. Then sketch the graph of your function.

Vertical asymptote: $x = 2$

Slant asymptote: $y = x + 1$

Zero of the function: $x = -2$

98. **HOW DO YOU SEE IT?** For each statement below, determine which graph(s) show that the statement is incorrect.

(i)

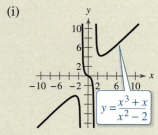

$$y = \frac{x^3 + x}{x^2 - 2}$$

(ii)

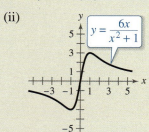

$$y = \frac{6x}{x^2 + 1}$$

(iii)

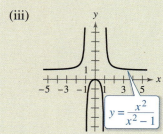

$$y = \frac{x^2}{x^2 - 1}$$

(iv)

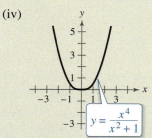

$$y = \frac{x^4}{x^2 + 1}$$

(a) The graph of every rational function has a vertical asymptote.

(b) The graph of every rational function has at least one vertical, horizontal, or slant asymptote.

(c) The graph of a rational function can have at most one vertical asymptote.

99. **Think About It** Is it possible for the graph of a rational function to have all three types of asymptotes? Why or why not?

Project: Department of Defense To work an extended application analyzing the total numbers of military personnel on active duty from 1984 through 2014, visit this text's website at *LarsonPrecalculus.com*. (*Source: U.S. Department of Defense*)

4.3 Conics

Conics have many real-life applications and are often used to model and solve engineering problems. For example, in Exercise 31 on page 338, you will use a parabola to model the cables of the Golden Gate Bridge.

■ **Recognize the four basic conics: circle, ellipse, parabola, and hyperbola.**
■ **Recognize, graph, and write equations of parabolas (vertex at origin).**
■ **Recognize, graph, and write equations of ellipses (center at origin).**
■ **Recognize, graph, and write equations of hyperbolas (center at origin).**

Introduction

The earliest basic descriptions of conic sections took place during the classical Greek period, 500 to 336 B.C. This early Greek study was largely concerned with the geometric properties of conics. It was not until the early 17th century that the broad applicability of conics became apparent and played a prominent role in the early development of calculus.

A **conic section** (or simply **conic**) is the intersection of a plane and a double-napped cone. Notice in Figure 4.16 that in the formation of the four basic conics, the intersecting plane does not pass through the vertex of the cone. When the plane does pass through the vertex, the resulting figure is a **degenerate conic,** as shown in Figure 4.17.

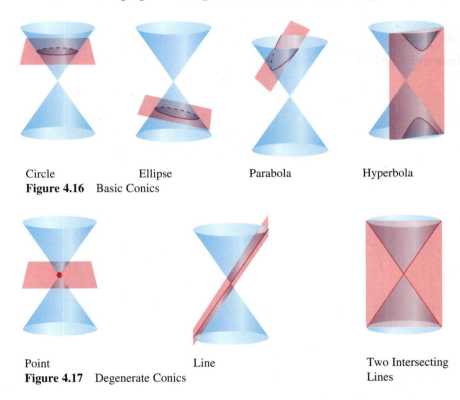

Circle   Ellipse   Parabola   Hyperbola

Figure 4.16   Basic Conics

Point   Line   Two Intersecting Lines

Figure 4.17   Degenerate Conics

There are several ways to approach the study of conics. You could begin by defining conics in terms of the intersections of planes and cones, as the Greeks did, or you could define them algebraically, in terms of the general second-degree equation

$$Ax^2 + Bxy + Cy^2 + Dx + Ey + F = 0.$$

However, you will study a third approach, in which each of the conics is defined as a *locus* (collection) of points satisfying a given geometric property. For example, in Section 1.1 you saw how the definition of a circle as *the collection of all points (x, y) that are equidistant from a fixed point (h, k)* led to the standard form of the equation of a circle

$$(x - h)^2 + (y - k)^2 = r^2. \qquad \text{Equation of a circle}$$

Recall that the center of a circle is at (h, k) and that the radius of the circle is r.

Parabolas

In Section 3.1, you learned that the graph of the quadratic function

$$f(x) = ax^2 + bx + c$$

is a parabola that opens upward or downward. The definition of a parabola below is more general in the sense that it is independent of the orientation of the parabola.

Definition of a Parabola

A **parabola** is the set of all points (x, y) in a plane that are equidistant from a fixed line, the **directrix,** and a fixed point, the **focus,** not on the line. (See figure.) The **vertex** is the midpoint between the focus and the directrix. The **axis** of the parabola is the line passing through the focus and the vertex.

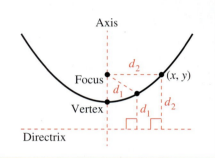

Note in the figure above that a parabola is symmetric with respect to its axis. The definition of a parabola can be used to derive the **standard form of the equation of a parabola** with vertex at $(0, 0)$ and directrix parallel to the x-axis or to the y-axis, stated below.

Standard Equation of a Parabola (Vertex at Origin)

The **standard form of the equation of a parabola** with vertex at $(0, 0)$ and directrix $y = -p$ is

$$x^2 = 4py, \quad p \neq 0. \qquad \text{Vertical axis}$$

For directrix $x = -p$, the equation is

$$y^2 = 4px, \quad p \neq 0. \qquad \text{Horizontal axis}$$

The focus is on the axis p units (directed distance) from the vertex.

For a proof of the standard form of the equation of a parabola, see Proofs in Mathematics on page 356.

Notice that a parabola can have a vertical or a horizontal axis, as shown in the figures below.

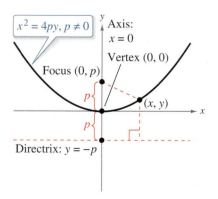

Parabola with vertical axis

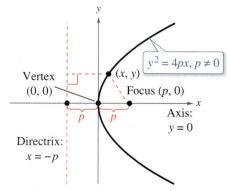

Parabola with horizontal axis

EXAMPLE 1 **Finding the Focus and Directrix of a Parabola**

Find the focus and directrix of the parabola $y = -2x^2$. Then sketch the parabola.

Solution   The squared term in the equation involves x, so you know that the axis is vertical, and the equation is of the form

$x^2 = 4py.$ Standard form, vertical axis

Write the original equation in this form.

$-2x^2 = y$ Write original equation.

$x^2 = -\dfrac{1}{2}y$ Divide each side by -2.

$x^2 = 4\left(-\dfrac{1}{8}\right)y$ Write in standard form.

So, $p = -\frac{1}{8}$. Because p is negative, the parabola opens downward. The focus of the parabola is

$(0, p) = \left(0, -\frac{1}{8}\right)$ Focus

and the directrix of the parabola is

$y = -p = -\left(-\frac{1}{8}\right) = \frac{1}{8}.$ Directrix

Figure 4.18 shows the parabola.

✓ **Checkpoint**))) Audio-video solution in English & Spanish at LarsonPrecalculus.com

Find the focus and directrix of the parabola $y = \frac{1}{4}x^2$. Then sketch the parabola.

EXAMPLE 2 **Finding the Standard Equation of a Parabola**

See LarsonPrecalculus.com for an interactive version of this type of example.

Find the standard form of the equation of the parabola with vertex at the origin and focus at $(2, 0)$.

Solution   The axis of the parabola is horizontal, passing through $(0, 0)$ and $(2, 0)$, as shown in Figure 4.19. So, the equation is of the form

$y^2 = 4px.$ Standard form, horizontal axis

The focus is $p = 2$ units from the vertex, so the standard form of the equation is

$y^2 = 4(2)x$ Standard form

$y^2 = 8x.$ Simplify.

✓ **Checkpoint**))) Audio-video solution in English & Spanish at LarsonPrecalculus.com

Find the standard form of the equation of the parabola with vertex at the origin and focus at $\left(0, \frac{3}{8}\right)$.

Parabolas occur in a wide variety of applications. For example, a parabolic reflector can be formed by revolving a parabola about its axis. The resulting surface has the property that all incoming rays parallel to the axis reflect through the focus of the parabola. This is the principle behind the construction of the parabolic mirrors used in reflecting telescopes. Conversely, the light rays emanating from the focus of a parabolic reflector used in a flashlight are all parallel to one another, as shown in Figure 4.20.

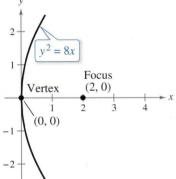

Figure 4.18

Figure 4.19

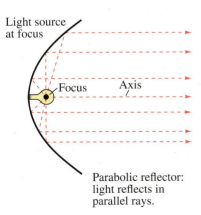

Figure 4.20

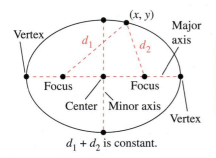

Vertex

d_1 d_2 Major axis

(x, y)

Focus   Focus

Center   Minor axis

Vertex

$d_1 + d_2$ is constant.

Figure 4.21

Figure 4.22

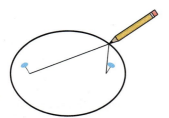

Each planet in our solar system moves in an elliptical orbit with the sun at one of the foci.

Ellipses

Definition of an Ellipse

An **ellipse** is the set of all points (x, y) in a plane, the sum of whose distances from two distinct fixed points (**foci**) is constant. See Figure 4.21.

The line through the foci intersects the ellipse at two points (**vertices**). The chord joining the vertices is the **major axis,** and its midpoint is the **center** of the ellipse. The chord perpendicular to the major axis at the center is the **minor axis.** (See Figure 4.21.)

To visualize the definition of an ellipse, imagine two thumbtacks placed at the foci, as shown in Figure 4.22. When the ends of a fixed length of string are fastened to the thumbtacks and the string is drawn taut with a pencil, the path traced by the pencil will be an ellipse.

The standard form of the equation of an ellipse takes one of two forms, depending on whether the major axis is horizontal or vertical.

Standard Equation of an Ellipse (Center at Origin)

The **standard form of the equation of an ellipse** centered at the origin with major and minor axes of lengths $2a$ and $2b$, respectively, where $0 < b < a$, is

$$\frac{x^2}{a^2} + \frac{y^2}{b^2} = 1 \qquad \text{Major axis is horizontal.}$$

$$\frac{x^2}{b^2} + \frac{y^2}{a^2} = 1. \qquad \text{Major axis is vertical.}$$

The vertices and foci lie on the major axis, a and c units, respectively, from the center. Moreover, a, b, and c are related by the equation $c^2 = a^2 - b^2$. See the figures below.

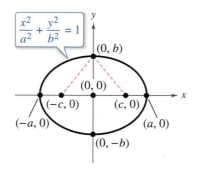

Major axis is horizontal;
minor axis is vertical.

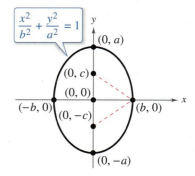

Major axis is vertical;
minor axis is horizontal.

To show that $c^2 = a^2 - b^2$, use either of the figures above and the fact that the sum of the distances from a point on the ellipse to the two foci is constant. For example, using the figure on the left above, you have

Sum of distances from $(0, b)$ to foci = sum of distances from $(a, 0)$ to foci

$$2\sqrt{b^2 + c^2} = (a + c) + (a - c)$$
$$\sqrt{b^2 + c^2} = a$$
$$c^2 = a^2 - b^2.$$

Figure 4.23

EXAMPLE 3 Finding the Standard Equation of an Ellipse

Find the standard form of the equation of the ellipse shown in Figure 4.23.

Solution From Figure 4.23, the foci occur at $(-2, 0)$ and $(2, 0)$. So, the center of the ellipse is $(0, 0)$, the major axis is horizontal, and the equation is of the form

$$\frac{x^2}{a^2} + \frac{y^2}{b^2} = 1.$$ Standard form, horizontal major axis

Also from Figure 4.23, the length of the major axis is $2a = 6$. This implies that $a = 3$. Moreover, the distance from the center to either focus is $c = 2$. Finally,

$$b^2 = a^2 - c^2 = 3^2 - 2^2 = 9 - 4 = 5.$$

Substituting $a^2 = 3^2$ and $b^2 = 5 = \left(\sqrt{5}\right)^2$ yields the standard form of the equation.

$$\frac{x^2}{3^2} + \frac{y^2}{\left(\sqrt{5}\right)^2} = 1$$ Standard form

This equation simplifies to

$$\frac{x^2}{9} + \frac{y^2}{5} = 1.$$

✓ **Checkpoint** *Audio-video solution in English & Spanish at LarsonPrecalculus.com*

Find the standard form of the equation of the ellipse that has a major axis of length 10 and foci at $(0, -3)$ and $(0, 3)$.

EXAMPLE 4 Sketching an Ellipse

Sketch the ellipse $4x^2 + y^2 = 36$ and identify the vertices.

Algebraic Solution

$$4x^2 + y^2 = 36$$ Write original equation.

$$\frac{4x^2}{36} + \frac{y^2}{36} = \frac{36}{36}$$ Divide each side by 36.

$$\frac{x^2}{3^2} + \frac{y^2}{6^2} = 1$$ Write in standard form.

The denominator of the y^2-term is greater than the denominator of the x^2-term, so the major axis is vertical. Moreover, $a^2 = 6^2$, so the endpoints of the major axis (the vertices) lie six units *up* and *down* from the center $(0, 0)$ at $(0, 6)$ and $(0, -6)$. Similarly, the denominator of the x^2-term is $b^2 = 3^2$, so the endpoints of the minor axis (the *co-vertices*) lie three units to the *right* and *left* of the center at $(3, 0)$ and $(-3, 0)$. A sketch of the ellipse is at the right.

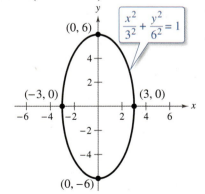

Graphical Solution

Solve the equation of the ellipse for y.

$$4x^2 + y^2 = 36$$

$$y^2 = 36 - 4x^2$$

$$y = \pm\sqrt{36 - 4x^2}$$

Then, use a graphing utility to graph

$$y_1 = \sqrt{36 - 4x^2} \quad \text{and} \quad y_2 = -\sqrt{36 - 4x^2}$$

in the same viewing window, as shown in the figure below. Be sure to use a square setting.

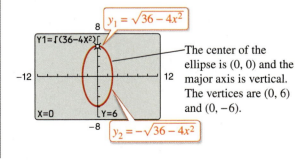

The center of the ellipse is $(0, 0)$ and the major axis is vertical. The vertices are $(0, 6)$ and $(0, -6)$.

✓ **Checkpoint** *Audio-video solution in English & Spanish at LarsonPrecalculus.com*

Sketch the ellipse $x^2 + 9y^2 = 81$ and identify the vertices.

Ellipses have many practical and aesthetic uses. For example, machine gears, supporting arches, and acoustic designs often involve elliptical shapes. The orbits of satellites and planets are also ellipses.

It was difficult for early astronomers to detect that the orbits of the planets are ellipses because the foci of the planetary orbits are relatively close to their centers, and so the orbits are nearly circular. You can measure the "ovalness" of an ellipse by using the concept of **eccentricity.**

Definition of Eccentricity

The **eccentricity** e of an ellipse is the ratio $e = \dfrac{c}{a}$.

Note that $0 < e < 1$ for *every* ellipse.

To see how this ratio describes the shape of an ellipse, note that because the foci of an ellipse are located along the major axis between the vertices and the center, it follows that $0 < c < a$. For an ellipse that is nearly circular, the foci are close to the center and the ratio c/a is close to 0, as shown in Figure 4.24. On the other hand, for an elongated ellipse, the foci are close to the vertices and the ratio c/a is close to 1, as shown in Figure 4.25.

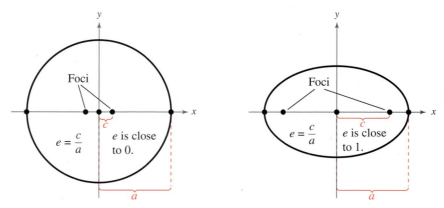

Figure 4.24   **Figure 4.25**

The orbit of the moon has an eccentricity of

$e \approx 0.0549.$ Eccentricity of the moon's orbit

The eccentricities of the eight planetary orbits are listed below.

Planet	Orbit Eccentricity, e
Mercury	0.2056
Venus	0.0067
Earth	0.0167
Mars	0.0935
Jupiter	0.0489
Saturn	0.0565
Uranus	0.0457
Neptune	0.0113

Hyperbolas

The definition of a **hyperbola** is similar to that of an ellipse. For an ellipse, the *sum* of the distances between the foci and a point on the ellipse is constant, whereas for a hyperbola, the absolute value of the *difference* of the distances between the foci and a point on the hyperbola is constant.

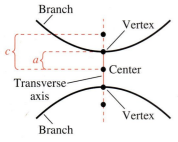

$|d_2 - d_1|$ is constant.

(a)

Definition of a Hyperbola

A **hyperbola** is the set of all points (x, y) in a plane for which the absolute value of the difference of the distances from two distinct fixed points (**foci**) is constant. See Figure 4.26(a).

The graph of a hyperbola has two disconnected parts (**branches**). The line through the foci intersects the hyperbola at two points (**vertices**). The line segment connecting the vertices is the **transverse axis,** and its midpoint is the **center** of the hyperbola. See Figure 4.26(b).

(b)

Figure 4.26

Standard Equation of a Hyperbola (Center at Origin)

The **standard form of the equation of a hyperbola** with center at the origin is

$$\frac{x^2}{a^2} - \frac{y^2}{b^2} = 1 \qquad \text{Transverse axis is horizontal.}$$

$$\frac{y^2}{a^2} - \frac{x^2}{b^2} = 1 \qquad \text{Transverse axis is vertical.}$$

where $a > 0$ and $b > 0$. The vertices and foci are, respectively, a and c units from the center. Moreover, a, b, and c are related by the equation $c^2 = a^2 + b^2$. See the figures below.

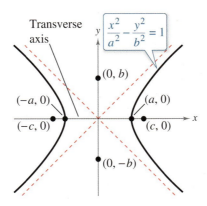

Transverse axis is horizontal

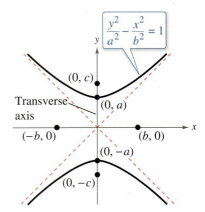

Transverse axis is vertical

Notice that the relationship among a, b, and c is different for hyperbolas than for ellipses.

For an ellipse:

$$c^2 = a^2 - b^2$$

For a hyperbola:

$$c^2 = a^2 + b^2$$

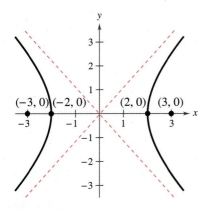

Figure 4.27

EXAMPLE 5 **Finding the Standard Equation of a Hyperbola**

Find the standard form of the equation of the hyperbola with foci at $(-3, 0)$ and $(3, 0)$ and vertices at $(-2, 0)$ and $(2, 0)$, as shown in Figure 4.27.

Solution From the graph, $c = 3$, because the foci are three units from the center $(0, 0)$. Moreover, $a = 2$ because the vertices are two units from the center. So, it follows that

$$b^2 = c^2 - a^2$$
$$= 3^2 - 2^2$$
$$= 9 - 4$$
$$= 5.$$

The transverse axis is horizontal, so the equation is of the form

$$\frac{x^2}{a^2} - \frac{y^2}{b^2} = 1. \qquad \text{Standard form, horizontal transverse axis}$$

Substitute

$$a^2 = 2^2 \quad \text{and} \quad b^2 = 5 = \left(\sqrt{5}\right)^2$$

to obtain

$$\frac{x^2}{2^2} - \frac{y^2}{\left(\sqrt{5}\right)^2} = 1 \qquad \text{Standard form}$$

$$\frac{x^2}{4} - \frac{y^2}{5} = 1. \qquad \text{Simplify.}$$

✓ **Checkpoint** 🔊))) *Audio-video solution in English & Spanish at LarsonPrecalculus.com*

Find the standard form of the equation of the hyperbola with foci at $(0, -6)$ and $(0, 6)$ and vertices at $(0, -3)$ and $(0, 3)$. ▮

An important aid in sketching the graph of a hyperbola is the determination of its *asymptotes,* as shown in the figures below. Every hyperbola has two asymptotes that intersect at the center of the hyperbola. Furthermore, the asymptotes pass through the corners of a rectangle of dimensions $2a$ by $2b$. The line segment of length $2b$ joining $(0, b)$ and $(0, -b)$ [or $(-b, 0)$ and $(b, 0)$] is the **conjugate axis** of the hyperbola.

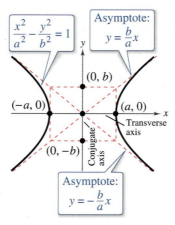

Transverse axis is horizontal;
conjugate axis is vertical.

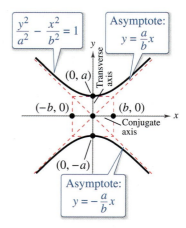

Transverse axis is vertical;
conjugate axis is horizontal.

Asymptotes of a Hyperbola (Center at Origin)

The asymptotes of a hyperbola with center at $(0, 0)$ are

$$y = \frac{b}{a}x \quad \text{and} \quad y = -\frac{b}{a}x \qquad \text{Transverse axis is horizontal.}$$

$$y = \frac{a}{b}x \quad \text{and} \quad y = -\frac{a}{b}x. \qquad \text{Transverse axis is vertical.}$$

EXAMPLE 6   Sketching a Hyperbola

Sketch the hyperbola $x^2 - \dfrac{y^2}{4} = 4$.

Algebraic Solution

$$x^2 - \frac{y^2}{4} = 4 \qquad \text{Write original equation.}$$

$$\frac{x^2}{4} - \frac{y^2}{16} = 1 \qquad \text{Divide each side by 4.}$$

$$\frac{x^2}{2^2} - \frac{y^2}{4^2} = 1 \qquad \text{Write in standard form.}$$

The x^2-term is positive, so the transverse axis is horizontal and the vertices occur at $(-2, 0)$ and $(2, 0)$. Moreover, the endpoints of the conjugate axis occur at $(0, -4)$ and $(0, 4)$. Use these four points to sketch the rectangle shown in Figure 4.28. Finally, draw the asymptotes

$$y = 2x$$

and

$$y = -2x$$

through the corners of this rectangle and complete the sketch, as shown in Figure 4.29.

Graphical Solution

Solve the equation of the hyperbola for y.

$$x^2 - \frac{y^2}{4} = 4$$

$$x^2 - 4 = \frac{y^2}{4}$$

$$4x^2 - 16 = y^2$$

$$\pm\sqrt{4x^2 - 16} = y$$

Then use a graphing utility to graph

$$y_1 = \sqrt{4x^2 - 16}$$

and

$$y_2 = -\sqrt{4x^2 - 16}$$

in the same viewing window, as shown in the figure below. Be sure to use a square setting.

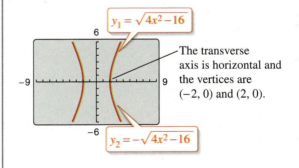

The transverse axis is horizontal and the vertices are $(-2, 0)$ and $(2, 0)$.

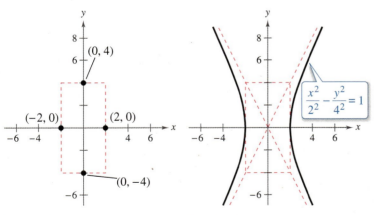

Figure 4.28   **Figure 4.29**

✓ **Checkpoint** 🔊))) *Audio-video solution in English & Spanish at LarsonPrecalculus.com*

Sketch the hyperbola $2x^2 - \dfrac{y^2}{2} = 2$.

EXAMPLE 7 **Finding the Standard Equation of a Hyperbola**

Find the standard form of the equation of the hyperbola that has vertices at $(0, -3)$ and $(0, 3)$ and asymptotes $y = -2x$ and $y = 2x$, as shown in the figure.

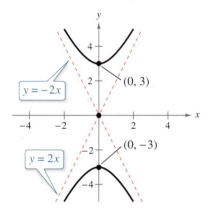

Solution

The transverse axis is vertical, so the asymptotes are of the forms

$$y = \frac{a}{b}x \quad \text{and} \quad y = -\frac{a}{b}x.$$

You are given that the asymptotes are $y = 2x$ and $y = -2x$, so

$$\frac{a}{b} = 2.$$

Substitute $a = 3$ and solve for b to obtain $b = \frac{3}{2}$. Finally, write the equation of the hyperbola in standard form.

$$\frac{y^2}{3^2} - \frac{x^2}{\left(\frac{3}{2}\right)^2} = 1 \qquad \text{\textcolor{red}{Standard form}}$$

$$\frac{y^2}{9} - \frac{x^2}{\frac{9}{4}} = 1 \qquad \text{\textcolor{red}{Simplify.}}$$

✓ **Checkpoint** 🔊))) *Audio-video solution in English & Spanish at LarsonPrecalculus.com*

Find the standard form of the equation of the hyperbola that has vertices at $(-1, 0)$ and $(1, 0)$ and asymptotes $y = -4x$ and $y = 4x$. ■

Summarize (Section 4.3)

1. List the four basic conic sections *(page 327)*.

2. State the definition of a parabola and the standard form of the equation of a parabola with its vertex at the origin *(page 328)*. For examples involving parabolas, see Examples 1 and 2.

3. State the definition of an ellipse and the standard form of the equation of an ellipse with its center at the origin *(page 330)*. For examples involving ellipses, see Examples 3 and 4.

4. State the definition of a hyperbola and the standard form of the equation of a hyperbola with its center at the origin *(page 333)*. For examples involving hyperbolas, see Examples 5–7.

4.3   Exercises

See **CalcChat.com** for tutorial help and worked-out solutions to odd-numbered exercises.

Vocabulary: Fill in the blanks.

1. A _____ is the intersection of a plane and a double-napped cone.
2. The equation $(x - h)^2 + (y - k)^2 = r^2$ is the standard form of the equation of a _____ with center _____ and radius _____.
3. A _____ is the set of all points (x, y) in a plane that are equidistant from a fixed line, called the _____, and a fixed point, called the _____, not on the line.
4. The line that passes through the focus and the vertex of a parabola is the _____ of the parabola.
5. An _____ is the set of all points (x, y) in a plane, the sum of whose distances from two distinct fixed points, called _____, is constant.
6. The chord joining the vertices of an ellipse is the _____ _____, and its midpoint is the _____ of the ellipse.
7. A _____ is the set of all points (x, y) in a plane for which the absolute value of the difference of the distances from two distinct fixed points, called _____, is constant.
8. The line segment connecting the vertices of a hyperbola is the _____ _____, and its midpoint is the _____ of the hyperbola.

Skills and Applications

Matching   In Exercises 9–14, match the equation with its graph. [The graphs are labeled (a)–(f).]

(a)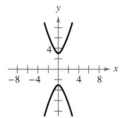

(b)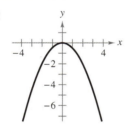

(c)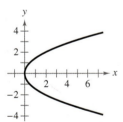

(d)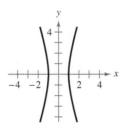

(e)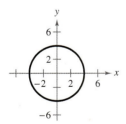

(f)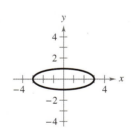

9. $x^2 = -2y$
10. $y^2 = 2x$
11. $\dfrac{x^2}{9} + y^2 = 1$
12. $x^2 - \dfrac{y^2}{9} = 1$
13. $y^2 - 9x^2 = 9$
14. $x^2 + y^2 = 16$

 Finding the Focus and Directrix of a Parabola   In Exercises 15–20, find the focus and directrix of the parabola. Then sketch the parabola.

15. $y = \frac{1}{2}x^2$
16. $y = -4x^2$
17. $y^2 = -6x$
18. $y^2 = 3x$
19. $x^2 + 12y = 0$
20. $x + y^2 = 0$

 Finding the Standard Equation of a Parabola   In Exercises 21–26, find the standard form of the equation of the parabola with the given characteristic(s) and vertex at the origin.

21. Focus: $(3, 0)$
22. Focus: $\left(0, \frac{1}{2}\right)$
23. Directrix: $y = 2$
24. Directrix: $x = -4$
25. Passes through the point $(-4, 6)$; horizontal axis
26. Passes through the point $\left(2, -\frac{1}{4}\right)$; vertical axis

Finding the Standard Equation of a Parabola   In Exercises 27 and 28, find the standard form of the equation of the parabola and determine the coordinates of the focus.

27.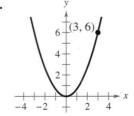

28.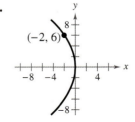

29. Flashlight The light bulb in a flashlight is at the focus of the parabolic reflector, 1.5 centimeters from the vertex of the reflector (see figure). Write an equation for a cross section of the flashlight's reflector with its focus on the positive x-axis and its vertex at the origin.

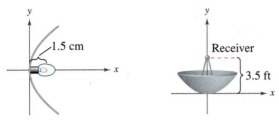

| Figure for 29 | Figure for 30 |

30. Satellite Dish The receiver of a parabolic satellite dish is at the focus of the parabola (see figure). Write an equation for a cross section of the satellite dish.

· · **31. Suspension Bridge** · · · · · · · · · · · ·

Each cable of the Golden Gate Bridge is suspended (in the shape of a parabola) between two towers that are 1280 meters apart. The top of each tower is 152 meters above the roadway. The cables touch the roadway at the midpoint between the towers.

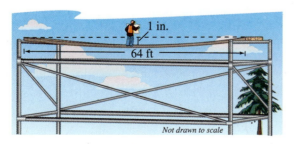

(a) Sketch the bridge on a rectangular coordinate system with the cables touching the roadway at the origin. Label the coordinates of the known points.

(b) Write an equation that models the cables.

32. Beam Deflection A simply supported beam (see figure) is 64 feet long and has a load at the center. The deflection of the beam at its center is 1 inch. The shape of the deflected beam is parabolic.

(a) Write an equation of the parabola. (Assume that the origin is at the center of the beam.)

(b) How far from the center of the beam is the deflection $\frac{1}{2}$ inch?

Finding the Standard Equation of an Ellipse In Exercises 33–42, find the standard form of the equation of the ellipse with the given characteristics and center at the origin.

33.

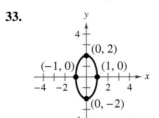

34.

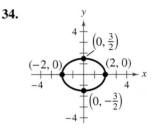

35. Vertices: $(\pm 5, 0)$; foci: $(\pm 2, 0)$

36. Vertices: $(0, \pm 8)$; foci: $(0, \pm 4)$

37. Foci: $(\pm 5, 0)$; major axis of length 14

38. Foci: $(\pm 2, 0)$; major axis of length 10

39. Vertices: $(\pm 9, 0)$; minor axis of length 6

40. Vertices: $(0, \pm 10)$; minor axis of length 2

41. Passes through the points $(0, 14)$ and $(-7, 0)$

42. Passes through the points $(0, 4)$ and $(2, 0)$

Sketching an Ellipse In Exercises 43–52, find the vertices and eccentricity of the ellipse. Then sketch the ellipse.

43. $\dfrac{x^2}{25} + \dfrac{y^2}{16} = 1$

44. $\dfrac{x^2}{121} + \dfrac{y^2}{144} = 1$

45. $\dfrac{x^2}{25/9} + \dfrac{y^2}{16/9} = 1$

46. $\dfrac{x^2}{4} + \dfrac{y^2}{1/4} = 1$

47. $7x^2 + 36y^2 = 252$

48. $16x^2 + 7y^2 = 448$

49. $9x^2 + 5y^2 = 45$

50. $5x^2 + 4y^2 = 20$

51. $4x^2 + y^2 = 1$

52. $x^2 + 9y^2 = 1$

Using Eccentricity In Exercises 53 and 54, find the standard form of the equation of the ellipse with the given vertices, eccentricity e, and center at the origin.

53. Vertices: $(\pm 5, 0)$; $e = \frac{4}{5}$ **54.** Vertices: $(0, \pm 8)$; $e = \frac{1}{2}$

55. Architecture A mason is building a semielliptical fireplace arch that has a height of 2 feet at the center and a width of 6 feet along the base (see figure). The mason draws the semiellipse on the wall by the method shown in Figure 4.22 on page 330. Find the positions of the thumbtacks and the length of the string.

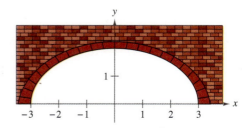

56. Architecture A semielliptical arch over a tunnel for a one-way road through a mountain has a major axis of 50 feet and a height at the center of 10 feet.

(a) Sketch the arch of the tunnel on a rectangular coordinate system with the center of the road entering the tunnel at the origin. Label the coordinates of the known points.

(b) Find an equation of the semielliptical arch over the tunnel.

(c) You are driving a moving truck that has a width of 8 feet and a height of 9 feet. Will the moving truck clear the opening of the arch?

57. Architecture Repeat Exercise 56 for a semielliptical arch with a major axis of 40 feet and a height at the center of 15 feet. The dimensions of the truck are 10 feet wide by 14 feet high.

58. Geometry A line segment through a focus of an ellipse with endpoints on the ellipse and perpendicular to the major axis is called a **latus rectum** of the ellipse. An ellipse has two latera recta. Knowing the length of the latera recta is helpful in sketching an ellipse because it yields other points on the curve (see figure). Show that the length of each latus rectum is $2b^2/a$.

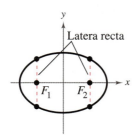

Using Latera Recta In Exercises 59–62, sketch the ellipse using the latera recta (see Exercise 58).

59. $\dfrac{x^2}{4} + \dfrac{y^2}{1} = 1$ **60.** $\dfrac{x^2}{9} + \dfrac{y^2}{16} = 1$

61. $9x^2 + 4y^2 = 36$ **62.** $3x^2 + 6y^2 = 30$

Finding the Standard Equation of a Hyperbola In Exercises 63–70, find the standard form of the equation of the hyperbola with the given characteristics and center at the origin.

63. Vertices: $(0, \pm 2)$; foci: $(0, \pm 6)$

64. Vertices: $(\pm 4, 0)$; foci: $(\pm 5, 0)$

65. Vertices: $(\pm 1, 0)$; asymptotes: $y = \pm 3x$

66. Vertices: $(0, \pm 3)$; asymptotes: $y = \pm 3x$

67. Foci: $(\pm 10, 0)$; asymptotes: $y = \pm\frac{3}{4}x$

68. Foci: $(0, \pm 8)$; asymptotes: $y = \pm 4x$

69. Vertices: $(0, \pm 3)$; passes through the point $(-2, 5)$

70. Vertices: $(\pm 2, 0)$; passes through the point $\left(3, \sqrt{3}\right)$

Sketching a Hyperbola In Exercises 71–80, find the vertices of the hyperbola. Then sketch the hyperbola using the asymptotes as an aid.

71. $\dfrac{x^2}{25} - \dfrac{y^2}{25} = 1$ **72.** $\dfrac{x^2}{9} - \dfrac{y^2}{16} = 1$

73. $\dfrac{1}{36}y^2 - \dfrac{1}{100}x^2 = 1$ **74.** $\dfrac{1}{144}x^2 - \dfrac{1}{169}y^2 = 1$

75. $2y^2 - \dfrac{x^2}{2} = 2$ **76.** $\dfrac{y^2}{3} - 3x^2 = 3$

77. $25y^2 - 9x^2 = 225$ **78.** $4x^2 - 36y^2 = 144$

79. $9x^2 - y^2 = 1$ **80.** $4y^2 - x^2 = 1$

81. Art A cross section of a sculpture can be modeled by a hyperbola (see figure).

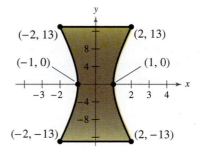

(a) Write an equation that models the curved sides of the sculpture.

(b) Each unit on the coordinate plane represents 1 foot. Find the width of the sculpture at a height of 18 feet.

82. Optics A hyperbolic mirror (used in some telescopes) has the property that a light ray directed at focus A is reflected to focus B. Find the vertex of the mirror when its mount at the top edge of the mirror has coordinates $(24, 24)$.

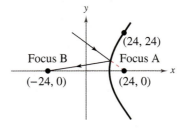

83. Aeronautics When an airplane travels faster than the speed of sound, the sound waves form a cone behind the airplane. If the airplane is flying parallel to the ground, then the sound waves intersect the ground in a hyperbola with the airplane directly above its center, and a sonic boom is heard along the hyperbola. You hear a sonic boom that is audible along a hyperbola with the equation $(x^2/100) - (y^2/4) = 1$, where x and y are measured in miles. What is the shortest horizontal distance you could be from the airplane?

84. Navigation Long-distance radio navigation for aircraft and ships uses synchronized pulses transmitted by widely separated transmitting stations. These pulses travel at the speed of light (186,000 miles per second). The difference in the times of arrival of these pulses at an aircraft or ship is constant on a hyperbola having the transmitting stations as foci.

Assume that two stations 300 miles apart are positioned on a rectangular coordinate system with coordinates $(-150, 0)$ and $(150, 0)$ and that a ship is traveling on a path with coordinates $(x, 75)$, as shown in the figure. Find the x-coordinate of the position of the ship when the time difference between the pulses from the transmitting stations is 1000 microseconds (0.001 second).

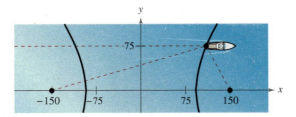

Exploration

True or False? In Exercises 85–88, determine whether the statement is true or false. Justify your answer.

85. The equation $x^2 - y^2 = 144$ represents a circle.

86. The major axis of the ellipse

$$y^2 + 16x^2 = 64$$

is vertical.

87. It is possible for a parabola to intersect its directrix.

88. When the vertex and focus of a parabola are on a horizontal line, the directrix of the parabola is vertical.

89. Area of an Ellipse Consider the ellipse

$$\frac{x^2}{a^2} + \frac{y^2}{b^2} = 1, \quad a + b = 20.$$

(a) The area of the ellipse is given by $A = \pi ab$. Write the area of the ellipse as a function of a.

(b) Find the equation of an ellipse with an area of 264 square centimeters.

(c) Complete the table using your equation from part (a), and make a conjecture about the shape of the ellipse with maximum area.

a	8	9	10	11	12	13
A						

(d) Use a graphing utility to graph the area function and use the graph to support your conjecture in part (c).

90. **HOW DO YOU SEE IT?** In parts (a)–(d), describe how a plane could intersect the double-napped cone to form each conic section (see figure).

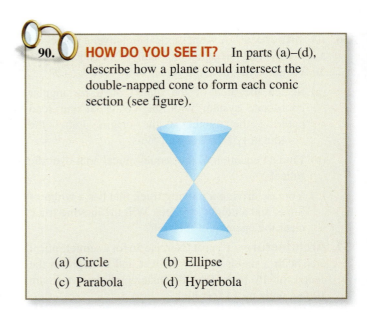

(a) Circle   (b) Ellipse

(c) Parabola   (d) Hyperbola

91. Writing Explain how to use a graphing utility to check your graph in Exercise 43. What equation(s) would you enter into the graphing utility?

92. Think About It How can you tell whether an ellipse is a circle from the equation?

93. Think About It Is the graph of $x^2 - 4y^4 = 4$ a hyperbola? Explain.

94. Degenerate Conic The graph of $x^2 - y^2 = 0$ is a degenerate conic. Sketch this graph and identify the degenerate conic.

95. Think About It Which part of the graph of the ellipse $4x^2 + 9y^2 = 36$ does each equation represent? Answer without graphing the equations.

(a) $x = -\frac{3}{2}\sqrt{4 - y^2}$

(b) $y = \frac{2}{3}\sqrt{9 - x^2}$

96. Writing Write a paragraph discussing the changes in the shape and orientation of the graph of the ellipse

$$\frac{x^2}{a^2} + \frac{y^2}{4^2} = 1$$

as a increases from 1 to 8.

97. Drawing Ellipses Use two thumbtacks, a string, and a pencil to draw an ellipse, as shown in Figure 4.22 on page 330. Vary the length of the string and the distance between the thumbtacks. Explain how to obtain ellipses that are almost circular. Explain how to obtain ellipses that are long and narrow.

98. Using a Definition Use the definition of an ellipse to derive the standard form of the equation of an ellipse. (*Hint:* The sum of the distances from a point (x, y) to the foci is $2a$.)

99. Using a Definition Use the definition of a hyperbola to derive the standard form of the equation of a hyperbola. (*Hint:* The absolute value of the difference of the distances from a point (x, y) to the foci is $2a$.)

4.4 Translations of Conics

In some real-life applications, it is convenient to use conics whose centers or vertices are not at the origin. For example, in Exercise 41 on page 348, you will use a parabola whose vertex is not at the origin to model the path of a satellite as it escapes Earth's gravity.

▷

REMARK Consider the equation of the ellipse

$$\frac{(x-h)^2}{a^2} + \frac{(y-k)^2}{b^2} = 1.$$

If you let $a = b = r$, then the equation can be rewritten as

$$(x-h)^2 + (y-k)^2 = r^2$$

which is the standard form of the equation of a circle with radius r. Geometrically, when $a = b$ for an ellipse, the major and minor axes are of equal length, and so the graph is a circle.

■ Recognize equations of conics that are shifted vertically or horizontally in the plane.
■ Write and graph equations of conics that are shifted vertically or horizontally in the plane.

Vertical and Horizontal Shifts of Conics

In Section 4.3, you studied conic sections whose graphs are in *standard position*, that is, whose centers or vertices are at the origin. In this section, you will study the equations of conic sections that are shifted vertically or horizontally in the plane.

Standard Forms of Equations of Conics

Circle: Center $= (h, k)$; radius $= r$

$$(x-h)^2 + (y-k)^2 = r^2$$

Ellipse: Center $= (h, k)$

Major axis length $= 2a$; minor axis length $= 2b$

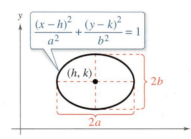
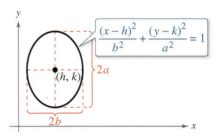

Hyperbola: Center $= (h, k)$

Transverse axis length $= 2a$; conjugate axis length $= 2b$

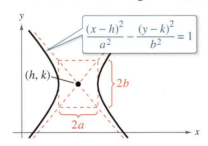
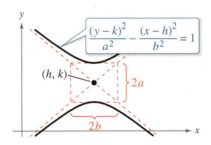

Parabola: Vertex $= (h, k)$

Directed distance from vertex to focus $= p$

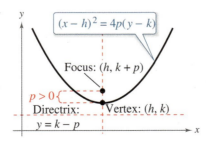
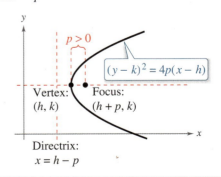

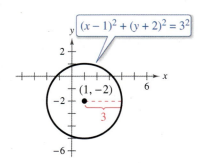

Figure 4.30 Circle

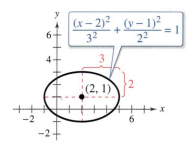

Figure 4.31 Ellipse

EXAMPLE 1 **Equations of Conic Sections**

Identify each conic. Then describe the translation of the conic from standard position.

a. $(x - 1)^2 + (y + 2)^2 = 3^2$ **b.** $\dfrac{(x - 2)^2}{3^2} + \dfrac{(y - 1)^2}{2^2} = 1$

c. $\dfrac{(x - 3)^2}{1^2} - \dfrac{(y - 2)^2}{3^2} = 1$ **d.** $(x - 2)^2 = 4(-1)(y - 3)$

Solution

a. The graph of $(x - 1)^2 + (y + 2)^2 = 3^2$ is a *circle* whose center is the point $(1, -2)$ and whose radius is 3, as shown in Figure 4.30. The circle is shifted one unit to the right and two units down from standard position.

b. The graph of

$$\frac{(x - 2)^2}{3^2} + \frac{(y - 1)^2}{2^2} = 1$$

is an *ellipse* whose center is the point $(2, 1)$. The major axis is horizontal and of length $2(3) = 6$, and the minor axis is vertical and of length $2(2) = 4$, as shown in Figure 4.31. The ellipse is shifted two units to the right and one unit up from standard position.

c. The graph of

$$\frac{(x - 3)^2}{1^2} - \frac{(y - 2)^2}{3^2} = 1$$

is a *hyperbola* whose center is the point $(3, 2)$ The transverse axis is horizontal and of length $2(1) = 2$, and the conjugate axis is vertical and of length $2(3) = 6$, as shown in Figure 4.32. The hyperbola is shifted three units to the right and two units up from standard position.

d. The graph of $(x - 2)^2 = 4(-1)(y - 3)$ is a *parabola* whose vertex is the point $(2, 3)$. The axis of the parabola is vertical. Moreover, $p = -1$, so the focus lies one unit below the vertex, and the directrix is $y = k - p = 3 - (-1) = 3 + 1 = 4$, as shown in Figure 4.33. The parabola is shifted two units to the right and three units up from standard position.

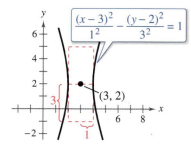

Figure 4.32 Hyperbola

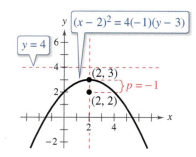

Figure 4.33 Parabola

✓ *Checkpoint*))) *Audio-video solution in English & Spanish at LarsonPrecalculus.com*

Identify each conic. Then describe the translation of the conic from standard position.

a. $\dfrac{(x + 1)^2}{3^2} + \dfrac{(y - 2)^2}{5^2} = 1$ **b.** $(x + 1)^2 + (y - 1)^2 = 2^2$

c. $(x + 4)^2 = 4(2)(y - 3)$ **d.** $\dfrac{(x - 3)^2}{2^2} - \dfrac{(y - 1)^2}{1^2} = 1$

Equations of Conics in Standard Form

EXAMPLE 2 Finding Characteristics of a Parabola

Find the vertex, focus, and directrix of the parabola $x^2 - 2x + 4y - 3 = 0$. Then sketch the parabola.

Solution

To write the equation in standard form, begin by completing the square.

$x^2 - 2x + 4y - 3 = 0$	Write original equation.
$x^2 - 2x = -4y + 3$	Isolate x-terms on one side of the equation.
$x^2 - 2x + 1 = -4y + 3 + 1$	Complete the square.
$(x - 1)^2 = -4y + 4$	Write in completed square form.
$(x - 1)^2 = 4(-1)(y - 1)$	Write in standard form, $(x - h)^2 = 4p(y - k)$.

From this standard form, it follows that $h = 1$, $k = 1$, and $p = -1$. The axis is vertical and p is negative, so the parabola opens downward.

The vertex is

$$(h, k) = (1, 1) \qquad \text{Vertex}$$

the focus is

$$(h, k + p) = (1, 0) \qquad \text{Focus}$$

and the directrix is

$$y = k - p = 2. \qquad \text{Directrix}$$

Figure 4.34 shows the parabola.

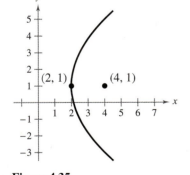

Figure 4.34

✓ *Checkpoint* *Audio-video solution in English & Spanish at LarsonPrecalculus.com*

Repeat Example 2 for the parabola $y^2 - 6y + 4x + 17 = 0$.

Recall that for a parabola, p is the directed distance from the vertex to the focus. In Example 2, the axis of the parabola is vertical and $p = -1$, so the focus is one unit below the vertex.

EXAMPLE 3 Finding the Standard Equation of a Parabola

Find the standard form of the equation of the parabola whose vertex is $(2, 1)$ and whose focus is $(4, 1)$, as shown in Figure 4.35.

Solution

The vertex and the focus lie on a horizontal line and the focus lies two units to the right of the vertex, so it follows that the axis of the parabola is horizontal and $p = 2$. So the standard form of the parabola is

$(y - k)^2 = 4p(x - h)$	Standard form, horizontal axis
$(y - 1)^2 = 4(2)(x - 2)$	Substitute.
$(y - 1)^2 = 8(x - 2).$	Simplify.

Figure 4.35

✓ *Checkpoint* *Audio-video solution in English & Spanish at LarsonPrecalculus.com*

Find the standard form of the equation of the parabola whose vertex is $(-1, 1)$ and whose directrix is $y = 0$.

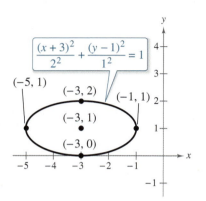

$$\frac{(x + 3)^2}{2^2} + \frac{(y - 1)^2}{1^2} = 1$$

Figure 4.36

| EXAMPLE 4 | **Sketching an Ellipse** |

Sketch the ellipse $x^2 + 4y^2 + 6x - 8y + 9 = 0$.

Solution

To write the equation in standard form, begin by completing the square.

$$x^2 + 4y^2 + 6x - 8y + 9 = 0 \qquad \text{Write original equation.}$$

$$\left(x^2 + 6x + \boxed{}\right) + \left(4y^2 - 8y + \boxed{}\right) = -9 \qquad \text{Group terms.}$$

$$\left(x^2 + 6x + \boxed{}\right) + 4\left(y^2 - 2y + \boxed{}\right) = -9 \qquad \text{Factor 4 out of } y\text{-terms.}$$

$$(x^2 + 6x + 9) + 4(y^2 - 2y + 1) = -9 + 9 + 4(1) \qquad \text{Complete the squares.}$$

$$(x + 3)^2 + 4(y - 1)^2 = 4 \qquad \text{Write in completed square form.}$$

$$\frac{(x + 3)^2}{4} + \frac{(y - 1)^2}{1} = 1 \qquad \text{Divide each side by 4.}$$

$$\frac{(x + 3)^2}{2^2} + \frac{(y - 1)^2}{1^2} = 1 \qquad \frac{(x - h)^2}{a^2} + \frac{(y - k)^2}{b^2} = 1$$

From this standard form, it follows that the center is $(h, k) = (-3, 1)$. The denominator of the x-term is $a^2 = 2^2$, so the endpoints of the major axis lie two units to the right and left of the center at $(-1, 1)$ and $(-5, 1)$. Similarly, the denominator of the y-term is $b^2 = 1^2$, so the endpoints of the minor axis lie one unit up and down from the center at $(-3, 2)$ and $(-3, 0)$. Figure 4.36 shows the ellipse.

✓ *Checkpoint*))) *Audio-video solution in English & Spanish at LarsonPrecalculus.com*

Sketch the ellipse $9x^2 + 4y^2 - 36x + 24y + 36 = 0$.

| EXAMPLE 5 | **Finding the Standard Equation of an Ellipse** |

See LarsonPrecalculus.com for an interactive version of this type of example.

Find the standard form of the equation of the ellipse whose vertices are $(2, -2)$ and $(2, 4)$, and whose minor axis length is 4, as shown in Figure 4.37.

Solution

The center of the ellipse lies at the midpoint of its vertices. So, the center is

$$(h, k) = \left(\frac{2 + 2}{2}, \frac{4 + (-2)}{2}\right) = (2, 1). \qquad \text{Center}$$

The vertices lie on a vertical line and are six units apart, so the major axis is vertical and has a length of $2a = 6$, which implies that $a = 3$. Moreover, the minor axis has a length of 4, so $2b = 4$, which implies that $b = 2$. The standard form of the ellipse is

$$\frac{(x - h)^2}{b^2} + \frac{(y - k)^2}{a^2} = 1 \qquad \text{Standard form, vertical major axis}$$

$$\frac{(x - 2)^2}{2^2} + \frac{(y - 1)^2}{3^2} = 1 \qquad \text{Substitute.}$$

$$\frac{(x - 2)^2}{4} + \frac{(y - 1)^2}{9} = 1. \qquad \text{Simplify.}$$

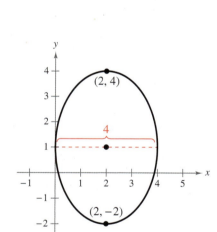

Figure 4.37

✓ *Checkpoint*))) *Audio-video solution in English & Spanish at LarsonPrecalculus.com*

Find the standard form of the equation of the ellipse whose vertices are $(3, 0)$ and $(3, 10)$, and whose minor axis length is 6.

EXAMPLE 6   **Sketching a Hyperbola**

Sketch the hyperbola

$$y^2 - 4x^2 + 4y + 24x - 41 = 0.$$

Solution

To write the equation in standard form, begin by completing the square.

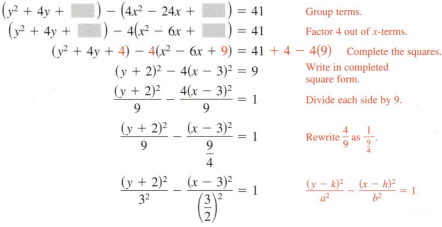

$$y^2 - 4x^2 + 4y + 24x - 41 = 0 \qquad \text{Write original equation.}$$

$$\left(y^2 + 4y + \quad\right) - \left(4x^2 - 24x + \quad\right) = 41 \qquad \text{Group terms.}$$

$$\left(y^2 + 4y + \quad\right) - 4\left(x^2 - 6x + \quad\right) = 41 \qquad \text{Factor 4 out of } x\text{-terms.}$$

$$(y^2 + 4y + 4) - 4(x^2 - 6x + 9) = 41 + 4 - 4(9) \qquad \text{Complete the squares.}$$

$$(y + 2)^2 - 4(x - 3)^2 = 9 \qquad \text{Write in completed square form.}$$

$$\frac{(y + 2)^2}{9} - \frac{4(x - 3)^2}{9} = 1 \qquad \text{Divide each side by 9.}$$

$$\frac{(y + 2)^2}{9} - \frac{(x - 3)^2}{\frac{9}{4}} = 1 \qquad \text{Rewrite } \frac{4}{9} \text{ as } \frac{1}{\frac{9}{4}}.$$

$$\frac{(y + 2)^2}{3^2} - \frac{(x - 3)^2}{\left(\frac{3}{2}\right)^2} = 1 \qquad \frac{(y - k)^2}{a^2} - \frac{(x - h)^2}{b^2} = 1$$

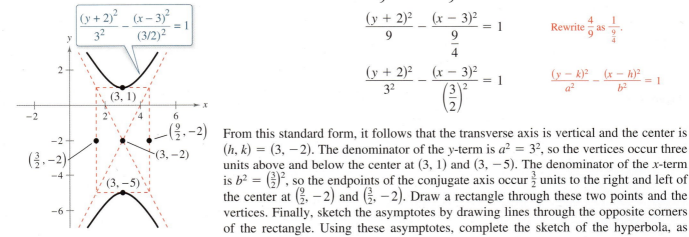

$$\frac{(y + 2)^2}{3^2} - \frac{(x - 3)^2}{(3/2)^2} = 1$$

(3, 1)

$\left(\frac{9}{2}, -2\right)$

$\left(\frac{3}{2}, -2\right)$

(3, −2)

(3, −5)

Figure 4.38

From this standard form, it follows that the transverse axis is vertical and the center is $(h, k) = (3, -2)$. The denominator of the y-term is $a^2 = 3^2$, so the vertices occur three units above and below the center at $(3, 1)$ and $(3, -5)$. The denominator of the x-term is $b^2 = \left(\frac{3}{2}\right)^2$, so the endpoints of the conjugate axis occur $\frac{3}{2}$ units to the right and left of the center at $\left(\frac{9}{2}, -2\right)$ and $\left(\frac{3}{2}, -2\right)$. Draw a rectangle through these two points and the vertices. Finally, sketch the asymptotes by drawing lines through the opposite corners of the rectangle. Using these asymptotes, complete the sketch of the hyperbola, as shown in Figure 4.38.

✓ **Checkpoint** 🔊))) *Audio-video solution in English & Spanish at LarsonPrecalculus.com*

Sketch the hyperbola

$$9x^2 - y^2 - 18x - 6y - 9 = 0.$$

To find the foci in Example 6, first find c. Recall from Section 4.3 that

$$c^2 = a^2 + b^2.$$

So,

$$c^2 = 9 + \frac{9}{4}$$

$$c^2 = \frac{45}{4}$$

$$c = \frac{3\sqrt{5}}{2}$$

The transverse axis is vertical, so the foci lie c units above and below the center.

$$\left(3, -2 + \frac{3}{2}\sqrt{5}\right) \quad \text{and} \quad \left(3, -2 - \frac{3}{2}\sqrt{5}\right) \qquad \text{Foci}$$

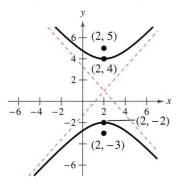

Figure 4.39

Finding the Standard Equation of a Hyperbola

Find the standard form of the equation of the hyperbola whose vertices are $(2, -2)$ and $(2, 4)$, and whose foci are $(2, -3)$ and $(2, 5)$, as shown in Figure 4.39.

Solution   The center of the hyperbola lies at the midpoint of its vertices. So, the center is $(h, k) = (2, 1)$. The vertices lie on a vertical line and are six units apart, so the transverse axis is vertical and has a length of $2a = 6$, which implies that $a = 3$. Moreover, the foci are four units from the center, so $c = 4$, and

$$b^2 = c^2 - a^2 = 4^2 - 3^2 = 16 - 9 = 7 = \left(\sqrt{7}\right)^2.$$

The transverse axis is vertical, so the standard form of the equation is

$$\frac{(y - k)^2}{a^2} - \frac{(x - h)^2}{b^2} = 1 \qquad \text{Standard form, vertical transverse axis}$$

$$\frac{(y - 1)^2}{3^2} - \frac{(x - 2)^2}{\left(\sqrt{7}\right)^2} = 1 \qquad \text{Substitute.}$$

$$\frac{(y - 1)^2}{9} - \frac{(x - 2)^2}{7} = 1. \qquad \text{Simplify.}$$

✓ *Checkpoint*))) *Audio-video solution in English & Spanish at LarsonPrecalculus.com*

Find the standard form of the equation of the hyperbola whose vertices are $(3, -1)$ and $(5, -1)$, and whose foci are $(1, -1)$ and $(7, -1)$. ■

An interesting application of conic sections involves the orbits of comets in our solar system. Comets can have elliptical, parabolic, or hyperbolic orbits. The center of the sun is a focus of each of these orbits, and each orbit has a vertex at the point where the comet is closest to the sun, as shown in Figure 4.40. Undoubtedly, many comets with parabolic or hyperbolic orbits have not been identified. You get to see such comets only *once*. Comets with elliptical orbits, such as Halley's comet, are the only ones that remain in our solar system.

If p is the distance between the vertex and the focus (in meters), and v is the speed of the comet at the vertex (in meters per second), then the type of orbit is determined as follows.

1. Elliptical: $v < \sqrt{2GM/p}$

2. Parabolic: $v = \sqrt{2GM/p}$

3. Hyperbolic: $v > \sqrt{2GM/p}$

In each of the above, $M = 1.989 \times 10^{30}$ kilograms (the mass of the sun) and $G \approx 6.67 \times 10^{-11}$ cubic meter per kilogram-second squared (the universal gravitational constant).

Figure 4.40

Summarize   (Section 4.4)

1. List the standard forms of the equations of conics *(page 341)*. For an example of identifying conics and describing translations of conics, see Example 1.

2. Explain how to write and graph equations of conics that are shifted vertically or horizontally in the plane *(pages 343–346)*. For examples of writing and graphing equations of conics that are shifted vertically or horizontally in the plane, see Examples 2–7.

4.4 Exercises

See **CalcChat.com** for tutorial help and worked-out solutions to odd-numbered exercises.

Vocabulary: Match the description of the conic with its standard equation. The equations are labeled (a)–(f).

(a) $\dfrac{(x-h)^2}{a^2} + \dfrac{(y-k)^2}{b^2} = 1$ 　　(b) $\dfrac{(x-h)^2}{a^2} - \dfrac{(y-k)^2}{b^2} = 1$ 　　(c) $\dfrac{(y-k)^2}{a^2} - \dfrac{(x-h)^2}{b^2} = 1$

(d) $\dfrac{(x-h)^2}{b^2} + \dfrac{(y-k)^2}{a^2} = 1$ 　　(e) $(x-h)^2 = 4p(y-k)$ 　　(f) $(y-k)^2 = 4p(x-h)$

1. Hyperbola with horizontal transverse axis 　　**2.** Ellipse with vertical major axis 　　**3.** Parabola with vertical axis

4. Hyperbola with vertical transverse axis 　　**5.** Ellipse with horizontal major axis 　　**6.** Parabola with horizontal axis

Skills and Applications

 Equations of Conic Sections In Exercises 7–14, identify the conic. Then describe the translation of the conic from standard position.

7. $(x+2)^2 + (y-1)^2 = 4$ 　　**8.** $(y-1)^2 = 4(2)(x+2)$

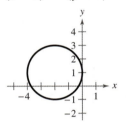

9. $\dfrac{(y+3)^2}{4} - (x-1)^2 = 1$

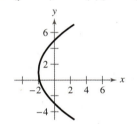

10. $\dfrac{(x-2)^2}{9} + \dfrac{(y+1)^2}{4} = 1$

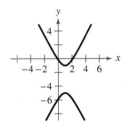

11. $(x+1)^2 = 4(-1)(y-2)$

12. $(x+1)^2 + (y-3)^2 = 6$

13. $\dfrac{(x+4)^2}{9} + \dfrac{(y+2)^2}{16} = 1$

14. $\dfrac{(x+2)^2}{4} - \dfrac{(y-3)^2}{9} = 1$

 Finding Characteristics of a Circle In Exercises 15–20, find the center and radius of the circle.

15. $x^2 + y^2 = 49$

16. $x^2 + y^2 = 1$

17. $(x-4)^2 + (y-5)^2 = 36$

18. $(x+8)^2 + (y+1)^2 = 144$

19. $(x-1)^2 + y^2 = 10$

20. $x^2 + (y+12)^2 = 24$

 Writing the Equation of a Circle in Standard Form In Exercises 21–26, write the equation of the circle in standard form, and then find its center and radius.

21. $x^2 + y^2 - 8y = 0$

22. $x^2 + y^2 - 10x + 16 = 0$

23. $x^2 + y^2 - 2x + 6y + 9 = 0$

24. $2x^2 + 2y^2 - 2x - 2y - 7 = 0$

25. $4x^2 + 4y^2 + 12x - 24y + 41 = 0$

26. $9x^2 + 9y^2 + 54x - 36y + 17 = 0$

 Finding Characteristics of a Parabola In Exercises 27–34, find the vertex, focus, and directrix of the parabola. Then sketch the parabola.

27. $(x-1)^2 + 8(y+2) = 0$

28. $(x+2) + (y-4)^2 = 0$

29. $\left(y+\tfrac{1}{2}\right)^2 = 2(x-5)$

30. $\left(x+\tfrac{1}{2}\right)^2 = 4(y-3)$

31. $y = \dfrac{x^2 - 2x + 5}{4}$

32. $\dfrac{y^2 + 14y + 19}{6} = x$

33. $8x + y^2 + 1 = 2y - 4$

34. $-12 - x^2 = 12y + 4x$

Finding the Standard Equation of a Parabola In Exercises 35–40, find the standard form of the equation of the parabola with the given characteristics.

35. Vertex: $(3, 2)$; focus: $(1, 2)$

36. Vertex: $(-1, 2)$; focus: $(-1, 0)$

37. Vertex: $(0, 4)$; directrix: $y = 2$

38. Vertex: $(-2, 1)$; directrix: $x = 1$

39. Focus: $(4, 4)$; directrix: $x = -4$

40. Focus: $(0, 0)$; directrix: $y = 4$

• • 41. Satellite Orbit • • • • • • • • • • •

A satellite in a 100-mile-high circular orbit around Earth has a velocity of approximately 17,500 miles per hour (see figure). When this velocity is multiplied by $\sqrt{2}$, the satellite has the minimum velocity necessary to escape Earth's gravity and follow a parabolic path with the center of Earth as the focus.

(a) Find the escape velocity of the satellite.

(b) Find the standard form of the equation that represents the parabolic path. (Assume that the radius of Earth is 4000 miles.)

Circular orbit
Parabolic path
4100 miles
Not drawn to scale

42. Fluid Flow Water flowing from a horizontal pipe 48 feet above the ground has the shape of a parabola whose vertex $(0, 48)$ is at the end of the pipe (see figure). The water strikes the ocean at the point $\left(10\sqrt{3}, 0\right)$.

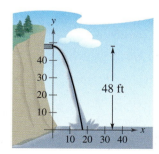

48 ft

(a) Find the standard form of the equation that represents the path of the water.

(b) The pipe is moved 12 feet up. Find an equation that represents the path of the water. Find the coordinates of the point where the water strikes the ocean.

Projectile Motion In Exercises 43 and 44, consider the path of an object projected horizontally with a velocity of v feet per second at a height of s feet, where the model for the path is $x^2 = (-v^2/16)(y - s)$. In this model (in which air resistance is disregarded), y is the height (in feet) of the projectile and x is the horizontal distance (in feet) the projectile travels.

43. A ball is thrown from the top of a 100-foot tower with a velocity of 28 feet per second.

(a) Find the equation that represents the parabolic path.

(b) How far does the ball travel horizontally before it strikes the ground?

44. A cargo plane is flying at an altitude of 500 feet and a speed of 135 miles per hour. A supply crate is dropped from the plane. How many *feet* will the crate travel horizontally before it hits the ground?

Sketching an Ellipse In Exercises 45–52, find the center, foci, and vertices of the ellipse. Then sketch the ellipse.

45. $\dfrac{(x - 1)^2}{9} + \dfrac{(y - 5)^2}{25} = 1$

46. $\dfrac{(x - 6)^2}{4} + \dfrac{(y + 7)^2}{16} = 1$

47. $(x + 2)^2 + \dfrac{(y + 4)^2}{1/4} = 1$

48. $\dfrac{(x - 3)^2}{25/9} + (y - 8)^2 = 1$

49. $9x^2 + 25y^2 - 36x - 50y + 52 = 0$

50. $16x^2 + 25y^2 - 32x + 50y + 16 = 0$

51. $25x^2 + 4y^2 + 50x - 75 = 0$

52. $9x^2 - 4y^2 - 36x + 8y + 31 = 0$

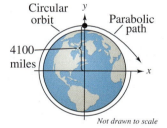

Finding the Standard Equation of an Ellipse In Exercises 53–64, find the standard form of the equation of the ellipse with the given characteristics.

53. Vertices: $(3, -3), (3, 3)$; minor axis of length 2

54. Vertices: $(-2, 3), (6, 3)$; minor axis of length 6

55. Foci: $(0, 0), (4, 0)$; major axis of length 8

56. Foci: $(0, 0), (0, 8)$; major axis of length 16

57. Center: $(1, 4)$; $a = 2c$; vertices: $(1, 0), (1, 8)$

58. Center: $(3, 2)$; $a = 3c$; foci: $(1, 2), (5, 2)$

59. Vertices: $(-3, 0), (7, 0)$; foci: $(0, 0), (4, 0)$

60. Vertices: $(2, 0), (2, 4)$; foci: $(2, 1), (2, 3)$

61. Foci: $(-3, -3), (-3, -1)$; eccentricity: $\frac{1}{3}$

62. Foci: $(7, -6), (7, -4)$; eccentricity: $\frac{1}{2}$

63. Vertices: $(2, -1), (2, 3)$; eccentricity: $\sqrt{3}/2$

64. Vertices: $\left(-\frac{7}{2}, 0\right), \left(\frac{3}{2}, 0\right)$; eccentricity: $\frac{3}{5}$

65. Planetary Motion   The dwarf planet Pluto moves in an elliptical orbit with the sun at one of the foci, as shown in the figure. The length of half of the major axis, a, is 3.67×10^9 miles, and the eccentricity is 0.249. Find the least distance (*perihelion*) and the greatest distance (*aphelion*) of Pluto from the center of the sun.

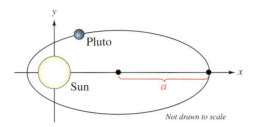

Not drawn to scale

66. Australian Football   In Australia, football by Australian Rules is played on elliptical fields. The field can be a maximum of 155 meters wide and a maximum of 185 meters long. Let the center of a field of maximum size be represented by the point $(0, 77.5)$. Find the standard form of the equation that represents this field.   (*Source: Australian Football League*)

 Sketching a Hyperbola   In Exercises 67–74, find the center, foci, and vertices of the hyperbola. Then sketch the hyperbola using the asymptotes as an aid.

67. $\dfrac{(x - 2)^2}{16} - \dfrac{(y + 1)^2}{9} = 1$

68. $\dfrac{(x - 1)^2}{144} - \dfrac{(y - 4)^2}{25} = 1$

69. $(y + 6)^2 - (x - 2)^2 = 1$

70. $\dfrac{(y - 1)^2}{1/4} - \dfrac{(x + 3)^2}{1/9} = 1$

71. $x^2 - 9y^2 + 2x - 54y - 85 = 0$

72. $16y^2 - x^2 + 2x + 64y + 62 = 0$

73. $9y^2 - x^2 - 36y - 6x + 18 = 0$

74. $x^2 - 9y^2 + 36y - 72 = 0$

 Finding the Standard Equation of a Hyperbola   In Exercises 75–82, find the standard form of the equation of the hyperbola with the given characteristics.

75. Vertices: $(0, 2), (0, 0)$; foci: $(0, 3), (0, -1)$

76. Vertices: $(1, 2), (5, 2)$; foci: $(0, 2), (6, 2)$

77. Foci: $(-8, 5), (0, 5)$; transverse axis of length 4

78. Foci: $(7, -5), (7, 3)$; conjugate axis of length 2

79. Vertices: $(2, 3), (2, -3)$; passes through the point $(0, 5)$

80. Vertices: $(-2, 1), (2, 1)$; passes through the point $(4, 3)$

81. Vertices: $(0, 2), (6, 2)$; asymptotes: $y = \frac{2}{3}x, \ y = 4 - \frac{2}{3}x$

82. Vertices: $(3, 0), (3, 4)$; asymptotes: $y = \frac{2}{3}x, \ y = 4 - \frac{2}{3}x$

Identifying a Conic   In Exercises 83–90, identify the conic by writing its equation in standard form. Then sketch its graph and describe the translation from standard position.

83. $y^2 - x^2 + 4y = 0$

84. $x^2 + y^2 - 6x + 4y + 9 = 0$

85. $16y^2 + 128x + 8y - 7 = 0$

86. $4x^2 - y^2 - 4x - 3 = 0$

87. $9x^2 + 16y^2 + 36x + 128y + 148 = 0$

88. $25x^2 - 10x - 200y - 119 = 0$

89. $16x^2 + 16y^2 - 16x + 24y - 3 = 0$

90. $4x^2 + 3y^2 + 8x - 24y + 51 = 0$

Exploration

True or False?   In Exercises 91–93, determine whether the statement is true or false. Justify your answer.

91. The conic represented by the equation $3x^2 + 2y^2 - 18x - 16y + 58 = 0$ is an ellipse.

92. The graphs of $x^2 + 10y - 10x + 5 = 0$ and $x^2 + 16y^2 + 10x - 32y - 23 = 0$ do not intersect.

93. A hyperbola can have vertices $(-9, 4)$ and $(-3, 4)$ and foci $(-6, 9)$ and $(-6, -1)$.

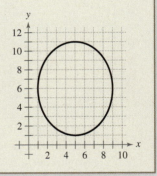

94.   **HOW DO YOU SEE IT?**   Consider the ellipse shown.

(a) Find the center, vertices, and foci of the ellipse.

(b) Find the standard form of the equation of the ellipse.

95. Alternate Form of the Equation of an Ellipse
Consider the ellipse $(x^2/a^2) + (y^2/b^2) = 1$.

(a) Show that the equation of the ellipse can be written as

$$\frac{(x - h)^2}{a^2} + \frac{(y - k)^2}{a^2(1 - e^2)} = 1$$

where e is the eccentricity.

(b) Use a graphing utility to graph the ellipse

$$\frac{(x - 2)^2}{4} + \frac{(y - 3)^2}{4(1 - e^2)} = 1$$

for $e = 0.95, 0.75, 0.5, 0.25,$ and 0. Make a conjecture about the change in the shape of the ellipse as e approaches 0.

Chapter Summary

What Did You Learn?	Explanation/Examples	Review Exercises
Section 4.1 Find domains of rational functions *(p. 310)*.	A rational function is a quotient of polynomial functions. It can be written in the form $f(x) = N(x)/D(x)$, where $N(x)$ and $D(x)$ are polynomials and $D(x)$ is not the zero polynomial. The domain of a rational function of x includes all real numbers except x-values that make the denominator zero.	1–4
Find vertical and horizontal asymptotes of graphs of rational functions *(p. 311)*.	The line $x = a$ is a vertical asymptote of the graph of f when $f(x) \to \infty$ or $f(x) \to -\infty$ as $x \to a$, either from the right or from the left. The line $y = b$ is a horizontal asymptote of the graph of f when $f(x) \to b$ as $x \to \infty$ or $x \to -\infty$.	5–10
Use rational functions to model and solve real-life problems *(p. 313)*.	A rational function can be used to model the cost of removing a given percent of the smokestack pollutants at a utility company that burns coal. (See Example 4.)	11, 12
Section 4.2 Sketch the graphs of rational functions *(p. 318)*.	Let $f(x) = N(x)/D(x)$, where $N(x)$ and $D(x)$ are polynomials and $D(x)$ is not the zero polynomial. **1.** Simplify f, if possible. List any restrictions on the domain of f that are not implied by the simplified function. **2.** Find and plot the y-intercept (if any) by evaluating $f(0)$. **3.** Find the zeros of the numerator (if any). Then plot the corresponding x-intercepts. **4.** Find the zeros of the denominator (if any). Then sketch the corresponding vertical asymptotes. **5.** Find and sketch the horizontal asymptote (if any). **6.** Plot at least one point *between* and one point *beyond* each x-intercept and vertical asymptote. **7.** Use smooth curves to complete the graph between and beyond the vertical asymptotes.	13–24
Sketch graphs of rational functions that have slant asymptotes *(p. 321)*.	Consider a rational function whose denominator is of degree 1 or greater. If the degree of the numerator is exactly *one more* than the degree of the denominator, then the graph of the function has a slant asymptote.	25–30
Use graphs of rational functions to model and solve real-life problems *(p. 322)*.	The graph of a rational function can be used to model the printed area of a rectangular page and to find the page dimensions that use the least amount of paper. (See Example 6.)	31–34
Section 4.3 Recognize the four basic conics: circle, ellipse, parabola, and hyperbola *(p. 327)*.	A conic section (or simply conic) is the intersection of a plane and a double-napped cone. Circle Ellipse Parabola Hyperbola	35–42

What Did You Learn?	**Explanation/Examples**	**Review Exercises**
Section 4.3 — Recognize, graph, and write equations of parabolas (vertex at origin) *(p. 328)*.	The standard form of the equation of a parabola with vertex at $(0, 0)$ and directrix $y = -p$ is $$x^2 = 4py, \quad p \neq 0. \qquad \text{Vertical axis}$$ For directrix $x = -p$, the equation is $$y^2 = 4px, \quad p \neq 0. \qquad \text{Horizontal axis}$$ The focus is on the axis p units (directed distance) from the vertex.	43–50
Recognize, graph, and write equations of ellipses (center at origin) *(p. 330)*.	The standard form of the equation of an ellipse centered at the origin with major and minor axes of lengths $2a$ and $2b$, respectively, where $0 < b < a$, is $$\frac{x^2}{a^2} + \frac{y^2}{b^2} = 1 \qquad \text{Major axis is horizontal.}$$ $$\frac{x^2}{b^2} + \frac{y^2}{a^2} = 1. \qquad \text{Major axis is vertical.}$$ The vertices and foci lie on the major axis, a and c units, respectively, from the center. Moreover, a, b, and c are related by the equation $c^2 = a^2 - b^2$.	51–58
Recognize, graph, and write equations of hyperbolas (center at origin) *(p. 333)*.	The standard form of the equation of a hyperbola with center at the origin is $$\frac{x^2}{a^2} - \frac{y^2}{b^2} = 1 \qquad \text{Transverse axis is horizontal.}$$ $$\frac{y^2}{a^2} - \frac{x^2}{b^2} = 1 \qquad \text{Transverse axis is vertical.}$$ where $a > 0$ and $b > 0$. The vertices and foci are, respectively, a and c units from the center. Moreover, a, b, and c are related by the equation $c^2 = a^2 + b^2$.	59–62
Section 4.4 — Recognize equations of conics that are shifted vertically or horizontally in the plane *(p. 341)*, and write and graph equations of conics that are shifted vertically or horizontally in the plane *(p. 343)*.	**Circle:** The graph of $(x - 2)^2 + (y + 1)^2 = 5^2$ is a circle whose center is the point $(2, -1)$ and whose radius is 5. The graph is shifted two units to the right and one unit down from standard position. **Parabola:** The graph of $(x - 1)^2 = 4(-1)(y - 6)$ is a parabola whose vertex is the point $(1, 6)$. The axis of the parabola is vertical. Moreover, $p = -1$, so the focus lies one unit below the vertex, and the directrix is $y = 6 - (-1) = 7$. The graph is shifted one unit to the right and six units up from standard position. **Ellipse:** The graph of $\dfrac{(x - 1)^2}{4^2} + \dfrac{(y - 2)^2}{3^2} = 1$ is an ellipse whose center is the point $(1, 2)$. The major axis is horizontal and of length $2(4) = 8$, and the minor axis is vertical and of length $2(3) = 6$. The graph is shifted one unit to the right and two units up from standard position. **Hyperbola:** The graph of $\dfrac{(x - 5)^2}{1^2} - \dfrac{(y - 4)^2}{2^2} = 1$ is a hyperbola whose center is the point $(5, 4)$. The transverse axis is horizontal and of length $2(1) = 2$, and the conjugate axis is vertical and of length $2(2) = 4$. The graph is shifted five units to the right and four units up from standard position.	63–87

Review Exercises

See **CalcChat.com** for tutorial help and worked-out solutions to odd-numbered exercises.

4.1 Finding the Domain of a Rational Function
In Exercises 1–4, find the domain of the function and discuss the behavior of f near any excluded x-values.

1. $f(x) = \dfrac{3x}{x + 10}$

2. $f(x) = \dfrac{4x^3}{2 + 5x}$

3. $f(x) = \dfrac{8}{x^2 - 10x + 24}$

4. $f(x) = \dfrac{x^2 + x - 2}{x^2 - 4x + 4}$

Finding Vertical and Horizontal Asymptotes In Exercises 5–10, find all vertical and horizontal asymptotes of the graph of the function.

5. $f(x) = \dfrac{6x^2}{x + 3}$

6. $f(x) = \dfrac{2x^2 + 5x - 3}{x^2 + 2}$

7. $g(x) = \dfrac{x^2}{x^2 - 4}$

8. $g(x) = \dfrac{x + 1}{x^2 - 1}$

9. $h(x) = \dfrac{5x + 20}{x^2 - 2x - 24}$

10. $h(x) = \dfrac{x^3 - 4x^2}{x^2 + 3x + 2}$

11. Average Cost The cost C (in dollars) of producing x units of a product is given by $C = 0.5x + 500$ and the average cost per unit $\overline{C}$ is given by

$$\overline{C} = \frac{C}{x} = \frac{0.5x + 500}{x}, \quad x > 0.$$

Determine the average cost per unit as x increases without bound.

12. Seizure of Illegal Drugs The cost C (in millions of dollars) for the federal government to seize $p\%$ of an illegal drug as it enters the country is given by

$$C = \frac{528p}{100 - p}, \quad 0 \le p < 100.$$

(a) Use a graphing utility to graph the cost function.

(b) Find the costs of seizing 25%, 50%, and 75% of the drug.

(c) According to the model, is it possible to seize 100% of the drug? Explain.

4.2 Sketching the Graph of a Rational Function
In Exercises 13–24, (a) state the domain of the function, (b) identify all intercepts, (c) find any vertical or horizontal asymptotes, and (d) plot additional solution points as needed to sketch the graph of the rational function.

13. $f(x) = \dfrac{-3}{2x^2}$

14. $f(x) = \dfrac{4}{x}$

15. $g(x) = \dfrac{2 + x}{1 - x}$

16. $h(x) = \dfrac{x - 4}{x - 7}$

17. $p(x) = \dfrac{5x^2}{4x^2 + 1}$

18. $f(x) = \dfrac{-8x}{x^2 + 4}$

19. $f(x) = \dfrac{x}{x^2 - 16}$

20. $h(x) = \dfrac{9}{(x - 3)^2}$

21. $f(x) = \dfrac{-6x^2}{x^2 + 1}$

22. $y = \dfrac{2x^2}{x^2 - 4}$

23. $f(x) = \dfrac{6x^2 - 11x + 3}{3x^2 - x}$

24. $f(x) = \dfrac{6x^2 - 7x + 2}{4x^2 - 1}$

A Rational Function with a Slant Asymptote In Exercises 25–30, (a) state the domain of the function, (b) identify all intercepts, (c) find any vertical or slant asymptotes, and (d) plot additional solution points as needed to sketch the graph of the rational function.

25. $f(x) = \dfrac{2x^3}{x^2 + 1}$

26. $f(x) = \dfrac{x^2 + 1}{x + 1}$

27. $f(x) = \dfrac{x^2 + 3x - 10}{x + 2}$

28. $f(x) = \dfrac{x^3}{2x^2 - 18}$

29. $f(x) = \dfrac{3x^3 - 2x^2 - 3x + 2}{3x^2 - x - 4}$

30. $f(x) = \dfrac{3x^3 - 4x^2 - 12x + 16}{3x^2 + 5x - 2}$

31. Minimum Cost The cost C (in dollars) of producing x units of a product is given by $C = 100{,}000 + 0.9x$ and the average cost per unit $\overline{C}$ is given by

$$\overline{C} = \frac{C}{x} = \frac{100{,}000 + 0.9x}{x}, \quad x > 0.$$

(a) Sketch the graph of the average cost function.

(b) Find the average costs when $x = 1000$, $x = 10{,}000$, and $x = 100{,}000$.

(c) By increasing the level of production, what is the least average cost per unit you can obtain? Explain.

32. Page Design A page that is x inches wide and y inches high contains 30 square inches of print. The top and bottom margins are each 2 inches deep and the margins on each side are 2 inches wide.

(a) Show that the total area A of the page is

$$A = \frac{2x(2x + 7)}{x - 4}.$$

(b) Determine the domain of the function based on the physical constraints of the problem.

(c) Use a graphing utility to graph the area function and approximate the dimensions of the page that use the least amount of paper. Verify your answer numerically using the *table* feature of the graphing utility.

33. Biology A parks and wildlife commission releases 80,000 fish into a lake. After t years, the population N of the fish (in thousands) is given by

$$N = \frac{20(3t + 4)}{0.05t + 1}, \quad t \geq 0.$$

(a) Sketch the graph of the function.

(b) Find the population when $t = 5$, $t = 10$, and $t = 25$.

(c) What is the maximum number of fish in the lake as time increases? Explain.

34. Medicine The concentration C of a chemical in the bloodstream t hours after injection into muscle tissue is given by $C(t) = (2t + 1)/(t^2 + 4)$, $t > 0$.

(a) Determine the horizontal asymptote of the graph of the function and interpret its meaning in the context of the problem.

(b) Use a graphing utility to graph the function and approximate the time when the bloodstream concentration is greatest.

4.3 Identifying a Conic In Exercises 35–42, identify the conic.

35. $y^2 = -10x$

36. $x^2 + \frac{y^2}{16} = 1$

37. $\frac{y^2}{12} - \frac{x^2}{9} = 1$

38. $x^2 + y^2 = 20$

39. $4x^2 + 18y^2 = 36$

40. $8x^2 - y^2 = 48$

41. $3x^2 + 3y^2 = 75$

42. $5y = x^2$

Finding the Standard Equation of a Parabola In Exercises 43–48, find the standard form of the equation of the parabola with the given characteristic(s) and vertex at the origin.

43. Focus: $(-6, 0)$

44. Focus: $(0, 7)$

45. Directrix: $y = -3$

46. Directrix: $x = 3$

47. Passes through the point $(3, 6)$; horizontal axis

48. Passes through the point $(4, -2)$; vertical axis

49. Satellite Dish A cross section of a large parabolic satellite dish is modeled by $y = x^2/200$, $-100 \leq x \leq 100$ (see figure). The receiving and transmitting equipment is positioned at the focus. Find the coordinates of the focus.

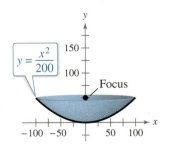

50. Suspension Bridge Each cable of a suspension bridge is suspended (in the shape of a parabola) between two towers (see figure).

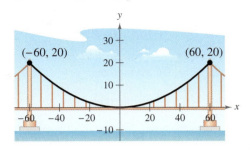

(a) Find the coordinates of the focus.

(b) Write an equation that models the cables.

Finding the Standard Equation of an Ellipse In Exercises 51–56, find the standard form of the equation of the ellipse with the given characteristics and center at the origin.

51. 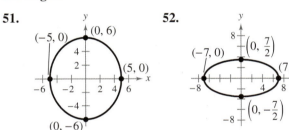 **52.**

53. Vertices: $(0, \pm 7)$; foci: $(0, \pm 6)$

54. Foci: $(\pm 3, 0)$; major axis of length 12

55. Foci: $(\pm 14, 0)$; minor axis of length 10

56. Passes through the points $(0, -6)$ and $(3, 0)$

57. Architecture A semielliptical archway is formed over the entrance to an estate. The arch is set on pillars that are 10 feet apart and has a height (atop the pillars) of 4 feet (see figure). Describe the location of the foci.

58. Wading Pool You are building a wading pool that is in the shape of an ellipse. An equation for the elliptical shape of the pool, measured in feet, is

$$\frac{x^2}{324} + \frac{y^2}{196} = 1.$$

Find the longest distance across the pool, the shortest distance, and the distance between the foci.

Finding the Standard Equation of a Hyperbola In Exercises 59–62, find the standard form of the equation of the hyperbola with the given characteristics and center at the origin.

59. Vertices: $(0, \pm 1)$; foci: $(0, \pm 5)$

60. Vertices: $(\pm 4, 0)$; foci: $(\pm 6, 0)$

61. Vertices: $(\pm 1, 0)$; asymptotes: $y = \pm 2x$

62. Vertices: $(0, \pm 2)$; asymptotes: $y = \pm \dfrac{2}{\sqrt{5}}x$

4.4 **Finding the Standard Equation of a Parabola** In Exercises 63–66, find the standard form of the equation of the parabola with the given characteristics.

63. Vertex: $(4, 2)$; focus: $(4, 0)$

64. Vertex: $(2, 0)$; focus: $(0, 0)$

65. Vertex: $(8, -8)$; directrix: $x = 1$

66. Focus: $(5, 6)$; directrix: $y = 0$

Finding the Standard Equation of an Ellipse In Exercises 67–72, find the standard form of the equation of the ellipse with the given characteristics.

67. Vertices: $(0, 2)$, $(4, 2)$; minor axis of length 2

68. Vertices: $(5, 0)$, $(5, 12)$; minor axis of length 10

69. Vertices: $(2, -2)$, $(2, 8)$; foci: $(2, 0)$, $(2, 6)$

70. Vertices: $(-9, -4)$, $(11, -4)$; foci: $(-5, -4)$, $(7, -4)$

71. Foci: $(0, -4)$, $(0, 0)$; eccentricity: $\frac{2}{3}$

72. Foci: $(5, 1)$, $(11, 1)$; eccentricity: $\frac{3}{5}$

Finding the Standard Equation of a Hyperbola In Exercises 73–76, find the standard form of the equation of the hyperbola with the given characteristics.

73. Vertices: $(-10, 3)$, $(6, 3)$; foci: $(-12, 3)$, $(8, 3)$

74. Vertices: $(2, -2)$, $(2, 2)$; foci: $(2, -4)$, $(2, 4)$

75. Vertices: $(3, -4)$, $(3, 4)$; passes through the point $(4, 6)$

76. Vertices: $(\pm 6, 7)$;

asymptotes: $y = -\frac{1}{2}x + 7$, $y = \frac{1}{2}x + 7$

Identifying a Conic In Exercises 77–84, identify the conic by writing its equation in standard form. Then sketch its graph and describe the translation from standard position.

77. $x^2 - 6x + 2y + 9 = 0$

78. $y^2 - 12y - 8x + 20 = 0$

79. $x^2 + y^2 - 2x - 4y + 1 = 0$

80. $16x^2 + 16y^2 - 16x + 24y - 3 = 0$

81. $x^2 + 9y^2 + 10x - 18y + 25 = 0$

82. $4x^2 + y^2 - 16x + 15 = 0$

83. $9x^2 - y^2 - 72x + 8y + 119 = 0$

84. $y^2 - 9x^2 + 10y + 18x + 7 = 0$

85. Architecture A parabolic archway is 12 meters high at the vertex. At a height of 10 meters, the width of the archway is 8 meters (see figure). How wide is the archway at ground level?

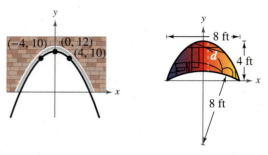

Figure for 85   Figure for 86

86. Architecture A church window is bounded above by a parabola and below by the arc of a circle (see figure).

(a) Find equations for the parabola and the circle.

(b) Complete the table by filling in the vertical distance d between the circle and the parabola for each given value of x.

x	0	1	2	3	4
d					

87. Running Path Let $(0, 0)$ represent a water fountain located in a city park. Each day you run through the park along a path given by

$$x^2 + y^2 - 200x - 52{,}500 = 0$$

where x and y are measured in meters.

(a) What type of conic is your path? Explain.

(b) Write the equation of the path in standard form. Sketch the graph of the equation.

(c) After you run, you walk to the water fountain. When you stop running at $(-100, 150)$, how far must you walk for a drink of water?

Exploration

True or False? In Exercises 88–90, determine whether the statement is true or false. Justify your answer.

88. The domain of a rational function can never be the set of all real numbers.

89. The graph of the equation

$$Ax^2 + Bxy + Cy^2 + Dx + Ey + F = 0$$

can be a single point.

90. If two ellipses have the same major axis length and same eccentricity, then they have the same minor axis length.

Chapter Test

See CalcChat.com for tutorial help and worked-out solutions to odd-numbered exercises.

Take this test as you would take a test in class. When you are finished, check your work against the answers given in the back of the book.

In Exercises 1–3, find the domain of the function and identify any asymptotes of the graph of the function.

1. $y = \dfrac{3x}{x + 1}$

2. $f(x) = \dfrac{3 - x^2}{3 + x^2}$

3. $g(x) = \dfrac{x - 4}{x^2 - 9x + 20}$

In Exercises 4–9, identify any intercepts and asymptotes of the graph of the function. Then sketch the graph of the function.

4. $h(x) = \dfrac{3}{x^2} - 1$

5. $g(x) = \dfrac{x^2 + 2}{x - 1}$

6. $f(x) = \dfrac{x + 1}{x^2 + x - 12}$

7. $f(x) = \dfrac{2x^2 - 5x - 12}{x^2 - 16}$

8. $f(x) = \dfrac{2x^2 + 9}{5x^2 + 9}$

9. $g(x) = \dfrac{2x^3 + 3x^2 - 8x - 12}{x^2 - x - 2}$

10. A rectangular page contains 36 square inches of print. The margins at the top and bottom of the page are each 2 inches deep. The margins on each side are 1 inch wide. What should the dimensions of the page be to use the least amount of paper?

11. A triangle is formed by the coordinate axes and a line through the point $(2, 1)$, as shown in the figure. The value of y is given by

$$y = 1 + \dfrac{2}{x - 2}.$$

(a) Write the area A of the triangle as a function of x. Determine the domain of the function in the context of the problem.

(b) Sketch the graph of the area function. Estimate the minimum area of the triangle from the graph.

In Exercises 12–17, sketch the conic and identify the center, vertices, and foci, if applicable.

12. $x^2 - \dfrac{y^2}{4} = 1$

13. $4y^2 - 5x^2 = 80$

14. $y^2 - 4x = 0$

15. $x^2 + y^2 - 10x + 4y + 4 = 0$

16. $x^2 - 10x - 2y + 19 = 0$

17. $x^2 + 3y^2 - 2x + 36y + 100 = 0$

18. Find the standard form of the equation of the ellipse with vertices $(0, 2)$ and $(8, 2)$ and minor axis of length 4. Then find the eccentricity of the ellipse.

19. Find the standard form of the equation of the hyperbola with vertices $(0, \pm 3)$ and asymptotes $y = \pm \frac{3}{2}x$.

20. A parabolic archway is 16 meters high at the vertex. At a height of 14 meters, the width of the archway is 12 meters, as shown in the figure. How wide is the archway at ground level?

21. The moon orbits Earth in an elliptical path with the center of Earth at one focus, as shown in the figure. The major and minor axes of the orbit have lengths of 768,800 kilometers and 767,641 kilometers, respectively. Find the least distance (perigee) and the greatest distance (apogee) from the center of the moon to the center of Earth.

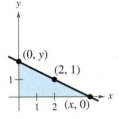

Figure for 11

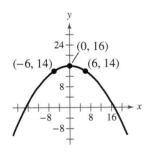

Figure for 20

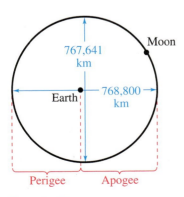

Figure for 21

Proofs in Mathematics ■ ■ ■ ■ ■ ■ ■ ■ ■ ■ ■ ■ ■ ■ ■ ■ ■

The definition of a parabola can be used to derive the standard form of the equation of a parabola whose vertex is at $(0, 0)$, and whose directrix is parallel to the x-axis or to the y-axis.

PARABOLIC PATHS

There are many natural occurrences of parabolas in real life. For example, Italian astronomer and mathematician Galileo Galilei discovered in the 17th century that an object projected upward and obliquely to the pull of gravity travels in a parabolic path. Examples of this include the path of a jumping dolphin and the path of water molecules from a drinking water fountain.

Standard Equation of a Parabola (Vertex at Origin) *(p. 328)*

The **standard form of the equation of a parabola** with vertex at $(0, 0)$ and directrix $y = -p$ is

$$x^2 = 4py, \quad p \neq 0. \qquad \text{Vertical axis}$$

For directrix $x = -p$, the equation is

$$y^2 = 4px, \quad p \neq 0. \qquad \text{Horizontal axis}$$

The focus is on the axis p units (directed distance) from the vertex.

Proof

For the first case, assume that the directrix $(y = -p)$ is parallel to the x-axis. In the figure, $p > 0$, and p is the directed distance from the vertex to the focus, so the focus must lie above the vertex. By the definition of a parabola, the point (x, y) is equidistant from $(0, p)$ and $y = -p$. Apply the Distance Formula to obtain

$$\sqrt{(x - 0)^2 + (y - p)^2} = y + p \qquad \text{Distance Formula}$$
$$x^2 + (y - p)^2 = (y + p)^2 \qquad \text{Square each side.}$$
$$x^2 + y^2 - 2py + p^2 = y^2 + 2py + p^2 \qquad \text{Expand.}$$
$$x^2 = 4py. \qquad \text{Simplify.}$$

A proof of the second case is similar to the proof of the first case. Assume that the directrix $(x = -p)$ is parallel to the y-axis. In the figure, $p > 0$, and p is the directed distance from the vertex to the focus, so the focus must lie to the right of the vertex. By the definition of a parabola, the point (x, y) is equidistant from $(p, 0)$ and $x = -p$. Apply the Distance Formula to obtain

$$\sqrt{(x - p)^2 + (y - 0)^2} = x + p \qquad \text{Distance Formula}$$
$$(x - p)^2 + y^2 = (x + p)^2 \qquad \text{Square each side.}$$
$$x^2 - 2px + p^2 + y^2 = x^2 + 2px + p^2 \qquad \text{Expand.}$$
$$y^2 = 4px. \qquad \text{Simplify.}$$

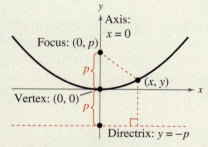

Parabola with vertical axis

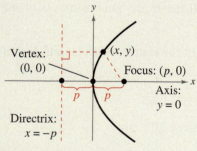

Parabola with horizontal axis

356

P.S. Problem Solving ■ ■ ■ ■ ■ ■ ■ ■ ■ ■ ■ ■ ■ ■ ■

1. Matching Match the graph of the rational function

$$f(x) = \frac{ax + b}{cx + d}$$

with the given conditions.

(a)
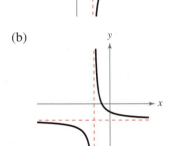

(b)
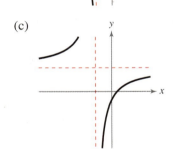

(c)

(d)
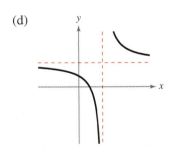

(i) $a > 0$ (ii) $a > 0$ (iii) $a < 0$ (iv) $a > 0$
    $b < 0$ $b > 0$ $b > 0$ $b < 0$
    $c > 0$ $c < 0$ $c > 0$ $c > 0$
    $d < 0$ $d < 0$ $d < 0$ $d > 0$

2. Effects of Values on a Graph Consider the function $f(x) = (ax)/(x - b)^2$.

(a) Determine the effect on the graph of f when $b \neq 0$ and a is varied. Consider cases in which a is positive and a is negative.

(b) Determine the effect on the graph of f when $a \neq 0$ and b is varied.

3. Distinct Vision The endpoints of the interval over which distinct vision is possible are called the *near point* and *far point* of the eye (see figure). With increasing age, these points normally change. The table shows the approximate near points y (in inches) for various ages x (in years).

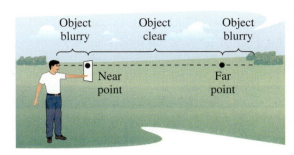

Age, x	Near Point, y
16	3.0
32	4.7
44	9.8
50	19.7
60	39.4

(a) Use the *regression* feature of a graphing utility to find a quadratic model for the data. Use the graphing utility to plot the data and graph the model in the same viewing window.

(b) Find a rational model for the data. Take the reciprocals of the near points to generate the points $(x, 1/y)$. Use the *regression* feature of the graphing utility to find a linear model for the data. The resulting line has the form

$$\frac{1}{y} = ax + b.$$

Solve for y. Use the graphing utility to plot the original data and graph the model in the same viewing window.

(c) Use the *table* feature of the graphing utility to construct a table showing the predicted near point based on each model for each of the ages in the original table. How well do the models fit the original data?

(d) Use both models to estimate the near point for a person who is 25 years old. Which model is a better fit?

(e) Do you think either model can be used to predict the near point for a person who is 70 years old? Explain.

4. Statuary Hall Statuary Hall is an elliptical room in the United States Capitol in Washington, D.C. The room is also called the Whispering Gallery because a person standing at one focus of the room can hear even a whisper spoken by a person standing at the other focus. This occurs because any sound emitted from one focus of an ellipse reflects off the side of the ellipse to the other focus. Statuary Hall is 46 feet wide and 97 feet long.

(a) Find an equation that models the shape of the room.

(b) How far apart are the two foci?

(c) What is the area of the floor of the room? (The area of an ellipse is $A = \pi ab$.)

5. Property of a Hyperbola Use the figure to show that

$$|d_2 - d_1| = 2a.$$

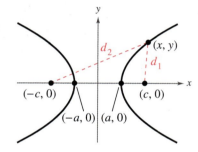

6. Finding the Equation of a Hyperbola Find an equation of a hyperbola such that for any point on the hyperbola, the difference between its distances from the points $(2, 2)$ and $(10, 2)$ is 6.

7. Tour Boat A tour boat travels between two islands that are 12 miles apart (see figure). For each trip between the islands, there is enough fuel for a 20-mile trip.

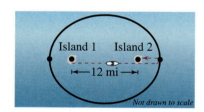

(a) Explain why the region in which the boat can travel is bounded by an ellipse.

(b) Let $(0, 0)$ represent the center of the ellipse. Find the coordinates of the center of each island.

(c) The boat travels from Island 1, past Island 2 to one vertex of the ellipse, and then to Island 2 (see figure). How many miles does the boat travel? Use your answer to find the coordinates of the vertex.

(d) Use the results of parts (b) and (c) to write an equation of the ellipse that bounds the region in which the boat can travel.

8. Car Headlight The filament of a light bulb is a thin wire that glows when electricity passes through it. The filament of a car headlight is at the focus of a parabolic reflector, which sends light out in a straight beam. Given that the filament is 1.5 inches from the vertex, find an equation for the cross section of the reflector. The reflector is 7 inches wide. How deep is it?

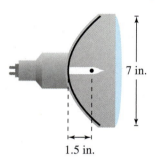

9. Analyzing Parabolas Consider the parabola $x^2 = 4py$.

(a) Use a graphing utility to graph the parabola for $p = 1$, $p = 2$, $p = 3$, and $p = 4$. Describe the effect on the graph when p increases.

(b) Locate the focus for each parabola in part (a).

(c) For each parabola in part (a), find the length of the chord passing through the focus and parallel to the directrix. How can you determine the length of this chord directly from the standard form of the equation of the parabola?

(d) Explain how the result of part (c) can be used as an aid when sketching parabolas.

10. Tangent Line Let (x_1, y_1) be the coordinates of a point on the parabola $x^2 = 4py$. The equation of the line that just touches the parabola at the point (x_1, y_1), called a *tangent line*, is

$$y - y_1 = \frac{x_1}{2p}(x - x_1).$$

(a) What is the slope of the tangent line?

(b) For each parabola in Exercise 9, find the equations of the tangent lines at the endpoints of the chord. Use a graphing utility to graph the parabola and tangent lines.

11. Proof Prove that the graph of the equation

$$Ax^2 + Cy^2 + Dx + Ey + F = 0$$

is one of the following (except in degenerate cases).

Conic	Condition
(a) Circle	$A = C, A \neq 0$
(b) Parabola	$A = 0$ or $C = 0$, but not both
(c) Ellipse	$AC > 0, A \neq C$
(d) Hyperbola	$AC < 0$

5 Exponential and Logarithmic Functions

Beaver Population *(Exercise 83, page 396)*

Earthquakes
(Example 6, page 404)

Sound Intensity *(Exercises 79–82, page 386)*

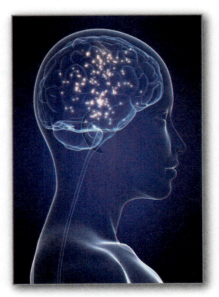

Human Memory Model
(Exercise 83, page 380)

Nuclear Reactor Accident *(Example 9, page 367)*

Clockwise from top left, Alexander Kuguchin/Shutterstock.com; Somjin Klong-ugkara/Shutterstock.com;
Sebastian Kaulitzki/Shutterstock.com; Fotokon/Shutterstock.com; Titima Ongkantong/Shutterstock.com

5.1 Exponential Functions and Their Graphs

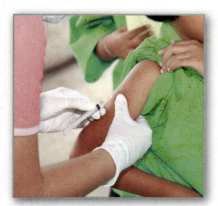

- ■ Recognize and evaluate exponential functions with base *a*.
- ■ Graph exponential functions and use the One-to-One Property.
- ■ Recognize, evaluate, and graph exponential functions with base *e*.
- ■ Use exponential functions to model and solve real-life problems.

Exponential Functions

So far, this text has dealt mainly with **algebraic functions,** which include polynomial functions and rational functions. In this chapter, you will study two types of nonalgebraic functions—*exponential functions* and *logarithmic functions*. These functions are examples of **transcendental functions.** This section will focus on exponential functions.

Exponential functions can help you model and solve real-life problems. For example, in Exercise 66 on page 370, you will use an exponential function to model the concentration of a drug in the bloodstream.

> **Definition of Exponential Function**
>
> The **exponential function** f **with base** a is denoted by
>
> $$f(x) = a^x$$
>
> where $a > 0$, $a \neq 1$, and x is any real number.

The base a of an exponential function cannot be 1 because $a = 1$ yields $f(x) = 1^x = 1$. This is a constant function, not an exponential function.

You have evaluated a^x for integer and rational values of x. For example, you know that $4^3 = 64$ and $4^{1/2} = 2$. However, to evaluate 4^x for any real number x, you need to interpret forms with *irrational* exponents. For the purposes of this text, it is sufficient to think of $a^{\sqrt{2}}$ (where $\sqrt{2} \approx 1.41421356$) as the number that has the successively closer approximations

$$a^{1.4}, a^{1.41}, a^{1.414}, a^{1.4142}, a^{1.41421}, \ldots .$$

EXAMPLE 1   Evaluating Exponential Functions

Use a calculator to evaluate each function at the given value of x.

Function	Value
a. $f(x) = 2^x$	$x = -3.1$
b. $f(x) = 2^{-x}$	$x = \pi$
c. $f(x) = 0.6^x$	$x = \frac{3}{2}$

Solution

Function Value	Calculator Keystrokes	Display
a. $f(-3.1) = 2^{-3.1}$	2 [∧] [(−)] 3.1 [ENTER]	0.1166291
b. $f(\pi) = 2^{-\pi}$	2 [∧] [(−)] π [ENTER]	0.1133147
c. $f\left(\frac{3}{2}\right) = (0.6)^{3/2}$	.6 [∧] [(] 3 [÷] 2 [)] [ENTER]	0.4647580

✔ **Checkpoint**))) *Audio-video solution in English & Spanish at LarsonPrecalculus.com*

Use a calculator to evaluate $f(x) = 8^{-x}$ at $x = \sqrt{2}$. ■

When evaluating exponential functions with a calculator, it may be necessary to enclose fractional exponents in parentheses. Some calculators do not correctly interpret an exponent that consists of an expression unless parentheses are used.

Graphs of Exponential Functions

The graphs of all exponential functions have similar characteristics, as shown in Examples 2, 3, and 5.

EXAMPLE 2 **Graphs of $y = a^x$**

▷ **ALGEBRA HELP** To review the techniques for sketching the graph of an equation, see Section 1.1.

In the same coordinate plane, sketch the graph of each function.

a. $f(x) = 2^x$ **b.** $g(x) = 4^x$

Solution Begin by constructing a table of values.

x	-3	-2	-1	0	1	2
2^x	$\frac{1}{8}$	$\frac{1}{4}$	$\frac{1}{2}$	1	2	4
4^x	$\frac{1}{64}$	$\frac{1}{16}$	$\frac{1}{4}$	1	4	16

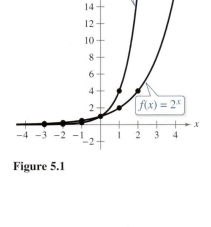

Figure 5.1

To sketch the graph of each function, plot the points from the table and connect them with a smooth curve, as shown in Figure 5.1. Note that both graphs are increasing. Moreover, the graph of $g(x) = 4^x$ is increasing more rapidly than the graph of $f(x) = 2^x$.

✓ **Checkpoint** *Audio-video solution in English & Spanish at LarsonPrecalculus.com*

In the same coordinate plane, sketch the graph of each function.

a. $f(x) = 3^x$ **b.** $g(x) = 9^x$

The table in Example 2 was evaluated by hand for integer values of x. You can also evaluate $f(x)$ and $g(x)$ for noninteger values of x by using a calculator.

EXAMPLE 3 **Graphs of $y = a^{-x}$**

In the same coordinate plane, sketch the graph of each function.

a. $F(x) = 2^{-x}$ **b.** $G(x) = 4^{-x}$

Solution Begin by constructing a table of values.

x	-2	-1	0	1	2	3
2^{-x}	4	2	1	$\frac{1}{2}$	$\frac{1}{4}$	$\frac{1}{8}$
4^{-x}	16	4	1	$\frac{1}{4}$	$\frac{1}{16}$	$\frac{1}{64}$

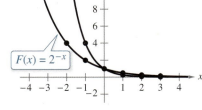

Figure 5.2

To sketch the graph of each function, plot the points from the table and connect them with a smooth curve, as shown in Figure 5.2. Note that both graphs are decreasing. Moreover, the graph of $G(x) = 4^{-x}$ is decreasing more rapidly than the graph of $F(x) = 2^{-x}$.

✓ **Checkpoint** *Audio-video solution in English & Spanish at LarsonPrecalculus.com*

In the same coordinate plane, sketch the graph of each function.

a. $f(x) = 3^{-x}$ **b.** $g(x) = 9^{-x}$

Note that it is possible to use one of the properties of exponents to rewrite the functions in Example 3 with positive exponents.

$$F(x) = 2^{-x} = \frac{1}{2^x} = \left(\frac{1}{2}\right)^x \quad \text{and} \quad G(x) = 4^{-x} = \frac{1}{4^x} = \left(\frac{1}{4}\right)^x$$

Comparing the functions in Examples 2 and 3, observe that

$$F(x) = 2^{-x} = f(-x) \quad \text{and} \quad G(x) = 4^{-x} = g(-x).$$

Consequently, the graph of F is a reflection (in the y-axis) of the graph of f. The graphs of G and g have the same relationship. The graphs in Figures 5.1 and 5.2 are typical of the exponential functions $y = a^x$ and $y = a^{-x}$. They have one y-intercept and one horizontal asymptote (the x-axis), and they are continuous. Here is a summary of the basic characteristics of the graphs of these exponential functions.

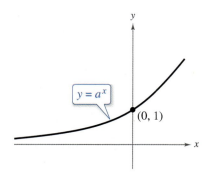

Graph of $y = a^x$, $a > 1$
- Domain: $(-\infty, \infty)$
- Range: $(0, \infty)$
- y-intercept: $(0, 1)$
- Increasing
- x-axis is a horizontal asymptote $(a^x \to 0$ as $x \to -\infty)$.
- Continuous

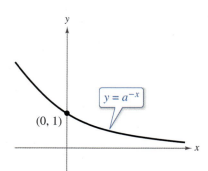

Graph of $y = a^{-x}$, $a > 1$
- Domain: $(-\infty, \infty)$
- Range: $(0, \infty)$
- y-intercept: $(0, 1)$
- Decreasing
- x-axis is a horizontal asymptote $(a^{-x} \to 0$ as $x \to \infty)$.
- Continuous

Notice that the graph of an exponential function is always increasing or always decreasing, so the graph passes the Horizontal Line Test. Therefore, an exponential function is a one-to-one function. You can use the following **One-to-One Property** to solve simple exponential equations.

For $a > 0$ and $a \neq 1$, $a^x = a^y$ if and only if $x = y$.   One-to-One Property

EXAMPLE 4   **Using the One-to-One Property**

a. $9 = 3^{x+1}$ Original equation

$3^2 = 3^{x+1}$ $9 = 3^2$

$2 = x + 1$ One-to-One Property

$1 = x$ Solve for x.

b. $\left(\frac{1}{2}\right)^x = 8$ Original equation

$2^{-x} = 2^3$ $\left(\frac{1}{2}\right)^x = 2^{-x}, 8 = 2^3$

$x = -3$ One-to-One Property

✓ *Checkpoint* *Audio-video solution in English & Spanish at LarsonPrecalculus.com*

Use the One-to-One Property to solve the equation for x.

a. $8 = 2^{2x-1}$ **b.** $\left(\frac{1}{3}\right)^{-x} = 27$

In Example 5, notice how the graph of $y = a^x$ can be used to sketch the graphs of functions of the form $f(x) = b \pm a^{x+c}$.

▷ **ALGEBRA HELP**   To review the techniques for transforming the graph of a function, see Section 2.5.

EXAMPLE 5

Transformations of Graphs of Exponential Functions

See LarsonPrecalculus.com for an interactive version of this type of example.

Describe the transformation of the graph of $f(x) = 3^x$ that yields each graph.

a.
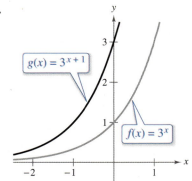

b.
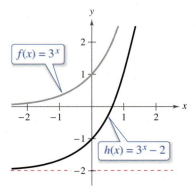

c.
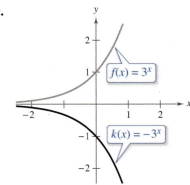

d.
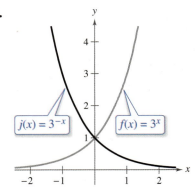

Solution

a. Because $g(x) = 3^{x+1} = f(x + 1)$, the graph of g is obtained by shifting the graph of f one unit to the *left*.

b. Because $h(x) = 3^x - 2 = f(x) - 2$, the graph of h is obtained by shifting the graph of f *down* two units.

c. Because $k(x) = -3^x = -f(x)$, the graph of k is obtained by *reflecting* the graph of f in the x-axis.

d. Because $j(x) = 3^{-x} = f(-x)$, the graph of j is obtained by *reflecting* the graph of f in the y-axis.

✓ *Checkpoint* 🔊 *Audio-video solution in English & Spanish at LarsonPrecalculus.com*

Describe the transformation of the graph of $f(x) = 4^x$ that yields the graph of each function.

a. $g(x) = 4^{x-2}$ b. $h(x) = 4^x + 3$ c. $k(x) = 4^{-x} - 3$

Note how each transformation in Example 5 affects the y-intercept and the horizontal asymptote.

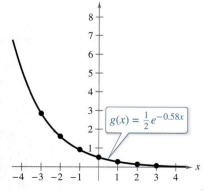

Figure 5.3

The Natural Base e

In many applications, the most convenient choice for a base is the irrational number

$$e \approx 2.718281828 \ldots .$$

This number is called the **natural base**. The function $f(x) = e^x$ is called the **natural exponential function.** Figure 5.3 shows its graph. Be sure you see that for the exponential function $f(x) = e^x$, e is the constant $2.718281828 \ldots$, whereas x is the variable.

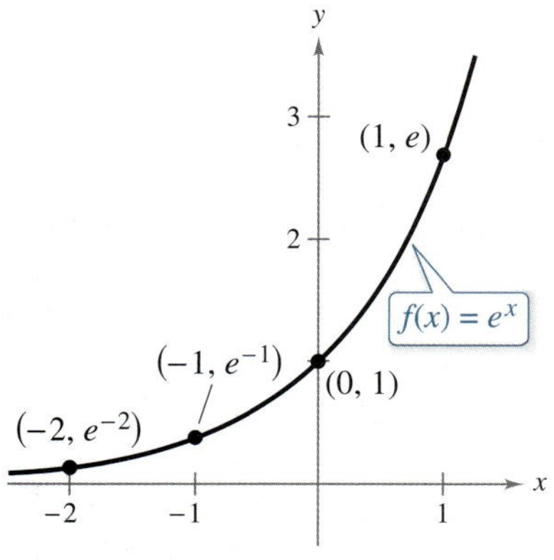

EXAMPLE 6   **Evaluating the Natural Exponential Function**

Use a calculator to evaluate the function $f(x) = e^x$ at each value of x.

a. $x = -2$ **b.** $x = -1$

c. $x = 0.25$ **d.** $x = -0.3$

Solution

Function Value	Calculator Keystrokes	Display
a. $f(-2) = e^{-2}$	$\boxed{e^x}$ $\boxed{(-)}$ 2 $\boxed{\text{ENTER}}$	0.1353353
b. $f(-1) = e^{-1}$	$\boxed{e^x}$ $\boxed{(-)}$ 1 $\boxed{\text{ENTER}}$	0.3678794
c. $f(0.25) = e^{0.25}$	$\boxed{e^x}$ 0.25 $\boxed{\text{ENTER}}$	1.2840254
d. $f(-0.3) = e^{-0.3}$	$\boxed{e^x}$ $\boxed{(-)}$ 0.3 $\boxed{\text{ENTER}}$	0.7408182

✓ *Checkpoint* ◀))) Audio-video solution in English & Spanish at LarsonPrecalculus.com

Use a calculator to evaluate the function $f(x) = e^x$ at each value of x.

a. $x = 0.3$

b. $x = -1.2$

c. $x = 6.2$

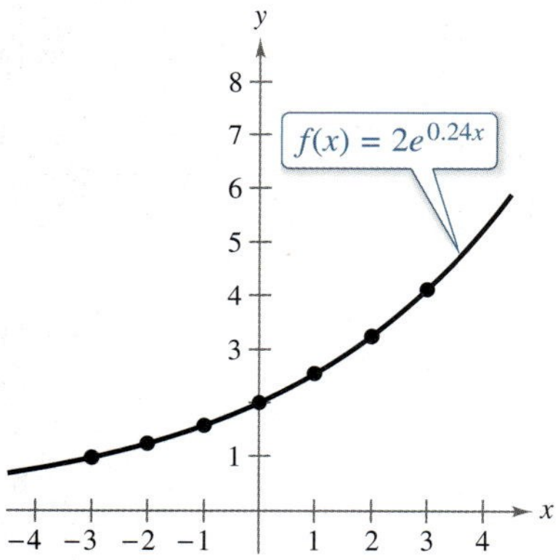

Figure 5.4

EXAMPLE 7   **Graphing Natural Exponential Functions**

Sketch the graph of each natural exponential function.

a. $f(x) = 2e^{0.24x}$

b. $g(x) = \frac{1}{2}e^{-0.58x}$

Solution   Begin by using a graphing utility to construct a table of values.

x	-3	-2	-1	0	1	2	3
$f(x)$	0.974	1.238	1.573	2.000	2.542	3.232	4.109
$g(x)$	2.849	1.595	0.893	0.500	0.280	0.157	0.088

To graph each function, plot the points from the table and connect them with a smooth curve, as shown in Figures 5.4 and 5.5. Note that the graph in Figure 5.4 is increasing, whereas the graph in Figure 5.5 is decreasing.

✓ *Checkpoint* ◀))) Audio-video solution in English & Spanish at LarsonPrecalculus.com

Sketch the graph of $f(x) = 5e^{0.17x}$.

Figure 5.5

Applications

One of the most familiar examples of exponential growth is an investment earning *continuously compounded interest*. Recall from page 95 in Section 1.3 that the formula for *interest compounded n times per year* is

$$A = P\left(1 + \frac{r}{n}\right)^{nt}.$$

In this formula, A is the balance in the account, P is the principal (or original deposit), r is the annual interest rate (in decimal form), n is the number of compoundings per year, and t is the time in years. Exponential functions can be used to *develop* this formula and show how it leads to continuous compounding.

Consider a principal P invested at an annual interest rate r, compounded once per year. When the interest is added to the principal at the end of the first year, the new balance P_1 is

$$\begin{aligned} P_1 &= P + Pr \\ &= P(1 + r). \end{aligned}$$

This pattern of multiplying the balance by $1 + r$ repeats each successive year, as shown here.

Year	Balance After Each Compounding
0	$P = P$
1	$P_1 = P(1 + r)$
2	$P_2 = P_1(1 + r) = P(1 + r)(1 + r) = P(1 + r)^2$
3	$P_3 = P_2(1 + r) = P(1 + r)^2(1 + r) = P(1 + r)^3$
$\vdots$	$\vdots$
t	$P_t = P(1 + r)^t$

To accommodate more frequent (quarterly, monthly, or daily) compounding of interest, let n be the number of compoundings per year and let t be the number of years. Then the rate per compounding is r/n, and the account balance after t years is

$$A = P\left(1 + \frac{r}{n}\right)^{nt}. \qquad \text{Amount (balance) with } n \text{ compoundings per year}$$

When the number of compoundings n increases without bound, the process approaches what is called **continuous compounding.** In the formula for n compoundings per year, let $m = n/r$. This yields a new expression.

m	$\left(1 + \dfrac{1}{m}\right)^m$
1	2
10	2.59374246
100	2.704813829
1,000	2.716923932
10,000	2.718145927
100,000	2.718268237
1,000,000	2.718280469
10,000,000	2.718281693
$\downarrow$	$\downarrow$
∞	e

$$A = P\left(1 + \frac{r}{n}\right)^{nt} \qquad \text{Amount with } n \text{ compoundings per year}$$

$$= P\left(1 + \frac{r}{mr}\right)^{mrt} \qquad \text{Substitute } mr \text{ for } n.$$

$$= P\left(1 + \frac{1}{m}\right)^{mrt} \qquad \text{Simplify.}$$

$$= P\left[\left(1 + \frac{1}{m}\right)^m\right]^{rt} \qquad \text{Property of exponents}$$

As m increases without bound (that is, as $m \to \infty$), the table at the left shows that $[1 + (1/m)]^m \to e$. This allows you to conclude that the formula for continuous compounding is

$$A = Pe^{rt}. \qquad \text{Substitute } e \text{ for } [1 + (1/m)]^m.$$

•• REMARK Be sure you see that, when using the formulas for compound interest, you must write the annual interest rate in decimal form. For example, you must write 6% as 0.06.

> **Formulas for Compound Interest**
>
> After t years, the balance A in an account with principal P and annual interest rate r (in decimal form) is given by one of these two formulas.
>
> **1.** For n compoundings per year: $A = P\left(1 + \dfrac{r}{n}\right)^{nt}$
>
> **2.** For continuous compounding: $A = Pe^{rt}$

EXAMPLE 8 **Compound Interest**

You invest \$12,000 at an annual rate of 3%. Find the balance after 5 years for each type of compounding.

a. Quarterly

b. Monthly

c. Continuous

Solution

a. For quarterly compounding, use $n = 4$ to find the balance after 5 years.

$$A = P\left(1 + \frac{r}{n}\right)^{nt} \qquad \text{Formula for compound interest}$$

$$= 12{,}000\left(1 + \frac{0.03}{4}\right)^{4(5)} \qquad \text{Substitute for } P, r, n, \text{ and } t.$$

$$\approx 13{,}934.21 \qquad \text{Use a calculator.}$$

b. For monthly compounding, use $n = 12$ to find the balance after 5 years.

$$A = P\left(1 + \frac{r}{n}\right)^{nt} \qquad \text{Formula for compound interest}$$

$$= 12{,}000\left(1 + \frac{0.03}{12}\right)^{12(5)} \qquad \text{Substitute for } P, r, n, \text{ and } t.$$

$$\approx \$13{,}939.40 \qquad \text{Use a calculator.}$$

c. Use the formula for continuous compounding to find the balance after 5 years.

$$A = Pe^{rt} \qquad \text{Formula for continuous compounding}$$

$$= 12{,}000e^{0.03(5)} \qquad \text{Substitute for } P, r, \text{ and } t.$$

$$\approx \$13{,}942.01 \qquad \text{Use a calculator.}$$

✓ **Checkpoint**))) *Audio-video solution in English & Spanish at LarsonPrecalculus.com*

You invest \$6000 at an annual rate of 4%. Find the balance after 7 years for each type of compounding.

a. Quarterly   **b.** Monthly   **c.** Continuous

In Example 8, note that continuous compounding yields more than quarterly and monthly compounding. This is typical of the two types of compounding. That is, for a given principal, interest rate, and time, continuous compounding will always yield a larger balance than compounding n times per year.

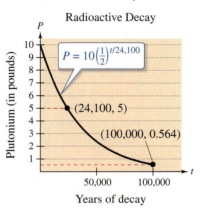

EXAMPLE 9   **Radioactive Decay**

In 1986, a nuclear reactor accident occurred in Chernobyl in what was then the Soviet Union. The explosion spread highly toxic radioactive chemicals, such as plutonium $\left(^{239}\text{Pu}\right)$, over hundreds of square miles, and the government evacuated the city and the surrounding area. To see why the city is now uninhabited, consider the model

$$P = 10\left(\frac{1}{2}\right)^{t/24,100}$$

which represents the amount of plutonium P that remains (from an initial amount of 10 pounds) after t years. Sketch the graph of this function over the interval from $t = 0$ to $t = 100,000$, where $t = 0$ represents 1986. How much of the 10 pounds will remain in the year 2020? How much of the 10 pounds will remain after 100,000 years?

The International Atomic Energy Authority ranks nuclear incidents and accidents by severity using a scale from 1 to 7 called the International Nuclear and Radiological Event Scale (INES). A level 7 ranking is the most severe. To date, the Chernobyl accident and an accident at Japan's Fukushima Daiichi power plant in 2011 are the only two disasters in history to be given an INES level 7 ranking.

Solution   The graph of this function is shown in the figure at the right. Note from this graph that plutonium has a *half-life* of about 24,100 years. That is, after 24,100 years, *half* of the original amount will remain. After another 24,100 years, one-quarter of the original amount will remain, and so on. In the year 2020 ($t = 34$), there will still be

$$P = 10\left(\frac{1}{2}\right)^{34/24,100}$$

$$\approx 10\left(\frac{1}{2}\right)^{0.0014108}$$

$$\approx 9.990 \text{ pounds}$$

of plutonium remaining. After 100,000 years, there will still be

$$P = 10\left(\frac{1}{2}\right)^{100,000/24,100}$$

$$\approx 0.564 \text{ pound}$$

of plutonium remaining.

Radioactive Decay

Plutonium (in pounds)

$P = 10\left(\frac{1}{2}\right)^{t/24,100}$

(24,100, 5)

(100,000, 0.564)

Years of decay

 Checkpoint))) *Audio-video solution in English & Spanish at LarsonPrecalculus.com*

In Example 9, how much of the 10 pounds will remain in the year 2089? How much of the 10 pounds will remain after 125,000 years?

Summarize   (Section 5.1)

1. State the definition of the exponential function f with base a *(page 360)*. For an example of evaluating exponential functions, see Example 1.

2. Describe the basic characteristics of the graphs of the exponential functions $y = a^x$ and $y = a^{-x}$, $a > 1$ *(page 362)*. For examples of graphing exponential functions, see Examples 2, 3, and 5.

3. State the definitions of the natural base and the natural exponential function *(page 364)*. For examples of evaluating and graphing natural exponential functions, see Examples 6 and 7.

4. Describe real-life applications involving exponential functions *(pages 366 and 367, Examples 8 and 9)*.

5.1 Exercises

See **CalcChat.com** for tutorial help and worked-out solutions to odd-numbered exercises.

Vocabulary: Fill in the blanks.

1. Polynomial and rational functions are examples of _____ functions.
2. Exponential and logarithmic functions are examples of nonalgebraic functions, also called _____ functions.
3. The _____ Property can be used to solve simple exponential equations.
4. The exponential function $f(x) = e^x$ is called the _____ _____ function, and the base e is called the _____ base.
5. To find the amount A in an account after t years with principal P and an annual interest rate r (in decimal form) compounded n times per year, use the formula _____.
6. To find the amount A in an account after t years with principal P and an annual interest rate r (in decimal form) compounded continuously, use the formula _____.

Skills and Applications

 Evaluating an Exponential Function In Exercises 7–12, evaluate the function at the given value of x. Round your result to three decimal places.

Function	Value
7. $f(x) = 0.9^x$	$x = 1.4$
8. $f(x) = 4.7^x$	$x = -\pi$
9. $f(x) = 3^x$	$x = \frac{2}{5}$
10. $f(x) = \left(\frac{2}{3}\right)^{5x}$	$x = \frac{3}{10}$
11. $f(x) = 5000(2^x)$	$x = -1.5$
12. $f(x) = 200(1.2)^{12x}$	$x = 24$

Matching an Exponential Function with Its Graph In Exercises 13–16, match the exponential function with its graph. [The graphs are labeled (a), (b), (c), and (d).]

(a)

(b)

(c)
(d)

13. $f(x) = 2^x$
14. $f(x) = 2^x + 1$
15. $f(x) = 2^{-x}$
16. $f(x) = 2^{x-2}$

 Graphing an Exponential Function In Exercises 17–24, use a graphing utility to construct a table of values for the function. Then sketch the graph of the function.

17. $f(x) = 7^x$
18. $f(x) = 7^{-x}$
19. $f(x) = \left(\frac{1}{4}\right)^{-x}$
20. $f(x) = \left(\frac{1}{4}\right)^x$
21. $f(x) = 4^{x-1}$
22. $f(x) = 4^{x+1}$
23. $f(x) = 2^{x+1} + 3$
24. $f(x) = 3^{x-2} + 1$

 Using the One-to-One Property In Exercises 25–28, use the One-to-One Property to solve the equation for x.

25. $3^{x+1} = 27$
26. $2^{x-2} = 64$
27. $\left(\frac{1}{2}\right)^x = 32$
28. $5^{x-2} = \frac{1}{125}$

 Transformations of the Graph of an Exponential Function In Exercises 29–32, describe the transformation(s) of the graph of f that yield(s) the graph of g.

29. $f(x) = 3^x$, $g(x) = 3^x + 1$
30. $f(x) = \left(\frac{7}{2}\right)^x$, $g(x) = -\left(\frac{7}{2}\right)^{-x}$
31. $f(x) = 10^x$, $g(x) = 10^{-x+3}$
32. $f(x) = 0.3^x$, $g(x) = -0.3^x + 5$

 Evaluating a Natural Exponential Function In Exercises 33–36, evaluate the function at the given value of x. Round your result to three decimal places.

Function	Value
33. $f(x) = e^x$	$x = 1.9$
34. $f(x) = 1.5e^{x/2}$	$x = 240$
35. $f(x) = 5000e^{0.06x}$	$x = 6$
36. $f(x) = 250e^{0.05x}$	$x = 20$

Graphing a Natural Exponential Function In Exercises 37–40, use a graphing utility to construct a table of values for the function. Then sketch the graph of the function.

37. $f(x) = 3e^{x+4}$ **38.** $f(x) = 2e^{-1.5x}$

39. $f(x) = 2e^{x-2} + 4$ **40.** $f(x) = 2 + e^{x-5}$

Graphing a Natural Exponential Function In Exercises 41–44, use a graphing utility to graph the exponential function.

41. $s(t) = 2e^{0.5t}$ **42.** $s(t) = 3e^{-0.2t}$

43. $g(x) = 1 + e^{-x}$ **44.** $h(x) = e^{x-2}$

Using the One-to-One Property In Exercises 45–48, use the One-to-One Property to solve the equation for x.

45. $e^{3x+2} = e^3$ **46.** $e^{2x-1} = e^4$

47. $e^{x^2-3} = e^{2x}$ **48.** $e^{x^2+6} = e^{5x}$

Compound Interest In Exercises 49–52, complete the table by finding the balance A when P dollars is invested at rate r for t years and compounded n times per year.

n	1	2	4	12	365	Continuous
A						

49. $P = \$1500$, $r = 2\%$, $t = 10$ years

50. $P = \$2500$, $r = 3.5\%$, $t = 10$ years

51. $P = \$2500$, $r = 4\%$, $t = 20$ years

52. $P = \$1000$, $r = 6\%$, $t = 40$ years

Compound Interest In Exercises 53–56, complete the table by finding the balance A when \$12,000 is invested at rate r for t years, compounded continuously.

t	10	20	30	40	50
A					

53. $r = 4\%$ **54.** $r = 6\%$

55. $r = 6.5\%$ **56.** $r = 3.5\%$

57. Trust Fund On the day of a child's birth, a parent deposits \$30,000 in a trust fund that pays 5% interest, compounded continuously. Determine the balance in this account on the child's 25th birthday.

58. Trust Fund A philanthropist deposits \$5000 in a trust fund that pays 7.5% interest, compounded continuously. The balance will be given to the college from which the philanthropist graduated after the money has earned interest for 50 years. How much will the college receive?

59. Inflation Assuming that the annual rate of inflation averages 4% over the next 10 years, the approximate costs C of goods or services during any year in that decade can be modeled by $C(t) = P(1.04)^t$, where t is the time in years and P is the present cost. The price of an oil change for your car is presently \$29.88. Estimate the price 10 years from now.

60. Computer Virus The number V of computers infected by a virus increases according to the model $V(t) = 100e^{4.6052t}$, where t is the time in hours. Find the number of computers infected after (a) 1 hour, (b) 1.5 hours, and (c) 2 hours.

61. Population Growth The projected population of the United States for the years 2025 through 2055 can be modeled by $P = 307.58e^{0.0052t}$, where P is the population (in millions) and t is the time (in years), with $t = 25$ corresponding to 2025. (*Source: U.S. Census Bureau*)

(a) Use a graphing utility to graph the function for the years 2025 through 2055.

(b) Use the *table* feature of the graphing utility to create a table of values for the same time period as in part (a).

(c) According to the model, during what year will the population of the United States exceed 430 million?

62. Population The population P (in millions) of Italy from 2003 through 2015 can be approximated by the model $P = 57.59e^{0.0051t}$, where t represents the year, with $t = 3$ corresponding to 2003. (*Source: U.S. Census Bureau*)

(a) According to the model, is the population of Italy increasing or decreasing? Explain.

(b) Find the populations of Italy in 2003 and 2015.

(c) Use the model to predict the populations of Italy in 2020 and 2025.

63. Radioactive Decay Let Q represent a mass (in grams) of radioactive plutonium (^{239}Pu), whose half-life is 24,100 years. The quantity of plutonium present after t years is $Q = 16\left(\frac{1}{2}\right)^{t/24,100}$.

(a) Determine the initial quantity (when $t = 0$).

(b) Determine the quantity present after 75,000 years.

(c) Use a graphing utility to graph the function over the interval $t = 0$ to $t = 150,000$.

64. Radioactive Decay Let Q represent a mass (in grams) of carbon (^{14}C), whose half-life is 5715 years. The quantity of carbon 14 present after t years is $Q = 10\left(\frac{1}{2}\right)^{t/5715}$.

(a) Determine the initial quantity (when $t = 0$).

(b) Determine the quantity present after 2000 years.

(c) Sketch the graph of the function over the interval $t = 0$ to $t = 10,000$.

65. Depreciation The value of a wheelchair conversion van that originally cost $49,810 depreciates so that each year it is worth $\frac{7}{8}$ of its value for the previous year.

(a) Find a model for $V(t)$, the value of the van after t years.

(b) Determine the value of the van 4 years after it was purchased.

66. Chemistry

Immediately following an injection, the concentration of a drug in the bloodstream is 300 milligrams per milliliter. After t hours, the concentration is 75% of the level of the previous hour.

(a) Find a model for $C(t)$, the concentration of the drug after t hours.

(b) Determine the concentration of the drug after 8 hours.

Exploration

True or False? **In Exercises 67 and 68, determine whether the statement is true or false. Justify your answer.**

67. The line $y = -2$ is an asymptote for the graph of $f(x) = 10^x - 2$.

68. $e = \dfrac{271{,}801}{99{,}990}$

Think About It **In Exercises 69–72, use properties of exponents to determine which functions (if any) are the same.**

69. $f(x) = 3^{x-2}$
$g(x) = 3^x - 9$
$h(x) = \frac{1}{9}(3^x)$

70. $f(x) = 4^x + 12$
$g(x) = 2^{2x+6}$
$h(x) = 64(4^x)$

71. $f(x) = 16(4^{-x})$
$g(x) = \left(\frac{1}{4}\right)^{x-2}$
$h(x) = 16(2^{-2x})$

72. $f(x) = e^{-x} + 3$
$g(x) = e^{3-x}$
$h(x) = -e^{x-3}$

73. Solving Inequalities Graph the functions $y = 3^x$ and $y = 4^x$ and use the graphs to solve each inequality.

(a) $4^x < 3^x$
(b) $4^x > 3^x$

74. Using Technology Use a graphing utility to graph each function. Use the graph to find where the function is increasing and decreasing, and approximate any relative maximum or minimum values.

(a) $f(x) = x^2 e^{-x}$

(b) $g(x) = x 2^{3-x}$

75. Graphical Reasoning Use a graphing utility to graph $y_1 = [1 + (1/x)]^x$ and $y_2 = e$ in the same viewing window. Using the *trace* feature, explain what happens to the graph of y_1 as x increases.

76. Graphical Reasoning Use a graphing utility to graph

$$f(x) = \left(1 + \frac{0.5}{x}\right)^x \quad \text{and} \quad g(x) = e^{0.5}$$

in the same viewing window. What is the relationship between f and g as x increases and decreases without bound?

77. Comparing Graphs Use a graphing utility to graph each pair of functions in the same viewing window. Describe any similarities and differences in the graphs.

(a) $y_1 = 2^x,\ y_2 = x^2$
(b) $y_1 = 3^x,\ y_2 = x^3$

78. HOW DO YOU SEE IT? The figure shows the graphs of $y = 2^x$, $y = e^x$, $y = 10^x$, $y = 2^{-x}$, $y = e^{-x}$, and $y = 10^{-x}$. Match each function with its graph. [The graphs are labeled (a) through (f).] Explain your reasoning.

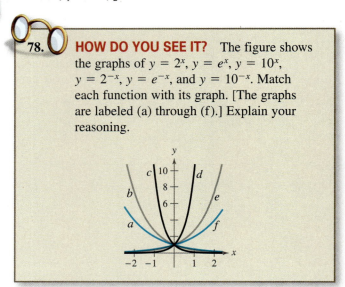

79. Think About It Which functions are exponential?

(a) $f(x) = 3x$
(b) $g(x) = 3x^2$
(c) $h(x) = 3^x$
(d) $k(x) = 2^{-x}$

80. Compound Interest Use the formula

$$A = P\left(1 + \frac{r}{n}\right)^{nt}$$

to calculate the balance A of an investment when $P = \$3000$, $r = 6\%$, and $t = 10$ years, and compounding is done (a) by the day, (b) by the hour, (c) by the minute, and (d) by the second. Does increasing the number of compoundings per year result in unlimited growth of the balance? Explain.

Project: Population per Square Mile To work an extended application analyzing the population per square mile of the United States, visit this text's website at *LarsonPrecalculus.com*. (*Source: U.S. Census Bureau*)

5.2 Logarithmic Functions and Their Graphs

- Recognize and evaluate logarithmic functions with base *a*.
- Graph logarithmic functions.
- Recognize, evaluate, and graph natural logarithmic functions.
- Use logarithmic functions to model and solve real-life problems.

Logarithmic Functions

In Section 5.1, you learned that the exponential function $f(x) = a^x$ is one-to-one. It follows that $f(x) = a^x$ must have an inverse function. This inverse function is the **logarithmic function with base *a*.**

Logarithmic functions can often model scientific observations. For example, in Exercise 83 on page 380, you will use a logarithmic function that models human memory.

> **Definition of Logarithmic Function with Base *a***
>
> For $x > 0$, $a > 0$, and $a \neq 1$,
>
> $\qquad y = \log_a x$ if and only if $x = a^y$.
>
> The function
>
> $\qquad f(x) = \log_a x$ Read as "log base *a* of *x*."
>
> is the **logarithmic function with base *a*.**

The equations $y = \log_a x$ and $x = a^y$ are equivalent. For example, $2 = \log_3 9$ is equivalent to $9 = 3^2$, and $5^3 = 125$ is equivalent to $\log_5 125 = 3$.

When evaluating logarithms, remember that *a logarithm is an exponent.* This means that $\log_a x$ is the exponent to which *a* must be raised to obtain *x*. For example, $\log_2 8 = 3$ because 2 raised to the third power is 8.

EXAMPLE 1 **Evaluating Logarithms**

Evaluate each logarithm at the given value of *x*.

a. $f(x) = \log_2 x$, $x = 32$ **b.** $f(x) = \log_3 x$, $x = 1$

c. $f(x) = \log_4 x$, $x = 2$ **d.** $f(x) = \log_{10} x$, $x = \frac{1}{100}$

Solution

a. $f(32) = \log_2 32 = 5$ because $2^5 = 32$.

b. $f(1) = \log_3 1 = 0$ because $3^0 = 1$.

c. $f(2) = \log_4 2 = \frac{1}{2}$ because $4^{1/2} = \sqrt{4} = 2$.

d. $f\left(\dfrac{1}{100}\right) = \log_{10} \dfrac{1}{100} = -2$ because $10^{-2} = \dfrac{1}{10^2} = \dfrac{1}{100}$.

✓ **Checkpoint** 🔊)) *Audio-video solution in English & Spanish at LarsonPrecalculus.com*

Evaluate each logarithm at the given value of *x*.

a. $f(x) = \log_6 x, x = 1$ **b.** $f(x) = \log_5 x, x = \frac{1}{125}$ **c.** $f(x) = \log_7 x, x = 343$ ■

The logarithmic function with base 10 is called the **common logarithmic function.** It is denoted by $\log_{10}$ or simply log. On most calculators, it is denoted by [LOG]. Example 2 shows how to use a calculator to evaluate common logarithmic functions. You will learn how to use a calculator to calculate logarithms with any base in Section 5.3.

EXAMPLE 2   **Evaluating Common Logarithms on a Calculator**

Use a calculator to evaluate the function $f(x) = \log x$ at each value of x.

a. $x = 10$ **b.** $x = \frac{1}{3}$ **c.** $x = -2$

Solution

Function Value	Calculator Keystrokes	Display
a. $f(10) = \log 10$	LOG 10 ENTER	1
b. $f\left(\frac{1}{3}\right) = \log \frac{1}{3}$	LOG (1 ÷ 3) ENTER	-0.4771213
c. $f(-2) = \log(-2)$	LOG (-) 2 ENTER	ERROR

Note that the calculator displays an error message (or a complex number) when you try to evaluate $\log(-2)$. This occurs because there is no real number power to which 10 can be raised to obtain -2.

✓ *Checkpoint* *Audio-video solution in English & Spanish at LarsonPrecalculus.com*

Use a calculator to evaluate the function $f(x) = \log x$ at each value of x.

a. $x = 275$ **b.** $x = -\frac{1}{2}$ **c.** $x = \frac{1}{2}$

The definition of the logarithmic function with base a leads to several properties.

Properties of Logarithms

1. $\log_a 1 = 0$ because $a^0 = 1$.

2. $\log_a a = 1$ because $a^1 = a$.

3. $\log_a a^x = x$ and $a^{\log_a x} = x$ Inverse Properties

4. If $\log_a x = \log_a y$, then $x = y$.   One-to-One Property

---

EXAMPLE 3   **Using Properties of Logarithms**

a. Simplify $\log_4 1$.   **b.** Simplify $\log_{\sqrt{7}} \sqrt{7}$.   **c.** Simplify $6^{\log_6 20}$.

Solution

a. $\log_4 1 = 0$ Property 1

b. $\log_{\sqrt{7}} \sqrt{7} = 1$ Property 2

c. $6^{\log_6 20} = 20$ Property 3 (Inverse Property)

✓ *Checkpoint* *Audio-video solution in English & Spanish at LarsonPrecalculus.com*

a. Simplify $\log_9 9$.   **b.** Simplify $20^{\log_{20} 3}$.   **c.** Simplify $\log_{\sqrt{3}} 1$.

EXAMPLE 4   **Using the One-to-One Property**

a. $\log_3 x = \log_3 12$ Original equation

       $x = 12$ One-to-One Property

b. $\log(2x + 1) = \log 3x$ ⟹ $2x + 1 = 3x$ ⟹ $1 = x$

c. $\log_4(x^2 - 6) = \log_4 10$ ⟹ $x^2 - 6 = 10$ ⟹ $x^2 = 16$ ⟹ $x = \pm 4$

✓ *Checkpoint*  *Audio-video solution in English & Spanish at LarsonPrecalculus.com*

Solve $\log_5(x^2 + 3) = \log_5 12$ for x.

Graphs of Logarithmic Functions

To sketch the graph of $y = \log_a x$, use the fact that the graphs of inverse functions are reflections of each other in the line $y = x$.

EXAMPLE 5 **Graphing Exponential and Logarithmic Functions**

In the same coordinate plane, sketch the graph of each function.

a. $f(x) = 2^x$ **b.** $g(x) = \log_2 x$

Solution

a. For $f(x) = 2^x$, construct a table of values. By plotting these points and connecting them with a smooth curve, you obtain the graph shown in Figure 5.6.

x	-2	-1	0	1	2	3
$f(x) = 2^x$	$\frac{1}{4}$	$\frac{1}{2}$	1	2	4	8

b. Because $g(x) = \log_2 x$ is the inverse function of $f(x) = 2^x$, the graph of g is obtained by plotting the points $(f(x), x)$ and connecting them with a smooth curve. The graph of g is a reflection of the graph of f in the line $y = x$, as shown in Figure 5.6.

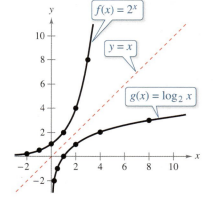

Figure 5.6

✓ *Checkpoint* *Audio-video solution in English & Spanish at LarsonPrecalculus.com*

In the same coordinate plane, sketch the graphs of (a) $f(x) = 8^x$ and (b) $g(x) = \log_8 x$.

EXAMPLE 6 **Sketching the Graph of a Logarithmic Function**

Sketch the graph of $f(x) = \log x$. Identify the vertical asymptote.

Solution Begin by constructing a table of values. Note that some of the values can be obtained without a calculator by using the properties of logarithms. Others require a calculator.

x	Without calculator				With calculator		
	$\frac{1}{100}$	$\frac{1}{10}$	1	10	2	5	8
$f(x) = \log x$	-2	-1	0	1	0.301	0.699	0.903

Next, plot the points and connect them with a smooth curve, as shown in the figure below. The vertical asymptote is $x = 0$ (y-axis).

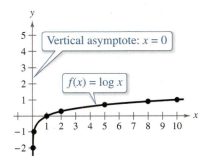

✓ *Checkpoint* *Audio-video solution in English & Spanish at LarsonPrecalculus.com*

Sketch the graph of $f(x) = \log_3 x$ by constructing a table of values without using a calculator. Identify the vertical asymptote.

The graph in Example 6 is typical for functions of the form $f(x) = \log_a x$, $a > 1$. They have one x-intercept and one vertical asymptote. Notice how slowly the graph rises for $x > 1$. Here are the basic characteristics of logarithmic graphs.

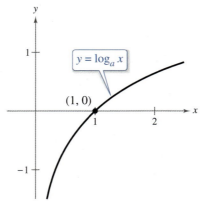

Graph of $y = \log_a x$, $a > 1$
- Domain: $(0, \infty)$
- Range: $(-\infty, \infty)$
- x-intercept: $(1, 0)$
- Increasing
- One-to-one, therefore has an inverse function
- y-axis is a vertical asymptote
  $(\log_a x \rightarrow -\infty$ as $x \rightarrow 0^+)$.
- Continuous
- Reflection of graph of $y = a^x$ in the line $y = x$

Some basic characteristics of the graph of $f(x) = a^x$ are listed below to illustrate the inverse relation between $f(x) = a^x$ and $g(x) = \log_a x$.

- Domain: $(-\infty, \infty)$
- Range: $(0, \infty)$
- y-intercept: $(0, 1)$
- x-axis is a horizontal asymptote $(a^x \rightarrow 0$ as $x \rightarrow -\infty)$.

The next example uses the graph of $y = \log_a x$ to sketch the graphs of functions of the form $f(x) = b \pm \log_a(x + c)$.

EXAMPLE 7   Shifting Graphs of Logarithmic Functions

See LarsonPrecalculus.com for an interactive version of this type of example.

Use the graph of $f(x) = \log x$ to sketch the graph of each function.

a. $g(x) = \log(x - 1)$ **b.** $h(x) = 2 + \log x$

Solution

a. Because $g(x) = \log(x - 1) = f(x - 1)$, the graph of g can be obtained by shifting the graph of f one unit to the right, as shown in Figure 5.7.

b. Because $h(x) = 2 + \log x = 2 + f(x)$, the graph of h can be obtained by shifting the graph of f two units up, as shown in Figure 5.8.

> •• **REMARK** Notice that the vertical transformation in Figure 5.8 keeps the y-axis as the vertical asymptote, but the horizontal transformation in Figure 5.7 yields a new vertical asymptote of $x = 1$.

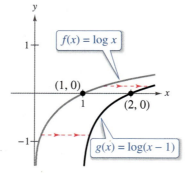

Figure 5.7

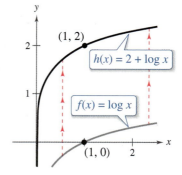

Figure 5.8

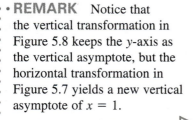

▷ **ALGEBRA HELP** To review the techniques for shifting, reflecting, and stretching graphs, see Section 2.5.

✓ *Checkpoint* ◀))) Audio-video solution in English & Spanish at LarsonPrecalculus.com

Use the graph of $f(x) = \log_3 x$ to sketch the graph of each function.

a. $g(x) = -1 + \log_3 x$ **b.** $h(x) = \log_3(x + 3)$

The Natural Logarithmic Function

By looking back at the graph of the natural exponential function introduced on page 364 in Section 5.1, you will see that $f(x) = e^x$ is one-to-one and so has an inverse function. This inverse function is called the **natural logarithmic function** and is denoted by the special symbol ln x, read as "the natural log of x" or "el en of x."

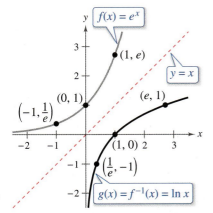

Reflection of graph of $f(x) = e^x$ in the line $y = x$
Figure 5.9

> ### The Natural Logarithmic Function
>
> The function
> $$f(x) = \log_e x = \ln x, \quad x > 0$$
> is called the **natural logarithmic function.**

The equations $y = \ln x$ and $x = e^y$ are equivalent. Note that the natural logarithm ln x is written without a base. The base is understood to be e.

Because the functions $f(x) = e^x$ and $g(x) = \ln x$ are inverse functions of each other, their graphs are reflections of each other in the line $y = x$, as shown in Figure 5.9.

EXAMPLE 8 **Evaluating the Natural Logarithmic Function**

Use a calculator to evaluate the function $f(x) = \ln x$ at each value of x.

a. $x = 2$

b. $x = 0.3$

c. $x = -1$

d. $x = 1 + \sqrt{2}$

▷ **TECHNOLOGY** On most calculators, the natural logarithm is denoted by ⌊LN⌋ as illustrated in Example 8.

Solution

Function Value	Calculator Keystrokes	Display
a. $f(2) = \ln 2$	⌊LN⌋ 2 ⌊ENTER⌋	0.6931472
b. $f(0.3) = \ln 0.3$	⌊LN⌋ .3 ⌊ENTER⌋	−1.2039728
c. $f(-1) = \ln(-1)$	⌊LN⌋ ⌊(−)⌋ 1 ⌊ENTER⌋	ERROR
d. $f(1 + \sqrt{2}) = \ln(1 + \sqrt{2})$	⌊LN⌋ ⌊(⌋ 1 ⌊+⌋ ⌊√⌋ 2 ⌊)⌋ ⌊ENTER⌋	0.8813736

✓ **Checkpoint** *Audio-video solution in English & Spanish at LarsonPrecalculus.com*

Use a calculator to evaluate the function $f(x) = \ln x$ at each value of x.

a. $x = 0.01$ **b.** $x = 4$

c. $x = \sqrt{3} + 2$ **d.** $x = \sqrt{3} - 2$

• • **REMARK** In Example 8(c), be sure you see that $\ln(-1)$ gives an error message on most calculators. This occurs because the domain of ln x is the set of *positive real numbers* (see Figure 5.9). So, $\ln(-1)$ is undefined.

The properties of logarithms on page 372 are also valid for natural logarithms.

> ### Properties of Natural Logarithms
>
> **1.** $\ln 1 = 0$ because $e^0 = 1$.
>
> **2.** $\ln e = 1$ because $e^1 = e$.
>
> **3.** $\ln e^x = x$ and $e^{\ln x} = x$ Inverse Properties
>
> **4.** If $\ln x = \ln y$, then $x = y$.   One-to-One Property

> **EXAMPLE 9** Using Properties of Natural Logarithms

Use the properties of natural logarithms to simplify each expression.

a. $\ln \dfrac{1}{e}$ **b.** $e^{\ln 5}$ **c.** $\dfrac{\ln 1}{3}$ **d.** $2 \ln e$

Solution

a. $\ln \dfrac{1}{e} = \ln e^{-1} = -1$ Property 3 (Inverse Property)

b. $e^{\ln 5} = 5$ Property 3 (Inverse Property)

c. $\dfrac{\ln 1}{3} = \dfrac{0}{3} = 0$ Property 1

d. $2 \ln e = 2(1) = 2$ Property 2

✓ *Checkpoint*  *Audio-video solution in English & Spanish at LarsonPrecalculus.com*

Use the properties of natural logarithms to simplify each expression.

a. $\ln e^{1/3}$ **b.** $5 \ln 1$ **c.** $\frac{3}{4} \ln e$ **d.** $e^{\ln 7}$

> **EXAMPLE 10** Finding the Domains of Logarithmic Functions

Find the domain of each function.

a. $f(x) = \ln(x - 2)$ **b.** $g(x) = \ln(2 - x)$ **c.** $h(x) = \ln x^2$

Solution

a. Because $\ln(x - 2)$ is defined only when

$$x - 2 > 0$$

it follows that the domain of f is $(2, \infty)$, as shown in Figure 5.10.

b. Because $\ln(2 - x)$ is defined only when

$$2 - x > 0$$

it follows that the domain of g is $(-\infty, 2)$, as shown in Figure 5.11.

c. Because $\ln x^2$ is defined only when

$$x^2 > 0$$

it follows that the domain of h is all real numbers except $x = 0$, as shown in Figure 5.12.

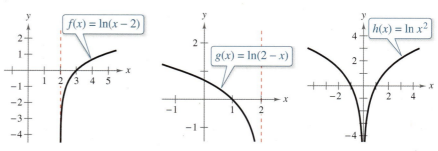

Figure 5.10   Figure 5.11   Figure 5.12

✓ *Checkpoint*  *Audio-video solution in English & Spanish at LarsonPrecalculus.com*

Find the domain of $f(x) = \ln(x + 3)$.

Application

EXAMPLE 11 **Human Memory Model**

Students participating in a psychology experiment attended several lectures on a subject and took an exam. Every month for a year after the exam, the students took a retest to see how much of the material they remembered. The average scores for the group are given by the *human memory model* $f(t) = 75 - 6\ln(t + 1)$, $0 \le t \le 12$, where t is the time in months.

a. What was the average score on the original exam ($t = 0$)?

b. What was the average score at the end of $t = 2$ months?

c. What was the average score at the end of $t = 6$ months?

Algebraic Solution

a. The original average score was

$$f(0) = 75 - 6\ln(0 + 1) \qquad \text{Substitute 0 for } t.$$

$$= 75 - 6\ln 1 \qquad \text{Simplify.}$$

$$= 75 - 6(0) \qquad \text{Property of natural logarithms}$$

$$= 75. \qquad \text{Solution}$$

b. After 2 months, the average score was

$$f(2) = 75 - 6\ln(2 + 1) \qquad \text{Substitute 2 for } t.$$

$$= 75 - 6\ln 3 \qquad \text{Simplify.}$$

$$\approx 75 - 6(1.0986) \qquad \text{Use a calculator.}$$

$$\approx 68.41. \qquad \text{Solution}$$

c. After 6 months, the average score was

$$f(6) = 75 - 6\ln(6 + 1) \qquad \text{Substitute 6 for } t.$$

$$= 75 - 6\ln 7 \qquad \text{Simplify.}$$

$$\approx 75 - 6(1.9459) \qquad \text{Use a calculator.}$$

$$\approx 63.32. \qquad \text{Solution}$$

Graphical Solution

a.

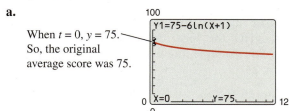

When $t = 0$, $y = 75$. So, the original average score was 75.

b.

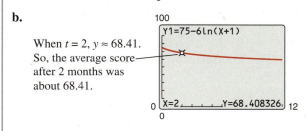

When $t = 2$, $y \approx 68.41$. So, the average score after 2 months was about 68.41.

c.

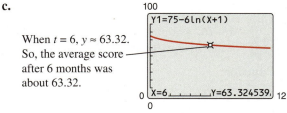

When $t = 6$, $y \approx 63.32$. So, the average score after 6 months was about 63.32.

✓ *Checkpoint* 🔊))) *Audio-video solution in English & Spanish at LarsonPrecalculus.com*

In Example 11, find the average score at the end of (a) $t = 1$ month, (b) $t = 9$ months, and (c) $t = 12$ months. ◼

Summarize (Section 5.2)

1. State the definition of the logarithmic function with base a *(page 371)* and make a list of the properties of logarithms *(page 372)*. For examples of evaluating logarithmic functions and using the properties of logarithms, see Examples 1–4.

2. Explain how to graph a logarithmic function *(pages 373 and 374)*. For examples of graphing logarithmic functions, see Examples 5–7.

3. State the definition of the natural logarithmic function and make a list of the properties of natural logarithms *(page 375)*. For examples of evaluating natural logarithmic functions and using the properties of natural logarithms, see Examples 8 and 9.

4. Describe a real-life application that uses a logarithmic function to model and solve a problem *(page 377, Example 11)*.

5.2 Exercises

See **CalcChat.com** for tutorial help and worked-out solutions to odd-numbered exercises.

Vocabulary: Fill in the blanks.

1. The inverse function of the exponential function $f(x) = a^x$ is the _____ function with base a.
2. The common logarithmic function has base _____.
3. The logarithmic function $f(x) = \ln x$ is the _____ logarithmic function and has base _____.
4. The Inverse Properties of logarithms state that $\log_a a^x = x$ and _____.
5. The One-to-One Property of natural logarithms states that if $\ln x = \ln y$, then _____.
6. The domain of the natural logarithmic function is the set of _____ _____ _____.

Skills and Applications

Writing an Exponential Equation In Exercises 7–10, write the logarithmic equation in exponential form. For example, the exponential form of $\log_5 25 = 2$ is $5^2 = 25$.

7. $\log_4 16 = 2$ **8.** $\log_9 \frac{1}{81} = -2$

9. $\log_{12} 12 = 1$ **10.** $\log_{32} 4 = \frac{2}{5}$

Writing a Logarithmic Equation In Exercises 11–14, write the exponential equation in logarithmic form. For example, the logarithmic form of $2^3 = 8$ is $\log_2 8 = 3$.

11. $5^3 = 125$ **12.** $9^{3/2} = 27$

13. $4^{-3} = \frac{1}{64}$ **14.** $24^0 = 1$

 Evaluating a Logarithm In Exercises 15–20, evaluate the logarithm at the given value of x without using a calculator.

Function	Value
15. $f(x) = \log_2 x$	$x = 64$
16. $f(x) = \log_{25} x$	$x = 5$
17. $f(x) = \log_8 x$	$x = 1$
18. $f(x) = \log x$	$x = 10$
19. $g(x) = \log_a x$	$x = a^{-2}$
20. $g(x) = \log_b x$	$x = \sqrt{b}$

Evaluating a Common Logarithm on a Calculator In Exercises 21–24, use a calculator to evaluate $f(x) = \log x$ at the given value of x. Round your result to three decimal places.

21. $x = \frac{7}{8}$ **22.** $x = \frac{1}{500}$

23. $x = 12.5$ **24.** $x = 96.75$

 Using Properties of Logarithms In Exercises 25–28, use the properties of logarithms to simplify the expression.

25. $\log_8 8$ **26.** $\log_\pi \pi^2$

27. $\log_{7.5} 1$ **28.** $5^{\log_5 3}$

 Using the One-to-One Property In Exercises 29–32, use the One-to-One Property to solve the equation for x.

29. $\log_5(x + 1) = \log_5 6$ **30.** $\log_2(x - 3) = \log_2 9$

31. $\log 11 = \log(x^2 + 7)$ **32.** $\log(x^2 + 6x) = \log 27$

 Graphing Exponential and Logarithmic Functions In Exercises 33–36, sketch the graphs of f and g in the same coordinate plane.

33. $f(x) = 7^x$, $g(x) = \log_7 x$
34. $f(x) = 5^x$, $g(x) = \log_5 x$
35. $f(x) = 6^x$, $g(x) = \log_6 x$
36. $f(x) = 10^x$, $g(x) = \log x$

Matching a Logarithmic Function with Its Graph In Exercises 37–40, use the graph of $g(x) = \log_3 x$ to match the given function with its graph. Then describe the relationship between the graphs of f and g. [The graphs are labeled (a), (b), (c), and (d).]

(a)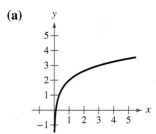

(b)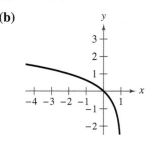

(c)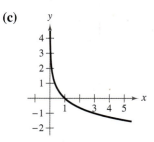

(d)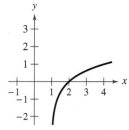

37. $f(x) = \log_3 x + 2$ **38.** $f(x) = \log_3(x - 1)$

39. $f(x) = \log_3(1 - x)$ **40.** $f(x) = -\log_3 x$

Sketching the Graph of a Logarithmic Function In Exercises 41–48, find the domain, x-intercept, and vertical asymptote of the logarithmic function and sketch its graph.

41. $f(x) = \log_4 x$ **42.** $g(x) = \log_6 x$

43. $y = \log_3 x + 1$

44. $h(x) = \log_4(x - 3)$

45. $f(x) = -\log_6(x + 2)$

46. $y = \log_5(x - 1) + 4$

47. $y = \log \dfrac{x}{7}$

48. $y = \log(-2x)$

Writing a Natural Exponential Equation In Exercises 49–52, write the logarithmic equation in exponential form.

49. $\ln \frac{1}{2} = -0.693 \ldots$ **50.** $\ln 7 = 1.945 \ldots$

51. $\ln 250 = 5.521 \ldots$ **52.** $\ln 1 = 0$

Writing a Natural Logarithmic Equation In Exercises 53–56, write the exponential equation in logarithmic form.

53. $e^2 = 7.3890 \ldots$ **54.** $e^{-3/4} = 0.4723 \ldots$

55. $e^{-4x} = \frac{1}{2}$ **56.** $e^{2x} = 3$

 Evaluating a Logarithmic Function In Exercises 57–60, use a calculator to evaluate the function at the given value of x. Round your result to three decimal places.

Function	Value
57. $f(x) = \ln x$	$x = 18.42$
58. $f(x) = 3 \ln x$	$x = 0.74$
59. $g(x) = 8 \ln x$	$x = \sqrt{5}$
60. $g(x) = -\ln x$	$x = \frac{1}{2}$

 Using Properties of Natural Logarithms In Exercises 61–66, use the properties of natural logarithms to simplify the expression.

61. $e^{\ln 4}$ **62.** $\ln \dfrac{1}{e^2}$

63. $2.5 \ln 1$ **64.** $\dfrac{\ln e}{\pi}$

65. $\ln e^{\ln e}$ **66.** $e^{\ln(1/e)}$

Graphing a Natural Logarithmic Function In Exercises 67–70, find the domain, x-intercept, and vertical asymptote of the logarithmic function and sketch its graph.

67. $f(x) = \ln(x - 4)$ **68.** $h(x) = \ln(x + 5)$

69. $g(x) = \ln(-x)$ **70.** $f(x) = \ln(3 - x)$

Graphing a Natural Logarithmic Function In Exercises 71–74, use a graphing utility to graph the function. Be sure to use an appropriate viewing window.

71. $f(x) = \ln(x - 1)$ **72.** $f(x) = \ln(x + 2)$

73. $f(x) = -\ln x + 8$ **74.** $f(x) = 3 \ln x - 1$

Using the One-to-One Property In Exercises 75–78, use the One-to-One Property to solve the equation for x.

75. $\ln(x + 4) = \ln 12$ **76.** $\ln(x - 7) = \ln 7$

77. $\ln(x^2 - x) = \ln 6$ **78.** $\ln(x^2 - 2) = \ln 23$

79. Monthly Payment The model

$$t = 16.625 \ln \frac{x}{x - 750}, \quad x > 750$$

approximates the length of a home mortgage of \$150,000 at 6% in terms of the monthly payment. In the model, t is the length of the mortgage in years and x is the monthly payment in dollars.

(a) Approximate the lengths of a \$150,000 mortgage at 6% when the monthly payment is \$897.72 and when the monthly payment is \$1659.24.

(b) Approximate the total amounts paid over the term of the mortgage with a monthly payment of \$897.72 and with a monthly payment of \$1659.24. What amount of the total is interest costs in each case?

(c) What is the vertical asymptote for the model? Interpret its meaning in the context of the problem.

80. Telephone Service The percent P of households in the United States with wireless-only telephone service from 2005 through 2014 can be approximated by the model

$$P = -3.42 + 1.297t \ln t, \quad 5 \le t \le 14$$

where t represents the year, with $t = 5$ corresponding to 2005. *(Source: National Center for Health Statistics)*

(a) Approximate the percents of households with wireless-only telephone service in 2008 and 2012.

(b) Use a graphing utility to graph the function.

(c) Can the model be used to predict the percent of households with wireless-only telephone service in 2020? in 2030? Explain.

81. Population The time t (in years) for the world population to double when it is increasing at a continuous rate r (in decimal form) is given by $t = (\ln 2)/r$.

(a) Complete the table and interpret your results.

r	0.005	0.010	0.015	0.020	0.025	0.030
t						

(b) Use a graphing utility to graph the function.

82. Compound Interest A principal P, invested at $5\frac{1}{2}\%$ and compounded continuously, increases to an amount K times the original principal after t years, where $t = (\ln K)/0.055$.

(a) Complete the table and interpret your results.

K	1	2	4	6	8	10	12
t							

(b) Sketch a graph of the function.

83. Human Memory Model

Students in a mathematics class took an exam and then took a retest monthly with an equivalent exam. The average scores for the class are given by the human memory model

$f(t) = 80 - 17 \log(t + 1), \quad 0 \le t \le 12$

where t is the time in months.

(a) Use a graphing utility to graph the model over the specified domain.

(b) What was the average score on the original exam ($t = 0$)?

(c) What was the average score after 4 months?

(d) What was the average score after 10 months?

84. Sound Intensity The relationship between the number of decibels β and the intensity of a sound I (in watts per square meter) is

$$\beta = 10 \log \frac{I}{10^{-12}}.$$

(a) Determine the number of decibels of a sound with an intensity of 1 watt per square meter.

(b) Determine the number of decibels of a sound with an intensity of 10^{-2} watt per square meter.

(c) The intensity of the sound in part (a) is 100 times as great as that in part (b). Is the number of decibels 100 times as great? Explain.

Exploration

True or False? In Exercises 85 and 86, determine whether the statement is true or false. Justify your answer.

85. The graph of $f(x) = \log_6 x$ is a reflection of the graph of $g(x) = 6^x$ in the x-axis.

86. The graph of $f(x) = \ln(-x)$ is a reflection of the graph of $h(x) = e^{-x}$ in the line $y = -x$.

87. Graphical Reasoning Use a graphing utility to graph f and g in the same viewing window and determine which is increasing at the greater rate as x approaches $+\infty$. What can you conclude about the rate of growth of the natural logarithmic function?

(a) $f(x) = \ln x, \quad g(x) = \sqrt{x}$

(b) $f(x) = \ln x, \quad g(x) = \sqrt[4]{x}$

88. **HOW DO YOU SEE IT?** The figure shows the graphs of $f(x) = 3^x$ and $g(x) = \log_3 x$. [The graphs are labeled m and n.]

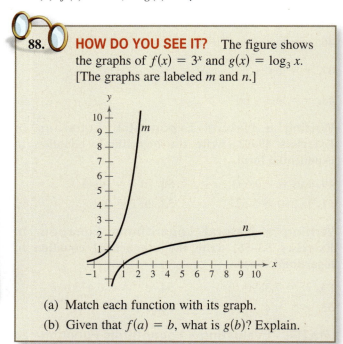

(a) Match each function with its graph.

(b) Given that $f(a) = b$, what is $g(b)$? Explain.

Error Analysis In Exercises 89 and 90, describe the error.

89.

x	1	2	8
y	0	1	3

From the table, you can conclude that y is an exponential function of x. ✗

90.

x	1	2	5
y	2	4	32

From the table, you can conclude that y is a logarithmic function of x. ✗

91. Numerical Analysis

(a) Complete the table for the function $f(x) = (\ln x)/x$.

x	1	5	10	10^2	10^4	10^6
$f(x)$						

(b) Use the table in part (a) to determine what value $f(x)$ approaches as x increases without bound.

(c) Use a graphing utility to confirm the result of part (b).

92. Writing Explain why $\log_a x$ is defined only for $0 < a < 1$ and $a > 1$.

5.3 Properties of Logarithms

- Use the change-of-base formula to rewrite and evaluate logarithmic expressions.
- Use properties of logarithms to evaluate or rewrite logarithmic expressions.
- Use properties of logarithms to expand or condense logarithmic expressions.
- Use logarithmic functions to model and solve real-life problems.

Change of Base

Most calculators have only two types of log keys, $\boxed{\text{LOG}}$ for common logarithms (base 10) and $\boxed{\text{LN}}$ for natural logarithms (base e). Although common logarithms and natural logarithms are the most frequently used, you may occasionally need to evaluate logarithms with other bases. To do this, use the **change-of-base formula.**

Logarithmic functions have many real-life applications. For example, in Exercises 79–82 on page 386, you will use a logarithmic function that models the relationship between the number of decibels and the intensity of a sound.

Change-of-Base Formula

Let a, b, and x be positive real numbers such that $a \neq 1$ and $b \neq 1$. Then $\log_a x$ can be converted to a different base as follows.

Base b	**Base 10**	**Base e**
$\log_a x = \dfrac{\log_b x}{\log_b a}$	$\log_a x = \dfrac{\log x}{\log a}$	$\log_a x = \dfrac{\ln x}{\ln a}$

One way to look at the change-of-base formula is that logarithms with base a are *constant multiples* of logarithms with base b. The constant multiplier is

$$\frac{1}{\log_b a}.$$

EXAMPLE 1 Changing Bases Using Common Logarithms

$$\log_4 25 = \frac{\log 25}{\log 4} \qquad \log_a x = \frac{\log x}{\log a}$$

$$\approx \frac{1.39794}{0.60206} \qquad \text{Use a calculator.}$$

$$\approx 2.3219 \qquad \text{Simplify.}$$

✓ **Checkpoint** 🔊))) *Audio-video solution in English & Spanish at LarsonPrecalculus.com*

Evaluate $\log_2 12$ using the change-of-base formula and common logarithms.

EXAMPLE 2 Changing Bases Using Natural Logarithms

$$\log_4 25 = \frac{\ln 25}{\ln 4} \qquad \log_a x = \frac{\ln x}{\ln a}$$

$$\approx \frac{3.21888}{1.38629} \qquad \text{Use a calculator.}$$

$$\approx 2.3219 \qquad \text{Simplify.}$$

✓ **Checkpoint** 🔊))) *Audio-video solution in English & Spanish at LarsonPrecalculus.com*

Evaluate $\log_2 12$ using the change-of-base formula and natural logarithms.

Properties of Logarithms

You know from the preceding section that the logarithmic function with base a is the *inverse function* of the exponential function with base a. So, it makes sense that the properties of exponents have corresponding properties involving logarithms. For example, the exponential property $a^m a^n = a^{m+n}$ has the corresponding logarithmic property $\log_a(uv) = \log_a u + \log_a v$.

•• **REMARK** There is no property that can be used to rewrite $\log_a(u \pm v)$. Specifically, $\log_a(u + v)$ is *not* equal to $\log_a u + \log_a v$.

Properties of Logarithms

Let a be a positive number such that $a \neq 1$, let n be a real number, and let u and v be positive real numbers.

	Logarithm with Base a	**Natural Logarithm**
1. Product Property:	$\log_a(uv) = \log_a u + \log_a v$	$\ln(uv) = \ln u + \ln v$
2. Quotient Property:	$\log_a \dfrac{u}{v} = \log_a u - \log_a v$	$\ln \dfrac{u}{v} = \ln u - \ln v$
3. Power Property:	$\log_a u^n = n \log_a u$	$\ln u^n = n \ln u$

For proofs of the properties listed above, see Proofs in Mathematics on page 418.

EXAMPLE 3 **Using Properties of Logarithms**

Write each logarithm in terms of $\ln 2$ and $\ln 3$.

a. $\ln 6$ **b.** $\ln \dfrac{2}{27}$

Solution

a. $\ln 6 = \ln(2 \cdot 3)$ Rewrite 6 as $2 \cdot 3$.

$\qquad = \ln 2 + \ln 3$ Product Property

b. $\ln \dfrac{2}{27} = \ln 2 - \ln 27$ Quotient Property

$\qquad = \ln 2 - \ln 3^3$ Rewrite 27 as 3^3.

$\qquad = \ln 2 - 3 \ln 3$ Power Property

✓ **Checkpoint**   *Audio-video solution in English & Spanish at LarsonPrecalculus.com*

Write each logarithm in terms of $\log 3$ and $\log 5$.

a. $\log 75$ **b.** $\log \dfrac{9}{125}$

EXAMPLE 4 **Using Properties of Logarithms**

Find the exact value of $\log_5 \sqrt[3]{5}$ without using a calculator.

Solution

$\log_5 \sqrt[3]{5} = \log_5 5^{1/3} = \frac{1}{3} \log_5 5 = \frac{1}{3}(1) = \frac{1}{3}$

✓ **Checkpoint**  *Audio-video solution in English & Spanish at LarsonPrecalculus.com*

Find the exact value of $\ln e^6 - \ln e^2$ without using a calculator.

Rewriting Logarithmic Expressions

The properties of logarithms are useful for rewriting logarithmic expressions in forms that simplify the operations of algebra. This is true because these properties convert complicated products, quotients, and exponential forms into simpler sums, differences, and products, respectively.

EXAMPLE 5 **Expanding Logarithmic Expressions**

Expand each logarithmic expression.

a. $\log_4 5x^3y$ **b.** $\ln \dfrac{\sqrt{3x - 5}}{7}$

Solution

a. $\log_4 5x^3y = \log_4 5 + \log_4 x^3 + \log_4 y$ Product Property

$\qquad\qquad\ = \log_4 5 + 3 \log_4 x + \log_4 y$ Power Property

> **ALGEBRA HELP** To review rewriting radicals and rational exponents, see Section P.2.

b. $\ln \dfrac{\sqrt{3x - 5}}{7} = \ln \dfrac{(3x - 5)^{1/2}}{7}$ Rewrite using rational exponent.

$\qquad\qquad\ = \ln(3x - 5)^{1/2} - \ln 7$ Quotient Property

$\qquad\qquad\ = \dfrac{1}{2} \ln(3x - 5) - \ln 7$ Power Property

✓ **Checkpoint**  *Audio-video solution in English & Spanish at LarsonPrecalculus.com*

Expand the expression $\log_3 \dfrac{4x^2}{\sqrt{y}}$.

Example 5 uses the properties of logarithms to *expand* logarithmic expressions. Example 6 reverses this procedure and uses the properties of logarithms to *condense* logarithmic expressions.

EXAMPLE 6 **Condensing Logarithmic Expressions**

See LarsonPrecalculus.com for an interactive version of this type of example.

Condense each logarithmic expression.

a. $\frac{1}{2} \log x + 3 \log(x + 1)$ **b.** $2 \ln(x + 2) - \ln x$ **c.** $\frac{1}{3}[\log_2 x + \log_2(x + 1)]$

Solution

a. $\frac{1}{2} \log x + 3 \log(x + 1) = \log x^{1/2} + \log(x + 1)^3$ Power Property

$\qquad\qquad\qquad\qquad\qquad = \log\left[\sqrt{x}\,(x + 1)^3\right]$ Product Property

b. $2 \ln(x + 2) - \ln x = \ln(x + 2)^2 - \ln x$ Power Property

$\qquad\qquad\qquad\quad = \ln \dfrac{(x + 2)^2}{x}$ Quotient Property

c. $\frac{1}{3}[\log_2 x + \log_2(x + 1)] = \frac{1}{3} \log_2[x(x + 1)]$ Product Property

$\qquad\qquad\qquad\qquad\quad = \log_2[x(x + 1)]^{1/3}$ Power Property

$\qquad\qquad\qquad\qquad\quad = \log_2 \sqrt[3]{x(x + 1)}$ Rewrite with a radical.

✓ **Checkpoint**  *Audio-video solution in English & Spanish at LarsonPrecalculus.com*

Condense the expression $2[\log(x + 3) - 2 \log(x - 2)]$.

Application

One way to determine a possible relationship between the x- and y-values of a set of nonlinear data is to take the natural logarithm of each x-value and each y-value. If the plotted points $(\ln x, \ln y)$ lie on a line, then x and y are related by the equation $\ln y = m \ln x$, where m is the slope of the line.

EXAMPLE 7 Finding a Mathematical Model

The table shows the mean distance x from the sun and the period y (the time it takes a planet to orbit the sun, in years) for each of the six planets that are closest to the sun. In the table, the mean distance is given in astronomical units (where one astronomical unit is defined as Earth's mean distance from the sun). The points from the table are plotted in Figure 5.13. Find an equation that relates y and x.

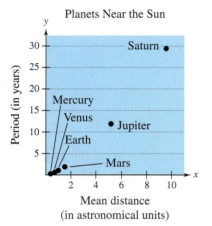

Planets Near the Sun

Figure 5.13

Planet	Mean Distance, x	Period, y
Mercury	0.387	0.241
Venus	0.723	0.615
Earth	1.000	1.000
Mars	1.524	1.881
Jupiter	5.203	11.862
Saturn	9.537	29.457

Spreadsheet at LarsonPrecalculus.com

Planet	$\ln x$	$\ln y$
Mercury	-0.949	-1.423
Venus	-0.324	-0.486
Earth	0.000	0.000
Mars	0.421	0.632
Jupiter	1.649	2.473
Saturn	2.255	3.383

Solution From Figure 5.13, it is not clear how to find an equation that relates y and x. To solve this problem, make a table of values giving the natural logarithms of all x- and y-values of the data (see the table at the left). Plot each point $(\ln x, \ln y)$. These points appear to lie on a line (see Figure 5.14). Choose two points to determine the slope of the line. Using the points $(0.421, 0.632)$ and $(0, 0)$, the slope of the line is

$$ m = \frac{0.632 - 0}{0.421 - 0} \approx 1.5 = \frac{3}{2}. $$

By the point-slope form, the equation of the line is $Y = \frac{3}{2} X$, where $Y = \ln y$ and $X = \ln x$. So, an equation that relates y and x is $\ln y = \frac{3}{2} \ln x$.

✓ **Checkpoint**   *Audio-video solution in English & Spanish at LarsonPrecalculus.com*

Find a logarithmic equation that relates y and x for the following ordered pairs.

$(0.37, 0.51), (1.00, 1.00), (2.72, 1.95), (7.39, 3.79), (20.09, 7.39)$

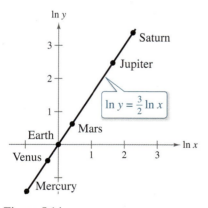

Figure 5.14

$\ln y = \frac{3}{2} \ln x$

Summarize (Section 5.3)

1. State the change-of-base formula *(page 381)*. For examples of using the change-of-base formula to rewrite and evaluate logarithmic expressions, see Examples 1 and 2.

2. Make a list of the properties of logarithms *(page 382)*. For examples of using the properties of logarithms to evaluate or rewrite logarithmic expressions, see Examples 3 and 4.

3. Explain how to use the properties of logarithms to expand or condense logarithmic expressions *(page 383)*. For examples of expanding and condensing logarithmic expressions, see Examples 5 and 6.

4. Describe an example of how to use a logarithmic function to model and solve a real-life problem *(page 384, Example 7)*.

5.3 Exercises

See **CalcChat.com** for tutorial help and worked-out solutions to odd-numbered exercises.

Vocabulary

In Exercises 1–3, fill in the blanks.

1. To evaluate a logarithm to any base, use the _____ formula.

2. The change-of-base formula for base e is $\log_a x =$ _____.

3. When you consider $\log_a x$ to be a constant multiple of $\log_b x$, the constant multiplier is _____.

4. Name the property of logarithms illustrated by each statement.

  (a) $\ln(uv) = \ln u + \ln v$ (b) $\log_a u^n = n \log_a u$ (c) $\ln \dfrac{u}{v} = \ln u - \ln v$

Skills and Applications

 Changing Bases In Exercises 5–8, rewrite the logarithm as a ratio of (a) common logarithms and (b) natural logarithms.

5. $\log_5 16$

6. $\log_{1/5} 4$

7. $\log_x \frac{3}{10}$

8. $\log_{2.6} x$

 Using the Change-of-Base Formula In Exercises 9–12, evaluate the logarithm using the change-of-base formula. Round your result to three decimal places.

9. $\log_3 17$

10. $\log_{0.4} 12$

11. $\log_\pi 0.5$

12. $\log_{2/3} 0.125$

 Using Properties of Logarithms In Exercises 13–18, use the properties of logarithms to write the logarithm in terms of $\log_3 5$ and $\log_3 7$.

13. $\log_3 35$

14. $\log_3 \frac{5}{7}$

15. $\log_3 \frac{7}{25}$

16. $\log_3 175$

17. $\log_3 \frac{21}{5}$

18. $\log_3 \frac{45}{49}$

 Using Properties of Logarithms In Exercises 19–32, find the exact value of the logarithmic expression without using a calculator. (If this is not possible, state the reason.)

19. $\log_3 9$

20. $\log_5 \frac{1}{125}$

21. $\log_6 \sqrt[3]{\frac{1}{6}}$

22. $\log_2 \sqrt[4]{8}$

23. $\log_2(-2)$

24. $\log_3(-27)$

25. $\ln \sqrt[4]{e^3}$

26. $\ln\left(1/\sqrt{e}\right)$

27. $\ln e^2 + \ln e^5$

28. $2 \ln e^6 - \ln e^5$

29. $\log_5 75 - \log_5 3$

30. $\log_4 2 + \log_4 32$

31. $\log_4 8$

32. $\log_8 16$

Using Properties of Logarithms In Exercises 33–40, approximate the logarithm using the properties of logarithms, given $\log_b 2 \approx 0.3562$, $\log_b 3 \approx 0.5646$, and $\log_b 5 \approx 0.8271$.

33. $\log_b 10$

34. $\log_b \frac{2}{3}$

35. $\log_b 0.04$

36. $\log_b \sqrt{2}$

37. $\log_b 45$

38. $\log_b(3b^2)$

39. $\log_b(2b)^{-2}$

40. $\log_b \sqrt[3]{3b}$

 Expanding a Logarithmic Expression In Exercises 41–60, use the properties of logarithms to expand the expression as a sum, difference, and/or constant multiple of logarithms. (Assume all variables are positive.)

41. $\ln 7x$

42. $\log_3 13z$

43. $\log_8 x^4$

44. $\ln(xy)^3$

45. $\log_5 \frac{5}{x}$

46. $\log_6 \frac{w^2}{v}$

47. $\ln \sqrt{z}$

48. $\ln \sqrt[3]{t}$

49. $\ln xyz^2$

50. $\log_4 11b^2c$

51. $\ln z(z-1)^2, \quad z > 1$

52. $\ln \dfrac{x^2 - 1}{x^3}, \quad x > 1$

53. $\log_2 \dfrac{\sqrt{a^2 - 4}}{7}, \quad a > 2$

54. $\ln \dfrac{3}{\sqrt{x^2 + 1}}$

55. $\log_5 \dfrac{x^2}{y^2 z^3}$

56. $\log_{10} \dfrac{xy^4}{z^5}$

57. $\ln \sqrt[3]{\dfrac{yz}{x^2}}$

58. $\log_2 x^4 \sqrt{\dfrac{y}{z^3}}$

59. $\ln \sqrt[4]{x^3(x^2 + 3)}$

60. $\ln \sqrt{x^2(x + 2)}$

Condensing a Logarithmic Expression
In Exercises 61–76, condense the expression to the logarithm of a single quantity.

61. $\ln 3 + \ln x$

62. $\log_5 8 - \log_5 t$

63. $\frac{2}{3} \log_7 (z - 2)$

64. $-4 \ln 3x$

65. $\log_3 5x - 4 \log_3 x$

66. $2 \log_2 x + 4 \log_2 y$

67. $\log x + 2 \log(x + 1)$

68. $2 \ln 8 - 5 \ln(z - 4)$

69. $\log x - 2 \log y + 3 \log z$

70. $3 \log_3 x + \frac{1}{4} \log_3 y - 4 \log_3 z$

71. $\ln x - [\ln(x + 1) + \ln(x - 1)]$

72. $4[\ln z + \ln(z + 5)] - 2 \ln(z - 5)$

73. $\frac{1}{2}[2 \ln(x + 3) + \ln x - \ln(x^2 - 1)]$

74. $2[3 \ln x - \ln(x + 1) - \ln(x - 1)]$

75. $\frac{1}{3}[\log_8 y + 2 \log_8(y + 4)] - \log_8(y - 1)$

76. $\frac{1}{2}[\log_4(x + 1) + 2 \log_4(x - 1)] + 6 \log_4 x$

Comparing Logarithmic Quantities
In Exercises 77 and 78, determine which (if any) of the logarithmic expressions are equal. Justify your answer.

77. $\dfrac{\log_2 32}{\log_2 4}$, $\quad \log_2 \dfrac{32}{4}$, $\quad \log_2 32 - \log_2 4$

78. $\log_7 \sqrt{70}$, $\quad \log_7 35$, $\quad \frac{1}{2} + \log_7 \sqrt{10}$

Sound Intensity

In Exercises 79–82, use the following information. The relationship between the number of decibels β and the intensity of a sound I (in watts per square meter) is

$$\beta = 10 \log \frac{I}{10^{-12}}.$$

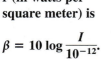

79. Use the properties of logarithms to write the formula in a simpler form. Then determine the number of decibels of a sound with an intensity of 10^{-6} watt per square meter.

80. Find the difference in loudness between an average office with an intensity of 1.26×10^{-7} watt per square meter and a broadcast studio with an intensity of 3.16×10^{-10} watt per square meter.

81. Find the difference in loudness between a vacuum cleaner with an intensity of 10^{-4} watt per square meter and rustling leaves with an intensity of 10^{-11} watt per square meter.

82. You and your roommate are playing your stereos at the same time and at the same intensity. How much louder is the music when both stereos are playing compared with just one stereo playing?

Curve Fitting
In Exercises 83–86, find a logarithmic equation that relates y and x.

83.

x	1	2	3	4	5	6
y	1	1.189	1.316	1.414	1.495	1.565

84.

x	1	2	3	4	5	6
y	1	0.630	0.481	0.397	0.342	0.303

85.

x	1	2	3	4	5	6
y	2.5	2.102	1.9	1.768	1.672	1.597

86.

x	1	2	3	4	5	6
y	0.5	2.828	7.794	16	27.951	44.091

87. Stride Frequency of Animals Four-legged animals run with two different types of motion: trotting and galloping. An animal that is trotting has at least one foot on the ground at all times, whereas an animal that is galloping has all four feet off the ground at some point in its stride. The number of strides per minute at which an animal breaks from a trot to a gallop depends on the weight of the animal. Use the table to find a logarithmic equation that relates an animal's weight x (in pounds) and its lowest stride frequency while galloping y (in strides per minute).

Weight, x	Stride Frequency, y
25	191.5
35	182.7
50	173.8
75	164.2
500	125.9
1000	114.2

Spreadsheet at LarsonPrecalculus.com

88. Nail Length The approximate lengths and diameters (in inches) of bright common wire nails are shown in the table. Find a logarithmic equation that relates the diameter y of a bright common wire nail to its length x.

Length, x	Diameter, y
2	0.113
3	0.148
4	0.192
5	0.225
6	0.262

89. Comparing Models A cup of water at an initial temperature of 78°C is placed in a room at a constant temperature of 21°C. The temperature of the water is measured every 5 minutes during a half-hour period. The results are recorded as ordered pairs of the form (t, T), where t is the time (in minutes) and T is the temperature (in degrees Celsius).

$(0, 78.0°)$, $(5, 66.0°)$, $(10, 57.5°)$, $(15, 51.2°)$,
$(20, 46.3°)$, $(25, 42.4°)$, $(30, 39.6°)$

(a) Subtract the room temperature from each of the temperatures in the ordered pairs. Use a graphing utility to plot the data points (t, T) and $(t, T - 21)$.

(b) An exponential model for the data $(t, T - 21)$ is $T - 21 = 54.4(0.964)^t$. Solve for T and graph the model. Compare the result with the plot of the original data.

(c) Use the graphing utility to plot the points $(t, \ln(T - 21))$ and observe that the points appear to be linear. Use the *regression* feature of the graphing utility to fit a line to these data. This resulting line has the form $\ln(T - 21) = at + b$, which is equivalent to $e^{\ln(T-21)} = e^{at+b}$. Solve for T, and verify that the result is equivalent to the model in part (b).

(d) Fit a rational model to the data. Take the reciprocals of the y-coordinates of the revised data points to generate the points

$$\left(t, \frac{1}{T - 21}\right).$$

Use the graphing utility to graph these points and observe that they appear to be linear. Use the *regression* feature of the graphing utility to fit a line to these data. The resulting line has the form

$$\frac{1}{T - 21} = at + b.$$

Solve for T, and use the graphing utility to graph the rational function and the original data points.

90. Writing Write a short paragraph explaining why the transformations of the data in Exercise 89 were necessary to obtain the models. Why did taking the logarithms of the temperatures lead to a linear scatter plot? Why did taking the reciprocals of the temperatures lead to a linear scatter plot?

Exploration

True or False? In Exercises 91–96, determine whether the statement is true or false given that $f(x) = \ln x$. Justify your answer.

91. $f(0) = 0$

92. $f(ax) = f(a) + f(x)$, $a > 0$, $x > 0$

93. $f(x - 2) = f(x) - f(2)$, $x > 2$

94. $\sqrt{f(x)} = \frac{1}{2}f(x)$

95. If $f(u) = 2f(v)$, then $v = u^2$.

96. If $f(x) < 0$, then $0 < x < 1$.

Using the Change-of-Base Formula In Exercises 97–100, use the change-of-base formula to rewrite the logarithm as a ratio of logarithms. Then use a graphing utility to graph the ratio.

97. $f(x) = \log_2 x$

98. $f(x) = \log_{1/2} x$

99. $f(x) = \log_{1/4} x$

100. $f(x) = \log_{11.8} x$

Error Analysis In Exercises 101 and 102, describe the error.

101. $(\ln e)^2 = 2(\ln e) = 2(1) = 2$ ✗

102. $\log_2 8 = \log_2(4 + 4)$
$= \log_2 4 + \log_2 4$
$= \log_2 2^2 + \log_2 2^2$
$= 2 + 2$
$= 4$

103. Graphical Reasoning Use a graphing utility to graph the functions $y_1 = \ln x - \ln(x - 3)$ and $y_2 = \ln \dfrac{x}{x - 3}$ in the same viewing window. Does the graphing utility show the functions with the same domain? If not, explain why some numbers are in the domain of one function but not the other.

104. **HOW DO YOU SEE IT?** The figure shows the graphs of $y = \ln x$, $y = \ln x^2$, $y = \ln 2x$, and $y = \ln 2$. Match each function with its graph. (The graphs are labeled A through D.) Explain.

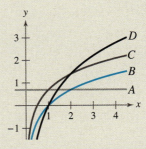

105. Think About It For which integers between 1 and 20 can you approximate natural logarithms, given the values $\ln 2 \approx 0.6931$, $\ln 3 \approx 1.0986$, and $\ln 5 \approx 1.6094$? Approximate these logarithms. (Do not use a calculator.)

5.4 Exponential and Logarithmic Equations

Exponential and logarithmic equations have many life science applications. For example, Exercise 83 on page 396 uses an exponential function to model the beaver population in a given area.

■ Solve simple exponential and logarithmic equations.
■ Solve more complicated exponential equations.
■ Solve more complicated logarithmic equations.
■ Use exponential and logarithmic equations to model and solve real-life problems.

Introduction

So far in this chapter, you have studied the definitions, graphs, and properties of exponential and logarithmic functions. In this section, you will study procedures for *solving equations* involving exponential and logarithmic expressions.

There are two basic strategies for solving exponential or logarithmic equations. The first is based on the One-to-One Properties and was used to solve simple exponential and logarithmic equations in Sections 5.1 and 5.2. The second is based on the Inverse Properties. For $a > 0$ and $a \neq 1$, the properties below are true for all x and y for which $\log_a x$ and $\log_a y$ are defined.

One-to-One Properties	**Inverse Properties**
$a^x = a^y$ if and only if $x = y$.	$a^{\log_a x} = x$
$\log_a x = \log_a y$ if and only if $x = y$.	$\log_a a^x = x$

EXAMPLE 1   Solving Simple Equations

Original Equation	Rewritten Equation	Solution	Property
a. $2^x = 32$	$2^x = 2^5$	$x = 5$	One-to-One
b. $\ln x - \ln 3 = 0$	$\ln x = \ln 3$	$x = 3$	One-to-One
c. $\left(\frac{1}{3}\right)^x = 9$	$3^{-x} = 3^2$	$x = -2$	One-to-One
d. $e^x = 7$	$\ln e^x = \ln 7$	$x = \ln 7$	Inverse
e. $\ln x = -3$	$e^{\ln x} = e^{-3}$	$x = e^{-3}$	Inverse
f. $\log x = -1$	$10^{\log x} = 10^{-1}$	$x = 10^{-1} = \frac{1}{10}$	Inverse
g. $\log_3 x = 4$	$3^{\log_3 x} = 3^4$	$x = 81$	Inverse

✓ **Checkpoint** 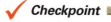 *Audio-video solution in English & Spanish at LarsonPrecalculus.com*

Solve each equation for x.

a. $2^x = 512$ **b.** $\log_6 x = 3$ **c.** $5 - e^x = 0$ **d.** $9^x = \frac{1}{3}$ ■

Strategies for Solving Exponential and Logarithmic Equations

1. Rewrite the original equation in a form that allows the use of the One-to-One Properties of exponential or logarithmic functions.

2. Rewrite an *exponential* equation in logarithmic form and apply the Inverse Property of logarithmic functions.

3. Rewrite a *logarithmic* equation in exponential form and apply the Inverse Property of exponential functions.

Solving Exponential Equations

> **EXAMPLE 2** Solving Exponential Equations

Solve each equation and approximate the result to three decimal places, if necessary.

a. $e^{-x^2} = e^{-3x-4}$ **b.** $3(2^x) = 42$

Solution

a.

$e^{-x^2} = e^{-3x-4}$	Write original equation.
$-x^2 = -3x - 4$	One-to-One Property
$x^2 - 3x - 4 = 0$	Write in general form.
$(x + 1)(x - 4) = 0$	Factor.
$x + 1 = 0 \implies x = -1$	Set 1st factor equal to 0.
$x - 4 = 0 \implies x = 4$	Set 2nd factor equal to 0.

> **REMARK**
> Another way to solve Example 2(b) is by taking the natural log of each side and then applying the Power Property.
>
> $$3(2^x) = 42$$
> $$2^x = 14$$
> $$\ln 2^x = \ln 14$$
> $$x \ln 2 = \ln 14$$
> $$x = \frac{\ln 14}{\ln 2} \approx 3.807$$
>
> Notice that you obtain the same result as in Example 2(b).

The solutions are $x = -1$ and $x = 4$. Check these in the original equation.

b.

$3(2^x) = 42$	Write original equation.
$2^x = 14$	Divide each side by 3.
$\log_2 2^x = \log_2 14$	Take log (base 2) of each side.
$x = \log_2 14$	Inverse Property
$x = \frac{\ln 14}{\ln 2} \approx 3.807$	Change-of-base formula

The solution is $x = \log_2 14 \approx 3.807$. Check this in the original equation.

✓ *Checkpoint* *Audio-video solution in English & Spanish at LarsonPrecalculus.com*

Solve each equation and approximate the result to three decimal places, if necessary.

a. $e^{2x} = e^{x^2-8}$ **b.** $2(5^x) = 32$

   In Example 2(b), the exact solution is $x = \log_2 14$, and the approximate solution is $x \approx 3.807$. An exact answer is preferred when the solution is an intermediate step in a larger problem. For a final answer, an approximate solution is more practical.

> **EXAMPLE 3** Solving an Exponential Equation

Solve $e^x + 5 = 60$ and approximate the result to three decimal places.

Solution

$e^x + 5 = 60$	Write original equation.
$e^x = 55$	Subtract 5 from each side.
$\ln e^x = \ln 55$	Take natural log of each side.
$x = \ln 55 \approx 4.007$	Inverse Property

> **REMARK** Remember that the natural logarithmic function has a base of e.

The solution is $x = \ln 55 \approx 4.007$. Check this in the original equation.

✓ *Checkpoint* *Audio-video solution in English & Spanish at LarsonPrecalculus.com*

Solve $e^x - 7 = 23$ and approximate the result to three decimal places.

EXAMPLE 4   **Solving an Exponential Equation**

Solve $2(3^{2t-5}) - 4 = 11$ and approximate the result to three decimal places.

Solution

$$2(3^{2t-5}) - 4 = 11 \qquad \text{Write original equation.}$$

$$2(3^{2t-5}) = 15 \qquad \text{Add 4 to each side.}$$

$$3^{2t-5} = \frac{15}{2} \qquad \text{Divide each side by 2.}$$

$$\log_3 3^{2t-5} = \log_3 \frac{15}{2} \qquad \text{Take log (base 3) of each side.}$$

$$2t - 5 = \log_3 \frac{15}{2} \qquad \text{Inverse Property}$$

$$2t = 5 + \log_3 7.5 \qquad \text{Add 5 to each side.}$$

$$t = \frac{5}{2} + \frac{1}{2}\log_3 7.5 \qquad \text{Divide each side by 2.}$$

$$t \approx 3.417 \qquad \text{Use a calculator.}$$

• • **REMARK** Remember that
to evaluate a logarithm such as
$\log_3 7.5$, you need to use the
change-of-base formula.

$$\log_3 7.5 = \frac{\ln 7.5}{\ln 3} \approx 1.834$$

The solution is $t = \frac{5}{2} + \frac{1}{2}\log_3 7.5 \approx 3.417$. Check this in the original equation.

✓ *Checkpoint*)) *Audio-video solution in English & Spanish at LarsonPrecalculus.com*

Solve $6(2^{t+5}) + 4 = 11$ and approximate the result to three decimal places.

When an equation involves two or more exponential expressions, you can still use a procedure similar to that demonstrated in Examples 2, 3, and 4. However, it may include additional algebraic techniques.

EXAMPLE 5   **Solving an Exponential Equation of Quadratic Type**

Solve $e^{2x} - 3e^x + 2 = 0$.

Algebraic Solution

$$e^{2x} - 3e^x + 2 = 0 \qquad \text{Write original equation.}$$

$$(e^x)^2 - 3e^x + 2 = 0 \qquad \text{Write in quadratic form.}$$

$$(e^x - 2)(e^x - 1) = 0 \qquad \text{Factor.}$$

$$e^x - 2 = 0 \qquad \text{Set 1st factor equal to 0.}$$

$$x = \ln 2 \qquad \text{Solve for } x.$$

$$e^x - 1 = 0 \qquad \text{Set 2nd factor equal to 0.}$$

$$x = 0 \qquad \text{Solve for } x.$$

The solutions are $x = \ln 2 \approx 0.693$ and $x = 0$. Check these in the original equation.

Graphical Solution

Use a graphing utility to graph $y = e^{2x} - 3e^x + 2$ and then find the zeros.

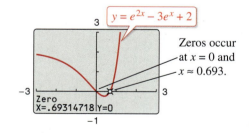

So, the solutions are $x = 0$ and $x \approx 0.693$.

✓ *Checkpoint*)) *Audio-video solution in English & Spanish at LarsonPrecalculus.com*

Solve $e^{2x} - 7e^x + 12 = 0$.

Solving Logarithmic Equations

To solve a logarithmic equation, write it in exponential form. This procedure is called *exponentiating* each side of an equation.

$\ln x = 3$ — Logarithmic form

$e^{\ln x} = e^3$ — Exponentiate each side.

$x = e^3$ — Exponential form

EXAMPLE 6   Solving Logarithmic Equations

a. $\ln x = 2$ — Original equation

$e^{\ln x} = e^2$ — Exponentiate each side.

$x = e^2$ — Inverse Property

b. $\log_3(5x - 1) = \log_3(x + 7)$ — Original equation

$5x - 1 = x + 7$ — One-to-One Property

$x = 2$ — Solve for x.

c. $\log_6(3x + 14) - \log_6 5 = \log_6 2x$ — Original equation

$\log_6\left(\dfrac{3x + 14}{5}\right) = \log_6 2x$ — Quotient Property of Logarithms

$\dfrac{3x + 14}{5} = 2x$ — One-to-One Property

$3x + 14 = 10x$ — Multiply each side by 5.

$x = 2$ — Solve for x.

✓ **Checkpoint** Audio-video solution in English & Spanish at LarsonPrecalculus.com

Solve each equation.

a. $\ln x = \frac{2}{3}$ **b.** $\log_2(2x - 3) = \log_2(x + 4)$ **c.** $\log 4x - \log(12 + x) = \log 2$

EXAMPLE 7   Solving a Logarithmic Equation

Solve $5 + 2 \ln x = 4$ and approximate the result to three decimal places.

Algebraic Solution

$5 + 2 \ln x = 4$ — Write original equation.

$2 \ln x = -1$ — Subtract 5 from each side.

$\ln x = -\dfrac{1}{2}$ — Divide each side by 2.

$e^{\ln x} = e^{-1/2}$ — Exponentiate each side.

$x = e^{-1/2}$ — Inverse Property

$x \approx 0.607$ — Use a calculator.

Graphical Solution

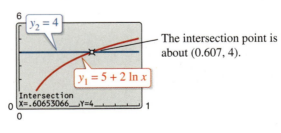

The intersection point is about $(0.607, 4)$.

So, the solution is $x \approx 0.607$.

✓ **Checkpoint** Audio-video solution in English & Spanish at LarsonPrecalculus.com

Solve $7 + 3 \ln x = 5$ and approximate the result to three decimal places.

EXAMPLE 8 **Solving a Logarithmic Equation**

Solve $2 \log_5 3x = 4$.

Solution

$$2 \log_5 3x = 4 \qquad \text{Write original equation.}$$

$$\log_5 3x = 2 \qquad \text{Divide each side by 2.}$$

$$5^{\log_5 3x} = 5^2 \qquad \text{Exponentiate each side (base 5).}$$

$$3x = 25 \qquad \text{Inverse Property}$$

$$x = \frac{25}{3} \qquad \text{Divide each side by 3.}$$

The solution is $x = \frac{25}{3}$. Check this in the original equation.

✓ *Checkpoint*  Audio-video solution in English & Spanish at LarsonPrecalculus.com

Solve $3 \log_4 6x = 9$.

The domain of a logarithmic function generally does not include all real numbers, so you should be sure to check for extraneous solutions of logarithmic equations.

EXAMPLE 9 **Checking for Extraneous Solutions**

Solve

$$\log 5x + \log(x - 1) = 2.$$

Algebraic Solution

$$\log 5x + \log(x - 1) = 2 \qquad \text{Write original equation.}$$

$$\log[5x(x - 1)] = 2 \qquad \text{Product Property of Logarithms}$$

$$10^{\log(5x^2 - 5x)} = 10^2 \qquad \text{Exponentiate each side (base 10).}$$

$$5x^2 - 5x = 100 \qquad \text{Inverse Property}$$

$$x^2 - x - 20 = 0 \qquad \text{Write in general form.}$$

$$(x - 5)(x + 4) = 0 \qquad \text{Factor.}$$

$$x - 5 = 0 \qquad \text{Set 1st factor equal to 0.}$$

$$x = 5 \qquad \text{Solve for } x.$$

$$x + 4 = 0 \qquad \text{Set 2nd factor equal to 0.}$$

$$x = -4 \qquad \text{Solve for } x.$$

The solutions appear to be $x = 5$ and $x = -4$. However, when you check these in the original equation, you can see that $x = 5$ is the only solution.

Graphical Solution

First, rewrite the original equation as

$$\log 5x + \log(x - 1) - 2 = 0.$$

Then use a graphing utility to graph the equation

$$y = \log 5x + \log(x - 1) - 2$$

and find the zero(s).

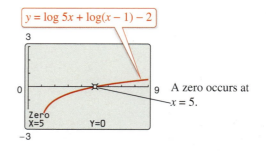

A zero occurs at $x = 5$.

So, the solution is $x = 5$.

✓ *Checkpoint*  Audio-video solution in English & Spanish at LarsonPrecalculus.com

Solve $\log x + \log(x - 9) = 1$.

In Example 9, the domain of $\log 5x$ is $x > 0$ and the domain of $\log(x - 1)$ is $x > 1$, so the domain of the original equation is $x > 1$. This means that the solution $x = -4$ is extraneous. The graphical solution verifies this conclusion.

Applications

EXAMPLE 10 Doubling an Investment

See LarsonPrecalculus.com for an interactive version of this type of example.

You invest $500 at an annual interest rate of 6.75%, compounded continuously. How long will it take your money to double?

Solution   Using the formula for continuous compounding, the balance is

$$A = Pe^{rt}$$

$$A = 500e^{0.0675t}.$$

To find the time required for the balance to double, let $A = 1000$ and solve the resulting equation for t.

$500e^{0.0675t} = 1000$	Let $A = 1000$.
$e^{0.0675t} = 2$	Divide each side by 500.
$\ln e^{0.0675t} = \ln 2$	Take natural log of each side.
$0.0675t = \ln 2$	Inverse Property
$t = \dfrac{\ln 2}{0.0675}$	Divide each side by 0.0675.
$t \approx 10.27$	Use a calculator.

The balance in the account will double after approximately 10.27 years. This result is demonstrated graphically below.

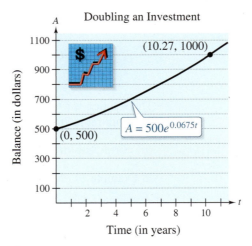

Doubling an Investment

$A = 500e^{0.0675t}$

(10.27, 1000)

(0, 500)

Balance (in dollars)

Time (in years)

✓ **Checkpoint** ◀))) *Audio-video solution in English & Spanish at LarsonPrecalculus.com*

You invest $500 at an annual interest rate of 5.25%, compounded continuously. How long will it take your money to double? Compare your result with that of Example 10.

In Example 10, an approximate answer of 10.27 years is given. Within the context of the problem, the exact solution

$$t = \frac{\ln 2}{0.0675}$$

does not make sense as an answer.

EXAMPLE 11 **Retail Sales**

The retail sales y (in billions of dollars) of e-commerce companies in the United States from 2009 through 2014 can be modeled by

$$y = -614 + 342.2 \ln t, \quad 9 \le t \le 14$$

where t represents the year, with $t = 9$ corresponding to 2009 (see figure). During which year did the sales reach \$240 billion? *(Source: U.S. Census Bureau)*

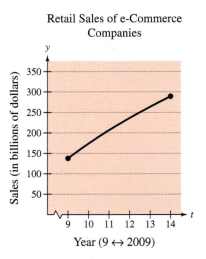

Retail Sales of e-Commerce Companies

Year (9 ↔ 2009)

Solution

$-614 + 342.2 \ln t = y$	Write original equation.
$-614 + 342.2 \ln t = 240$	Substitute 240 for y.
$342.2 \ln t = 854$	Add 614 to each side.
$\ln t = \dfrac{854}{342.2}$	Divide each side by 342.2.
$e^{\ln t} = e^{854/342.2}$	Exponentiate each side.
$t = e^{854/342.2}$	Inverse Property
$t \approx 12$	Use a calculator.

The solution is $t \approx 12$. Because $t = 9$ represents 2009, it follows that the sales reached \$240 billion in 2012.

✓ *Checkpoint*  *Audio-video solution in English & Spanish at LarsonPrecalculus.com*

In Example 11, during which year did the sales reach \$180 billion?

Summarize (Section 5.4)

1. State the One-to-One Properties and the Inverse Properties that are used to solve simple exponential and logarithmic equations *(page 388)*. For an example of solving simple exponential and logarithmic equations, see Example 1.

2. Describe strategies for solving exponential equations *(pages 389 and 390)*. For examples of solving exponential equations, see Examples 2–5.

3. Describe strategies for solving logarithmic equations *(pages 391 and 392)*. For examples of solving logarithmic equations, see Examples 6–9.

4. Describe examples of how to use exponential and logarithmic equations to model and solve real-life problems *(pages 393 and 394, Examples 10 and 11)*.

5.4 Exercises

See CalcChat.com for tutorial help and worked-out solutions to odd-numbered exercises.

Vocabulary: Fill in the blanks.

1. To solve exponential and logarithmic equations, you can use the One-to-One and Inverse Properties below.

(a) $a^x = a^y$ if and only if _____.

(b) $\log_a x = \log_a y$ if and only if _____.

(c) $a^{\log_a x} =$ _____

(d) $\log_a a^x =$ _____

2. An _____ solution does not satisfy the original equation.

Skills and Applications

Determining Solutions In Exercises 3–6, determine whether each x-value is a solution (or an approximate solution) of the equation.

3. $4^{2x-7} = 64$
  (a) $x = 5$
  (b) $x = 2$
  (c) $x = \frac{1}{2}(\log_4 64 + 7)$

4. $4e^{x-1} = 60$
  (a) $x = 1 + \ln 15$
  (b) $x \approx 1.708$
  (c) $x = \ln 16$

5. $\log_2(x + 3) = 10$
  (a) $x = 1021$
  (b) $x = 17$
  (c) $x = 10^2 - 3$

6. $\ln(2x + 3) = 5.8$
  (a) $x = \frac{1}{2}(-3 + \ln 5.8)$
  (b) $x = \frac{1}{2}(-3 + e^{5.8})$
  (c) $x \approx 163.650$

 Solving a Simple Equation In Exercises 7–16, solve for x.

7. $4^x = 16$

8. $\left(\frac{1}{2}\right)^x = 32$

9. $\ln x - \ln 2 = 0$

10. $\log x - \log 10 = 0$

11. $e^x = 2$

12. $e^x = \frac{1}{3}$

13. $\ln x = -1$

14. $\log x = -2$

15. $\log_4 x = 3$

16. $\log_5 x = \frac{1}{2}$

Approximating a Point of Intersection In Exercises 17 and 18, approximate the point of intersection of the graphs of f and g. Then solve the equation $f(x) = g(x)$ algebraically to verify your approximation.

17. $f(x) = 2^x,\ g(x) = 8$

18. $f(x) = \log_3 x,\ g(x) = 2$

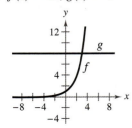

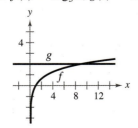

 Solving an Exponential Equation In Exercises 19–46, solve the exponential equation algebraically. Approximate the result to three decimal places, if necessary.

19. $e^x = e^{x^2 - 2}$

20. $e^{x^2 - 3} = e^{x-2}$

21. $4(3^x) = 20$

22. $4e^x = 91$

23. $e^x - 8 = 31$

24. $5^x + 8 = 26$

25. $3^{2x} = 80$

26. $4^{-3t} = 0.10$

27. $3^{2-x} = 400$

28. $7^{-3-x} = 242$

29. $8(10^{3x}) = 12$

30. $8(3^{6-x}) = 40$

31. $e^{3x} = 12$

32. $500e^{-2x} = 125$

33. $7 - 2e^x = 5$

34. $-14 + 3e^x = 11$

35. $6(2^{3x-1}) - 7 = 9$

36. $8(4^{6-2x}) + 13 = 41$

37. $3^x = 2^{x-1}$

38. $e^{x+1} = 2^{x+2}$

39. $4^x = 5^{x^2}$

40. $3^{x^2} = 7^{6-x}$

41. $e^{2x} - 4e^x - 5 = 0$

42. $e^{2x} - 5e^x + 6 = 0$

43. $\dfrac{1}{1 - e^x} = 5$

44. $\dfrac{100}{1 + e^{2x}} = 1$

45. $\left(1 + \dfrac{0.065}{365}\right)^{365t} = 4$

46. $\left(1 + \dfrac{0.10}{12}\right)^{12t} = 2$

 Solving a Logarithmic Equation In Exercises 47–62, solve the logarithmic equation algebraically. Approximate the result to three decimal places, if necessary.

47. $\ln x = -3$

48. $\ln x - 7 = 0$

49. $2.1 = \ln 6x$

50. $\log 3z = 2$

51. $3 - 4\ln x = 11$

52. $3 + 8\ln x = 7$

53. $6\log_3 0.5x = 11$

54. $4\log(x - 6) = 11$

55. $\ln x - \ln(x + 1) = 2$

56. $\ln x + \ln(x + 1) = 1$

57. $\ln(x + 5) = \ln(x - 1) - \ln(x + 1)$

58. $\ln(x + 1) - \ln(x - 2) = \ln x$

59. $\log(3x + 4) = \log(x - 10)$

60. $\log_2 x + \log_2(x + 2) = \log_2(x + 6)$

61. $\log_4 x - \log_4(x - 1) = \frac{1}{2}$

62. $\log 8x - \log\left(1 + \sqrt{x}\right) = 2$

Using Technology In Exercises 63–70, use a graphing utility to graphically solve the equation. Approximate the result to three decimal places. Verify your result algebraically.

63. $5^x = 212$

64. $6e^{1-x} = 25$

65. $8e^{-2x/3} = 11$

66. $e^{0.09t} = 3$

67. $3 - \ln x = 0$

68. $10 - 4\ln(x - 2) = 0$

69. $2\ln(x + 3) = 3$

70. $\ln(x + 1) = 2 - \ln x$

Compound Interest In Exercises 71 and 72, you invest $2500 in an account at interest rate r, compounded continuously. Find the time required for the amount to (a) double and (b) triple.

71. $r = 0.025$

72. $r = 0.0375$

Algebra of Calculus In Exercises 73–80, solve the equation algebraically. Round your result to three decimal places, if necessary. Verify your answer using a graphing utility.

73. $2x^2e^{2x} + 2xe^{2x} = 0$

74. $-x^2e^{-x} + 2xe^{-x} = 0$

75. $-xe^{-x} + e^{-x} = 0$

76. $e^{-2x} - 2xe^{-2x} = 0$

77. $\dfrac{1 + \ln x}{2} = 0$

78. $\dfrac{1 - \ln x}{x^2} = 0$

79. $2x\ln x + x = 0$

80. $2x\ln\left(\dfrac{1}{x}\right) - x = 0$

81. Average Heights The percent m of American males between the ages of 20 and 29 who are under x inches tall is modeled by

$$m(x) = \frac{100}{1 + e^{-0.5536(x - 69.51)}}, \quad 64 \le x \le 78$$

and the percent f of American females between the ages of 20 and 29 who are under x inches tall is modeled by

$$f(x) = \frac{100}{1 + e^{-0.5834(x - 64.49)}}, \quad 60 \le x \le 78.$$

(Source: U.S. National Center for Health Statistics)

(a) Use the graph to determine any horizontal asymptotes of the graphs of the functions. Interpret the meaning in the context of the problem.

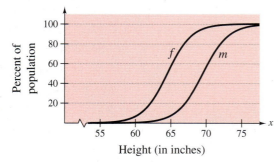

Height (in inches)

(b) What is the average height of each sex?

82. Demand The demand equation for a smartphone is

$$p = 5000\left(1 - \frac{4}{4 + e^{-0.002x}}\right).$$

Find the demand x for each price.

(a) $p = \$169$

(b) $p = \$299$

83. Ecology

The number N of beavers in a given area after x years can be approximated by

$$N = 5.5 \cdot 10^{0.23x}, \quad 0 \le x \le 10.$$

Use the model to approximate how many years it will take for the beaver population to reach 78.

84. Ecology The number N of trees of a given species per acre is approximated by the model

$$N = 3500(10^{-0.12x}), \quad 3 \le x \le 30$$

where x is the average diameter of the trees (in inches) 4.5 feet above the ground. Use the model to approximate the average diameter of the trees in a test plot when $N = 22$.

85. Population The population P (in thousands) of Alaska in the years 2005 through 2015 can be modeled by

$$P = 75\ln t + 540, \quad 5 \le t \le 15$$

where t represents the year, with $t = 5$ corresponding to 2005. During which year did the population of Alaska exceed 720 thousand? *(Source: U.S. Census Bureau)*

86. Population The population P (in thousands) of Montana in the years 2005 through 2015 can be modeled by

$$P = 81\ln t + 807, \quad 5 \le t \le 15$$

where t represents the year, with $t = 5$ corresponding to 2005. During which year did the population of Montana exceed 965 thousand? *(Source: U.S. Census Bureau)*

87. Temperature An object at a temperature of 80°C is placed in a room at 20°C. The temperature of the object is given by

$$T = 20 + 60e^{-0.06m}$$

where m represents the number of minutes after the object is placed in the room. How long does it take the object to reach a temperature of 70°C?

88. Temperature An object at a temperature of 160°C was removed from a furnace and placed in a room at 20°C. The temperature T of the object was measured each hour h and recorded in the table. A model for the data is

$T = 20 + 140e^{-0.68h}$.

Hour, h	Temperature, T
0	160°
1	90°
2	56°
3	38°
4	29°
5	24°

DATA

Spreadsheet at LarsonPrecalculus.com

(a) The figure below shows the graph of the model. Use the graph to identify the horizontal asymptote of the model and interpret the asymptote in the context of the problem.

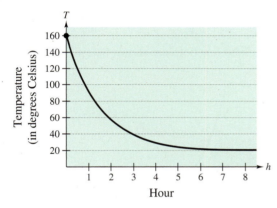

Hour

(b) Use the model to approximate the time it took for the object to reach a temperature of 100°C.

Exploration

True or False?   In Exercises 89–92, rewrite each verbal statement as an equation. Then decide whether the statement is true or false. Justify your answer.

89. The logarithm of the product of two numbers is equal to the sum of the logarithms of the numbers.

90. The logarithm of the sum of two numbers is equal to the product of the logarithms of the numbers.

91. The logarithm of the difference of two numbers is equal to the difference of the logarithms of the numbers.

92. The logarithm of the quotient of two numbers is equal to the difference of the logarithms of the numbers.

93. Think About It Is it possible for a logarithmic equation to have more than one extraneous solution? Explain.

94. **HOW DO YOU SEE IT?** Solving $\log_3 x + \log_3(x - 8) = 2$ algebraically, the solutions appear to be $x = 9$ and $x = -1$. Use the graph of

$y = \log_3 x + \log_3(x - 8) - 2$

to determine whether each value is an actual solution of the equation. Explain.

(9, 0)

95. Finance You are investing P dollars at an annual interest rate of r, compounded continuously, for t years. Which change below results in the highest value of the investment? Explain.

(a) Double the amount you invest.

(b) Double your interest rate.

(c) Double the number of years.

96. Think About It Are the times required for the investments in Exercises 71 and 72 to quadruple twice as long as the times for them to double? Give a reason for your answer and verify your answer algebraically.

97. Effective Yield The *effective yield* of an investment plan is the percent increase in the balance after 1 year. Find the effective yield for each investment plan. Which investment plan has the greatest effective yield? Which investment plan will have the highest balance after 5 years?

(a) 7% annual interest rate, compounded annually

(b) 7% annual interest rate, compounded continuously

(c) 7% annual interest rate, compounded quarterly

(d) 7.25% annual interest rate, compounded quarterly

98. Graphical Reasoning Let $f(x) = \log_a x$ and $g(x) = a^x$, where $a > 1$.

(a) Let $a = 1.2$ and use a graphing utility to graph the two functions in the same viewing window. What do you observe? Approximate any points of intersection of the two graphs.

(b) Determine the value(s) of a for which the two graphs have one point of intersection.

(c) Determine the value(s) of a for which the two graphs have two points of intersection.

5.5 Exponential and Logarithmic Models

■ Recognize the five most common types of models involving exponential and logarithmic functions.
■ Use exponential growth and decay functions to model and solve real-life problems.
■ Use Gaussian functions to model and solve real-life problems.
■ Use logistic growth functions to model and solve real-life problems.
■ Use logarithmic functions to model and solve real-life problems.

Exponential growth and decay models can often represent populations. For example, in Exercise 30 on page 406, you will use exponential growth and decay models to compare the populations of several countries.

Introduction

The five most common types of mathematical models involving exponential functions and logarithmic functions are listed below.

1. **Exponential growth model:** $y = ae^{bx}, \quad b > 0$

2. **Exponential decay model:** $y = ae^{-bx}, \quad b > 0$

3. **Gaussian model:** $y = ae^{-(x-b)^2/c}$

4. **Logistic growth model:** $y = \dfrac{a}{1 + be^{-rx}}$

5. **Logarithmic models:** $y = a + b \ln x, \quad y = a + b \log x$

The basic shapes of the graphs of these functions are shown below.

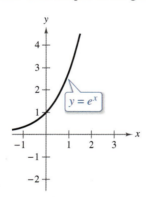

Exponential growth model

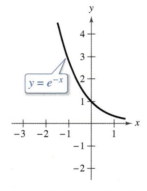

Exponential decay model

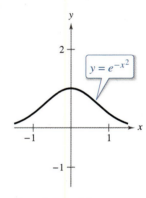

Gaussian model

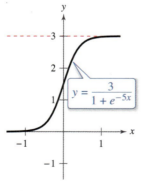

Logistic growth model

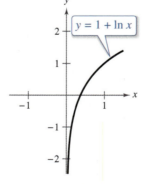

Natural logarithmic model

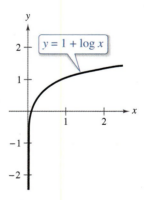

Common logarithmic model

You often gain insight into a situation modeled by an exponential or logarithmic function by identifying and interpreting the asymptotes of the graph of the function. Identify the asymptote(s) of the graph of each function shown above.

Exponential Growth and Decay

EXAMPLE 1 Online Advertising

The amounts S (in billions of dollars) spent in the United States on mobile online advertising in the years 2010 through 2014 are shown in the table. A scatter plot of the data is shown at the right. *(Source: IAB/Price Waterhouse Coopers)*

Year	2010	2011	2012	2013	2014
Advertising Spending	0.6	1.6	3.4	7.1	12.5

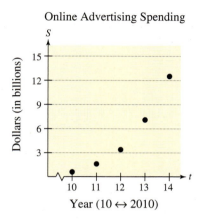

Online Advertising Spending

Year (10 ↔ 2010)

An exponential growth model that approximates the data is

$S = 0.00036e^{0.7563t}, \quad 10 \le t \le 14$

where t represents the year, with $t = 10$ corresponding to 2010. Compare the values found using the model with the amounts shown in the table. According to this model, in what year will the amount spent on mobile online advertising be approximately $65 billion?

Algebraic Solution

The table compares the actual amounts with the values found using the model.

Year	2010	2011	2012	2013	2014
Advertising Spending	0.6	1.6	3.4	7.1	12.5
Model	0.7	1.5	3.1	6.7	14.3

To find when the amount spent on mobile online advertising is about $65 billion, let $S = 65$ in the model and solve for t.

$0.00036e^{0.7563t} = S$	Write original model.
$0.00036e^{0.7563t} = 65$	Substitute 65 for S.
$e^{0.7563t} \approx 180{,}556$	Divide each side by 0.00036.
$\ln e^{0.7563t} \approx \ln 180{,}556$	Take natural log of each side.
$0.7563t \approx 12.1038$	Inverse Property
$t \approx 16$	Divide each side by 0.7563.

According to the model, the amount spent on mobile online advertising will be about $65 billion in 2016.

Graphical Solution

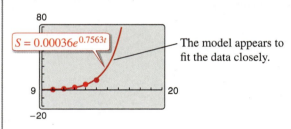

The model appears to fit the data closely.

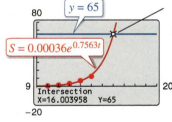

The intersection point of the model and the line $y = 65$ is about (16, 65). So, according to the model, the amount spent on mobile online advertising will be about $65 billion in 2016.

✓ **Checkpoint** 🔊))) *Audio-video solution in English & Spanish at LarsonPrecalculus.com*

In Example 1, in what year will the amount spent on mobile online advertising be about $300 billion?

▷ **TECHNOLOGY** Some graphing utilities have an *exponential regression* feature that can help you find exponential models to represent data. If you have such a graphing utility, use it to find an exponential model for the data given in Example 1. How does your model compare with the model given in Example 1?

In Example 1, the exponential growth model is given. Sometimes you must find such a model. One technique for doing this is shown in Example 2.

EXAMPLE 2 Modeling Population Growth

In a research experiment, a population of fruit flies is increasing according to the law of exponential growth. After 2 days there are 100 flies, and after 4 days there are 300 flies. How many flies will there be after 5 days?

Solution Let y be the number of flies at time t (in days). From the given information, you know that $y = 100$ when $t = 2$ and $y = 300$ when $t = 4$. Substituting this information into the model $y = ae^{bt}$ produces

$$100 = ae^{2b} \quad \text{and} \quad 300 = ae^{4b}.$$

To solve for b, solve for a in the first equation.

$$100 = ae^{2b} \qquad \text{Write first equation.}$$

$$\frac{100}{e^{2b}} = a \qquad \text{Solve for } a.$$

Then substitute the result into the second equation.

$$300 = ae^{4b} \qquad \text{Write second equation.}$$

$$300 = \left(\frac{100}{e^{2b}}\right)e^{4b} \qquad \text{Substitute } \frac{100}{e^{2b}} \text{ for } a.$$

$$300 = 100e^{2b} \qquad \text{Simplify.}$$

$$\frac{300}{100} = e^{2b} \qquad \text{Divide each side by 100.}$$

$$\ln 3 = 2b \qquad \text{Take natural log of each side.}$$

$$\frac{1}{2}\ln 3 = b \qquad \text{Solve for } b.$$

Now substitute $\frac{1}{2}\ln 3$ for b in the expression you found for a.

$$a = \frac{100}{e^{2[(1/2)\ln 3]}} \qquad \text{Substitute } \tfrac{1}{2}\ln 3 \text{ for } b.$$

$$= \frac{100}{e^{\ln 3}} \qquad \text{Simplify.}$$

$$= \frac{100}{3} \qquad \text{Inverse Property}$$

$$\approx 33.33 \qquad \text{Divide.}$$

So, with $a \approx 33.33$ and $b = \frac{1}{2}\ln 3 \approx 0.5493$, the exponential growth model is

$$y = 33.33e^{0.5493t}$$

as shown in Figure 5.15. After 5 days, the population will be

$$y = 33.33e^{0.5493(5)}$$

$$\approx 520 \text{ flies.}$$

✓ **Checkpoint** ◀))) *Audio-video solution in English & Spanish at LarsonPrecalculus.com*

The number of bacteria in a culture is increasing according to the law of exponential growth. After 1 hour there are 100 bacteria, and after 2 hours there are 200 bacteria. How many bacteria will there be after 3 hours?

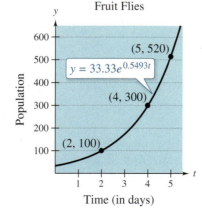

Fruit Flies

$y = 33.33e^{0.5493t}$

(5, 520)

(4, 300)

(2, 100)

Figure 5.15

In living organic material, the ratio of the number of radioactive carbon isotopes (carbon-14) to the number of nonradioactive carbon isotopes (carbon-12) is about 1 to 10^{12}. When organic material dies, its carbon-12 content remains fixed, whereas its radioactive carbon-14 begins to decay with a half-life of about 5700 years. To estimate the age (the number of years since death) of organic material, scientists use the formula

$$R = \frac{1}{10^{12}}e^{-t/8223} \qquad \text{Carbon dating model}$$

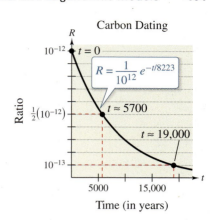

Carbon Dating

where R represents the ratio of carbon-14 to carbon-12 of organic material t years after death. The graph of R is shown at the right. Note that R decreases as t increases.

EXAMPLE 3   Carbon Dating

Estimate the age of a newly discovered fossil for which the ratio of carbon-14 to carbon-12 is $R = \dfrac{1}{10^{13}}$.

Algebraic Solution

In the carbon dating model, substitute the given value of R to obtain the following.

$\dfrac{1}{10^{12}}e^{-t/8223} = R$ Write original model.

$\dfrac{e^{-t/8223}}{10^{12}} = \dfrac{1}{10^{13}}$ Substitute $\dfrac{1}{10^{13}}$ for R.

$e^{-t/8223} = \dfrac{1}{10}$ Multiply each side by 10^{12}.

$\ln e^{-t/8223} = \ln \dfrac{1}{10}$ Take natural log of each side.

$-\dfrac{t}{8223} \approx -2.3026$ Inverse Property

$t \approx 18{,}934$ Multiply each side by -8223.

So, to the nearest thousand years, the age of the fossil is about 19,000 years.

Graphical Solution

Use a graphing utility to graph

$$y_1 = \frac{1}{10^{12}}e^{-x/8223} \quad \text{and} \quad y_2 = \frac{1}{10^{13}}$$

in the same viewing window.

Use the *intersect* feature to estimate that $x \approx 18{,}934$ when $y = 1/10^{13}$.

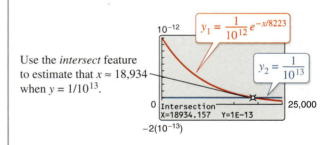

So, to the nearest thousand years, the age of the fossil is about 19,000 years.

✓ **Checkpoint** 🔊))) *Audio-video solution in English & Spanish at LarsonPrecalculus.com*

Estimate the age of a newly discovered fossil for which the ratio of carbon-14 to carbon-12 is $R = 1/10^{14}$.

The value of b in the exponential decay model $y = ae^{-bt}$ determines the *decay* of radioactive isotopes. For example, to find how much of an initial 10 grams of ^{226}Ra isotope with a half-life of 1599 years is left after 500 years, substitute this information into the model $y = ae^{-bt}$.

$$\frac{1}{2}(10) = 10e^{-b(1599)} \implies \ln\frac{1}{2} = -1599b \implies b = -\frac{\ln\frac{1}{2}}{1599}$$

Using the value of b found above and $a = 10$, the amount left is

$$y = 10e^{-[-\ln(1/2)/1599](500)} \approx 8.05 \text{ grams.}$$

Gaussian Models

As mentioned at the beginning of this section, Gaussian models are of the form

$$y = ae^{-(x-b)^2/c}.$$

This type of model is commonly used in probability and statistics to represent populations that are **normally distributed.** For *standard* normal distributions, the model takes the form

$$y = \frac{1}{\sqrt{2\pi}}e^{-x^2/2}.$$

The graph of a Gaussian model is called a **bell-shaped curve.** Use a graphing utility to graph the standard normal distribution curve. Can you see why it is called a bell-shaped curve?

The **average value** of a population can be found from the bell-shaped curve by observing where the maximum y-value of the function occurs. The x-value corresponding to the maximum y-value of the function represents the average value of the independent variable—in this case, x.

EXAMPLE 4 **SAT Scores**

See LarsonPrecalculus.com for an interactive version of this type of example.

In 2015, the SAT mathematics scores for college-bound seniors in the United States roughly followed the normal distribution

$$y = 0.0033e^{-(x-511)^2/28,800}, \quad 200 \le x \le 800$$

where x is the SAT score for mathematics. Use a graphing utility to graph this function and estimate the average SAT mathematics score. *(Source: The College Board)*

Solution The graph of the function is shown below. On this bell-shaped curve, the maximum value of the curve corresponds to the average score. Using the *maximum* feature of the graphing utility, you find that the average mathematics score for college-bound seniors in 2015 was about 511.

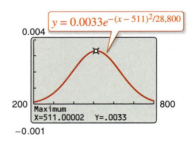

✓ Checkpoint ◀))) *Audio-video solution in English & Spanish at LarsonPrecalculus.com*

In 2015, the SAT critical reading scores for college-bound seniors in the United States roughly followed the normal distribution

$$y = 0.0034e^{-(x-495)^2/26,912}, \quad 200 \le x \le 800$$

where x is the SAT score for critical reading. Use a graphing utility to graph this function and estimate the average SAT critical reading score. *(Source: The College Board)*

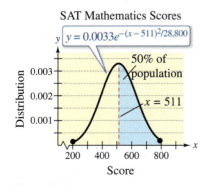

SAT Mathematics Scores

Figure 5.16

In Example 4, note that 50% of the seniors who took the test earned scores greater than 511 (see Figure 5.16).

Logistic Growth Models

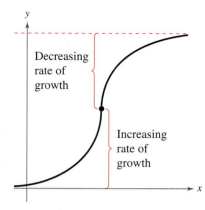

Decreasing rate of growth

Increasing rate of growth

Figure 5.17

Some populations initially have rapid growth, followed by a declining rate of growth, as illustrated by the graph in Figure 5.17. One model for describing this type of growth pattern is the **logistic curve** given by the function

$$y = \frac{a}{1 + be^{-rx}}$$

where y is the population size and x is the time. An example is a bacteria culture that is initially allowed to grow under ideal conditions and then under less favorable conditions that inhibit growth. A logistic growth curve is also called a **sigmoidal curve.**

EXAMPLE 5 **Spread of a Virus**

On a college campus of 5000 students, one student returns from vacation with a contagious and long-lasting flu virus. The spread of the virus is modeled by

$$y = \frac{5000}{1 + 4999e^{-0.8t}}, \quad t \geq 0$$

where y is the total number of students infected after t days. The college will cancel classes when 40% or more of the students are infected.

a. How many students are infected after 5 days?

b. After how many days will the college cancel classes?

Algebraic Solution

a. After 5 days, the number of students infected is

$$y = \frac{5000}{1 + 4999e^{-0.8(5)}} = \frac{5000}{1 + 4999e^{-4}} \approx 54.$$

b. The college will cancel classes when the number of infected students is $(0.40)(5000) = 2000$.

$$2000 = \frac{5000}{1 + 4999e^{-0.8t}}$$

$$1 + 4999e^{-0.8t} = 2.5$$

$$e^{-0.8t} = \frac{1.5}{4999}$$

$$-0.8t = \ln\frac{1.5}{4999}$$

$$t = -\frac{1}{0.8}\ln\frac{1.5}{4999}$$

$$t \approx 10.14$$

So, after about 10 days, at least 40% of the students will be infected, and the college will cancel classes.

Graphical Solution

a.

Use the *value* feature to estimate that $y \approx 54$ when $x = 5$. So, after 5 days, about 54 students are infected.

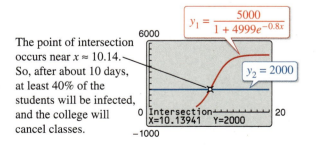

$$y = \frac{5000}{1 + 4999e^{-0.8x}}$$

b. The college will cancel classes when the number of infected students is $(0.40)(5000) = 2000$. Use a graphing utility to graph

$$y_1 = \frac{5000}{1 + 4999e^{-0.8x}} \quad \text{and} \quad y_2 = 2000$$

in the same viewing window. Use the *intersect* feature of the graphing utility to find the point of intersection of the graphs.

The point of intersection occurs near $x \approx 10.14$. So, after about 10 days, at least 40% of the students will be infected, and the college will cancel classes.

$$y_1 = \frac{5000}{1 + 4999e^{-0.8x}}$$

$y_2 = 2000$

 ✓ *Checkpoint* ◀))) *Audio-video solution in English & Spanish at LarsonPrecalculus.com*

In Example 5, after how many days are 250 students infected?

Logarithmic Models

On the Richter scale, the magnitude R of an earthquake of intensity I is given by

$$R = \log \frac{I}{I_0}$$

where $I_0 = 1$ is the minimum intensity used for comparison. (Intensity is a measure of the wave energy of an earthquake.)

 **EXAMPLE 6**   **Magnitudes of Earthquakes**

Find the intensity of each earthquake.

a. Piedmont, California, in 2015: $R = 4.0$ **b.** Nepal in 2015: $R = 7.8$

Solution

a. Because $I_0 = 1$ and $R = 4.0$, you have

$$4.0 = \log \frac{I}{1}$$ Substitute 1 for I_0 and 4.0 for R.

$$10^{4.0} = 10^{\log I}$$ Exponentiate each side.

$$10^{4.0} = I$$ Inverse Property

$$10,000 = I.$$ Simplify.

b. For $R = 7.8$, you have

$$7.8 = \log \frac{I}{1}$$ Substitute 1 for I_0 and 7.8 for R.

$$10^{7.8} = 10^{\log I}$$ Exponentiate each side.

$$10^{7.8} = I$$ Inverse Property

$$63,000,000 \approx I.$$ Use a calculator.

Note that an increase of 3.8 units on the Richter scale (from 4.0 to 7.8) represents an increase in intensity by a factor of $10^{7.8}/10^4 \approx 63,000,000/10,000 = 6300$. In other words, the intensity of the earthquake in Nepal was about 6300 times as great as that of the earthquake in Piedmont, California.

 Checkpoint *Audio-video solution in English & Spanish at LarsonPrecalculus.com*

Find the intensities of earthquakes whose magnitudes are (a) $R = 6.0$ and (b) $R = 7.9$.

On April 25, 2015, an earthquake of magnitude 7.8 struck in Nepal. The city of Kathmandu took extensive damage, including the collapse of the 203-foot Dharahara Tower, built by Nepal's first prime minister in 1832.

Summarize   (Section 5.5)

1. State the five most common types of models involving exponential and logarithmic functions (*page 398*).

2. Describe examples of real-life applications that use exponential growth and decay functions (*pages 399–401, Examples 1–3*).

3. Describe an example of a real-life application that uses a Gaussian function (*page 402, Example 4*).

4. Describe an example of a real-life application that uses a logistic growth function (*page 403, Example 5*).

5. Describe an example of a real-life application that uses a logarithmic function (*page 404, Example 6*).

5.5 Exercises

Vocabulary: Fill in the blanks.

1. An exponential growth model has the form _____, and an exponential decay model has the form _____.
2. A logarithmic model has the form _____ or _____.
3. In probability and statistics, Gaussian models commonly represent populations that are _____ _____.
4. A logistic growth model has the form _____.

Skills and Applications

Solving for a Variable In Exercises 5 and 6, (a) solve for *P* and (b) solve for *t*.

5. $A = Pe^{rt}$

6. $A = P\left(1 + \dfrac{r}{n}\right)^{nt}$

 Compound Interest In Exercises 7–12, find the missing values assuming continuously compounded interest.

	Initial Investment	Annual % Rate	Time to Double	Amount After 10 Years
7.	$1000	3.5%		
8.	$750	$10\frac{1}{2}$%		
9.	$750		$7\frac{3}{4}$ yr	
10.	$500			$1505.00
11.		4.5%		$10,000.00
12.			12 yr	$2000.00

Compound Interest In Exercises 13 and 14, determine the principal *P* that must be invested at rate *r*, compounded monthly, so that $500,000 will be available for retirement in *t* years.

13. $r = 5\%,\ t = 10$

14. $r = 3\frac{1}{2}\%,\ t = 15$

Compound Interest In Exercises 15 and 16, determine the time necessary for *P* dollars to double when it is invested at interest rate *r* compounded (a) annually, (b) monthly, (c) daily, and (d) continuously.

15. $r = 10\%$

16. $r = 6.5\%$

17. **Compound Interest** Complete the table for the time *t* (in years) necessary for *P* dollars to triple when it is invested at an interest rate *r* compounded (a) continuously and (b) annually.

r	2%	4%	6%	8%	10%	12%
t						

18. **Modeling Data** Draw scatter plots of the data in Exercise 17. Use the *regression* feature of a graphing utility to find models for the data.

19. **Comparing Models** If $1 is invested over a 10-year period, then the balance *A* after *t* years is given by either $A = 1 + 0.075[\![t]\!]$ or $A = e^{0.07t}$ depending on whether the interest is simple interest at $7\frac{1}{2}$% or continuous compound interest at 7%. Graph each function on the same set of axes. Which grows at a greater rate? (Remember that $[\![t]\!]$ is the greatest integer function discussed in Section 2.4.)

20. **Comparing Models** If $1 is invested over a 10-year period, then the balance *A* after *t* years is given by either $A = 1 + 0.06[\![t]\!]$ or $A = [1 + (0.055/365)]^{[\![365t]\!]}$ depending on whether the interest is simple interest at 6% or compound interest at $5\frac{1}{2}$% compounded daily. Use a graphing utility to graph each function in the same viewing window. Which grows at a greater rate?

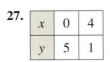 **Radioactive Decay** In Exercises 21–24, find the missing value for the radioactive isotope.

	Isotope	Half-life (years)	Initial Quantity	Amount After 1000 Years
21.	^{226}Ra	1599	10 g	
22.	^{14}C	5715	6.5 g	
23.	^{14}C	5715		2 g
24.	^{239}Pu	24,100		0.4 g

Finding an Exponential Model In Exercises 25–28, find the exponential model that fits the points shown in the graph or table.

25.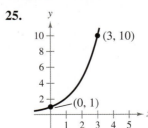

26.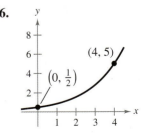

27.

x	0	4
y	5	1

28.

x	0	3
y	1	$\frac{1}{4}$

29. Population The populations P (in thousands) of Horry County, South Carolina, from 1971 through 2014 can be modeled by

$$P = 76.6e^{0.0313t}$$

where t represents the year, with $t = 1$ corresponding to 1971.   *(Source: U.S. Census Bureau)*

(a) Use the model to complete the table.

Year	Population
1980	
1990	
2000	
2010	

(b) According to the model, when will the population of Horry County reach 360,000?

(c) Do you think the model is valid for long-term predictions of the population? Explain.

30. Population

The table shows the mid-year populations (in millions) of five countries in 2015 and the projected populations (in millions) for the year 2025.   *(Source: U.S. Census Bureau)*

Country	2015	2025
Bulgaria	7.2	6.7
Canada	35.1	37.6
China	1367.5	1407.0
United Kingdom	64.1	67.2
United States	321.4	347.3

(a) Find the exponential growth or decay model $y = ae^{bt}$ or $y = ae^{-bt}$ for the population of each country by letting $t = 15$ correspond to 2015. Use the model to predict the population of each country in 2035.

(b) You can see that the populations of the United States and the United Kingdom are growing at different rates. What constant in the equation $y = ae^{bt}$ gives the growth rate? Discuss the relationship between the different growth rates and the magnitude of the constant.

31. Website Growth The number y of hits a new website receives each month can be modeled by $y = 4080e^{kt}$, where t represents the number of months the website has been operating. In the website's third month, there were 10,000 hits. Find the value of k, and use this value to predict the number of hits the website will receive after 24 months.

32. Population The population P (in thousands) of Tallahassee, Florida, from 2000 through 2014 can be modeled by $P = 150.9e^{kt}$, where t represents the year, with $t = 0$ corresponding to 2000. In 2005, the population of Tallahassee was about 163,075. *(Source: U.S. Census Bureau)*

(a) Find the value of k. Is the population increasing or decreasing? Explain.

(b) Use the model to predict the populations of Tallahassee in 2020 and 2025. Are the results reasonable? Explain.

(c) According to the model, during what year will the population reach 200,000?

33. Bacteria Growth The number of bacteria in a culture is increasing according to the law of exponential growth. After 3 hours there are 100 bacteria, and after 5 hours there are 400 bacteria. How many bacteria will there be after 6 hours?

34. Bacteria Growth The number of bacteria in a culture is increasing according to the law of exponential growth. The initial population is 250 bacteria, and the population after 10 hours is double the population after 1 hour. How many bacteria will there be after 6 hours?

35. Depreciation A laptop computer that costs $575 new has a book value of $275 after 2 years.

(a) Find the linear model $V = mt + b$.

(b) Find the exponential model $V = ae^{kt}$.

(c) Use a graphing utility to graph the two models in the same viewing window. Which model depreciates faster in the first 2 years?

(d) Find the book values of the computer after 1 year and after 3 years using each model.

(e) Explain the advantages and disadvantages of using each model to a buyer and a seller.

36. Learning Curve The management at a plastics factory has found that the maximum number of units a worker can produce in a day is 30. The learning curve for the number N of units produced per day after a new employee has worked t days is modeled by $N = 30(1 - e^{kt})$. After 20 days on the job, a new employee produces 19 units.

(a) Find the learning curve for this employee. (*Hint:* First, find the value of k.)

(b) How many days does the model predict will pass before this employee is producing 25 units per day?

37. Carbon Dating The ratio of carbon-14 to carbon-12 in a piece of wood discovered in a cave is $R = 1/8^{14}$. Estimate the age of the piece of wood.

38. Carbon Dating The ratio of carbon-14 to carbon-12 in a piece of paper buried in a tomb is $R = 1/13^{11}$. Estimate the age of the piece of paper.

39. IQ Scores The IQ scores for a sample of students at a small college roughly follow the normal distribution

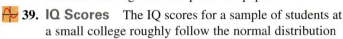

$$y = 0.0266e^{-(x-100)^2/450}, \quad 70 \le x \le 115$$

where x is the IQ score.

(a) Use a graphing utility to graph the function.

(b) From the graph in part (a), estimate the average IQ score of a student.

40. Education The amount of time (in hours per week) a student utilizes a math-tutoring center roughly follows the normal distribution

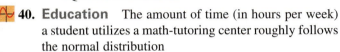

$$y = 0.7979e^{-(x-5.4)^2/0.5}, \quad 4 \le x \le 7$$

where x is the number of hours.

(a) Use a graphing utility to graph the function.

(b) From the graph in part (a), estimate the average number of hours per week a student uses the tutoring center.

41. Cell Sites A cell site is a site where electronic communications equipment is placed in a cellular network for the use of mobile phones. The numbers y of cell sites from 1985 through 2014 can be modeled by

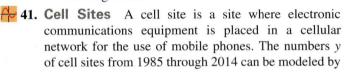

$$y = \frac{320{,}110}{1 + 374e^{-0.252t}}$$

where t represents the year, with $t = 5$ corresponding to 1985. *(Source: CTIA-The Wireless Association)*

(a) Use the model to find the numbers of cell sites in the years 1998, 2003, and 2006.

(b) Use a graphing utility to graph the function.

(c) Use the graph to determine the year in which the number of cell sites reached 270,000.

(d) Confirm your answer to part (c) algebraically.

42. Population The population P (in thousands) of a city from 2000 through 2016 can be modeled by

$$P = \frac{2632}{1 + 0.083e^{0.050t}}$$

where t represents the year, with $t = 0$ corresponding to 2000.

(a) Use the model to find the populations of the city in the years 2000, 2005, 2010, and 2015.

(b) Use a graphing utility to graph the function.

(c) Use the graph to determine the year in which the population reached 2.2 million.

(d) Confirm your answer to part (c) algebraically.

43. Population Growth A conservation organization released 100 animals of an endangered species into a game preserve. The preserve has a carrying capacity of 1000 animals. The growth of the pack is modeled by the logistic curve

$$p(t) = \frac{1000}{1 + 9e^{-0.1656t}}$$

where t is measured in months (see figure).

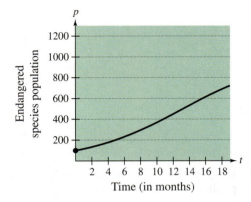

(a) Estimate the population after 5 months.

(b) After how many months is the population 500?

(c) Use a graphing utility to graph the function. Use the graph to determine the horizontal asymptotes, and interpret the meaning of the asymptotes in the context of the problem.

44. Sales After discontinuing all advertising for a tool kit in 2010, the manufacturer noted that sales began to drop according to the model

$$S = \frac{500{,}000}{1 + 0.1e^{kt}}$$

where S represents the number of units sold and t represents the year, with $t = 0$ corresponding to 2010 (see figure). In 2014, 300,000 units were sold.

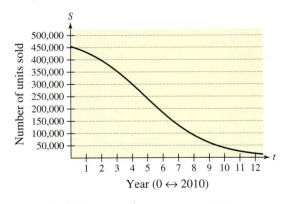

(a) Use the graph to estimate sales in 2020.

(b) Complete the model by solving for k.

(c) Use the model to estimate sales in 2020. Compare your results with that of part (a).

Geology In Exercises 45 and 46, use the Richter scale

$$R = \log \frac{I}{I_0}$$

for measuring the magnitude R of an earthquake.

45. Find the intensity I of an earthquake measuring R on the Richter scale (let $I_0 = 1$).

(a) Peru in 2015: $R = 7.6$

(b) Pakistan in 2015: $R = 5.6$

(c) Indonesia in 2015: $R = 6.6$

46. Find the magnitude R of each earthquake of intensity I (let $I_0 = 1$).

(a) $I = 199{,}500{,}000$

(b) $I = 48{,}275{,}000$

(c) $I = 17{,}000$

Intensity of Sound In Exercises 47–50, use the following information for determining sound intensity. The number of decibels β of a sound with an intensity of I watts per square meter is given by $\beta = 10 \log(I/I_0)$, where I_0 is an intensity of 10^{-12} watt per square meter, corresponding roughly to the faintest sound that can be heard by the human ear. In Exercises 47 and 48, find the number of decibels β of the sound.

47. (a) $I = 10^{-10}$ watt per m² (quiet room)

(b) $I = 10^{-5}$ watt per m² (busy street corner)

(c) $I = 10^{-8}$ watt per m² (quiet radio)

(d) $I = 10^{-3}$ watt per m² (loud car horn)

48. (a) $I = 10^{-11}$ watt per m² (rustle of leaves)

(b) $I = 10^2$ watt per m² (jet at 30 meters)

(c) $I = 10^{-4}$ watt per m² (door slamming)

(d) $I = 10^{-6}$ watt per m² (normal conversation)

49. Due to the installation of noise suppression materials, the noise level in an auditorium decreased from 93 to 80 decibels. Find the percent decrease in the intensity of the noise as a result of the installation of these materials.

50. Due to the installation of a muffler, the noise level of an engine decreased from 88 to 72 decibels. Find the percent decrease in the intensity of the noise as a result of the installation of the muffler.

pH Levels In Exercises 51–56, use the acidity model $pH = -\log[H^+]$, where acidity (pH) is a measure of the hydrogen ion concentration $[H^+]$ (measured in moles of hydrogen per liter) of a solution.

51. Find the pH when $[H^+] = 2.3 \times 10^{-5}$.

52. Find the pH when $[H^+] = 1.13 \times 10^{-5}$.

53. Compute $[H^+]$ for a solution in which pH $= 5.8$.

54. Compute $[H^+]$ for a solution in which pH $= 3.2$.

55. Apple juice has a pH of 2.9 and drinking water has a pH of 8.0. The hydrogen ion concentration of the apple juice is how many times the concentration of drinking water?

56. The pH of a solution decreases by one unit. By what factor does the hydrogen ion concentration increase?

57. Forensics At 8:30 A.M., a coroner went to the home of a person who had died during the night. In order to estimate the time of death, the coroner took the person's temperature twice. At 9:00 A.M. the temperature was 85.7°F, and at 11:00 A.M. the temperature was 82.8°F. From these two temperatures, the coroner was able to determine that the time elapsed since death and the body temperature were related by the formula

$$t = -10 \ln \frac{T - 70}{98.6 - 70}$$

where t is the time in hours elapsed since the person died and T is the temperature (in degrees Fahrenheit) of the person's body. (This formula comes from a general cooling principle called *Newton's Law of Cooling*. It uses the assumptions that the person had a normal body temperature of 98.6°F at death and that the room temperature was a constant 70°F.) Use the formula to estimate the time of death of the person.

58. Home Mortgage A $120,000 home mortgage for 30 years at $7\frac{1}{2}\%$ has a monthly payment of $839.06. Part of the monthly payment covers the interest charge on the unpaid balance, and the remainder of the payment reduces the principal. The amount paid toward the interest is

$$u = M - \left(M - \frac{Pr}{12}\right)\left(1 + \frac{r}{12}\right)^{12t}$$

and the amount paid toward the reduction of the principal is

$$v = \left(M - \frac{Pr}{12}\right)\left(1 + \frac{r}{12}\right)^{12t}.$$

In these formulas, P is the amount of the mortgage, r is the interest rate (in decimal form), M is the monthly payment, and t is the time in years.

(a) Use a graphing utility to graph each function in the same viewing window. (The viewing window should show all 30 years of mortgage payments.)

(b) In the early years of the mortgage, is the greater part of the monthly payment paid toward the interest or the principal? Approximate the time when the monthly payment is evenly divided between interest and principal reduction.

(c) Repeat parts (a) and (b) for a repayment period of 20 years ($M = \$966.71$). What can you conclude?

59. Home Mortgage The total interest u paid on a home mortgage of P dollars at interest rate r (in decimal form) for t years is

$$u = P\left[\frac{rt}{1 - \left(\dfrac{1}{1 + r/12}\right)^{12t}} - 1\right].$$

Consider a $120,000 home mortgage at $7\frac{1}{2}\%$.

(a) Use a graphing utility to graph the total interest function.

(b) Approximate the length of the mortgage for which the total interest paid is the same as the size of the mortgage. Is it possible that some people are paying twice as much in interest charges as the size of the mortgage?

60. Car Speed The table shows the time t (in seconds) required for a car to attain a speed of s miles per hour from a standing start.

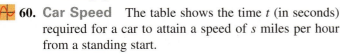

Speed, s	Time, t
30	3.4
40	5.0
50	7.0
60	9.3
70	12.0
80	15.8
90	20.0

Spreadsheet at LarsonPrecalculus.com

Two models for these data are given below.

$t_1 = 40.757 + 0.556s - 15.817 \ln s$

$t_2 = 1.2259 + 0.0023s^2$

(a) Use the *regression* feature of a graphing utility to find a linear model t_3 and an exponential model t_4 for the data.

(b) Use the graphing utility to graph the data and each model in the same viewing window.

(c) Create a table comparing the data with estimates obtained from each model.

(d) Use the results of part (c) to find the sum of the absolute values of the differences between the data and the estimated values found using each model. Based on the four sums, which model do you think best fits the data? Explain.

Exploration

True or False? In Exercises 61–64, determine whether the statement is true or false. Justify your answer.

61. The domain of a logistic growth function cannot be the set of real numbers.

62. A logistic growth function will always have an x-intercept.

63. The graph of $f(x) = \dfrac{4}{1 + 6e^{-2x}} + 5$ is the graph of $g(x) = \dfrac{4}{1 + 6e^{-2x}}$ shifted to the right five units.

64. The graph of a Gaussian model will never have an x-intercept.

65. Writing Use your school's library, the Internet, or some other reference source to write a paper describing John Napier's work with logarithms.

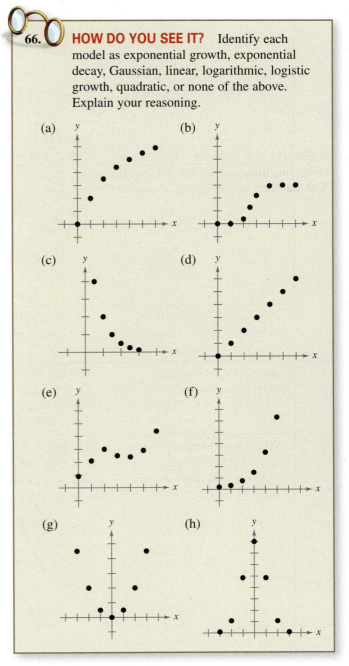

66. **HOW DO YOU SEE IT?** Identify each model as exponential growth, exponential decay, Gaussian, linear, logarithmic, logistic growth, quadratic, or none of the above. Explain your reasoning.

(a) (b) (c) (d) (e) (f) (g) (h)

Project: Sales per Share To work an extended application analyzing the sales per share for Kohl's Corporation from 1999 through 2014, visit this text's website at *LarsonPrecalculus.com*. (*Source: Kohl's Corporation*)

Chapter Summary

What Did You Learn?	Explanation/Examples	Review Exercises
Section 5.1 Recognize and evaluate exponential functions with base a (p. 360).	The exponential function f with base a is denoted by $f(x) = a^x$, where $a > 0$, $a \neq 1$, and x is any real number.	1–6
Graph exponential functions and use a One-to-One Property (p. 361).	**One-to-One Property:** For $a > 0$ and $a \neq 1$, $a^x = a^y$ if and only if $x = y$.	7–20
Recognize, evaluate, and graph exponential functions with base e (p. 364).	The function $f(x) = e^x$ is called the natural exponential function.	21–28
Use exponential functions to model and solve real-life problems (p. 365).	Exponential functions are used in compound interest formulas (see Example 8) and in radioactive decay models (see Example 9).	29–32
Section 5.2 Recognize and evaluate logarithmic functions with base a (p. 371).	For $x > 0$, $a > 0$, and $a \neq 1$, $y = \log_a x$ if and only if $x = a^y$. The function $f(x) = \log_a x$ is the logarithmic function with base a. The logarithmic function with base 10 is called the common logarithmic function. It is denoted by $\log_{10}$ or log.	33–44
Graph logarithmic functions (p. 373), and recognize, evaluate, and graph natural logarithmic functions (p. 375).	The graph of $g(x) = \log_a x$ is a reflection of the graph of $f(x) = a^x$ in the line $y = x$. The function $g(x) = \ln x$, $x > 0$, is called the natural logarithmic function. Its graph is a reflection of the graph of $f(x) = e^x$ in the line $y = x$.	45–56
Use logarithmic functions to model and solve real-life problems (p. 377).	A logarithmic function can model human memory. (See Example 11.)	57, 58

	What Did You Learn?	**Explanation/Examples**	**Review Exercises**
Section 5.3	Use the change-of-base formula to rewrite and evaluate logarithmic expressions *(p. 381).*	Let a, b, and x be positive real numbers such that $a \neq 1$ and $b \neq 1$. Then $\log_a x$ can be converted to a different base as follows. **Base b** $\qquad\qquad$ **Base 10** $\qquad\qquad$ **Base e** $\log_a x = \dfrac{\log_b x}{\log_b a} \qquad \log_a x = \dfrac{\log x}{\log a} \qquad \log_a x = \dfrac{\ln x}{\ln a}$	59–62
	Use properties of logarithms to evaluate, rewrite, expand, or condense logarithmic expressions *(pp. 382–383).*	Let a be a positive number such that $a \neq 1$, let n be a real number, and let u and v be positive real numbers. **1. Product Property:** $\log_a(uv) = \log_a u + \log_a v$ $\ln(uv) = \ln u + \ln v$ **2. Quotient Property:** $\log_a(u/v) = \log_a u - \log_a v$ $\ln(u/v) = \ln u - \ln v$ **3. Power Property:** $\log_a u^n = n \log_a u, \ \ln u^n = n \ln u$	63–78
	Use logarithmic functions to model and solve real-life problems *(p. 384).*	Logarithmic functions can help you find an equation that relates the periods of several planets and their distances from the sun. (See Example 7.)	79, 80
Section 5.4	Solve simple exponential and logarithmic equations *(p. 388).*	One-to-One Properties and Inverse Properties of exponential or logarithmic functions are used to solve exponential or logarithmic equations.	81–86
	Solve more complicated exponential equations *(p. 389)* and logarithmic equations *(p. 391).*	To solve more complicated equations, rewrite the equations to allow the use of the One-to-One Properties or Inverse Properties of exponential or logarithmic functions. (See Examples 2–9.)	87–102
	Use exponential and logarithmic equations to model and solve real-life problems *(p. 393).*	Exponential and logarithmic equations can help you determine how long it will take to double an investment (see Example 10) and find the year in which an industry had a given amount of sales (see Example 11).	103, 104
Section 5.5	Recognize the five most common types of models involving exponential and logarithmic functions *(p. 398).*	**1. Exponential growth model:** $y = ae^{bx}, \ \ b > 0$ **2. Exponential decay model:** $y = ae^{-bx}, \ \ b > 0$ **3. Gaussian model:** $y = ae^{-(x-b)^2/c}$ **4. Logistic growth model:** $y = \dfrac{a}{1 + be^{-rx}}$ **5. Logarithmic models:** $y = a + b \ln x, \ y = a + b \log x$	105–110
	Use exponential growth and decay functions to model and solve real-life problems *(p. 399).*	An exponential growth function can help you model a population of fruit flies (see Example 2), and an exponential decay function can help you estimate the age of a fossil (see Example 3).	111, 112
	Use Gaussian functions *(p. 402)*, logistic growth functions *(p. 403)*, and logarithmic functions *(p. 404)* to model and solve real-life problems.	A Gaussian function can help you model SAT mathematics scores for college-bound seniors. (See Example 4.) A logistic growth function can help you model the spread of a flu virus. (See Example 5.) A logarithmic function can help you find the intensity of an earthquake given its magnitude. (See Example 6.)	113–115

5.1 **Evaluating an Exponential Function** In Exercises 1–6, evaluate the function at the given value of x. Round your result to three decimal places.

1. $f(x) = 0.3^x$, $x = 1.5$ **2.** $f(x) = 30^x$, $x = \sqrt{3}$

3. $f(x) = 2^x$, $x = \frac{2}{3}$ **4.** $f(x) = \left(\frac{1}{2}\right)^{2x}$, $x = \pi$

5. $f(x) = 7(0.2^x)$, $x = -\sqrt{11}$

6. $f(x) = -14(5^x)$, $x = -0.8$

Graphing an Exponential Function In Exercises 7–12, use a graphing utility to construct a table of values for the function. Then sketch the graph of the function.

7. $f(x) = 4^{-x} + 4$ **8.** $f(x) = 2.65^{x-1}$

9. $f(x) = 5^{x-2} + 4$ **10.** $f(x) = 2^{x-6} - 5$

11. $f(x) = \left(\frac{1}{2}\right)^{-x} + 3$ **12.** $f(x) = \left(\frac{1}{8}\right)^{x+2} - 5$

Using a One-to-One Property In Exercises 13–16, use a One-to-One Property to solve the equation for x.

13. $\left(\frac{1}{3}\right)^{x-3} = 9$ **14.** $3^{x+3} = \frac{1}{81}$

15. $e^{3x-5} = e^7$ **16.** $e^{8-2x} = e^{-3}$

Transforming the Graph of an Exponential Function In Exercises 17–20, describe the transformation of the graph of f that yields the graph of g.

17. $f(x) = 5^x$, $g(x) = 5^x + 1$

18. $f(x) = 6^x$, $g(x) = 6^{x+1}$

19. $f(x) = 3^x$, $g(x) = 1 - 3^x$

20. $f(x) = \left(\frac{1}{2}\right)^x$, $g(x) = -\left(\frac{1}{2}\right)^{x+2}$

Evaluating the Natural Exponential Function In Exercises 21–24, evaluate $f(x) = e^x$ at the given value of x. Round your result to three decimal places.

21. $x = 3.4$ **22.** $x = -2.5$

23. $x = \frac{3}{5}$ **24.** $x = \frac{2}{7}$

Graphing a Natural Exponential Function In Exercises 25–28, use a graphing utility to construct a table of values for the function. Then sketch the graph of the function.

25. $h(x) = e^{-x/2}$ **26.** $h(x) = 2 - e^{-x/2}$

27. $f(x) = e^{x+2}$ **28.** $s(t) = 4e^{t-1}$

29. Waiting Times The average time between new posts on a message board is 3 minutes. The probability F of waiting less than t minutes until the next post is approximated by the model $F(t) = 1 - e^{-t/3}$. A message has just been posted. Find the probability that the next post will be within (a) 1 minute, (b) 2 minutes, and (c) 5 minutes.

30. Depreciation After t years, the value V of a car that originally cost \$23,970 is given by $V(t) = 23{,}970\left(\frac{3}{4}\right)^t$.

(a) Use a graphing utility to graph the function.

(b) Find the value of the car 2 years after it was purchased.

(c) According to the model, when does the car depreciate most rapidly? Is this realistic? Explain.

(d) According to the model, when will the car have no value?

Compound Interest In Exercises 31 and 32, complete the table by finding the balance A when P dollars is invested at rate r for t years and compounded n times per year.

n	1	2	4	12	365	Continuous
A						

31. $P = \$5000$, $r = 3\%$, $t = 10$ years

32. $P = \$4500$, $r = 2.5\%$, $t = 30$ years

5.2 **Writing a Logarithmic Equation** In Exercises 33–36, write the exponential equation in logarithmic form. For example, the logarithmic form of $2^3 = 8$ is $\log_2 8 = 3$.

33. $3^3 = 27$ **34.** $25^{3/2} = 125$

35. $e^{0.8} = 2.2255\ldots$ **36.** $e^0 = 1$

Evaluating a Logarithm In Exercises 37–40, evaluate the logarithm at the given value of x without using a calculator.

37. $f(x) = \log x$, $x = 1000$ **38.** $g(x) = \log_9 x$, $x = 3$

39. $g(x) = \log_2 x$, $x = \frac{1}{4}$ **40.** $f(x) = \log_3 x$, $x = \frac{1}{81}$

Using a One-to-One Property In Exercises 41–44, use a One-to-One Property to solve the equation for x.

41. $\log_4(x + 7) = \log_4 14$ **42.** $\log_8(3x - 10) = \log_8 5$

43. $\ln(x + 9) = \ln 4$ **44.** $\log(3x - 2) = \log 7$

Sketching the Graph of a Logarithmic Function In Exercises 45–48, find the domain, x-intercept, and vertical asymptote of the logarithmic function and sketch its graph.

45. $g(x) = \log_7 x$

46. $f(x) = \log \dfrac{x}{3}$

47. $f(x) = 4 - \log(x + 5)$

48. $f(x) = \log(x - 3) + 1$

Evaluating a Logarithmic Function In Exercises 49–52, use a calculator to evaluate the function at the given value of x. Round your result to three decimal places, if necessary.

49. $f(x) = \ln x$, $x = 22.6$ **50.** $f(x) = \ln x$, $x = e^{-12}$

51. $f(x) = \frac{1}{2}\ln x$, $x = \sqrt{e}$

52. $f(x) = 5\ln x$, $x = 0.98$

Graphing a Natural Logarithmic Function In Exercises 53–56, find the domain, x-intercept, and vertical asymptote of the logarithmic function and sketch its graph.

53. $f(x) = \ln x + 6$ **54.** $f(x) = \ln x - 5$

55. $h(x) = \ln(x - 6)$ **56.** $f(x) = \ln(x + 4)$

57. Astronomy The formula $M = m - 5\log(d/10)$ gives the distance d (in parsecs) from Earth to a star with apparent magnitude m and absolute magnitude M. The star Rasalhague has an apparent magnitude of 2.08 and an absolute magnitude of 1.3. Find the distance from Earth to Rasalhague.

58. Snow Removal The number of miles s of roads cleared of snow is approximated by the model

$$s = 25 - \frac{13\ln(h/12)}{\ln 3}, \quad 2 \le h \le 15$$

where h is the depth (in inches) of the snow. Use this model to find s when $h = 10$ inches.

5.3 **Using the Change-of-Base Formula** In Exercises 59–62, evaluate the logarithm using the change-of-base formula (a) with common logarithms and (b) with natural logarithms. Round your results to three decimal places.

59. $\log_2 6$ **60.** $\log_{12} 200$

61. $\log_{1/2} 5$ **62.** $\log_4 0.75$

Using Properties of Logarithms In Exercises 63–66, use the properties of logarithms to write the logarithm in terms of $\log_2 3$ and $\log_2 5$.

63. $\log_2 \frac{5}{3}$ **64.** $\log_2 45$

65. $\log_2 \frac{9}{5}$ **66.** $\log_2 \frac{20}{9}$

Expanding a Logarithmic Expression In Exercises 67–72, use the properties of logarithms to expand the expression as a sum, difference, and/or constant multiple of logarithms. (Assume all variables are positive.)

67. $\log 7x^2$ **68.** $\log 11x^3$

69. $\log_3 \dfrac{9}{\sqrt{x}}$ **70.** $\log_7 \dfrac{\sqrt[3]{x}}{19}$

71. $\ln x^2 y^2 z$ **72.** $\ln\left(\dfrac{y - 1}{3}\right)^2$, $y > 1$

Condensing a Logarithmic Expression In Exercises 73–78, condense the expression to the logarithm of a single quantity.

73. $\ln 7 + \ln x$

74. $\log_2 y - \log_2 3$

75. $\log x - \frac{1}{2}\log y$

76. $3\ln x + 2\ln(x + 1)$

77. $\frac{1}{2}\log_3 x - 2\log_3(y + 8)$

78. $5\ln(x - 2) - \ln(x + 2) - 3\ln x$

79. Climb Rate The time t (in minutes) for a small plane to climb to an altitude of h feet is modeled by

$$t = 50\log[18{,}000/(18{,}000 - h)]$$

where 18,000 feet is the plane's absolute ceiling.

(a) Determine the domain of the function in the context of the problem.

(b) Use a graphing utility to graph the function and identify any asymptotes.

(c) As the plane approaches its absolute ceiling, what can be said about the time required to increase its altitude?

(d) Find the time it takes for the plane to climb to an altitude of 4000 feet.

80. Human Memory Model Students in a learning theory study took an exam and then retested monthly for 6 months with an equivalent exam. The data obtained in the study are given by the ordered pairs (t, s), where t is the time (in months) after the initial exam and s is the average score for the class. Use the data to find a logarithmic equation that relates t and s.

$(1, 84.2)$, $(2, 78.4)$, $(3, 72.1)$,
$(4, 68.5)$, $(5, 67.1)$, $(6, 65.3)$

5.4 **Solving a Simple Equation** In Exercises 81–86, solve for x.

81. $5^x = 125$

82. $6^x = \frac{1}{216}$

83. $e^x = 3$

84. $\log x - \log 5 = 0$

85. $\ln x = 4$

86. $\ln x = -1.6$

Solving an Exponential Equation In Exercises 87–90, solve the exponential equation algebraically. Approximate the result to three decimal places.

87. $e^{4x} = e^{x^2 + 3}$

88. $e^{3x} = 25$

89. $2^x - 3 = 29$

90. $e^{2x} - 6e^x + 8 = 0$

Solving a Logarithmic Equation In Exercises 91–98, solve the logarithmic equation algebraically. Approximate the result to three decimal places.

91. $\ln 3x = 8.2$

92. $4 \ln 3x = 15$

93. $\ln x + \ln(x - 3) = 1$

94. $\ln(x + 2) - \ln x = 2$

95. $\log_8(x - 1) = \log_8(x - 2) - \log_8(x + 2)$

96. $\log_6(x + 2) - \log_6 x = \log_6(x + 5)$

97. $\log(1 - x) = -1$

98. $\log(-x - 4) = 2$

Using Technology In Exercises 99–102, use a graphing utility to graphically solve the equation. Approximate the result to three decimal places. Verify your result algebraically.

99. $25e^{-0.3x} = 12$

100. $2 = 5 - e^{x+7}$

101. $2 \ln(x + 3) - 3 = 0$

102. $2 \ln x - \ln(3x - 1) = 0$

103. Compound Interest You deposit $8500 in an account that pays 1.5% interest, compounded continuously. How long will it take for the money to triple?

104. Meteorology The speed of the wind S (in miles per hour) near the center of a tornado and the distance d (in miles) the tornado travels are related by the model $S = 93 \log d + 65$. On March 18, 1925, a large tornado struck portions of Missouri, Illinois, and Indiana with a wind speed at the center of about 283 miles per hour. Approximate the distance traveled by this tornado.

5.5 Matching a Function with Its Graph In Exercises 105–110, match the function with its graph. [The graphs are labeled (a), (b), (c), (d), (e), and (f).]

(a)

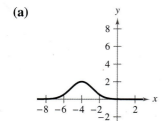

(b)

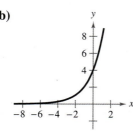

(c)

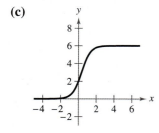

(d)

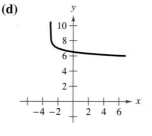

(e)

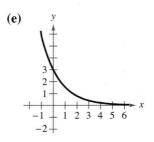

(f)

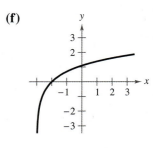

105. $y = 3e^{-2x/3}$

106. $y = 4e^{2x/3}$

107. $y = \ln(x + 3)$

108. $y = 7 - \log(x + 3)$

109. $y = 2e^{-(x+4)^2/3}$

110. $y = \dfrac{6}{1 + 2e^{-2x}}$

111. Finding an Exponential Model Find the exponential model $y = ae^{bx}$ that fits the points $(0, 2)$ and $(4, 3)$.

112. Wildlife Population A species of bat is in danger of becoming extinct. Five years ago, the total population of the species was 2000. Two years ago, the total population of the species was 1400. What was the total population of the species one year ago?

113. Test Scores The test scores for a biology test follow the normal distribution

$$y = 0.0499e^{-(x-71)^2/128}, \quad 40 \le x \le 100$$

where x is the test score. Use a graphing utility to graph the equation and estimate the average test score.

114. Typing Speed In a typing class, the average number N of words per minute typed after t weeks of lessons is

$$N = 157/(1 + 5.4e^{-0.12t}).$$

Find the time necessary to type (a) 50 words per minute and (b) 75 words per minute.

115. Sound Intensity The relationship between the number of decibels β and the intensity of a sound I (in watts per square meter) is

$$\beta = 10 \log(I/10^{-12}).$$

Find the intensity I for each decibel level β.

(a) $\beta = 60$ (b) $\beta = 135$ (c) $\beta = 1$

Exploration

116. Graph of an Exponential Function Consider the graph of $y = e^{kt}$. Describe the characteristics of the graph when k is positive and when k is negative.

True or False? In Exercises 117 and 118, determine whether the equation is true or false. Justify your answer.

117. $\log_b b^{2x} = 2x$

118. $\ln(x + y) = \ln x + \ln y$

Take this test as you would take a test in class. When you are finished, check your work against the answers given in the back of the book.

In Exercises 1–4, evaluate the expression. Round your result to three decimal places.

1. $0.7^{2.5}$ **2.** $3^{-\pi}$ **3.** $e^{-7/10}$ **4.** $e^{3.1}$

 In Exercises 5–7, use a graphing utility to construct a table of values for the function. Then sketch the graph of the function.

5. $f(x) = 10^{-x}$ **6.** $f(x) = -6^{x-2}$ **7.** $f(x) = 1 - e^{2x}$

8. Evaluate (a) $\log_7 7^{-0.89}$ and (b) $4.6 \ln e^2$.

In Exercises 9–11, find the domain, x-intercept, and vertical asymptote of the logarithmic function and sketch its graph.

9. $f(x) = 4 + \log x$ **10.** $f(x) = \ln(x - 4)$ **11.** $f(x) = 1 + \ln(x + 6)$

In Exercises 12–14, evaluate the logarithm using the change-of-base formula. Round your result to three decimal places.

12. $\log_5 35$ **13.** $\log_{16} 0.63$ **14.** $\log_{3/4} 24$

In Exercises 15–17, use the properties of logarithms to expand the expression as a sum, difference, and/or constant multiple of logarithms. (Assume all variables are positive.)

15. $\log_2 3a^4$ **16.** $\ln \dfrac{\sqrt{x}}{7}$ **17.** $\log \dfrac{10x^2}{y^3}$

In Exercises 18–20, condense the expression to the logarithm of a single quantity.

18. $\log_3 13 + \log_3 y$ **19.** $4 \ln x - 4 \ln y$

20. $3 \ln x - \ln(x + 3) + 2 \ln y$

In Exercises 21–26, solve the equation algebraically. Approximate the result to three decimal places, if necessary.

21. $5^x = \dfrac{1}{25}$ **22.** $3e^{-5x} = 132$

23. $\dfrac{1025}{8 + e^{4x}} = 5$ **24.** $\ln x = \dfrac{1}{2}$

25. $18 + 4 \ln x = 7$ **26.** $\log x + \log(x - 15) = 2$

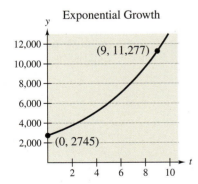

Figure for 27

27. Find the exponential growth model that fits the points shown in the graph.

28. The half-life of radioactive actinium (^{227}Ac) is 21.77 years. What percent of a present amount of radioactive actinium will remain after 19 years?

29. A model that can predict a child's height H (in centimeters) based on the child's age is $H = 70.228 + 5.104x + 9.222 \ln x, \frac{1}{4} \leq x \leq 6$, where x is the child's age in years. (*Source: Snapshots of Applications in Mathematics*)

   (a) Construct a table of values for the model. Then sketch the graph of the model.

   (b) Use the graph from part (a) to predict the height of a four-year-old child. Then confirm your prediction algebraically.

Cumulative Test for Chapters 3–5

See CalcChat.com for tutorial help and worked-out solutions to odd-numbered exercises.

Take this test as you would take a test in class. When you are finished, check your work against the answers given in the back of the book.

1. Write the standard form of the quadratic function whose graph is a parabola with vertex at $(-8, 5)$ and that passes through the point $(-4, -7)$.

In Exercises 2–4, sketch the graph of the function.

2. $h(x) = -x^2 + 10x - 21$ 3. $f(t) = -\frac{1}{2}(t - 1)^2(t + 2)^2$ 4. $g(s) = s^3 - 3s^2$

In Exercises 5 and 6, find all the zeros of the function.

5. $f(x) = x^3 + 2x^2 + 4x + 8$ 6. $f(x) = x^4 + 4x^3 - 21x^2$

7. Use long division to divide: $\dfrac{6x^3 - 4x^2}{2x^2 + 1}$.

8. Use synthetic division to divide $3x^4 + 2x^2 - 5x + 3$ by $x - 2$.

9. Use a graphing utility to approximate (to three decimal places) the real zero of the function $g(x) = x^3 + 3x^2 - 6$.

10. Find a polynomial function with real coefficients that has -5, -2, and $2 + \sqrt{3}i$ as its zeros. (There are many correct answers.)

In Exercises 11 and 12, state the domain of the function and find any asymptotes. Sketch the graph of the function.

11. $f(x) = \dfrac{2x}{x - 3}$

12. $f(x) = \dfrac{4x^2}{x - 5}$

In Exercises 13–15, sketch the graph of the rational function. Identify all intercepts and find any asymptotes.

13. $f(x) = \dfrac{2x}{x^2 + 2x - 3}$

14. $f(x) = \dfrac{x^2 - 4}{x^2 + x - 2}$

15. $f(x) = \dfrac{x^3 - 2x^2 - 9x + 18}{x^2 + 4x + 3}$

In Exercises 16 and 17, sketch the conic.

16. $\dfrac{(x + 3)^2}{16} - \dfrac{(y + 4)^2}{25} = 1$ 17. $\dfrac{(x - 2)^2}{4} + \dfrac{(y + 1)^2}{9} = 1$

18. Find the standard form of the equation of the parabola shown in the figure below.

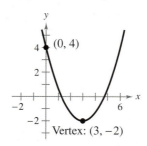

Vertex: $(3, -2)$

19. Find the standard form of the equation of the hyperbola whose vertices are $(-1, -5)$ and $(-1, 1)$, and whose foci are $(-1, -7)$ and $(-1, 3)$.

In Exercises 20 and 21, describe the transformations of the graph of f that yield the graph of g.

20. $f(x) = \left(\frac{2}{5}\right)^x$, $g(x) = -\left(\frac{2}{5}\right)^{-x+3}$

21. $f(x) = 2.2^x$, $g(x) = -2.2^x + 4$

In Exercises 22–25, use a calculator to evaluate the expression. Round your result to three decimal places.

22. $\log 98$

23. $\log \frac{6}{7}$

24. $\ln \sqrt{31}$

25. $\ln\left(\sqrt{30} - 4\right)$

In Exercises 26–28, evaluate the logarithm using the change-of-base formula. Round your result to three decimal places.

26. $\log_5 4.3$

27. $\log_3 0.149$

28. $\log_{1/2} 17$

29. Use the properties of logarithms to expand $\ln\left(\dfrac{x^2 - 25}{x^4}\right)$, where $x > 5$.

30. Condense $2 \ln x - \frac{1}{2} \ln(x + 5)$ to the logarithm of a single quantity.

In Exercises 31–36, solve the equation algebraically. Approximate the result to three decimal places.

31. $6e^{2x} = 72$

32. $4^{x-5} + 21 = 30$

33. $e^{2x} - 13e^x + 42 = 0$

34. $\log_2 x + \log_2 5 = 6$

35. $\ln 4x - \ln 2 = 8$

36. $\ln \sqrt{x + 2} = 3$

37. On the day a grandchild is born, a grandparent deposits $2500 in a fund earning 7.5% interest, compounded continuously. Determine the balance in the account on the grandchild's 25th birthday.

38. The number N of bacteria in a culture is given by the model $N = 175e^{kt}$, where t is the time in hours. Given that $N = 420$ when $t = 8$, estimate the time required for the population to double in size.

39. The population P (in millions) of Texas from 2001 through 2014 can be approximated by the model $P = 20.913e^{0.0184t}$, where t represents the year, with $t = 1$ corresponding to 2001. According to this model, when will the population reach 32 million? *(Source: U.S. Census Bureau)*

40. The population p of a species of bird t years after it is introduced into a new habitat is given by

$$p = \frac{1200}{1 + 3e^{-t/5}}.$$

(a) Determine the population size that was introduced into the habitat.

(b) Determine the population after 5 years.

(c) After how many years will the population be 800?

Proofs in Mathematics ■ ■ ■ ■ ■ ■ ■ ■ ■ ■ ■ ■ ■ ■

Each of the three properties of logarithms listed below can be proved by using properties of exponential functions.

Properties of Logarithms (p. 382)

Let a be a positive number such that $a \neq 1$, let n be a real number, and let u and v be positive real numbers.

	Logarithm with Base a	**Natural Logarithm**
1. **Product Property:**	$\log_a(uv) = \log_a u + \log_a v$	$\ln(uv) = \ln u + \ln v$
2. **Quotient Property:**	$\log_a \dfrac{u}{v} = \log_a u - \log_a v$	$\ln \dfrac{u}{v} = \ln u - \ln v$
3. **Power Property:**	$\log_a u^n = n \log_a u$	$\ln u^n = n \ln u$

Proof

Let

$$x = \log_a u \quad \text{and} \quad y = \log_a v.$$

The corresponding exponential forms of these two equations are

$$a^x = u \quad \text{and} \quad a^y = v.$$

To prove the Product Property, multiply u and v to obtain

$$uv = a^x a^y$$
$$= a^{x+y}.$$

The corresponding logarithmic form of $uv = a^{x+y}$ is $\log_a(uv) = x + y$. So,

$$\log_a(uv) = \log_a u + \log_a v.$$

To prove the Quotient Property, divide u by v to obtain

$$\frac{u}{v} = \frac{a^x}{a^y}$$
$$= a^{x-y}.$$

The corresponding logarithmic form of $\dfrac{u}{v} = a^{x-y}$ is $\log_a \dfrac{u}{v} = x - y$. So,

$$\log_a \frac{u}{v} = \log_a u - \log_a v.$$

To prove the Power Property, substitute a^x for u in the expression $\log_a u^n$.

$$\log_a u^n = \log_a (a^x)^n \qquad \text{Substitute } a^x \text{ for } u.$$
$$= \log_a a^{nx} \qquad \text{Property of Exponents}$$
$$= nx \qquad \text{Inverse Property}$$
$$= n \log_a u \qquad \text{Substitute } \log_a u \text{ for } x.$$

So, $\log_a u^n = n \log_a u$.

P.S. Problem Solving ■ ■ ■ ■ ■ ■ ■ ■ ■ ■ ■ ■ ■ 📕

1. **Graphical Reasoning** Graph the exponential function $y = a^x$ for $a = 0.5$, 1.2, and 2.0. Which of these curves intersects the line $y = x$? Determine all positive numbers a for which the curve $y = a^x$ intersects the line $y = x$.

2. **Graphical Reasoning** Use a graphing utility to graph each of the functions $y_1 = e^x$, $y_2 = x^2$, $y_3 = x^3$, $y_4 = \sqrt{x}$, and $y_5 = |x|$. Which function increases at the greatest rate as x approaches ∞?

3. **Conjecture** Use the result of Exercise 2 to make a conjecture about the rate of growth of $y_1 = e^x$ and $y = x^n$, where n is a natural number and x approaches ∞.

4. **Implication of "Growing Exponentially"** Use the results of Exercises 2 and 3 to describe what is implied when it is stated that a quantity is growing exponentially.

5. **Exponential Function** Given the exponential function

   $$f(x) = a^x$$

   show that
   (a) $f(u + v) = f(u) \cdot f(v)$ and (b) $f(2x) = [f(x)]^2$.

6. **Hyperbolic Functions** Given that

   $$f(x) = \frac{e^x + e^{-x}}{2} \quad \text{and} \quad g(x) = \frac{e^x - e^{-x}}{2}$$

   show that

   $$[f(x)]^2 - [g(x)]^2 = 1.$$

7. **Graphical Reasoning** Use a graphing utility to compare the graph of the function $y = e^x$ with the graph of each function. [$n!$ (read "n factorial") is defined as $n! = 1 \cdot 2 \cdot 3 \cdots (n - 1) \cdot n$.]

   (a) $y_1 = 1 + \dfrac{x}{1!}$

   (b) $y_2 = 1 + \dfrac{x}{1!} + \dfrac{x^2}{2!}$

   (c) $y_3 = 1 + \dfrac{x}{1!} + \dfrac{x^2}{2!} + \dfrac{x^3}{3!}$

8. **Identifying a Pattern** Identify the pattern of successive polynomials given in Exercise 7. Extend the pattern one more term and compare the graph of the resulting polynomial function with the graph of $y = e^x$. What do you think this pattern implies?

9. **Finding an Inverse Function** Graph the function

   $$f(x) = e^x - e^{-x}.$$

   From the graph, the function appears to be one-to-one. Assume that f has an inverse function and find $f^{-1}(x)$.

10. **Finding a Pattern for an Inverse Function** Find a pattern for $f^{-1}(x)$ when

    $$f(x) = \frac{a^x + 1}{a^x - 1}$$

    where $a > 0$, $a \neq 1$.

11. **Determining the Equation of a Graph** Determine whether the graph represents equation (a), (b), or (c). Explain your reasoning.

    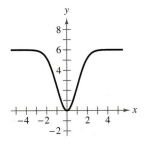

    (a) $y = 6e^{-x^2/2}$

    (b) $y = \dfrac{6}{1 + e^{-x/2}}$

    (c) $y = 6(1 - e^{-x^2/2})$

12. **Simple and Compound Interest** You have two options for investing \$500. The first earns 7% interest compounded annually, and the second earns 7% simple interest. The figure shows the growth of each investment over a 30-year period.

    (a) Determine which graph represents each type of investment. Explain your reasoning.

    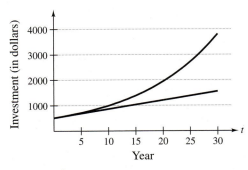

    (b) Verify your answer in part (a) by finding the equations that model the investment growth and by graphing the models.

    (c) Which option would you choose? Explain.

13. **Radioactive Decay** Two different samples of radioactive isotopes are decaying. The isotopes have initial amounts of c_1 and c_2 and half-lives of k_1 and k_2, respectively. Find an expression for the time t required for the samples to decay to equal amounts.

14. Bacteria Decay   A lab culture initially contains 500 bacteria. Two hours later, the number of bacteria decreases to 200. Find the exponential decay model of the form

$$B = B_0 a^{kt}$$

that approximates the number of bacteria B in the culture after t hours.

15. Colonial Population   The table shows the colonial population estimates of the American colonies for each decade from 1700 through 1780.   *(Source: U.S. Census Bureau)*

Year	Population
1700	250,900
1710	331,700
1720	466,200
1730	629,400
1740	905,600
1750	1,170,800
1760	1,593,600
1770	2,148,100
1780	2,780,400

DATA — Spreadsheet at LarsonPrecalculus.com

Let y represent the population in the year t, with $t = 0$ corresponding to 1700.

(a) Use the *regression* feature of a graphing utility to find an exponential model for the data.

(b) Use the *regression* feature of the graphing utility to find a quadratic model for the data.

(c) Use the graphing utility to plot the data and the models from parts (a) and (b) in the same viewing window.

(d) Which model is a better fit for the data? Would you use this model to predict the population of the United States in 2020? Explain your reasoning.

16. Ratio of Logarithms   Show that

$$\frac{\log_a x}{\log_{a/b} x} = 1 + \log_a \frac{1}{b}.$$

17. Solving a Logarithmic Equation   Solve

$(\ln x)^2 = \ln x^2.$

18. Graphical Reasoning   Use a graphing utility to compare the graph of each function with the graph of $y = \ln x$.

(a) $y_1 = x - 1$

(b) $y_2 = (x - 1) - \frac{1}{2}(x - 1)^2$

(c) $y_3 = (x - 1) - \frac{1}{2}(x - 1)^2 + \frac{1}{3}(x - 1)^3$

19. Identifying a Pattern   Identify the pattern of successive polynomials given in Exercise 18. Extend the pattern one more term and compare the graph of the resulting polynomial function with the graph of $y = \ln x$. What do you think the pattern implies?

20. Finding Slope and y-Intercept   Take the natural log of each side of each equation below.

$$y = ab^x, \quad y = ax^b$$

(a) What are the slope and y-intercept of the line relating x and $\ln y$ for $y = ab^x$?

(b) What are the slope and y-intercept of the line relating $\ln x$ and $\ln y$ for $y = ax^b$?

Ventilation Rate   In Exercises 21 and 22, use the model

$$y = 80.4 - 11 \ln x, \quad 100 \le x \le 1500$$

which approximates the minimum required ventilation rate in terms of the air space per child in a public school classroom. In the model, x is the air space (in cubic feet) per child and y is the ventilation rate (in cubic feet per minute) per child.

21. Use a graphing utility to graph the model and approximate the required ventilation rate when there are 300 cubic feet of air space per child.

22. In a classroom designed for 30 students, the air conditioning system can move 450 cubic feet of air per minute.

(a) Determine the ventilation rate per child in a full classroom.

(b) Estimate the air space required per child.

(c) Determine the minimum number of square feet of floor space required for the room when the ceiling height is 30 feet.

Using Technology   In Exercises 23–26, (a) use a graphing utility to create a scatter plot of the data, (b) decide whether the data could best be modeled by a linear model, an exponential model, or a logarithmic model, (c) explain why you chose the model you did in part (b), (d) use the *regression* feature of the graphing utility to find the model you chose in part (b) for the data and graph the model with the scatter plot, and (e) determine how well the model you chose fits the data.

23. $(1, 2.0), (1.5, 3.5), (2, 4.0), (4, 5.8), (6, 7.0), (8, 7.8)$

24. $(1, 4.4), (1.5, 4.7), (2, 5.5), (4, 9.9), (6, 18.1), (8, 33.0)$

25. $(1, 7.5), (1.5, 7.0), (2, 6.8), (4, 5.0), (6, 3.5), (8, 2.0)$

26. $(1, 5.0), (1.5, 6.0), (2, 6.4), (4, 7.8), (6, 8.6), (8, 9.0)$

6 Systems of Equations and Inequalities

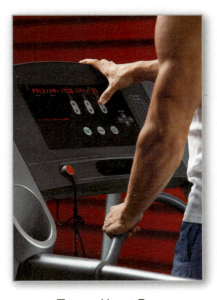

Target Heart Rate
(Exercise 68, page 472)

Thermodynamics *(Exercise 60, page 463)*

Global Positioning System *(page 449)*

Fuel Mixture
(Exercise 50, page 441)

Environmental Science *(Exercise 62, page 431)*

6.1 Linear and Nonlinear Systems of Equations

Graphs of systems of equations can help you solve real-life problems. For example, in Exercise 62 on page 431, you will use the graph of a system of equations to compare the consumption of wind energy and the consumption of geothermal energy.

■ Use the method of substitution to solve systems of linear equations in two variables.
■ Use the method of substitution to solve systems of nonlinear equations in two variables.
■ Use a graphical method to solve systems of equations in two variables.
■ Use systems of equations to model and solve real-life problems.

The Method of Substitution

Up to this point in the text, most problems have involved either a function of one variable or a single equation in two variables. However, many problems in science, business, and engineering involve two or more equations in two or more variables. To solve such a problem, you need to find the solutions of a **system of equations.** Here is an example of a system of two equations in two variables, x and y.

$$\begin{cases} 2x + y = 5 & \text{Equation 1} \\ 3x - 2y = 4 & \text{Equation 2} \end{cases}$$

A **solution** of this system is an ordered pair that satisfies each equation in the system. Finding the set of all solutions is called **solving the system of equations.** For example, the ordered pair $(2, 1)$ is a solution of this system. To check this, substitute 2 for x and 1 for y in *each* equation.

Check (2, 1) in Equation 1 and Equation 2:

$$2x + y = 5 \qquad \text{Write Equation 1.}$$
$$2(2) + 1 \overset{?}{=} 5 \qquad \text{Substitute 2 for } x \text{ and 1 for } y.$$
$$4 + 1 = 5 \qquad \text{Solution checks in Equation 1.} \checkmark$$
$$3x - 2y = 4 \qquad \text{Write Equation 2.}$$
$$3(2) - 2(1) \overset{?}{=} 4 \qquad \text{Substitute 2 for } x \text{ and 1 for } y.$$
$$6 - 2 = 4 \qquad \text{Solution checks in Equation 2.} \checkmark$$

In this chapter, you will study four ways to solve systems of equations, beginning with the **method of substitution.**

Method	Section	Type of System
1. Substitution	6.1	Linear or nonlinear, two variables
2. Graphical method	6.1	Linear or nonlinear, two variables
3. Elimination	6.2	Linear, two variables
4. Gaussian elimination	6.3	Linear, three or more variables

Method of Substitution

1. *Solve* one of the equations for one variable in terms of the other.
2. *Substitute* the expression found in Step 1 into the other equation to obtain an equation in one variable.
3. *Solve* the equation obtained in Step 2.
4. *Back-substitute* the value obtained in Step 3 into the expression obtained in Step 1 to find the value of the other variable.
5. *Check* that the solution satisfies *each* of the original equations.

EXAMPLE 1 **Solving a System of Equations by Substitution**

Solve the system of equations.

$$\begin{cases} x + y = 4 & \text{Equation 1} \\ x - y = 2 & \text{Equation 2} \end{cases}$$

Solution Begin by solving for y in Equation 1.

$$y = 4 - x \qquad\qquad \text{Solve for } y \text{ in Equation 1.}$$

Next, substitute this expression for y into Equation 2 and solve the resulting single-variable equation for x.

$x - y = 2$	Write Equation 2.
$x - (4 - x) = 2$	Substitute $4 - x$ for y.
$x - 4 + x = 2$	Distributive Property
$2x = 6$	Combine like terms.
$x = 3$	Divide each side by 2.

Finally, solve for y by *back-substituting* $x = 3$ into the equation $y = 4 - x$.

$y = 4 - x$	Write revised Equation 1.
$y = 4 - 3$	Substitute 3 for x.
$y = 1$	Solve for y.

So, the solution of the system is the ordered pair

$$(3, 1).$$

Check this solution as follows.

Check

Substitute $(3, 1)$ into Equation 1:

$x + y = 4$	Write Equation 1.
$3 + 1 \stackrel{?}{=} 4$	Substitute for x and y.
$4 = 4$	Solution checks in Equation 1. ✔

Substitute $(3, 1)$ into Equation 2:

$x - y = 2$	Write Equation 2.
$3 - 1 \stackrel{?}{=} 2$	Substitute for x and y.
$2 = 2$	Solution checks in Equation 2. ✔

The point $(3, 1)$ satisfies both equations in the system. This confirms that $(3, 1)$ is a solution of the system of equations.

✔ *Checkpoint* *Audio-video solution in English & Spanish at LarsonPrecalculus.com*

Solve the system of equations.

$$\begin{cases} x - y = 0 \\ 5x - 3y = 6 \end{cases}$$

The term *back-substitution* implies that you work *backwards*. First you solve for one of the variables, and then you substitute that value *back* into one of the equations in the system to find the value of the other variable.

REMARK Many steps are required to solve a system of equations, so there are many possible ways to make errors in arithmetic. You should always check your solution by substituting it into *each* equation in the original system.

EXAMPLE 2 **Solving a System by Substitution**

A total of $12,000 is invested in two funds paying 5% and 3% simple interest. The total annual interest is $500. How much is invested at each rate?

Solution

Recall that the formula for simple interest is $I = Prt$, where P is the principal, r is the annual interest rate (in decimal form), and t is the time.

Verbal model:

Amount in 5% fund	+	Amount in 3% fund	=	Total investment

Interest for 5% fund	+	Interest for 3% fund	=	Total interest

Labels:
Amount in 5% fund $= x$ (dollars)
Interest for 5% fund $= 0.05x$ (dollars)
Amount in 3% fund $= y$ (dollars)
Interest for 3% fund $= 0.03y$ (dollars)
Total investment $= 12,000$ (dollars)
Total interest $= 500$ (dollars)

System:
$$\begin{cases} x + y = 12{,}000 & \text{Equation 1} \\ 0.05x + 0.03y = 500 & \text{Equation 2} \end{cases}$$

•• **REMARK** In Example 2, note that you would obtain the same solution by first solving for y in Equation 1 or solving for x or y in Equation 2.

To begin, it is convenient to multiply each side of Equation 2 by 100. This eliminates the need to work with decimals.

$$100(0.05x + 0.03y) = 100(500)$$ Multiply each side of Equation 2 by 100.

$$5x + 3y = 50{,}000$$ Revised Equation 2.

To solve this system, you can solve for x in Equation 1.

$$x = 12{,}000 - y$$ Revised Equation 1.

Then, substitute this expression for x into revised Equation 2 and solve for y.

$$5(12{,}000 - y) + 3y = 50{,}000$$ Substitute $12{,}000 - y$ for x in revised Equation 2.

$$60{,}000 - 5y + 3y = 50{,}000$$ Distributive Property

$$-2y = -10{,}000$$ Combine like terms.

$$y = 5000$$ Divide each side by -2.

Next, back-substitute $y = 5000$ to solve for x.

$$x = 12{,}000 - y$$ Write revised Equation 1.

$$x = 12{,}000 - 5000$$ Substitute 5000 for y.

$$x = 7000$$ Subtract.

▷ **TECHNOLOGY** One way to check the answers you obtain in this section is to use a graphing utility. For instance, graph the two equations in Example 2

$$y_1 = 12{,}000 - x$$

$$y_2 = \frac{500 - 0.05x}{0.03}$$

and find the point of intersection. Does this point agree with the solution obtained at the right?

The solution is $(7000, 5000)$. So, $7000 is invested at 5% and $5000 is invested at 3%. Check this in the original system.

✓ *Checkpoint*))) *Audio-video solution in English & Spanish at LarsonPrecalculus.com*

A total of $25,000 is invested in two funds paying 6.5% and 8.5% simple interest. The total annual interest is $2000. How much is invested at each rate?

Nonlinear Systems of Equations

The equations in Examples 1 and 2 are linear. The method of substitution can also be used to solve systems in which one or both of the equations are nonlinear.

EXAMPLE 3 **Substitution: Two-Solution Case**

Solve the system of equations.

$$\begin{cases} 3x^2 + 4x - y = 7 & \text{Equation 1} \\ 2x - y = -1 & \text{Equation 2} \end{cases}$$

Solution Begin by solving for y in Equation 2 to obtain $y = 2x + 1$. Next, substitute this expression for y into Equation 1 and solve for x.

$$3x^2 + 4x - (2x + 1) = 7 \qquad \text{Substitute } 2x + 1 \text{ for } y \text{ in Equation 1.}$$

$$3x^2 + 2x - 1 = 7 \qquad \text{Simplify.}$$

$$3x^2 + 2x - 8 = 0 \qquad \text{Write in general form.}$$

$$(3x - 4)(x + 2) = 0 \qquad \text{Factor.}$$

$$x = \frac{4}{3}, -2 \qquad \text{Solve for } x.$$

▷ **ALGEBRA HELP** To review the techniques for factoring, see Section P.4.

Back-substituting these values of x to solve for the corresponding values of y produces the solutions $\left(\frac{4}{3}, \frac{11}{3}\right)$ and $(-2, -3)$. Check these in the original system.

✓ *Checkpoint*  *Audio-video solution in English & Spanish at LarsonPrecalculus.com*

Solve the system of equations.

$$\begin{cases} -2x + y = 5 \\ x^2 - y + 3x = 1 \end{cases}$$

EXAMPLE 4 **Substitution: No-Real-Solution Case**

Solve the system of equations.

$$\begin{cases} -x + y = 4 & \text{Equation 1} \\ x^2 + y = 3 & \text{Equation 2} \end{cases}$$

Solution Begin by solving for y in Equation 1 to obtain $y = x + 4$. Next, substitute this expression for y into Equation 2 and solve for x.

$$x^2 + (x + 4) = 3 \qquad \text{Substitute } x + 4 \text{ for } y \text{ in Equation 2.}$$

$$x^2 + x + 1 = 0 \qquad \text{Write in general form.}$$

$$x = \frac{-1 \pm \sqrt{-3}}{2} \qquad \text{Use the Quadratic Formula.}$$

Because the discriminant is negative, the equation

$$x^2 + x + 1 = 0$$

has no real solution. So, the original system has no real solution.

✓ *Checkpoint*  *Audio-video solution in English & Spanish at LarsonPrecalculus.com*

Solve the system of equations.

$$\begin{cases} 2x - y = -3 \\ 2x^2 + 4x - y^2 = 0 \end{cases}$$

▷ **TECHNOLOGY** Most graphing utilities have built-in features for approximating the point(s) of intersection of two graphs. Use a graphing utility to find the points of intersection of the graphs in Figures 6.1 through 6.3. Be sure to adjust the viewing window so that you see all the points of intersection.

Graphical Method for Finding Solutions

Notice from Examples 2, 3, and 4 that a system of two equations in two variables can have exactly one solution, more than one solution, or no solution. A **graphical method** helps you to gain insight about the number and location(s) of the solution(s) of a system of equations. When you graph each of the equations in the same coordinate plane, the solutions of the system correspond to the **points of intersection** of the graphs. For example, the two equations in Figure 6.1 graph as two lines with a *single point* of intersection; the two equations in Figure 6.2 graph as a parabola and a line with *two points* of intersection; and the two equations in Figure 6.3 graph as a parabola and a line with *no points* of intersection.

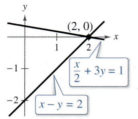

One intersection point
Figure 6.1

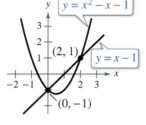

Two intersection points
Figure 6.2

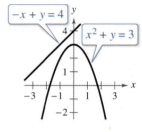

No intersection points
Figure 6.3

EXAMPLE 5   **Solving a System of Equations Graphically**

See LarsonPrecalculus.com for an interactive version of this type of example.

Solve the system of equations.

$$\begin{cases} y = \ln x & \text{Equation 1} \\ x + y = 1 & \text{Equation 2} \end{cases}$$

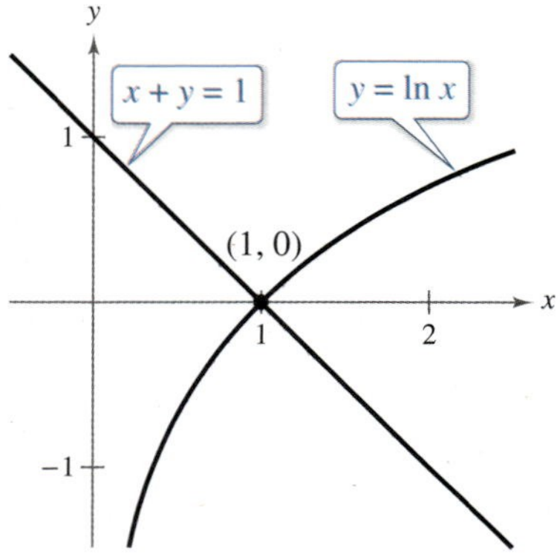

Solution   There is only one point of intersection of the graphs of the two equations, and $(1, 0)$ is the solution point (see Figure 6.4). Check this solution as follows.

Check (1, 0) in Equation 1:

$y = \ln x$	Write Equation 1.
$0 = \ln 1$	Substitute for x and y.
$0 = 0$	Solution checks in Equation 1. ✓

Check (1, 0) in Equation 2:

$x + y = 1$	Write Equation 2.
$1 + 0 = 1$	Substitute for x and y.
$1 = 1$	Solution checks in Equation 2. ✓

✓ *Checkpoint* 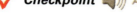 *Audio-video solution in English & Spanish at LarsonPrecalculus.com*

Solve the system of equations.

$$\begin{cases} y = 3 - \log x \\ -2x + y = 1 \end{cases}$$

Example 5 shows the benefit of solving systems of equations in two variables graphically. Note that using the substitution method in Example 5 produces $x + \ln x = 1$. It is difficult to solve this equation for x using standard algebraic techniques.

Figure 6.4

Applications

The total cost C of producing x units of a product typically has two components—the initial cost and the cost per unit. When enough units have been sold so that the total revenue R equals the total cost C, the sales are said to have reached the **break-even point.** The break-even point corresponds to the point of intersection of the cost and revenue curves.

EXAMPLE 6   **Break-Even Analysis**

A shoe company invests $300,000 in equipment to produce a new line of athletic footwear. Each pair of shoes costs $15 to produce and sells for $70. How many pairs of shoes must the company sell to break even?

Solution

The total cost of producing x units is

Total cost	=	Cost per unit	·	Number of units	+	Initial cost

$$C = 15x + 300,000. \qquad \text{Equation 1}$$

The revenue obtained by selling x units is

Total revenue	=	Price per unit	·	Number of units

$$R = 70x. \qquad \text{Equation 2}$$

The break-even point occurs when $R = C$, so you have $C = 70x$. This gives you the system of equations below to solve.

$$\begin{cases} C = 15x + 300,000 \\ C = 70x \end{cases}$$

Solve by substitution.

$$70x = 15x + 300,000 \qquad \text{Substitute } 70x \text{ for } C \text{ in Equation 1.}$$

$$55x = 300,000 \qquad \text{Subtract } 15x \text{ from each side.}$$

$$x \approx 5455 \qquad \text{Divide each side by 55.}$$

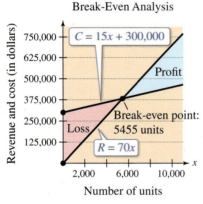

Break-Even Analysis

Figure 6.5

The company must sell about 5455 pairs of shoes to break even. Note in Figure 6.5 that revenue less than the break-even point corresponds to an overall loss, whereas revenue greater than the break-even point corresponds to a profit. Verify the break-even point using the *intersect* feature or the *zoom* and *trace* features of a graphing utility.

✓ *Checkpoint* 🔊))) *Audio-video solution in English & Spanish at LarsonPrecalculus.com*

In Example 6, each pair of shoes costs $12 to produce. How many pairs of shoes must the company sell to break even?

Another way to view the solution in Example 6 is to consider the profit function

$$P = R - C.$$

The break-even point occurs when the profit P is 0, which is the same as saying that

$$R = C.$$

EXAMPLE 7 **Movie Ticket Sales**

Two new movies, a comedy and a drama, are released in the same week. In the first six weeks, the weekly ticket sales S (in millions of dollars) decrease for the comedy and increase for the drama according to the models

$$\begin{cases} S = 60 - 8x & \text{Comedy (Equation 1)} \\ S = 10 + 4.5x & \text{Drama (Equation 2)} \end{cases}$$

where x represents the time (in weeks), with $x = 1$ corresponding to the first week of release. According to the models, in what week are the ticket sales of the two movies equal?

Algebraic Solution

Both equations are already solved for S in terms of x, so substitute the expression for S from Equation 2 into Equation 1 and solve for x.

$10 + 4.5x = 60 - 8x$ Substitute for S in Equation 1.

$4.5x + 8x = 60 - 10$ Add $8x$ and -10 to each side.

$12.5x = 50$ Combine like terms.

$x = 4$ Divide each side by 12.5.

According to the models, the weekly ticket sales for the two movies are equal in the fourth week.

Numerical Solution

Create a table of values for each model.

Number of Weeks, x	1	2	3	4	5	6
Sales, S (comedy)	52	44	36	28	20	12
Sales, S (drama)	14.5	19	23.5	28	32.5	37

According to the table, the weekly ticket sales for the two movies are equal in the fourth week.

✓ *Checkpoint* *Audio-video solution in English & Spanish at LarsonPrecalculus.com*

Two new movies, an animated movie and a horror movie, are released in the same week. In the first eight weeks, the weekly ticket sales S (in millions of dollars) decrease for the animated movie and increase for the horror movie according to the models

$$\begin{cases} S = 108 - 9.4x & \text{Animated} \\ S = 16 + 9x & \text{Horror} \end{cases}$$

where x represents the time (in weeks), with $x = 1$ corresponding to the first week of release. According to the models, in what week are the ticket sales of the two movies equal?

Summarize (Section 6.1)

1. Explain how to use the method of substitution to solve a system of linear equations in two variables *(page 422)*. For examples of using the method of substitution to solve systems of linear equations in two variables, see Examples 1 and 2.

2. Explain how to use the method of substitution to solve a system of nonlinear equations in two variables *(page 425)*. For examples of using the method of substitution to solve systems of nonlinear equations in two variables, see Examples 3 and 4.

3. Explain how to use a graphical method to solve a system of equations in two variables *(page 426)*. For an example of using a graphical approach to solve a system of equations in two variables, see Example 5.

4. Describe examples of how to use systems of equations to model and solve real-life problems *(pages 427 and 428, Examples 6 and 7)*.

6.1 Exercises

See **CalcChat.com** for tutorial help and worked-out solutions to odd-numbered exercises.

Vocabulary: Fill in the blanks.

1. A _____ of a system of equations is an ordered pair that satisfies each equation in the system.

2. The first step in solving a system of equations by the method of _____ is to solve one of the equations for one variable in terms of the other.

3. Graphically, solutions of a system of two equations correspond to the _____ of _____ of the graphs of the two equations.

4. In business applications, the total revenue equals the total cost at the _____ point.

Skills and Applications

Checking Solutions In Exercises 5 and 6, determine whether each ordered pair is a solution of the system.

5. $\begin{cases} 2x - y = 4 \\ 8x + y = -9 \end{cases}$ (a) $(0, -4)$ (b) $(3, -1)$ (c) $\left(\frac{3}{2}, -1\right)$ (d) $\left(-\frac{1}{2}, -5\right)$

6. $\begin{cases} 4x^2 + y = 3 \\ -x - y = 11 \end{cases}$ (a) $(2, -13)$ (b) $(1, -2)$ (c) $\left(-\frac{3}{2}, -\frac{31}{3}\right)$ (d) $\left(-\frac{7}{4}, -\frac{37}{4}\right)$

Solving a System by Substitution In Exercises 7–14, solve the system by the method of substitution. Check your solution(s) graphically.

7. $\begin{cases} 2x + y = 6 \\ -x + y = 0 \end{cases}$

8. $\begin{cases} x - 4y = -11 \\ x + 3y = 3 \end{cases}$

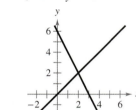

9. $\begin{cases} x - y = -4 \\ x^2 - y = -2 \end{cases}$

10. $\begin{cases} 3x + y = 2 \\ x^3 - 2 + y = 0 \end{cases}$

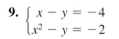

11. $\begin{cases} x^2 + y = 0 \\ x^2 - 4x - y = 0 \end{cases}$

12. $\begin{cases} x + y = 0 \\ x^3 - 5x - y = 0 \end{cases}$

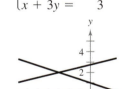
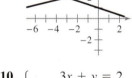

13. $\begin{cases} y = x^3 - 3x^2 + 1 \\ y = x^2 - 3x + 1 \end{cases}$

14. $\begin{cases} -\frac{1}{2}x + y = -\frac{5}{2} \\ x^2 + y^2 = 25 \end{cases}$

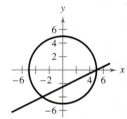

Solving a System by Substitution In Exercises 15–24, solve the system by the method of substitution.

15. $\begin{cases} x - y = 2 \\ 6x - 5y = 16 \end{cases}$

16. $\begin{cases} 2x + y = 9 \\ 3x - 5y = 20 \end{cases}$

17. $\begin{cases} 2x - y + 2 = 0 \\ 4x + y - 5 = 0 \end{cases}$

18. $\begin{cases} 6x - 3y - 4 = 0 \\ x + 2y - 4 = 0 \end{cases}$

19. $\begin{cases} 1.5x + 0.8y = 2.3 \\ 0.3x - 0.2y = 0.1 \end{cases}$

20. $\begin{cases} 0.5x + y = -3.5 \\ x - 3.2y = 3.4 \end{cases}$

21. $\begin{cases} \frac{1}{5}x + \frac{1}{2}y = 8 \\ x + y = 20 \end{cases}$

22. $\begin{cases} \frac{1}{2}x + \frac{3}{4}y = 10 \\ \frac{3}{4}x - y = 4 \end{cases}$

23. $\begin{cases} 6x + 5y = -3 \\ -x - \frac{5}{6}y = -7 \end{cases}$

24. $\begin{cases} -\frac{2}{3}x + y = 2 \\ 2x - 3y = 6 \end{cases}$

Solving a System by Substitution In Exercises 25–28, the given amount of annual interest is earned from a total of $12,000 invested in two funds paying simple interest. Write and solve a system of equations to find the amount invested at each given rate.

	Annual Interest	Rate 1	Rate 2
25.	$500	2%	6%
26.	$630	4%	7%
27.	$396	2.8%	3.8%
28.	$254	1.75%	2.25%

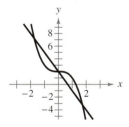
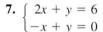
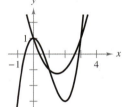
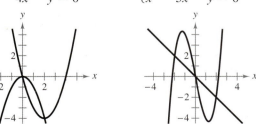

Solving a System with a Nonlinear Equation In Exercises 29–32, solve the system by the method of substitution.

29. $\begin{cases} x^2 - y = 0 \\ 2x + y = 0 \end{cases}$
30. $\begin{cases} x - 2y = 0 \\ 3x - y^2 = 0 \end{cases}$

31. $\begin{cases} x - y = -1 \\ x^2 - y = -4 \end{cases}$
32. $\begin{cases} y = -x \\ y = x^3 + 3x^2 + 2x \end{cases}$

Solving a System of Equations Graphically In Exercises 33–42, solve the system graphically.

33. $\begin{cases} -x + 2y = -2 \\ 3x + y = 20 \end{cases}$
34. $\begin{cases} x + y = 0 \\ 2x - 7y = -18 \end{cases}$

35. $\begin{cases} x - 3y = -3 \\ 5x + 3y = -6 \end{cases}$
36. $\begin{cases} -x + 2y = -7 \\ x - y = 2 \end{cases}$

37. $\begin{cases} x + y = 4 \\ x^2 + y^2 - 4x = 0 \end{cases}$
38. $\begin{cases} -x + y = 3 \\ x^2 - 6x - 27 + y^2 = 0 \end{cases}$

39. $\begin{cases} 3x - 2y = 0 \\ x^2 - y^2 = 4 \end{cases}$
40. $\begin{cases} 2x - y + 3 = 0 \\ x^2 + y^2 - 4x = 0 \end{cases}$

41. $\begin{cases} x^2 + y^2 = 25 \\ 3x^2 - 16y = 0 \end{cases}$
42. $\begin{cases} x^2 + y^2 = 25 \\ (x - 8)^2 + y^2 = 25 \end{cases}$

Using Technology In Exercises 43–46, use a graphing utility to solve the system of equations graphically. Round your solution(s) to two decimal places, if necessary.

43. $\begin{cases} y = e^x \\ x - y + 1 = 0 \end{cases}$
44. $\begin{cases} y = -4e^{-x} \\ y + 3x + 8 = 0 \end{cases}$

45. $\begin{cases} y + 2 = \ln(x - 1) \\ 3y + 2x = 9 \end{cases}$
46. $\begin{cases} x^2 + y^2 = 4 \\ 2x^2 - y = 2 \end{cases}$

Choosing a Solution Method In Exercises 47–54, solve the system graphically or algebraically. Explain your choice of method.

47. $\begin{cases} y = 2x \\ y = x^2 + 1 \end{cases}$
48. $\begin{cases} x^2 + y^2 = 9 \\ x - y = -3 \end{cases}$

49. $\begin{cases} x - 2y = 4 \\ x^2 - y = 0 \end{cases}$
50. $\begin{cases} y = x^3 - 2x^2 + x - 1 \\ y = -x^2 + 3x - 1 \end{cases}$

51. $\begin{cases} y - e^{-x} = 1 \\ y - \ln x = 3 \end{cases}$
52. $\begin{cases} x^2 + y = 4 \\ e^x - y = 0 \end{cases}$

53. $\begin{cases} xy - 1 = 0 \\ 2x - 4y + 7 = 0 \end{cases}$
54. $\begin{cases} x - 2y = 1 \\ y = \sqrt{x - 1} \end{cases}$

Break-Even Analysis In Exercises 55 and 56, use the equations for the total cost C and total revenue R to find the number x of units a company must sell to break even. (Round to the nearest whole unit.)

55. $C = 8650x + 250{,}000, \quad R = 9502x$

56. $C = 5.5\sqrt{x} + 10{,}000, \quad R = 4.22x$

57. **Break-Even Analysis** A small software development company invests $16,000 to produce a software package that will sell for $55.95. Each unit costs $9.45 to produce.

(a) How many units must the company sell to break even?

(b) How many units must the company sell to make a profit of $100,000?

58. **Choice of Two Jobs** You receive two sales job offers. One company offers a straight commission of 6% of sales. The other company offers a salary of $500 per week plus 3% of sales. How much would you have to sell per week in order to make the straight commission job offer better?

59. **DVD Rentals** Two new DVDs, a horror film and a comedy film, are released in the same week. The weekly number N of rentals decreases for the horror film and increases for the comedy film according to the models

$$\begin{cases} N = 360 - 24x & \text{Horror film} \\ N = 24 + 18x & \text{Comedy film} \end{cases}$$

where x represents the time (in weeks), with $x = 1$ corresponding to the first week of release.

(a) After how many weeks will the numbers of DVDs rented for the two films be equal?

(b) Use a table to solve the system of equations numerically. Compare your result with that of part (a).

60. **Supply and Demand** The supply and demand curves for a business dealing with wheat are

Supply: $p = 1.45 + 0.00014x^2$

Demand: $p = (2.388 - 0.07x)^2$

where p is the price (in dollars) per bushel and x is the quantity in bushels per day. Use a graphing utility to graph the supply and demand equations and find the market equilibrium. (The market equilibrium is the point of intersection of the graphs for $x > 0$.)

61. **Error Analysis** Describe the error in solving the system of equations.

$$\begin{cases} x^2 + 2x - y = 3 \\ 2x - y = 2 \end{cases}$$

$$x^2 + 2x - (-2x + 2) = 3$$
$$x^2 + 4x - 2 = 3$$
$$x^2 + 4x - 5 = 0$$
$$(x + 5)(x - 1) = 0$$
$$x = -5, 1$$

When $x = -5$, $y = -2(-5) + 2 = -8$, and when $x = 1$, $y = -2(1) + 2 = 0$.
So, the solutions are $(-5, -8)$ and $(1, 0)$.

62. Environmental Science

The table shows the consumption C (in trillions of Btus) of geothermal energy and wind energy in the United States from 2004 through 2014. *(Source: U.S. Energy Information Administration)*

DATA	Year	Geothermal, C	Wind, C
	2004	178	142
	2005	181	178
	2006	181	264
	2007	186	341
	2008	192	546
	2009	200	721
	2010	208	923
	2011	212	1168
	2012	212	1340
	2013	214	1601
	2014	215	1733

Spreadsheet at LarsonPrecalculus.com

(a) Use a graphing utility to find a cubic model for the geothermal energy consumption data and a cubic model for the wind energy consumption data. Let t represent the year, with $t = 4$ corresponding to 2004.

(b) Use the graphing utility to graph the data and the two models in the same viewing window.

(c) Use the graph from part (b) to approximate the point of intersection of the graphs of the models. Interpret your answer in the context of the problem.

(d) Describe the behavior of each model. Do you think the models can accurately predict the consumption of geothermal energy and wind energy in the United States for future years? Explain.

(e) Use your school's library, the Internet, or some other reference source to research the advantages and disadvantages of using renewable energy.

Geometry In Exercises 63 and 64, use a system of equations to find the dimensions of the rectangle meeting the specified conditions.

63. The perimeter is 56 meters and the length is 4 meters greater than the width.

64. The perimeter is 42 inches and the width is three-fourths the length.

65. Geometry What are the dimensions of a rectangular tract of land when its perimeter is 44 kilometers and its area is 120 square kilometers?

66. Geometry What are the dimensions of a right triangle with a two-inch hypotenuse and an area of 1 square inch?

Exploration

True or False? In Exercises 67 and 68, determine whether the statement is true or false. Justify your answer.

67. In order to solve a system of equations by substitution, you must always solve for y in one of the two equations and then back-substitute.

68. If the graph of a system consists of a parabola and a circle, then the system can have at most two solutions.

69. Think About It When solving a system of equations by substitution, how do you recognize that the system has no solution?

70. **HOW DO YOU SEE IT?** The cost C of producing x units and the revenue R obtained by selling x units are shown in the figure.

(a) Estimate the point of intersection. What does this point represent?

(b) Use the figure to identify the x-values that correspond to (i) an overall loss and (ii) a profit. Explain.

71. Think About It Consider the system of equations

$$\begin{cases} ax + by = c \\ dx + ey = f \end{cases}.$$

(a) Find values for a, b, c, d, e, and f so that the system has one distinct solution. (There is more than one correct answer.)

(b) Explain how to solve the system in part (a) by the method of substitution and graphically.

(c) Write a brief paragraph describing any advantages of the method of substitution over the graphical method of solving a system of equations.

6.2 Two-Variable Linear Systems

Systems of equations in two variables can help you model and solve mixture problems. For example, in Exercise 50 on page 441, you will write, graph, and solve a system of equations to find the numbers of gallons of 87- and 92-octane gasoline that must be mixed to obtain 500 gallons of 89-octane gasoline.

■ Use the method of elimination to solve systems of linear equations in two variables.
■ Interpret graphically the numbers of solutions of systems of linear equations in two variables.
■ Use systems of linear equations in two variables to model and solve real-life problems.

The Method of Elimination

In Section 6.1, you studied two methods for solving a system of equations: substitution and graphing. Now, you will study the **method of elimination.** The key step in this method is to obtain, for one of the variables, coefficients that differ only in sign so that *adding* the equations eliminates the variable.

$$
\begin{array}{ll}
3x + 5y = 7 & \text{Equation 1} \\
\underline{-3x - 2y = -1} & \text{Equation 2} \\
3y = 6 & \text{Add equations.}
\end{array}
$$

Note that by adding the two equations, you eliminate the x-terms and obtain a single equation in y. Solving this equation for y produces $y = 2$, which you can back-substitute into one of the original equations to solve for x.

EXAMPLE 1 **Solving a System of Equations by Elimination**

Solve the system of linear equations.

$$
\begin{cases}
3x + 2y = 4 & \text{Equation 1} \\
5x - 2y = 12 & \text{Equation 2}
\end{cases}
$$

Solution The coefficients of y differ only in sign, so eliminate the y-terms by adding the two equations.

$$
\begin{array}{ll}
3x + 2y = 4 & \text{Write Equation 1.} \\
\underline{5x - 2y = 12} & \text{Write Equation 2.} \\
8x = 16 & \text{Add equations.} \\
x = 2 & \text{Solve for } x.
\end{array}
$$

Solve for y by back-substituting $x = 2$ into Equation 1.

$$
\begin{array}{ll}
3x + 2y = 4 & \text{Write Equation 1.} \\
3(2) + 2y = 4 & \text{Substitute 2 for } x. \\
y = -1 & \text{Solve for } y.
\end{array}
$$

The solution is $(2, -1)$. Check this in the original system, as follows.

Check

$$
\begin{array}{ll}
3(2) + 2(-1) = 4 & \text{Solution checks in Equation 1. } ✓ \\
5(2) - 2(-1) = 12 & \text{Solution checks in Equation 2. } ✓
\end{array}
$$

✓ *Checkpoint* ◀))) *Audio-video solution in English & Spanish at LarsonPrecalculus.com*

Solve the system of linear equations.

$$
\begin{cases}
2x + y = 4 \\
2x - y = -1
\end{cases}
$$

Method of Elimination

To use the **method of elimination** to solve a system of two linear equations in x and y, perform the following steps.

1. *Obtain coefficients* for x (or y) that differ only in sign by multiplying all terms of one or both equations by suitably chosen constants.

2. *Add* the equations to eliminate one variable.

3. *Solve* the equation obtained in Step 2.

4. *Back-substitute* the value obtained in Step 3 into either of the original equations and solve for the other variable.

5. *Check* that the solution satisfies *each* of the original equations.

EXAMPLE 2   **Solving a System of Equations by Elimination**

Solve the system of linear equations.

$$\begin{cases} 2x - 4y = -7 & \text{Equation 1} \\ 5x + y = -1 & \text{Equation 2} \end{cases}$$

Solution   To obtain coefficients for y that differ only in sign, multiply Equation 2 by 4.

$$
\begin{aligned}
2x - 4y &= -7 \quad\Longrightarrow\quad & 2x - 4y &= -7 & &\text{Write Equation 1.}\\
5x + y &= -1 \quad\Longrightarrow\quad & 20x + 4y &= -4 & &\text{Multiply Equation 2 by 4.}\\
& & 22x &= -11 & &\text{Add equations.}\\
& & x &= -\tfrac{1}{2} & &\text{Solve for } x.
\end{aligned}
$$

Solve for y by back-substituting $x = -\frac{1}{2}$ into Equation 1.

$$
\begin{aligned}
2x - 4y &= -7 & &\text{Write Equation 1.}\\
2\left(-\tfrac{1}{2}\right) - 4y &= -7 & &\text{Substitute } -\tfrac{1}{2} \text{ for } x.\\
-4y &= -6 & &\text{Simplify.}\\
y &= \tfrac{3}{2} & &\text{Solve for } y.
\end{aligned}
$$

The solution is $\left(-\frac{1}{2}, \frac{3}{2}\right)$. Check this in the original system, as follows.

Check

$$
\begin{aligned}
2x - 4y &= -7 & &\text{Write Equation 1.}\\
2\left(-\tfrac{1}{2}\right) - 4\left(\tfrac{3}{2}\right) &\overset{?}{=} -7 & &\text{Substitute for } x \text{ and } y.\\
-1 - 6 &= -7 & &\text{Solution checks in Equation 1. } \checkmark\\
5x + y &= -1 & &\text{Write Equation 2.}\\
5\left(-\tfrac{1}{2}\right) + \tfrac{3}{2} &\overset{?}{=} -1 & &\text{Substitute for } x \text{ and } y.\\
-\tfrac{5}{2} + \tfrac{3}{2} &= -1 & &\text{Solution checks in Equation 2. } \checkmark
\end{aligned}
$$

✓ *Checkpoint* ◀))) *Audio-video solution in English & Spanish at LarsonPrecalculus.com*

Solve the system of linear equations.

$$\begin{cases} 2x + 3y = 17 \\ 5x - y = 17 \end{cases}$$

In Example 2, the two systems of linear equations (the original system and the system obtained by multiplying Equation 2 by a constant)

$$\begin{cases} 2x - 4y = -7 \\ 5x + y = -1 \end{cases} \quad \text{and} \quad \begin{cases} 2x - 4y = -7 \\ 20x + 4y = -4 \end{cases}$$

are **equivalent systems** because they have the same solution set. The operations that can be performed on a system of linear equations to produce an equivalent system are (1) interchange two equations, (2) multiply one of the equations by a nonzero constant, and (3) add a multiple of one equation to another equation to replace the latter equation. You will study these operations in more depth in Section 6.3.

EXAMPLE 3 **Solving a System of Linear Equations**

Solve the system of linear equations.

$$\begin{cases} 5x + 3y = 9 & \text{Equation 1} \\ 2x - 4y = 14 & \text{Equation 2} \end{cases}$$

Algebraic Solution

To obtain coefficients of y that differ only in sign, multiply Equation 1 by 4 and multiply Equation 2 by 3.

$$5x + 3y = 9 \implies 20x + 12y = 36 \quad \text{Multiply Equation 1 by 4.}$$

$$\underline{2x - 4y = 14 \implies 6x - 12y = 42} \quad \text{Multiply Equation 2 by 3.}$$

$$26x = 78 \quad \text{Add equations.}$$

$$x = 3 \quad \text{Solve for } x.$$

Solve for y by back-substituting $x = 3$ into Equation 2.

$$2x - 4y = 14 \quad \text{Write Equation 2.}$$

$$2(3) - 4y = 14 \quad \text{Substitute 3 for } x.$$

$$-4y = 8 \quad \text{Simplify.}$$

$$y = -2 \quad \text{Solve for } y.$$

The solution is $(3, -2)$. Check this in the original system, as shown below.

Check

$$5x + 3y = 9 \quad \text{Write Equation 1.}$$

$$5(3) + 3(-2) \stackrel{?}{=} 9 \quad \text{Substitute 3 for } x \text{ and } -2 \text{ for } y.$$

$$15 - 6 = 9 \quad \text{Solution checks in Equation 1. } ✓$$

$$2x - 4y = 14 \quad \text{Write Equation 2.}$$

$$2(3) - 4(-2) \stackrel{?}{=} 14 \quad \text{Substitute 3 for } x \text{ and } -2 \text{ for } y.$$

$$6 + 8 = 14 \quad \text{Solution checks in Equation 2. } ✓$$

Graphical Solution

Solve each equation for y and use a graphing utility to graph the equations in the same viewing window.

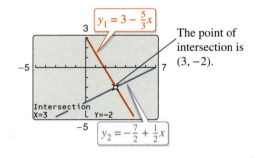

$y_1 = 3 - \frac{5}{3}x$

The point of intersection is $(3, -2)$.

$y_2 = -\frac{7}{2} + \frac{1}{2}x$

Intersection X=3 Y=-2

From the graph, the solution is $(3, -2)$. Check this in the original system, as shown below.

✓ **Checkpoint** 🔊))) *Audio-video solution in English & Spanish at LarsonPrecalculus.com*

Solve the system of linear equations.

$$\begin{cases} 3x + 2y = 7 \\ 2x + 5y = 1 \end{cases}$$

▷ TECHNOLOGY The
solution of the linear system

$$\begin{cases} ax + by = c \\ dx + ey = f \end{cases}$$

is given by

$$x = \frac{ce - bf}{ae - bd}$$

and

$$y = \frac{af - cd}{ae - bd}.$$

If $ae - bd = 0$, then the
system does not have a
unique solution. A graphing
utility program for solving
such a system is available at
CengageBrain.com. Use this
program, called "Systems of
Linear Equations," to solve the
system in Example 4.

Example 4 illustrates a strategy for solving a system of linear equations that has decimal coefficients.

EXAMPLE 4 **A Linear System Having Decimal Coefficients**

Solve the system of linear equations.

$$\begin{cases} 0.02x - 0.05y = -0.38 & \text{Equation 1} \\ 0.03x + 0.04y = 1.04 & \text{Equation 2} \end{cases}$$

Solution The coefficients in this system have two decimal places, so multiply each equation by 100 to produce a system in which the coefficients are all integers.

$$\begin{cases} 2x - 5y = -38 & \text{Revised Equation 1} \\ 3x + 4y = 104 & \text{Revised Equation 2} \end{cases}$$

Now, to obtain coefficients that differ only in sign, multiply revised Equation 1 by 3 and multiply revised Equation 2 by -2.

$$2x - 5y = -38 \implies 6x - 15y = -114 \qquad \text{Multiply revised Equation 1 by 3.}$$
$$3x + 4y = 104 \implies -6x - 8y = -208 \qquad \text{Multiply revised Equation 2 by } -2.$$
$$-23y = -322 \qquad \text{Add equations.}$$
$$y = \frac{-322}{-23} \qquad \text{Divide each side by } -23.$$
$$y = 14 \qquad \text{Simplify.}$$

Solve for x by back-substituting $y = 14$ into revised Equation 2.

$$3x + 4y = 104 \qquad \text{Write revised Equation 2.}$$
$$3x + 4(14) = 104 \qquad \text{Substitute 14 for } y.$$
$$3x = 48 \qquad \text{Simplify.}$$
$$x = 16 \qquad \text{Solve for } x.$$

The solution is

$$(16, 14).$$

Check this in the original system, as follows.

Check

$$0.02x - 0.05y = -0.38 \qquad \text{Write Equation 1.}$$
$$0.02(16) - 0.05(14) \overset{?}{=} -0.38 \qquad \text{Substitute for } x \text{ and } y.$$
$$0.32 - 0.70 = -0.38 \qquad \text{Solution checks in Equation 1.} ✓$$
$$0.03x + 0.04y = 1.04 \qquad \text{Write Equation 2.}$$
$$0.03(16) + 0.04(14) \overset{?}{=} 1.04 \qquad \text{Substitute for } x \text{ and } y.$$
$$0.48 + 0.56 = 1.04 \qquad \text{Solution checks in Equation 2.} ✓$$

✓ *Checkpoint*  *Audio-video solution in English & Spanish at LarsonPrecalculus.com*

Solve the system of linear equations.

$$\begin{cases} 0.03x + 0.04y = 0.75 \\ 0.02x + 0.06y = 0.90 \end{cases}$$

Graphical Interpretation of Solutions

It is possible for a system of equations to have exactly one solution, two or more solutions, or no solution. In a system of *linear* equations, however, if the system has two different solutions, then it must have an *infinite* number of solutions. To see why this is true, consider the following graphical interpretation of a system of two linear equations in two variables.

Graphical Interpretations of Solutions

For a system of two linear equations in two variables, the number of solutions is one of the following.

Number of Solutions	Graphical Interpretation	Slopes of Lines
1. Exactly one solution	The two lines intersect at one point.	The slopes of the two lines are not equal.
2. Infinitely many solutions	The two lines coincide (are identical).	The slopes of the two lines are equal.
3. No solution	The two lines are parallel.	The slopes of the two lines are equal.

A system of linear equations is **consistent** when it has at least one solution. A system is **inconsistent** when it has no solution.

EXAMPLE 5   **Recognizing Graphs of Linear Systems**

See LarsonPrecalculus.com for an interactive version of this type of example.

Match each system of linear equations with its graph. Describe the number of solutions and state whether the system is consistent or inconsistent.

a. $\begin{cases} 2x - 3y = 3 \\ -4x + 6y = 6 \end{cases}$ **b.** $\begin{cases} 2x - 3y = 3 \\ x + 2y = 5 \end{cases}$ **c.** $\begin{cases} 2x - 3y = 3 \\ -4x + 6y = -6 \end{cases}$

i. 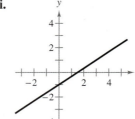   **ii.** 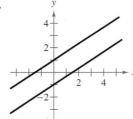   **iii.**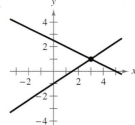

• • **REMARK**   When solving a system of linear equations graphically, it helps to begin by writing the equations in slope-intercept form, so you can compare the slopes and *y*-intercepts of their graphs. Do this for the systems in Example 5.

Solution

a. The graph of system (a) is a pair of parallel lines (ii). The lines have no point of intersection, so the system has no solution. The system is inconsistent.

b. The graph of system (b) is a pair of intersecting lines (iii). The lines have one point of intersection, so the system has exactly one solution. The system is consistent.

c. The graph of system (c) is a pair of lines that coincide (i). The lines have infinitely many points of intersection, so the system has infinitely many solutions. The system is consistent.

✓ **Checkpoint** *Audio-video solution in English & Spanish at LarsonPrecalculus.com*

Sketch the graph of the system of linear equations. Then describe the number of solutions and state whether the system is consistent or inconsistent.

$$\begin{cases} -2x + 3y = 6 \\ 4x - 6y = -9 \end{cases}$$

In Examples 6 and 7, note how the method of elimination is used to determine that a system of linear equations has no solution or infinitely many solutions.

EXAMPLE 6 **Method of Elimination: No-Solution Case**

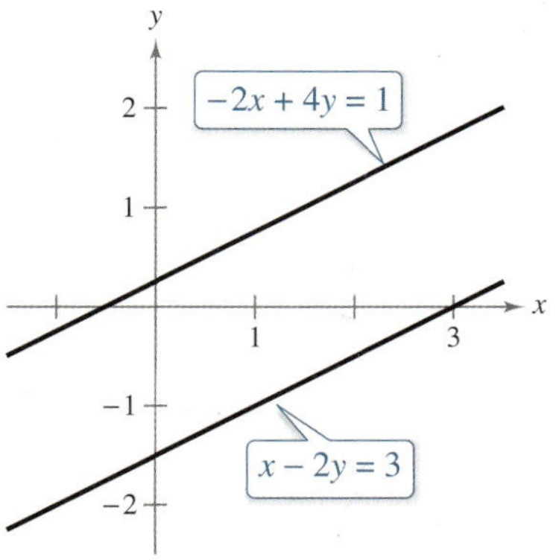

Solve the system of linear equations.

$$\begin{cases} x - 2y = 3 & \text{Equation 1} \\ -2x + 4y = 1 & \text{Equation 2} \end{cases}$$

Solution To obtain coefficients that differ only in sign, multiply Equation 1 by 2.

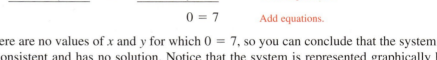

$x - 2y = 3$ ⟹	$2x - 4y = 6$	Multiply Equation 1 by 2.
$-2x + 4y = 1$ ⟹	$-2x + 4y = 1$	Write Equation 2.
	$0 = 7$	Add equations.

There are no values of x and y for which $0 = 7$, so you can conclude that the system is inconsistent and has no solution. Notice that the system is represented graphically by two parallel lines with no point of intersection, as shown in Figure 6.6.

✓ *Checkpoint* 🔊))) *Audio-video solution in English & Spanish at LarsonPrecalculus.com*

Solve the system of linear equations.

$$\begin{cases} 6x - 5y = 3 \\ -12x + 10y = 5 \end{cases}$$

In Example 6, note that the occurrence of a false statement, such as $0 = 7$, indicates that the system has no solution. In the next example, note that the occurrence of a statement that is true for all values of the variables, such as $0 = 0$, indicates that the system has infinitely many solutions.

EXAMPLE 7 **Method of Elimination: Infinitely-Many-Solutions Case**

Solve the system of linear equations.

$$\begin{cases} 2x - y = 1 & \text{Equation 1} \\ 4x - 2y = 2 & \text{Equation 2} \end{cases}$$

Solution To obtain coefficients that differ only in sign, multiply Equation 1 by -2.

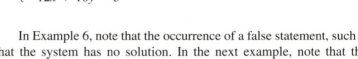

$2x - y = 1$ ⟹	$-4x + 2y = -2$	Multiply Equation 1 by -2.
$4x - 2y = 2$ ⟹	$4x - 2y = 2$	Write Equation 2.
	$0 = 0$	Add equations.

The two equations are equivalent (have the same solution set), so the system has infinitely many solutions. The solution set consists of all points (x, y) lying on the line $2x - y = 1$, as shown in Figure 6.7. Letting $x = a$, where a is any real number, you find that $y = 2a - 1$. So, the solutions of the system are all ordered pairs of the form $(a, 2a - 1)$.

✓ *Checkpoint* 🔊))) *Audio-video solution in English & Spanish at LarsonPrecalculus.com*

Solve the system of linear equations.

$$\begin{cases} \dfrac{1}{2}x - \dfrac{1}{8}y = -\dfrac{3}{8} \\ -4x + y = 3 \end{cases}$$

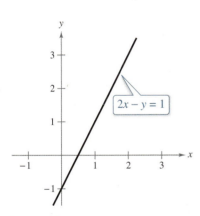

Figure 6.6

Figure 6.7

Applications

At this point, you may be asking the question "How can I tell whether I can solve an application problem using a system of linear equations?" To answer this question, start with the following considerations.

1. Does the problem involve more than one unknown quantity?

2. Are there two (or more) equations or conditions to be satisfied?

When the answer to one or both of these questions is "yes," the appropriate model for the problem may be a system of linear equations.

EXAMPLE 8   **An Application of a Linear System**

An airplane flying into a headwind travels the 2000-mile flying distance between Tallahassee, Florida, and Los Angeles, California, in 4 hours and 24 minutes. On the return flight, the airplane travels this distance in 4 hours. Find the airspeed of the plane and the speed of the wind, assuming that both remain constant.

Solution   The two unknown quantities are the airspeed of the plane and the speed of the wind. Let r_1 be the airspeed of the plane and r_2 be the speed of the wind (see Figure 6.8).

$$r_1 - r_2 = \text{speed of the plane against the wind}$$

$$r_1 + r_2 = \text{speed of the plane with the wind}$$

Use the formula

$$\text{distance} = (\text{rate})(\text{time})$$

to write equations involving these two speeds.

$$2000 = (r_1 - r_2)\left(4 + \frac{24}{60}\right)$$

$$2000 = (r_1 + r_2)(4)$$

These two equations simplify as follows.

$$\begin{cases} 5000 = 11r_1 - 11r_2 & \text{Equation 1} \\ 500 = r_1 + r_2 & \text{Equation 2} \end{cases}$$

To solve this system by elimination, multiply Equation 2 by 11.

$$5000 = 11r_1 - 11r_2 \quad \Longrightarrow \quad 5000 = 11r_1 - 11r_2 \qquad \text{Write Equation 1.}$$

$$\underline{500 = r_1 + r_2} \quad \Longrightarrow \quad \underline{5500 = 11r_1 + 11r_2} \qquad \text{Multiply Equation 2 by 11.}$$

$$10{,}500 = 22r_1 \qquad \text{Add equations.}$$

So,

$$r_1 = \frac{10{,}500}{22} = \frac{5250}{11} \approx 477.27 \text{ miles per hour} \qquad \text{Airspeed of plane}$$

and

$$r_2 = 500 - \frac{5250}{11} = \frac{250}{11} \approx 22.73 \text{ miles per hour.} \qquad \text{Speed of wind}$$

Check this solution in the original statement of the problem.

✓ **Checkpoint** ◀))) *Audio-video solution in English & Spanish at LarsonPrecalculus.com*

In Example 8, the return flight takes 4 hours and 6 minutes. Find the airspeed of the plane and the speed of the wind, assuming that both remain constant.

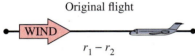
Original flight

$r_1 - r_2$

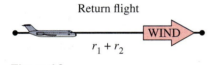
Return flight

$r_1 + r_2$

Figure 6.8

In a free market, the demands for many products are related to the prices of the products. As the prices decrease, the quantities demanded by consumers increase and the quantities that producers are able or willing to supply decrease. The **equilibrium point** is the price p and number of units x that satisfy both the demand and supply equations.

EXAMPLE 9   **Finding the Equilibrium Point**

The demand and supply equations for a video game console are

$$\begin{cases} p = 180 - 0.00001x & \text{Demand equation} \\ p = 90 + 0.00002x & \text{Supply equation} \end{cases}$$

where p is the price per unit (in dollars) and x is the number of units. Find the equilibrium point for this market.

Solution   Because p is written in terms of x, begin by substituting the expression for p from the supply equation into the demand equation.

$$p = 180 - 0.00001x \qquad \text{Write demand equation.}$$
$$90 + 0.00002x = 180 - 0.00001x \qquad \text{Substitute } 90 + 0.00002x \text{ for } p.$$
$$0.00003x = 90 \qquad \text{Combine like terms.}$$
$$x = 3{,}000{,}000 \qquad \text{Solve for } x.$$

So, the equilibrium point occurs when the demand and supply are each 3 million units. (See Figure 6.9.) To find the price that corresponds to this x-value, back-substitute $x = 3{,}000{,}000$ into either of the original equations. Using the demand equation produces the following.

$$p = 180 - 0.00001(3{,}000{,}000)$$
$$= 180 - 30$$
$$= \$150$$

The equilibrium point is $(3{,}000{,}000, 150)$. Check this by substituting $(3{,}000{,}000, 150)$ into the demand and supply equations.

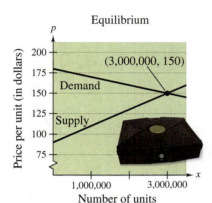

Equilibrium

(3,000,000, 150)

Demand

Supply

Price per unit (in dollars)

Number of units

Figure 6.9

✓ *Checkpoint* ◀))) *Audio-video solution in English & Spanish at LarsonPrecalculus.com*

The demand and supply equations for a flat-screen television are

$$\begin{cases} p = 567 - 0.00002x & \text{Demand equation} \\ p = 492 + 0.00003x & \text{Supply equation} \end{cases}$$

where p is the price per unit (in dollars) and x is the number of units. Find the equilibrium point for this market.

Summarize   (Section 6.2)

1. Explain how to use the method of elimination to solve a system of linear equations in two variables *(page 432)*. For examples of using the method of elimination to solve systems of linear equations in two variables, see Examples 1–4.

2. Explain how to interpret graphically the numbers of solutions of systems of linear equations in two variables *(page 436)*. For examples of interpreting graphically the numbers of solutions of systems of linear equations in two variables, see Examples 5–7.

3. Describe real-life applications of systems of linear equations in two variables *(pages 438 and 439, Examples 8 and 9)*.

6.2 Exercises

See **CalcChat.com** for tutorial help and worked-out solutions to odd-numbered exercises.

Vocabulary: Fill in the blanks.

1. The first step in solving a system of equations by the method of _____ is to obtain coefficients for x (or y) that differ only in sign.

2. Two systems of equations that have the same solution set are _____ systems.

3. A system of linear equations that has at least one solution is _____, whereas a system of linear equations that has no solution is _____.

4. In business applications, the _____ _____ (x, p) is the price p and the number of units x that satisfy both the demand and supply equations.

Skills and Applications

 Solving a System by Elimination In Exercises 5–12, solve the system by the method of elimination. Label each line with its equation. To print an enlarged copy of the graph, go to *MathGraphs.com*.

5. $\begin{cases} 2x + y = 7 \\ x - y = -4 \end{cases}$

6. $\begin{cases} x + 3y = 1 \\ -x + 2y = 4 \end{cases}$

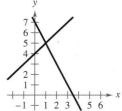

7. $\begin{cases} x + y = 0 \\ 3x + 2y = 1 \end{cases}$

8. $\begin{cases} \frac{1}{2}x - y = -2 \\ x + \frac{1}{3}y = 3 \end{cases}$

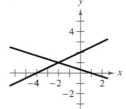

9. $\begin{cases} x - y = 2 \\ -2x + 2y = 5 \end{cases}$

10. $\begin{cases} 3x + 2y = 3 \\ 6x + 4y = 14 \end{cases}$

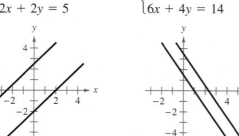

11. $\begin{cases} 3x - 2y = 5 \\ -6x + 4y = -10 \end{cases}$

12. $\begin{cases} 9x - 3y = -15 \\ -3x + y = 5 \end{cases}$

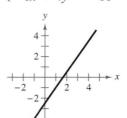 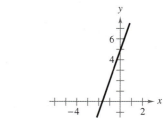

 **Solving a System by Elimination** In Exercises 13–30, solve the system by the method of elimination and check any solutions algebraically.

13. $\begin{cases} x + 2y = 6 \\ x - 2y = 2 \end{cases}$

14. $\begin{cases} 3x - 5y = 8 \\ 2x + 5y = 22 \end{cases}$

15. $\begin{cases} 5x + 3y = 6 \\ 3x - y = 5 \end{cases}$

16. $\begin{cases} x + 5y = 10 \\ 3x - 10y = -5 \end{cases}$

17. $\begin{cases} 2u + 3v = -1 \\ 7u + 15v = 4 \end{cases}$

18. $\begin{cases} 2r + 4s = 5 \\ 16r + 50s = 55 \end{cases}$

19. $\begin{cases} 3x + 2y = 10 \\ 2x + 5y = 3 \end{cases}$

20. $\begin{cases} 3x + 11y = 4 \\ -2x - 5y = 9 \end{cases}$

21. $\begin{cases} 4b + 3m = 3 \\ 3b + 11m = 13 \end{cases}$

22. $\begin{cases} 2x + 5y = 8 \\ 5x + 8y = 10 \end{cases}$

23. $\begin{cases} 0.2x - 0.5y = -27.8 \\ 0.3x + 0.4y = 68.7 \end{cases}$

24. $\begin{cases} 0.5x - 0.3y = 6.5 \\ 0.7x + 0.2y = 6.0 \end{cases}$

25. $\begin{cases} 3x + 2y = 4 \\ 9x + 6y = 3 \end{cases}$

26. $\begin{cases} -6x + 4y = 7 \\ 15x - 10y = -16 \end{cases}$

27. $\begin{cases} -5x + 6y = -3 \\ 20x - 24y = 12 \end{cases}$

28. $\begin{cases} 7x + 8y = 6 \\ -14x - 16y = -12 \end{cases}$

29. $\begin{cases} \dfrac{x + 3}{4} + \dfrac{y - 1}{3} = 1 \\ 2x - y = 12 \end{cases}$

30. $\begin{cases} \dfrac{x - 1}{2} + \dfrac{y + 2}{3} = 4 \\ x - 2y = 5 \end{cases}$

Matching a System with Its Graph In Exercises 31–34, match the system of linear equations with its graph. Describe the number of solutions and state whether the system is consistent or inconsistent. [The graphs are labeled (a), (b), (c) and (d).]

(a)
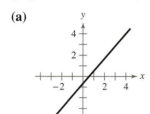

(b)
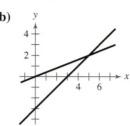

(c)
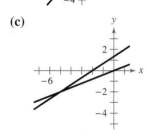

(d)
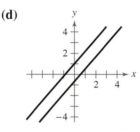

31. $\begin{cases} -7x + 6y = -4 \\ 14x - 12y = 8 \end{cases}$

32. $\begin{cases} 2x - 5y = 0 \\ 2x - 3y = -4 \end{cases}$

33. $\begin{cases} 7x - 6y = -6 \\ -7x + 6y = -4 \end{cases}$

34. $\begin{cases} 2x - 5y = 0 \\ x - y = 3 \end{cases}$

Choosing a Solution Method In Exercises 35–40, use any method to solve the system. Explain your choice of method.

35. $\begin{cases} 3x - 5y = 7 \\ 2x + y = 9 \end{cases}$

36. $\begin{cases} -x + 3y = 17 \\ 4x + 3y = 7 \end{cases}$

37. $\begin{cases} -2x + 8y = 20 \\ y = x - 5 \end{cases}$

38. $\begin{cases} -5x + 9y = 13 \\ y = x - 4 \end{cases}$

39. $\begin{cases} y = -2x - 17 \\ y = 2 - 3x \end{cases}$

40. $\begin{cases} y = -3x - 8 \\ y = 15 - 2x \end{cases}$

41. Airplane Speed An airplane flying into a headwind travels the 1800-mile flying distance between Indianapolis, Indiana, and Phoenix, Arizona, in 3 hours. On the return flight, the airplane travels this distance in 2 hours and 30 minutes. Find the airspeed of the plane and the speed of the wind, assuming that both remain constant.

42. Airplane Speed Two planes start from Los Angeles International Airport and fly in opposite directions. The second plane starts $\frac{1}{2}$ hour after the first plane, but its speed is 80 kilometers per hour faster. Find the speed of each plane when 2 hours after the first plane departs the planes are 3200 kilometers apart.

43. Nutrition Two cheeseburgers and one small order of fries contain a total of 1420 calories. Three cheeseburgers and two small orders of fries contain a total of 2290 calories. Find the caloric content of each item.

44. Nutrition One eight-ounce glass of apple juice and one eight-ounce glass of orange juice contain a total of 179.2 milligrams of vitamin C. Two eight-ounce glasses of apple juice and three eight-ounce glasses of orange juice contain a total of 442.1 milligrams of vitamin C. How much vitamin C is in an eight-ounce glass of each type of juice?

 Finding the Equilibrium Point In Exercises 45–48, find the equilibrium point of the demand and supply equations.

Demand	Supply
45. $p = 500 - 0.4x$	$p = 380 + 0.1x$
46. $p = 100 - 0.05x$	$p = 25 + 0.1x$
47. $p = 140 - 0.00002x$	$p = 80 + 0.00001x$
48. $p = 400 - 0.0002x$	$p = 225 + 0.0005x$

49. Chemistry Thirty liters of a 40% acid solution is obtained by mixing a 25% solution with a 50% solution.

(a) Write a system of equations in which one equation represents the total amount of final mixture required and the other represents the percent of acid in the final mixture. Let x and y represent the amounts of the 25% and 50% solutions, respectively.

(b) Use a graphing utility to graph the two equations in part (a) in the same viewing window. As the amount of the 25% solution increases, how does the amount of the 50% solution change?

(c) How much of each solution is required to obtain the specified concentration of the final mixture?

50. Fuel Mixture

Five hundred gallons of 89-octane gasoline is obtained by mixing 87-octane gasoline with 92-octane gasoline.

(a) Write a system of equations in which one equation represents the total amount of final mixture required and the other represents the amounts of 87- and 92-octane gasoline in the final mixture. Let x and y represent the numbers of gallons of 87- and 92-octane gasoline, respectively.

(b) Use a graphing utility to graph the two equations in part (a) in the same viewing window. As the amount of 87-octane gasoline increases, how does the amount of 92-octane gasoline change?

(c) How much of each type of gasoline is required to obtain the 500 gallons of 89-octane gasoline?

51. Investment Portfolio A total of $24,000 is invested in two corporate bonds that pay 3.5% and 5% simple interest. The investor wants an annual interest income of $930 from the investments. What amount should be invested in the 3.5% bond?

52. Investment Portfolio A total of $32,000 is invested in two municipal bonds that pay 5.75% and 6.25% simple interest. The investor wants an annual interest income of $1900 from the investments. What amount should be invested in the 5.75% bond?

53. Pharmacology The numbers of prescriptions P (in thousands) filled at two pharmacies from 2012 through 2016 are shown in the table.

Year	Pharmacy A	Pharmacy B
2012	19.2	20.4
2013	19.6	20.8
2014	20.0	21.1
2015	20.6	21.5
2016	21.3	22.0

(a) Use a graphing utility to create a scatter plot of the data for pharmacy A and find a linear model. Let t represent the year, with $t = 12$ corresponding to 2012. Repeat the procedure for pharmacy B.

(b) Assume that the models in part (a) can be used to represent future years. Will the number of prescriptions filled at pharmacy A ever exceed the number of prescriptions filled at pharmacy B? If so, when?

54. Daily Sales A store manager wants to know the demand for a product as a function of the price. The table shows the daily sales y for different prices x of the product.

Price, x	Demand, y
$1.00	45
$1.20	37
$1.50	23

(a) Find the least squares regression line $y = ax + b$ for the data by solving the system

$$\begin{cases} 3.00b + 3.70a = 105.00 \\ 3.70b + 4.69a = 123.90 \end{cases}$$

for a and b. Use a graphing utility to confirm the result.

(b) Use the linear model from part (a) to predict the demand when the price is $1.75.

Fitting a Line to Data One way to find the least squares regression line $y = ax + b$ for a set of points

$$(x_1, y_1), (x_2, y_2), \ldots, (x_n, y_n)$$

is by solving the system below for a and b.

$$\begin{cases} nb + \left(\sum_{i=1}^{n} x_i\right)a = \left(\sum_{i=1}^{n} y_i\right) \\ \left(\sum_{i=1}^{n} x_i\right)b + \left(\sum_{i=1}^{n} x_i^2\right)a = \left(\sum_{i=1}^{n} x_i y_i\right) \end{cases}$$

In Exercises 55 and 56, the sums have been evaluated. Solve the simplified system for a and b to find the least squares regression line for the points. Use a graphing utility to confirm the result. (*Note:* The symbol Σ is used to denote a sum of the terms of a sequence. You will learn how to use this notation in Section 8.1.)

55. $\begin{cases} 5b + 10a = 20.2 \\ 10b + 30a = 50.1 \end{cases}$

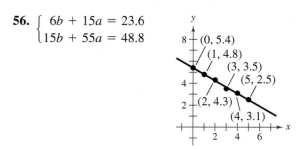

56. $\begin{cases} 6b + 15a = 23.6 \\ 15b + 55a = 48.8 \end{cases}$

57. Agriculture An agricultural scientist used four test plots to determine the relationship between wheat yield y (in bushels per acre) and the amount of fertilizer x (in hundreds of pounds per acre). The table shows the results.

Fertilizer, x	1.0	1.5	2.0	2.5
Yield, y	32	41	48	53

(a) Find the least squares regression line $y = ax + b$ for the data by solving the system for a and b.

$$\begin{cases} 4b + 7a = 174 \\ 7b + 13.5a = 322 \end{cases}$$

(b) Use the linear model from part (a) to estimate the yield for a fertilizer application of 160 pounds per acre.

58. Gross Domestic Product   The table shows the total gross domestic products y (in billions of dollars) of the United States for the years 2009 through 2015. (*Source: U.S. Office of Management and Budget*)

Year	GDP, y
2009	14,414.6
2010	14,798.5
2011	15,379.2
2012	16,027.2
2013	16,498.1
2014	17,183.5
2015	17,803.4

DATA

Spreadsheet at LarsonPrecalculus.com

(a) Find the least squares regression line $y = at + b$ for the data, where t represents the year with $t = 9$ corresponding to 2009, by solving the system

$$\begin{cases} 7b + 84a = 112{,}104.5 \\ 84b + 1036a = 1{,}361{,}309.3 \end{cases}$$

for a and b. Use the *regression* feature of a graphing utility to confirm the result.

(b) Use the linear model to create a table of estimated values of y. Compare the estimated values with the actual data.

(c) Use the linear model to estimate the gross domestic product for 2016.

(d) Use the Internet, your school's library, or some other reference source to find the total national outlay for 2016. How does this value compare with your answer in part (c)?

(e) Is the linear model valid for long-term predictions of gross domestic products? Explain.

Exploration

True or False?   In Exercises 59 and 60, determine whether the statement is true or false. Justify your answer.

59. If two lines do not have exactly one point of intersection, then they must be parallel.

60. Solving a system of equations graphically will always give an exact solution.

Finding the Value of a Constant   In Exercises 61 and 62, find the value of k such that the system of linear equations is inconsistent.

61. $\begin{cases} 4x - 8y = -3 \\ 2x + ky = 16 \end{cases}$ **62.** $\begin{cases} 15x + 3y = 6 \\ -10x + ky = 9 \end{cases}$

63. Writing   Briefly explain whether it is possible for a consistent system of linear equations to have exactly two solutions.

64. Think About It   Give examples of systems of linear equations that have (a) no solution and (b) infinitely many solutions.

65. Comparing Methods   Use the method of substitution to solve the system in Example 1. Do you prefer the method of substitution or the method of elimination? Explain.

66. HOW DO YOU SEE IT?   Use the graphs of the two equations shown below.

(a) Describe the graphs of the two equations.

(b) Can you conclude that the system of equations whose graphs are shown is inconsistent? Explain.

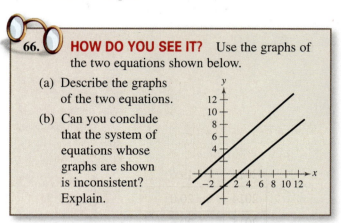

Think About It   In Exercises 67 and 68, the graphs of the two equations appear to be parallel. Yet, when you solve the system algebraically, you find that the system does have a solution. Find the solution and explain why it does not appear on the portion of the graph shown.

67. $\begin{cases} 100y - x = 200 \\ 99y - x = -198 \end{cases}$

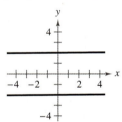

68. $\begin{cases} 21x - 20y = 0 \\ 13x - 12y = 120 \end{cases}$

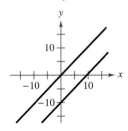

Project: College Expenses   To work an extended application analyzing the average undergraduate tuition, room, and board charges at private degree-granting institutions in the United States from 1993 through 2013, visit this text's website at *LarsonPrecalculus.com*. (*Source: U.S. Department of Education*)

6.3 Multivariable Linear Systems

■ Use back-substitution to solve linear systems in row-echelon form.
■ Use Gaussian elimination to solve systems of linear equations.
■ Solve nonsquare systems of linear equations.
■ Use systems of linear equations in three or more variables to model and solve real-life problems.

Row-Echelon Form and Back-Substitution

Systems of linear equations in three or more variables can help you model and solve real-life problems. For example, in Exercise 70 on page 455, you will use a system of linear equations in three variables to analyze the reproductive rates of deer in a wildlife preserve.

The method of elimination can be applied to a system of linear equations in more than two variables. In fact, this method adapts to computer use for solving linear systems with dozens of variables.

When using the method of elimination to solve a system of linear equations, the goal is to rewrite the system in a form to which back-substitution can be applied. To see how this works, consider the following two systems of linear equations.

$$\begin{cases} x - 2y + 3z = 9 \\ -x + 3y \quad\quad = -4 \\ 2x - 5y + 5z = 17 \end{cases}$$ System of three linear equations in three variables (See Example 3.)

$$\begin{cases} x - 2y + 3z = 9 \\ y + 3z = 5 \\ z = 2 \end{cases}$$ Equivalent system in row-echelon form (See Example 1.)

The second system is in **row-echelon form,** which means that it has a "stair-step" pattern with leading coefficients of 1. In comparing the two systems, notice that the row-echelon form can readily be solved using back-substitution.

EXAMPLE 1   Using Back-Substitution in Row-Echelon Form

Solve the system of linear equations.

$$\begin{cases} x - 2y + 3z = 9 & \text{Equation 1} \\ y + 3z = 5 & \text{Equation 2} \\ z = 2 & \text{Equation 3} \end{cases}$$

Solution   From Equation 3, you know the value of z. To solve for y, back-substitute $z = 2$ into Equation 2.

$$y + 3(2) = 5 \qquad \text{Substitute 2 for } z.$$
$$y = -1 \qquad \text{Solve for } y.$$

To solve for x, back-substitute $y = -1$ and $z = 2$ into Equation 1.

$$x - 2(-1) + 3(2) = 9 \qquad \text{Substitute } -1 \text{ for } y \text{ and 2 for } z.$$
$$x = 1 \qquad \text{Solve for } x.$$

The solution is $x = 1$, $y = -1$, and $z = 2$, which can be written as the **ordered triple** $(1, -1, 2)$. Check this in the original system of equations.

✓ *Checkpoint* ◀))) *Audio-video solution in English & Spanish at LarsonPrecalculus.com*

Solve the system of linear equations.

$$\begin{cases} x - y + 5z = 22 \\ y + 3z = 6 \\ z = 3 \end{cases}$$

Gaussian Elimination

Recall from Section 6.2 that two systems of equations are *equivalent* when they have the same solution set. To solve a system that is not in row-echelon form, first convert it to an *equivalent* system that is in row-echelon form by using one or more of the row operations given below.

> ### Operations That Produce Equivalent Systems
>
> Each of the following **row operations** on a system of linear equations produces an *equivalent* system of linear equations.
>
> 1. Interchange two equations.
>
> 2. Multiply one of the equations by a nonzero constant.
>
> 3. Add a multiple of one equation to another equation to replace the latter equation.

To see how to use row operations, take another look at the method of elimination, as applied to a system of two linear equations.

EXAMPLE 2   **Using Row Operations to Solve a System**

Solve the system of linear equations.

$$\begin{cases} 3x - 2y = -1 \\ x - y = 0 \end{cases}$$

Solution   Two strategies seem reasonable: eliminate the variable x or eliminate the variable y. The following steps show one way to eliminate x in the second equation. Start by interchanging the equations to obtain a leading coefficient of 1 in the first equation.

$$\begin{cases} x - y = 0 \\ 3x - 2y = -1 \end{cases}$$ Interchange the two equations in the system.

$$-3x + 3y = 0$$ Multiply the first equation by -3.

$$\begin{array}{r} -3x + 3y = 0 \\ \underline{3x - 2y = -1} \\ y = -1 \end{array}$$ Add the multiple of the first equation to the second equation to obtain a new second equation.

$$\begin{cases} x - y = 0 \\ y = -1 \end{cases}$$ New system in row-echelon form

Now back-substitute $y = -1$ into the first equation of the new system in row-echelon form and solve for x.

$$x - (-1) = 0$$ Substitute -1 for y.

$$x = -1$$ Solve for x.

The solution is $(-1, -1)$. Check this in the original system.

✓ *Checkpoint* *Audio-video solution in English & Spanish at LarsonPrecalculus.com*

Solve the system of linear equations.

$$\begin{cases} 2x + y = 3 \\ x + 2y = 3 \end{cases}$$

Historically, one of the most influential Chinese mathematics books was the *Chui-chang suan-shu* or *Nine Chapters on the Mathematical Art*, a compilation of ancient Chinese problems published in 263 A.D. Chapter Eight of the *Nine Chapters* contained solutions of systems of linear equations such as the system below.

$$\begin{cases} 3x + 2y + z = 39 \\ 2x + 3y + z = 34 \\ x + 2y + 3z = 26 \end{cases}$$

This system was solved by performing column operations on a matrix, using the same strategies as Gaussian elimination. Matrices (plural for matrix) are discussed in the next chapter.

Rewriting a system of linear equations in row-echelon form usually involves a chain of equivalent systems, each of which is obtained by using one of the three basic row operations listed on the previous page. This process is called **Gaussian elimination,** after the German mathematician Carl Friedrich Gauss (1777–1855).

EXAMPLE 3 **Using Gaussian Elimination to Solve a System**

Solve the system of linear equations.

$$\begin{cases} x - 2y + 3z = 9 & \text{Equation 1} \\ -x + 3y = -4 & \text{Equation 2} \\ 2x - 5y + 5z = 17 & \text{Equation 3} \end{cases}$$

Solution The leading coefficient of the first equation is 1, so begin by keeping the x in the upper left position and eliminating the other x-terms from the first column.

> **REMARK** Arithmetic errors are common when performing row operations. You should note the operation performed in each step to make checking your work easier.

$$\begin{array}{ll} x - 2y + 3z = 9 & \text{Write Equation 1.} \\ \underline{-x + 3y = -4} & \text{Write Equation 2.} \\ y + 3z = 5 & \text{Add.} \end{array}$$

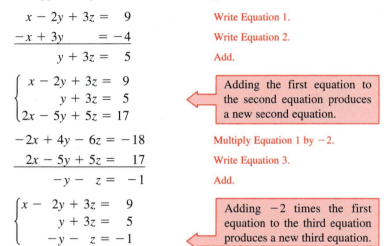

$$\begin{cases} x - 2y + 3z = 9 \\ y + 3z = 5 \\ 2x - 5y + 5z = 17 \end{cases}$$

Adding the first equation to the second equation produces a new second equation.

$$\begin{array}{ll} -2x + 4y - 6z = -18 & \text{Multiply Equation 1 by } -2. \\ \underline{2x - 5y + 5z = 17} & \text{Write Equation 3.} \\ -y - z = -1 & \text{Add.} \end{array}$$

$$\begin{cases} x - 2y + 3z = 9 \\ y + 3z = 5 \\ -y - z = -1 \end{cases}$$

Adding -2 times the first equation to the third equation produces a new third equation.

Now that all but the first x have been eliminated from the first column, begin to work on the second column. (You need to eliminate y from the third equation.)

$$\begin{array}{ll} y + 3z = 5 & \text{Write second equation} \\ \underline{-y - z = -1} & \text{Write third equation.} \\ 2z = 4 & \text{Add.} \end{array}$$

$$\begin{cases} x - 2y + 3z = 9 \\ y + 3z = 5 \\ 2z = 4 \end{cases}$$

Adding the second equation to the third equation produces a new third equation.

Finally, you need a coefficient of 1 for z in the third equation.

$$\begin{cases} x - 2y + 3z = 9 \\ y + 3z = 5 \\ z = 2 \end{cases}$$

Multiplying the third equation by $\frac{1}{2}$ produces a new third equation.

This is the same system that was solved in Example 1. So, as in that example, the solution is $(1, -1, 2)$.

✓ *Checkpoint*  *Audio-video solution in English & Spanish at LarsonPrecalculus.com*

Solve the system of linear equations.

$$\begin{cases} x + y + z = 6 \\ 2x - y + z = 3 \\ 3x + y - z = 2 \end{cases}$$

The next example involves an inconsistent system—one that has no solution. The key to recognizing an inconsistent system is that at some stage in the elimination process, you obtain a false statement such as $0 = -2$.

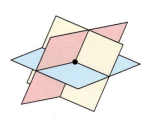

Solution: one point
Figure 6.10

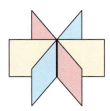

Solution: one line
Figure 6.11

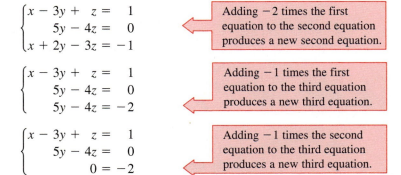

EXAMPLE 4 **An Inconsistent System**

Solve the system of linear equations.

$$\begin{cases} x - 3y + z = 1 \\ 2x - y - 2z = 2 \\ x + 2y - 3z = -1 \end{cases}$$

Solution

$$\begin{cases} x - 3y + z = 1 \\ 5y - 4z = 0 \\ x + 2y - 3z = -1 \end{cases}$$ Adding -2 times the first equation to the second equation produces a new second equation.

$$\begin{cases} x - 3y + z = 1 \\ 5y - 4z = 0 \\ 5y - 4z = -2 \end{cases}$$ Adding -1 times the first equation to the third equation produces a new third equation.

$$\begin{cases} x - 3y + z = 1 \\ 5y - 4z = 0 \\ 0 = -2 \end{cases}$$ Adding -1 times the second equation to the third equation produces a new third equation.

You obtain the false statement $0 = -2$, so this system is inconsistent and has no solution. Moreover, this system is equivalent to the original system, so the original system also has no solution.

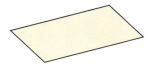

✓ *Checkpoint*  *Audio-video solution in English & Spanish at LarsonPrecalculus.com*

Solve the system of linear equations.

$$\begin{cases} x + y - 2z = 3 \\ 3x - 2y + 4z = 1 \\ 2x - 3y + 6z = 8 \end{cases}$$

Note that the graph of a linear equation in three variables is a plane. As with a system of linear equations in two variables, the number of solutions of a system of linear equations in more than two variables must fall into one of three categories.

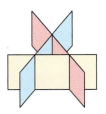

Solution: one plane
Figure 6.12

The Number of Solutions of a Linear System

For a system of linear equations, exactly one of the following is true.

1. There is exactly one solution.

2. There are infinitely many solutions.

3. There is no solution.

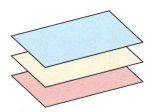

Solution: none
Figure 6.13

Solution: none
Figure 6.14

In Section 6.2, you learned that the graph of a system of two linear equations in two variables is a pair of lines that intersect, coincide, or are parallel. Similarly, the graph of a system of three linear equations in three variables consists of three planes in space. These planes must intersect in one point (exactly one solution, see Figure 6.10), intersect in a line or a plane (infinitely many solutions, see Figures 6.11 and 6.12), or have no points common to all three planes (no solution, see Figures 6.13 and 6.14).

EXAMPLE 5   A System with Infinitely Many Solutions

Solve the system of linear equations.

$$\begin{cases} x + y - 3z = -1 & \text{Equation 1} \\ y - z = 0 & \text{Equation 2} \\ -x + 2y = 1 & \text{Equation 3} \end{cases}$$

Solution

$$\begin{cases} x + y - 3z = -1 \\ y - z = 0 \\ 3y - 3z = 0 \end{cases}$$

> Adding the first equation to the third equation produces a new third equation.

$$\begin{cases} x + y - 3z = -1 \\ y - z = 0 \\ 0 = 0 \end{cases}$$

> Adding -3 times the second equation to the third equation produces a new third equation.

You have $0 = 0$, so Equation 3 depends on Equations 1 and 2 in the sense that it gives no additional information about the variables. So, the original system is equivalent to

$$\begin{cases} x + y - 3z = -1 \\ y - z = 0 \end{cases}.$$

In the second equation, solve for y in terms of z to obtain $y = z$. Back-substituting for y in the first equation yields $x = 2z - 1$. So, the system has infinitely many solutions consisting of all real values of x, y, and z for which

$$y = z \quad \text{and} \quad x = 2z - 1.$$

Letting $z = a$, where a is any real number, the solutions of the original system are all ordered triples of the form

$$(2a - 1, a, a).$$

✓ **Checkpoint** ◀)) *Audio-video solution in English & Spanish at LarsonPrecalculus.com*

Solve the system of linear equations.

$$\begin{cases} x + 2y - 7z = -4 \\ 2x + 3y + z = 5 \\ 3x + 7y - 36z = -25 \end{cases}$$

In Example 5, there are other ways to write the same infinite set of solutions. For instance, letting $x = b$, the solutions could have been written as

$$\left(b, \tfrac{1}{2}(b + 1), \tfrac{1}{2}(b + 1)\right).$$ *b is a real number.*

To convince yourself that this form produces the same set of solutions, consider the following.

Substitution	Solution	
$a = 0$	$(2(0) - 1, 0, 0) = (-1, 0, 0)$	Same solution
$b = -1$	$\left(-1, \tfrac{1}{2}(-1 + 1), \tfrac{1}{2}(-1 + 1)\right) = (-1, 0, 0)$	
$a = 1$	$(2(1) - 1, 1, 1) = (1, 1, 1)$	Same solution
$b = 1$	$\left(1, \tfrac{1}{2}(1 + 1), \tfrac{1}{2}(1 + 1)\right) = (1, 1, 1)$	
$a = 2$	$(2(2) - 1, 2, 2) = (3, 2, 2)$	Same solution
$b = 3$	$\left(3, \tfrac{1}{2}(3 + 1), \tfrac{1}{2}(3 + 1)\right) = (3, 2, 2)$	

Nonsquare Systems

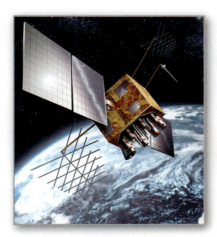

The Global Positioning System (GPS) is a network of 24 satellites originally developed by the U.S. military as a navigational tool. Civilian applications of GPS receivers include determining directions, locating vessels lost at sea, and monitoring earthquakes. A GPS receiver works by using satellite readings to calculate its location. In a simplified mathematical model, nonsquare systems of three linear equations in four variables (three dimensions and time) determine the coordinates of the receiver as a function of time.

So far, each system of linear equations you have looked at has been *square*, which means that the number of equations is equal to the number of variables. In a **nonsquare** system, the number of equations differs from the number of variables. A system of linear equations cannot have a unique solution unless there are at least as many equations as there are variables in the system.

EXAMPLE 6 **A System with Fewer Equations than Variables**

See LarsonPrecalculus.com for an interactive version of this type of example.

Solve the system of linear equations.

$$\begin{cases} x - 2y + z = 2 & \text{Equation 1} \\ 2x - y - z = 1 & \text{Equation 2} \end{cases}$$

Solution   The system has three variables and only two equations, so the system does not have a unique solution. Begin by rewriting the system in row-echelon form.

$$\begin{cases} x - 2y + z = 2 \\ \quad\quad 3y - 3z = -3 \end{cases}$$

> Adding -2 times the first equation to the second equation produces a new second equation.

$$\begin{cases} x - 2y + z = 2 \\ \quad\quad y - z = -1 \end{cases}$$

> Multiplying the second equation by $\frac{1}{3}$ produces a new second equation.

Solve the new second equation for y in terms of z to obtain

$$y = z - 1.$$

Solve for x by back-substituting $y = z - 1$ into Equation 1.

$x - 2y + z = 2$	Write Equation 1.
$x - 2(z - 1) + z = 2$	Substitute $z - 1$ for y.
$x - 2z + 2 + z = 2$	Distributive Property
$x = z$	Solve for x.

Finally, by letting $z = a$, where a is any real number, you have the solution

$$x = a, \quad y = a - 1, \quad \text{and} \quad z = a.$$

So, the solution set of the system consists of all ordered triples of the form $(a, a - 1, a)$, where a is a real number.

✓ **Checkpoint**  *Audio-video solution in English & Spanish at LarsonPrecalculus.com*

Solve the system of linear equations.

$$\begin{cases} x - y + 4z = 3 \\ 4x \quad\quad - z = 0 \end{cases}$$

In Example 6, choose several values of a to obtain different solutions of the system, such as

$$(1, 0, 1), \quad (2, 1, 2), \quad \text{and} \quad (3, 2, 3).$$

Then check each ordered triple in the original system to verify that it is a solution of the system.

Applications

Figure 6.15

EXAMPLE 7 **Modeling Vertical Motion**

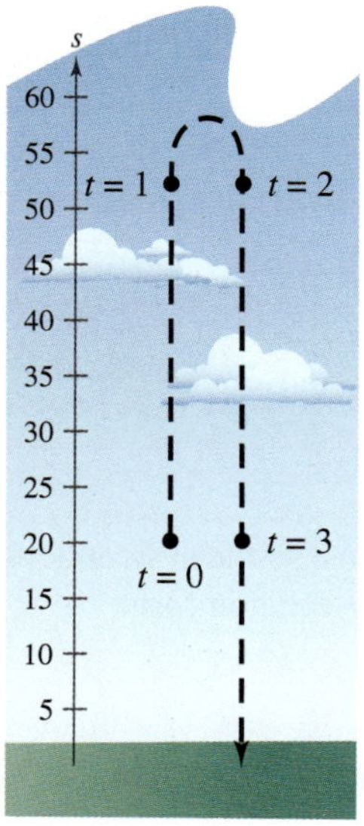

The height at time t of an object that is moving in a (vertical) line with constant acceleration a is given by the **position equation**

$$s = \tfrac{1}{2}at^2 + v_0t + s_0$$

where s is the height in feet, a is the acceleration in feet per second squared, t is the time in seconds, v_0 is the initial velocity (at $t = 0$), and s_0 is the initial height. Find the values of a, v_0, and s_0 when $s = 52$ at $t = 1$, $s = 52$ at $t = 2$, and $s = 20$ at $t = 3$, and interpret the result. (See Figure 6.15.)

Solution Substitute the three sets of values for t and s into the position equation to obtain three linear equations in a, v_0, and s_0.

When $t = 1$: $\tfrac{1}{2}a(1)^2 + v_0(1) + s_0 = 52$ $\Longrightarrow$ $a + 2v_0 + 2s_0 = 104$

When $t = 2$: $\tfrac{1}{2}a(2)^2 + v_0(2) + s_0 = 52$ $\Longrightarrow$ $2a + 2v_0 + s_0 = 52$

When $t = 3$: $\tfrac{1}{2}a(3)^2 + v_0(3) + s_0 = 20$ $\Longrightarrow$ $9a + 6v_0 + 2s_0 = 40$

Solve the resulting system of linear equations using Gaussian elimination.

$$\begin{cases} a + 2v_0 + 2s_0 = 104 \\ 2a + 2v_0 + s_0 = 52 \\ 9a + 6v_0 + 2s_0 = 40 \end{cases}$$

$$\begin{cases} a + 2v_0 + 2s_0 = 104 \\ -2v_0 - 3s_0 = -156 \\ 9a + 6v_0 + 2s_0 = 40 \end{cases}$$

> Adding -2 times the first equation to the second equation produces a new second equation.

$$\begin{cases} a + 2v_0 + 2s_0 = 104 \\ -2v_0 - 3s_0 = -156 \\ -12v_0 - 16s_0 = -896 \end{cases}$$

> Adding -9 times the first equation to the third equation produces a new third equation.

$$\begin{cases} a + 2v_0 + 2s_0 = 104 \\ -2v_0 - 3s_0 = -156 \\ 2s_0 = 40 \end{cases}$$

> Adding -6 times the second equation to the third equation produces a new third equation.

$$\begin{cases} a + 2v_0 + 2s_0 = 104 \\ v_0 + \tfrac{3}{2}s_0 = 78 \\ s_0 = 20 \end{cases}$$

> Multiplying the second equation by $-\tfrac{1}{2}$ produces a new second equation and multiplying the third equation by $\tfrac{1}{2}$ produces a new third equation.

Using back-substitution, the solution of this system is

$$a = -32, \quad v_0 = 48, \quad \text{and} \quad s_0 = 20.$$

So, the position equation for the object is $s = -16t^2 + 48t + 20$, which implies that the object was thrown upward at a velocity of 48 feet per second from a height of 20 feet.

✓ **Checkpoint** 🔊))) *Audio-video solution in English & Spanish at LarsonPrecalculus.com*

Use the position equation

$$s = \tfrac{1}{2}at^2 + v_0t + s_0$$

from Example 7 to find the values of a, v_0, and s_0 when $s = 104$ at $t = 1$, $s = 76$ at $t = 2$, and $s = 16$ at $t = 3$, and interpret the result. ∎

EXAMPLE 8   **Data Analysis: Curve-Fitting**

Find a quadratic equation $y = ax^2 + bx + c$ whose graph passes through the points $(-1, 3)$, $(1, 1)$, and $(2, 6)$.

Solution   Because the graph of $y = ax^2 + bx + c$ passes through the points $(-1, 3)$, $(1, 1)$, and $(2, 6)$, you can write the following.

When $x = -1$, $y = 3$: $a(-1)^2 + b(-1) + c = 3$

When $x = 1$, $y = 1$: $a(1)^2 + b(1) + c = 1$

When $x = 2$, $y = 6$: $a(2)^2 + b(2) + c = 6$

This yields the following system of linear equations.

$$\begin{cases} a - b + c = 3 \\ a + b + c = 1 \\ 4a + 2b + c = 6 \end{cases}$$

The solution of this system is $a = 2$, $b = -1$, and $c = 0$. So, the equation of the parabola is

$$y = 2x^2 - x$$

as shown below.

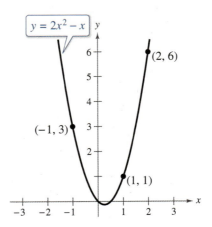

$y = 2x^2 - x$

 **Checkpoint**))) *Audio-video solution in English & Spanish at LarsonPrecalculus.com*

Find a quadratic equation $y = ax^2 + bx + c$ whose graph passes through the points $(0, 0)$, $(3, -3)$, and $(6, 0)$.

Summarize   (Section 6.3)

1. Explain what row-echelon form is *(page 444)*. For an example of solving a linear system in row-echelon form, see Example 1.

2. Describe the process of Gaussian elimination *(pages 445 and 446)*. For examples of using Gaussian elimination to solve systems of linear equations, see Examples 2–5.

3. Explain the difference between a square system of linear equations and a nonsquare system of linear equations *(page 449)*. For an example of solving a nonsquare system of linear equations, see Example 6.

4. Describe examples of how to use systems of linear equations in three or more variables to model and solve real-life problems *(pages 450 and 451, Examples 7 and 8)*.

6.3 Exercises

See **CalcChat.com** for tutorial help and worked-out solutions to odd-numbered exercises.

Vocabulary: Fill in the blanks.

1. A system of equations in _____ form has a "stair-step" pattern with leading coefficients of 1.

2. A solution of a system of three linear equations in three variables can be written as an _____ _____, which has the form (x, y, z).

3. The process used to write a system of linear equations in row-echelon form is called _____ elimination.

4. Interchanging two equations of a system of linear equations is a _____ _____ that produces an equivalent system.

5. In a _____ system, the number of equations differs from the number of variables in the system.

6. The equation $s = \frac{1}{2}at^2 + v_0t + s_0$ is called the _____ equation, and it models the height s of an object at time t that is moving in a vertical line with a constant acceleration a.

Skills and Applications

Checking Solutions In Exercises 7–10, determine whether each ordered triple is a solution of the system of equations.

7. $\begin{cases} 6x - y + z = -1 \\ 4x \quad\quad - 3z = -19 \\ \quad\quad 2y + 5z = 25 \end{cases}$
   (a) $(0, 3, 1)$ (b) $(-3, 0, 5)$
   (c) $(0, -1, 4)$ (d) $(-1, 0, 5)$

8. $\begin{cases} 3x + 4y - z = 17 \\ 5x - y + 2z = -2 \\ 2x - 3y + 7z = -21 \end{cases}$
   (a) $(3, -1, 2)$ (b) $(1, 3, -2)$
   (c) $(1, 5, 6)$ (d) $(1, -2, 2)$

9. $\begin{cases} 4x + y - z = 0 \\ -8x - 6y + z = -\frac{7}{4} \\ 3x - y \quad\quad = -\frac{9}{4} \end{cases}$
   (a) $\left(\frac{1}{2}, -\frac{3}{4}, -\frac{7}{4}\right)$ (b) $\left(\frac{3}{2}, -\frac{2}{5}, \frac{3}{5}\right)$
   (c) $\left(-\frac{1}{2}, \frac{3}{4}, -\frac{5}{4}\right)$ (d) $\left(-\frac{1}{2}, \frac{1}{6}, -\frac{3}{4}\right)$

10. $\begin{cases} -4x - y - 8z = -6 \\ \quad\quad y + z = 0 \\ 4x - 7y \quad\quad = 6 \end{cases}$
    (a) $(-2, -2, 2)$ (b) $\left(-\frac{33}{2}, -10, 10\right)$
    (c) $\left(\frac{1}{8}, -\frac{1}{2}, \frac{1}{2}\right)$ (d) $\left(-\frac{1}{2}, -2, 1\right)$

 Using Back-Substitution in Row-Echelon Form In Exercises 11–16, use back-substitution to solve the system of linear equations.

11. $\begin{cases} x - y + 5z = 37 \\ \quad y + 2z = 6 \\ \quad\quad z = 8 \end{cases}$

12. $\begin{cases} x - 2y + 2z = 20 \\ \quad y - z = 8 \\ \quad\quad z = -1 \end{cases}$

13. $\begin{cases} x + y - 3z = 7 \\ \quad y + z = 12 \\ \quad\quad z = 2 \end{cases}$

14. $\begin{cases} x - y + 2z = 22 \\ \quad y - 8z = 13 \\ \quad\quad z = -3 \end{cases}$

15. $\begin{cases} x - 2y + z = -\frac{1}{4} \\ \quad y - z = -4 \\ \quad\quad z = 11 \end{cases}$

16. $\begin{cases} x \quad - 8z = \frac{1}{2} \\ \quad y - 5z = 22 \\ \quad\quad z = -4 \end{cases}$

Performing Row Operations In Exercises 17 and 18, perform the row operation and write the equivalent system.

17. Add Equation 1 to Equation 2.

   $\begin{cases} x - 2y + 3z = 5 & \text{Equation 1} \\ -x + 3y - 5z = 4 & \text{Equation 2} \\ 2x \quad\quad - 3z = 0 & \text{Equation 3} \end{cases}$

   What did this operation accomplish?

18. Add -2 times Equation 1 to Equation 3.

   $\begin{cases} x - 2y + 3z = 5 & \text{Equation 1} \\ -x + 3y - 5z = 4 & \text{Equation 2} \\ 2x \quad\quad - 3z = 0 & \text{Equation 3} \end{cases}$

   What did this operation accomplish?

 Solving a System of Linear Equations In Exercises 19–22, solve the system of linear equations and check any solutions algebraically.

19. $\begin{cases} -2x + 3y = 10 \\ x + y = 0 \end{cases}$

20. $\begin{cases} 2x - y = 0 \\ x - y = 7 \end{cases}$

21. $\begin{cases} 3x - y = 9 \\ x - 2y = -2 \end{cases}$

22. $\begin{cases} x + 2y = 1 \\ 5x - 4y = -23 \end{cases}$

Solving a System of Linear Equations
In Exercises 23–40, solve the system of linear equations and check any solutions algebraically.

23. $\begin{cases} x + y + z = 7 \\ 2x - y + z = 9 \\ 3x \quad\;\; - z = 10 \end{cases}$ 24. $\begin{cases} x + y + z = 5 \\ x - 2y + 4z = 13 \\ 3y + 4z = 13 \end{cases}$

25. $\begin{cases} 2x + 4y - z = 7 \\ 2x - 4y + 2z = -6 \\ x + 4y + z = 0 \end{cases}$ 26. $\begin{cases} 2x + 4y + z = 1 \\ x - 2y - 3z = 2 \\ x + y - z = -1 \end{cases}$

27. $\begin{cases} 2x + y - z = 7 \\ x - 2y + 2z = -9 \\ 3x - y + z = 5 \end{cases}$ 28. $\begin{cases} 5x - 3y + 2z = 3 \\ 2x + 4y - z = 7 \\ x - 11y + 4z = 3 \end{cases}$

29. $\begin{cases} 3x - 5y + 5z = 1 \\ 2x - 2y + 3z = 0 \\ 7x - y + 3z = 0 \end{cases}$ 30. $\begin{cases} 2x + y + 3z = 1 \\ 2x + 6y + 8z = 3 \\ 6x + 8y + 18z = 5 \end{cases}$

31. $\begin{cases} 2x + 3y = 0 \\ 4x + 3y - z = 0 \\ 8x + 3y + 3z = 0 \end{cases}$ 32. $\begin{cases} 4x + 3y + 17z = 0 \\ 5x + 4y + 22z = 0 \\ 4x + 2y + 19z = 0 \end{cases}$

33. $\begin{cases} x \quad\;\; + 4z = 1 \\ x + y + 10z = 10 \\ 2x - y + 2z = -5 \end{cases}$ 34. $\begin{cases} 2x - 2y - 6z = -4 \\ -3x + 2y + 6z = 1 \\ x - y - 5z = -3 \end{cases}$

35. $\begin{cases} 3x - 3y + 6z = 6 \\ x + 2y - z = 5 \\ 5x - 8y + 13z = 7 \end{cases}$ 36. $\begin{cases} x \quad\;\; + 2z = 5 \\ 3x - y - z = 1 \\ 6x - y + 5z = 16 \end{cases}$

37. $\begin{cases} x + 2y - 7z = -4 \\ 2x + y + z = 13 \\ 3x + 9y - 36z = -33 \end{cases}$

38. $\begin{cases} 2x + y - 3z = 4 \\ 4x \quad\;\; + 2z = 10 \\ -2x + 3y - 13z = -8 \end{cases}$

39. $\begin{cases} x \quad\qquad\;\; + 3w = 4 \\ 2y - z - w = 0 \\ 3y \quad\;\; - 2w = 1 \\ 2x - y + 4z = 5 \end{cases}$

40. $\begin{cases} x + y + z + w = 6 \\ 2x + 3y - w = 0 \\ -3x + 4y + z + 2w = 4 \\ x + 2y - z + w = 0 \end{cases}$

Solving a Nonsquare System In Exercises 41–44, solve the system of linear equations and check any solutions algebraically.

41. $\begin{cases} x - 2y + 5z = 2 \\ 4x \quad\;\; - z = 0 \end{cases}$ 42. $\begin{cases} x - 3y + 2z = 18 \\ 5x - 13y + 12z = 80 \end{cases}$

43. $\begin{cases} 2x - 3y + z = -2 \\ -4x + 9y = 7 \end{cases}$ 44. $\begin{cases} 2x + 3y + 3z = 7 \\ 4x + 18y + 15z = 44 \end{cases}$

Modeling Vertical Motion In Exercises 45 and 46, an object moving vertically is at the given heights at the specified times. Find the position equation $s = \frac{1}{2}at^2 + v_0 t + s_0$ for the object.

45. At $t = 1$ second, $s = 128$ feet
At $t = 2$ seconds, $s = 80$ feet
At $t = 3$ seconds, $s = 0$ feet

46. At $t = 1$ second, $s = 132$ feet
At $t = 2$ seconds, $s = 100$ feet
At $t = 3$ seconds, $s = 36$ feet

Finding the Equation of a Parabola In Exercises 47–52, find the equation

$$y = ax^2 + bx + c$$

of the parabola that passes through the points. To verify your result, use a graphing utility to plot the points and graph the parabola.

47. $(0, 0), (2, -2), (4, 0)$ 48. $(0, 3), (1, 4), (2, 3)$

49. $(2, 0), (3, -1), (4, 0)$ 50. $(1, 3), (2, 2), (3, -3)$

51. $\left(\frac{1}{2}, 1\right), (1, 3), (2, 13)$

52. $(-2, -3), (-1, 0), \left(\frac{1}{2}, -3\right)$

Finding the Equation of a Circle In Exercises 53–56, find the equation

$$x^2 + y^2 + Dx + Ey + F = 0$$

of the circle that passes through the points. To verify your result, use a graphing utility to plot the points and graph the circle.

53. $(0, 0), (5, 5), (10, 0)$ 54. $(0, 0), (0, 6), (3, 3)$

55. $(-3, -1), (2, 4), (-6, 8)$

56. $(0, 0), (0, -2), (3, 0)$

57. **Error Analysis** Describe the error.

The system

$$\begin{cases} x - 2y + 3x = 12 \\ y + 3z = 5 \\ 2z = 4 \end{cases}$$

is in row-echelon form.

58. **Agriculture** A mixture of 5 pounds of fertilizer A, 13 pounds of fertilizer B, and 4 pounds of fertilizer C provides the optimal nutrients for a plant. Commercial brand X contains equal parts of fertilizer B and fertilizer C. Commercial brand Y contains one part of fertilizer A and two parts of fertilizer B. Commercial brand Z contains two parts of fertilizer A, five parts of fertilizer B, and two parts of fertilizer C. How much of each fertilizer brand is needed to obtain the desired mixture?

59. Finance To expand its clothing line, a small corporation borrowed \$775,000 from three different lenders. The money was borrowed at 8%, 9%, and 10% simple interest. How much was borrowed at each rate when the annual interest owed was \$67,500 and the amount borrowed at 8% was four times the amount borrowed at 10%?

60. Advertising A health insurance company advertises on television, on radio, and in the local newspaper. The marketing department has an advertising budget of \$42,000 per month. A television ad costs \$1000, a radio ad costs \$200, and a newspaper ad costs \$500. The department wants to run 60 ads per month and have as many television ads as radio and newspaper ads combined. How many of each type of ad can the department run each month?

Geometry **In Exercises 61 and 62, find the values of**
x, *y*, and *z* in the figure.

61.

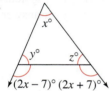

62.

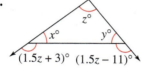

63. Geometry The perimeter of a triangle is 180 feet. The longest side of the triangle is 9 feet shorter than twice the shortest side. The sum of the lengths of the two shorter sides is 30 feet more than the length of the longest side. Find the lengths of the sides of the triangle.

64. Chemistry A chemist needs 10 liters of a 25% acid solution. The solution is to be mixed from three solutions whose concentrations are 10%, 20%, and 50%. How many liters of each solution will satisfy each condition?

(a) Use 2 liters of the 50% solution.

(b) Use as little as possible of the 50% solution.

(c) Use as much as possible of the 50% solution.

65. Electrical Network Applying Kirchhoff's Laws to the electrical network in the figure, the currents I_1, I_2, and I_3, are the solution of the system

$$\begin{cases} I_1 - I_2 + I_3 = 0 \\ 3I_1 + 2I_2 = 7. \\ 2I_2 + 4I_3 = 8 \end{cases}$$

Find the currents.

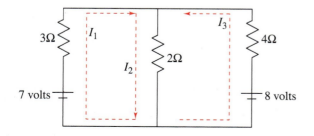

66. Pulley System A system of pulleys is loaded with 128-pound and 32-pound weights (see figure). The tensions t_1 and t_2 in the ropes and the acceleration a of the 32-pound weight are found by solving the system

$$\begin{cases} t_1 - 2t_2 = 0 \\ t_1 - 2a = 128 \\ t_2 + a = 32 \end{cases}$$

where t_1 and t_2 are in pounds and a is in feet per second squared. Solve this system.

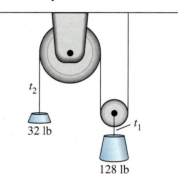

Fitting a Parabola One way to find the least squares regression parabola $y = ax^2 + bx + c$ for a set of points

$(x_1, y_1), (x_2, y_2), \ldots, (x_n, y_n)$

is by solving the system below for a, b, and c.

$$\begin{cases} nc + \left(\sum_{i=1}^{n} x_i\right)b + \left(\sum_{i=1}^{n} x_i^2\right)a = \sum_{i=1}^{n} y_i \\ \left(\sum_{i=1}^{n} x_i\right)c + \left(\sum_{i=1}^{n} x_i^2\right)b + \left(\sum_{i=1}^{n} x_i^3\right)a = \sum_{i=1}^{n} x_i y_i \\ \left(\sum_{i=1}^{n} x_i^2\right)c + \left(\sum_{i=1}^{n} x_i^3\right)b + \left(\sum_{i=1}^{n} x_i^4\right)a = \sum_{i=1}^{n} x_i^2 y_i \end{cases}$$

In Exercises 67 and 68, the sums have been evaluated. Solve the simplified system for *a*, *b*, and *c* to find the least squares regression parabola for the points. Use a graphing utility to confirm the result. *(Note:* The symbol Σ is used to denote a sum of the terms of a sequence. You will learn how to use this notation in Section 8.1.)

67.
$$\begin{cases} 4c + 9b + 29a = 20 \\ 9c + 29b + 99a = 70 \\ 29c + 99b + 353a = 254 \end{cases}$$

68.
$$\begin{cases} 4c + 40a = 19 \\ 40b = -12 \\ 40c + 544a = 160 \end{cases}$$

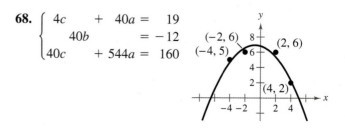

69. Stopping Distance In testing a new automobile braking system, engineers recorded the speed x (in miles per hour) and the stopping distance y (in feet). The table shows the results.

Speed, x	30	40	50	60	70
Stopping Distance, y	75	118	175	240	315

(a) Find the least squares regression parabola $y = ax^2 + bc + c$ for the data by solving the system.

$$\begin{cases} 5c + 250b + 13{,}500a = 923 \\ 250c + 13{,}500b + 775{,}000a = 52{,}170 \\ 13{,}500c + 775{,}000b + 46{,}590{,}000a = 3{,}101{,}300 \end{cases}$$

(b) Use a graphing utility to graph the model you found in part (a) and the data in the same viewing window. How well does the model fit the data? Explain.

(c) Use the model to estimate the stopping distance when the speed is 75 miles per hour.

70. Wildlife

A wildlife management team studied the reproductive rates of deer in four tracts of a wildlife preserve. In each tract, the number of females x and the percent of females y that had offspring the following year were recorded. The table shows the results.

Number, x	100	120	140	160
Percent, y	75	68	55	30

(a) Find the least squares regression parabola $y = ax^2 + bx + c$ for the data by solving the system.

$$\begin{cases} 4c + 520b + 69{,}600a = 228 \\ 520c + 69{,}600b + 9{,}568{,}000a = 28{,}160 \\ 69{,}600c + 9{,}568{,}000b + 1{,}346{,}880{,}000a = 3{,}575{,}200 \end{cases}$$

(b) Use a graphing utility to graph the model you found in part (a) and the data in the same viewing window. How well does the model fit the data? Explain.

(c) Use the model to estimate the percent of females that had offspring when there were 170 females.

(d) Use the model to estimate the number of females when 40% of the females had offspring.

Advanced Applications In Exercises 71 and 72, find values of x, y, and λ that satisfy the system. These systems arise in certain optimization problems in calculus, and λ is called a Lagrange multiplier.

71. $\begin{cases} 2x - 2x\lambda = 0 \\ -2y + \lambda = 0 \\ y - x^2 = 0 \end{cases}$

72. $\begin{cases} 2 + 2y + 2\lambda = 0 \\ 2x + 1 + \lambda = 0 \\ 2x + y - 100 = 0 \end{cases}$

Exploration

True or False? In Exercises 73 and 74, determine whether the statement is true or false. Justify your answer.

73. Every nonsquare system of linear equations has a unique solution.

74. If a system of three linear equations is inconsistent, then there are no points common to the graphs of all three equations of the system.

75. Think About It Are the following two systems of equations equivalent? Give reasons for your answer.

$$\begin{cases} x + 3y - z = 6 \\ 2x - y + 2z = 1 \\ 3x + 2y - z = 2 \end{cases} \qquad \begin{cases} x + 3y - z = 6 \\ -7y + 4z = 1 \\ -7y - 4z = -16 \end{cases}$$

76. HOW DO YOU SEE IT? The number of sides x and the combined number of sides and diagonals y for each of three regular polygons are shown below. Write a system of linear equations to find an equation of the form $y = ax^2 + bx + c$ that represents the relationship between x and y for the three polygons.

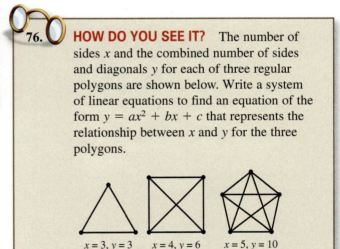

$x = 3, y = 3$ $x = 4, y = 6$ $x = 5, y = 10$

Finding Systems of Linear Equations In Exercises 77–80, find two systems of linear equations that have the ordered triple as a solution. (There are many correct answers.)

77. $(2, 0, -1)$

78. $(-5, 3, -2)$

79. $\left(\tfrac{1}{2}, -3, 0\right)$

80. $\left(4, \tfrac{2}{5}, \tfrac{1}{2}\right)$

Project: Earnings per Share To work an extended application analyzing the earnings per share for Wal-Mart Stores, Inc., from 2001 through 2015, visit this text's website at *LarsonPrecalculus.com*. (Source: Wal-Mart Stores, Inc.)

6.4 Partial Fractions

Partial fractions can help you analyze the behavior of a rational function. For example, in Exercise 60 on page 463, you will use partial fractions to analyze the exhaust temperatures of a diesel engine.

■ Recognize partial fraction decompositions of rational expressions.
■ Find partial fraction decompositions of rational expressions.

Introduction

In this section, you will learn to write a rational expression as the sum of two or more simpler rational expressions. For example, the rational expression

$$\frac{x + 7}{x^2 - x - 6}$$

can be written as the sum of two fractions with first-degree denominators. That is,

Partial fraction decomposition
of $\dfrac{x + 7}{x^2 - x - 6}$

$$\frac{x + 7}{x^2 - x - 6} = \overbrace{\frac{2}{x - 3}}^{} + \overbrace{\frac{-1}{x + 2}}^{}.$$

Partial fraction   Partial fraction

Each fraction on the right side of the equation is a **partial fraction,** and together they make up the **partial fraction decomposition** of the left side.

▷ **ALGEBRA HELP** To review how to find the degree of a polynomial (such as $x - 3$ and $x + 2$), see Section P.3.

· · **REMARK** Section P.5 shows you how to combine expressions such as

$$\frac{1}{x - 2} + \frac{-1}{x + 3} = \frac{5}{(x - 2)(x + 3)}.$$

The method of partial fraction decomposition shows you how to reverse this process and write

$$\frac{5}{(x - 2)(x + 3)} = \frac{1}{x - 2} + \frac{-1}{x + 3}.$$

Decomposition of $N(x)/D(x)$ into Partial Fractions

1. *Divide when improper:* When $N(x)/D(x)$ is an improper fraction [degree of $N(x) \geq$ degree of $D(x)$], divide the denominator into the numerator to obtain

$$\frac{N(x)}{D(x)} = (\text{polynomial}) + \frac{N_1(x)}{D(x)}$$

and apply Steps 2, 3, and 4 to the proper rational expression

$$\frac{N_1(x)}{D(x)}.$$

Note that $N_1(x)$ is the remainder from the division of $N(x)$ by $D(x)$.

2. *Factor the denominator:* Completely factor the denominator into factors of the form

$$(px + q)^m \quad \text{and} \quad (ax^2 + bx + c)^n$$

where $(ax^2 + bx + c)$ is irreducible.

3. *Linear factors:* For *each* factor of the form $(px + q)^m$, the partial fraction decomposition must include the following sum of m fractions.

$$\frac{A_1}{(px + q)} + \frac{A_2}{(px + q)^2} + \cdots + \frac{A_m}{(px + q)^m}$$

4. *Quadratic factors:* For *each* factor of the form $(ax^2 + bx + c)^n$, the partial fraction decomposition must include the following sum of n fractions.

$$\frac{B_1x + C_1}{ax^2 + bx + c} + \frac{B_2x + C_2}{(ax^2 + bx + c)^2} + \cdots + \frac{B_nx + C_n}{(ax^2 + bx + c)^n}$$

Partial Fraction Decomposition

The examples in this section demonstrate algebraic techniques for determining the constants in the numerators of partial fractions. Note that the techniques vary slightly, depending on the type of factors of the denominator: linear or quadratic, distinct or repeated.

EXAMPLE 1   **Distinct Linear Factors**

Write the partial fraction decomposition of

$$\frac{x + 7}{x^2 - x - 6}.$$

Solution   The expression is proper, so begin by factoring the denominator.

$$x^2 - x - 6 = (x - 3)(x + 2)$$

Include one partial fraction with a constant numerator for each linear factor of the denominator. Write the form of the decomposition with A and B as the unknown constants.

$$\frac{x + 7}{x^2 - x - 6} = \frac{A}{x - 3} + \frac{B}{x + 2} \qquad \text{\color{red}Write form of decomposition.}$$

Multiply each side of this equation by the least common denominator, $(x - 3)(x + 2)$, to obtain the **basic equation.**

$$x + 7 = A(x + 2) + B(x - 3) \qquad \text{\color{red}Basic equation}$$

This equation is true for all x, so substitute any *convenient* values of x that will help determine the constants A and B. Values of x that are especially convenient are those that make the factors $(x + 2)$ and $(x - 3)$ equal to zero. For example, to solve for B, let $x = -2$.

$$-2 + 7 = A(-2 + 2) + B(-2 - 3) \qquad \text{\color{red}Substitute } -2 \text{ for } x.$$
$$5 = A(0) + B(-5)$$
$$5 = -5B$$
$$-1 = B$$

To solve for A, let $x = 3$.

$$3 + 7 = A(3 + 2) + B(3 - 3) \qquad \text{\color{red}Substitute 3 for } x.$$
$$10 = A(5) + B(0)$$
$$10 = 5A$$
$$2 = A$$

So, the partial fraction decomposition is

$$\frac{x + 7}{x^2 - x - 6} = \frac{2}{x - 3} + \frac{-1}{x + 2}.$$

Check this result by combining the two partial fractions on the right side of the equation, or by using your graphing utility.

✓ *Checkpoint*  *Audio-video solution in English & Spanish at LarsonPrecalculus.com*

Write the partial fraction decomposition of

$$\frac{x + 5}{2x^2 - x - 1}.$$

▷ **TECHNOLOGY** To use a graphing utility to check the decomposition found in Example 1, graph

$$y_1 = \frac{x + 7}{x^2 - x - 6}$$

and

$$y_2 = \frac{2}{x - 3} + \frac{-1}{x + 2}$$

in the same viewing window. The graphs should be identical, as shown below.

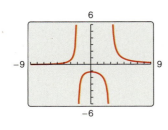

EXAMPLE 2 **Repeated Linear Factors**

Write the partial fraction decomposition of $\dfrac{x^4 + 2x^3 + 6x^2 + 20x + 6}{x^3 + 2x^2 + x}$.

▷ **ALGEBRA HELP** To review long division of polynomials, see Section 3.3. To review factoring of polynomials, see Section P.4.

Solution This rational expression is improper, so begin by dividing the numerator by the denominator.

$$
\begin{array}{r}
x \\
x^3 + 2x^2 + x \overline{)\, x^4 + 2x^3 + 6x^2 + 20x + 6} \\
\underline{x^4 + 2x^3 + x^2 } \\
5x^2 + 20x + 6
\end{array}
$$

The result is

$$x + \frac{5x^2 + 20x + 6}{x^3 + 2x^2 + x}.$$

The denominator of the remainder factors as

$$x^3 + 2x^2 + x = x(x^2 + 2x + 1) = x(x + 1)^2$$

so include a partial fraction with a constant numerator for each power of x and $(x + 1)$.

$$\frac{5x^2 + 20x + 6}{x(x + 1)^2} = \frac{A}{x} + \frac{B}{x + 1} + \frac{C}{(x + 1)^2} \qquad \textcolor{red}{\text{Write form of decomposition.}}$$

Multiply each side by the LCD, $x(x + 1)^2$, to obtain the basic equation.

$$5x^2 + 20x + 6 = A(x + 1)^2 + Bx(x + 1) + Cx \qquad \textcolor{red}{\text{Basic equation}}$$

▷ **REMARK** To obtain the basic equation, be sure to multiply *each* fraction by the LCD.

Let $x = -1$ to eliminate the A- and B-terms.

$$5(-1)^2 + 20(-1) + 6 = A(-1 + 1)^2 + B(-1)(-1 + 1) + C(-1)$$
$$5 - 20 + 6 = 0 + 0 - C$$
$$C = 9$$

Let $x = 0$ to eliminate the B- and C-terms.

$$5(0)^2 + 20(0) + 6 = A(0 + 1)^2 + B(0)(0 + 1) + C(0)$$
$$6 = A(1) + 0 + 0$$
$$6 = A$$

You have exhausted the most convenient values of x, but you can now use the known values of A and C to find the value of B. So, let $x = 1$, $A = 6$, and $C = 9$.

$$5(1)^2 + 20(1) + 6 = 6(1 + 1)^2 + B(1)(1 + 1) + 9(1)$$
$$31 = 6(4) + 2B + 9$$
$$-2 = 2B$$
$$-1 = B$$

So, the partial fraction decomposition is

$$\frac{x^4 + 2x^3 + 6x^2 + 20x + 6}{x^3 + 2x^2 + x} = x + \frac{6}{x} + \frac{-1}{x + 1} + \frac{9}{(x + 1)^2}.$$

✓ *Checkpoint* 🔊))) *Audio-video solution in English & Spanish at LarsonPrecalculus.com*

Write the partial fraction decomposition of $\dfrac{x^4 + x^3 + x + 4}{x^3 + x^2}$.

The procedure used to solve for the constants $A, B, \ldots$ in Examples 1 and 2 works well when the factors of the denominator are linear. When the denominator contains irreducible quadratic factors, a better process is to write the right side of the basic equation in polynomial form, *equate the coefficients* of like terms to form a system of equations, and solve the resulting system for the constants.

EXAMPLE 3 Distinct Linear and Quadratic Factors

Write the partial fraction decomposition of

$$\frac{3x^2 + 4x + 4}{x^3 + 4x}.$$

Solution This expression is proper, so begin by factoring the denominator. The denominator factors as

$$x^3 + 4x = x(x^2 + 4)$$

so when writing the form of the decomposition, include one partial fraction with a constant numerator and one partial fraction with a linear numerator.

$$\frac{3x^2 + 4x + 4}{x^3 + 4x} = \frac{A}{x} + \frac{Bx + C}{x^2 + 4} \qquad \text{Write form of decomposition.}$$

Multiply each side by the LCD, $x(x^2 + 4)$, to obtain the basic equation.

$$3x^2 + 4x + 4 = A(x^2 + 4) + (Bx + C)x \qquad \text{Basic equation}$$

Expand this basic equation and collect like terms.

$$3x^2 + 4x + 4 = Ax^2 + 4A + Bx^2 + Cx$$
$$= (A + B)x^2 + Cx + 4A \qquad \text{Polynomial form}$$

Use the fact that two polynomials are equal if and only if the coefficients of like terms are equal to write a system of linear equations.

$$3x^2 + 4x + 4 = (A + B)x^2 + Cx + 4A \qquad \text{Equate coefficients of like terms.}$$

$$\begin{cases} A + B = 3 & \text{Equation 1} \\ C = 4 & \text{Equation 2} \\ 4A = 4 & \text{Equation 3} \end{cases}$$

From Equation 3 and Equation 2, you have

$$A = 1 \quad \text{and} \quad C = 4.$$

Back-substituting $A = 1$ into Equation 1 yields

$$1 + B = 3 \implies B = 2.$$

So, the partial fraction decomposition is

$$\frac{3x^2 + 4x + 4}{x^3 + 4x} = \frac{1}{x} + \frac{2x + 4}{x^2 + 4}.$$

✓ **Checkpoint** Audio-video solution in English & Spanish at LarsonPrecalculus.com

Write the partial fraction decomposition of

$$\frac{2x^2 - 5}{x^3 + x}.$$

Johann Bernoulli (1667–1748), a Swiss mathematician, introduced the method of partial fractions and was instrumental in the early development of calculus. Bernoulli was a professor at the University of Basel and taught many outstanding students, including the renowned Leonhard Euler.

The next example shows how to find the partial fraction decomposition of a rational expression whose denominator has a *repeated* quadratic factor.

> **EXAMPLE 4**   **Repeated Quadratic Factors**

See LarsonPrecalculus.com for an interactive version of this type of example.

Write the partial fraction decomposition of

$$\frac{8x^3 + 13x}{(x^2 + 2)^2}.$$

Solution   Include one partial fraction with a linear numerator for each power of $(x^2 + 2)$.

$$\frac{8x^3 + 13x}{(x^2 + 2)^2} = \frac{Ax + B}{x^2 + 2} + \frac{Cx + D}{(x^2 + 2)^2} \qquad \text{Write form of decomposition.}$$

Multiply each side by the LCD, $(x^2 + 2)^2$, to obtain the basic equation.

$$8x^3 + 13x = (Ax + B)(x^2 + 2) + Cx + D \qquad \text{Basic equation}$$

$$= Ax^3 + 2Ax + Bx^2 + 2B + Cx + D$$

$$= Ax^3 + Bx^2 + (2A + C)x + (2B + D) \qquad \text{Polynomial form}$$

Equate coefficients of like terms on opposite sides of the equation to write a system of linear equations.

$$8x^3 + 0x^2 + 13x + 0 = Ax^3 + Bx^2 + (2A + C)x + (2B + D)$$

$$\begin{cases} A & = 8 & \text{Equation 1} \\ \quad B & = 0 & \text{Equation 2} \\ 2A + \quad C & = 13 & \text{Equation 3} \\ \quad 2B + D & = 0 & \text{Equation 4} \end{cases}$$

Use the values $A = 8$ and $B = 0$ to obtain the values of C and D.

$$2(8) + C = 13 \qquad \text{Substitute 8 for } A \text{ in Equation 3.}$$

$$C = -3$$

$$2(0) + D = 0 \qquad \text{Substitute 0 for } B \text{ in Equation 4.}$$

$$D = 0$$

So, using

$$A = 8, \quad B = 0, \quad C = -3, \quad \text{and} \quad D = 0$$

the partial fraction decomposition is

$$\frac{8x^3 + 13x}{(x^2 + 2)^2} = \frac{8x}{x^2 + 2} + \frac{-3x}{(x^2 + 2)^2}.$$

Check this result by combining the two partial fractions on the right side of the equation, or by using your graphing utility.

✓ *Checkpoint* ))) *Audio-video solution in English & Spanish at LarsonPrecalculus.com*

Write the partial fraction decomposition of $\dfrac{x^3 + 3x^2 - 2x + 7}{(x^2 + 4)^2}$. ■

> **EXAMPLE 5** **Repeated Linear and Quadratic Factors**

Write the partial fraction decomposition of $\dfrac{x + 5}{x^2(x^2 + 1)^2}$.

Solution Include one partial fraction with a constant numerator for each power of x and one partial fraction with a linear numerator for each power of $(x^2 + 1)$.

$$\frac{x + 5}{x^2(x^2 + 1)^2} = \frac{A}{x} + \frac{B}{x^2} + \frac{Cx + D}{x^2 + 1} + \frac{Ex + F}{(x^2 + 1)^2} \qquad \textcolor{red}{\text{Write form of decomposition.}}$$

Multiply each side by the LCD, $x^2(x^2 + 1)^2$, to obtain the basic equation.

$$x + 5 = Ax(x^2 + 1)^2 + B(x^2 + 1)^2 + (Cx + D)x^2(x^2 + 1) + (Ex + F)x^2 \qquad \textcolor{red}{\substack{\text{Basic} \\ \text{equation}}}$$

$$= (A + C)x^5 + (B + D)x^4 + (2A + C + E)x^3 + (2B + D + F)x^2 + Ax + B$$

Write and solve the system of equations formed by equating coefficients on opposite sides of the equation to show that $A = 1$, $B = 5$, $C = -1$, $D = -5$, $E = -1$, and $F = -5$, and that the partial fraction decomposition is

$$\frac{x + 5}{x^2(x^2 + 1)^2} = \frac{1}{x} + \frac{5}{x^2} - \frac{x + 5}{x^2 + 1} - \frac{x + 5}{(x^2 + 1)^2}.$$

✓ *Checkpoint*  *Audio-video solution in English & Spanish at LarsonPrecalculus.com*

Write the partial fraction decomposition of $\dfrac{4x - 8}{x^2(x^2 + 2)^2}$.

Guidelines for Solving the Basic Equation

Linear Factors

1. Substitute the *zeros* of the distinct linear factors into the basic equation.
2. For repeated linear factors, use the coefficients determined in Step 1 to rewrite the basic equation. Then substitute *other* convenient values of x and solve for the remaining coefficients.

Quadratic Factors

1. Expand the basic equation.
2. Collect terms according to powers of x.
3. Equate the coefficients of like terms to obtain a system of equations involving the constants, $A, B, C, \ldots$
4. Use the system of linear equations to solve for $A, B, C, \ldots$

Keep in mind that for *improper* rational expressions, you must first divide before applying partial fraction decomposition.

Summarize (Section 6.4)

1. Explain what is meant by the partial fraction decomposition of a rational expression *(page 456)*.
2. Explain how to find the partial fraction decomposition of a rational expression *(pages 456–461)*. For examples of finding partial fraction decompositions of rational expressions, see Examples 1–5.

6.4 Exercises

See **CalcChat.com** for tutorial help and worked-out solutions to odd-numbered exercises.

Vocabulary: Fill in the blanks.

1. The result of writing a rational expression as the sum of two or more simpler rational expressions is called the _____ _____ _____.

2. If the degree of the numerator of a rational expression is greater than or equal to the degree of the denominator, then the fraction is _____.

3. Each fraction on the right side of the equation $\dfrac{x-1}{x^2-8x+15} = \dfrac{-1}{x-3} + \dfrac{2}{x-5}$ is a _____ _____.

4. You obtain the _____ _____ by multiplying each side of the partial fraction decomposition form by the least common denominator.

Skills and Applications

Matching In Exercises 5–8, match the rational expression with the form of its decomposition. [The decompositions are labeled (a), (b), (c), and (d).]

(a) $\dfrac{A}{x} + \dfrac{B}{x+2} + \dfrac{C}{x-2}$

(b) $\dfrac{A}{x} + \dfrac{B}{x-4}$

(c) $\dfrac{A}{x} + \dfrac{B}{x^2} + \dfrac{C}{x-4}$

(d) $\dfrac{A}{x} + \dfrac{B}{x-4} + \dfrac{C}{(x-4)^2}$

5. $\dfrac{3x-1}{x(x-4)}$

6. $\dfrac{3x-1}{x^2(x-4)}$

7. $\dfrac{3x-1}{x(x-4)^2}$

8. $\dfrac{3x-1}{x(x^2-4)}$

 Writing the Form of the Decomposition In Exercises 9–16, write the form of the partial fraction decomposition of the rational expression. Do not solve for the constants.

9. $\dfrac{3}{x^2-2x}$

10. $\dfrac{x-2}{x^2+4x+3}$

11. $\dfrac{6x+5}{(x+2)^4}$

12. $\dfrac{5x^2+3}{x^2(x-4)^2}$

13. $\dfrac{2x-3}{x^3+10x}$

14. $\dfrac{x-1}{x(x^2+1)^2}$

15. $\dfrac{8x}{x^2(x^2+3)^2}$

16. $\dfrac{x^2-9}{x^3(x^2+2)^2}$

 Writing the Partial Fraction Decomposition In Exercises 17–42, write the partial fraction decomposition of the rational expression. Check your result algebraically.

17. $\dfrac{1}{x^2+x}$

18. $\dfrac{3}{x^2-3x}$

19. $\dfrac{3}{x^2+x-2}$

20. $\dfrac{x+1}{x^2-x-6}$

21. $\dfrac{1}{x^2-1}$

22. $\dfrac{1}{4x^2-9}$

23. $\dfrac{x^2+12x+12}{x^3-4x}$

24. $\dfrac{x+2}{x(x^2-9)}$

25. $\dfrac{3x}{(x-3)^2}$

26. $\dfrac{2x-3}{(x-1)^2}$

27. $\dfrac{4x^2+2x-1}{x^2(x+1)}$

28. $\dfrac{6x^2+1}{x^2(x-1)^2}$

29. $\dfrac{x^2+2x+3}{x^3+x}$

30. $\dfrac{2x}{x^3-1}$

31. $\dfrac{x}{x^3-x^2-2x+2}$

32. $\dfrac{x+6}{x^3-3x^2-4x+12}$

33. $\dfrac{x}{16x^4-1}$

34. $\dfrac{3}{x^4+x}$

35. $\dfrac{x^2+5}{(x+1)(x^2-2x+3)}$

36. $\dfrac{x^2-4x+7}{(x+1)(x^2-2x+3)}$

37. $\dfrac{2x^2+x+8}{(x^2+4)^2}$

38. $\dfrac{3x^2+1}{(x^2+2)^2}$

39. $\dfrac{5x^2-2}{(x^2+3)^3}$

40. $\dfrac{x^2-4x+6}{(x^2+4)^3}$

41. $\dfrac{8x-12}{x^2(x^2+2)^2}$

42. $\dfrac{x+1}{x^3(x^2+1)^2}$

 Improper Rational Expression Decomposition In Exercises 43–50, write the partial fraction decomposition of the improper rational expression.

43. $\dfrac{x^2-x}{x^2+x+1}$

44. $\dfrac{x^2-4x}{x^2+x+6}$

45. $\dfrac{2x^3-x^2+x+5}{x^2+3x+2}$

46. $\dfrac{x^3+2x^2-x+1}{x^2+3x-4}$

47. $\dfrac{x^4}{(x-1)^3}$

48. $\dfrac{16x^4}{(2x-1)^3}$

49. $\dfrac{x^4+2x^3+4x^2+8x+2}{x^3+2x^2+x}$

50. $\dfrac{2x^4+8x^3+7x^2-7x-12}{x^3+4x^2+4x}$

Writing the Partial Fraction Decomposition In Exercises 51–58, write the partial fraction decomposition of the rational expression. Use a graphing utility to check your result.

51. $\dfrac{5 - x}{2x^2 + x - 1}$

52. $\dfrac{4x^2 - 1}{2x(x + 1)^2}$

53. $\dfrac{3x^2 - 7x - 2}{x^3 - x}$

54. $\dfrac{3x + 6}{x^3 + 2x}$

55. $\dfrac{x^2 + x + 2}{(x^2 + 2)^2}$

56. $\dfrac{x^3}{(x + 2)^2(x - 2)^2}$

57. $\dfrac{2x^3 - 4x^2 - 15x + 5}{x^2 - 2x - 8}$

58. $\dfrac{x^3 - x + 3}{x^2 + x - 2}$

59. **Environmental Science** The predicted cost C (in thousands of dollars) for a company to remove $p\%$ of a chemical from its waste water is given by the model

$$C = \frac{120p}{10{,}000 - p^2}, \quad 0 \le p < 100.$$

Write the partial fraction decomposition for the rational function. Verify your result by using a graphing utility to create a table comparing the original function with the partial fractions.

60. **Thermodynamics**

The magnitude of the range R of exhaust temperatures (in degrees Fahrenheit) in an experimental diesel engine is approximated by the model

$$R = \frac{5000(4 - 3x)}{(11 - 7x)(7 - 4x)}, \quad 0 < x \le 1$$

where x is the relative load (in foot-pounds).

(a) Write the partial fraction decomposition of the equation.

(b) The decomposition in part (a) is the difference of two fractions. The absolute values of the terms give the expected maximum and minimum temperatures of the exhaust gases for different loads.

Ymax = |1st term| Ymin = |2nd term|

Write the equations for Ymax and Ymin.

(c) Use a graphing utility to graph each equation from part (b) in the same viewing window.

(d) Determine the expected maximum and minimum temperatures for a relative load of 0.5.

Exploration

True or False? In Exercises 61–63, determine whether the statement is true or false. Justify your answer.

61. For the rational expression $\dfrac{x}{(x + 10)(x - 10)^2}$, the partial fraction decomposition is of the form

$$\frac{A}{x + 10} + \frac{B}{(x - 10)^2}.$$

62. When writing the partial fraction decomposition of the expression $\dfrac{x^3 + x - 2}{x^2 - 5x - 14}$, the first step is to divide the numerator by the denominator.

63. In the partial fraction decomposition of a rational expression, the denominators of each partial fraction always have a lower degree than the denominator of the original expression.

64. **HOW DO YOU SEE IT?** Identify the graph of the rational function and the graph representing each partial fraction of its partial fraction decomposition. Then state any relationship between the vertical asymptotes of the graph of the rational function and the vertical asymptotes of the graphs representing the partial fractions of the decomposition. To print an enlarged copy of the graph, go to *MathGraphs.com*.

(a) $y = \dfrac{x - 12}{x(x - 4)}$

$\qquad = \dfrac{3}{x} - \dfrac{2}{x - 4}$

(b) $y = \dfrac{2(4x - 3)}{x^2 - 9}$

$\qquad = \dfrac{3}{x - 3} + \dfrac{5}{x + 3}$

65. **Error Analysis** Describe the error in writing the basic equation for the partial fraction decomposition of the rational expression.

$$\frac{x^2 + 1}{x(x - 1)} = \frac{A}{x} + \frac{B}{x - 1}$$

$$x^2 + 1 = A(x - 1) + Bx$$

66. **Writing** Describe two ways of solving for the constants in a partial fraction decomposition.

6.5 Systems of Inequalities

- Sketch the graphs of inequalities in two variables.
- Solve systems of inequalities.
- Use systems of inequalities in two variables to model and solve real-life problems.

Systems of inequalities in two variables can help you model and solve real-life problems. For example, in Exercise 68 on page 472, you will use a system of inequalities to analyze a person's recommended target heart rate during exercise.

The Graph of an Inequality

The statements

$$3x - 2y < 6 \quad \text{and} \quad 2x^2 + 3y^2 \geq 6$$

are inequalities in two variables. An ordered pair (a, b) is a **solution of an inequality** in x and y when the inequality is true after a and b are substituted for x and y, respectively. The **graph of an inequality** is the collection of all solutions of the inequality. To sketch the graph of an inequality, begin by sketching the graph of the *corresponding equation*. The graph of the equation will usually separate the plane into two or more regions. In each such region, one of the following must be true.

1. *All* points in the region are solutions of the inequality.

2. *No* point in the region is a solution of the inequality.

So, you can determine whether the points in an entire region satisfy the inequality by testing *one* point in the region.

Sketching the Graph of an Inequality in Two Variables

1. Replace the inequality sign by an equal sign and sketch the graph of the equation. (Use a dashed line for $<$ or $>$ and a solid line for $\leq$ or $\geq$.)

2. Test one point in each of the regions formed by the graph in Step 1. If the point satisfies the inequality, then shade the entire region to denote that every point in the region satisfies the inequality.

• • **REMARK** Be careful when you are sketching the graph of an inequality in two variables. A dashed line means that the points on the line or curve *are not* solutions of the inequality. A solid line means that the points on the line or curve *are* solutions of the inequality.

▷

EXAMPLE 1 Sketching the Graph of an Inequality

See LarsonPrecalculus.com for an interactive version of this type of example.

Sketch the graph of $y \geq x^2 - 1$.

Solution Begin by graphing the corresponding equation $y = x^2 - 1$, as shown at the right. Test a point *above* the parabola, such as $(0, 0)$, and a point *below* the parabola, such as $(0, -2)$.

$$(0, 0): \quad 0 \overset{?}{\geq} 0^2 - 1$$

$$0 \geq -1 \qquad (0, 0) \text{ is a solution.}$$

$$(0, -2): \quad -2 \overset{?}{\geq} 0^2 - 1$$

$$-2 \not\geq -1 \qquad (0, -2) \text{ is not a solution.}$$

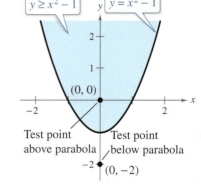

The points that satisfy the inequality $y \geq x^2 - 1$ are those lying above (or on) the parabola, as shown by the shaded region in the figure.

✓ *Checkpoint* 🔊))) *Audio-video solution in English & Spanish at LarsonPrecalculus.com*

Sketch the graph of $(x + 2)^2 + (y - 2)^2 < 16$.

The inequality in Example 1 is a nonlinear inequality in two variables. Many of the examples in this section involve **linear inequalities** such as $ax + by < c$ (where a and b are not both zero). The graph of a linear inequality is a half-plane lying on one side of the line $ax + by = c$.

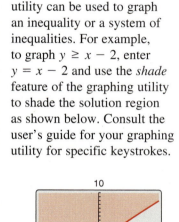

> **EXAMPLE 2**   **Sketching the Graph of a Linear Inequality**

Sketch the graph of each linear inequality.

a. $x > -2$ **b.** $y \le 3$

Solution

a. The graph of the corresponding equation $x = -2$ is a vertical line. The points that satisfy the inequality $x > -2$ are those lying to the right of this line, as shown in Figure 6.16.

b. The graph of the corresponding equation $y = 3$ is a horizontal line. The points that satisfy the inequality $y \le 3$ are those lying below (or on) this line, as shown in Figure 6.17.

> TECHNOLOGY A graphing utility can be used to graph an inequality or a system of inequalities. For example, to graph $y \ge x - 2$, enter $y = x - 2$ and use the *shade* feature of the graphing utility to shade the solution region as shown below. Consult the user's guide for your graphing utility for specific keystrokes.

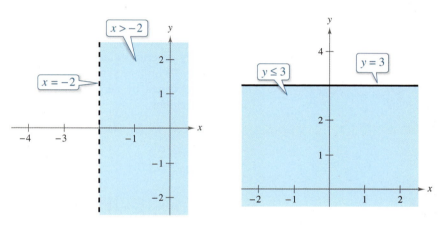

Figure 6.16   **Figure 6.17**

✓ *Checkpoint*  *Audio-video solution in English & Spanish at LarsonPrecalculus.com*

Sketch the graph of $x \ge 3$.

> **EXAMPLE 3**   **Sketching the Graph of a Linear Inequality**

Sketch the graph of $x - y < 2$.

Solution   The graph of the corresponding equation $x - y = 2$ is a line, as shown in Figure 6.18. The origin $(0, 0)$ satisfies the inequality, so the graph consists of the half-plane lying above the line. (Check a point below the line. Regardless of which point you choose, you will find that it does not satisfy the inequality.)

✓ *Checkpoint* *Audio-video solution in English & Spanish at LarsonPrecalculus.com*

Sketch the graph of $x + y > -2$.   ■

To graph a linear inequality, it sometimes helps to write the inequality in slope-intercept form. For example, writing $x - y < 2$ as

$$y > x - 2$$

helps you to see that the solution points lie *above* the line $x - y = 2$ (or $y = x - 2$), as shown in Figure 6.18.

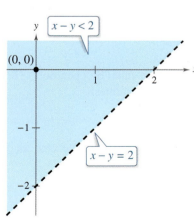

Figure 6.18

Systems of Inequalities

Many practical problems in business, science, and engineering involve systems of linear inequalities. A **solution** of a system of inequalities in x and y is a point (x, y) that satisfies each inequality in the system.

To sketch the graph of a system of inequalities in two variables, first sketch the graph of each individual inequality (on the same coordinate system) and then find the region that is *common* to every graph in the system. This region represents the **solution set** of the system. For a system of *linear inequalities,* it is helpful to find the vertices of the solution region.

EXAMPLE 4   Solving a System of Inequalities

Sketch the graph of the solution set of the system of inequalities. Label the vertices of the region.

$$\begin{cases} x - y < 2 \\ \quad x > -2 \\ \quad y \le 3 \end{cases}$$

Solution   The graphs of these inequalities are shown in Figures 6.18, 6.16, and 6.17, respectively, on page 465. The triangular region common to all three graphs can be found by superimposing the graphs on the same coordinate system, as shown in Figure 6.19. To find the vertices of the region, solve the three systems of corresponding equations obtained by taking *pairs* of equations representing the boundaries of the individual regions.

Vertex A: $(-2, -4)$

$$\begin{cases} x - y = 2 \\ \quad x = -2 \end{cases}$$

Vertex B: $(5, 3)$

$$\begin{cases} x - y = 2 \\ \quad y = 3 \end{cases}$$

Vertex C: $(-2, 3)$

$$\begin{cases} x = -2 \\ y = 3 \end{cases}$$

• • **REMARK** Using a different colored pencil to shade the solution of each inequality in a system will make identifying the solution of the system of inequalities easier.

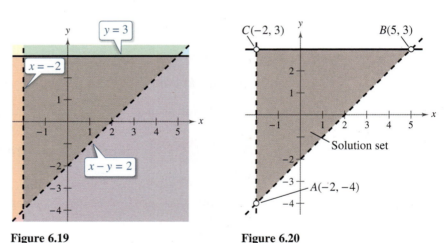

Figure 6.19

Figure 6.20

Note in Figure 6.20 that the vertices of the region are represented by open dots. This means that the vertices *are not* solutions of the system of inequalities.

✓ **Checkpoint** 🔊))) *Audio-video solution in English & Spanish at LarsonPrecalculus.com*

Sketch the graph of the solution set of the system of inequalities. Label the vertices of the region.

$$\begin{cases} \quad x + y \ge 1 \\ -x + y \ge 1 \\ \qquad y \le 2 \end{cases}$$

For the triangular region shown in Example 4, each pair of boundary lines intersects at a vertex of the region. With more complicated regions, two boundary lines can sometimes intersect at a point that is not a vertex of the region, as shown below. As you sketch the graph of a solution set, use your sketch along with the inequalities of the system to determine which points of intersection are actually vertices of the region.

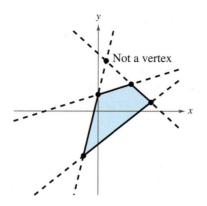

EXAMPLE 5   Solving a System of Inequalities

Sketch the graph of the solution set of the system of inequalities.

$$\begin{cases} x^2 - y \le 1 & \text{Inequality 1} \\ -x + y \le 1 & \text{Inequality 2} \end{cases}$$

Solution The points that satisfy the inequality

$$x^2 - y \le 1 \qquad \text{Inequality 1}$$

are the points lying above (or on) the parabola

$$y = x^2 - 1. \qquad \text{Parabola}$$

The points satisfying the inequality

$$-x + y \le 1 \qquad \text{Inequality 2}$$

are the points lying below (or on) the line

$$y = x + 1. \qquad \text{Line}$$

To find the points of intersection of the parabola and the line, solve the system of corresponding equations.

$$\begin{cases} x^2 - y = 1 \\ -x + y = 1 \end{cases}$$

Using the method of substitution, you find that the solutions are $(-1, 0)$ and $(2, 3)$. These points are both solutions of the original system, so they are represented by closed dots in the graph of the solution region shown at the right.

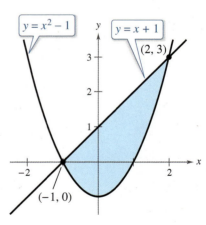

✓ **Checkpoint** ◀))) *Audio-video solution in English & Spanish at LarsonPrecalculus.com*

Sketch the graph of the solution set of the system of inequalities.

$$\begin{cases} x - y^2 > 0 \\ x + y < 2 \end{cases}$$

When solving a system of inequalities, be aware that the system might have no solution *or* its graph might be an unbounded region in the plane. Examples 6 and 7 show these two possibilities.

EXAMPLE 6 **A System with No Solution**

Sketch the solution set of the system of inequalities.

$$\begin{cases} x + y > 3 \\ x + y < -1 \end{cases}$$

Solution It should be clear from the way it is written that the system has no solution, because the quantity $(x + y)$ cannot be both less than -1 and greater than 3. The graph of the inequality $x + y > 3$ is the half-plane lying above the line $x + y = 3$, and the graph of the inequality $x + y < -1$ is the half-plane lying below the line $x + y = -1$, as shown below. These two half-planes have no points in common. So, the system of inequalities has no solution.

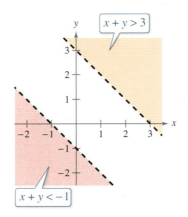

✓ **Checkpoint** 🔊)) *Audio-video solution in English & Spanish at LarsonPrecalculus.com*

Sketch the solution set of the system of inequalities.

$$\begin{cases} 2x - y < -3 \\ 2x - y > 1 \end{cases}$$

EXAMPLE 7 **An Unbounded Solution Set**

Sketch the solution set of the system of inequalities.

$$\begin{cases} x + y < 3 \\ x + 2y > 3 \end{cases}$$

Solution The graph of the inequality $x + y < 3$ is the half-plane that lies below the line $x + y = 3$. The graph of the inequality $x + 2y > 3$ is the half-plane that lies above the line $x + 2y = 3$. The intersection of these two half-planes is an *infinite wedge* that has a vertex at $(3, 0)$, as shown in Figure 6.21. So, the solution set of the system of inequalities is unbounded.

✓ **Checkpoint** 🔊)) *Audio-video solution in English & Spanish at LarsonPrecalculus.com*

Sketch the solution set of the system of inequalities.

$$\begin{cases} x^2 - y < 0 \\ x - y < -2 \end{cases}$$

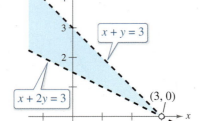

Figure 6.21

Applications

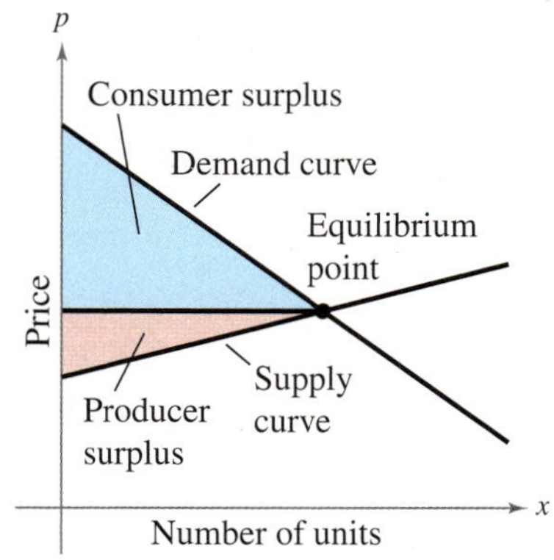

Example 9 in Section 6.2 discussed the equilibrium point for a system of demand and supply equations. The next example discusses two related concepts that economists call *consumer surplus* and *producer surplus*. As shown in Figure 6.22, the **consumer surplus** is the area of the region formed by the demand curve, the horizontal line passing through the equilibrium point, and the *p*-axis. Similarly, the **producer surplus** is the area of the region formed by the supply curve, the horizontal line passing through the equilibrium point, and the *p*-axis. The consumer surplus is a measure of the amount that consumers would have been willing to pay *above* what they actually paid, whereas the producer surplus is a measure of the amount that producers would have been willing to receive *below* what they actually received.

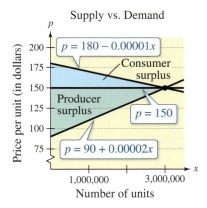

Figure 6.22

EXAMPLE 8   Consumer Surplus and Producer Surplus

The demand and supply equations for a new type of video game console are

$$\begin{cases} p = 180 - 0.00001x & \text{Demand equation} \\ p = 90 + 0.00002x & \text{Supply equation} \end{cases}$$

where *p* is the price per unit (in dollars) and *x* is the number of units. Find the consumer surplus and producer surplus for these two equations.

Solution   Begin by finding the equilibrium point (when supply and demand are equal) by solving the equation

$$90 + 0.00002x = 180 - 0.00001x.$$

In Example 9 in Section 6.2, you saw that the solution is $x = 3{,}000{,}000$ units, which corresponds to a price of $p = \$150$. So, the consumer surplus and producer surplus are the areas of the solution sets of the following systems of inequalities.

Consumer Surplus	Producer Surplus
$\begin{cases} p \le 180 - 0.00001x \\ p \ge 150 \\ x \ge 0 \end{cases}$	$\begin{cases} p \ge 90 + 0.00002x \\ p \le 150 \\ x \ge 0 \end{cases}$

In other words, the consumer and producer surpluses are the areas of the shaded triangles shown in Figure 6.23.

$$\begin{aligned} \text{Consumer surplus} &= \tfrac{1}{2}(\text{base})(\text{height}) \\ &= \tfrac{1}{2}(3{,}000{,}000)(30) \\ &= \$45{,}000{,}000 \\ \text{Producer surplus} &= \tfrac{1}{2}(\text{base})(\text{height}) \\ &= \tfrac{1}{2}(3{,}000{,}000)(60) \\ &= \$90{,}000{,}000 \end{aligned}$$

✓ **Checkpoint**  *Audio-video solution in English & Spanish at LarsonPrecalculus.com*

The demand and supply equations for a flat-screen television are

$$\begin{cases} p = 567 - 0.00002x & \text{Demand equation} \\ p = 492 + 0.00003x & \text{Supply equation} \end{cases}$$

where *p* is the price per unit (in dollars) and *x* is the number of units. Find the consumer surplus and producer surplus for these two equations.

Figure 6.23

EXAMPLE 9 **Nutrition**

The liquid portion of a diet is to provide at least 300 calories, 36 units of vitamin A, and 90 units of vitamin C. A cup of dietary drink X provides 60 calories, 12 units of vitamin A, and 10 units of vitamin C. A cup of dietary drink Y provides 60 calories, 6 units of vitamin A, and 30 units of vitamin C. Write a system of linear inequalities that describes how many cups of each drink must be consumed each day to meet or exceed the minimum daily requirements for calories and vitamins.

Solution   Begin by letting x represent the number of cups of dietary drink X and y represent the number of cups of dietary drink Y. To meet or exceed the minimum daily requirements, the following inequalities must be satisfied.

$$\begin{cases} 60x + 60y \geq 300 & \text{Calories} \\ 12x + 6y \geq 36 & \text{Vitamin A} \\ 10x + 30y \geq 90 & \text{Vitamin C} \\ x \geq 0 \\ y \geq 0 \end{cases}$$

The last two inequalities are included because x and y cannot be negative. The graph of this system of inequalities is shown below. (More is said about this application in Example 6 in Section 6.6.)

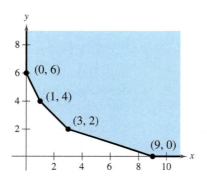

 Checkpoint ◀))) *Audio-video solution in English & Spanish at LarsonPrecalculus.com*

A public aquarium is adding coral nutrients to a large reef tank. A bottle of brand X nutrients contains 8 units of nutrient A, 1 unit of nutrient B, and 2 units of nutrient C. A bottle of brand Y nutrients contains 2 units of nutrient A, 1 unit of nutrient B, and 7 units of nutrient C. The minimum amounts of nutrients A, B, and C that need to be added to the tank are 16 units, 5 units, and 20 units, respectively. Set up a system of linear inequalities that describes how many bottles of each brand must be added to meet or exceed the needs. ◼

Summarize   (Section 6.5)

1. Explain how to sketch the graph of an inequality in two variables *(page 464)*. For examples of sketching the graphs of inequalities in two variables, see Examples 1–3.

2. Explain how to solve a system of inequalities *(page 466)*. For examples of solving systems of inequalities, see Examples 4–7.

3. Describe examples of how to use systems of inequalities in two variables to model and solve real-life problems *(pages 469 and 470, Examples 8 and 9)*.

6.5 Exercises

See CalcChat.com for tutorial help and worked-out solutions to odd-numbered exercises.

Vocabulary: Fill in the blanks.

1. An ordered pair (a, b) is a _____ of an inequality in x and y when the inequality is true after a and b are substituted for x and y, respectively.

2. The _____ of an inequality is the collection of all solutions of the inequality.

3. A _____ of a system of inequalities in x and y is a point (x, y) that satisfies each inequality in the system.

4. The _____ _____ of a system of inequalities in two variables is represented by the region that is common to every graph in the system.

Skills and Applications

 Graphing an Inequality In Exercises 5–18, sketch the graph of the inequality.

5. $y < 5 - x^2$
6. $y^2 - x < 0$
7. $x \geq 6$
8. $x < -4$
9. $y > -7$
10. $10 \geq y$
11. $y < 2 - x$
12. $y > 4x - 3$
13. $2y - x \geq 4$
14. $5x + 3y \geq -15$
15. $x^2 + (y - 3)^2 < 4$
16. $(x + 2)^2 + y^2 > 9$
17. $y > -\dfrac{2}{x^2 + 1}$
18. $y \leq \dfrac{3}{x^2 + x + 1}$

 Graphing an Inequality In Exercises 19–26, use a graphing utility to graph the inequality.

19. $y \geq -\ln(x - 1)$
20. $y < \ln(x + 3) - 1$
21. $y < 2^x$
22. $y \geq 3^{-x} - 2$
23. $y \leq 2 - \frac{1}{5}x$
24. $y > -2.4x + 3.3$
25. $\frac{2}{3}y + 2x^2 - 5 \geq 0$
26. $-\frac{1}{6}x^2 - \frac{2}{7}y < -\frac{1}{3}$

Writing an Inequality In Exercises 27–30, write an inequality for the shaded region shown in the figure.

27.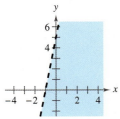
28.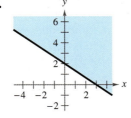
29.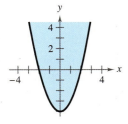
30.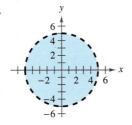

 Solving a System of Inequalities In Exercises 31–38, sketch the graph of the solution set of the system of inequalities. Label the vertices of the region.

31. $\begin{cases} x + y \leq 1 \\ -x + y \leq 1 \\ y \geq 0 \end{cases}$
32. $\begin{cases} 3x + 4y < 12 \\ x > 0 \\ y > 0 \end{cases}$
33. $\begin{cases} -3x + 2y < 6 \\ x - 4y > -2 \\ 2x + y < 3 \end{cases}$
34. $\begin{cases} x - 7y > -36 \\ 5x + 2y > 5 \\ 6x - 5y > 6 \end{cases}$
35. $\begin{cases} 2x + y > 2 \\ 6x + 3y < 2 \end{cases}$
36. $\begin{cases} x - 2y < -6 \\ 5x - 3y > -9 \end{cases}$
37. $\begin{cases} 2x - 3y > 7 \\ 5x + y < 9 \end{cases}$
38. $\begin{cases} 4x - 6y > 2 \\ -2x + 3y \geq 5 \end{cases}$

Solving a System of Inequalities In Exercises 39–44, sketch the graph of the solution set of the system.

39. $\begin{cases} x^2 + y \leq 7 \\ x \geq -2 \\ y \geq 0 \end{cases}$
40. $\begin{cases} 4x^2 + y \geq 2 \\ x \leq 1 \\ y \leq 1 \end{cases}$
41. $\begin{cases} x - y^2 > 0 \\ x - y > 2 \end{cases}$
42. $\begin{cases} x^2 + y^2 \leq 25 \\ 4x - 3y \leq 0 \end{cases}$
43. $\begin{cases} 3x + 4 \geq y^2 \\ x - y < 0 \end{cases}$
44. $\begin{cases} x < 2y - y^2 \\ 0 < x + y \end{cases}$

Solving a System of Inequalities In Exercises 45–50, use a graphing utility to graph the solution set of the system of inequalities.

45. $\begin{cases} y \leq \sqrt{3x} + 1 \\ y \geq x^2 + 1 \end{cases}$
46. $\begin{cases} y < 2\sqrt{x} - 1 \\ y \geq x^2 - 1 \end{cases}$
47. $\begin{cases} y < -x^2 + 2x + 3 \\ y > x^2 - 4x + 3 \end{cases}$
48. $\begin{cases} y \geq x^4 - 2x^2 + 1 \\ y \leq 1 - x^2 \end{cases}$
49. $\begin{cases} x^2y \geq 1 \\ 0 < x \leq 4 \\ y \leq 4 \end{cases}$
50. $\begin{cases} y \leq e^{-x^2/2} \\ y \geq 0 \\ -2 \leq x \leq 2 \end{cases}$

Writing a System of Inequalities In Exercises 51–58, write a system of inequalities that describes the region.

51.

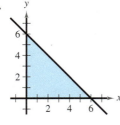

52.

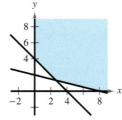

53.

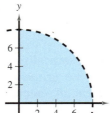

54.

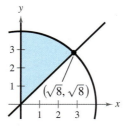

$(\sqrt{8}, \sqrt{8})$

55. Rectangle: vertices at $(4, 3)$, $(9, 3)$, $(9, 9)$, $(4, 9)$

56. Parallelogram: vertices at $(0, 0)$, $(4, 0)$, $(1, 4)$, $(5, 4)$

57. Triangle: vertices at $(0, 0)$, $(6, 0)$, $(1, 5)$

58. Triangle: vertices at $(-1, 0)$, $(1, 0)$, $(0, 1)$

Consumer Surplus and Producer Surplus In Exercises 59–62, (a) graph the systems of inequalities representing the consumer surplus and producer surplus for the supply and demand equations and (b) find the consumer surplus and producer surplus.

Demand	Supply
59. $p = 50 - 0.5x$	$p = 0.125x$
60. $p = 100 - 0.05x$	$p = 25 + 0.1x$
61. $p = 140 - 0.00002x$	$p = 80 + 0.00001x$
62. $p = 400 - 0.0002x$	$p = 225 + 0.0005x$

63. Investment Analysis A person plans to invest up to $20,000 in two different interest-bearing accounts. Each account must contain at least $5000. The amount in one account is to be at least twice the amount in the other account. Write and graph a system of inequalities that describes the various amounts that can be deposited in each account.

64. Ticket Sales For a concert event, there are $30 reserved seat tickets and $20 general admission tickets. There are 2000 reserved seats available, and fire regulations limit the number of paid ticket holders to 3000. The promoter must take in at least $75,000 in ticket sales. Write and graph a system of inequalities that describes the different numbers of tickets that can be sold.

65. Production A furniture company produces tables and chairs. Each table requires 1 hour in the assembly center and $1\frac{1}{3}$ hours in the finishing center. Each chair requires $1\frac{1}{2}$ hours in the assembly center and $1\frac{1}{2}$ hours in the finishing center. The assembly center is available 12 hours per day, and the finishing center is available 15 hours per day. Write and graph a system of inequalities that describes all possible production levels.

66. Inventory A store sells two models of laptop computers. The store stocks at least twice as many units of model A as of model B. The costs to the store for the two models are $800 and $1200, respectively. The management does not want more than $20,000 in computer inventory at any one time, and it wants at least four model A laptop computers and two model B laptop computers in inventory at all times. Write and graph a system of inequalities that describes all possible inventory levels.

67. Nutrition A dietician prescribes a special dietary plan using two different foods. Each ounce of food X contains 180 milligrams of calcium, 6 milligrams of iron, and 220 milligrams of magnesium. Each ounce of food Y contains 100 milligrams of calcium, 1 milligram of iron, and 40 milligrams of magnesium. The minimum daily requirements of the diet are 1000 milligrams of calcium, 18 milligrams of iron, and 400 milligrams of magnesium.

(a) Write and graph a system of inequalities that describes the different amounts of food X and food Y that can be prescribed.

(b) Find two solutions of the system and interpret their meanings in the context of the problem.

68. Target Heart Rate

One formula for a person's maximum heart rate is $220 - x$, where x is the person's age in years for $20 \le x \le 70$. The American Heart Association recommends that when a person exercises, the person should strive for a heart rate that is at least 50% of the maximum and at most 85% of the maximum. *(Source: American Heart Association)*

(a) Write and graph a system of inequalities that describes the exercise target heart rate region.

(b) Find two solutions of the system and interpret their meanings in the context of the problem.

69. Shipping A warehouse supervisor has instructions to ship at least 50 bags of gravel that weigh 55 pounds each and at least 40 bags of stone that weigh 70 pounds each. The maximum weight capacity of the truck being used is 7500 pounds.

(a) Write and graph a system that describes the numbers of bags of stone and gravel that can be shipped.

(b) Find two solutions of the system and interpret their meanings in the context of the problem.

70. Physical Fitness Facility A physical fitness facility is constructing an indoor running track with space for exercise equipment inside the track (see figure). The track must be at least 125 meters long, and the exercise space must have an area of at least 500 square meters.

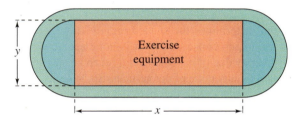

(a) Write and graph a system of inequalities that describes the requirements of the facility.

(b) Find two solutions of the system and interpret their meanings in the context of the problem.

Exploration

True or False? **In Exercises 71 and 72, determine whether the statement is true or false. Justify your answer.**

71. The area of the figure described by the system

$$\begin{cases} x \geq -3 \\ x \leq 6 \\ y \leq 5 \\ y \geq -6 \end{cases}$$

is 99 square units.

72. The graph shows the solution of the system

$$\begin{cases} y \leq 6 \\ -4x - 9y > 6. \\ 3x + y^2 \geq 2 \end{cases}$$

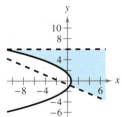

73. Think About It After graphing the boundary line of the inequality $x + y < 3$, explain how to determine the region that you need to shade.

74. **HOW DO YOU SEE IT?** The graph of the solution of the inequality $x + 2y < 6$ is shown in the figure. Describe how the solution set would change for each inequality.

(a) $x + 2y \leq 6$

(b) $x + 2y > 6$

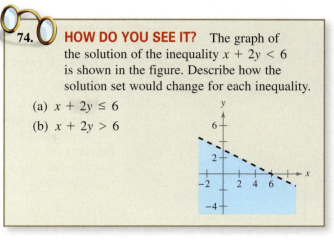

75. Matching Match the system of inequalities with the graph of its solution. [The graphs are labeled (i), (ii), (iii), and (iv).]

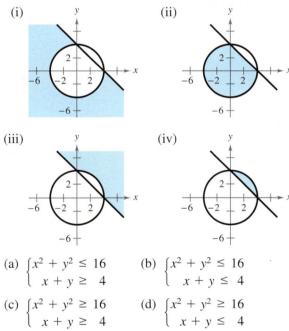

(a) $\begin{cases} x^2 + y^2 \leq 16 \\ x + y \geq 4 \end{cases}$ (b) $\begin{cases} x^2 + y^2 \leq 16 \\ x + y \leq 4 \end{cases}$

(c) $\begin{cases} x^2 + y^2 \geq 16 \\ x + y \geq 4 \end{cases}$ (d) $\begin{cases} x^2 + y^2 \geq 16 \\ x + y \leq 4 \end{cases}$

76. Graphical Reasoning Two concentric circles have radii x and y, where $y > x$. The area between the circles is at least 10 square units.

(a) Write a system of inequalities that describes the constraints on the circles.

(b) Use a graphing utility to graph the system of inequalities in part (a). Graph the line $y = x$ in the same viewing window.

(c) Identify the graph of the line in relation to the boundary of the inequality. Explain its meaning in the context of the problem.

6.6 Linear Programming

■ Solve linear programming problems.
■ Use linear programming to model and solve real-life problems.

Linear programming is often used to make real-life decisions. For example, in Exercise 43 on page 482, you will use linear programming to determine the optimal acreage and yield for two fruit crops.

Linear Programming: A Graphical Approach

Many applications in business and economics involve a process called **optimization,** in which you find the minimum or maximum value of a quantity. In this section, you will study an optimization strategy called **linear programming.**

A two-dimensional linear programming problem consists of a linear **objective function** and a system of linear inequalities called **constraints.** The objective function gives the quantity to be maximized (or minimized), and the constraints determine the set of **feasible solutions.** For example, one such problem is to maximize the value of

$$z = ax + by \qquad \text{\textcolor{red}{Objective function}}$$

subject to a set of constraints that determines the shaded region shown below. Every point in the shaded region satisfies each constraint, so it is not clear how you should find the point that yields a maximum value of z. Fortunately, it can be shown that when there is an optimal solution, it must occur at one of the vertices. So, *to find the maximum value of z, evaluate z at each of the vertices* and compare the resulting z-values.

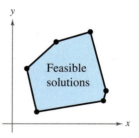

Optimal Solution of a Linear Programming Problem

If a linear programming problem has an optimal solution, then it must occur at a vertex of the set of feasible solutions.

A linear programming problem can include hundreds, and sometimes even thousands, of variables. However, in this section, you will solve linear programming problems that involve only two variables. The guidelines for solving a linear programming problem in two variables are listed below.

Solving a Linear Programming Problem

1. Sketch the region corresponding to the system of constraints. (The points inside or on the boundary of the region are *feasible solutions*.)

2. Find the vertices of the region.

3. Evaluate the objective function at each of the vertices and select the values of the variables that optimize the objective function. For a bounded region, both a minimum and a maximum value will exist. (For an unbounded region, *if* an optimal solution exists, then it will occur at a vertex.)

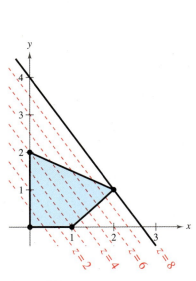

Figure 6.24

EXAMPLE 1 **Solving a Linear Programming Problem**

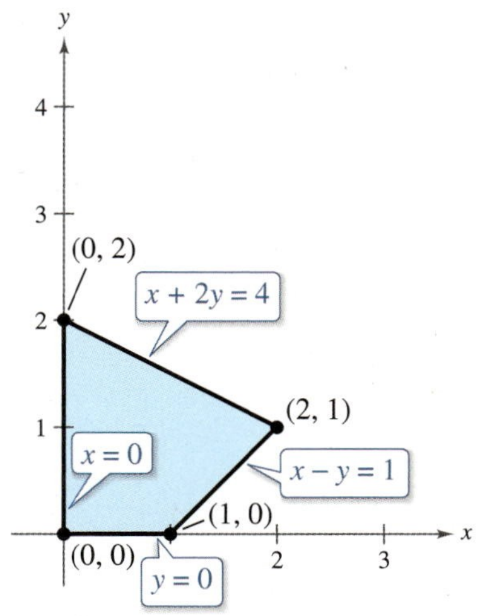

Find the maximum value of

$$z = 3x + 2y \qquad \text{Objective function}$$

subject to the following constraints.

$$\left.\begin{array}{r} x \geq 0 \\ y \geq 0 \\ x + 2y \leq 4 \\ x - y \leq 1 \end{array}\right\} \quad \text{Constraints}$$

Solution The constraints form the region shown in Figure 6.24. At the four vertices of this region, the objective function has the following values.

At $(0, 0)$: $z = 3(0) + 2(0) = 0$

At $(0, 2)$: $z = 3(0) + 2(2) = 4$

At $(2, 1)$: $z = 3(2) + 2(1) = 8$ Maximum value of z

At $(1, 0)$: $z = 3(1) + 2(0) = 3$

So, the maximum value of z is 8, and this occurs when $x = 2$ and $y = 1$.

✓ **Checkpoint** 🔊)) *Audio-video solution in English & Spanish at LarsonPrecalculus.com*

Find the maximum value of

$$z = 4x + 5y$$

subject to the following constraints.

$$\begin{array}{r} x \geq 0 \\ y \geq 0 \\ x + y \leq 6 \end{array}$$

In Example 1, consider some of the *interior* points in the region. You will see that the corresponding values of z are less than 8. Here are some examples.

At $(1, 1)$: $z = 3(1) + 2(1) = 5$

At $\left(\frac{1}{2}, \frac{3}{2}\right)$: $z = 3\left(\frac{1}{2}\right) + 2\left(\frac{3}{2}\right) = \frac{9}{2}$

At $\left(\frac{3}{2}, 1\right)$: $z = 3\left(\frac{3}{2}\right) + 2(1) = \frac{13}{2}$

To see why the maximum value of the objective function in Example 1 must occur at a vertex, consider writing the objective function in slope-intercept form.

$$y = -\frac{3}{2}x + \frac{z}{2} \qquad \text{Family of lines}$$

Notice that the y-intercept

$$b = \frac{z}{2}$$

varies according to the value of z. This equation represents a family of lines, each of slope $-\frac{3}{2}$. Of these infinitely many lines, you want the one that has the largest z-value while still intersecting the region determined by the constraints. In other words, of all the lines whose slope is $-\frac{3}{2}$, you want the one that has the largest y-intercept *and* intersects the region, as shown in Figure 6.25. Notice from the graph that this line will pass through one point of the region, the vertex $(2, 1)$.

Figure 6.25

The next example shows that the same basic procedure can be used to solve a problem in which the objective function is to be *minimized*.

EXAMPLE 2   **Minimizing an Objective Function**

See LarsonPrecalculus.com for an interactive version of this type of example.

Find the minimum value of

$$z = 5x + 7y \qquad \text{Objective function}$$

where $x \geq 0$ and $y \geq 0$, subject to the following constraints.

$$\left.\begin{array}{rcl} 2x + 3y & \geq & 6 \\ 3x - y & \leq & 15 \\ -x + y & \leq & 4 \\ 2x + 5y & \leq & 27 \end{array}\right\} \text{Constraints}$$

y

5 — (1, 5)
4 — (0, 4)   (6, 3)
3 —
2 — (0, 2)
1 —   (3, 0)   (5, 0)
   1 2 3 4 5 6   x

Figure 6.26

Solution   Figure 6.26 shows the region bounded by the constraints. Evaluate the objective function at each vertex.

At $(0, 2)$: $z = 5(0) + 7(2) = 14 \qquad$ Minimum value of z

At $(0, 4)$: $z = 5(0) + 7(4) = 28$

At $(1, 5)$: $z = 5(1) + 7(5) = 40$

At $(6, 3)$: $z = 5(6) + 7(3) = 51$

At $(5, 0)$: $z = 5(5) + 7(0) = 25$

At $(3, 0)$: $z = 5(3) + 7(0) = 15$

The minimum value of z is 14, and this occurs when $x = 0$ and $y = 2$.

✓ **Checkpoint** 🔊))) *Audio-video solution in English & Spanish at LarsonPrecalculus.com*

Find the minimum value of

$$z = 12x + 8y$$

where $x \geq 0$ and $y \geq 0$, subject to the following constraints.

$$\begin{array}{rcr} 5x + 6y & \leq & 420 \\ -x + 6y & \leq & 240 \\ -2x + y & \geq & -100 \end{array}$$

EXAMPLE 3   **Maximizing an Objective Function**

Find the maximum value of

$$z = 5x + 7y \qquad \text{Objective function}$$

where $x \geq 0$ and $y \geq 0$, subject to the constraints given in Example 2.

Solution   Using the values of z found at the vertices in Example 2, you can conclude that the maximum value of z is 51, which occurs when $x = 6$ and $y = 3$.

✓ **Checkpoint** 🔊))) *Audio-video solution in English & Spanish at LarsonPrecalculus.com*

Find the maximum value of

$$z = 12x + 8y$$

where $x \geq 0$ and $y \geq 0$, subject to the constraints given in the Checkpoint with Example 2.

George Dantzig (1914–2005) was the first to propose the simplex method for linear programming in 1947. This technique defined the steps needed to find the optimal solution of a complex multivariable problem.

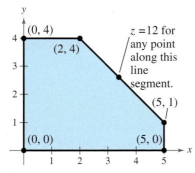

Figure 6.27

It is possible for the maximum (or minimum) value in a linear programming problem to occur at *two* different vertices. For example, at the vertices of the region shown in Figure 6.27, the objective function

$$z = 2x + 2y \qquad \text{Objective function}$$

has the following values.

At $(0, 0)$: $z = 2(0) + 2(0) = 0$

At $(0, 4)$: $z = 2(0) + 2(4) = 8$

At $(2, 4)$: $z = 2(2) + 2(4) = 12 \qquad$ Maximum value of z

At $(5, 1)$: $z = 2(5) + 2(1) = 12 \qquad$ Maximum value of z

At $(5, 0)$: $z = 2(5) + 2(0) = 10$

In this case, the objective function has a maximum value not only at the vertices $(2, 4)$ and $(5, 1)$; it also has a maximum value (of 12) at *any point on the line segment connecting these two vertices*. Note that the objective function in slope-intercept form $y = -x + \frac{1}{2}z$ has the same slope as the line through the vertices $(2, 4)$ and $(5, 1)$.

Some linear programming problems have no optimal solution. This can occur when the region determined by the constraints is *unbounded*. Example 4 illustrates such a problem.

EXAMPLE 4   **An Unbounded Region**

Find the maximum value of

$$z = 4x + 2y \qquad \text{Objective function}$$

where $x \geq 0$ and $y \geq 0$, subject to the following constraints.

$$\left. \begin{array}{r} x + 2y \geq 4 \\ 3x + y \geq 7 \\ -x + 2y \leq 7 \end{array} \right\} \text{Constraints}$$

Solution   Figure 6.28 shows the region determined by the constraints. For this unbounded region, there is no maximum value of z. To see this, note that the point $(x, 0)$ lies in the region for all values of $x \geq 4$. Substituting this point into the objective function, you get

$$z = 4(x) + 2(0) = 4x.$$

You can choose values of x to obtain values of z that are as large as you want. So, there is no maximum value of z. However, there *is* a minimum value of z.

At $(1, 4)$: $z = 4(1) + 2(4) = 12$

At $(2, 1)$: $z = 4(2) + 2(1) = 10 \qquad$ Minimum value of z

At $(4, 0)$: $z = 4(4) + 2(0) = 16$

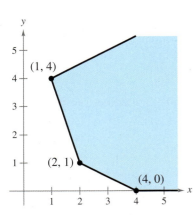

Figure 6.28

So, the minimum value of z is 10, and this occurs when $x = 2$ and $y = 1$.

✓ *Checkpoint*  Audio-video solution in English & Spanish at LarsonPrecalculus.com

Find the minimum value of

$$z = 3x + 7y$$

where $x \geq 0$ and $y \geq 0$, subject to the following constraints.

$$\begin{array}{r} x + y \geq 8 \\ 3x + 5y \geq 30 \end{array}$$

Applications

Example 5 shows how linear programming can help you find the maximum profit in a business application.

EXAMPLE 5 **Optimal Profit**

A candy manufacturer wants to maximize the combined profit for two types of boxed chocolates. A box of chocolate-covered creams yields a profit of $1.50 per box, and a box of chocolate-covered nuts yields a profit of $2.00 per box. Market tests and available resources indicate the following constraints.

1. The combined production level must not exceed 1200 boxes per month.

2. The demand for chocolate-covered nuts is no more than half the demand for chocolate-covered creams.

3. The production level for chocolate-covered creams must be less than or equal to 600 boxes plus three times the production level for chocolate-covered nuts.

What is the maximum monthly profit? How many boxes of each type are produced per month to yield the maximum profit?

Solution Let x be the number of boxes of chocolate-covered creams and let y be the number of boxes of chocolate-covered nuts. Then, the objective function (for the combined profit) is

$$P = 1.5x + 2y. \qquad \text{Objective function}$$

The three constraints yield the following linear inequalities.

1. $x + y \le 1200$ $\implies$ $x + y \le 1200$

2. $y \le \frac{1}{2}x$ $\implies$ $-x + 2y \le 0$

3. $x \le 600 + 3y$ $\implies$ $x - 3y \le 600$

Neither x nor y can be negative, so there are two additional constraints of

$$x \ge 0 \quad \text{and} \quad y \ge 0.$$

Figure 6.29 shows the region determined by the constraints. To find the maximum monthly profit, evaluate P at the vertices of the region.

At $(0, 0)$: $\quad P = 1.5(0) \quad + 2(0) \quad = \quad 0$

At $(800, 400)$: $\quad P = 1.5(800) \quad + 2(400) = 2000 \qquad \text{Maximum profit}$

At $(1050, 150)$: $\quad P = 1.5(1050) + 2(150) = 1875$

At $(600, 0)$: $\quad P = 1.5(600) \quad + 2(0) \quad = \quad 900$

So, the maximum monthly profit is $2000, and it occurs when the monthly production consists of 800 boxes of chocolate-covered creams and 400 boxes of chocolate-covered nuts.

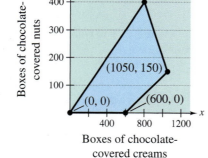

Maximum Monthly Profit

Boxes of chocolate-covered nuts

Boxes of chocolate-covered creams

Figure 6.29

✓ **Checkpoint** ◀))) *Audio-video solution in English & Spanish at LarsonPrecalculus.com*

In Example 5, the candy manufacturer improves the production of chocolate-covered creams so that the profit is $2.50 per box. The constraints do not change. What is the maximum monthly profit? How many boxes of each type are produced per month to yield the maximum profit? ◼

Example 6 shows how linear programming can help you find the optimal cost in a real-life application.

Optimal Cost

The liquid portion of a daily diet is to provide at least 300 calories, 36 units of vitamin A, and 90 units of vitamin C. Two dietary drinks (drink X and drink Y) will be used to meet these requirements. Information about one cup of each drink is shown below. How many cups of each dietary drink must be consumed each day to satisfy the daily requirements at the minimum possible cost?

	Cost	Calories	Vitamin A	Vitamin C
Drink X	$0.72	60	12 units	10 units
Drink Y	$0.90	60	6 units	30 units

Solution As in Example 9 in Section 6.5, let x be the number of cups of dietary drink X and let y be the number of cups of dietary drink Y.

$$\left. \begin{array}{ll} \text{For calories:} & 60x + 60y \geq 300 \\ \text{For vitamin A:} & 12x + 6y \geq 36 \\ \text{For vitamin C:} & 10x + 30y \geq 90 \\ & x \geq 0 \\ & y \geq 0 \end{array} \right\} \text{Constraints}$$

The cost C is given by

$$C = 0.72x + 0.90y. \qquad \text{Objective function}$$

Figure 6.30 shows the graph of the region corresponding to the constraints. You want to incur as little cost as possible, so you want to determine the *minimum* cost. Evaluate C at each vertex of the region.

At $(0, 6)$: $C = 0.72(0) + 0.90(6) = 5.40$

At $(1, 4)$: $C = 0.72(1) + 0.90(4) = 4.32$

At $(3, 2)$: $C = 0.72(3) + 0.90(2) = 3.96$ Minimum value of C

At $(9, 0)$: $C = 0.72(9) + 0.90(0) = 6.48$

You find that the minimum cost is $3.96 per day, and this occurs when 3 cups of drink X and 2 cups of drink Y are consumed each day.

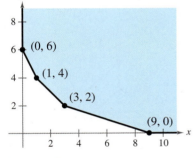

Figure 6.30

✓ **Checkpoint** 🔊⟍⟍ *Audio-video solution in English & Spanish at LarsonPrecalculus.com*

A public aquarium is adding coral nutrients to a large reef tank. The minimum amounts of nutrients A, B, and C that need to be added to the tank are 16 units, 5 units, and 20 units, respectively. Information about each bottle of brand X and brand Y additives is shown below. How many bottles of each brand must be added to satisfy the needs of the reef tank at the minimum possible cost?

	Cost	Nutrient A	Nutrient B	Nutrient C
Brand X	$15	8 units	1 unit	2 units
Brand Y	$30	2 units	1 unit	7 units

Summarize   (Section 6.6)

1. State the guidelines for solving a linear programming problem *(page 474)*. For examples of solving linear programming problems, see Examples 1–4.

2. Describe examples of real-life applications of linear programming *(pages 478 and 479, Examples 5 and 6)*.

6.6 Exercises

See **CalcChat.com** for tutorial help and worked-out solutions to odd-numbered exercises.

Vocabulary: Fill in the blanks.

1. In the process called _____, you find the maximum or minimum value of a quantity.
2. One type of optimization strategy is called _____ _____.
3. The _____ function of a linear programming problem gives the quantity to be maximized (or minimized).
4. The _____ of a linear programming problem determine the set of _____ _____.
5. The feasible solutions are _____ or _____ the boundary of the region corresponding to a system of constraints.
6. If a linear programming problem has a solution, then it must occur at a _____ of the set of feasible solutions.

Skills and Applications

Solving a Linear Programming Problem In Exercises 7–12, use the graph of the region corresponding to the system of constraints to find the minimum and maximum values of the objective function subject to the constraints. Identify the points where the optimal values occur.

7. Objective function:

$z = 4x + 3y$

Constraints:

$$x \geq 0$$
$$y \geq 0$$
$$x + y \leq 5$$

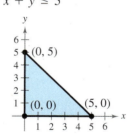

8. Objective function:

$z = 2x + 8y$

Constraints:

$$x \geq 0$$
$$y \geq 0$$
$$2x + y \leq 4$$

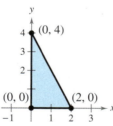

9. Objective function:

$z = 2x + 5y$

Constraints:

$$y \geq 0$$
$$5x + y \geq 5$$
$$x + 3y \leq 15$$
$$4x + y \leq 16$$

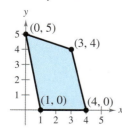

10. Objective function:

$z = 4x + 5y$

Constraints:

$$x \geq 0$$
$$2x + 3y \geq 6$$
$$3x - y \leq 9$$
$$x + 4y \leq 16$$

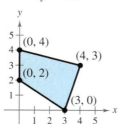

11. Objective function:

$z = 10x + 7y$

Constraints:

$$0 \leq x \leq 60$$
$$y \leq 45$$
$$5x + 6y \leq 420$$
$$x + 3y \geq 60$$

12. Objective function:

$z = 40x + 45y$

Constraints:

$$x \geq 0$$
$$y \geq 0$$
$$8x + 9y \leq 7200$$
$$8x + 9y \geq 3600$$

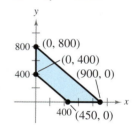

Solving a Linear Programming Problem In Exercises 13–16, sketch the region corresponding to the system of constraints. Then find the minimum and maximum values of the objective function (if possible) and the points where they occur, subject to the constraints.

13. Objective function:

$z = 3x + 2y$

Constraints:

$$x \geq 0$$
$$y \geq 0$$
$$3x + 2y \leq 24$$
$$4x + y \geq 12$$

14. Objective function:

$z = 5x + \frac{1}{2}y$

Constraints:

$$x \geq 0$$
$$y \geq 0$$
$$\frac{1}{2}x + y \leq 8$$
$$x + \frac{1}{2}y \geq 4$$

15. Objective function:

$z = 4x + 5y$

Constraints:

$$x \geq 0$$
$$y \geq 0$$
$$x + y \geq 8$$
$$3x + 5y \geq 30$$

16. Objective function:

$z = 5x + 4y$

Constraints:

$$x \geq 0$$
$$y \geq 0$$
$$2x + 2y \geq 10$$
$$x + 2y \geq 6$$

Using Technology In Exercises 17–20, use a graphing utility to graph the region corresponding to the system of constraints. Then find the minimum and maximum values of the objective function and the points where they occur, subject to the constraints.

17. Objective function:

$z = 3x + y$

Constraints:

$x \geq 0$

$y \geq 0$

$x + 4y \leq 60$

$3x + 2y \geq 48$

18. Objective function:

$z = 6x + 3y$

Constraints:

$x \geq 0$

$y \geq 0$

$2x + 3y \leq 60$

$2x + y \leq 28$

$4x + y \leq 48$

19. Objective function:

$z = x$

Constraints:

(See Exercise 17.)

20. Objective function:

$z = y$

Constraints:

(See Exercise 18.)

Finding Minimum and Maximum Values In Exercises 21–24, find the minimum and maximum values of the objective function and the points where they occur, subject to the constraints $x \geq 0$, $y \geq 0$, $x + 4y \leq 20$, $x + y \leq 18$, and $2x + 2y \leq 21$.

21. $z = x + 5y$

22. $z = 2x + 4y$

23. $z = 4x + 5y$

24. $z = 4x + y$

Finding Minimum and Maximum Values In Exercises 25–28, find the minimum and maximum values (if possible) of the objective function and the points where they occur, subject to the constraints $x \geq 0$, $3x + y \geq 15$, $-x + 4y \geq 8$, and $-2x + y \geq -19$.

25. $z = x + 2y$

26. $z = 5x + 3y$

27. $z = x - y$

28. $z = y - x$

Describing an Unusual Characteristic In Exercises 29–36, the linear programming problem has an unusual characteristic. Sketch a graph of the solution region for the problem and describe the unusual characteristic. Find the minimum and maximum values of the objective function (if possible) and the points where they occur.

29. Objective function:

$z = 2.5x + y$

Constraints:

$x \geq 0$

$y \geq 0$

$3x + 5y \leq 15$

$5x + 2y \leq 10$

30. Objective function:

$z = x + y$

Constraints:

$x \geq 0$

$y \geq 0$

$-x + y \leq 1$

$-x + 2y \leq 4$

31. Objective function:

$z = -x + 2y$

Constraints:

$x \geq 0$

$y \geq 0$

$x \leq 10$

$x + y \leq 7$

32. Objective function:

$z = x + y$

Constraints:

$x \geq 0$

$y \geq 0$

$-x + y \leq 0$

$-3x + y \geq 3$

33. Objective function:

$z = 3x + 4y$

Constraints:

$x \geq 0$

$y \geq 0$

$x + y \leq 1$

$2x + y \geq 4$

34. Objective function:

$z = x + 2y$

Constraints:

$x \geq 0$

$y \geq 0$

$x + 2y \leq 4$

$2x + y \leq 4$

35. Objective function:

$z = x + y$

Constraints:

$x \geq 9$

$0 \leq y \leq 7$

$-x + 3y \leq -6$

36. Objective function:

$z = 2x - y$

Constraints:

$0 \leq x \leq 9$

$0 \leq y \leq 11$

$5x + 2y \leq 67$

37. Optimal Profit A merchant plans to sell two models of MP3 players at prices of $225 and $250. The $225 model yields a profit of $30 per unit and the $250 model yields a profit of $31 per unit. The merchant estimates that the total monthly demand will not exceed 275 units. The merchant does not want to invest more than $63,000 in inventory for these products. What is the optimal inventory level for each model? What is the optimal profit?

38. Optimal Profit A manufacturer produces two models of elliptical cross-training exercise machines. The times for assembling, finishing, and packaging model X are 3 hours, 3 hours, and 0.8 hour, respectively. The times for model Y are 4 hours, 2.5 hours, and 0.4 hour. The total times available for assembling, finishing, and packaging are 6000 hours, 4200 hours, and 950 hours, respectively. The profits per unit are $300 for model X and $375 for model Y. What is the optimal production level for each model? What is the optimal profit?

39. Optimal Cost A public aquarium is adding coral nutrients to a large reef tank. The minimum amounts of nutrients A, B, and C that need to be added to the tank are 30 units, 16 units, and 24 units, respectively. Information about each bottle of brand X and brand Y additives is shown below. How many bottles of each brand must be added to satisfy the needs of the reef tank at the minimum possible cost?

	Cost	Nutrient A	Nutrient B	Nutrient C
Brand X	$25	3 units	3 units	7 units
Brand Y	$15	9 units	2 units	2 units

40. Optimal Labor A manufacturer has two different factories that produce three grades of steel: structural steel, rail steel, and pipe steel. They must produce 32 tons of structural, 26 tons of rail, and 30 tons of pipe steel to fill an order. The table shows the number of employees at each factory and the amounts of steel they produce hourly. How many hours should each factory operate to fill the orders at the minimum labor (in employee-hours)? What is the minimum labor?

	Factory X	Factory Y
Employees	120	80
Structural steel	2	5
Rail steel	8	2
Pipe steel	3	3

41. Optimal Revenue An accounting firm has 780 hours of staff time and 272 hours of reviewing time available each week. The firm charges $1600 for an audit and $250 for a tax return. Each audit requires 60 hours of staff time and 16 hours of review time. Each tax return requires 10 hours of staff time and 4 hours of review time. What numbers of audits and tax returns will yield an optimal revenue? What is the optimal revenue?

42. Optimal Revenue The accounting firm in Exercise 41 lowers its charge for an audit to $1400. What numbers of audits and tax returns will yield an optimal revenue? What is the optimal revenue?

43. Agriculture

A fruit grower raises crops A and B. The yield is 300 bushels per acre for crop A and 500 bushels per acre for crop B. Research and available resources indicate the following constraints.

• The fruit grower has 150 acres of land available.

• It takes 1 day to trim the trees on an acre of crop A and 2 days to trim an acre of crop B, and there are 240 days per year available for trimming.

• It takes 0.3 day to pick an acre of crop A and 0.1 day to pick an acre of crop B, and there are 30 days per year available for picking.

What is the optimal acreage for each fruit? What is the optimal yield?

44. Optimal Profit In Exercise 43, the profit is $185 per acre for crop A and $245 per acre for crop B. What is the optimal profit?

45. Media Selection A company budgets a maximum of $1,000,000 for national advertising of an allergy medication. Each TV ad costs $100,000 and each one-page newspaper ad costs $20,000. Each TV ad is expected to be viewed by 20 million viewers, and each newspaper ad is expected to be seen by 5 million readers. The company's marketing department recommends that at most 80% of the budget be spent on TV ads. What is the optimal amount that should be spent on each type of ad? What is the optimal total audience?

46. Investment Portfolio An investor has up to $450,000 to invest in two types of investments. Type A pays 6% annually and type B pays 10% annually. To have a well-balanced portfolio, the investor imposes the following conditions. At least one-half of the total portfolio is to be allocated to type A investments and at least one-fourth of the portfolio is to be allocated to type B investments. What is the optimal amount that should be invested in each type of investment? What is the optimal return?

Exploration

True or False? In Exercises 47–49, determine whether the statement is true or false. Justify your answer.

47. If an objective function has a maximum value at the vertices $(4, 7)$ and $(8, 3)$, then it also has a maximum value at the points $(4.5, 6.5)$ and $(7.8, 3.2)$.

48. If an objective function has a minimum value at the vertex $(20, 0)$, then it also has a minimum value at $(0, 0)$.

49. If the constraint region of a linear programming problem lies in Quadrant I and is unbounded, the objective function cannot have a maximum value.

50. HOW DO YOU SEE IT? Using the constraint region shown below, determine which of the following objective functions has (a) a maximum at vertex A, (b) a maximum at vertex B, (c) a maximum at vertex C, and (d) a minimum at vertex C.

(i) $z = 2x + y$

(ii) $z = 2x - y$

(iii) $z = -x + 2y$

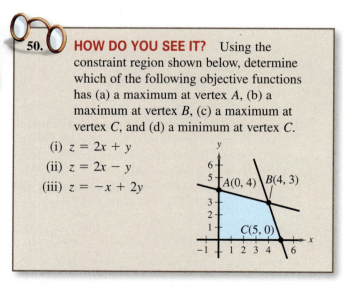

51. Think About It A linear programming problem has an objective function $z = 3x + 5y$ and an infinite number of optimal solutions that lie on the line segment connecting two points. What is the slope between the points?

Chapter Summary

What Did You Learn?	Explanation/Examples	Review Exercises
Section 6.1 Use the method of substitution to solve systems of linear equations in two variables (*p. 422*).	**Method of Substitution** 1. *Solve* one of the equations for one variable in terms of the other. 2. *Substitute* the expression found in Step 1 into the other equation to obtain an equation in one variable. 3. *Solve* the equation obtained in Step 2. 4. *Back-substitute* the value obtained in Step 3 into the expression obtained in Step 1 to find the value of the other variable. 5. *Check* that the solution satisfies *each* of the original equations.	1–6
Use the method of substitution to solve systems of nonlinear equations in two variables (*p. 425*).	The method of substitution (see steps above) can be used to solve systems in which one or both of the equations are nonlinear. (See Examples 3 and 4.)	7–10
Use a graphical method to solve systems of equations in two variables (*p. 426*).	One intersection point Two intersection points No intersection points	11–18
Use systems of equations to model and solve real-life problems (*p. 427*).	A system of equations can help you find the break-even point for a company. (See Example 6.)	19–22
Section 6.2 Use the method of elimination to solve systems of linear equations in two variables (*p. 432*).	**Method of Elimination** 1. *Obtain coefficients* for x (or y) that differ only in sign. 2. *Add* the equations to eliminate one variable. 3. *Solve* the equation obtained in Step 2. 4. *Back-substitute* the value obtained in Step 3 into either of the original equations and solve for the other variable. 5. *Check* that the solution satisfies *each* of the original equations.	23–28
Interpret graphically the numbers of solutions of systems of linear equations in two variables (*p. 436*).	Exactly one solution Infinitely many solutions No solution	29–32
Use systems of linear equations in two variables to model and solve real-life problems (*p. 438*).	A system of linear equations in two variables can help you find the equilibrium point for a market. (See Example 9.)	33, 34

What Did You Learn?	**Explanation/Examples**	**Review Exercises**
Section 6.3 Use back-substitution to solve linear systems in row-echelon form *(p. 444)*.	$$\begin{cases} x - 2y + 3z = 9 \\ -x + 3y\quad\quad = -4 \\ 2x - 5y + 5z = 17 \end{cases} \Rightarrow \begin{cases} x - 2y + 3z = 9 \\ \quad\quad y + 3z = 5 \\ \quad\quad\quad\quad z = 2 \end{cases}$$ **Row-Echelon Form**	35, 36
Use Gaussian elimination to solve systems of linear equations *(p. 445)*.	To produce an equivalent system of linear equations, use one or more of the following row operations. (1) Interchange two equations. (2) Multiply one equation by a nonzero constant. (3) Add a multiple of one of the equations to another equation to replace the latter equation.	37–42
Solve nonsquare systems of linear equations *(p. 449)*.	In a nonsquare system, the number of equations differs from the number of variables. A system of linear equations cannot have a unique solution unless there are at least as many equations as there are variables in the system.	43, 44
Use systems of linear equations in three or more variables to model and solve real-life problems *(p. 450)*.	A system of linear equations in three variables can help you find the position equation of an object that is moving in a (vertical) line with constant acceleration. (See Example 7.)	45–54
Section 6.4 Recognize partial fraction decompositions of rational expressions *(p. 456)*.	$$\frac{9}{x^3 - 6x^2} = \frac{9}{x^2(x-6)} = \frac{A}{x} + \frac{B}{x^2} + \frac{C}{x-6}$$	55–58
Find partial fraction decompositions of rational expressions *(p. 457)*.	The techniques used for determining the constants in the numerators of partial fractions vary slightly, depending on the type of factors of the denominator: linear or quadratic, distinct or repeated.	59–66
Section 6.5 Sketch the graphs of inequalities in two variables *(p. 464)*, and solve systems of inequalities *(p. 466)*.	A solution of a system of inequalities in x and y is a point (x, y) that satisfies each inequality in the system. $$\begin{cases} x^2 + y \leq 5 \\ \quad x \geq -1 \\ \quad y \geq 0 \end{cases}$$	67–80
Use systems of inequalities in two variables to model and solve real-life problems *(p. 469)*.	A system of inequalities in two variables can help you find the consumer surplus and producer surplus for given demand and supply equations. (See Example 8.)	81–86
Section 6.6 Solve linear programming problems *(p. 474)*.	To solve a linear programming problem, (1) sketch the region corresponding to the system of constraints, (2) find the vertices of the region, and (3) evaluate the objective function at each of the vertices and select the values of the variables that optimize the objective function.	87–90
Use linear programming to model and solve real-life problems *(p. 478)*.	Linear programming can help you find the maximum profit in business applications. (See Example 5.)	91, 92

Review Exercises   See CalcChat.com for tutorial help and worked-out solutions to odd-numbered exercises.

6.1 **Solving a System by Substitution** In Exercises 1–10, solve the system by the method of substitution.

1. $\begin{cases} x + y = 2 \\ x - y = 0 \end{cases}$

2. $\begin{cases} 2x - 3y = 3 \\ x - y = 0 \end{cases}$

3. $\begin{cases} 4x - y - 1 = 0 \\ 8x + y - 17 = 0 \end{cases}$

4. $\begin{cases} 10x + 6y + 14 = 0 \\ x + 9y + 7 = 0 \end{cases}$

5. $\begin{cases} 0.5x + y = 0.75 \\ 1.25x - 4.5y = -2.5 \end{cases}$

6. $\begin{cases} -x + \frac{2}{5}y = \frac{3}{5} \\ -x + \frac{1}{5}y = -\frac{4}{5} \end{cases}$

7. $\begin{cases} x^2 - y^2 = 9 \\ x - y = 1 \end{cases}$

8. $\begin{cases} x^2 + y^2 = 169 \\ 3x + 2y = 39 \end{cases}$

9. $\begin{cases} y = 2x^2 \\ y = x^4 - 2x^2 \end{cases}$

10. $\begin{cases} x = y + 3 \\ x = y^2 + 1 \end{cases}$

Solving a System of Equations Graphically In Exercises 11–14, solve the system graphically.

11. $\begin{cases} 2x - y = 10 \\ x + 5y = -6 \end{cases}$

12. $\begin{cases} 8x - 3y = -3 \\ 2x + 5y = 28 \end{cases}$

13. $\begin{cases} y = 2x^2 - 4x + 1 \\ y = x^2 - 4x + 3 \end{cases}$

14. $\begin{cases} y^2 - 2y + x = 0 \\ x + y = 0 \end{cases}$

Using Technology In Exercises 15–18, use a graphing utility to solve the systems of equations. Round your solution(s) to two decimal places.

15. $\begin{cases} y = -2e^{-x} \\ 2e^x + y = 0 \end{cases}$

16. $\begin{cases} x^2 + y^2 = 100 \\ 2x - 3y = -12 \end{cases}$

17. $\begin{cases} y = 2 + \log x \\ y = \frac{3}{4}x + 5 \end{cases}$

18. $\begin{cases} y = \ln(x - 1) - 3 \\ y = 4 - \frac{1}{2}x \end{cases}$

19. **Body Mass Index** Body Mass Index (BMI) is a measure of body fat based on height and weight. The 85th percentile BMI for females, ages 9 to 20, increases more slowly than that for males of the same age range. Models that represent the 85th percentile BMI for males and females, ages 9 to 20, are

$\begin{cases} B = 0.78a + 11.7 & \text{Males} \\ B = 0.68a + 13.5 & \text{Females} \end{cases}$

where B is the BMI (kg/m^2) and a represents the age, with $a = 9$ corresponding to 9 years old. Use a graphing utility to determine when the BMI for males exceeds the BMI for females. *(Source: Centers for Disease Control and Prevention)*

20. **Choice of Two Jobs** You receive two sales job offers. One company offers an annual salary of $55,000 plus a year-end bonus of 1.5% of your total sales. The other company offers an annual salary of $52,000 plus a year-end bonus of 2% of your total sales. How much would you have to sell to make the second job offer better?

21. **Geometry** The perimeter of a rectangle is 68 feet and its width is $\frac{8}{9}$ times its length. Use a system of equations to find the dimensions of the rectangle.

22. **Geometry** The perimeter of a rectangle is 40 inches. The area of the rectangle is 96 square inches. Use a system of equations to find the dimensions of the rectangle.

6.2 **Solving a System by Elimination** In Exercises 23–28, solve the system by the method of elimination and check any solutions algebraically.

23. $\begin{cases} 2x - y = 2 \\ 6x + 8y = 39 \end{cases}$

24. $\begin{cases} 12x + 42y = -17 \\ 30x - 18y = 19 \end{cases}$

25. $\begin{cases} 3x - 2y = 0 \\ 3x + 2y = 0 \end{cases}$

26. $\begin{cases} 7x + 12y = 63 \\ 2x + 3y = 15 \end{cases}$

27. $\begin{cases} 1.25x - 2y = 3.5 \\ 5x - 8y = 14 \end{cases}$

28. $\begin{cases} 1.5x + 2.5y = 8.5 \\ 6x + 10y = 24 \end{cases}$

Matching a System with Its Graph In Exercises 29–32, match the system of linear equations with its graph. Describe the number of solutions and state whether the system is consistent or inconsistent. [The graphs are labeled (a), (b), (c), and (d).]

(a)

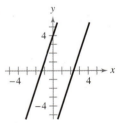

(b)

(c)

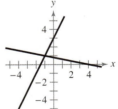

(d)

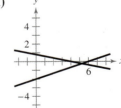
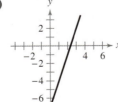

29. $\begin{cases} x + 5y = 4 \\ x - 3y = 6 \end{cases}$

30. $\begin{cases} -3x + y = -7 \\ 9x - 3y = 21 \end{cases}$

31. $\begin{cases} 3x - y = 7 \\ -6x + 2y = 8 \end{cases}$

32. $\begin{cases} 2x - y = -3 \\ x + 5y = 4 \end{cases}$

Finding the Equilibrium Point In Exercises 33 and 34, find the equilibrium point of the demand and supply equations.

Demand	Supply
33. $p = 43 - 0.0002x$	$p = 22 + 0.00001x$
34. $p = 120 - 0.0001x$	$p = 45 + 0.0002x$

6.3 Using Back-Substitution in Row-Echelon Form In Exercises 35 and 36, use back-substitution to solve the system of linear equations.

35. $\begin{cases} x - 4y + 3z = 3 \\ \quad\ y - z = 1 \\ \quad\quad\ z = -5 \end{cases}$

36. $\begin{cases} x - 7y + 8z = 85 \\ \quad\ y - 9z = -35 \\ \quad\quad\ z = 3 \end{cases}$

Solving a System of Linear Equations In Exercises 37–42, solve the system of linear equations and check any solutions algebraically.

37. $\begin{cases} 4x - 3y - 2z = -65 \\ \quad\quad 8y - 7z = -14 \\ 4x \quad\quad - 2z = -44 \end{cases}$

38. $\begin{cases} 5x \quad\quad - 7z = 9 \\ \quad\ 3y - 8z = -4 \\ 5x - 3y \quad = 20 \end{cases}$

39. $\begin{cases} x + 2y + 6z = 4 \\ -3x + 2y - z = -4 \\ 4x \quad\quad + 2z = 16 \end{cases}$

40. $\begin{cases} x - 2y + z = -6 \\ 2x - 3y \quad = -7 \\ -x + 3y - 3z = 11 \end{cases}$

41. $\begin{cases} 2x \quad\quad + 6z = -9 \\ 3x - 2y + 11z = -16 \\ 3x - y + 7z = -11 \end{cases}$

42. $\begin{cases} x \quad\quad\quad + 4w = 1 \\ \quad 3y + z - w = 4 \\ \quad 2y \quad - 3w = 2 \\ 4x - y + 2z \quad = 5 \end{cases}$

Solving a Nonsquare System In Exercises 43 and 44, solve the system of linear equations and check any solutions algebraically.

43. $\begin{cases} 5x - 12y + 7z = 16 \\ 3x - 7y + 4z = 9 \end{cases}$

44. $\begin{cases} 2x + 5y - 19z = 34 \\ 3x + 8y - 31z = 54 \end{cases}$

Finding the Equation of a Parabola In Exercises 45 and 46, find the equation of the parabola

$$y = ax^2 + bx + c$$

that passes through the points. To verify your result, use a graphing utility to plot the points and graph the parabola.

45.
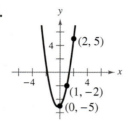

46.
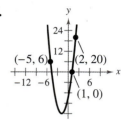

Finding the Equation of a Circle In Exercises 47 and 48, find the equation of the circle

$$x^2 + y^2 + Dx + Ey + F = 0$$

that passes through the points. To verify your result, use a graphing utility to plot the points and graph the circle.

47.
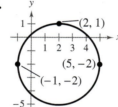

48.
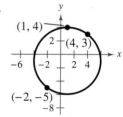

49. **Agriculture** A mixture of 6 gallons of chemical A, 8 gallons of chemical B, and 13 gallons of chemical C is required to kill a destructive crop insect. Commercial spray X contains one, two, and two parts, respectively, of these chemicals. Commercial spray Y contains only chemical C. Commercial spray Z contains chemicals A, B, and C in equal amounts. How much of each type of commercial spray gives the desired mixture?

50. **Sports** The Old Course at St Andrews Links in St Andrews, Scotland, is one of the oldest golf courses in the world. It is an 18-hole course that consists of par-3 holes, par-4 holes, and par-5 holes. There are seven times as many par-4 holes as par-5 holes, and the sum of the numbers of par-3 and par-5 holes is four. Find the numbers of par-3, par-4, and par-5 holes on the course. (Source: St Andrews Links Trust)

51. **Investment** An inheritance of $40,000 is divided among three investments yielding $3500 in interest per year. The interest rates for the three investments are 7%, 9%, and 11% simple interest. Find the amount placed in each investment when the second and third amounts are $3000 and $5000 less than the first, respectively.

52. Investment An amount of $46,000 is divided among three investments yielding $3020 in interest per year. The interest rates for the three investments are 5%, 7%, and 8% simple interest. Find the amount placed in each investment when the second and third amounts are $2000 and $3000 less than the first, respectively.

Modeling Vertical Motion In Exercises 53 and 54, an object moving vertically is at the given heights at the specified times. Find the position equation

$$s = \tfrac{1}{2}at^2 + v_0t + s_0$$

for the object.

53. At $t = 1$ second, $s = 134$ feet

At $t = 2$ seconds, $s = 86$ feet

At $t = 3$ seconds, $s = 6$ feet

54. At $t = 1$ second, $s = 184$ feet

At $t = 2$ seconds, $s = 116$ feet

At $t = 3$ seconds, $s = 16$ feet

6.4 **Writing the Form of the Decomposition** In Exercises 55–58, write the form of the partial fraction decomposition of the rational expression. Do not solve for the constants.

55. $\dfrac{3}{x^2 + 20x}$

56. $\dfrac{x - 8}{x^2 - 3x - 28}$

57. $\dfrac{3x - 4}{x^3 - 5x^2}$

58. $\dfrac{x - 2}{x(x^2 + 2)^2}$

Writing the Partial Fraction Decomposition In Exercises 59–66, write the partial fraction decomposition of the rational expression. Check your result algebraically.

59. $\dfrac{4 - x}{x^2 + 6x + 8}$

60. $\dfrac{-x}{x^2 + 3x + 2}$

61. $\dfrac{x^2}{x^2 + 2x - 15}$

62. $\dfrac{9}{x^2 - 9}$

63. $\dfrac{x^2 + 2x}{x^3 - x^2 + x - 1}$

64. $\dfrac{4x}{3(x - 1)^2}$

65. $\dfrac{3x^2 + 4x}{(x^2 + 1)^2}$

66. $\dfrac{4x^2}{(x - 1)(x^2 + 1)}$

6.5 **Graphing an Inequality** In Exercises 67–72, sketch the graph of the inequality.

67. $y \geq 5$

68. $x < -3$

69. $y \leq 5 - 2x$

70. $3y - x \geq 7$

71. $(x - 1)^2 + (y - 3)^2 < 16$

72. $x^2 + (y + 5)^2 > 1$

Solving a System of Inequalities In Exercises 73–76, sketch the graph of the solution set of the system of inequalities. Label the vertices of the region.

73. $\begin{cases} x + 2y \leq 2 \\ -x + 2y \leq 2 \\ \quad\quad y \geq 0 \end{cases}$

74. $\begin{cases} 2x + 3y < 6 \\ x \quad\quad > 0 \\ \quad\quad y > 0 \end{cases}$

75. $\begin{cases} 2x - y < -1 \\ -3x + 2y > 4 \\ \quad\quad y > 0 \end{cases}$

76. $\begin{cases} 3x - 2y > -4 \\ 6x - y < 5 \\ \quad\quad y < 1 \end{cases}$

Solving a System of Inequalities In Exercises 77–80, sketch the graph of the solution set of the system of inequalities.

77. $\begin{cases} y < x + 1 \\ y > x^2 - 1 \end{cases}$

78. $\begin{cases} y \leq 6 - 2x - x^2 \\ y \geq x + 6 \end{cases}$

79. $\begin{cases} x^2 + y^2 > 4 \\ x^2 + y^2 \leq 9 \end{cases}$

80. $\begin{cases} x^2 + y^2 \leq 169 \\ x + y \leq 7 \end{cases}$

81. Geometry Write a system of inequalities to describe the region of a rectangle with vertices at $(3, 1)$, $(7, 1)$, $(7, 10)$, and $(3, 10)$.

82. Geometry Write a system of inequalities that describes the triangular region with vertices $(0, 5)$, $(5, 0)$, and $(0, 0)$.

Consumer Surplus and Producer Surplus In Exercises 83 and 84, (a) graph the systems of inequalities representing the consumer surplus and producer surplus for the supply and demand equations and (b) find the consumer surplus and producer surplus.

Demand	Supply
83. $p = 160 - 0.0001x$	$p = 70 + 0.0002x$
84. $p = 130 - 0.0002x$	$p = 30 + 0.0003x$

85. Inventory Costs A warehouse operator has 24,000 square feet of floor space in which to store two products. Each unit of product I requires 20 square feet of floor space and costs $12 per day to store. Each unit of product II requires 30 square feet of floor space and costs $8 per day to store. The total storage cost per day cannot exceed $12,400. Write and graph a system that describes all possible inventory levels.

86. Nutrition A dietician prescribes a special dietary plan using two different foods. Each ounce of food X contains 200 milligrams of calcium, 3 milligrams of iron, and 100 milligrams of magnesium. Each ounce of food Y contains 150 milligrams of calcium, 2 milligrams of iron, and 80 milligrams of magnesium. The minimum daily requirements of the diet are 800 milligrams of calcium, 10 milligrams of iron, and 200 milligrams of magnesium.

(a) Write and graph a system of inequalities that describes the different amounts of food X and food Y that can be prescribed.

(b) Find two solutions to the system and interpret their meanings in the context of the problem.

6.6 Solving a Linear Programming Problem In Exercises 87–90, sketch the region corresponding to the system of constraints. Then find the minimum and maximum values of the objective function (if possible) and the points where they occur, subject to the constraints.

87. Objective function:

$z = 3x + 4y$

Constraints:

$x \geq 0$
$y \geq 0$
$2x + 5y \leq 50$
$4x + y \leq 28$

88. Objective function:

$z = 10x + 7y$

Constraints:

$x \geq 0$
$y \geq 0$
$2x + y \geq 100$
$x + y \geq 75$

89. Objective function:

$z = 1.75x + 2.25y$

Constraints:

$x \geq 0$
$y \geq 0$
$2x + y \geq 25$
$3x + 2y \geq 45$

90. Objective function:

$z = 50x + 70y$

Constraints:

$x \geq 0$
$y \geq 0$
$x + 2y \leq 1500$
$5x + 2y \leq 3500$

91. Optimal Revenue A student is working part time as a hairdresser to pay college expenses. The student may work no more than 24 hours per week. Haircuts cost $25 and require an average of 20 minutes, and permanents cost $70 and require an average of 1 hour and 10 minutes. How many haircuts and/or permanents will yield an optimal revenue? What is the optimal revenue?

92. Optimal Profit A manufacturer produces two models of bicycles. The table shows the times (in hours) required for assembling, painting, and packaging each model.

Process	Hours, Model A	Hours, Model B
Assembling	2	2.5
Painting	4	1
Packaging	1	0.75

The total times available for assembling, painting, and packaging are 4000 hours, 4800 hours, and 1500 hours, respectively. The profits per unit are $45 for model A and $50 for model B. What is the optimal production level for each model? What is the optimal profit?

Exploration

True or False? In Exercises 93 and 94, determine whether the statement is true or false. Justify your answer.

93. The system

$$\begin{cases} y \leq 2 \\ y \leq -2 \\ y \leq 4x - 10 \\ y \leq -4x + 26 \end{cases}$$

represents a region in the shape of an isosceles trapezoid.

94. For the rational expression $\dfrac{2x + 3}{x^2(x + 2)^2}$, the partial fraction decomposition is of the form $\dfrac{Ax + B}{x^2} + \dfrac{Cx + D}{(x + 2)^2}$.

Writing a System of Linear Equations In Exercises 95–98, write a system of linear equations that has the ordered pair as a solution. (There are many correct answers.)

95. $(-8, 10)$

96. $(5, -4)$

97. $\left(\frac{4}{3}, 3\right)$

98. $\left(-2, \frac{11}{5}\right)$

Writing a System of Linear Equations In Exercises 99–102, write a system of linear equations that has the ordered triple as a solution. (There are many correct answers.)

99. $(4, -1, 3)$

100. $(-3, 5, 6)$

101. $\left(5, \frac{3}{2}, 2\right)$

102. $\left(-\frac{1}{2}, -2, -\frac{3}{4}\right)$

103. Writing Explain what is meant by an inconsistent system of linear equations.

104. Graphical Reasoning How can you tell graphically that a system of linear equations in two variables has no solution? Give an example.

Chapter Test

See **CalcChat.com** for tutorial help and worked-out solutions to odd-numbered exercises.

Take this test as you would take a test in class. When you are finished, check your work against the answers given in the back of the book.

In Exercises 1–3, solve the system of equations by the method of substitution.

1. $\begin{cases} x + y = -9 \\ 5x - 8y = 20 \end{cases}$

2. $\begin{cases} y = x + 1 \\ y = (x - 1)^3 \end{cases}$

3. $\begin{cases} 2x - y^2 = 0 \\ x - y = 4 \end{cases}$

In Exercises 4–6, solve the system of equations graphically.

4. $\begin{cases} 3x - 6y = 0 \\ 2x + 5y = 18 \end{cases}$

5. $\begin{cases} y = 9 - x^2 \\ y = x + 3 \end{cases}$

6. $\begin{cases} y - \ln x = 4 \\ 7x - 2y - 5 = -6 \end{cases}$

In Exercises 7 and 8, solve the system of equations by the method of elimination.

7. $\begin{cases} 3x + 4y = -26 \\ 7x - 5y = 11 \end{cases}$

8. $\begin{cases} 1.4x - y = 17 \\ 0.8x + 6y = -10 \end{cases}$

In Exercises 9 and 10, solve the system of linear equations and check any solutions algebraically.

9. $\begin{cases} x - 2y + 3z = 11 \\ 2x \quad - z = 3 \\ \quad 3y + z = -8 \end{cases}$

10. $\begin{cases} 3x + 2y + z = 17 \\ -x + y + z = 4 \\ x - y - z = 3 \end{cases}$

In Exercises 11–14, write the partial fraction decomposition of the rational expression. Check your result algebraically.

11. $\dfrac{2x + 5}{x^2 - x - 2}$

12. $\dfrac{3x^2 - 2x + 4}{x^2(2 - x)}$

13. $\dfrac{x^4 + 5}{x^3 - x}$

14. $\dfrac{x^2 - 4}{x^3 + 2x}$

In Exercises 15–17, sketch the graph of the solution set of the system of inequalities.

15. $\begin{cases} 2x + y \le 4 \\ 2x - y \ge 0 \\ x \ge 0 \end{cases}$

16. $\begin{cases} y < -x^2 + x + 4 \\ y > 4x \end{cases}$

17. $\begin{cases} x^2 + y^2 \le 36 \\ x \ge 2 \\ y \ge -4 \end{cases}$

18. Find the minimum and maximum values of the objective function $z = 20x + 12y$ and the points where they occur, subject to the following constraints.

$$\left. \begin{array}{r} x \ge 0 \\ y \ge 0 \\ x + 4y \le 32 \\ 3x + 2y \le 36 \end{array} \right\} \text{Constraints}$$

19. A total of \$50,000 is invested in two funds that pay 4% and 5.5% simple interest. The yearly interest is \$2390. How much is invested at each rate?

20. Find the equation of the parabola $y = ax^2 + bx + c$ that passes through the points $(0, 6)$, $(-2, 2)$, and $\left(3, \frac{9}{2}\right)$.

21. A manufacturer produces two models of television stands. The table at the left shows the times (in hours) required for assembling, staining, and packaging the two models. The total times available for assembling, staining, and packaging are 3750 hours, 8950 hours, and 2650 hours, respectively. The profits per unit are \$30 for model I and \$40 for model II. What is the optimal inventory level for each model? What is the optimal profit?

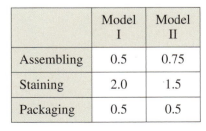

	Model I	Model II
Assembling	0.5	0.75
Staining	2.0	1.5
Packaging	0.5	0.5

Table for 21

Proofs in Mathematics ■ ■ ■ ■ ■ ■ ■ ■ ■ ■ ■ ■ ■ ■ ■

An **indirect proof** can be useful in proving statements of the form "*p* implies *q*." Recall that the conditional statement $p \rightarrow q$ is false only when *p* is true and *q* is false. To prove a conditional statement indirectly, assume that *p* is true and *q* is false. If this assumption leads to an impossibility, then you have proved that the conditional statement is true. An indirect proof is also called a **proof by contradiction.**

An indirect proof can be used to prove the conditional statement

"If *a* is a positive integer and a^2 is divisible by 2, then *a* is divisible by 2."

The proof is as follows.

Proof

First, assume that *p*, "*a* is a positive integer and a^2 is divisible by 2," is true and *q*, "*a* is divisible by 2," is false. This means that *a* is not divisible by 2. If so, then *a* is odd and can be written as $a = 2n + 1$, where *n* is an integer.

$a = 2n + 1$ Definition of an odd integer

$a^2 = 4n^2 + 4n + 1$ Square each side.

$a^2 = 2(2n^2 + 2n) + 1$ Distributive Property

So, by the definition of an odd integer, a^2 is odd. This contradicts the assumption, and you can conclude that *a* is divisible by 2. ■

EXAMPLE   Using an Indirect Proof

Use an indirect proof to prove that $\sqrt{2}$ is an irrational number.

Solution Begin by assuming that $\sqrt{2}$ is *not* an irrational number. Then $\sqrt{2}$ can be written as the quotient of two integers *a* and *b* ($b \neq 0$) that have no common factors.

$\sqrt{2} = \dfrac{a}{b}$ Assume that $\sqrt{2}$ is a rational number.

$2 = \dfrac{a^2}{b^2}$ Square each side.

$2b^2 = a^2$ Multiply each side by b^2.

This implies that 2 is a factor of a^2. So, 2 is also a factor of *a*, and *a* can be written as $2c$, where *c* is an integer.

$2b^2 = (2c)^2$ Substitute $2c$ for *a*.

$2b^2 = 4c^2$ Simplify.

$b^2 = 2c^2$ Divide each side by 2.

This implies that 2 is a factor of b^2 and also a factor of *b*. So, 2 is a factor of both *a* and *b*. This contradicts the assumption that *a* and *b* have no common factors. So, you can conclude that $\sqrt{2}$ is an irrational number. ■

P.S. Problem Solving

1. **Geometry**   A theorem from geometry states that if a triangle is inscribed in a circle such that one side of the triangle is a diameter of the circle, then the triangle is a right triangle. Show that this theorem is true for the circle

   $$x^2 + y^2 = 100$$

   and the triangle formed by the lines

   $$y = 0$$

   $$y = \tfrac{1}{2}x + 5$$

   and

   $$y = -2x + 20.$$

2. **Finding Values of Constants**   Find values of k_1 and k_2 such that the system of equations has an infinite number of solutions.

   $$\begin{cases} 3x - 5y = 8 \\ 2x + k_1 y = k_2 \end{cases}$$

3. **Finding Conditions on Constants**   Under what condition(s) will the system of equations in x and y have exactly one solution?

   $$\begin{cases} ax + by = e \\ cx + dy = f \end{cases}$$

4. **Finding Values of Constants**   Find values of a, b, and c (if possible) such that the system of linear equations has (a) a unique solution, (b) no solution, and (c) an infinite number of solutions.

   $$\begin{cases} x + y \quad\;\;\; = 2 \\ \quad\; y + z = 2 \\ x \quad\;\;\; + z = 2 \\ ax + by + cz = 0 \end{cases}$$

5. **Graphical Analysis**   Graph the lines determined by each system of linear equations. Then use Gaussian elimination to solve each system. At each step of the elimination process, graph the corresponding lines. How do the graphs at the different steps compare?

   (a) $\begin{cases} x - 4y = -3 \\ 5x - 6y = 13 \end{cases}$

   (b) $\begin{cases} 2x - 3y = 7 \\ -4x + 6y = -14 \end{cases}$

6. **Maximum Numbers of Solutions**   A system of two equations in two variables has a finite number of solutions. Determine the maximum number of solutions of the system satisfying each condition.

   (a) Both equations are linear.

   (b) One equation is linear and the other is quadratic.

   (c) Both equations are quadratic.

7. **Vietnam Veterans Memorial**   The Vietnam Veterans Memorial (or "The Wall") in Washington, D.C., was designed by Maya Ying Lin when she was a student at Yale University. This monument has two vertical, triangular sections of black granite with a common side (see figure). The bottom of each section is level with the ground. The tops of the two sections can be approximately modeled by the equations

   $$-2x + 50y = 505$$

   and

   $$2x + 50y = 505$$

   when the x-axis is superimposed at the base of the wall. Each unit in the coordinate system represents 1 foot. How high is the memorial at the point where the two sections meet? How long is each section?

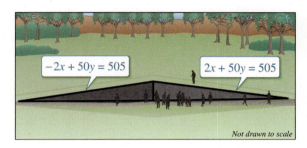

   $-2x + 50y = 505$ $2x + 50y = 505$

   Not drawn to scale

8. **Finding Atomic Weights**   Weights of atoms and molecules are measured in atomic mass units (u). A molecule of C_2H_6 (ethane) is made up of two carbon atoms and six hydrogen atoms and weighs 30.069 u. A molecule of C_3H_8 (propane) is made up of three carbon atoms and eight hydrogen atoms and weighs 44.096 u. Find the weights of a carbon atom and a hydrogen atom.

9. **DVD Connector Cables**   Connecting a DVD player to a television set requires a cable with special connectors at both ends. You buy a six-foot cable for $15.50 and a three-foot cable for $10.25. Assuming that the cost of a cable is the sum of the cost of the two connectors and the cost of the cable itself, what is the cost of a four-foot cable?

10. **Distance**   A hotel 35 miles from an airport runs a shuttle service to and from the airport. The 9:00 A.M. bus leaves for the airport traveling at 30 miles per hour. The 9:15 A.M. bus leaves for the airport traveling at 40 miles per hour.

    (a) Write a system of linear equations that represents distance as a function of time for the buses.

    (b) Graph and solve the system.

    (c) How far from the airport will the 9:15 A.M. bus catch up to the 9:00 A.M. bus?

11. Systems with Rational Expressions Solve each system of equations by letting $X = 1/x$, $Y = 1/y$, and $Z = 1/z$.

(a) $\begin{cases} \dfrac{12}{x} - \dfrac{12}{y} = 7 \\[2mm] \dfrac{3}{x} + \dfrac{4}{y} = 0 \end{cases}$

(b) $\begin{cases} \dfrac{2}{x} + \dfrac{1}{y} - \dfrac{3}{z} = 4 \\[2mm] \dfrac{4}{x} \phantom{+\dfrac{1}{y}} + \dfrac{2}{z} = 10 \\[2mm] -\dfrac{2}{x} + \dfrac{3}{y} - \dfrac{13}{z} = -8 \end{cases}$

12. Finding Values of Constants For what values of a, b, and c does the linear system have $(-1, 2, -3)$ as its only solution?

$\begin{cases} x + 2y - 3z = a & \text{\textcolor{red}{Equation 1}} \\ -x - y + z = b & \text{\textcolor{red}{Equation 2}} \\ 2x + 3y - 2z = c & \text{\textcolor{red}{Equation 3}} \end{cases}$

13. System of Linear Equations The following system has one solution: $x = 1$, $y = -1$, and $z = 2$.

$\begin{cases} 4x - 2y + 5z = 16 & \text{\textcolor{red}{Equation 1}} \\ x + y = 0 & \text{\textcolor{red}{Equation 2}} \\ -x - 3y + 2z = 6 & \text{\textcolor{red}{Equation 3}} \end{cases}$

Solve each system of two equations that consists of (a) Equation 1 and Equation 2, (b) Equation 1 and Equation 3, and (c) Equation 2 and Equation 3. (d) How many solutions does each of these systems have?

14. System of Linear Equations Solve the system of linear equations algebraically.

$\begin{cases} x_1 - x_2 + 2x_3 + 2x_4 + 6x_5 = 6 \\ 3x_1 - 2x_2 + 4x_3 + 4x_4 + 12x_5 = 14 \\ -x_2 - x_3 - x_4 - 3x_5 = -3 \\ 2x_1 - 2x_2 + 4x_3 + 5x_4 + 15x_5 = 10 \\ 2x_1 - 2x_2 + 4x_3 + 4x_4 + 13x_5 = 13 \end{cases}$

15. Biology Each day, an average adult moose can process about 32 kilograms of terrestrial vegetation (twigs and leaves) and aquatic vegetation. From this food, it needs to obtain about 1.9 grams of sodium and 11,000 calories of energy. Aquatic vegetation has about 0.15 gram of sodium per kilogram and about 193 calories of energy per kilogram, whereas terrestrial vegetation has minimal sodium and about four times as much energy as aquatic vegetation. Write and graph a system of inequalities that describes the amounts t and a of terrestrial and aquatic vegetation, respectively, for the daily diet of an average adult moose. *(Source: Biology by Numbers)*

16. Height and Weight For a healthy person who is 4 feet 10 inches tall, the recommended minimum weight is about 91 pounds and increases by about 3.6 pounds for each additional inch of height. The recommended maximum weight is about 115 pounds and increases by about 4.5 pounds for each additional inch of height. *(Source: National Institutes of Health)*

(a) Let x be the number of inches by which a person's height exceeds 4 feet 10 inches and let y be the person's weight (in pounds). Write a system of inequalities that describes the possible values of x and y for a healthy person.

(b) Use a graphing utility to graph the system of inequalities from part (a).

(c) What is the recommended weight range for a healthy person who is 6 feet tall?

17. Cholesterol Cholesterol in human blood is necessary, but too much can lead to health problems. There are three main types of cholesterol: HDL (high-density lipoproteins), LDL (low-density lipoproteins), and VLDL (very low-density lipoproteins). HDL is considered "good" cholesterol; LDL and VLDL are considered "bad" cholesterol.

A standard fasting cholesterol blood test measures total cholesterol, HDL cholesterol, and triglycerides. These numbers are used to estimate LDL and VLDL, which are difficult to measure directly. Your doctor recommends that your combined LDL/VLDL cholesterol level be less than 130 milligrams per deciliter, your HDL cholesterol level be at least 60 milligrams per deciliter, and your total cholesterol level be no more than 200 milligrams per deciliter.

(a) Write a system of linear inequalities for the recommended cholesterol levels. Let x represent the HDL cholesterol level, and let y represent the combined LDL/VLDL cholesterol level.

(b) Graph the system of inequalities from part (a). Label any vertices of the solution region.

(c) Is the following set of cholesterol levels within the recommendations? Explain.

LDL/VLDL: 120 milligrams per deciliter

HDL: 90 milligrams per deciliter

Total: 210 milligrams per deciliter

(d) Give an example of cholesterol levels in which the LDL/VLDL cholesterol level is too high but the HDL cholesterol level is acceptable.

(e) Another recommendation is that the ratio of total cholesterol to HDL cholesterol be less than 4 (that is, less than 4 to 1). Identify a point in the solution region from part (b) that meets this recommendation, and explain why it meets the recommendation.

7 Matrices and Determinants

Sudoku *(page 534)*

Data Encryption *(page 545)*

Beam Deflection *(page 522)*

Flight Crew Scheduling
(page 515)

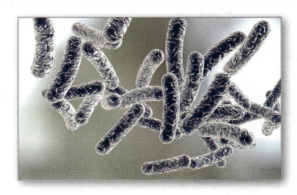

Waterborne Disease *(Exercise 93, page 506)*

7.1 Matrices and Systems of Equations

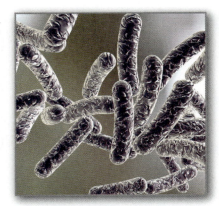

- Write matrices and determine their dimensions.
- Perform elementary row operations on matrices.
- Use matrices and Gaussian elimination to solve systems of linear equations.
- Use matrices and Gauss-Jordan elimination to solve systems of linear equations.

Matrices

Matrices can help you solve real-life problems that are represented by systems of equations. For example, in Exercise 93 on page 506, you will use a matrix to find a model for the numbers of new cases of a waterborne disease in a small city.

In this section, you will study a streamlined technique for solving systems of linear equations. This technique involves the use of a rectangular array of numbers called a **matrix.** The plural of matrix is *matrices.*

> ### Definition of Matrix
>
> If m and n are positive integers, then an $m \times n$ (read "m by n") matrix is a rectangular array
>
> $$\begin{array}{c} \quad\;\; \text{Column 1} \;\; \text{Column 2} \;\; \text{Column 3} \quad \cdots \quad \text{Column } n \\ \begin{array}{r} \text{Row 1} \\ \text{Row 2} \\ \text{Row 3} \\ \vdots \\ \text{Row } m \end{array} \begin{bmatrix} a_{11} & a_{12} & a_{13} & \cdots & a_{1n} \\ a_{21} & a_{22} & a_{23} & \cdots & a_{2n} \\ a_{31} & a_{32} & a_{33} & \cdots & a_{3n} \\ \vdots & \vdots & \vdots & & \vdots \\ a_{m1} & a_{m2} & a_{m3} & \cdots & a_{mn} \end{bmatrix} \end{array}$$
>
> in which each **entry** a_{ij} of the matrix is a number. An $m \times n$ matrix has m rows and n columns.

The entry in the ith row and jth column of a matrix is denoted by the *double subscript* notation a_{ij}. For example, a_{23} refers to the entry in the second row, third column. A matrix having m rows and n columns is said to be of **dimension** $m \times n$. If $m = n$, then the matrix is **square** of dimension $m \times m$ (or $n \times n$). For a square matrix, the entries $a_{11}, a_{22}, a_{33}, \ldots$ are the **main diagonal** entries. A matrix with only one row is called a **row matrix,** and a matrix with only one column is called a **column matrix.**

EXAMPLE 1 Dimensions of Matrices

Determine the dimension of each matrix.

a. $[2]$ **b.** $\begin{bmatrix} 1 & -3 & 0 & \frac{1}{2} \end{bmatrix}$ **c.** $\begin{bmatrix} 0 & 0 \\ 0 & 0 \end{bmatrix}$ **d.** $\begin{bmatrix} 5 & 0 \\ 2 & -2 \\ -7 & 4 \end{bmatrix}$

Solution

a. This matrix has *one* row and *one* column. The dimension of the matrix is 1×1.

b. This matrix has *one* row and *four* columns. The dimension of the matrix is 1×4.

c. This matrix has *two* rows and *two* columns. The dimension of the matrix is 2×2.

d. This matrix has *three* rows and *two* columns. The dimension of the matrix is 3×2.

✓ **Checkpoint** 🔊))) *Audio-video solution in English & Spanish at LarsonPrecalculus.com*

Determine the dimension of the matrix $\begin{bmatrix} 14 & 7 & 10 \\ -2 & -3 & -8 \end{bmatrix}$.

A matrix derived from a system of linear equations (each written in standard form with the constant term on the right) is the **augmented matrix** of the system. Moreover, the matrix derived from the coefficients of the system (but not including the constant terms) is the **coefficient matrix** of the system.

$$\text{System:} \begin{cases} x - 4y + 3z = 5 \\ -x + 3y - z = -3 \\ 2x - 4z = 6 \end{cases}$$

$$\text{Augmented matrix:} \begin{bmatrix} 1 & -4 & 3 & \vdots & 5 \\ -1 & 3 & -1 & \vdots & -3 \\ 2 & 0 & -4 & \vdots & 6 \end{bmatrix} \qquad \text{Coefficient matrix:} \begin{bmatrix} 1 & -4 & 3 \\ -1 & 3 & -1 \\ 2 & 0 & -4 \end{bmatrix}$$

> **REMARK** The vertical dots in an augmented matrix separate the coefficients of the linear system from the constant terms.

Note the use of 0 for the coefficient of the missing y-variable in the third equation, and also note the fourth column of constant terms in the augmented matrix.

When forming either the coefficient matrix or the augmented matrix of a system, you should begin by vertically aligning the variables in the equations and using zeros for the coefficients of the missing variables.

EXAMPLE 2   Writing an Augmented Matrix

Write the augmented matrix for the system of linear equations.

$$\begin{cases} x + 3y - w = 9 \\ -y + 4z + 2w = -2 \\ x - 5z - 6w = 0 \\ 2x + 4y - 3z = 4 \end{cases}$$

What is the dimension of the augmented matrix?

Solution

Begin by rewriting the linear system and aligning the variables.

$$\begin{cases} x + 3y - w = 9 \\ -y + 4z + 2w = -2 \\ x - 5z - 6w = 0 \\ 2x + 4y - 3z = 4 \end{cases}$$

Next, use the coefficients and constant terms as the matrix entries. Include zeros for the coefficients of the missing variables.

$$\begin{matrix} R_1 \\ R_2 \\ R_3 \\ R_4 \end{matrix} \begin{bmatrix} 1 & 3 & 0 & -1 & \vdots & 9 \\ 0 & -1 & 4 & 2 & \vdots & -2 \\ 1 & 0 & -5 & -6 & \vdots & 0 \\ 2 & 4 & -3 & 0 & \vdots & 4 \end{bmatrix}$$

The augmented matrix has four rows and five columns, so it is a 4×5 matrix. The notation R_n is used to designate each row in the matrix. For example, Row 1 is represented by R_1.

✓ *Checkpoint*))) *Audio-video solution in English & Spanish at LarsonPrecalculus.com*

Write the augmented matrix for the system of linear equations. What is the dimension of the augmented matrix?

$$\begin{cases} x + y + z = 2 \\ 2x - y + 3z = -1 \\ -x + 2y - z = 4 \end{cases}$$

Elementary Row Operations

In Section 6.3, you studied three operations that can be used on a system of linear equations to produce an equivalent system.

1. Interchange two equations.

2. Multiply an equation by a nonzero constant.

3. Add a multiple of an equation to another equation.

In matrix terminology, these three operations correspond to **elementary row operations.** An elementary row operation on an augmented matrix of a given system of linear equations produces a new augmented matrix corresponding to a new (but equivalent) system of linear equations. Two matrices are **row-equivalent** when one can be obtained from the other by a sequence of elementary row operations.

> **Elementary Row Operations**
>
> | Operation | Notation |
> |---|---|
> | **1.** Interchange two rows. | $R_a \leftrightarrow R_b$ |
> | **2.** Multiply a row by a nonzero constant. | $cR_a \quad (c \neq 0)$ |
> | **3.** Add a multiple of a row to another row. | $cR_a + R_b$ |

EXAMPLE 3 **Elementary Row Operations**

a. Interchange the first and second rows of the original matrix.

Original Matrix	New Row-Equivalent Matrix
$\begin{bmatrix} 0 & 1 & 3 & 4 \\ -1 & 2 & 0 & 3 \\ 2 & -3 & 4 & 1 \end{bmatrix}$	$\begin{matrix} R_2 \\ R_1 \end{matrix} \begin{bmatrix} -1 & 2 & 0 & 3 \\ 0 & 1 & 3 & 4 \\ 2 & -3 & 4 & 1 \end{bmatrix}$

b. Multiply the first row of the original matrix by $\frac{1}{2}$.

Original Matrix	New Row-Equivalent Matrix
$\begin{bmatrix} 2 & -4 & 6 & -2 \\ 1 & 3 & -3 & 0 \\ 5 & -2 & 1 & 2 \end{bmatrix}$	$\frac{1}{2}R_1 \rightarrow \begin{bmatrix} 1 & -2 & 3 & -1 \\ 1 & 3 & -3 & 0 \\ 5 & -2 & 1 & 2 \end{bmatrix}$

c. Add -2 times the first row of the original matrix to the third row.

Original Matrix	New Row-Equivalent Matrix
$\begin{bmatrix} 1 & 2 & -4 & 3 \\ 0 & 3 & -2 & -1 \\ 2 & 1 & 5 & -2 \end{bmatrix}$	$-2R_1 + R_3 \rightarrow \begin{bmatrix} 1 & 2 & -4 & 3 \\ 0 & 3 & -2 & -1 \\ 0 & -3 & 13 & -8 \end{bmatrix}$

Note that the elementary row operation is written beside the row that is *changed*.

✓ **Checkpoint**  *Audio-video solution in English & Spanish at LarsonPrecalculus.com*

Identify the elementary row operation performed to obtain the new row-equivalent matrix.

Original Matrix	New Row-Equivalent Matrix
$\begin{bmatrix} 1 & 0 & 2 \\ 3 & 1 & 7 \\ 2 & -6 & 14 \end{bmatrix}$	$\begin{bmatrix} 1 & 0 & 2 \\ 0 & 1 & 1 \\ 2 & -6 & 14 \end{bmatrix}$

Gaussian Elimination with Back-Substitution

In Example 3 in Section 6.3, you used Gaussian elimination with back-substitution to solve a system of linear equations. The next example demonstrates the matrix version of Gaussian elimination. The two methods are essentially the same. The basic difference is that with matrices you do not need to keep writing the variables.

EXAMPLE 4   **Comparing Linear Systems and Matrix Operations**

Linear System

$$\begin{cases} x - 2y + 3z = 9 \\ -x + 3y \quad\quad = -4 \\ 2x - 5y + 5z = 17 \end{cases}$$

Associated Augmented Matrix

$$\begin{bmatrix} 1 & -2 & 3 & \vdots & 9 \\ -1 & 3 & 0 & \vdots & -4 \\ 2 & -5 & 5 & \vdots & 17 \end{bmatrix}$$

Add the first equation to the second equation.

$$\begin{cases} x - 2y + 3z = 9 \\ y + 3z = 5 \\ 2x - 5y + 5z = 17 \end{cases}$$

Add the first row to the second row: $R_1 + R_2$.

$$R_1 + R_2 \rightarrow \begin{bmatrix} 1 & -2 & 3 & \vdots & 9 \\ 0 & 1 & 3 & \vdots & 5 \\ 2 & -5 & 5 & \vdots & 17 \end{bmatrix}$$

Add -2 times the first equation to the third equation.

$$\begin{cases} x - 2y + 3z = 9 \\ y + 3z = 5 \\ -y - z = -1 \end{cases}$$

Add -2 times the first row to the third row: $-2R_1 + R_3$.

$$-2R_1 + R_3 \rightarrow \begin{bmatrix} 1 & -2 & 3 & \vdots & 9 \\ 0 & 1 & 3 & \vdots & 5 \\ 0 & -1 & -1 & \vdots & -1 \end{bmatrix}$$

Add the second equation to the third equation.

$$\begin{cases} x - 2y + 3z = 9 \\ y + 3z = 5 \\ 2z = 4 \end{cases}$$

Add the second row to the third row: $R_2 + R_3$.

$$R_2 + R_3 \rightarrow \begin{bmatrix} 1 & -2 & 3 & \vdots & 9 \\ 0 & 1 & 3 & \vdots & 5 \\ 0 & 0 & 2 & \vdots & 4 \end{bmatrix}$$

Multiply the third equation by $\frac{1}{2}$.

$$\begin{cases} x - 2y + 3z = 9 \\ y + 3z = 5 \\ z = 2 \end{cases}$$

Multiply the third row by $\frac{1}{2}$: $\frac{1}{2}R_3$.

$$\frac{1}{2}R_3 \rightarrow \begin{bmatrix} 1 & -2 & 3 & \vdots & 9 \\ 0 & 1 & 3 & \vdots & 5 \\ 0 & 0 & 1 & \vdots & 2 \end{bmatrix}$$

At this point, use back-substitution to find x and y.

$$y + 3(2) = 5 \qquad \text{Substitute 2 for } z.$$
$$y = -1 \qquad \text{Solve for } y.$$
$$x - 2(-1) + 3(2) = 9 \qquad \text{Substitute } -1 \text{ for } y \text{ and 2 for } z.$$
$$x = 1 \qquad \text{Solve for } x.$$

The solution is $(1, -1, 2)$.

• • **REMARK** Remember that you should check a solution by substituting the values of x, y, and z into each equation of the original system. For example, check the solution to Example 4 as shown below.

Equation 1:
$1 - 2(-1) + 3(2) = 9$ ✔

Equation 2:
$-1 + 3(-1) = -4$ ✔

Equation 3:
$2(1) - 5(-1) + 5(2) = 17$ ✔

✓ *Checkpoint* ◀))) *Audio-video solution in English & Spanish at LarsonPrecalculus.com*

Compare solving the linear system below to solving it using its associated augmented matrix.

$$\begin{cases} 2x + y - z = -3 \\ 4x - 2y + 2z = -2 \\ -6x + 5y + 4z = 10 \end{cases}$$

The last matrix in Example 4 is in *row-echelon form*. The term *echelon* refers to the stair-step pattern formed by the nonzero entries of the matrix. The row-echelon form and *reduced row-echelon form* of matrices are described below.

Row-Echelon Form and Reduced Row-Echelon Form

A matrix in **row-echelon form** has the following properties.

1. Any rows consisting entirely of zeros occur at the bottom of the matrix.

2. For each row that does not consist entirely of zeros, the first nonzero entry is 1 (called a **leading 1**).

3. For two successive (nonzero) rows, the leading 1 in the higher row is farther to the left than the leading 1 in the lower row.

A matrix in *row-echelon form* is in **reduced row-echelon form** when every column that has a leading 1 has zeros in every position above and below its leading 1.

It is worth noting that the row-echelon form of a matrix is not unique. That is, two different sequences of elementary row operations may yield different row-echelon forms. The *reduced* row-echelon form of a matrix, however, is unique.

EXAMPLE 5   Row-Echelon Form

Determine whether each matrix is in row-echelon form. If it is, determine whether it is in reduced row-echelon form.

a. $\begin{bmatrix} 1 & 2 & -1 & 4 \\ 0 & 1 & 0 & 3 \\ 0 & 0 & 1 & -2 \end{bmatrix}$ **b.** $\begin{bmatrix} 1 & 2 & -1 & 2 \\ 0 & 0 & 0 & 0 \\ 0 & 1 & 2 & -4 \end{bmatrix}$

c. $\begin{bmatrix} 1 & -5 & 2 & -1 & 3 \\ 0 & 0 & 1 & 3 & -2 \\ 0 & 0 & 0 & 1 & 4 \\ 0 & 0 & 0 & 0 & 1 \end{bmatrix}$ **d.** $\begin{bmatrix} 1 & 0 & 0 & -1 \\ 0 & 1 & 0 & 2 \\ 0 & 0 & 1 & 3 \\ 0 & 0 & 0 & 0 \end{bmatrix}$

e. $\begin{bmatrix} 1 & 2 & -3 & 4 \\ 0 & 2 & 1 & -1 \\ 0 & 0 & 1 & -3 \end{bmatrix}$ **f.** $\begin{bmatrix} 0 & 1 & 0 & 5 \\ 0 & 0 & 1 & 3 \\ 0 & 0 & 0 & 0 \end{bmatrix}$

Solution   The matrices in (a), (c), (d), and (f) are in row-echelon form. The matrices in (d) and (f) are in *reduced* row-echelon form because every column that has a leading 1 has zeros in every position above and below its leading 1. The matrix in (b) is not in row-echelon form because a row of all zeros occurs above a row that is not all zeros. The matrix in (e) is not in row-echelon form because the first nonzero entry in Row 2 is not a leading 1.

✓ *Checkpoint* ◀))) *Audio-video solution in English & Spanish at LarsonPrecalculus.com*

Determine whether the matrix is in row-echelon form. If it is, determine whether it is in reduced row-echelon form.

$$\begin{bmatrix} 1 & 0 & -2 & 4 \\ 0 & 1 & 11 & 3 \\ 0 & 0 & 0 & 0 \end{bmatrix}$$

Every matrix is row-equivalent to a matrix in row-echelon form. For instance, in Example 5, you can change the matrix in part (e) to row-echelon form by multiplying its second row by $\frac{1}{2}$.

Gaussian elimination with back-substitution works well for solving systems of linear equations by hand or with a computer. For this algorithm, the order in which the elementary row operations are performed is important. You should operate from left to right by columns, using elementary row operations to obtain zeros in all entries directly below the leading 1's.

EXAMPLE 6   **Gaussian Elimination with Back-Substitution**

Solve the system

$$\begin{cases} y + z - 2w = -3 \\ x + 2y - z = 2 \\ 2x + 4y + z - 3w = -2 \\ x - 4y - 7z - w = -19 \end{cases}$$

Solution

$$\begin{bmatrix} 0 & 1 & 1 & -2 & \vdots & -3 \\ 1 & 2 & -1 & 0 & \vdots & 2 \\ 2 & 4 & 1 & -3 & \vdots & -2 \\ 1 & -4 & -7 & -1 & \vdots & -19 \end{bmatrix}$$ Write augmented matrix.

$$\begin{matrix} R_2 \\ R_1 \end{matrix} \begin{bmatrix} 1 & 2 & -1 & 0 & \vdots & 2 \\ 0 & 1 & 1 & -2 & \vdots & -3 \\ 2 & 4 & 1 & -3 & \vdots & -2 \\ 1 & -4 & -7 & -1 & \vdots & -19 \end{bmatrix}$$ Interchange R_1 and R_2 so first column has leading 1 in upper left corner.

$$\begin{matrix} \\ \\ -2R_1 + R_3 \to \\ -R_1 + R_4 \to \end{matrix} \begin{bmatrix} 1 & 2 & -1 & 0 & \vdots & 2 \\ 0 & 1 & 1 & -2 & \vdots & -3 \\ 0 & 0 & 3 & -3 & \vdots & -6 \\ 0 & -6 & -6 & -1 & \vdots & -21 \end{bmatrix}$$ Perform operations on R_3 and R_4 so first column has zeros below its leading 1.

$$\begin{matrix} \\ \\ \\ 6R_2 + R_4 \to \end{matrix} \begin{bmatrix} 1 & 2 & -1 & 0 & \vdots & 2 \\ 0 & 1 & 1 & -2 & \vdots & -3 \\ 0 & 0 & 3 & -3 & \vdots & -6 \\ 0 & 0 & 0 & -13 & \vdots & -39 \end{bmatrix}$$ Perform operations on R_4 so second column has zeros below its leading 1.

$$\begin{matrix} \\ \\ \frac{1}{3}R_3 \to \\ -\frac{1}{13}R_4 \to \end{matrix} \begin{bmatrix} 1 & 2 & -1 & 0 & \vdots & 2 \\ 0 & 1 & 1 & -2 & \vdots & -3 \\ 0 & 0 & 1 & -1 & \vdots & -2 \\ 0 & 0 & 0 & 1 & \vdots & 3 \end{bmatrix}$$ Perform operations on R_3 and R_4 so third and fourth columns have leading 1's.

The matrix is now in row-echelon form, and the corresponding system is

$$\begin{cases} x + 2y - z = 2 \\ y + z - 2w = -3 \\ z - w = -2 \\ w = 3 \end{cases}$$

Using back-substitution, the solution is $(-1, 2, 1, 3)$.

REMARK Note that the order of the variables in the system of equations is x, y, z, and w. The coordinates of the solution are given in this order.

✓ *Checkpoint* ◀))) *Audio-video solution in English & Spanish at LarsonPrecalculus.com*

Solve the system

$$\begin{cases} -3x + 5y + 3z = -19 \\ 3x + 4y + 4z = 8 \\ 4x - 8y - 6z = 26 \end{cases}$$

The steps below summarize the procedure used in Example 6.

> **Gaussian Elimination with Back-Substitution**
> 1. Write the augmented matrix of the system of linear equations.
> 2. Use elementary row operations to rewrite the augmented matrix in row-echelon form.
> 3. Write the system of linear equations corresponding to the matrix in row-echelon form and use back-substitution to find the solution.

When solving a system of linear equations, remember that it is possible for the system to have no solution. If, in the elimination process, you obtain a row of all zeros except for the last entry, then the system has no solution, or is *inconsistent*.

EXAMPLE 7 **A System with No Solution**

Solve the system $\begin{cases} x - y + 2z = 4 \\ x \quad\ \ + z = 6 \\ 2x - 3y + 5z = 4 \\ 3x + 2y - z = 1 \end{cases}$.

Solution

$$\begin{bmatrix} 1 & -1 & 2 & \vdots & 4 \\ 1 & 0 & 1 & \vdots & 6 \\ 2 & -3 & 5 & \vdots & 4 \\ 3 & 2 & -1 & \vdots & 1 \end{bmatrix}$$ Write augmented matrix.

$$\begin{matrix} \\ -R_1 + R_2 \rightarrow \\ -2R_1 + R_3 \rightarrow \\ -3R_1 + R_4 \rightarrow \end{matrix} \begin{bmatrix} 1 & -1 & 2 & \vdots & 4 \\ 0 & 1 & -1 & \vdots & 2 \\ 0 & -1 & 1 & \vdots & -4 \\ 0 & 5 & -7 & \vdots & -11 \end{bmatrix}$$ Perform row operations.

$$\begin{matrix} \\ \\ R_2 + R_3 \rightarrow \\ \\ \end{matrix} \begin{bmatrix} 1 & -1 & 2 & \vdots & 4 \\ 0 & 1 & -1 & \vdots & 2 \\ 0 & 0 & 0 & \vdots & -2 \\ 0 & 5 & -7 & \vdots & -11 \end{bmatrix}$$ Perform row operations.

Note that the third row of this matrix consists entirely of zeros except for the last entry. This means that the original system of linear equations is inconsistent. You can see why this is true by converting back to a system of linear equations.

$$\begin{cases} x - y + 2z = 4 \\ y - z = 2 \\ 0 = -2 \\ 5y - 7z = -11 \end{cases}$$

The third equation is not possible, so the system has no solution.

✓ *Checkpoint* ◀))) *Audio-video solution in English & Spanish at LarsonPrecalculus.com*

Solve the system $\begin{cases} x + y + z = 1 \\ x + 2y + 2z = 2 \\ x - y - z = 1 \end{cases}$.

Gauss-Jordan Elimination

With Gaussian elimination, elementary row operations are applied to a matrix to obtain a (row-equivalent) row-echelon form of the matrix. A second method of elimination, called **Gauss-Jordan elimination,** after Carl Friedrich Gauss and Wilhelm Jordan (1842–1899), continues the reduction process until the *reduced* row-echelon form is obtained. This procedure is demonstrated in Example 8.

EXAMPLE 8   Gauss-Jordan Elimination

See LarsonPrecalculus.com for an interactive version of this type of example.

▷ **TECHNOLOGY** For a demonstration of a graphical approach to Gauss-Jordan elimination on a 2×3 matrix, see the program called "Visualizing Row Operations," available at *CengageBrain.com.*

Use Gauss-Jordan elimination to solve the system $\begin{cases} x - 2y + 3z = 9 \\ -x + 3y \quad\;\; = -4. \\ 2x - 5y + 5z = 17 \end{cases}$

Solution   In Example 4, Gaussian elimination was used to obtain the row-echelon form of the linear system above.

$$\begin{bmatrix} 1 & -2 & 3 & \vdots & 9 \\ 0 & 1 & 3 & \vdots & 5 \\ 0 & 0 & 1 & \vdots & 2 \end{bmatrix}$$

Now, rather than using back-substitution, apply elementary row operations until you obtain zeros above each of the leading 1's.

$$\begin{matrix} 2R_2 + R_1 \rightarrow \end{matrix} \begin{bmatrix} 1 & 0 & 9 & \vdots & 19 \\ 0 & 1 & 3 & \vdots & 5 \\ 0 & 0 & 1 & \vdots & 2 \end{bmatrix}$$

Perform operations on R_1 so second column has a zero above its leading 1.

$$\begin{matrix} -9R_3 + R_1 \rightarrow \\ -3R_3 + R_2 \rightarrow \end{matrix} \begin{bmatrix} 1 & 0 & 0 & \vdots & 1 \\ 0 & 1 & 0 & \vdots & -1 \\ 0 & 0 & 1 & \vdots & 2 \end{bmatrix}$$

Perform operations on R_1 and R_2 so third column has zeros above its leading 1.

The matrix is now in reduced row-echelon form. Converting back to a system of linear equations, you have

$$\begin{cases} x = 1 \\ y = -1. \\ z = 2 \end{cases}$$

•• **REMARK**   The advantage of using Gauss-Jordan elimination to solve a system of linear equations is that the solution of the system is easily found without using back-substitution, as illustrated in Example 8. ▷

So, the solution is $(1, -1, 2)$.

✓ *Checkpoint* *Audio-video solution in English & Spanish at LarsonPrecalculus.com*

Use Gauss-Jordan elimination to solve the system $\begin{cases} -3x + 7y + 2z = 1 \\ -5x + 3y - 5z = -8. \\ 2x - 2y - 3z = 15 \end{cases}$

The elimination procedures described in this section sometimes result in fractional coefficients. For example, consider the system

$$\begin{cases} 2x - 5y + 5z = 17 \\ 3x - 2y + 3z = 11. \\ -3x + 3y \quad\;\; = -6 \end{cases}$$

Multiplying the first row by $\frac{1}{2}$ to produce a leading 1 results in fractional coefficients. You can sometimes avoid fractions by judiciously choosing the order in which you apply elementary row operations.

EXAMPLE 9   **A System with an Infinite Number of Solutions**

Solve the system $\begin{cases} 2x + 4y - 2z = 0 \\ 3x + 5y \qquad = 1 \end{cases}$.

Solution

$$\begin{bmatrix} 2 & 4 & -2 & \vdots & 0 \\ 3 & 5 & 0 & \vdots & 1 \end{bmatrix}$$

$$\frac{1}{2}R_1 \rightarrow \begin{bmatrix} 1 & 2 & -1 & \vdots & 0 \\ 3 & 5 & 0 & \vdots & 1 \end{bmatrix}$$

$$-3R_1 + R_2 \rightarrow \begin{bmatrix} 1 & 2 & -1 & \vdots & 0 \\ 0 & -1 & 3 & \vdots & 1 \end{bmatrix}$$

$$-R_2 \rightarrow \begin{bmatrix} 1 & 2 & -1 & \vdots & 0 \\ 0 & 1 & -3 & \vdots & -1 \end{bmatrix}$$

$$-2R_2 + R_1 \rightarrow \begin{bmatrix} 1 & 0 & 5 & \vdots & 2 \\ 0 & 1 & -3 & \vdots & -1 \end{bmatrix}$$

The corresponding system of equations is

$$\begin{cases} x + 5z = 2 \\ y - 3z = -1 \end{cases}.$$

Solving for x and y in terms of z, you have

$$x = -5z + 2 \quad \text{and} \quad y = 3z - 1.$$

To write a solution of the system that does not use any of the three variables of the system, let a represent any real number and let $z = a$. Substitute a for z in the equations for x and y.

$$x = -5z + 2 = -5a + 2 \quad \text{and} \quad y = 3z - 1 = 3a - 1$$

So, the solution set can be written as an ordered triple of the form

$$(-5a + 2, 3a - 1, a)$$

where a is any real number. Remember that a solution set of this form represents an infinite number of solutions. Substitute values for a to obtain a few solutions. Then check each solution in the original system of equations.

✓ *Checkpoint*   *Audio-video solution in English & Spanish at LarsonPrecalculus.com*

Solve the system $\begin{cases} 2x - 6y + 6z = 46 \\ 2x - 3y \qquad = 31 \end{cases}$.

Summarize   (Section 7.1)

1. State the definition of a matrix *(page 494)*. For examples of writing matrices and determining their dimensions, see Examples 1 and 2.

2. List the elementary row operations *(page 496)*. For an example of performing elementary row operations, see Example 3.

3. Explain how to use matrices and Gaussian elimination to solve systems of linear equations *(page 497)*. For examples of using Gaussian elimination, see Examples 4, 6, and 7.

4. Explain how to use matrices and Gauss-Jordan elimination to solve systems of linear equations *(page 501)*. For examples of using Gauss-Jordan elimination, see Examples 8 and 9.

7.1 Exercises

See CalcChat.com for tutorial help and worked-out solutions to odd-numbered exercises.

Vocabulary: Fill in the blanks.

1. A matrix is _____ when the number of rows equals the number of columns.
2. For a square matrix, the entries $a_{11}, a_{22}, a_{33}, \ldots$ are the _____ _____ entries.
3. A matrix derived from a system of linear equations (each written in standard form with the constant term on the right) is the _____ matrix of the system.
4. A matrix derived from the coefficients of a system of linear equations (but not including the constant terms) is the _____ matrix of the system.
5. Two matrices are _____ when one can be obtained from the other by a sequence of elementary row operations.
6. A matrix in row-echelon form is in _____ _____ _____ when every column that has a leading 1 has zeros in every position above and below its leading 1.

Skills and Applications

Dimension of a Matrix In Exercises 7–14, determine the dimension of the matrix.

7. $\begin{bmatrix} 7 & 0 \end{bmatrix}$

8. $\begin{bmatrix} 5 & -3 & 8 & 7 \end{bmatrix}$

9. $\begin{bmatrix} 2 \\ 36 \\ 3 \end{bmatrix}$

10. $\begin{bmatrix} -3 & 7 & 15 & 0 \\ 0 & 0 & 3 & 3 \\ 1 & 1 & 6 & 7 \end{bmatrix}$

11. $\begin{bmatrix} 33 & 45 \\ -9 & 20 \end{bmatrix}$

12. $\begin{bmatrix} -7 & 6 & 4 \\ 0 & -5 & 1 \end{bmatrix}$

13. $\begin{bmatrix} 1 & 6 & -1 \\ 8 & 0 & 3 \\ 3 & -9 & 9 \end{bmatrix}$

14. $\begin{bmatrix} 3 & -1 \\ 4 & 1 \\ -5 & 9 \end{bmatrix}$

Writing an Augmented Matrix In Exercises 15–20, write the augmented matrix for the system of linear equations.

15. $\begin{cases} 2x - y = 7 \\ x + y = 2 \end{cases}$

16. $\begin{cases} 5x + 2y = 13 \\ -3x + 4y = -24 \end{cases}$

17. $\begin{cases} x - y + 2z = 2 \\ 4x - 3y + z = -1 \\ 2x + y = 0 \end{cases}$

18. $\begin{cases} -2x - 4y + z = 13 \\ 6x - 7z = 22 \\ 3x - y + z = 9 \end{cases}$

19. $\begin{cases} 3x - 5y + 2z = 12 \\ 12x - 7z = 10 \end{cases}$

20. $\begin{cases} 9x + y - 3z = 21 \\ -15y + 13z = -8 \end{cases}$

Writing a System of Equations In Exercises 21–26, write the system of linear equations represented by the augmented matrix. (Use variables x, y, z, and w, if applicable.)

21. $\begin{bmatrix} 1 & 1 & \vdots & 3 \\ 5 & -3 & \vdots & -1 \end{bmatrix}$

22. $\begin{bmatrix} 5 & 2 & \vdots & 9 \\ 3 & -8 & \vdots & 0 \end{bmatrix}$

23. $\begin{bmatrix} 2 & 0 & 5 & \vdots & -12 \\ 0 & 1 & -2 & \vdots & 7 \\ 6 & 3 & 0 & \vdots & 2 \end{bmatrix}$

24. $\begin{bmatrix} 4 & -5 & -1 & \vdots & 18 \\ -11 & 0 & 6 & \vdots & 25 \\ 3 & 8 & 0 & \vdots & -29 \end{bmatrix}$

25. $\begin{bmatrix} 9 & 12 & 3 & 0 & \vdots & 0 \\ -2 & 18 & 5 & 2 & \vdots & 10 \\ 1 & 7 & -8 & 0 & \vdots & -4 \\ 3 & 0 & 2 & 0 & \vdots & -10 \end{bmatrix}$

26. $\begin{bmatrix} 6 & 2 & -1 & -5 & \vdots & -25 \\ -1 & 0 & 7 & 3 & \vdots & 7 \\ 4 & -1 & -10 & 6 & \vdots & 23 \\ 0 & 8 & 1 & -11 & \vdots & -21 \end{bmatrix}$

Identifying an Elementary Row Operation In Exercises 27–30, identify the elementary row operation(s) performed to obtain the new row-equivalent matrix.

Original Matrix	New Row-Equivalent Matrix

27. $\begin{bmatrix} -2 & 5 & 1 \\ 3 & -1 & -8 \end{bmatrix}$ $\qquad$ $\begin{bmatrix} 13 & 0 & -39 \\ 3 & -1 & -8 \end{bmatrix}$

28. $\begin{bmatrix} 3 & -1 & -4 \\ -4 & 3 & 7 \end{bmatrix}$ $\qquad$ $\begin{bmatrix} 3 & -1 & -4 \\ 5 & 0 & -5 \end{bmatrix}$

29. $\begin{bmatrix} 0 & -1 & -5 & 5 \\ -1 & 3 & -7 & 6 \\ 4 & -5 & 1 & 3 \end{bmatrix}$ $\qquad$ $\begin{bmatrix} -1 & 3 & -7 & 6 \\ 0 & -1 & -5 & 5 \\ 0 & 7 & -27 & 27 \end{bmatrix}$

30. $\begin{bmatrix} -1 & -2 & 3 & -2 \\ 2 & -5 & 1 & -7 \\ 5 & 4 & -7 & 6 \end{bmatrix}$ $\qquad$ $\begin{bmatrix} -1 & -2 & 3 & -2 \\ 0 & -9 & 7 & -11 \\ 0 & -6 & 8 & -4 \end{bmatrix}$

Elementary Row Operations In Exercises 31–38, fill in the blank(s) using elementary row operations to form a row-equivalent matrix.

31. $\begin{bmatrix} 3 & 6 & 8 \\ 4 & -3 & 6 \end{bmatrix}$

$\begin{bmatrix} 1 & & \frac{8}{3} \\ 4 & -3 & 6 \end{bmatrix}$

32. $\begin{bmatrix} 1 & 4 & 3 \\ 2 & 10 & 5 \end{bmatrix}$

$\begin{bmatrix} 1 & 4 & 3 \\ 0 & & -1 \end{bmatrix}$

33. $\begin{bmatrix} 1 & 1 & 1 \\ 5 & -2 & 4 \end{bmatrix}$

$\begin{bmatrix} 1 & 1 & 1 \\ 0 & & -1 \end{bmatrix}$

34. $\begin{bmatrix} -3 & 3 & 12 \\ 18 & -8 & 4 \end{bmatrix}$

$\begin{bmatrix} 1 & -1 & \\ 18 & -8 & 4 \end{bmatrix}$

35. $\begin{bmatrix} 1 & 5 & 4 & -1 \\ 0 & 1 & -2 & 2 \\ 0 & 0 & 1 & -7 \end{bmatrix}$

$\begin{bmatrix} 1 & 0 & & \\ 0 & 1 & -2 & 2 \\ 0 & 0 & 1 & -7 \end{bmatrix}$

36. $\begin{bmatrix} 1 & 0 & 6 & 1 \\ 0 & -1 & 0 & 7 \\ 0 & 0 & -1 & 3 \end{bmatrix}$

$\begin{bmatrix} 1 & 0 & 6 & 1 \\ 0 & 1 & 0 & \\ 0 & 0 & 1 & \end{bmatrix}$

37. $\begin{bmatrix} 1 & 1 & 4 & -1 \\ 3 & 8 & 10 & 3 \\ -2 & 1 & 12 & 6 \end{bmatrix}$

$\begin{bmatrix} 1 & 1 & 4 & -1 \\ 0 & 5 & & \\ 0 & 3 & & \end{bmatrix}$

$\begin{bmatrix} 1 & 1 & 4 & -1 \\ 0 & 1 & -\frac{2}{5} & \frac{6}{5} \\ 0 & 3 & & \end{bmatrix}$

38. $\begin{bmatrix} 2 & 4 & 8 & 3 \\ 1 & -1 & -3 & 2 \\ 2 & 6 & 4 & 9 \end{bmatrix}$

$\begin{bmatrix} 1 & & & \\ 1 & -1 & -3 & 2 \\ 2 & 6 & 4 & 9 \end{bmatrix}$

$\begin{bmatrix} 1 & 2 & 4 & \frac{3}{2} \\ 0 & & -7 & \frac{1}{2} \\ 0 & 2 & & \end{bmatrix}$

Comparing Linear Systems and Matrix Operations In Exercises 39 and 40, (a) perform the row operations to solve the augmented matrix, (b) write and solve the system of linear equations (in variables x, y, and z, if applicable) represented by the augmented matrix, and (c) compare the two solution methods. Which do you prefer?

39. $\begin{bmatrix} -3 & 4 & \vdots & 22 \\ 6 & -4 & \vdots & -28 \end{bmatrix}$

(i) Add R_2 to R_1.

(ii) Add -2 times R_1 to R_2.

(iii) Multiply R_2 by $-\frac{1}{4}$.

(iv) Multiply R_1 by $\frac{1}{3}$.

40. $\begin{bmatrix} 7 & 13 & 1 & \vdots & -4 \\ -3 & -5 & -1 & \vdots & -4 \\ 3 & 6 & 1 & \vdots & -2 \end{bmatrix}$

(i) Add R_2 to R_1.

(ii) Multiply R_1 by $\frac{1}{4}$.

(iii) Add R_3 to R_2.

(iv) Add -3 times R_1 to R_3.

(v) Add -2 times R_2 to R_1.

 Row-Echelon Form In Exercises 41–44, determine whether the matrix is in row-echelon form. If it is, determine whether it is in reduced row-echelon form.

41. $\begin{bmatrix} 1 & 0 & 0 & 0 \\ 0 & 1 & 1 & 5 \\ 0 & 0 & 0 & 0 \end{bmatrix}$

42. $\begin{bmatrix} 1 & 0 & 0 & 5 \\ 0 & 1 & 0 & 3 \\ 0 & 1 & 0 & 0 \end{bmatrix}$

43. $\begin{bmatrix} 1 & 0 & 0 & 1 \\ 0 & 1 & 0 & -1 \\ 0 & 0 & 0 & 2 \end{bmatrix}$

44. $\begin{bmatrix} 1 & 0 & 0 & 5 \\ 0 & 1 & -2 & 3 \\ 0 & 0 & 1 & 0 \end{bmatrix}$

Writing a Matrix in Row-Echelon Form In Exercises 45–48, write the matrix in row-echelon form. (Remember that the row-echelon form of a matrix is not unique.)

45. $\begin{bmatrix} 1 & 1 & 0 & 5 \\ -2 & -1 & 2 & -10 \\ 3 & 6 & 7 & 14 \end{bmatrix}$

46. $\begin{bmatrix} 1 & 2 & -1 & 3 \\ 3 & 7 & -5 & 14 \\ -2 & -1 & -3 & 8 \end{bmatrix}$

47. $\begin{bmatrix} 1 & -1 & -1 & 1 \\ 5 & -4 & 1 & 8 \\ -6 & 8 & 18 & 0 \end{bmatrix}$

48. $\begin{bmatrix} 1 & -3 & 0 & -7 \\ -3 & 10 & 1 & 23 \\ 4 & -10 & 2 & -24 \end{bmatrix}$

Using a Graphing Utility In Exercises 49–54, use the matrix capabilities of a graphing utility to write the matrix in reduced row-echelon form.

49. $\begin{bmatrix} -1 & 2 & 1 \\ 3 & 4 & 9 \\ 2 & 1 & -2 \end{bmatrix}$

50. $\begin{bmatrix} 1 & 3 & 2 \\ 5 & 15 & 9 \\ 2 & 6 & 10 \end{bmatrix}$

51. $\begin{bmatrix} 1 & 2 & 3 & -5 \\ 1 & 2 & 4 & -9 \\ -2 & -4 & -4 & 3 \\ 4 & 8 & 11 & -14 \end{bmatrix}$

52. $\begin{bmatrix} -2 & 3 & -1 & -2 \\ 4 & -2 & 5 & 8 \\ 1 & 5 & -2 & 0 \\ 3 & 8 & -10 & -30 \end{bmatrix}$

53. $\begin{bmatrix} -3 & 5 & 1 & 12 \\ 1 & -1 & 1 & 4 \end{bmatrix}$

54. $\begin{bmatrix} 5 & 1 & 2 & 4 \\ -1 & 5 & 10 & -32 \end{bmatrix}$

Using Back-Substitution In Exercises 55–58, write the system of linear equations represented by the augmented matrix. Then use back-substitution to solve the system. (Use variables x, y, and z, if applicable.)

55. $\begin{bmatrix} 1 & -2 & \vdots & 4 \\ 0 & 1 & \vdots & -1 \end{bmatrix}$

56. $\begin{bmatrix} 1 & 5 & \vdots & 0 \\ 0 & 1 & \vdots & 6 \end{bmatrix}$

57. $\begin{bmatrix} 1 & -1 & 2 & \vdots & 4 \\ 0 & 1 & -1 & \vdots & 2 \\ 0 & 0 & 1 & \vdots & -2 \end{bmatrix}$

58. $\begin{bmatrix} 1 & 2 & -2 & \vdots & -1 \\ 0 & 1 & 1 & \vdots & 9 \\ 0 & 0 & 1 & \vdots & -3 \end{bmatrix}$

Gaussian Elimination with Back-Substitution In Exercises 59–68, use matrices to solve the system of linear equations, if possible. Use Gaussian elimination with back-substitution.

59. $\begin{cases} x + 2y = 7 \\ -x + y = 8 \end{cases}$ **60.** $\begin{cases} 2x + 6y = 16 \\ 2x + 3y = 7 \end{cases}$

61. $\begin{cases} 3x - 2y = -27 \\ x + 3y = 13 \end{cases}$ **62.** $\begin{cases} -x + y = 4 \\ 2x - 4y = -34 \end{cases}$

63. $\begin{cases} x + 2y - 3z = -28 \\ 4y + 2z = 0 \\ -x + y - z = -5 \end{cases}$

64. $\begin{cases} 3x - 2y + z = 15 \\ -x + y + 2z = -10 \\ x - y - 4z = 14 \end{cases}$

65. $\begin{cases} -3x + 2y = -22 \\ 3x + 4y = 4 \\ 4x - 8y = 32 \end{cases}$ **66.** $\begin{cases} x + 2y = 0 \\ x + y = 6 \\ 3x - 2y = 8 \end{cases}$

67. $\begin{cases} 3x + 2y - z + w = 0 \\ x - y + 4z + 2w = 25 \\ -2x + y + 2z - w = 2 \\ x + y + z + w = 6 \end{cases}$

68. $\begin{cases} x - 4y + 3z - 2w = 9 \\ 3x - 2y + z - 4w = -13 \\ -4x + 3y - 2z + w = -4 \\ -2x + y - 4z + 3w = -10 \end{cases}$

Interpreting Reduced Row-Echelon Form In Exercises 69 and 70, an augmented matrix that represents a system of linear equations (in variables x, y, and z, if applicable) has been reduced using Gauss-Jordan elimination. Write the solution represented by the augmented matrix.

69. $\begin{bmatrix} 1 & 0 & \vdots & 3 \\ 0 & 1 & \vdots & -4 \end{bmatrix}$

70. $\begin{bmatrix} 1 & 0 & 0 & \vdots & 5 \\ 0 & 1 & 0 & \vdots & -3 \\ 0 & 0 & 1 & \vdots & 0 \end{bmatrix}$

Gauss-Jordan Elimination In Exercises 71–78, use matrices to solve the system of linear equations, if possible. Use Gauss-Jordan elimination.

71. $\begin{cases} -2x + 6y = -22 \\ x + 2y = -9 \end{cases}$ **72.** $\begin{cases} 5x - 5y = -5 \\ -2x - 3y = 7 \end{cases}$

73. $\begin{cases} x + 2y + z = 8 \\ 3x + 7y + 6z = 26 \end{cases}$ **74.** $\begin{cases} x + y + 4z = 5 \\ 2x + y - z = 9 \end{cases}$

75. $\begin{cases} x - 3z = -2 \\ 3x + y - 2z = 5 \\ 2x + 2y + z = 4 \end{cases}$ **76.** $\begin{cases} 2x - y + 3z = 24 \\ 2y - z = 14 \\ 7x - 5y = 6 \end{cases}$

77. $\begin{cases} -x + y - z = -14 \\ 2x - y + z = 21 \\ 3x + 2y + z = 19 \end{cases}$

78. $\begin{cases} 2x + 2y - z = 2 \\ x - 3y + z = -28 \\ -x + y = 14 \end{cases}$

Using a Graphing Utility In Exercises 79–84, use the matrix capabilities of a graphing utility to write the augmented matrix corresponding to the system of linear equations in reduced row-echelon form. Then solve the system.

79. $\begin{cases} 3x + 3y + 12z = 6 \\ x + y + 4z = 2 \\ 2x + 5y + 20z = 10 \\ -x + 2y + 8z = 4 \end{cases}$

80. $\begin{cases} 2x + 10y + 2z = 6 \\ x + 5y + 2z = 6 \\ x + 5y + z = 3 \\ -3x - 15y - 3z = -9 \end{cases}$

81. $\begin{cases} 2x + y - z + 2w = -6 \\ 3x + 4y + w = 1 \\ x + 5y + 2z + 6w = -3 \\ 5x + 2y - z - w = 3 \end{cases}$

82. $\begin{cases} x + 2y + 2z + 4w = 11 \\ 3x + 6y + 5z + 12w = 30 \\ x + 3y - 3z + 2w = -5 \\ 6x - y - z + w = -9 \end{cases}$

83. $\begin{cases} x + y + z + w = 0 \\ 2x + 3y + z - 2w = 0 \\ 3x + 5y + z = 0 \end{cases}$

84. $\begin{cases} x + 2y + z + 3w = 0 \\ x - y + w = 0 \\ y - z + 2w = 0 \end{cases}$

85. Error Analysis Describe the error.

The matrix

$$\begin{bmatrix} 3 \\ 0 \\ 8 \\ -1 \end{bmatrix}$$

has four rows and one column, so the dimension of the matrix is 1×4.

86. Error Analysis Describe the error.

The matrix

$$\begin{bmatrix} 1 & 2 & 7 & 2 \\ 0 & 1 & 4 & 0 \\ 0 & 0 & 1 & \frac{1}{3} \end{bmatrix}$$

is in reduced row-echelon form.

Curve Fitting In Exercises 87–92, use a system of linear equations to find the quadratic function $f(x) = ax^2 + bx + c$ that satisfies the given conditions. Solve the system using matrices.

87. $f(1) = 1$, $f(2) = -1$, $f(3) = -5$
88. $f(1) = 2$, $f(2) = 9$, $f(3) = 20$
89. $f(-2) = -15$, $f(-1) = 7$, $f(1) = -3$
90. $f(-2) = -3$, $f(1) = -3$, $f(2) = -11$
91. $f(1) = 8$, $f(2) = 13$, $f(3) = 20$
92. $f(1) = 9$, $f(2) = 8$, $f(3) = 5$

93. Waterborne Disease

From 2005 through 2016, the numbers of new cases of a waterborne disease in a small city increased in a pattern that was approximately linear (see figure). Find the least squares regression line

$y = at + b$

for the data shown in the figure by solving the system below using matrices. Let t represent the year, with $t = 0$ corresponding to 2005.

$$\begin{cases} 12b + 66a = 831 \\ 66b + 506a = 5643 \end{cases}$$

Use the result to predict the number of new cases of the waterborne disease in 2020. Is the estimate reasonable? Explain.

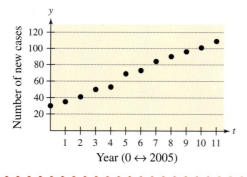

Year ($0 \leftrightarrow 2005$)

94. **Museum** A natural history museum borrows $2,000,000 at simple annual interest to purchase new exhibits. Some of the money is borrowed at 7%, some at 8.5%, and some at 9.5%. Use a system of linear equations to determine how much is borrowed at each rate given that the total annual interest is $169,750 and the amount borrowed at 8.5% is four times the amount borrowed at 9.5%. Solve the system of linear equations using matrices.

95. **Breeding Facility** A city zoo borrows $2,000,000 at simple annual interest to construct a breeding facility. Some of the money is borrowed at 8%, some at 9%, and some at 12%. Use a system of linear equations to determine how much is borrowed at each rate given that the total annual interest is $186,000 and the amount borrowed at 8% is twice the amount borrowed at 12%. Solve the system of linear equations using matrices.

96. **Mathematical Modeling** A video of the path of a ball thrown by a baseball player was analyzed with a grid covering the TV screen. The video was paused three times, and the position of the ball was measured each time. The coordinates obtained are shown in the table. (x and y are measured in feet.)

Horizontal Distance, x	0	15	30
Height, y	5.0	9.6	12.4

(a) Use a system of equations to find the equation of the parabola $y = ax^2 + bx + c$ that passes through the three points. Solve the system using matrices.

(b) Use a graphing utility to graph the parabola.

(c) Graphically approximate the maximum height of the ball and the point at which the ball struck the ground.

(d) Analytically find the maximum height of the ball and the point at which the ball struck the ground.

(e) Compare your results from parts (c) and (d).

Exploration

True or False? In Exercises 97 and 98, determine whether the statement is true or false. Justify your answer.

97. $\begin{bmatrix} 5 & 0 & -2 & 7 \\ -1 & 3 & -6 & 0 \end{bmatrix}$ is a 4×2 matrix.

98. The method of Gaussian elimination reduces a matrix until a reduced row-echelon form is obtained.

99. **Think About It** What is the relationship between the three elementary row operations performed on an augmented matrix and the operations that lead to equivalent systems of equations?

100. **HOW DO YOU SEE IT?** Determine whether the matrix below is in row-echelon form, reduced row-echelon form, or neither when it satisfies the given conditions.

$$\begin{bmatrix} 1 & b \\ c & 1 \end{bmatrix}$$

(a) $b = 0$, $c = 0$ (b) $b \neq 0$, $c = 0$

(c) $b = 0$, $c \neq 0$ (d) $b \neq 0$, $c \neq 0$

7.2 Operations with Matrices

- Determine whether two matrices are equal.
- Add and subtract matrices, and multiply matrices by scalars.
- Multiply two matrices.
- Use matrix operations to model and solve real-life problems.

Matrix operations have many practical applications. For example, in Exercise 74 on page 520, you will use matrix multiplication to analyze the calories burned by individuals of different body weights while performing different types of exercises.

Equality of Matrices

In Section 7.1, you used matrices to solve systems of linear equations. There is a rich mathematical theory of matrices, and its applications are numerous. This section and the next two sections introduce some fundamental concepts of matrix theory. It is standard mathematical convention to represent matrices in any of the three ways listed below.

Representation of Matrices

1. A matrix can be denoted by an uppercase letter such as A, B, or C.

2. A matrix can be denoted by a representative element enclosed in brackets, such as $[a_{ij}]$, $[b_{ij}]$, or $[c_{ij}]$.

3. A matrix can be denoted by a rectangular array of numbers such as

$$A = [a_{ij}] = \begin{bmatrix} a_{11} & a_{12} & a_{13} & \cdots & a_{1n} \\ a_{21} & a_{22} & a_{23} & \cdots & a_{2n} \\ a_{31} & a_{32} & a_{33} & \cdots & a_{3n} \\ \vdots & \vdots & \vdots & & \vdots \\ a_{m1} & a_{m2} & a_{m3} & \cdots & a_{mn} \end{bmatrix}.$$

Two matrices $A = [a_{ij}]$ and $B = [b_{ij}]$ are **equal** when they have the same dimension $(m \times n)$ and $a_{ij} = b_{ij}$ for $1 \le i \le m$ and $1 \le j \le n$. In other words, two matrices are equal when their corresponding entries are equal.

EXAMPLE 1 **Equality of Matrices**

Solve for a_{11}, a_{12}, a_{21}, and a_{22} in the matrix equation $\begin{bmatrix} a_{11} & a_{12} \\ a_{21} & a_{22} \end{bmatrix} = \begin{bmatrix} 2 & -1 \\ -3 & 0 \end{bmatrix}$.

Solution Two matrices are equal when their corresponding entries are equal, so

$$a_{11} = 2, \quad a_{12} = -1, \quad a_{21} = -3, \quad \text{and} \quad a_{22} = 0.$$

✓ *Checkpoint* ◀))) *Audio-video solution in English & Spanish at LarsonPrecalculus.com*

Solve for a_{11}, a_{12}, a_{21}, and a_{22} in the matrix equation $\begin{bmatrix} a_{11} & a_{12} \\ a_{21} & a_{22} \end{bmatrix} = \begin{bmatrix} 6 & 3 \\ -2 & 4 \end{bmatrix}$. ■

Be sure you see that for two matrices to be equal, they must have the same dimension *and* their corresponding entries must be equal. For example,

$$\begin{bmatrix} 2 & -1 \\ \sqrt{4} & \frac{1}{2} \end{bmatrix} = \begin{bmatrix} 2 & -1 \\ 2 & 0.5 \end{bmatrix} \quad \text{but} \quad \begin{bmatrix} 2 & -1 & 0 \\ 3 & 4 & 0 \end{bmatrix} \ne \begin{bmatrix} 2 & -1 \\ 3 & 4 \end{bmatrix}.$$

Matrix Addition and Scalar Multiplication

Two basic matrix operations are matrix addition and scalar multiplication. With matrix addition, you add two matrices (of the same dimension) by adding their corresponding entries.

Definition of Matrix Addition

If $A = [a_{ij}]$ and $B = [b_{ij}]$ are matrices of dimension $m \times n$, then their sum is the $m \times n$ matrix

$$A + B = [a_{ij} + b_{ij}].$$

The sum of two matrices of different dimensions is undefined.

EXAMPLE 2 **Addition of Matrices**

a. $\begin{bmatrix} -1 & 2 \\ 0 & 1 \end{bmatrix} + \begin{bmatrix} 1 & 3 \\ -1 & 2 \end{bmatrix} = \begin{bmatrix} -1+1 & 2+3 \\ 0+(-1) & 1+2 \end{bmatrix} = \begin{bmatrix} 0 & 5 \\ -1 & 3 \end{bmatrix}$

b. $\begin{bmatrix} 0 & 1 & -2 \\ 1 & 2 & 3 \end{bmatrix} + \begin{bmatrix} 0 & 0 & 0 \\ 0 & 0 & 0 \end{bmatrix} = \begin{bmatrix} 0 & 1 & -2 \\ 1 & 2 & 3 \end{bmatrix}$

c. $\begin{bmatrix} 1 \\ -3 \\ -2 \end{bmatrix} + \begin{bmatrix} -1 \\ 3 \\ 2 \end{bmatrix} = \begin{bmatrix} 0 \\ 0 \\ 0 \end{bmatrix}$

d. The sum of

$$A = \begin{bmatrix} 2 & 1 & 0 \\ 4 & 0 & -1 \\ 3 & -2 & 2 \end{bmatrix} \quad \text{and} \quad B = \begin{bmatrix} 0 & 1 \\ -1 & 3 \\ 2 & 4 \end{bmatrix}$$

is undefined because A is of dimension 3×3 and B is of dimension 3×2.

✓ *Checkpoint* *Audio-video solution in English & Spanish at LarsonPrecalculus.com*

Find each sum, if possible.

a. $\begin{bmatrix} 4 & -1 \\ 2 & -3 \end{bmatrix} + \begin{bmatrix} 2 & -1 \\ 0 & 6 \end{bmatrix}$ **b.** $\begin{bmatrix} 2 & -1 \\ 3 & 4 \\ 0 & -2 \end{bmatrix} + \begin{bmatrix} -2 & 1 \\ -3 & -4 \\ 0 & 2 \end{bmatrix}$

c. $\begin{bmatrix} 3 & 9 & 6 \\ 0 & 4 & -2 \\ 1 & -1 & 0 \end{bmatrix} + \begin{bmatrix} 3 & 9 & 6 \\ 0 & 2 & -4 \end{bmatrix}$ **d.** $\begin{bmatrix} 1 \\ -1 \\ 1 \end{bmatrix} + \begin{bmatrix} -1 \\ 1 \\ 1 \end{bmatrix}$

In operations with matrices, numbers are usually referred to as **scalars.** In this text, scalars will always be real numbers. To multiply a matrix A by a scalar c, multiply each entry in A by c.

Definition of Scalar Multiplication

If $A = [a_{ij}]$ is an $m \times n$ matrix and c is a scalar, then the **scalar multiple** of A by c is the $m \times n$ matrix

$$cA = [ca_{ij}].$$

The symbol $-A$ represents the **negation** of A, which is the scalar product $(-1)A$. Moreover, if A and B are of the same dimension, then $A - B$ represents the sum of A and $(-1)B$. That is,

$$A - B = A + (-1)B. \qquad \text{Subtraction of matrices}$$

EXAMPLE 3 **Operations with Matrices**

For the matrices below, find (a) $3A$, (b) $-B$, and (c) $3A - B$.

$$A = \begin{bmatrix} 2 & 2 & 4 \\ -3 & 0 & -1 \\ 2 & 1 & 2 \end{bmatrix} \quad \text{and} \quad B = \begin{bmatrix} 2 & 0 & 0 \\ 1 & -4 & 3 \\ -1 & 3 & 2 \end{bmatrix}$$

Solution

a. $3A = 3\begin{bmatrix} 2 & 2 & 4 \\ -3 & 0 & -1 \\ 2 & 1 & 2 \end{bmatrix}$ Scalar multiplication

$$= \begin{bmatrix} 3(2) & 3(2) & 3(4) \\ 3(-3) & 3(0) & 3(-1) \\ 3(2) & 3(1) & 3(2) \end{bmatrix} \qquad \text{Multiply each entry by 3.}$$

$$= \begin{bmatrix} 6 & 6 & 12 \\ -9 & 0 & -3 \\ 6 & 3 & 6 \end{bmatrix} \qquad \text{Simplify.}$$

b. $-B = (-1)\begin{bmatrix} 2 & 0 & 0 \\ 1 & -4 & 3 \\ -1 & 3 & 2 \end{bmatrix}$ Definition of negation

$$= \begin{bmatrix} -2 & 0 & 0 \\ -1 & 4 & -3 \\ 1 & -3 & -2 \end{bmatrix} \qquad \text{Multiply each entry by } -1.$$

c. $3A - B = \begin{bmatrix} 6 & 6 & 12 \\ -9 & 0 & -3 \\ 6 & 3 & 6 \end{bmatrix} + \begin{bmatrix} -2 & 0 & 0 \\ -1 & 4 & -3 \\ 1 & -3 & -2 \end{bmatrix}$ $3A - B = 3A + (-1)B$

$$= \begin{bmatrix} 4 & 6 & 12 \\ -10 & 4 & -6 \\ 7 & 0 & 4 \end{bmatrix} \qquad \text{Add corresponding entries.}$$

REMARK The order of operations for matrix expressions is similar to that for real numbers. As shown in Example 3(c), you perform scalar multiplication before matrix addition and subtraction.

✓ *Checkpoint* ◄))) Audio-video solution in English & Spanish at LarsonPrecalculus.com

For the matrices below, find (a) $A - B$, (b) $3A$, and (c) $3A - 2B$.

$$A = \begin{bmatrix} 4 & -1 \\ 0 & 4 \\ -3 & 8 \end{bmatrix} \quad \text{and} \quad B = \begin{bmatrix} 0 & 4 \\ -1 & 3 \\ 1 & 7 \end{bmatrix}$$

It is often convenient to rewrite the scalar multiple cA by factoring c out of every entry in the matrix. The example below shows factoring the scalar $\frac{1}{2}$ out of a matrix.

$$\begin{bmatrix} \frac{1}{2} & -\frac{3}{2} \\ \frac{5}{2} & \frac{1}{2} \end{bmatrix} = \begin{bmatrix} \frac{1}{2}(1) & \frac{1}{2}(-3) \\ \frac{1}{2}(5) & \frac{1}{2}(1) \end{bmatrix} = \frac{1}{2}\begin{bmatrix} 1 & -3 \\ 5 & 1 \end{bmatrix}$$

▷ **ALGEBRA HELP**   To review
the properties of addition and
multiplication of real numbers
(and other properties of real
numbers), see Section P.1.

The properties of matrix addition and scalar multiplication are similar to those of addition and multiplication of real numbers.

Properties of Matrix Addition and Scalar Multiplication

Let A, B, and C be $m \times n$ matrices and let c and d be scalars.

1. $A + B = B + A$ Commutative Property of Matrix Addition

2. $A + (B + C) = (A + B) + C$ Associative Property of Matrix Addition

3. $(cd)A = c(dA)$ Associative Property of Scalar Multiplication

4. $1A = A$ Scalar Identity Property

5. $c(A + B) = cA + cB$ Distributive Property

6. $(c + d)A = cA + dA$ Distributive Property

Note that the Associative Property of Matrix Addition allows you to write expressions such as $A + B + C$ without ambiguity because the same sum occurs no matter how the matrices are grouped. This same reasoning applies to sums of four or more matrices.

▷ **TECHNOLOGY**   Most
graphing utilities can perform
matrix operations. Consult the
user's guide for your graphing
utility for specific keystrokes.
Use a graphing utility to find
the sum of the matrices

$$A = \begin{bmatrix} 2 & -3 \\ -1 & 0 \end{bmatrix}$$

and

$$B = \begin{bmatrix} -1 & 4 \\ 2 & -5 \end{bmatrix}.$$

EXAMPLE 4   **Addition of More than Two Matrices**

$$\begin{bmatrix} 1 \\ 2 \\ -3 \end{bmatrix} + \begin{bmatrix} -1 \\ -1 \\ 2 \end{bmatrix} + \begin{bmatrix} 0 \\ 1 \\ 4 \end{bmatrix} + \begin{bmatrix} 2 \\ -3 \\ -2 \end{bmatrix} = \begin{bmatrix} 2 \\ -1 \\ 1 \end{bmatrix} \qquad \text{Add corresponding entries.}$$

✓ **Checkpoint** *Audio-video solution in English & Spanish at LarsonPrecalculus.com*

Evaluate the expression.

$$\begin{bmatrix} 3 & -8 \\ 0 & 2 \end{bmatrix} + \begin{bmatrix} -2 & 3 \\ 6 & -5 \end{bmatrix} + \begin{bmatrix} 0 & 7 \\ 4 & -1 \end{bmatrix}$$

EXAMPLE 5   **Evaluating an Expression**

$$3\left(\begin{bmatrix} -2 & 0 \\ 4 & 1 \end{bmatrix} + \begin{bmatrix} 4 & -2 \\ 3 & 7 \end{bmatrix} \right) = 3\begin{bmatrix} -2 & 0 \\ 4 & 1 \end{bmatrix} + 3\begin{bmatrix} 4 & -2 \\ 3 & 7 \end{bmatrix}$$

$$= \begin{bmatrix} -6 & 0 \\ 12 & 3 \end{bmatrix} + \begin{bmatrix} 12 & -6 \\ 9 & 21 \end{bmatrix}$$

$$= \begin{bmatrix} 6 & -6 \\ 21 & 24 \end{bmatrix}$$

✓ **Checkpoint** *Audio-video solution in English & Spanish at LarsonPrecalculus.com*

Evaluate the expression.

$$2\left(\begin{bmatrix} 1 & 3 \\ -2 & 2 \end{bmatrix} + \begin{bmatrix} -4 & 0 \\ -3 & 1 \end{bmatrix} \right)$$

In Example 5, you could add the two matrices first and then multiply the resulting matrix by 3. The result would be the same.

One important property of addition of real numbers is that the number 0 is the additive identity. That is, $c + 0 = c$ for any real number c. For matrices, a similar property holds. That is, if A is an $m \times n$ matrix and O is the $m \times n$ **zero matrix** consisting entirely of zeros, then

$$A + O = A.$$

In other words, O is the **additive identity** for the set of all $m \times n$ matrices. For example, the matrices below are the additive identities for the sets of all 2×3 and 2×2 matrices.

$$O = \begin{bmatrix} 0 & 0 & 0 \\ 0 & 0 & 0 \end{bmatrix} \quad \text{and} \quad O = \begin{bmatrix} 0 & 0 \\ 0 & 0 \end{bmatrix}.$$

2×3 zero matrix $\qquad$ 2×2 zero matrix

The algebra of real numbers and the algebra of matrices have many similarities. For example, compare the solutions below.

Real Numbers (Solve for x.)	$m \times n$ **Matrices** (Solve for X.)
$x + a = b$	$X + A = B$
$x + a + (-a) = b + (-a)$	$X + A + (-A) = B + (-A)$
$x + 0 = b - a$	$X + O = B - A$
$x = b - a$	$X = B - A$

The algebra of real numbers and the algebra of matrices also have important differences (see Example 9 and Exercises 77–82).

> **REMARK** When you solve for X in a matrix equation, you are solving for a *matrix X* that makes the equation true.

EXAMPLE 6 Solving a Matrix Equation

Solve for X in the equation $3X + A = B$, where

$$A = \begin{bmatrix} 1 & -2 \\ 0 & 3 \end{bmatrix} \quad \text{and} \quad B = \begin{bmatrix} -3 & 4 \\ 2 & 1 \end{bmatrix}.$$

Solution Begin by solving the matrix equation for X.

$$3X + A = B$$
$$3X = B - A$$
$$X = \tfrac{1}{3}(B - A)$$

Now, substituting the matrices A and B, you have

$$X = \tfrac{1}{3}\left(\begin{bmatrix} -3 & 4 \\ 2 & 1 \end{bmatrix} - \begin{bmatrix} 1 & -2 \\ 0 & 3 \end{bmatrix}\right) \qquad \text{Substitute the matrices.}$$

$$= \tfrac{1}{3}\begin{bmatrix} -4 & 6 \\ 2 & -2 \end{bmatrix} \qquad \text{Subtract matrix } A \text{ from matrix } B.$$

$$= \begin{bmatrix} -\tfrac{4}{3} & 2 \\ \tfrac{2}{3} & -\tfrac{2}{3} \end{bmatrix} \qquad \text{Multiply the resulting matrix by } \tfrac{1}{3}.$$

✓ *Checkpoint* ◀))) *Audio-video solution in English & Spanish at LarsonPrecalculus.com*

Solve for X in the equation $2X - A = B$, where

$$A = \begin{bmatrix} 6 & 1 \\ 0 & 3 \end{bmatrix} \quad \text{and} \quad B = \begin{bmatrix} 4 & -1 \\ -2 & 5 \end{bmatrix}.$$ ∎

Matrix Multiplication

Another basic matrix operation is **matrix multiplication.** At first glance, the definition may seem unusual. You will see later, however, that this definition of the product of two matrices has many practical applications.

> ### Definition of Matrix Multiplication
>
> If $A = [a_{ij}]$ is an $m \times n$ matrix and $B = [b_{ij}]$ is an $n \times p$ matrix, then the product AB is an $m \times p$ matrix given by $AB = [c_{ij}]$, where
> $$c_{ij} = a_{i1}b_{1j} + a_{i2}b_{2j} + a_{i3}b_{3j} + \cdots + a_{in}b_{nj}.$$

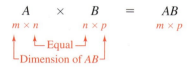

$$
\begin{array}{ccccc}
A & \times & B & = & AB \\
m \times n & & n \times p & & m \times p
\end{array}
$$
Equal
Dimension of AB

The definition of matrix multiplication uses a *row-by-column* multiplication, where the entry in the ith row and jth column of the product AB is obtained by multiplying the entries in the ith row of A by the corresponding entries in the jth column of B and then adding the results. So, for the product of two matrices to be defined, the number of columns of the first matrix must equal the number of rows of the second matrix. That is, the middle two indices must be the same. The outside two indices give the dimension of the product, as shown at the left. The general pattern for matrix multiplication is shown below.

$$
\begin{bmatrix}
a_{11} & a_{12} & a_{13} & \cdots & a_{1n} \\
a_{21} & a_{22} & a_{23} & \cdots & a_{2n} \\
a_{31} & a_{32} & a_{33} & \cdots & a_{3n} \\
\vdots & \vdots & \vdots & & \vdots \\
a_{i1} & a_{i2} & a_{i3} & \cdots & a_{in} \\
\vdots & \vdots & \vdots & & \vdots \\
a_{m1} & a_{m2} & a_{m3} & \cdots & a_{mn}
\end{bmatrix}
\begin{bmatrix}
b_{11} & b_{12} & \cdots & b_{1j} & \cdots & b_{1p} \\
b_{21} & b_{22} & \cdots & b_{2j} & \cdots & b_{2p} \\
b_{31} & b_{32} & \cdots & b_{3j} & \cdots & b_{3p} \\
\vdots & \vdots & & \vdots & & \vdots \\
b_{n1} & b_{n2} & \cdots & b_{nj} & \cdots & b_{np}
\end{bmatrix}
=
\begin{bmatrix}
c_{11} & c_{12} & \cdots & c_{1j} & \cdots & c_{1p} \\
c_{21} & c_{22} & \cdots & c_{2j} & \cdots & c_{2p} \\
\vdots & \vdots & & & & \vdots \\
c_{i1} & c_{i2} & \cdots & c_{ij} & \cdots & c_{ip} \\
\vdots & \vdots & & & & \vdots \\
c_{m1} & c_{m2} & \cdots & c_{mj} & \cdots & c_{mp}
\end{bmatrix}
$$

$$a_{i1}b_{1j} + a_{i2}b_{2j} + a_{i3}b_{3j} + \cdots + a_{in}b_{nj} = c_{ij}$$

EXAMPLE 7   **Finding the Product of Two Matrices**

Find the product AB, where $A = \begin{bmatrix} -1 & 3 \\ 4 & -2 \\ 5 & 0 \end{bmatrix}$ and $B = \begin{bmatrix} -3 & 2 \\ -4 & 1 \end{bmatrix}$.

Solution   To find the entries of the product, multiply each row of A by each column of B.

$$
AB = \begin{bmatrix} -1 & 3 \\ 4 & -2 \\ 5 & 0 \end{bmatrix} \begin{bmatrix} -3 & 2 \\ -4 & 1 \end{bmatrix}
$$

$$
= \begin{bmatrix}
(-1)(-3) + 3(-4) & (-1)(2) + 3(1) \\
4(-3) + (-2)(-4) & 4(2) + (-2)(1) \\
5(-3) + 0(-4) & 5(2) + 0(1)
\end{bmatrix}
$$

$$
= \begin{bmatrix} -9 & 1 \\ -4 & 6 \\ -15 & 10 \end{bmatrix}
$$

· · REMARK In Example 7, the product AB is defined because the number of columns of A is equal to the number of rows of B. Also, note that the product AB has dimension 3×2.

✓ *Checkpoint* 🔊))) *Audio-video solution in English & Spanish at LarsonPrecalculus.com*

Find the product AB, where $A = \begin{bmatrix} -1 & 4 \\ 2 & 0 \\ 1 & 2 \end{bmatrix}$ and $B = \begin{bmatrix} 1 & -2 \\ 0 & 7 \end{bmatrix}$.

EXAMPLE 8 **Finding the Product of Two Matrices**

Find the product AB, where $A = \begin{bmatrix} 1 & 0 & 3 \\ 2 & -1 & -2 \end{bmatrix}$ and $B = \begin{bmatrix} -2 & 4 \\ 1 & 0 \\ -1 & 1 \end{bmatrix}$.

Solution Note that the dimension of A is 2×3 and the dimension of B is 3×2. So, the product AB has dimension 2×2.

$$AB = \begin{bmatrix} 1 & 0 & 3 \\ 2 & -1 & -2 \end{bmatrix} \begin{bmatrix} -2 & 4 \\ 1 & 0 \\ -1 & 1 \end{bmatrix}$$

$$= \begin{bmatrix} 1(-2) + 0(1) + 3(-1) & 1(4) + 0(0) + 3(1) \\ 2(-2) + (-1)(1) + (-2)(-1) & 2(4) + (-1)(0) + (-2)(1) \end{bmatrix}$$

$$= \begin{bmatrix} -5 & 7 \\ -3 & 6 \end{bmatrix}$$

✓ **Checkpoint** ◀))) *Audio-video solution in English & Spanish at LarsonPrecalculus.com*

Find the product AB, where $A = \begin{bmatrix} 0 & 4 & -3 \\ 2 & 1 & 7 \\ 3 & -2 & 1 \end{bmatrix}$ and $B = \begin{bmatrix} -2 & 0 \\ 0 & -4 \\ 1 & 2 \end{bmatrix}$.

EXAMPLE 9 **Matrix Multiplication**

See LarsonPrecalculus.com for an interactive version of this type of example.

a. $\begin{bmatrix} 3 & 4 \\ -2 & 5 \end{bmatrix} \begin{bmatrix} 1 & 0 \\ 0 & 1 \end{bmatrix} = \begin{bmatrix} 3 & 4 \\ -2 & 5 \end{bmatrix}$

    2×2 2×2 2×2

b. $\begin{bmatrix} 6 & 2 & 0 \\ 3 & -1 & 2 \\ 1 & 4 & 6 \end{bmatrix} \begin{bmatrix} 1 \\ 2 \\ -3 \end{bmatrix} = \begin{bmatrix} 10 \\ -5 \\ -9 \end{bmatrix}$

    3×3 3×1 3×1

c. $\begin{bmatrix} 1 & -2 & -3 \end{bmatrix} \begin{bmatrix} 2 \\ -1 \\ 1 \end{bmatrix} = \begin{bmatrix} 1 \end{bmatrix}$

    1×3 3×1 1×1

> **REMARK** In Examples 9(c) and 9(d), note that the two products are different. Even when both AB and BA are defined, matrix multiplication is not, in general, commutative. That is, for most matrices, $AB \neq BA$. This is one way in which the algebra of real numbers and the algebra of matrices differ.

d. $\begin{bmatrix} 2 \\ -1 \\ 1 \end{bmatrix} \begin{bmatrix} 1 & -2 & -3 \end{bmatrix} = \begin{bmatrix} 2 & -4 & -6 \\ -1 & 2 & 3 \\ 1 & -2 & -3 \end{bmatrix}$

    3×1 1×3 3×3

e. The product $\begin{bmatrix} -2 & 1 \\ 1 & -3 \\ 1 & 4 \end{bmatrix} \begin{bmatrix} -2 & 3 & 1 & 4 \\ 0 & 1 & -1 & 2 \\ 2 & -1 & 0 & 1 \end{bmatrix}$ is not defined.

    3×2 3×4

✓ **Checkpoint** ◀))) *Audio-video solution in English & Spanish at LarsonPrecalculus.com*

Find each product, if possible.

a. $\begin{bmatrix} 1 \\ -3 \end{bmatrix} \begin{bmatrix} 3 & -1 \end{bmatrix}$ **b.** $\begin{bmatrix} 3 & -1 \end{bmatrix} \begin{bmatrix} 1 \\ -3 \end{bmatrix}$ **c.** $\begin{bmatrix} 3 & 1 & 2 \\ 7 & 0 & -2 \end{bmatrix} \begin{bmatrix} 6 & 4 \\ 2 & -1 \end{bmatrix}$ ■

EXAMPLE 10 **Squaring a Matrix**

Find A^2, where $A = \begin{bmatrix} 3 & 1 \\ -1 & 2 \end{bmatrix}$. (*Note:* $A^2 = AA$.)

Solution

$$A^2 = \begin{bmatrix} 3 & 1 \\ -1 & 2 \end{bmatrix}\begin{bmatrix} 3 & 1 \\ -1 & 2 \end{bmatrix}$$

$$= \begin{bmatrix} 8 & 5 \\ -5 & 3 \end{bmatrix}$$

✓ **Checkpoint** 🔊))) *Audio-video solution in English & Spanish at LarsonPrecalculus.com*

Find A^2, where $A = \begin{bmatrix} 2 & 1 \\ 3 & -2 \end{bmatrix}$.

Properties of Matrix Multiplication

Let A, B, and C be matrices and let c be a scalar.

1. $A(BC) = (AB)C$ Associative Property of Matrix Multiplication

2. $A(B + C) = AB + AC$ Left Distributive Property

3. $(A + B)C = AC + BC$ Right Distributive Property

4. $c(AB) = (cA)B = A(cB)$ Associative Property of Scalar Multiplication

---

Definition of the Identity Matrix

The $n \times n$ matrix that consists of 1's on its main diagonal and 0's elsewhere is called the **identity matrix of dimension $n \times n$** and is denoted by

$$I_n = \begin{bmatrix} 1 & 0 & 0 & \ldots & 0 \\ 0 & 1 & 0 & \ldots & 0 \\ 0 & 0 & 1 & \ldots & 0 \\ \vdots & \vdots & \vdots & & \vdots \\ 0 & 0 & 0 & \ldots & 1 \end{bmatrix}. \qquad \text{Identity matrix}$$

Note that an identity matrix must be *square*. When the dimension is understood to be $n \times n$, you can denote I_n simply by I.

If A is an $n \times n$ matrix, then the identity matrix has the property that $AI_n = A$ and $I_nA = A$. For example,

$$\begin{bmatrix} 3 & -2 & 5 \\ 1 & 0 & 4 \\ -1 & 2 & -3 \end{bmatrix}\begin{bmatrix} 1 & 0 & 0 \\ 0 & 1 & 0 \\ 0 & 0 & 1 \end{bmatrix} = \begin{bmatrix} 3 & -2 & 5 \\ 1 & 0 & 4 \\ -1 & 2 & -3 \end{bmatrix} \qquad AI = A$$

and

$$\begin{bmatrix} 1 & 0 & 0 \\ 0 & 1 & 0 \\ 0 & 0 & 1 \end{bmatrix}\begin{bmatrix} 3 & -2 & 5 \\ 1 & 0 & 4 \\ -1 & 2 & -3 \end{bmatrix} = \begin{bmatrix} 3 & -2 & 5 \\ 1 & 0 & 4 \\ -1 & 2 & -3 \end{bmatrix}. \qquad IA = A$$

Many real-life applications of linear systems involve enormous numbers of equations and variables. For example, a flight crew scheduling problem for American Airlines required the manipulation of matrices with 837 rows and 12,753,313 columns. *(Source: Very Large-Scale Linear Programming. A Case Study in Combining Interior Point and Simplex Methods, Bixby, Robert E., et al., Operations Research, 40, no. 5)*

Applications

Matrix multiplication can be used to represent a system of linear equations. Note how the system below can be written as the matrix equation $AX = B$, where A is the *coefficient matrix* of the system and X and B are column matrices. The column matrix B is also called a *constant matrix*. Its entries are the constant terms in the system of equations.

System

$$\begin{cases} a_{11}x_1 + a_{12}x_2 + a_{13}x_3 = b_1 \\ a_{21}x_1 + a_{22}x_2 + a_{23}x_3 = b_2 \\ a_{31}x_1 + a_{32}x_2 + a_{33}x_3 = b_3 \end{cases}$$

Matrix Equation $AX = B$

$$\underbrace{\begin{bmatrix} a_{11} & a_{12} & a_{13} \\ a_{21} & a_{22} & a_{23} \\ a_{31} & a_{32} & a_{33} \end{bmatrix}}_{A} \times \underbrace{\begin{bmatrix} x_1 \\ x_2 \\ x_3 \end{bmatrix}}_{X} = \underbrace{\begin{bmatrix} b_1 \\ b_2 \\ b_3 \end{bmatrix}}_{B}$$

In Example 11, $[A \;\vdots\; B]$ represents the augmented matrix formed when you *adjoin* matrix B to matrix A. Also, $[I \;\vdots\; X]$ represents the reduced row-echelon form of the augmented matrix that yields the solution of the system.

EXAMPLE 11   Solving a System of Linear Equations

For the system of linear equations, (a) write the system as a matrix equation, $AX = B$, and (b) use Gauss-Jordan elimination on $[A \;\vdots\; B]$ to solve for the matrix X.

$$\begin{cases} x_1 - 2x_2 + x_3 = -4 \\ x_2 + 2x_3 = 4 \\ 2x_1 + 3x_2 - 2x_3 = 2 \end{cases}$$

Solution

a. In matrix form, $AX = B$, the system is

$$\begin{bmatrix} 1 & -2 & 1 \\ 0 & 1 & 2 \\ 2 & 3 & -2 \end{bmatrix} \begin{bmatrix} x_1 \\ x_2 \\ x_3 \end{bmatrix} = \begin{bmatrix} -4 \\ 4 \\ 2 \end{bmatrix}.$$

b. Form the augmented matrix by adjoining matrix B to matrix A.

$$[A \;\vdots\; B] = \begin{bmatrix} 1 & -2 & 1 & \vdots & -4 \\ 0 & 1 & 2 & \vdots & 4 \\ 2 & 3 & -2 & \vdots & 2 \end{bmatrix}$$

Using Gauss-Jordan elimination, rewrite this matrix as

$$[I \;\vdots\; X] = \begin{bmatrix} 1 & 0 & 0 & \vdots & -1 \\ 0 & 1 & 0 & \vdots & 2 \\ 0 & 0 & 1 & \vdots & 1 \end{bmatrix}.$$

So, the solution of the matrix equation is

$$X = \begin{bmatrix} x_1 \\ x_2 \\ x_3 \end{bmatrix} = \begin{bmatrix} -1 \\ 2 \\ 1 \end{bmatrix}.$$

 Checkpoint *Audio-video solution in English & Spanish at LarsonPrecalculus.com*

For the system of linear equations, (a) write the system as a matrix equation, $AX = B$, and (b) use Gauss-Jordan elimination on $[A \;\vdots\; B]$ to solve for the matrix X.

$$\begin{cases} -2x_1 - 3x_2 = -4 \\ 6x_1 + x_2 = -36 \end{cases}$$

EXAMPLE 12   **Softball Team Expenses**

Two softball teams submit equipment lists to their sponsors.

Equipment	Women's Team	Men's Team
Bats	12	15
Balls	45	38
Gloves	15	17

Each bat costs $80, each ball costs $4, and each glove costs $90. Use matrices to find the total cost of equipment for each team.

Solution   Write the equipment lists E and the costs per item C in matrix form as

$$E = \begin{bmatrix} 12 & 15 \\ 45 & 38 \\ 15 & 17 \end{bmatrix}$$

and

$$C = \begin{bmatrix} 80 & 4 & 90 \end{bmatrix}.$$

· · REMARK Notice in Example 12 that it is not possible to find the total cost using the product EC because EC is not defined. That is, the number of columns of E (2 columns) does not equal the number of rows of C (1 row).

To find the total cost of equipment for each team, use the product CE because the number of columns of C (3 columns) equals the number of rows of E (3 rows). So, the total cost of equipment for each team is given by

$$CE = \begin{bmatrix} 80 & 4 & 90 \end{bmatrix} \begin{bmatrix} 12 & 15 \\ 45 & 38 \\ 15 & 17 \end{bmatrix}$$

$$= \begin{bmatrix} 80(12) + 4(45) + 90(15) & 80(15) + 4(38) + 90(17) \end{bmatrix}$$

$$= \begin{bmatrix} 2490 & 2882 \end{bmatrix}.$$

The total cost of equipment for the women's team is $2490 and the total cost of equipment for the men's team is $2882.

✓ *Checkpoint*   Audio-video solution in English & Spanish at LarsonPrecalculus.com

Repeat Example 12 when each bat costs $100, each ball costs $3, and each glove costs $65.

Summarize (Section 7.2)

1. State the conditions under which two matrices are equal *(page 507)*. For an example involving matrix equality, see Example 1.

2. Explain how to add matrices *(page 508)*. For an example of matrix addition, see Example 2.

3. Explain how to multiply a matrix by a scalar *(page 508)*. For an example of scalar multiplication, see Example 3.

4. List the properties of matrix addition and scalar multiplication *(page 510)*. For examples of using these properties, see Examples 4–6.

5. Explain how to multiply two matrices *(page 512)*. For examples of matrix multiplication, see Examples 7–10.

6. Describe real-life applications of matrix operations *(pages 515 and 516, Examples 11 and 12)*.

7.2 Exercises

See **CalcChat.com** for tutorial help and worked-out solutions to odd-numbered exercises.

Vocabulary: Fill in the blanks.

1. Two matrices are _____ when their corresponding entries are equal.

2. When performing matrix operations, real numbers are usually referred to as _____.

3. A matrix consisting entirely of zeros is called a _____ matrix and is denoted by _____.

4. The $n \times n$ matrix that consists of 1's on its main diagonal and 0's elsewhere is called the _____ matrix of dimension $n \times n$.

Skills and Applications

Equality of Matrices In Exercises 5–8, solve for x and y.

5. $\begin{bmatrix} x & -2 \\ 7 & 23 \end{bmatrix} = \begin{bmatrix} -4 & -2 \\ 7 & y \end{bmatrix}$ 6. $\begin{bmatrix} -5 & x \\ 3y & 8 \end{bmatrix} = \begin{bmatrix} -5 & 13 \\ 12 & 8 \end{bmatrix}$

7. $\begin{bmatrix} 16 & 4 & x & 4 \\ 0 & 2 & 4 & 0 \end{bmatrix} = \begin{bmatrix} 16 & 4 & 2x+1 & 4 \\ 0 & 2 & 3y-5 & 0 \end{bmatrix}$

8. $\begin{bmatrix} x+2 & 8 & -3 \\ 1 & 18 & -8 \\ 7 & -2 & y+2 \end{bmatrix} = \begin{bmatrix} 2x+6 & 8 & -3 \\ 1 & 18 & -8 \\ 7 & -2 & x \end{bmatrix}$

Operations with Matrices In Exercises 9–16, if possible, find (a) $A + B$, (b) $A - B$, (c) $3A$, and (d) $3A - 2B$.

9. $A = \begin{bmatrix} 1 & -1 \\ 2 & -1 \end{bmatrix}, \quad B = \begin{bmatrix} 2 & -1 \\ -1 & 8 \end{bmatrix}$

10. $A = \begin{bmatrix} 1 & 2 \\ 2 & 1 \end{bmatrix}, \quad B = \begin{bmatrix} -3 & -2 \\ 4 & 2 \end{bmatrix}$

11. $A = \begin{bmatrix} 6 & 0 & 3 \\ -1 & -4 & 0 \end{bmatrix}, \quad B = \begin{bmatrix} 8 & -1 \\ 4 & -3 \end{bmatrix}$

12. $A = \begin{bmatrix} 3 \\ 2 \\ -1 \end{bmatrix}, \quad B = \begin{bmatrix} -4 & 6 & 2 \end{bmatrix}$

13. $A = \begin{bmatrix} 8 & -1 \\ 2 & 3 \\ -4 & 5 \end{bmatrix}, \quad B = \begin{bmatrix} 1 & 6 \\ -1 & -5 \\ 1 & 10 \end{bmatrix}$

14. $A = \begin{bmatrix} 1 & -1 & 3 \\ 0 & 6 & 9 \end{bmatrix}, \quad B = \begin{bmatrix} -2 & 0 & -5 \\ -3 & 4 & -7 \end{bmatrix}$

15. $A = \begin{bmatrix} 4 & 5 & -1 & 3 & 4 \\ 1 & 2 & -2 & -1 & 0 \end{bmatrix},$

$B = \begin{bmatrix} 1 & 0 & -1 & 1 & 0 \\ -6 & 8 & 2 & -3 & -7 \end{bmatrix}$

16. $A = \begin{bmatrix} -1 & 4 & 0 \\ 3 & -2 & 2 \\ 5 & 4 & -1 \\ 0 & 8 & -6 \\ -4 & -1 & 0 \end{bmatrix}, \quad B = \begin{bmatrix} -3 & 5 & 1 \\ 2 & -4 & -7 \\ 10 & -9 & -1 \\ 3 & 2 & -4 \\ 0 & 1 & -2 \end{bmatrix}$

Evaluating an Expression In Exercises 17–22, evaluate the expression.

17. $\begin{bmatrix} -5 & 0 \\ 3 & -6 \end{bmatrix} + \begin{bmatrix} 7 & 1 \\ -2 & -1 \end{bmatrix} + \begin{bmatrix} -10 & -8 \\ 14 & 6 \end{bmatrix}$

18. $\begin{bmatrix} 6 & 8 \\ -1 & 0 \end{bmatrix} + \begin{bmatrix} 0 & 5 \\ -3 & -1 \end{bmatrix} + \begin{bmatrix} -11 & -7 \\ 2 & -1 \end{bmatrix}$

19. $4\left(\begin{bmatrix} -4 & 0 & 1 \\ 0 & 2 & 3 \end{bmatrix} - \begin{bmatrix} 2 & 1 & -2 \\ 3 & -6 & 0 \end{bmatrix} \right)$

20. $\frac{1}{2}\left(\begin{bmatrix} 5 & -2 & 4 & 0 \end{bmatrix} + \begin{bmatrix} 14 & 6 & -18 & 9 \end{bmatrix} \right)$

21. $-3\left(\begin{bmatrix} 0 & -3 \\ 7 & 2 \end{bmatrix} + \begin{bmatrix} -6 & 3 \\ 8 & 1 \end{bmatrix} \right) - 2\begin{bmatrix} 4 & -4 \\ 7 & -9 \end{bmatrix}$

22. $-1\begin{bmatrix} 4 & 11 \\ -2 & -1 \\ 9 & 3 \end{bmatrix} + \frac{1}{6}\left(\begin{bmatrix} -5 & -1 \\ 3 & 4 \\ 0 & 13 \end{bmatrix} + \begin{bmatrix} 7 & 5 \\ -9 & -1 \\ 6 & -1 \end{bmatrix} \right)$

Operations with Matrices In Exercises 23–26, use the matrix capabilities of a graphing utility to evaluate the expression.

23. $\frac{11}{25}\begin{bmatrix} 2 & 5 \\ -1 & -4 \end{bmatrix} + 6\begin{bmatrix} -3 & 0 \\ 2 & 2 \end{bmatrix}$

24. $55\left(\begin{bmatrix} 14 & -11 \\ -22 & 19 \end{bmatrix} - \begin{bmatrix} -8 & 20 \\ 13 & 6 \end{bmatrix} \right)$

25. $-2\begin{bmatrix} 1.23 & 4.19 & -3.85 \\ 7.21 & -2.60 & 6.54 \end{bmatrix} - \begin{bmatrix} 8.35 & -3.02 & 7.30 \\ -0.38 & -5.49 & 1.68 \end{bmatrix}$

26. $-1\begin{bmatrix} 10 & 15 \\ -20 & 10 \\ 12 & 4 \end{bmatrix} + \frac{1}{8}\left(\begin{bmatrix} -13 & 11 \\ 7 & 0 \\ 6 & 9 \end{bmatrix} + \begin{bmatrix} -3 & 13 \\ -3 & 8 \\ -14 & 15 \end{bmatrix} \right)$

Solving a Matrix Equation In Exercises 27–34, solve for X in the equation, where

$A = \begin{bmatrix} -2 & 1 & 3 \\ -1 & 0 & 4 \end{bmatrix}$ and $B = \begin{bmatrix} 0 & 2 & -4 \\ 3 & 0 & 1 \end{bmatrix}$.

27. $X = 2A + 2B$ 28. $X = 3A - 2B$

29. $2X = 2A - B$ 30. $2X = A + B$

31. $2X + 3A = B$ 32. $3X - 4A = 2B$

33. $4B = -2X - 2A$ 34. $5A = 6B - 3X$

Finding the Product of Two Matrices
In Exercises 35–40, if possible, find AB and state the dimension of the result.

35. $A = \begin{bmatrix} -1 & 6 \\ -4 & 5 \\ 0 & 3 \end{bmatrix}$, $B = \begin{bmatrix} 2 & 3 \\ 0 & 9 \end{bmatrix}$

36. $A = \begin{bmatrix} 0 & -1 & 2 \\ 6 & 0 & 3 \\ 7 & -1 & 8 \end{bmatrix}$, $B = \begin{bmatrix} 2 & -1 \\ 4 & -5 \\ 1 & 6 \end{bmatrix}$

37. $A = \begin{bmatrix} 2 & 1 \\ -3 & 4 \\ 1 & 6 \end{bmatrix}$, $B = \begin{bmatrix} 0 & -1 & 0 \\ 4 & 0 & 2 \\ 8 & -1 & 7 \end{bmatrix}$

38. $A = \begin{bmatrix} 1 & 0 & 3 & -2 \\ 6 & 13 & 8 & -17 \end{bmatrix}$, $B = \begin{bmatrix} 1 & 6 \\ 4 & 2 \end{bmatrix}$

39. $A = \begin{bmatrix} 5 & 0 & 0 \\ 0 & -8 & 0 \\ 0 & 0 & 7 \end{bmatrix}$, $B = \begin{bmatrix} \frac{1}{5} & 0 & 0 \\ 0 & -\frac{1}{8} & 0 \\ 0 & 0 & \frac{1}{2} \end{bmatrix}$

40. $A = \begin{bmatrix} 0 & 0 & 5 \\ 0 & 0 & -3 \\ 0 & 0 & 4 \end{bmatrix}$, $B = \begin{bmatrix} 6 & -11 & 4 \\ 8 & 16 & 4 \\ 0 & 0 & 0 \end{bmatrix}$

Finding the Product of Two Matrices **In Exercises 41–46, use the matrix capabilities of a graphing utility to find AB, if possible.**

41. $A = \begin{bmatrix} 7 & 5 & -4 \\ -2 & 5 & 1 \\ 10 & -4 & -7 \end{bmatrix}$, $B = \begin{bmatrix} 2 & -2 & 3 \\ 8 & 1 & 4 \\ -4 & 2 & -8 \end{bmatrix}$

42. $A = \begin{bmatrix} 11 & -12 & 4 \\ 14 & 10 & 12 \\ 6 & -2 & 9 \end{bmatrix}$, $B = \begin{bmatrix} 12 & 10 \\ -5 & 12 \\ 15 & 16 \end{bmatrix}$

43. $A = \begin{bmatrix} -3 & 2 \\ 5 & -4 \\ 1 & -7 \end{bmatrix}$, $B = \begin{bmatrix} -1 & 4 & 0 \\ 3 & -2 & 2 \\ 5 & 4 & -1 \\ 0 & 8 & -6 \\ 4 & -1 & 0 \end{bmatrix}$

44. $A = \begin{bmatrix} -1 & 4 & 0 \\ 3 & -2 & 2 \\ 5 & 4 & -1 \\ 0 & 8 & -6 \\ 4 & -1 & 0 \end{bmatrix}$, $B = \begin{bmatrix} -3 & 2 \\ 5 & -4 \\ 1 & -7 \end{bmatrix}$

45. $A = \begin{bmatrix} -3 & 8 & -6 & 8 \\ -12 & 15 & 9 & 6 \\ 5 & -1 & 1 & 5 \end{bmatrix}$, $B = \begin{bmatrix} 3 & 1 & 6 \\ 24 & 15 & 14 \\ 16 & 10 & 21 \\ 8 & -4 & 10 \end{bmatrix}$

46. $A = \begin{bmatrix} -2 & 4 & 8 \\ 21 & 5 & 6 \\ 13 & 2 & 6 \end{bmatrix}$, $B = \begin{bmatrix} 2 & 0 \\ -7 & 15 \\ 32 & 14 \\ 0.5 & 1.6 \end{bmatrix}$

Operations with Matrices **In Exercises 47–54, if possible, find (a) AB, (b) BA, and (c) A^2.**

47. $A = \begin{bmatrix} 1 & 2 \\ 4 & 2 \end{bmatrix}$, $B = \begin{bmatrix} 2 & -1 \\ -1 & 8 \end{bmatrix}$

48. $A = \begin{bmatrix} 6 & 3 \\ -2 & -4 \end{bmatrix}$, $B = \begin{bmatrix} -2 & 0 \\ 2 & 4 \end{bmatrix}$

49. $A = \begin{bmatrix} 5 & -9 & 0 \\ 3 & 0 & -8 \\ -1 & 4 & 11 \end{bmatrix}$, $B = \begin{bmatrix} 1 & 0 & 0 \\ 0 & 1 & 0 \\ 0 & 0 & 1 \end{bmatrix}$

50. $A = \begin{bmatrix} 2 & -2 \\ -3 & 0 \\ 7 & 6 \end{bmatrix}$, $B = \begin{bmatrix} 1 & 0 \\ 0 & 1 \end{bmatrix}$

51. $A = \begin{bmatrix} -4 & -1 \\ 2 & 12 \end{bmatrix}$, $B = \begin{bmatrix} -6 \\ 5 \end{bmatrix}$

52. $A = \begin{bmatrix} 1 & 3 & -2 \\ -5 & 10 & 1 \end{bmatrix}$, $B = \begin{bmatrix} 3 \\ 3 \\ 3 \end{bmatrix}$

53. $A = \begin{bmatrix} 7 \\ 8 \\ -1 \end{bmatrix}$, $B = \begin{bmatrix} 1 & 1 & 2 \end{bmatrix}$

54. $A = \begin{bmatrix} 3 & 2 & 1 & 4 \end{bmatrix}$, $B = \begin{bmatrix} 2 \\ 3 \\ 0 \\ 1 \end{bmatrix}$

Operations with Matrices **In Exercises 55–60, evaluate the expression. Use the matrix capabilities of a graphing utility to verify your answer.**

55. $\begin{bmatrix} 3 & 1 \\ 0 & -2 \end{bmatrix} \begin{bmatrix} 1 & 0 \\ -2 & 2 \end{bmatrix} \begin{bmatrix} 1 & 0 \\ 2 & 4 \end{bmatrix}$

56. $-3\left(\begin{bmatrix} 6 & 5 & -1 \\ 1 & -2 & 0 \end{bmatrix} \begin{bmatrix} 0 & 3 \\ -1 & -3 \\ 4 & 1 \end{bmatrix} \right)$

57. $\begin{bmatrix} 0 & 2 & -2 \\ 4 & 1 & 2 \end{bmatrix} \left(\begin{bmatrix} 4 & 0 \\ 0 & -1 \\ -1 & 2 \end{bmatrix} + \begin{bmatrix} -2 & 3 \\ -3 & 5 \\ 0 & -3 \end{bmatrix} \right)$

58. $\left(\begin{bmatrix} 4 & 0 \\ 0 & -1 \\ -1 & 2 \end{bmatrix} + \begin{bmatrix} -2 & 3 \\ -3 & 5 \\ 0 & -3 \end{bmatrix} \right) \begin{bmatrix} 0 & 2 & -2 \\ 4 & 1 & 2 \end{bmatrix}$

59. $\begin{bmatrix} 3 \\ -1 \\ 5 \\ 7 \end{bmatrix} \left(\begin{bmatrix} 5 & -6 \end{bmatrix} + \begin{bmatrix} 7 & -1 \end{bmatrix} + \begin{bmatrix} -8 & 9 \end{bmatrix} \right)$

60. $\begin{bmatrix} 3 & -1 & 5 & 7 \end{bmatrix} \left(\begin{bmatrix} 5 \\ -6 \\ 6 \\ -5 \end{bmatrix} + \begin{bmatrix} 7 \\ -1 \\ 1 \\ -7 \end{bmatrix} + \begin{bmatrix} -8 \\ 9 \\ -9 \\ 8 \end{bmatrix} \right)$

Solving a System of Linear Equations
In Exercises 61–66, (a) write the system of linear equations as a matrix equation, $AX = B$, and (b) use Gauss-Jordan elimination on $[A \vdots B]$ to solve for the matrix X.

61. $\begin{cases} 2x_1 + 3x_2 = 5 \\ x_1 + 4x_2 = 10 \end{cases}$ **62.** $\begin{cases} -2x_1 - 3x_2 = -4 \\ 6x_1 + x_2 = -36 \end{cases}$

63. $\begin{cases} x_1 - 2x_2 + 3x_3 = 9 \\ -x_1 + 3x_2 - x_3 = -6 \\ 2x_1 - 5x_2 + 5x_3 = 17 \end{cases}$

64. $\begin{cases} x_1 + x_2 - 3x_3 = -1 \\ -x_1 + 2x_2 = 1 \\ x_1 - x_2 + x_3 = 2 \end{cases}$

65. $\begin{cases} x_1 - 5x_2 + 2x_3 = -20 \\ -3x_1 + x_2 - x_3 = 8 \\ -2x_2 + 5x_3 = -16 \end{cases}$

66. $\begin{cases} x_1 - x_2 + 4x_3 = 17 \\ x_1 + 3x_2 = -11 \\ -6x_2 + 5x_3 = 40 \end{cases}$

67. Manufacturing A corporation has four factories that manufacture sport utility vehicles and pickup trucks. The production levels are represented by A.

$$A = \begin{matrix} & \overbrace{\begin{matrix} 1 & 2 & 3 & 4 \end{matrix}}^{\text{Factory}} & \\ \begin{bmatrix} 100 & 90 & 70 & 30 \\ 40 & 20 & 60 & 60 \end{bmatrix} & \begin{matrix} \text{SUV} \\ \text{Pickup} \end{matrix} \left.\begin{matrix} \\ \end{matrix}\right\} \begin{matrix} \text{Vehicle} \\ \text{Type} \end{matrix} \end{matrix}$$

Find the production levels when production increases by 10%.

68. Vacation Packages A travel agent identifies four resorts with special all-inclusive packages. The current rates for two types of rooms (double and quadruple occupancy) at the four resorts are represented by A.

$$A = \begin{matrix} \overbrace{\begin{matrix} \text{Resort} & \text{Resort} & \text{Resort} & \text{Resort} \\ w & x & y & z \end{matrix}}^{} \\ \begin{bmatrix} 615 & 670 & 740 & 990 \\ 995 & 1030 & 1180 & 1105 \end{bmatrix} \begin{matrix} \text{Double} \\ \text{Quadruple} \end{matrix} \left.\begin{matrix} \\ \end{matrix}\right\} \text{Occupancy} \end{matrix}$$

The rates are expected to increase by no more than 12% by next season. Find the maximum rate per package per resort.

69. Agriculture A farmer grows apples and peaches. Each crop is shipped to three different outlets. The shipment levels are represented by A.

$$A = \begin{matrix} & \overbrace{\begin{matrix} 1 & 2 & 3 \end{matrix}}^{\text{Outlet}} & \\ \begin{bmatrix} 125 & 100 & 75 \\ 100 & 175 & 125 \end{bmatrix} & \begin{matrix} \text{Apples} \\ \text{Peaches} \end{matrix} \left.\begin{matrix} \\ \end{matrix}\right\} \text{Crop} \end{matrix}$$

The profits per unit are represented by the matrix $B = [\$3.50 \quad \$6.00]$. Compute BA and interpret the result.

70. Revenue An electronics manufacturer produces three models of high-definition televisions, which are shipped to two warehouses. The shipment levels are represented by A.

$$A = \begin{matrix} & \overbrace{\begin{matrix} 1 & 2 \end{matrix}}^{\text{Warehouse}} & \\ \begin{bmatrix} 5,000 & 4,000 \\ 6,000 & 10,000 \\ 8,000 & 5,000 \end{bmatrix} & \begin{matrix} \text{A} \\ \text{B} \\ \text{C} \end{matrix} \left.\begin{matrix} \\ \\ \end{matrix}\right\} \text{Model} \end{matrix}$$

The prices per unit are represented by the matrix

$$B = [\$699.95 \quad \$899.95 \quad \$1099.95].$$

Compute BA and interpret the result.

71. Labor and Wages A company has two factories that manufacture three sizes of boats. The numbers of hours of labor required to manufacture each size are represented by S.

$$S = \begin{matrix} & \overbrace{\begin{matrix} \text{Cutting} & \text{Assembly} & \text{Packaging} \end{matrix}}^{\text{Department}} & \\ \begin{bmatrix} 1.0 & 0.5 & 0.2 \\ 1.6 & 1.0 & 0.2 \\ 2.5 & 2.0 & 1.4 \end{bmatrix} & \begin{matrix} \text{Small} \\ \text{Medium} \\ \text{Large} \end{matrix} \left.\begin{matrix} \\ \\ \end{matrix}\right\} \text{Boat size} \end{matrix}$$

The wages of the workers are represented by T.

$$T = \begin{matrix} & \overbrace{\begin{matrix} \text{A} & \text{B} \end{matrix}}^{\text{Factory}} & \\ \begin{bmatrix} \$15 & \$13 \\ \$12 & \$11 \\ \$11 & \$10 \end{bmatrix} & \begin{matrix} \text{Cutting} \\ \text{Assembly} \\ \text{Packaging} \end{matrix} \left.\begin{matrix} \\ \\ \end{matrix}\right\} \text{Department} \end{matrix}$$

Compute ST and interpret the result.

72. Profit At a local store, the numbers of gallons of skim milk, 2% milk, and whole milk sold over the weekend are represented by A.

$$A = \begin{matrix} \overbrace{\begin{matrix} \text{Skim} & \text{2\%} & \text{Whole} \\ \text{milk} & \text{milk} & \text{milk} \end{matrix}}^{} \\ \begin{bmatrix} 40 & 64 & 52 \\ 60 & 82 & 76 \\ 76 & 96 & 84 \end{bmatrix} \begin{matrix} \text{Friday} \\ \text{Saturday} \\ \text{Sunday} \end{matrix} \end{matrix}$$

The selling prices per gallon and the profits per gallon for the three types of milk are represented by B.

$$B = \begin{matrix} \overbrace{\begin{matrix} \text{Selling} & \\ \text{price} & \text{Profit} \end{matrix}}^{} \\ \begin{bmatrix} \$3.45 & \$1.20 \\ \$3.65 & \$1.30 \\ \$3.85 & \$1.45 \end{bmatrix} \begin{matrix} \text{Skim milk} \\ \text{2\% milk} \\ \text{Whole milk} \end{matrix} \end{matrix}$$

(a) Compute AB and interpret the result.

(b) Find the store's total profit from milk sales for the weekend.

73. Voting Preferences The matrix

$$P = \begin{bmatrix} 0.6 & 0.1 & 0.1 \\ 0.2 & 0.7 & 0.1 \\ 0.2 & 0.2 & 0.8 \end{bmatrix} \begin{matrix} R \\ D \\ I \end{matrix} \Big\} \text{To}$$

From
R   D   I

is called a *stochastic matrix*. Each entry p_{ij} $(i \neq j)$ represents the proportion of the voting population that changes from party i to party j, and p_{ii} represents the proportion that remains loyal to the party from one election to the next. Compute and interpret P^2.

• • **74. Exercise** • • • • • • • • • • •

The numbers of calories burned by individuals of different body weights while performing different types of exercises for a one-hour time period are represented by A.

Calories burned
130-lb   155-lb
person   person

$$A = \begin{bmatrix} 472 & 563 \\ 590 & 704 \\ 177 & 211 \end{bmatrix} \begin{matrix} \text{Basketball} \\ \text{Jumping rope} \\ \text{Weight lifting} \end{matrix}$$

(a) A 130-pound person and a 155-pound person play basketball for 2 hours, jump rope for 15 minutes, and lift weights for 30 minutes. Organize the times spent exercising in a matrix B.

(b) Compute BA and interpret the result.

Exploration

True or False? **In Exercises 75 and 76, determine whether the statement is true or false. Justify your answer.**

75. Two matrices can be added only when they have the same dimension.

76. Matrix multiplication is commutative.

Think About It **In Exercises 77–80, use the matrices**

$$A = \begin{bmatrix} 2 & -1 \\ 1 & 3 \end{bmatrix} \quad \text{and} \quad B = \begin{bmatrix} -1 & 1 \\ 0 & -2 \end{bmatrix}.$$

77. Show that $(A + B)^2 \neq A^2 + 2AB + B^2$.

78. Show that $(A - B)^2 \neq A^2 - 2AB + B^2$.

79. Show that $(A + B)(A - B) \neq A^2 - B^2$.

80. Show that $(A + B)^2 = A^2 + AB + BA + B^2$.

81. Think About It If a, b, and c are real numbers such that $c \neq 0$ and $ac = bc$, then $a = b$. However, if A, B, and C are nonzero matrices such that $AC = BC$, then A is *not necessarily* equal to B. Illustrate this using the following matrices.

$$A = \begin{bmatrix} 0 & 1 \\ 0 & 1 \end{bmatrix}, \quad B = \begin{bmatrix} 1 & 0 \\ 1 & 0 \end{bmatrix}, \quad C = \begin{bmatrix} 2 & 3 \\ 2 & 3 \end{bmatrix}$$

82. Think About It If a and b are real numbers such that $ab = 0$, then $a = 0$ or $b = 0$. However, if A and B are matrices such that $AB = O$, it is *not necessarily* true that $A = O$ or $B = O$. Illustrate this using the following matrices.

$$A = \begin{bmatrix} 3 & 3 \\ 4 & 4 \end{bmatrix}, \quad B = \begin{bmatrix} 1 & -1 \\ -1 & 1 \end{bmatrix}$$

83. Finding Matrices Find two matrices A and B such that $AB = BA$.

84. HOW DO YOU SEE IT? A corporation has three factories that manufacture acoustic guitars and electric guitars. The production levels are represented by A.

Factory
A   B   C

$$A = \begin{bmatrix} 70 & 50 & 25 \\ 35 & 100 & 70 \end{bmatrix} \begin{matrix} \text{Acoustic} \\ \text{Electric} \end{matrix} \Big\} \text{Guitar type}$$

(a) Interpret the value of a_{22}.

(b) How could you find the production levels when production increases by 20%?

(c) Each acoustic guitar sells for $80 and each electric guitar sells for $120. How could you use matrices to find the total sales value of the guitars produced at each factory?

85. Conjecture Let A and B be unequal diagonal matrices of the same dimension. (A **diagonal matrix** is a square matrix in which each entry not on the main diagonal is zero.) Determine the products AB for several pairs of such matrices. Make a conjecture about a rule that can be used to calculate AB without using row-by-column multiplication.

86. Matrices with Complex Entries Let $i = \sqrt{-1}$ and let

$$A = \begin{bmatrix} i & 0 \\ 0 & i \end{bmatrix} \quad \text{and} \quad B = \begin{bmatrix} 0 & -i \\ i & 0 \end{bmatrix}.$$

(a) Find A^2, A^3, and A^4. Identify any similarities with i^2, i^3, and i^4.

(b) Find and identify B^2.

7.3 The Inverse of a Square Matrix

Inverse matrices are used to model and solve real-life problems. For example, in Exercises 59–62 on page 528, you will use an inverse matrix to find the currents in a circuit.

■ Verify that two matrices are inverses of each other.
■ Use Gauss-Jordan elimination to find the inverses of matrices.
■ Use a formula to find the inverses of 2 × 2 matrices.
■ Use inverse matrices to solve systems of linear equations.

The Inverse of a Matrix

This section further develops the algebra of matrices. To begin, consider the real number equation $ax = b$. To solve this equation for x, multiply each side of the equation by a^{-1} (provided that $a \neq 0$).

$$ax = b$$
$$(a^{-1}a)x = a^{-1}b$$
$$(1)x = a^{-1}b$$
$$x = a^{-1}b$$

The number a^{-1} is called the *multiplicative inverse* of a because $a^{-1}a = 1$. The multiplicative **inverse of a matrix** is defined in a similar way.

Definition of the Inverse of a Square Matrix

Let A be an $n \times n$ matrix and let I_n be the $n \times n$ identity matrix. If there exists a matrix A^{-1} such that

$$AA^{-1} = I_n = A^{-1}A$$

then A^{-1} is the **inverse** of A. The symbol A^{-1} is read as "A inverse."

EXAMPLE 1   The Inverse of a Matrix

Show that $B = \begin{bmatrix} 1 & -2 \\ 1 & -1 \end{bmatrix}$ is the inverse of $A = \begin{bmatrix} -1 & 2 \\ -1 & 1 \end{bmatrix}$.

Solution   To show that B is the inverse of A, show that $AB = I = BA$.

$$AB = \begin{bmatrix} -1 & 2 \\ -1 & 1 \end{bmatrix}\begin{bmatrix} 1 & -2 \\ 1 & -1 \end{bmatrix} = \begin{bmatrix} -1+2 & 2-2 \\ -1+1 & 2-1 \end{bmatrix} = \begin{bmatrix} 1 & 0 \\ 0 & 1 \end{bmatrix}$$

$$BA = \begin{bmatrix} 1 & -2 \\ 1 & -1 \end{bmatrix}\begin{bmatrix} -1 & 2 \\ -1 & 1 \end{bmatrix} = \begin{bmatrix} -1+2 & 2-2 \\ -1+1 & 2-1 \end{bmatrix} = \begin{bmatrix} 1 & 0 \\ 0 & 1 \end{bmatrix}$$

So, B is the inverse of A because $AB = I = BA$. This is an example of a square matrix that has an inverse. Note that not all square matrices have inverses.

✓ *Checkpoint*))) *Audio-video solution in English & Spanish at LarsonPrecalculus.com*

Show that $B = \begin{bmatrix} -1 & -1 \\ -3 & -2 \end{bmatrix}$ is the inverse of $A = \begin{bmatrix} 2 & -1 \\ -3 & 1 \end{bmatrix}$. ■

Recall that it is not always true that $AB = BA$, even when both products are defined. However, if A and B are both square matrices and $AB = I_n$, then it can be shown that $BA = I_n$. So, in Example 1, you need only to check that $AB = I_2$.

One real-life application of inverse matrices is in the study of beam deflection. In a simply supported elastic beam subjected to multiple forces, deflection **d** is related to force **w** by the matrix equation

$$\mathbf{d} = F\mathbf{w}$$

where F is a *flexibility matrix* whose entries depend on the material of the beam. The inverse of the flexibility matrix, F^{-1}, is the *stiffness matrix*.

Finding Inverse Matrices

If a matrix A has an inverse, then A is **invertible** (or **nonsingular**); otherwise, A is **singular.** A nonsquare matrix cannot have an inverse. To see this, note that when A is of dimension $m \times n$ and B is of dimension $n \times m$ (where $m \neq n$), the products AB and BA are of different dimensions and so cannot be equal to each other. Not all square matrices have inverses (see the matrix at the bottom of page 524). When a matrix does have an inverse, however, that inverse is unique. Example 2 shows how to use a system of equations to find the inverse of a matrix.

EXAMPLE 2 **Finding the Inverse of a Matrix**

Find the inverse of $A = \begin{bmatrix} 1 & 4 \\ -1 & -3 \end{bmatrix}$.

Solution To find the inverse of A, solve the matrix equation $AX = I$ for X.

$$\underset{A}{\begin{bmatrix} 1 & 4 \\ -1 & -3 \end{bmatrix}} \underset{X}{\begin{bmatrix} x_{11} & x_{12} \\ x_{21} & x_{22} \end{bmatrix}} \underset{=}{=} \underset{I}{\begin{bmatrix} 1 & 0 \\ 0 & 1 \end{bmatrix}}$$ Write matrix equation.

$$\begin{bmatrix} x_{11} + 4x_{21} & x_{12} + 4x_{22} \\ -x_{11} - 3x_{21} & -x_{12} - 3x_{22} \end{bmatrix} = \begin{bmatrix} 1 & 0 \\ 0 & 1 \end{bmatrix}$$ Multiply.

Equating corresponding entries, you obtain two systems of linear equations.

$$\begin{cases} x_{11} + 4x_{21} = 1 \\ -x_{11} - 3x_{21} = 0 \end{cases} \qquad \begin{cases} x_{12} + 4x_{22} = 0 \\ -x_{12} - 3x_{22} = 1 \end{cases}$$

Solve the first system using elementary row operations to determine that

$$x_{11} = -3 \quad \text{and} \quad x_{21} = 1.$$

Solve the second system to determine that

$$x_{12} = -4 \quad \text{and} \quad x_{22} = 1.$$

So, the inverse of A is

$$X = A^{-1}$$

$$= \begin{bmatrix} -3 & -4 \\ 1 & 1 \end{bmatrix}.$$

Use matrix multiplication to check this result in two ways.

Check

$$AA^{-1} = \begin{bmatrix} 1 & 4 \\ -1 & -3 \end{bmatrix} \begin{bmatrix} -3 & -4 \\ 1 & 1 \end{bmatrix}$$

$$= \begin{bmatrix} 1 & 0 \\ 0 & 1 \end{bmatrix} \; ✓$$

$$A^{-1}A = \begin{bmatrix} -3 & -4 \\ 1 & 1 \end{bmatrix} \begin{bmatrix} 1 & 4 \\ -1 & -3 \end{bmatrix}$$

$$= \begin{bmatrix} 1 & 0 \\ 0 & 1 \end{bmatrix} \; ✓$$

✓ *Checkpoint* 🔊⟩⟩⟩ *Audio-video solution in English & Spanish at LarsonPrecalculus.com*

Find the inverse of $A = \begin{bmatrix} 1 & -2 \\ -1 & 3 \end{bmatrix}$.

In Example 2, note that the two systems of linear equations have the *same coefficient matrix A*. Rather than solve the two systems represented by

$$\begin{bmatrix} 1 & 4 & \vdots & 1 \\ -1 & -3 & \vdots & 0 \end{bmatrix}$$

and

$$\begin{bmatrix} 1 & 4 & \vdots & 0 \\ -1 & -3 & \vdots & 1 \end{bmatrix}$$

separately, you can solve them *simultaneously* by *adjoining* the identity matrix to the coefficient matrix to obtain

$$\begin{matrix} A & & I \end{matrix}$$
$$\begin{bmatrix} 1 & 4 & \vdots & 1 & 0 \\ -1 & -3 & \vdots & 0 & 1 \end{bmatrix}.$$

This "doubly augmented" matrix can be represented as

$[A \; \vdots \; I].$

By applying Gauss-Jordan elimination to this matrix, you can solve *both* systems with a single elimination process.

$$\begin{bmatrix} 1 & 4 & \vdots & 1 & 0 \\ -1 & -3 & \vdots & 0 & 1 \end{bmatrix}$$

$$R_1 + R_2 \rightarrow \begin{bmatrix} 1 & 4 & \vdots & 1 & 0 \\ 0 & 1 & \vdots & 1 & 1 \end{bmatrix}$$

$$-4R_2 + R_1 \rightarrow \begin{bmatrix} 1 & 0 & \vdots & -3 & -4 \\ 0 & 1 & \vdots & 1 & 1 \end{bmatrix}$$

So, from the "doubly augmented" matrix $[A \; \vdots \; I]$, you obtain the matrix $[I \; \vdots \; A^{-1}]$.

$$\begin{matrix} A & & I \end{matrix}$$
$$\begin{bmatrix} 1 & 4 & \vdots & 1 & 0 \\ -1 & -3 & \vdots & 0 & 1 \end{bmatrix} \Rightarrow \begin{matrix} I & & A^{-1} \end{matrix} \begin{bmatrix} 1 & 0 & \vdots & -3 & -4 \\ 0 & 1 & \vdots & 1 & 1 \end{bmatrix}$$

This procedure (or algorithm) works for any square matrix that has an inverse.

▷ **TECHNOLOGY** Most graphing utilities can find the inverse of a square matrix. To do so, you may have to use the inverse key $\boxed{x^{-1}}$. Consult the user's guide for your graphing utility for specific keystrokes.

Finding an Inverse Matrix

Let A be a square matrix of dimension $n \times n$.

1. Write the $n \times 2n$ matrix that consists of the given matrix A on the left and the $n \times n$ identity matrix I on the right to obtain

$[A \; \vdots \; I].$

2. If possible, row reduce A to I using elementary row operations on the *entire* matrix

$[A \; \vdots \; I].$

The result will be the matrix

$[I \; \vdots \; A^{-1}].$

If this is not possible, then A is not invertible.

3. Check your work by multiplying to see that

$AA^{-1} = I = A^{-1}A.$

EXAMPLE 3 **Finding the Inverse of a Matrix**

Find the inverse of

$$A = \begin{bmatrix} 1 & -1 & 0 \\ 1 & 0 & -1 \\ 6 & -2 & -3 \end{bmatrix}.$$

Solution Begin by adjoining the identity matrix to A to form the matrix

$$[A \vdots I] = \begin{bmatrix} 1 & -1 & 0 & \vdots & 1 & 0 & 0 \\ 1 & 0 & -1 & \vdots & 0 & 1 & 0 \\ 6 & -2 & -3 & \vdots & 0 & 0 & 1 \end{bmatrix}.$$

Use elementary row operations to obtain the form $[I \vdots A^{-1}]$.

$$\begin{array}{c} \\ -R_1 + R_2 \rightarrow \\ -6R_1 + R_3 \rightarrow \end{array} \begin{bmatrix} 1 & -1 & 0 & \vdots & 1 & 0 & 0 \\ 0 & 1 & -1 & \vdots & -1 & 1 & 0 \\ 0 & 4 & -3 & \vdots & -6 & 0 & 1 \end{bmatrix}$$

$$\begin{array}{c} R_2 + R_1 \rightarrow \\ \\ -4R_2 + R_3 \rightarrow \end{array} \begin{bmatrix} 1 & 0 & -1 & \vdots & 0 & 1 & 0 \\ 0 & 1 & -1 & \vdots & -1 & 1 & 0 \\ 0 & 0 & 1 & \vdots & -2 & -4 & 1 \end{bmatrix}$$

$$\begin{array}{c} R_3 + R_1 \rightarrow \\ R_3 + R_2 \rightarrow \\ \end{array} \begin{bmatrix} 1 & 0 & 0 & \vdots & -2 & -3 & 1 \\ 0 & 1 & 0 & \vdots & -3 & -3 & 1 \\ 0 & 0 & 1 & \vdots & -2 & -4 & 1 \end{bmatrix} = [I \vdots A^{-1}]$$

So, the matrix A is invertible and its inverse is

$$A^{-1} = \begin{bmatrix} -2 & -3 & 1 \\ -3 & -3 & 1 \\ -2 & -4 & 1 \end{bmatrix}.$$

Check

$$AA^{-1} = \begin{bmatrix} 1 & -1 & 0 \\ 1 & 0 & -1 \\ 6 & -2 & -3 \end{bmatrix} \begin{bmatrix} -2 & -3 & 1 \\ -3 & -3 & 1 \\ -2 & -4 & 1 \end{bmatrix} = \begin{bmatrix} 1 & 0 & 0 \\ 0 & 1 & 0 \\ 0 & 0 & 1 \end{bmatrix} = I$$

> • • • • • • • • • • • • • • • • ▷
> •
> •**REMARK** Be sure to check
> your solution because it is not
> uncommon to make arithmetic
> errors when using elementary
> row operations.

✓ *Checkpoint*  *Audio-video solution in English & Spanish at LarsonPrecalculus.com*

Find the inverse of

$$A = \begin{bmatrix} 1 & -2 & -1 \\ 0 & -1 & 2 \\ 1 & -2 & 0 \end{bmatrix}.$$

The process shown in Example 3 applies to any $n \times n$ matrix A. When using this algorithm, if the matrix A does not reduce to the identity matrix, then A does not have an inverse. For example, the matrix below has no inverse.

$$A = \begin{bmatrix} 1 & 2 & 0 \\ 3 & -1 & 2 \\ -2 & 3 & -2 \end{bmatrix}$$

To confirm that this matrix has no inverse, adjoin the identity matrix to A to form $[A \vdots I]$ and try to apply Gauss-Jordan elimination to the matrix. You will find that it is impossible to obtain the identity matrix I on the left. So, A is not invertible.

The Inverse of a 2 × 2 Matrix

Using Gauss-Jordan elimination to find the inverse of a matrix works well (even as a computer technique) for matrices of dimension 3×3 or greater. For 2×2 matrices, however, many people prefer to use a formula for the inverse rather than Gauss-Jordan elimination. This simple formula, which works *only* for 2×2 matrices, is explained as follows. A 2×2 matrix A given by

$$A = \begin{bmatrix} a & b \\ c & d \end{bmatrix}$$

is invertible if and only if

$$ad - bc \neq 0.$$

Moreover, if $ad - bc \neq 0$, then the inverse is given by

$$A^{-1} = \frac{1}{ad - bc} \begin{bmatrix} d & -b \\ -c & a \end{bmatrix}. \qquad \text{\color{red}Formula for the inverse of a 2 × 2 matrix}$$

The denominator

$$ad - bc$$

is the **determinant** of the 2×2 matrix A. You will study determinants in the next section.

EXAMPLE 4   **Finding the Inverse of a 2 × 2 Matrix**

See LarsonPrecalculus.com for an interactive version of this type of example.

If possible, find the inverse of each matrix.

a. $A = \begin{bmatrix} 3 & -1 \\ -2 & 2 \end{bmatrix}$ **b.** $B = \begin{bmatrix} 3 & -1 \\ -6 & 2 \end{bmatrix}$

Solution

a. The determinant of a matrix A is

$$ad - bc = 3(2) - (-1)(-2) = 4.$$

This quantity is not zero, so the matrix is invertible. The inverse is formed by interchanging the entries on the main diagonal, changing the signs of the other two entries, and multiplying by the scalar $\frac{1}{4}$.

$$A^{-1} = \frac{1}{ad - bc} \begin{bmatrix} d & -b \\ -c & a \end{bmatrix} \qquad \text{\color{red}Formula for the inverse of a 2 × 2 matrix}$$

$$= \frac{1}{4} \begin{bmatrix} 2 & 1 \\ 2 & 3 \end{bmatrix} \qquad \text{\color{red}Substitute for } a, b, c, d, \text{ and the determinant.}$$

$$= \begin{bmatrix} \dfrac{1}{2} & \dfrac{1}{4} \\ \dfrac{1}{2} & \dfrac{3}{4} \end{bmatrix} \qquad \text{\color{red}Multiply by the scalar } \frac{1}{4}.$$

b. The determinant of matrix B is

$$ad - bc = 3(2) - (-1)(-6) = 0.$$

Because $ad - bc = 0$, B is not invertible.

✓ **Checkpoint** 🔊)) *Audio-video solution in English & Spanish at LarsonPrecalculus.com*

If possible, find the inverse of $A = \begin{bmatrix} 5 & -1 \\ 3 & 4 \end{bmatrix}$.

Systems of Linear Equations

You know that a system of linear equations can have exactly one solution, infinitely many solutions, or no solution. If the coefficient matrix A of a *square* system (a system that has the same number of equations as variables) is invertible, then the system has a unique solution, which can be found using an inverse matrix as follows.

> ### A System of Equations with a Unique Solution
>
> If A is an invertible matrix, then the system of linear equations represented by $AX = B$ has a unique solution given by $X = A^{-1}B$.

EXAMPLE 5 **Solving a System Using an Inverse Matrix**

Use an inverse matrix to solve the system

$$\begin{cases} x + y + z = 10{,}000 \\ 0.06x + 0.075y + 0.095z = 730. \\ x - 2z = 0 \end{cases}$$

> ▷ **TECHNOLOGY** On most graphing utilities, to solve a linear system that has an invertible coefficient matrix, you can use the formula $X = A^{-1}B$. That is, enter the $n \times n$ coefficient matrix $[A]$ and the $n \times 1$ column matrix $[B]$. The solution matrix X is given by
>
> $$[A]^{-1}[B].$$

Solution Begin by writing the system in the matrix form $AX = B$.

$$\begin{bmatrix} 1 & 1 & 1 \\ 0.06 & 0.075 & 0.095 \\ 1 & 0 & -2 \end{bmatrix} \begin{bmatrix} x \\ y \\ z \end{bmatrix} = \begin{bmatrix} 10{,}000 \\ 730 \\ 0 \end{bmatrix}$$

Then, use Gauss-Jordan elimination to find A^{-1}.

$$A^{-1} = \begin{bmatrix} 15 & -200 & -2 \\ -21.5 & 300 & 3.5 \\ 7.5 & -100 & -1.5 \end{bmatrix}$$

Finally, multiply B by A^{-1} on the left to obtain the solution.

$$X = A^{-1}B = \begin{bmatrix} 15 & -200 & -2 \\ -21.5 & 300 & 3.5 \\ 7.5 & -100 & -1.5 \end{bmatrix} \begin{bmatrix} 10{,}000 \\ 730 \\ 0 \end{bmatrix} = \begin{bmatrix} 4000 \\ 4000 \\ 2000 \end{bmatrix}$$

The solution of the system is $x = 4000$, $y = 4000$, and $z = 2000$, or $(4000, 4000, 2000)$.

✓ *Checkpoint* *Audio-video solution in English & Spanish at LarsonPrecalculus.com*

Use an inverse matrix to solve the system $\begin{cases} 2x + 3y + z = -1 \\ 3x + 3y + z = 1. \\ 2x + 4y + z = -2 \end{cases}$

> ## Summarize (Section 7.3)
>
> 1. State the definition of the inverse of a square matrix *(page 521)*. For an example of how to show that a matrix is the inverse of another matrix, see Example 1.
> 2. Explain how to find an inverse matrix *(pages 522 and 523)*. For examples of finding inverse matrices, see Examples 2 and 3.
> 3. State the formula for the inverse of a 2×2 matrix *(page 525)*. For an example of using this formula to find an inverse matrix, see Example 4.
> 4. Explain how to use an inverse matrix to solve a system of linear equations *(page 526)*. For an example of using an inverse matrix to solve a system of linear equations, see Example 5.

7.3 Exercises

See **CalcChat.com** for tutorial help and worked-out solutions to odd-numbered exercises.

Vocabulary: Fill in the blanks.

1. If there exists an $n \times n$ matrix A^{-1} such that $AA^{-1} = I_n = A^{-1}A$, then A^{-1} is the _____ of A.

2. A matrix that has an inverse is invertible or _____. A matrix that does not have an inverse is _____.

3. A 2×2 matrix is invertible if and only if its _____ is not zero.

4. If A is an invertible matrix, then the system of linear equations represented by $AX = B$ has a unique solution given by $X =$ _____.

Skills and Applications

The Inverse of a Matrix In Exercises 5–12, show that B is the inverse of A.

5. $A = \begin{bmatrix} 2 & 1 \\ 5 & 3 \end{bmatrix}$, $B = \begin{bmatrix} 3 & -1 \\ -5 & 2 \end{bmatrix}$

6. $A = \begin{bmatrix} 1 & -1 \\ -1 & 2 \end{bmatrix}$, $B = \begin{bmatrix} 2 & 1 \\ 1 & 1 \end{bmatrix}$

7. $A = \begin{bmatrix} 3 & 2 \\ 1 & 4 \end{bmatrix}$, $B = \dfrac{1}{10}\begin{bmatrix} 4 & -2 \\ -1 & 3 \end{bmatrix}$

8. $A = \begin{bmatrix} 1 & -1 \\ 2 & 3 \end{bmatrix}$, $B = \dfrac{1}{5}\begin{bmatrix} 3 & 1 \\ -2 & 1 \end{bmatrix}$

9. $A = \begin{bmatrix} 2 & -17 & 11 \\ -1 & 11 & -7 \\ 0 & 3 & -2 \end{bmatrix}$, $B = \begin{bmatrix} 1 & 1 & 2 \\ 2 & 4 & -3 \\ 3 & 6 & -5 \end{bmatrix}$

10. $A = \begin{bmatrix} -4 & 1 & 5 \\ -1 & 2 & 4 \\ 0 & -1 & -1 \end{bmatrix}$,

    $B = \dfrac{1}{4}\begin{bmatrix} -2 & 4 & 6 \\ 1 & -4 & -11 \\ -1 & 4 & 7 \end{bmatrix}$

11. $A = \begin{bmatrix} 2 & 0 & 2 & 1 \\ 3 & 0 & 0 & 1 \\ -1 & 1 & -2 & 1 \\ 3 & -1 & 1 & 0 \end{bmatrix}$,

    $B = \dfrac{1}{3}\begin{bmatrix} -1 & 3 & -2 & -2 \\ -2 & 9 & -7 & -10 \\ 1 & 0 & -1 & -1 \\ 3 & -6 & 6 & 6 \end{bmatrix}$

12. $A = \begin{bmatrix} -1 & 1 & 0 & -1 \\ 1 & -1 & 1 & 0 \\ -1 & 1 & 2 & 0 \\ 0 & -1 & 1 & 1 \end{bmatrix}$,

    $B = \dfrac{1}{3}\begin{bmatrix} -3 & 1 & 1 & -3 \\ -3 & -1 & 2 & -3 \\ 0 & 1 & 1 & 0 \\ -3 & -2 & 1 & 0 \end{bmatrix}$

Finding the Inverse of a Matrix In Exercises 13–24, find the inverse of the matrix, if possible.

13. $\begin{bmatrix} 2 & 1 \\ 5 & 3 \end{bmatrix}$

14. $\begin{bmatrix} 1 & 2 \\ 3 & 7 \end{bmatrix}$

15. $\begin{bmatrix} 1 & -2 \\ 2 & -3 \end{bmatrix}$

16. $\begin{bmatrix} -7 & 33 \\ 4 & -19 \end{bmatrix}$

17. $\begin{bmatrix} 3 & 1 \\ 4 & 2 \end{bmatrix}$

18. $\begin{bmatrix} 4 & -1 \\ -3 & 1 \end{bmatrix}$

19. $\begin{bmatrix} 1 & 1 & 1 \\ 3 & 5 & 4 \\ 3 & 6 & 5 \end{bmatrix}$

20. $\begin{bmatrix} 1 & 2 & 2 \\ 3 & 7 & 9 \\ -1 & -4 & -7 \end{bmatrix}$

21. $\begin{bmatrix} -5 & 0 & 0 \\ 2 & 0 & 0 \\ -1 & 5 & 7 \end{bmatrix}$

22. $\begin{bmatrix} 1 & 0 & 0 \\ 3 & 0 & 0 \\ 2 & 5 & 5 \end{bmatrix}$

23. $\begin{bmatrix} -8 & 0 & 0 & 0 \\ 0 & 1 & 0 & 0 \\ 0 & 0 & 4 & 0 \\ 0 & 0 & 0 & -5 \end{bmatrix}$

24. $\begin{bmatrix} 1 & 3 & -2 & 0 \\ 0 & 2 & 4 & 6 \\ 0 & 0 & -2 & 1 \\ 0 & 0 & 0 & 5 \end{bmatrix}$

Finding the Inverse of a Matrix In Exercises 25–32, use the matrix capabilities of a graphing utility to find the inverse of the matrix, if possible.

25. $\begin{bmatrix} 1 & 2 & -1 \\ 3 & 7 & -10 \\ -5 & -7 & -15 \end{bmatrix}$

26. $\begin{bmatrix} 10 & 5 & -7 \\ -5 & 1 & 4 \\ 3 & 2 & -2 \end{bmatrix}$

27. $\begin{bmatrix} -\frac{1}{2} & \frac{3}{4} & \frac{1}{4} \\ 1 & 0 & -\frac{3}{2} \\ 0 & -1 & \frac{1}{2} \end{bmatrix}$

28. $\begin{bmatrix} -\frac{5}{6} & \frac{1}{3} & \frac{11}{6} \\ 0 & \frac{2}{3} & 2 \\ 1 & -\frac{1}{2} & -\frac{5}{2} \end{bmatrix}$

29. $\begin{bmatrix} 0.1 & 0.2 & 0.3 \\ -0.3 & 0.2 & 0.2 \\ 0.5 & 0.4 & 0.4 \end{bmatrix}$

30. $\begin{bmatrix} 0.6 & 0 & -0.3 \\ 0.7 & -1 & 0.2 \\ 1 & 0 & -0.9 \end{bmatrix}$

31. $\begin{bmatrix} -1 & 0 & 1 & 0 \\ 0 & 2 & 0 & -1 \\ 2 & 0 & -1 & 0 \\ 0 & -1 & 0 & 1 \end{bmatrix}$

32. $\begin{bmatrix} 1 & -2 & -1 & -2 \\ 3 & -5 & -2 & -3 \\ 2 & -5 & -2 & -5 \\ -1 & 4 & 4 & 11 \end{bmatrix}$

Finding the Inverse of a 2 × 2 Matrix
In Exercises 33–38, use the formula on page 525 to find the inverse of the 2 × 2 matrix, if possible.

33. $\begin{bmatrix} 2 & 3 \\ -1 & 5 \end{bmatrix}$

34. $\begin{bmatrix} 1 & -2 \\ -3 & 2 \end{bmatrix}$

35. $\begin{bmatrix} -4 & -6 \\ 2 & 3 \end{bmatrix}$

36. $\begin{bmatrix} -12 & 3 \\ 5 & -2 \end{bmatrix}$

37. $\begin{bmatrix} 0.5 & 0.3 \\ 1.5 & 0.6 \end{bmatrix}$

38. $\begin{bmatrix} -1.25 & 0.625 \\ 0.16 & 0.32 \end{bmatrix}$

Solving a System Using an Inverse Matrix
In Exercises 39–42, use the inverse matrix found in Exercise 15 to solve the system of linear equations.

39. $\begin{cases} x - 2y = 5 \\ 2x - 3y = 10 \end{cases}$

40. $\begin{cases} x - 2y = 0 \\ 2x - 3y = 3 \end{cases}$

41. $\begin{cases} x - 2y = 4 \\ 2x - 3y = 2 \end{cases}$

42. $\begin{cases} x - 2y = 1 \\ 2x - 3y = -2 \end{cases}$

Solving a System Using an Inverse Matrix
In Exercises 43 and 44, use the inverse matrix found in Exercise 19 to solve the system of linear equations.

43. $\begin{cases} x + y + z = 0 \\ 3x + 5y + 4z = 5 \\ 3x + 6y + 5z = 2 \end{cases}$

44. $\begin{cases} x + y + z = -1 \\ 3x + 5y + 4z = 2 \\ 3x + 6y + 5z = 0 \end{cases}$

Solving a System Using an Inverse Matrix
In Exercises 45 and 46, use the inverse matrix found in Exercise 32 to solve the system of linear equations.

45. $\begin{cases} x_1 - 2x_2 - x_3 - 2x_4 = 0 \\ 3x_1 - 5x_2 - 2x_3 - 3x_4 = 1 \\ 2x_1 - 5x_2 - 2x_3 - 5x_4 = -1 \\ -x_1 + 4x_2 + 4x_3 + 11x_4 = 2 \end{cases}$

46. $\begin{cases} x_1 - 2x_2 - x_3 - 2x_4 = 1 \\ 3x_1 - 5x_2 - 2x_3 - 3x_4 = -2 \\ 2x_1 - 5x_2 - 2x_3 - 5x_4 = 0 \\ -x_1 + 4x_2 + 4x_3 + 11x_4 = -3 \end{cases}$

Solving a System Using an Inverse Matrix
In Exercises 47–54, use an inverse matrix to solve the system of linear equations, if possible.

47. $\begin{cases} 5x + 4y = -1 \\ 2x + 5y = 3 \end{cases}$

48. $\begin{cases} 18x + 12y = 13 \\ 30x + 24y = 23 \end{cases}$

49. $\begin{cases} -0.4x + 0.8y = 1.6 \\ 2x - 4y = 5 \end{cases}$

50. $\begin{cases} 0.2x - 0.6y = 2.4 \\ -x + 1.4y = -8.8 \end{cases}$

51. $\begin{cases} 2.3x - 1.9y = 6 \\ 1.5x + 0.75y = -12 \end{cases}$

52. $\begin{cases} 5.1x - 3.4y = -20 \\ 0.9x - 0.6y = -51 \end{cases}$

53. $\begin{cases} 4x - y + z = -5 \\ 2x + 2y + 3z = 10 \\ 5x - 2y + 6z = 1 \end{cases}$

54. $\begin{cases} 4x - 2y + 3z = -2 \\ 2x + 2y + 5z = 16 \\ 8x - 5y - 2z = 4 \end{cases}$

Using a Graphing Utility
In Exercises 55 and 56, use the matrix capabilities of a graphing utility to solve the system of linear equations, if possible.

55. $\begin{cases} 5x - 3y + 2z = 2 \\ 2x + 2y - 3z = 3 \\ x - 7y + 7z = -4 \end{cases}$

56. $\begin{cases} 2x + 3y + 5z = 4 \\ 3x + 5y + 9z = 7 \\ 5x + 9y + 16z = 13 \end{cases}$

Investment Portfolio
In Exercises 57 and 58, you invest in AAA-rated bonds, A-rated bonds, and B-rated bonds. The average yields are 4.5% on AAA bonds, 5% on A bonds, and 9% on B bonds. You invest twice as much in B bonds as in A bonds. Let x, y, and z represent the amounts invested in AAA, A, and B bonds, respectively.

$\begin{cases} x + y + z = \text{(total investment)} \\ 0.045x + 0.05y + 0.09z = \text{(annual return)} \\ 2y - z = 0 \end{cases}$

Use the inverse of the coefficient matrix of this system to find the amount invested in each type of bond for the given total investment and annual return.

	Total Investment	Annual Return
57.	\$10,000	\$650
58.	\$12,000	\$835

Circuit Analysis

In Exercises 59–62, consider the circuit shown in the figure. The currents I_1, I_2, and I_3 (in amperes) are the solution of the system

$\begin{cases} 2I_1 + 4I_3 = E_1 \\ I_2 + 4I_3 = E_2 \\ I_1 + I_2 - I_3 = 0 \end{cases}$

where E_1 and E_2 are voltages. Use the inverse of the coefficient matrix of this system to find the unknown currents for the given voltages.

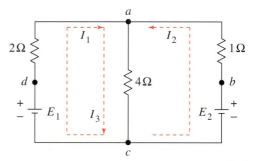

59. $E_1 = 15$ volts, $E_2 = 17$ volts

60. $E_1 = 10$ volts, $E_2 = 10$ volts

61. $E_1 = 28$ volts, $E_2 = 21$ volts

62. $E_1 = 24$ volts, $E_2 = 23$ volts

Raw Materials In Exercises 63 and 64, find the numbers of bags of potting soil that a company can produce for seedlings, general potting, and hardwood plants with the given amounts of raw materials. The raw materials used in one bag of each type of potting soil are shown below.

	Sand	Loam	Peat Moss
Seedlings	2 units	1 unit	1 unit
General	1 unit	2 units	1 unit
Hardwoods	2 units	2 units	2 units

63. 500 units of sand
500 units of loam
400 units of peat moss

64. 500 units of sand
750 units of loam
450 units of peat moss

65. Floral Design A florist is creating 10 centerpieces. Roses cost $2.50 each, lilies cost $4 each, and irises cost $2 each. The customer has a budget of $300 allocated for the centerpieces and wants each centerpiece to contain 12 flowers, with twice as many roses as the number of irises and lilies combined.

(a) Write a system of linear equations that represents the situation. Then write a matrix equation that corresponds to your system.

(b) Solve your system of linear equations using an inverse matrix. Find the number of flowers of each type that the florist can use to create the 10 centerpieces.

66. International Travel The table shows the numbers of visitors y (in thousands) to the United States from China from 2012 through 2014. *(Source: U.S. Department of Commerce)*

Year	Visitors, y (in thousands)
2012	1474
2013	1807
2014	2188

(a) The data can be modeled by the quadratic function $y = at^2 + bt + c$. Write a system of linear equations for the data. Let t represent the year, with $t = 12$ corresponding to 2012.

(b) Use the matrix capabilities of a graphing utility to find the inverse of the coefficient matrix of the system from part (a).

(c) Use the result of part (b) to solve the system and write the model $y = at^2 + bt + c$.

(d) Use the graphing utility to graph the model with the data.

Exploration

True or False? In Exercises 67 and 68, determine whether the statement is true or false. Justify your answer.

67. Multiplication of an invertible matrix and its inverse is commutative.

68. When the product of two square matrices is the identity matrix, the matrices are inverses of one another.

69. Writing Explain how to determine whether the inverse of a 2×2 matrix exists, as well as how to find the inverse when it exists.

70. Writing Explain how to write a system of three linear equations in three variables as a matrix equation $AX = B$, as well as how to solve the system using an inverse matrix.

Think About It In Exercises 71 and 72, find the value of k that makes the matrix singular.

71. $\begin{bmatrix} 4 & 3 \\ -2 & k \end{bmatrix}$

72. $\begin{bmatrix} 2k+1 & 3 \\ -7 & 1 \end{bmatrix}$

73. Conjecture Consider matrices of the form

$$A = \begin{bmatrix} a_{11} & 0 & 0 & 0 & \cdots & 0 \\ 0 & a_{22} & 0 & 0 & \cdots & 0 \\ 0 & 0 & a_{33} & 0 & \cdots & 0 \\ \vdots & \vdots & \vdots & \vdots & & \vdots \\ 0 & 0 & 0 & 0 & \cdots & a_{nn} \end{bmatrix}.$$

(a) Write a 2×2 matrix and a 3×3 matrix in the form of A. Find the inverse of each.

(b) Use the result of part (a) to make a conjecture about the inverses of matrices in the form of A.

74. **HOW DO YOU SEE IT?** Consider the matrix

$$A = \begin{bmatrix} x & y \\ 0 & z \end{bmatrix}.$$

Use the determinant of A to state the conditions for which (a) A^{-1} exists and (b) $A^{-1} = A$.

75. Verifying a Formula Verify that the inverse of an invertible 2×2 matrix

$$A = \begin{bmatrix} a & b \\ c & d \end{bmatrix}$$

is given by $A^{-1} = \dfrac{1}{ad - bc} \begin{bmatrix} d & -b \\ -c & a \end{bmatrix}.$

Project: Consumer Credit To work an extended application analyzing the outstanding consumer credit in the United States, visit this text's website at *LarsonPrecalculus.com*. *(Source: Board of Governors of the Federal Reserve System)*

7.4 The Determinant of a Square Matrix

- Find the determinants of 2 × 2 matrices.
- Find minors and cofactors of square matrices.
- Find the determinants of square matrices.

Determinants are often used in other branches of mathematics. For example, the types of determinants in Exercises 87–92 on page 537 occur when changes of variables are made in calculus.

The Determinant of a 2 × 2 Matrix

Every *square* matrix can be associated with a real number called its **determinant.** Determinants have many uses, and several will be discussed in this section and the next section. Historically, the use of determinants arose from special number patterns that occur when systems of linear equations are solved. For example, the system

$$\begin{cases} a_1 x + b_1 y = c_1 \\ a_2 x + b_2 y = c_2 \end{cases}$$

has a solution

$$x = \frac{c_1 b_2 - c_2 b_1}{a_1 b_2 - a_2 b_1} \quad \text{and} \quad y = \frac{a_1 c_2 - a_2 c_1}{a_1 b_2 - a_2 b_1}$$

provided that $a_1 b_2 - a_2 b_1 \neq 0$. Note that the denominators of the two fractions are the same. This denominator is called the *determinant* of the coefficient matrix of the system.

Coefficient Matrix **Determinant**

$$A = \begin{bmatrix} a_1 & b_1 \\ a_2 & b_2 \end{bmatrix} \qquad \det(A) = a_1 b_2 - a_2 b_1$$

The determinant of matrix A can also be denoted by vertical bars on both sides of the matrix, as shown in the definition below.

Definition of the Determinant of a 2 × 2 Matrix

The **determinant** of the matrix

$$A = \begin{bmatrix} a_1 & b_1 \\ a_2 & b_2 \end{bmatrix}$$

is given by

$$\det(A) = |A| = \begin{vmatrix} a_1 & b_1 \\ a_2 & b_2 \end{vmatrix} = a_1 b_2 - a_2 b_1.$$

In this text, $\det(A)$ and $|A|$ are used interchangeably to represent the determinant of A. Although vertical bars are also used to denote the absolute value of a real number, the context will show which use is intended.

A convenient method for remembering the formula for the determinant of a 2 × 2 matrix is shown below.

$$\det(A) = \begin{vmatrix} a_1 & b_1 \\ a_2 & b_2 \end{vmatrix} = a_1 b_2 - a_2 b_1$$

Note that the determinant is the difference of the products of the two diagonals of the matrix.

In Example 1, you will see that the determinant of a matrix can be positive, zero, or negative.

EXAMPLE 1 **The Determinant of a 2 × 2 Matrix**

Find the determinant of each matrix.

a. $A = \begin{bmatrix} 2 & -3 \\ 1 & 2 \end{bmatrix}$

b. $B = \begin{bmatrix} 2 & 1 \\ 4 & 2 \end{bmatrix}$

c. $C = \begin{bmatrix} 0 & \frac{3}{2} \\ 2 & 4 \end{bmatrix}$

Solution

a. $\det(A) = \begin{vmatrix} 2 & -3 \\ 1 & 2 \end{vmatrix} = 2(2) - 1(-3) = 4 + 3 = 7$

b. $\det(B) = \begin{vmatrix} 2 & 1 \\ 4 & 2 \end{vmatrix} = 2(2) - 4(1) = 4 - 4 = 0$

c. $\det(C) = \begin{vmatrix} 0 & \frac{3}{2} \\ 2 & 4 \end{vmatrix} = 0(4) - 2\left(\frac{3}{2}\right) = 0 - 3 = -3$

✓ *Checkpoint* 🔊)) *Audio-video solution in English & Spanish at LarsonPrecalculus.com*

Find the determinant of each matrix.

a. $A = \begin{bmatrix} 1 & 2 \\ 3 & -1 \end{bmatrix}$

b. $B = \begin{bmatrix} 5 & 0 \\ -4 & 2 \end{bmatrix}$

c. $C = \begin{bmatrix} 3 & 6 \\ 2 & 4 \end{bmatrix}$

The determinant of a matrix of dimension 1×1 is defined simply as the entry of the matrix. For example, if $A = [-2]$, then $\det(A) = -2$.

▷ **TECHNOLOGY** Most graphing utilities can find the determinant of a matrix. For example, to find the determinant of

$$A = \begin{bmatrix} 2.4 & 0.8 \\ -0.6 & -3.2 \end{bmatrix}$$

use the *matrix editor* to enter the matrix as $[A]$ and then choose the *determinant* feature. The result is -7.2, as shown below.

```
[A]
        ┌ 2.4    .8 ┐
        └ -.6  -3.2 ┘
det([A])
                -7.2
```

Consult the user's guide for your graphing utility for specific keystrokes.

Minors and Cofactors

To define the determinant of a square matrix of dimension 3×3 or greater, it is helpful to introduce the concepts of **minors** and **cofactors.**

Sign Pattern for Cofactors

$$\begin{bmatrix} + & - & + \\ - & + & - \\ + & - & + \end{bmatrix}$$

3×3 matrix

$$\begin{bmatrix} + & - & + & - \\ - & + & - & + \\ + & - & + & - \\ - & + & - & + \end{bmatrix}$$

4×4 matrix

$$\begin{bmatrix} + & - & + & - & + & \cdots \\ - & + & - & + & - & \cdots \\ + & - & + & - & + & \cdots \\ - & + & - & + & - & \cdots \\ + & - & + & - & + & \cdots \\ \vdots & \vdots & \vdots & \vdots & \vdots \end{bmatrix}$$

$n \times n$ matrix

Minors and Cofactors of a Square Matrix

If A is a square matrix, then the **minor** M_{ij} of the entry a_{ij} is the determinant of the matrix obtained by deleting the ith row and jth column of A. The **cofactor** C_{ij} of the entry a_{ij} is

$$C_{ij} = (-1)^{i+j}M_{ij}.$$

In the sign pattern for cofactors at the left, notice that *odd* positions (where $i + j$ is odd) have negative signs and *even* positions (where $i + j$ is even) have positive signs.

EXAMPLE 2   Finding the Minors and Cofactors of a Matrix

Find all the minors and cofactors of

$$A = \begin{bmatrix} 0 & 2 & 1 \\ 3 & -1 & 2 \\ 4 & 0 & 1 \end{bmatrix}.$$

Solution To find the minor M_{11}, delete the first row and first column of A and find the determinant of the resulting matrix.

$$\begin{bmatrix} 0 & 2 & 1 \\ 3 & -1 & 2 \\ 4 & 0 & 1 \end{bmatrix}, \quad M_{11} = \begin{vmatrix} -1 & 2 \\ 0 & 1 \end{vmatrix} = -1(1) - 0(2) = -1$$

Similarly, to find M_{12}, delete the first row and second column.

$$\begin{bmatrix} 0 & 2 & 1 \\ 3 & -1 & 2 \\ 4 & 0 & 1 \end{bmatrix}, \quad M_{12} = \begin{vmatrix} 3 & 2 \\ 4 & 1 \end{vmatrix} = 3(1) - 4(2) = -5$$

Continuing this pattern, you obtain the minors.

$$\begin{array}{lll} M_{11} = -1 & M_{12} = -5 & M_{13} = 4 \\ M_{21} = 2 & M_{22} = -4 & M_{23} = -8 \\ M_{31} = 5 & M_{32} = -3 & M_{33} = -6 \end{array}$$

Now, to find the cofactors, combine these minors with the checkerboard pattern of signs for a 3×3 matrix shown at the upper left.

$$\begin{array}{lll} C_{11} = -1 & C_{12} = 5 & C_{13} = 4 \\ C_{21} = -2 & C_{22} = -4 & C_{23} = 8 \\ C_{31} = 5 & C_{32} = 3 & C_{33} = -6 \end{array}$$

✓ **Checkpoint**  *Audio-video solution in English & Spanish at LarsonPrecalculus.com*

Find all the minors and cofactors of

$$A = \begin{bmatrix} 1 & 2 & 3 \\ 0 & -1 & 5 \\ 2 & 1 & 4 \end{bmatrix}.$$

The Determinant of a Square Matrix

The definition below is *inductive* because it uses determinants of matrices of dimension $(n - 1) \times (n - 1)$ to define determinants of matrices of dimension $n \times n$.

Determinant of a Square Matrix

If A is a square matrix (of dimension 2×2 or greater), then the determinant of A is the sum of the entries in any row (or column) of A multiplied by their respective cofactors. For example, expanding along the first row yields

$$|A| = a_{11}C_{11} + a_{12}C_{12} + \cdots + a_{1n}C_{1n}.$$

Applying this definition to find a determinant is called **expanding by cofactors.**

Verify that for a 2×2 matrix

$$A = \begin{bmatrix} a_1 & b_1 \\ a_2 & b_2 \end{bmatrix}$$

this definition of the determinant yields

$$|A| = a_1 b_2 - a_2 b_1$$

as previously defined.

EXAMPLE 3 **The Determinant of a 3 × 3 Matrix**

See LarsonPrecalculus.com for an interactive version of this type of example.

Find the determinant of $A = \begin{bmatrix} 0 & 2 & 1 \\ 3 & -1 & 2 \\ 4 & 0 & 1 \end{bmatrix}$.

Solution Note that this is the same matrix used in Example 2. There you found that the cofactors of the entries in the first row are

$$C_{11} = -1, \quad C_{12} = 5, \quad \text{and} \quad C_{13} = 4.$$

Use the definition of the determinant of a square matrix to expand along the first row.

$$|A| = a_{11}C_{11} + a_{12}C_{12} + a_{13}C_{13} \qquad \text{First-row expansion}$$
$$= 0(-1) + 2(5) + 1(4)$$
$$= 14$$

✓ *Checkpoint*))) *Audio-video solution in English & Spanish at LarsonPrecalculus.com*

Find the determinant of $A = \begin{bmatrix} 3 & 4 & -2 \\ 3 & 5 & 0 \\ -1 & 4 & 1 \end{bmatrix}$.

In Example 3, it was efficient to expand by cofactors along the first row, but any row or column can be used. For example, expanding along the second row gives the same result.

$$|A| = a_{21}C_{21} + a_{22}C_{22} + a_{23}C_{23} \qquad \text{Second-row expansion}$$
$$= 3(-2) + (-1)(-4) + 2(8)$$
$$= 14$$

The goal of Sudoku is to fill in a 9×9 grid so that each column, row, and 3×3 sub-grid contains all the numbers 1 through 9 without repetition. When solved correctly, no two rows or two columns are the same. Note that when a matrix has two rows or two columns that are the same, the determinant is zero.

When expanding by cofactors, you do not need to find cofactors of zero entries, because zero times its cofactor is zero. So, the row (or column) containing the most zeros is usually the best choice for expansion by cofactors. This is demonstrated in the next example.

EXAMPLE 4 **The Determinant of a 4 × 4 Matrix**

Find the determinant of $A = \begin{bmatrix} 1 & -2 & 3 & 0 \\ -1 & 1 & 0 & 2 \\ 0 & 2 & 0 & 3 \\ 3 & 4 & 0 & 2 \end{bmatrix}$.

Solution Notice that three of the entries in the third column are zeros. So, to eliminate some of the work in the expansion, expand along the third column.

$$|A| = 3(C_{13}) + 0(C_{23}) + 0(C_{33}) + 0(C_{43})$$

The cofactors C_{23}, C_{33}, and C_{43} have zero coefficients, so the only cofactor you need to find is C_{13}. Start by deleting the first row and third column of A to form the determinant that gives the minor M_{13}.

$$C_{13} = (-1)^{1+3} \begin{vmatrix} -1 & 1 & 2 \\ 0 & 2 & 3 \\ 3 & 4 & 2 \end{vmatrix} \qquad \textcolor{red}{\text{Delete 1st row and 3rd column.}}$$

$$= \begin{vmatrix} -1 & 1 & 2 \\ 0 & 2 & 3 \\ 3 & 4 & 2 \end{vmatrix} \qquad \textcolor{red}{\text{Simplify.}}$$

Now, expand by cofactors along the second row.

$$C_{13} = 0(-1)^3 \begin{vmatrix} 1 & 2 \\ 4 & 2 \end{vmatrix} + 2(-1)^4 \begin{vmatrix} -1 & 2 \\ 3 & 2 \end{vmatrix} + 3(-1)^5 \begin{vmatrix} -1 & 1 \\ 3 & 4 \end{vmatrix}$$

$$= 0 + 2(1)(-8) + 3(-1)(-7)$$

$$= 5$$

So, $|A| = 3C_{13} = 3(5) = 15$.

✓ *Checkpoint*))) *Audio-video solution in English & Spanish at LarsonPrecalculus.com*

Find the determinant of $A = \begin{bmatrix} 2 & 6 & -4 & 2 \\ 2 & -2 & 3 & 6 \\ 1 & 5 & 0 & 1 \\ 3 & 1 & 0 & -5 \end{bmatrix}$.

Summarize (Section 7.4)

1. State the definition of the determinant of a 2×2 matrix *(page 530)*. For an example of finding the determinants of 2×2 matrices, see Example 1.

2. State the definitions of minors and cofactors of a square matrix *(page 532)*. For an example of finding the minors and cofactors of a square matrix, see Example 2.

3. State the definition of the determinant of a square matrix using expanding by cofactors *(page 533)*. For examples of finding determinants using expanding by cofactors, see Examples 3 and 4.

7.4 Exercises

See **CalcChat.com** for tutorial help and worked-out solutions to odd-numbered exercises.

Vocabulary: Fill in the blanks.

1. Both $\det(A)$ and $|A|$ represent the _____ of the matrix A.

2. The _____ M_{ij} of the entry a_{ij} is the determinant of the matrix obtained by deleting the ith row and jth column of the square matrix A.

3. The _____ C_{ij} of the entry a_{ij} of the square matrix A is given by $(-1)^{i+j}M_{ij}$.

4. The method of finding the determinant of a matrix of dimension 2×2 or greater is called _____ by _____.

Skills and Applications

Finding the Determinant of a Matrix In Exercises 5–22, find the determinant of the matrix.

5. $[4]$

6. $[-10]$

7. $\begin{bmatrix} 8 & 4 \\ 2 & 3 \end{bmatrix}$

8. $\begin{bmatrix} -9 & 0 \\ 6 & -2 \end{bmatrix}$

9. $\begin{bmatrix} 6 & -3 \\ -5 & 2 \end{bmatrix}$

10. $\begin{bmatrix} 3 & -3 \\ 4 & -8 \end{bmatrix}$

11. $\begin{bmatrix} -7 & 0 \\ 3 & 0 \end{bmatrix}$

12. $\begin{bmatrix} 4 & -3 \\ 0 & 0 \end{bmatrix}$

13. $\begin{bmatrix} 2 & 6 \\ 0 & 3 \end{bmatrix}$

14. $\begin{bmatrix} 2 & -3 \\ -6 & 9 \end{bmatrix}$

15. $\begin{bmatrix} -3 & -2 \\ -6 & -4 \end{bmatrix}$

16. $\begin{bmatrix} 4 & 7 \\ -2 & 5 \end{bmatrix}$

17. $\begin{bmatrix} -2 & -7 \\ -3 & 1 \end{bmatrix}$

18. $\begin{bmatrix} 2 & -5 \\ -4 & -1 \end{bmatrix}$

19. $\begin{bmatrix} -7 & 6 \\ 0.5 & 3 \end{bmatrix}$

20. $\begin{bmatrix} 0 & 2.5 \\ -3 & 2 \end{bmatrix}$

21. $\begin{bmatrix} -\frac{1}{2} & \frac{1}{3} \\ -6 & \frac{1}{3} \end{bmatrix}$

22. $\begin{bmatrix} \frac{2}{3} & -\frac{4}{3} \\ -1 & \frac{1}{3} \end{bmatrix}$

Using a Graphing Utility In Exercises 23–28, use the matrix capabilities of a graphing utility to find the determinant of the matrix.

23. $\begin{bmatrix} 3 & 4 \\ -2 & 1 \end{bmatrix}$

24. $\begin{bmatrix} 5 & -9 \\ 7 & 16 \end{bmatrix}$

25. $\begin{bmatrix} 19 & 20 \\ 43 & -56 \end{bmatrix}$

26. $\begin{bmatrix} 101 & 197 \\ -253 & 172 \end{bmatrix}$

27. $\begin{bmatrix} \frac{1}{10} & \frac{1}{5} \\ -\frac{3}{10} & \frac{1}{5} \end{bmatrix}$

28. $\begin{bmatrix} 0.1 & 0.1 \\ 7.5 & 6.2 \end{bmatrix}$

Finding the Minors and Cofactors of a Matrix In Exercises 29–34, find all the **(a) minors** and **(b) cofactors** of the matrix.

29. $\begin{bmatrix} 4 & 5 \\ 3 & -6 \end{bmatrix}$

30. $\begin{bmatrix} 0 & 10 \\ 3 & -4 \end{bmatrix}$

31. $\begin{bmatrix} 4 & 0 & 2 \\ -3 & 2 & 1 \\ 1 & -1 & 1 \end{bmatrix}$

32. $\begin{bmatrix} 1 & -1 & 0 \\ 3 & 2 & 5 \\ 4 & -6 & 4 \end{bmatrix}$

33. $\begin{bmatrix} -4 & 6 & 3 \\ 7 & -2 & 8 \\ 1 & 0 & -5 \end{bmatrix}$

34. $\begin{bmatrix} -2 & 9 & 4 \\ 7 & -6 & 0 \\ 6 & 7 & -6 \end{bmatrix}$

Finding the Determinant of a Matrix In Exercises 35–44, find the determinant of the matrix. Expand by cofactors using the indicated row or column.

35. $\begin{bmatrix} 2 & 5 \\ 6 & -3 \end{bmatrix}$

  (a) Row 1

  (b) Column 1

36. $\begin{bmatrix} 7 & -1 \\ -4 & 10 \end{bmatrix}$

  (a) Row 2

  (b) Column 2

37. $\begin{bmatrix} 5 & 0 & -3 \\ 0 & 12 & 4 \\ 1 & 6 & 3 \end{bmatrix}$

  (a) Row 2

  (b) Column 2

38. $\begin{bmatrix} 3 & -2 & 5 \\ 1 & 0 & 3 \\ 0 & 4 & -1 \end{bmatrix}$

  (a) Row 3

  (b) Column 1

39. $\begin{bmatrix} -3 & 2 & 1 \\ 4 & 5 & 6 \\ 2 & -3 & 1 \end{bmatrix}$

  (a) Row 1

  (b) Column 2

40. $\begin{bmatrix} -3 & 4 & 2 \\ 6 & 3 & 1 \\ 4 & -7 & -8 \end{bmatrix}$

  (a) Row 2

  (b) Column 3

41. $\begin{bmatrix} 6 & 0 & -3 & 5 \\ 4 & 0 & 6 & -8 \\ -1 & 0 & 7 & 4 \\ 8 & 0 & 0 & 2 \end{bmatrix}$

  (a) Row 4

  (b) Column 2

42. $\begin{bmatrix} 10 & 8 & 3 & -7 \\ 4 & 0 & 5 & -6 \\ 0 & 3 & 2 & 7 \\ 0 & 0 & 0 & 0 \end{bmatrix}$

  (a) Row 4

  (b) Column 1

43. $\begin{bmatrix} -2 & 4 & 7 & 1 \\ 3 & 0 & 0 & 0 \\ 8 & 5 & 10 & 5 \\ 6 & 0 & 5 & 0 \end{bmatrix}$

  (a) Row 2

  (b) Column 4

44. $\begin{bmatrix} 7 & 0 & 0 & -6 \\ 6 & 0 & 1 & -2 \\ 1 & -2 & 3 & 2 \\ -3 & 0 & -1 & 4 \end{bmatrix}$

  (a) Row 1

  (b) Column 2

Finding the Determinant of a Matrix
In Exercises 45–58, find the determinant of the matrix. Expand by cofactors using the row or column that appears to make the computations easiest.

45. $\begin{bmatrix} -1 & 8 & -3 \\ 0 & 3 & -6 \\ 0 & 0 & 3 \end{bmatrix}$ 46. $\begin{bmatrix} 1 & 0 & 0 \\ -1 & -1 & 0 \\ 4 & 11 & 5 \end{bmatrix}$

47. $\begin{bmatrix} 6 & 3 & -7 \\ 0 & 0 & 0 \\ 4 & -6 & 3 \end{bmatrix}$ 48. $\begin{bmatrix} 0 & 1 & 2 \\ 3 & 1 & 0 \\ -2 & 0 & 3 \end{bmatrix}$

49. $\begin{bmatrix} 2 & -1 & 0 \\ 4 & 2 & 1 \\ 4 & 2 & 1 \end{bmatrix}$ 50. $\begin{bmatrix} -2 & 2 & 3 \\ 1 & -1 & 0 \\ 0 & 1 & 4 \end{bmatrix}$

51. $\begin{bmatrix} 1 & 4 & -2 \\ 3 & 2 & 0 \\ -1 & 4 & 3 \end{bmatrix}$ 52. $\begin{bmatrix} 2 & -1 & 3 \\ -4 & 2 & -6 \\ 1 & 0 & 2 \end{bmatrix}$

53. $\begin{bmatrix} 2 & 6 & 0 & 2 \\ 2 & 7 & 3 & 6 \\ 1 & 0 & 0 & 1 \\ 3 & 7 & 0 & 7 \end{bmatrix}$ 54. $\begin{bmatrix} 1 & 4 & 3 & 2 \\ -5 & 6 & 2 & 1 \\ 0 & 0 & 0 & 0 \\ 3 & -2 & 1 & 5 \end{bmatrix}$

55. $\begin{bmatrix} 5 & 3 & 0 & 6 \\ 4 & 6 & 4 & 12 \\ 0 & 2 & -3 & 4 \\ 0 & 1 & -2 & 2 \end{bmatrix}$ 56. $\begin{bmatrix} 3 & 6 & -5 & 4 \\ -2 & 0 & 6 & 0 \\ 1 & 1 & 2 & 2 \\ 0 & 3 & -1 & -1 \end{bmatrix}$

57. $\begin{bmatrix} 3 & 2 & 4 & -1 & 5 \\ -2 & 0 & 1 & 3 & 2 \\ 1 & 0 & 0 & 4 & 0 \\ 6 & 0 & 2 & -1 & 0 \\ 3 & 0 & 5 & 1 & 0 \end{bmatrix}$

58. $\begin{bmatrix} 5 & 2 & 0 & 0 & -2 \\ 0 & 1 & 4 & 3 & \frac{1}{2} \\ 0 & 0 & 2 & 6 & 3 \\ 0 & 0 & 3 & \frac{3}{2} & 1 \\ 0 & 0 & 0 & 0 & 2 \end{bmatrix}$

Using a Graphing Utility In Exercises 59–62, use the matrix capabilities of a graphing utility to find the determinant.

59. $\begin{vmatrix} 3 & 8 & -7 \\ 0 & -5 & 4 \\ 8 & 1 & 6 \end{vmatrix}$ 60. $\begin{vmatrix} 5 & -8 & 0 \\ 9 & 7 & 4 \\ -8 & 7 & 1 \end{vmatrix}$

61. $\begin{vmatrix} 1 & -1 & 8 & 4 \\ 2 & 6 & 0 & -4 \\ 2 & 0 & 2 & 6 \\ 0 & 2 & 8 & 0 \end{vmatrix}$ 62. $\begin{vmatrix} 0 & -3 & 8 & 2 \\ 8 & 1 & -1 & 6 \\ -4 & 6 & 0 & 9 \\ -7 & 0 & 0 & 14 \end{vmatrix}$

The Determinant of a Matrix Product In Exercises 63–68, find (a) $|A|$, (b) $|B|$, (c) AB, and (d) $|AB|$.

63. $A = \begin{bmatrix} -1 & 0 \\ 0 & 3 \end{bmatrix}$, $B = \begin{bmatrix} 2 & 0 \\ 0 & -1 \end{bmatrix}$

64. $A = \begin{bmatrix} -2 & 1 \\ 4 & -2 \end{bmatrix}$, $B = \begin{bmatrix} 1 & 2 \\ 0 & -1 \end{bmatrix}$

65. $A = \begin{bmatrix} 4 & 0 \\ 3 & -2 \end{bmatrix}$, $B = \begin{bmatrix} -1 & 1 \\ -2 & 2 \end{bmatrix}$

66. $A = \begin{bmatrix} 5 & 4 \\ 3 & -1 \end{bmatrix}$, $B = \begin{bmatrix} 0 & 6 \\ 1 & -2 \end{bmatrix}$

67. $A = \begin{bmatrix} -1 & 2 & 1 \\ 1 & 0 & 1 \\ 0 & 1 & 0 \end{bmatrix}$, $B = \begin{bmatrix} -1 & 0 & 0 \\ 0 & 2 & 0 \\ 0 & 0 & 3 \end{bmatrix}$

68. $A = \begin{bmatrix} 2 & 0 & 1 \\ 1 & -1 & 2 \\ 3 & 1 & 0 \end{bmatrix}$, $B = \begin{bmatrix} 2 & -1 & 4 \\ 0 & 1 & 3 \\ 3 & -2 & 1 \end{bmatrix}$

Creating a Matrix In Exercises 69–74, create a matrix A with the given characteristics. (There are many correct answers.)

69. Dimension: 2×2, $|A| = 3$
70. Dimension: 2×2, $|A| = -5$
71. Dimension: 3×3, $|A| = -1$
72. Dimension: 3×3, $|A| = 4$
73. Dimension: 2×2, $|A| = 0$, $A \neq O$
74. Dimension: 3×3, $|A| = 0$, $A \neq O$

 Verifying an Equation In Exercises 75–80, find the determinant(s) to verify the equation.

75. $\begin{vmatrix} w & x \\ y & z \end{vmatrix} = -\begin{vmatrix} y & z \\ w & x \end{vmatrix}$ 76. $\begin{vmatrix} w & cx \\ y & cz \end{vmatrix} = c\begin{vmatrix} w & x \\ y & z \end{vmatrix}$

77. $\begin{vmatrix} w & x \\ y & z \end{vmatrix} = \begin{vmatrix} w & x + cw \\ y & z + cy \end{vmatrix}$ 78. $\begin{vmatrix} w & x \\ cw & cx \end{vmatrix} = 0$

79. $\begin{vmatrix} 1 & x & x^2 \\ 1 & y & y^2 \\ 1 & z & z^2 \end{vmatrix} = (y - x)(z - x)(z - y)$

80. $\begin{vmatrix} a + b & a & a \\ a & a + b & a \\ a & a & a + b \end{vmatrix} = b^2(3a + b)$

Solving an Equation In Exercises 81–86, solve for x.

81. $\begin{vmatrix} x & 2 \\ 1 & x \end{vmatrix} = 2$ 82. $\begin{vmatrix} x & 4 \\ -1 & x \end{vmatrix} = 20$

83. $\begin{vmatrix} x + 1 & 2 \\ -1 & x \end{vmatrix} = 4$ 84. $\begin{vmatrix} x - 2 & -1 \\ -3 & x \end{vmatrix} = 0$

85. $\begin{vmatrix} x + 3 & 2 \\ 1 & x + 2 \end{vmatrix} = 0$ 86. $\begin{vmatrix} x + 4 & -2 \\ 7 & x - 5 \end{vmatrix} = 0$

• • **Entries Involving Expressions** • • • • • • • • •

In Exercises 87–92, find the determinant in which the entries are functions. Determinants of this type occur when changes of variables are made in calculus.

Let $x = Ce^{\lambda t}$

87. $\begin{vmatrix} 4u & -1 \\ -1 & 2v \end{vmatrix}$

88. $\begin{vmatrix} 3x^2 & -3y^2 \\ 1 & 1 \end{vmatrix}$

89. $\begin{vmatrix} e^{2x} & e^{3x} \\ 2e^{2x} & 3e^{3x} \end{vmatrix}$

90. $\begin{vmatrix} e^{-x} & xe^{-x} \\ -e^{-x} & (1-x)e^{-x} \end{vmatrix}$

91. $\begin{vmatrix} x & \ln x \\ 1 & 1/x \end{vmatrix}$

92. $\begin{vmatrix} x & x \ln x \\ 1 & 1 + \ln x \end{vmatrix}$

Exploration

True or False? In Exercises 93 and 94, determine whether the statement is true or false. Justify your answer.

93. If a square matrix has an entire row of zeros, then the determinant of the matrix is zero.

94. If the rows of a 2×2 matrix are the same, then the determinant of the matrix is zero.

95. Think About It Find square matrices A and B such that $|A + B| \neq |A| + |B|$.

96. Conjecture Consider square matrices in which the entries are consecutive integers. An example of such a matrix is

$$\begin{bmatrix} 4 & 5 & 6 \\ 7 & 8 & 9 \\ 10 & 11 & 12 \end{bmatrix}.$$

(a) Use the matrix capabilities of a graphing utility to find the determinants of four matrices of this type. Make a conjecture based on the results.

(b) Verify your conjecture.

97. Error Analysis Describe the error.

$$\begin{vmatrix} 1 & 1 & 4 \\ 3 & 2 & 0 \\ 2 & 1 & 3 \end{vmatrix} = 3(1)\begin{vmatrix} 1 & 4 \\ 1 & 3 \end{vmatrix} + 2(-1)\begin{vmatrix} 1 & 4 \\ 2 & 3 \end{vmatrix}$$
$$+ 0(1)\begin{vmatrix} 1 & 1 \\ 2 & 1 \end{vmatrix}$$
$$= 3(-1) - 2(-5) + 0$$
$$= 7$$

98. Think About It Let A be a 3×3 matrix such that $|A| = 5$. Is it possible to find $|2A|$? Explain.

Properties of Determinants In Exercises 99–101, explain why each equation is an example of the given property of determinants (A and B are square matrices). Use a graphing utility to verify the results.

99. If B is obtained from A by interchanging two rows of A or interchanging two columns of A, then $|B| = -|A|$.

(a) $\begin{vmatrix} 1 & 3 & 4 \\ -7 & 2 & -5 \\ 6 & 1 & 2 \end{vmatrix} = -\begin{vmatrix} 1 & 4 & 3 \\ -7 & -5 & 2 \\ 6 & 2 & 1 \end{vmatrix}$

(b) $\begin{vmatrix} 1 & 3 & 4 \\ -2 & 2 & 0 \\ 1 & 6 & 2 \end{vmatrix} = -\begin{vmatrix} 1 & 6 & 2 \\ -2 & 2 & 0 \\ 1 & 3 & 4 \end{vmatrix}$

100. If B is obtained from A by adding a multiple of a row of A to another row of A or by adding a multiple of a column of A to another column of A, then $|B| = |A|$.

(a) $\begin{vmatrix} 1 & -3 \\ 5 & 2 \end{vmatrix} = \begin{vmatrix} 1 & -3 \\ 0 & 17 \end{vmatrix}$

(b) $\begin{vmatrix} 5 & 4 & 2 \\ 2 & -3 & 4 \\ 7 & 6 & 3 \end{vmatrix} = \begin{vmatrix} 1 & 10 & -6 \\ 2 & -3 & 4 \\ 7 & 6 & 3 \end{vmatrix}$

101. If B is obtained from A by multiplying a row by a nonzero constant c or by multiplying a column by a nonzero constant c, then $|B| = c|A|$.

(a) $\begin{vmatrix} 5 & 10 \\ 2 & -3 \end{vmatrix} = 5\begin{vmatrix} 1 & 2 \\ 2 & -3 \end{vmatrix}$

(b) $\begin{vmatrix} 1 & 8 & -3 \\ 3 & -12 & 6 \\ 7 & 4 & 9 \end{vmatrix} = 12\begin{vmatrix} 1 & 2 & -1 \\ 3 & -3 & 2 \\ 7 & 1 & 3 \end{vmatrix}$

102. HOW DO YOU SEE IT? Explain why the determinant of each matrix is equal to zero.

(a) $\begin{bmatrix} 2 & -4 & 5 \\ 1 & -2 & 3 \\ 0 & 0 & 0 \end{bmatrix}$

(b) $\begin{bmatrix} 4 & -4 & 5 & 7 \\ 2 & -2 & 3 & 1 \\ 4 & -4 & 5 & 7 \\ 6 & 1 & -3 & -3 \end{bmatrix}$

103. Conjecture A **diagonal matrix** is a square matrix in which each entry not on the main diagonal is zero. Find the determinant of each diagonal matrix. Make a conjecture based on your results.

(a) $\begin{bmatrix} 7 & 0 \\ 0 & 4 \end{bmatrix}$ (b) $\begin{bmatrix} -1 & 0 & 0 \\ 0 & 5 & 0 \\ 0 & 0 & 2 \end{bmatrix}$ (c) $\begin{bmatrix} 2 & 0 & 0 & 0 \\ 0 & -2 & 0 & 0 \\ 0 & 0 & 1 & 0 \\ 0 & 0 & 0 & 3 \end{bmatrix}$

7.5 Applications of Matrices and Determinants

■ **Use Cramer's Rule to solve systems of linear equations.**
■ **Use determinants to find areas of triangles.**
■ **Use determinants to test for collinear points and find equations of lines passing through two points.**
■ **Use 2 × 2 matrices to perform transformations in the plane and find areas of parallelograms.**
■ **Use matrices to encode and decode messages.**

Cramer's Rule

So far, you have studied four methods for solving a system of linear equations: substitution, graphing, elimination with equations, and elimination with matrices. In this section, you will study one more method, **Cramer's Rule,** named after the Swiss mathematician Gabriel Cramer (1704–1752). This rule uses determinants to write the solution of a system of linear equations. To see how Cramer's Rule works, consider the system described at the beginning of Section 7.4, which is shown below.

$$\begin{cases} a_1x + b_1y = c_1 \\ a_2x + b_2y = c_2 \end{cases}$$

This system has a solution

$$x = \frac{c_1b_2 - c_2b_1}{a_1b_2 - a_2b_1} \quad \text{and} \quad y = \frac{a_1c_2 - a_2c_1}{a_1b_2 - a_2b_1}$$

provided that

$$a_1b_2 - a_2b_1 \neq 0.$$

Each numerator and denominator in this solution can be expressed as a determinant.

$$x = \frac{c_1b_2 - c_2b_1}{a_1b_2 - a_2b_1} = \frac{\begin{vmatrix} c_1 & b_1 \\ c_2 & b_2 \end{vmatrix}}{\begin{vmatrix} a_1 & b_1 \\ a_2 & b_2 \end{vmatrix}} \qquad y = \frac{a_1c_2 - a_2c_1}{a_1b_2 - a_2b_1} = \frac{\begin{vmatrix} a_1 & c_1 \\ a_2 & c_2 \end{vmatrix}}{\begin{vmatrix} a_1 & b_1 \\ a_2 & b_2 \end{vmatrix}}$$

Relative to the original system, the denominators for x and y are the determinant of the *coefficient* matrix of the system. This determinant is denoted by D. The numerators for x and y are denoted by D_x and D_y, respectively, and are formed by using the column of constants as replacements for the coefficients of x and y.

Coefficient Matrix	D	D_x	D_y
$\begin{bmatrix} a_1 & b_1 \\ a_2 & b_2 \end{bmatrix}$	$\begin{vmatrix} a_1 & b_1 \\ a_2 & b_2 \end{vmatrix}$	$\begin{vmatrix} c_1 & b_1 \\ c_2 & b_2 \end{vmatrix}$	$\begin{vmatrix} a_1 & c_1 \\ a_2 & c_2 \end{vmatrix}$

For example, given the system

$$\begin{cases} 2x - 5y = 3 \\ -4x + 3y = 8 \end{cases}$$

the coefficient matrix, D, D_x, and D_y are as follows.

Coefficient Matrix	D	D_x	D_y
$\begin{bmatrix} 2 & -5 \\ -4 & 3 \end{bmatrix}$	$\begin{vmatrix} 2 & -5 \\ -4 & 3 \end{vmatrix}$	$\begin{vmatrix} 3 & -5 \\ 8 & 3 \end{vmatrix}$	$\begin{vmatrix} 2 & 3 \\ -4 & 8 \end{vmatrix}$

Determinants have many applications in real life. For example, in Exercise 21 on page 548, you will use a determinant to find the area of a region of forest infested with gypsy moths.

Cramer's Rule generalizes to systems of n equations in n variables. The value of each variable is given as the quotient of two determinants. The denominator is the determinant of the coefficient matrix, and the numerator is the determinant of the matrix formed by replacing the column in the coefficient matrix corresponding to the variable being solved for with the column representing the constants. For example, the solution for x_3 in the system below is shown.

$$\begin{cases} a_{11}x_1 + a_{12}x_2 + a_{13}x_3 = b_1 \\ a_{21}x_1 + a_{22}x_2 + a_{23}x_3 = b_2 \\ a_{31}x_1 + a_{32}x_2 + a_{33}x_3 = b_3 \end{cases} \qquad x_3 = \frac{|A_3|}{|A|} = \frac{\begin{vmatrix} a_{11} & a_{12} & b_1 \\ a_{21} & a_{22} & b_2 \\ a_{31} & a_{32} & b_3 \end{vmatrix}}{\begin{vmatrix} a_{11} & a_{12} & a_{13} \\ a_{21} & a_{22} & a_{23} \\ a_{31} & a_{32} & a_{33} \end{vmatrix}}$$

Cramer's Rule

If a system of n linear equations in n variables has a coefficient matrix A with a nonzero determinant $|A|$, then the solution of the system is

$$x_1 = \frac{|A_1|}{|A|}, \quad x_2 = \frac{|A_2|}{|A|}, \quad \dots, \quad x_n = \frac{|A_n|}{|A|}$$

where the ith column of A_i is the column of constants in the system of equations. If the determinant of the coefficient matrix is zero, then the system has either no solution or infinitely many solutions.

EXAMPLE 1 Using Cramer's Rule for a 2 × 2 System

Use Cramer's Rule (if possible) to solve the system

$$\begin{cases} 4x - 2y = 10 \\ 3x - 5y = 11 \end{cases}.$$

Solution To begin, find the determinant of the coefficient matrix.

$$D = \begin{vmatrix} 4 & -2 \\ 3 & -5 \end{vmatrix} = -20 - (-6) = -14$$

This determinant is not zero, so you can apply Cramer's Rule.

$$x = \frac{D_x}{D} = \frac{\begin{vmatrix} 10 & -2 \\ 11 & -5 \end{vmatrix}}{-14} = \frac{-50 - (-22)}{-14} = \frac{-28}{-14} = 2$$

$$y = \frac{D_y}{D} = \frac{\begin{vmatrix} 4 & 10 \\ 3 & 11 \end{vmatrix}}{-14} = \frac{44 - 30}{-14} = \frac{14}{-14} = -1$$

The solution is $(2, -1)$. Check this in the original system.

✓ **Checkpoint** 🔊)) *Audio-video solution in English & Spanish at LarsonPrecalculus.com*

Use Cramer's Rule (if possible) to solve the system

$$\begin{cases} 3x + 4y = 1 \\ 5x + 3y = 9 \end{cases}.$$

EXAMPLE 2   **Using Cramer's Rule for a 3 × 3 System**

Use Cramer's Rule (if possible) to solve the system $\begin{cases} -x + 2y - 3z = 1 \\ 2x \quad\quad + \ z = 0. \\ 3x - 4y + 4z = 2 \end{cases}$

Solution   To find the determinant of the coefficient matrix

$$\begin{bmatrix} -1 & 2 & -3 \\ 2 & 0 & 1 \\ 3 & -4 & 4 \end{bmatrix}$$

expand along the second row.

$$D = 2(-1)^3 \begin{vmatrix} 2 & -3 \\ -4 & 4 \end{vmatrix} + 0(-1)^4 \begin{vmatrix} -1 & -3 \\ 3 & 4 \end{vmatrix} + 1(-1)^5 \begin{vmatrix} -1 & 2 \\ 3 & -4 \end{vmatrix}$$

$$= -2(-4) + 0 - 1(-2)$$

$$= 10$$

This determinant is not zero, so you can apply Cramer's Rule.

$$x = \frac{D_x}{D} = \frac{\begin{vmatrix} 1 & 2 & -3 \\ 0 & 0 & 1 \\ 2 & -4 & 4 \end{vmatrix}}{10} = \frac{8}{10} = \frac{4}{5}$$

$$y = \frac{D_y}{D} = \frac{\begin{vmatrix} -1 & 1 & -3 \\ 2 & 0 & 1 \\ 3 & 2 & 4 \end{vmatrix}}{10} = \frac{-15}{10} = -\frac{3}{2}$$

$$z = \frac{D_z}{D} = \frac{\begin{vmatrix} -1 & 2 & 1 \\ 2 & 0 & 0 \\ 3 & -4 & 2 \end{vmatrix}}{10} = \frac{-16}{10} = -\frac{8}{5}$$

The solution is

$\left(\frac{4}{5}, -\frac{3}{2}, -\frac{8}{5}\right).$

Check this in the original system.

✓ *Checkpoint* ◀))) *Audio-video solution in English & Spanish at LarsonPrecalculus.com*

Use Cramer's Rule (if possible) to solve the system $\begin{cases} 4x - \ y + \ z = 12 \\ 2x + 2y + 3z = \ 1. \\ 5x - 2y + 6z = 22 \end{cases}$

Remember that Cramer's Rule does not apply when the determinant of the coefficient matrix is zero. This would create division by zero, which is undefined. For example, consider the system of linear equations below.

$$\begin{cases} -x \quad\quad + \ z = \ 4 \\ 2x - y + \ z = -3 \\ \quad\quad y - 3z = \ 1 \end{cases}$$

The determinant of the coefficient matrix is zero, so you cannot apply Cramer's Rule.

Area of a Triangle

Another application of matrices and determinants is finding the area of a triangle whose vertices are given as three points in a coordinate plane.

Area of a Triangle

The area of a triangle with vertices (x_1, y_1), (x_2, y_2), and (x_3, y_3) is

$$\text{Area} = \pm\frac{1}{2}\begin{vmatrix} x_1 & y_1 & 1 \\ x_2 & y_2 & 1 \\ x_3 & y_3 & 1 \end{vmatrix}$$

where you choose the sign $(\pm)$ so that the area is positive.

EXAMPLE 3 **Finding the Area of a Triangle**

See LarsonPrecalculus.com for an interactive version of this type of example.

Find the area of the triangle whose vertices are $(1, 0)$, $(2, 2)$, and $(4, 3)$, as shown below.

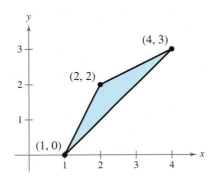

Solution Letting $(x_1, y_1) = (1, 0)$, $(x_2, y_2) = (2, 2)$, and $(x_3, y_3) = (4, 3)$, you have

$$\begin{vmatrix} x_1 & y_1 & 1 \\ x_2 & y_2 & 1 \\ x_3 & y_3 & 1 \end{vmatrix} = \begin{vmatrix} 1 & 0 & 1 \\ 2 & 2 & 1 \\ 4 & 3 & 1 \end{vmatrix}$$

$$= 1(-1)^2\begin{vmatrix} 2 & 1 \\ 3 & 1 \end{vmatrix} + 0(-1)^3\begin{vmatrix} 2 & 1 \\ 4 & 1 \end{vmatrix} + 1(-1)^4\begin{vmatrix} 2 & 2 \\ 4 & 3 \end{vmatrix}$$

$$= -3.$$

Using this value, the area of the triangle is

$$\text{Area} = -\frac{1}{2}\begin{vmatrix} 1 & 0 & 1 \\ 2 & 2 & 1 \\ 4 & 3 & 1 \end{vmatrix} \qquad \text{Choose } (-) \text{ so that the area is positive.}$$

$$= -\frac{1}{2}(-3)$$

$$= \frac{3}{2} \text{ square units.}$$

✓ *Checkpoint* 🔊))) *Audio-video solution in English & Spanish at LarsonPrecalculus.com*

Find the area of the triangle whose vertices are $(0, 0)$, $(4, 1)$, and $(2, 5)$.

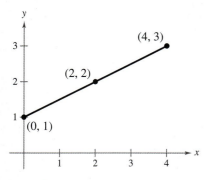

Figure 7.1

Lines in a Plane

In Example 3, what would have happened if the three points were collinear (lying on the same line)? The answer is that the determinant would have been zero. Consider, for example, the three collinear points $(0, 1)$, $(2, 2)$, and $(4, 3)$, as shown in Figure 7.1. The area of the "triangle" that has these three points as vertices is

$$\frac{1}{2}\begin{vmatrix} 0 & 1 & 1 \\ 2 & 2 & 1 \\ 4 & 3 & 1 \end{vmatrix} = \frac{1}{2}\left[0(-1)^2\begin{vmatrix} 2 & 1 \\ 3 & 1 \end{vmatrix} + 1(-1)^3\begin{vmatrix} 2 & 1 \\ 4 & 1 \end{vmatrix} + 1(-1)^4\begin{vmatrix} 2 & 2 \\ 4 & 3 \end{vmatrix}\right]$$

$$= \frac{1}{2}[0 - 1(-2) + 1(-2)]$$

$$= 0.$$

A generalization of this result is below.

Test for Collinear Points

Three points

$$(x_1, y_1), \quad (x_2, y_2), \quad \text{and} \quad (x_3, y_3)$$

are collinear (lie on the same line) if and only if

$$\begin{vmatrix} x_1 & y_1 & 1 \\ x_2 & y_2 & 1 \\ x_3 & y_3 & 1 \end{vmatrix} = 0.$$

EXAMPLE 4   **Testing for Collinear Points**

Determine whether the points

$$(-2, -2), \quad (1, 1), \quad \text{and} \quad (7, 5)$$

are collinear. (See Figure 7.2.)

Solution   Letting $(x_1, y_1) = (-2, -2)$, $(x_2, y_2) = (1, 1)$, and $(x_3, y_3) = (7, 5)$, you have

$$\begin{vmatrix} x_1 & y_1 & 1 \\ x_2 & y_2 & 1 \\ x_3 & y_3 & 1 \end{vmatrix} = \begin{vmatrix} -2 & -2 & 1 \\ 1 & 1 & 1 \\ 7 & 5 & 1 \end{vmatrix}$$

$$= -2(-1)^2\begin{vmatrix} 1 & 1 \\ 5 & 1 \end{vmatrix} + (-2)(-1)^3\begin{vmatrix} 1 & 1 \\ 7 & 1 \end{vmatrix} + 1(-1)^4\begin{vmatrix} 1 & 1 \\ 7 & 5 \end{vmatrix}$$

$$= -2(-4) + 2(-6) + 1(-2)$$

$$= -6.$$

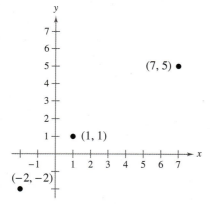

Figure 7.2

The value of this determinant is *not* zero, so the three points are not collinear. Note that the area of the triangle with vertices at these points is $\left(-\frac{1}{2}\right)(-6) = 3$ square units.

✓ *Checkpoint* 🔊))) *Audio-video solution in English & Spanish at LarsonPrecalculus.com*

Determine whether the points

$$(-2, 4), \quad (3, -1), \quad \text{and} \quad (6, -4)$$

are collinear.

The test for collinear points can be adapted for another use. Given two points on a rectangular coordinate system, you can find an equation of the line passing through the two points.

Two-Point Form of the Equation of a Line

An equation of the line passing through the distinct points (x_1, y_1) and (x_2, y_2) is given by

$$\begin{vmatrix} x & y & 1 \\ x_1 & y_1 & 1 \\ x_2 & y_2 & 1 \end{vmatrix} = 0.$$

EXAMPLE 5   **Finding an Equation of a Line**

Find an equation of the line passing through the points $(2, 4)$ and $(-1, 3)$, as shown in the figure.

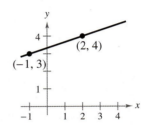

Solution   Let $(x_1, y_1) = (2, 4)$ and $(x_2, y_2) = (-1, 3)$. Applying the determinant formula for the equation of a line produces

$$\begin{vmatrix} x & y & 1 \\ 2 & 4 & 1 \\ -1 & 3 & 1 \end{vmatrix} = 0.$$

Evaluate this determinant to find an equation of the line.

$$x(-1)^2\begin{vmatrix} 4 & 1 \\ 3 & 1 \end{vmatrix} + y(-1)^3\begin{vmatrix} 2 & 1 \\ -1 & 1 \end{vmatrix} + 1(-1)^4\begin{vmatrix} 2 & 4 \\ -1 & 3 \end{vmatrix} = 0$$

$$x(1) - y(3) + (1)(10) = 0$$

$$x - 3y + 10 = 0$$

✓ *Checkpoint* 🔊))) *Audio-video solution in English & Spanish at LarsonPrecalculus.com*

Find an equation of the line passing through the points $(-3, -1)$ and $(3, 5)$. ∎

Note that this method of finding an equation of a line works for all lines, including horizontal and vertical lines. For example, an equation of the vertical line passing through $(2, 0)$ and $(2, 2)$ is

$$\begin{vmatrix} x & y & 1 \\ 2 & 0 & 1 \\ 2 & 2 & 1 \end{vmatrix} = 0$$

$$-2x + 4 = 0$$

$$x = 2.$$

Further Applications of 2 × 2 Matrices

You can use transformation matrices to transform figures in the coordinate plane. Several transformations and their corresponding transformation matrices are listed below.

Transformation Matrices

Reflection in the y-axis	Reflection in the x-axis	Horizontal stretch ($k > 1$) or shrink ($0 < k < 1$)	Vertical stretch ($k > 1$) or shrink ($0 < k < 1$)
$\begin{bmatrix} -1 & 0 \\ 0 & 1 \end{bmatrix}$	$\begin{bmatrix} 1 & 0 \\ 0 & -1 \end{bmatrix}$	$\begin{bmatrix} k & 0 \\ 0 & 1 \end{bmatrix}$	$\begin{bmatrix} 1 & 0 \\ 0 & k \end{bmatrix}$

EXAMPLE 6 **Transforming a Square**

To find the image of the square whose vertices are $(0, 0)$, $(2, 0)$, $(0, 2)$, and $(2, 2)$ after a reflection in the y-axis, first write the vertices as column matrices. Then multiply each column matrix by the appropriate transformation matrix on the left.

$$\begin{bmatrix} -1 & 0 \\ 0 & 1 \end{bmatrix}\begin{bmatrix} 0 \\ 0 \end{bmatrix} = \begin{bmatrix} 0 \\ 0 \end{bmatrix} \qquad \begin{bmatrix} -1 & 0 \\ 0 & 1 \end{bmatrix}\begin{bmatrix} 2 \\ 0 \end{bmatrix} = \begin{bmatrix} -2 \\ 0 \end{bmatrix}$$

$$\begin{bmatrix} -1 & 0 \\ 0 & 1 \end{bmatrix}\begin{bmatrix} 0 \\ 2 \end{bmatrix} = \begin{bmatrix} 0 \\ 2 \end{bmatrix} \qquad \begin{bmatrix} -1 & 0 \\ 0 & 1 \end{bmatrix}\begin{bmatrix} 2 \\ 2 \end{bmatrix} = \begin{bmatrix} -2 \\ 2 \end{bmatrix}$$

So, the vertices of the image are $(0, 0)$, $(-2, 0)$, $(0, 2)$, and $(-2, 2)$. Figure 7.3 shows a sketch of the square and its image.

✓ *Checkpoint* 🔊)) *Audio-video solution in English & Spanish at LarsonPrecalculus.com*

Find the image of the square in Example 6 after a vertical stretch by a factor of $k = 2$. ■

You can find the area of a parallelogram using the determinant of a 2 × 2 matrix.

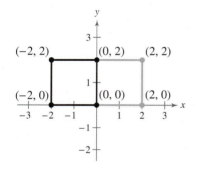

Figure 7.3

Area of a Parallelogram

The area of a parallelogram with vertices $(0, 0)$, (a, b), (c, d), and $(a + c, b + d)$ is

$$\text{Area} = |\det(A)| \qquad \text{\color{red}{$|\det(A)|$ is the absolute value of the determinant.}}$$

where $A = \begin{bmatrix} a & b \\ c & d \end{bmatrix}$.

⬦ **· ·REMARK** For an informal *proof without words* of this formula, see Proofs in Mathematics on page 558.

EXAMPLE 7 **Finding the Area of a Parallelogram**

To find the area of the parallelogram shown in Figure 7.4 using the formula above, let $(a, b) = (2, 0)$ and $(c, d) = (1, 3)$. Then

$$A = \begin{bmatrix} 2 & 0 \\ 1 & 3 \end{bmatrix}$$

and the area of the parallelogram is

$$\text{Area} = |\det(A)| = |6| = 6 \text{ square units.}$$

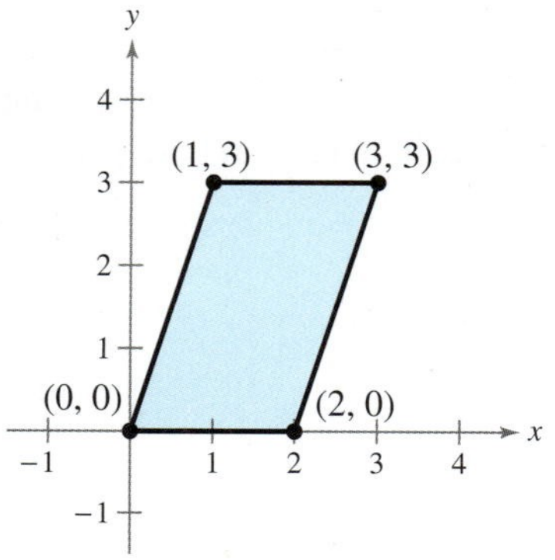

✓ *Checkpoint* 🔊)) *Audio-video solution in English & Spanish at LarsonPrecalculus.com*

Find the area of the parallelogram with vertices $(0, 0)$, $(5, 5)$, $(2, 4)$, and $(7, 9)$. ■

Figure 7.4

Information security is of the utmost importance when conducting business online, and can include the use of data *encryption*. This is the process of encoding information so that the only way to decode it, apart from an "exhaustion attack," is to use a *key*. Data encryption technology uses algorithms based on the material presented here, but on a much more sophisticated level.

Cryptography

A **cryptogram** is a message written according to a secret code. (The Greek word *kryptos* means "hidden.") Matrix multiplication can be used to encode and decode messages. To begin, assign a number to each letter in the alphabet (with 0 assigned to a blank space), as listed below.

0 = _	9 = I	18 = R
1 = A	10 = J	19 = S
2 = B	11 = K	20 = T
3 = C	12 = L	21 = U
4 = D	13 = M	22 = V
5 = E	14 = N	23 = W
6 = F	15 = O	24 = X
7 = G	16 = P	25 = Y
8 = H	17 = Q	26 = Z

Then convert the message to numbers and partition the numbers into **uncoded row matrices,** each having n entries, as demonstrated in Example 8.

EXAMPLE 8   **Forming Uncoded Row Matrices**

Write the uncoded 1×3 row matrices for the message

MEET ME MONDAY.

Solution   Partitioning the message (including blank spaces, but ignoring punctuation) into groups of three produces the uncoded row matrices below.

$$[13 \quad 5 \quad 5] \quad [20 \quad 0 \quad 13] \quad [5 \quad 0 \quad 13] \quad [15 \quad 14 \quad 4] \quad [1 \quad 25 \quad 0]$$
$$\text{M} \quad \text{E} \quad \text{E} \quad \text{T} \qquad \text{M} \quad \text{E} \qquad \text{M} \quad \text{O} \quad \text{N} \quad \text{D} \quad \text{A} \quad \text{Y}$$

Note the use of a blank space to fill out the last uncoded row matrix.

✓ *Checkpoint* 🔊 *Audio-video solution in English & Spanish at LarsonPrecalculus.com*

Write the uncoded 1×3 row matrices for the message

OWLS ARE NOCTURNAL.   ■

To encode a message, create an $n \times n$ invertible matrix A, called an **encoding matrix,** such as

$$A = \begin{bmatrix} 1 & -2 & 2 \\ -1 & 1 & 3 \\ 1 & -1 & -4 \end{bmatrix}.$$

Multiply the uncoded row matrices by A (on the right) to obtain the **coded row matrices.** Here is an example.

Uncoded Matrix	Encoding Matrix A	Coded Matrix

$$[13 \quad 5 \quad 5] \begin{bmatrix} 1 & -2 & 2 \\ -1 & 1 & 3 \\ 1 & -1 & -4 \end{bmatrix} = [13 \quad -26 \quad 21]$$

HISTORICAL NOTE

During World War II, Navajo soldiers created a code using their native language to send messages between battalions. The soldiers assigned native words to represent characters in the English alphabet, and they created a number of expressions for important military terms, such as *iron-fish* to mean *submarine*. Without the Navajo Code Talkers, the Second World War might have had a very different outcome.

EXAMPLE 9 **Encoding a Message**

Use the invertible matrix below to encode the message MEET ME MONDAY.

$$A = \begin{bmatrix} 1 & -2 & 2 \\ -1 & 1 & 3 \\ 1 & -1 & -4 \end{bmatrix}$$

Solution Obtain the coded row matrices by multiplying each of the uncoded row matrices found in Example 8 by the matrix A.

Uncoded Matrix	Encoding Matrix A	Coded Matrix

$$\begin{bmatrix} 13 & 5 & 5 \end{bmatrix} \begin{bmatrix} 1 & -2 & 2 \\ -1 & 1 & 3 \\ 1 & -1 & -4 \end{bmatrix} = \begin{bmatrix} 13 & -26 & 21 \end{bmatrix}$$

$$\begin{bmatrix} 20 & 0 & 13 \end{bmatrix} \begin{bmatrix} 1 & -2 & 2 \\ -1 & 1 & 3 \\ 1 & -1 & -4 \end{bmatrix} = \begin{bmatrix} 33 & -53 & -12 \end{bmatrix}$$

$$\begin{bmatrix} 5 & 0 & 13 \end{bmatrix} \begin{bmatrix} 1 & -2 & 2 \\ -1 & 1 & 3 \\ 1 & -1 & -4 \end{bmatrix} = \begin{bmatrix} 18 & -23 & -42 \end{bmatrix}$$

$$\begin{bmatrix} 15 & 14 & 4 \end{bmatrix} \begin{bmatrix} 1 & -2 & 2 \\ -1 & 1 & 3 \\ 1 & -1 & -4 \end{bmatrix} = \begin{bmatrix} 5 & -20 & 56 \end{bmatrix}$$

$$\begin{bmatrix} 1 & 25 & 0 \end{bmatrix} \begin{bmatrix} 1 & -2 & 2 \\ -1 & 1 & 3 \\ 1 & -1 & -4 \end{bmatrix} = \begin{bmatrix} -24 & 23 & 77 \end{bmatrix}$$

So, the sequence of coded row matrices is

$$\begin{bmatrix} 13 & -26 & 21 \end{bmatrix}\begin{bmatrix} 33 & -53 & -12 \end{bmatrix}\begin{bmatrix} 18 & -23 & -42 \end{bmatrix}\begin{bmatrix} 5 & -20 & 56 \end{bmatrix}\begin{bmatrix} -24 & 23 & 77 \end{bmatrix}.$$

Finally, removing the matrix notation produces the cryptogram

$$13 \quad -26 \quad 21 \quad 33 \quad -53 \quad -12 \quad 18 \quad -23 \quad -42 \quad 5 \quad -20 \quad 56 \quad -24 \quad 23 \quad 77.$$

✓ *Checkpoint*   *Audio-video solution in English & Spanish at LarsonPrecalculus.com*

Use the invertible matrix below to encode the message OWLS ARE NOCTURNAL.

$$A = \begin{bmatrix} 1 & -1 & 0 \\ 1 & 0 & -1 \\ 6 & -2 & -3 \end{bmatrix}$$

If you do not know the encoding matrix A, decoding a cryptogram such as the one found in Example 9 can be difficult. But if you know the encoding matrix A, decoding is straightforward. You just multiply the coded row matrices by A^{-1} (on the right) to obtain the uncoded row matrices. Here is an example.

$$\underbrace{\begin{bmatrix} 13 & -26 & 21 \end{bmatrix}}_{\text{Coded}} \underbrace{\begin{bmatrix} -1 & -10 & -8 \\ -1 & -6 & -5 \\ 0 & -1 & -1 \end{bmatrix}}_{A^{-1}} = \underbrace{\begin{bmatrix} 13 & 5 & 5 \end{bmatrix}}_{\text{Uncoded}}$$

EXAMPLE 10   **Decoding a Message**

Use the inverse of A in Example 9 to decode the cryptogram

13 −26 21 33 −53 −12 18 −23 −42 5 −20 56 −24 23 77.

Solution   Find the decoding matrix A^{-1}, partition the message into groups of three to form the coded row matrices and multiply each coded row matrix by A^{-1} (on the right).

Coded Matrix	Decoding Matrix A^{-1}	Decoded Matrix

$$[13 \; -26 \; \; 21]\begin{bmatrix} -1 & -10 & -8 \\ -1 & -6 & -5 \\ 0 & -1 & -1 \end{bmatrix} = [13 \quad 5 \quad 5]$$

$$[33 \; -53 \; -12]\begin{bmatrix} -1 & -10 & -8 \\ -1 & -6 & -5 \\ 0 & -1 & -1 \end{bmatrix} = [20 \quad 0 \quad 13]$$

$$[18 \; -23 \; -42]\begin{bmatrix} -1 & -10 & -8 \\ -1 & -6 & -5 \\ 0 & -1 & -1 \end{bmatrix} = [5 \quad 0 \quad 13]$$

$$[5 \; -20 \; \; 56]\begin{bmatrix} -1 & -10 & -8 \\ -1 & -6 & -5 \\ 0 & -1 & -1 \end{bmatrix} = [15 \quad 14 \quad 4]$$

$$[-24 \; \; 23 \; \; 77]\begin{bmatrix} -1 & -10 & -8 \\ -1 & -6 & -5 \\ 0 & -1 & -1 \end{bmatrix} = [1 \quad 25 \quad 0]$$

So, the message is

$$[13 \quad 5 \quad 5] \; [20 \quad 0 \quad 13] \; [5 \quad 0 \quad 13] \; [15 \quad 14 \quad 4] \; [1 \quad 25 \quad 0].$$

M E E T   M E   M O N D A Y

✓ *Checkpoint*  *Audio-video solution in English & Spanish at LarsonPrecalculus.com*

Use the inverse of A in the Checkpoint with Example 9 to decode the cryptogram

110 −39 −59 25 −21 −3 23 −18 −5 47 −20 −24
149 −56 −75 87 −38 −37.

Summarize   (Section 7.5)

1. Explain how to use Cramer's Rule to solve systems of linear equations *(page 539)*. For examples of using Cramer's Rule, see Examples 1 and 2.

2. State the formula for finding the area of a triangle using a determinant *(page 541)*. For an example of using this formula to find the area of a triangle, see Example 3.

3. Explain how to use determinants to test for collinear points *(page 542)* and find equations of lines passing through two points *(page 543)*. For examples of these applications, see Examples 4 and 5.

4. Explain how to use 2×2 matrices to perform transformations in the plane and find areas of parallelograms *(page 544)*. For examples of these applications, see Examples 6 and 7.

5. Explain how to use matrices to encode and decode messages *(pages 545–547)*. For examples involving encoding and decoding messages, see Examples 8–10.

7.5   Exercises

See **CalcChat.com** for tutorial help and worked-out solutions to odd-numbered exercises.

Vocabulary: Fill in the blanks.

1. The method of using determinants to solve a system of linear equations is called _____ _____.
2. Three points are _____ when they lie on the same line.
3. The area A of a triangle with vertices (x_1, y_1), (x_2, y_2), and (x_3, y_3) is given by _____.
4. A message written according to a secret code is a _____.
5. To encode a message, create an invertible matrix A and multiply the _____ row matrices by A (on the right) to obtain the _____ row matrices.
6. A message encoded using an invertible matrix A can be decoded by multiplying the coded row matrices by _____ (on the right).

Skills and Applications

Using Cramer's Rule In Exercises 7–14, use Cramer's Rule (if possible) to solve the system of equations.

7. $\begin{cases} -5x + 9y = -14 \\ 3x - 7y = 10 \end{cases}$

8. $\begin{cases} 4x - 3y = -10 \\ 6x + 9y = 12 \end{cases}$

9. $\begin{cases} 3x + 2y = -2 \\ 6x + 4y = 4 \end{cases}$

10. $\begin{cases} 12x - 7y = -4 \\ -11x + 8y = 10 \end{cases}$

11. $\begin{cases} 4x - y + z = -5 \\ 2x + 2y + 3z = 10 \\ 5x - 2y + 6z = 1 \end{cases}$

12. $\begin{cases} 4x - 2y + 3z = -2 \\ 2x + 2y + 5z = 16 \\ 8x - 5y - 2z = 4 \end{cases}$

13. $\begin{cases} x + 2y + 3z = -3 \\ -2x + y - z = 6 \\ 3x - 3y + 2z = -11 \end{cases}$

14. $\begin{cases} 5x - 4y + z = -14 \\ -x + 2y - 2z = 10 \\ 3x + y + z = 1 \end{cases}$

Finding the Area of a Triangle In Exercises 15–18, use a determinant to find the area of the triangle with the given vertices.

15.
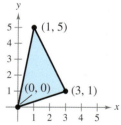

16.
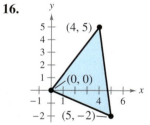

17. $(0, 4), (-2 -3), (2, -3)$
18. $(-2, 1), (1, 6), (3, -1)$

Finding a Coordinate In Exercises 19 and 20, find a value of y such that the triangle with the given vertices has an area of 4 square units.

19. $(-5, 1), (0, 2), (-2, y)$
20. $(-4, 2), (-3, 5), (-1, y)$

21. Area of Infestation

A large region of forest is infested with gypsy moths. The region is triangular, as shown in the figure. From vertex A, the distances to the other vertices are 25 miles south

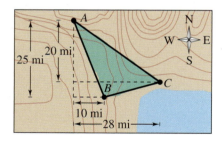

and 10 miles east (for vertex B), and 20 miles south and 28 miles east (for vertex C). Use a graphing utility to find the area (in square miles) of the region.

22. Botany A botanist is studying the plants growing in the triangular region shown in the figure. Starting at vertex A, the botanist walks 65 feet east and 50 feet north to vertex B, and then walks 85 feet west and 30 feet north to vertex C. Use a graphing utility to find the area (in square feet) of the region.

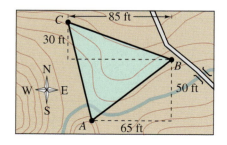

Testing for Collinear Points In Exercises 23–28, use a determinant to determine whether the points are collinear.

23. $(2, -6), (0, -2), (3, -8)$

24. $(3, -5), (6, 1), (4, 2)$

25. $\left(2, -\frac{1}{2}\right), (-4, 4), (6, -3)$

26. $(0, 1), \left(-2, \frac{7}{2}\right), \left(1, -\frac{1}{4}\right)$

27. $(0, 2), (1, 2.4), (-1, 1.6)$

28. $(3, 7), (4, 9.5), (-1, -5)$

Finding a Coordinate In Exercises 29 and 30, find the value of y such that the points are collinear.

29. $(2, -5), (4, y), (5, -2)$ **30.** $(-6, 2), (-5, y), (-3, 5)$

Finding an Equation of a Line In Exercises 31–36, use a determinant to find an equation of the line passing through the points.

31. $(0, 0), (5, 3)$

32. $(0, 0), (-2, 2)$

33. $(-4, 3), (2, 1)$

34. $(10, 7), (-2, -7)$

35. $\left(-\frac{1}{2}, 3\right), \left(\frac{5}{2}, 1\right)$

36. $\left(\frac{2}{3}, 4\right), (6, 12)$

Transforming a Square In Exercises 37–40, use matrices to find the vertices of the image of the square with the given vertices after the given transformation. Then sketch the square and its image.

37. $(0, 0), (0, 3), (3, 0), (3, 3)$; horizontal stretch, $k = 2$

38. $(1, 2), (3, 2), (1, 4), (3, 4)$; reflection in the x-axis

39. $(4, 3), (5, 3), (4, 4), (5, 4)$; reflection in the y-axis

40. $(1, 1), (3, 2), (0, 3), (2, 4)$; vertical shrink, $k = \frac{1}{2}$

Finding the Area of a Parallelogram In Exercises 41–44, use a determinant to find the area of the parallelogram with the given vertices.

41. $(0, 0), (1, 0), (2, 2), (3, 2)$

42. $(0, 0), (3, 0), (4, 1), (7, 1)$

43. $(0, 0), (-2, 0), (3, 5), (1, 5)$

44. $(0, 0), (0, 8), (8, -6), (8, 2)$

Encoding a Message In Exercises 45 and 46, (a) write the uncoded 1×2 row matrices for the message, and then (b) encode the message using the encoding matrix.

Message

45. COME HOME SOON

46. HELP IS ON THE WAY

Encoding Matrix

$\begin{bmatrix} 1 & 2 \\ 3 & 5 \end{bmatrix}$

$\begin{bmatrix} -2 & 3 \\ -1 & 1 \end{bmatrix}$

Encoding a Message In Exercises 47 and 48, (a) write the uncoded 1×3 row matrices for the message, and then (b) encode the message using the encoding matrix.

Message

47. CALL ME TOMORROW

48. PLEASE SEND MONEY

Encoding Matrix

$\begin{bmatrix} 1 & -1 & 0 \\ 1 & 0 & -1 \\ -6 & 2 & 3 \end{bmatrix}$

$\begin{bmatrix} 4 & 2 & 1 \\ -3 & -3 & -1 \\ 3 & 2 & 1 \end{bmatrix}$

Encoding a Message In Exercises 49–52, write a cryptogram for the message using the matrix

$$A = \begin{bmatrix} 1 & 2 & 2 \\ 3 & 7 & 9 \\ -1 & -4 & -7 \end{bmatrix}.$$

49. LANDING SUCCESSFUL

50. ICEBERG DEAD AHEAD

51. HAPPY BIRTHDAY

52. OPERATION OVERLOAD

Decoding a Message In Exercises 53–56, use A^{-1} to decode the cryptogram.

53. $A = \begin{bmatrix} 1 & 2 \\ 3 & 5 \end{bmatrix}$

11 21 64 112 25 50 29 53 23 46 40
75 55 92

54. $A = \begin{bmatrix} 2 & 3 \\ 3 & 4 \end{bmatrix}$

85 120 6 8 10 15 84 117 42 56 90
125 60 80 30 45 19 26

55. $A = \begin{bmatrix} 1 & -1 & 0 \\ 1 & 0 & -1 \\ -6 & 2 & 3 \end{bmatrix}$

9 -1 -9 38 -19 -19 28 -9 -19
-80 25 41 -64 21 31 9 -5 -4

56. $A = \begin{bmatrix} 3 & -4 & 2 \\ 0 & 2 & 1 \\ 4 & -5 & 3 \end{bmatrix}$

112 -140 83 19 -25 13 72 -76 61 95
-118 71 20 21 38 35 -23 36 42 -48 32

Decoding a Message In Exercises 57 and 58, decode the cryptogram by using the inverse of A in Exercises 49–52.

57. 20 17 -15 -12 -56 -104 1 -25 -65
62 143 181

58. 13 -9 -59 61 112 106 -17 -73
-131 11 24 29 65 144 172

59. Decoding a Message The cryptogram below was encoded with a 2×2 matrix.

8 21 -15 -10 -13 -13 5 10 5 25
5 19 -1 6 20 40 -18 -18 1 16

The last word of the message is _RON. What is the message?

60. Decoding a Message The cryptogram below was encoded with a 2×2 matrix.

5 2 25 11 -2 -7 -15 -15 32 14
-8 -13 38 19 -19 -19 37 16

The last word of the message is _SUE. What is the message?

61. Circuit Analysis Consider the circuit shown in the figure. The currents I_1, I_2, and I_3 (in amperes) are the solution of the system

$$\begin{cases} 4I_1 & + 8I_3 = 2 \\ 2I_2 + 8I_3 = 6. \\ I_1 + I_2 - I_3 = 0 \end{cases}$$

Use Cramer's Rule to find the three currents.

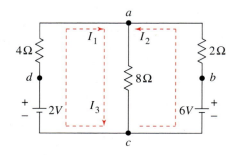

62. Pulley System A system of pulleys is loaded with 192-pound and 64-pound weights (see figure). The tensions t_1 and t_2 in the ropes and the acceleration a of the 64-pound weight are found by solving the system of equations

$$\begin{cases} t_1 - 2t_2 & = 0 \\ t_1 & - 3a = 192 \\ t_2 + 2a = 64 \end{cases}$$

where t_1 and t_2 are measured in pounds and a is in feet per second squared. Use Cramer's Rule to find t_1, t_2, and a.

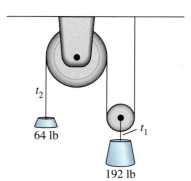

Exploration

True or False? **In Exercises 63 and 64, determine whether the statement is true or false. Justify your answer.**

63. In Cramer's Rule, the numerator is the determinant of the coefficient matrix.

64. Cramer's Rule cannot be used to solve a system of linear equations when the determinant of the coefficient matrix is zero.

65. Error Analysis Describe the error.

Consider the system

$$\begin{cases} 2x - 3y = 0 \\ 4x - 6y = 0 \end{cases}.$$

The determinant of the coefficient matrix is

$$D = \begin{vmatrix} 2 & -3 \\ 4 & -6 \end{vmatrix}$$

$$= -12 - (-12)$$

$$= 0$$

so the system has no solution.

66. 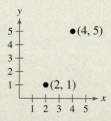 **HOW DO YOU SEE IT?** At this point in the text, you know several methods for finding an equation of a line that passes through two given points. Briefly describe the methods that can be used to find an equation of the line that passes through the two points shown. Discuss the advantages and disadvantages of each method.

67. Finding the Area of a Triangle Use a determinant to find the area of the triangle whose vertices are $(3, -1)$, $(7, -1)$, and $(7, 5)$. Confirm your answer by plotting the points in a coordinate plane and using the formula

Area $= \frac{1}{2}$(base)(height).

68. Writing Use your school's library, the Internet, or some other reference source to research a few current real-life uses of cryptography. Write a short summary of these uses. Include a description of how messages are encoded and decoded in each case.

Chapter Summary

What Did You Learn?	Explanation/Examples	Review Exercises

Section 7.1

What Did You Learn?	Explanation/Examples	Review Exercises
Write matrices and determine their dimensions (p. 494).	$\begin{bmatrix} -1 & 1 \\ 4 & 7 \end{bmatrix}$ $\quad$ $\begin{bmatrix} -2 & 3 & 0 \end{bmatrix}$ $\quad$ $\begin{bmatrix} 4 & -3 \\ 5 & 0 \\ -2 & 1 \end{bmatrix}$ $\quad$ $\begin{bmatrix} 8 \\ -8 \end{bmatrix}$ $\quad 2 \times 2 \qquad\qquad 1 \times 3 \qquad\qquad 3 \times 2 \qquad 2 \times 1$	1–8
Perform elementary row operations on matrices (p. 496).	**Elementary Row Operations** **1.** Interchange two rows. **2.** Multiply a row by a nonzero constant. **3.** Add a multiple of a row to another row.	9, 10
Use matrices and Gaussian elimination to solve systems of linear equations (p. 497).	**Gaussian Elimination with Back-Substitution** **1.** Write the augmented matrix of the system of linear equations. **2.** Use elementary row operations to rewrite the augmented matrix in row-echelon form. **3.** Write the system of linear equations corresponding to the matrix in row-echelon form and use back-substitution to find the solution.	11–26
Use matrices and Gauss-Jordan elimination to solve systems of linear equations (p. 501).	Gauss-Jordan elimination continues the reduction process on a matrix in row-echelon form until the *reduced* row-echelon form is obtained. (See Example 8.)	27–32

Section 7.2

What Did You Learn?	Explanation/Examples	Review Exercises
Determine whether two matrices are equal (p. 507).	Two matrices are equal when their corresponding entries are equal.	33–36
Add and subtract matrices and multiply matrices by scalars (p. 508).	**Definition of Matrix Addition** If $A = [a_{ij}]$ and $B = [b_{ij}]$ are matrices of dimension $m \times n$, then their sum is the $m \times n$ matrix $A + B = [a_{ij} + b_{ij}]$. **Definition of Scalar Multiplication** If $A = [a_{ij}]$ is an $m \times n$ matrix and c is a scalar, then the scalar multiple of A by c is the $m \times n$ matrix $cA = [ca_{ij}]$.	37–50
Multiply two matrices (p. 512).	**Definition of Matrix Multiplication** If $A = [a_{ij}]$ is an $m \times n$ matrix and $B = [b_{ij}]$ is an $n \times p$ matrix, then the product AB is an $m \times p$ matrix given by $$AB = [c_{ij}]$$ where $c_{ij} = a_{i1}b_{1j} + a_{i2}b_{2j} + a_{i3}b_{3j} + \cdots + a_{in}b_{nj}$.	51–60
Use matrix operations to model and solve real-life problems (p. 515).	Matrix operations can be used to find the total cost of equipment for two softball teams. (See Example 12.)	61, 62

Section 7.3

What Did You Learn?	Explanation/Examples	Review Exercises
Verify that two matrices are inverses of each other (p. 521).	**Definition of the Inverse of a Square Matrix** Let A be an $n \times n$ matrix and let I_n be the $n \times n$ identity matrix. If there exists a matrix A^{-1} such that $$AA^{-1} = I_n = A^{-1}A$$ then A^{-1} is the inverse of A.	63–66

What Did You Learn?	Explanation/Examples	Review Exercises

Section 7.3

Use Gauss-Jordan elimination to find the inverses of matrices (*p. 523*).	**Finding an Inverse Matrix** Let A be a square matrix of dimension $n \times n$. **1.** Write the $n \times 2n$ matrix that consists of the given matrix A on the left and the $n \times n$ identity matrix I on the right to obtain $[A \vdots I]$. **2.** If possible, row reduce A to I using elementary row operations on the *entire* matrix $[A \vdots I]$. The result will be the matrix $[I \vdots A^{-1}]$. If this is not possible, then A is not invertible. **3.** Check your work by multiplying to see that $AA^{-1} = I = A^{-1}A$.	67–72
Use a formula to find the inverses of 2×2 matrices (*p. 525*).	If $A = \begin{bmatrix} a & b \\ c & d \end{bmatrix}$ and $ad - bc \neq 0$, then $A^{-1} = \dfrac{1}{ad - bc} \begin{bmatrix} d & -b \\ -c & a \end{bmatrix}.$	73–76
Use inverse matrices to solve systems of linear equations (*p. 526*).	If A is an invertible matrix, then the system of linear equations represented by $AX = B$ has a unique solution given by $X = A^{-1}B$.	77–90

Section 7.4

Find the determinants of 2×2 matrices (*p. 530*).	The determinant of the matrix $A = \begin{bmatrix} a_1 & b_1 \\ a_2 & b_2 \end{bmatrix}$ is given by $\det(A) = |A| = \begin{vmatrix} a_1 & b_1 \\ a_2 & b_2 \end{vmatrix} = a_1 b_2 - a_2 b_1.$	91–94
Find minors and cofactors of square matrices (*p. 532*).	If A is a square matrix, then the minor M_{ij} of the entry a_{ij} is the determinant of the matrix obtained by deleting the ith row and jth column of A. The cofactor C_{ij} of the entry a_{ij} is $C_{ij} = (-1)^{i+j} M_{ij}$.	95–98
Find the determinants of square matrices (*p. 533*).	If A is a square matrix (of dimension 2×2 or greater), then the determinant of A is the sum of the entries in any row (or column) of A multiplied by their respective cofactors.	99–104

Section 7.5

Use Cramer's Rule to solve systems of linear equations (*p. 539*).	Cramer's Rule uses determinants to write the solution of a system of linear equations.	105–108
Use determinants to find areas of triangles (*p. 541*), test for collinear points (*p. 542*), and find equations of lines passing through two points (*p. 543*).	The area of a triangle with vertices (x_1, y_1), (x_2, y_2), and (x_3, y_3) is $\text{Area} = \pm\dfrac{1}{2}\begin{vmatrix} x_1 & y_1 & 1 \\ x_2 & y_2 & 1 \\ x_3 & y_3 & 1 \end{vmatrix}$ where you choose the sign ($\pm$) so that the area is positive.	109–116
Use 2×2 matrices to perform transformations in the plane and find areas of parallelograms (*p. 544*).	The area of a parallelogram with vertices $(0, 0)$, (a, b), (c, d), and $(a + c, b + d)$ is $\text{Area} = |\det(A)|,$ where $A = \begin{bmatrix} a & b \\ c & d \end{bmatrix}.$	117, 118
Use matrices to encode and decode messages (*p. 545*).	The inverse of a matrix can be used to decode a cryptogram. (See Example 10.)	119, 120

Review Exercises
See **CalcChat.com** for tutorial help and worked-out solutions to odd-numbered exercises.

7.1 **Dimension of a Matrix** In Exercises 1–4, determine the dimension of the matrix.

1. $\begin{bmatrix} -1 & 3 \end{bmatrix}$

2. $\begin{bmatrix} 3 & 1 \\ 5 & -2 \end{bmatrix}$

3. $\begin{bmatrix} 2 & 1 & 0 & 4 & -1 \\ 6 & 2 & 1 & 8 & 0 \end{bmatrix}$

4. $\begin{bmatrix} 5 \end{bmatrix}$

Writing an Augmented Matrix In Exercises 5 and 6, write the augmented matrix for the system of linear equations.

5. $\begin{cases} 3x - 10y = 15 \\ 5x + 4y = 22 \end{cases}$

6. $\begin{cases} 8x - 7y + 4z = 12 \\ 3x - 5y + 2z = 20 \end{cases}$

Writing a System of Equations In Exercises 7 and 8, write the system of linear equations represented by the augmented matrix. (Use variables x, y, z, and w, if applicable.)

7. $\begin{bmatrix} 1 & 0 & 2 & \vdots & -8 \\ 2 & -2 & 3 & \vdots & 12 \\ 4 & 7 & 1 & \vdots & 3 \end{bmatrix}$

8. $\begin{bmatrix} 2 & 10 & 8 & 5 & \vdots & -1 \\ -3 & 4 & 0 & 9 & \vdots & 2 \end{bmatrix}$

Writing a Matrix in Row-Echelon Form In Exercises 9 and 10, write the matrix in row-echelon form. (Remember that the row-echelon form of a matrix is not unique.)

9. $\begin{bmatrix} 0 & 1 & 1 \\ 1 & 2 & 3 \\ 2 & 2 & 2 \end{bmatrix}$

10. $\begin{bmatrix} 4 & 8 & 16 \\ 3 & -1 & 2 \\ -2 & 10 & 12 \end{bmatrix}$

Using Back-Substitution In Exercises 11–14, write the system of linear equations represented by the augmented matrix. Then use back-substitution to solve the system. (Use variables x, y, and z, if applicable.)

11. $\begin{bmatrix} 1 & 2 & 3 & \vdots & 9 \\ 0 & 1 & -2 & \vdots & 2 \\ 0 & 0 & 1 & \vdots & -1 \end{bmatrix}$

12. $\begin{bmatrix} 1 & 3 & -9 & \vdots & 4 \\ 0 & 1 & -1 & \vdots & 10 \\ 0 & 0 & 1 & \vdots & -2 \end{bmatrix}$

13. $\begin{bmatrix} 1 & 3 & 4 & \vdots & 1 \\ 0 & 1 & 2 & \vdots & 3 \\ 0 & 0 & 1 & \vdots & 4 \end{bmatrix}$

14. $\begin{bmatrix} 1 & -8 & 0 & \vdots & -2 \\ 0 & 1 & -1 & \vdots & -7 \\ 0 & 0 & 1 & \vdots & 1 \end{bmatrix}$

Gaussian Elimination with Back-Substitution In Exercises 15–26, use matrices to solve the system of linear equations, if possible. Use Gaussian elimination with back-substitution.

15. $\begin{cases} 5x + 4y = 2 \\ -x + y = -22 \end{cases}$

16. $\begin{cases} 2x - 5y = 2 \\ 3x - 7y = 1 \end{cases}$

17. $\begin{cases} 0.3x - 0.1y = -0.13 \\ 0.2x - 0.3y = -0.25 \end{cases}$

18. $\begin{cases} 0.2x - 0.1y = 0.07 \\ 0.4x - 0.5y = -0.01 \end{cases}$

19. $\begin{cases} -x + 2y = 3 \\ 2x - 4y = 6 \end{cases}$

20. $\begin{cases} -x + 2y = 3 \\ 2x - 4y = -6 \end{cases}$

21. $\begin{cases} x - 2y + z = 7 \\ 2x + y - 2z = -4 \\ -x + 3y + 2z = -3 \end{cases}$

22. $\begin{cases} x - 2y + z = 4 \\ 2x + y - 2z = -24 \\ -x + 3y + 2z = 20 \end{cases}$

23. $\begin{cases} 2x + y + 2z = 4 \\ 2x + 2y = 5 \\ 2x - y + 6z = 2 \end{cases}$

24. $\begin{cases} x + 2y + 6z = 1 \\ 2x + 5y + 15z = 4 \\ 3x + y + 3z = -6 \end{cases}$

25. $\begin{cases} 2x + 3y + z = 10 \\ 2x - 3y - 3z = 22 \\ 4x - 2y + 3z = -2 \end{cases}$

26. $\begin{cases} 2x + 3y + 3z = 3 \\ 6x + 6y + 12z = 13 \\ 12x + 9y - z = 2 \end{cases}$

Gauss-Jordan Elimination In Exercises 27–30, use matrices to solve the system of linear equations, if possible. Use Gauss-Jordan elimination.

27. $\begin{cases} x + 2y - z = 3 \\ x - y - z = -3 \\ 2x + y + 3z = 10 \end{cases}$

28. $\begin{cases} x - 3y + z = 2 \\ 3x - y - z = -6 \\ -x + y - 3z = -2 \end{cases}$

29. $\begin{cases} -x + y + 2z = 1 \\ 2x + 3y + z = -2 \\ 5x + 4y + 2z = 4 \end{cases}$

30. $\begin{cases} 4x + 4y + 4z = 5 \\ 4x - 2y - 8z = 1 \\ 5x + 3y + 8z = 6 \end{cases}$

Using a Graphing Utility In Exercises 31 and 32, use the matrix capabilities of a graphing utility to write the augmented matrix corresponding to the system of linear equations in reduced row-echelon form. Then solve the system, if possible.

31. $\begin{cases} 3x - y + 5z - 2w = -44 \\ x + 6y + 4z - w = 1 \\ 5x - y + z + 3w = -15 \\ 4y - z - 8w = 58 \end{cases}$

32. $\begin{cases} 4x + 12y + 2z = 20 \\ x + 6y + 4z = 12 \\ x + 6y + z = 8 \\ -2x - 10y - 2z = -10 \end{cases}$

7.2 **Equality of Matrices** In Exercises 33–36, solve for x and y.

33. $\begin{bmatrix} -1 & x \\ y & 9 \end{bmatrix} = \begin{bmatrix} -1 & 12 \\ 11 & 9 \end{bmatrix}$

34. $\begin{bmatrix} -1 & 0 \\ x & 5 \\ -4 & -3 \end{bmatrix} = \begin{bmatrix} -1 & 0 \\ 8 & 5 \\ -4 & y \end{bmatrix}$

35. $\begin{bmatrix} x+3 & -4 & 44 \\ 0 & -3 & 2 \\ -2 & y+5 & 6 \end{bmatrix} = \begin{bmatrix} 5x-1 & -4 & 44 \\ 0 & -3 & 2 \\ -2 & 16 & 6 \end{bmatrix}$

36. $\begin{bmatrix} -9 & 4 & 2 & -5 \\ 0 & -3 & 7 & 2y \\ 6 & -1 & 1 & 0 \end{bmatrix} = \begin{bmatrix} -9 & 4 & x-10 & -5 \\ 0 & -3 & 7 & -6 \\ 6 & -1 & 1 & 0 \end{bmatrix}$

Operations with Matrices In Exercises 37–42, if possible, find (a) $A + B$, (b) $A - B$, (c) $4A$, and (d) $2A + 2B$.

37. $A = \begin{bmatrix} 2 & -2 \\ 3 & 5 \end{bmatrix}$, $B = \begin{bmatrix} -3 & 10 \\ 12 & 8 \end{bmatrix}$

38. $A = \begin{bmatrix} -1 \\ 3 \\ 7 \end{bmatrix}$, $B = \begin{bmatrix} 3 \\ 1 \\ -7 \end{bmatrix}$

39. $A = \begin{bmatrix} 4 & 3 \\ -6 & 1 \\ 10 & 1 \end{bmatrix}$, $B = \begin{bmatrix} 3 & 11 \\ 15 & 25 \\ 20 & 29 \end{bmatrix}$

40. $A = \begin{bmatrix} 5 & 4 \\ -7 & 2 \\ 11 & 2 \end{bmatrix}$, $B = \begin{bmatrix} 0 & 3 \\ 4 & 12 \\ 20 & 40 \end{bmatrix}$

41. $A = \begin{bmatrix} -2 & 3 & -2 \\ 3 & -2 & 3 \\ -2 & 3 & -2 \end{bmatrix}$, $B = \begin{bmatrix} 3 & -2 & 3 \\ -2 & 3 & -2 \\ 3 & -2 & 3 \end{bmatrix}$

42. $A = \begin{bmatrix} 6 & -5 & 7 \end{bmatrix}$, $B = \begin{bmatrix} -1 \\ 4 \\ 8 \end{bmatrix}$

Evaluating an Expression In Exercises 43–46, evaluate the expression.

43. $\begin{bmatrix} 7 & 3 \\ -1 & 5 \end{bmatrix} + \begin{bmatrix} 10 & -20 \\ 14 & -3 \end{bmatrix} + \begin{bmatrix} 5 & 0 \\ 1 & 9 \end{bmatrix}$

44. $\begin{bmatrix} -11 & -7 \\ 16 & -2 \\ 19 & 1 \end{bmatrix} - \begin{bmatrix} 6 & 0 \\ 8 & -4 \\ -2 & 10 \end{bmatrix} + \begin{bmatrix} -3 & 1 \\ 2 & 28 \\ 12 & -2 \end{bmatrix}$

45. $-2\left(\begin{bmatrix} 1 & 2 \\ 5 & -4 \\ 6 & 0 \end{bmatrix} + \begin{bmatrix} 7 & 1 \\ 1 & 2 \\ 1 & 4 \end{bmatrix} \right)$

46. $5\left(\begin{bmatrix} 8 & -1 & 8 \\ -2 & 4 & 12 \\ 0 & -6 & 0 \end{bmatrix} - \begin{bmatrix} -2 & 0 & -4 \\ 3 & -1 & 1 \\ 6 & 12 & -8 \end{bmatrix} \right)$

Solving a Matrix Equation In Exercises 47–50, solve for X in the equation, where

$$A = \begin{bmatrix} -4 & 0 \\ 1 & -5 \\ -3 & 2 \end{bmatrix} \text{ and } B = \begin{bmatrix} 1 & 2 \\ -2 & 1 \\ 4 & 4 \end{bmatrix}.$$

47. $X = 2A - 3B$

48. $6X = 4A + 3B$

49. $3X + 2A = B$

50. $2A - 5B = 3X$

Finding the Product of Two Matrices In Exercises 51–54, if possible, find AB and state the dimension of the result.

51. $A = \begin{bmatrix} 2 & -2 \\ 3 & 5 \end{bmatrix}$, $B = \begin{bmatrix} -3 & 10 \\ 12 & 8 \end{bmatrix}$

52. $A = \begin{bmatrix} 5 & 4 \\ -7 & 2 \\ 11 & 2 \end{bmatrix}$, $B = \begin{bmatrix} 4 & 12 \\ 20 & 40 \\ 15 & 30 \end{bmatrix}$

53. $A = \begin{bmatrix} 5 & 4 \\ -7 & 2 \\ 11 & 2 \end{bmatrix}$, $B = \begin{bmatrix} 4 & 12 \\ 20 & 40 \end{bmatrix}$

54. $A = \begin{bmatrix} 6 & -5 & 7 \end{bmatrix}$, $B = \begin{bmatrix} -1 \\ 4 \\ 8 \end{bmatrix}$

Finding the Product of Two Matrices In Exercises 55–58, use the matrix capabilities of a graphing utility to find AB, if possible.

55. $A = \begin{bmatrix} 4 & 1 \\ 11 & -7 \\ 12 & 3 \end{bmatrix}$, $B = \begin{bmatrix} 3 & -5 & 6 \\ 2 & -2 & -2 \end{bmatrix}$

56. $A = \begin{bmatrix} -2 & 3 & 10 \\ 4 & -2 & 2 \end{bmatrix}$, $B = \begin{bmatrix} 1 & 1 \\ -5 & 2 \\ 3 & 2 \end{bmatrix}$

57. $A = \begin{bmatrix} 1 & 2 & -1 \\ 0 & 4 & -2 \\ 1 & 1 & 3 \end{bmatrix}$, $B = \begin{bmatrix} 1 & -1 & 2 \end{bmatrix}$

58. $A = \begin{bmatrix} 4 & -2 & 6 \end{bmatrix}$, $B = \begin{bmatrix} -2 & 1 \\ 0 & -3 \\ 2 & 0 \end{bmatrix}$

Operations with Matrices In Exercises 59 and 60, if possible, find (a) AB, (b) BA, and (c) A^2.

59. $A = \begin{bmatrix} 1 & 3 \\ 4 & 1 \end{bmatrix}$, $B = \begin{bmatrix} 5 & -1 \\ -2 & 0 \end{bmatrix}$

60. $A = \begin{bmatrix} 2 & 3 \\ 8 & -1 \\ 0 & 2 \end{bmatrix}$, $B = \begin{bmatrix} 4 \\ 1 \end{bmatrix}$

61. Manufacturing A tire corporation has three factories that manufacture two models of tires. The production levels are represented by A.

$$\begin{matrix} & \text{Factory} \\ & \begin{matrix} 1 & 2 & 3 \end{matrix} \\ A = & \begin{bmatrix} 80 & 120 & 140 \\ 40 & 100 & 80 \end{bmatrix} \begin{matrix} A \\ B \end{matrix} \end{matrix} \right\} \text{Model}$$

Find the production levels when production decreases by 5%.

62. Cell Phone Charges The pay-as-you-go charges (per minute) of two cell phone companies for calls inside the coverage area, regional roaming calls, and calls outside the coverage area are represented by C.

$$\begin{matrix} & \text{Company} \\ & \begin{matrix} \text{A} & \text{B} \end{matrix} \\ C = & \begin{bmatrix} \$0.07 & \$0.095 \\ \$0.10 & \$0.08 \\ \$0.28 & \$0.25 \end{bmatrix} \begin{matrix} \text{Inside} \\ \text{Regional Roaming} \\ \text{Outside} \end{matrix} \end{matrix} \right\} \text{Coverage area}$$

The numbers of minutes you plan to use in the coverage areas per month are represented by the matrix

$$T = \begin{bmatrix} 120 & 80 & 20 \end{bmatrix}.$$

Compute TC and interpret the result.

7.3 The Inverse of a Matrix In Exercises 63–66, show that B is the inverse of A.

63. $A = \begin{bmatrix} -4 & -1 \\ 7 & 2 \end{bmatrix}$, $B = \begin{bmatrix} -2 & -1 \\ 7 & 4 \end{bmatrix}$

64. $A = \begin{bmatrix} 5 & -1 \\ 11 & -2 \end{bmatrix}$, $B = \begin{bmatrix} -2 & 1 \\ -11 & 5 \end{bmatrix}$

65. $A = \begin{bmatrix} 1 & 1 & 0 \\ 1 & 0 & 1 \\ 6 & 2 & 3 \end{bmatrix}$, $B = \begin{bmatrix} -2 & -3 & 1 \\ 3 & 3 & -1 \\ 2 & 4 & -1 \end{bmatrix}$

66. $A = \begin{bmatrix} 1 & -1 & 0 \\ -1 & 0 & -1 \\ 8 & -4 & 2 \end{bmatrix}$,

$B = \begin{bmatrix} -2 & 1 & \frac{1}{2} \\ -3 & 1 & \frac{1}{2} \\ 2 & -2 & -\frac{1}{2} \end{bmatrix}$

Finding the Inverse of a Matrix In Exercises 67–70, find the inverse of the matrix, if possible.

67. $\begin{bmatrix} -6 & 5 \\ -5 & 4 \end{bmatrix}$

68. $\begin{bmatrix} 3 & 4 \\ 6 & 8 \end{bmatrix}$

69. $\begin{bmatrix} 2 & 0 & 3 \\ -1 & 1 & 1 \\ 2 & -2 & 1 \end{bmatrix}$

70. $\begin{bmatrix} 0 & -2 & 1 \\ -5 & -2 & -3 \\ 7 & 3 & 4 \end{bmatrix}$

Finding the Inverse of a Matrix In Exercises 71 and 72, use the matrix capabilities of a graphing utility to find the inverse of the matrix, if possible.

71. $\begin{bmatrix} -1 & -2 & -2 \\ 3 & 7 & 9 \\ 1 & 4 & 7 \end{bmatrix}$

72. $\begin{bmatrix} 8 & 0 & 2 & 8 \\ 4 & -2 & 0 & -2 \\ 1 & 2 & 1 & 4 \\ -1 & 4 & 1 & 1 \end{bmatrix}$

Finding the Inverse of a 2 × 2 Matrix In Exercises 73–76, use the formula on page 525 to find the inverse of the 2 × 2 matrix, if possible.

73. $\begin{bmatrix} -7 & 2 \\ -8 & 2 \end{bmatrix}$

74. $\begin{bmatrix} 10 & 4 \\ 7 & 3 \end{bmatrix}$

75. $\begin{bmatrix} -12 & 6 \\ 10 & -5 \end{bmatrix}$

76. $\begin{bmatrix} -18 & -15 \\ -6 & -5 \end{bmatrix}$

Solving a System Using an Inverse Matrix In Exercises 77–86, use an inverse matrix to solve the system of linear equations, if possible.

77. $\begin{cases} -x + 4y = 8 \\ 2x - 7y = -5 \end{cases}$

78. $\begin{cases} 5x - y = 13 \\ -9x + 2y = -24 \end{cases}$

79. $\begin{cases} -3x + 10y = 8 \\ 5x - 17y = -13 \end{cases}$

80. $\begin{cases} 4x - 2y = -10 \\ -19x + 9y = 47 \end{cases}$

81. $\begin{cases} \frac{1}{2}x + \frac{1}{3}y = 2 \\ -3x + 2y = 0 \end{cases}$

82. $\begin{cases} -\frac{5}{6}x + \frac{3}{8}y = -2 \\ 4x - 3y = 0 \end{cases}$

83. $\begin{cases} 0.3x + 0.7y = 10.2 \\ 0.4x + 0.6y = 7.6 \end{cases}$

84. $\begin{cases} 3.5x - 4.5y = 8 \\ 2.5x - 7.5y = 25 \end{cases}$

85. $\begin{cases} 3x + 2y - z = 6 \\ x - y + 2z = -1 \\ 5x + y + z = 7 \end{cases}$

86. $\begin{cases} 4x + 5y - 6z = -6 \\ 3x + 2y + 2z = 8 \\ 2x + y + z = 3 \end{cases}$

Using a Graphing Utility In Exercises 87–90, use the matrix capabilities of a graphing utility to solve the system of linear equations, if possible.

87. $\begin{cases} x + 2y = -1 \\ 3x + 4y = -5 \end{cases}$

88. $\begin{cases} x + 3y = 23 \\ -6x + 2y = -18 \end{cases}$

89. $\begin{cases} \frac{6}{5}x - \frac{4}{7}y = \frac{6}{5} \\ -\frac{12}{5}x + \frac{12}{7}y = -\frac{17}{15} \end{cases}$

90. $\begin{cases} 5x + 10y = 7 \\ 2x + y = -98 \end{cases}$

7.4 Finding the Determinant of a Matrix In Exercises 91–94, find the determinant of the matrix.

91. $\begin{bmatrix} 2 & 5 \\ -4 & 3 \end{bmatrix}$

92. $\begin{bmatrix} -3 & 1 \\ 5 & -2 \end{bmatrix}$

93. $\begin{bmatrix} 10 & -2 \\ 18 & 8 \end{bmatrix}$

94. $\begin{bmatrix} -30 & 10 \\ 5 & 2 \end{bmatrix}$

Finding the Minors and Cofactors of a Matrix In Exercises 95–98, find all the (a) minors and (b) cofactors of the matrix.

95. $\begin{bmatrix} 2 & -1 \\ 7 & 4 \end{bmatrix}$ **96.** $\begin{bmatrix} 3 & 6 \\ 5 & -4 \end{bmatrix}$

97. $\begin{bmatrix} 3 & 2 & -1 \\ -2 & 5 & 0 \\ 1 & 8 & 6 \end{bmatrix}$ **98.** $\begin{bmatrix} 8 & 3 & 4 \\ 6 & 5 & -9 \\ -4 & 1 & 2 \end{bmatrix}$

Finding the Determinant of a Matrix In Exercises 99–104, find the determinant of the matrix. Expand by cofactors using the row or column that appears to make the computations easiest.

99. $\begin{bmatrix} -2 & 0 & 0 \\ 2 & -1 & 0 \\ -1 & 1 & -3 \end{bmatrix}$ **100.** $\begin{bmatrix} 0 & 1 & -2 \\ 0 & 1 & 2 \\ -1 & -1 & 3 \end{bmatrix}$

101. $\begin{bmatrix} 4 & 1 & -1 \\ 2 & 3 & 2 \\ 1 & -1 & 0 \end{bmatrix}$ **102.** $\begin{bmatrix} -1 & -2 & 1 \\ 2 & 3 & 0 \\ -5 & -1 & 3 \end{bmatrix}$

103. $\begin{bmatrix} -2 & 4 & 1 \\ -6 & 0 & 2 \\ 5 & 3 & 4 \end{bmatrix}$ **104.** $\begin{bmatrix} 1 & 1 & 4 \\ -4 & 1 & 2 \\ 0 & 1 & -1 \end{bmatrix}$

7.5 **Using Cramer's Rule** In Exercises 105–108, use Cramer's Rule (if possible) to solve the system of equations.

105. $\begin{cases} 5x - 2y = 6 \\ -11x + 3y = -23 \end{cases}$ **106.** $\begin{cases} 3x + 8y = -7 \\ 9x - 5y = 37 \end{cases}$

107. $\begin{cases} -2x + 3y - 5z = -11 \\ 4x - y + z = -3 \\ -x - 4y + 6z = 15 \end{cases}$

108. $\begin{cases} 5x - 2y + z = 15 \\ 3x - 3y - z = -7 \\ 2x - y - 7z = -3 \end{cases}$

Finding the Area of a Triangle In Exercises 109 and 110, use a determinant to find the area of the triangle with the given vertices.

109.

110.

Testing for Collinear Points In Exercises 111 and 112, use a determinant to determine whether the points are collinear.

111. $(-1, 7), (3, -9), (-3, 15)$

112. $(0, -5), (-2, -6), (8, -1)$

Finding an Equation of a Line In Exercises 113–116, use a determinant to find an equation of the line passing through the points.

113. $(-4, 0), (4, 4)$ **114.** $(2, 5), (6, -1)$

115. $\left(-\frac{5}{2}, 3\right), \left(\frac{7}{2}, 1\right)$ **116.** $(-0.8, 0.2), (0.7, 3.2)$

Finding the Area of a Parallelogram In Exercises 117 and 118, use a determinant to find the area of the parallelogram with the given vertices.

117. $(0, 0), (2, 0), (1, 4), (3, 4)$

118. $(0, 0), (-3, 0), (1, 3), (-2, 3)$

Decoding a Message In Exercises 119 and 120, decode the cryptogram using the inverse of the matrix

$$A = \begin{bmatrix} -5 & 4 & -3 \\ 10 & -7 & 6 \\ 8 & -6 & 5 \end{bmatrix}.$$

119. $-5 \ 11 \ -2 \ 370 \ -265 \ 225 \ -57 \ 48 \ -33 \ 32$
$-15 \ 20 \ 245 \ -171 \ 147$

120. $145 \ -105 \ 92 \ 264 \ -188 \ 160 \ 23 \ -16 \ 15$
$129 \ -84 \ 78 \ -9 \ 8 \ -5 \ 159 \ -118 \ 100 \ 219$
$-152 \ 133 \ 370 \ -265 \ 225 \ -105 \ 84 \ -63$

Exploration

True or False? In Exercises 121 and 122, determine whether the statement is true or false. Justify your answer.

121. It is possible to find the determinant of a 4×5 matrix.

122. $\begin{vmatrix} a_{11} & a_{12} & a_{13} \\ a_{21} & a_{22} & a_{23} \\ a_{31} + c_1 & a_{32} + c_2 & a_{33} + c_3 \end{vmatrix}$

$= \begin{vmatrix} a_{11} & a_{12} & a_{13} \\ a_{21} & a_{22} & a_{23} \\ a_{31} & a_{32} & a_{33} \end{vmatrix} + \begin{vmatrix} a_{11} & a_{12} & a_{13} \\ a_{21} & a_{22} & a_{23} \\ c_1 & c_2 & c_3 \end{vmatrix}$

123. Writing What is the cofactor of an entry of a matrix? How are cofactors used to find the determinant of the matrix?

124. Think About It Three people are solving a system of equations using an augmented matrix. Each person writes the matrix in row-echelon form. Their reduced matrices are shown below.

$$\begin{bmatrix} 1 & 2 & \vdots & 3 \\ 0 & 1 & \vdots & 1 \end{bmatrix}$$

$$\begin{bmatrix} 1 & 0 & \vdots & 1 \\ 0 & 1 & \vdots & 1 \end{bmatrix}$$

$$\begin{bmatrix} 1 & 2 & \vdots & 3 \\ 0 & 0 & \vdots & 0 \end{bmatrix}$$

Can all three be right? Explain.

Take this test as you would take a test in class. When you are finished, check your work against the answers given in the back of the book.

In Exercises 1 and 2, write the matrix in reduced row-echelon form.

1. $\begin{bmatrix} 1 & -1 & 5 \\ 6 & 2 & 3 \\ 5 & 3 & -3 \end{bmatrix}$

2. $\begin{bmatrix} 1 & 0 & -1 & 2 \\ -1 & 1 & 1 & -3 \\ 1 & 1 & -1 & 1 \\ 3 & 2 & -3 & 4 \end{bmatrix}$

3. Write the augmented matrix for the system of equations and solve the system.

$$\begin{cases} 4x + 3y - 2z = 14 \\ -x - y + 2z = -5 \\ 3x + y - 4z = 8 \end{cases}$$

4. If possible, find (a) $A - B$, (b) $3C$, (c) $3A - 2B$, (d) BC, and (e) C^2.

$$A = \begin{bmatrix} 6 & 5 \\ -5 & -5 \end{bmatrix}, \quad B = \begin{bmatrix} 5 & 0 \\ -5 & -1 \end{bmatrix}, \quad C = \begin{bmatrix} 2 & -1 & 4 \\ 0 & 6 & -3 \end{bmatrix}$$

In Exercises 5 and 6, find the inverse of the matrix, if possible.

5. $\begin{bmatrix} -4 & 3 \\ 5 & -2 \end{bmatrix}$

6. $\begin{bmatrix} -2 & 4 & -6 \\ 2 & 1 & 0 \\ 4 & -2 & 5 \end{bmatrix}$

7. Use the result of Exercise 5 to solve the system.

$$\begin{cases} -4x + 3y = 6 \\ 5x - 2y = 24 \end{cases}$$

In Exercises 8–10, find the determinant of the matrix.

8. $\begin{bmatrix} -6 & 4 \\ 10 & 12 \end{bmatrix}$

9. $\begin{bmatrix} \frac{5}{2} & -\frac{3}{8} \\ -8 & \frac{6}{5} \end{bmatrix}$

10. $\begin{bmatrix} 6 & -7 & 2 \\ 3 & -2 & 0 \\ 1 & 5 & 1 \end{bmatrix}$

In Exercises 11 and 12, use Cramer's Rule (if possible) to solve the system of equations.

11. $\begin{cases} 7x + 6y = 9 \\ -2x - 11y = -49 \end{cases}$

12. $\begin{cases} 6x - y + 2z = -4 \\ -2x + 3y - z = 10 \\ 4x - 4y + z = -18 \end{cases}$

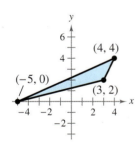

Figure for 13

13. Use a determinant to find the area of the triangle at the left.

14. Write the uncoded 1×3 row matrices for the message KNOCK ON WOOD. Then encode the message using the encoding matrix A below.

$$A = \begin{bmatrix} 1 & -1 & 0 \\ 1 & 0 & -1 \\ 6 & -2 & -3 \end{bmatrix}$$

15. One hundred liters of a 50% solution is obtained by mixing a 60% solution with a 20% solution. Use a system of linear equations to determine how many liters of each solution are required to obtain the desired mixture. Solve the system using matrices.

Proofs in Mathematics

A proof without words is a picture or diagram that gives a visual understanding of why a theorem or statement is true. It can also provide a starting point for writing a formal proof.

In Section 7.5 (page 544), you learned that the area of a parallelogram with vertices $(0, 0)$, (a, b), (c, d), and $(a + c, b + d)$ is the absolute value of the determinant of the matrix A, where

$$A = \begin{bmatrix} a & b \\ c & d \end{bmatrix}.$$

The color-coded visual proof below shows this for a case in which the determinant is positive. Also shown is a brief explanation of why this proof works.

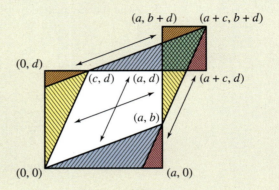

$$\begin{vmatrix} a & b \\ c & d \end{vmatrix} = ad - bc = \|\square\| - \|\square\| = \|\square\|$$

Area of $\square$ = Area of orange $\triangle$ + Area of yellow $\triangle$ + Area of blue $\triangle$
   + Area of pink $\triangle$ + Area of white quadrilateral

Area of $\square$ = Area of orange $\triangle$ + Area of pink $\triangle$ + Area of green quadrilateral

Area of $\square$ = Area of white quadrilateral + Area of blue $\triangle$ + Area of yellow $\triangle$
   − Area of green quadrilateral

   = Area of $\square$ − Area of $\square$

The formula in Section 7.5 is a generalization, taking into consideration the possibility that the coordinates could yield a negative determinant. Area is always positive, which is the reason the formula uses absolute value. Verify the formula using values of a, b, c, and d that produce a negative determinant.

From "Proof Without Words: A 2 × 2 Determinant Is the Area of a Parallelogram" by Solomon W. Golomb, *Mathematics Magazine*, Vol. 58, No. 2, pg. 107.

P.S. Problem Solving ▪ ■ ▪ ■ ▪ ■ ▪ ■ ▪ ■ ▪ ■ ▪ ■

1. Multiplying by a Transformation Matrix The columns of matrix T show the coordinates of the vertices of a triangle. Matrix A is a transformation matrix.

$$A = \begin{bmatrix} 0 & -1 \\ 1 & 0 \end{bmatrix} \qquad T = \begin{bmatrix} 1 & 2 & 3 \\ 1 & 4 & 2 \end{bmatrix}$$

(a) Find AT and AAT. Then sketch the original triangle and the two images of the triangle. What transformation does A represent?

(b) Given the triangle determined by AAT, describe the transformation that produces the triangle determined by AT and then the triangle determined by T.

2. Population The matrices show the male and female populations in the United States in 2011 and 2014. The male and female populations are separated into three age groups. *(Source: U.S. Census Bureau)*

2011

	0–19	20–64	65+
Male	42,376,825	92,983,543	17,934,267
Female	40,463,751	94,530,885	23,432,361

2014

	0–19	20–64	65+
Male	41,969,399	94,615,796	20,351,292
Female	40,166,203	95,862,447	25,891,919

(a) The total population in 2011 was 311,721,632 and the total population in 2014 was 318,857,056. Rewrite the matrices to give the information as percents of the total population.

(b) Write a matrix that gives the change in the percent of the population for each gender and age group from 2011 to 2014.

(c) Based on the result of part (b), which gender(s) and age group(s) had percents that decreased from 2011 to 2014?

3. Determining Whether Matrices are Idempotent A square matrix is **idempotent** when $A^2 = A$. Determine whether each matrix is idempotent.

(a) $\begin{bmatrix} 1 & 0 \\ 0 & 0 \end{bmatrix}$

(b) $\begin{bmatrix} 0 & 1 \\ 1 & 0 \end{bmatrix}$

(c) $\begin{bmatrix} 2 & 3 \\ -1 & -2 \end{bmatrix}$

(d) $\begin{bmatrix} 2 & 3 \\ 1 & 2 \end{bmatrix}$

(e) $\begin{bmatrix} 0 & 0 & 1 \\ 0 & 1 & 0 \\ 1 & 0 & 0 \end{bmatrix}$

(f) $\begin{bmatrix} 0 & 1 & 0 \\ 1 & 0 & 0 \\ 0 & 0 & 1 \end{bmatrix}$

4. Finding a Matrix Find a singular 2×2 matrix satisfying $A^2 = A$.

5. Quadratic Matrix Equation Let

$$A = \begin{bmatrix} 1 & 2 \\ -2 & 1 \end{bmatrix}.$$

(a) Show that $A^2 - 2A + 5I = O$, where I is the identity matrix of dimension 2×2.

(b) Show that $A^{-1} = \frac{1}{5}(2I - A)$.

(c) Show that for any square matrix satisfying

$$A^2 - 2A + 5I = O$$

the inverse of A is given by

$$A^{-1} = \frac{1}{5}(2I - A).$$

6. Satellite Television Two competing companies offer satellite television to a city with 100,000 households. Gold Satellite System has 25,000 subscribers and Galaxy Satellite Network has 30,000 subscribers. (The other 45,000 households do not subscribe.) The matrix shows the percent changes in satellite subscriptions each year.

	Percent Changes		
	From Gold	From Galaxy	From Non-subscriber
Percent Changes To Gold	0.70	0.15	0.15
To Galaxy	0.20	0.80	0.15
To Nonsubscriber	0.10	0.05	0.70

(a) Find the number of subscribers each company will have in 1 year using matrix multiplication. Explain how you obtained your answer.

(b) Find the number of subscribers each company will have in 2 years using matrix multiplication. Explain how you obtained your answer.

(c) Find the number of subscribers each company will have in 3 years using matrix multiplication. Explain how you obtained your answer.

(d) What is happening to the number of subscribers to each company? What is happening to the number of nonsubscribers?

7. The Transpose of a Matrix The **transpose** of a matrix, denoted A^T, is formed by writing its rows as columns. Find the transpose of each matrix and verify that $(AB)^T = B^T A^T$.

$$A = \begin{bmatrix} -1 & 1 & -2 \\ 2 & 0 & 1 \end{bmatrix}, \quad B = \begin{bmatrix} -3 & 0 \\ 1 & 2 \\ 1 & -1 \end{bmatrix}$$

8. Finding a Value Find x such that the matrix is equal to its own inverse.

$$A = \begin{bmatrix} 3 & x \\ -2 & -3 \end{bmatrix}$$

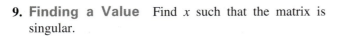

9. Finding a Value Find x such that the matrix is singular.

$$A = \begin{bmatrix} 4 & x \\ -2 & -3 \end{bmatrix}$$

10. Verifying an Equation Verify the following equation.

$$\begin{vmatrix} 1 & 1 & 1 \\ a & b & c \\ a^2 & b^2 & c^2 \end{vmatrix} = (a - b)(b - c)(c - a)$$

11. Verifying an Equation Verify the following equation.

$$\begin{vmatrix} 1 & 1 & 1 \\ a & b & c \\ a^3 & b^3 & c^3 \end{vmatrix} = (a - b)(b - c)(c - a)(a + b + c)$$

12. Verifying an Equation Verify the following equation.

$$\begin{vmatrix} x & 0 & c \\ -1 & x & b \\ 0 & -1 & a \end{vmatrix} = ax^2 + bx + c$$

13. Finding a Matrix Find a 4×4 matrix whose determinant is equal to $ax^3 + bx^2 + cx + d$. (*Hint:* Use the equation in Exercise 12 as a model.)

14. Finding the Determinant of a Matrix Let A be an $n \times n$ matrix each of whose rows sum to zero. Find $|A|$.

15. Finding Atomic Masses The table shows the masses (in atomic mass units) of three compounds. Use a linear system and Cramer's Rule to find the atomic masses of sulfur (S), nitrogen (N), and fluorine (F).

Compound	Formula	Mass
Tetrasulfur tetranitride	S_4N_4	184
Sulfur hexafluoride	SF_6	146
Dinitrogen tetrafluoride	N_2F_4	104

16. Finding the Costs of Items A walkway lighting package includes a transformer, a certain length of wire, and a certain number of lights on the wire. The price of each lighting package depends on the length of wire and the number of lights on the wire. Use the information below to find the cost of a transformer, the cost per foot of wire, and the cost of a light. Assume that the cost of each item is the same in each lighting package.

- A package that contains a transformer, 25 feet of wire, and 5 lights costs $20.
- A package that contains a transformer, 50 feet of wire, and 15 lights costs $35.
- A package that contains a transformer, 100 feet of wire, and 20 lights costs $50.

17. Decoding a Message Use the inverse of A to decode the cryptogram.

$$A = \begin{bmatrix} 1 & -2 & 2 \\ 1 & 1 & -3 \\ 1 & -1 & 4 \end{bmatrix}$$

23 13 -34 31 -34 63 25 -17 61
24 14 -37 41 -17 -8 20 -29 40 38
-56 116 13 -11 1 22 -3 -6 41
-53 85 28 -32 16

18. Decoding a Message A code breaker intercepts the encoded message below.

45 -35 38 -30 18 -18 35 -30 81 -60
42 -28 75 -55 2 -2 22 -21 15 -10

Let $A^{-1} = \begin{bmatrix} w & x \\ y & z \end{bmatrix}$.

(a) You know that

$$[45 \;\; -35]A^{-1} = [10 \;\; 15]$$

$$[38 \;\; -30]A^{-1} = [8 \;\; 14]$$

where A^{-1} is the inverse of the encoding matrix A. Write and solve two systems of equations to find w, x, y, and z.

(b) Decode the message.

19. Conjecture Let

$$A = \begin{bmatrix} 6 & 4 & 1 \\ 0 & 2 & 3 \\ 1 & 1 & 2 \end{bmatrix}.$$

Use a graphing utility to find A^{-1}. Compare $|A^{-1}|$ with $|A|$. Make a conjecture about the determinant of the inverse of a matrix.

20. Conjecture Consider matrices of the form

$$A = \begin{bmatrix} 0 & a_{12} & a_{13} & a_{14} & \cdots & a_{1n} \\ 0 & 0 & a_{23} & a_{24} & \cdots & a_{2n} \\ 0 & 0 & 0 & a_{34} & \cdots & a_{3n} \\ \vdots & \vdots & \vdots & \vdots & & \vdots \\ 0 & 0 & 0 & 0 & \cdots & a_{(n-1)n} \\ 0 & 0 & 0 & 0 & \cdots & 0 \end{bmatrix}.$$

(a) Write a 2×2 matrix and a 3×3 matrix in the form of A.

(b) Use a graphing utility to raise each of the matrices to higher powers. Describe the result.

(c) Use the result of part (b) to make a conjecture about powers of A when A is a 4×4 matrix. Use the graphing utility to test your conjecture.

(d) Use the results of parts (b) and (c) to make a conjecture about powers of A when A is an $n \times n$ matrix.

8 Sequences, Series, and Probability

Horse Racing *(Example 6, page 611)*

Tossing Dice
(Example 3, page 620)

Electricity *(Exercise 86, page 607)*

Dominoes *(page 591)*

Physical Activity *(Exercise 98, page 571)*

8.1 Sequences and Series

Sequences and series model many real-life situations over time. For example, in Exercise 98 on page 571, a sequence models the percent of United States adults who met federal physical activity guidelines from 2007 through 2014.

- Use sequence notation to write the terms of sequences.
- Use factorial notation.
- Use summation notation to write sums.
- Find the sums of series.
- Use sequences and series to model and solve real-life problems.

Sequences

In mathematics, the word *sequence* is used in much the same way as in ordinary English. Saying that a collection is listed in *sequence* means that it is ordered so that it has a first member, a second member, a third member, and so on. Two examples are 1, 2, 3, 4, . . . and 1, 3, 5, 7, ...

Mathematically, you can think of a sequence as a *function* whose domain is the set of positive integers. Rather than using function notation, however, sequences are usually written using subscript notation, as shown in the following definition.

> ### Definition of Sequence
>
> An **infinite sequence** is a function whose domain is the set of positive integers. The function values
>
> $$a_1, a_2, a_3, a_4, \ldots , a_n, \ldots$$
>
> are the **terms** of the sequence. When the domain of the function consists of the first n positive integers only, the sequence is a **finite sequence.**

On occasion, it is convenient to begin subscripting a sequence with 0 instead of 1 so that the terms of the sequence become

$$a_0, a_1, a_2, a_3, \ldots .$$

When this is the case, the domain includes 0.

REMARK The subscripts of a sequence make up the domain of the sequence and serve to identify the positions of terms within the sequence. For example, a_4 is the fourth term of the sequence, and a_n is the nth term of the sequence. Any variable can be a subscript. The most commonly used variable subscripts in sequence and series notation are $i, j, k,$ and n.

EXAMPLE 1 Writing the Terms of a Sequence

a. The first four terms of the sequence given by $a_n = 3n - 2$ are

$$a_1 = 3(1) - 2 = 1 \qquad \text{1st term}$$
$$a_2 = 3(2) - 2 = 4 \qquad \text{2nd term}$$
$$a_3 = 3(3) - 2 = 7 \qquad \text{3rd term}$$
$$a_4 = 3(4) - 2 = 10. \qquad \text{4th term}$$

b. The first four terms of the sequence given by $a_n = 3 + (-1)^n$ are

$$a_1 = 3 + (-1)^1 = 3 - 1 = 2 \qquad \text{1st term}$$
$$a_2 = 3 + (-1)^2 = 3 + 1 = 4 \qquad \text{2nd term}$$
$$a_3 = 3 + (-1)^3 = 3 - 1 = 2 \qquad \text{3rd term}$$
$$a_4 = 3 + (-1)^4 = 3 + 1 = 4. \qquad \text{4th term}$$

✓ **Checkpoint** *Audio-video solution in English & Spanish at LarsonPrecalculus.com*

Write the first four terms of the sequence given by $a_n = 2n + 1$.

· · REMARK Write the first four terms of the sequence given by

$$a_n = \frac{(-1)^{n+1}}{2n+1}.$$

Are they the same as the first four terms of the sequence in Example 2? If not, then how do they differ? ▷

EXAMPLE 2   A Sequence Whose Terms Alternate in Sign

Write the first four terms of the sequence given by $a_n = \dfrac{(-1)^n}{2n+1}$.

Solution   The first four terms of the sequence are as follows.

$$a_1 = \frac{(-1)^1}{2(1)+1} = \frac{-1}{2+1} = -\frac{1}{3} \qquad \text{1st term}$$

$$a_2 = \frac{(-1)^2}{2(2)+1} = \frac{1}{4+1} = \frac{1}{5} \qquad \text{2nd term}$$

$$a_3 = \frac{(-1)^3}{2(3)+1} = \frac{-1}{6+1} = -\frac{1}{7} \qquad \text{3rd term}$$

$$a_4 = \frac{(-1)^4}{2(4)+1} = \frac{1}{8+1} = \frac{1}{9} \qquad \text{4th term}$$

✓ **Checkpoint**   *Audio-video solution in English & Spanish at LarsonPrecalculus.com*

Write the first four terms of the sequence given by $a_n = \dfrac{2+(-1)^n}{n}$. ■

Simply listing the first few terms is not sufficient to define a unique sequence—the *n*th term *must be given*. To see this, consider the following sequences, both of which have the same first three terms.

$$\frac{1}{2}, \frac{1}{4}, \frac{1}{8}, \frac{1}{16}, \cdots \frac{1}{2^n}, \cdots$$

$$\frac{1}{2}, \frac{1}{4}, \frac{1}{8}, \frac{1}{15}, \cdots \frac{6}{(n+1)(n^2-n+6)}, \cdots$$

▷ **TECHNOLOGY**
To graph a sequence using a graphing utility, set the mode to *sequence* and *dot* and enter the expression for a_n. The graph of the sequence in Example 3(a) is shown below. To identify the terms, use the *trace* feature or *value* feature.

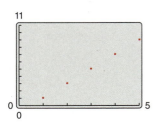

EXAMPLE 3   Finding the *n*th Term of a Sequence

Write an expression for the apparent *n*th term (a_n) of each sequence.

a. $1, 3, 5, 7, \ldots$ **b.** $2, -5, 10, -17, \ldots$

Solution

a.   *n*: 1 2 3 4 . . . *n*
        Terms: 1 3 5 7 . . . a_n

   Apparent pattern: Each term is 1 less than twice *n*. So, the apparent *n*th term is

   $$a_n = 2n - 1.$$

b.   *n*: 1   2   3   4 . . . *n*
        Terms: 2   -5   10   -17 . . . a_n

   Apparent pattern: The absolute value of each term is 1 more than the square of *n*, and the terms have alternating signs, with those in the even positions being negative. So, the apparent *n*th term is

   $$a_n = (-1)^{n+1}(n^2 + 1).$$

✓ **Checkpoint**  *Audio-video solution in English & Spanish at LarsonPrecalculus.com*

Write an expression for the apparent *n*th term (a_n) of each sequence.

a. $1, 5, 9, 13, \ldots$ **b.** $2, -4, 6, -8, \ldots$ ■

Some sequences are defined **recursively.** To define a sequence recursively, you need to be given one or more of the first few terms. All other terms of the sequence are then defined using previous terms.

EXAMPLE 4 A Recursive Sequence

Write the first five terms of the sequence defined recursively as

$$a_1 = 3$$
$$a_k = 2a_{k-1} + 1, \quad \text{where } k \geq 2.$$

Solution

$a_1 = 3$	1st term is given.
$a_2 = 2a_{2-1} + 1 = 2a_1 + 1 = 2(3) + 1 = 7$	Use recursion formula.
$a_3 = 2a_{3-1} + 1 = 2a_2 + 1 = 2(7) + 1 = 15$	Use recursion formula.
$a_4 = 2a_{4-1} + 1 = 2a_3 + 1 = 2(15) + 1 = 31$	Use recursion formula.
$a_5 = 2a_{5-1} + 1 = 2a_4 + 1 = 2(31) + 1 = 63$	Use recursion formula.

✓ *Checkpoint* ◀))) *Audio-video solution in English & Spanish at LarsonPrecalculus.com*

Write the first five terms of the sequence defined recursively as

$$a_1 = 6$$
$$a_{k+1} = a_k + 1, \quad \text{where} \quad k \geq 1.$$

In the next example, you will study a well-known recursive sequence, the Fibonacci sequence.

EXAMPLE 5 The Fibonacci Sequence: A Recursive Sequence

The Fibonacci sequence is defined recursively, as follows.

$$a_0 = 1$$
$$a_1 = 1$$
$$a_k = a_{k-2} + a_{k-1}, \quad \text{where} \quad k \geq 2$$

Write the first six terms of this sequence.

Solution

$a_0 = 1$	0th term is given.
$a_1 = 1$	1st term is given.
$a_2 = a_{2-2} + a_{2-1} = a_0 + a_1 = 1 + 1 = 2$	Use recursion formula.
$a_3 = a_{3-2} + a_{3-1} = a_1 + a_2 = 1 + 2 = 3$	Use recursion formula.
$a_4 = a_{4-2} + a_{4-1} = a_2 + a_3 = 2 + 3 = 5$	Use recursion formula.
$a_5 = a_{5-2} + a_{5-1} = a_3 + a_4 = 3 + 5 = 8$	Use recursion formula.

✓ *Checkpoint* ◀))) *Audio-video solution in English & Spanish at LarsonPrecalculus.com*

Write the first five terms of the sequence defined recursively as

$$a_0 = 1, \quad a_1 = 3, \quad a_k = a_{k-2} + a_{k-1}, \quad \text{where} \quad k \geq 2.$$

Factorial Notation

Many sequences involve terms defined using special products called **factorials.**

> **Definition of Factorial**
>
> If n is a positive integer, then n **factorial** is defined as
>
> $$n! = 1 \cdot 2 \cdot 3 \cdot 4 \cdots (n-1) \cdot n.$$
>
> As a special case, zero factorial is defined as $0! = 1$.

· · **REMARK** The value of n does not have to be very large before the value of $n!$ becomes extremely large. For example, $10! = 3,628,800$.
· · · · · · · · · · · · · · · ▷

Notice that $1! = 1, 2! = 1 \cdot 2 = 2, 3! = 1 \cdot 2 \cdot 3 = 6$, and $4! = 1 \cdot 2 \cdot 3 \cdot 4 = 24$. Factorials follow the same conventions for order of operations as exponents. So,

$$2n! = 2(n!) = 2(1 \cdot 2 \cdot 3 \cdot 4 \cdots n), \quad \text{whereas} \quad (2n)! = 1 \cdot 2 \cdot 3 \cdot 4 \cdots 2n.$$

EXAMPLE 6 Writing the Terms of a Sequence Involving Factorials

Write the first five terms of the sequence given by $a_n = \dfrac{2^n}{n!}$. Begin with $n = 0$.

Algebraic Solution

$$a_0 = \frac{2^0}{0!} = \frac{1}{1} = 1 \qquad \text{0th term}$$

$$a_1 = \frac{2^1}{1!} = \frac{2}{1} = 2 \qquad \text{1st term}$$

$$a_2 = \frac{2^2}{2!} = \frac{4}{2} = 2 \qquad \text{2nd term}$$

$$a_3 = \frac{2^3}{3!} = \frac{8}{6} = \frac{4}{3} \qquad \text{3rd term}$$

$$a_4 = \frac{2^4}{4!} = \frac{16}{24} = \frac{2}{3} \qquad \text{4th term}$$

Graphical Solution

Using a graphing utility set to *dot* and *sequence* modes, enter the expression for a_n. Next, graph the sequence. Use the graph to estimate the first five terms.

$$u_0 = 1$$
$$u_1 = 2$$
$$u_2 = 2$$
$$u_3 \approx 1.333 \approx \tfrac{4}{3}$$
$$u_4 \approx 0.667 \approx \tfrac{2}{3}$$

Use the *trace* feature to approximate the first five terms.

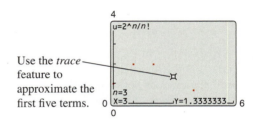

✓ *Checkpoint* 🔊))) *Audio-video solution in English & Spanish at LarsonPrecalculus.com*

Write the first five terms of the sequence given by $a_n = \dfrac{3^n + 1}{n!}$. Begin with $n = 0$. ■

▷ **ALGEBRA HELP** Here is another way to simplify the expression in Example 7(a).

$$\frac{8!}{2! \cdot 6!} = \frac{8 \cdot 7 \cdot 6\!\!\!/}{2 \cdot 1 \cdot 6\!\!\!/} = 28$$

When fractions involve factorials, you can often divide out common factors.

EXAMPLE 7 Simplifying Factorial Expressions

a. $\dfrac{8!}{2! \cdot 6!} = \dfrac{1 \cdot 2 \cdot 3 \cdot 4 \cdot 5 \cdot 6 \cdot 7 \cdot 8}{1 \cdot 2 \cdot 1 \cdot 2 \cdot 3 \cdot 4 \cdot 5 \cdot 6} = \dfrac{7 \cdot 8}{2} = 28$

b. $\dfrac{n!}{(n-1)!} = \dfrac{1 \cdot 2 \cdot 3 \cdots (n-1) \cdot n}{1 \cdot 2 \cdot 3 \cdots (n-1)} = n$

✓ *Checkpoint* 🔊))) *Audio-video solution in English & Spanish at LarsonPrecalculus.com*

Simplify the factorial expression $\dfrac{4!(n+1)!}{3!n!}$. ■

•• REMARK Summation
notation is an instruction to
add the terms of a sequence.
Note that the upper limit of
summation tells you the last
term of the sum. Summation
notation helps you generate the
terms of the sequence prior to
finding the sum.

•• REMARK In Example 8,
note that the lower limit of a
summation does not have to be
1 and the index of summation
does not have to be the letter i.
For example, in part (b), the
lower limit of summation is 3
and the index of summation is k.

Summation Notation

A convenient notation for the sum of the terms of a finite sequence is called **summation notation** or **sigma notation.** It involves the use of the uppercase Greek letter sigma, written as Σ.

Definition of Summation Notation

The sum of the first n terms of a sequence is represented by

$$\sum_{i=1}^{n} a_i = a_1 + a_2 + a_3 + a_4 + \cdots + a_n$$

where i is the **index of summation,** n is the **upper limit of summation,** and 1 is the **lower limit of summation.**

EXAMPLE 8 **Summation Notation for a Sum**

a. $\displaystyle\sum_{i=1}^{5} 3i = 3(1) + 3(2) + 3(3) + 3(4) + 3(5)$

$$= 45$$

b. $\displaystyle\sum_{k=3}^{6} (1 + k^2) = (1 + 3^2) + (1 + 4^2) + (1 + 5^2) + (1 + 6^2)$

$$= 10 + 17 + 26 + 37$$

$$= 90$$

c. $\displaystyle\sum_{i=0}^{8} \frac{1}{i!} = \frac{1}{0!} + \frac{1}{1!} + \frac{1}{2!} + \frac{1}{3!} + \frac{1}{4!} + \frac{1}{5!} + \frac{1}{6!} + \frac{1}{7!} + \frac{1}{8!}$

$$= 1 + 1 + \frac{1}{2} + \frac{1}{6} + \frac{1}{24} + \frac{1}{120} + \frac{1}{720} + \frac{1}{5040} + \frac{1}{40,320}$$

$$\approx 2.71828$$

For this summation, note that the sum is very close to the irrational number

$$e \approx 2.718281828.$$

It can be shown that as more terms of the sequence whose nth term is $1/n!$ are added, the sum becomes closer and closer to e.

✓ *Checkpoint*  *Audio-video solution in English & Spanish at LarsonPrecalculus.com*

Find the sum $\displaystyle\sum_{i=1}^{4} (4i + 1)$. ∎

Properties of Sums

1. $\displaystyle\sum_{i=1}^{n} c = cn,$ c is a constant. 2. $\displaystyle\sum_{i=1}^{n} ca_i = c \sum_{i=1}^{n} a_i,$ c is a constant.

3. $\displaystyle\sum_{i=1}^{n} (a_i + b_i) = \sum_{i=1}^{n} a_i + \sum_{i=1}^{n} b_i$ 4. $\displaystyle\sum_{i=1}^{n} (a_i - b_i) = \sum_{i=1}^{n} a_i - \sum_{i=1}^{n} b_i$

For proofs of these properties, see Proofs in Mathematics on page 638.

Series

Many applications involve the sum of the terms of a finite or infinite sequence. Such a sum is called a **series.**

Definition of Series

Consider the infinite sequence $a_1, a_2, a_3, \ldots, a_i, \ldots$.

1. The sum of the first n terms of the sequence is called a **finite series** or the **nth partial sum** of the sequence and is denoted by

$$a_1 + a_2 + a_3 + \cdots + a_n = \sum_{i=1}^{n} a_i.$$

2. The sum of all the terms of the infinite sequence is called an **infinite series** and is denoted by

$$a_1 + a_2 + a_3 + \cdots + a_i + \cdots = \sum_{i=1}^{\infty} a_i.$$

EXAMPLE 9   **Finding the Sum of a Series**

See LarsonPrecalculus.com for an interactive version of this type of example.

For the series

$$\sum_{i=1}^{\infty} \frac{3}{10^i}$$

find (a) the third partial sum and (b) the sum.

Solution

a. The third partial sum is

$$\sum_{i=1}^{3} \frac{3}{10^i} = \frac{3}{10^1} + \frac{3}{10^2} + \frac{3}{10^3}$$

$$= 0.3 + 0.03 + 0.003$$

$$= 0.333.$$

b. The sum of the series is

$$\sum_{i=1}^{\infty} \frac{3}{10^i} = \frac{3}{10^1} + \frac{3}{10^2} + \frac{3}{10^3} + \frac{3}{10^4} + \frac{3}{10^5} + \cdots$$

$$= 0.3 + 0.03 + 0.003 + 0.0003 + 0.00003 + \cdots$$

$$= 0.33333\ldots$$

$$= \frac{1}{3}.$$

✓ *Checkpoint* ◀))) *Audio-video solution in English & Spanish at LarsonPrecalculus.com*

For the series

$$\sum_{i=1}^{\infty} \frac{5}{10^i}$$

find (a) the fourth partial sum and (b) the sum.

Notice in Example 9(b) that the sum of an infinite series can be a finite number.

Application

Sequences have many applications in business and science. Example 10 illustrates one such application.

EXAMPLE 10 **Compound Interest**

An investor deposits $5000 in an account that earns 3% interest compounded quarterly. The balance in the account after n quarters is given by

$$A_n = 5000\left(1 + \frac{0.03}{4}\right)^n, \quad n = 0, 1, 2, \ldots .$$

a. Write the first three terms of the sequence.

b. Find the balance in the account after 10 years by computing the 40th term of the sequence.

Solution

a. The first three terms of the sequence are as follows.

$$A_0 = 5000\left(1 + \frac{0.03}{4}\right)^0 = \$5000.00 \qquad \text{Original deposit}$$

$$A_1 = 5000\left(1 + \frac{0.03}{4}\right)^1 = \$5037.50 \qquad \text{First-quarter balance}$$

$$A_2 = 5000\left(1 + \frac{0.03}{4}\right)^2 \approx \$5075.28 \qquad \text{Second-quarter balance}$$

b. The 40th term of the sequence is

$$A_{40} = 5000\left(1 + \frac{0.03}{4}\right)^{40} \approx \$6741.74. \qquad \text{Ten-year balance}$$

✓ **Checkpoint**  *Audio-video solution in English & Spanish at LarsonPrecalculus.com*

An investor deposits $1000 in an account that earns 3% interest compounded monthly. The balance in the account after n months is given by

$$A_n = 1000\left(1 + \frac{0.03}{12}\right)^n, \quad n = 0, 1, 2, \ldots .$$

a. Write the first three terms of the sequence.

b. Find the balance in the account after four years by computing the 48th term of the sequence.

Summarize (Section 8.1)

1. State the definition of a sequence *(page 562)*. For examples of writing the terms of sequences, see Examples 1–5.

2. State the definition of a factorial *(page 565)*. For examples of using factorial notation, see Examples 6 and 7.

3. State the definition of summation notation *(page 566)*. For an example of using summation notation, see Example 8.

4. State the definition of a series *(page 567)*. For an example of finding the sum of a series, see Example 9.

5. Describe an example of how to use a sequence to model and solve a real-life problem *(page 568, Example 10)*.

8.1 Exercises

See **CalcChat.com** for tutorial help and worked-out solutions to odd-numbered exercises.

Vocabulary: Fill in the blanks.

1. An _____ _____ is a function whose domain is the set of positive integers.

2. A sequence is a _____ sequence when the domain of the function consists only of the first n positive integers.

3. When you are given one or more of the first few terms of a sequence, and all other terms of the sequence are defined using previous terms, the sequence is defined _____.

4. If n is a positive integer, then n _____ is defined as $n! = 1 \cdot 2 \cdot 3 \cdot 4 \cdots (n-1) \cdot n$.

5. For the sum $\sum_{i=1}^{n} a_i$, i is the _____ of summation, n is the _____ limit of summation, and 1 is the _____ limit of summation.

6. The sum of the terms of a finite or infinite sequence is called a _____.

Skills and Applications

 Writing the Terms of a Sequence In Exercises 7–22, write the first five terms of the sequence. (Assume that n begins with 1.)

7. $a_n = 4n - 7$
8. $a_n = -2n + 8$

9. $a_n = (-1)^{n+1} + 4$
10. $a_n = 1 - (-1)^n$

11. $a_n = (-2)^n$
12. $a_n = \left(\frac{1}{2}\right)^n$

13. $a_n = \frac{2}{3}$
14. $a_n = 6(-1)^{n+1}$

15. $a_n = \frac{1}{3}n^3$
16. $a_n = \frac{1}{n^2}$

17. $a_n = \frac{n}{n+2}$
18. $a_n = \frac{6n}{3n^2 - 1}$

19. $a_n = n(n-1)(n-2)$
20. $a_n = n(n^2 - 6)$

21. $a_n = (-1)^n \left(\frac{n}{n+1}\right)$

22. $a_n = \frac{(-1)^{n+1}}{n^2 + 1}$

Finding a Term of a Sequence In Exercises 23–26, find the missing term of the sequence.

23. $a_n = (-1)^n(3n - 2)$
    $a_{25} = $ ▨

24. $a_n = (-1)^{n-1}[n(n-1)]$
    $a_{16} = $ ▨

25. $a_n = \frac{4n}{2n^2 - 3}$
    $a_{11} = $ ▨

26. $a_n = \frac{4n^2 - n + 3}{n(n-1)(n+2)}$
    $a_{13} = $ ▨

 Graphing the Terms of a Sequence In Exercises 27–32, use a graphing utility to graph the first 10 terms of the sequence. (Assume that n begins with 1.)

27. $a_n = \frac{2}{3}n$
28. $a_n = 3n + 3(-1)^n$

29. $a_n = 16(-0.5)^{n-1}$
30. $a_n = 8(0.75)^{n-1}$

31. $a_n = \frac{2n}{n+1}$
32. $a_n = \frac{3n^2}{n^2 + 1}$

Matching a Sequence with a Graph In Exercises 33–36, match the sequence with the graph of its first 10 terms. [The graphs are labeled (a), (b), (c), and (d).]

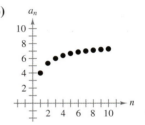

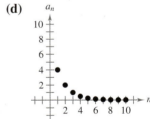

33. $a_n = \frac{8}{n+1}$
34. $a_n = \frac{8n}{n+1}$

35. $a_n = 4(0.5)^{n-1}$
36. $a_n = n\left(2 - \frac{n}{10}\right)$

 Finding the nth Term of a Sequence In Exercises 37–50, write an expression for the apparent nth term (a_n) of the sequence. (Assume that n begins with 1.)

37. $3, 7, 11, 15, 19, \ldots$
38. $0, 3, 8, 15, 24, \ldots$

39. $3, 10, 29, 66, 127, \ldots$
40. $91, 82, 73, 64, 55, \ldots$

41. $1, -1, 1, -1, 1, \ldots$
42. $1, 3, 1, 3, 1, \ldots$

43. $-\frac{2}{3}, \frac{3}{4}, -\frac{4}{5}, \frac{5}{6}, -\frac{6}{7}, \ldots$
44. $\frac{1}{2}, -\frac{1}{4}, \frac{1}{8}, -\frac{1}{16}, \ldots$

45. $\frac{2}{3}, \frac{3}{5}, \frac{4}{7}, \frac{5}{9}, \frac{6}{9}, \ldots$
46. $\frac{1}{3}, \frac{2}{9}, \frac{4}{27}, \frac{8}{81}, \ldots$

47. $1, \frac{1}{2}, \frac{1}{6}, \frac{1}{24}, \frac{1}{120}, \ldots$
48. $2, 3, 7, 25, 121, \ldots$

49. $\frac{1}{1}, \frac{3}{1}, \frac{9}{2}, \frac{27}{6}, \frac{81}{24}, \ldots$
50. $\frac{2}{1}, \frac{6}{3}, \frac{24}{7}, \frac{120}{15}, \frac{720}{31}, \ldots$

Writing the Terms of a Recursive Sequence In Exercises 51–56, write the first five terms of the sequence defined recursively.

51. $a_1 = 28, \quad a_{k+1} = a_k - 4$

52. $a_1 = 3, \quad a_{k+1} = 2(a_k - 1)$

53. $a_1 = 81, \quad a_{k+1} = \frac{1}{3}a_k$

54. $a_1 = 14, \quad a_{k+1} = (-2)a_k$

55. $a_0 = 1, \quad a_1 = 2, \quad a_k = a_{k-2} + \frac{1}{2}a_{k-1}$

56. $a_0 = -1, \quad a_1 = 1, \quad a_k = a_{k-2} + a_{k-1}$

Fibonacci Sequence In Exercises 57 and 58, use the Fibonacci sequence. (See Example 5.)

57. Write the first 12 terms of the Fibonacci sequence whose nth term is a_n and the first 10 terms of the sequence given by

$$b_n = \frac{a_{n+1}}{a_n}, \quad n \geq 1.$$

58. Using the definition for b_n in Exercise 57, show that b_n can be defined recursively by

$$b_n = 1 + \frac{1}{b_{n-1}}.$$

Writing the Terms of a Sequence Involving Factorials In Exercises 59–62, write the first five terms of the sequence. (Assume that n begins with 0.)

59. $a_n = \dfrac{5}{n!}$

60. $a_n = \dfrac{1}{(n+1)!}$

61. $a_n = \dfrac{(-1)^n(n+3)!}{n!}$

62. $a_n = \dfrac{(-1)^{2n+1}}{(2n+1)!}$

Simplifying a Factorial Expression In Exercises 63–66, simplify the factorial expression.

63. $\dfrac{4!}{6!}$

64. $\dfrac{12!}{4! \cdot 8!}$

65. $\dfrac{(n+1)!}{n!}$

66. $\dfrac{(2n-1)!}{(2n+1)!}$

Finding a Sum In Exercises 67–74, find the sum.

67. $\displaystyle\sum_{i=0}^{4} 3i^2$

68. $\displaystyle\sum_{k=1}^{4} 10$

69. $\displaystyle\sum_{j=3}^{5} \frac{1}{j^2 - 3}$

70. $\displaystyle\sum_{i=1}^{5} (2i - 1)$

71. $\displaystyle\sum_{k=2}^{5} (k+1)^2(k-3)$

72. $\displaystyle\sum_{i=1}^{4} [(i-1)^2 + (i+1)^3]$

73. $\displaystyle\sum_{i=1}^{4} \frac{i!}{2^i}$

74. $\displaystyle\sum_{j=0}^{5} \frac{(-1)^j}{j!}$

 Finding a Sum In Exercises 75–78, use a graphing utility to find the sum.

75. $\displaystyle\sum_{k=0}^{4} \frac{(-1)^k}{k!}$

76. $\displaystyle\sum_{k=0}^{4} \frac{(-1)^k}{k+1}$

77. $\displaystyle\sum_{n=0}^{25} \frac{1}{4^n}$

78. $\displaystyle\sum_{n=0}^{10} \frac{n!}{2^n}$

Using Sigma Notation to Write a Sum In Exercises 79–88, use sigma notation to write the sum.

79. $\dfrac{1}{3(1)} + \dfrac{1}{3(2)} + \dfrac{1}{3(3)} + \cdots + \dfrac{1}{3(9)}$

80. $\dfrac{5}{1+1} + \dfrac{5}{1+2} + \dfrac{5}{1+3} + \cdots + \dfrac{5}{1+15}$

81. $\left[2\left(\frac{1}{8}\right) + 3\right] + \left[2\left(\frac{2}{8}\right) + 3\right] + \cdots + \left[2\left(\frac{8}{8}\right) + 3\right]$

82. $\left[1 - \left(\frac{1}{6}\right)^2\right] + \left[1 - \left(\frac{2}{6}\right)^2\right] + \cdots + \left[1 - \left(\frac{6}{6}\right)^2\right]$

83. $3 - 9 + 27 - 81 + 243 - 729$

84. $1 - \frac{1}{2} + \frac{1}{4} - \frac{1}{8} + \cdots - \frac{1}{128}$

85. $\dfrac{1^2}{2} + \dfrac{2^2}{6} + \dfrac{3^2}{24} + \dfrac{4^2}{120} + \cdots + \dfrac{7^2}{40{,}320}$

86. $\dfrac{1}{1 \cdot 3} + \dfrac{1}{2 \cdot 4} + \dfrac{1}{3 \cdot 5} + \cdots + \dfrac{1}{10 \cdot 12}$

87. $\frac{1}{4} + \frac{3}{8} + \frac{7}{16} + \frac{15}{32} + \frac{31}{64}$

88. $\frac{1}{2} + \frac{2}{4} + \frac{6}{8} + \frac{24}{16} + \frac{120}{32} + \frac{720}{64}$

Finding a Partial Sum of a Series In Exercises 89–92, find the (a) third, (b) fourth, and (c) fifth partial sums of the series.

89. $\displaystyle\sum_{i=1}^{\infty} \left(\frac{1}{2}\right)^i$

90. $\displaystyle\sum_{i=1}^{\infty} 2\left(\frac{1}{3}\right)^i$

91. $\displaystyle\sum_{n=1}^{\infty} 4\left(-\frac{1}{2}\right)^n$

92. $\displaystyle\sum_{n=1}^{\infty} 5\left(-\frac{1}{4}\right)^n$

Finding the Sum of an Infinite Series In Exercises 93–96, find the sum of the infinite series.

93. $\displaystyle\sum_{i=1}^{\infty} \frac{6}{10^i}$

94. $\displaystyle\sum_{k=1}^{\infty} \left(\frac{1}{10}\right)^k$

95. $\displaystyle\sum_{k=1}^{\infty} 7\left(\frac{1}{10}\right)^k$

96. $\displaystyle\sum_{i=1}^{\infty} \frac{2}{10^i}$

97. Compound Interest   An investor deposits $10,000 in an account that earns 3.5% interest compounded quarterly. The balance in the account after n quarters is given by

$$A_n = 10{,}000\left(1 + \frac{0.035}{4}\right)^n, \quad n = 1, 2, 3, \ldots.$$

(a) Write the first eight terms of the sequence.

(b) Find the balance in the account after 10 years by computing the 40th term of the sequence.

(c) Is the balance after 20 years twice the balance after 10 years? Explain.

98. Physical Activity

The percent p_n of United States adults who met federal physical activity guidelines from 2007 through 2014 can be approximated by

$$p_n = 0.0061n^3 - 0.419n^2 + 7.85n + 4.9,$$

$$n = 7, 8, \ldots, 14$$

where n is the year, with $n = 7$ corresponding to 2007. (*Source: National Center for Health Statistics*)

(a) Write the terms of this finite sequence. Use a graphing utility to construct a bar graph that represents the sequence.

(b) What can you conclude from the bar graph in part (a)?

Exploration

True or False?   In Exercises 99 and 100, determine whether the statement is true or false. Justify your answer.

99. $\displaystyle\sum_{i=1}^{4}(i^2 + 2i) = \sum_{i=1}^{4}i^2 + 2\sum_{i=1}^{4}i$

100. $\displaystyle\sum_{j=1}^{4}2^j = \sum_{j=3}^{6}2^{j-2}$

Arithmetic Mean   In Exercises 101–103, use the following definition of the arithmetic mean $\bar{x}$ of a set of n measurements $x_1, x_2, x_3, \ldots, x_n$.

$$\bar{x} = \frac{1}{n}\sum_{i=1}^{n}x_i$$

101. Find the arithmetic mean of the six checking account balances $327.15, $785.69, $433.04, $265.38, $604.12, and $590.30. Use the statistical capabilities of a graphing utility to verify your result.

102. Proof   Prove that $\displaystyle\sum_{i=1}^{n}(x_i - \bar{x}) = 0.$

103. Proof   Prove that

$$\sum_{i=1}^{n}(x_i - \bar{x})^2 = \sum_{i=1}^{n}x_i^2 - \frac{1}{n}\left(\sum_{i=1}^{n}x_i\right)^2.$$

104.   HOW DO YOU SEE IT?   The graph represents the first 10 terms of a sequence. Complete each expression for the apparent nth term (a_n) of the sequence. Which expressions are appropriate to represent the cost a_n to buy n MP3 songs at a cost of $1 per song? Explain.

(a) $a_n = 1 \cdot \boxed{}$

(b) $a_n = \dfrac{\boxed{}!}{(n-1)!}$

(c) $a_n = \displaystyle\sum_{k=1}^{n}\boxed{}$

Error Analysis   In Exercises 105 and 106, describe the error in finding the sum.

105. $\displaystyle\sum_{k=1}^{4}(3 + 2k^2) = \sum_{k=1}^{4}3 + \sum_{k=1}^{4}2k^2$

$$= 3 + (2 + 8 + 18 + 32)$$

$$= 63$$

106. $\displaystyle\sum_{n=0}^{3}(-1)^n n! = (-1)(1) + (1)(2) + (-1)(6)$

$$= -5$$

107. Cube   A $3 \times 3 \times 3$ cube is made up of 27 unit cubes (a unit cube has a length, width, and height of 1 unit), and only the faces of each cube that are visible are painted blue, as shown in the figure.

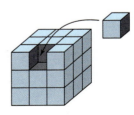

(a) Determine how many unit cubes of the $3 \times 3 \times 3$ cube have 0 blue faces, 1 blue face, 2 blue faces, and 3 blue faces.

(b) Repeat part (a) for a $4 \times 4 \times 4$ cube, a $5 \times 5 \times 5$ cube, and a $6 \times 6 \times 6$ cube.

(c) Write formulas you could use to repeat part (a) for an $n \times n \times n$ cube.

8.2 Arithmetic Sequences and Partial Sums

Arithmetic sequences have many real-life applications. For example, in Exercise 73 on page 579, you will use an arithmetic sequence to determine how far an object falls in 7 seconds when dropped from the top of the Willis Tower in Chicago.

- Recognize, write, and find the *n*th terms of arithmetic sequences.
- Find *n*th partial sums of arithmetic sequences.
- Use arithmetic sequences to model and solve real-life problems.

Arithmetic Sequences

A sequence whose consecutive terms have a common difference is an **arithmetic sequence.**

Definition of Arithmetic Sequence

A sequence is **arithmetic** when the differences between consecutive terms are the same. So, the sequence

$$a_1, a_2, a_3, a_4, \ldots, a_n, \ldots$$

is arithmetic when there is a number d such that

$$a_2 - a_1 = a_3 - a_2 = a_4 - a_3 = \cdots = d.$$

The number d is the **common difference** of the arithmetic sequence.

EXAMPLE 1   **Examples of Arithmetic Sequences**

a. The sequence whose *n*th term is $4n + 3$ is arithmetic. The common difference between consecutive terms is 4.

$$7, 11, 15, 19, \ldots, 4n + 3, \ldots \qquad \text{Begin with } n = 1.$$
$$11 - 7 = 4$$

b. The sequence whose *n*th term is $7 - 5n$ is arithmetic. The common difference between consecutive terms is -5.

$$2, -3, -8, -13, \ldots, 7 - 5n, \ldots \qquad \text{Begin with } n = 1.$$
$$-3 - 2 = -5$$

c. The sequence whose *n*th term is $\frac{1}{4}(n + 3)$ is arithmetic. The common difference between consecutive terms is $\frac{1}{4}$.

$$1, \frac{5}{4}, \frac{3}{2}, \frac{7}{4}, \ldots, \frac{n + 3}{4}, \ldots \qquad \text{Begin with } n = 1.$$
$$\tfrac{5}{4} - 1 = \tfrac{1}{4}$$

✓ *Checkpoint* 🔊 *Audio-video solution in English & Spanish at LarsonPrecalculus.com*

Write the first four terms of the arithmetic sequence whose *n*th term is $3n - 1$. Then find the common difference between consecutive terms. ■

The sequence 1, 4, 9, 16, . . . , whose *n*th term is n^2, is *not* arithmetic. The difference between the first two terms is

$$a_2 - a_1 = 4 - 1 = 3$$

but the difference between the second and third terms is

$$a_3 - a_2 = 9 - 4 = 5.$$

The nth term of an arithmetic sequence can be derived from the pattern below.

$a_1 = a_1$ 1st term

$a_2 = a_1 + d$ 2nd term

$a_3 = a_1 + 2d$ 3rd term

$a_4 = a_1 + 3d$ 4th term

$a_5 = a_1 + 4d$ 5th term

       1 less

$\vdots$

$a_n = a_1 + (n - 1)d$ nth term

       1 less

The following definition summarizes this result.

The *n*th Term of an Arithmetic Sequence

The nth term of an arithmetic sequence has the form

$$a_n = a_1 + (n - 1)d$$

where d is the common difference between consecutive terms of the sequence and a_1 is the first term.

EXAMPLE 2   **Finding the *n*th Term**

Find a formula for the nth term of the arithmetic sequence whose common difference is 3 and whose first term is 2.

Solution You know that the formula for the nth term is of the form $a_n = a_1 + (n - 1)d$. Moreover, the common difference is $d = 3$ and the first term is $a_1 = 2$, so the formula must have the form

$$a_n = 2 + 3(n - 1). \qquad \text{Substitute 2 for } a_1 \text{ and 3 for } d.$$

So, the formula for the nth term is $a_n = 3n - 1$.

✓ *Checkpoint* ◀))) *Audio-video solution in English & Spanish at LarsonPrecalculus.com*

Find a formula for the nth term of the arithmetic sequence whose common difference is 5 and whose first term is -1.

The sequence in Example 2 is as follows.

$$2, 5, 8, 11, 14, \ldots, 3n - 1, \ldots$$

The figure below shows a graph of the first 15 terms of this sequence. Notice that the points lie on a line. This makes sense because a_n is a linear function of n. In other words, the terms "arithmetic" and "linear" are closely connected.

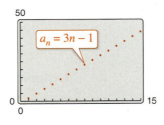

•• REMARK Another way to find a_1 in Example 3 is to use the definition of the nth term of an arithmetic sequence, as shown below.

$$a_n = a_1 + (n-1)d$$

$$a_4 = a_1 + (4-1)d$$

$$20 = a_1 + (4-1)5$$

$$20 = a_1 + 15$$

$$5 = a_1$$

EXAMPLE 3 **Writing the Terms of an Arithmetic Sequence**

The 4th term of an arithmetic sequence is 20, and the 13th term is 65. Write the first 11 terms of this sequence.

Solution You know that $a_4 = 20$ and $a_{13} = 65$. So, you must add the common difference d nine times to the 4th term to obtain the 13th term. Therefore, the 4th and 13th terms of the sequence are related by

$$a_{13} = a_4 + 9d. \qquad \text{a_4 and a_{13} are nine terms apart.}$$

Using $a_4 = 20$ and $a_{13} = 65$, you have $65 = 20 + 9d$. Solve for d to find that the common difference is $d = 5$. Use the common difference with the known term a_4 to write the other terms of the sequence.

a_1	a_2	a_3	a_4	a_5	a_6	a_7	a_8	a_9	a_{10}	$a_{11} \cdots$
5	10	15	20	25	30	35	40	45	50	$55 \cdots$

✓ *Checkpoint* ◀))) *Audio-video solution in English & Spanish at LarsonPrecalculus.com*

The 8th term of an arithmetic sequence is 25, and the 12th term is 41. Write the first 11 terms of this sequence.

When you know the nth term of an arithmetic sequence *and* you know the common difference of the sequence, you can find the $(n + 1)$th term by using the *recursion formula*

$$a_{n+1} = a_n + d. \qquad \text{Recursion formula}$$

With this formula, you can find any term of an arithmetic sequence, *provided* that you know the preceding term. For example, when you know the first term, you can find the second term. Then, knowing the second term, you can find the third term, and so on.

EXAMPLE 4 **Using a Recursion Formula**

Find the ninth term of the arithmetic sequence whose first two terms are 2 and 9.

Solution The common difference between consecutive terms of this sequence is

$$d = 9 - 2 = 7.$$

There are two ways to find the ninth term. One way is to write the first nine terms (by repeatedly adding 7).

$$2, 9, 16, 23, 30, 37, 44, 51, 58$$

Another way to find the ninth term is to first find a formula for the nth term. The common difference is $d = 7$ and the first term is $a_1 = 2$, so the formula must have the form

$$a_n = 2 + 7(n - 1). \qquad \text{Substitute 2 for a_1 and 7 for d.}$$

Therefore, a formula for the nth term is

$$a_n = 7n - 5$$

which implies that the ninth term is

$$a_9 = 7(9) - 5$$

$$= 58.$$

✓ *Checkpoint* ◀))) *Audio-video solution in English & Spanish at LarsonPrecalculus.com*

Find the 10th term of the arithmetic sequence that begins with 7 and 15.

The Sum of a Finite Arithmetic Sequence

There is a formula for the *sum* of a finite arithmetic sequence.

•••**REMARK** Note that this formula works only for *arithmetic* sequences.

> **The Sum of a Finite Arithmetic Sequence**
>
> The sum of a finite arithmetic sequence with n terms is given by $S_n = \dfrac{n}{2}(a_1 + a_n)$.

For a proof of this formula, see Proofs in Mathematics on page 639.

EXAMPLE 5   **Sum of a Finite Arithmetic Sequence**

Find the sum: $1 + 3 + 5 + 7 + 9 + 11 + 13 + 15 + 17 + 19$.

Solution   To begin, notice that the sequence is arithmetic (with a common difference of 2). Moreover, the sequence has 10 terms. So, the sum of the sequence is

$$S_n = \frac{n}{2}(a_1 + a_n) \qquad \text{\color{red}Sum of a finite arithmetic sequence}$$

$$= \frac{10}{2}(1 + 19) \qquad \text{\color{red}Substitute 10 for } n, \text{ 1 for } a_1, \text{ and 19 for } a_n.$$

$$= 5(20) = 100. \qquad \text{\color{red}Simplify.}$$

✓ *Checkpoint*))) *Audio-video solution in English & Spanish at LarsonPrecalculus.com*

Find the sum: $40 + 37 + 34 + 31 + 28 + 25 + 22$.

EXAMPLE 6   **Sum of a Finite Arithmetic Sequence**

Find the sum of the integers (a) from 1 to 100 and (b) from 1 to N.

Solution

a. The integers from 1 to 100 form an arithmetic sequence that has 100 terms. So, use the formula for the sum of a finite arithmetic sequence.

$$S_n = 1 + 2 + 3 + 4 + 5 + 6 + \cdots + 99 + 100$$

$$= \frac{n}{2}(a_1 + a_n) \qquad \text{\color{red}Sum of a finite arithmetic sequence}$$

$$= \frac{100}{2}(1 + 100) \qquad \text{\color{red}Substitute 100 for } n, \text{ 1 for } a_1, \text{ and 100 for } a_n.$$

$$= 50(101) = 5050 \qquad \text{\color{red}Simplify.}$$

b. $S_n = 1 + 2 + 3 + 4 + \cdots + N$

$$= \frac{n}{2}(a_1 + a_n) \qquad \text{\color{red}Sum of a finite arithmetic sequence}$$

$$= \frac{N}{2}(1 + N) \qquad \text{\color{red}Substitute } N \text{ for } n, \text{ 1 for } a_1, \text{ and } N \text{ for } a_n.$$

✓ *Checkpoint* ))) *Audio-video solution in English & Spanish at LarsonPrecalculus.com*

Find the sum of the integers (a) from 1 to 35 and (b) from 1 to $2N$. ■

A teacher of Carl Friedrich Gauss (1777–1855) asked him to add all the integers from 1 to 100. When Gauss returned with the correct answer after only a few moments, the teacher could only look at him in astounded silence. This is what Gauss did:

$$\begin{array}{rcccccc} S_n = & 1 & + & 2 & + & 3 & + \cdots + & 100 \\ S_n = & 100 & + & 99 & + & 98 & + \cdots + & 1 \\ \hline 2S_n = & 101 & + & 101 & + & 101 & + \cdots + & 101 \end{array}$$

$$S_n = \frac{100 \times 101}{2} = 5050.$$

Recall that the sum of the first n terms of an infinite sequence is the *nth partial sum*. The nth partial sum of an arithmetic sequence can be found by using the formula for the sum of a finite arithmetic sequence.

EXAMPLE 7 **Partial Sum of an Arithmetic Sequence**

Find the 150th partial sum of the arithmetic sequence

$$5, 16, 27, 38, 49, \ldots.$$

Solution For this arithmetic sequence, $a_1 = 5$ and $d = 16 - 5 = 11$. So,

$$a_n = 5 + 11(n - 1)$$

and the nth term is

$$a_n = 11n - 6.$$

Therefore, $a_{150} = 11(150) - 6 = 1644$, and the sum of the first 150 terms is

$$S_{150} = \frac{n}{2}(a_1 + a_{150}) \qquad \textcolor{red}{\text{nth partial sum formula}}$$

$$= \frac{150}{2}(5 + 1644) \qquad \textcolor{red}{\text{Substitute 150 for n, 5 for a_1, and 1644 for a_{150}.}}$$

$$= 75(1649) \qquad \textcolor{red}{\text{Simplify.}}$$

$$= 123{,}675. \qquad \textcolor{red}{\text{nth partial sum}}$$

✓ *Checkpoint*  *Audio-video solution in English & Spanish at LarsonPrecalculus.com*

Find the 120th partial sum of the arithmetic sequence

$$6, 12, 18, 24, 30, \ldots.$$

EXAMPLE 8 **Partial Sum of an Arithmetic Sequence**

Find the 16th partial sum of the arithmetic sequence

$$100, 95, 90, 85, 80, \ldots.$$

Solution For this arithmetic sequence, $a_1 = 100$ and $d = 95 - 100 = -5$. So,

$$a_n = 100 + (-5)(n - 1)$$

and the nth term is

$$a_n = -5n + 105.$$

Therefore, $a_{16} = -5(16) + 105 = 25$, and the sum of the first 16 terms is

$$S_{16} = \frac{n}{2}(a_1 + a_{16}) \qquad \textcolor{red}{\text{nth partial sum formula}}$$

$$= \frac{16}{2}(100 + 25) \qquad \textcolor{red}{\text{Substitute 16 for n, 100 for a_1, and 25 for a_{16}.}}$$

$$= 8(125) \qquad \textcolor{red}{\text{Simplify.}}$$

$$= 1000. \qquad \textcolor{red}{\text{nth partial sum}}$$

✓ *Checkpoint*  *Audio-video solution in English & Spanish at LarsonPrecalculus.com*

Find the 30th partial sum of the arithmetic sequence

$$78, 76, 74, 72, 70, \ldots.$$

Application

EXAMPLE 9   **Total Sales**

See LarsonPrecalculus.com for an interactive version of this type of example.

A small business sells $10,000 worth of skin care products during its first year. The owner of the business has set a goal of increasing annual sales by $7500 each year for 9 years. Assuming that this goal is met, find the total sales during the first 10 years this business is in operation.

Solution   When the goal is met the annual sales form an arithmetic sequence with

$$a_1 = 10,000 \quad \text{and} \quad d = 7500.$$

So,

$$a_n = 10,000 + 7500(n - 1)$$

and the nth term of the sequence is

$$a_n = 7500n + 2500.$$

Therefore, the 10th term of the sequence is

$$a_{10} = 7500(10) + 2500$$

$$= 77,500. \qquad \text{See figure.}$$

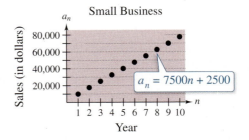

The sum of the first 10 terms of the sequence is

$$S_{10} = \frac{n}{2}(a_1 + a_{10}) \qquad \textit{n}\text{th partial sum formula}$$

$$= \frac{10}{2}(10,000 + 77,500) \qquad \text{Substitute 10 for } n, 10,000 \text{ for } a_1, \text{ and } 77,500 \text{ for } a_{10}.$$

$$= 5(87,500) \qquad \text{Simplify.}$$

$$= 437,500. \qquad \text{Multiply.}$$

So, the total sales for the first 10 years will be $437,500.

 Checkpoint 🔊))) *Audio-video solution in English & Spanish at LarsonPrecalculus.com*

A company sells $160,000 worth of printing paper during its first year. The sales manager has set a goal of increasing annual sales of printing paper by $20,000 each year for 9 years. Assuming that this goal is met, find the total sales of printing paper during the first 10 years this company is in operation.

Summarize   (Section 8.2)

1. State the definition of an arithmetic sequence *(page 572)*, and state the formula for the nth term of an arithmetic sequence *(page 573)*. For examples of recognizing, writing, and finding the nth terms of arithmetic sequences, see Examples 1–4.

2. State the formula for the sum of a finite arithmetic sequence and explain how to use it to find the nth partial sum of an arithmetic sequence *(pages 575 and 576)*. For examples of finding sums of arithmetic sequences, see Examples 5–8.

3. Describe an example of how to use an arithmetic sequence to model and solve a real-life problem *(page 577, Example 9)*.

8.2 Exercises

See **CalcChat.com** for tutorial help and worked-out solutions to odd-numbered exercises.

Vocabulary: Fill in the blanks.

1. A sequence is _____ when the differences between consecutive terms are the same. This difference is the _____ difference.

2. The nth term of an arithmetic sequence has the form $a_n = $ _____.

3. When you know the nth term of an arithmetic sequence *and* you know the common difference of the sequence, you can find the $(n + 1)$th term by using the _____ formula $a_{n+1} = a_n + d$.

4. The formula $S_n = \dfrac{n}{2}(a_1 + a_n)$ gives the sum of a _____ _____ _____ with n terms.

Skills and Applications

 Determining Whether a Sequence Is Arithmetic In Exercises 5–12, determine whether the sequence is arithmetic. If so, find the common difference.

5. $1, 2, 4, 8, 16, \ldots$ 6. $4, 9, 14, 19, 24, \ldots$

7. $10, 8, 6, 4, 2, \ldots$ 8. $80, 40, 20, 10, 5, \ldots$

9. $\frac{5}{4}, \frac{3}{2}, \frac{7}{4}, 2, \frac{9}{4}, \ldots$ 10. $6.6, 5.9, 5.2, 4.5, 3.8, \ldots$

11. $1^2, 2^2, 3^2, 4^2, 5^2, \ldots$

12. $\ln 1, \ln 2, \ln 4, \ln 8, \ln 16, \ldots$

Writing the Terms of a Sequence In Exercises 13–20, write the first five terms of the sequence. Determine whether the sequence is arithmetic. If so, find the common difference. (Assume that n begins with 1.)

13. $a_n = 5 + 3n$ 14. $a_n = 100 - 3n$

15. $a_n = 3 - 4(n - 2)$ 16. $a_n = 1 + (n - 1)n$

17. $a_n = (-1)^n$ 18. $a_n = n - (-1)^n$

19. $a_n = (2^n)n$ 20. $a_n = \dfrac{3(-1)^n}{n}$

 **Finding the nth Term** In Exercises 21–30, find a formula for a_n for the arithmetic sequence.

21. $a_1 = 1, d = 3$ 22. $a_1 = 15, d = 4$

23. $a_1 = 100, d = -8$ 24. $a_1 = 0, d = -\frac{2}{3}$

25. $4, \frac{3}{2}, -1, -\frac{7}{2}, \ldots$ 26. $10, 5, 0, -5, -10, \ldots$

27. $a_1 = 5, a_4 = 15$ 28. $a_1 = -4, a_5 = 16$

29. $a_3 = 94, a_6 = 103$ 30. $a_5 = 190, a_{10} = 115$

 Writing the Terms of an Arithmetic Sequence In Exercises 31–36, write the first five terms of the arithmetic sequence.

31. $a_1 = 5, d = 6$ 32. $a_1 = 5, d = -\frac{3}{4}$

33. $a_1 = 2, a_{12} = -64$ 34. $a_4 = 16, a_{10} = 46$

35. $a_8 = 26, a_{12} = 42$ 36. $a_3 = 19, a_{15} = -1.7$

Writing the Terms of an Arithmetic Sequence In Exercises 37–40, write the first five terms of the arithmetic sequence defined recursively.

37. $a_1 = 15, a_{n+1} = a_n + 4$

38. $a_1 = 200, a_{n+1} = a_n - 10$

39. $a_5 = 7, a_{n+1} = a_n - 2$

40. $a_3 = 0.5, a_{n+1} = a_n + 0.75$

 Using a Recursion Formula In Exercises 41–44, the first two terms of the arithmetic sequence are given. Find the missing term.

41. $a_1 = 5, a_2 = -1, a_{10} = $ ▨

42. $a_1 = 3, a_2 = 13, a_9 = $ ▨

43. $a_1 = \dfrac{1}{8}, a_2 = \dfrac{3}{4}, a_7 = $ ▨

44. $a_1 = -0.7, a_2 = -13.8, a_8 = $ ▨

 Sum of a Finite Arithmetic Sequence In Exercises 45–50, find the sum of the finite arithmetic sequence.

45. $2 + 4 + 6 + 8 + 10 + 12 + 14 + 16 + 18 + 20$

46. $1 + 4 + 7 + 10 + 13 + 16 + 19$

47. $-1 + (-3) + (-5) + (-7) + (-9)$

48. $-5 + (-3) + (-1) + 1 + 3 + 5$

49. Sum of the first 100 positive odd integers

50. Sum of the integers from -100 to 30

 Partial Sum of an Arithmetic Sequence In Exercises 51–54, find the nth partial sum of the arithmetic sequence for the given value of n.

51. $8, 20, 32, 44, \ldots, \quad n = 50$

52. $-6, -2, 2, 6, \ldots, \quad n = 100$

53. $0, -9, -18, -27, \ldots, \quad n = 40$

54. $75, 70, 65, 60, \ldots, \quad n = 25$

Finding a Sum In Exercises 55–60, find the partial sum.

55. $\displaystyle\sum_{n=1}^{50} n$

56. $\displaystyle\sum_{n=51}^{100} 7n$

57. $\displaystyle\sum_{n=1}^{500} (n + 8)$

58. $\displaystyle\sum_{n=1}^{250} (1000 - n)$

59. $\displaystyle\sum_{n=1}^{100} (-6n + 20)$

60. $\displaystyle\sum_{n=1}^{75} (12n - 9)$

Matching an Arithmetic Sequence with Its Graph In Exercises 61–64, match the arithmetic sequence with its graph. [The graphs are labeled (a)–(d).]

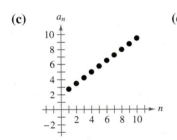

61. $a_n = -\frac{3}{4}n + 8$

62. $a_n = 3n - 5$

63. $a_n = 2 + \frac{3}{4}n$

64. $a_n = 25 - 3n$

Graphing the Terms of a Sequence In Exercises 65–68, use a graphing utility to graph the first 10 terms of the sequence. (Assume that n begins with 1.)

65. $a_n = 15 - \frac{3}{2}n$

66. $a_n = -5 + 2n$

67. $a_n = 0.2n + 3$

68. $a_n = -0.3n + 8$

Job Offer In Exercises 69 and 70, consider a job offer with the given starting salary and annual raise. (a) Determine the salary during the sixth year of employment. (b) Determine the total compensation from the company through six full years of employment.

	Starting Salary	Annual Raise
69.	$32,500	$1500
70.	$36,800	$1750

71. Seating Capacity Determine the seating capacity of an auditorium with 36 rows of seats when there are 15 seats in the first row, 18 seats in the second row, 21 seats in the third row, and so on.

72. Brick Pattern A triangular brick wall is made by cutting some bricks in half to use in the first column of every other row (see figure). The wall has 28 rows. The top row is one-half brick wide and the bottom row is 14 bricks wide. How many bricks are in the finished wall?

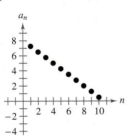

73. Falling Object

An object with negligible air resistance is dropped from the top of the Willis Tower in Chicago at a height of 1451 feet. During the first second of fall, the object falls 16 feet; during the second second, it falls 48 feet; during the third second, it falls 80 feet; during the fourth second, it falls 112 feet. Assuming this pattern continues, how many feet does the object fall in the first 7 seconds after it is dropped?

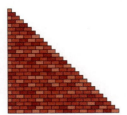

74. Prize Money A county fair is holding a baked goods competition in which the top eight bakers receive cash prizes. First place receives $200, second place receives $175, third place receives $150, and so on.

(a) Write the nth term (a_n) of a sequence that represents the cash prize received in terms of the place n the baked good is awarded.

(b) Find the total amount of prize money awarded at the competition.

75. Total Sales An entrepreneur sells $15,000 worth of sports memorabilia during one year and sets a goal of increasing annual sales by $5000 each year for the next 9 years. Assuming that the entrepreneur meets this goal, find the total sales during the first 10 years of this business. What kinds of economic factors could prevent the business from meeting its goals?

76. Borrowing Money You borrow $5000 from your parents to purchase a used car. The arrangements of the loan are such that you make payments of $250 per month toward the balance plus 1% interest on the unpaid balance from the previous month.

(a) Find the first year's monthly payments and the unpaid balance after each month.

(b) Find the total amount of interest paid over the term of the loan.

77. Business The table shows the net numbers of new stores opened by H&M from 2011 through 2015. (*Source: H&M Hennes & Mauritz AB*)

DATA	Year	New Stores
	2011	266
	2012	304
	2013	356
	2014	379
	2015	413

Spreadsheet at LarsonPrecalculus.com

(a) Construct a bar graph showing the annual net numbers of new stores opened by H&M from 2011 through 2015.

(b) Find the nth term (a_n) of an arithmetic sequence that approximates the data. Let n represent the year, with $n = 1$ corresponding to 2011. (*Hint:* Use the average change per year for d.)

(c) Use a graphing utility to graph the terms of the finite sequence you found in part (b).

(d) Use summation notation to represent the *total* number of new stores opened from 2011 through 2015. Use this sum to approximate the total number of new stores opened during these years.

78. Business In Exercise 77, there are a total number of 2206 stores at the end of 2010. Write the terms of a sequence that represents the total number of stores at the end of each year from 2011 through 2015. Is the sequence approximately arithmetic? Explain.

Exploration

True or False? In Exercises 79 and 80, determine whether the statement is true or false. Justify your answer.

79. Given an arithmetic sequence for which only the first two terms are known, it is possible to find the nth term.

80. When the first term, the nth term, and n are known for an arithmetic sequence, you have enough information to find the nth partial sum of the sequence.

81. Comparing Graphs of a Sequence and a Line

(a) Graph the first 10 terms of the arithmetic sequence $a_n = 2 + 3n$.

(b) Graph the equation of the line $y = 3x + 2$.

(c) Discuss any differences between the graph of $a_n = 2 + 3n$ and the graph of $y = 3x + 2$.

(d) Compare the slope of the line in part (b) with the common difference of the sequence in part (a). What can you conclude about the slope of a line and the common difference of an arithmetic sequence?

82. Writing Describe two ways to use the first two terms of an arithmetic sequence to find the 13th term.

Finding the Terms of a Sequence In Exercises 83 and 84, find the first 10 terms of the sequence.

83. $a_1 = x,\, d = 2x$

84. $a_1 = -y,\, d = 5y$

85. Error Analysis Describe the error in finding the sum of the first 50 odd integers.

$$S_n = \frac{n}{2}(a_1 + a_n) = \frac{50}{2}(1 + 101) = 2550 \quad \times$$

86. 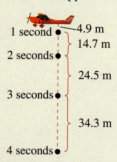 **HOW DO YOU SEE IT?** A steel ball with negligible air resistance is dropped from an airplane. The figure shows the distance that the ball falls during each of the first four seconds after it is dropped.

1 second •⎴ 4.9 m
⎱ 14.7 m
2 seconds •
⎱ 24.5 m
3 seconds •
⎱ 34.3 m
4 seconds •

(a) Describe a pattern in the distances shown. Explain why the distances form a finite arithmetic sequence.

(b) Assume the pattern described in part (a) continues. Describe the steps and formulas involved in using the sum of a finite sequence to find the total distance the ball falls in n seconds, where n is a whole number.

87. Pattern Recognition

(a) Compute the following sums of consecutive positive odd integers.

$1 + 3 = $ ▪

$1 + 3 + 5 = $ ▪

$1 + 3 + 5 + 7 = $ ▪

$1 + 3 + 5 + 7 + 9 = $ ▪

$1 + 3 + 5 + 7 + 9 + 11 = $ ▪

(b) Use the sums in part (a) to make a conjecture about the sums of consecutive positive odd integers. Check your conjecture for the sum

$1 + 3 + 5 + 7 + 9 + 11 + 13 = $ ▪.

(c) Verify your conjecture algebraically.

Project: Net Sales To work an extended application analyzing the net sales for Dollar Tree from 2001 through 2014, visit the textbook's website at *LarsonPrecalculus.com*. (*Source: Dollar Tree, Inc.*)

8.3 Geometric Sequences and Series

- Recognize, write, and find the *n*th terms of geometric sequences.
- Find the sum of a finite geometric sequence.
- Find the sum of an infinite geometric series.
- Use geometric sequences to model and solve real-life problems.

Geometric Sequences

In Section 8.2, you learned that a sequence whose consecutive terms have a common *difference* is an arithmetic sequence. In this section, you will study another important type of sequence called a **geometric sequence**. Consecutive terms of a geometric sequence have a common *ratio*.

Geometric sequences can help you model and solve real-life problems. For example, in Exercise 84 on page 588, you will use a geometric sequence to model the population of Argentina from 2009 through 2015.

REMARK Be sure you understand that a sequence such as $1, 4, 9, 16, \ldots$, whose *n*th term is n^2, is *not* geometric. The ratio of the second term to the first term is

$$\frac{a_2}{a_1} = \frac{4}{1} = 4$$

but the ratio of the third term to the second term is

$$\frac{a_3}{a_2} = \frac{9}{4}.$$

> ### Definition of Geometric Sequence
>
> A sequence is **geometric** when the ratios of consecutive terms are the same. So, the sequence $a_1, a_2, a_3, a_4, \ldots, a_n, \ldots$ is geometric when there is a number r such that
>
> $$\frac{a_2}{a_1} = \frac{a_3}{a_2} = \frac{a_4}{a_3} = \cdots = r, \quad r \neq 0.$$
>
> The number r is the **common ratio** of the geometric sequence.

EXAMPLE 1 **Examples of Geometric Sequences**

a. The sequence whose *n*th term is 2^n is geometric. The common ratio of consecutive terms is 2.

$$2, 4, 8, 16, \ldots, 2^n, \ldots \qquad \text{Begin with } n = 1.$$
$$\underbrace{\qquad}_{\frac{4}{2} = 2}$$

b. The sequence whose *n*th term is $4(3^n)$ is geometric. The common ratio of consecutive terms is 3.

$$12, 36, 108, 324, \ldots, 4(3^n), \ldots \qquad \text{Begin with } n = 1.$$
$$\underbrace{\qquad}_{\frac{36}{12} = 3}$$

c. The sequence whose *n*th term is $\left(-\frac{1}{3}\right)^n$ is geometric. The common ratio of consecutive terms is $-\frac{1}{3}$.

$$-\frac{1}{3}, \frac{1}{9}, -\frac{1}{27}, \frac{1}{81}, \ldots, \left(-\frac{1}{3}\right)^n, \ldots \qquad \text{Begin with } n = 1.$$
$$\underbrace{\qquad}_{\frac{1/9}{-1/3} = -\frac{1}{3}}$$

✓ *Checkpoint* 🔊)) *Audio-video solution in English & Spanish at LarsonPrecalculus.com*

Write the first four terms of the geometric sequence whose *n*th term is $6(-2)^n$. Then find the common ratio of the consecutive terms. ■

In Example 1, notice that each of the geometric sequences has an *n*th term that is of the form ar^n, where the common ratio of the sequence is r. A geometric sequence may be thought of as an exponential function whose domain is the set of natural numbers.

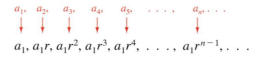

The *n*th Term of a Geometric Sequence

The *n*th term of a geometric sequence has the form

$$a_n = a_1 r^{n-1}$$

where r is the common ratio of consecutive terms of the sequence. So, every geometric sequence can be written in the form below.

$$a_1, \quad a_2, \quad a_3, \quad a_4, \quad a_5, \quad \ldots, \quad a_n, \ldots$$
$$a_1, a_1 r, a_1 r^2, a_1 r^3, a_1 r^4, \ldots, a_1 r^{n-1}, \ldots$$

When you know the *n*th term of a geometric sequence, multiply by r to find the $(n + 1)$th term. That is, $a_{n+1} = a_n r$.

EXAMPLE 2   **Writing the Terms of a Geometric Sequence**

Write the first five terms of the geometric sequence whose first term is $a_1 = 3$ and whose common ratio is $r = 2$. Then graph the terms on a set of coordinate axes.

Solution   Starting with 3, repeatedly multiply by 2 to obtain the terms below.

$a_1 = 3$	1st term	$a_4 = 3(2^3) = 24$	4th term
$a_2 = 3(2^1) = 6$	2nd term	$a_5 = 3(2^4) = 48$	5th term
$a_3 = 3(2^2) = 12$	3rd term		

Figure 8.1 shows the graph of the first five terms of this geometric sequence.

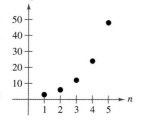

Figure 8.1

✓ *Checkpoint*  *Audio-video solution in English & Spanish at LarsonPrecalculus.com*

Write the first five terms of the geometric sequence whose first term is $a_1 = 2$ and whose common ratio is $r = 4$. Then graph the terms on a set of coordinate axes.

EXAMPLE 3   **Finding a Term of a Geometric Sequence**

Find the 15th term of the geometric sequence whose first term is 20 and whose common ratio is 1.05.

Algebraic Solution

$a_n = a_1 r^{n-1}$ Formula for *n*th term of a geometric sequence

$a_{15} = 20(1.05)^{15-1}$ Substitute 20 for a_1, 1.05 for r, and 15 for n.

≈ 39.60 Use a calculator.

Numerical Solution

For this sequence, $r = 1.05$ and $a_1 = 20$. So, $a_n = 20(1.05)^{n-1}$. Use a graphing utility to create a table that shows the terms of the sequence.

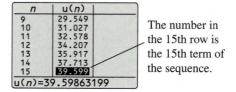

The number in the 15th row is the 15th term of the sequence.

So, $a_{15} \approx 39.60$.

✓ *Checkpoint*  *Audio-video solution in English & Spanish at LarsonPrecalculus.com*

Find the 12th term of the geometric sequence whose first term is 14 and whose common ratio is 1.2.

| EXAMPLE 4 | **Writing the *n*th Term of a Geometric Sequence** |

Find a formula for the *n*th term of the geometric sequence

   5, 15, 45, ...

What is the 12th term of the sequence?

Solution The common ratio of this sequence is $r = 15/5 = 3$. The first term is $a_1 = 5$, so the formula for the *n*th term is

$$a_n = a_1 r^{n-1}$$
$$= 5(3)^{n-1}.$$

Use the formula for a_n to find the 12th term of the sequence.

$$a_{12} = 5(3)^{12-1}$$ Substitute 12 for *n*.

$$= 5(177,147)$$ Use a calculator.

$$= 885,735.$$ Multiply.

✓ *Checkpoint* ◀))) *Audio-video solution in English & Spanish at LarsonPrecalculus.com*

Find a formula for the *n*th term of the geometric sequence

   4, 20, 100, ...

What is the 12th term of the sequence?

When you know *any* two terms of a geometric sequence, you can use that information to find *any other* term of the sequence.

▷ **ALGEBRA HELP**
Remember that *r* is the common ratio of consecutive terms of a geometric sequence. So, in Example 5

$$a_{10} = a_1 r^9$$
$$= a_1 \cdot r \cdot r \cdot r \cdot r^6$$
$$= a_1 \cdot \frac{a_2}{a_1} \cdot \frac{a_3}{a_2} \cdot \frac{a_4}{a_3} \cdot r^6$$
$$= a_4 r^6.$$

| EXAMPLE 5 | **Finding a Term of a Geometric Sequence** |

The 4th term of a geometric sequence is 125, and the 10th term is 125/64. Find the 14th term. (Assume that the terms of the sequence are positive.)

Solution The 10th term is related to the 4th term by the equation

$$a_{10} = a_4 r^6.$$ Multiply fourth term by r^{10-4}.

Use $a_{10} = 125/64$ and $a_4 = 125$ to solve for *r*.

$$\frac{125}{64} = 125r^6$$ Substitute $\frac{125}{64}$ for a_{10} and 125 for a_4.

$$\frac{1}{64} = r^6$$ Divide each side by 125.

$$\frac{1}{2} = r$$ Take the sixth root of each side.

Multiply the 10th term by $r^{14-10} = r^4$ to obtain the 14th term.

$$a_{14} = a_{10}r^4 = \frac{125}{64}\left(\frac{1}{2}\right)^4 = \frac{125}{64}\left(\frac{1}{16}\right) = \frac{125}{1024}$$

✓ *Checkpoint* ◀))) *Audio-video solution in English & Spanish at LarsonPrecalculus.com*

The second term of a geometric sequence is 6, and the fifth term is 81/4. Find the eighth term. (Assume that the terms of the sequence are positive.)

The Sum of a Finite Geometric Sequence

The formula for the sum of a *finite* geometric sequence is as follows.

The Sum of a Finite Geometric Sequence

The sum of the finite geometric sequence

$$a_1, a_1 r, a_1 r^2, a_1 r^3, a_1 r^4, \ldots, a_1 r^{n-1}$$

with common ratio $r \neq 1$ is given by $S_n = \sum_{i=1}^{n} a_1 r^{i-1} = a_1 \left(\dfrac{1 - r^n}{1 - r} \right)$.

For a proof of this formula for the sum of a finite geometric sequence, see Proofs in Mathematics on page 639.

EXAMPLE 6 **Sum of a Finite Geometric Sequence**

Find the sum $\displaystyle\sum_{i=1}^{12} 4(0.3)^{i-1}$.

Solution You have

$$\sum_{i=1}^{12} 4(0.3)^{i-1} = 4(0.3)^0 + 4(0.3)^1 + 4(0.3)^2 + \cdots + 4(0.3)^{11}.$$

Using $a_1 = 4$, $r = 0.3$, and $n = 12$, apply the formula for the sum of a finite geometric sequence.

$$S_n = a_1 \left(\frac{1 - r^n}{1 - r} \right) \qquad \text{Sum of a finite geometric sequence}$$

$$\sum_{i=1}^{12} 4(0.3)^{i-1} = 4 \left[\frac{1 - (0.3)^{12}}{1 - 0.3} \right] \qquad \text{Substitute 4 for } a_1, 0.3 \text{ for } r, \text{ and 12 for } n.$$

$$\approx 5.714 \qquad \text{Use a calculator.}$$

✓ *Checkpoint* 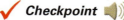 *Audio-video solution in English & Spanish at LarsonPrecalculus.com*

Find the sum $\displaystyle\sum_{i=1}^{10} 2(0.25)^{i-1}$.

When using the formula for the sum of a finite geometric sequence, make sure that the sum is of the form

$$\sum_{i=1}^{n} a_1 r^{i-1}. \qquad \text{Exponent for } r \text{ is } i - 1.$$

For a sum that is not of this form, you must rewrite the sum before applying the formula. For example, the sum $\displaystyle\sum_{i=1}^{12} 4(0.3)^i$ is evaluated as follows.

$$\sum_{i=1}^{12} 4(0.3)^i = \sum_{i=1}^{12} 4[(0.3)(0.3)^{i-1}] \qquad \text{Property of exponents}$$

$$= \sum_{i=1}^{12} 4(0.3)(0.3)^{i-1} \qquad \text{Associative Property}$$

$$= 4(0.3) \left[\frac{1 - (0.3)^{12}}{1 - 0.3} \right] \qquad a_1 = 4(0.3), r = 0.3, n = 12$$

$$\approx 1.714$$

Geometric Series

The sum of the terms of an infinite geometric *sequence* is called an **infinite geometric series** or simply a **geometric series.**

The formula for the sum of a *finite geometric sequence* can, depending on the value of r, be extended to produce a formula for the sum of an *infinite geometric series.* Specifically, if the common ratio r has the property that $|r| < 1$, then it can be shown that r^n approaches zero as n increases without bound. Consequently,

$$a_1\left(\frac{1 - r^n}{1 - r}\right) \rightarrow a_1\left(\frac{1 - 0}{1 - r}\right) \quad \text{as} \quad n \rightarrow \infty.$$

The following summarizes this result.

The Sum of an Infinite Geometric Series

If $|r| < 1$, then the infinite geometric series

$$a_1 + a_1 r + a_1 r^2 + a_1 r^3 + \cdots + a_1 r^{n-1} + \cdots$$

has the sum

$$S = \sum_{i=0}^{\infty} a_1 r^i = \frac{a_1}{1 - r}.$$

Note that when $|r| \geq 1$, the series does not have a sum.

EXAMPLE 7   **Finding the Sum of an Infinite Geometric Series**

Find each sum.

a. $\displaystyle\sum_{n=0}^{\infty} 4(0.6)^n$

b. $3 + 0.3 + 0.03 + 0.003 + \cdots$

Solution

a. $\displaystyle\sum_{n=0}^{\infty} 4(0.6)^n = 4 + 4(0.6) + 4(0.6)^2 + 4(0.6)^3 + \cdots + 4.(0.6)^n + \cdots$

$$= \frac{4}{1 - 0.6} \qquad\qquad \frac{a_1}{1 - r}$$

$$= 10$$

b. $3 + 0.3 + 0.03 + 0.003 + \cdots = 3 + 3(0.1) + 3(0.1)^2 + 3(0.1)^3 + \cdots$

$$= \frac{3}{1 - 0.1} \qquad\qquad \frac{a_1}{1 - r}$$

$$= \frac{10}{3}$$

$$\approx 3.33$$

✓ *Checkpoint* *Audio-video solution in English & Spanish at LarsonPrecalculus.com*

Find each sum.

a. $\displaystyle\sum_{n=0}^{\infty} 5(0.5)^n$

b. $5 + 1 + 0.2 + 0.04 + \cdots$

Application

EXAMPLE 8 **Increasing Annuity**

See LarsonPrecalculus.com for an interactive version of this type of example.

An investor deposits $50 on the first day of each month in an account that pays 3% interest, compounded monthly. What is the balance at the end of 2 years? (This type of investment plan is called an **increasing annuity.**)

Solution To find the balance in the account after 24 months, consider each of the 24 deposits separately. The first deposit will gain interest for 24 months, and its balance will be

$$A_{24} = 50\left(1 + \frac{0.03}{12}\right)^{24}$$

$$= 50(1.0025)^{24}.$$

The second deposit will gain interest for 23 months, and its balance will be

$$A_{23} = 50\left(1 + \frac{0.03}{12}\right)^{23}$$

$$= 50(1.0025)^{23}.$$

The last deposit will gain interest for only 1 month, and its balance will be

$$A_1 = 50\left(1 + \frac{0.03}{12}\right)^{1}$$

$$= 50(1.0025).$$

The total balance in the annuity will be the sum of the balances of the 24 deposits. Using the formula for the sum of a finite geometric sequence, with $A_1 = 50(1.0025)$, $r = 1.0025$, and $n = 24$, you have

$$S_n = A_1\left(\frac{1 - r^n}{1 - r}\right) \qquad \text{Sum of a finite geometric sequence}$$

$$S_{24} = 50(1.0025)\left[\frac{1 - (1.0025)^{24}}{1 - 1.0025}\right] \qquad \text{Substitute } 50(1.0025) \text{ for } A_1, 1.0025 \text{ for } r, \text{ and } 24 \text{ for } n.$$

$$\approx \$1238.23. \qquad \text{Use a calculator.}$$

✓ *Checkpoint* *Audio-video solution in English & Spanish at LarsonPrecalculus.com*

An investor deposits $70 on the first day of each month in an account that pays 2% interest, compounded monthly. What is the balance at the end of 4 years? ■

> **REMARK** Recall from Section 5.1 that the formula for compound interest (for n compoundings per year) is
>
> $$A = P\left(1 + \frac{r}{n}\right)^{nt}.$$
>
> So, in Example 8, $50 is the principal P, 0.03 is the annual interest rate r, 12 is the number n of compoundings per year, and 2 is the time t in years. When you substitute these values into the formula, you obtain
>
> $$A = 50\left(1 + \frac{0.03}{12}\right)^{12(2)}$$
>
> $$= 50\left(1 + \frac{0.03}{12}\right)^{24}.$$

Summarize (Section 8.3)

1. State the definition of a geometric sequence *(page 581)* and state the formula for the nth term of a geometric sequence *(page 582)*. For examples of recognizing, writing, and finding the nth terms of geometric sequences, see Examples 1–5.

2. State the formula for the sum of a finite geometric sequence *(page 584)*. For an example of finding the sum of a finite geometric sequence, see Example 6.

3. State the formula for the sum of an infinite geometric series *(page 585)*. For an example of finding the sums of infinite geometric series, see Example 7.

4. Describe an example of how to use a geometric sequence to model and solve a real-life problem *(page 586, Example 8)*.

8.3 Exercises

See CalcChat.com for tutorial help and worked-out solutions to odd-numbered exercises.

Vocabulary: Fill in the blanks.

1. A sequence is _____ when the ratios of consecutive terms are the same. This ratio is the _____ ratio.
2. The _____ term of a geometric sequence has the form $a_n =$ _____.
3. The sum of a finite geometric sequence with common ratio $r \neq 1$ is given by $S_n =$ _____.
4. The sum of the terms of an infinite geometric sequence is called a _____ _____.

Skills and Applications

 Determining Whether a Sequence Is Geometric In Exercises 5–12, determine whether the sequence is geometric. If so, find the common ratio.

5. $3, 6, 12, 24, \ldots$
6. $5, 10, 15, 20, \ldots$
7. $\frac{1}{27}, \frac{1}{9}, \frac{1}{3}, 1, \ldots$
8. $27, -9, 3, -1, \ldots$
9. $1, \frac{1}{2}, \frac{1}{3}, \frac{1}{4}, \ldots$
10. $5, 1, 0.2, 0.04, \ldots$
11. $1, -\sqrt{7}, 7, -7\sqrt{7}, \ldots$
12. $2, \dfrac{4}{\sqrt{3}}, \dfrac{8}{3}, \dfrac{16}{3\sqrt{3}}, \ldots$

 Writing the Terms of a Geometric Sequence In Exercises 13–22, write the first five terms of the geometric sequence.

13. $a_1 = 4, r - 3$
14. $a_1 = 7, r = 4$
15. $a_1 = 1, r = \frac{1}{2}$
16. $a_1 = 6, r = -\frac{1}{4}$
17. $a_1 = 1, r = e$
18. $a_1 = 2, r = \pi$
19. $a_1 = 3, r = \sqrt{5}$
20. $a_1 = 4, r = -1/\sqrt{2}$
21. $a_1 = 2, r = 3x$
22. $a_1 = 4, r = x/5$

 Finding a Term of a Geometric Sequence In Exercises 23–32, write an expression for the nth term of the geometric sequence. Then find the missing term.

23. $a_1 = 4, r = \frac{1}{2}, a_{10} = $ ▨
24. $a_1 = 5, r = \frac{7}{2}, a_8 = $ ▨
25. $a_1 = 6, r = -\frac{1}{3}, a_{12} = $ ▨
26. $a_1 = 64, r = -\frac{1}{4}, a_{10} = $ ▨
27. $a_1 = 100, r = e^x, a_9 = $ ▨
28. $a_1 = 1, r = e^{-x}, a_4 = $ ▨
29. $a_1 = 1, r = \sqrt{2}, a_{12} = $ ▨
30. $a_1 = 1, r = \sqrt{3}, a_8 = $ ▨
31. $a_1 = 500, r = 1.02, a_{40} = $ ▨
32. $a_1 = 1000, r = 1.005, a_{60} = $ ▨

 Writing the nth Term of a Geometric Sequence In Exercises 33–38, find a formula for the nth term of the sequence.

33. $64, 32, 16, \ldots$
34. $81, 27, 9, \ldots$
35. $9, 18, 36, \ldots$
36. $5, -10, 20, \ldots$
37. $6, -9, \frac{27}{2}, \ldots$
38. $80, -40, 20, \ldots$

 Finding a Term of a Geometric Sequence In Exercises 39–46, find the specified term of the geometric sequence.

39. 8th term: $6, 18, 54, \ldots$
40. 7th term: $5, 20, 80, \ldots$
41. 9th term: $\frac{1}{3}, -\frac{1}{6}, \frac{1}{12}, \ldots$
42. 8th term: $\frac{3}{2}, -1, \frac{2}{3}, \ldots$
43. a_3: $a_1 = 16, a_4 = \frac{27}{4}$
44. a_1: $a_2 = 3, a_5 = \frac{3}{64}$
45. a_6: $a_4 = -18, a_7 = \frac{2}{3}$
46. a_5: $a_2 = 2, a_3 = -\sqrt{2}$

Matching a Geometric Sequence with Its Graph
In Exercises 47–50, match the geometric sequence with its graph. [The graphs are labeled (a), (b), (c), and (d).]

(a)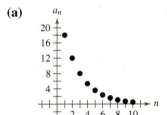
(b)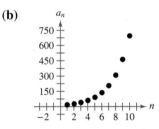

(c)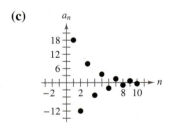
(d)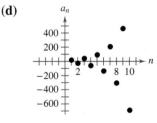

47. $a_n = 18\left(\frac{2}{3}\right)^{n-1}$
48. $a_n = 18\left(-\frac{2}{3}\right)^{n-1}$
49. $a_n = 18\left(\frac{3}{2}\right)^{n-1}$
50. $a_n = 18\left(-\frac{3}{2}\right)^{n-1}$

⚡ Graphing the Terms of a Sequence In Exercises 51–54, use a graphing utility to graph the first 10 terms of the sequence.

51. $a_n = 14(1.4)^{n-1}$
52. $a_n = 18(0.7)^{n-1}$
53. $a_n = 8(-0.3)^{n-1}$
54. $a_n = 11(-1.9)^{n-1}$

Sum of a Finite Geometric Sequence
In Exercises 55–64, find the sum of the finite geometric sequence.

55. $\sum_{n=1}^{7} 4^{n-1}$

56. $\sum_{n=1}^{10} \left(\frac{3}{2}\right)^{n-1}$

57. $\sum_{n=1}^{6} (-7)^{n-1}$

58. $\sum_{n=1}^{8} 5\left(-\frac{5}{2}\right)^{n-1}$

59. $\sum_{n=0}^{20} 3\left(\frac{3}{2}\right)^{n}$

60. $\sum_{n=0}^{40} 5\left(\frac{3}{5}\right)^{n}$

61. $\sum_{n=0}^{5} 200(1.05)^{n}$

62. $\sum_{n=0}^{6} 500(1.04)^{n}$

63. $\sum_{n=0}^{40} 2\left(-\frac{1}{4}\right)^{n}$

64. $\sum_{n=0}^{50} 10\left(\frac{2}{3}\right)^{n-1}$

Using Summation Notation   In Exercises 65–68, use summation notation to write the sum.

65. $10 + 30 + 90 + \cdots + 7290$

66. $15 - 3 + \frac{3}{5} - \cdots - \frac{3}{625}$

67. $0.1 + 0.4 + 1.6 + \cdots + 102.4$

68. $32 + 24 + 18 + 13.5 + 10.125$

Sum of an Infinite Geometric Series
In Exercises 69–78, find the sum of the infinite geometric series.

69. $\sum_{n=0}^{\infty} \left(\frac{1}{2}\right)^{n}$

70. $\sum_{n=0}^{\infty} 2\left(\frac{3}{4}\right)^{n}$

71. $\sum_{n=0}^{\infty} \left(-\frac{1}{2}\right)^{n}$

72. $\sum_{n=0}^{\infty} 2\left(-\frac{2}{3}\right)^{n}$

73. $\sum_{n=0}^{\infty} (0.8)^{n}$

74. $\sum_{n=0}^{\infty} 4(0.2)^{n}$

75. $8 + 6 + \frac{9}{2} + \frac{27}{8} + \cdots$

76. $9 + 6 + 4 + \frac{8}{3} + \cdots$

77. $\frac{1}{9} - \frac{1}{3} + 1 - 3 + \cdots$

78. $-\frac{125}{36} + \frac{25}{6} - 5 + 6 - \cdots$

Writing a Repeating Decimal as a Rational Number   In Exercises 79 and 80, find the rational number representation of the repeating decimal.

79. $0.\overline{36}$

80. $0.3\overline{18}$

Graphical Reasoning   In Exercises 81 and 82, use a graphing utility to graph the function. Identify the horizontal asymptote of the graph and determine its relationship to the sum.

81. $f(x) = 6\left[\frac{1 - (0.5)^{x}}{1 - (0.5)}\right]$, $\sum_{n=0}^{\infty} 6\left(\frac{1}{2}\right)^{n}$

82. $f(x) = 2\left[\frac{1 - (0.8)^{x}}{1 - (0.8)}\right]$, $\sum_{n=0}^{\infty} 2\left(\frac{4}{5}\right)^{n}$

83. Depreciation   A tool and die company buys a machine for \$175,000 and it depreciates at a rate of 30% per year. (In other words, at the end of each year the depreciated value is 70% of what it was at the beginning of the year.) Find the depreciated value of the machine after 5 full years.

84. Population

The table shows the mid-year populations of Argentina (in millions) from 2009 through 2015. (*Source: U.S. Census Bureau*)

DATA	Year	Population
	2009	40.9
	2010	41.3
	2011	41.8
	2012	42.2
	2013	42.6
	2014	43.0
	2015	43.4

Spreadsheet at LarsonPrecalculus.com

(a) Use the *exponential regression* feature of a graphing utility to find the nth term (a_n) of a geometric sequence that models the data. Let n represent the year, with $n = 9$ corresponding to 2009.

(b) Use the sequence from part (a) to describe the rate at which the population of Argentina is growing.

(c) Use the sequence from part (a) to predict the population of Argentina in 2025. The U.S. Census Bureau predicts the population of Argentina will be 47.2 million in 2025. How does this value compare with your prediction?

(d) Use the sequence from part (a) to predict when the population of Argentina will reach 50.0 million.

85. Annuity   An investor deposits P dollars on the first day of each month in an account with an annual interest rate r, compounded monthly. The balance A after t years is

$$A = P\left(1 + \frac{r}{12}\right) + \cdots + P\left(1 + \frac{r}{12}\right)^{12t}.$$

Show that the balance is

$$A = P\left[\left(1 + \frac{r}{12}\right)^{12t} - 1\right]\left(1 + \frac{12}{r}\right).$$

86. Annuity An investor deposits $100 on the first day of each month in an account that pays 2% interest, compounded monthly. The balance A in the account at the end of 5 years is

$$A = 100\left(1 + \frac{0.02}{12}\right)^1 + \cdots + 100\left(1 + \frac{0.02}{12}\right)^{60}.$$

Use the result of Exercise 85 to find A.

Multiplier Effect **In Exercises 87 and 88, use the following information.** A state government gives property owners a tax rebate with the anticipation that each property owner will spend approximately $p\%$ of the rebate, and in turn each recipient of this amount will spend $p\%$ of what he or she receives, and so on. Economists refer to this exchange of money and its circulation within the economy as the "multiplier effect." The multiplier effect operates on the idea that the expenditures of one individual become the income of another individual. For the given tax rebate, find the total amount of spending that results, assuming that this effect continues without end.

	Tax rebate	$p\%$
87.	$400	75%
88.	$600	72.5%

89. Geometry The sides of a square are 27 inches in length. New squares are formed by dividing the original square into nine squares. The center square is then shaded (see figure). This process is repeated three more times. Determine the total area of the shaded region.

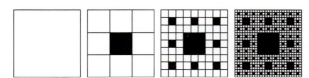

∫ 90. Distance A ball is dropped from a height of 6 feet and begins bouncing as shown in the figure. The height of each bounce is three-fourths the height of the previous bounce. Find the total vertical distance the ball travels before coming to rest.

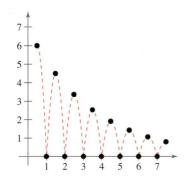

91. Salary An investment firm has a job opening with a salary of $45,000 for the first year. During the next 39 years, there is a 5% raise each year. Find the total compensation over the 40-year period.

92. HOW DO YOU SEE IT? Use the figures shown below.

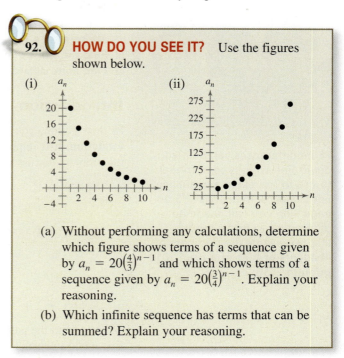

(a) Without performing any calculations, determine which figure shows terms of a sequence given by $a_n = 20\left(\frac{4}{3}\right)^{n-1}$ and which shows terms of a sequence given by $a_n = 20\left(\frac{3}{4}\right)^{n-1}$. Explain your reasoning.

(b) Which infinite sequence has terms that can be summed? Explain your reasoning.

Exploration

True or False? **In Exercises 93 and 94, determine whether the statement is true or false. Justify your answer.**

93. A sequence is geometric when the ratios of consecutive differences of consecutive terms are the same.

94. To find the nth term of a geometric sequence, multiply its common ratio by the first term of the sequence raised to the $(n - 1)$th power.

95. Graphical Reasoning Consider the graph of

$$y = \frac{1 - r^x}{1 - r}.$$

(a) Use a graphing utility to graph y for $r = \frac{1}{2}, \frac{2}{3}$, and $\frac{4}{5}$. What happens as $x \to \infty$?

(b) Use the graphing utility to graph y for $r = 1.5, 2$, and 3. What happens as $x \to \infty$?

96. Writing Write a brief paragraph explaining why the terms of a geometric sequence decrease in magnitude when $-1 < r < 1$.

Project: Population To work an extended application analyzing the population of Delaware, visit this text's website at *LarsonPrecalculus.com*. *(Source: U.S. Census Bureau)*

8.4 Mathematical Induction

- ■ Use mathematical induction to prove statements involving a positive integer n.
- ■ Use pattern recognition and mathematical induction to write a formula for the nth term of a sequence.
- ■ Find the sums of powers of integers.
- ■ Find finite differences of sequences.

Finite differences can help you determine what type of model to use to represent a sequence. For example, in Exercise 75 on page 599, you will use finite differences to find a model that represents the numbers of residents of Alabama from 2010 through 2015.

Introduction

In this section, you will study a form of mathematical proof called **mathematical induction.** It is important that you see the logical need for it, so take a closer look at the problem discussed in Example 5 in Section 8.2.

$$S_1 = 1 = 1^2$$

$$S_2 = 1 + 3 = 2^2$$

$$S_3 = 1 + 3 + 5 = 3^2$$

$$S_4 = 1 + 3 + 5 + 7 = 4^2$$

$$S_5 = 1 + 3 + 5 + 7 + 9 = 5^2$$

$$S_6 = 1 + 3 + 5 + 7 + 9 + 11 = 6^2$$

Judging from the pattern formed by these first six sums, it appears that the sum of the first n odd integers is

$$S_n = 1 + 3 + 5 + 7 + 9 + 11 + \cdots + (2n - 1) = n^2.$$

Although this particular formula *is* valid, it is important for you to see that recognizing a pattern and then simply *jumping to the conclusion* that the pattern must be true for all values of n is *not* a logically valid method of proof. There are many examples in which a pattern appears to be developing for small values of n, but then at some point the pattern fails. One of the most famous cases of this was the conjecture by the French mathematician Pierre de Fermat (1601–1665), who speculated that all numbers of the form

$$F_n = 2^{2^n} + 1, \quad n = 0, 1, 2, \ldots$$

are prime. For $n = 0, 1, 2, 3,$ and 4, the conjecture is true.

$$F_0 = 3$$

$$F_1 = 5$$

$$F_2 = 17$$

$$F_3 = 257$$

$$F_4 = 65,537$$

The size of the next Fermat number $(F_5 = 4,294,967,297)$ is so great that it was difficult for Fermat to determine whether it was prime or not. However, another well-known mathematician, Leonhard Euler (1707–1783), later found the factorization

$$F_5 = 4,294,967,297 = 641(6,700,417)$$

which proved that F_5 is not prime and therefore Fermat's conjecture was false.

Just because a rule, pattern, or formula seems to work for several values of n, you cannot simply decide that it is valid for all values of n without going through a *legitimate proof*. Mathematical induction is one method of proof.

REMARK It is important to recognize that in order to prove a statement by induction, both parts of the Principle of Mathematical Induction are necessary.

> **The Principle of Mathematical Induction**
>
> Let P_n be a statement involving the positive integer n. If
>
> **1.** P_1 is true, and
>
> **2.** for every positive integer k, the truth of P_k implies the truth of P_{k+1}
>
> then the statement P_n must be true for all positive integers n.

To apply the Principle of Mathematical Induction, you need to be able to determine the statement P_{k+1} for a given statement P_k. To determine P_{k+1}, substitute the quantity $k + 1$ for k in the statement P_k.

EXAMPLE 1   A Preliminary Example

Find the statement P_{k+1} for each given statement P_k.

a. P_k: $S_k = \dfrac{k^2(k + 1)^2}{4}$

b. P_k: $S_k = 1 + 5 + 9 + \cdots + [4(k - 1) - 3] + (4k - 3)$

c. P_k: $k + 3 < 5k^2$

d. P_k: $3^k \geq 2k + 1$

Solution

a. P_{k+1}: $S_{k+1} = \dfrac{(k + 1)^2(k + 1 + 1)^2}{4}$ Replace k with $k + 1$.

$= \dfrac{(k + 1)^2(k + 2)^2}{4}$ Simplify.

b. P_{k+1}: $S_{k+1} = 1 + 5 + 9 + \cdots + \{4[(k + 1) - 1] - 3\} + [4(k + 1) - 3]$

$= 1 + 5 + 9 + \cdots + (4k - 3) + (4k + 1)$

c. P_{k+1}: $(k + 1) + 3 < 5(k + 1)^2$

$k + 4 < 5(k^2 + 2k + 1)$

d. P_{k+1}: $3^{k+1} \geq 2(k + 1) + 1$

$3^{k+1} \geq 2k + 3$

✓ **Checkpoint**  *Audio-video solution in English & Spanish at LarsonPrecalculus.com*

Find the statement P_{k+1} for each given statement P_k.

a. P_k: $S_k = \dfrac{6}{k(k + 3)}$ **b.** P_k: $k + 2 \leq 3(k - 1)^2$ **c.** P_k: $2^{4k-2} + 1 > 5k$ ■

An unending line of dominoes can illustrate how the Principle of Mathematical Induction works.

A well-known illustration of how the Principle of Mathematical Induction works is an unending line of dominoes. It is clear that you could not knock down an infinite number of dominoes *one domino* at a time. However, if it were true that each domino would knock down the next one as it fell, then you could knock them all down by pushing the first one and starting a chain reaction. Mathematical induction works in the same way. If the truth of P_k implies the truth of P_{k+1} and if P_1 is true, then the chain reaction proceeds as follows: P_1 implies P_2, P_2 implies P_3, P_3 implies P_4, and so on.

When using mathematical induction to prove a *summation* formula (such as the one in Example 2), it is helpful to think of S_{k+1} as $S_{k+1} = S_k + a_{k+1}$, where a_{k+1} is the $(k + 1)$th term of the original sum.

EXAMPLE 2 Using Mathematical Induction

Use mathematical induction to prove the formula

$$S_n = 1 + 3 + 5 + 7 + \cdots + (2n - 1)$$
$$= n^2$$

for all integers $n \geq 1$.

Solution Mathematical induction consists of two distinct parts.

1. First, you must show that the formula is true when $n = 1$. When $n = 1$, the formula is valid, because

$$S_1 = 1$$
$$= 1^2.$$

2. The second part of mathematical induction has two steps. The first step is to *assume* that the formula is valid for some integer k. The second step is to use this assumption to prove that the formula is valid for the *next* integer, $k + 1$. Assuming that the formula

$$S_k = 1 + 3 + 5 + 7 + \cdots + (2k - 1) \qquad \textcolor{red}{a_k = 2k - 1}$$
$$= k^2$$

is true, you must show that the formula $S_{k+1} = (k + 1)^2$ is true.

$$S_{k+1} = 1 + 3 + 5 + 7 + \cdots + (2k - 1)$$
$$\qquad\qquad\qquad + [2(k + 1) - 1] \qquad \textcolor{red}{S_{k+1} = S_k + a_{k+1}}$$
$$= [1 + 3 + 5 + 7 + \cdots + (2k - 1)] + (2k + 2 - 1)$$
$$= S_k + (2k + 1) \qquad\qquad\qquad \textcolor{red}{\text{Group terms to form } S_k.}$$
$$= k^2 + 2k + 1 \qquad\qquad\qquad \textcolor{red}{\text{By assumption}}$$
$$= (k + 1)^2 \qquad\qquad\qquad\qquad \textcolor{red}{S_k \text{ implies } S_{k+1}.}$$

Combining the results of parts (1) and (2), you can conclude by mathematical induction that the formula is valid for all integers $n \geq 1$.

✓ *Checkpoint*  *Audio-video solution in English & Spanish at LarsonPrecalculus.com*

Use mathematical induction to prove the formula

$$S_n = 5 + 7 + 9 + 11 + \cdots + (2n + 3) = n(n + 4)$$

for all integers $n \geq 1$.

It occasionally happens that a statement involving natural numbers is not true for the first $k - 1$ positive integers but is true for all values of $n \geq k$. In these instances, you use a slight variation of the Principle of Mathematical Induction in which you verify P_k rather than P_1. This variation is called the *Extended Principle of Mathematical Induction*. To see the validity of this, note in the unending line of dominoes discussed on page 591 that all but the first $k - 1$ dominoes can be knocked down by knocking over the kth domino. This suggests that you can prove a statement P_n to be true for $n \geq k$ by showing that P_k is true and that P_k implies P_{k+1}. In Exercises 25–28 of this section, you will apply the Extended Principle of Mathematical Induction.

EXAMPLE 3 **Using Mathematical Induction**

Use mathematical induction to prove the formula

$$S_n = 1^2 + 2^2 + 3^2 + 4^2 + \cdots + n^2$$

$$= \frac{n(n + 1)(2n + 1)}{6}$$

for all integers $n \geq 1$.

Solution

1. When $n = 1$, the formula is valid, because

$$S_1 = 1^2$$

$$= \frac{1(2)(3)}{6}.$$

2. Assuming that the formula

$$S_k = 1^2 + 2^2 + 3^2 + 4^2 + \cdots + k^2 \qquad \color{red}{a_k = k^2}$$

$$= \frac{k(k + 1)(2k + 1)}{6}$$

is true, you must show that

$$S_{k+1} = \frac{(k + 1)(k + 1 + 1)[2(k + 1) + 1]}{6}$$

$$= \frac{(k + 1)(k + 2)(2k + 3)}{6}$$

is true.

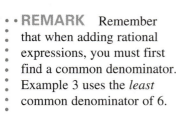

•• **REMARK** Remember that when adding rational expressions, you must first find a common denominator. Example 3 uses the *least* common denominator of 6.

$$S_{k+1} = 1^2 + 2^2 + 3^2 + 4^2 + \cdots + k^2 + (k + 1)^2 \qquad \color{red}{S_{k+1} = S_k + a_{k+1}}$$

$$= S_k + (k + 1)^2 \qquad \color{red}{\text{Group terms to form } S_k.}$$

$$= \frac{k(k + 1)(2k + 1)}{6} + (k + 1)^2 \qquad \color{red}{\text{By assumption}}$$

$$= \frac{k(k + 1)(2k + 1) + 6(k + 1)^2}{6} \qquad \color{red}{\text{Combine fractions.}}$$

$$= \frac{(k + 1)[k(2k + 1) + 6(k + 1)]}{6} \qquad \color{red}{\text{Factor.}}$$

$$= \frac{(k + 1)(2k^2 + 7k + 6)}{6} \qquad \color{red}{\text{Simplify.}}$$

$$= \frac{(k + 1)(k + 2)(2k + 3)}{6} \qquad \color{red}{S_k \text{ implies } S_{k+1}.}$$

Combining the results of parts (1) and (2), you can conclude by mathematical induction that the formula is valid for all integers $n \geq 1$.

✓ *Checkpoint* ◄))) *Audio-video solution in English & Spanish at LarsonPrecalculus.com*

Use mathematical induction to prove the formula

$$S_n = 1(1 - 1) + 2(2 - 1) + 3(3 - 1) + \cdots + n(n - 1) = \frac{n(n - 1)(n + 1)}{3}$$

for all integers $n \geq 1$.

When proving a formula using mathematical induction, the only statement that you *need* to verify is P_1. As a check, however, it is a good idea to try verifying some of the other statements. For instance, in Example 3, try verifying P_2 and P_3.

EXAMPLE 4   Proving an Inequality

Prove that

$$n < 2^n$$

for all integers $n \geq 1$.

Solution

1. For $n = 1$, the statement is true because

$$1 < 2^1.$$

2. Assuming that $k < 2^k$, you need to show that $k + 1 < 2^{k+1}$. To do this, use the fact that $2^{n+1} = 2(2^n) = 2^n + 2^n$. Then for $n = k$, you have

$$2^k + 2^k > 2^k + 1 > k + 1. \qquad \text{By assumption}$$

It follows that

$$2^{k+1} > 2^k + 1 > k + 1 \quad \text{or} \quad k + 1 < 2^{k+1}.$$

Combining the results of parts (1) and (2), you can conclude by mathematical induction that $n < 2^n$ for all integers $n \geq 1$.

✓ *Checkpoint* *Audio-video solution in English & Spanish at LarsonPrecalculus.com*

Prove that

$$n! \geq n$$

for all integers $n \geq 1$.

EXAMPLE 5   Proving a Property

Prove that 3 is a factor of $4^n - 1$ for all integers $n \geq 1$.

Solution

1. For $n = 1$, the statement is true because

$$4^1 - 1 = 3.$$

So, 3 is a factor.

2. Assuming that 3 is a factor of $4^k - 1$, you must show that 3 is a factor of $4^{k+1} - 1$. To do this, write the following.

$$4^{k+1} - 1 = 4^{k+1} - 4^k + 4^k - 1 \qquad \text{Subtract and add } 4^k.$$

$$= 4^k(4 - 1) + (4^k - 1) \qquad \text{Regroup terms.}$$

$$= 4^k \cdot 3 + (4^k - 1) \qquad \text{Simplify.}$$

Because 3 is a factor of $4^k \cdot 3$ and 3 is also a factor of $4^k - 1$, it follows that 3 is a factor of $4^{k+1} - 1$. Combining the results of parts (1) and (2), you can conclude by mathematical induction that 3 is a factor of $4^n - 1$ for all integers $n \geq 1$.

✓ *Checkpoint*  *Audio-video solution in English & Spanish at LarsonPrecalculus.com*

Prove that 2 is a factor of $3^n + 1$ for all integers $n \geq 1$.

Pattern Recognition

Although choosing a formula on the basis of a few observations does *not* guarantee the validity of the formula, pattern recognition *is* important. Once you have a pattern or formula that you think works, try using mathematical induction to prove your formula.

Finding a Formula for the *n*th Term of a Sequence

To find a formula for the *n*th term of a sequence, consider these guidelines.

1. Calculate the first several terms of the sequence. It is often a good idea to write the terms in both simplified and factored forms.

2. Try to find a recognizable pattern for the terms and write a formula for the *n*th term of the sequence. This is your *hypothesis* or *conjecture*. You might compute one or two more terms in the sequence to test your hypothesis.

3. Use mathematical induction to prove your hypothesis.

EXAMPLE 6   **Finding a Formula for a Finite Sum**

Find a formula for the finite sum and prove its validity.

$$\frac{1}{1 \cdot 2} + \frac{1}{2 \cdot 3} + \frac{1}{3 \cdot 4} + \frac{1}{4 \cdot 5} + \cdots + \frac{1}{n(n + 1)}$$

Solution   Begin by writing the first few sums.

$$S_1 = \frac{1}{1 \cdot 2} = \frac{1}{2} = \frac{1}{1 + 1}$$

$$S_2 = \frac{1}{1 \cdot 2} + \frac{1}{2 \cdot 3} = \frac{4}{6} = \frac{2}{3} = \frac{2}{2 + 1}$$

$$S_3 = \frac{1}{1 \cdot 2} + \frac{1}{2 \cdot 3} + \frac{1}{3 \cdot 4} = \frac{9}{12} = \frac{3}{4} = \frac{3}{3 + 1}$$

From this sequence, it appears that the formula for the *k*th sum is

$$S_k = \frac{1}{1 \cdot 2} + \frac{1}{2 \cdot 3} + \frac{1}{3 \cdot 4} + \frac{1}{4 \cdot 5} + \cdots + \frac{1}{k(k + 1)} = \frac{k}{k + 1}.$$

To prove the validity of this hypothesis, use mathematical induction. Note that you have already verified the formula for $n = 1$, so begin by assuming that the formula is valid for $n = k$ and trying to show that it is valid for $n = k + 1$.

$$S_{k+1} = \left[\frac{1}{1 \cdot 2} + \frac{1}{2 \cdot 3} + \frac{1}{3 \cdot 4} + \frac{1}{4 \cdot 5} + \cdots + \frac{1}{k(k + 1)} \right] + \frac{1}{(k + 1)(k + 2)}$$

$$= \frac{k}{k + 1} + \frac{1}{(k + 1)(k + 2)} \qquad \text{By assumption}$$

$$= \frac{k(k + 2) + 1}{(k + 1)(k + 2)} = \frac{k^2 + 2k + 1}{(k + 1)(k + 2)} = \frac{(k + 1)^2}{(k + 1)(k + 2)} = \frac{k + 1}{k + 2}$$

So, by mathematical induction the hypothesis is valid.

✓ *Checkpoint* ◀))) *Audio-video solution in English & Spanish at LarsonPrecalculus.com*

Find a formula for the finite sum and prove its validity.

$$3 + 7 + 11 + 15 + \cdots + 4n - 1$$

Sums of Powers of Integers

The formula in Example 3 is one of a collection of useful summation formulas. This and other formulas dealing with the sums of various powers of the first n positive integers are as follows.

Sums of Powers of Integers

1. $1 + 2 + 3 + 4 + \cdots + n = \dfrac{n(n + 1)}{2}$

2. $1^2 + 2^2 + 3^2 + 4^2 + \cdots + n^2 = \dfrac{n(n + 1)(2n + 1)}{6}$

3. $1^3 + 2^3 + 3^3 + 4^3 + \cdots + n^3 = \dfrac{n^2(n + 1)^2}{4}$

4. $1^4 + 2^4 + 3^4 + 4^4 + \cdots + n^4 = \dfrac{n(n + 1)(2n + 1)(3n^2 + 3n - 1)}{30}$

5. $1^5 + 2^5 + 3^5 + 4^5 + \cdots + n^5 = \dfrac{n^2(n + 1)^2(2n^2 + 2n - 1)}{12}$

EXAMPLE 7 **Finding Sums**

Find each sum.

a. $\displaystyle\sum_{i=1}^{7} i^3 = 1^3 + 2^3 + 3^3 + 4^3 + 5^3 + 6^3 + 7^3$　　**b.** $\displaystyle\sum_{i=1}^{4} (6i - 4i^2)$

Solution

a. Using the formula for the sum of the cubes of the first n positive integers, you obtain

$$\sum_{i=1}^{7} i^3 = 1^3 + 2^3 + 3^3 + 4^3 + 5^3 + 6^3 + 7^3$$

$$= \frac{7^2(7 + 1)^2}{4} \qquad \text{Formula 3}$$

$$= \frac{49(64)}{4}$$

$$= 784.$$

b. $\displaystyle\sum_{i=1}^{4} (6i - 4i^2) = \sum_{i=1}^{4} 6i - \sum_{i=1}^{4} 4i^2$

$$= 6 \sum_{i=1}^{4} i - 4 \sum_{i=1}^{4} i^2$$

$$= 6\left[\frac{4(4 + 1)}{2}\right] - 4\left[\frac{4(4 + 1)(2 \cdot 4 + 1)}{6}\right] \qquad \text{Formulas 1 and 2}$$

$$= 6(10) - 4(30)$$

$$= -60$$

✓ **Checkpoint** 🔊))) *Audio-video solution in English & Spanish at LarsonPrecalculus.com*

Find each sum.

a. $\displaystyle\sum_{i=1}^{20} i$　　**b.** $\displaystyle\sum_{i=1}^{5} (2i^2 + 3i^3)$

Finite Differences

The **first differences** of a sequence are found by subtracting consecutive terms. The **second differences** are found by subtracting consecutive first differences. The first and second differences of the sequence 3, 5, 8, 12, 17, 23, . . . are shown below.

n: 1   2   3   4   5   6

a_n: 3   5   8   12   17   23

First differences: 2   3   4   5   6

Second differences: 1   1   1   1

For this sequence, the second differences are all the same. When this happens, the sequence has a perfect *quadratic* model. When the first differences are all the same, the sequence has a perfect *linear* model. That is, the sequence is arithmetic.

EXAMPLE 8   **Finding a Quadratic Model**

See LarsonPrecalculus.com for an interactive version of this type of example.

Find the quadratic model for the sequence 3, 5, 8, 12, 17, 23, ...

Solution   You know from the second differences shown above that the model is quadratic and has the form $a_n = an^2 + bn + c$. By substituting 1, 2, and 3 for n, you obtain a system of three linear equations in three variables.

$a_1 = a(1)^2 + b(1) + c = 3$ Substitute 1 for n.

$a_2 = a(2)^2 + b(2) + c = 5$ Substitute 2 for n.

$a_3 = a(3)^2 + b(3) + c = 8$ Substitute 3 for n.

You now have a system of three equations in a, b, and c.

$$\begin{cases} a + b + c = 3 & \text{Equation 1} \\ 4a + 2b + c = 5 & \text{Equation 2} \\ 9a + 3b + c = 8 & \text{Equation 3} \end{cases}$$

Using the techniques discussed in Chapter 6, you find that the solution of the system is $a = \frac{1}{2}$, $b = \frac{1}{2}$, and $c = 2$. So, the quadratic model is $a_n = \frac{1}{2}n^2 + \frac{1}{2}n + 2$. Check the values of a_1, a_2, and a_3 in this model.

✓ **Checkpoint** 🔊)) *Audio-video solution in English & Spanish at LarsonPrecalculus.com*

Find a quadratic model for the sequence -2, 0, 4, 10, 18, 28, ... ◼

Summarize   (Section 8.4)

1. State the Principle of Mathematical Induction *(page 591)*. For examples of using mathematical induction to prove statements involving a positive integer n, see Examples 2–5.

2. Explain how to use pattern recognition and mathematical induction to write a formula for the nth term of a sequence *(page 595)*. For an example of using pattern recognition and mathematical induction to write a formula for the nth term of a sequence, see Example 6.

3. State the formulas for the sums of powers of integers *(page 596)*. For an example of finding sums of powers of integers, see Example 7.

4. Explain how to find finite differences of sequences *(page 597)*. For an example of using finite differences to find a quadratic model, see Example 8.

8.4 Exercises

See **CalcChat.com** for tutorial help and worked-out solutions to odd-numbered exercises.

Vocabulary: Fill in the blanks.

1. The first step in proving a formula by _____ _____ is to show that the formula is true when $n = 1$.

2. To find the _____ differences of a sequence, subtract consecutive terms.

3. A sequence is an _____ sequence when the first differences are all the same nonzero number.

4. If the _____ differences of a sequence are all the same nonzero number, then the sequence has a perfect quadratic model.

Skills and Applications

 Finding P_{k+1} **Given** P_k **In Exercises 5–10, find the statement** P_{k+1} **for the given statement** P_k.

5. $P_k = \dfrac{5}{k(k+1)}$ 6. $P_k = \dfrac{1}{2(k+2)}$

7. $P_k = k^2(k+3)^2$ 8. $P_k = \frac{1}{3}k(2k+1)$

9. $P_k = \dfrac{3}{(k+2)(k+3)}$ 10. $P_k = \dfrac{k^2}{2(k+1)^2}$

 Using Mathematical Induction In Exercises 11–24, use mathematical induction to prove the formula for all integers $n \geq 1$.

11. $2 + 4 + 6 + 8 + \cdots + 2n = n(n+1)$

12. $6 + 12 + 18 + 24 + \cdots + 6n = 3n(n+1)$

13. $2 + 7 + 12 + 17 + \cdots + (5n-3) = \dfrac{n}{2}(5n-1)$

14. $1 + 4 + 7 + 10 + \cdots + (3n-2) = \dfrac{n}{2}(3n-1)$

15. $1 + 2 + 2^2 + 2^3 + \cdots + 2^{n-1} = 2^n - 1$

16. $2(1 + 3 + 3^2 + 3^3 + \cdots + 3^{n-1}) = 3^n - 1$

17. $1 + 2 + 3 + 4 + \cdots + n = \dfrac{n(n+1)}{2}$

18. $1^3 + 2^3 + 3^3 + 4^3 + \cdots + n^3 = \dfrac{n^2(n+1)^2}{4}$

19. $1^2 + 3^2 + 5^2 + \cdots + (2n-1)^2 = \dfrac{n(2n-1)(2n+1)}{3}$

20. $\left(1 + \dfrac{1}{1}\right)\left(1 + \dfrac{1}{2}\right)\left(1 + \dfrac{1}{3}\right)\cdots\left(1 + \dfrac{1}{n}\right) = n + 1$

21. $\displaystyle\sum_{i=1}^{n} i^5 = \dfrac{n^2(n+1)^2(2n^2+2n-1)}{12}$

22. $\displaystyle\sum_{i=1}^{n} i^4 = \dfrac{n(n+1)(2n+1)(3n^2+3n-1)}{30}$

23. $\displaystyle\sum_{i=1}^{n} i(i+1) = \dfrac{n(n+1)(n+2)}{3}$

24. $\displaystyle\sum_{i=1}^{n} \dfrac{1}{(2i-1)(2i+1)} = \dfrac{n}{2n+1}$

 Proving an Inequality In Exercises 25–30, use mathematical induction to prove the inequality for the specified integer values of n.

25. $n! > 2^n, \quad n \geq 4$ 26. $\left(\frac{4}{3}\right)^n > n, \quad n \geq 7$

27. $\dfrac{1}{\sqrt{1}} + \dfrac{1}{\sqrt{2}} + \dfrac{1}{\sqrt{3}} + \cdots + \dfrac{1}{\sqrt{n}} > \sqrt{n}, \quad n \geq 2$

28. $2n^2 > (n+1)^2, \quad n \geq 3$

29. $\left(\dfrac{x}{y}\right)^{n+1} < \left(\dfrac{x}{y}\right)^n, \quad n \geq 1 \text{ and } 0 < x < y$

30. $(1+a)^n \geq na, \quad n \geq 1 \text{ and } a > 0$

 Proving a Property In Exercises 31–40, use mathematical induction to prove the property for all integers $n \geq 1$.

31. A factor of $n^3 + 3n^2 + 2n$ is 3.

32. A factor of $n^4 - n + 4$ is 2.

33. A factor of $2^{2n+1} + 1$ is 3.

34. A factor of $2^{2n-1} + 3^{2n-1}$ is 5.

35. $(ab)^n = a^n b^n$ 36. $\left(\dfrac{a}{b}\right)^n = \dfrac{a^n}{b^n}$

37. If $x_1 \neq 0, x_2 \neq 0, \ldots, x_n \neq 0$, then
$(x_1 x_2 x_3 \cdots x_n)^{-1} = x_1^{-1} x_2^{-1} x_3^{-1} \cdots x_n^{-1}$.

38. If $x_1 > 0, x_2 > 0, \ldots, x_n > 0$, then
$\ln(x_1 x_2 \cdots x_n) = \ln x_1 + \ln x_2 + \cdots + \ln x_n$.

39. $x(y_1 + y_2 + \cdots + y_n) = xy_1 + xy_2 + \cdots + xy_n$

40. $(a + bi)^n$ and $(a - bi)^n$ are complex conjugates.

 Finding a Formula for a Finite Sum In Exercises 41–44, find a formula for the sum of the first n **terms of the sequence. Prove the validity of your formula.**

41. $1, 5, 9, 13, \ldots$ 42. $3, -\frac{9}{2}, \frac{27}{4}, -\frac{81}{8}, \ldots$

43. $\dfrac{1}{4}, \dfrac{1}{12}, \dfrac{1}{24}, \dfrac{1}{40}, \ldots, \dfrac{1}{2n(n+1)}, \ldots$

44. $\dfrac{1}{2 \cdot 3}, \dfrac{1}{3 \cdot 4}, \dfrac{1}{4 \cdot 5}, \dfrac{1}{5 \cdot 6}, \ldots, \dfrac{1}{(n+1)(n+2)}, \ldots$

Finding a Sum In Exercises 45–54, find the sum using the formulas for the sums of powers of integers.

45. $\displaystyle\sum_{n=1}^{15} n$

46. $\displaystyle\sum_{n=1}^{30} n$

47. $\displaystyle\sum_{n=1}^{6} n^2$

48. $\displaystyle\sum_{n=1}^{10} n^3$

49. $\displaystyle\sum_{n=1}^{5} n^4$

50. $\displaystyle\sum_{n=1}^{8} n^5$

51. $\displaystyle\sum_{n=1}^{6} (n^2 - n)$

52. $\displaystyle\sum_{n=1}^{20} (n^3 - n)$

53. $\displaystyle\sum_{i=1}^{6} (6i - 8i^3)$

54. $\displaystyle\sum_{j=1}^{10} \left(3 - \tfrac{1}{2}j + \tfrac{1}{2}j^2\right)$

Finding a Linear or Quadratic Model In Exercises 55–60, decide whether the sequence can be represented perfectly by a linear or a quadratic model. Then find the model.

55. 5, 14, 23, 32, 41, 50, . . .

56. 3, 9, 15, 21, 27, 33, . . .

57. 4, 10, 20, 34, 52, 74, . . .

58. 0, 9, 24, 45, 72, 105, . . .

59. −1, 11, 31, 59, 95, 139, . . .

60. −2, 13, 38, 73, 118, 173, . . .

Linear Model, Quadratic Model, or Neither? In Exercises 61–68, write the first six terms of the sequence beginning with the term a_1. Then calculate the first and second differences of the sequence. State whether the sequence has a perfect linear model, a perfect quadratic model, or neither.

61. $a_1 = 0$
    $a_n = a_{n-1} + 3$

62. $a_1 = 2$
    $a_n = a_{n-1} + 2$

63. $a_1 = 4$
    $a_n = a_{n-1} + 3n$

64. $a_1 = 3$
    $a_n = 2a_{n-1}$

65. $a_1 = 3$
    $a_n = a_{n-1} + n^2$

66. $a_1 = 0$
    $a_n = a_{n-1} - 2n$

67. $a_1 = 5$
    $a_n = 4n - a_{n-1}$

68. $a_1 = -2$
    $a_n = a_{n-1} + 4n$

Finding a Quadratic Model In Exercises 69–74, find the quadratic model for the sequence with the given terms.

69. $a_0 = 3, a_1 = 3, a_4 = 15$

70. $a_0 = 7, a_1 = 6, a_3 = 10$

71. $a_0 = -1, a_2 = 5, a_4 = 15$

72. $a_0 = 3, a_2 = -3, a_6 = 21$

73. $a_1 = 0, a_2 = 7, a_4 = 27$

74. $a_0 = -7, a_2 = -3, a_6 = -43$

75. Residents

The table shows the numbers a_n (in thousands) of residents of Alabama from 2010 through 2015. (*Source: U.S. Census Bureau*)

Year	Number of Residents, a_n
2010	4785
2011	4801
2012	4816
2013	4831
2014	4846
2015	4859

Spreadsheet at LarsonPrecalculus.com

(a) Find the first differences of the data shown in the table. Then find a linear model that approximates the data. Let n represent the year, with $n = 10$ corresponding to 2010.

(b) Use a graphing utility to find a linear model for the data. Compare this model with the model from part (a).

(c) Use the models found in parts (a) and (b) to predict the number of residents in 2021. How do these values compare?

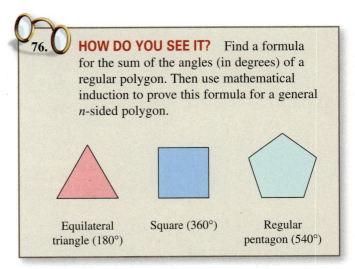

76. HOW DO YOU SEE IT? Find a formula for the sum of the angles (in degrees) of a regular polygon. Then use mathematical induction to prove this formula for a general n-sided polygon.

Equilateral triangle (180°) Square (360°) Regular pentagon (540°)

Exploration

True or False? In Exercises 77 and 78, determine whether the statement is true or false. Justify your answer.

77. If the statement P_k is true and P_k implies P_{k+1}, then P_1 is also true.

78. A sequence with n terms has $n - 1$ second differences.

8.5   The Binomial Theorem

Binomial coefficients have many applications in real life. For example, in Exercise 86 on page 607, you will use binomial coefficients to write the expansion of a model that represents the average prices of residential electricity in the United States.

■ Use the Binomial Theorem to find binomial coefficients.
■ Use Pascal's Triangle to find binomial coefficients.
■ Use binomial coefficients to write binomial expansions.

Binomial Coefficients

Recall that a *binomial* is a polynomial that has two terms. In this section, you will study a formula that provides a quick method of finding the terms that result from raising a binomial to a power, or **expanding a binomial.** To begin, look at the expansion of

$$(x + y)^n$$

for several values of n.

$$(x + y)^0 = 1$$
$$(x + y)^1 = x + y$$
$$(x + y)^2 = x^2 + 2xy + y^2$$
$$(x + y)^3 = x^3 + 3x^2y + 3xy^2 + y^3$$
$$(x + y)^4 = x^4 + 4x^3y + 6x^2y^2 + 4xy^3 + y^4$$
$$(x + y)^5 = x^5 + 5x^4y + 10x^3y^2 + 10x^2y^3 + 5xy^4 + y^5$$

There are several observations you can make about these expansions.

1. In each expansion, there are $n + 1$ terms.

2. In each expansion, x and y have symmetric roles. The powers of x decrease by 1 in successive terms, whereas the powers of y increase by 1.

3. The sum of the powers of each term is n. For example, in the expansion of $(x + y)^5$, the sum of the powers of each term is 5.

$$4 + 1 = 5 \quad 3 + 2 = 5$$
$$(x + y)^5 = x^5 + 5x^4y^1 + 10x^3y^2 + 10x^2y^3 + 5x^1y^4 + y^5$$

4. The coefficients increase and then decrease in a symmetric pattern.

The coefficients of a binomial expansion are called **binomial coefficients.** To find them, you can use the **Binomial Theorem.**

The Binomial Theorem

In the expansion of $(x + y)^n$

$$(x + y)^n = x^n + nx^{n-1}y + \cdots + {}_nC_r x^{n-r}y^r + \cdots + nxy^{n-1} + y^n$$

the coefficient of $x^{n-r}y^r$ is

$${}_nC_r = \frac{n!}{(n - r)!r!}.$$

The symbol $\binom{n}{r}$ is often used in place of ${}_nC_r$ to denote binomial coefficients.

For a proof of the Binomial Theorem, see Proofs in Mathematics on page 640.

▷ **TECHNOLOGY**
Most graphing utilities can evaluate $_nC_r$. If yours can, use it to check Example 1.

EXAMPLE 1 **Finding Binomial Coefficients**

Find each binomial coefficient.

a. $_8C_2$ **b.** $\begin{pmatrix} 10 \\ 3 \end{pmatrix}$ **c.** $_7C_0$ **d.** $\begin{pmatrix} 8 \\ 8 \end{pmatrix}$

Solution

a. $_8C_2 = \dfrac{8!}{6! \cdot 2!} = \dfrac{(8 \cdot 7) \cdot \cancel{6!}}{\cancel{6!} \cdot 2!} = \dfrac{8 \cdot 7}{2 \cdot 1} = 28$

b. $\begin{pmatrix} 10 \\ 3 \end{pmatrix} = \dfrac{10!}{7! \cdot 3!} = \dfrac{(10 \cdot 9 \cdot 8) \cdot \cancel{7!}}{\cancel{7!} \cdot 3!} = \dfrac{10 \cdot 9 \cdot 8}{3 \cdot 2 \cdot 1} = 120$

c. $_7C_0 = \dfrac{\cancel{7!}}{\cancel{7!} \cdot 0!} = 1$ **d.** $\begin{pmatrix} 8 \\ 8 \end{pmatrix} = \dfrac{\cancel{8!}}{0! \cdot \cancel{8!}} = 1$

✓ *Checkpoint* *Audio-video solution in English & Spanish at LarsonPrecalculus.com*

Find each binomial coefficient.

a. $\begin{pmatrix} 11 \\ 5 \end{pmatrix}$ **b.** $_9C_2$ **c.** $\begin{pmatrix} 5 \\ 0 \end{pmatrix}$ **d.** $_{15}C_{15}$

When $r \neq 0$ and $r \neq n$, as in parts (a) and (b) above, there is a pattern for evaluating binomial coefficients that works because there will always be factorial terms that divide out of the expression.

$$_8C_2 = \overbrace{\dfrac{8 \cdot 7}{2 \cdot 1}}^{\text{2 factors}} \quad \text{and} \quad \begin{pmatrix} 10 \\ 3 \end{pmatrix} = \overbrace{\dfrac{10 \cdot 9 \cdot 8}{3 \cdot 2 \cdot 1}}^{\text{3 factors}}$$

2 factors 3 factors

EXAMPLE 2 **Finding Binomial Coefficients**

a. $_7C_3 = \dfrac{7 \cdot \cancel{6} \cdot 5}{\cancel{3} \cdot \cancel{2} \cdot 1} = 35$

b. $\begin{pmatrix} 7 \\ 4 \end{pmatrix} = \dfrac{7 \cdot \cancel{6} \cdot 5 \cdot \cancel{4}}{\cancel{4} \cdot \cancel{3} \cdot \cancel{2} \cdot 1} = 35$

c. $_{12}C_1 = \dfrac{12}{1} = 12$

d. $\begin{pmatrix} 12 \\ 11 \end{pmatrix} = \dfrac{12 \cdot \cancel{11} \cdot \cancel{10} \cdot \cancel{9} \cdot \cancel{8} \cdot \cancel{7} \cdot \cancel{6} \cdot \cancel{5} \cdot \cancel{4} \cdot \cancel{3} \cdot \cancel{2}}{\cancel{11} \cdot \cancel{10} \cdot \cancel{9} \cdot \cancel{8} \cdot \cancel{7} \cdot \cancel{6} \cdot \cancel{5} \cdot \cancel{4} \cdot \cancel{3} \cdot \cancel{2} \cdot 1} = \dfrac{12}{1} = 12$

✓ *Checkpoint* *Audio-video solution in English & Spanish at LarsonPrecalculus.com*

Find each binomial coefficient.

a. $_7C_5$ **b.** $\begin{pmatrix} 7 \\ 2 \end{pmatrix}$ **c.** $_{14}C_{13}$ **d.** $\begin{pmatrix} 14 \\ 1 \end{pmatrix}$

•• **REMARK** The property $_nC_r = {_nC_{n-r}}$ produces the symmetric pattern of binomial coefficients identified earlier.

In Example 2, it is not a coincidence that the results in parts (a) and (b) are the same and that the results in parts (c) and (d) are the same. In general, it is true that $_nC_r = {_nC_{n-r}}$, for all integers r and n, where $0 \leq r \leq n$.

Pascal's Triangle

There is a convenient way to remember the pattern for binomial coefficients. By arranging the coefficients in a triangular pattern, you obtain the array shown below, which is called **Pascal's Triangle.** This triangle is named after the French mathematician Blaise Pascal (1623–1662).

$$
\begin{array}{ccccccccccccccccc}
 & & & & & & & & 1 & & & & & & & & \\
 & & & & & & & 1 & & 1 & & & & & & & \\
 & & & & & & 1 & & 2 & & 1 & & & & & & \\
 & & & & & 1 & & 3 & & 3 & & 1 & & & & & \\
 & & & & 1 & & 4 & & 6 & & 4 & & 1 & & & & \\
 & & & 1 & & 5 & & 10 & & 10 & & 5 & & 1 & & & 4 + 6 = 10 \\
 & & 1 & & 6 & & 15 & & 20 & & 15 & & 6 & & 1 & & \\
 & 1 & & 7 & & 21 & & 35 & & 35 & & 21 & & 7 & & 1 & 15 + 6 = 21
\end{array}
$$

Note the pattern in Pascal's Triangle. The first and last numbers in each row are 1. Each other number in a row is the sum of the two numbers immediately above it. Pascal noticed that numbers in this triangle are precisely the same numbers as the coefficients of binomial expansions.

$$(x + y)^0 = 1 \qquad \text{0th row}$$
$$(x + y)^1 = 1x + 1y \qquad \text{1st row}$$
$$(x + y)^2 = 1x^2 + 2xy + 1y^2 \qquad \text{2nd row}$$
$$(x + y)^3 = 1x^3 + 3x^2y + 3xy^2 + 1y^3 \qquad \text{3rd row}$$
$$(x + y)^4 = 1x^4 + 4x^3y + 6x^2y^2 + 4xy^3 + 1y^4 \qquad \vdots$$
$$(x + y)^5 = 1x^5 + 5x^4y + 10x^3y^2 + 10x^2y^3 + 5xy^4 + 1y^5$$
$$(x + y)^6 = 1x^6 + 6x^5y + 15x^4y^2 + 20x^3y^3 + 15x^2y^4 + 6xy^5 + 1y^6$$
$$(x + y)^7 = 1x^7 + 7x^6y + 21x^5y^2 + 35x^4y^3 + 35x^3y^4 + 21x^2y^5 + 7xy^6 + 1y^7$$

The top row in Pascal's Triangle is called the *zeroth row* because it corresponds to the binomial expansion $(x + y)^0 = 1$. Similarly, the next row is called the *first row* because it corresponds to the binomial expansion

$$(x + y)^1 = 1x + 1y.$$

In general, the *nth row* in Pascal's Triangle gives the coefficients of $(x + y)^n$.

EXAMPLE 3   Using Pascal's Triangle

Use the seventh row of Pascal's Triangle to find the binomial coefficients.

$$_8C_0, \; _8C_1, \; _8C_2, \; _8C_3, \; _8C_4, \; _8C_5, \; _8C_6, \; _8C_7, \; _8C_8$$

Solution

$$
\begin{array}{ccccccccccccccccc}
 & 1 & & 7 & & 21 & & 35 & & 35 & & 21 & & 7 & & 1 & \\
1 & & 8 & & 28 & & 56 & & 70 & & 56 & & 28 & & 8 & & 1 \\
_8C_0 & & _8C_1 & & _8C_2 & & _8C_3 & & _8C_4 & & _8C_5 & & _8C_6 & & _8C_7 & & _8C_8
\end{array}
$$

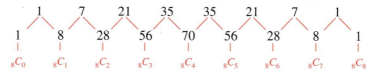

✓ **Checkpoint** 🔊))) *Audio-video solution in English & Spanish at LarsonPrecalculus.com*

Use the eighth row of Pascal's Triangle to find the binomial coefficients.

$$_9C_0, \; _9C_1, \; _9C_2, \; _9C_3, \; _9C_4, \; _9C_5, \; _9C_6, \; _9C_7, \; _9C_8, \; _9C_9$$

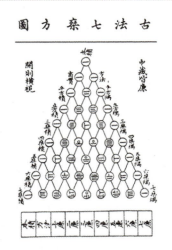

PRECIOUS MIRROR OF THE FOUR ELEMENTS

Eastern cultures were familiar with "Pascal's" Triangle and forms of the Binomial Theorem prior to the Western "discovery" of the theorem. A Chinese text entitled *Precious Mirror of the Four Elements* contains a triangle of binomial expansions through the eighth power.

Binomial Expansions

The formula for binomial coefficients and Pascal's Triangle give you a systematic way to write the coefficients of a binomial expansion, as demonstrated in the next four examples.

EXAMPLE 4   Expanding a Binomial

Write the expansion of the expression

$(x + 1)^3$.

Solution   The binomial coefficients from the third row of Pascal's Triangle are

$1, 3, 3, 1$.

So, the expansion is

$$(x + 1)^3 = (1)x^3 + (3)x^2(1) + (3)x(1^2) + (1)(1^3)$$
$$= x^3 + 3x^2 + 3x + 1.$$

✓ *Checkpoint* 🔊))) *Audio-video solution in English & Spanish at LarsonPrecalculus.com*

Write the expansion of the expression

$(x + 2)^4$.

To expand binomials representing *differences* rather than sums, you alternate signs. Here are two examples.

$$(x - 1)^2 = x^2 - 2x + 1$$
$$(x - 1)^3 = x^3 - 3x^2 + 3x - 1$$

▷ **ALGEBRA HELP**   The solutions to Example 5 use the property of exponents

$(ab)^m = a^m b^m$.

For instance, in Example 5(a),

$(2x)^4 = 2^4 x^4 = 16x^4$.

To review properties of exponents, see Section P.2.

EXAMPLE 5   Expanding a Binomial

See LarsonPrecalculus.com for an interactive version of this type of example.

Write the expansion of each expression.

a. $(2x - 3)^4$

b. $(x - 2y)^4$

Solution   The binomial coefficients from the fourth row of Pascal's Triangle are

$1, 4, 6, 4, 1$.

The expansions are given below.

a. $(2x - 3)^4 = (1)(2x)^4 - (4)(2x)^3(3) + (6)(2x)^2(3^2) - (4)(2x)(3^3) + (1)(3^4)$
$$= 16x^4 - 96x^3 + 216x^2 - 216x + 81$$

b. $(x - 2y)^4 = (1)x^4 - (4)x^3(2y) + (6)x^2(2y)^2 - (4)x(2y)^3 + (1)(2y)^4$
$$= x^4 - 8x^3 y + 24x^2 y^2 - 32xy^3 + 16y^4$$

✓ *Checkpoint* 🔊))) *Audio-video solution in English & Spanish at LarsonPrecalculus.com*

Write the expansion of each expression.

a. $(y - 2)^4$

b. $(2x - y)^5$

▷ TECHNOLOGY Use a graphing utility to check the expansion in Example 6. Graph the original binomial expression and the expansion in the same viewing window. The graphs should coincide, as shown in the figure below.

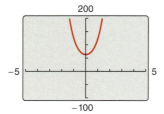

EXAMPLE 6 **Expanding a Binomial**

Write the expansion of $(x^2 + 4)^3$.

Solution Use the third row of Pascal's Triangle.

$(x^2 + 4)^3 = (1)(x^2)^3 + (3)(x^2)^2(4) + (3)x^2(4^2) + (1)(4^3)$

$= x^6 + 12x^4 + 48x^2 + 64$

✓ *Checkpoint* 🔊))) *Audio-video solution in English & Spanish at LarsonPrecalculus.com*

Write the expansion of $(5 + y^2)^3$.

Sometimes you will need to find a specific term in a binomial expansion. Instead of writing the entire expansion, use the fact that, from the Binomial Theorem, the $(r + 1)$th term is $_nC_r x^{n-r} y^r$.

EXAMPLE 7 **Finding a Term or Coefficient**

a. Find the sixth term of $(a + 2b)^8$.

b. Find the coefficient of the term a^6b^5 in the expansion of $(3a - 2b)^{11}$.

Solution

a. Remember that the formula is for the $(r + 1)$th term, so r is one less than the number of the term you need. So, to find the sixth term in this binomial expansion, use $r = 5$, $n = 8$, $x = a$, and $y = 2b$.

$_nC_r x^{n-r} y^r = {}_8C_5 a^3(2b)^5$

$= 56a^3(32b^5)$

$= 1792a^3b^5$

b. In this case, $n = 11$, $r = 5$, $x = 3a$, and $y = -2b$. Substitute these values to obtain

$_nC_r x^{n-r} y^r = {}_{11}C_5(3a)^6(-2b)^5$

$= (462)(729a^6)(-32b^5)$

$= -10,777,536a^6b^5.$

So, the coefficient is $-10,777,536$.

✓ *Checkpoint* 🔊))) *Audio-video solution in English & Spanish at LarsonPrecalculus.com*

a. Find the fifth term of $(a + 2b)^8$.

b. Find the coefficient of the term a^4b^7 in the expansion of $(3a - 2b)^{11}$.

Summarize (Section 8.5)

1. State the Binomial Theorem *(page 600)*. For examples of using the Binomial Theorem to find binomial coefficients, see Examples 1 and 2.

2. Explain how to use Pascal's Triangle to find binomial coefficients *(page 602)*. For an example of using Pascal's Triangle to find binomial coefficients, see Example 3.

3. Explain how to use binomial coefficients to write a binomial expansion *(page 603)*. For examples of using binomial coefficients to write binomial expansions, see Examples 4–6.

8.5 Exercises

See **CalcChat.com** for tutorial help and worked-out solutions to odd-numbered exercises.

Vocabulary: Fill in the blanks.

1. When you find the terms that result from raising a binomial to a power, you are _____ the binomial.
2. The coefficients of a binomial expansion are called _____ _____.
3. To find binomial coefficients, you can use the _____ _____ or _____ _____.
4. The symbol used to denote a binomial coefficient is _____ or _____.

Skills and Applications

 Finding a Binomial Coefficient In Exercises 5–12, find the binomial coefficient.

5. $_5C_3$

6. $_7C_6$

7. $_{12}C_0$

8. $_{20}C_{20}$

9. $\binom{10}{4}$

10. $\binom{10}{6}$

11. $\binom{100}{98}$

12. $\binom{100}{2}$

 Using Pascal's Triangle In Exercises 13–16, evaluate using Pascal's Triangle.

13. $_6C_3$

14. $_4C_2$

15. $\binom{5}{1}$

16. $\binom{7}{4}$

 Expanding a Binomial In Exercises 17–24, use the Binomial Theorem to write the expansion of the expression.

17. $(x + 1)^6$

18. $(x + 1)^4$

19. $(y - 3)^3$

20. $(y - 2)^5$

21. $(r + 3s)^3$

22. $(x + 2y)^4$

23. $(3a - 4b)^5$

24. $(2x - 5y)^5$

 Expanding an Expression In Exercises 25–38, expand the expression by using Pascal's Triangle to determine the coefficients.

25. $(a + 6)^4$

26. $(a + 5)^5$

27. $(y - 1)^6$

28. $(y - 4)^4$

29. $(3 - 2z)^4$

30. $(3v + 2)^6$

31. $(x + 2y)^5$

32. $(2t - s)^5$

33. $(x^2 + y^2)^4$

34. $(x^2 + y^2)^6$

35. $\left(\dfrac{1}{x} + y\right)^5$

36. $\left(\dfrac{1}{x} + 2y\right)^6$

37. $2(x - 3)^4 + 5(x - 3)^2$

38. $(4x - 1)^3 - 2(4x - 1)^4$

 Finding a Term In Exercises 39–46, find the specified nth term in the expansion of the binomial.

39. $(x + y)^{10}$, $n = 4$

40. $(x - y)^6$, $n = 2$

41. $(x - 6y)^5$, $n = 3$

42. $(x + 2z)^7$, $n = 4$

43. $(4x + 3y)^9$, $n = 8$

44. $(5a + 6b)^5$, $n = 5$

45. $(10x - 3y)^{12}$, $n = 10$

46. $(7x + 2y)^{15}$, $n = 7$

Finding a Coefficient In Exercises 47–54, find the coefficient a of the term in the expansion of the binomial.

Binomial	Term
47. $(x + 2)^6$	ax^3
48. $(x - 2)^6$	ax^3
49. $(4x - y)^{10}$	ax^2y^8
50. $(x - 2y)^{10}$	ax^8y^2
51. $(2x - 5y)^9$	ax^4y^5
52. $(3x + 4y)^8$	ax^6y^2
53. $(x^2 + y)^{10}$	ax^8y^6
54. $(z^2 - t)^{10}$	az^4t^8

Expanding an Expression In Exercises 55–60, use the Binomial Theorem to write the expansion of the expression.

55. $\left(\sqrt{x} + 5\right)^3$

56. $\left(2\sqrt{t} - 1\right)^3$

57. $(x^{2/3} - y^{1/3})^3$

58. $(u^{3/5} + 2)^5$

59. $\left(3\sqrt{t} + \sqrt[4]{t}\right)^4$

60. $(x^{3/4} - 2x^{5/4})^4$

𝑓 Simplifying a Difference Quotient In Exercises 61–66, simplify the difference quotient, using the Binomial Theorem if necessary.

$$\frac{f(x + h) - f(x)}{h} \qquad \text{Difference quotient}$$

61. $f(x) = x^3$

62. $f(x) = x^4$

63. $f(x) = x^6$

64. $f(x) = x^7$

65. $f(x) = \sqrt{x}$

66. $f(x) = \dfrac{1}{x}$

Expanding a Complex Number In Exercises 67–72, use the Binomial Theorem to expand the complex number. Simplify your result.

67. $(1 + i)^4$

68. $(2 - i)^5$

69. $(2 - 3i)^6$

70. $\left(5 + \sqrt{-9}\right)^3$

71. $\left(-\dfrac{1}{2} + \dfrac{\sqrt{3}}{2}i\right)^3$

72. $\left(5 - \sqrt{3}i\right)^4$

Approximation In Exercises 73–76, use the Binomial Theorem to approximate the quantity accurate to three decimal places. For example, in Exercise 73, use the expansion

$$(1.02)^8 = (1 + 0.02)^8$$

$$= 1 + 8(0.02) + 28(0.02)^2 + \cdots + (0.02)^8.$$

73. $(1.02)^8$

74. $(2.005)^{10}$

75. $(2.99)^{12}$

76. $(1.98)^9$

Probability In Exercises 77–80, consider n independent trials of an experiment in which each trial has two possible outcomes: "success" or "failure." The probability of a success on each trial is p, and the probability of a failure is $q = 1 - p$. In this context, the term $_nC_k p^k q^{n-k}$ in the expansion of $(p + q)^n$ gives the probability of k successes in the n trials of the experiment.

77. You toss a fair coin seven times. To find the probability of obtaining four heads, evaluate the term

$$_7C_4\left(\tfrac{1}{2}\right)^4\left(\tfrac{1}{2}\right)^3$$

in the expansion of $\left(\tfrac{1}{2} + \tfrac{1}{2}\right)^7$.

78. The probability of a baseball player getting a hit during any given time at bat is $\tfrac{1}{4}$. To find the probability that the player gets three hits during the next 10 times at bat, evaluate the term

$$_{10}C_3\left(\tfrac{1}{4}\right)^3\left(\tfrac{3}{4}\right)^7$$

in the expansion of $\left(\tfrac{1}{4} + \tfrac{3}{4}\right)^{10}$.

79. The probability of a sales representative making a sale with any one customer is $\tfrac{1}{3}$. The sales representative makes eight contacts a day. To find the probability of making four sales, evaluate the term

$$_8C_4\left(\tfrac{1}{3}\right)^4\left(\tfrac{2}{3}\right)^4$$

in the expansion of $\left(\tfrac{1}{3} + \tfrac{2}{3}\right)^8$.

80. To find the probability that the sales representative in Exercise 79 makes four sales when the probability of a sale with any one customer is $\tfrac{1}{2}$, evaluate the term

$$_8C_4\left(\tfrac{1}{2}\right)^4\left(\tfrac{1}{2}\right)^4$$

in the expansion of $\left(\tfrac{1}{2} + \tfrac{1}{2}\right)^8$.

Graphical Reasoning In Exercises 81 and 82, use a graphing utility to graph f and g in the same viewing window. What is the relationship between the two graphs? Use the Binomial Theorem to write the polynomial function g in standard form.

81. $f(x) = x^3 - 4x$

$\quad g(x) = f(x + 4)$

82. $f(x) = -x^4 + 4x^2 - 1$

$\quad g(x) = f(x - 3)$

83. Finding a Pattern Describe the pattern formed by the sums of the numbers along the diagonal line segments shown in Pascal's Triangle (see figure).

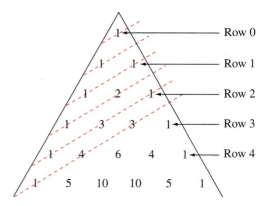

84. Error Analysis Describe the error.

$$(x - 3)^3 = {}_3C_0 x^3 + {}_3C_1 x^2(3) + {}_3C_2 x(3)^2$$
$$+ {}_3C_3(3)^3$$
$$= 1x^3 + 3x^2(3) + 3x(3)^2 + 1(3)^3$$
$$= x^3 + 9x^2 + 27x + 27$$

85. Child Support The amounts $f(t)$ (in billions of dollars) of child support collected in the United States from 2005 through 2014 can be approximated by the model

$$f(t) = -0.056t^2 + 1.62t + 16.4, \quad 5 \le t \le 14$$

where t represents the year, with $t = 5$ corresponding to 2005. (*Source: U.S. Department of Health and Human Services*)

(a) You want to adjust the model so that $t = 5$ corresponds to 2010 rather than 2005. To do this, you shift the graph of f five units *to the left* to obtain $g(t) = f(t + 5)$. Use binomial coefficients to write $g(t)$ in standard form.

(b) Use a graphing utility to graph f and g in the same viewing window.

(c) Use the graphs to estimate when the child support collections exceeded $27 billion.

86. Electricity

The table shows the average prices $f(t)$ (in cents per kilowatt-hour) of residential electricity in the United States from 2007 through 2014. *(Source: U.S. Energy Information Administration)*

DATA	Year	Average Price, $f(t)$
	2007	10.65
	2008	11.26
	2009	11.51
	2010	11.54
	2011	11.72
	2012	11.88
	2013	12.13
	2014	12.52

Spreadsheet at LarsonPrecalculus.com

(a) Use the *regression* feature of a graphing utility to find a cubic model for the data. Let t represent the year, with $t = 7$ corresponding to 2007.

(b) Use the graphing utility to plot the data and the model in the same viewing window.

(c) You want to adjust the model so that $t = 7$ corresponds to 2012 rather than 2007. To do this, you shift the graph of f five units *to the left* to obtain $g(t) = f(t + 5)$. Use binomial coefficients to write $g(t)$ in standard form.

(d) Use the graphing utility to graph g in the same viewing window as f.

(e) Use both models to predict the average price in 2015. Do you obtain the same answer?

(f) Do your answers to part (e) seem reasonable? Explain.

(g) What factors do you think contributed to the change in the average price?

Exploration

True or False? In Exercises 87 and 88, determine whether the statement is true or false. Justify your answer.

87. The Binomial Theorem could be used to produce each row of Pascal's Triangle.

88. A binomial that represents a difference cannot always be accurately expanded using the Binomial Theorem.

89. Writing Explain how to form the rows of Pascal's Triangle.

90. Forming Rows of Pascal's Triangle Form rows 8–10 of Pascal's Triangle.

91. Graphical Reasoning Use a graphing utility to graph the functions in the same viewing window. Which two functions have identical graphs, and why?

$f(x) = (1 - x)^3$

$g(x) = 1 - x^3$

$h(x) = 1 + 3x + 3x^2 + x^3$

$k(x) = 1 - 3x + 3x^2 - x^3$

$p(x) = 1 + 3x - 3x^2 + x^3$

92. HOW DO YOU SEE IT? The expansions of $(x + y)^4$, $(x + y)^5$, and $(x + y)^6$ are shown below.

$(x + y)^4 = 1x^4 + 4x^3y + 6x^2y^2 + 4xy^3 + 1y^4$

$(x + y)^5 = 1x^5 + 5x^4y + 10x^3y^2 + 10x^2y^3 + 5xy^4 + 1y^5$

$(x + y)^6 = 1x^6 + 6x^5y + 15x^4y^2 + 20x^3y^3 + 15x^2y^4 + 6xy^5 + 1y^6$

(a) Explain how the exponent of a binomial is related to the number of terms in its expansion.

(b) How many terms are in the expansion of $(x + y)^n$?

Proof In Exercises 93–96, prove the property for all integers r and n, where $0 \le r \le n$.

93. $_nC_r = {_nC_{n-r}}$

94. $_nC_0 - {_nC_1} + {_nC_2} - \cdots \pm {_nC_n} = 0$

95. $_{n+1}C_r = {_nC_r} + {_nC_{r-1}}$

96. The sum of the numbers in the nth row of Pascal's Triangle is 2^n.

97. Binomial Coefficients and Pascal's Triangle Complete the table. What characteristic of Pascal's Triangle does this table illustrate?

n	r	$_nC_r$	$_nC_{n-r}$
9	5		
7	1		
12	4		
6	0		
10	7		

8.6   Counting Principles

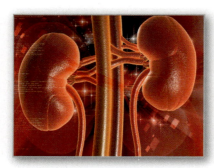

■ Solve simple counting problems.
■ Use the Fundamental Counting Principle to solve counting problems.
■ Use permutations to solve counting problems.
■ Use combinations to solve counting problems.

Simple Counting Problems

This section and Section 8.7 present a brief introduction to some of the basic counting principles and their applications to probability. In Section 8.7, you will see that much of probability has to do with counting the number of ways an event can occur. The two examples below describe simple counting problems.

Counting principles are useful for helping you solve counting problems that occur in real life. For example, in Exercise 39 on page 616, you will use counting principles to determine the number of possible orders there are for best match, second-best match, and third-best match kidney donors.

EXAMPLE 1   **Selecting Pairs of Numbers at Random**

You place eight pieces of paper, numbered from 1 to 8, in a box. You draw one piece of paper at random from the box, record its number, and *replace* the paper in the box. Then, you draw a second piece of paper at random from the box and record its number. Finally, you add the two numbers. How many different ways can you obtain a sum of 12?

Solution   To solve this problem, count the different ways to obtain a sum of 12 using two numbers from 1 to 8.

First number	4	5	6	7	8
Second number	8	7	6	5	4

So, a sum of 12 can occur in five different ways.

✓ *Checkpoint*   *Audio-video solution in English & Spanish at LarsonPrecalculus.com*

In Example 1, how many different ways can you obtain a sum of 14?

EXAMPLE 2   **Selecting Pairs of Numbers at Random**

You place eight pieces of paper, numbered from 1 to 8, in a box. You draw one piece of paper at random from the box, record its number, and *do not* replace the paper in the box. Then, you draw a second piece of paper at random from the box and record its number. Finally, you add the two numbers. How many different ways can you obtain a sum of 12?

Solution   To solve this problem, count the different ways to obtain a sum of 12 using two *different* numbers from 1 to 8.

First number	4	5	7	8
Second number	8	7	5	4

So, a sum of 12 can occur in four different ways.

✓ *Checkpoint*   *Audio-video solution in English & Spanish at LarsonPrecalculus.com*

In Example 2, how many different ways can you obtain a sum of 14?   ■

Notice the difference between the counting problems in Examples 1 and 2. The random selection in Example 1 occurs **with replacement,** whereas the random selection in Example 2 occurs **without replacement,** which eliminates the possibility of choosing two 6's.

The Fundamental Counting Principle

Examples 1 and 2 describe simple counting problems and *list* each possible way that an event can occur. When it is possible, this is always the best way to solve a counting problem. However, some events can occur in so many different ways that it is not feasible to write the entire list. In such cases, you must rely on formulas and counting principles. The most important of these is the **Fundamental Counting Principle.**

> **Fundamental Counting Principle**
>
> Let E_1 and E_2 be two events. The first event E_1 can occur in m_1 different ways. After E_1 has occurred, E_2 can occur in m_2 different ways. The number of ways the two events can occur is $m_1 \cdot m_2$.

The Fundamental Counting Principle can be extended to three or more events. For example, the number of ways that three events E_1, E_2, and E_3 can occur is

$$m_1 \cdot m_2 \cdot m_3.$$

EXAMPLE 3 **Using the Fundamental Counting Principle**

How many different pairs of letters from the English alphabet are possible?

Solution There are two events in this situation. The first event is the choice of the first letter, and the second event is the choice of the second letter. The English alphabet contains 26 letters, so it follows that the number of two-letter pairs is

$$26 \cdot 26 = 676.$$

✓ *Checkpoint* *Audio-video solution in English & Spanish at LarsonPrecalculus.com*

A combination lock will open when you select the right choice of three numbers (from 1 to 30, inclusive). How many different lock combinations are possible?

EXAMPLE 4 **Using the Fundamental Counting Principle**

Telephone numbers in the United States have 10 digits. The first three digits are the *area code* and the next seven digits are the *local telephone number*. How many different telephone numbers are possible within each area code? (Note that a local telephone number cannot begin with 0 or 1.)

Solution The first digit of a local telephone number cannot be 0 or 1, so there are only eight choices for the first digit. For each of the other six digits, there are 10 choices.

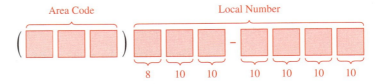

So, the number of telephone numbers that are possible within each area code is

$$8 \cdot 10 \cdot 10 \cdot 10 \cdot 10 \cdot 10 \cdot 10 = 8{,}000{,}000.$$

✓ *Checkpoint* *Audio-video solution in English & Spanish at LarsonPrecalculus.com*

A product's catalog number is made up of one letter from the English alphabet followed by a five-digit number. How many different catalog numbers are possible?

Permutations

One important application of the Fundamental Counting Principle is in determining the number of ways that *n* elements can be arranged (in order). An ordering of *n* elements is called a **permutation** of the elements.

Definition of a Permutation

A **permutation** of *n* different elements is an ordering of the elements such that one element is first, one is second, one is third, and so on.

EXAMPLE 5   **Finding the Number of Permutations**

How many permutations of the letters

A,   B,   C,   D,   E,   and   F

are possible?

Solution   Consider the reasoning below.

First position: Any of the *six* letters
Second position: Any of the remaining *five* letters
Third position: Any of the remaining *four* letters
Fourth position: Any of the remaining *three* letters
Fifth position: Either of the remaining *two* letters
Sixth position: The *one* remaining letter

So, the numbers of choices for the six positions are as shown in the figure.

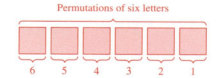

Permutations of six letters

| 6 | 5 | 4 | 3 | 2 | 1 |

The total number of permutations of the six letters is

$$6! = 6 \cdot 5 \cdot 4 \cdot 3 \cdot 2 \cdot 1 = 720.$$

✓ *Checkpoint* ◀))) *Audio-video solution in English & Spanish at LarsonPrecalculus.com*

How many permutations of the letters

W,   X,   Y,   and   Z

are possible?

Generalizing the result in Example 5, the number of permutations of *n* different elements is *n*!.

Number of Permutations of *n* Elements

The number of permutations of *n* elements is

$$n \cdot (n-1) \cdots 4 \cdot 3 \cdot 2 \cdot 1 = n!.$$

In other words, there are *n*! different ways of ordering *n* elements.

As of 2015, twelve thoroughbred racehorses hold the title of Triple Crown winner for winning the Kentucky Derby, the Preakness Stakes, and the Belmont Stakes in the same year. Fifty-two horses have won two out of the three races.

It is useful, on occasion, to order a *subset* of a collection of elements rather than the entire collection. For example, you may want to order r elements out of a collection of n elements. Such an ordering is called a **permutation of n elements taken r at a time.** The next example demonstrates this ordering.

EXAMPLE 6   **Counting Horse Race Finishes**

Eight horses are running in a race. In how many different ways can these horses come in first, second, and third? (Assume that there are no ties.)

Solution   Here are the different possibilities.

Win (first position): *Eight* choices

Place (second position): *Seven* choices

Show (third position): *Six* choices

The numbers of choices for the three positions are as shown in the figure.

Different orders of horses

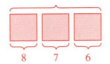

8   7   6

So, using the Fundamental Counting Principle, there are

$$8 \cdot 7 \cdot 6 = 336$$

different ways in which the eight horses can come in first, second, and third.

✓ *Checkpoint* 🔊))) *Audio-video solution in English & Spanish at LarsonPrecalculus.com*

A coin club has five members. In how many different ways can there be a president and a vice-president?

Generalizing the result in Example 6 gives the formula below.

▷ **TECHNOLOGY**

Most graphing utilities can evaluate $_nP_r$. If yours can, use it to evaluate several permutations. Check your results algebraically by hand.

Permutations of n Elements Taken r at a Time

The number of permutations of n elements taken r at a time is

$$_nP_r = \frac{n!}{(n-r)!} = n(n-1)(n-2) \cdots (n-r+1).$$

Using this formula, rework Example 6 to find that the number of permutations of eight horses taken three at a time is

$$_8P_3 = \frac{8!}{(8-3)!}$$

$$= \frac{8!}{5!}$$

$$= \frac{8 \cdot 7 \cdot 6 \cdot 5!}{5!}$$

$$= 336$$

which is the same answer obtained in the example.

Remember that for permutations, order is important. For example, to find the possible permutations of the letters A, B, C, and D taken three at a time, count A, B, D and B, A, D as different because the *order* of the elements is different.

Consider, however, the possible permutations of the letters A, A, B, and C. The total number of permutations of the four letters is $_4P_4 = 4!$. However, not all of these arrangements are *distinguishable* because there are two A's in the list. To find the number of distinguishable permutations, use the formula below.

Distinguishable Permutations

Consider a set of n objects that has n_1 of one kind of object, n_2 of a second kind, n_3 of a third kind, and so on, with

$$n = n_1 + n_2 + n_3 + \cdots + n_k.$$

The number of **distinguishable permutations** of the n objects is

$$\frac{n!}{n_1! \cdot n_2! \cdot n_3! \cdot \cdots \cdot n_k!}.$$

EXAMPLE 7 **Distinguishable Permutations**

See LarsonPrecalculus.com for an interactive version of this type of example.

In how many distinguishable ways can the letters in BANANA be written?

Solution This word has six letters, of which three are A's, two are N's, and one is a B. So, the number of distinguishable ways the letters can be written is

$$\frac{n!}{n_1! \cdot n_2! \cdot n_3!} = \frac{6!}{3! \cdot 2! \cdot 1!}$$

$$= \frac{6 \cdot 5 \cdot 4 \cdot 3!}{3! \cdot 2!}$$

$$= 60.$$

The 60 different distinguishable permutations are as listed below.

AAABNN	AAANBN	AAANNB	AABANN
AABNAN	AABNNA	AANABN	AANANB
AANBAN	AANBNA	AANNAB	AANNBA
ABAANN	ABANAN	ABANNA	ABNAAN
ABNANA	ABNNAA	ANAABN	ANAANB
ANABAN	ANABNA	ANANAB	ANANBA
ANBAAN	ANBANA	ANBNAA	ANNAAB
ANNABA	ANNBAA	BAAANN	BAANAN
BAANNA	BANAAN	BANANA	BANNAA
BNAAAN	BNAANA	BNANAA	BNNAAA
NAAABN	NAAANB	NAABAN	NAABNA
NAANAB	NAANBA	NABAAN	NABANA
NABNAA	NANAAB	NANABA	NANBAA
NBAAAN	NBAANA	NBANAA	NBNAAA
NNAAAB	NNAABA	NNABAA	NNBAAA

✓ **Checkpoint** ◀))) *Audio-video solution in English & Spanish at LarsonPrecalculus.com*

In how many distinguishable ways can the letters in MITOSIS be written?

Combinations

When you count the number of possible permutations of a set of elements, order is important. As a final topic in this section, you will look at a method of selecting subsets of a larger set in which order is *not* important. Such subsets are called **combinations of *n* elements taken *r* at a time.** For example, the combinations

$$\{A, B, C\} \quad \text{and} \quad \{B, A, C\}$$

are equivalent because both sets contain the same three elements, and the order in which the elements are listed is not important. So, you would count only one of the two sets. Another example of how a combination occurs is in a card game in which players are free to reorder the cards after they have been dealt.

EXAMPLE 8 **Combinations of *n* Elements Taken *r* at a Time**

In how many different ways can three letters be chosen from the letters

A, B, C, D, and E?

(The order of the three letters is not important.)

Solution The subsets listed below represent the different combinations of three letters that can be chosen from the five letters.

$$\{A, B, C\} \qquad \{A, B, D\}$$
$$\{A, B, E\} \qquad \{A, C, D\}$$
$$\{A, C, E\} \qquad \{A, D, E\}$$
$$\{B, C, D\} \qquad \{B, C, E\}$$
$$\{B, D, E\} \qquad \{C, D, E\}$$

So, when order is not important, there are 10 different ways that three letters can be chosen from five letters.

✓ *Checkpoint*  *Audio-video solution in English & Spanish at LarsonPrecalculus.com*

In how many different ways can two letters be chosen from the letters A, B, C, D, E, F, and G? (The order of the two letters is not important.)

Combinations of *n* Elements Taken *r* at a Time

The number of combinations of *n* elements taken *r* at a time is

$$_nC_r = \frac{n!}{(n-r)!r!}$$

which is equivalent to $_nC_r = \dfrac{_nP_r}{r!}$.

····REMARK Note that the formula for $_nC_r$ is the same one given for binomial coefficients.

To see how to use this formula, rework the counting problem in Example 8. In that problem, you want to find the number of combinations of five elements taken three at a time. So, $n = 5$, $r = 3$, and the number of combinations is

$$_5C_3 = \frac{5!}{2!3!} = \frac{5 \cdot \overset{2}{\cancel{4}} \cdot \cancel{3!}}{\cancel{2} \cdot 1 \cdot \cancel{3!}} = 10$$

which is the same answer obtained in the example.

Ranks and suits in a standard deck of playing cards
Figure 8.2

EXAMPLE 9   **Counting Card Hands**

A standard poker hand consists of five cards dealt from a deck of 52 (see Figure 8.2). How many different poker hands are possible? (Order is not important.)

Solution   To determine the number of different poker hands, find the number of combinations of 52 elements taken five at a time.

$$
\begin{aligned}
{}_{52}C_5 &= \frac{52!}{(52-5)!5!} \\[4pt]
&= \frac{52!}{47!5!} \\[4pt]
&= \frac{52 \cdot 51 \cdot 50 \cdot 49 \cdot 48 \cdot \cancel{47!}}{\cancel{47!} \cdot 5 \cdot 4 \cdot 3 \cdot 2 \cdot 1} \\[4pt]
&= 2{,}598{,}960
\end{aligned}
$$

✓ *Checkpoint*))) *Audio-video solution in English & Spanish at LarsonPrecalculus.com*

In three-card poker, a hand consists of three cards dealt from a deck of 52. How many different three-card poker hands are possible? (Order is not important.)

EXAMPLE 10   **Forming a Team**

You are forming a 12-member swim team from 10 girls and 15 boys. The team must consist of five girls and seven boys. How many different 12-member teams are possible?

Solution   There are ${}_{10}C_5$ ways of choosing five girls. There are ${}_{15}C_7$ ways of choosing seven boys. By the Fundamental Counting Principle, there are ${}_{10}C_5 \cdot {}_{15}C_7$ ways of choosing five girls and seven boys.

$$
{}_{10}C_5 \cdot {}_{15}C_7 = \frac{10!}{5! \cdot 5!} \cdot \frac{15!}{8! \cdot 7!} = 252 \cdot 6435 = 1{,}621{,}620
$$

So, there are 1,621,620 12-member swim teams possible.

✓ *Checkpoint*))) *Audio-video solution in English & Spanish at LarsonPrecalculus.com*

In Example 10, the team must consist of six boys and six girls. How many different 12-member teams are possible?

Summarize   **(Section 8.6)**

1. Explain how to solve a simple counting problem *(page 608)*. For examples of solving simple counting problems, see Examples 1 and 2.

2. State the Fundamental Counting Principle *(page 609)*. For examples of using the Fundamental Counting Principle to solve counting problems, see Examples 3 and 4.

3. Explain how to find the number of permutations of *n* elements *(page 610)*, the number of permutations of *n* elements taken *r* at a time *(page 611)*, and the number of distinguishable permutations *(page 612)*. For examples of using permutations to solve counting problems, see Examples 5–7.

4. Explain how to find the number of combinations of *n* elements taken *r* at a time *(page 613)*. For examples of using combinations to solve counting problems, see Examples 8–10.

8.6 Exercises

See **CalcChat.com** for tutorial help and worked-out solutions to odd-numbered exercises.

Vocabulary: Fill in the blanks.

1. The _____ _____ _____ states that when there are m_1 different ways for one event to occur and m_2 different ways for a second event to occur, there are $m_1 \cdot m_2$ ways for both events to occur.

2. An ordering of n elements is a _____ of the elements.

3. The number of permutations of n elements taken r at a time is given by _____.

4. The number of _____ _____ of n objects is given by $\dfrac{n!}{n_1! \cdot n_2! \cdot n_3! \cdot \, \cdots \, \cdot n_k!}$.

5. When selecting subsets of a larger set in which order is not important, you are finding the number of _____ of n elements taken r at a time.

6. The number of combinations of n elements taken r at a time is given by _____.

Skills and Applications

Random Selection In Exercises 7–14, determine the number of ways a computer can randomly generate one or more such integers from 1 through 12.

7. An odd integer
8. An even integer
9. A prime integer
10. An integer that is greater than 9
11. An integer that is divisible by 4
12. An integer that is divisible by 3
13. Two *distinct* integers whose sum is 9
14. Two *distinct* integers whose sum is 8

15. **Entertainment Systems** A customer can choose one of three amplifiers, one of two compact disc players, and one of five speaker models for an entertainment system. Determine the number of possible system configurations.

16. **Job Applicants** A small college needs two additional faculty members: a chemist and a statistician. There are five applicants for the chemistry position and three applicants for the statistics position. In how many ways can the college fill these positions?

17. **Course Schedule** A college student is preparing a course schedule for the next semester. The student may select one of two mathematics courses, one of three science courses, and one of five courses from the social sciences. How many schedules are possible?

18. **Physiology** In a physiology class, a student must dissect three different specimens. The student can select one of nine earthworms, one of four frogs, and one of seven fetal pigs. In how many ways can the student select the specimens?

19. **True-False Exam** In how many ways can you answer a six-question true-false exam? (Assume that you do not omit any questions.)

20. **True-False Exam** In how many ways can you answer a 12-question true-false exam? (Assume that you do not omit any questions.)

21. **License Plate Numbers** In the state of Pennsylvania, each standard automobile license plate number consists of three letters followed by a four-digit number. How many distinct license plate numbers are possible in Pennsylvania?

22. **License Plate Numbers** In a certain state, each automobile license plate number consists of two letters followed by a four-digit number. To avoid confusion between "O" and "zero" and between "I" and "one," the letters "O" and "I" are not used. How many distinct license plate numbers are possible in this state?

23. **Three-Digit Numbers** How many three-digit numbers are possible under each condition?
   (a) The leading digit cannot be zero.
   (b) The leading digit cannot be zero and no repetition of digits is allowed.
   (c) The leading digit cannot be zero and the number must be a multiple of 5.
   (d) The number is at least 400.

24. **Four-Digit Numbers** How many four-digit numbers are possible under each condition?
   (a) The leading digit cannot be zero.
   (b) The leading digit cannot be zero and no repetition of digits is allowed.
   (c) The leading digit cannot be zero and the number must be less than 5000.
   (d) The leading digit cannot be zero and the number must be even.

25. **Combination Lock** A combination lock will open when you select the right choice of three numbers (from 1 to 40, inclusive). How many different lock combinations are possible?

26. Combination Lock A combination lock will open when you select the right choice of three numbers (from 1 to 50, inclusive). How many different lock combinations are possible?

27. Concert Seats Four couples reserve seats in one row for a concert. In how many different ways can they sit when

(a) there are no seating restrictions?

(b) the two members of each couple wish to sit together?

28. Single File In how many orders can four girls and four boys walk through a doorway single file when

(a) there are no restrictions?

(b) the girls walk through before the boys?

29. Posing for a Photograph In how many ways can five children posing for a photograph line up in a row?

30. Riding in a Car In how many ways can six people sit in a six-passenger car?

 Evaluating $_nP_r$ In Exercises 31–34, evaluate $_nP_r$.

31. $_5P_2$ **32.** $_6P_6$ **33.** $_{12}P_2$ **34.** $_6P_5$

Evaluating $_nP_r$ In Exercises 35–38, use a graphing utility to evaluate $_nP_r$.

35. $_{15}P_3$ **36.** $_{100}P_4$ **37.** $_{50}P_4$ **38.** $_{10}P_5$

39. Kidney Donors

A patient with end-stage kidney disease has nine family members who are potential kidney donors. How many possible orders are there for a best match, a second-best match, and a third-best match?

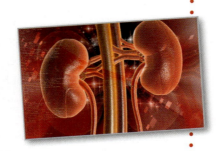

40. Choosing Officers From a pool of 12 candidates, the offices of president, vice-president, secretary, and treasurer need to be filled. In how many different ways can the offices be filled?

41. Batting Order A baseball coach is creating a nine-player batting order by selecting from a team of 15 players. How many different batting orders are possible?

42. Athletics Eight sprinters qualify for the finals in the 100-meter dash at the NCAA national track meet. In how many ways can the sprinters come in first, second, and third? (Assume there are no ties.)

 Number of Distinguishable Permutations In Exercises 43–46, find the number of distinguishable permutations of the group of letters.

43. A, A, G, E, E, E, M   **44.** B, B, B, T, T, T, T, T

45. A, L, G, E, B, R, A   **46.** M, I, S, S, I, S, S, I, P, P, I

47. Writing Permutations Write all permutations of the letters A, B, C, and D.

48. Writing Permutations Write all permutations of the letters A, B, C, and D when letters B and C must remain between A and D.

 Evaluating $_nC_r$ In Exercises 49–52, evaluate $_nC_r$ using the formula from this section.

49. $_6C_4$ **50.** $_5C_4$ **51.** $_9C_9$ **52.** $_{12}C_0$

Evaluating $_nC_r$ In Exercises 53–56, use a graphing utility to evaluate $_nC_r$.

53. $_{16}C_2$ **54.** $_{17}C_5$ **55.** $_{20}C_6$ **56.** $_{50}C_8$

57. Writing Combinations Write all combinations of two letters that can be formed from the letters A, B, C, D, E, and F. (Order is not important.)

58. Forming an Experimental Group To conduct an experiment, researchers randomly select five students from a class of 20. How many different groups of five students are possible?

59. Jury Selection In how many different ways can a jury of 12 people be randomly selected from a group of 40 people?

60. Committee Members A U.S. Senate Committee has 14 members. Assuming party affiliation is not a factor in selection, how many different committees are possible from the 100 U.S. senators?

61. Lottery Choices In the Massachusetts Mass Cash game, a player randomly chooses five distinct numbers from 1 to 35. In how many ways can a player select the five numbers?

62. Lottery Choices In the Louisiana Lotto game, a player randomly chooses six distinct numbers from 1 to 40. In how many ways can a player select the six numbers?

63. Defective Units A shipment of 25 television sets contains three defective units. In how many ways can a vending company purchase four of these units and receive (a) all good units, (b) two good units, and (c) at least two good units?

64. Interpersonal Relationships The complexity of interpersonal relationships increases dramatically as the size of a group increases. Determine the numbers of different two-person relationships in groups of people of sizes (a) 3, (b) 8, (c) 12, and (d) 20.

65. Poker Hand You are dealt five cards from a standard deck of 52 playing cards. In how many ways can you get (a) a full house and (b) a five-card combination containing two jacks and three aces? (A full house consists of three of one kind and two of another. For example, A-A-A-5-5 and K-K-K-10-10 are full houses.)

66. Job Applicants An employer interviews 12 people for four openings at a company. Five of the 12 people are women. All 12 applicants are qualified. In how many ways can the employer fill the four positions when (a) the selection is random and (b) exactly two selections are women?

67. Forming a Committee A local college is forming a six-member research committee with one administrator, three faculty members, and two students. There are seven administrators, 12 faculty members, and 20 students in contention for the committee. How many six-member committees are possible?

68. Law Enforcement A police department uses computer imaging to create digital photographs of alleged perpetrators from eyewitness accounts. One software package contains 195 hairlines, 99 sets of eyes and eyebrows, 89 noses, 105 mouths, and 74 chin and cheek structures.

  (a) Find the possible number of different faces that the software could create.

  (b) An eyewitness can clearly recall the hairline and eyes and eyebrows of a suspect. How many different faces are possible with this information?

Geometry In Exercises 69–72, find the number of diagonals of the polygon. (A *diagonal* is a line segment connecting any two nonadjacent vertices of a polygon.)

69. Pentagon

70. Hexagon

71. Octagon

72. Decagon (10 sides)

73. Geometry Three points that are not collinear determine three lines. How many lines are determined by nine points, no three of which are collinear?

74. Lottery Powerball is a lottery game that is operated by the Multi-State Lottery Association and is played in 44 states, Washington D.C., Puerto Rico, and the U.S. Virgin Islands. The game is played by drawing five white balls out of a drum of 69 white balls (numbered 1–69) and one red powerball out of a drum of 26 red balls (numbered 1–26). The jackpot is won by matching all five white balls in any order and the red powerball.

  (a) Find the possible number of winning Powerball numbers.

  (b) Find the possible number of winning Powerball numbers when you win the jackpot by matching all five white balls in order and the red powerball.

Solving an Equation In Exercises 75–82, solve for n.

75. $4 \cdot {}_{n+1}P_2 = {}_{n+2}P_3$

76. $5 \cdot {}_{n-1}P_1 = {}_nP_2$

77. ${}_{n+1}P_3 = 4 \cdot {}_nP_2$

78. ${}_{n+2}P_3 = 6 \cdot {}_{n+2}P_1$

79. $14 \cdot {}_nP_3 = {}_{n+2}P_4$

80. ${}_nP_5 = 18 \cdot {}_{n-2}P_4$

81. ${}_nP_4 = 10 \cdot {}_{n-1}P_3$

82. ${}_nP_6 = 12 \cdot {}_{n-1}P_5$

Exploration

True or False? In Exercises 83 and 84, determine whether the statement is true or false. Justify your answer.

83. The number of letter pairs that can be formed in any order from any two of the first 13 letters in the alphabet (A–M) is an example of a permutation.

84. The number of permutations of n elements can be determined by using the Fundamental Counting Principle.

85. Think About It Without calculating, determine which of the following is greater. Explain.

  (a) The number of combinations of 10 elements taken six at a time

  (b) The number of permutations of 10 elements taken six at a time

86. HOW DO YOU SEE IT? Without calculating, determine whether the value of ${}_nP_r$ is greater than the value of ${}_nC_r$ for the values of n and r given in the table. Complete the table using yes (Y) or no (N). Is the value of ${}_nP_r$ always greater than the value of ${}_nC_r$? Explain.

n \ r	0	1	2	3	4	5	6	7
1								
2								
3								
4								
5								
6								
7								

Proof In Exercises 87–90, prove the identity.

87. ${}_nP_{n-1} = {}_nP_n$

88. ${}_nC_n = {}_nC_0$

89. ${}_nC_{n-1} = {}_nC_1$

90. ${}_nC_r = \dfrac{{}_nP_r}{r!}$

91. Think About It Can your graphing utility evaluate ${}_{100}P_{80}$? If not, explain why.

8.7 Probability

Probability applies to many real-life applications. For example, in Exercise 59 on page 628, you will find probabilities that relate to a communication network and an independent backup system for a space vehicle.

- Find probabilities of events.
- Find probabilities of mutually exclusive events.
- Find probabilities of independent events.
- Find the probability of the complement of an event.

The Probability of an Event

Any happening for which the result is uncertain is an **experiment.** The possible results of the experiment are **outcomes,** the set of all possible outcomes of the experiment is the **sample space** of the experiment, and any subcollection of a sample space is an **event.**

For example, when you toss a six-sided die, the numbers 1 through 6 can represent the sample space. For this experiment, each of the outcomes is *equally likely*.

To describe sample spaces in such a way that each outcome is equally likely, you must sometimes distinguish between or among various outcomes in ways that appear artificial. Example 1 illustrates such a situation.

EXAMPLE 1   **Finding a Sample Space**

Find the sample space for each experiment.

a. You toss one coin.

b. You toss two coins.

c. You toss three coins.

Solution

a. The coin will land either heads up (denoted by H) or tails up (denoted by T), so the sample space is

$$S = \{H, T\}.$$

b. Either coin can land heads up or tails up, so the possible outcomes are as follows.

$HH =$ heads up on both coins

$HT =$ heads up on the first coin and tails up on the second coin

$TH =$ tails up on the first coin and heads up on the second coin

$TT =$ tails up on both coins

So, the sample space is

$$S = \{HH, HT, TH, TT\}.$$

Note that this list distinguishes between the two cases HT and TH, even though these two outcomes appear to be similar.

c. Using notation similar to that used in part (b), the sample space is

$$S = \{HHH, HHT, HTH, HTT, THH, THT, TTH, TTT\}.$$

Note that this list distinguishes among the cases HHT, HTH, and THH, and among the cases HTT, THT, and TTH.

 Checkpoint *Audio-video solution in English & Spanish at LarsonPrecalculus.com*

Find the sample space for the experiment.

You toss a coin twice and a six-sided die once.

To find the probability of an event, count the number of outcomes in the event and in the sample space. The *number of equally likely outcomes* in event E is denoted by $n(E)$, and the number of equally likely outcomes in the sample space S is denoted by $n(S)$. The probability that event E will occur is given by $n(E)/n(S)$.

The Probability of an Event

If an event E has $n(E)$ equally likely outcomes and its sample space S has $n(S)$ equally likely outcomes, then the **probability** of event E is

$$P(E) = \frac{n(E)}{n(S)}.$$

The number of outcomes in an event must be less than or equal to the number of outcomes in the sample space, so the probability of an event must be a number between 0 and 1, inclusive. That is,

$$0 \le P(E) \le 1$$

as shown in the figure. If $P(E) = 0$, then event E *cannot occur,* and E is an **impossible event.** If $P(E) = 1$, then event E *must occur,* and E is a **certain event.**

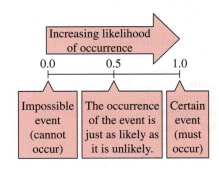

EXAMPLE 2 **Finding the Probability of an Event**

See LarsonPrecalculus.com for an interactive version of this type of example.

a. You toss two coins. What is the probability that both land heads up?

b. You draw one card at random from a standard deck of 52 playing cards. What is the probability that it is an ace?

Solution

a. Using the results of Example 1(b), let

$$E = \{HH\} \quad \text{and} \quad S = \{HH, HT, TH, TT\}.$$

The probability of getting two heads is

$$P(E) = \frac{n(E)}{n(S)} = \frac{1}{4}.$$

b. The deck has four aces (one in each suit), so the probability of drawing an ace is

$$P(E) = \frac{n(E)}{n(S)} = \frac{4}{52} = \frac{1}{13}.$$

REMARK
You can write a probability as a fraction, a decimal, or a percent. For instance, in Example 2(a), the probability of getting two heads can be written as $\frac{1}{4}$, 0.25, or 25%.

✓ *Checkpoint* 🔊)) *Audio-video solution in English & Spanish at LarsonPrecalculus.com*

a. You toss three coins. What is the probability that all three land tails up?

b. You draw one card at random from a standard deck of 52 playing cards. What is the probability that it is a diamond?

In some cases, the number of outcomes in the sample space may not be given. In these cases, either write out the sample space or use the counting principles discussed in Section 8.6. Example 3 on the next page uses the Fundamental Counting Principle.

As shown in Example 3, when you toss two six-sided dice, the probability of rolling a total of 7 is $\frac{1}{6}$.

EXAMPLE 3 **Finding the Probability of an Event**

You toss two six-sided dice. What is the probability that the total of the two dice is 7?

Solution There are six possible outcomes on each die, so by the Fundamental Counting Principle, there are 6 · 6 or 36 different outcomes when you toss two dice. To find the probability of rolling a total of 7, you must first count the number of ways in which this can occur.

First Die	Second Die
1	6
2	5
3	4
4	3
5	2
6	1

So, a total of 7 can be rolled in six ways, which means that the probability of rolling a total of 7 is

$$P(E) = \frac{n(E)}{n(S)} = \frac{6}{36} = \frac{1}{6}.$$

✓ *Checkpoint*)) *Audio-video solution in English & Spanish at LarsonPrecalculus.com*

You toss two six-sided dice. What is the probability that the total of the two dice is 5?

EXAMPLE 4 **Finding the Probability of an Event**

Twelve-sided dice, as shown in Figure 8.3, can be constructed (in the shape of regular dodecahedrons) such that each of the numbers from 1 to 6 occurs twice on each die. Show that these dice can be used in any game requiring ordinary six-sided dice without changing the probabilities of the various events.

Solution For an ordinary six-sided die, each of the numbers

   1, 2, 3, 4, 5, and 6

occurs once, so the probability of rolling any one of these numbers is

$$P(E) = \frac{n(E)}{n(S)} = \frac{1}{6}.$$

For one of the 12-sided dice, each number occurs twice, so the probability of rolling each number is

$$P(E) = \frac{n(E)}{n(S)} = \frac{2}{12} = \frac{1}{6}.$$

Figure 8.3

✓ *Checkpoint*)) *Audio-video solution in English & Spanish at LarsonPrecalculus.com*

Show that the probability of drawing a club at random from a standard deck of 52 playing cards is the same as the probability of drawing the ace of hearts at random from a set of four cards consisting of the aces of hearts, diamonds, clubs, and spades.

EXAMPLE 5 **Random Selection**

The figure shows the numbers of degree-granting postsecondary institutions in various regions of the United States in 2015. What is the probability that an institution selected at random is in one of the three southern regions? *(Source: National Center for Education Statistics)*

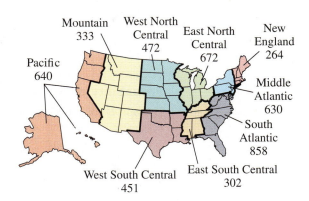

Mountain 333
West North Central 472
East North Central 672
New England 264
Pacific 640
Middle Atlantic 630
South Atlantic 858
West South Central 451
East South Central 302

Solution From the figure, the total number of institutions is 4622. There are $858 + 302 + 451 = 1611$ institutions in the three southern regions, so the probability that the institution is in one of these regions is

$$P(E) = \frac{n(E)}{n(S)} = \frac{1611}{4622} \approx 0.349.$$

 Checkpoint  *Audio-video solution in English & Spanish at LarsonPrecalculus.com*

In Example 5, what is the probability that an institution selected at random is in the Pacific region?

EXAMPLE 6 **Finding the Probability of Winning a Lottery**

In Arizona's The Pick game, a player chooses six different numbers from 1 to 44. If these six numbers match the six numbers drawn (in any order), the player wins (or shares) the top prize. What is the probability of winning the top prize when the player buys one ticket?

Solution To find the number of outcomes in the sample space, use the formula for the number of combinations of 44 numbers taken six at a time.

$$n(S) = {}_{44}C_6$$
$$= \frac{44 \cdot 43 \cdot 42 \cdot 41 \cdot 40 \cdot 39}{6 \cdot 5 \cdot 4 \cdot 3 \cdot 2 \cdot 1}$$
$$= 7,059,052$$

When a player buys one ticket, the probability of winning is

$$P(E) = \frac{1}{7,059,052}.$$

 Checkpoint *Audio-video solution in English & Spanish at LarsonPrecalculus.com*

In Pennsylvania's Cash 5 game, a player chooses five different numbers from 1 to 43. If these five numbers match the five numbers drawn (in any order), the player wins (or shares) the top prize. What is the probability of winning the top prize when the player buys one ticket?

Mutually Exclusive Events

Two events A and B (from the same sample space) are **mutually exclusive** when A and B have no outcomes in common. In the terminology of sets, the intersection of A and B is the empty set, which implies that

$$P(A \cap B) = 0.$$

For example, when you toss two dice, the event A of rolling a total of 6 and the event B of rolling a total of 9 are mutually exclusive. To find the probability that one or the other of two mutually exclusive events will occur, *add* their individual probabilities.

Probability of the Union of Two Events

If A and B are events in the same sample space, then the probability of A or B occurring is given by

$$P(A \cup B) = P(A) + P(B) - P(A \cap B).$$

If A and B are mutually exclusive, then

$$P(A \cup B) = P(A) + P(B).$$

EXAMPLE 7   **Probability of a Union of Events**

You draw one card at random from a standard deck of 52 playing cards. What is the probability that the card is either a heart or a face card?

Solution   The deck has 13 hearts, so the probability of drawing a heart (event A) is

$$P(A) = \frac{13}{52}.$$

Similarly, the deck has 12 face cards, so the probability of drawing a face card (event B) is

$$P(B) = \frac{12}{52}.$$

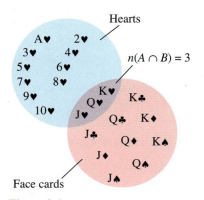

Hearts

$n(A \cap B) = 3$

Face cards

Figure 8.4

Three of the cards are hearts *and* face cards (see Figure 8.4), so it follows that

$$P(A \cap B) = \frac{3}{52}.$$

Finally, applying the formula for the probability of the union of two events, the probability of drawing either a heart or a face card is

$$P(A \cup B) = P(A) + P(B) - P(A \cap B)$$

$$= \frac{13}{52} + \frac{12}{52} - \frac{3}{52}$$

$$= \frac{22}{52}$$

$$\approx 0.423.$$

✓ *Checkpoint*)) *Audio-video solution in English & Spanish at LarsonPrecalculus.com*

You draw one card at random from a standard deck of 52 playing cards. What is the probability that the card is either an ace or a spade?

EXAMPLE 8   **Probability of Mutually Exclusive Events**

The human resources department of a company has compiled data showing the number of years of service for each employee. The table shows the results.

Years of Service	Number of Employees
0–4	157
5–9	89
10–14	74
15–19	63
20–24	42
25–29	38
30–34	35
35–39	21
40–44	8
45 or more	2

DATA — Spreadsheet at LarsonPrecalculus.com

a. What is the probability that an employee chosen at random has 4 or fewer years of service?

b. What is the probability that an employee chosen at random has 9 or fewer years of service?

Solution

a. To begin, add the number of employees to find that the total is 529. Next, let event A represent choosing an employee with 0 to 4 years of service. Then the probability of choosing an employee who has 4 or fewer years of service is

$$P(A) = \frac{157}{529}$$

$$\approx 0.297.$$

b. Let event B represent choosing an employee with 5 to 9 years of service. Then

$$P(B) = \frac{89}{529}.$$

Event A from part (a) and event B have no outcomes in common, so these two events are mutually exclusive and

$$P(A \cup B) = P(A) + P(B)$$

$$= \frac{157}{529} + \frac{89}{529}$$

$$= \frac{246}{529}$$

$$\approx 0.465.$$

So, the probability of choosing an employee who has 9 or fewer years of service is about 0.465.

 Checkpoint *Audio-video solution in English & Spanish at LarsonPrecalculus.com*

In Example 8, what is the probability that an employee chosen at random has 30 or more years of service?

Independent Events

Two events are **independent** when the occurrence of one has no effect on the occurrence of the other. For example, rolling a total of 12 with two six-sided dice has no effect on the outcome of future rolls of the dice. To find the probability that two independent events will occur, *multiply* the probabilities of each.

Probability of Independent Events

If A and B are independent events, then the probability that both A and B will occur is

$$P(A \text{ and } B) = P(A) \cdot P(B).$$

This rule can be extended to any number of independent events.

EXAMPLE 9 **Probability of Independent Events**

A random number generator selects three integers from 1 to 20. What is the probability that all three numbers are less than or equal to 5?

Solution Let event A represent selecting a number from 1 to 5. Then the probability of selecting a number from 1 to 5 is

$$P(A) = \frac{5}{20} = \frac{1}{4}.$$

So, the probability that all three numbers are less than or equal to 5 is

$$P(A) \cdot P(A) \cdot P(A) = \left(\frac{1}{4}\right)\left(\frac{1}{4}\right)\left(\frac{1}{4}\right)$$

$$= \frac{1}{64}.$$

✓ *Checkpoint* 🔊))) *Audio-video solution in English & Spanish at LarsonPrecalculus.com*

A random number generator selects two integers from 1 to 30. What is the probability that both numbers are less than 12?

EXAMPLE 10 **Probability of Independent Events**

In 2015, approximately 65% of Americans expected much of the workforce to be automated within 50 years. In a survey, researchers selected 10 people at random from the population. What is the probability that all 10 people expected much of the workforce to be automated within 50 years? *(Source: Pew Research Center)*

Solution Let event A represent selecting a person who expected much of the workforce to be automated within 50 years. The probability of event A is 0.65. Each of the 10 occurrences of event A is an independent event, so the probability that all 10 people expected much of the workforce to be automated within 50 years is

$$[P(A)]^{10} = (0.65)^{10}$$

$$\approx 0.013.$$

✓ *Checkpoint* 🔊))) *Audio-video solution in English & Spanish at LarsonPrecalculus.com*

In Example 10, researchers selected five people at random from the population. What is the probability that all five people expected much of the workforce to be automated within 50 years?

The Complement of an Event

The **complement of an event** A is the collection of all outcomes in the sample space that are *not* in A. The complement of event A is denoted by A'. Because $P(A \text{ or } A') = 1$ and A and A' are mutually exclusive, it follows that $P(A) + P(A') = 1$. So, the probability of A' is

$$P(A') = 1 - P(A).$$

Probability of a Complement

Let A be an event and let A' be its complement. If the probability of A is $P(A)$, then the probability of the complement is

$$P(A') = 1 - P(A).$$

For example, if the probability of *winning* a game is $P(A) = \frac{1}{4}$, then the probability of *losing* the game is $P(A') = 1 - \frac{1}{4} = \frac{3}{4}$.

EXAMPLE 11   Probability of a Complement

A manufacturer has determined that a machine averages one faulty unit for every 1000 it produces. What is the probability that an order of 200 units will have one or more faulty units?

Solution   To solve this problem as stated, you would need to find the probabilities of having exactly one faulty unit, exactly two faulty units, exactly three faulty units, and so on. However, using complements, it is much less tedious to find the probability that all units are perfect and then subtract this value from 1. The probability that any given unit is perfect is 999/1000, so the probability that all 200 units are perfect is

$$P(A) = \left(\frac{999}{1000}\right)^{200} \approx 0.819$$

and the probability that at least one unit is faulty is

$$P(A') = 1 - P(A) \approx 1 - 0.819 = 0.181.$$

✓ *Checkpoint* ◀)) *Audio-video solution in English & Spanish at LarsonPrecalculus.com*

A manufacturer has determined that a machine averages one faulty unit for every 500 it produces. What is the probability that an order of 300 units will have one or more faulty units?

Summarize   (Section 8.7)

1. State the definition of the probability of an event *(page 619)*. For examples of finding the probabilities of events, see Examples 2–6.

2. State the definition of mutually exclusive events and explain how to find the probability of the union of two events *(page 622)*. For examples of finding the probabilities of the unions of two events, see Examples 7 and 8.

3. State the definition of, and explain how to find the probability of, independent events *(page 624)*. For examples of finding the probabilities of independent events, see Examples 9 and 10.

4. State the definition of, and explain how to find the probability of, the complement of an event *(page 625)*. For an example of finding the probability of the complement of an event, see Example 11.

8.7 Exercises

See **CalcChat.com** for tutorial help and worked-out solutions to odd-numbered exercises.

Vocabulary

In Exercises 1–7, fill in the blanks.

1. An _____ is any happening for which the result is uncertain, and the possible results are called _____.

2. The set of all possible outcomes of an experiment is the _____ _____.

3. The formula for the _____ of an event is $P(E) = \dfrac{n(E)}{n(S)}$, where $n(E)$ is the number of equally likely outcomes in the event and $n(S)$ is the number of equally likely outcomes in the sample space.

4. If $P(E) = 0$, then E is an _____ event, and if $P(E) = 1$, then E is a _____ event.

5. Two events A and B (from the same sample space) are _____ _____ when A and B have no outcomes in common.

6. Two events are _____ when the occurrence of one has no effect on the occurrence of the other.

7. The _____ of an event A is the collection of all outcomes in the sample space that are not in A.

8. Match the probability formula with the correct probability name.
   (a) Probability of the union of two events
   (b) Probability of mutually exclusive events
   (c) Probability of independent events
   (d) Probability of a complement

   (i) $P(A \cup B) = P(A) + P(B)$
   (ii) $P(A') = 1 - P(A)$
   (iii) $P(A \cup B) = P(A) + P(B) - P(A \cap B)$
   (iv) $P(A \text{ and } B) = P(A) \cdot P(B)$

Skills and Applications

Finding a Sample Space In Exercises 9–14, find the sample space for the experiment.

9. You toss a coin and a six-sided die.

10. You toss a six-sided die twice and record the sum.

11. A taste tester ranks three varieties of yogurt, A, B, and C, according to preference.

12. You select two marbles (without replacement) from a bag containing two red marbles, two blue marbles, and one yellow marble. You record the color of each marble.

13. Two county supervisors are selected from five supervisors, A, B, C, D, and E, to study a recycling plan.

14. A sales representative visits three homes per day. In each home, there may be a sale (denote by S) or there may be no sale (denote by F).

Tossing a Coin In Exercises 15–20, find the probability for the experiment of tossing a coin three times.

15. The probability of getting exactly one tail

16. The probability of getting exactly two tails

17. The probability of getting a head on the first toss

18. The probability of getting a tail on the last toss

19. The probability of getting at least one head

20. The probability of getting at least two heads

Drawing a Card In Exercises 21–24, find the probability for the experiment of drawing a card at random from a standard deck of 52 playing cards.

21. The card is a face card.

22. The card is not a face card.

23. The card is a red face card.

24. The card is a 9 or lower. (Aces are low.)

Tossing a Die In Exercises 25–30, find the probability for the experiment of tossing a six-sided die twice.

25. The sum is 6.

26. The sum is at least 8.

27. The sum is less than 11.

28. The sum is 2, 3, or 12.

29. The sum is odd and no more than 7.

30. The sum is odd or prime.

Drawing Marbles In Exercises 31–34, find the probability for the experiment of drawing two marbles at random (without replacement) from a bag containing one green, two yellow, and three red marbles.

31. Both marbles are red.   32. Both marbles are yellow.

33. Neither marble is yellow.

34. The marbles are different colors.

35. Unemployment In 2015, there were approximately 8.3 million unemployed workers in the United States. The circle graph shows the age profile of these unemployed workers. *(Source: U.S. Bureau of Labor Statistics)*

Ages of Unemployed Workers

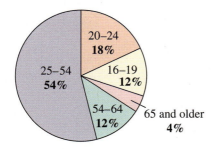

(a) Estimate the number of unemployed workers in the 16–19 age group.

(b) What is the probability that a person selected at random from the population of unemployed workers is in the 20–24 age group?

(c) What is the probability that a person selected at random from the population of unemployed workers is in the 25–54 age group?

(d) What is the probability that a person selected at random from the population of unemployed workers is 55 or older?

36. Political Poll An independent polling organization interviewed 100 college students to determine their political party affiliations and whether they favor a balanced-budget amendment to the Constitution. The table lists the results of the study. In the table, *D* represents Democrat and *R* represents Republican.

	Favor	Not Favor	Unsure	Total
D	23	25	7	55
R	32	9	4	45
Total	55	34	11	100

Find the probability that a person selected at random from the sample is as described.

(a) A person who does not favor the amendment

(b) A Republican

(c) A Democrat who favors the amendment

37. Education In a high school graduating class of 128 students, 52 are on the honor roll. Of these, 48 are going on to college. Of the 76 students not on the honor roll, 56 are going on to college. What is the probability that a student selected at random from the class is (a) going to college, (b) not going to college, and (c) not going to college and on the honor roll?

38. Alumni Association A college sends a survey to members of the class of 2016. Of the 1254 people who graduated that year, 672 are women, of whom 124 went on to graduate school. Of the 582 male graduates, 198 went on to graduate school. Find the probability that a class of 2016 alumnus selected at random is as described.

(a) Female

(b) Male

(c) Female and did not attend graduate school

39. Winning an Election Three people are running for president of a class. The results of a poll show that the first candidate has an estimated 37% chance of winning and the second candidate has an estimated 44% chance of winning. What is the probability that the third candidate will win?

40. Payroll Error The employees of a company work in six departments: 31 are in sales, 54 are in research, 42 are in marketing, 20 are in engineering, 47 are in finance, and 58 are in production. The payroll clerk loses one employee's paycheck. What is the probability that the employee works in the research department?

41. Exam Questions A class receives a list of 20 study problems, from which 10 will be part of an upcoming exam. A student knows how to solve 15 of the problems. Find the probability that the student will be able to answer (a) all 10 questions on the exam, (b) exactly eight questions on the exam, and (c) at least nine questions on the exam.

42. Payroll Error A payroll clerk addresses five paychecks and envelopes to five different people and randomly inserts the paychecks into the envelopes. Find the probability of each event.

(a) Exactly one paycheck is inserted in the correct envelope.

(b) At least one paycheck is inserted in the correct envelope.

43. Game Show On a game show, you are given five digits to arrange in the proper order to form the price of a car. If you are correct, you win the car. What is the probability of winning, given the following conditions?

(a) You guess the position of each digit.

(b) You know the first digit and guess the positions of the other digits.

44. Card Game The deck for a card game contains 108 cards. Twenty-five each are red, yellow, blue, and green, and eight are wild cards. Each player is randomly dealt a seven-card hand.

(a) What is the probability that a hand will contain exactly two wild cards?

(b) What is the probability that a hand will contain two wild cards, two red cards, and three blue cards?

45. Drawing a Card You draw one card at random from a standard deck of 52 playing cards. Find the probability that (a) the card is an even-numbered card, (b) the card is a heart or a diamond, and (c) the card is a nine or a face card.

46. Drawing Cards You draw five cards at random from a standard deck of 52 playing cards. What is the probability that the hand drawn is a full house? (A full house consists of three of one kind and two of another.)

47. Shipment A shipment of 12 microwave ovens contains three defective units. A vending company purchases four units at random. What is the probability that (a) all four units are good, (b) exactly two units are good, and (c) at least two units are good?

48. PIN Code ATM personal identification number (PIN) codes typically consist of four-digit sequences of numbers. Find the probability that if you forget your PIN, you can guess the correct sequence (a) at random and (b) when you recall the first two digits.

49. Random Number Generator A random number generator selects two integers from 1 through 40. What is the probability that (a) both numbers are even, (b) one number is even and one number is odd, (c) both numbers are less than 30, and (d) the same number is selected twice?

50. Flexible Work Hours In a recent survey, people were asked whether they would prefer to work flexible hours—even when it meant slower career advancement—so they could spend more time with their families. The figure shows the results of the survey. What is the probability that three people chosen at random would prefer flexible work hours?

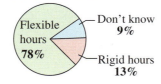

Flexible Work Hours

Flexible hours **78%** — Don't know **9%** — Rigid hours **13%**

Probability of a Complement In Exercises 51–54, you are given the probability that an event *will* happen. Find the probability that the event *will not* happen.

51. $P(E) = 0.73$ **52.** $P(E) = 0.28$

53. $P(E) = \dfrac{1}{5}$ **54.** $P(E) = \dfrac{2}{7}$

Probability of a Complement In Exercises 55–58, you are given the probability that an event *will not* happen. Find the probability that the event *will* happen.

55. $P(E') = 0.29$ **56.** $P(E') = 0.89$

57. $P(E') = \dfrac{14}{25}$ **58.** $P(E') = \dfrac{79}{100}$

• 59. Backup System •
A space vehicle has an independent backup system for one of its communication networks. The probability that either system will function satisfactorily during a flight is 0.985. What is the probability that during a given flight (a) both systems function satisfactorily, (b) both systems fail, and (c) at least one system functions satisfactorily?

60. Backup Vehicle A fire department keeps two rescue vehicles. Due to the demand on the vehicles and the chance of mechanical failure, the probability that a specific vehicle is available when needed is 90%. The availability of one vehicle is independent of the availability of the other. Find the probability that (a) both vehicles are available at a given time, (b) neither vehicle is available at a given time, and (c) at least one vehicle is available at a given time.

61. Roulette American roulette is a game in which a wheel turns on a spindle and is divided into 38 pockets. Thirty-six of the pockets are numbered 1–36, of which half are red and half are black. Two of the pockets are green and are numbered 0 and 00 (see figure). The dealer spins the wheel and a small ball in opposite directions. As the ball slows to a stop, it has an equal probability of landing in any of the numbered pockets.

(a) Find the probability of landing in the number 00 pocket.

(b) Find the probability of landing in a red pocket.

(c) Find the probability of landing in a green pocket or a black pocket.

(d) Find the probability of landing in the number 14 pocket on two consecutive spins.

(e) Find the probability of landing in a red pocket on three consecutive spins.

62. A Boy or a Girl? Assume that the probability of the birth of a child of a particular sex is 50%. In a family with four children, find the probability of each event.

(a) All the children are boys.

(b) All the children are the same sex.

(c) There is at least one boy.

63. Geometry You and a friend agree to meet at your favorite restaurant between 5:00 P.M. and 6:00 P.M. The one who arrives first will wait 15 minutes for the other, and then will leave (see figure). What is the probability that the two of you will actually meet, assuming that your arrival times are random within the hour?

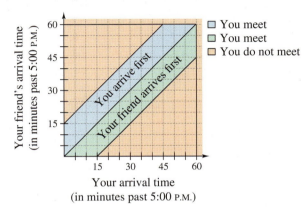

Your arrival time
(in minutes past 5:00 P.M.)

64. Estimating π You drop a coin of diameter d onto a paper that contains a grid of squares d units on a side (see figure).

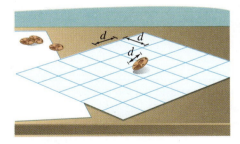

(a) Find the probability that the coin covers a vertex of one of the squares on the grid.

(b) Perform the experiment 100 times and use the results to approximate π.

Exploration

True or False? In Exercises 65 and 66, determine whether the statement is true or false. Justify your answer.

65. If A and B are independent events with nonzero probabilities, then A can occur when B occurs.

66. Rolling a number less than 3 on a normal six-sided die has a probability of $\frac{1}{3}$. The complement of this event is rolling a number greater than 3, which has a probability of $\frac{1}{2}$.

67. Pattern Recognition Consider a group of n people.

(a) Explain why the pattern below gives the probabilities that the n people have distinct birthdays.

$$n = 2: \frac{365}{365} \cdot \frac{364}{365} = \frac{365 \cdot 364}{365^2}$$

$$n = 3: \frac{365}{365} \cdot \frac{364}{365} \cdot \frac{363}{365} = \frac{365 \cdot 364 \cdot 363}{365^3}$$

(b) Use the pattern in part (a) to write an expression for the probability that $n = 4$ people have distinct birthdays.

(c) Let P_n be the probability that the n people have distinct birthdays. Verify that this probability can be obtained recursively by

$$P_1 = 1 \quad \text{and} \quad P_n = \frac{365 - (n-1)}{365} P_{n-1}.$$

(d) Explain why $Q_n = 1 - P_n$ gives the probability that at least two people in a group of n people have the same birthday.

(e) Use the results of parts (c) and (d) to complete the table.

n	10	15	20	23	30	40	50
P_n							
Q_n							

(f) How many people must be in a group so that the probability of at least two of them having the same birthday is greater than $\frac{1}{2}$? Explain.

68. **HOW DO YOU SEE IT?** The circle graphs show the percents of undergraduate students by class level at two colleges. A student is chosen at random from the combined undergraduate population of the two colleges. The probability that the student is a freshman, sophomore, or junior is 81%. Which college has a greater number of undergraduate students? Explain.

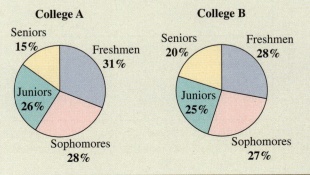

Chapter Summary

	What Did You Learn?	Explanation/Examples	Review Exercises
Section 8.1	Use sequence notation to write the terms of sequences (p. 562).	$a_n = 7n - 4$; $a_1 = 7(1) - 4 = 3$, $a_2 = 7(2) - 4 = 10$, $a_3 = 7(3) - 4 = 17$, $a_4 = 7(4) - 4 = 24$	1–8
	Use factorial notation (p. 565).	If n is a positive integer, then $n! = 1 \cdot 2 \cdot 3 \cdot 4 \cdots (n-1) \cdot n.$	9–12
	Use summation notation to write sums (p. 566).	The sum of the first n terms of a sequence is represented by $$\sum_{i=1}^{n} a_i = a_1 + a_2 + a_3 + a_4 + \cdots + a_n.$$	13–16
	Find the sums of series (p. 567).	$$\sum_{i=1}^{\infty} \frac{8}{10^i} = \frac{8}{10^1} + \frac{8}{10^2} + \frac{8}{10^3} + \frac{8}{10^4} + \frac{8}{10^5} + \cdots$$ $$= 0.8 + 0.08 + 0.008 + 0.0008 + 0.00008 + \cdots$$ $$= 0.88888 \ldots = \frac{8}{9}$$	17, 18
	Use sequences and series to model and solve real-life problems (p. 568).	A sequence can help you model the balance after n compoundings in an account that earns compound interest. (See Example 10.)	19, 20
Section 8.2	Recognize, write, and find the nth terms of arithmetic sequences (p. 572).	$a_n = 9n + 5$; $a_1 = 9(1) + 5 = 14$, $a_2 = 9(2) + 5 = 23$, $a_3 = 9(3) + 5 = 32$, $a_4 = 9(4) + 5 = 41$; common difference: $d = 9$	21–30
	Find nth partial sums of arithmetic sequences (p. 575).	The sum of a finite arithmetic sequence with n terms is given by $S_n = (n/2)(a_1 + a_n).$	31–36
	Use arithmetic sequences to model and solve real-life problems (p. 577).	An arithmetic sequence can help you find the total sales of a small business. (See Example 9.)	37, 38
Section 8.3	Recognize, write, and find the nth terms of geometric sequences (p. 581).	$a_n = 3(4^n)$; $a_1 = 3(4^1) = 12$, $a_2 = 3(4^2) = 48$, $a_3 = 3(4^3) = 192$, $a_4 = 3(4^4) = 768$; common ratio: $r = 4$	39–50
	Find the sum of a finite geometric sequence (p. 584).	The sum of the finite geometric sequence $a_1, a_1 r, a_1 r^2, \ldots, a_1 r^{n-1}$ with common ratio $r \neq 1$ is given by $S_n = \sum_{i=1}^{n} a_1 r^{i-1} = a_1 \left(\frac{1 - r^n}{1 - r} \right).$	51–58
	Find the sum of an infinite geometric series (p. 585).	If $|r| < 1$, then the infinite geometric series $a_1 + a_1 r + a_1 r^2 + \cdots + a_1 r^{n-1} + \cdots$ has the sum $S = \sum_{i=0}^{\infty} a_1 r^i = \frac{a_1}{1 - r}.$	59–62
	Use geometric sequences to model and solve real-life problems (p. 586).	A finite geometric sequence can help you find the balance of an increasing annuity at the end of two years. (See Example 8.)	63, 64

What Did You Learn?	**Explanation/Examples**	**Review Exercises**
Section 8.4		
Use mathematical induction to prove statements involving a positive integer n *(p. 590)*.	Let P_n be a statement involving the positive integer n. If (1) P_1 is true, and (2) for every positive integer k, the truth of P_k implies the truth of P_{k+1}, then the statement P_n must be true for all positive integers n.	65–68
Use pattern recognition and mathematical induction to write a formula for the nth term of a sequence *(p. 595)*.	To find a formula for the nth term of a sequence, (1) calculate the first several terms of the sequence, (2) try to find a pattern for the terms and write a formula for the nth term of the sequence (hypothesis), and (3) use mathematical induction to prove your hypothesis.	69–72
Find the sums of powers of integers *(p. 596)*.	$$\sum_{i=1}^{8} i^2 = \frac{n(n+1)(2n+1)}{6} = \frac{8(8+1)(16+1)}{6} = 204$$	73, 74
Find finite differences of sequences *(p. 597)*.	The first differences of a sequence are found by subtracting consecutive terms. The second differences are found by subtracting consecutive first differences.	75, 76
Section 8.5		
Use the Binomial Theorem to find binomial coefficients *(p. 600)*.	**The Binomial Theorem:** In the expansion of $(x+y)^n = x^n + nx^{n-1}y + \cdots + {}_nC_r x^{n-r}y^r + \cdots + nxy^{n-1} + y^n$, the coefficient of $x^{n-r}y^r$ is ${}_nC_r = \dfrac{n!}{(n-r)!r!}$.	77, 78
Use Pascal's Triangle to find binomial coefficients *(p. 602)*.	First several rows of Pascal's Triangle: $\begin{array}{ccccccccc} & & & & 1 & & & & \\ & & & 1 & & 1 & & & \\ & & 1 & & 2 & & 1 & & \\ & 1 & & 3 & & 3 & & 1 & \\ 1 & & 4 & & 6 & & 4 & & 1 \end{array}$	79, 80
Use binomial coefficients to write binomial expansions *(p. 603)*.	$(x+1)^3 = x^3 + 3x^2 + 3x + 1$ $(x-1)^4 = x^4 - 4x^3 + 6x^2 - 4x + 1$	81–84
Section 8.6		
Solve simple counting problems *(p. 608)*.	A computer randomly generates an integer from 1 through 15. The computer can generate an integer that is divisible by 3 in 5 ways (3, 6, 9, 12, and 15).	85, 86
Use the Fundamental Counting Principle to solve counting problems *(p. 609)*.	**Fundamental Counting Principle:** Let E_1 and E_2 be two events. The first event E_1 can occur in m_1 different ways. After E_1 has occurred, E_2 can occur in m_2 different ways. The number of ways the two events can occur is $m_1 \cdot m_2$.	87, 88
Use permutations *(p. 610)* and combinations *(p. 613)* to solve counting problems.	The number of permutations of n elements taken r at a time is ${}_nP_r = n!/(n-r)!$. The number of combinations of n elements taken r at a time is ${}_nC_r = n!/[(n-r)!r!]$, or ${}_nC_r = {}_nP_r/r!$.	89–92
Section 8.7		
Find probabilities of events *(p. 619)*.	If an event E has $n(E)$ equally likely outcomes and its sample space S has $n(S)$ equally likely outcomes, then the probability of event E is $P(E) = n(E)/n(S)$.	93, 94
Find probabilities of mutually exclusive events *(p. 622)* and independent events *(p. 624)*, and find the probability of the complement of an event *(p. 625)*.	If A and B are events in the same sample space, then the probability of A or B occurring is $P(A \cup B) = P(A) + P(B) - P(A \cap B)$. If A and B are mutually exclusive, then $P(A \cup B) = P(A) + P(B)$. If A and B are independent, then $P(A \text{ and } B) = P(A) \cdot P(B)$. The probability of the complement of A is $P(A') = 1 - P(A)$.	95–100

Review Exercises   See CalcChat.com for tutorial help and worked-out solutions to odd-numbered exercises.

8.1 **Writing the Terms of a Sequence** In Exercises 1–4, write the first five terms of the sequence. (Assume that n begins with 1.)

1. $a_n = 3 + \dfrac{12}{n}$

2. $a_n = \dfrac{(-1)^n 5n}{2n - 1}$

3. $a_n = \dfrac{120}{n!}$

4. $a_n = (n + 1)(n + 2)$

Finding the nth Term of a Sequence In Exercises 5–8, write an expression for the apparent nth term (a_n) of the sequence. (Assume that n begins with 1.)

5. $-2, 2, -2, 2, -2, \ldots$

6. $-1, 2, 7, 14, 23, \ldots$

7. $4, 2, \frac{4}{3}, 1, \frac{4}{5}, \ldots$

8. $1, -\frac{1}{2}, \frac{1}{3}, -\frac{1}{4}, \frac{1}{5}, \ldots$

Simplifying a Factorial Expression In Exercises 9–12, simplify the factorial expression.

9. $\dfrac{3!}{5!}$

10. $\dfrac{7!}{3! \cdot 4!}$

11. $\dfrac{(n - 1)!}{(n + 1)!}$

12. $\dfrac{n!}{(n + 2)!}$

Finding a Sum In Exercises 13 and 14, find the sum.

13. $\displaystyle\sum_{j=1}^{4} \dfrac{6}{j^2}$

14. $\displaystyle\sum_{k=1}^{10} 2k^3$

Using Sigma Notation to Write a Sum In Exercises 15 and 16, use sigma notation to write the sum.

15. $\dfrac{1}{2(1)} + \dfrac{1}{2(2)} + \dfrac{1}{2(3)} + \cdots + \dfrac{1}{2(20)}$

16. $\dfrac{1}{2} + \dfrac{2}{3} + \dfrac{3}{4} + \cdots + \dfrac{9}{10}$

Finding the Sum of an Infinite Series In Exercises 17 and 18, find the sum of the infinite series.

17. $\displaystyle\sum_{i=1}^{\infty} \dfrac{4}{10^i}$

18. $\displaystyle\sum_{k=1}^{\infty} 8\left(\dfrac{1}{10}\right)^k$

19. **Compound Interest** An investor deposits $10,000 in an account that earns 2.25% interest compounded monthly. The balance in the account after n months is given by

$$A_n = 10{,}000\left(1 + \dfrac{0.0225}{12}\right)^n, \quad n = 1, 2, 3, \ldots.$$

(a) Write the first 10 terms of the sequence.

(b) Find the balance in the account after 10 years by computing the 120th term of the sequence.

20. **Population** The population a_n (in thousands) of Miami, Florida, from 2010 through 2014 can be approximated by

$$a_n = -0.34n^2 + 14.8n + 288, \quad n = 10, 11, \ldots, 14$$

where n is the year with $n = 10$ corresponding to 2010. Write the terms of this finite sequence. Use a graphing utility to construct a bar graph that represents the sequence. *(Source: U.S. Census Bureau)*

8.2 **Determining Whether a Sequence Is Arithmetic** In Exercises 21–24, determine whether the sequence is arithmetic. If so, find the common difference.

21. $5, -1, -7, -13, -19, \ldots$

22. $0, 1, 3, 6, 10, \ldots$

23. $\frac{1}{8}, \frac{1}{4}, \frac{1}{2}, 1, 2, \ldots$

24. $1, \frac{15}{16}, \frac{7}{8}, \frac{13}{16}, \frac{3}{4}, \ldots$

Finding the nth Term In Exercises 25–28, find a formula for a_n for the arithmetic sequence.

25. $a_1 = 7, d = 12$

26. $a_1 = 34, d = -4$

27. $a_3 = 96, a_7 = 24$

28. $a_7 = 8, a_{13} = 6$

Writing the Terms of an Arithmetic Sequence In Exercises 29 and 30, write the first five terms of the arithmetic sequence.

29. $a_1 = 4, d = 17$

30. $a_1 = 25, a_{n+1} = a_n + 3$

31. **Sum of a Finite Arithmetic Sequence** Find the sum of the first 100 positive multiples of 9.

32. **Sum of a Finite Arithmetic Sequence** Find the sum of the integers from 30 to 80.

Finding a Sum In Exercises 33–36, find the sum.

33. $\displaystyle\sum_{j=1}^{10} (2j - 3)$

34. $\displaystyle\sum_{j=1}^{8} (20 - 3j)$

35. $\displaystyle\sum_{k=1}^{11} \left(\dfrac{2}{3}k + 4\right)$

36. $\displaystyle\sum_{k=1}^{25} \left(\dfrac{3k + 1}{4}\right)$

37. **Job Offer** The starting salary for a job is $43,800 with a guaranteed increase of $1950 per year. Determine (a) the salary during the fifth year and (b) the total compensation through five full years of employment.

38. **Baling Hay** In the first two trips baling hay around a large field, a farmer obtains 123 bales and 112 bales, respectively. Each round gets shorter, so the farmer estimates that the same pattern will continue. Estimate the total number of bales made after the farmer takes another six trips around the field.

8.3 **Determining Whether a Sequence Is Geometric**   In Exercises 39–42, determine whether the sequence is geometric. If so, find the common ratio.

39. 2, 6, 18, 54, 162, . . .   **40.** 48, -24, 12, -6, . . .

41. $\frac{1}{5}, -\frac{3}{5}, \frac{9}{5}, -\frac{27}{5}, \ldots$ **42.** $\frac{1}{4}, \frac{2}{5}, \frac{3}{6}, \frac{4}{7}, \ldots$

Writing the Terms of a Geometric Sequence   In Exercises 43–46, write the first five terms of the geometric sequence.

43. $a_1 = 2$, $r = 15$

44. $a_1 = 6$, $r = -\frac{1}{3}$

45. $a_1 = 9$, $a_3 = 4$

46. $a_1 = 2$, $a_3 = 12$

Finding a Term of a Geometric Sequence   In Exercises 47–50, write an expression for the nth term of the geometric sequence. Then find the 10th term of the sequence.

47. $a_1 = 100$, $r = 1.05$

48. $a_1 = 5$, $r = 0.2$

49. $a_1 = 18$, $a_2 = -9$

50. $a_3 = 6$, $a_4 = 1$

Sum of a Finite Geometric Sequence   In Exercises 51–58, find the sum of the finite geometric sequence.

51. $\displaystyle\sum_{i=1}^{7} 2^{i-1}$ **52.** $\displaystyle\sum_{i=1}^{5} 3^{i-1}$

53. $\displaystyle\sum_{i=1}^{4} \left(\frac{1}{2}\right)^{i}$ **54.** $\displaystyle\sum_{i=1}^{6} \left(\frac{1}{3}\right)^{i-1}$

55. $\displaystyle\sum_{i=1}^{5} (2)^{i-1}$ **56.** $\displaystyle\sum_{i=1}^{4} 6(3)^{i}$

57. $\displaystyle\sum_{i=1}^{5} 10(0.6)^{i-1}$ **58.** $\displaystyle\sum_{i=1}^{4} 20(0.2)^{i-1}$

Sum of an Infinite Geometric Sequence   In Exercises 59–62, find the sum of the infinite geometric series.

59. $\displaystyle\sum_{i=0}^{\infty} \left(\frac{7}{8}\right)^{i}$ **60.** $\displaystyle\sum_{i=0}^{\infty} (0.5)^{i}$

61. $\displaystyle\sum_{k=1}^{\infty} 4\left(\frac{2}{3}\right)^{k-1}$ **62.** $\displaystyle\sum_{k=1}^{\infty} 1.3\left(\frac{1}{10}\right)^{k-1}$

63. Depreciation   A paper manufacturer buys a machine for $120,000. It depreciates at a rate of 30% per year. (In other words, at the end of each year the depreciated value is 70% of what it was at the beginning of the year.)

   (a) Find the formula for the nth term of a geometric sequence that gives the value of the machine t full years after it is purchased.

   (b) Find the depreciated value of the machine after 5 full years.

64. Annuity   An investor deposits $800 in an account on the first day of each month for 10 years. The account pays 3%, compounded monthly. What is the balance at the end of 10 years?

8.4 **Using Mathematical Induction**   In Exercises 65–68, use mathematical induction to prove the formula for all integers $n \geq 1$.

65. $3 + 5 + 7 + \cdots + (2n + 1) = n(n + 2)$

66. $1 + \frac{3}{2} + 2 + \frac{5}{2} + \cdots + \frac{1}{2}(n + 1) = \frac{n}{4}(n + 3)$

67. $\displaystyle\sum_{i=0}^{n-1} ar^i = \frac{a(1 - r^n)}{1 - r}$

68. $\displaystyle\sum_{k=0}^{n-1} (a + kd) = \frac{n}{2}[2a + (n - 1)d]$

Finding a Formula for a Finite Sum   In Exercises 69–72, find a formula for the sum of the first n terms of the sequence. Prove the validity of your formula.

69. 9, 13, 17, 21, . . .

70. 68, 60, 52, 44, . . .

71. $1, \frac{3}{5}, \frac{9}{25}, \frac{27}{125}, \ldots$

72. $12, -1, \frac{1}{12}, -\frac{1}{144}, \ldots$

Finding a Sum   In Exercises 73 and 74, find the sum using the formulas for the sums of powers of integers.

73. $\displaystyle\sum_{n=1}^{75} n$ **74.** $\displaystyle\sum_{n=1}^{6} (n^5 - n^2)$

Linear Model, Quadratic Model, or Neither?   In Exercises 75 and 76, write the first five terms of the sequence beginning with the term a_1. Then calculate the first and second differences of the sequence. State whether the sequence has a perfect linear model, a perfect quadratic model, or neither.

75. $a_1 = 5$

   $a_n = a_{n-1} + 5$

76. $a_1 = -3$

   $a_n = a_{n-1} - 2n$

8.5 **Finding a Binomial Coefficient**   In Exercises 77 and 78, find the binomial coefficient.

77. $_6C_4$ **78.** $_{12}C_3$

Using Pascal's Triangle   In Exercises 79 and 80, evaluate using Pascal's Triangle.

79. $\dbinom{7}{2}$ **80.** $\dbinom{10}{4}$

Expanding a Bionomial   In Exercises 81–84, use the Binomial Theorem to write the expansion of the expression.

81. $(x + 4)^4$ **82.** $(5 + 2z)^4$

83. $(4 - 5x)^3$ **84.** $(a - 3b)^5$

8.6 **Random Selection** **In Exercises 85 and 86, determine the number of ways a computer can generate the sum using randomly selected integers from 1 through 14.**

85. Two *distinct* integers whose sum is 7

86. Two *distinct* integers whose sum is 12

87. Telephone Numbers All of the landline telephone numbers in a small town use the same three-digit prefix. How many different telephone numbers are possible by changing only the last four digits?

88. Course Schedule A college student is preparing a course schedule for the next semester. The student may select one of three mathematics courses, one of four science courses, and one of six history courses. How many schedules are possible?

89. Genetics A geneticist is using gel electrophoresis to analyze five DNA samples. The geneticist treats each sample with a different restriction enzyme and then injects it into one of five wells formed in a bed of gel. In how many orders can the geneticist inject the five samples into the wells?

90. Race There are 10 bicyclists entered in a race. In how many different ways can the top three places be decided?

91. Jury Selection In how many different ways can a jury of 12 people be randomly selected from a group of 32 people?

92. Menu Choices A local sandwich shop offers five different breads, four different meats, three different cheeses, and six different vegetables. A customer can choose one bread, one or no meat, one or no cheese, and up to three vegetables. Find the total number of combinations of sandwiches possible.

8.7

93. Apparel A drawer contains six white socks, two blue socks, and two gray socks.

(a) What is the probability of randomly selecting one blue sock?

(b) What is the probability of randomly selecting one white sock?

94. Bookshelf Order A child returns a five-volume set of books to a bookshelf. The child is not able to read, and so cannot distinguish one volume from another. What is the probability that the child shelves the books in the correct order?

95. Students by Class At a university, 31% of the students are freshmen, 26% are sophomores, 25% are juniors, and 18% are seniors. One student receives a cash scholarship randomly by lottery. Find the probability that the scholarship winner is as described.

(a) A junior or senior

(b) A freshman, sophomore, or junior

96. Opinion Poll In a survey, a sample of college students, faculty members, and administrators were asked whether they favor a proposed increase in the annual activity fee to enhance student life on campus. The table lists the results of the survey.

	Students	Faculty	Admin.	Total
Favor	237	37	18	292
Oppose	163	38	7	208
Total	400	75	25	500

Find the probability that a person selected at random from the sample is as described.

(a) A person who opposes the proposal

(b) A student

(c) A faculty member who favors the proposal

97. Tossing a Die You toss a six-sided die four times. What is the probability of getting four 5's?

98. Tossing a Die You toss a six-sided die six times. What is the probability of getting each number exactly once?

99. Drawing a Card You draw one card at random from a standard deck of 52 playing cards. What is the probability that the card is not a club?

100. Tossing a Coin You toss a coin five times. What is the probability of getting at least one tail?

Exploration

True or False? **In Exercises 101–104, determine whether the statement is true or false. Justify your answer.**

101. $\dfrac{(n+2)!}{n!} = \dfrac{n+2}{n}$

102. $\displaystyle\sum_{i=1}^{5}(i^3 + 2i) = \sum_{i=1}^{5} i^3 + \sum_{i=1}^{5} 2i$

103. $\displaystyle\sum_{k=1}^{8} 3k = 3\sum_{k=1}^{8} k$

104. $\displaystyle\sum_{j=1}^{6} 2^j = \sum_{j=3}^{8} 2^{j-2}$

105. Think About It An infinite sequence beginning with a_1 is a function. What is the domain of the function?

106. Think About It How do the two sequences differ?

(a) $a_n = \dfrac{(-1)^n}{n}$ (b) $a_n = \dfrac{(-1)^{n+1}}{n}$

107. Writing Explain what is meant by a recursion formula.

108. Writing Write a brief paragraph explaining how to identify the graph of an arithmetic sequence and the graph of a geometric sequence.

Chapter Test

See **CalcChat.com** for tutorial help and worked-out solutions to odd-numbered exercises.

Take this test as you would take a test in class. When you are finished, check your work against the answers given in the back of the book.

1. Write the first five terms of the sequence $a_n = \dfrac{(-1)^n}{3n + 2}$. (Assume that n begins with 1.)

2. Write an expression for the apparent nth term (a_n) of the sequence. (Assume that n begins with 1.)

   $$\frac{3}{1!}, \frac{4}{2!}, \frac{5}{3!}, \frac{6}{4!}, \frac{7}{5!}, \cdots$$

3. Write the next three terms of the series. Then find the seventh partial sum of the series.

   $$8 + 21 + 34 + 47 + \cdots$$

4. The 5th term of an arithmetic sequence is 45, and the 12th term is 24. Find the nth term.

5. The second term of a geometric sequence is 14, and the sixth term is 224. Find the nth term. (Assume that the terms of the sequence are positive.)

In Exercises 6–9, find the sum.

6. $\displaystyle\sum_{i=1}^{50} (2i^2 + 5)$

7. $\displaystyle\sum_{n=1}^{9} (12n - 7)$

8. $\displaystyle\sum_{i=1}^{\infty} 4\left(\frac{1}{2}\right)^i$

9. $\displaystyle\sum_{n=1}^{\infty} \left(-\frac{1}{3}\right)^n$

10. Use mathematical induction to prove the formula for all integers $n \geq 1$.

    $$5 + 10 + 15 + \cdots + 5n = \frac{5n(n + 1)}{2}$$

11. Use the Binomial Theorem to write the expansion of $(x + 6y)^4$.

12. Expand $3(x - 2)^5 + 4(x - 2)^3$ by using Pascal's Triangle to determine the coefficients.

13. Find the coefficient of the term $a^4 b^3$ in the expansion of $(3a - 2b)^7$.

In Exercises 14 and 15, evaluate each expression.

14. (a) $_9P_2$ (b) $_{70}P_3$ 15. (a) $_{11}C_4$ (b) $_{66}C_4$

16. How many distinct license plate numbers consisting of one letter followed by a three-digit number are possible?

17. Eight people are going for a ride in a boat that seats eight people. One person will drive, and only three of the remaining people are willing to ride in the two bow seats. How many seating arrangements are possible?

18. You attend a karaoke night and hope to hear your favorite song. The karaoke song book has 300 different songs (your favorite song is among them). Assuming that the singers are equally likely to pick any song and no song repeats, what is the probability that your favorite song is one of the 20 that you hear that night?

19. You and three of your friends are at a party. Names of all of the 30 guests are placed in a hat and drawn randomly to award four door prizes. Each guest can win only one prize. What is the probability that you and your friends win all four prizes?

20. The weather report calls for a 90% chance of snow. According to this report, what is the probability that it will *not* snow?

Cumulative Test for Chapters 6–8

See CalcChat.com for tutorial help and worked-out solutions to odd-numbered exercises.

Take this test as you would take a test in class. When you are finished, check your work against the answers given in the back of the book.

In Exercises 1–4, solve the system by the specified method.

1. Substitution

$$\begin{cases} y = 3 - x^2 \\ 2(y - 2) = x - 1 \end{cases}$$

2. Elimination

$$\begin{cases} x + 3y = -6 \\ 2x + 4y = -10 \end{cases}$$

3. Gaussian Elimination

$$\begin{cases} -2x + 4y - z = -16 \\ x - 2y + 2z = 5 \\ x - 3y - z = 13 \end{cases}$$

4. Gauss-Jordan Elimination

$$\begin{cases} x + 3y - 2z = -7 \\ -2x + y - z = -5 \\ 4x + y + z = 3 \end{cases}$$

5. A custom-blend bird seed is made by mixing two types of bird seeds costing $0.75 per pound and $1.25 per pound. How many pounds of each type of seed mixture are used to make 200 pounds of custom-blend bird seed costing $0.95 per pound?

6. Find a quadratic equation $y = ax^2 + bx + c$ whose graph passes through the points $(0, 6)$, $(2, 3)$, and $(4, 2)$.

7. Write the partial fraction decomposition of the rational expression $\dfrac{2x^2 - x - 6}{x^3 + 2x}$.

In Exercises 8 and 9, sketch the graph of the solution set of the system of inequalities. Label the vertices of the region.

8. $\begin{cases} 2x + y \geq -3 \\ x - 3y \leq 2 \end{cases}$

9. $\begin{cases} x - y > 6 \\ 5x + 2y < 10 \end{cases}$

10. Sketch the region corresponding to the system of constraints. Then find the minimum and maximum values of the objective function $z = 3x + 2y$ and the points where they occur, subject to the constraints.

$$\begin{aligned} x + 4y &\leq 20 \\ 2x + y &\leq 12 \\ x &\geq 0 \\ y &\geq 0 \end{aligned}$$

$$\begin{cases} -x + 2y - z = 9 \\ 2x - y + 2z = -9 \\ 3x + 3y - 4z = 7 \end{cases}$$

System for 11 and 12

In Exercises 11 and 12, use the system of linear equations shown at the left.

11. Write the augmented matrix for the system.

12. Solve the system using the matrix found in Exercise 11 and Gauss-Jordan elimination.

In Exercises 13–18, perform the operation(s) using the matrices below, if possible.

$$A = \begin{bmatrix} -1 & 3 \\ 6 & 2 \end{bmatrix}, \qquad B = \begin{bmatrix} -2 & 5 \\ 0 & -1 \end{bmatrix}, \qquad C = \begin{bmatrix} 4 & 0 & 1 \\ -3 & 2 & -1 \end{bmatrix}$$

13. $A + B$

14. $2A - 5B$

15. AC

16. CB

17. A^2

18. $BA - B^2$

19. Find the inverse of the matrix, if possible: $\begin{bmatrix} 1 & 2 & -1 \\ 3 & 7 & -10 \\ -5 & -7 & -15 \end{bmatrix}$.

$$\begin{bmatrix} 7 & 1 & 0 \\ -2 & 4 & -1 \\ 3 & 8 & 5 \end{bmatrix}$$

Matrix for 20

	Gym shoes	Jogging shoes	Walking shoes
14–17	0.079	0.064	0.029
18–24	0.050	0.060	0.022
25–34	0.103	0.259	0.085

Age group (bracket for 14–17, 18–24, 25–34)

Matrix for 22

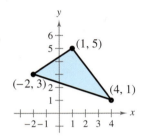

Figure for 25

20. Find the determinant of the matrix shown at the left.

21. Use matrices to find the vertices of the image of the square with vertices $(0, 2)$, $(0, 5)$, $(3, 2)$, and $(3, 5)$ after a reflection in the x-axis.

22. The matrix at the left shows the percents (in decimal form) of the total amounts spent on three types of footwear in a recent year. The total amounts (in millions of dollars) spent by the age groups on the three types of footwear were $479.88 (14–17 age group), $365.88 (18–24 age group), and $1248.89 (25–34 age group). How many dollars worth of gym shoes, jogging shoes, and walking shoes were sold that year? *(Source: National Sporting Goods Association)*

In Exercises 23 and 24, use Cramer's Rule to solve the system of equations.

23. $\begin{cases} 8x - 3y = -52 \\ 3x + 5y = 5 \end{cases}$

24. $\begin{cases} 5x + 4y + 3z = 7 \\ -3x - 8y + 7z = -9 \\ 7x - 5y - 6z = -53 \end{cases}$

25. Use a determinant to find the area of the triangle shown at the left.

26. Write the first five terms of the sequence $a_n = \dfrac{(-1)^{n+1}}{2n + 3}$. (Assume that n begins with 1.)

27. Write an expression for the apparent nth term (a_n) of the sequence.

$$\frac{2!}{4}, \frac{3!}{5}, \frac{4!}{6}, \frac{5!}{7}, \frac{6!}{8}, \dots$$

28. Find the 16th partial sum of the arithmetic sequence $6, 18, 30, 42, \dots$.

29. The sixth term of an arithmetic sequence is 20.6, and the ninth term is 30.2.

(a) Find the 20th term.

(b) Find the nth term.

30. Write the first five terms of the sequence $a_n = 3(2)^{n-1}$. (Assume that n begins with 1.)

31. Find the sum: $\displaystyle\sum_{i=0}^{\infty} 1.9\left(\frac{1}{10}\right)^{i-1}$.

32. Use mathematical induction to prove the inequality

$$(n + 1)! > 2^n, \quad n \geq 2.$$

33. Use the Binomial Theorem to write the expansion of $(w - 9)^4$.

In Exercises 34–37, evaluate the expression.

34. $_{14}P_3$

35. $_{25}P_2$

36. $\dbinom{8}{4}$

37. $_{11}C_6$

In Exercises 38 and 39, find the number of distinguishable permutations of the group of letters.

38. B, A, S, K, E, T, B, A, L, L

39. A, N, T, A, R, C, T, I, C, A

40. There are 10 applicants for three sales positions at a department store. All of the applicants are qualified. In how many ways can the department store fill the three positions?

41. On a game show, a contestant is given the digits 3, 4, and 5 to arrange in the proper order to form the price of an appliance. If the contestant is correct, he or she wins the appliance. What is the probability of winning when the contestant knows that the price is at least $400?

Proofs in Mathematics ■ ■ ■ ■ ■ ■ ■ ■ ■ ■ ■ ■ ■ ■ ■ ■

Properties of Sums *(p. 566)*

1. $\displaystyle\sum_{i=1}^{n} c = cn,$ c is a constant.

2. $\displaystyle\sum_{i=1}^{n} ca_i = c\sum_{i=1}^{n} a_i,$ c is a constant.

3. $\displaystyle\sum_{i=1}^{n} (a_i + b_i) = \sum_{i=1}^{n} a_i + \sum_{i=1}^{n} b_i$

4. $\displaystyle\sum_{i=1}^{n} (a_i - b_i) = \sum_{i=1}^{n} a_i - \sum_{i=1}^{n} b_i$

INFINITE SERIES

People considered the study of infinite series a novelty in the fourteenth century. Logician Richard Suiseth, whose nickname was Calculator, solved this problem.

If throughout the first half of a given time interval a variation continues at a certain intensity; throughout the next quarter of the interval at double the intensity; throughout the following eighth at triple the intensity and so ad infinitum; The average intensity for the whole interval will be the intensity of the variation during the second subinterval (or double the intensity).

This is the same as saying that the sum of the infinite series

$$\frac{1}{2} + \frac{2}{4} + \frac{3}{8} + \cdots$$
$$+ \frac{n}{2^n} + \cdots$$

is 2.

Proof

Each of these properties follows directly from the properties of real numbers.

1. $\displaystyle\sum_{i=1}^{n} c = c + c + c + \cdots + c = cn$ n terms

The proof of Property 2 uses the Distributive Property.

2. $\displaystyle\sum_{i=1}^{n} ca_i = ca_1 + ca_2 + ca_3 + \cdots + ca_n$

$$= c(a_1 + a_2 + a_3 + \cdots + a_n)$$

$$= c\sum_{i=1}^{n} a_i$$

The proof of Property 3 uses the Commutative and Associative Properties of Addition.

3. $\displaystyle\sum_{i=1}^{n} (a_i + b_i) = (a_1 + b_1) + (a_2 + b_2) + (a_3 + b_3) + \cdots + (a_n + b_n)$

$$= (a_1 + a_2 + a_3 + \cdots + a_n) + (b_1 + b_2 + b_3 + \cdots + b_n)$$

$$= \sum_{i=1}^{n} a_i + \sum_{i=1}^{n} b_i$$

The proof of Property 4 uses the Commutative and Associative Properties of Addition and the Distributive Property.

4. $\displaystyle\sum_{i=1}^{n} (a_i - b_i) = (a_1 - b_1) + (a_2 - b_2) + (a_3 - b_3) + \cdots + (a_n - b_n)$

$$= (a_1 + a_2 + a_3 + \cdots + a_n) + (-b_1 - b_2 - b_3 - \cdots - b_n)$$

$$= (a_1 + a_2 + a_3 + \cdots + a_n) - (b_1 + b_2 + b_3 + \cdots + b_n)$$

$$= \sum_{i=1}^{n} a_i - \sum_{i=1}^{n} b_i$$

The Sum of a Finite Arithmetic Sequence   *(p. 575)*

The sum of a finite arithmetic sequence with n terms is given by

$$S_n = \frac{n}{2}(a_1 + a_n).$$

Proof

Begin by generating the terms of the arithmetic sequence in two ways. In the first way, repeatedly add d to the first term.

$$S_n = a_1 + a_2 + a_3 + \cdots + a_{n-2} + a_{n-1} + a_n$$
$$= a_1 + [a_1 + d] + [a_1 + 2d] + \cdots + [a_1 + (n-1)d]$$

In the second way, repeatedly subtract d from the nth term.

$$S_n = a_n + a_{n-1} + a_{n-2} + \cdots + a_3 + a_2 + a_1$$
$$= a_n + [a_n - d] + [a_n - 2d] + \cdots + [a_n - (n-1)d]$$

Add these two versions of S_n. The multiples of d sum to zero and you obtain the formula.

$$2S_n = (a_1 + a_n) + (a_1 + a_n) + (a_1 + a_n) + \cdots + (a_1 + a_n) \qquad n \text{ terms}$$
$$2S_n = n(a_1 + a_n)$$
$$S_n = \frac{n}{2}(a_1 + a_n)$$

The Sum of a Finite Geometric Sequence   *(p. 584)*

The sum of the finite geometric sequence

$$a_1, a_1r, a_1r^2, a_1r^3, a_1r^4, \ldots, a_1r^{n-1}$$

with common ratio $r \neq 1$ is given by $S_n = \displaystyle\sum_{i=1}^{n} a_1 r^{i-1} = a_1\left(\frac{1 - r^n}{1 - r}\right).$

Proof

$$S_n = a_1 + a_1r + a_1r^2 + \cdots + a_1r^{n-2} + a_1r^{n-1}$$
$$rS_n = a_1r + a_1r^2 + a_1r^3 + \cdots + a_1r^{n-1} + a_1r^n \qquad \text{Multiply by } r.$$

Subtracting the second equation from the first yields

$$S_n - rS_n = a_1 - a_1r^n.$$

So, $S_n(1 - r) = a_1(1 - r^n)$, and, because $r \neq 1$, you have $S_n = a_1\left(\dfrac{1 - r^n}{1 - r}\right).$

The Binomial Theorem *(p. 600)*

In the expansion of $(x + y)^n$

$$(x + y)^n = x^n + nx^{n-1}y + \cdots + {}_nC_r x^{n-r}y^r + \cdots + nxy^{n-1} + y^n$$

the coefficient of $x^{n-r}y^r$ is

$${}_nC_r = \frac{n!}{(n-r)!r!}.$$

Proof

Use mathematical induction. The steps are straightforward but look a little messy, so only an outline of the proof is given below.

1. For $n = 1$, you have $(x + y)^1 = x^1 + y^1 = {}_1C_0 x + {}_1C_1 y$, and the formula is valid.

2. Assuming that the formula is true for $n = k$, the coefficient of $x^{k-r}y^r$ is

$${}_kC_r = \frac{k!}{(k-r)!r!} = \frac{k(k-1)(k-2)\cdots(k-r+1)}{r!}.$$

To show that the formula is true for $n = k + 1$, look at the coefficient of $x^{k+1-r}y^r$ in the expansion of

$$(x + y)^{k+1} = (x + y)^k(x + y).$$

On the right-hand side, the term involving $x^{k+1-r}y^r$ is the sum of two products.

$$({}_kC_r x^{k-r}y^r)(x) + ({}_kC_{r-1} x^{k+1-r}y^{r-1})(y)$$

$$= \left[\frac{k!}{(k-r)!r!} + \frac{k!}{(k+1-r)!(r-1)!}\right]x^{k+1-r}y^r$$

$$= \left[\frac{(k+1-r)k!}{(k+1-r)!r!} + \frac{k!r}{(k+1-r)!r!}\right]x^{k+1-r}y^r$$

$$= \left[\frac{k!(k+1-r+r)}{(k+1-r)!r!}\right]x^{k+1-r}y^r$$

$$= \left[\frac{(k+1)!}{(k+1-r)!r!}\right]x^{k+1-r}y^r$$

$$= {}_{k+1}C_r x^{k+1-r}y^r$$

So, by mathematical induction, the Binomial Theorem is valid for all positive integers n.

P.S. Problem Solving

 1. Decreasing Sequence Consider the sequence

$$a_n = \frac{n+1}{n^2+1}.$$

(a) Use a graphing utility to graph the first 10 terms of the sequence.

(b) Use the graph from part (a) to estimate the value of a_n as n approaches infinity.

(c) Complete the table.

n	1	10	100	1000	10,000
a_n					

(d) Use the table from part (c) to determine (if possible) the value of a_n as n approaches infinity.

 2. Alternating Sequence Consider the sequence

$$a_n = 3 + (-1)^n.$$

(a) Use a graphing utility to graph the first 10 terms of the sequence.

(b) Use the graph from part (a) to describe the behavior of the graph of the sequence.

(c) Complete the table.

n	1	10	101	1000	10,001
a_n					

(d) Use the table from part (c) to determine (if possible) the value of a_n as n approaches infinity.

3. Greek Mythology Can the Greek hero Achilles, running at 20 feet per second, ever catch a tortoise, starting 20 feet ahead of Achilles and running at 10 feet per second? The Greek mathematician Zeno said no. When Achilles runs 20 feet, the tortoise will be 10 feet ahead. Then, when Achilles runs 10 feet, the tortoise will be 5 feet ahead. Achilles will keep cutting the distance in half but will never catch the tortoise. The table shows Zeno's reasoning. In the table, both the distances and the times required to achieve them form infinite geometric series. Using the table, show that both series have finite sums. What do these sums represent?

Distance (in feet)	Time (in seconds)
20	1
10	0.5
5	0.25
2.5	0.125
1.25	0.0625
0.625	0.03125

DATA

Spreadsheet at LarsonPrecalculus.com

4. Conjecture Let $x_0 = 1$ and consider the sequence x_n given by

$$x_n = \frac{1}{2}x_{n-1} + \frac{1}{x_{n-1}}, \quad n = 1, 2, \dots$$

Use a graphing utility to compute the first 10 terms of the sequence and make a conjecture about the value of x_n as n approaches infinity.

5. Operations on an Arithmetic Sequence Determine whether each operation results in an arithmetic sequence when performed on an arithmetic sequence. If so, state the common difference.

(a) A constant C is added to each term.

(b) Each term is multiplied by a nonzero constant C.

(c) Each term is squared.

6. Sequences of Powers The following sequence of perfect squares is not arithmetic.

1, 4, 9, 16, 25, 36, 49, 64, 81, . . .

The related sequence formed from the first differences of this sequence, however, is arithmetic.

(a) Write the first eight terms of the related arithmetic sequence described above. What is the nth term of this sequence?

(b) Explain how to find an arithmetic sequence that is related to the following sequence of perfect cubes.

1, 8, 27, 64, 125, 216, 343, 512, 729, . . .

(c) Write the first seven terms of the related arithmetic sequence in part (b) and find the nth term of the sequence.

(d) Explain how to find an arithmetic sequence that is related to the following sequence of perfect fourth powers.

1, 16, 81, 256, 625, 1296, 2401, 4096, 6561, . . .

(e) Write the first six terms of the related arithmetic sequence in part (d) and find the nth term of the sequence.

7. Piecewise-Defined Sequence A sequence can be defined using a piecewise formula. An example of a piecewise-defined sequence is given below.

$$a_1 = 7, \ a_n = \begin{cases} \frac{1}{2}a_{n-1}, & \text{when } a_{n-1} \text{ is even.} \\ 3a_{n-1} + 1, & \text{when } a_{n-1} \text{ is odd.} \end{cases}$$

(a) Write the first 20 terms of the sequence.

(b) Write the first 10 terms of the sequences for which $a_1 = 4$, $a_1 = 5$, and $a_1 = 12$ (using a_n as defined above). What conclusion can you make about the behavior of each sequence?

8. Fibonacci Sequence Let $f_1, f_2, \ldots, f_n, \ldots$ be the Fibonacci sequence.

(a) Use mathematical induction to prove that
$$f_1 + f_2 + \cdots + f_n = f_{n+2} - 1.$$

(b) Find the sum of the first 20 terms of the Fibonacci sequence.

9. Pentagonal Numbers The numbers 1, 5, 12, 22, 35, 51, . . . are called pentagonal numbers because they represent the numbers of dots in the sequence of figures shown below. Use mathematical induction to prove that the nth pentagonal number P_n is given by
$$P_n = \frac{n(3n - 1)}{2}.$$

10. Think About It What conclusion can be drawn about the sequence of statements P_n for each situation?

(a) P_3 is true and P_k implies P_{k+1}.

(b) $P_1, P_2, P_3, \ldots, P_{50}$ are all true.

(c) $P_1, P_2,$ and P_3 are all true, but the truth of P_k does not imply that P_{k+1} is true.

(d) P_2 is true and P_{2k} implies P_{2k+2}.

11. Sierpinski Triangle Recall that a *fractal* is a geometric figure that consists of a pattern that is repeated infinitely on a smaller and smaller scale. One well-known fractal is the *Sierpinski Triangle*. In the first stage, the midpoints of the three sides are used to create the vertices of a new triangle, which is then removed, leaving three triangles. The figure below shows the first two stages. Note that each remaining triangle is similar to the original triangle. Assume that the length of each side of the original triangle is one unit. Write a formula that describes the side length of the triangles generated in the nth stage. Write a formula for the area of the triangles generated in the nth stage.

12. Job Offer You work for a company that pays $0.01 the first day, $0.02 the second day, $0.04 the third day, and so on. If the daily wage keeps doubling, what will your total income be for working 30 days?

13. Multiple Choice A multiple choice question has five possible answers. You know that the answer is not B or D, but you are not sure about answers A, C, and E. What is the probability that you will get the right answer when you take a guess?

14. Throwing a Dart You throw a dart at the circular target shown below. The dart is equally likely to hit any point inside the target. What is the probability that it hits the region outside the triangle?

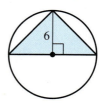

15. Odds The odds in favor of an event occurring is the ratio of the probability that the event will occur to the probability that the event will not occur. The reciprocal of this ratio represents the odds against the event occurring.

(a) A bag contains three blue marbles and seven yellow marbles. What are the odds in favor of choosing a blue marble? What are the odds against choosing a blue marble?

(b) Six of the marbles in a bag are red. The odds against choosing a red marble are 4 to 1. How many marbles are in the bag?

(c) Write a formula for converting the odds in favor of an event to the probability of the event.

(d) Write a formula for converting the probability of an event to the odds in favor of the event.

16. Expected Value An event A has n possible outcomes, which have the values $x_1, x_2, \ldots, x_n$. The probabilities of the n outcomes occurring are $p_1, p_2, \ldots, p_n$. The **expected value** V of an event A is the sum of the products of the outcomes' probabilities and their values,
$$V = p_1 x_1 + p_2 x_2 + \cdots + p_n x_n.$$

(a) To win California's Super Lotto Plus game, you must match five different numbers chosen from the numbers 1 to 47, plus one MEGA number chosen from the numbers 1 to 27. You purchase a ticket for $1. If the jackpot for the next drawing is $12,000,000, what is the expected value of the ticket?

(b) You are playing a dice game in which you need to score 60 points to win. On each turn, you toss two six-sided dice. Your score for the turn is 0 when the dice do not show the same number. Your score for the turn is the product of the numbers on the dice when they do show the same number. What is the expected value of each turn? How many turns will it take on average to score 60 points?

Appendix A   Errors and the Algebra of Calculus

- Avoid common algebraic errors.
- Recognize and use algebraic techniques that are common in calculus.

Algebraic Errors to Avoid

This section contains five lists of common algebraic errors: errors involving parentheses, errors involving fractions, errors involving exponents, errors involving radicals, and errors involving dividing out common factors. Many of these errors occur because they seem to be the *easiest* things to do. For example, students often believe that the operations of subtraction and division are commutative and associative. The examples below illustrate the fact that subtraction and division are neither commutative nor associative.

Not commutative

$$4 - 3 \neq 3 - 4$$
$$15 \div 5 \neq 5 \div 15$$

Not associative

$$8 - (6 - 2) \neq (8 - 6) - 2$$
$$20 \div (4 \div 2) \neq (20 \div 4) \div 2$$

Errors Involving Parentheses

Potential Error	Correct Form	Comment
$a - (x - b) = a - x - b$ ✗	$a - (x - b) = a - x + b$	Distribute negative sign to each term in parentheses.
$(a + b)^2 = a^2 + b^2$ ✗	$(a + b)^2 = a^2 + 2ab + b^2$	Remember the middle term when squaring binomials.
$\left(\frac{1}{2}a\right)\left(\frac{1}{2}b\right) = \frac{1}{2}(ab)$ ✗	$\left(\frac{1}{2}a\right)\left(\frac{1}{2}b\right) = \frac{1}{4}(ab) = \frac{ab}{4}$	$\frac{1}{2}$ occurs twice as a factor.
$(3x + 6)^2 = 3(x + 2)^2$ ✗	$(3x + 6)^2 = [3(x + 2)]^2$ $= 3^2(x + 2)^2$	When factoring, raise all factors to the power.

Errors Involving Fractions

Potential Error	Correct Form	Comment
$\dfrac{2}{x + 4} = \dfrac{2}{x} + \dfrac{2}{4}$ ✗	Leave as $\dfrac{2}{x + 4}$.	The fraction is already in simplest form.
$\dfrac{\left(\dfrac{x}{a}\right)}{b} = \dfrac{bx}{a}$ ✗	$\dfrac{\left(\dfrac{x}{a}\right)}{b} = \left(\dfrac{x}{a}\right)\left(\dfrac{1}{b}\right) = \dfrac{x}{ab}$	Multiply by the reciprocal when dividing fractions.
$\dfrac{1}{a} + \dfrac{1}{b} = \dfrac{1}{a + b}$ ✗	$\dfrac{1}{a} + \dfrac{1}{b} = \dfrac{b + a}{ab}$	Use the property for adding fractions with unlike denominators.
$\dfrac{1}{3x} = \dfrac{1}{3}x$ ✗	$\dfrac{1}{3x} = \dfrac{1}{3} \cdot \dfrac{1}{x}$	Use the property for multiplying fractions.
$(1/3)x = \dfrac{1}{3x}$ ✗	$(1/3)x = \dfrac{1}{3} \cdot x = \dfrac{x}{3}$	Be careful when expressing fractions in the form $1/a$.
$(1/x) + 2 = \dfrac{1}{x + 2}$ ✗	$(1/x) + 2 = \dfrac{1}{x} + 2 = \dfrac{1 + 2x}{x}$	Be careful when expressing fractions in the form $1/a$. Be sure to find a common denominator before adding fractions.

Errors Involving Exponents

Potential Error	Correct Form	Comment
$(x^2)^3 = x^5$ ✗	$(x^2)^3 = x^{2 \cdot 3} = x^6$	Multiply exponents when raising a power to a power.
$x^2 \cdot x^3 = x^6$ ✗	$x^2 \cdot x^3 = x^{2+3} = x^5$	Add exponents when multiplying powers with like bases.
$(2x)^3 = 2x^3$ ✗	$(2x)^3 = 2^3 x^3 = 8x^3$	Raise each factor to the power.
$\dfrac{1}{x^2 - x^3} = x^{-2} - x^{-3}$ ✗	Leave as $\dfrac{1}{x^2 - x^3}$.	Do not move term-by-term from denominator to numerator.

Errors Involving Radicals

Potential Error	Correct Form	Comment
$\sqrt{5x} = 5\sqrt{x}$ ✗	$\sqrt{5x} = \sqrt{5}\sqrt{x}$	Radicals apply to every factor inside the radical.
$\sqrt{x^2 + a^2} = x + a$ ✗	Leave as $\sqrt{x^2 + a^2}$.	Do not apply radicals term-by-term when adding or subtracting terms.
$\sqrt{-x + a} = -\sqrt{x - a}$ ✗	Leave as $\sqrt{-x + a}$.	Do not factor negative signs out of square roots.

Errors Involving Dividing Out

Potential Error	Correct Form	Comment
$\dfrac{a + bx}{a} = 1 + bx$ ✗	$\dfrac{a + bx}{a} = \dfrac{a}{a} + \dfrac{bx}{a} = 1 + \dfrac{b}{a}x$	Divide out common factors, not common terms.
$\dfrac{a + ax}{a} = a + x$ ✗	$\dfrac{a + ax}{a} = \dfrac{a(1 + x)}{a} = 1 + x$	Factor before dividing out common factors.
$1 + \dfrac{x}{2x} = 1 + \dfrac{1}{x}$ ✗	$1 + \dfrac{x}{2x} = 1 + \dfrac{1}{2} = \dfrac{3}{2}$	Divide out common factors.

A good way to avoid errors is to *work slowly, write neatly,* and *think about each step.* Each time you write a step, ask yourself why the step is algebraically legitimate. For example, the step below is legitimate because *dividing the numerator and denominator by the same nonzero number produces an equivalent fraction.*

$$\frac{2x}{6} = \frac{2 \cdot x}{2 \cdot 3} = \frac{x}{3}$$

EXAMPLE 1 **Describing and Correcting an Error**

Describe and correct the error. $\dfrac{1}{2x} + \dfrac{1}{3x} = \dfrac{1}{5x}$ ✗

Solution Use the property for adding fractions with unlike denominators.

$$\frac{1}{2x} + \frac{1}{3x} = \frac{3x + 2x}{6x^2} = \frac{5x}{6x^2} = \frac{5}{6x}$$

✓ *Checkpoint*   *Audio-video solution in English & Spanish at LarsonPrecalculus.com*

Describe and correct the error. $\sqrt{x^2 + 4} = x + 2$ ✗

Some Algebra of Calculus

In calculus it is often necessary to rewrite a simplified algebraic expression. See the following lists, which are from a standard calculus text.

Unusual Factoring

Expression	Useful Calculus Form	Comment
$\dfrac{5x^4}{8}$	$\dfrac{5}{8}x^4$	Write with fractional coefficient.
$\dfrac{x^2 + 3x}{-6}$	$-\dfrac{1}{6}(x^2 + 3x)$	Write with fractional coefficient.
$2x^2 - x - 3$	$2\left(x^2 - \dfrac{x}{2} - \dfrac{3}{2}\right)$	Factor out the leading coefficient.
$\dfrac{x}{2}(x + 1)^{-1/2} + (x + 1)^{1/2}$	$\dfrac{(x + 1)^{-1/2}}{2}[x + 2(x + 1)]$	Factor out the fractional coefficient and the variable expression with the lesser exponent.

Writing with Negative Exponents

Expression	Useful Calculus Form	Comment
$\dfrac{9}{5x^3}$	$\dfrac{9}{5}x^{-3}$	Move the factor to the numerator and change the sign of the exponent.
$\dfrac{7}{\sqrt{2x - 3}}$	$7(2x - 3)^{-1/2}$	Move the factor to the numerator and change the sign of the exponent.

Writing a Fraction as a Sum

Expression	Useful Calculus Form	Comment
$\dfrac{x + 2x^2 + 1}{\sqrt{x}}$	$x^{1/2} + 2x^{3/2} + x^{-1/2}$	Divide each term of the numerator by $x^{1/2}$.
$\dfrac{1 + x}{x^2 + 1}$	$\dfrac{1}{x^2 + 1} + \dfrac{x}{x^2 + 1}$	Rewrite the fraction as a sum of fractions.
$\dfrac{2x}{x^2 + 2x + 1}$	$\dfrac{2x + 2 - 2}{x^2 + 2x + 1}$	Add and subtract the same term.
	$= \dfrac{2x + 2}{x^2 + 2x + 1} - \dfrac{2}{(x + 1)^2}$	Rewrite the fraction as a difference of fractions.
$\dfrac{x^2 - 2}{x + 1}$	$x - 1 - \dfrac{1}{x + 1}$	Use polynomial long division. (See Section 3.3.)
$\dfrac{x + 7}{x^2 - x - 6}$	$\dfrac{2}{x - 3} - \dfrac{1}{x + 2}$	Use the method of partial fractions. (See Section 6.4.)

Inserting Factors and Terms

Expression	Useful Calculus Form	Comment
$(2x - 1)^3$	$\dfrac{1}{2}(2x - 1)^3(2)$	Multiply and divide by 2.
$7x^2(4x^3 - 5)^{1/2}$	$\dfrac{7}{12}(4x^3 - 5)^{1/2}(12x^2)$	Multiply and divide by 12.
$\dfrac{4x^2}{9} - 4y^2 = 1$	$\dfrac{x^2}{9/4} - \dfrac{y^2}{1/4} = 1$	Write with fractional denominators.
$\dfrac{x}{x + 1}$	$\dfrac{x + 1 - 1}{x + 1} = 1 - \dfrac{1}{x + 1}$	Add and subtract the same term.

The next five examples demonstrate many of the steps in the preceding lists.

EXAMPLE 2   **Factors Involving Negative Exponents**

Factor $x(x + 1)^{-1/2} + (x + 1)^{1/2}$.

Solution   When multiplying powers with like bases, you add exponents. When factoring, you are undoing multiplication, and so you *subtract* exponents.

$$x(x + 1)^{-1/2} + (x + 1)^{1/2} = (x + 1)^{-1/2}[x(x + 1)^0 + (x + 1)^1]$$
$$= (x + 1)^{-1/2}[x + (x + 1)]$$
$$= (x + 1)^{-1/2}(2x + 1)$$

✓ *Checkpoint* *Audio-video solution in English & Spanish at LarsonPrecalculus.com*

Factor $x(x - 2)^{-1/2} + 6(x - 2)^{1/2}$.

Another way to simplify the expression in Example 2 is to multiply the expression by a fractional form of 1 and then use the Distributive Property.

$$\left[x(x + 1)^{-1/2} + (x + 1)^{1/2}\right] \cdot \frac{(x + 1)^{1/2}}{(x + 1)^{1/2}} = \frac{x(x + 1)^0 + (x + 1)^1}{(x + 1)^{1/2}}$$
$$= \frac{2x + 1}{\sqrt{x + 1}}$$

EXAMPLE 3   **Inserting Factors in an Expression**

Insert the required factor: $\dfrac{x + 2}{(x^2 + 4x - 3)^2} = (\;\rule{1.2em}{0.8em}\;)\dfrac{1}{(x^2 + 4x - 3)^2}(2x + 4)$.

Solution   The expression on the right side of the equation is twice the expression on the left side. To make both sides equal, insert a factor of $\frac{1}{2}$.

$$\frac{x + 2}{(x^2 + 4x - 3)^2} = \left(\frac{1}{2}\right)\frac{1}{(x^2 + 4x - 3)^2}(2x + 4)$$

✓ *Checkpoint* *Audio-video solution in English & Spanish at LarsonPrecalculus.com*

Insert the required factor: $\dfrac{6x - 3}{(x^2 - x + 4)^2} = (\;\rule{1.2em}{0.8em}\;)\dfrac{1}{(x^2 - x + 4)^2}(2x - 1)$.

Answers to Odd-Numbered Exercises and Tests

Chapter P

Section P.1 *(page 12)*

1. irrational   **3.** absolute value   **5.** terms
7. (a) $5, 1, 2$ (b) $0, 5, 1, 2$ (c) $-9, 5, 0, 1, -4, 2, -11$
   (d) $-9, -\frac{7}{2}, 5, \frac{2}{3}, 0, 1, -4, 2, -11$ (e) $\sqrt{2}$
9. (a) 1 (b) 1 (c) $-13, 1, -6$
   (d) $2.01, 0.\overline{6}, -13, 1, -6$ (e) $0.010110111\ldots$

11. (a) 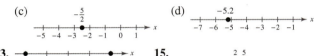   (b)

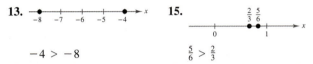

   (c)   (d)

13.   **15.**

   $-4 > -8$ $\frac{5}{6} > \frac{2}{3}$

17. (a) $x \le 5$ denotes the set of all real numbers less than or
   equal to 5.
   (b)   (c) Unbounded

19. (a) $-2 < x < 2$ denotes the set of all real numbers greater
   than -2 and less than 2.
   (b)   (c) Bounded

21. (a) $[4, \infty)$ denotes the set of all real numbers greater than or
   equal to 4.
   (b)   (c) Unbounded

23. (a) $[-5, 2)$ denotes the set of all real numbers greater than or
   equal to -5 and less than 2.
   (b)   (c) Bounded

Inequality	Interval
25. $y \ge 0$	$[0, \infty)$
27. $10 \le t \le 22$	$[10, 22]$

29. 10 **31.** 5 **33.** -1 **35.** 25 **37.** -1
39. $|-4| = |4|$ **41.** $-|-6| < |-6|$ **43.** 51 **45.** $\frac{5}{2}$
47. $|x - 5| \le 3$ **49.** 2524.0 billion; 458.5 billion
51. 2450.0 billion; 1087.0 billion
53. $7x$ and 4 are the terms; 7 is the coefficient.
55. $6x^3$ and $-5x$ are the terms; 6 and -5 are the coefficients.
57. $3\sqrt{3}x^2$ and 1 are the terms; $3\sqrt{3}$ is the coefficient.
59. (a) -10 (b) -6 **61.** (a) 2 (b) 6
63. (a) Division by 0 is undefined.   (b) 0
65. Multiplicative Inverse Property
67. Associative and Commutative Properties of Multiplication
69. $\dfrac{5x}{12}$ **71.** $\dfrac{x}{4}$

73. False. Zero is nonnegative, but not positive.
75. True. The product of two negative numbers is positive.

77. (a)

n	0.0001	0.01	1	100	10,000
$\dfrac{5}{n}$	50,000	500	5	0.05	0.0005

   (b) (i) The value of $5/n$ approaches infinity as n approaches 0.
       (ii) The value of $5/n$ approaches 0 as n increases without
           bound.

Section P.2 *(page 24)*

1. exponent; base   **3.** square root   **5.** like radicals
7. rationalizing   **9.** (a) 625 (b) $\frac{1}{25}$
11. (a) 5184 (b) $-\frac{3}{5}$ **13.** (a) $\frac{16}{3}$ (b) 1
15. -24 **17.** 6 **19.** -48 **21.** (a) $125z^3$ (b) $5x^6$
23. (a) $24y^2$ (b) $-3z^7$ **25.** (a) $\dfrac{5184}{y^7}$ (b) 1
27. (a) 1 (b) $\dfrac{1}{4x^4}$ **29.** (a) $\dfrac{125x^9}{y^{12}}$ (b) $\dfrac{b^5}{a^5}$
31. 1.02504×10^4 **33.** 0.000314
35. $9,460,000,000,000$ km
37. (a) 6.8×10^5 (b) 6.0×10^4 **39.** (a) 3 (b) $\frac{3}{2}$
41. (a) 2 (b) $2x$ **43.** (a) $2\sqrt{5}$ (b) $4\sqrt[3]{2}$
45. (a) $6x\sqrt{2x}$ (b) $3y^2\sqrt{6x}$
47. (a) $2x\sqrt[3]{2x^2}$ (b) $\dfrac{5|x|\sqrt{3}}{y^2}$
49. (a) $29|x|\sqrt{5}$ (b) $44\sqrt{3x}$ **51.** $\dfrac{\sqrt{3}}{3}$ **53.** $\dfrac{\sqrt{14}+2}{2}$
55. $\dfrac{2}{3\left(\sqrt{5}-\sqrt{3}\right)}$ **57.** $64^{1/3}$ **59.** $\dfrac{3}{\sqrt[3]{x^2}}$
61. (a) $\frac{1}{8}$ (b) $\frac{27}{8}$ **63.** (a) $\sqrt{3}$ (b) $\sqrt[3]{(x+1)^2}$
65. (a) $2\sqrt[4]{2}$ (b) $\sqrt[8]{2x}$ **67.** (a) $x - 1$ (b) $\dfrac{1}{x-1}$

69.

h	0	1	2	3	4	5	6
t	0	2.93	5.48	7.67	9.53	11.08	12.32

h	7	8	9	10	11	12
t	13.29	14.00	14.50	14.80	14.93	14.96

71. False. When $x = 0$, the expressions are not equal.
73. False. For instance, $(3 + 5)^2 = 8^2 = 64 \ne 34 = 3^2 + 5^2$.

Section P.3 *(page 31)*

1. n; a_n; a_0 **3.** like terms
5. (a) $7x$ (b) Degree: 1; Leading coefficient: 7
   (c) Monomial
7. (a) $-\frac{1}{2}x^5 + 14x$ (b) Degree: 5; Leading coefficient: $-\frac{1}{2}$
   (c) Binomial
9. (a) $-4x^5 + 6x^4 + 1$
   (b) Degree: 5; Leading coefficient: -4 (c) Trinomial
11. Polynomial: $-3x^3 + 2x + 8$
13. Not a polynomial because it includes a term with a negative
   exponent

15. Polynomial: $-y^4 + y^3 + y^2$ **17.** $-2x - 10$
19. $7t^3 - 5t - 1$ **21.** $-8.3x^3 + 0.3x^2 - 23$ **23.** $12z + 8$
25. $3x^3 - 6x^2 + 3x$ **27.** $-15z^2 + 5z$ **29.** $-4.5t^3 - 15t$
31. $-0.2x^2 - 34x$ **33.** $x^2 + 12x + 35$ **35.** $6x^2 - 7x - 5$
37. $x^4 + 2x^2 + x + 2$ **39.** $x^2 - 100$ **41.** $x^2 - 4y^2$
43. $4x^2 + 12x + 9$ **45.** $16x^6 - 24x^3 + 9$
47. $x^3 + 9x^2 + 27x + 27$ **49.** $8x^3 - 12x^2y + 6xy^2 - y^3$
51. $\frac{1}{25}x^2 - 9$ **53.** $36x^2 - 9y^2$ **55.** $\frac{1}{16}x^2 - \frac{5}{2}x + 25$
57. $x^2 + 2xy + y^2 - 6x - 6y + 9$ **59.** $m^2 - n^2 - 6m + 9$
61. $u^4 - 16$ **63.** $-3x^3 - 3x^2 + 14$
65. $y^4 - 3y^3 - 19y^2 + 42y - 20$ **67.** $x - y$
69. $x^2 - 2\sqrt{5}x + 5$
71. (a) $P = 42x - 35,000$ (b) \$175,000
73. (a) 25%   (b) $0.25N^2 + 0.5Na + 0.25a^2$
 (c) 25%; Answers will vary.
75. $3x^2 + 8x$ **77.** $15x^2 + 18x + 3$
79. (a) $V = 4x^3 - 88x^2 + 468x$
 (b)

x (cm)	1	2	3
V (cm³)	384	616	720

81. (a) Approximations will vary.
 (b) The difference of the safe loads decreases in magnitude as the span increases.
83. False. $(4x^2 + 1)(3x + 1) = 12x^3 + 4x^2 + 3x + 1$
85. False. $(4x + 3) + (-4x + 6) = 4x + 3 - 4x + 6$
$$= 3 + 6$$
$$= 9$$
87. $m + n$
89. The middle term was omitted when squaring the binomial.
$(x - 3)^2 = x^2 - 6x + 9 \neq x^2 + 9$
91. $-x^3 + 8x^2 + 2x + 7$

Section P.4   *(page 39)*

1. factoring   **3.** perfect square trinomial   **5.** $2x(x^2 - 3)$
7. $(x - 5)(3x + 8)$ **9.** $(x + 9)(x - 9)$
11. $(5y + 2)(5y - 2)$ **13.** $(8 + 3z)(8 - 3z)$
15. $(x + 1)(x - 3)$ **17.** $(3u - 1)(3u + 1)(9u^2 + 1)$
19. $(x - 2)^2$ **21.** $(5z - 3)^2$ **23.** $(2y - 3)^2$
25. $(x - 2)(x^2 + 2x + 4)$ **27.** $(2t - 1)(4t^2 + 2t + 1)$
29. $(3x + 2)(9x^2 - 6x + 4)$ **31.** $(u + 3v)(u^2 - 3uv + 9v^2)$
33. $(x + 2)(x - 1)$ **35.** $(s - 3)(s - 2)$
37. $(3x - 2)(x + 4)$ **39.** $(5x + 1)(x + 6)$
41. $-(5y - 2)(y + 2)$ **43.** $(x - 1)(x^2 + 2)$
45. $(2x - 1)(x^2 - 3)$ **47.** $(3x^3 - 2)(x^2 + 2)$
49. $(x + 3)(2x + 3)$ **51.** $(2x + 3)(3x - 5)$
53. $6(x + 3)(x - 3)$ **55.** $x^2(x - 1)$ **57.** $(1 - 2x)^2$
59. $-2x(x + 1)(x - 2)$ **61.** $(x + 3)(x + 1)(x - 3)(x - 1)$
63. $(x - 2)(x + 2)(2x + 1)$ **65.** $(3x + 1)(5x + 1)$
67. $-(x - 2)(x + 1)(x - 8)$ **69.** $(x + 1)^2(2x + 1)(11x + 6)$
71. $\left(4x + \frac{1}{3}\right)\left(4x - \frac{1}{3}\right)$ **73.** $\left(z + \frac{1}{2}\right)^2$
75. $\left(y + \frac{2}{3}\right)\left(y^2 - \frac{2}{3}y + \frac{4}{9}\right)$

77.

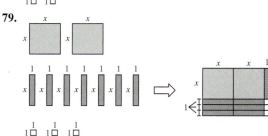

79.

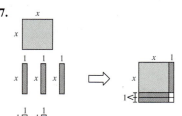

81. $4\pi(r + 1)$
83. (a) $\pi h(R + r)(R - r)$ (b) $V = 2\pi\left[\left(\frac{R + r}{2}\right)(R - r)\right]h$
85. $-14, 14, -2, 2$ **87.** Two possible answers: $2, -12$
89. True. $a^2 - b^2 = (a + b)(a - b)$
91. 3 should be factored out of the second binomial to yield $9(x + 2)(x - 3)$.
93. $(x^n + y^n)(x^n - y^n)$
95. Answers will vary. *Sample answer:* $x^2 - 3$
97. $(u + v)(u - v)(u^2 + uv + v^2)(u^2 - uv + v^2)$
 $(x - 1)(x + 1)(x^2 + x + 1)(x^2 - x + 1)$
 $(x - 2)(x + 2)(x^2 - 2x + 4)(x^2 + 2x + 4)$

Section P.5   *(page 48)*

1. domain   **3.** complex   **5.** All real numbers x
7. All real numbers x such that $x \neq 3$
9. All real numbers x such that $x \neq -\frac{2}{3}$
11. All real numbers x such that $x \neq -4, -2$
13. All real numbers x such that $x \geq 7$
15. All real numbers x such that $x > 3$
17. $\frac{3x}{2}$, $x \neq 0$ **19.** $-\frac{1}{2}$, $x \neq 5$ **21.** $y - 4$, $y \neq -4$
23. $\frac{3y}{4}$, $y \neq -\frac{2}{3}$ **25.** $\frac{x - 1}{x + 3}$, $x \neq -5$
27. $-\frac{x + 1}{x + 5}$, $x \neq 2$ **29.** $\frac{1}{x + 1}$, $x \neq \pm 4$
31. When simplifying fractions, only common factors can be divided out, not terms.
33. $\frac{1}{5(x - 2)}$, $x \neq 1$ **35.** $-\frac{(x + 2)^2}{6}$, $x \neq \pm 2$
37. $\frac{x - y}{x(x + y)^2}$, $x \neq -2y$ **39.** $\frac{3}{x + 2}$ **41.** $\frac{3x^2 + 3x + 1}{(x + 1)(3x + 2)}$
43. $\frac{3 - 2x}{2(x + 2)}$ **45.** $\frac{2 - x}{x^2 + 1}$, $x \neq 0$
47. The minus sign should be distributed to each term in the numerator of the second fraction to yield $x + 4 - 3x + 8$.
49. $\frac{1}{2}$, $x \neq 2$ **51.** $x(x + 1)$, $x \neq -1, 0$
53. $\frac{2x - 1}{2x}$, $x > 0$ **55.** $\frac{x^2 + (x^2 + 3)^7}{(x^2 + 3)^4}$

57. $\dfrac{2x^3 - 2x^2 - 5}{(x - 1)^{1/2}}$ **59.** $\dfrac{3x - 1}{3}, \; x \neq 0$

61. $\dfrac{-1}{x(x + h)}, \; h \neq 0$ **63.** $\dfrac{-1}{(x - 4)(x + h - 4)}, \; h \neq 0$

65. $\dfrac{1}{\sqrt{x + 2} + \sqrt{x}}$ **67.** $\dfrac{1}{\sqrt{t + 3} + \sqrt{3}}, \; t \neq 0$

69. $\dfrac{1}{\sqrt{x + h + 1} + \sqrt{x + 1}}, \; h \neq 0$

71. (a)

t	0	2	4	6	8	10	12
T	75	55.9	48.3	45	43.3	42.3	41.7

t	14	16	18	20	22
T	41.3	41.1	40.9	40.7	40.6

(b) The model is approaching a T-value of 40.

73. $\dfrac{x}{2(2x + 1)}, \; x \neq 0$

75. (a)

Year, t	Online Banking	Mobile Banking
11	79.1	17.9
12	80.9	24.0
13	83.1	29.6
14	86.0	34.8

(b) The values from the models are close to the actual data.

(c) Ratio $= \dfrac{0.0313t^3 - 0.661t^2 - 2.23t + 47}{0.0208t^3 - 0.494t^2 + 2.97t - 70.5}$

(d)

Year	Ratio
2011	0.2267
2012	0.2977
2013	0.3578
2014	0.4061

Answers will vary.

77. $\dfrac{R_1 R_2}{R_2 + R_1}$

79. False. In order for the simplified expression to be equivalent to the original expression, the domain of the simplified expression needs to be restricted. If n is even, $x \neq -1, 1$. If n is odd, $x \neq 1$.

Section P.6   *(page 57)*

1. Cartesian   **3.** Distance Formula

5.

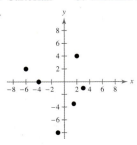

7. $(-3, 4)$ **9.** Quadrant IV   **11.** Quadrant II

13. Quadrant II or IV

15.

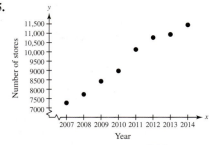

17. 13   **19.** $\sqrt{61}$ **21.** $\dfrac{\sqrt{277}}{6}$

23. (a) 5, 12, 13   (b) $5^2 + 12^2 = 13^2$

25. $\left(\sqrt{5}\right)^2 + \left(\sqrt{45}\right)^2 = \left(\sqrt{50}\right)^2$

27. Distances between the points: $\sqrt{29}, \sqrt{58}, \sqrt{29}$

29. (a)

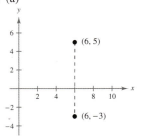

31. (a)

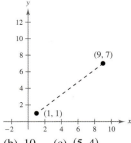

(b) 8   (c) $(6, 1)$ (b) 10   (c) $(5, 4)$

33. (a)

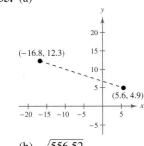

35. (a)

(b) $2\sqrt{10}$ (b) $\sqrt{556.52}$

(c) $(2, 3)$ (c) $(-5.6, 8.6)$

37. $30\sqrt{41} \approx 192$ km   **39.** \$40,560.5 million

41. $(0, 1), (4, 2), (1, 4)$ **43.** $(-3, 6), (2, 10), (2, 4), (-3, 4)$

45. (a) 2000–2010   (b) 53.7%; 40.8%   (c) \$10.21

(d) Answers will vary.

47. True. Because $x < 0$ and $y > 0$, $2x < 0$ and $-3y < 0$, which is located in Quadrant III.

49. True. Two sides of the triangle have lengths of $\sqrt{149}$, and the third side has a length of $\sqrt{18}$.

51. *Sample answer:* When the x-values are much larger or smaller that the y-values.

53. $(2x_m - x_1, 2y_m - y_1)$

55. $\left(\dfrac{3x_1 + x_2}{4}, \dfrac{3y_1 + y_2}{4}\right), \left(\dfrac{x_1 + x_2}{2}, \dfrac{y_1 + y_2}{2}\right),$

$\left(\dfrac{x_1 + 3x_2}{4}, \dfrac{y_1 + 3y_2}{4}\right)$

CHAPTER P

57. Use the Midpoint Formula to prove that the diagonals of the parallelogram bisect each other.

$$\left(\frac{b+a}{2}, \frac{c+0}{2}\right) = \left(\frac{a+b}{2}, \frac{c}{2}\right)$$

$$\left(\frac{a+b+0}{2}, \frac{c+0}{2}\right) = \left(\frac{a+b}{2}, \frac{c}{2}\right)$$

59. (a)

	First Set	Second Set
Distance A to B	3	$\sqrt{10}$
Distance B to C	5	$\sqrt{10}$
Distance A to C	4	$\sqrt{40}$
	Right triangle	Isosceles triangle

(b)

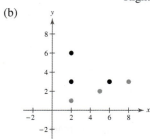

The first set of points is not collinear. The second set of points is collinear.

(c) A set of three points is collinear when the sum of two distances among the points is exactly equal to the third distance.

Review Exercises   (page 62)

1. (a) 11   (b) 11   (c) 11, -14
   (d) 11, -14, $-\frac{8}{9}, \frac{5}{2}, 0.4$ (e) $\sqrt{6}$

3.

$\frac{5}{4} > \frac{7}{8}$

5. (a) $x \geq 6$ denotes the set of all real numbers greater than or equal to 6.

(b)

(c) Unbounded

7. (a) $[-3, 4)$ denotes the set of all real numbers greater than or equal to -3 and less than 4.

(b)

(c) Bounded

9. 122   **11.** $|x - 7| \geq 4$ **13.** (a) -7 (b) -19

15. (a) -1 (b) -3 **17.** Additive Identity Property

19. Associative Property of Addition

21. Commutative Property of Addition

23. 0   **25.** 32   **27.** $\frac{47x}{60}$ **29.** $\frac{x}{2}$

31. (a) $192x^{11}$ (b) $\frac{y^5}{2}, \ y \neq 0$ **33.** (a) $-8z^3$ (b) $\frac{1}{y^2}$

35. (a) a^2b^2 (b) $3a^3b^2$ **37.** (a) $\frac{1}{625a^4}$ (b) $64x^2$

39. 2.744×10^8 **41.** 484,000,000

43. (a) 9   (b) 343   **45.** (a) 216   (b) 32

47. (a) $(2x + 1)\sqrt{3x}$ (b) $2x\sqrt{3x}$

49. Radicals cannot be combined by addition or subtraction unless the index and the radicand are the same.

51. $\frac{\sqrt{3}}{4}$ **53.** $2 + \sqrt{3}$ **55.** $\frac{3}{\sqrt{7} - 1}$

57. 64   **59.** $6x^{9/10}$

61. $-11x^2 + 3$; Degree: 2; Leading coefficient: -11

63. $-12x^2 - 4$; Degree: 2; Leading coefficient: -12

65. $-3x^2 - 7x + 1$ **67.** $2x^3 - 10x^2 + 12x$

69. $15x^2 - 27x - 6$ **71.** $36x^2 - 25$ **73.** $4x^2 - 12x + 9$

75. $x^4 - 6x^3 - 4x^2 - 37x - 10$

77. $2500r^2 + 5000r + 2500$ **79.** $x^2 + 28x + 192$

81. $x(x + 1)(x - 1)$ **83.** $(5x + 7)(5x - 7)$

85. $(x - 4)(x^2 + 4x + 16)$ **87.** $(x + 10)(2x + 1)$

89. $(x - 1)(x^2 + 2)$

91. All real numbers x such that $x \neq -1$

93. All real numbers x such that $x \geq -2$

95. $\frac{x - 8}{15}, \ x \neq -8$ **97.** $\frac{x - 4}{(x + 3)(x + 4)}, \ x \neq 3$

99. $-\frac{4}{x - 3}$ **101.** $\frac{3ax^2}{(a^2 - x)(a - x)}, \ x \neq 0$

103. $\frac{-1}{2x(x + h)}, \ h \neq 0$

105.

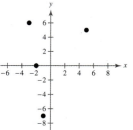

107. Quadrant IV

109. (a)

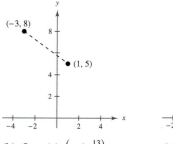

111. (a)

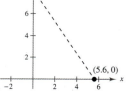

(b) 5   (c) $\left(-1, \frac{13}{2}\right)$ (b) $\sqrt{98.6}$ (c) $(2.8, 4.1)$

113. $(0, 0), (2, 0), (0, -5), (2, -5)$

115. \$6.45 billion

117. False. There is also a cross-product term when a binomial sum is squared.
   $(x + a)^2 = x^2 + 2ax + a^2$

Chapter Test   (page 65)

1. $-\frac{10}{3} < -\frac{5}{3}$ **2.** 3   **3.** Additive Identity Property

4. (a) $-\frac{27}{125}$ (b) $\frac{8}{729}$ (c) $\frac{5}{49}$ (d) $\frac{1}{64}$

5. (a) 25   (b) $\frac{3\sqrt{6}}{2}$ (c) 1.8×10^5 (d) 9.6×10^5

6. (a) $12z^8$ (b) $(u - 2)^{-7}$ (c) $\frac{3x^2}{y^2}$

7. (a) $15z\sqrt{2z}$ (b) $4x^{14/15}$ (c) $\frac{2\sqrt[3]{2v}}{v^2}$

8. $-2x^5 - x^4 + 3x^3 + 3$; Degree: 5; Leading coefficient: -2

9. $2x^2 - 3x - 5$ **10.** $x^2 - 5$ **11.** $5, x \neq 4$

12. $x^2(x - 1), \ x \neq 0, 1$

13. (a) $x^2(2x + 1)(x - 2)$ (b) $(x - 2)(x + 2)^2$

14. (a) $4\sqrt[3]{4}$ (b) $-4(1 + \sqrt{2})$

15. All real numbers x except $x = 1$

16. $\dfrac{4}{y + 4}, \ y \neq 2$ **17.** \$545

18.

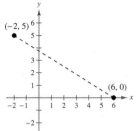

Midpoint: $\left(2, \frac{5}{2}\right)$

Distance: $\sqrt{89}$

19. $\frac{5}{2}x^2 + 3x + \frac{1}{2}$

Problem Solving   *(page 67)*

1. (a) Men's: 1,150,347 mm³; 696,910 mm³

Women's: 696,910 mm³; 448,921 mm³

(b) Men's: 1.04×10^{-5} kg/mm³; 6.31×10^{-6} kg/mm³

Women's: 8.91×10^{-6} kg/mm³; 5.74×10^{-6} kg/mm³

(c) No. Iron has a greater density than cork.

3. Answers will vary.

5. Man: 2,812,834,080 beats; Woman: 2,989,556,640 beats

7. $r \approx 0.28$ **9.** $SA = 10x^2 + 4x - 8$; 376 in.²

11. (a) $(2, -1), (3, 0)$ (b) $\left(-\frac{4}{3}, -2\right), \left(-\frac{2}{3}, -1\right)$

13. About 0.101 lb; 1.616 oz

Chapter 1

Section 1.1   *(page 78)*

1. solution or solution point   **3.** intercepts

5. circle; (h, k); r **7.** (a) Yes   (b) Yes

9. (a) Yes   (b) No   **11.** (a) No   (b) Yes

13. (a) No   (b) Yes

15.

x	-1	0	1	2	$\frac{5}{2}$
y	7	5	3	1	0
(x, y)	$(-1, 7)$	$(0, 5)$	$(1, 3)$	$(2, 1)$	$\left(\frac{5}{2}, 0\right)$

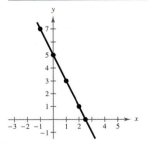

17.

x	-1	0	1	2	3
y	4	0	-2	-2	0
(x, y)	$(-1, 4)$	$(0, 0)$	$(1, -2)$	$(2, -2)$	$(3, 0)$

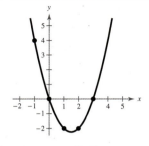

19. x-intercept: $(3, 0)$ **21.** x-intercept: $(-2, 0)$

y-intercept: $(0, 9)$ y-intercept: $(0, 2)$

23. x-intercept: $(1, 0)$

y-intercept: $(0, 2)$

25. y-axis symmetry   **27.** Origin symmetry

29. Origin symmetry   **31.** x-axis symmetry

33. 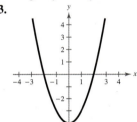   **35.**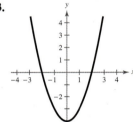

37. No symmetry   **39.** No symmetry

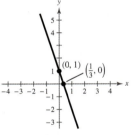   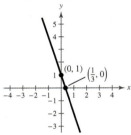

x-intercept: $\left(\frac{1}{3}, 0\right)$ x-intercepts: $(0, 0), (2, 0)$

y-intercept: $(0, 1)$ y-intercept: $(0, 0)$

41. No symmetry   **43.** No symmetry

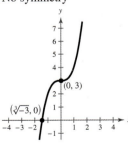   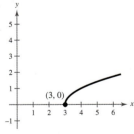

x-intercept: $\left(\sqrt[3]{-3}, 0\right)$ x-intercept: $(3, 0)$

y-intercept: $(0, 3)$ y-intercept: None

45. No symmetry

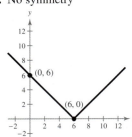

x-intercept: $(6, 0)$
y-intercept: $(0, 6)$

47. x-axis symmetry

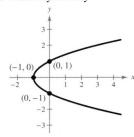

x-intercept: $(-1, 0)$
y-intercepts: $(0, \pm 1)$

49.

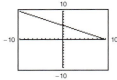

Intercepts: $(10, 0), (0, 5)$

51.

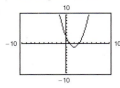

Intercepts:
$(3, 0), (1, 0), (0, 3)$

53.

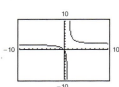

Intercept: $(0, 0)$

55.

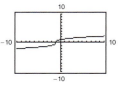

Intercepts: $(-1, 0), (0, 1)$

57.

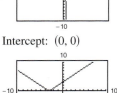

Intercepts: $(-3, 0), (0, 3)$

59. $x^2 + y^2 = 9$ **61.** $(x + 4)^2 + (y - 5)^2 = 4$

63. $(x - 3)^2 + (y - 8)^2 = 169$

65. $(x + 3)^2 + (y + 3)^2 = 61$

67. Center: $(0, 0)$; Radius: 5

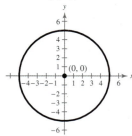

69. Center: $(1, -3)$; Radius: 3

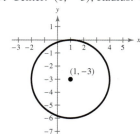

71. Center: $\left(\frac{1}{2}, \frac{1}{2}\right)$; Radius: $\frac{3}{2}$

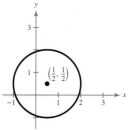

73.

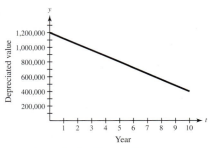

75. (a)

(b) Answers will vary.

(c)

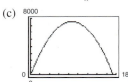

(d) $x = 86\frac{2}{3}, y = 86\frac{2}{3}$

(e) A regulation NFL playing field is 120 yards long and $53\frac{1}{3}$ yards wide. The actual area is 6400 square yards.

77. (a)

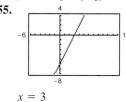

The model fits the data well;
Each data value is close to
the graph of the model.

(b) 74.7 yr (c) 1964

(d) $(0, 63.6)$; In 1940, the life expectancy of a child (at birth) was 63.6 years.

(e) Answers will vary.

79. False. $y = x$ is symmetric with respect to the origin.

81. True. *Sample answer:* Depending on the center and radius, the graph could intersect one, both, or neither axis.

83. (a) $a = 1, b = 0$ (b) $a = 0, b = 1$

Section 1.2 *(page 87)*

1. equation **3.** $ax + b = 0$ **5.** rational **7.** Identity

9. Conditional equation **11.** Contradiction **13.** Identity

15. 4 **17.** -9 **19.** 12 **21.** 1 **23.** No solution

25. 9 **27.** $-\frac{96}{23}$ **29.** -4 **31.** 4 **33.** 3 **35.** 0

37. No solution; The variable is divided out.

39. No solution; The solution is extraneous. **41.** 5

43. No solution; The solution is extraneous.

45. x-intercept: $\left(\frac{12}{5}, 0\right)$ **47.** x-intercept: $\left(-\frac{1}{2}, 0\right)$
y-intercept: $(0, 12)$ y-intercept: $(0, -3)$

49. x-intercept: $(5, 0)$ **51.** x-intercept: $(1.6, 0)$
y-intercept: $\left(0, \frac{10}{3}\right)$ y-intercept: $(0, -0.3)$

53. x-intercept: $(-20, 0)$
y-intercept: $\left(0, \frac{8}{3}\right)$

55.

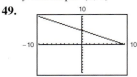

$x = 3$

57.

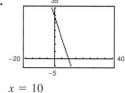

$x = 10$

59.

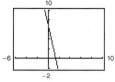

$x = \frac{7}{5}$

61. 138.889   **63.** 19.993   **65.** $h = 10$ ft   **67.** 63.7 in.

69. (a) About $(0, 295)$

(b) $(0, 293.4)$; In 2000, the population of Raleigh was about 293,400.

(c) 2022; Answers will vary.

71. 20,000 mi

73. False. $x(3 - x) = 10$

$3x - x^2 = 10$

The equation cannot be written in the form $ax + b = 0$.

75. False. The equation is an identity.

77. False. The equation is a contradiction.

79. (a)

x	-1	0	1	2	3	4
$3.2x - 5.8$	-9	-5.8	-2.6	0.6	3.8	7

(b) $1 < x < 2$. The expression changes from negative to positive in this interval.

(c)

x	1.5	1.6	1.7	1.8	1.9	2
$3.2x - 5.8$	-1	-0.68	-0.36	-0.04	0.28	0.6

(d) $1.8 < x < 1.9$. To improve accuracy, evaluate the expression in this interval and determine where the sign changes.

(e) $8.16 < x < 8.17$

81. (a)

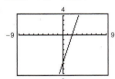

(b) $(2, 0)$

(c) The x-intercept is the solution of the equation $3x - 6 = 0$.

83. (a) $\left(-\frac{b}{a}, 0\right)$ (b) $(0, b)$

(c) x-intercept: $(-2, 0)$

y-intercept: $(0, 10)$

Section 1.3   *(page 97)*

1. mathematical modeling   **3.** A number increased by 2

5. A number divided by 6

7. A number decreased by 2 then divided by 3

9. The product of -2 and a number increased by 5

11. The product of 2 less than a number and 3 is then divided by the same number.

13. $n + (n + 1) = 2n + 1$ **15.** $(2n - 1)(2n + 1) = 4n^2 - 1$

17. $55t$ **19.** $0.20x$ **21.** $6x$ **23.** $2500 + 40x$

25. $d = 0.30L$ **27.** $N = \dfrac{p}{100} \cdot 672$ **29.** $4x + 8x = 12x$

31. 262, 263   **33.** 37, 185   **35.** $-5, -4$

37. First salesperson: \$516.89; Second salesperson: \$608.11

39. \$49,000

41. (a)

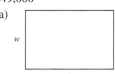

(b) $l = 1.5w$; $P = 5w$

(c) 7.5 m $\times$ 5 m

43. 97   **45.** 5 h   **47.** About 8.33 min   **49.** 945 ft

51. (a)

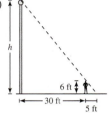

(b) 42 ft

53. \$4000 at $4\frac{1}{2}\%$, \$8000 at 5%

55. Red maple: \$25,000; Dogwood: \$15,000

57. About 1.09 gal   **59.** $\dfrac{2A}{b}$ **61.** $\dfrac{S}{1 + R}$ **63.** $\dfrac{A - P}{Pt}$

65. 37°C   **67.** 80.6°F   **69.** $\sqrt[3]{\dfrac{4.47}{\pi}} \approx 1.12$ in.

71. (a) $x = 6$ feet from the 50-pound child

(b) $x = 3\frac{2}{3}$ feet from the person

73. False. The expression should be $\dfrac{z^3 - 8}{z^2 - 9}$.

75. True.

Circle: $A = \pi r^2$

$= \pi(2)^2$

≈ 12.56 in.^{2}

Square: $A = s^2$

$= 4^2$

$= 16$ in.^{2}

Section 1.4   *(page 110)*

1. quadratic equation

3. factoring; square roots; completing; square; Quadratic Formula

5. position equation   **7.** $0, -\frac{1}{2}$ **9.** $3, -\frac{1}{2}$ **11.** -5

13. $\pm\frac{3}{4}$ **15.** $-\frac{3}{2}, 11$ **17.** $-\frac{20}{3}, -4$ **19.** ± 7

21. $\pm\sqrt{19} \approx 4.36$ **23.** $\pm 3\sqrt{3} \approx \pm 5.20$

25. $-3, 11$ **27.** $-2 \pm \sqrt{14} \approx 1.74, -5.74$

29. $\dfrac{1 \pm 3\sqrt{2}}{2} \approx 2.62, -1.62$ **31.** 2   **33.** $4, -8$

35. $-2 \pm \sqrt{2}$ **37.** $1 \pm \dfrac{\sqrt{2}}{2}$ **39.** $1 \pm 2\sqrt{2}$

41. $\dfrac{-5 \pm \sqrt{89}}{4}$ **43.** $\dfrac{1}{(x - 1)^2 + 4}$ **45.** $\dfrac{4}{(x + 5)^2 + 49}$

47. $\dfrac{1}{\sqrt{4 - (x - 1)^2}}$ **49.** $\dfrac{1}{\sqrt{16 - (x - 2)^2}}$

51. (a)

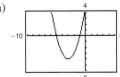

(b) and (c) $-1, -5$

(d) The answers are the same.

53. (a)

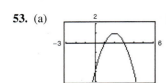

(b) and (c) 3, 1

(d) The answers are the same.

55. (a)

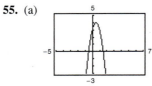

(b) and (c) $-\frac{1}{2}, \frac{3}{2}$

(d) The answers are the same.

57. (a)

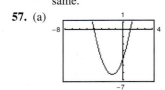

(b) and (c) 1, -4

(d) The answers are the same.

59. One repeated real solution   **61.** No real solution

63. Two real solutions   **65.** No real solution

67. Two real solutions   **69.** $\frac{1}{2}, -1$ **71.** $\frac{1}{4}, -\frac{3}{4}$

73. $-4 \pm 2\sqrt{5}$ **75.** $\frac{7}{4} \pm \frac{\sqrt{41}}{4}$ **77.** $1 \pm \sqrt{3}$

79. $-6 \pm 2\sqrt{5}$ **81.** $-\frac{3}{4} \pm \frac{\sqrt{41}}{4}$ **83.** $\frac{2}{7}$ **85.** $2 \pm \frac{\sqrt{6}}{2}$

87. $6 \pm \sqrt{11}$ **89.** $-\frac{3}{8} \pm \frac{\sqrt{265}}{8}$ **91.** 0.976, -0.643

93. $-1.107, 1.853$ **95.** $-0.290, -2.200$ **97.** $1 \pm \sqrt{2}$

99. $-10, 6$ **101.** $\frac{1}{2} \pm \sqrt{3}$ **103.** $\frac{3}{4} \pm \frac{\sqrt{97}}{4}$

105. (a) $w(w + 14) = 1632$ (b) $w = 34$ ft, $l = 48$ ft

107. 6 in. $\times$ 6 in. $\times$ 3 in.

109. (a) About 16.51 ft $\times$ 15.51 ft   (b) 63,897.6 lb

111. (a) $s = -16t^2 + 984$ (b) 728 ft   (c) About 7.84 sec

113. (a) About 5.86 sec   (b) About 0.2 mi

115. (a)

t	8	9	10	11
D	8.98	10.71	12.30	13.75

t	12	13	14
D	15.06	16.22	17.24

The public debt reached $13 trillion sometime in 2010.

(b) $t \approx 10.47$ (2010)

(c) $19.125 trillion; Answers will vary.

117. 4:00 P.M.   **119.** (a) $x^2 + 15^2 = l^2$ (b) $30\sqrt{6} \approx 73.5$ ft

121. True. $b^2 - 4ac = 61 > 0$ **123.** $x^2 - 4x = 0$

125. $x^2 - 22x + 112 = 0$ **127.** $x^2 - 2x - 1 = 0$

129. Yes, the vertex of the parabola would be on the x-axis.

Section 1.5   *(page 119)*

1. real   **3.** pure imaginary   **5.** principal square

7. $a = 9, b = 8$ **9.** $a = 8, b = 4$ **11.** $2 + 5i$

13. $1 - 2\sqrt{3}i$ **15.** $2\sqrt{10}i$ **17.** 23   **19.** $-1 - 6i$

21. $0.2i$ **23.** $7 + 4i$ **25.** 1   **27.** $3 - 3\sqrt{2}i$

29. $-14 + 20i$ **31.** $5 + i$ **33.** $108 + 12i$ **35.** 11

37. $-13 + 84i$ **39.** $9 - 2i, 85$ **41.** $-1 + \sqrt{5}i, 6$

43. $-2\sqrt{5}i, 20$ **45.** $\sqrt{6}, 6$ **47.** $\frac{8}{41} + \frac{10}{41}i$

49. $\frac{12}{13} + \frac{5}{13}i$ **51.** $-4 - 9i$ **53.** $-\frac{120}{1681} - \frac{27}{1681}i$

55. $-\frac{1}{2} - \frac{5}{2}i$ **57.** $\frac{62}{949} + \frac{297}{949}i$ **59.** $-2\sqrt{3}$ **61.** -15

63. $7\sqrt{2}i$ **65.** $\left(21 + 5\sqrt{2}\right) + \left(7\sqrt{5} - 3\sqrt{10}\right)i$ **67.** $1 \pm i$

69. $-2 \pm \frac{1}{2}i$ **71.** $-2 \pm \frac{\sqrt{5}}{2}i$ **73.** $2 \pm \sqrt{2}i$

75. $\frac{5}{7} \pm \frac{5\sqrt{13}}{7}i$ **77.** $-1 + 6i$ **79.** $-14i$

81. $-432\sqrt{2}i$ **83.** i **85.** 81

87. (a) $z_1 = 9 + 16i, z_2 = 20 - 10i$

(b) $z = \dfrac{11{,}240}{877} + \dfrac{4630}{877}i$

89. False. *Sample answer:* $(1 + i) + (3 + i) = 4 + 2i$

91. True.
$$x^4 - x^2 + 14 = 56$$
$$\left(-i\sqrt{6}\right)^4 - \left(-i\sqrt{6}\right)^2 + 14 \stackrel{?}{=} 56$$
$$36 + 6 + 14 \stackrel{?}{=} 56$$
$$56 = 56$$

93. $i, -1, -i, 1, i, -1, -i, 1$; The pattern repeats the first four results. Divide the exponent by 4.

When the remainder is 1, the result is i.

When the remainder is 2, the result is -1.

When the remainder is 3, the result is $-i$.

When the remainder is 0, the result is 1.

95. $\sqrt{-6}\sqrt{-6} = \sqrt{6}i\sqrt{6}i = 6i^2 = -6$ **97.** Proof

Section 1.6   *(page 128)*

1. polynomial   **3.** radical   **5.** $0, \pm 3$ **7.** $-3, 0$

9. $\pm 3, \pm 3i$ **11.** $-8, 4 \pm 4\sqrt{3}i$ **13.** $3, 1, -1$

15. $\pm 1, \frac{1}{2} \pm \frac{\sqrt{3}}{2}i$ **17.** $\pm\sqrt{3}, \pm 1$ **19.** $\pm\frac{1}{2}, \pm 4$

21. $1, -2, 1 \pm \sqrt{3}i, -\frac{1}{2} \pm \frac{\sqrt{3}i}{2}$ **23.** $-\frac{1}{5}, -\frac{1}{3}$

25. $-\frac{2}{3}, -4$ **27.** $\frac{1}{4}$ **29.** $-\frac{64}{27}$ **31.** 20   **33.** 17

35. $-\frac{55}{2}$ **37.** 1   **39.** 4   **41.** 3   **43.** $\frac{7}{4}$

45. 9   **47.** $\pm\sqrt{14}$ **49.** 1   **51.** $2, -\frac{3}{2}$

53. $\dfrac{-3 \pm \sqrt{21}}{6}$ **55.** $-\frac{1}{3}, 5$ **57.** $\dfrac{1 \pm \sqrt{31}}{3}$ **59.** $8, -3$

61. $2\sqrt{6}, -6$ **63.** $3, \dfrac{-1 - \sqrt{17}}{2}$

65. (a)

(b) and (c) 0, 3, -1

(d) The x-intercepts and the solutions are the same.

67. (a)

(b) and (c) $\pm 3, \pm 1$

(d) The x-intercepts and the solutions are the same.

69. (a)

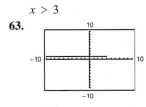

(b) and (c) 5, 6

(d) The x-intercepts and the solutions are the same.

71. (a)

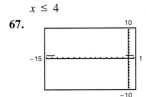

(b) and (c) -1

(d) The x-intercept and the solution are the same.

73. (a)

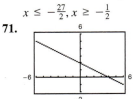

(b) and (c) 1, -3

(d) The x-intercepts and the solutions are the same.

75. ± 1.038 **77.** $-1.143, 0.968$ **79.** 16.756

81. $-2.280, -0.320$ **83.** $x^2 - 3x - 28 = 0$

85. $21x^2 + 31x - 42 = 0$ **87.** $x^3 - 4x^2 - 3x + 12 = 0$

89. $x^2 + 1 = 0$ **91.** $x^4 - 1 = 0$ **93.** 34 students

95. 191.5 mi/h   **97.** About 1.5%   **99.** 26,250 passengers

101. (a) About 15 lb/in.^{2} (b) Answers will vary.

103. 500 units   **105.** 3 h   **107.** $U = \dfrac{kd^2}{2}$

109. False. See Example 7 on page 125.   **111.** $x = -2, 8$

113. The quadratic equation was not written in general form. As a result, the substitutions in the Quadratic Formula are incorrect.

115. $a = 9, b = 9$ **117.** $a = 4, b = 24$

Section 1.7 *(page 137)*

1. solution set   **3.** double   **5.** $-2 \le x < 6$; Bounded

7. $-1 \le x \le 5$; Bounded   **9.** $x > 11$; Unbounded

11. $x < -2$; Unbounded

13. $x < 3$

15. $x < \dfrac{3}{2}$

17. $x \ge 12$

19. $x > 2$

21. $x \ge 1$

23. $x < 5$

25. $x \ge 4$

27. $x \ge 2$

29. $x \ge -4$

31. $-1 < x < 3$

33. $-7 < x \le -\dfrac{1}{3}$

35. $-\dfrac{9}{2} < x < \dfrac{15}{2}$

37. $-5 \le x < 1$

39. $-\dfrac{3}{4} < x < -\dfrac{1}{4}$

41. $10.5 \le x \le 13.5$

43. $-5 < x < 5$

45. $x < -2, x > 2$

47. No solution

49. $14 \le x \le 26$

51. $x \le -1, x \ge 8$

53. $x \le -5, x \ge 11$

55. $4 < x < 5$

57. $x \le -\dfrac{29}{2}, x \ge -\dfrac{11}{2}$

59.

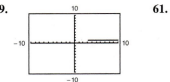

$x > 3$

61.

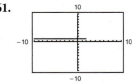

$x \le 2$

63.

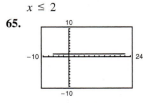

$x \le 4$

65.

$-6 \le x \le 22$

67.

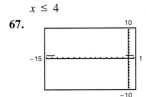

$x \le -\dfrac{27}{2}, x \ge -\dfrac{1}{2}$

69.

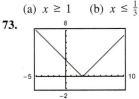

(a) $x \ge 1$ (b) $x \le \dfrac{1}{3}$

71.

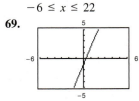

(a) $-2 \le x \le 4$

(b) $x \le 4$

73.

(a) $1 \le x \le 5$

(b) $x \le -1, x \ge 7$

75. All real numbers less than eight units from 10

77. $|x| \le 3$ **79.** $|x - 7| \ge 3$ **81.** $|x - 7| \ge 3$

83. $|x + 3| < 4$ **85.** $7.25 \le P \le 7.75$ **87.** $r < 0.08$

89. $100 \le r \le 170$ **91.** More than 6 units per hour

93. Greater than 10%   **95.** $x \ge 36$ **97.** $87 \le x \le 210$

99. (a)

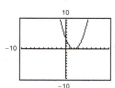

(b) $x \geq 2.9$

(c) $x \geq 2.908$

(d) Answers will vary.

101. (a) $4.34 \leq t \leq 6.56$ (Between 2004 and 2006)

(b) $t > 20.95$ (2020)

103. $13.7 < t < 17.5$ **105.** \$0.28

107. $106.864 \text{ in.}^2 \leq \text{area} \leq 109.464 \text{ in.}^2$

109. True by the Addition of a Constant Property of Inequalities.

111. False. If $-10 \leq x \leq 8$, then $10 \geq -x$ and $-x \geq -8$.

113. *Sample answer:* $x < x + 1$

115. *Sample answer:* $a = 1, b = 5, c = 5$

Section 1.8 *(page 147)*

1. positive; negative **3.** zeros; undefined values

5. (a) No (b) Yes (c) Yes (d) No

7. (a) Yes (b) No (c) No (d) Yes

9. $-3, 6$ **11.** $4, 5$

13. $(-2, 0)$

15. $(-3, 3)$

17. $[-7, 3]$

19. $(-\infty, -4] \cup [-2, \infty)$

21. $(-3, 2)$

23. $(-3, 1)$

25. $\left(-\infty, -\frac{4}{3}\right) \cup (5, \infty)$

27. $(-1, 1) \cup (3, \infty)$

29. $(-\infty, -3) \cup (3, 7)$

31. $(-\infty, 0) \cup \left(0, \frac{3}{2}\right)$

33. $[-2, 0] \cup [2, \infty)$

35. $[-2, \infty)$

37. The solution set consists of the single real number $\frac{1}{2}$.

39. The solution set is empty.

41. $(-\infty, 0) \cup \left(\frac{1}{4}, \infty\right)$

43. $(-7, 1)$

45. $(-5, 3) \cup (11, \infty)$

47. $\left(-\frac{3}{4}, 3\right) \cup [6, \infty)$

49. $(-3, -2] \cup [0, 3)$

51. $(-\infty, -1) \cup (1, \infty)$

53.

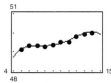

(a) $x \leq -1, x \geq 3$

(b) $0 \leq x \leq 2$

55.

(a) $-2 \leq x \leq 0,$

$2 \leq x \leq \infty$

(b) $x \leq 4$

57.

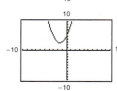

(a) $0 \leq x < 2$

(b) $2 < x \leq 4$

59.

(a) $|x| \geq 2$

(b) $-\infty < x < \infty$

61. $(-3.89, 3.89)$ **63.** $(-0.13, 25.13)$ **65.** $(2.26, 2.39)$

67. (a) $t = 10$ sec (b) 4 sec $< t < 6$ sec

69. $40,000 \leq x \leq 50,000; \$50.00 \leq p \leq \$55.00$

71. $[-2, 2]$ **73.** $(-\infty, 4] \cup [5, \infty)$ **75.** $(-5, 0] \cup (7, \infty)$

77. (a) and (c)

(b) $N = -0.001231t^4 + 0.04723t^3 - 0.6452t^2$
$+ 3.783t + 41.21$

(d) 2017

(e) *Sample answer:* No. For $t > 15$, the model rapidly decreases.

79. $13.8 \text{ m} \leq L \leq 36.2 \text{ m}$ **81.** $R_1 \geq 2$ ohms

83. False. There are four test intervals.

85.

For part (b), the y-values that are less than or equal to 0 occur only at $x = -1$.

For part (c), there are no y-values that are less than 0.

For part (d), the y-values that are greater than 0 occur for all values of x except 2.

87. (a) $(-\infty, -6] \cup [6, \infty)$

(b) When $a > 0$ and $c > 0, b \leq -2\sqrt{ac}$ or $b \geq 2\sqrt{ac}$.

89. (a) $\left(-\infty, -2\sqrt{30}\right] \cup \left[2\sqrt{30}, \infty\right)$

(b) When $a > 0$ and $c > 0, b \leq -2\sqrt{ac}$ or $b \geq 2\sqrt{ac}$.

Review Exercises *(page 152)*

1.

x	-2	-1	0	1	2
y	9	5	1	-3	-7

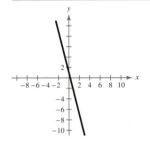

3. x-intercepts: $(1, 0), (5, 0)$
    y-intercept: $(0, 5)$

5. No symmetry

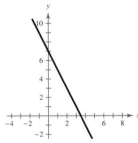

7. x-axis symmetry

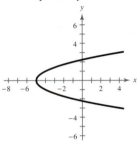

9. y-axis symmetry

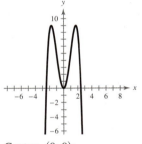

11. Origin symmetry

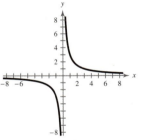

13. Center: $(0, 0)$;
    Radius: 3

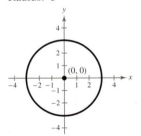

15. Center: $(-2, 0)$;
    Radius: 4

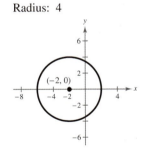

17. $(x - 2)^2 + (y + 3)^2 = 13$

19. (a)   (b) 2010

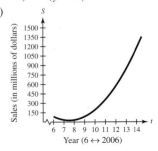

21. Identity   **23.** Conditional equation   **25.** 5
27. 6   **29.** -30 **31.** 13
33. x-intercept: $\left(\frac{1}{3}, 0\right)$ **35.** x-intercept: $(4, 0)$
    y-intercept: $(0, -1)$ y-intercept: $(0, -8)$
37. x-intercept: $\left(\frac{4}{3}, 0\right)$
    y-intercept: $\left(0, \frac{2}{3}\right)$
39. $h = 10$ in.   **41.** 2013: \$1.53 billion; 2014: \$2.22 billion
43. \$90,000   **45.** $\frac{20}{7}$ L ≈ 2.857 L   **47.** $\frac{3V}{\pi r^2}$ **49.** $-\frac{5}{2}, 3$
51. $\pm\sqrt{2}$ **53.** $-8, -18$ **55.** $-6 \pm \sqrt{11}$
57. $-\frac{5}{4} \pm \frac{\sqrt{241}}{4}$
59. (a) $x = 0, 20$
    (b)   (c) $x = 10$

61. $4 + 3i$ **63.** $-1 + 3i$ **65.** $-3 - 3i$ **67.** $15 + 6i$
69. 50   **71.** $\frac{4}{5} + \frac{8}{5}i$ **73.** $\frac{17}{26} + \frac{7}{26}i$ **75.** $\frac{21}{13} - \frac{1}{13}i$
77. $1 \pm 3i$ **79.** $-\frac{1}{2} \pm \frac{\sqrt{6}}{2}i$ **81.** $0, \frac{12}{5}$ **83.** $7, \pm2i$
85. $-1, 2, \frac{1}{2} \pm \frac{\sqrt{3}}{2}i, -1 \pm \sqrt{3}i$ **87.** No solution
89. $-124, 126$ **91.** $\pm\sqrt{10}$ **93.** $-5, 15$ **95.** $1, 3$
97. 143,203 units   **99.** $-7 < x \leq 2$; Bounded
101. $x \leq -10$; Unbounded
103. $x < -18$ **105.** $x \geq \frac{32}{15}$

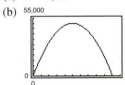

107. $-11 < x < -1$

109. $x > 37$ units
111. $(-3, 9)$ **113.** $(-\infty, -3) \cup (0, 3)$

115. $[-5, -1) \cup (1, \infty)$

117. 4.9%
119. False. $\sqrt{-18}\sqrt{-2} = (3\sqrt{2}i)(\sqrt{2}i) = 6i^2 = -6$
    and $\sqrt{(-18)(-2)} = \sqrt{36} = 6$
121. Some solutions to certain types of equations may be extraneous solutions, which do not satisfy the original equations. So, checking is crucial.

CHAPTER 1

Chapter Test   *(page 155)*

1. No symmetry

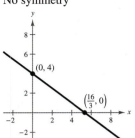

2. *y*-axis symmetry

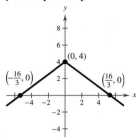

3. No symmetry

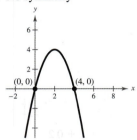

4. Origin symmetry

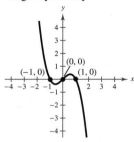

5. No symmetry

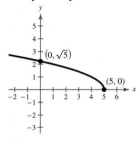

6. *x*-axis symmetry

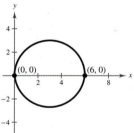

7. $\frac{128}{11}$ **8.** $-3, 5$

9. No solution. The variable is divided out.

10. $\pm\sqrt{2}, \pm\sqrt{3}i$ **11.** 4   **12.** $-2, \frac{8}{3}$

13. $-\frac{11}{2} \le x < 3$

14. $x < -6$ or $0 < x < 4$

15. $x < -4$ or $x > \frac{3}{2}$

16. $x \le -5$ or $x \ge \frac{5}{3}$

17. (a) -14 (b) $19 + 17i$ **18.** $\frac{8}{5} - \frac{16}{5}i$ **19.** $\frac{3}{2} \pm \frac{1}{2}i$

20. (a)

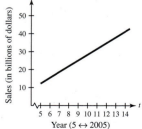

(b) and (c) $50.8 billion

21. 4.774 in.   **22.** $93\frac{3}{4}$ km/h   **23.** $a = 80, b = 20$

Problem Solving   *(page 157)*

1.

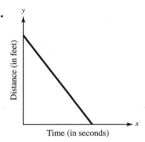

3. (a) Answers will vary.

(b)

a	4	7	10	13	16
A	64π	91π	100π	91π	64π

(c) $\dfrac{10\pi + 10\sqrt{\pi(\pi - 3)}}{\pi} \approx 12.12$ or

$\dfrac{10\pi - 10\sqrt{\pi(\pi - 3)}}{\pi} \approx 7.88$

(d)

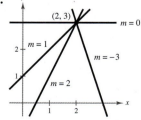

(e) 0, 20; They represent the minimum and maximum values of a.

(f) 100π; $a = 10$; $b = 10$

5. (a) 88.4 mi/h

(b) No, the maximum wind speed that the library can survive is 125 miles per hour.

(c) Answers will vary.

7. (a) $mn = 9$, $m + n = 0$

(b) Answers will vary. m and n are imaginary numbers.

9. (a) *Sample answers:* 5, 12, 13; 8, 15, 17

(b) Yes; yes; yes

(c) The product of the three numbers in a Pythagorean Triple is divisible by 60.

11. Proof

13. (a) $\frac{1}{2} - \frac{1}{2}i$ (b) $\frac{3}{10} + \frac{1}{10}i$ (c) $-\frac{1}{34} - \frac{2}{17}i$

15. (a) Yes   (b) No   (c) Yes

Chapter 2

Section 2.1   *(page 169)*

1. linear   **3.** point-slope   **5.** perpendicular

7. Linear extrapolation   **9.** (a) L_2 (b) L_3 (c) L_1

11.

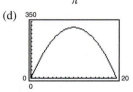

13. $\frac{3}{2}$

15. $m = 5$

y-intercept: $(0, 3)$

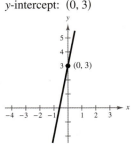

17. $m = -\frac{3}{4}$

y-intercept: $(0, -1)$

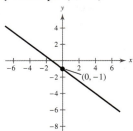

19. $m = 0$

y-intercept: $(0, 5)$

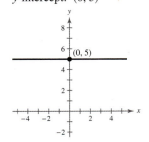

21. m is undefined.

y-intercept: none

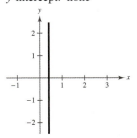

23. $m = \frac{7}{6}$

y-intercept: $(0, -5)$

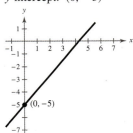

25. $m = -\frac{3}{2}$ **27.** $m = 2$ **29.** $m = 0$

31. m is undefined.   **33.** $m = 0.15$

35. $(-1, 7), (0, 7), (4, 7)$ **37.** $(-4, 6), (-3, 8), (-2, 10)$

39. $(-2, 7), \left(0, \frac{19}{3}\right), (1, 6)$ **41.** $(-4, -5), (-4, 0), (-4, 2)$

43. $y = 3x - 2$ **45.** $y = -2x$

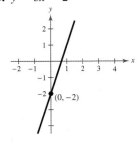

47. $y = -\frac{1}{3}x + \frac{4}{3}$ **49.** $y = -\frac{1}{2}x - 2$

51. $y = \frac{5}{2}$

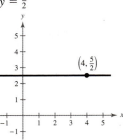

53. $y = 5x + 27.3$

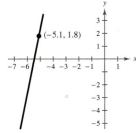

55. $y = -\frac{3}{5}x + 2$

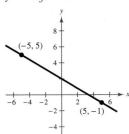

57. $x = -7$

59. $y = -\frac{1}{2}x + \frac{3}{2}$

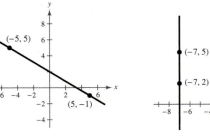

61. $y = 0.4x + 0.2$

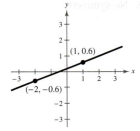

63. $y = -1$

65. Parallel   **67.** Neither   **69.** Perpendicular

71. Parallel   **73.** (a) $y = 2x - 3$ (b) $y = -\frac{1}{2}x + 2$

75. (a) $y = -\frac{3}{4}x + \frac{3}{8}$ (b) $y = \frac{4}{3}x + \frac{127}{72}$

77. (a) $y = 4$ (b) $x = -2$

79. (a) $y = x + 4.3$ (b) $y = -x + 9.3$

81. $5x + 3y - 15 = 0$ **83.** $12x + 3y + 2 = 0$

85. $x + y - 3 = 0$

87. (a) Sales increasing 135 units/yr

(b) No change in sales

(c) Sales decreasing 40 units/yr

89. 12 ft   **91.** $V(t) = -150t + 5400$, $16 \le t \le 21$

93. C-intercept: fixed initial cost; Slope: cost of producing an additional laptop bag

95. $V(t) = -175t + 875$, $0 \le t \le 5$

97. $F = 1.8C + 32$ or $C = \frac{5}{9}F - \frac{160}{9}$

99. (a) $C = 21t + 42,000$ (b) $R = 45t$

(c) $P = 24t - 42,000$ (d) 1750 h

CHAPTER 2

101. False. The slope with the greatest magnitude corresponds to the steepest line.

103. Find the slopes of the lines containing each two points and use the relationship $m_1 = -\dfrac{1}{m_2}$.

105. The scale on the y-axis is unknown, so the slopes of the lines cannot be determined.

107. No. The slopes of two perpendicular lines have opposite signs (assume that neither line is vertical or horizontal).

109. The line $y = 4x$ rises most quickly, and the line $y = -4x$ falls most quickly. The greater the magnitude of the slope (the absolute value of the slope), the faster the line rises or falls.

111. $3x - 2y - 1 = 0$ **113.** $80x + 12y + 139 = 0$

Section 2.2 (page 182)

1. domain; range; function   **3.** implied domain

5. Function   **7.** Not a function

9. (a) Function

(b) Not a function, because the element 1 in A corresponds to two elements, -2 and 1, in B.

(c) Function

(d) Not a function, because not every element in A is matched with an element in B.

11. Not a function   **13.** Function   **15.** Function

17. Function   **19.** (a) -2 (b) -14 (c) $3x + 1$

21. (a) 15   (b) $4t^2 - 19t + 27$ (c) $4t^2 - 3t - 10$

23. (a) 1   (b) 2.5   (c) $3 - 2|x|$

25. (a) $-\dfrac{1}{9}$ (b) Undefined   (c) $\dfrac{1}{y^2 + 6y}$

27. (a) 1   (b) -1 (c) $\dfrac{|x - 1|}{x - 1}$

29. (a) -1 (b) 2   (c) 6

31.

x	-2	-1	0	1	2
$f(x)$	1	4	5	4	1

33.

x	-2	-1	0	1	2
$f(x)$	5	$\frac{9}{2}$	4	1	0

35. 5   **37.** $\frac{4}{3}$ **39.** ± 9 **41.** $0, \pm 1$ **43.** $-1, 2$

45. $0, \pm 2$ **47.** All real numbers x

49. All real numbers y such that $y \ge -6$

51. All real numbers x except $x = 0, -2$

53. All real numbers s such that $s \ge 1$ except $s = 4$

55. All real numbers x such that $x > 0$

57. (a) The maximum volume is 1024 cubic centimeters.

(b)

Yes, V is a function of x.

(c) $V = x(24 - 2x)^2,\ \ 0 < x < 12$

59. $A = \dfrac{P^2}{16}$ **61.** No, the ball will be at a height of 18.5 feet.

63. $A = \dfrac{x^2}{2(x - 2)},\ \ x > 2$

65. 2008: 67.36%
2009: 70.13%
2010: 72.90%
2011: 75.67%
2012: 79.30%
2013: 81.25%
2014: 83.20%

67. (a) $C = 12.30x + 98{,}000$ (b) $R = 17.98x$

(c) $P = 5.68x - 98{,}000$

69. (a)

(b) $h = \sqrt{d^2 - 3000^2},\ \ d \ge 3000$

71. (a) $R = \dfrac{240n - n^2}{20},\ \ n \ge 80$

(b)

n	90	100	110	120	130	140	150
$R(n)$	\$675	\$700	\$715	\$720	\$715	\$700	\$675

The revenue is maximum when 120 people take the trip.

73. $2 + h,\ \ h \ne 0$ **75.** $3x^2 + 3xh + h^2 + 3,\ \ h \ne 0$

77. $-\dfrac{x + 3}{9x^2},\ \ x \ne 3$ **79.** $\dfrac{\sqrt{5x} - 5}{x - 5}$

81. $g(x) = cx^2;\ c = -2$ **83.** $r(x) = \dfrac{c}{x};\ c = 32$

85. False. A function is a special type of relation.

87. False. The range is $[-1, \infty)$.

89. The domain of $f(x)$ includes $x = 1$ and the domain of $g(x)$ does not because you cannot divide by 0. So, the functions do not have the same domain.

91. No; x is the independent variable, f is the name of the function.

93. (a) Yes. The amount you pay in sales tax will increase as the price of the item purchased increases.

(b) No. The length of time that you study will not necessarily determine how well you do on an exam.

Section 2.3 (page 194)

1. Vertical Line Test   **3.** decreasing

5. average rate of change; secant

7. Domain: $(-2, 2]$; range: $[-1, 8]$

(a) -1 (b) 0   (c) -1 (d) 8

9. Domain: $(-\infty, \infty)$; range: $(-2, \infty)$

(a) 0   (b) 1   (c) 2   (d) 3

11. Function   **13.** Not a function   **15.** -6 **17.** $-\frac{5}{2}, 6$

19. -3 **21.** $0, \pm\sqrt{6}$ **23.** $\pm 3, 4$ **25.** $\frac{1}{2}$

27. (a) 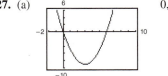   0, 6

(b) 0, 6

29. (a) 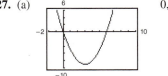   −5.5

(b) $-\frac{11}{2}$

31. (a)   0.3333

(b) $\frac{1}{3}$

33. Decreasing on $(-\infty, \infty)$

35. Increasing on $(1, \infty)$; Decreasing on $(-\infty, -1)$

37. Increasing on $(1, \infty)$; Decreasing on $(-\infty, -1)$
Constant on $(-1, 1)$

39. Increasing on $(-\infty, -1)$, $(0, \infty)$; Decreasing on $(-1, 0)$

41.
Constant on $(-\infty, \infty)$

43.
Decreasing on $(-\infty, 0)$
Increasing on $(0, \infty)$

45.
Decreasing on $(-\infty, 1)$

47.
Increasing on $(0, \infty)$

49. Relative minimum: $(-1.5, -2.25)$

51. Relative maximum: $(0, 15)$
Relative minimum: $(4, -17)$

53. Relative minimum: $(0.33, -0.38)$

55.
$(-\infty, 4]$

57.
$[-3, 3]$

59.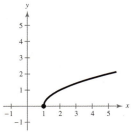

$[1, \infty)$

61. -2 **63.** -1

65. (a)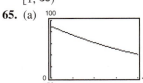

(b) About -6.14; The amount the U.S. federal government spent on research and development for defense decreased by about \$6.14 billion each year from 2010 to 2014.

67. (a) $s = -16t^2 + 64t + 6$

(b) 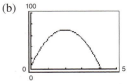   (c) 16 ft/sec

(d) The slope of the secant line is positive.

(e) $y = 16t + 6$

(f)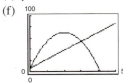

69. (a) $s = -16t^2 + 120t$

(b) 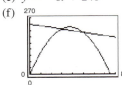   (c) -8 ft/sec

(d) The slope of the secant line is negative.

(e) $y = -8t + 240$

(f)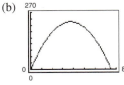

71. Even; y-axis symmetry   **73.** Neither; no symmetry

75. Neither; no symmetry

77.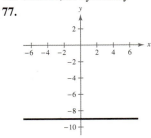

Even

CHAPTER 2

79.
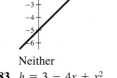

81.
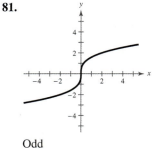

Neither   Odd

83. $h = 3 - 4x + x^2$ **85.** $L = 2 - \sqrt[3]{2y}$

87. The negative symbol should be divided out of each term, which yields $f(-x) = -(2x^3 + 5)$. So, the function is neither even nor odd.

89. (a) Ten thousands   (b) Ten millions   (c) Tens
    (d) Ones

91. False. The function $f(x) = \sqrt{x^2 + 1}$ has a domain of all real numbers.

93. True. A graph that is symmetric with respect to the y-axis cannot be increasing on its entire domain.

95. (a) $\left(\frac{5}{3}, -7\right)$ (b) $\left(\frac{5}{3}, 7\right)$

97. (a)
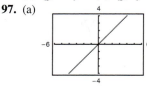
(b)
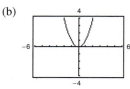

(c)
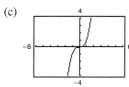
(d)
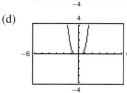

(e)
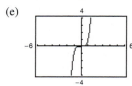
(f)
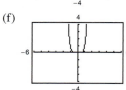

All the graphs pass through the origin. The graphs of the odd powers of x are symmetric with respect to the origin, and the graphs of the even powers are symmetric with respect to the y-axis. As the powers increase, the graphs become flatter in the interval $-1 < x < 1$.

99. (a) Even. The graph is a reflection in the x-axis.
    (b) Even. The graph is a reflection in the y-axis.
    (c) Even. The graph is a downward shift of f.
    (d) Neither. The graph is a right shift of f.

Section 2.4   (page 203)

1. Greatest integer function   **3.** Reciprocal function
5. Square root function   **7.** Absolute value function
9. Linear function

11. (a) $f(x) = -2x + 6$
    (b)

13. (a) $f(x) = \frac{2}{3}x - 2$
    (b)
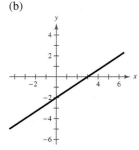

15.
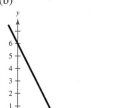

17.
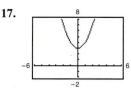

19.
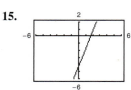

21.
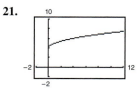

23.
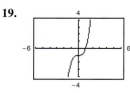

25.
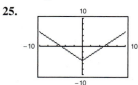

27. (a) 2   (b) 2   (c) -4 (d) 3
29. (a) 1   (b) -4 (c) 3   (d) 2

31.
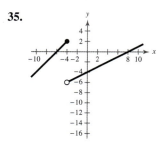

33.
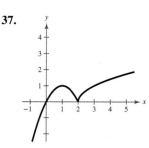

35.
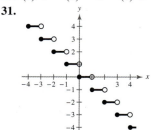

37.
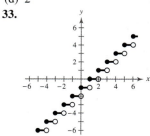

39.
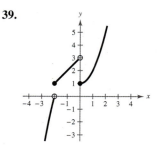

41. (a)

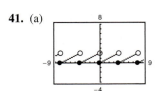

(b) Domain: $(-\infty, \infty)$

     Range: $[0, 2)$

43. (a) $W(30) = 420;\ W(40) = 560;$

      $W(45) = 665;\ W(50) = 770$

(b) $W(h) = \begin{cases} 14h, & 0 < h \le 36 \\ 21(h - 36) + 504, & h > 36 \end{cases}$

(c) $W(h) = \begin{cases} 16h, & 0 < h \le 40 \\ 24(h - 40) + 640, & h > 40 \end{cases}$

45.

Interval	Input Pipe	Drain Pipe 1	Drain Pipe 2
$[0, 5]$	Open	Closed	Closed
$[5, 10]$	Open	Open	Closed
$[10, 20]$	Closed	Closed	Closed
$[20, 30]$	Closed	Closed	Open
$[30, 40]$	Open	Open	Open
$[40, 45]$	Open	Closed	Open
$[45, 50]$	Open	Open	Open
$[50, 60]$	Open	Open	Closed

47. $f(t) = \begin{cases} t, & 0 \le t \le 2 \\ 2t - 2, & 2 < t \le 8 \\ \frac{1}{2}t + 10, & 8 < t \le 9 \end{cases}$

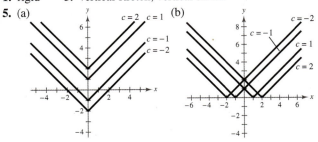

Total accumulation $= 14.5$ in.

49. False. A piecewise-defined function is a function that is defined by two or more equations over a specified domain. That domain may or may not include x- and y-intercepts.

Section 2.5   *(page 210)*

1. rigid   **3.** vertical stretch; vertical shrink

5. (a)   (b)

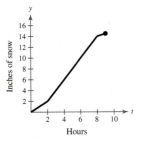

7. (a)   (b)

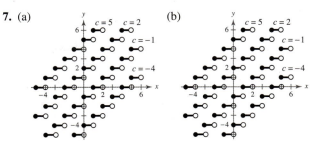

9. (a)   (b)

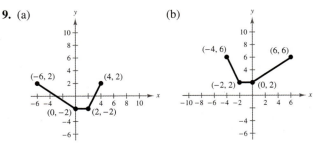

(c)   (d)

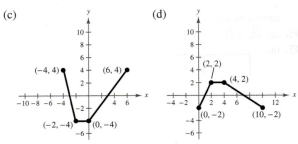

(e)   (f)

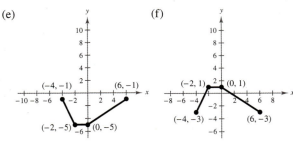

(g)

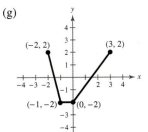

11. (a) $y = x^2 - 1$ (b) $y = -(x + 1)^2 + 1$

13. (a) $y = -|x + 3|$ (b) $y = |x - 2| - 4$

15. Right shift of $y = x^3;\ y = (x - 2)^3$

17. Reflection in the x-axis of $y = x^2;\ y = -x^2$

19. Reflection in the x-axis and upward shift of $y = \sqrt{x}$; $y = 1 - \sqrt{x}$

CHAPTER 2

21. (a) $f(x) = x^2$
(b) Upward shift of six units
(c)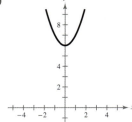
(d) $g(x) = f(x) + 6$

23. (a) $f(x) = x^3$
(b) Reflection in the x-axis and a right shift of two units
(c)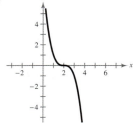
(d) $g(x) = -f(x - 2)$

25. (a) $f(x) = x^2$
(b) Reflection in the x-axis, a left shift of one unit, and a downward shift of three units
(c)
(d) $g(x) = -3 - f(x + 1)$

27. (a) $f(x) = |x|$
(b) Right shift of one unit and an upward shift of two units
(c)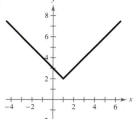
(d) $g(x) = f(x - 1) + 2$

29. (a) $f(x) = \sqrt{x}$ (b) Vertical stretch
(c)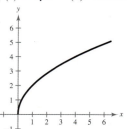
(d) $g(x) = 2f(x)$

31. (a) $f(x) = [\![x]\!]$
(b) Vertical stretch and a downward shift of one unit
(c)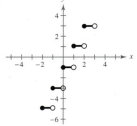
(d) $g(x) = 2f(x) - 1$

33. (a) $f(x) = |x|$
(b) Horizontal shrink
(c)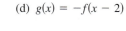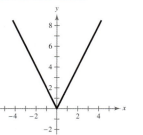
(d) $g(x) = f(2x)$

35. (a) $f(x) = x^2$
(b) Reflection in the x-axis, a vertical stretch, and an upward shift of one unit
(c)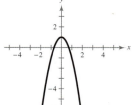
(d) $g(x) = -2f(x) + 1$

37. (a) $f(x) = |x|$
(b) Vertical stretch, a right shift of one unit, and an upward shift of two units
(c)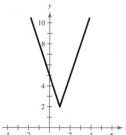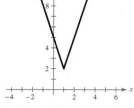
(d) $g(x) = 3f(x - 1) + 2$

39. $g(x) = (x - 3)^2 - 7$ **41.** $g(x) = (x - 13)^3$
43. $g(x) = -|x| - 12$ **45.** $g(x) = -\sqrt{-x + 6}$
47. (a) $y = -3x^2$ (b) $y = 4x^2 + 3$
49. (a) $y = -\frac{1}{2}|x|$ (b) $y = 3|x| - 3$
51. Vertical stretch of $y = x^3$; $y = 2x^3$
53. Reflection in the x-axis and vertical shrink of $y = x^2$; $y = -\frac{1}{2}x^2$
55. Reflection in the y-axis and vertical shrink of $y = \sqrt{x}$; $y = \frac{1}{2}\sqrt{-x}$
57. $y = -(x - 2)^3 + 2$ **59.** $y = -\sqrt{x} - 3$

61. (a)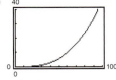

(b) $H\left(\dfrac{x}{1.6}\right) = 0.00001132x^3$; Horizontal stretch

63. False. The graph of $y = f(-x)$ is a reflection of the graph of $f(x)$ in the y-axis.

65. True. $|-x| = |x|$

67. $(-2, 0), (-1, 1), (0, 2)$

69. The equation should be $g(x) = (x - 1)^3$.

71. (a) $g(t) = \frac{3}{4}f(t)$ (b) $g(t) = f(t) + 10{,}000$

(c) $g(t) = f(t - 2)$

Section 2.6 *(page 219)*

1. addition; subtraction; multiplication; division

3.

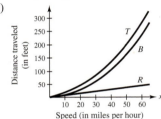

5. (a) $2x$ (b) 4 (c) $x^2 - 4$

(d) $\dfrac{x + 2}{x - 2}$; all real numbers x except $x = 2$

7. (a) $x^2 + 4x - 5$ (b) $x^2 - 4x + 5$ (c) $4x^3 - 5x^2$

(d) $\dfrac{x^2}{4x - 5}$; all real numbers x except $x = \dfrac{5}{4}$

9. (a) $x^2 + 6 + \sqrt{1 - x}$ (b) $x^2 + 6 - \sqrt{1 - x}$

(c) $(x^2 + 6)\sqrt{1 - x}$

(d) $\dfrac{(x^2 + 6)\sqrt{1 - x}}{1 - x}$; all real numbers x such that $x < 1$

11. (a) $\dfrac{x^4 + x^3 + x}{x + 1}$ (b) $\dfrac{-x^4 - x^3 + x}{x + 1}$ (c) $\dfrac{x^4}{x + 1}$

(d) $\dfrac{1}{x^2(x + 1)}$; all real numbers x except $x = 0, -1$

13. 7 **15.** 5 **17.** $-9t^2 + 3t + 5$ **19.** 306

21. $\frac{8}{23}$ **23.** -9

25.

27.

$f(x), g(x)$

$f(x), f(x)$

29. (a) $x + 5$ (b) $x + 5$ (c) $x - 6$

31. (a) $(x - 1)^2$ (b) $x^2 - 1$ (c) $x - 2$

33. (a) x (b) x (c) $x^9 + 3x^6 + 3x^3 + 2$

35. (a) $\sqrt{x^2 + 4}$ (b) $x + 4$

Domains of f and $g \circ f$: all real numbers x such that $x \geq -4$

Domains of g and $f \circ g$: all real numbers x

37. (a) x^2 (b) x^2

Domains of $f, g, f \circ g$, and $g \circ f$: all real numbers x

39. (a) $|x + 6|$ (b) $|x| + 6$

Domains of $f, g, f \circ g$, and $g \circ f$: all real numbers x

41. (a) $\dfrac{1}{x + 3}$ (b) $\dfrac{1}{x} + 3$

Domains of f and $g \circ f$: all real numbers x except $x = 0$

Domain of g: all real numbers x

Domain of $f \circ g$: all real numbers x except $x = -3$

43. (a)   (b)

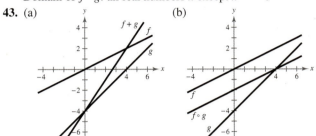

45. (a) 3 (b) 0 **47.** (a) 0 (b) 4

49. $f(x) = x^2, g(x) = 2x + 1$ **51.** $f(x) = \sqrt[3]{x}, g(x) = x^2 - 4$

53. $f(x) = \dfrac{1}{x}, g(x) = x + 2$ **55.** $f(x) = \dfrac{x + 3}{4 + x}, g(x) = -x^2$

57. (a) $T = \frac{3}{4}x + \frac{1}{15}x^2$

(b)

(c) The braking function $B(x)$; As x increases, $B(x)$ increases at a faster rate than $R(x)$.

59. (a) $c(t) = \dfrac{b(t) - d(t)}{p(t)} \times 100$

(b) $c(16)$ is the percent change in the population due to births and deaths in the year 2016.

61. (a) $r(x) = \dfrac{x}{2}$ (b) $A(r) = \pi r^2$

(c) $(A \circ r)(x) = \pi\left(\dfrac{x}{2}\right)^2$;

$(A \circ r)(x)$ represents the area of the circular base of the tank on the square foundation with side length x.

63. $g(f(x))$ represents 3 percent of an amount over $500{,}000$.

65. False. $(f \circ g)(x) = 6x + 1$ and $(g \circ f)(x) = 6x + 6$

67. (a) $O(M(Y)) = 2\left(6 + \frac{1}{2}Y\right) = 12 + Y$

(b) Middle child is 8 years old; youngest child is 4 years old.

69. Proof

71. (a) *Sample answer:* $f(x) = x + 1, g(x) = x + 3$

(b) *Sample answer:* $f(x) = x^2, g(x) = x^3$

73. (a) Proof

(b) $\frac{1}{2}[f(x) + f(-x)] + \frac{1}{2}[f(x) - f(-x)]$

$= \frac{1}{2}[f(x) + f(-x) + f(x) - f(-x)]$

$= \frac{1}{2}[2f(x)]$

$= f(x)$

(c) $f(x) = (x^2 + 1) + (-2x)$

$k(x) = \dfrac{-1}{(x + 1)(x - 1)} + \dfrac{x}{(x + 1)(x - 1)}$

CHAPTER 2

Section 2.7   *(page 228)*

1. inverse   **3.** range; domain   **5.** one-to-one

7. $f^{-1}(x) = \dfrac{1}{6}x$ **9.** $f^{-1}(x) = \dfrac{x-1}{3}$

11. $f^{-1}(x) = \sqrt{x+4}$ **13.** $f^{-1}(x) = \sqrt[3]{x-1}$

15. $f(g(x)) = f(4x+9) = \dfrac{(4x+9)-9}{4} = x$

$g(f(x)) = g\left(\dfrac{x-9}{4}\right) = 4\left(\dfrac{x-9}{4}\right) + 9 = x$

17. $f(g(x)) = f(\sqrt[3]{4x}) = \dfrac{(\sqrt[3]{4x})^3}{4} = x$

$g(f(x)) = g\left(\dfrac{x^3}{4}\right) = \sqrt[3]{4\left(\dfrac{x^3}{4}\right)} = x$

19.

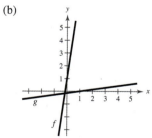

21. (a) $f(g(x)) = f(x+5) = (x+5) - 5 = x$
$g(f(x)) = g(x-5) = (x-5) + 5 = x$

(b)

23. (a) $f(g(x)) = f\left(\dfrac{x-1}{7}\right) = 7\left(\dfrac{x-1}{7}\right) + 1 = x$

$g(f(x)) = g(7x+1) = \dfrac{(7x+1)-1}{7} = x$

(b)

25. (a) $f(g(x)) = f(\sqrt[3]{x}) = (\sqrt[3]{x})^3 = x$
$g(f(x)) = g(x^3) = \sqrt[3]{(x^3)} = x$

(b)

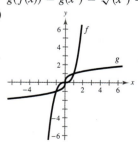

27. (a) $f(g(x)) = f(x^2 - 5) = \sqrt{(x^2 - 5) + 5} = x,\ x \geq 0$
$g(f(x)) = g(\sqrt{x+5}) = (\sqrt{x+5})^2 - 5 = x$

(b)

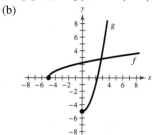

29. (a) $f(g(x)) = f\left(\dfrac{1}{x}\right) = \dfrac{1}{(1/x)} = x$

$g(f(x)) = g\left(\dfrac{1}{x}\right) = \dfrac{1}{(1/x)} = x$

(b)

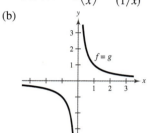

31. (a) $f(g(x)) = f\left(-\dfrac{5x+1}{x-1}\right) = \dfrac{-\left(\dfrac{5x+1}{x-1}\right) - 1}{-\left(\dfrac{5x+1}{x-1}\right) + 5}$

$= \dfrac{-5x - 1 - x + 1}{-5x - 1 + 5x - 5} = x$

$g(f(x)) = g\left(\dfrac{x-1}{x+5}\right) = \dfrac{-5\left(\dfrac{x-1}{x+5}\right) - 1}{\dfrac{x-1}{x+5} - 1}$

$= \dfrac{-5x + 5 - x - 5}{x - 1 - x - 5} = x$

(b)

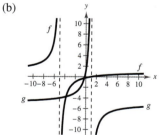

33. No

35.

x	3	5	7	9	11	13
$f^{-1}(x)$	-1	0	1	2	3	4

37. Yes   **39.** No

41.

The function does not have an inverse function.

43.

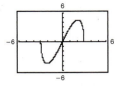

The function does not have an inverse function.

45. (a) $f^{-1}(x) = \sqrt[5]{x + 2}$

(b)

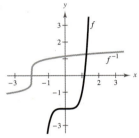

(c) The graph of f^{-1} is the reflection of the graph of f in the line $y = x$.

(d) The domains and ranges of f and f^{-1} are all real numbers x.

47. (a) $f^{-1}(x) = \sqrt{4 - x^2}$, $0 \le x \le 2$

(b)

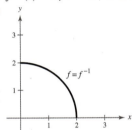

(c) The graph of f^{-1} is the same as the graph of f.

(d) The domains and ranges of f and f^{-1} are all real numbers x such that $0 \le x \le 2$.

49. (a) $f^{-1}(x) = \dfrac{4}{x}$

(b)

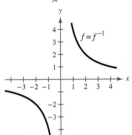

(c) The graph of f^{-1} is the same as the graph of f.

(d) The domains and ranges of f and f^{-1} are all real numbers x except $x = 0$.

51. (a) $f^{-1}(x) = \dfrac{2x + 1}{x - 1}$

(b)

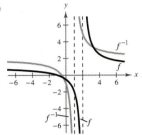

(c) The graph of f^{-1} is the reflection of the graph of f in the line $y = x$.

(d) The domain of f and the range of f^{-1} are all real numbers x except $x = 2$. The domain of f^{-1} and the range of f are all real numbers x except $x = 1$.

53. (a) $f^{-1}(x) = x^3 + 1$

(b)

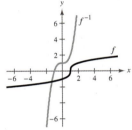

(c) The graph of f^{-1} is the reflection of the graph of f in the line $y = x$.

(d) The domains and ranges of f and f^{-1} are all real numbers x.

55. No inverse function   **57.** $g^{-1}(x) = 6x - 1$

59. No inverse function   **61.** $f^{-1}(x) = \sqrt{x} - 3$

63. No inverse function   **65.** No inverse function

67. $f^{-1}(x) = \dfrac{x^2 - 3}{2}$, $x \ge 0$ **69.** $f^{-1}(x) = \dfrac{5x - 4}{6 - 4x}$

71. $f^{-1}(x) = x - 2$

The domain of f and the range of f^{-1} are all real numbers x such that $x \ge -2$. The domain of f^{-1} and the range of f are all real numbers x such that $x \ge 0$.

73. $f^{-1}(x) = \sqrt{x} - 6$

The domain of f and the range of f^{-1} are all real numbers x such that $x \ge -6$. The domain of f^{-1} and the range of f are all real numbers x such that $x \ge 0$.

75. $f^{-1}(x) = \dfrac{\sqrt{-2(x - 5)}}{2}$

The domain of f and the range of f^{-1} are all real numbers x such that $x \ge 0$. The domain of f^{-1} and the range of f are all real numbers x such that $x \le 5$.

77. $f^{-1}(x) = x + 3$

The domain of f and the range of f^{-1} are all real numbers x such that $x \ge 4$. The domain of f^{-1} and the range of f are all real numbers x such that $x \ge 1$.

79. 32   **81.** 472   **83.** $2\sqrt[3]{x} + 3$

85. $\dfrac{x + 1}{2}$ **87.** $\dfrac{x + 1}{2}$

89. (a) $y = \dfrac{x - 10}{0.75}$

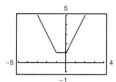

x = hourly wage; y = number of units produced

(b) 19 units

91. False. $f(x) = x^2$ has no inverse function.

93.

x	1	3	4	6
y	1	2	6	7

x	1	2	6	7
$f^{-1}(x)$	1	3	4	6

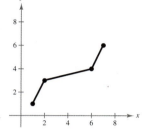

95. Proof   **97.** $k = \frac{1}{4}$

99.

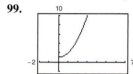

There is an inverse function $f^{-1}(x) = \sqrt{x - 1}$ because the domain of f is equal to the range of f^{-1} and the range of f is equal to the domain of f^{-1}.

101. This situation could be represented by a one-to-one function if the runner does not stop to rest. The inverse function would represent the time in hours for a given number of miles completed.

Review Exercises   *(page 233)*

1. $m = -\frac{1}{2}$
   y-intercept: $(0, 1)$

3. $m = 0$
   y-intercept: $(0, 1)$

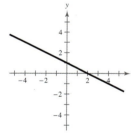

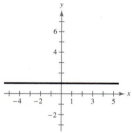

5. $m = -1$

7. $y = \frac{1}{3}x - 7$

9. $y = \frac{1}{2}x + 7$

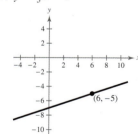

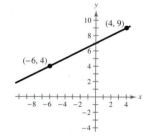

11. (a) $y = \frac{5}{4}x - \frac{23}{4}$ (b) $y = -\frac{4}{5}x + \frac{2}{5}$ **13.** $S = 0.80L$

15. Not a function   **17.** Function

19. (a) 16   (b) $(t + 1)^{4/3}$ (c) 81   (d) $x^{4/3}$

21. All real numbers x such that $-5 \le x \le 5$

23. 16 ft/sec   **25.** $4x + 2h + 3$, $h \ne 0$ **27.** Function

29. $-1, \frac{1}{5}$ **31.** $-\frac{1}{2}$

33.

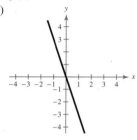

Increasing on $(0, \infty)$
Decreasing on $(-\infty, -1)$
Constant on $(-1, 0)$

35. Relative maximum: $(1, 2)$ **37.** 4

39. Neither even nor odd; no symmetry

41. Odd; origin symmetry

43. (a) $f(x) = -3x$

(b)

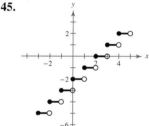

45.

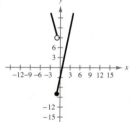

47.

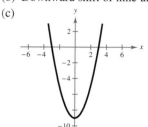

49. (a) $f(x) = x^2$

(b) Downward shift of nine units

(c)

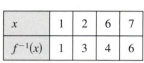

(d) $h(x) = f(x) - 9$

51. (a) $f(x) = \sqrt{x}$

(b) Reflection in the x-axis and an upward shift of four units

(c)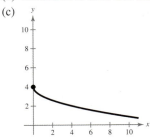

(d) $h(x) = -f(x) + 4$

53. (a) $f(x) = x^2$

(b) Reflection in the x-axis, a left shift of two units, and an upward shift of three units

(c)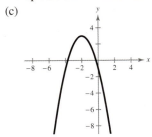

(d) $h(x) = -f(x + 2) + 3$

55. (a) $f(x) = [\![x]\!]$

(b) Reflection in the x-axis and an upward shift of six units

(c)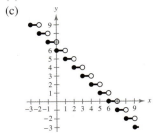

(d) $h(x) = -f(x) + 6$

57. (a) $f(x) = [\![x]\!]$

(b) Vertical stretch and a right shift of nine units

(c)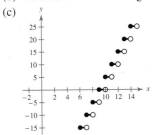

(d) $h(x) = 5f(x - 9)$

59. (a) $x^2 + 2x + 2$ (b) $x^2 - 2x + 4$

(c) $2x^3 - x^2 + 6x - 3$

(d) $\dfrac{x^2 + 3}{2x - 1}$; all real numbers x except $x = \dfrac{1}{2}$

61. (a) $x - \dfrac{8}{3}$ (b) $x - 8$

Domains of $f, g, f \circ g$, and $g \circ f$: all real numbers x

63. $(f \circ g)(x) = 0.95x - 100$; $(f \circ g)(x)$ represents the 5% discount before the $100 rebate.

65. $f^{-1}(x) = 5x + 4$

$$f(f^{-1}(x)) = \dfrac{5x + 4 - 4}{5} = x$$

$$f^{-1}(f(x)) = 5\left(\dfrac{x - 4}{5}\right) + 4 = x$$

67.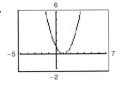

The function does not have an inverse function.

69. (a) $f^{-1}(x) = 2x + 6$

(b)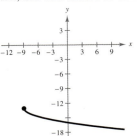

(c) The graphs are reflections of each other in the line $y = x$.

(d) Both f and f^{-1} have domains and ranges that are all real numbers x.

71. $x > 4$; $f^{-1}(x) = \sqrt{\dfrac{x}{2} + 4}, \ x \neq 0$

73. False. The graph is reflected in the x-axis, shifted 9 units to the left, and then shifted 13 units down.

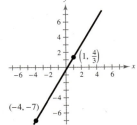

Chapter Test   *(page 235)*

1. $y = -4x - 3$

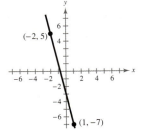

2. $y = \dfrac{5}{3}x - \dfrac{1}{3}$

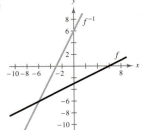

3. (a) $y = -\dfrac{5}{2}x + 4$ (b) $y = \dfrac{2}{5}x + 4$

4. (a) -9 (b) 1 (c) $|x - 4| - 15$

5. (a) $-\dfrac{1}{8}$ (b) $-\dfrac{1}{28}$ (c) $\dfrac{\sqrt{x}}{x^2 - 18x}$

6. (a) All real numbers x such that $x \neq 0, \dfrac{1}{2}$ (b) 5

7. (a) All real numbers x such that $x \leq 3$ (b) -97

8. (a)

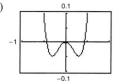

(b) Increasing on $(-0.31, 0)$, $(0.31, \infty)$
Decreasing on $(-\infty, -0.31)$, $(0, 0.31)$

(c) Even

9. (a)

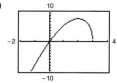

(b) Increasing on $(-\infty, 2)$
Decreasing on $(2, 3)$

(c) Neither

10. (a)

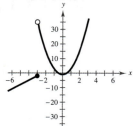

(b) Increasing on $(-5, \infty)$
Decreasing on $(-\infty, -5)$

(c) Neither

11. Relative maximum: $(0.82, 0.09)$
Relative minimum: $(-0.82, -2.09)$

12. -3

13.

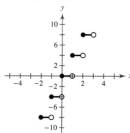

14. (a) $f(x) = [\![x]\!]$

(b) Vertical stretch

(c)

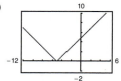

15. (a) $f(x) = \sqrt{x}$

(b) Reflection in the x-axis, left shift of five units, and an upward shift of eight units

(c)

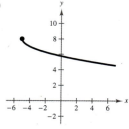

16. (a) $f(x) = x^3$

(b) Vertical stretch, a reflection in the x-axis, a right shift of five units, and an upward shift of three units

(c)

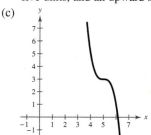

17. (a) $2x^2 - 4x - 2$ (b) $4x^2 + 4x - 12$

(c) $-3x^4 - 12x^3 + 22x^2 + 28x - 35$

(d) $\dfrac{3x^2 - 7}{-x^2 - 4x + 5}$

(e) $3x^4 + 24x^3 + 18x^2 - 120x + 68$

(f) $-9x^4 + 30x^2 - 16$

18. (a) $\dfrac{1 + 2x^{3/2}}{x}$ (b) $\dfrac{1 - 2x^{3/2}}{x}$ (c) $\dfrac{2\sqrt{x}}{x}$

(d) $\dfrac{1}{2x^{3/2}}$ (e) $\dfrac{\sqrt{x}}{2x}$ (f) $\dfrac{2\sqrt{x}}{x}$

19. $f^{-1}(x) = \sqrt[3]{x - 8}$ **20.** No inverse function

21. $f^{-1}(x) = \left(\frac{1}{3}x\right)^{2/3}, \quad x \geq 0$ **22.** $153

Cumulative Test for Chapters P–2   *(page 236)*

1. $\dfrac{4x^3}{15y^5}, \quad x \neq 0$ **2.** $2|x|y\sqrt{5y}$ **3.** $5x - 6$

4. $x^3 - x^2 - 5x + 6$ **5.** $\dfrac{s - 19}{(s - 3)(s + 5)}$

6. $(x + 7)(5 - x)$ **7.** $x(x + 1)(1 - 6x)$

8. $2(3x + 2)(9x^2 - 6x + 4)$

9. $4x^2 + 5x + 1$ **10.** $\frac{5}{2}x^2 + 2x + \frac{1}{2}$

11.

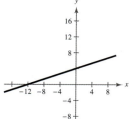

12.

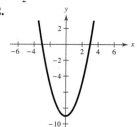

13.

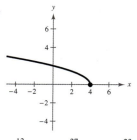

14. $-\dfrac{13}{3}$ **15.** $\dfrac{27}{5}$ **16.** $\dfrac{23}{6}$ **17.** $1, 3$ **18.** $-1, 3$

19. $-\dfrac{3}{2} \pm \dfrac{\sqrt{69}}{6}$ **20.** $\dfrac{-5 \pm \sqrt{97}}{6}$ **21.** ± 6

22. ± 8 **23.** $0, -12, \pm 2i$ **24.** $0, 3$ **25.** No solution

26. 6 **27.** $-3, 5$ **28.** No solution

29. $[-7, 5]$

30. $\left(-\infty, -\frac{4}{3}\right) \cup \left(-\frac{1}{3}, \infty\right)$

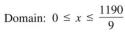

31. $\left(-\infty, -\frac{7}{5}\right] \cup [-1, \infty)$

32. $\left(-\frac{1}{4}, \frac{3}{2}\right)$

33. $y = 2x + 2$

34. For some values of x there correspond two values of y.

35. (a) $\dfrac{3}{2}$ (b) Division by 0 is undefined.   (c) $\dfrac{s + 2}{s}$

36. Neither   **37.** Odd   **38.** Even

39. (a) Vertical shrink   (b) Upward shift of two units
(c) Left shift of two units

40. (a) $4x - 3$ (b) $-2x - 5$ (c) $3x^2 - 11x - 4$
(d) $\dfrac{x - 4}{3x + 1}$; Domain: all real numbers x except $x = -\dfrac{1}{3}$

41. (a) $\sqrt{x - 1} + x^2 + 1$ (b) $\sqrt{x - 1} - x^2 - 1$
(c) $x^2\sqrt{x - 1} + \sqrt{x - 1}$
(d) $\dfrac{\sqrt{x - 1}}{x^2 + 1}$; Domain: all real numbers x such that $x \geq 1$

42. (a) $2x + 12$ (b) $\sqrt{2x^2 + 6}$
Domain of $f \circ g$: all real numbers x such that $x \geq -6$
Domain of $g \circ f$: all real numbers x

43. (a) $|x| - 2$ (b) $|x - 2|$
Domains of $f \circ g$ and $g \circ f$: all real numbers x

44. $h(x)^{-1} = \frac{1}{3}(x + 4)$ **45.** $n = 9$

46. (a) $R(n) = -0.05n^2 + 13n, \quad n \geq 60$
(b) 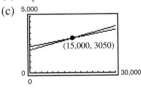   130 people

47. 4; Answers will vary.

Problem Solving   (page 239)

1. (a) $W_1 = 2000 + 0.07S$ (b) $W_2 = 2300 + 0.05S$
(c)

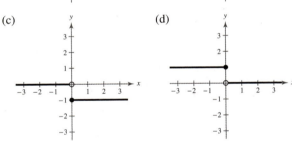

Both jobs pay the same monthly salary when sales equal $15,000.
(d) No. Job 1 would pay $3400 and job 2 would pay $3300.

3. (a) The function will be even.
(b) The function will be odd.
(c) The function will be neither even nor odd.

5. $f(x) = a_{2n}x^{2n} + a_{2n-2}x^{2n-2} + \cdots + a_2x^2 + a_0$
$f(-x) = a_{2n}(-x)^{2n} + a_{2n-2}(-x)^{2n-2}$
$\qquad\qquad + \cdots + a_2(-x)^2 + a_0$
$\qquad = f(x)$

7. (a) $81\frac{2}{3}$ h
(b) $25\frac{5}{7}$ mi/h
(c) $y = \dfrac{-180}{7}x + 3400$
Domain: $0 \leq x \leq \dfrac{1190}{9}$
Range: $0 \leq y \leq 3400$

(d)

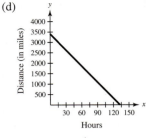

9. (a) $(f \circ g)(x) = 4x + 24$ (b) $(f \circ g)^{-1}(x) = \frac{1}{4}x - 6$
(c) $f^{-1}(x) = \frac{1}{4}x$; $g^{-1}(x) = x - 6$
(d) $(g^{-1} \circ f^{-1})(x) = \frac{1}{4}x - 6$; They are the same.
(e) $(f \circ g)(x) = 8x^3 + 1$; $(f \circ g)^{-1}(x) = \frac{1}{2}\sqrt[3]{x - 1}$;
$f^{-1}(x) = \sqrt[3]{x - 1}$; $g^{-1}(x) = \frac{1}{2}x$;
$(g^{-1} \circ f^{-1})(x) = \frac{1}{2}\sqrt[3]{x - 1}$
(f) Answers will vary.
(g) $(f \circ g)^{-1}(x) = (g^{-1} \circ f^{-1})(x)$

11. (a)   (b)

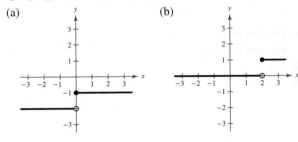

(c)   (d)

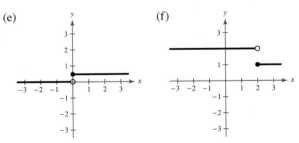

(e)   (f)

13. Proof

CHAPTER 2

15. (a)

x	-4	-2	0	4
$f(f^{-1}(x))$	-4	-2	0	4

(b)

x	-3	-2	0	1
$(f + f^{-1})(x)$	5	1	-3	-5

(c)

x	-3	-2	0	1
$(f \cdot f^{-1})(x)$	4	0	2	6

(d)

x	-4	-3	0	4
$|f^{-1}(x)|$	2	1	1	3

Chapter 3

Section 3.1   *(page 248)*

1. polynomial   **3.** quadratic; parabola
5. b   **6.** a   **7.** c   **8.** d
9. (a)

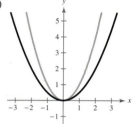

Vertical shrink

(b)

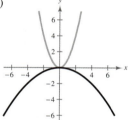

Vertical shrink and a reflection in the x-axis

(c)

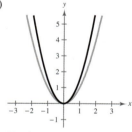

Vertical stretch

(d)

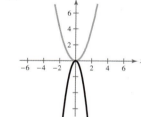

Vertical stretch and a reflection in the x-axis

11. (a)

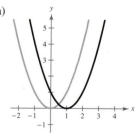

Right shift of one unit

(b)

Horizontal shrink and an upward shift of one unit

(c)

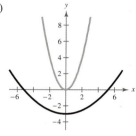

Horizontal stretch and a downward shift of three units

(d)

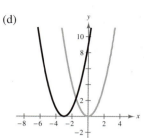

Left shift of three units

13. $f(x) = (x - 3)^2 - 9$ **15.** $h(x) = (x - 4)^2$

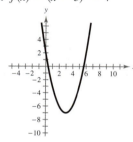

Vertex: $(3, -9)$
Axis of symmetry: $x = 3$
x-intercepts: $(0, 0), (6, 0)$

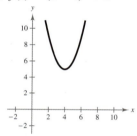

Vertex: $(4, 0)$
Axis of symmetry: $x = 4$
x-intercept: $(4, 0)$

17. $f(x) = (x - 3)^2 - 7$ **19.** $f(x) = (x - 4)^2 + 5$

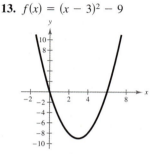

Vertex: $(3, -7)$
Axis of symmetry: $x = 3$
x-intercepts: $\left(3 \pm \sqrt{7}, 0\right)$

Vertex: $(4, 5)$
Axis of symmetry: $x = 4$
No x-intercept

21. $f(x) = \left(x - \frac{1}{2}\right)^2 + 1$ **23.** $f(x) = -(x - 1)^2 + 6$

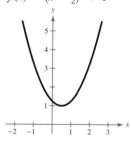

Vertex: $\left(\frac{1}{2}, 1\right)$
Axis of symmetry: $x = \frac{1}{2}$
No x-intercept

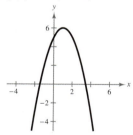

Vertex: $(1, 6)$
Axis of symmetry: $x = 1$
x-intercepts: $\left(1 \pm \sqrt{6}, 0\right)$

25. $h(x) = 4\left(x - \frac{1}{2}\right)^2 + 20$

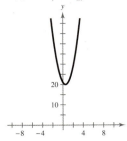

Vertex: $\left(\frac{1}{2}, 20\right)$

Axis of symmetry: $x = \frac{1}{2}$

No x-intercept

27.

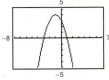

Vertex: $(-1, 4)$

Axis of symmetry: $x = -1$

x-intercepts: $(1, 0), (-3, 0)$

29.

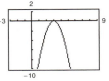

Vertex: $(-4, -5)$

Axis of symmetry: $x = -4$

x-intercepts: $\left(-4 \pm \sqrt{5}, 0\right)$

31.

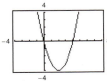

Vertex: $(3, 0)$

Axis of symmetry: $x = 3$

x-intercept: $(3, 0)$

33.

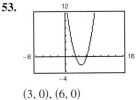

Vertex: $(-2, -3)$

Axis of symmetry: $x = -2$

x-intercepts: $\left(-2 \pm \sqrt{6}, 0\right)$

35. $f(x) = (x + 2)^2 - 1$ **37.** $f(x) = (x + 2)^2 + 5$

39. $f(x) = 4(x - 1)^2 - 2$ **41.** $f(x) = \frac{3}{4}(x - 5)^2 + 12$

43. $f(x) = -\frac{24}{49}\left(x + \frac{1}{4}\right)^2 + \frac{3}{2}$ **45.** $f(x) = -\frac{16}{3}\left(x + \frac{5}{2}\right)^2$

47. $(-1, 0), (3, 0)$ **49.** $\left(-3, 0\right), \left(\frac{1}{2}, 0\right)$

51.

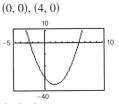

$(0, 0), (4, 0)$

53.

$(3, 0), (6, 0)$

55.

$\left(-\frac{5}{2}, 0\right), (6, 0)$

57. $f(x) = x^2 - 9$
     $g(x) = -x^2 + 9$

59. $f(x) = x^2 - 3x - 4$
     $g(x) = -x^2 + 3x + 4$

61. $f(x) = 2x^2 + 7x + 3$
     $g(x) = -2x^2 - 7x - 3$

63. 55, 55   **65.** 12, 6   **67.** 16 ft   **69.** 20 fixtures

71. (a) \$14,000,000; \$14,375,000; \$13,500,000

    (b) \$24; \$14,400,000; Answers will vary.

73. (a) $A = \dfrac{8x(50 - x)}{3}$

    (b) $x = 25$ ft, $y = 33\frac{1}{3}$ ft

75. True. The equation has no real solutions, so the graph has no x-intercepts.

77. $b = \pm 20$ **79.** $f(x) = a\left(x + \dfrac{b}{2a}\right)^2 + \dfrac{4ac - b^2}{4a}$

81. Proof

Section 3.2   *(page 260)*

1. continuous   **3.** $n; n - 1$ **5.** touches; crosses

7. standard   **9.** c   **10.** f   **11.** a   **12.** e

13. d   **14.** b

15. (a)

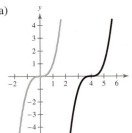

(b)

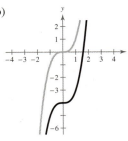

(c)

(d)

17. (a)

(b)

(c)

(d)

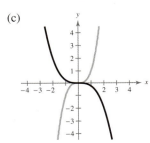

(e)

(f)

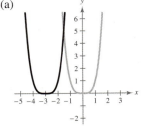
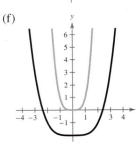

19. Falls to the left, rises to the right
21. Falls to the left and to the right
23. Rises to the left, falls to the right
25. Rises to the left and to the right
27. Rises to the left, falls to the right

29. 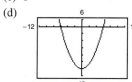   **31.**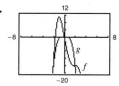

33. (a) ± 6 **35.** (a) 3
    (b) Odd multiplicity   (b) Even multiplicity
    (c) 1   (c) 1
    (d) 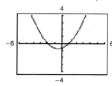   (d)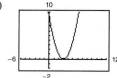

37. (a) $-2, 1$ **39.** (a) $0, 1 \pm \sqrt{2}$
    (b) Odd multiplicity   (b) Odd multiplicity
    (c) 1   (c) 2
    (d) 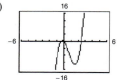   (d)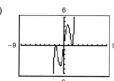

41. (a) $0, 2 \pm \sqrt{3}$ **43.** (a) $0, \pm\sqrt{3}$
    (b) Odd multiplicity   (b) 0, odd multiplicity;
    (c) 2 $\pm\sqrt{3}$, even multiplicity
    (d) 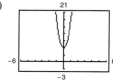   (c) 4
                    (d)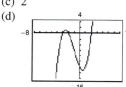

45. (a) No real zero   **47.** (a) $\pm 2, -3$
    (b) No multiplicity   (b) Odd multiplicity
    (c) 1   (c) 2
    (d) 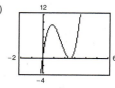   (d)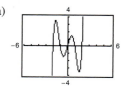

49. (a) 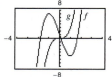   **51.** (a) (image)
    (b) and (c) $0, \frac{5}{2}$ (b) and (c) $0, \pm 1, \pm 2$
    (d) The answers are the   (d) The answers are the
        same.   same.

53. $f(x) = x^2 - 7x$ **55.** $f(x) = x^3 + 6x^2 + 8x$
57. $f(x) = x^4 - 4x^3 - 9x^2 + 36x$ **59.** $f(x) = x^2 - 2x - 1$
61. $f(x) = x^3 - 6x^2 + 7x + 2$ **63.** $f(x) = x^2 + 6x + 9$

65. $f(x) = x^3 + 4x^2 - 5x$
67. $f(x) = x^4 + x^3 - 15x^2 + 23x - 10$ **69.** $f(x) = x^5 - 3x^3$
71. (a) Rises to the left and to the right
    (b) No zeros   (c) Answers will vary.
    (d)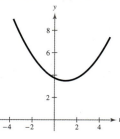

73. (a) Falls to the left, rises to the right
    (b) $0, 5, -5$ (c) Answers will vary.
    (d)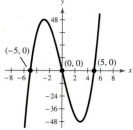

75. (a) Rises to the left and to the right
    (b) $-2, 2$ (c) Answers will vary.
    (d)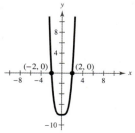

77. (a) Falls to the left, rises to the right
    (b) $0, 2, 3$ (c) Answers will vary.
    (d)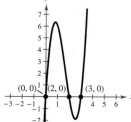

79. (a) Rises to the left, falls to the right
    (b) $-5, 0$ (c) Answers will vary.
    (d)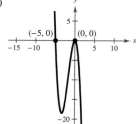

81. (a) Falls to the left, rises to the right
  (b) $-2, 0$
  (c) Answers will vary.
  (d)

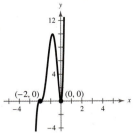

83. (a) Falls to the left and to the right
  (b) ± 2
  (c) Answers will vary.
  (d)

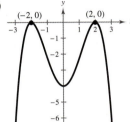

85.

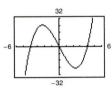

Zeros: $0, \pm 4$,
odd multiplicity

87.

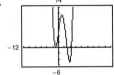

Zeros: -1,
even multiplicity;
$3, \frac{9}{2}$, odd multiplicity

89. (a) $[-1, 0], [1, 2], [2, 3]$ (b) $-0.879, 1.347, 2.532$
91. (a) $[-2, -1], [0, 1]$ (b) $-1.585, 0.779$
93. (a) $V(x) = x(36 - 2x)^2$
  (b) Domain: $0 < x < 18$
  (c)

6 in. × 24 in. × 24 in.

  (d)

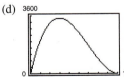

$x = 6$; The results are the same.
95. (a) Relative maximum: $(4.44, 1512.60)$
    Relative minimum: $(11.97, 189.37)$
  (b) Increasing: $(3, 4.44), (11.97, 16)$
    Decreasing: $(4.44, 11.97)$
  (c) Answers will vary.
97. $x \approx 200$
99. True. A polynomial function falls to the right only when the leading coefficient is negative.

101. False. The graph falls to the left and to the right or the graph rises to the left and to the right.
103. Answers will vary. *Sample answers:*
  $a_4 < 0$ $\qquad\qquad\qquad$ $a_4 > 0$

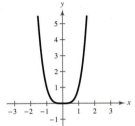

105.

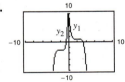

  (a) Upward shift of two units; Even
  (b) Left shift of two units; Neither
  (c) Reflection in the y-axis; Even
  (d) Reflection in the x-axis; Even
  (e) Horizontal stretch; Even   (f) Vertical shrink; Even
  (g) $g(x) = x^3, x \geq 0$; Neither   (h) $g(x) = x^{16}$; Even
107.

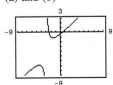

  (a) y_1 is decreasing, y_2 is increasing.
  (b) Yes; a; If $a > 0$, then the graph is increasing, and if $a < 0$, then the graph is decreasing.
  (c)

No; f is not strictly increasing or strictly decreasing, so f cannot be written in the form $f(x) = a(x - h)^5 + k$.

Section 3.3 *(page 270)*

1. $f(x)$: dividend; $d(x)$: divisor;
   $q(x)$: quotient; $r(x)$: remainder
3. improper   **5.** Factor   **7.** Answers will vary.
9. (a) and (b)

  (c) Answers will vary

11. $2x + 4$, $x \neq -3$ **13.** $x^2 - 3x + 1$, $x \neq -\frac{5}{4}$

15. $x^3 + 3x^2 - 1$, $x \neq -2$ **17.** $6 - \dfrac{1}{x+1}$

19. $x - \dfrac{x+9}{x^2+1}$ **21.** $2x - 8 + \dfrac{x-1}{x^2+1}$

23. $x + 3 + \dfrac{6x^2 - 8x + 3}{(x-1)^3}$ **25.** $2x^2 - 2x + 6$, $x \neq 4$

27. $6x^2 + 25x + 74 + \dfrac{248}{x-3}$ **29.** $4x^2 - 9$, $x \neq -2$

31. $-x^2 + 10x - 25$, $x \neq -10$ **33.** $x^2 + x + 4 + \dfrac{21}{x-4}$

35. $10x^3 + 10x^2 + 60x + 360 + \dfrac{1360}{x-6}$

37. $x^2 - 8x + 64$, $x \neq -8$

39. $-3x^3 - 6x^2 - 12x - 24 - \dfrac{48}{x-2}$

41. $-x^3 - 6x^2 - 36x - 36 - \dfrac{216}{x-6}$

43. $4x^2 + 14x - 30$, $x \neq -\frac{1}{2}$

45. $f(x) = (x-3)(x^2 + 2x - 4) - 5$, $f(3) = -5$

47. $f(x) = \left(x + \frac{2}{3}\right)(15x^3 - 6x + 4) + \frac{34}{3}$, $f\left(-\frac{2}{3}\right) = \frac{34}{3}$

49. $f(x) = \left(x - 1 + \sqrt{3}\right)\left[-4x^2 + \left(2 + 4\sqrt{3}\right)x + \left(2 + 2\sqrt{3}\right)\right]$, $f\left(1 - \sqrt{3}\right) = 0$

51. (a) -2 (b) 1 (c) 36 (d) 5

53. (a) -35 (b) $-\frac{5}{8}$ (c) -10 (d) -211

55. $(x+3)(x+2)(x+1)$; Solutions: $-3, -2, -1$

57. $(2x-1)(x-5)(x-2)$; Solutions: $\frac{1}{2}, 5, 2$

59. $\left(x + \sqrt{3}\right)\left(x - \sqrt{3}\right)(x+2)$; Solutions: $-\sqrt{3}, \sqrt{3}, -2$

61. $(x-1)\left(x - 1 - \sqrt{3}\right)\left(x - 1 + \sqrt{3}\right)$; Solutions: $1, 1 + \sqrt{3}, 1 - \sqrt{3}$

63. (a) Answers will vary.   (b) $2x - 1$
(c) $f(x) = (2x-1)(x+2)(x-1)$ (d) $\frac{1}{2}, -2, 1$
(e)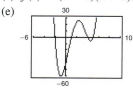

65. (a) Answers will vary.   (b) $(x-4)(x-1)$
(c) $f(x) = (x-5)(x+2)(x-4)(x-1)$ (d) $5, -2, 4, 1$
(e)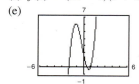

67. (a) Answers will vary.   (b) $x + 7$
(c) $f(x) = (x+7)(2x+1)(3x-2)$ (d) $-7, -\frac{1}{2}, \frac{2}{3}$
(e)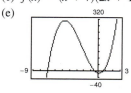

69. (a) Answers will vary.   (b) $x - \sqrt{5}$
(c) $f(x) = \left(x - \sqrt{5}\right)\left(x + \sqrt{5}\right)(2x - 1)$ (d) $\pm\sqrt{5}, \frac{1}{2}$
(e)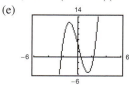

71. (a) $2, \pm 2.236$ (b) 2
(c) $f(x) = (x-2)\left(x - \sqrt{5}\right)\left(x + \sqrt{5}\right)$

73. (a) $-2, 0.268, 3.732$ (b) -2
(c) $h(t) = (t+2)\left[t - \left(2 + \sqrt{3}\right)\right]\left[t - \left(2 - \sqrt{3}\right)\right]$

75. (a) $0, 3, 4, \pm 1.414$ (b) 0
(c) $h(x) = x(x-4)(x-3)\left(x + \sqrt{2}\right)\left(x - \sqrt{2}\right)$

77. $x^2 - 7x - 8$, $x \neq -8$

79. $x^2 + 3x$, $x \neq -2, -1$

81. (a) 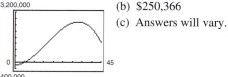   (b) $\$250,366$
(c) Answers will vary.

83. False. $-\frac{4}{7}$ is a zero of f.

85. True. The degree of the numerator is greater than the degree of the denominator.

87. $x^{2n} + 6x^n + 9$, $x^n \neq -3$ **89.** $k = -1$, not 1.

91. $c = -210$ **93.** $k = 7$

Section 3.4   *(page 283)*

1. Fundamental Theorem of Algebra   **3.** Rational Zero

5. linear; quadratic; quadratic   **7.** Descartes's Rule of Signs

9. 3   **11.** 5   **13.** 2   **15.** $\pm 1, \pm 2$

17. $\pm 1, \pm 3, \pm 5, \pm 9, \pm 15, \pm 45, \pm\frac{1}{2}, \pm\frac{3}{2}, \pm\frac{5}{2}, \pm\frac{9}{2}, \pm\frac{15}{2}, \pm\frac{45}{2}$

19. $-2, -1, 3$ **21.** No rational zeros   **23.** $-6, -1$

25. $-1, \frac{1}{2}$ **27.** $-2, 3, \pm\frac{2}{3}$ **29.** $1, \frac{3}{5} \pm \dfrac{\sqrt{19}}{5}$

31. $-3, 1, -2 \pm \sqrt{6}$

33. (a) $\pm 1, \pm 2, \pm 4$
(b) 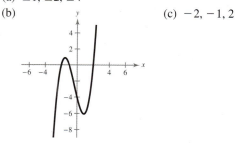   (c) $-2, -1, 2$

35. (a) $\pm 1, \pm 3, \pm\frac{1}{2}, \pm\frac{3}{2}, \pm\frac{1}{4}, \pm\frac{3}{4}$
(b) 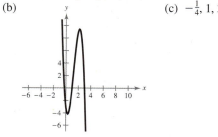   (c) $-\frac{1}{4}, 1, 3$

37. (a) $\pm 1, \pm 2, \pm 4, \pm 8, \pm\frac{1}{2}$

(b)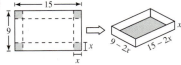

(c) $-\frac{1}{2}, 1, 2, 4$

39. (a) $\pm 1, \pm 3, \pm\frac{1}{2}, \pm\frac{3}{2}, \pm\frac{1}{4}, \pm\frac{3}{4}, \pm\frac{1}{8}, \pm\frac{3}{8}, \pm\frac{1}{16}, \pm\frac{3}{16}, \pm\frac{1}{32}, \pm\frac{3}{32}$

(b)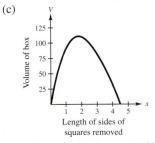

(c) $1, \frac{3}{4}, -\frac{1}{8}$

41. $f(x) = x^3 - x^2 + 25x - 25$

43. $f(x) = x^4 - 6x^3 + 14x^2 - 16x + 8$

45. $f(x) = 3x^4 - 17x^3 + 25x^2 + 23x - 22$

47. $f(x) = 2x^4 + 2x^3 - 2x^2 + 2x - 4$

49. $f(x) = x^3 + x^2 - 2x + 12$

51. (a) $(x^2 + 4)(x^2 - 2)$

(b) $(x^2 + 4)(x + \sqrt{2})(x - \sqrt{2})$

(c) $(x + 2i)(x - 2i)(x + \sqrt{2})(x - \sqrt{2})$

53. (a) $(x^2 - 6)(x^2 - 2x + 3)$

(b) $(x + \sqrt{6})(x - \sqrt{6})(x^2 - 2x + 3)$

(c) $(x + \sqrt{6})(x - \sqrt{6})(x - 1 - \sqrt{2}i)(x - 1 + \sqrt{2}i)$

55. $\pm 2i, 1$ **57.** $2, 3 \pm 2i$ **59.** $1, 3, 1 \pm \sqrt{2}i$

61. $(x + 6i)(x - 6i); \pm 6i$

63. $(x - 1 - 4i)(x - 1 + 4i); 1 \pm 4i$

65. $(x - 2)(x + 2)(x - 2i)(x + 2i); \pm 2, \pm 2i$

67. $(z - 1 + i)(z - 1 - i); 1 \pm i$

69. $(x + 1)(x - 2 + i)(x - 2 - i); -1, 2 \pm i$

71. $(x - 2)^2(x + 2i)(x - 2i); 2, \pm 2i$

73. $-10, -7 \pm 5i$ **75.** $-\frac{3}{4}, 1 \pm \frac{1}{2}i$ **77.** $-2, -\frac{1}{2}, \pm i$

79. One positive real zero, no negative real zeros

81. No positive real zeros, one negative real zero

83. Two or no positive real zeros, two or no negative real zeros

85. Two or no positive real zeros, one negative real zero

87–89. Answers will vary.

91. $\frac{3}{4}, \pm\frac{1}{2}$ **93.** $-\frac{3}{4}$ **95.** $\pm 2, \pm\frac{3}{2}$ **97.** $\pm 1, \frac{1}{4}$

99. d   **100.** a   **101.** b   **102.** c

103. (a)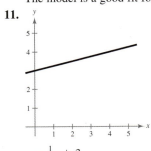

(b) $V(x) = x(9 - 2x)(15 - 2x)$

Domain: $0 < x < \frac{9}{2}$

(c)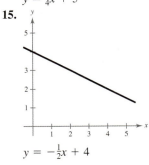

Length of sides of
squares removed

$1.82 \text{ cm} \times 5.36 \text{ cm} \times 11.36 \text{ cm}$

(d) $\frac{1}{2}, \frac{7}{2}, 8$; 8 is not in the domain of V.

105. (a) $V(x) = x^3 + 9x^2 + 26x + 24 = 120$

(b) $4 \text{ ft} \times 5 \text{ ft} \times 6 \text{ ft}$

107. False. The most complex zeros it can have is two, and the Linear Factorization Theorem guarantees that there are three linear factors, so one zero must be real.

109. r_1, r_2, r_3 **111.** $5 + r_1, 5 + r_2, 5 + r_3$

113. The zeros cannot be determined.

115. Answers will vary. There are infinitely many possible functions for f. *Sample equation and graph:*

$f(x) = -2x^3 + 3x^2 + 11x - 6$

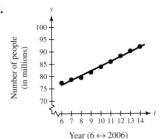

117. $f(x) = x^3 - 3x^2 + 4x - 2$

119. The function should be

$f(x) = (x + 2)(x - 3.5)(x + i)(x - i)$.

121. $f(x) = x^4 + 5x^2 + 4$

123. (a) $x^2 + b$ (b) $x^2 - 2ax + a^2 + b^2$

Section 3.5   *(page 294)*

1. variation; regression   **3.** least squares regression

5. directly proportional   **7.** combined

9.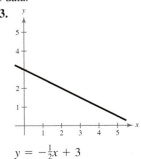

Year (6 $\leftrightarrow$ 2006)

The model is a good fit for the data.

11.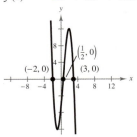

$y = \frac{1}{4}x + 3$

13.

$y = -\frac{1}{2}x + 3$

15.

$y = -\frac{1}{2}x + 4$

17. (a) and (b)

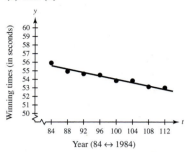

Sample answer: $y = -0.1t + 64$
(c) $y = -0.097t + 63.72$ (d) The models are similar.

19. $y = 7x$ **21.** $y = \frac{1}{5}x$ **23.** $y = 2\pi x$

25.

x	2	4	6	8	10
$y = x^2$	4	16	36	64	100

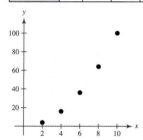

27.

x	2	4	6	8	10
$y = \frac{1}{2}x^3$	4	32	108	256	500

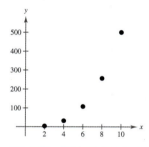

29.

x	2	4	6	8	10
$y = \dfrac{2}{x}$	1	$\frac{1}{2}$	$\frac{1}{3}$	$\frac{1}{4}$	$\frac{1}{5}$

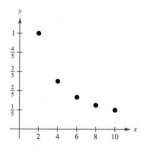

31.

x	2	4	6	8	10
$y = \dfrac{10}{x^2}$	$\frac{5}{2}$	$\frac{5}{8}$	$\frac{5}{18}$	$\frac{5}{32}$	$\frac{1}{10}$

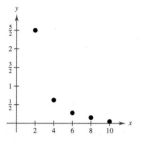

33. Inversely; Answers will vary.   **35.** $y = \dfrac{5}{x}$

37. $y = -\dfrac{7}{10}x$ **39.** $A = kr^2$ **41.** $y = \dfrac{k}{x^2}$

43. $F = \dfrac{kg}{r^2}$ **45.** $R = k(T - T_e)$ **47.** $P = kVI$

49. y is directly proportional to the square of x.
51. A is jointly proportional to b and h.

53. $y = 18x$ **55.** $y = \dfrac{75}{x}$ **57.** $z = 2xy$ **59.** $P = \dfrac{18x}{y^2}$

61. $I = 0.035P$ **63.** Model: $y = \frac{33}{13}x$; 25.4 cm, 50.8 cm
65. $293\frac{1}{3}$N   **67.** About 39.47 lb   **69.** About 0.61 mi/h
71. (a)

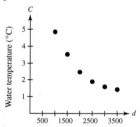

(b) Inverse variation
(c) About 4919.9
(d) 1640 m

73. (a) 200 Hz   (b) 50 Hz   (c) 100 Hz
75. True. If $y = k_1x$ and $x = k_2z$, then $y = k_1(k_2z) = (k_1k_2)z$.
77. π is a constant, not a variable.   **79.** Direct; $y = 2t$

Review Exercises   *(page 300)*

1. (a)   (b)

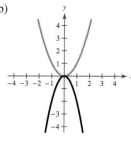

Vertical stretch   Vertical stretch and a
reflection in the x-axis

(c)

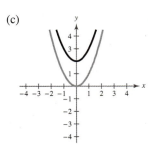

(d)

Upward shift of two units   Left shift of two units

3. $g(x) = (x - 1)^2 - 1$ **5.** $f(x) = (x - 3)^2 - 8$

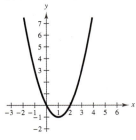

Vertex: $(1, -1)$
Axis of symmetry: $x = 1$
x-intercepts: $(0, 0), (2, 0)$

Vertex: $(3, -8)$
Axis of symmetry: $x = 3$
x-intercepts: $(3 \pm 2\sqrt{2}, 0)$

7. $f(x) = (x + 4)^2 - 6$ **9.** $h(x) = -(x - 2)^2 + 7$

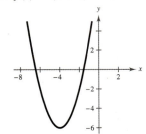

Vertex: $(-4, -6)$
Axis of symmetry: $x = -4$
x-intercepts: $(-4 \pm \sqrt{6}, 0)$

Vertex: $(2, 7)$
Axis of symmetry: $x = 2$
x-intercepts: $(2 \pm \sqrt{7}, 0)$

11. $h(x) = 4\left(x + \frac{1}{2}\right)^2 + 12$ **13.** $f(x) = \frac{1}{3}\left(x + \frac{5}{2}\right)^2 - \frac{41}{12}$

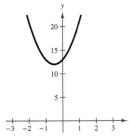

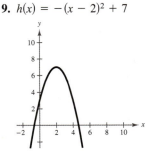

Vertex: $\left(-\frac{1}{2}, 12\right)$
Axis of symmetry: $x = -\frac{1}{2}$
No x-intercept

Vertex: $\left(-\frac{5}{2}, -\frac{41}{12}\right)$
Axis of symmetry: $x = -\frac{5}{2}$
x-intercepts:
$\left(\frac{\pm\sqrt{41} - 5}{2}, 0\right)$

15. $f(x) = -\frac{1}{2}(x - 4)^2 + 1$ **17.** $f(x) = -(x - 6)^2$
19. $f(x) = \frac{3}{4}(x - 2)^2 - \frac{5}{2}$

21. (a) $A = x\left(\dfrac{8 - x}{2}\right)$ (b) $0 < x < 8$

(c)

x	1	2	3	4	5	6
A	$\frac{7}{2}$	6	$\frac{15}{2}$	8	$\frac{15}{2}$	6

$x = 4, y = 2$

(d)

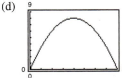

$x = 4, y = 2$

(e) $A = -\frac{1}{2}(x - 4)^2 + 8$; $x = 4, y = 2$

23. (a) $12,000; $13,750; $15,000
   (b) Maximum revenue at $40; $16,000; Any price greater or less than $40 per unit will not yield as much revenue.

25. 1091 units

27.

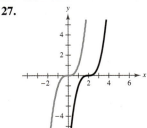

29.

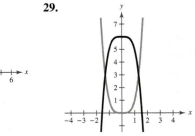

31.

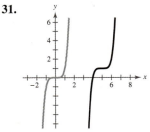

33. Falls to the left and to the right
35. Falls to the left, rises to the right
37. (a) $-8, \frac{4}{3}$ (b) Odd multiplicity   (c) 1
   (d)

39. (a) $0, \pm\sqrt{3}$ (b) Odd multiplicity   (c) 2
   (d)

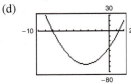

41. (a) ± 3 (b) Odd multiplicity   (c) 3
   (d)

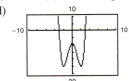

43. (a) Rises to the left, falls to the right (b) -1
(c) Answers will vary.
(d)

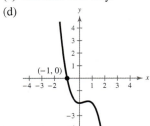

45. (a) Rises to the left and to the right (b) $-3, 0, 1$
(c) Answers will vary.
(d)

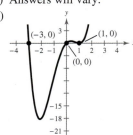

47. (a) $[-1, 0]$ (b) -0.900
49. (a) $[-1, 0], [1, 2]$ (b) $-0.200, 1.772$
51. $6x + 3 + \dfrac{17}{5x - 3}$ **53.** $5x + 4, \ x \neq \dfrac{5}{2} \pm \dfrac{\sqrt{29}}{2}$
55. $x^2 - 3x + 2 - \dfrac{1}{x^2 + 2}$ **57.** $2x^2 - 9x - 6, \ x \neq 8$
59. $x^3 - 3x^2 + 7x - 12 + \dfrac{36}{x + 3}$ **61.** (a) -421 (b) -9
63. (a) Yes (b) Yes (c) Yes (d) No
65. (a) Answers will vary. (b) $(x + 7), (x + 1)$
(c) $f(x) = (x + 7)(x + 1)(x - 4)$ (d) $-7, -1, 4$
(e)

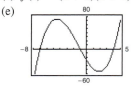

67. (a) Answers will vary. (b) $(x + 1), (x - 4)$
(c) $f(x) = (x + 1)(x - 4)(x + 2)(x - 3)$
(d) $-2, -1, 3, 4$
(e)

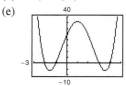

69. 1 **71.** 5 **73.** 3 **75.** $-6, -2, 5$ **77.** $-2, \frac{4}{3}$
79. $-4, 3$ **81.** $f(x) = 3x^4 - 14x^3 + 17x^2 - 42x + 24$
83. $2, \pm 4i$ **85.** $x(x - 1)(x + 5); \ 0, 1, -5$
87. $(x + 4)^2(x - 2 - 3i)(x - 2 + 3i); \ -4, 2 \pm 3i$
89. $16, \pm i$ **91.** $-3, 6, \pm 2i$
93. Two or no positive real zeros, one negative real zero
95. Answers will vary.
97. Radius: 1.82 in., height: 10.82 in.

99.

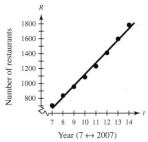

The model fits the data well.
101. $y = \frac{8}{5}x$; 8 km; 40 km **103.** A factor of 4
105. About 2 h, 26 min
107. True. The leading coefficient is negative and the degree is odd.
109. True. If y is directly proportional to x, then $y = kx$, so $x = (1/k)y$. Therefore, x is directly proportional to y.
111. Answers will vary. *Sample answer:*
A polynomial of degree $n > 0$ with real coefficients can be written as the product of linear and quadratic factors with real coefficients, where the quadratic factors have no real zeros.
Setting the factors equal to zero and solving for the variable can find the zeros of a polynomial function.
To solve an equation is to find all the values of the variable for which the equation is true.

Chapter Test *(page 304)*

1. (a)   (b)

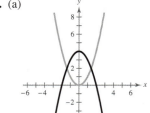

Reflection in the x-axis   Right shift of $\frac{3}{2}$ units
and an upward shift of
four units

2. Vertex: $(1, -4)$; Intercepts: $(3, 0), (-1, 0), (0, -3)$
3. $y = (x - 3)^2 - 6$
4. (a) 50 ft
(b) 5. Yes, changing the constant term results in a vertical shift of the graph, so the maximum height changes.
5. Rises to the left, falls to the right

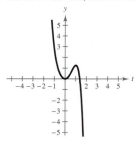

6. $3x + \dfrac{x - 1}{x^2 + 1}$ **7.** $2x^3 - 4x^2 + 5x - 6 + \dfrac{11}{x + 2}$
8. $(2x - 5)(x - \sqrt{3})(x + \sqrt{3})$; Zeros: $\pm\sqrt{3}, \frac{5}{2}$
9. $-2, \frac{3}{2}$ **10.** $\pm 1, -\frac{2}{3}$

11. $f(x) = x^4 - 2x^3 + 9x^2 - 18x$

12. $f(x) = x^4 - 6x^3 + 16x^2 - 18x + 7$

13. $1, -5, -\frac{2}{3}$ **14.** $-2, 4, -1 \pm \sqrt{2}\,i$

15. $v = 6\sqrt{s}$ **16.** $A = \frac{25}{6}xy$ **17.** $b = \frac{48}{a}$

18. $y = -232.8t + 8890$; The model fits the data well.

Problem Solving   *(page 307)*

1. (a) (i) $6, -2$ (ii) $0, -5$ (iii) $-5, 2$

(iv) 2 (v) $1 \pm \sqrt{7}$ (vi) $\dfrac{-3 \pm \sqrt{7}\,i}{2}$

(b) (i) 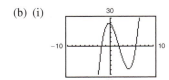   (ii)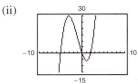

(iii) 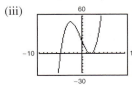   (iv)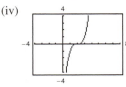

(v) 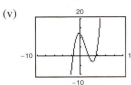   (vi)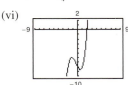

Graph (iii) touches the x-axis at $(2, 0)$, and all the other graphs pass through the x-axis at $(2, 0)$.

(c) (i) $(6, 0), (-2, 0)$ (ii) $(0, 0), (-5, 0)$

(iii) $(-5, 0)$ (iv) No other x-intercepts

(v) $(-1.6, 0), (3.6, 0)$ (vi) No other x-intercepts

(d) When the function has two real zeros, the results are the same. When the function has one real zero, the graph touches the x-axis at the zero. When there are no real zeros, there is no x-intercept.

3. (a) $l = \dfrac{600 - \pi r^2}{\pi r}$ (b) $V(r) = 300r - \dfrac{\pi}{2}r^3$

(c) About 1595.8 ft^3; $r \approx 8$ ft, $l \approx 16$ ft

5. (a)

y	1	2	3	4	5
$y^3 + y^2$	2	12	36	80	150

y	6	7	8	9	10
$y^3 + y^2$	252	392	576	810	1100

(b) (i) $x = 6$ (ii) $x = 6$ (iii) $x = 3$ (iv) $x = 10$

(v) $x = 6$ (vi) $x = 3$

(c) Answers will vary.

7. (a)

Function	Zeros	Sum of zeros	Product of zeros
$f_1(x)$	2, 3	5	6
$f_2(x)$	$-3, 1, 2$	0	-6
$f_3(x)$	$-3, 1, \pm 2i$	-2	-12
$f_4(x)$	$0, 2, -3, 2 \pm \sqrt{3}$	3	0

(b) The sum of the zeros is equal to the opposite of the coefficient of the $(n - 1)$th term.

(c) The product of the zeros is equal to the constant term when the function is of an even degree and to the opposite of the constant term when the function is of an odd degree.

9. (a) and (b) $y = -x^2 + 5x - 4$

11. (a) $A(x) = x\left(\dfrac{100 - x}{2}\right)$; Domain: $0 < x < 100$

(b) 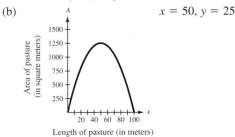 $x = 50, y = 25$

(c) $A(x) = -\frac{1}{2}(x - 50)^2 + 1250$; $x = 50, y = 25$

13. 2 in. $\times$ 2 in. $\times$ 5 in.

Chapter 4

Section 4.1   *(page 315)*

1. rational function   **3.** horizontal asymptote

5. Domain: all real numbers x except $x = 1$

$f(x) \to -\infty$ as $x \to 1^-$, $f(x) \to \infty$ as $x \to 1^+$

7. Domain: all real numbers x except $x = -2$

$f(x) \to \infty$ as $x \to -2^-$, $f(x) \to -\infty$ as $x \to -2^+$

9. Domain: all real numbers x except $x = \pm 1$

$f(x) \to \infty$ as $x \to -1^-$ and as $x \to 1^+$,

$f(x) \to -\infty$ as $x \to -1^+$ and as $x \to 1^-$

11. Domain: all real numbers x except $x = 1$

$f(x) \to \infty$ as $x \to 1^-$ and as $x \to 1^+$

13. Vertical asymptote: $x = 0$

Horizontal asymptote: $y = 0$

15. Vertical asymptote: $x = 5$

Horizontal asymptote: $x = -1$

17. Vertical asymptotes: $x = \pm 1$

19. Horizontal asymptote: $y = -4$

21. Vertical asymptote: $x = -4$

Horizontal asymptote: $y = 0$

23. Vertical asymptotes: $x = -2, x = 3$

Horizontal asymptote: $y = 1$

25. Vertical asymptote: $x = \frac{1}{2}$

Horizontal asymptote: $y = \frac{1}{2}$

27. Vertical asymptote: $x = 2$

Horizontal asymptote: $y = 2$

29. f   **30.** e   **31.** a   **32.** g
33. c   **34.** b   **35.** d   **36.** h
37. (a) Domain of f: all real numbers x except $x = -2$
    Domain of g: all real numbers x

(b) $x - 2$; Vertical asymptote: none
(c)

x	-4	-3	-2.5	-2	-1.5	-1	0
$f(x)$	-6	-5	-4.5	Undef.	-3.5	-3	-2
$g(x)$	-6	-5	-4.5	-4	-3.5	-3	-2

(d) The functions differ only at $x = -2$, where f is undefined.
39. (a) Domain of f: all real numbers x except $x = 0, \frac{1}{2}$
    Domain of g: all real numbers x except $x = 0$

(b) $\frac{1}{x}$; Vertical asymptote: $x = 0$
(c)

x	-1	-0.5	0	0.5	2	3	4
$f(x)$	-1	-2	Undef.	Undef.	$\frac{1}{2}$	$\frac{1}{3}$	$\frac{1}{4}$
$g(x)$	-1	-2	Undef.	2	$\frac{1}{2}$	$\frac{1}{3}$	$\frac{1}{4}$

(d) The functions differ only at $x = 0.5$ where f is undefined.
41. (a)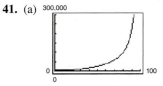

(b) \$4411.76; \$25,000; \$225,000
(c) No. The function is undefined at $p = 100$.
43. (a)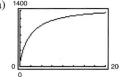

(b) 333 deer, 500 deer, 800 deer   (c) 1500 deer
45. (a)

n	1	2	3	4	5
P	0.50	0.74	0.82	0.86	0.89

n	6	7	8	9	10
P	0.91	0.92	0.93	0.94	0.95

P approaches 1 as n increases.
(b) 100%
47. False. Polynomials do not have vertical asymptotes.
49. True. A rational function has at most one horizontal asymptote.
51. (a) 4   (b) Less than   (c) Greater than
53. (a) 0   (b) Greater than   (c) Less than
55. $f(x) = \dfrac{2x^2}{x^2 + 1}$
57. The denominator could have a factor of $(x - c)$, so $x = c$ could be undefined for the rational function.

Section 4.2   *(page 323)*

1. slant asymptote
3.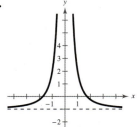
5.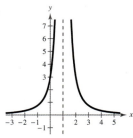
7.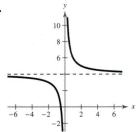
9.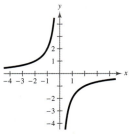
11.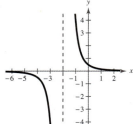
13.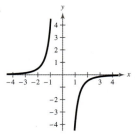
15. (a) Domain: all real numbers x except $x = -1$
(b) y-intercept: $(0, 1)$
(c) Vertical asymptote: $x = -1$
    Horizontal asymptote: $y = 0$
(d)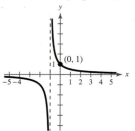

17. (a) Domain: all real numbers x except $x = -4$
(b) y-intercept: $\left(0, -\frac{1}{4}\right)$
(c) Vertical asymptote: $x = -4$
    Horizontal asymptote: $y = 0$
(d)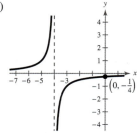

19. (a) Domain: all real numbers x except $x = -2$
(b) x-intercept: $\left(-\frac{3}{2}, 0\right)$
y-intercept: $\left(0, \frac{3}{2}\right)$
(c) Vertical asymptote: $x = -2$
Horizontal asymptote: $y = 2$
(d)

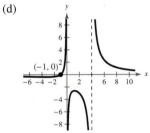

21. (a) Domain: all real numbers x except $x = -2$
(b) x-intercept: $\left(-\frac{5}{2}, 0\right)$
y-intercept: $\left(0, \frac{5}{2}\right)$
(c) Vertical asymptote: $x = -2$
Horizontal asymptote: $y = 2$
(d)

23. (a) Domain: all real numbers x
(b) Intercept: $(0, 0)$
(c) Horizontal asymptote: $y = 1$
(d)

25. (a) Domain: all real numbers x except $x = \pm 3$
(b) Intercept: $(0, 0)$
(c) Vertical asymptotes: $x = \pm 3$
Horizontal asymptote: $y = 1$
(d)

27. (a) Domain: all real numbers s
(b) Intercept: $(0, 0)$
(c) Horizontal asymptote: $y = 0$
(d)

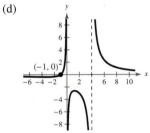

29. (a) Domain: all real numbers x except $x = 0, 4$
(b) x-intercept: $(-1, 0)$
(c) Vertical asymptotes: $x = 0, x = 4$
Horizontal asymptote: $y = 0$
(d)

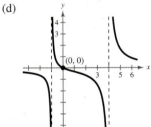

31. (a) Domain: all real numbers x except $x = 4, -1$
(b) Intercept: $(0, 0)$
(c) Vertical asymptotes: $x = -1, x = 4$
Horizontal asymptote: $y = 0$
(d)

33. (a) Domain: all real numbers x except $x = 3, -4$
(b) y-intercept: $\left(0, -\frac{5}{3}\right)$
(c) Vertical asymptote: $x = 3$
Horizontal asymptote: $y = 0$
(d)

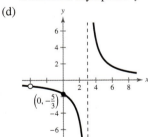

CHAPTER 4

35. (a) Domain: all real numbers t except $t = 1$

(b) t-intercept: $(-1, 0)$

   y-intercept: $(0, 1)$

(c) No asymptotes

(d)

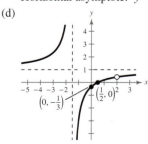

37. (a) Domain: all real numbers x except $x = \pm 2$

(b) x-intercepts: $(1, 0), (4, 0)$

   y-intercept: $(0, -1)$

(c) Vertical asymptotes: $x = \pm 2$

   Horizontal asymptote: $y = 1$

(d)

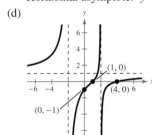

39. (a) Domain: all real numbers x except $x = -4, 1$

(b) y-intercept: $(0, -1)$

(c) Vertical asymptotes: $x = -4, x = 1$

   Horizontal asymptote: $y = 1$

(d)

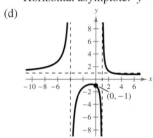

41. (a) Domain: all real numbers x except $x = -\frac{3}{2}, 2$

(b) x-intercept: $\left(\frac{1}{2}, 0\right)$

   y-intercept: $\left(0, -\frac{1}{3}\right)$

(c) Vertical asymptote: $x = -\frac{3}{2}$

   Horizontal asymptote: $y = 1$

(d)

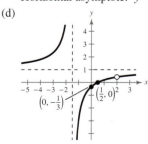

43. (a) Domain: all real numbers x except $x = \pm 1, 2$

(b) x-intercepts: $(3, 0), \left(-\frac{1}{2}, 0\right)$

   y-intercept: $\left(0, -\frac{3}{2}\right)$

(c) Vertical asymptotes: $x = 2, x = \pm 1$

   Horizontal asymptote: $y = 0$

(d)

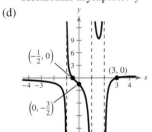

45. (a) Domain of f: all real numbers x except $x = -1$

   Domain of g: all real numbers x

(b)

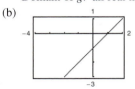

(c) Because there are only finitely many pixels, the graphing utility may not attempt to evaluate the function where it does not exist.

47. (a) Domain: all real numbers x except $x = 0$

(b) x-intercepts: $(\pm 2, 0)$

(c) Vertical asymptote: $x = 0$

   Slant asymptote: $y = x$

(d)

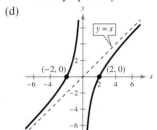

49. (a) Domain: all real numbers x except $x = 0$

(b) No intercepts

(c) Vertical asymptote: $x = 0$

   Slant asymptote: $y = 2x$

(d)

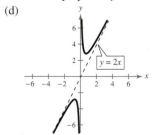

51. (a) Domain: all real numbers x except $x = 0$
(b) No intercepts
(c) Vertical asymptote: $x = 0$
  Slant asymptote: $y = x$
(d)

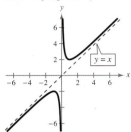

53. (a) Domain: all real numbers t except $t = -5$
(b) y-intercept: $\left(0, -\frac{1}{5}\right)$
(c) Vertical asymptote: $t = -5$
  Slant asymptote: $y = -t + 5$
(d)

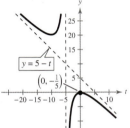

55. (a) Domain: all real numbers x except $x = \pm2$
(b) Intercept: $(0, 0)$
(c) Vertical asymptotes: $x = \pm2$
  Slant asymptote: $y = x$
(d)

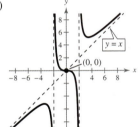

57. (a) Domain: all real numbers x except $x = 0, 1$
(b) No intercepts
(c) Vertical asymptote: $x = 0$
  Slant asymptote: $y = x + 1$
(d)

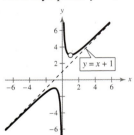

59. (a) Domain: all real numbers x except $x = 1$
(b) y-intercept: $(0, -1)$
(c) Vertical asymptote: $x = 1$
  Slant asymptote: $y = x$
(d)

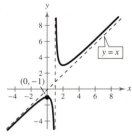

61. (a) Domain: all real numbers x except $x = -1, -2$
(b) y-intercept: $\left(0, \frac{1}{2}\right)$
  x-intercepts: $\left(\frac{1}{2}, 0\right), (1, 0)$
(c) Vertical asymptote: $x = -2$
  Slant asymptote: $y = 2x - 7$
(d)

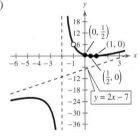

63.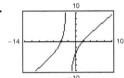

Domain: all real numbers
x except $x = -2$
Vertical asymptote: $x = -2$
Slant asymptote: $y = x$
$y = x$

65.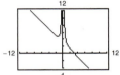

Domain: all real numbers
x except $x = 0$
Vertical asymptote: $x = 0$
Slant asymptote:
$y = -x + 3$
$y = -x + 3$

67. (a) $(-1, 0)$ (b) -1
69. (a) $(0, 0)$ (b) 0
71. (a) $(1, 0), (-1, 0)$ (b) ±1
73. (a) $(-1, 0)$ (b) -1
75. (a)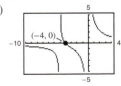

$(-4, 0)$

(b) -4

77. (a)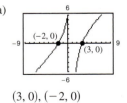

$(3, 0), (-2, 0)$

(b) $3, -2$

79. (a) $y = \dfrac{600}{x}$ (b) $(0, \infty)$

(c)

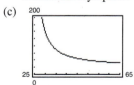

$17\frac{1}{7}$ m

81. (a) Answers will vary.   (b) $(4, \infty)$

(c) 200

11.75 in. $\times$ 5.87 in.

83.

Relative minimum: $(-2, -1)$
Relative maximum: $(0, 3)$

85.

Relative minimum: $(5.657, 9.314)$
Relative maximum:
$(-5.657, -13.314)$

87. 300

$x \approx 40.45$, or 4045 components

89. (a) Answers will vary.

(b) Vertical asymptote: $x = 25$
Horizontal asymptote: $y = 25$

(c) 200

(d)

x	30	35	40	45	50	55	60
y	150	87.5	66.7	56.3	50	45.8	42.9

(e) *Sample answer:* No. You might expect the average speed for the round trip to be the average of the average speeds for the two parts of the trip.

(f) No. At 20 miles per hour you would use more time in one direction than is required for the round trip at an average speed of 50 miles per hour.

91. False. There are two distinct branches of the graph.

93. True. The degree of the numerator is 1 more than the degree of the denominator.

95. $h(x) = \dfrac{6 - 2x}{3 - x} = 2$, $x \ne 3$, so the graph has a hole at $x = 3$.

97. Answers will vary. *Sample answer:* $f(x) = \dfrac{x^2 - x - 6}{x - 2}$

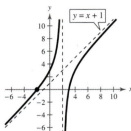

99. No. Given $f(x) = \dfrac{a_n x^n + \cdots + a_0}{b_m x^m + \cdots + b_0}$, when $n > m$, there is no horizontal asymptote, and n must be greater than m for a slant asymptote to occur.

Section 4.3   *(page 337)*

1. conic or conic section   **3.** parabola; directrix; focus

5. ellipse; foci   **7.** hyperbola; foci

9. b   **10.** c   **11.** f   **12.** d   **13.** a   **14.** e

15. Focus: $\left(0, \frac{1}{2}\right)$
Directrix: $y = -\frac{1}{2}$

17. Focus: $\left(-\frac{3}{2}, 0\right)$
Directrix: $x = \frac{3}{2}$

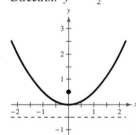

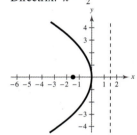

19. Focus: $(0, -3)$
Directrix: $y = 3$

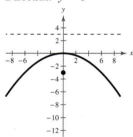

21. $y^2 = 12x$ **23.** $x^2 = -8y$ **25.** $y^2 = -9x$

27. $x^2 = \frac{3}{2}y$; Focus: $\left(0, \frac{3}{8}\right)$ **29.** $y^2 = 6x$

31. (a)

(−640, 152)   (640, 152)

(b) $y = \dfrac{19x^2}{51,200}$

33. $\dfrac{x^2}{1} + \dfrac{y^2}{4} = 1$ **35.** $\dfrac{x^2}{25} + \dfrac{y^2}{21} = 1$ **37.** $\dfrac{x^2}{49} + \dfrac{y^2}{24} = 1$

39. $\dfrac{x^2}{81} + \dfrac{y^2}{9} = 1$ **41.** $\dfrac{x^2}{49} + \dfrac{y^2}{196} = 1$

43. Vertices: $(\pm 5, 0)$
$e = \frac{3}{5}$

45. Vertices: $\left(\pm\frac{5}{3}, 0\right)$
$e = \frac{3}{5}$

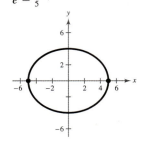

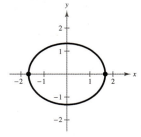

47. Vertices: $(\pm 6, 0)$
$e = \dfrac{\sqrt{29}}{6}$

49. Vertices: $(0, \pm 3)$
$e = \dfrac{2}{3}$

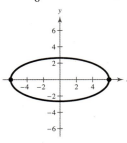

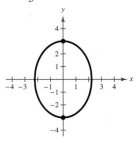

51. Vertices: $(0, \pm 1)$
$e = \dfrac{\sqrt{3}}{2}$

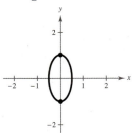

53. $\dfrac{x^2}{25} + \dfrac{y^2}{9} = 1$ **55.** $\left(\pm\sqrt{5}, 0\right)$; Length of string: 6 ft

57. (a)

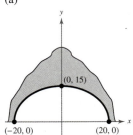

(b) $y = \frac{3}{4}\sqrt{400 - x^2}$
(c) Yes, with clearance of 0.52 foot.

59.

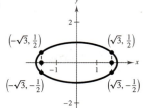

61.

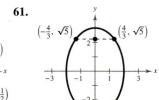

63. $\dfrac{y^2}{4} - \dfrac{x^2}{32} = 1$ **65.** $\dfrac{x^2}{1} - \dfrac{y^2}{9} = 1$

67. $\dfrac{x^2}{64} - \dfrac{y^2}{36} = 1$ **69.** $\dfrac{y^2}{9} - \dfrac{x^2}{9/4} = 1$

71. Vertices: $(\pm 5, 0)$ **73.** Vertices: $(0, \pm 6)$

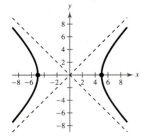

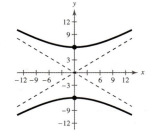

75. Vertices: $(0, \pm 1)$ **77.** Vertices: $(0, \pm 3)$

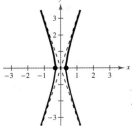

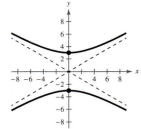

79. Vertices: $\left(\pm\frac{1}{3}, 0\right)$

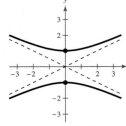

81. (a) $\dfrac{x^2}{1} - \dfrac{y^2}{169/3} = 1$ (b) 2.403 ft   **83.** 10 mi

85. False. The equation represents a hyperbola:
$$\dfrac{x^2}{144} - \dfrac{y^2}{144} = 1.$$

87. False. If the graph crossed the directrix, there would exist points nearer the directrix than the focus.

89. (a) $A = \pi a(20 - a)$ (b) $\dfrac{x^2}{196} + \dfrac{y^2}{36} = 1$

(c)

a	8	9	10	11	12	13
A	301.6	311.0	314.2	311.0	301.6	285.9

$a = 10$, circle

(d)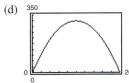

$a = 10$, circle

91. *Sample answer:* Solve the equation for y and graph the functions in the same viewing window; $y_1 = \sqrt{16 - 16x^2/25}$, $y_2 = -\sqrt{16 - 16x^2/25}$

93. No. If it were a hyperbola, the equation would have to be second degree.

95. (a) Left half of ellipse (portion to the left of the y-axis)
(b) Top half of ellipse (portion above the x-axis)

97–99. Answers will vary.

Section 4.4   *(page 347)*

1. b   **2.** d   **3.** e   **4.** c   **5.** a   **6.** f

7. Circle; shifted two units to the left and one unit up

9. Hyperbola; shifted one unit to the right and three units down

11. Parabola; shifted one unit to the left and two units up

13. Ellipse; shifted four units to the left and two units down

15. Center: $(0, 0)$
Radius: 7

17. Center: $(4, 5)$
Radius: 6

19. Center: $(1, 0)$
Radius: $\sqrt{10}$

21. $x^2 + (y - 4)^2 = 16$
Center: $(0, 4)$
Radius: 4

23. $(x - 1)^2 + (y + 3)^2 = 1$
Center: $(1, -3)$
Radius: 1

25. $\left(x + \tfrac{3}{2}\right)^2 + (y - 3)^2 = 1$
Center: $\left(-\tfrac{3}{2}, 3\right)$
Radius: 1

27. Vertex: $(1, -2)$
Focus: $(1, -4)$
Directrix: $y = 0$

29. Vertex: $\left(5, -\tfrac{1}{2}\right)$
Focus: $\left(\tfrac{11}{2}, -\tfrac{1}{2}\right)$
Directrix: $x = \tfrac{9}{2}$

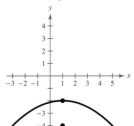
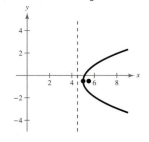

31. Vertex: $(1, 1)$
Focus: $(1, 2)$
Directrix: $y = 0$

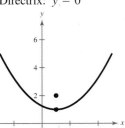

33. Vertex: $\left(-\tfrac{1}{2}, 1\right)$
Focus: $\left(-\tfrac{5}{2}, 1\right)$
Directrix: $x = \tfrac{3}{2}$

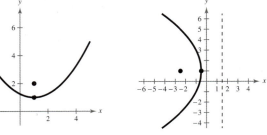

35. $(y - 2)^2 = -8(x - 3)$ **37.** $x^2 = 8(y - 4)$

39. $(y - 4)^2 = 16x$

41. (a) $17,500\sqrt{2}$ mi/h   (b) $x^2 = -16,400(y - 4100)$

43. (a) $x^2 = -49(y - 100)$ (b) 70 ft

45. Center: $(1, 5)$
Foci: $(1, 9), (1, 1)$
Vertices: $(1, 10), (1, 0)$

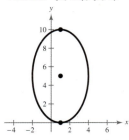

47. Center: $(-2, -4)$
Foci: $\left(-2 \pm \dfrac{\sqrt{3}}{2}, -4\right)$
Vertices: $(-3, -4), (-1, -4)$

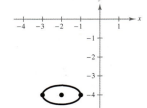

49. Center: $(2, 1)$
Foci: $\left(\tfrac{14}{5}, 1\right), \left(\tfrac{6}{5}, 1\right)$
Vertices: $(1, 1), (3, 1)$

51. Center: $(-1, 0)$
Foci: $\left(-1, \pm\sqrt{21}\right)$
Vertices: $(-1, \pm 5)$

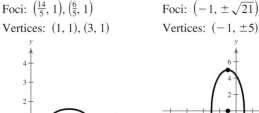

53. $\dfrac{(x - 3)^2}{1} + \dfrac{y^2}{9} = 1$ **55.** $\dfrac{(x - 2)^2}{16} + \dfrac{y^2}{12} = 1$

57. $\dfrac{(x - 1)^2}{12} + \dfrac{(y - 4)^2}{16} = 1$ **59.** $\dfrac{(x - 2)^2}{25} + \dfrac{y^2}{21} = 1$

61. $\dfrac{(x + 3)^2}{8} + \dfrac{(y + 2)^2}{9} = 1$ **63.** $(x - 2)^2 + \dfrac{(y - 1)^2}{4} = 1$

65. $2,756,170,000$ mi; $4,583,830,000$ mi

67. Center: $(2, -1)$
Foci: $(7, -1), (-3, -1)$
Vertices: $(6, -1), (-2, -1)$

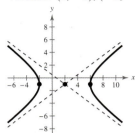

69. Center: $(2, -6)$
Foci: $(2, -6 \pm \sqrt{2})$
Vertices: $(2, -5), (2, -7)$

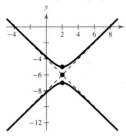

71. Center: $(-1, -3)$
Foci: $\left(-1 \pm \dfrac{5\sqrt{2}}{3}, -3\right)$
Vertices: $(-1 \pm \sqrt{5}, -3)$

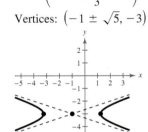

73. Center: $(-3, 2)$
Foci: $(-3, 2 \pm \sqrt{10})$
Vertices: $(-3, 1), (-3, 3)$

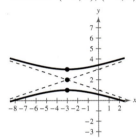

75. $\dfrac{(y-1)^2}{1} - \dfrac{x^2}{3} = 1$

77. $\dfrac{(x+4)^2}{4} - \dfrac{(y-5)^2}{12} = 1$

79. $\dfrac{y^2}{9} - \dfrac{4(x-2)^2}{9} = 1$

81. $\dfrac{(x-3)^2}{9} - \dfrac{(y-2)^2}{4} = 1$

83. $\dfrac{(y+2)^2}{4} - \dfrac{x^2}{4} = 1$

85. $\left(y + \dfrac{1}{4}\right)^2 = -8\left(x - \dfrac{1}{16}\right)$

Hyperbola

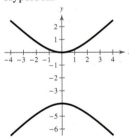

Shifted two units down

Parabola

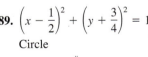
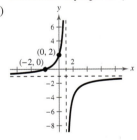

Shifted $\frac{1}{16}$ unit right and $\frac{1}{4}$ unit down

87. $\dfrac{(x+2)^2}{16} + \dfrac{(y+4)^2}{9} = 1$

Ellipse

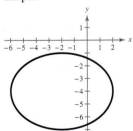

Shifted two units left and four units downs

89. $\left(x - \dfrac{1}{2}\right)^2 + \left(y + \dfrac{3}{4}\right)^2 = 1$

Circle

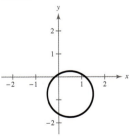

Shifted $\frac{1}{2}$ unit right and $\frac{3}{4}$ unit down

91. True. The conic is an ellipse.
93. False. The vertices and the foci must be collinear.
95. (a) Answers will vary.

(b) 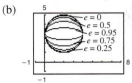   As e approaches 0, the ellipse becomes a circle.

Review Exercises   (page 352)

1. Domain: all real numbers x except $x = -10$
$f(x) \to \infty$ as $x \to -10^-$, $f(x) \to -\infty$ as $x \to -10^+$
3. Domain: all real numbers x except $x = 6, 4$
$f(x) \to \infty$ as $x \to 4^-$ and as $x \to 6^+$,
$f(x) \to -\infty$ as $x \to 4^+$ and as $x \to 6^-$
5. Vertical asymptote: $x = -3$
7. Vertical asymptotes: $x = \pm 2$
Horizontal asymptote: $y = 1$
9. Vertical asymptote: $x = 6$
Horizontal asymptote: $y = 0$
11. $\$0.50$ is the horizontal asymptote of the function.
13. (a) Domain: all real numbers x except $x = 0$
(b) No intercepts
(c) Vertical asymptote: $x = 0$
Horizontal asymptote: $y = 0$
(d)

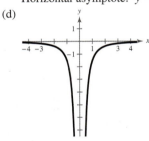

15. (a) Domain: all real numbers x except $x = 1$
(b) x-intercept: $(-2, 0)$
y-intercept: $(0, 2)$
(c) Vertical asymptote: $x = 1$
Horizontal asymptote: $y = -1$
(d)

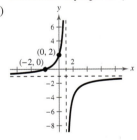

CHAPTER 4

17. (a) Domain: all real numbers x
(b) Intercept: $(0, 0)$
(c) Horizontal asymptote: $y = \frac{5}{4}$
(d)

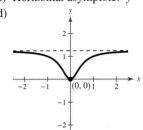

19. (a) Domain: all real numbers x except $x = \pm 4$
(b) Intercept: $(0, 0)$
(c) Vertical asymptotes: $x = \pm 4$
    Horizontal asymptote: $y = 0$
(d)

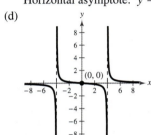

21. (a) Domain: all real numbers x
(b) Intercept: $(0, 0)$
(c) Horizontal asymptote: $y = -6$
(d)

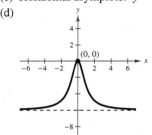

23. (a) Domain: all real numbers x except $x = 0, \frac{1}{3}$
(b) x-intercept: $\left(\frac{3}{2}, 0\right)$
(c) Vertical asymptote: $x = 0$
    Horizontal asymptote: $y = 2$
(d)

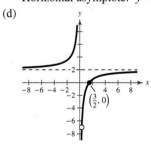

25. (a) Domain: all real numbers x
(b) Intercept: $(0, 0)$
(c) Slant asymptote: $y = 2x$
(d)

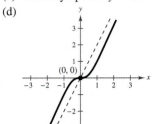

27. (a) Domain: all real numbers x except $x = -2$
(b) x-intercepts: $(2, 0), (-5, 0)$
    y-intercept: $(0, -5)$
(c) Vertical asymptote: $x = -2$
    Slant asymptote: $y = x + 1$
(d)

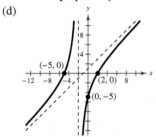

29. (a) Domain: all real numbers x except $x = \frac{4}{3}, -1$
(b) x-intercepts: $\left(\frac{2}{3}, 0\right), (1, 0)$
    y-intercept: $\left(0, -\frac{1}{2}\right)$
(c) Vertical asymptote: $x = \frac{4}{3}$
    Slant asymptote: $y = x - \frac{1}{3}$
(d)

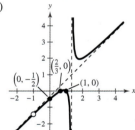

31. (a)

(b) $100.90, $10.90, $1.90
(c) $0.90 is the horizontal asymptote of the function.

33. (a)

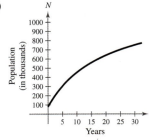

(b) $N(5) = 304,000$ fish;
$N(10) \approx 453,333$ fish;
$N(25) \approx 702,222$ fish

(c) $1,200,000$ fish; The graph has $N = 1200$ as a horizontal asymptote.

35. Parabola   **37.** Hyperbola   **39.** Ellipse   **41.** Circle

43. $y^2 = -24x$ **45.** $x^2 = 12y$ **47.** $y^2 = 12x$

49. $(0, 50)$ **51.** $\dfrac{x^2}{25} + \dfrac{y^2}{36} = 1$ **53.** $\dfrac{x^2}{13} + \dfrac{y^2}{49} = 1$

55. $\dfrac{x^2}{221} + \dfrac{y^2}{25} = 1$

57. The foci are 3 feet from the center and have the same height as the pillars.

59. $\dfrac{y^2}{1} - \dfrac{x^2}{24} = 1$ **61.** $\dfrac{x^2}{1} - \dfrac{y^2}{4} = 1$

63. $(x - 4)^2 = -8(y - 2)$ **65.** $(y + 8)^2 = 28(x - 8)$

67. $\dfrac{(x - 2)^2}{4} + \dfrac{(y - 2)^2}{1} = 1$ **69.** $\dfrac{(x - 2)^2}{16} + \dfrac{(y - 3)^2}{25} = 1$

71. $\dfrac{x^2}{5} + \dfrac{(y + 2)^2}{9} = 1$ **73.** $\dfrac{(x + 2)^2}{64} - \dfrac{(y - 3)^2}{36} = 1$

75. $\dfrac{y^2}{16} - \dfrac{(x - 3)^2}{4/5} = 1$

77. $(x - 3)^2 = -2y$ **79.** $(x - 1)^2 + (y - 2)^2 = 4$
    Parabola   Circle

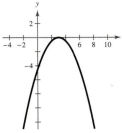

Shifted three units right

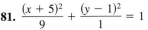

Shifted one unit right and two units up

81. $\dfrac{(x + 5)^2}{9} + \dfrac{(y - 1)^2}{1} = 1$ **83.** $\dfrac{(x - 4)^2}{1} - \dfrac{(y - 4)^2}{9} = 1$
    Ellipse   Hyperbola

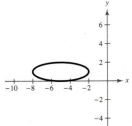

Shifted five units left and one unit up

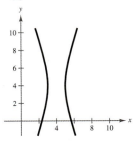

Shifted four units right and four units up

85. $8\sqrt{6}$ m

87. (a) Circle
    (b) $(x - 100)^2 + y^2 = 62,500$ (c) Approximately 180.28 m

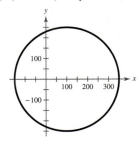

89. True. It could be a degenerate conic.

Chapter Test   *(page 355)*

1. Domain: all real numbers x except $x = -1$
    Vertical asymptote: $x = -1$
    Horizontal asymptote: $y = 3$

2. Domain: all real numbers x
    Vertical asymptote: none
    Horizontal asymptote: $y = -1$

3. Domain: all real numbers x except $x = 4, 5$
    Vertical asymptote: $x = 5$
    Horizontal asymptote: $y = 0$

4. x-intercepts: $\left(\pm\sqrt{3}, 0\right)$
    Vertical asymptote: $x = 0$
    Horizontal asymptote: $y = -1$

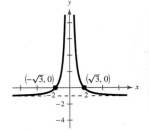

5. y-intercept: $(0, -2)$
    Vertical asymptote: $x = 1$
    Slant asymptote: $y = x + 1$

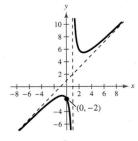

6. x-intercept: $(-1, 0)$
    y-intercept: $\left(0, -\dfrac{1}{12}\right)$
    Vertical asymptotes:
    $x = 3, x = -4$
    Horizontal asymptote: $y = 0$

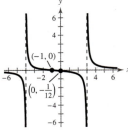

CHAPTER 4

7. x-intercept: $\left(-\frac{3}{2}, 0\right)$

y-intercept: $\left(0, \frac{3}{4}\right)$

Vertical asymptote: $x = -4$

Horizontal asymptote: $y = 2$

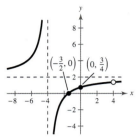

8. y-intercept: $(0, 1)$

Horizontal asymptote: $y = \frac{2}{5}$

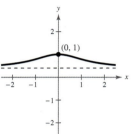

9. x-intercepts: $(-2, 0), \left(-\frac{3}{2}, 0\right)$

y-intercept: $(0, 6)$

Vertical asymptote: $x = -1$

Slant asymptote: $y = 2x + 5$

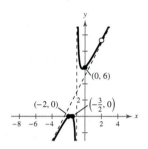

10. 6.24 in. $\times$ 12.49 in.

11. (a) $A = \dfrac{x^2}{2(x-2)}, x > 2$

(b)

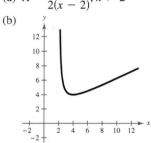

$A = 4$

12.

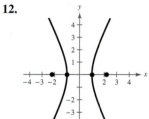

Center: $(0, 0)$

Vertices: $(\pm 1, 0)$

Foci: $\left(\pm\sqrt{5}, 0\right)$

13.

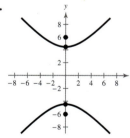

Center: $(0, 0)$

Vertices: $\left(0, \pm 2\sqrt{5}\right)$

Foci: $(0, \pm 6)$

14.

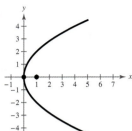

Vertex: $(0, 0)$

Focus: $(1, 0)$

15.

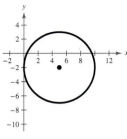

Center: $(5, -2)$

16.

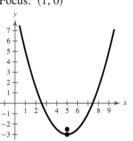

Vertex: $(5, -3)$

Focus: $\left(5, -\frac{5}{2}\right)$

17.

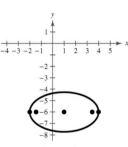

Center: $(1, -6)$

Vertices: $(4, -6), (-2, -6)$

Foci: $\left(1 \pm \sqrt{6}, -6\right)$

18. $\dfrac{(x-4)^2}{16} + \dfrac{(y-2)^2}{4} = 1; e = \dfrac{\sqrt{3}}{2}$

19. $\dfrac{y^2}{9} - \dfrac{x^2}{4} = 1$ **20.** 34 m

21. Least distance: About 363,301 km

Greatest distance: About 405,499 km

Problem Solving *(page 357)*

1. (a) iii   (b) ii   (c) iv   (d) i

3. (a) $y_1 \approx 0.031x^2 - 1.59x + 21.0$

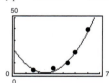

(b) $y_2 \approx \dfrac{1}{-0.007x + 0.44}$

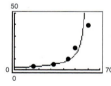

(c) The models are a good fit for the original data.

(d) $y_1(25) = 0.625; y_2(25) = 3.774$

The rational model is the better fit for the original data.

(e) The reciprocal model should not be used to predict the near point for a person who is 70 years old because a negative value is obtained. The quadratic model is a better fit.

5. Answers will vary.

7. (a) Answers will vary.

(b) Island 1: $(-6, 0)$; Island 2: $(6, 0)$

(c) 20 mi; Vertex: $(10, 0)$ (d) $\dfrac{x^2}{100} + \dfrac{y^2}{64} = 1$

9. (a)

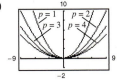

As p increases, the graph becomes wider.

(b) $p = 1$: focus $(0, 1)$
   $p = 2$: focus $(0, 2)$
   $p = 3$: focus $(0, 3)$
   $p = 4$: focus $(0, 4)$

(c) $p = 1$: 4 units
   $p = 2$: 8 units
   $p = 3$: 12 units
   $p = 4$: 16 units
   Length of chord $= 4|p|$ units

(d) Answers will vary.

11. Proof

Chapter 5

Section 5.1 *(page 368)*

1. algebraic   **3.** One-to-One   **5.** $A = P\left(1 + \dfrac{r}{n}\right)^{nt}$

7. 0.863   **9.** 1.552   **11.** 1767.767

13. d   **14.** c   **15.** a   **16.** b

17.

x	-2	-1	0	1	2
$f(x)$	0.020	0.143	1	7	49

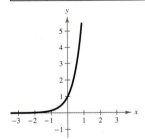

19.

x	-2	-1	0	1	2
$f(x)$	0.063	0.25	1	4	16

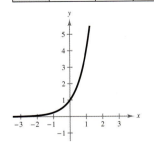

21.

x	-2	-1	0	1	2
$f(x)$	0.016	0.063	0.25	1	4

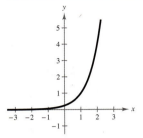

23.

x	-3	-2	-1	0	1
$f(x)$	3.25	3.5	4	5	7

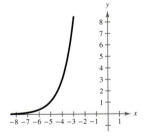

25. 2   **27.** -5 **29.** Shift the graph of f one unit up.

31. Reflect the graph of f in the y-axis and shift three units to the right.

33. 6.686   **35.** 7166.647

37.

x	-8	-7	-6	-5	-4
$f(x)$	0.055	0.149	0.406	1.104	3

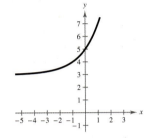

39.

x	-2	-1	0	1	2
$f(x)$	4.037	4.100	4.271	4.736	6

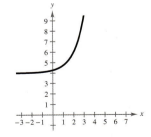

41.

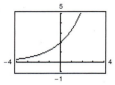

43.

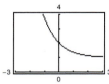

45. $\frac{1}{3}$ **47.** $3, -1$

49.

n	1	2	4	12
A	\$1828.49	\$1830.29	\$1831.19	\$1831.80

n	365	Continuous
A	\$1832.09	\$1832.10

51.

n	1	2	4	12
A	\$5477.81	\$5520.10	\$5541.79	\$5556.46

n	365	Continuous
A	\$5563.61	\$5563.85

53.

t	10	20	30
A	\$17,901.90	\$26,706.49	\$39,841.40

t	40	50
A	\$59,436.39	\$88,668.67

55.

t	10	20	30
A	\$22,986.49	\$44,031.56	\$84,344.25

t	40	50
A	\$161,564.86	\$309,484.08

57. \$104,710.29
59. \$44.23

61. (a)

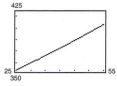

(b)

t	25	26	27	28
P (in millions)	350.281	352.107	353.943	355.788

t	29	30	31	32
P (in millions)	357.643	359.508	361.382	363.266

t	33	34	35	36
P (in millions)	365.160	367.064	368.977	370.901

t	37	38	39	40
P (in millions)	372.835	374.779	376.732	378.697

t	41	42	43	44
P (in millions)	380.671	382.656	384.651	386.656

t	45	46	47	48
P (in millions)	388.672	390.698	392.735	394.783

t	49	50	51	52
P (in millions)	396.841	398.910	400.989	403.080

t	53	54	55
P (in millions)	405.182	407.294	409.417

(c) 2064
63. (a) 16 g   (b) 1.85 g
(c)

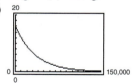

65. (a) $V(t) = 49{,}810\left(\frac{7}{8}\right)^t$ (b) \$29,197.71
67. True. As $x \to -\infty$, $f(x) \to -2$ but never reaches -2.
69. $f(x) = h(x)$
71. $f(x) = g(x) = h(x)$
73.

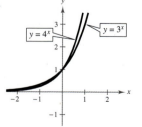

(a) $x < 0$ (b) $x > 0$

75.

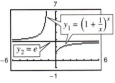

As the x-value increases, y_1 approaches the value of e.

77. (a)

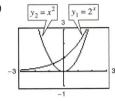

(b)

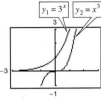

In both viewing windows, the constant raised to a variable power increases more rapidly than the variable raised to a constant power.

79. c, d

Section 5.2 *(page 378)*

1. logarithmic　　**3.** natural; e　　**5.** $x = y$　　**7.** $4^2 = 16$

9. $12^1 = 12$　　**11.** $\log_5 125 = 3$　　**13.** $\log_4 \frac{1}{64} = -3$

15. 6　　**17.** 0　　**19.** -2　　**21.** -0.058　　**23.** 1.097

25. 1　　**27.** 0　　**29.** 5　　**31.** ± 2

33.

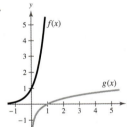

35.

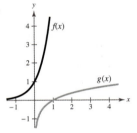

37. a; Upward shift of two units
38. d; Right shift of one unit
39. b; Reflection in the y-axis and a right shift of one unit
40. c; Reflection in the x-axis

41.

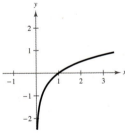

Domain: $(0, \infty)$
x-intercept: $(1, 0)$
Vertical asymptote: $x = 0$

43.

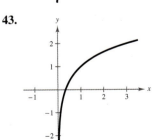

Domain: $(0, \infty)$
x-intercept: $\left(\frac{1}{3}, 0\right)$
Vertical asymptote: $x = 0$

45.

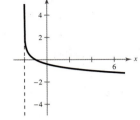

Domain: $(-2, \infty)$
x-intercept: $(-1, 0)$
Vertical asymptote: $x = -2$

47.

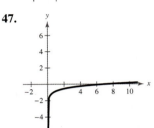

Domain: $(0, \infty)$
x-intercept: $(7, 0)$
Vertical asymptote: $x = 0$

49. $e^{-0.693\cdots} = \frac{1}{2}$　　**51.** $e^{5.521\cdots} = 250$

53. $\ln 7.3890\ldots = 2$　　**55.** $\ln \frac{1}{2} = -4x$　　**57.** 2.913

59. 6.438　　**61.** 4　　**63.** 0　　**65.** 1

67.

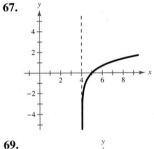

Domain: $(4, \infty)$
x-intercept: $(5, 0)$
Vertical asymptote: $x = 4$

69.

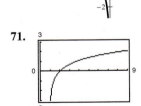

Domain: $(-\infty, 0)$
x-intercept: $(-1, 0)$
Vertical asymptote: $x = 0$

71.

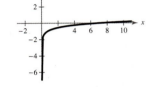

73.

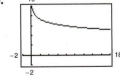

75. 8　　**77.** $-2, 3$

79. (a) 30 yr; 10 yr
(b) \$323,179; \$199,109; \$173,179; \$49,109
(c) $x = 750$; The monthly payment must be greater than \$750.

CHAPTER 5

81. (a)

r	0.005	0.010	0.015	0.020	0.025	0.030
t	138.6	69.3	46.2	34.7	27.7	23.1

As the rate of increase r increases, the time t in years for the population to double decreases.

(b)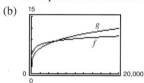

83. (a) 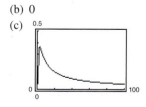   (b) 80   (c) 68.1   (d) 62.3

85. False. Reflecting $g(x)$ in the line $y = x$ will determine the graph of $f(x)$.

87. (a)

$g(x)$; The natural log function grows at a slower rate than the square root function.

(b)

$g(x)$; The natural log function grows at a slower rate than the fourth root function.

89. $y = \log_2 x$, so y is a logarithmic function of x.

91. (a)

x	1	5	10	10^2
f(x)	0	0.322	0.230	0.046

x	10^4	10^6
f(x)	0.00092	0.0000138

(b) 0

(c)

Section 5.3 *(page 385)*

1. change-of-base   **3.** $\dfrac{1}{\log_b a}$

5. (a) $\dfrac{\log 16}{\log 5}$ (b) $\dfrac{\ln 16}{\ln 5}$ **7.** (a) $\dfrac{\log \frac{3}{10}}{\log x}$ (b) $\dfrac{\ln \frac{3}{10}}{\ln x}$

9. 2.579   **11.** -0.606 **13.** $\log_3 5 + \log_3 7$

15. $\log_3 7 - 2 \log_3 5$ **17.** $1 + \log_3 7 - \log_3 5$

19. 2   **21.** $-\frac{1}{3}$ **23.** -2 is not in the domain of $\log_2 x$.
25. $\frac{3}{4}$ **27.** 7   **29.** 2   **31.** $\frac{3}{2}$ **33.** 1.1833
35. -1.6542 **37.** 1.9563   **39.** -2.7124
41. $\ln 7 + \ln x$ **43.** $4 \log_8 x$ **45.** $1 - \log_5 x$
47. $\frac{1}{2} \ln z$ **49.** $\ln x + \ln y + 2 \ln z$
51. $\ln z + 2 \ln(z - 1)$
53. $\frac{1}{2} \log_2 (a + 2) + \frac{1}{2} \log_2 (a - 2) - \log_2 7$
55. $2 \log_5 x - 2 \log_5 y - 3 \log_5 z$
57. $\frac{1}{3} \ln y + \frac{1}{3} \ln z - \frac{2}{3} \ln x$ **59.** $\frac{3}{4} \ln x + \frac{1}{4} \ln(x^2 + 3)$

61. $\ln 3x$ **63.** $\log_7 (z - 2)^{2/3}$ **65.** $\log_3 \dfrac{5}{x^3}$

67. $\log x(x + 1)^2$ **69.** $\log \dfrac{xz^3}{y^2}$ **71.** $\ln \dfrac{x}{(x + 1)(x - 1)}$

73. $\ln \sqrt{\dfrac{x(x + 3)^2}{x^2 - 1}}$ **75.** $\log_8 \dfrac{\sqrt[3]{y(y + 4)^2}}{y - 1}$

77. $\log_2 \frac{32}{4} = \log_2 32 - \log_2 4$; Property 2
79. $\beta = 10(\log I + 12)$; 60 dB   **81.** 70 dB
83. $\ln y = \frac{1}{4} \ln x$ **85.** $\ln y = -\frac{1}{4} \ln x + \ln \frac{5}{2}$
87. $\ln y = -0.14 \ln x + 5.7$
89. (a) and (b)   (c)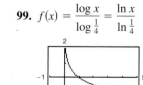

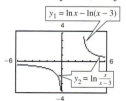

$T = 21 + e^{-0.037t + 3.997}$
The results are similar.

(d)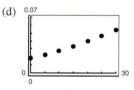

$$T = 21 + \dfrac{1}{0.001t + 0.016}$$

91. False; $\ln 1 = 0$ **93.** False; $\ln(x - 2) \neq \ln x - \ln 2$
95. False; $u = v^2$

97. $f(x) = \dfrac{\log x}{\log 2} = \dfrac{\ln x}{\ln 2}$ **99.** $f(x) = \dfrac{\log x}{\log \frac{1}{4}} = \dfrac{\ln x}{\ln \frac{1}{4}}$

101. The Power Property cannot be used because $\ln e$ is raised to the second power, not just e.

103.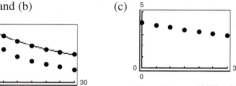

No; $\dfrac{x}{x - 3} > 0$ when $x < 0$.

105. $\ln 1 = 0$ $\ln 9 \approx 2.1972$
$\ln 2 \approx 0.6931$ $\ln 10 \approx 2.3025$
$\ln 3 \approx 1.0986$ $\ln 12 \approx 2.4848$
$\ln 4 \approx 1.3862$ $\ln 15 \approx 2.7080$
$\ln 5 \approx 1.6094$ $\ln 16 \approx 2.7724$
$\ln 6 \approx 1.7917$ $\ln 18 \approx 2.8903$
$\ln 8 \approx 2.0793$ $\ln 20 \approx 2.9956$

Section 5.4   (page 395)

1. (a) $x = y$ (b) $x = y$ (c) x (d) x
3. (a) Yes   (b) No   (c) Yes
5. (a) Yes   (b) No   (c) No   **7.** 2
9. 2   **11.** $\ln 2 \approx 0.693$ **13.** $e^{-1} \approx 0.368$
15. 64   **17.** $(3, 8)$ **19.** $2, -1$
21. $\dfrac{\ln 5}{\ln 3} \approx 1.465$ **23.** $\ln 39 \approx 3.664$
25. $\dfrac{\ln 80}{2 \ln 3} \approx 1.994$ **27.** $2 - \dfrac{\ln 400}{\ln 3} \approx -3.454$
29. $\dfrac{1}{3} \log \dfrac{3}{2} \approx 0.059$ **31.** $\dfrac{\ln 12}{3} \approx 0.828$
33. 0   **35.** $\dfrac{\ln \frac{8}{3}}{3 \ln 2} + \dfrac{1}{3} \approx 0.805$
37. $-\dfrac{\ln 2}{\ln 3 - \ln 2} \approx -1.710$ **39.** $0, \dfrac{\ln 4}{\ln 5} \approx 0.861$
41. $\ln 5 \approx 1.609$ **43.** $\ln \frac{4}{5} \approx -0.223$
45. $\dfrac{\ln 4}{365 \ln\left(1 + \dfrac{0.065}{365}\right)} \approx 21.330$ **47.** $e^{-3} \approx 0.050$
49. $\dfrac{e^{2.1}}{6} \approx 1.361$ **51.** $e^{-2} \approx 0.135$
53. $2(3^{11/6}) \approx 14.988$ **55.** No solution   **57.** No solution
59. No solution   **61.** 2   **63.** 3.328   **65.** -0.478
67. 20.086   **69.** 1.482   **71.** (a) 27.73 yr   (b) 43.94 yr
73. $-1, 0$ **75.** 1   **77.** $e^{-1} \approx 0.368$
79. $e^{-1/2} \approx 0.607$
81. (a) $y = 100$ and $y = 0$; The range falls between 0% and 100%.
    (b) Males: 69.51 in.   Females: 64.49 in.
83. 5 years   **85.** 2011   **87.** About 3.039 min
89. $\log_b uv = \log_b u + \log_b v$
    True by Property 1 in Section 5.3.
91. $\log_b(u - v) = \log_b u - \log_b v$
    False.
    $1.95 \approx \log(100 - 10) \neq \log 100 - \log 10 = 1$
93. Yes. See Exercise 57.
95. For $rt < \ln 2$ years, double the amount you invest. For $rt > \ln 2$ years, double your interest rate or double the number of years, because either of these will double the exponent in the exponential function.
97. (a) 7%   (b) 7.25%   (c) 7.19%   (d) 7.45%
    The investment plan with the greatest effective yield and the highest balance after 5 years is plan (d).

Section 5.5   (page 405)

1. $y = ae^{bx}$; $y = ae^{-bx}$ **3.** normally distributed
5. (a) $P = \dfrac{A}{e^{rt}}$ (b) $t = \dfrac{\ln\left(\dfrac{A}{P}\right)}{r}$
7. 19.8 yr; \$1419.07   **9.** 8.9438%; \$1834.37
11. \$6376.28; 15.4 yr   **13.** \$303,580.52
15. (a) 7.27 yr   (b) 6.96 yr   (c) 6.93 yr   (d) 6.93 yr
17. (a)

r	2%	4%	6%	8%	10%	12%
t	54.93	27.47	18.31	13.73	10.99	9.16

    (b)

r	2%	4%	6%	8%	10%	12%
t	55.48	28.01	18.85	14.27	11.53	9.69

19.

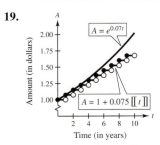

    Continuous compounding
21. 6.48 g   **23.** 2.26 g   **25.** $y = e^{0.7675x}$
27. $y = 5e^{-0.4024x}$
29. (a)

Year	Population
1980	104,752
1990	143,251
2000	195,899
2010	267,896

    (b) 2019
    (c) *Sample answer:* No; As t increases, the population increases rapidly.
31. $k = 0.2988$; About 5,309,734 hits
33. About 800 bacteria
35. (a) $V = -150t + 575$ (b) $V = 575e^{-0.3688t}$
    (c)

    The exponential model depreciates faster.
    (d) Linear model: \$425; \$125
        Exponential model: \$397.65; \$190.18
    (e) Answers will vary.
37. About 12,180 yr old

39. (a)
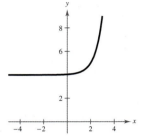

(b) 100

41. (a) 1998: 63,922 sites
2003: 149,805 sites
2006: 208,705 sites

(b)
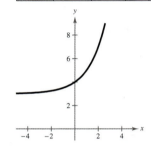

(c) and (d) 2010

43. (a) 203 animals   (b) 13 mo

(c)
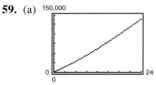

Horizontal asymptotes: $p = 0$, $p = 1000$. The population
size will approach 1000 as time increases.

45. (a) $10^{7.6} \approx 39,810,717$

(b) $10^{5.6} \approx 398,107$ (c) $10^{6.6} \approx 3,981,072$

47. (a) 20 dB   (b) 70 dB   (c) 40 dB   (d) 90 dB

49. 95%   **51.** 4.64   **53.** 1.58×10^{-6} moles/L

55. $10^{5.1}$ **57.** 3:00 A.M.

59. (a)
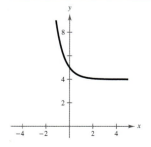
(b) $t \approx 21$ yr; Yes

61. False. The domain can be the set of real numbers for a logistic
growth function.

63. False. The graph of $f(x)$ is the graph of $g(x)$ shifted five
units up.

65. Answers will vary.

Review Exercises   *(page 412)*

1. 0.164   **3.** 1.587   **5.** 1456.529

7.

x	-1	0	1	2	3
$f(x)$	8	5	4.25	4.063	4.016

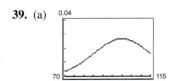

9.

x	-1	0	1	2	3
$f(x)$	4.008	4.04	4.2	5	9

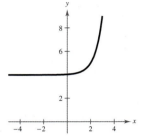

11.

x	-2	-1	0	1	2
$f(x)$	3.25	3.5	4	5	7

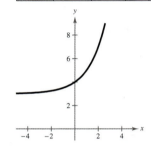

13. 1   **15.** 4   **17.** Shift the graph of f one unit up.

19. Reflect f in the x-axis and shift one unit up.

21. 29.964   **23.** 1.822

25.

x	-2	-1	0	1	2
$h(x)$	2.72	1.65	1	0.61	0.37

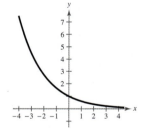

27.

x	-3	-2	-1	0	1
$f(x)$	0.37	1	2.72	7.39	20.09

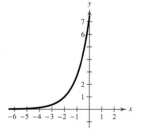

29. (a) 0.283   (b) 0.487   (c) 0.811

31.

n	1	2	4	12
A	\$6719.58	\$6734.28	\$6741.74	\$6746.77

n	365	Continuous
A	\$6749.21	\$6749.29

33. $\log_3 27 = 3$ **35.** $\ln 2.2255 \ldots = 0.8$

37. 3   **39.** -2 **41.** 7   **43.** -5

45. Domain: $(0, \infty)$ **47.** Domain: $(-5, \infty)$
x-intercept: $(1, 0)$ x-intercept: $(9995, 0)$
Vertical asymptote: $x = 0$ Vertical asymptote: $x = -5$

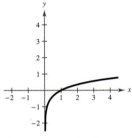

49. 3.118   **51.** 0.25

53. Domain: $(0, \infty)$ **55.** Domain: $(6, \infty)$
x-intercept: $(e^{-6}, 0)$ x-intercept: $(7, 0)$
Vertical asymptote: $x = 0$ Vertical asymptote: $x = 6$

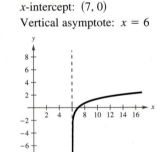

57. About 14.32 parsecs   **59.** (a) and (b) 2.585

61. (a) and (b) -2.322 **63.** $\log_2 5 - \log_2 3$

65. $2 \log_2 3 - \log_2 5$ **67.** $\log 7 + 2 \log x$

69. $2 - \frac{1}{2} \log_3 x$ **71.** $2 \ln x + 2 \ln y + \ln z$

73. $\ln 7x$ **75.** $\log \dfrac{x}{\sqrt{y}}$ **77.** $\log_3 \dfrac{\sqrt{x}}{(y + 8)^2}$

79. (a) $0 \le h < 18{,}000$
(b)

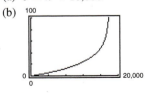

Vertical asymptote: $h = 18{,}000$
(c) The plane is climbing at a slower rate, so the time required increases.
(d) 5.46 min

81. 3   **83.** $\ln 3 \approx 1.099$ **85.** $e^4 \approx 54.598$ **87.** 1, 3

89. $\dfrac{\ln 32}{\ln 2} = 5$ **91.** $\dfrac{1}{3} e^{8.2} \approx 1213.650$

93. $\dfrac{3}{2} + \dfrac{\sqrt{9 + 4e}}{2} \approx 3.729$ **95.** No solution

97. 0.900   **99.** 2.447   **101.** 1.482   **103.** 73.2 yr

105. e   **106.** b   **107.** f   **108.** d   **109.** a   **110.** c

111. $y = 2e^{0.1014x}$

113.

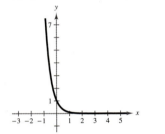

71

115. (a) 10^{-6} W/m²   (b) $10\sqrt{10}$ W/m²
(c) 1.259×10^{-12} W/m²

117. True by the inverse properties.

Chapter Test   *(page 415)*

1. 0.410   **2.** 0.032   **3.** 0.497   **4.** 22.198

5.

x	-1	$-\frac{1}{2}$	0	$\frac{1}{2}$	1
$f(x)$	10	3.162	1	0.316	0.1

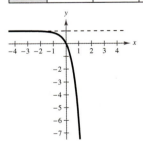

6.

x	-1	0	1	2	3
$f(x)$	-0.005	-0.028	-0.167	-1	-6

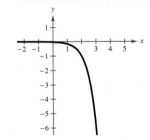

7.

x	-1	$-\frac{1}{2}$	0	$\frac{1}{2}$	1
$f(x)$	0.865	0.632	0	-1.718	-6.389

8. (a) -0.89 (b) 9.2

CHAPTER 5

9. Domain: $(0, \infty)$
x-intercept: $(10^{-4}, 0)$
Vertical asymptote: $x = 0$

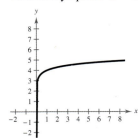

10. Domain: $(4, \infty)$
x-intercept: $(5, 0)$
Vertical asymptote: $x = 4$

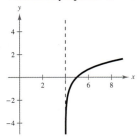

11. Domain: $(-6, \infty)$
x-intercept: $(e^{-1} - 6, 0)$
Vertical asymptote: $x = -6$

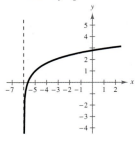

12. 2.209   **13.** -0.167 **14.** -11.047

15. $\log_2 3 + 4 \log_2 a$ **16.** $\frac{1}{2} \ln x - \ln 7$

17. $1 + 2 \log x - 3 \log y$ **18.** $\log_3 13y$ **19.** $\ln \dfrac{x^4}{y^4}$

20. $\ln \dfrac{x^3 y^2}{x + 3}$ **21.** -2 **22.** $\dfrac{\ln 44}{-5} \approx -0.757$

23. $\dfrac{\ln 197}{4} \approx 1.321$ **24.** $e^{1/2} \approx 1.649$

25. $e^{-11/4} \approx 0.0639$ **26.** 20

27. $y = 2745 e^{0.1570t}$ **28.** 55%

29. (a)

x	$\frac{1}{4}$	1	2	4	5	6
H	58.720	75.332	86.828	103.43	110.59	117.38

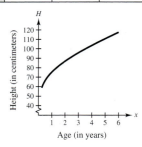

(b) 103 cm; 103.43 cm

Cumulative Test for Chapters 3–5   (page 416)

1. $y = -\frac{3}{4}(x + 8)^2 + 5$

2.

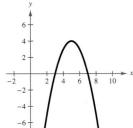

3.

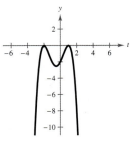

4.

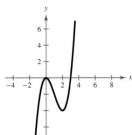

5. $-2, \pm 2i$ **6.** $-7, 0, 3$ **7.** $3x - 2 - \dfrac{3x - 2}{2x^2 + 1}$

8. $3x^3 + 6x^2 + 14x + 23 + \dfrac{49}{x - 2}$ **9.** 1.196

10. $f(x) = x^4 + 3x^3 - 11x^2 + 9x + 70$

11. Domain: all real numbers
x except $x = 3$
Vertical asymptote: $x = 3$
Horizontal asymptote:
$y = 2$

12. Domain: all real numbers
x except $x = 5$
Vertical asymptote: $x = 5$
Slant asymptote:
$y = 4x + 20$

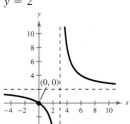

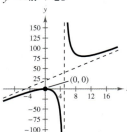

13. Intercept: $(0, 0)$
Vertical asymptotes:
$x = 1, x = -3$
Horizontal asymptote:
$y = 0$

14. y-intercept: $(0, 2)$
x-intercept: $(2, 0)$
Vertical asymptote: $x = 1$
Horizontal asymptote:
$y = 1$

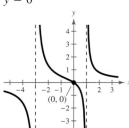

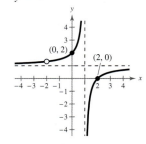

15. *y*-intercept: $(0, 6)$
x-intercepts: $(2, 0), (3, 0)$
Vertical asymptote: $x = -1$
Slant asymptote: $y = x - 6$

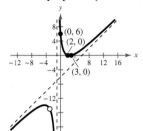

16. 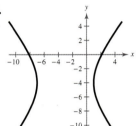 **17.**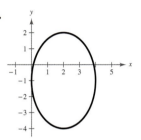

18. $(x - 3)^2 = \frac{3}{2}(y + 2)$ **19.** $\frac{(y + 2)^2}{9} - \frac{(x + 1)^2}{16} = 1$

20. Reflect *f* in the *x*-axis and *y*-axis, and shift three units to the right.
21. Reflect *f* in the *x*-axis, and shift four units up.
22. 1.991   **23.** -0.067 **24.** 1.717   **25.** 0.390
26. 0.906   **27.** -1.733 **28.** -4.087
29. $\ln(x + 5) + \ln(x - 5) - 4 \ln x$
30. $\ln \dfrac{x^2}{\sqrt{x + 5}}, x > 0$ **31.** $\dfrac{\ln 12}{2} \approx 1.242$
32. $\dfrac{\ln 9}{\ln 4} + 5 \approx 6.585$ **33.** $\ln 6 \approx 1.792$ or $\ln 7 \approx 1.946$
34. $\frac{64}{5} = 12.8$ **35.** $\frac{1}{2}e^8 \approx 1490.479$
36. $e^6 - 2 \approx 401.429$ **37.** \$16,302.05
38. 6.3 h   **39.** 2023
40. (a) 300   (b) 570   (c) About 9 yr

Problem Solving   *(page 419)*

1.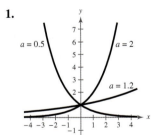

$y = 0.5^x$ and $y = 1.2^x$
$0 < a \le e^{1/e}$

3. As $x \to \infty$, the graph of e^x increases at a greater rate than the graph of x^n.
5. Answers will vary.

7. (a) 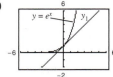   (b)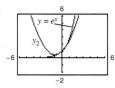
(c)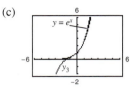

9.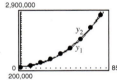

$f^{-1}(x) = \ln\left(\dfrac{x + \sqrt{x^2 + 4}}{2}\right)$

11. c   **13.** $t = \dfrac{\ln c_1 - \ln c_2}{\left(\dfrac{1}{k_2} - \dfrac{1}{k_1}\right)\ln \dfrac{1}{2}}$

15. (a) $y_1 = 252{,}606(1.0310)^t$
(b) $y_2 = 400.88t^2 - 1464.6t + 291{,}782$
(c)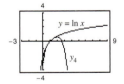
(d) The exponential model is a better fit. No, because the model is rapidly approaching infinity.
17. $1, e^2$
19. $y_4 = (x - 1) - \frac{1}{2}(x - 1)^2 + \frac{1}{3}(x - 1)^3 - \frac{1}{4}(x - 1)^4$

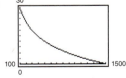

The pattern implies that
$\ln x = (x - 1) - \frac{1}{2}(x - 1)^2 + \frac{1}{3}(x - 1)^3 - \cdots$.
21.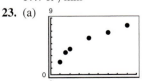

$17.7 \text{ ft}^3/\text{min}$
23. (a) 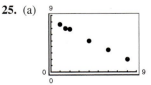   **25.** (a)

(b)–(e) Answers will vary.   (b)–(e) Answers will vary.

Chapter 6

Section 6.1   *(page 429)*

1. solution   **3.** points; intersection
5. (a) No   (b) No   (c) No   (d) Yes
7. $(2, 2)$ **9.** $(2, 6), (-1, 3)$ **11.** $(0, 0), (2, -4)$
13. $(0, 1), (1, -1), (3, 1)$ **15.** $(6, 4)$ **17.** $\left(\frac{1}{2}, 3\right)$
19. $(1, 1)$ **21.** $\left(\frac{20}{3}, \frac{40}{3}\right)$ **23.** No solution
25. \$5500 at 2%; \$6500 at 6%
27. \$6000 at 2.8%; \$6000 at 3.8%   **29.** $(-2, 4), (0, 0)$
31. No solution   **33.** $(6, 2)$ **35.** $\left(-\frac{3}{2}, \frac{1}{2}\right)$
37. $(2, 2), (4, 0)$ **39.** No solution   **41.** $(4, 3), (-4, 3)$
43. $(0, 1)$ **45.** $(5.31, -0.54)$ **47.** $(1, 2)$
49. No solution   **51.** $(0.287, 1.751)$ **53.** $\left(\frac{1}{2}, 2\right), \left(-4, -\frac{1}{4}\right)$
55. 293 units   **57.** (a) 344 units   (b) 2495 units
59. (a) 8 weeks

(b)

	1	2	3	4
$360 - 24x$	336	312	288	264
$24 + 18x$	42	60	78	96

	5	6	7	8
$360 - 24x$	240	216	192	168
$24 + 18x$	114	132	150	168

61. $y = 2x - 2$, not $-2x + 2$.   **63.** $12 \text{ m} \times 16 \text{ m}$
65. $10 \text{ km} \times 12 \text{ km}$
67. False. You can solve for either variable in either equation and then back-substitute.
69. *Sample answer:* After substituting, the resulting equation may be a contradiction or have imaginary solutions.
71. (a)–(c) Answers will vary.

Section 6.2   *(page 440)*

1. elimination   **3.** consistent; inconsistent
5. $(1, 5)$ **7.** $(1, -1)$

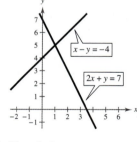
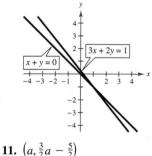

9. No solution   **11.** $\left(a, \frac{3}{2}a - \frac{5}{2}\right)$

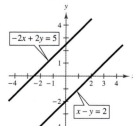

13. $(4, 1)$ **15.** $\left(\frac{3}{2}, -\frac{1}{2}\right)$ **17.** $\left(-3, \frac{5}{3}\right)$ **19.** $(4, -1)$
21. $\left(-\frac{6}{35}, \frac{43}{35}\right)$ **23.** $(101, 96)$ **25.** No solution
27. Infinitely many solutions: $\left(a, -\frac{1}{2} + \frac{5}{6}a\right)$ **29.** $(5, -2)$
31. a; infinitely many solutions; consistent
32. c; one solution; consistent
33. d; no solutions; inconsistent
34. b; one solution; consistent   **35.** $(4, 1)$ **37.** $(10, 5)$
39. $(19, -55)$ **41.** 660 mi/h; 60 mi/h
43. Cheeseburger: 550 calories; fries: 320 calories
45. $(240, 404)$ **47.** $(2{,}000{,}000, 100)$
49. (a) $\begin{cases} x + y = 30 \\ 0.25x + 0.5y = 12 \end{cases}$

(b)

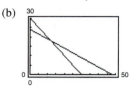

Decreases
(c) 25% solution: 12 L; 50% solution: 18 L
51. \$18,000
53. (a) Pharmacy A: $P = 0.52t + 12.9$

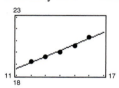

Pharmacy B: $P = 0.39t + 15.7$

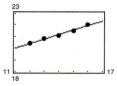

(b) Yes. 2021
55. $y = 0.97x + 2.1$
57. (a) $y = 14x + 19$ (b) 41.4 bushels/acre
59. False. Two lines that coincide have infinitely many points of intersection.
61. $k = -4$
63. No. Two lines will intersect only once or will coincide, and if they coincide the system will have infinitely many solutions.
65. Answers will vary.
67. $(39{,}600, 398)$. It is necessary to change the scale on the axes to see the point of intersection.

Section 6.3   *(page 452)*

1. row-echelon   **3.** Gaussian   **5.** nonsquare
7. (a) No   (b) No   (c) No   (d) Yes
9. (a) No   (b) No   (c) Yes   (d) No
11. $(-13, -10, 8)$ **13.** $(3, 10, 2)$ **15.** $\left(\frac{11}{4}, 7, 11\right)$
17. $\begin{cases} x - 2y + 3z = 5 \\ y - 2z = 9 \\ 2x - 3z = 0 \end{cases}$ First step in putting the system in row-echelon form.
19. $(-2, 2)$ **21.** $(4, 3)$ **23.** $(4, 1, 2)$ **25.** $\left(1, \frac{1}{2}, -3\right)$
27. No solution   **29.** $\left(\frac{1}{8}, -\frac{5}{8}, -\frac{1}{2}\right)$ **31.** $(0, 0, 0)$

33. No solution   **35.** $(-a + 3, a + 1, a)$
37. $(-3a + 10, 5a - 7, a)$ **39.** $(1, 1, 1, 1)$
41. $(2a, 21a - 1, 8a)$ **43.** $\left(-\frac{3}{2}a + \frac{1}{2}, -\frac{2}{3}a + 1, a\right)$
45. $s = -16t^2 + 144$ **47.** $y = \frac{1}{2}x^2 - 2x$
49. $y = x^2 - 6x + 8$ **51.** $y = 4x^2 - 2x + 1$
53. $x^2 + y^2 - 10x = 0$ **55.** $x^2 + y^2 + 6x - 8y = 0$
57. The leading coefficient of the third equation is not 1, so the system is not in row-echelon form.
59. $300,000 at 8%
$400,000 at 9%
$75,000 at 10%
61. $x = 60°, y = 67°, z = 53°$ **63.** 75 ft, 63 ft, 42 ft
65. $I_1 = 1, I_2 = 2, I_3 = 1$ **67.** $y = x^2 - x$
69. (a) $y = 0.0514x^2 + 0.8771x + 1.8857$

(b)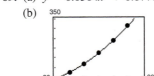

The model fits the data well.
(c) About 356 ft
71. $x = \pm\frac{\sqrt{2}}{2}, y = \frac{1}{2}, \lambda = 1$ or $x = 0, y = 0, \lambda = 0$
73. False. See Example 6 on page 449.
75. No. Answers will vary.   **77–79.** Answers will vary.

Section 6.4   *(page 462)*

1. partial fraction decomposition   **3.** partial fraction
5. b   **6.** c   **7.** d   **8.** a   **9.** $\dfrac{A}{x} + \dfrac{B}{x - 2}$
11. $\dfrac{A}{x + 2} + \dfrac{B}{(x + 2)^2} + \dfrac{C}{(x + 2)^3} + \dfrac{D}{(x + 2)^4}$
13. $\dfrac{A}{x} + \dfrac{Bx + C}{x^2 + 10}$ **15.** $\dfrac{A}{x} + \dfrac{B}{x^2} + \dfrac{Cx + D}{x^2 + 3} + \dfrac{Ex + F}{(x^2 + 3)^2}$
17. $\dfrac{1}{x} - \dfrac{1}{x + 1}$ **19.** $\dfrac{1}{x - 1} - \dfrac{1}{x + 2}$
21. $\dfrac{1}{2}\left(\dfrac{1}{x - 1} - \dfrac{1}{x + 1}\right)$ **23.** $-\dfrac{3}{x} - \dfrac{1}{x + 2} + \dfrac{5}{x - 2}$
25. $\dfrac{3}{x - 3} + \dfrac{9}{(x - 3)^2}$ **27.** $\dfrac{3}{x} - \dfrac{1}{x^2} + \dfrac{1}{x + 1}$
29. $\dfrac{3}{x} - \dfrac{2x - 2}{x^2 + 1}$ **31.** $-\dfrac{1}{x - 1} + \dfrac{x + 2}{x^2 - 2}$
33. $\dfrac{1}{8}\left(\dfrac{1}{2x + 1} + \dfrac{1}{2x - 1} - \dfrac{4x}{4x^2 + 1}\right)$
35. $\dfrac{1}{x + 1} + \dfrac{2}{x^2 - 2x + 3}$ **37.** $\dfrac{2}{x^2 + 4} + \dfrac{x}{(x^2 + 4)^2}$
39. $\dfrac{5}{(x^2 + 3)^2} - \dfrac{17}{(x^2 + 3)^3}$
41. $\dfrac{2}{x} - \dfrac{3}{x^2} - \dfrac{2x - 3}{x^2 + 2} - \dfrac{4x - 6}{(x^2 + 2)^2}$ **43.** $1 - \dfrac{2x + 1}{x^2 + x + 1}$
45. $2x - 7 + \dfrac{17}{x + 2} + \dfrac{1}{x + 1}$
47. $x + 3 + \dfrac{6}{x - 1} + \dfrac{4}{(x - 1)^2} + \dfrac{1}{(x - 1)^3}$
49. $x + \dfrac{2}{x} + \dfrac{1}{x + 1} + \dfrac{3}{(x + 1)^2}$ **51.** $\dfrac{3}{2x - 1} - \dfrac{2}{x + 1}$

53. $\dfrac{2}{x} + \dfrac{4}{x + 1} - \dfrac{3}{x - 1}$ **55.** $\dfrac{1}{x^2 + 2} + \dfrac{x}{(x^2 + 2)^2}$
57. $2x + \dfrac{1}{2}\left(\dfrac{3}{x - 4} - \dfrac{1}{x + 2}\right)$ **59.** $\dfrac{60}{100 - p} - \dfrac{60}{100 + p}$
61. False. The partial fraction decomposition is
$$\dfrac{A}{x + 10} + \dfrac{B}{x - 10} + \dfrac{C}{(x - 10)^2}.$$
63. False. The degrees could be equal.
65. The expression is improper, so first divide the denominator into the numerator to obtain $1 + \dfrac{x + 1}{x^2 - x}$.

Section 6.5   *(page 471)*

1. solution   **3.** solution

5. 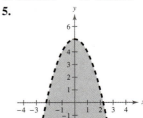   **7.**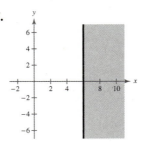

9. 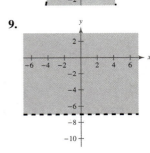   **11.**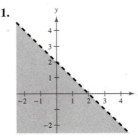

13. 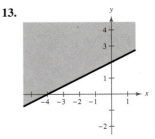   **15.**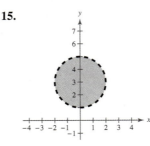

17.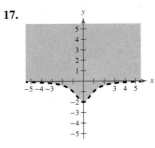

19. 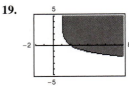   **21.**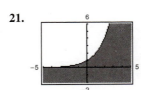

CHAPTER 6

23. 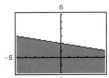   **25.**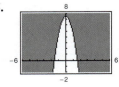

27. $y < 5x + 5$ **29.** $y \geq x^2 - 4$

31. 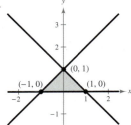   **33.**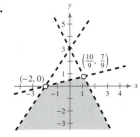

35. 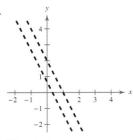   **37.**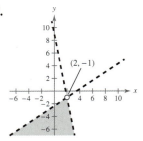

No solution

39. 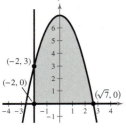   **41.**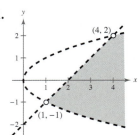

43. 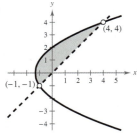   **45.**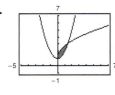

47. 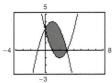   **49.**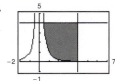

51. $\begin{cases} x \geq 0 \\ y \geq 0 \\ y \leq 6 - x \end{cases}$ **53.** $\begin{cases} x \geq 0 \\ y \geq 0 \\ x^2 + y^2 < 64 \end{cases}$ **55.** $\begin{cases} x \geq 4 \\ x \leq 9 \\ y \geq 3 \\ y \leq 9 \end{cases}$

57. $\begin{cases} y \geq 0 \\ y \leq 5x \\ y \leq -x + 6 \end{cases}$

59. (a) 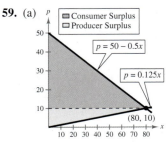   (b) Consumer surplus: $1600
Producer surplus: $400

61. (a) 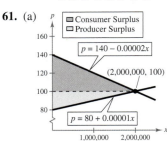   (b) Consumer surplus:
$40,000,000
Producer surplus:
$20,000,000

63. $\begin{cases} x + y \leq 20,000 \\ y \geq 2x \\ x \geq 5,000 \\ y \geq 5,000 \end{cases}$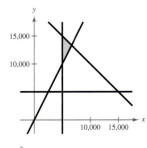

65. $\begin{cases} x + \frac{3}{2}y \leq 12 \\ \frac{4}{3}x + \frac{3}{2}y \leq 15 \\ x \geq 0 \\ y \geq 0 \end{cases}$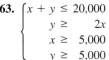

67. (a) $\begin{cases} 180x + 100y \geq 1000 \\ 6x + y \geq 18 \\ 220x + 40y \geq 400 \\ x \geq 0 \\ y \geq 0 \end{cases}$

(b) Answers will vary.

69. (a) $\begin{cases} x \geq 50 \\ y \geq 40 \\ 55x + 70y \leq 7500 \end{cases}$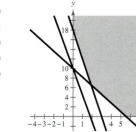

(b) Answers will vary.

71. True. The figure is a rectangle with a length of 9 units and a width of 11 units.

73. Test a point on each side of the line.
75. (a) iv   (b) ii   (c) iii   (d) i

Section 6.6   *(page 480)*

1. optimization   **3.** objective   **5.** inside; on
7. Minimum at $(0, 0)$: 0   **9.** Minimum at $(1, 0)$: 2
   Maximum at $(5, 0)$: 20   Maximum at $(3, 4)$: 26
11. Minimum at $(0, 20)$: 140
   Maximum at $(60, 20)$: 740

13. 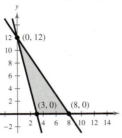   **15.**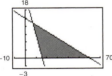

Minimum at $(3, 0)$: 9   Minimum at $(5, 3)$: 35
Maximum at any point on   No maximum
the line segment connecting
$(0, 12)$ and $(8, 0)$: 24

17. 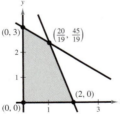   **19.**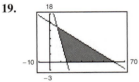

Minimum at $(7.2, 13.2)$: 34.8   Minimum at $(7.2, 13.2)$: 7.2
Maximum at $(60, 0)$: 180   Maximum at $(60, 0)$: 60
21. Minimum at $(0, 0)$: 0   **23.** Minimum at $(0, 0)$: 0
   Maximum at $(0, 5)$: 25   Maximum at $\left(\frac{22}{3}, \frac{19}{6}\right)$: $\frac{271}{6}$
25. Minimum at $(4, 3)$: 10   **27.** No minimum
   No maximum   Maximum at $(12, 5)$: 7

29.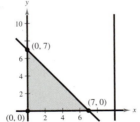

The maximum, 5, occurs at any point on the line segment
connecting $(2, 0)$ and $\left(\frac{20}{19}, \frac{45}{19}\right)$. Minimum at $(0, 0)$: 0

31.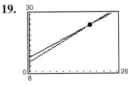

The constraint $x \le 10$ is extraneous. Minimum at $(7, 0)$: -7;
maximum at $(0, 7)$: 14

33.

The feasible set is empty.

35.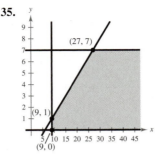

The solution region is
unbounded.
Minimum at $(9, 0)$: 9
No maximum

37. 230 units of the \$225 model
   45 units of the \$250 model
   Optimal profit: \$8295
39. 2 bottles of brand X   **41.** 13 audits
   5 bottles of brand Y   0 tax returns
                             Optimal revenue: \$20,800

43. 60 acres for crop A
   90 acres for crop B
   Optimal yield: 63,000 bushels
45. \$0 on TV ads
   \$1,000,000 on newspaper ads
   Optimal audience: 250 million people
47. True. The objective function has a maximum value at any
   point on the line segment connecting the two vertices.
49. False. See Exercise 27.   **51.** $-\frac{3}{5}$

Review Exercises   *(page 485)*

1. $(1, 1)$ **3.** $\left(\frac{3}{2}, 5\right)$ **5.** $(0.25, 0.625)$ **7.** $(5, 4)$
9. $(0, 0), (2, 8), (-2, 8)$ **11.** $(4, -2)$
13. $(1.41, -0.66), (-1.41, 10.66)$ **15.** $(0, -2)$
17. No solution
19.

The BMI for males exceeds the
BMI for females after age 18.

21. $16 \text{ ft} \times 18 \text{ ft}$ **23.** $\left(\frac{5}{2}, 3\right)$ **25.** $(0, 0)$ **27.** $\left(\frac{8}{5}a + \frac{14}{5}, a\right)$
29. d, one solution, consistent
30. c, infinitely many solutions, consistent
31. b, no solution, inconsistent
32. a, one solution, consistent   **33.** $(100,000, 23)$
35. $(2, -4, -5)$ **37.** $(-6, 7, 10)$ **39.** $\left(\frac{24}{5}, \frac{22}{5}, -\frac{8}{5}\right)$
41. $\left(-\frac{3}{4}, 0, -\frac{5}{4}\right)$ **43.** $(a - 4, a - 3, a)$
45. $y = 2x^2 + x - 5$ **47.** $x^2 + y^2 - 4x + 4y - 1 = 0$
49. 10 gal of spray X   **51.** \$16,000 at 7%
   5 gal of spray Y   \$13,000 at 9%
   12 gal of spray Z   \$11,000 at 11%
53. $s = -16t^2 + 150$ **55.** $\dfrac{A}{x} + \dfrac{B}{x + 20}$
57. $\dfrac{A}{x} + \dfrac{B}{x^2} + \dfrac{C}{x - 5}$ **59.** $\dfrac{3}{x + 2} - \dfrac{4}{x + 4}$
61. $1 - \dfrac{25}{8(x + 5)} + \dfrac{9}{8(x - 3)}$ **63.** $\dfrac{1}{2}\left(\dfrac{3}{x - 1} - \dfrac{x - 3}{x^2 + 1}\right)$

CHAPTER 6

65. $\dfrac{3}{x^2 + 1} + \dfrac{4x - 3}{(x^2 + 1)^2}$

67.

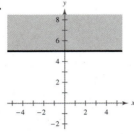

69.

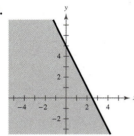

71.

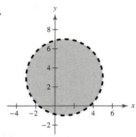

73.

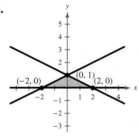

75.

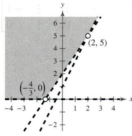

77.

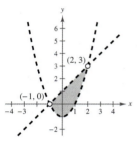

79.

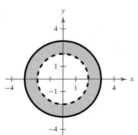

81. $\begin{cases} x \ge 3 \\ x \le 7 \\ y \ge 1 \\ y \le 10 \end{cases}$

83. (a)

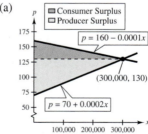

(b) Consumer surplus: $4,500,000
Producer surplus: $9,000,000

85. $\begin{cases} 20x + 30y \le 24{,}000 \\ 12x + 8y \le 12{,}400 \\ x \ge 0 \\ y \ge 0 \end{cases}$

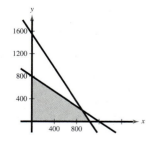

87.

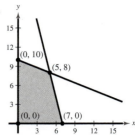

Minimum at $(0, 0)$: 0
Maximum at $(5, 8)$: 47

89.

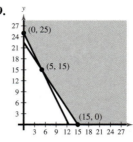

Minimum at $(15, 0)$: 26.25
No maximum

91. 72 haircuts
0 permanents
Optimal revenue: $1800

93. True. The nonparallel sides of the trapezoid are equal in length.

95. $\begin{cases} 4x + y = -22 \\ \frac{1}{2}x + y = 6 \end{cases}$ **97.** $\begin{cases} 3x + y = 7 \\ -6x + 3y = 1 \end{cases}$

99. $\begin{cases} x + y + z = 6 \\ x + y - z = 0 \\ x - y - z = 2 \end{cases}$ **101.** $\begin{cases} 2x + 2y - 3z = 7 \\ x - 2y + z = 4 \\ -x + 4y - z = -1 \end{cases}$

103. An inconsistent system of linear equations has no solution.

Chapter Test   *(page 489)*

1. $(-4, -5)$ **2.** $(0, -1), (1, 0), (2, 1)$ **3.** $(8, 4), (2, -2)$
4. $(4, 2)$ **5.** $(-3, 0), (2, 5)$ **6.** $(1, 4), (0.034, 0.619)$
7. $(-2, -5)$ **8.** $(10, -3)$ **9.** $(2, -3, 1)$

10. No solution   **11.** $-\dfrac{1}{x + 1} + \dfrac{3}{x - 2}$ **12.** $\dfrac{2}{x^2} + \dfrac{3}{2 - x}$

13. $x - \dfrac{5}{x} + \dfrac{3}{x + 1} + \dfrac{3}{x - 1}$ **14.** $-\dfrac{2}{x} + \dfrac{3x}{x^2 + 2}$

15.

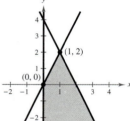

16.

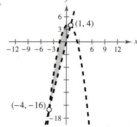

17.

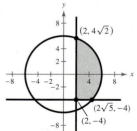

18. Minimum at $(0, 0)$: 0
Maximum at $(12, 0)$: 240

19. $24,000 in 4% fund
$26,000 in 5.5% fund

20. $y = -\frac{1}{2}x^2 + x + 6$

21. 900 units of model I
4400 units of model II
Optimal profit: $203,000

Problem Solving *(page 491)*

1.

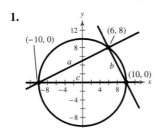

$a = 8\sqrt{5}, b = 4\sqrt{5}, c = 20$
$\left(8\sqrt{5}\right)^2 + \left(4\sqrt{5}\right)^2 = 20^2$
Therefore, the triangle is a right triangle.

3. $ad \neq bc$

5. (a)

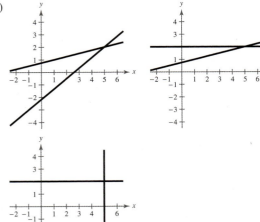

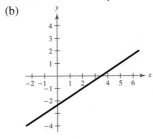

$(5, 2)$
Answers will vary.

(b)

$\left(\frac{3}{2}a + \frac{7}{2}, a\right)$
Answers will vary.

7. 10.1 ft; About 252.7 ft   **9.** $12.00

11. (a) $(3, -4)$ (b) $\left(\dfrac{2}{-a + 5}, \dfrac{1}{4a - 1}, \dfrac{1}{a}\right)$

13. (a) $\left(\dfrac{-5a + 16}{6}, \dfrac{5a - 16}{6}, a\right)$

(b) $\left(\dfrac{-11a + 36}{14}, \dfrac{13a - 40}{14}, a\right)$

(c) $(-a + 3, a - 3, a)$ (d) Infinitely many

15. $\begin{cases} a + \quad t \leq \quad 32 \\ 0.15a \quad\quad \geq \quad 1.9 \\ 193a + 772t \geq 11,000 \end{cases}$

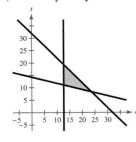

17. (a) $\begin{cases} 0 < y < 130 \\ x \geq 60 \\ x + y \leq 200 \end{cases}$ (b)

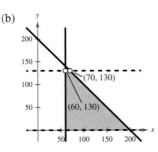

(70, 130)
(60, 130)

(c) No. The point $(90, 120)$ is not in the solution region.

(d) *Sample answer:* LDL/VLDL: 135 mg/dL; HDL: 65 mg/dL

(e) *Sample answer:* $(75, 90)$; $\dfrac{165}{75} = 2.2 < 4$

Chapter 7

Section 7.1 *(page 503)*

1. square   **3.** augmented   **5.** row-equivalent

7. 1×2 **9.** 3×1 **11.** 2×2 **13.** 3×3

15. $\begin{bmatrix} 2 & -1 & \vdots & 7 \\ 1 & 1 & \vdots & 2 \end{bmatrix}$ **17.** $\begin{bmatrix} 1 & -1 & 2 & \vdots & 2 \\ 4 & -3 & 1 & \vdots & -1 \\ 2 & 1 & 0 & \vdots & 0 \end{bmatrix}$

19. $\begin{bmatrix} 3 & -5 & 2 & \vdots & 12 \\ 12 & 0 & -7 & \vdots & 10 \end{bmatrix}$ **21.** $\begin{cases} x + \quad y = \quad 3 \\ 5x - 3y = -1 \end{cases}$

23. $\begin{cases} 2x \quad\quad + 5z = -12 \\ \quad y - 2z = \quad 7 \\ 6x + 3y \quad\quad = \quad 2 \end{cases}$

25. $\begin{cases} 9x + 12y + 3z \quad\quad = \quad 0 \\ -2x + 18y + 5z + 2w = \quad 10 \\ x + 7y - 8z \quad\quad = -4 \\ 3x \quad\quad + 2z \quad\quad = -10 \end{cases}$

27. Add 5 times Row 2 to Row 1.

29. Interchange Row 1 and Row 2.
Add 4 times new Row 1 to Row 3.

31. $\begin{bmatrix} 1 & 2 & \frac{8}{3} \\ 4 & -3 & 6 \end{bmatrix}$ **33.** $\begin{bmatrix} 1 & 1 & 1 \\ 0 & -7 & -1 \end{bmatrix}$

35. $\begin{bmatrix} 1 & 0 & 14 & -11 \\ 0 & 1 & -2 & 2 \\ 0 & 0 & 1 & -7 \end{bmatrix}$

37. $\begin{bmatrix} 1 & 1 & 4 & -1 \\ 0 & 5 & -2 & 6 \\ 0 & 3 & 20 & 4 \end{bmatrix}; \begin{bmatrix} 1 & 1 & 4 & -1 \\ 0 & 1 & -\frac{2}{5} & \frac{6}{5} \\ 0 & 3 & 20 & 4 \end{bmatrix}$

39. (a) (i) $\begin{bmatrix} 3 & 0 & \vdots & -6 \\ 6 & -4 & \vdots & -28 \end{bmatrix}$

(ii) $\begin{bmatrix} 3 & 0 & \vdots & -6 \\ 0 & -4 & \vdots & -16 \end{bmatrix}$

(iii) $\begin{bmatrix} 3 & 0 & \vdots & -6 \\ 0 & 1 & \vdots & 4 \end{bmatrix}$

(iv) $\begin{bmatrix} 1 & 0 & \vdots & -2 \\ 0 & 1 & \vdots & 4 \end{bmatrix}$

(b) $\begin{cases} -3x + 4y = \quad 22 \\ 6x - 4y = -28 \end{cases}$

Solution: $(-2, 4)$

(c) Answers will vary.

CHAPTER 7

41. Reduced row-echelon form   **43.** Not in row-echelon form

45. $\begin{bmatrix} 1 & 1 & 0 & 5 \\ 0 & 1 & 2 & 0 \\ 0 & 0 & 1 & -1 \end{bmatrix}$ **47.** $\begin{bmatrix} 1 & -1 & -1 & 1 \\ 0 & 1 & 6 & 3 \\ 0 & 0 & 0 & 0 \end{bmatrix}$

49. $\begin{bmatrix} 1 & 0 & 0 \\ 0 & 1 & 0 \\ 0 & 0 & 1 \end{bmatrix}$ **51.** $\begin{bmatrix} 1 & 2 & 0 & 0 \\ 0 & 0 & 1 & 0 \\ 0 & 0 & 0 & 1 \\ 0 & 0 & 0 & 0 \end{bmatrix}$

53. $\begin{bmatrix} 1 & 0 & 3 & 16 \\ 0 & 1 & 2 & 12 \end{bmatrix}$

55. $\begin{cases} x - 2y = 4 \\ \quad\;\; y = -1 \end{cases}$ **57.** $\begin{cases} x - y + 2z = 4 \\ \quad\; y - z = 2 \\ \quad\qquad z = -2 \end{cases}$

$(2, -1)$ $(8, 0, -2)$

59. $(-3, 5)$ **61.** $(-5, 6)$ **63.** $(-4, -3, 6)$

65. No solution   **67.** $(3, -2, 5, 0)$ **69.** $(3, -4)$

71. $(-1, -4)$ **73.** $(5a + 4, -3a + 2, a)$ **75.** $(4, -3, 2)$

77. $(7, -3, 4)$ **79.** $(0, 2 - 4a, a)$ **81.** $(1, 0, 4, -2)$

83. $(-2a, a, a, 0)$ **85.** The dimension is 4×1.

87. $f(x) = -x^2 + x + 1$ **89.** $f(x) = -9x^2 - 5x + 11$

91. $f(x) = x^2 + 2x + 5$

93. $y = 7.5t + 28$; About 141 cases; Yes, because the data values increase in a linear pattern.

95. \$1,200,000 at 8%
\$200,000 at 9%
\$600,000 at 12%

97. False. It is a 2×4 matrix.   **99.** They are the same.

Section 7.2   (page 517)

1. equal   **3.** zero; O **5.** $x = -4, y = 23$

7. $x = -1, y = 3$

9. (a) $\begin{bmatrix} 3 & -2 \\ 1 & 7 \end{bmatrix}$ (b) $\begin{bmatrix} -1 & 0 \\ 3 & -9 \end{bmatrix}$ (c) $\begin{bmatrix} 3 & -3 \\ 6 & -3 \end{bmatrix}$

(d) $\begin{bmatrix} -1 & -1 \\ 8 & -19 \end{bmatrix}$

11. (a), (b), and (d) Not possible   (c) $\begin{bmatrix} 18 & 0 & 9 \\ -3 & -12 & 0 \end{bmatrix}$

13. (a) $\begin{bmatrix} 9 & 5 \\ 1 & -2 \\ -3 & 15 \end{bmatrix}$ (b) $\begin{bmatrix} 7 & -7 \\ 3 & 8 \\ -5 & -5 \end{bmatrix}$ (c) $\begin{bmatrix} 24 & -3 \\ 6 & 9 \\ -12 & 15 \end{bmatrix}$

(d) $\begin{bmatrix} 22 & -15 \\ 8 & 19 \\ -14 & -5 \end{bmatrix}$

15. (a) $\begin{bmatrix} 5 & 5 & -2 & 4 & 4 \\ -5 & 10 & 0 & -4 & -7 \end{bmatrix}$

(b) $\begin{bmatrix} 3 & 5 & 0 & 2 & 4 \\ 7 & -6 & -4 & 2 & 7 \end{bmatrix}$

(c) $\begin{bmatrix} 12 & 15 & -3 & 9 & 12 \\ 3 & 6 & -6 & -3 & 0 \end{bmatrix}$

(d) $\begin{bmatrix} 10 & 15 & -1 & 7 & 12 \\ 15 & -10 & -10 & 3 & 14 \end{bmatrix}$

17. $\begin{bmatrix} -8 & -7 \\ 15 & -1 \end{bmatrix}$ **19.** $\begin{bmatrix} -24 & -4 & 12 \\ -12 & 32 & 12 \end{bmatrix}$

21. $\begin{bmatrix} 10 & 8 \\ -59 & 9 \end{bmatrix}$ **23.** $\begin{bmatrix} -17.12 & 2.2 \\ 11.56 & 10.24 \end{bmatrix}$

25. $\begin{bmatrix} -10.81 & -5.36 & 0.4 \\ -14.04 & 10.69 & -14.76 \end{bmatrix}$ **27.** $\begin{bmatrix} -4 & 6 & -2 \\ 4 & 0 & 10 \end{bmatrix}$

29. $\begin{bmatrix} -2 & 0 & 5 \\ -\frac{5}{2} & 0 & \frac{7}{2} \end{bmatrix}$ **31.** $\begin{bmatrix} 3 & -\frac{1}{2} & -\frac{13}{2} \\ 3 & 0 & -\frac{11}{2} \end{bmatrix}$

33. $\begin{bmatrix} 2 & -5 & 5 \\ -5 & 0 & -6 \end{bmatrix}$ **35.** $\begin{bmatrix} -2 & 51 \\ -8 & 33 \\ 0 & 27 \end{bmatrix}$ **37.** Not possible

3×2

39. $\begin{bmatrix} 1 & 0 & 0 \\ 0 & 1 & 0 \\ 0 & 0 & \frac{7}{2} \end{bmatrix}$ **41.** $\begin{bmatrix} 70 & -17 & 73 \\ 32 & 11 & 6 \\ 16 & -38 & 70 \end{bmatrix}$ **43.** Not possible

3×3

45. $\begin{bmatrix} 151 & 25 & 48 \\ 516 & 279 & 387 \\ 47 & -20 & 87 \end{bmatrix}$

47. (a) $\begin{bmatrix} 0 & 15 \\ 6 & 12 \end{bmatrix}$ (b) $\begin{bmatrix} -2 & 2 \\ 31 & 14 \end{bmatrix}$ (c) $\begin{bmatrix} 9 & 6 \\ 12 & 12 \end{bmatrix}$

49. (a) $\begin{bmatrix} 5 & -9 & 0 \\ 3 & 0 & -8 \\ -1 & 4 & 11 \end{bmatrix}$ (b) $\begin{bmatrix} 5 & -9 & 0 \\ 3 & 0 & -8 \\ -1 & 4 & 11 \end{bmatrix}$

(c) $\begin{bmatrix} -2 & -45 & 72 \\ 23 & -59 & -88 \\ -4 & 53 & 89 \end{bmatrix}$

51. (a) $\begin{bmatrix} 19 \\ 48 \end{bmatrix}$ (b) Not possible   (c) $\begin{bmatrix} 14 & -8 \\ 16 & 142 \end{bmatrix}$

53. (a) $\begin{bmatrix} 7 & 7 & 14 \\ 8 & 8 & 16 \\ -1 & -1 & -2 \end{bmatrix}$ (b) $\begin{bmatrix} 13 \end{bmatrix}$ (c) Not possible

55. $\begin{bmatrix} 5 & 8 \\ -4 & -16 \end{bmatrix}$ **57.** $\begin{bmatrix} -4 & 10 \\ 3 & 14 \end{bmatrix}$ **59.** $\begin{bmatrix} 12 & 6 \\ -4 & -2 \\ 20 & 10 \\ 28 & 14 \end{bmatrix}$

61. (a) $\begin{bmatrix} 2 & 3 \\ 1 & 4 \end{bmatrix}\begin{bmatrix} x_1 \\ x_2 \end{bmatrix} = \begin{bmatrix} 5 \\ 10 \end{bmatrix}$ (b) $\begin{bmatrix} -2 \\ 3 \end{bmatrix}$

63. (a) $\begin{bmatrix} 1 & -2 & 3 \\ -1 & 3 & -1 \\ 2 & -5 & 5 \end{bmatrix}\begin{bmatrix} x_1 \\ x_2 \\ x_3 \end{bmatrix} = \begin{bmatrix} 9 \\ -6 \\ 17 \end{bmatrix}$ (b) $\begin{bmatrix} 1 \\ -1 \\ 2 \end{bmatrix}$

65. (a) $\begin{bmatrix} 1 & -5 & 2 \\ -3 & 1 & -1 \\ 0 & -2 & 5 \end{bmatrix}\begin{bmatrix} x_1 \\ x_2 \\ x_3 \end{bmatrix} = \begin{bmatrix} -20 \\ 8 \\ -16 \end{bmatrix}$ (b) $\begin{bmatrix} -1 \\ 3 \\ -2 \end{bmatrix}$

67. $\begin{bmatrix} 110 & 99 & 77 & 33 \\ 44 & 22 & 66 & 66 \end{bmatrix}$

69. $\begin{bmatrix} \$1037.50 & \$1400 & \$1012.50 \end{bmatrix}$

The entries represent the profits from both crops at each of the three outlets.

71. $\begin{bmatrix} \$23.20 & \$20.50 \\ \$38.20 & \$33.80 \\ \$76.90 & \$68.50 \end{bmatrix}$

The entries represent the labor costs at each plant for each size of boat.

73. $\begin{bmatrix} 0.40 & 0.15 & 0.15 \\ 0.28 & 0.53 & 0.17 \\ 0.32 & 0.32 & 0.68 \end{bmatrix}$

P^2 gives the proportions of the voting population that changed parties or remained loyal to their parties from the first election to the third.

75. True. The sum of two matrices of different dimensions is undefined.

77. $\begin{bmatrix} 1 & 0 \\ 2 & 1 \end{bmatrix} \neq \begin{bmatrix} 0 & 0 \\ 3 & 2 \end{bmatrix}$ **79.** $\begin{bmatrix} 3 & -2 \\ 4 & 3 \end{bmatrix} \neq \begin{bmatrix} 2 & -2 \\ 5 & 4 \end{bmatrix}$

81. $AC = BC = \begin{bmatrix} 2 & 3 \\ 2 & 3 \end{bmatrix}$ **83.** Answers will vary.

85. AB is a diagonal matrix whose entries are the products of the corresponding entries of A and B.

Section 7.3 *(page 527)*

1. inverse   **3.** determinant   **5–11.** $AB = I$ and $BA = I$

13. $\begin{bmatrix} 3 & -1 \\ -5 & 2 \end{bmatrix}$ **15.** $\begin{bmatrix} -3 & 2 \\ -2 & 1 \end{bmatrix}$ **17.** $\begin{bmatrix} 1 & -\frac{1}{2} \\ -2 & \frac{3}{2} \end{bmatrix}$

19. $\begin{bmatrix} 1 & 1 & -1 \\ -3 & 2 & -1 \\ 3 & -3 & 2 \end{bmatrix}$ **21.** Not possible

23. $\begin{bmatrix} -\frac{1}{8} & 0 & 0 & 0 \\ 0 & 1 & 0 & 0 \\ 0 & 0 & \frac{1}{4} & 0 \\ 0 & 0 & 0 & -\frac{1}{5} \end{bmatrix}$ **25.** $\begin{bmatrix} -175 & 37 & -13 \\ 95 & -20 & 7 \\ 14 & -3 & 1 \end{bmatrix}$

27. $\begin{bmatrix} -12 & -5 & -9 \\ -4 & -2 & -4 \\ -8 & -4 & -6 \end{bmatrix}$ **29.** $\begin{bmatrix} 0 & -1.\overline{81} & 0.\overline{90} \\ -10 & 5 & 5 \\ 10 & -2.\overline{72} & -3.\overline{63} \end{bmatrix}$

31. $\begin{bmatrix} 1 & 0 & 1 & 0 \\ 0 & 1 & 0 & 1 \\ 2 & 0 & 1 & 0 \\ 0 & 1 & 0 & 2 \end{bmatrix}$ **33.** $\begin{bmatrix} \frac{5}{13} & -\frac{3}{13} \\ \frac{1}{13} & \frac{2}{13} \end{bmatrix}$

35. Not possible   **37.** $\begin{bmatrix} -4 & 2 \\ 10 & -\frac{10}{3} \end{bmatrix}$

39. $(5, 0)$ **41.** $(-8, -6)$ **43.** $(3, 8, -11)$
45. $(2, 1, 0, 0)$ **47.** $(-1, 1)$ **49.** No solution
51. $(-4, -8)$ **53.** $(-1, 3, 2)$ **55.** $\left(\frac{13}{16}, \frac{11}{16}, 0\right)$
57. $3684.21 in AAA-rated bonds
$2105.26 in A-rated bonds
$4210.53 in B-rated bonds
59. $I_1 = 0.5$ amp   **61.** $I_1 = 4$ amps
   $I_2 = 3$ amps $I_2 = 1$ amp
   $I_3 = 3.5$ amps $I_3 = 5$ amps
63. 100 bags of potting soil for seedlings
   100 bags of potting soil for general potting
   100 bags of potting soil for hardwood plants
65. (a) $\begin{cases} 2.5r + 4l + 2i = 300 \\ -r + 2l + 2i = 0 \\ r + l + i = 120 \end{cases}$

   $\begin{bmatrix} 2.5 & 4 & 2 \\ -1 & 2 & 2 \\ 1 & 1 & 1 \end{bmatrix} \begin{bmatrix} r \\ l \\ i \end{bmatrix} = \begin{bmatrix} 300 \\ 0 \\ 120 \end{bmatrix}$

   (b) 80 roses, 10 lilies, 30 irises

67. True. If B is the inverse of A, then $AB = I = BA$.
69. Answers will vary.   **71.** $k = -\frac{3}{2}$
73. (a) Answers will vary.

   (b) $A^{-1} = \begin{bmatrix} 1/a_{11} & 0 & 0 & \ldots & 0 \\ 0 & 1/a_{22} & 0 & \ldots & 0 \\ 0 & 0 & 1/a_{33} & \ldots & 0 \\ \vdots & \vdots & \vdots & & \vdots \\ 0 & 0 & 0 & \ldots & 1/a_{nn} \end{bmatrix}$

75. Answers will vary.

Section 7.4 *(page 535)*

1. determinant   **3.** cofactor   **5.** 4   **7.** 16   **9.** -3
11. 0   **13.** 6   **15.** 0   **17.** -23 **19.** -24
21. $\frac{11}{6}$ **23.** 11   **25.** -1924 **27.** 0.08
29. (a) $M_{11} = -6, M_{12} = 3, M_{21} = 5, M_{22} = 4$
   (b) $C_{11} = -6, C_{12} = -3, C_{21} = -5, C_{22} = 4$
31. (a) $M_{11} = 3, M_{12} = -4, M_{13} = 1, M_{21} = 2, M_{22} = 2$,
   $M_{23} = -4, M_{31} = -4, M_{32} = 10, M_{33} = 8$
   (b) $C_{11} = 3, C_{12} = 4, C_{13} = 1, C_{21} = -2, C_{22} = 2$,
   $C_{23} = 4, C_{31} = -4, C_{32} = -10, C_{33} = 8$
33. (a) $M_{11} = 10, M_{12} = -43, M_{13} = 2, M_{21} = -30, M_{22} = 17$,
   $M_{23} = -6, M_{31} = 54, M_{32} = -53, M_{33} = -34$
   (b) $C_{11} = 10, C_{12} = 43, C_{13} = 2, C_{21} = 30, C_{22} = 17$,
   $C_{23} = 6, C_{31} = 54, C_{32} = 53, C_{33} = -34$
35. (a) and (b) -36 **37.** (a) and (b) 96
39. (a) and (b) -75 **41.** (a) and (b) 0
43. (a) and (b) 225   **45.** -9 **47.** 0   **49.** 0
51. -58 **53.** 72   **55.** 0   **57.** 412   **59.** -126
61. -336

63. (a) -3 (b) -2 (c) $\begin{bmatrix} -2 & 0 \\ 0 & -3 \end{bmatrix}$ (d) 6

65. (a) -8 (b) 0   (c) $\begin{bmatrix} -4 & 4 \\ 1 & -1 \end{bmatrix}$ (d) 0

67. (a) 2   (b) -6 (c) $\begin{bmatrix} 1 & 4 & 3 \\ -1 & 0 & 3 \\ 0 & 2 & 0 \end{bmatrix}$ (d) -12

69. $A = \begin{bmatrix} 3 & 3 \\ 1 & 2 \end{bmatrix}$ **71.** $A = \begin{bmatrix} 4 & 2 & -1 \\ 2 & 1 & 0 \\ 1 & 1 & 3 \end{bmatrix}$

73. $A = \begin{bmatrix} 2 & 3 \\ 8 & 12 \end{bmatrix}$ **75–79.** Answers will vary.   **81.** ± 2

83. $-2, 1$ **85.** $-1, -4$ **87.** $8uv - 1$ **89.** e^{5x}
91. $1 - \ln x$
93. True. If an entire row is zero, then each cofactor in the expansion is multiplied by zero.
95. Answers will vary.
97. The signs of the cofactors should be $-, +, -$.
99. (a) Columns 2 and 3 of A were interchanged.
   $|A| = -115 = -|B|$
   (b) Rows 1 and 3 of A were interchanged.
   $|A| = -40 = -|B|$
101. (a) Multiply Row 1 by 5.
   (b) Multiply Column 2 by 4 and Column 3 by 3.
103. (a) 28   (b) -10 (c) -12
   The determinant of a diagonal matrix is the product of the entries on the main diagonal.

Section 7.5 *(page 548)*

1. Cramer's Rule   **3.** $A = \pm\dfrac{1}{2}\begin{vmatrix} x_1 & y_1 & 1 \\ x_2 & y_2 & 1 \\ x_3 & y_3 & 1 \end{vmatrix}$

5. uncoded; coded   **7.** $(1, -1)$ **9.** Not possible

11. $(-1, 3, 2)$ **13.** $(-2, 1, -1)$ **15.** 7   **17.** 14

19. $y = \frac{16}{5}$ or $y = 0$ **21.** 250 mi^2 **23.** Collinear

25. Not collinear   **27.** Collinear   **29.** $y = -3$

31. $3x - 5y = 0$ **33.** $x + 3y - 5 = 0$

35. $2x + 3y - 8 = 0$

37. $(0, 0), (0, 3), (6, 0), (6, 3)$

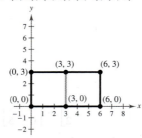

39. $(-4, 3), (-5, 3), (-4, 4), (-5, 4)$

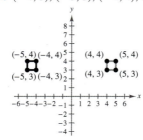

41. 2 square units   **43.** 10 square units

45. (a) Uncoded: $[3 \ \ 15], [13 \ \ 5], [0 \ \ 8], [15 \ \ 13], [5 \ \ 0],$
$[19 \ \ 15], [15 \ \ 14]$

(b) Encoded: 48 81 28 51 24 40 54 95 5
10 64 113 57 100

47. (a) Uncoded: $[3 \ \ 1 \ \ 12], [12 \ \ 0 \ \ 13], [5 \ \ 0 \ \ 20],$
$[15 \ \ 13 \ \ 15], [18 \ \ 18 \ \ 15], [23 \ \ 0 \ \ 0]$

(b) Encoded: -68 21 35 -66 14 39 -115
35 60 -62 15 32 -54 12 27 23 -23 0

49. 1 -25 -65 17 15 -9 -12 -62 -119
27 51 48 43 67 48 57 111 117

51. -5 -41 -87 91 207 257 11 -5 -41 40
80 84 76 177 227

53. HAPPY NEW YEAR   **55.** CLASS IS CANCELED

57. SEND PLANES   **59.** MEET ME TONIGHT RON

61. $I_1 = -0.5$ amp
$I_2 = 1$ amp
$I_3 = 0.5$ amp

63. False. The denominator is the determinant of the coefficient matrix.

65. The system has either no solutions or infinitely many solutions.

67. 12

Review Exercises *(page 553)*

1. 1×2 **3.** 2×5 **5.** $\begin{bmatrix} 3 & -10 & \vdots & 15 \\ 5 & 4 & \vdots & 22 \end{bmatrix}$

7. $\begin{cases} x \ \ \ \ \ + 2z = -8 \\ 2x - 2y + 3z = 12 \\ 4x + 7y + \ z = \ \ 3 \end{cases}$ **9.** $\begin{bmatrix} 1 & 2 & 3 \\ 0 & 1 & 1 \\ 0 & 0 & 1 \end{bmatrix}$

11. $\begin{cases} x + 2y + 3z = \ \ 9 \\ \ \ \ \ \ \ y - 2z = \ \ 2 \\ \ \ \ \ \ \ \ \ \ \ \ z = -1 \end{cases}$ **13.** $\begin{cases} x + 3y + 4z = 1 \\ \ \ \ \ \ \ y + 2z = 3 \\ \ \ \ \ \ \ \ \ \ \ \ z = 4 \end{cases}$

$(12, 0, -1)$ $(0, -5, 4)$

15. $(10, -12)$ **17.** $\left(-\frac{1}{5}, \frac{7}{10}\right)$ **19.** No solution

21. $(1, -2, 2)$ **23.** $\left(-2a + \frac{3}{2}, 2a + 1, a\right)$

25. $(5, 2, -6)$ **27.** $(1, 2, 2)$ **29.** $(2, -3, 3)$

31. $(2, 6, -10, -3)$ **33.** $x = 12, y = 11$

35. $x = 1, y = 11$

37. (a) $\begin{bmatrix} -1 & 8 \\ 15 & 13 \end{bmatrix}$ (b) $\begin{bmatrix} 5 & -12 \\ -9 & -3 \end{bmatrix}$

(c) $\begin{bmatrix} 8 & -8 \\ 12 & 20 \end{bmatrix}$ (d) $\begin{bmatrix} -2 & 16 \\ 30 & 26 \end{bmatrix}$

39. (a) $\begin{bmatrix} 7 & 14 \\ 9 & 26 \\ 30 & 30 \end{bmatrix}$ (b) $\begin{bmatrix} 1 & -8 \\ -21 & -24 \\ -10 & -28 \end{bmatrix}$

(c) $\begin{bmatrix} 16 & 12 \\ -24 & 4 \\ 40 & 4 \end{bmatrix}$ (d) $\begin{bmatrix} 14 & 28 \\ 18 & 52 \\ 60 & 60 \end{bmatrix}$

41. (a) $\begin{bmatrix} 1 & 1 & 1 \\ 1 & 1 & 1 \\ 1 & 1 & 1 \end{bmatrix}$ (b) $\begin{bmatrix} -5 & 5 & -5 \\ 5 & -5 & 5 \\ -5 & 5 & -5 \end{bmatrix}$

(c) $\begin{bmatrix} -8 & 12 & -8 \\ 12 & -8 & 12 \\ -8 & 12 & -8 \end{bmatrix}$ (d) $\begin{bmatrix} 2 & 2 & 2 \\ 2 & 2 & 2 \\ 2 & 2 & 2 \end{bmatrix}$

43. $\begin{bmatrix} 22 & -17 \\ 14 & 11 \end{bmatrix}$ **45.** $\begin{bmatrix} -16 & -6 \\ -12 & 4 \\ -14 & -8 \end{bmatrix}$ **47.** $\begin{bmatrix} -11 & -6 \\ 8 & -13 \\ -18 & -8 \end{bmatrix}$

49. $\begin{bmatrix} 3 & \frac{2}{3} \\ -\frac{4}{3} & \frac{11}{3} \\ \frac{10}{3} & 0 \end{bmatrix}$ **51.** $\begin{bmatrix} -30 & 4 \\ 51 & 70 \end{bmatrix}; 2 \times 2$

53. $\begin{bmatrix} 100 & 220 \\ 12 & -4 \\ 84 & 212 \end{bmatrix}; 3 \times 2$ **55.** $\begin{bmatrix} 14 & -22 & 22 \\ 19 & -41 & 80 \\ 42 & -66 & 66 \end{bmatrix}$

57. Not possible

59. (a) $\begin{bmatrix} -1 & -1 \\ 18 & -4 \end{bmatrix}$ (b) $\begin{bmatrix} 1 & 14 \\ -2 & -6 \end{bmatrix}$ (c) $\begin{bmatrix} 13 & 6 \\ 8 & 13 \end{bmatrix}$

61. $\begin{bmatrix} 76 & 114 & 133 \\ 38 & 95 & 76 \end{bmatrix}$ **63–65.** $AB = I$ and $BA = I$

67. $\begin{bmatrix} 4 & -5 \\ 5 & -6 \end{bmatrix}$ **69.** $\begin{bmatrix} \frac{1}{2} & -1 & -\frac{1}{2} \\ \frac{1}{2} & -\frac{2}{3} & -\frac{5}{6} \\ 0 & \frac{2}{3} & \frac{1}{3} \end{bmatrix}$

71. $\begin{bmatrix} 13 & 6 & -4 \\ -12 & -5 & 3 \\ 5 & 2 & -1 \end{bmatrix}$ **73.** $\begin{bmatrix} 1 & -1 \\ 4 & -\frac{7}{2} \end{bmatrix}$

75. Not possible   **77.** $(36, 11)$

79. $(-6, -1)$ **81.** $(2, 3)$ **83.** $(-8, 18)$

85. $(2, -1, -2)$ **87.** $(-3, 1)$ **89.** $\left(\frac{1}{6}, -\frac{7}{4}\right)$

91. 26   **93.** 116

95. (a) $M_{11} = 4, M_{12} = 7, M_{21} = -1, M_{22} = 2$

(b) $C_{11} = 4, C_{12} = -7, C_{21} = 1, C_{22} = 1$

97. (a) $M_{11} = 30, M_{12} = -12, M_{13} = -21, M_{21} = 20,$

$M_{22} = 19, M_{23} = 22, M_{31} = 5, M_{32} = -2, M_{33} = 19$

(b) $C_{11} = 30, C_{12} = 12, C_{13} = -21, C_{21} = -20,$

$C_{22} = 19, C_{23} = -22, C_{31} = 5, C_{32} = 2, C_{33} = 19$

99. -6 **101.** 15   **103.** 130   **105.** $(4, 7)$

107. $(-1, 4, 5)$ **109.** 16   **111.** Collinear

113. $x - 2y + 4 = 0$ **115.** $2x + 6y - 13 = 0$

117. 8 square units   **119.** SEE YOU FRIDAY

121. False. The matrix must be square.

123. If A is a square matrix, then the cofactor C_{ij} of the entry a_{ij} is $(-1)^{i+j}M_{ij}$, where M_{ij} is the determinant obtained by deleting the ith row and jth column of A. The determinant of A is the sum of the entries of any row or column of A multiplied by their respective cofactors.

Chapter Test   *(page 557)*

1. $\begin{bmatrix} 1 & 0 & 0 \\ 0 & 1 & 0 \\ 0 & 0 & 1 \end{bmatrix}$ **2.** $\begin{bmatrix} 1 & 0 & -1 & 2 \\ 0 & 1 & 0 & -1 \\ 0 & 0 & 0 & 0 \\ 0 & 0 & 0 & 0 \end{bmatrix}$

3. $\begin{bmatrix} 4 & 3 & -2 & \vdots & 14 \\ -1 & -1 & 2 & \vdots & -5 \\ 3 & 1 & -4 & \vdots & 8 \end{bmatrix}, \left(1, 3, -\frac{1}{2}\right)$

4. (a) $\begin{bmatrix} 1 & 5 \\ 0 & -4 \end{bmatrix}$

(b) $\begin{bmatrix} 6 & -3 & 12 \\ 0 & 18 & -9 \end{bmatrix}$

(c) $\begin{bmatrix} 8 & 15 \\ -5 & -13 \end{bmatrix}$

(d) $\begin{bmatrix} 10 & -5 & 20 \\ -10 & -1 & -17 \end{bmatrix}$

(e) Not possible

5. $\begin{bmatrix} \frac{2}{7} & \frac{3}{7} \\ \frac{5}{7} & \frac{4}{7} \end{bmatrix}$ **6.** $\begin{bmatrix} -\frac{5}{2} & 4 & -3 \\ 5 & -7 & 6 \\ 4 & -6 & 5 \end{bmatrix}$ **7.** $(12, 18)$

8. -112 **9.** 0   **10.** 43   **11.** $(-3, 5)$

12. $(-2, 4, 6)$ **13.** 7

14. Uncoded: $[11 \ 14 \ 15], [3 \ 11 \ 0], [15 \ 14 \ 0], [23 \ 15 \ 15],$
$[4 \ 0 \ 0]$

Encoded: $115 \ -41 \ -59 \ 14 \ -3 \ -11 \ 29 \ -15$
$-14 \ 128 \ -53 \ -60 \ 4 \ -4 \ 0$

15. 75 L of 60% solution, 25 L of 20% solution

Problem Solving   *(page 559)*

1. (a) $AT = \begin{bmatrix} -1 & -4 & -2 \\ 1 & 2 & 3 \end{bmatrix}, AAT = \begin{bmatrix} -1 & -2 & -3 \\ -1 & -4 & -2 \end{bmatrix}$

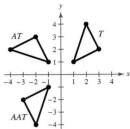

A represents a counterclockwise rotation.

(b) AAT is rotated clockwise 90° to obtain AT. AT is then rotated clockwise 90° to obtain T.

3. (a) Yes   (b) No   (c) No   (d) No   (e) No   (f) No

5. (a) $A^2 - 2A + 5I = \begin{bmatrix} 1 & 2 \\ -2 & 1 \end{bmatrix}^2 - 2\begin{bmatrix} 1 & 2 \\ -2 & 1 \end{bmatrix} + 5\begin{bmatrix} 1 & 0 \\ 0 & 1 \end{bmatrix}$

$= \begin{bmatrix} -3 & 4 \\ -4 & -3 \end{bmatrix} - \begin{bmatrix} 2 & 4 \\ -4 & 2 \end{bmatrix} + \begin{bmatrix} 5 & 0 \\ 0 & 5 \end{bmatrix}$

$= \begin{bmatrix} 0 & 0 \\ 0 & 0 \end{bmatrix}$

(b) $A^{-1} = \frac{1}{5}(2I - A)$

$\begin{bmatrix} 1 & 2 \\ -2 & 1 \end{bmatrix}^{-1} = \frac{1}{5}\left(2\begin{bmatrix} 1 & 0 \\ 0 & 1 \end{bmatrix} - \begin{bmatrix} 1 & 2 \\ -2 & 1 \end{bmatrix}\right)$

$\begin{bmatrix} \frac{1}{5} & -\frac{2}{5} \\ \frac{2}{5} & \frac{1}{5} \end{bmatrix} = \frac{1}{5}\left(\begin{bmatrix} 2 & 0 \\ 0 & 2 \end{bmatrix} - \begin{bmatrix} 1 & 2 \\ -2 & 1 \end{bmatrix}\right)$

$\begin{bmatrix} \frac{1}{5} & -\frac{2}{5} \\ \frac{2}{5} & \frac{1}{5} \end{bmatrix} = \frac{1}{5}\begin{bmatrix} 1 & -2 \\ 2 & 1 \end{bmatrix}$

$\begin{bmatrix} \frac{1}{5} & -\frac{2}{5} \\ \frac{2}{5} & \frac{1}{5} \end{bmatrix} = \begin{bmatrix} \frac{1}{5} & -\frac{2}{5} \\ \frac{2}{5} & \frac{1}{5} \end{bmatrix}$

(c) Answers will vary.

7. $A^T = \begin{bmatrix} -1 & 2 \\ 1 & 0 \\ -2 & 1 \end{bmatrix}, B^T = \begin{bmatrix} -3 & 1 & 1 \\ 0 & 2 & -1 \end{bmatrix}$ **9.** $x = 6$

$(AB)^T = \begin{bmatrix} 2 & -5 \\ 4 & -1 \end{bmatrix} = B^TA^T$

11. Answers will vary.   **13.** $\begin{vmatrix} x & 0 & 0 & d \\ -1 & x & 0 & c \\ 0 & -1 & x & b \\ 0 & 0 & -1 & a \end{vmatrix}$

15. Sulfur: 32 atomic mass units
Nitrogen: 14 atomic mass units
Fluorine: 19 atomic mass units

17. REMEMBER SEPTEMBER THE ELEVENTH

19. $A^{-1} = \begin{bmatrix} 0.0625 & -0.4375 & 0.625 \\ 0.1875 & 0.6875 & -1.125 \\ -0.125 & -0.125 & 0.75 \end{bmatrix}$

$|A^{-1}| = \frac{1}{16}, |A| = 16$

$|A^{-1}| = \frac{1}{|A|}$

CHAPTER 7

Chapter 8

Section 8.1　*(page 569)*

1. infinite sequence　**3.** recursively
5. index; upper; lower　**7.** $-3, 1, 5, 9, 13$
9. $5, 3, 5, 3, 5$　**11.** $-2, 4, -8, 16, -32$　**13.** $\frac{2}{3}, \frac{2}{3}, \frac{2}{3}, \frac{2}{3}, \frac{2}{3}$
15. $\frac{1}{3}, \frac{8}{3}, 9, \frac{64}{3}, \frac{125}{3}$　**17.** $\frac{1}{3}, \frac{1}{2}, \frac{3}{5}, \frac{2}{3}, \frac{5}{7}$　**19.** $0, 0, 6, 24, 60$
21. $-\frac{1}{2}, \frac{2}{3}, -\frac{3}{4}, \frac{4}{5}, -\frac{5}{6}$　**23.** -73　**25.** $\frac{44}{239}$
27. 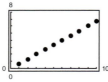　**29.**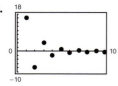

31.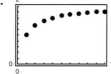

33. c　**34.** b　**35.** d　**36.** a　**37.** $a_n = 4n - 1$
39. $a_n = n^3 + 2$　**41.** $a_n = (-1)^{n+1}$
43. $a_n = \dfrac{(-1)^n(n+1)}{n+2}$　**45.** $a_n = \dfrac{n+1}{2n-1}$　**47.** $a_n = \dfrac{1}{n!}$
49. $a_n = \dfrac{3^{n-1}}{(n-1)!}$　**51.** $28, 24, 20, 16, 12$
53. $81, 27, 9, 3, 1$　**55.** $1, 2, 2, 3, \frac{7}{2}$
57. $1, 1, 2, 3, 5, 8, 13, 21, 34, 55, 89, 144$
$1, 2, \frac{3}{2}, \frac{5}{3}, \frac{8}{5}, \frac{13}{8}, \frac{21}{13}, \frac{34}{21}, \frac{55}{34}, \frac{89}{55}$
59. $5, 5, \frac{5}{2}, \frac{5}{6}, \frac{5}{24}$　**61.** $6, -24, 60, -120, 210$　**63.** $\frac{1}{30}$
65. $n + 1$　**67.** 90　**69.** $\frac{124}{429}$　**71.** 88　**73.** $\frac{13}{4}$
75. $\dfrac{3}{8}$　**77.** 1.33　**79.** $\displaystyle\sum_{i=1}^{9} \frac{1}{3i}$　**81.** $\displaystyle\sum_{i=1}^{8} \left[2\left(\frac{i}{8}\right) + 3 \right]$
83. $\displaystyle\sum_{i=1}^{6} (-1)^{i+1} 3^i$　**85.** $\displaystyle\sum_{i=1}^{7} \frac{i^2}{(i+1)!}$　**87.** $\displaystyle\sum_{i=1}^{5} \frac{2^i - 1}{2^{i+1}}$
89. (a) $\frac{7}{8}$　(b) $\frac{15}{16}$　(c) $\frac{31}{32}$
91. (a) $-\frac{3}{2}$　(b) $-\frac{5}{4}$　(c) $-\frac{11}{8}$　**93.** $\frac{2}{3}$　**95.** $\frac{7}{9}$
97. (a) $A_1 = \$10,087.50$, $A_2 \approx \$10,175.77$, $A_3 \approx \$10,264.80$,
$A_4 \approx \$10,354.62$, $A_5 \approx \$10,445.22$, $A_6 \approx \$10,536.62$,
$A_7 \approx \$10,628.81$, $A_8 \approx \$10,721.82$
(b) $\$14,169.09$
(c) No. $A_{80} \approx \$20,076.31 \neq 2A_{40} \approx \$28,338.18$
99. True by the Properties of Sums.　**101.** $\$500.95$
103. Proof　**105.** $\displaystyle\sum_{k=1}^{4} 3 = 3(4) = 12$
107. (a) 0 blue faces: 1
1 blue face: 6
2 blue faces: 12
3 blue faces: 8
(b)

Number of blue faces	0	1	2	3
$4 \times 4 \times 4$	8	24	24	8
$5 \times 5 \times 5$	27	54	36	8
$6 \times 6 \times 6$	64	96	48	8

(c) 0 blue faces: $(n-2)^3$
1 blue face: $6(n-2)^2$
2 blue faces: $12(n-2)$
3 blue faces: 8

Section 8.2　*(page 578)*

1. arithmetic; common　**3.** recursion　**5.** Not arithmetic
7. Arithmetic, $d = -2$　**9.** Arithmetic, $d = \frac{1}{4}$
11. Not arithmetic
13. $8, 11, 14, 17, 20$　　　**15.** $7, 3, -1, -5, -9$
Arithmetic, $d = 3$　　　　Arithmetic, $d = -4$
17. $-1, 1, -1, 1, -1$　　**19.** $2, 8, 24, 64, 160$
Not arithmetic　　　　　Not arithmetic
21. $a_n = 3n - 2$　**23.** $a_n = -8n + 108$
25. $a_n = -\frac{5}{2}n + \frac{13}{2}$　**27.** $a_n = \frac{10}{3}n + \frac{5}{3}$　**29.** $a_n = 3n + 85$
31. $5, 11, 17, 23, 29$　**33.** $2, -4, -10, -16, -22$
35. $-2, 2, 6, 10, 14$　**37.** $15, 19, 23, 27, 31$
39. $15, 13, 11, 9, 7$　**41.** -49　**43.** $\frac{31}{8}$　**45.** 110
47. -25　**49.** $10,000$　**51.** $15,100$　**53.** -7020
55. 1275　**57.** $129,250$　**59.** $-28,300$
61. b　**62.** d　**63.** c　**64.** a
65.

67.

69. (a) $\$40,000$　(b) $\$217,500$　**71.** 2430 seats
73. 784 ft　**75.** $\$375,000$; Answers will vary.
77. (a)

(b) $a_n = 229.25 + 36.75n$
(c)

(d) $\displaystyle\sum_{n=1}^{5} (229.25 + 36.75n)$; About 1698 stores
79. True. Given a_1 and a_2, $d = a_2 - a_1$ and $a_n = a_1 + (n-1)d$.

81. (a)

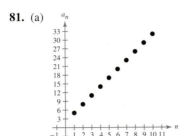

(b)

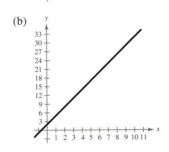

(c) The graph of $y = 3x + 2$ contains all points on the line. The graph of $a_n = 2 + 3n$ contains only points at the positive integers.

(d) The slope of the line and the common difference of the arithmetic sequence are equal.

83. $x, 3x, 5x, 7x, 9x, 11x, 13x, 15x, 17x, 19x$

85. When $n = 50$, $a_n = 2(50) - 1 = 99$.

87. (a) $4, 9, 16, 25, 36$ (b) $S_n = n^2; S_7 = 49 = 7^2$

   (c) $\dfrac{n}{2}[1 + (2n - 1)] = n^2$

Section 8.3   *(page 587)*

1. geometric; common   **3.** $a_1\left(\dfrac{1 - r^n}{1 - r}\right)$

5. Geometric, $r = 2$ **7.** Geometric, $r = 3$

9. Not geometric   **11.** Geometric, $r = -\sqrt{7}$

13. $4, 12, 36, 108, 324$ **15.** $1, \frac{1}{2}, \frac{1}{4}, \frac{1}{8}, \frac{1}{16}$ **17.** $1, e, e^2, e^3, e^4$

19. $3, 3\sqrt{5}, 15, 15\sqrt{5}, 75$ **21.** $2, 6x, 18x^2, 54x^3, 162x^4$

23. $a_n = 4\left(\dfrac{1}{2}\right)^{n-1}; \dfrac{1}{128}$ **25.** $a_n = 6\left(-\dfrac{1}{3}\right)^{n-1}; -\dfrac{2}{59,049}$

27. $a_n = 100e^{x(n-1)}; 100e^{8x}$ **29.** $a_n = \left(\sqrt{2}\right)^{n-1}; 32\sqrt{2}$

31. $a_n = 500(1.02)^{n-1}$; About 1082.372   **33.** $a_n = 64\left(\frac{1}{2}\right)^{n-1}$

35. $a_n = 9(2)^{n-1}$ **37.** $a_n = 6\left(-\frac{3}{2}\right)^{n-1}$ **39.** $13,122$

41. $\frac{1}{768}$ **43.** $a_3 = 9$ **45.** $a_6 = -2$

47. a   **48.** c   **49.** b   **50.** d

51.

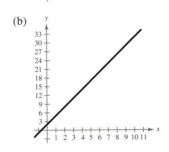

53.

55. 5461 **57.** $-14,706$ **59.** $29,921.311$ **61.** 1360.383

63. 1.600 **65.** $\displaystyle\sum_{n=1}^{7} 10(3)^{n-1}$ **67.** $\displaystyle\sum_{n=1}^{6} 0.1(4)^{n-1}$ **69.** 2

71. $\frac{2}{3}$ **73.** 5 **75.** 32 **77.** Undefined   **79.** $\frac{4}{11}$

81.

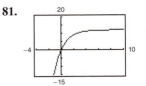

Horizontal asymptote: $y = 12$

Corresponds to the sum of the series

83. $29,412.25$ **85.** Answers will vary.   **87.** $1600

89. $273\frac{8}{9}$ in.^{2} **91.** $5,435,989.84$

93. False. A sequence is geometric when the ratios of consecutive terms are the same.

95. (a)

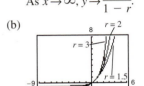

As $x \to \infty$, $y \to \dfrac{1}{1 - r}$.

(b)

As $x \to \infty$, $y \to \infty$.

Section 8.4   *(page 598)*

1. mathematical induction   **3.** arithmetic

5. $\dfrac{5}{(k + 1)(k + 2)}$ **7.** $(k + 1)^2(k + 4)^2$

9. $\dfrac{3}{(k + 3)(k + 4)}$ **11–39.** Proofs

41. $S_n = n(2n - 1)$; Proof   **43.** $S_n = \dfrac{n}{2(n + 1)}$; Proof

45. 120 **47.** 91 **49.** 979 **51.** 70 **53.** -3402

55. Linear; $a_n = 9n - 4$ **57.** Quadratic; $a_n = 2n^2 + 2$

59. Quadratic; $a_n = 4n^2 - 5$

61. $0, 3, 6, 9, 12, 15$

   First differences: $3, 3, 3, 3, 3$

   Second differences: $0, 0, 0, 0$

   Linear

63. $4, 10, 19, 31, 46, 64$

   First differences: $6, 9, 12, 15, 18$

   Second differences: $3, 3, 3, 3$

   Quadratic

65. $3, 7, 16, 32, 57, 93$

   First differences: $4, 9, 16, 25, 36$

   Second differences: $5, 7, 9, 11$

   Neither

67. $5, 3, 9, 7, 13, 11$

   First differences: $-2, 6, -2, 6, -2$

   Second differences: $8, -8, 8, -8$

   Neither

69. $a_n = n^2 - n + 3$ **71.** $a_n = \frac{1}{2}n^2 + 2n - 1$

73. $a_n = n^2 + 4n - 5$

75. (a) 16, 15, 15, 15, 13; *Sample answer:* $a_n = 15n + 4636$

    (b) $a_n \approx 14.9n + 4637$; The models are similar.

    (c) Part (a): 4,951,000,

        Part (b): 4,949,900;

        The values are similar.

77. False. P_1 must be proven to be true.

Section 8.5 *(page 605)*

1. expanding   **3.** Binomial Theorem; Pascal's Triangle

5. 10   **7.** 1   **9.** 210   **11.** 4950   **13.** 20   **15.** 5

17. $x^6 + 6x^5 + 15x^4 + 20x^3 + 15x^2 + 6x + 1$

19. $y^3 - 9y^2 + 27y - 27$ **21.** $r^3 + 9r^2s + 27rs^2 + 27s^3$

23. $243a^5 - 1620a^4b + 4320a^3b^2 - 5760a^2b^3$
       $+ 3840ab^4 - 1024b^5$

25. $a^4 + 24a^3 + 216a^2 + 864a + 1296$

27. $y^6 - 6y^5 + 15y^4 - 20y^3 + 15y^2 - 6y + 1$

29. $81 - 216z + 216z^2 - 96z^3 + 16z^4$

31. $x^5 + 10x^4y + 40x^3y^2 + 80x^2y^3 + 80xy^4 + 32y^5$

33. $x^8 + 4x^6y^2 + 6x^4y^4 + 4x^2y^6 + y^8$

35. $\dfrac{1}{x^5} + \dfrac{5y}{x^4} + \dfrac{10y^2}{x^3} + \dfrac{10y^3}{x^2} + \dfrac{5y^4}{x} + y^5$

37. $2x^4 - 24x^3 + 113x^2 - 246x + 207$ **39.** $120x^7y^3$

41. $360x^3y^2$ **43.** $1{,}259{,}712x^2y^7$ **45.** $-4{,}330{,}260{,}000y^9x^3$

47. 160   **49.** 720   **51.** $-6{,}300{,}000$ **53.** 210

55. $x^{3/2} + 15x + 75x^{1/2} + 125$

57. $x^2 - 3x^{4/3}y^{1/3} + 3x^{2/3}y^{2/3} - y$

59. $81t^2 + 108t^{7/4} + 54t^{3/2} + 12t^{5/4} + t$

61. $3x^2 + 3xh + h^2,\ h \neq 0$

63. $6x^5 + 15x^4h + 20x^3h^2 + 15x^2h^3 + 6xh^4 + h^5,\ h \neq 0$

65. $\dfrac{1}{\sqrt{x + h} + \sqrt{x}},\ h \neq 0$ **67.** -4 **69.** $2035 + 828i$

71. 1   **73.** 1.172   **75.** 510,568.785   **77.** 0.273

79. 0.171

81.

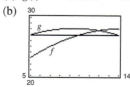

The graph of g is shifted four units to the left of the graph of f.
$g(x) = x^3 + 12x^2 + 44x + 48$

83. Fibonacci sequence

85. (a) $g(t) = -0.056t^2 + 1.06t + 23.1$

    (b)

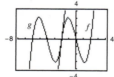

    (c) 2010

87. True. The coefficients from the Binomial Theorem can be used to find the numbers in Pascal's Triangle.

89. The first and last numbers in each row are 1. Every other number in each row is formed by adding the two numbers immediately above the number.

91.

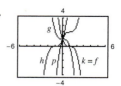

k, f; $k(x)$ is the expansion of $f(x)$.

93–95. Proofs

97.

n	r	$_nC_r$	$_nC_{n-r}$
9	5	126	126
7	1	7	7
12	4	495	495
6	0	1	1
10	7	120	120

This illustrates the symmetry of Pascal's Triangle.

Section 8.6 *(page 615)*

1. Fundamental Counting Principle   **3.** $_nP_r = \dfrac{n!}{(n-r)!}$

5. combinations   **7.** 6   **9.** 5   **11.** 3   **13.** 8

15. 30   **17.** 30   **19.** 64   **21.** 175,760,000

23. (a) 900   (b) 648   (c) 180   (d) 600

25. 64,000   **27.** (a) 40,320   (b) 384   **29.** 120

31. 20   **33.** 132   **35.** 2730   **37.** 5,527,200

39. 504   **41.** 1,816,214,400   **43.** 420   **45.** 2520

47. ABCD, ABDC, ACBD, ACDB, ADBC, ADCB, BACD, BADC, CABD, CADB, DABC, DACB, BCAD, BDAC, CBAD, CDAB, DBAC, DCAB, BCDA, BDCA, CBDA, CDBA, DBCA, DCBA

49. 15   **51.** 1   **53.** 120   **55.** 38,760

57. AB, AC, AD, AE, AF, BC, BD, BE, BF, CD, CE, CF, DE, DF, EF

59. 5,586,853,480   **61.** 324,632

63. (a) 7315   (b) 693   (c) 12,628

65. (a) 3744   (b) 24   **67.** 292,600   **69.** 5   **71.** 20

73. 36   **75.** $n = 2$ **77.** $n = 3$ **79.** $n = 5$ or $n = 6$

81. $n = 10$ **83.** False. It is an example of a combination.

85. $_{10}P_6 > {}_{10}C_6$. Changing the order of any of the six elements selected results in a different permutation but the same combination.

87–89. Proofs

91. No. For some calculators the number is too great.

Section 8.7 *(page 626)*

1. experiment; outcomes   **3.** probability

5. mutually exclusive   **7.** complement

9. $\{(H, 1), (H, 2), (H, 3), (H, 4), (H, 5), (H, 6),$
    $(T, 1), (T, 2), (T, 3), (T, 4), (T, 5), (T, 6)\}$

11. $\{ABC, ACB, BAC, BCA, CAB, CBA\}$

13. $\{AB, AC, AD, AE, BC, BD, BE, CD, CE, DE\}$

15. $\frac{3}{8}$ **17.** $\frac{1}{2}$ **19.** $\frac{7}{8}$ **21.** $\frac{3}{13}$ **23.** $\frac{3}{26}$ **25.** $\frac{5}{36}$

27. $\frac{11}{12}$ **29.** $\frac{1}{3}$ **31.** $\frac{1}{5}$ **33.** $\frac{2}{5}$

35. (a) 996,000   (b) $\frac{9}{50}$ (c) $\frac{27}{50}$ (d) $\frac{4}{25}$

37. (a) $\frac{13}{16}$ (b) $\frac{3}{16}$ (c) $\frac{1}{32}$ **39.** 19%

41. (a) $\frac{21}{1292}$ (b) $\frac{225}{646}$ (c) $\frac{49}{323}$ **43.** (a) $\frac{1}{120}$ (b) $\frac{1}{24}$

45. (a) $\frac{5}{13}$ (b) $\frac{1}{2}$ (c) $\frac{4}{13}$ **47.** (a) $\frac{14}{55}$ (b) $\frac{12}{55}$ (c) $\frac{54}{55}$

49. (a) $\frac{1}{4}$ (b) $\frac{1}{2}$ (c) $\frac{841}{1600}$ (d) $\frac{1}{40}$ **51.** 0.27   **53.** $\frac{4}{5}$

55. 0.71   **57.** $\frac{11}{25}$

59. (a) 0.9702   (b) 0.0002   (c) 0.9998

61. (a) $\frac{1}{38}$ (b) $\frac{9}{19}$ (c) $\frac{10}{19}$ (d) $\frac{1}{1444}$ (e) $\frac{729}{6859}$ **63.** $\frac{7}{16}$

65. True. Two events are independent when the occurrence of one has no effect on the occurrence of the other.

67. (a) As you consider successive people with distinct birthdays, the probabilities must decrease to take into account the birth dates already used. Because the birth dates of people are independent events, multiply the respective probabilities of distinct birthdays.
  (b) $\frac{365}{365} \cdot \frac{364}{365} \cdot \frac{363}{365} \cdot \frac{362}{365}$ (c) Answers will vary.
  (d) Q_n is the probability that the birthdays are *not* distinct, which is equivalent to at least two people having the same birthday.
  (e)

n	10	15	20	23	30	40	50
P_n	0.88	0.75	0.59	0.49	0.29	0.11	0.03
Q_n	0.12	0.25	0.41	0.51	0.71	0.89	0.97

  (f) 23; $Q_n > 0.5$ for $n \geq 23$.

Review Exercises   (page 632)

1. 15, 9, 7, 6, $\frac{27}{5}$ **3.** 120, 60, 20, 5, 1   **5.** $a_n = 2(-1)^n$

7. $a_n = \dfrac{4}{n}$ **9.** $\dfrac{1}{20}$ **11.** $\dfrac{1}{n(n+1)}$ **13.** $\dfrac{205}{24}$

15. $\displaystyle\sum_{k=1}^{20} \frac{1}{2k}$ **17.** $\dfrac{4}{9}$

19. (a) $A_1 = \$10,018.75$
  $A_2 \approx \$10,037.54$
  $A_3 \approx \$10,056.36$
  $A_4 \approx \$10,075.21$
  $A_5 \approx \$10,094.10$
  $A_6 \approx \$10,113.03$
  $A_7 \approx \$10,131.99$
  $A_8 \approx \$10,150.99$
  $A_9 \approx \$10,170.02$
  $A_{10} \approx \$10,189.09$
  (b) $\$12,520.59$

21. Arithmetic, $d = -6$ **23.** Not arithmetic

25. $a_n = 12n - 5$ **27.** $a_n = -18n + 150$

29. 4, 21, 38, 55, 72   **31.** 45,450   **33.** 80   **35.** 88

37. (a) $\$51,600$ (b) $\$238,500$ **39.** Geometric, $r = 3$

41. Geometric, $r = -3$ **43.** 2, 30, 450, 6750, 101,250

45. 9, 6, 4, $\frac{8}{3}$, $\frac{16}{9}$ or 9, -6, 4, $-\frac{8}{3}$, $\frac{16}{9}$

47. $a_n = 100(1.05)^{n-1}$; About 155.133

49. $a_n = 18\left(\frac{1}{2}\right)^{n-1}$; $-\frac{9}{256}$ **51.** 127   **53.** $\frac{15}{16}$

55. 31   **57.** 23.056   **59.** 8   **61.** 12

63. (a) $a_n = 120,000(0.7)^n$ (b) $\$20,168.40$

65–67. Proofs   **69.** $S_n = n(2n+7)$; Proof

71. $S_n = \frac{5}{2}\left[1 - \left(\frac{3}{5}\right)^n\right]$; Proof   **73.** 2850

75. 5, 10, 15, 20, 25
  First differences: 5, 5, 5, 5
  Second differences: 0, 0, 0
  Linear

77. 15   **79.** 21   **81.** $x^4 + 16x^3 + 96x^2 + 256x + 256$

83. $64 - 240x + 300x^2 - 125x^3$ **85.** 6   **87.** 10,000

89. 120   **91.** 225,792,840   **93.** (a) $\frac{1}{5}$ (b) $\frac{3}{5}$

95. (a) 43%   (b) 82%   **97.** $\frac{1}{1296}$ **99.** $\frac{3}{4}$

101. False. $\dfrac{(n+2)!}{n!} = \dfrac{(n+2)(n+1)n!}{n!} = (n+2)(n+1)$

103. True by the Properties of Sums.

105. The set of positive integers

107. Each term of the sequence is defined in terms of preceding terms.

Chapter Test   (page 635)

1. $-\dfrac{1}{5}, \dfrac{1}{8}, -\dfrac{1}{11}, \dfrac{1}{14}, -\dfrac{1}{17}$ **2.** $a_n = \dfrac{n+2}{n!}$

3. 60, 73, 86; 329   **4.** $a_n = -3n + 60$ **5.** $a_n = \frac{7}{2}(2)^n$

6. 86,100   **7.** 477   **8.** 4   **9.** $-\frac{1}{4}$ **10.** Proof

11. $x^4 + 24x^3y + 216x^2y^2 + 864xy^3 + 1296y^4$

12. $3x^5 - 30x^4 + 124x^3 - 264x^2 + 288x - 128$

13. $-22,680$ **14.** (a) 72   (b) 328,440

15. (a) 330   (b) 720,720   **16.** 26,000   **17.** 720

18. $\dfrac{1}{15}$ **19.** $\dfrac{1}{27,405}$ **20.** 10%

Cumulative Test for Chapters 6–8   (page 636)

1. $(1, 2), \left(-\frac{3}{2}, \frac{3}{4}\right)$ **2.** $(-3, -1)$ **3.** $(5, -2, -2)$

4. $(1, -2, 1)$ **5.** $\$0.75$ mixture: 120 lb; $\$1.25$ mixture: 80 lb

6. $y = \dfrac{1}{4}x^2 - 2x + 6$ **7.** $-\dfrac{3}{x} + \dfrac{5x - 1}{x^2 + 2}$

8.

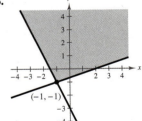

9.

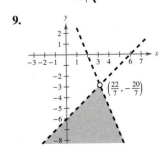

10.

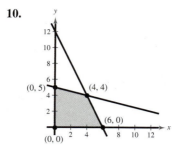

Maximum at $(4, 4)$: 20
Minimum at $(0, 0)$: 0

11. $\begin{bmatrix} -1 & 2 & -1 & \vdots & 9 \\ 2 & -1 & 2 & \vdots & -9 \\ 3 & 3 & -4 & \vdots & 7 \end{bmatrix}$ **12.** $(-2, 3, -1)$

13. $\begin{bmatrix} -3 & 8 \\ 6 & 1 \end{bmatrix}$ **14.** $\begin{bmatrix} 8 & -19 \\ 12 & 9 \end{bmatrix}$

15. $\begin{bmatrix} -13 & 6 & -4 \\ 18 & 4 & 4 \end{bmatrix}$ **16.** Not possible

17. $\begin{bmatrix} 19 & 3 \\ 6 & 22 \end{bmatrix}$ **18.** $\begin{bmatrix} 28 & 19 \\ -6 & -3 \end{bmatrix}$

19. $\begin{bmatrix} -175 & 37 & -13 \\ 95 & -20 & 7 \\ 14 & -3 & 1 \end{bmatrix}$ **20.** 203

21. $(0, -2), (3, -5), (0, -5)\ (3, -2)$

22. Gym shoes: \$2539 million
Jogging shoes: \$2362 million
Walking shoes: \$4418 million

23. $(-5, 4)$ **24.** $(-3, 4, 2)$ **25.** 9

26. $\frac{1}{5}, -\frac{1}{7}, \frac{1}{9}, -\frac{1}{11}, \frac{1}{13}$ **27.** $a_n = \dfrac{(n + 1)!}{n + 3}$

28. 1536   **29.** (a) 65.4   (b) $a_n = 3.2n + 1.4$

30. 3, 6, 12, 24, 48   **31.** $\frac{190}{9}$ **32.** Proof

33. $w^4 - 36w^3 + 486w^2 - 2916w + 6561$ **34.** 2184

35. 600   **36.** 70   **37.** 462   **38.** 453,600

39. 151,200   **40.** 720   **41.** $\frac{1}{4}$

Problem Solving   (page 641)

1. (a)

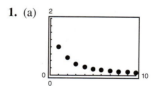

(b) 0

(c)

n	1	10	100	1000	10,000
a_n	1	0.1089	0.0101	0.0010	0.0001

(d) 0

3. $s_d = \dfrac{a_1}{1 - r} = \dfrac{20}{1 - \frac{1}{2}} = 40$

This represents the total distance Achilles ran.

$s_t = \dfrac{a_1}{1 - r} = \dfrac{1}{1 - \frac{1}{2}} = 2$

This represents the total amount of time Achilles ran.

5. (a) Arithmetic sequence, difference $= d$
(b) Arithmetic sequence, difference $= dC$
(c) Not an arithmetic sequence

7. (a) 7, 22, 11, 34, 17, 52, 26, 13, 40, 20, 10, 5, 16, 8, 4, 2, 1, 4, 2, 1
(b) $a_1 = 4$: 4, 2, 1, 4, 2, 1, 4, 2, 1, 4
    $a_1 = 5$: 5, 16, 8, 4, 2, 1, 4, 2, 1, 4
    $a_1 = 12$: 12, 6, 3, 10, 5, 16, 8, 4, 2, 1
    Eventually, the terms repeat: 4, 2, 1.

9. Proof   **11.** $S_n = \left(\dfrac{1}{2}\right)^{n-1}$; $A_n = \dfrac{\sqrt{3}}{4} S_n^2$ **13.** $\frac{1}{3}$

15. (a) 3 to 7; 7 to 3   (b) 30 marbles

(c) $P(E) = \dfrac{\text{odds in favor of } E}{\text{odds in favor of } E + 1}$

(d) Odds in favor of event $E = \dfrac{P(E)}{P(E')}$

Appendix A   *(page A6)*

1. numerator

3. The middle term needs to be included.
$(x + 3)^2 = x^2 + 6x + 9$

5. $\sqrt{x + 9}$ cannot be simplified.

7. Divide out common factors, not common terms.
$\dfrac{2x^2 + 1}{5x}$ cannot be simplified.

9. The exponent also applies to the coefficient. $(4x)^2 = 16x^2$

11. To add fractions, first find a common denominator.
$\dfrac{3}{x} + \dfrac{4}{y} = \dfrac{3y + 4x}{xy}$

13. $(x + 2)^{-1/2}(3x + 2)$

15. $2x(2x - 1)^{-1/2}[2x^2(2x - 1)^2 - 1]$ **17.** $5x + 3$

19. $2x^2 + x + 15$ **21.** $1 - 5x$ **23.** $3x - 1$ **25.** $\frac{1}{3}$

27. 2   **29–33.** Answers will vary.   **35.** $7(x + 3)^{-5}$

37. $2x^5(3x + 5)^{-4}$ **39.** $\frac{4}{3}x^{-1} + 4x^{-4} - 7x(2x)^{-1/3}$

41. $\dfrac{x}{3} + 2 + \dfrac{4}{x}$ **43.** $4x^{8/3} - 7x^{5/3} + \dfrac{1}{x^{1/3}}$

45. $\dfrac{3}{x^{1/2}} - 5x^{3/2} - x^{7/2}$ **47.** $\dfrac{-7x^2 - 4x + 9}{(x^2 - 3)^3(x + 1)^4}$

49. $\dfrac{27x^2 - 24x + 2}{(6x + 1)^4}$ **51.** $\dfrac{-1}{(x + 3)^{2/3}(x + 2)^{7/4}}$

53. $\dfrac{4x - 3}{(3x - 1)^{4/3}}$ **55.** $\dfrac{x}{x^2 + 4}$

57. $\dfrac{(3x - 2)^{1/2}(15x^2 - 4x + 45)}{2(x^2 + 5)^{1/2}}$

59. (a) Answers will vary.

(b)

x	-2	-1	$-\frac{1}{2}$	0	1	2	$\frac{5}{2}$
y_1	-8.7	-2.9	-1.1	0	2.9	8.7	12.5
y_2	-8.7	-2.9	-1.1	0	2.9	8.7	12.5

(c) Answers will vary.

61. You cannot move term-by-term from the denominator to the numerator.

Technology

Chapter P   *(page 52)*

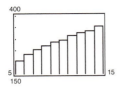

Chapter 1   *(page 122)*

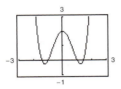

The *x*-intercepts occur at $x = \pm 1, \pm\sqrt{2}$.

Chapter 2   *(page 165)*

The lines appear perpendicular on the square setting.

(page 178)

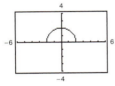

Domain: $[-2, 2]$

Domain: $(-\infty, -2] \cup [2, \infty)$
Yes, for -2 and 2.

(page 201)

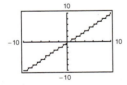

The graph in *dot* mode illustrates that the range is the set of all integers.

Chapter 5   *(page 399)*

$S = 0.00036(2.130)^t$

The exponential regression model has the same coefficient as the model in Example 1. However, the model given in Example 1 contains the natural exponential function.

Chapter 6   *(page 424)*

The point of intersection (7000, 5000) agrees with the solution.

(page 426)

$(2, 0); (0, -1), (2, 1);$ None

Chapter 7   *(page 510)*

$$\begin{bmatrix} 1 & 1 \\ 1 & -5 \end{bmatrix}$$

Checkpoints

Chapter P

Section P.1

1. (a) Natural numbers: $\left\{\frac{6}{3}, 8\right\}$ (b) Whole numbers: $\left\{\frac{6}{3}, 8\right\}$
   (c) Integers: $\left\{\frac{6}{3}, -1, 8, -22\right\}$
   (d) Rational numbers: $\left\{-\frac{1}{4}, \frac{6}{3}, -7.5, -1, 8, -22\right\}$
   (e) Irrational numbers: $\left\{-\pi, \frac{1}{2}\sqrt{2}\right\}$

2.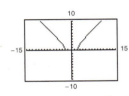

3. (a) $1 > -5$ (b) $\frac{3}{2} < 7$ (c) $-\frac{2}{3} > -\frac{3}{4}$

4. (a) The inequality $x > -3$ denotes all real numbers greater than -3.
   (b) The inequality $0 < x \le 4$ denotes all real numbers between 0 and 4, not including 0, but including 4.

5. The interval consists of all real numbers greater than or equal to -2 and less than 5.

6. $-2 \le x < 4$ 7. (a) 1   (b) $-\frac{3}{4}$ (c) $\frac{2}{3}$ (d) -0.7

8. (a) 1   (b) -1

9. (a) $|-3| < |4|$ (b) $-|-4| = -|4|$
   (c) $|-3| > -|-3|$

10. (a) 58   (b) 12   (c) 12

11. Terms: $-2x, 4$; Coefficients: $-2, 4$ 12. -5

13. (a) Commutative Property of Addition
    (b) Associative Property of Multiplication
    (c) Distributive Property

14. (a) $\frac{x}{10}$ (b) $\frac{x}{2}$

Section P.2

1. (a) -81 (b) 81   (c) 27   (d) $\frac{1}{27}$

2. (a) $-\frac{1}{16}$ (b) 64

3. (a) $-2x^2y^4$ (b) 1   (c) $-125z^5$ (d) $\frac{9x^4}{y^4}$

4. (a) $\frac{2}{a^2}$ (b) $\frac{b^5}{5a^4}$ (c) $\frac{10}{x}$ (d) $-2x^3$

5. 4.585×10^4 6. -0.002718 7. $864,000,000$

8. (a) -12 (b) Not a real number   (c) $\frac{5}{8}$ (d) $-\frac{2}{3}$

9. (a) 5   (b) 25   (c) x (d) $\sqrt[4]{x}$

10. (a) $4\sqrt{2}$ (b) $5\sqrt[3]{2}$ (c) $2a^2\sqrt{6a}$ (d) $-3x\sqrt[3]{5}$

11. (a) $9\sqrt{2}$ (b) $(3x - 2)\sqrt[3]{3x^2}$

12. (a) $\frac{5\sqrt{2}}{6}$ (b) $\frac{\sqrt[3]{5}}{5}$ 13. $2(\sqrt{6} + \sqrt{2})$ 14. $\frac{2}{3(2 + \sqrt{2})}$

15. (a) $27^{1/3}$ (b) $x^{3/2}y^{5/2}z^{1/2}$ (c) $3x^{5/3}$

16. (a) $\frac{1}{\sqrt{x^2 - 7}}$ (b) $-3\sqrt[3]{bc^2}$ (c) $\sqrt[4]{a^3}$ (d) $\sqrt[5]{x^4}$

17. (a) $\frac{1}{25}$ (b) $-12x^{5/3}y^{9/10}, x \ne 0, y \ne 0$
    (c) $\sqrt[4]{3}$ (d) $(3x + 2)^2, x \ne -\frac{2}{3}$

Section P.3

1. Standard form: $-7x^3 + 2x + 6$
   Degree: 3; Leading coefficient: -7
2. $2x^3 - x^2 + x + 6$　　**3.** $3x^2 - 16x + 5$
4. $x^4 + 2x^2 + 9$　　**5.** $9x^2 - 4$　　**6.** $x^2 + 20x + 100$
7. $64x^3 - 48x^2 + 12x - 1$　　**8.** $x^2 - 9y^2 - 4x + 4$
9. Volume $= 4x^3 - 44x^2 + 120x$
   $x = 2$: Volume $= 96$ in.^{3}
   $x = 3$: Volume $= 72$ in.^{3}

Section P.4

1. (a) $5x^2(x - 3)$　　(b) $-3(1 - 2x + 4x^3)$
   (c) $(x + 1)(x^2 - 2)$
2. $4(5 + y)(5 - y)$　　**3.** $(x - 1 + 3y^2)(x - 1 - 3y^2)$
4. $(3x - 5)^2$　　**5.** $(4x - 1)(16x^2 + 4x + 1)$
6. (a) $(x + 6)(x^2 - 6x + 36)$　　(b) $5(y + 3)(y^2 - 3y + 9)$
7. $(x + 3)(x - 2)$　　**8.** $(2x - 3)(x - 1)$
9. $(x^2 - 5)(x + 1)$　　**10.** $(2x - 3)(x + 4)$

Section P.5

1. (a) All nonnegative real numbers x
   (b) All real numbers x such that $x \geq -7$
   (c) All real numbers x such that $x \neq 0$
2. $\dfrac{4}{x - 6}, x \neq -3$　　**3.** $-\dfrac{3x + 2}{5 + x}, x \neq 1$
4. $\dfrac{5(x - 5)}{(x - 6)(x - 3)}, x \neq -3, x \neq -\dfrac{1}{3}, x \neq 0$
5. $x + 1, x \neq \pm 1$　　**6.** $\dfrac{x^2 + 1}{(2x - 1)(x + 2)}$
7. $\dfrac{7x^2 - 13x - 16}{x(x + 2)(x - 2)}$　　**8.** $\dfrac{3(x + 3)}{(x - 3)(x + 2)}$　　**9.** $-\dfrac{1}{(x - 1)^{4/3}}$
10. $\dfrac{2(x + 1)(x - 1)}{(x^2 - 2)^{3/2}}$　　**11.** $\dfrac{1}{\sqrt{9 + h} + 3}, h \neq 0$

Section P.6

1.
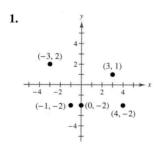

2.
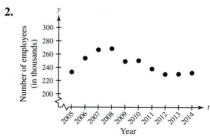

3. $\sqrt{37} \approx 6.08$
4. $d_1 = \sqrt{45}, d_2 = \sqrt{20}, d_3 = \sqrt{65}$
   $\left(\sqrt{45}\right)^2 + \left(\sqrt{20}\right)^2 = \left(\sqrt{65}\right)^2$

5. $(1, -1)$　　**6.** $\sqrt{709} \approx 27$ yd　　**7.** \$4.8 billion
8. $(1, 2), (1, -2), (-1, 0), (-1, -4)$

Chapter 1

Section 1.1

1. (a) No　　(b) Yes
2. (a)　　　　　　　　　　　(b)
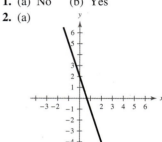
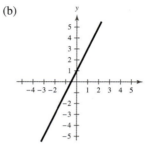

3. (a)　　　　　　　　　　　(b)
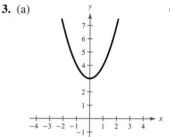
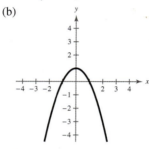

4. x-intercepts: $(0, 0), (-5, 0)$,　　**5.** x-axis symmetry
   y-intercept: $(0, 0)$
6.　　　　　　　　　　　7.
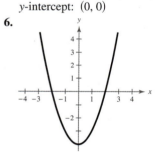
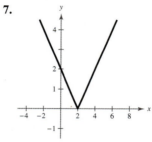

8. $(x + 3)^2 + (y + 5)^2 = 25$　　**9.** About 221 lb

Section 1.2

1. (a) -4　　(b) 8　　**2.** 26　　**3.** $-\dfrac{18}{5}$　　**4.** No solution
5. (a) x-intercept: $\left(-\dfrac{2}{3}, 0\right)$　　(b) x-intercept: $(3, 0)$
   y-intercept: $(0, -2)$　　　　y-intercept: $(0, 5)$
6. (a) $(0, 91.4)$; There were about 91,400 male participants in 2010.
   (b) 2020

Section 1.3

1. \$1100　　**2.** 20%　　**3.** \$64,000　　**4.** 14 ft by 42 ft
5. 1 hr 24 min　　**6.** About 122 ft
7. \$2375 at $2\frac{1}{2}$%, \$2625 at $3\frac{1}{2}$%
8. 24-inch television: \$10,000; 50-inch television: \$20,000
9. About 2.97 in.

Section 1.4

1. $-1, \dfrac{5}{2}$　　**2.** (a) $\pm 2\sqrt{3}$　　(b) $1 \pm \sqrt{10}$　　**3.** $2 \pm \sqrt{5}$

4. $1 \pm \dfrac{\sqrt{2}}{2}$ **5.** $\dfrac{5}{3} \pm \dfrac{\sqrt{31}}{3}$ **6.** $-\dfrac{1}{3} \pm \dfrac{\sqrt{31}}{3}$

7. 14 ft by 8 ft   **8.** 3.5 sec   **9.** 2013   **10.** About 27.02 ft

Section 1.5

1. (a) $12 - i$ (b) $-2 + 7i$ (c) i (d) 0
2. (a) $-15 + 10i$ (b) $18 - 6i$ (c) 41   (d) $12 + 16i$
3. (a) 45   (b) 29   **4.** $\dfrac{3}{5} + \dfrac{4}{5}i$ **5.** $-2\sqrt{7}$
6. $-\dfrac{7}{8} \pm \dfrac{\sqrt{23}}{8}i$

Section 1.6

1. $0, \pm\dfrac{2\sqrt{3}}{3}$ **2.** (a) $5, \pm\sqrt{2}$ (b) $0, -\dfrac{3}{2}, 6$
3. (a) $\pm2, \pm\sqrt{3}$ (b) $\pm\dfrac{1}{3}, \pm2$ **4.** -9 **5.** $-59, 69$
6. $-4, -1$ **7.** $-2, 6$ **8.** 32 students   **9.** 7%

Section 1.7

1. (a) $1 \le x \le 3$; Bounded   (b) $-1 < x < 6$; Bounded
(c) $x < 4$; Unbounded   (d) $x \ge 0$; Unbounded
2. $x \le 2$

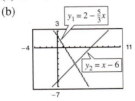

3. (a) $x < 3$
(b)

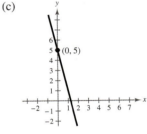

$y_1 > y_2$ for $x < 3$.
4. $(-3, 2)$ **5.** $[16, 24]$

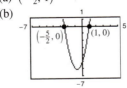

6. More than 68 hours
7. You might have been overcharged by as much as $0.16 or undercharged by as much as $0.15.

Section 1.8

1. $(-4, 5)$

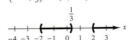

2. (a) $\left(-\dfrac{5}{2}, 1\right)$
(b)

The graph is below the x-axis when x is greater than $-\dfrac{5}{2}$ and less than 1. So, the solution set is $\left(-\dfrac{5}{2}, 1\right)$.

3. $\left(-2, \dfrac{1}{3}\right) \cup (2, \infty)$

4. (a) The solution set is empty.
(b) The solution set consists of the single real number $\{-2\}$.
(c) The solution set consists of all real numbers except $x = 3$.
(d) The solution set is all real numbers.

5. (a) $\left(-\infty, \dfrac{11}{4}\right] \cup (3, \infty)$ (b) $(-\infty, -17) \cup (6, \infty)$

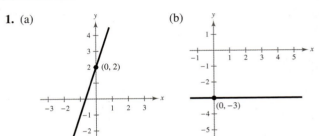

6. $180{,}000 \le x \le 300{,}000$ **7.** $(-\infty, 2] \cup [5, \infty)$

Chapter 2

Section 2.1

1. (a)   (b)

(c)

2. (a) 2   (b) $-\dfrac{3}{2}$ (c) Undefined   (d) 0
3. (a) $y = 2x - 13$ (b) $y = -\dfrac{2}{3}x + \dfrac{5}{3}$ (c) $y = 1$
4. (a) $y = \dfrac{5}{3}x + \dfrac{23}{3}$ (b) $y = -\dfrac{3}{5}x - \dfrac{7}{5}$ **5.** Yes
6. The y-intercept, $(0, 1500)$, tells you that the initial value of a copier at the time it is purchased is $1500. The slope, $m = -300$, tells you that the value of the copier decreases by $300 each year after it is purchased.
7. $y = -4125x + 24{,}750$ **8.** $y = 0.7t + 4.4$; $9.3 billion

Section 2.2

1. (a) Not a function   (b) Function
2. (a) Not a function   (b) Function
3. (a) -2 (b) -38 (c) $-3x^2 + 6x + 7$
4. $f(-2) = 5, f(2) = 1, f(3) = 2$ **5.** ±4 **6.** $-4, 3$
7. (a) $\{-2, -1, 0, 1, 2\}$
(b) All real numbers x except $x = 3$
(c) All real numbers r such that $r > 0$
(d) All real numbers x such that $x \ge 16$
8. (a) $S(r) = 10\pi r^2$ (b) $S(h) = \dfrac{5}{8}\pi h^2$ **9.** No
10. 2009: 772   2013: 1277
2010: 841   2014: 1437
2011: 910   2015: 1597
2012: 1117
11. $2x + h + 2, h \ne 0$

Section 2.3

1. (a) All real numbers x except $x = -3$
(b) $f(0) = 3; f(3) = -6$ (c) $(-\infty, 3]$
2. Function
3. (a) $x = -8, x = \dfrac{3}{2}$ (b) $t = 25$ (c) $x = \pm\sqrt{2}$

4.

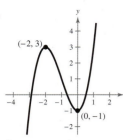

Increasing on $(-\infty, -2)$ and $(0, \infty)$

Decreasing on $(-2, 0)$

5. $(-0.88, 6.06)$ **6.** (a) -3 (b) 0

7. (a) 20 ft/sec   (b) $\frac{140}{3}$ ft/sec

8. (a) Neither; No symmetry   (b) Even; y-axis symmetry
(c) Odd; Origin symmetry

Section 2.4

1. $f(x) = -\frac{5}{2}x + 1$ **2.** $f\left(-\frac{3}{2}\right) = 0, f(1) = 3, f\left(-\frac{5}{2}\right) = -1$

3.

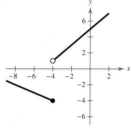

Section 2.5

1. (a)

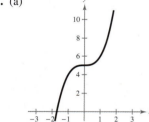

(b)

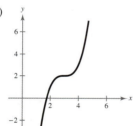

2. $j(x) = -(x + 3)^4$

3. (a) The graph of g is a reflection in the x-axis of the graph of f.
(b) The graph of h is a reflection in the y-axis of the graph of f.

4. (a) The graph of g is a vertical stretch of the graph of f.
(b) The graph of h is a vertical shrink of the graph of f.

5. (a) The graph of g is a horizontal shrink of the graph of f.
(b) The graph of h is a horizontal stretch of the graph of f.

Section 2.6

1. $x^2 - x + 1; 3$ **2.** $x^2 + x - 1; 11$ **3.** $x^2 - x^3; -18$

4. $\left(\dfrac{f}{g}\right)(x) = \dfrac{\sqrt{x - 3}}{\sqrt{16 - x^2}}$; Domain: $[3, 4)$

$\left(\dfrac{g}{f}\right)(x) = \dfrac{\sqrt{16 - x^2}}{\sqrt{x - 3}}$; Domain: $(3, 4]$

5. (a) $8x^2 + 7$ (b) $16x^2 + 80x + 101$ (c) 9

6. All real numbers x **7.** $f(x) = \frac{1}{5}\sqrt[3]{x}, g(x) = 8 - x$

8. (a) $(N \circ T)(t) = 32t^2 + 36t + 204$ (b) About 4.5 h

Section 2.7

1. $f^{-1}(x) = 5x, f(f^{-1}(x)) = \frac{1}{5}(5x) = x, f^{-1}(f(x)) = 5\left(\frac{1}{5}x\right) = x$

2. $g(x) = 7x + 4$

3.

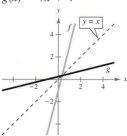

4.

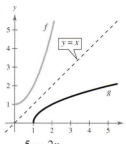

5. (a) Yes   (b) No   **6.** $f^{-1}(x) = \dfrac{5 - 2x}{x + 3}$

7. $f^{-1}(x) = x^3 - 10$

Chapter 3

Section 3.1

1. (a)

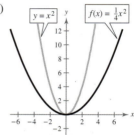

The graph of $f(x) = \frac{1}{4}x^2$ is broader than the graph of $y = x^2$.

(b)

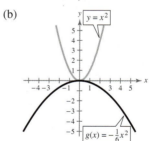

The graph of $g(x) = -\frac{1}{6}x^2$ is a reflection in the x-axis and is broader than the graph of $y = x^2$.

(c)

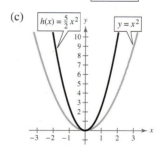

The graph of $h(x) = \frac{5}{2}x^2$ is narrower than the graph of $y = x^2$.

(d)

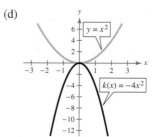

The graph of $k(x) = -4x^2$ is a reflection in the x-axis and is narrower than the graph of $y = x^2$.

2.

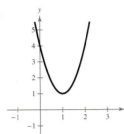

Vertex: $(1, 1)$
Axis: $x = 1$

3.

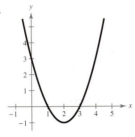

Vertex: $(2, -1)$
x-intercepts: $(1, 0), (3, 0)$

4. $y = (x + 4)^2 + 11$ **5.** About 39.7 ft

Section 3.2

1. (a)

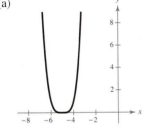

(b)

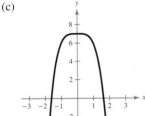

(c)

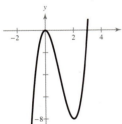

(d)

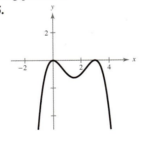

2. (a) Falls to the left, rises to the right
(b) Rises to the left, falls to the right

3. Real zeros: $x = 0, x = 6$; Turning points: 2

4.

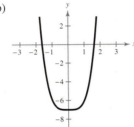

5.

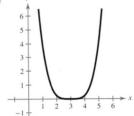

6. $x \approx 3.196$

Section 3.3

1. $(x + 4)(3x + 7)(3x - 7)$ **2.** $x^2 + x + 3, x \neq 3$

3. $2x^3 - x^2 + 3 - \dfrac{6}{3x + 1}$ **4.** $5x^2 - 2x + 3$

5. (a) 1   (b) 396   (c) $-\frac{13}{2}$ (d) -17

6. $f(-3) = 0, f(x) = (x + 3)(x - 5)(x + 2)$

Section 3.4

1. 4   **2.** No rational zeros   **3.** 5   **4.** $-3, \frac{1}{2}, 2$

5. $-1, \dfrac{-3 - 3\sqrt{17}}{4} \approx -3.8423, \dfrac{-3 + 3\sqrt{17}}{4} \approx 2.3423$

6. $f(x) = x^4 + 45x^2 - 196$

7. $f(x) = -x^4 - x^3 - 2x^2 - 4x + 8$ **8.** $\frac{2}{3}, \pm 4i$

9. $f(x) = (x - 1)(x + 1)(x - 3i)(x + 3i); \pm 1, \pm 3i$

10. No positive real zeros, three or no negative real zeros

11. $\frac{1}{2}$ **12.** 7 in. $\times$ 7 in. $\times$ 9 in.

Section 3.5

1.

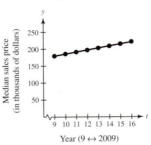

The model is a good fit for the data.

2.

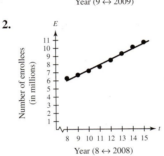

$E = 0.65t + 0.8$
The model is a good fit for the data.

3. $I = 0.075P$ **4.** 576 ft   **5.** 508 units

6. About 1314 ft   **7.** 14,000 joules

Chapter 4

Section 4.1

1. Domain: all real numbers x such that $x \neq 1$
$f(x)$ decreases without bound as x approaches 1 from the left and increases without bound as x approaches 1 from the right.

2. Vertical asymptotes: $x = \pm 1$
Horizontal asymptote: $y = 5$

3. Vertical asymptote: $x = -1$
Horizontal asymptote: $y = 3$

4. (a) $63.75 million; $208.64 million; $1020 million
(b) No. The function is undefined at $p = 100$.

5. (a) $8.40; $1.40; $0.80; $0.48
(b) $C = 0.40$; As the number of units increases, the average cost per unit gets closer to $0.40.

Section 4.2

1.

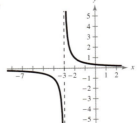

Domain: all real numbers x except $x = -3$

2.

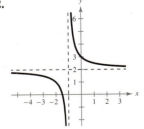

Domain: all real numbers x except $x = -1$

3.

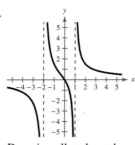

4.

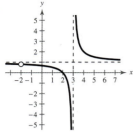

Domain: all real numbers
x except $x = -2, 1$

Domain: all real numbers
x except $x = -2, 3$

5.

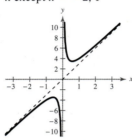

Domain: all real numbers x except $x = 0$

6. 12.9 in. by 6.5 in.

Section 4.3

1. Focus: $(0, 1)$
Directrix: $y = -1$

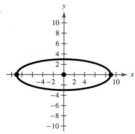

2. $x^2 = \dfrac{3}{2}y$ **3.** $\dfrac{x^2}{16} + \dfrac{y^2}{25} = 1$

4.

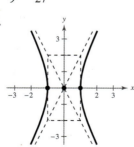

Vertices: $(-9, 0), (9, 0)$

5. $\dfrac{y^2}{9} - \dfrac{x^2}{27} = 1$

6.

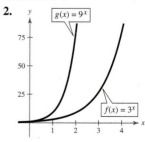

7. $x^2 - \dfrac{y^2}{16} = 1$

Section 4.4

1. (a) Ellipse centered at $(-1, 2)$; horizontal shift one unit to the left and vertical shift two units up
   (b) Circle centered at $(-1, 1)$; horizontal shift one unit to the left and vertical shift one unit up
   (c) Parabola with vertex at $(-4, 3)$; horizontal shift four units to the left and vertical shift three units up
   (d) Hyperbola centered at $(3, 1)$; horizontal shift three units to the right and vertical shift one unit up

2. Vertex: $(-2, 3)$
Focus: $(-3, 3)$
Directrix: $x = -1$

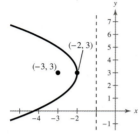

3. $(x + 1)^2 = 4(y - 1)$ **4.**

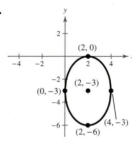

5. $\dfrac{(x - 3)^2}{9} + \dfrac{(y - 5)^2}{25} = 1$

6.

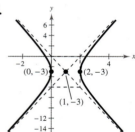

7. $\dfrac{(x - 4)^2}{1} - \dfrac{(y + 1)^2}{8} = 1$

Chapter 5

Section 5.1

1. 0.0528248

2.

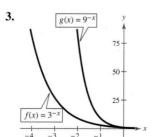

3.

4. (a) 2   (b) 3
5. (a) Shift the graph of f two units to the right.
   (b) Shift the graph of f three units up.
   (c) Reflect the graph of f in the y-axis and shift three units down.

6. (a) 1.3498588   (b) 0.3011942   (c) 492.7490411

7.

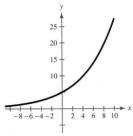

8. (a) $7927.75   (b) $7935.08   (c) $7938.78

9. About 9.970 lb; about 0.275 lb

Section 5.2

1. (a) 0   (b) -3 (c) 3

2. (a) 2.4393327   (b) Error or complex number
    (c) -0.3010300

3. (a) 1   (b) 3   (c) 0   **4.** ± 3

5.

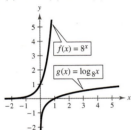

6.

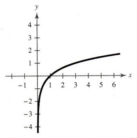

Vertical asymptote: $x = 0$

7. (a)

(b)

8. (a) -4.6051702 (b) 1.3862944   (c) 1.3169579
    (d) Error or complex number

9. (a) $\frac{1}{3}$ (b) 0   (c) $\frac{3}{4}$ (d) 7   **10.** $(-3, \infty)$

11. (a) 70.84   (b) 61.18   (c) 59.61

Section 5.3

1. 3.5850   **2.** 3.5850

3. (a) $\log 3 + 2 \log 5$ (b) $2 \log 3 - 3 \log 5$ **4.** 4

5. $\log_3 4 + 2 \log_3 x - \frac{1}{2} \log_3 y$ **6.** $\log \dfrac{(x + 3)^2}{(x - 2)^4}$

7. $\ln y = \frac{2}{3} \ln x$

Section 5.4

1. (a) 9   (b) 216   (c) $\ln 5$ (d) $-\frac{1}{2}$

2. (a) 4, -2 (b) 1.723   **3.** 3.401   **4.** -4.778

5. 1.099, 1.386   **6.** (a) $e^{2/3}$ (b) 7   (c) 12

7. 0.513   **8.** $\frac{32}{3}$ **9.** 10

10. About 13.2 years; It takes longer for your money to double at a lower interest rate.

11. 2010

Section 5.5

1. 2018   **2.** 400 bacteria   **3.** About 38,000 yr

4.

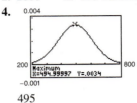

495

5. About 7 days   **6.** (a) 1,000,000   (b) About 80,000,000

Chapter 6

Section 6.1

1. (3, 3)   **2.** $6250 at 6.5%, $18,750 at 8.5%

3. $(-3, -1), (2, 9)$ **4.** No solution   **5.** (1, 3)

6. About 5172 pairs   **7.** 5 weeks

Section 6.2

1. $\left(\frac{3}{4}, \frac{5}{2}\right)$ **2.** (4, 3)   **3.** $(3, -1)$ **4.** (9, 12)

5.

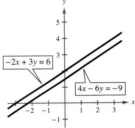

No solution; inconsistent

6. No solution   **7.** Infinitely many solutions: $(a, 4a + 3)$

8. About 471.18 mi/h; about 16.63 mi/h   **9.** (1,500,000, 537)

Section 6.3

1. $(4, -3, 3)$ **2.** (1, 1)   **3.** (1, 2, 3)   **4.** No solution

5. Infinitely many solutions: $(-23a + 22, 15a - 13, a)$

6. Infinitely many solutions: $\left(\frac{1}{4}a, \frac{17}{4}a - 3, a\right)$

7. $s = -16t^2 + 20t + 100$; The object was thrown upward at a velocity of 20 feet per second from a height of 100 feet.

8. $y = \frac{1}{3}x^2 - 2x$

Section 6.4

1. $-\dfrac{3}{2x + 1} + \dfrac{2}{x - 1}$ **2.** $x - \dfrac{3}{x} + \dfrac{4}{x^2} + \dfrac{3}{x + 1}$

3. $-\dfrac{5}{x} + \dfrac{7x}{x^2 + 1}$ **4.** $\dfrac{x + 3}{x^2 + 4} - \dfrac{6x + 5}{(x^2 + 4)^2}$

5. $\dfrac{1}{x} - \dfrac{2}{x^2} + \dfrac{2 - x}{x^2 + 2} + \dfrac{4 - 2x}{(x^2 + 2)^2}$

Section 6.5

1.

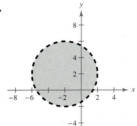

2.

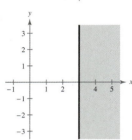

3.

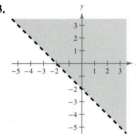

4.

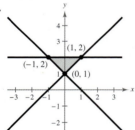

5.

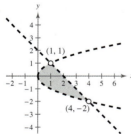

6.

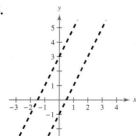

No solution

7.

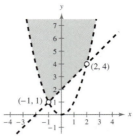

8. Consumer surplus: $22,500,000
Producer surplus: $33,750,000

9. $\begin{cases} 8x + 2y \geq 16 & \text{Nutrient A} \\ x + y \geq 5 & \text{Nutrient B} \\ 2x + 7y \geq 20 & \text{Nutrient C} \\ x \geq 0 \\ y \geq 0 \end{cases}$

Section 6.6

1. Maximum at $(0, 6)$: 30   **2.** Minimum at $(0, 0)$: 0
3. Maximum at $(60, 20)$: 880   **4.** Minimum at $(10, 0)$: 30
5. $2925; 1050 boxes of chocolate-covered creams, 150 boxes of chocolate-covered nuts
6. 3 bottles of brand X, 2 bottles of brand Y

Chapter 7

Section 7.1

1. 2×3 **2.** $\begin{bmatrix} 1 & 1 & 1 & \vdots & 2 \\ 2 & -1 & 3 & \vdots & -1 \\ -1 & 2 & -1 & \vdots & 4 \end{bmatrix}$; 3×4

3. Add -3 times Row 1 to Row 2.
4. Answers will vary. Solution: $(-1, 0, 1)$
5. Reduced row-echelon form   **6.** $(4, -2, 1)$
7. No solution   **8.** $(7, 4, -3)$ **9.** $(3a + 8, 2a - 5, a)$

Section 7.2

1. $a_{11} = 6, a_{12} = 3, a_{21} = -2, a_{22} = 4$

2. (a) $\begin{bmatrix} 6 & -2 \\ 2 & 3 \end{bmatrix}$ (b) $\begin{bmatrix} 0 & 0 \\ 0 & 0 \\ 0 & 0 \end{bmatrix}$

(c) Not possible   (d) $\begin{bmatrix} 0 \\ 0 \\ 2 \end{bmatrix}$

3. (a) $\begin{bmatrix} 4 & -5 \\ 1 & 1 \\ -4 & 1 \end{bmatrix}$ (b) $\begin{bmatrix} 12 & -3 \\ 0 & 12 \\ -9 & 24 \end{bmatrix}$ (c) $\begin{bmatrix} 12 & -11 \\ 2 & 6 \\ -11 & 10 \end{bmatrix}$

4. $\begin{bmatrix} 1 & 2 \\ 10 & -4 \end{bmatrix}$ **5.** $\begin{bmatrix} -6 & 6 \\ -10 & 6 \end{bmatrix}$ **6.** $\begin{bmatrix} 5 & 0 \\ -1 & 4 \end{bmatrix}$

7. $\begin{bmatrix} -1 & 30 \\ 2 & -4 \\ 1 & 12 \end{bmatrix}$ **8.** $\begin{bmatrix} -3 & -22 \\ 3 & 10 \\ -5 & 10 \end{bmatrix}$

9. (a) $\begin{bmatrix} 3 & -1 \\ -9 & 3 \end{bmatrix}$ (b) $[6]$ (c) Not possible

10. $\begin{bmatrix} 7 & 0 \\ 0 & 7 \end{bmatrix}$

11. (a) $\begin{bmatrix} -2 & -3 \\ 6 & 1 \end{bmatrix}\begin{bmatrix} x_1 \\ x_2 \end{bmatrix} = \begin{bmatrix} -4 \\ -36 \end{bmatrix}$

(b) $\begin{bmatrix} -7 \\ 6 \end{bmatrix}$

12. Total cost for women's team: $2310
Total cost for men's team: $2719

Section 7.3

1. $AB = I$ and $BA = I$ **2.** $\begin{bmatrix} 3 & 2 \\ 1 & 1 \end{bmatrix}$

3. $\begin{bmatrix} -4 & -2 & 5 \\ -2 & -1 & 2 \\ -1 & 0 & 1 \end{bmatrix}$ **4.** $\begin{bmatrix} \frac{4}{23} & \frac{1}{23} \\ -\frac{3}{23} & \frac{5}{23} \end{bmatrix}$ **5.** $(2, -1, -2)$

Section 7.4

1. (a) -7 (b) 10 (c) 0
2. $M_{11} = -9, M_{12} = -10, M_{13} = 2, M_{21} = 5, M_{22} = -2,$
$M_{23} = -3, M_{31} = 13, M_{32} = 5, M_{33} = -1$
$C_{11} = -9, C_{12} = 10, C_{13} = 2, C_{21} = -5, C_{22} = -2,$
$C_{23} = 3, C_{31} = 13, C_{32} = -5, C_{33} = -1$
3. -31 **4.** 704

Section 7.5

1. $(3, -2)$ **2.** $(2, -3, 1)$ **3.** 9 square units
4. Collinear   **5.** $x - y + 2 = 0$
6. $(0, 0), (2, 0), (0, 4), (2, 4)$ **7.** 10 square units
8. $[15 \quad 23 \quad 12][19 \quad 0 \quad 1][18 \quad 5 \quad 0][14 \quad 15 \quad 3]$
$[20 \quad 21 \quad 18][14 \quad 1 \quad 12]$
9. $110 \quad -39 \quad -59 \quad 25 \quad -21 \quad -3 \quad 23 \quad -18 \quad -5 \quad 47$
$-20 \quad -24 \quad 149 \quad -56 \quad -75 \quad 87 \quad -38 \quad -37$
10. OWLS ARE NOCTURNAL

Chapter 8

Section 8.1

1. 3, 5, 7, 9   **2.** $1, \frac{3}{2}, \frac{1}{3}, \frac{3}{4}$
3. (a) $a_n = 4n - 3$ (b) $a_n = (-1)^{n+1}(2n)$
4. 6, 7, 8, 9, 10   **5.** 1, 3, 4, 7, 11   **6.** $2, 4, 5, \frac{14}{3}, \frac{41}{12}$
7. $4(n + 1)$ **8.** 44   **9.** (a) 0.5555   (b) $\frac{5}{9}$
10. (a) $1000, $1002.50, $1005.01   (b) $1127.33

Section 8.2

1. 2, 5, 8, 11; $d = 3$ **2.** $a_n = 5n - 6$
3. $-3, 1, 5, 9, 13, 17, 21, 25, 29, 33, 37$ **4.** 79   **5.** 217
6. (a) 630   (b) $N(1 + 2N)$ **7.** 43,560   **8.** 1470
9. $2,500,000

Section 8.3

1. $-12, 24, -48, 96; r = -2$

2. 2, 8, 32, 128, 512

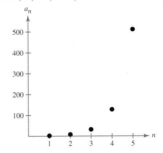

3. 104.02   **4.** $a_n = 4(5)^{n-1}$; 195,312,500   **5.** $\frac{2187}{32}$
6. 2.667   **7.** (a) 10   (b) 6.25   **8.** $3500.85

Section 8.4

1. (a) $\dfrac{6}{(k+1)(k+4)}$ (b) $k + 3 \le 3k^2$
(c) $2^{4k+2} + 1 > 5k + 5$
2–5. Proofs   **6.** $S_k = k(2k + 1)$; Proof
7. (a) 210   (b) 785   **8.** $a_n = n^2 - n - 2$

Section 8.5

1. (a) 462   (b) 36   (c) 1   (d) 1
2. (a) 21   (b) 21   (c) 14   (d) 14
3. 1, 9, 36, 84, 126, 126, 84, 36, 9, 1
4. $x^4 + 8x^3 + 24x^2 + 32x + 16$
5. (a) $y^4 - 8y^3 + 24y^2 - 32y + 16$
(b) $32x^5 - 80x^4y + 80x^3y^2 - 40x^2y^3 + 10xy^4 - y^5$
6. $125 + 75y^2 + 15y^4 + y^6$
7. (a) $1120a^4b^4$ (b) $-3,421,440$

Section 8.6

1. 3 ways   **2.** 2 ways   **3.** 27,000 combinations
4. 2,600,000 numbers   **5.** 24 permutations   **6.** 20 ways
7. 1260 ways   **8.** 21 ways
9. 22,100 three-card poker hands   **10.** 1,051,050 teams

Section 8.7

1. {HH1, HH2, HH3, HH4, HH5, HH6, HT1, HT2, HT3, HT4,
HT5, HT6, TH1, TH2, TH3, TH4, TH5, TH6, TT1, TT2, TT3
TT4, TT5, TT6}
2. (a) $\frac{1}{8}$ (b) $\frac{1}{4}$ **3.** $\frac{1}{9}$ **4.** Answers will vary.
5. $\dfrac{320}{2311} \approx 0.138$ **6.** $\dfrac{1}{962,598}$ **7.** $\dfrac{4}{13} \approx 0.308$
8. $\frac{66}{529} \approx 0.125$ **9.** $\frac{121}{900} \approx 0.134$ **10.** About 0.116
11. 0.452

Appendix A

1. Do not apply radicals term-by-term. Leave as $\sqrt{x^2 + 4}$.
2. $(x - 2)^{-1/2}(7x - 12)$ **3.** 3
4. Answers will vary.   **5.** $-6x(1 - 3x^2)^{-2} + x^{-1/3}$
6. (a) $x - 2 + \dfrac{5}{x^3}$ (b) $x^{3/2} - x^{1/2} + 5x^{-1/2}$

Index